READINGS FROM THE TREATISE ON GEOCHEMISTRY

Editors

H. D. Holland
Harvard University, Cambridge, MA, USA
University of Pennsylvania, Philadelphia, PA

K. K. Turekian
Yale University, New Haven, CT, USA

ELSEVIER

AMSTERDAM • BOSTON • HEIDELBERG • LONDON • NEW YORK • OXFORD
PARIS • SAN DIEGO • SAN FRANCISCO • SINGAPORE • SYDNEY • TOKYO

Elsevier Ltd.
Radarweg 29, 1043 NX Amsterdam, the Netherlands

First edition 2010

British Library Cataloguing in Publication Data
A catalogue record for this book is available from the British Library

Library of Congress Control Number: 2009937759

ISBN: 978-0-12-381391-6

CONTENTS

Introduction

These readings from the updated first edition of the *Treatise on Geochemistry* were chosen to serve as supplements for students in General Geochemistry courses, and to introduce professionals to the field of geochemistry. They are only selections, but they span the entire range of geochemistry that is represented in the *Treatise.*

We start with a discussion of the origin of the elements that is basic to any further exploration of the geochemistry of the Solar System and the Earth in particular. This introduction is followed by an essay on the use of isotopic studies of meteorites and mantle-derived rocks to arrive at the history of the Earth's formation and its separation into the crust, mantle, and core. This essay is followed by chapters on the composition of these three major parts of the Earth, and on the processes associated with plate tectonics that appear to have altered significant portions of the mantle over time. These chapters are followed by an essay on the emplacement of the atmosphere and oceans and on the mechanisms by which volatiles move from the Earth's interior to its surface.

Earth surface processes transport and redistribute the elements and their compounds. These processes are tracked in chapters on soil formation and on the chemistry of rivers and groundwater. The oceans modify the geochemistry of many chemical species. Their effect is illustrated by chapters on the marine $CaCO_3$ cycle and on the effects of marine hydrothermal systems on the chemistry and the biology of the oceans. Three chapters are devoted to sedimentary rocks, which carry an impressive record of changes in the composition of the atmosphere and of seawater, and in the Earth's surface environments during its long history. The evolution of these environments has been influenced strongly by the biosphere. Chapters are therefore devoted to the biogeochemistry of primary organic production in the oceans and to the carbon cycle as a whole.

The cycles of the elements at the Earth's surface, together with many other parts of the Earth system are strongly affected by humanity. The final chapter of the *Readings* deals with the environmental geochemistry of radioactive contaminants. It serves as an example of the human impact and as a warning of potential disasters.

The chapters in this volume were chosen from the large array in the *Treatise on Geochemistry.* They are only fragments but we hope that they will convey the wide sweep and scope of the field that is geochemistry.

H.D. Holland and K.K. Turekian
Executive Editors
September 2009

CONTRIBUTORS

R Amundson
University of California, Berkeley, CA, USA

W S Broecker
Lamont-Doherty Earth Observatory, Columbia University, Palisades, NY

C R Bryan
Sandia National Laboratories, Albuquerque, NM, USA

F H Chapelle
US Geological Survey, Columbia, SC, USA

P G Falkowski
Rutgers University, New Brunswick, NJ, USA

S Gao
China University of Geosciences, Wuhan, People's Republic of China and Northwest University, Xi'an, People's Republic of China

C R German
Southampton Oceanography Centre, Southampton, UK

A N Halliday
University of Oxford, UK

A Heger
University of Chicago, IL, USA

A W Hofmann
Max-Plank-Institut für Chemie, Mainz, Germany and Lamont-Doherty Earth Observatory, Columbia University, Palisades, NY, USA

H D Holland
Harvard University, Cambridge, MA, USA
University of Pennsylvania, Philadelphia, PA

R A Houghton
Woods Hole Research Center, MA, USA

T W Lyons
University of California, Riverside, CA, USA

F T Mackenzie
University of Hawaii, Honolulu, HI, USA

W F McDonough
University of Maryland, College Park, USA

M Meybeck
University of Paris VI, CNRS, Paris, France

D Porcelli
University of Oxford, UK

R L Rudnick
University of Maryland, College Park, MD, USA

B B Sageman
Northwestern University, Evanston, IL, USA

M D Siegel
Sandia National Laboratories, Albuquerque, NM, USA

J W Truran Jr.
University of Chicago, IL, USA

K K Turekian
Yale University, New Haven, CT, USA

J Veizer
Ruhr University, Bochum, Germany and University of Ottawa, ON, Canada

K L Von Damm[†]
University of New Hampshire, Durham, NH, USA

[†] Deceased.

1
Origin of the Elements

J. W. Truran, Jr. and A. Heger
University of Chicago, IL, USA

1.1 INTRODUCTION

Nucleosynthesis is the study of the nuclear processes responsible for the formation of the elements which constitute the baryonic matter of the Universe. The elements of which the Universe is composed indeed have a quite complicated nucleosynthesis history, which extends from the first three minutes of the Big Bang through to the present. Contemporary nucleosynthesis theory associates the production of certain elements/isotopes or groups of elements with a number of specific astrophysical settings, the most significant of which are: (i) the cosmological Big Bang, (ii) stars, and (iii) supernovae.

Cosmological nucleosynthesis studies predict that the conditions characterizing the Big Bang are consistent with the synthesis only of the lightest elements: ^1H, ^2H, ^3He, ^4He, and ^7Li (Burles *et al.*, 2001; Cyburt *et al.*, 2002). These contributions define the primordial compositions both of galaxies and of the first stars formed therein. Within galaxies, stars and supernovae play the dominant role both in synthesizing the elements from carbon to uranium and in returning heavy-element-enriched matter to the interstellar gas from which new stars are formed. The mass fraction of our solar system (formed ~4.6 Gyr ago) in the form of heavy elements is ~1.8%, and stars formed today in our galaxy can be a factor 2 or 3 more enriched (Edvardsson *et al.*, 1993). It is the processes of nucleosynthesis operating in stars and supernovae that we will review in this chapter. We will confine our attention to three broad categories of stellar and supernova site with which specific nucleosynthesis products are understood to be identified: (i) intermediate mass stars, (ii) massive stars and associated type II supernovae, and (iii) type Ia supernovae. The first two of these sites are the straightforward consequence of the evolution of single stars, while type Ia supernovae are understood to result from binary stellar evolution.

Stellar nucleosynthesis resulting from the evolution of single stars is a strong function of stellar mass (Woosley *et al.*, 2002). Following phases of hydrogen and helium burning, all stars consist of a carbon–oxygen core. In the mass range of the so-called "intermediate mass" stars ($1 \lesssim M/M_{\odot} \lesssim 10$), the temperatures realized in their degenerate cores never reach levels at which carbon ignition can occur. Substantial element production occurs in such stars during the asymptotic giant branch (AGB) phase of evolution, accompanied by significant mass loss, and they evolve to white dwarfs of carbon–oxygen (or, less commonly, oxygen–neon) composition. In contrast, the increased pressures that are experienced in the cores of stars of masses $M \gtrsim 10 M_{\odot}$ yield higher core temperatures that enable subsequent phases of carbon, neon, oxygen, and silicon burning to proceed. Collapse of an iron core devoid of further nuclear energy then gives rise to a type II supernova and the formation of a neutron star or black hole remnant (Heger *et al.*, 2003). The ejecta of type IIs contain the ashes of nuclear burning of the entire life of the star, but are also modified by the explosion itself. They are the source of most material (by mass) heavier than helium.

Observations reveal that binary stellar systems comprise roughly half of all stars in our galaxy. Single star evolution, as noted above, can leave in its wake compact stellar remnants: white dwarfs, neutron stars, and black holes. Indeed, we have evidence for the occurrence of all three types of condensed remnant in binaries. In close binary systems, mass transfer can take place from an evolving companion onto a compact object. This naturally gives rise to a variety of interesting phenomena: classical novae (involving hydrogen thermonuclear runaways in accreted shells on white dwarfs (Gehrz *et al.*, 1998)), X-ray bursts (hydrogen/helium thermonuclear runaways on neutron stars (Strohmayer and Bildsten, 2003)), and X-ray binaries (accretion onto black holes). For some range of conditions, accretion onto carbon–oxygen white dwarfs will permit growth of the CO core to the Chandrasekhar limit $M_{Ch} = 1.4 M_{\odot}$, and a thermonuclear runaway in to core leads to a type Ia supernova.

In this chapter, we will review the characteristics of thermonuclear processing in the three environments we have identified: (i) intermediate-mass stars; (ii) massive stars and type II supernovae; and (iii) type Ia supernovae. This will be followed by a brief discussion of galactic chemical evolution, which illustrates how the contributions from each of these environments are first introduced into the interstellar media of galaxies. Reviews of nucleosynthesis processes include those by Arnett (1995), Trimble (1975), Truran (1984), Wallerstein *et al.* (1997), and Woosley *et al.* (2002). An overview of galactic chemical evolution is presented by Tinsley (1980).

1.2 ABUNDANCES AND NUCLEOSYNTHESIS

The ultimate goal of nucleosynthesis theory is, of course, to explain the composition of the Universe, as reflected, for example, in the stellar and gas components of galaxies. Significant progress has been achieved in this regard as a consequence of a wealth of new information of cosmic abundances—spectroscopic properties of stars in our galaxy and of gas clouds and galaxies at high redshifts—pouring in from new ground- and space-based observatories. Given that, it remains true that the most significant clues to nucleosynthesis are those provided by our detailed knowledge of the elemental and isotopic composition of solar system matter. The mass fractions of the stable isotopes in the solar are displayed in Figure 1. Key features that reflect the nature of the nuclear processes by which the heavy elements are formed include: (i) the large abundances of ^{12}C and ^{16}O, the main products of stellar helium burning; (ii) the dominance of the α-particle nuclei through calcium (^{20}Ne, ^{24}Mg, ^{28}Si, ^{32}S, ^{36}Ar, and ^{40}Ca); (iii) the "nuclear statistical equilibrium" peak at the position of ^{56}Fe; and (iv) the abundance peaks in the region past iron at the neutron closed shell positions (zirconium, barium, and lead), confirming the occurrence of processes of neutron-capture synthesis. The solar system abundance patterns associated specifically with the slow (s-process) and fast (r-process) processes of neutron capture synthesis are shown in Figure 2.

It is important here to call attention to the revised determinations of the oxygen and carbon abundances in the Sun. Allende Prieto *et al.* (2001) derived an accurate oxygen abundance for the Sun of $\log \varepsilon(O) = 8.69 \pm 0.05$ dex, a value approximately a factor of 2 below that quoted by Anders and Grevesse (1989). Subsequently, Allende Prieto *et al.* (2002) determined the solar carbon abundance to be $\log \varepsilon(C) = 8.39 \pm 0.04$ dex, and the ratio $C/O = 0.5 \pm 0.07$. The bottom line here is a reduction in the abundances of the two most abundant heavy elements in the Sun, relative to hydrogen and helium, by a factor ~ 2. The implications of these results for stellar evolution, nucleosynthesis, the formation of carbon stars, and galactic chemical evolution remain to be explored.

Guided by early compilations of the "cosmic abundances" as reflected in solar system material (e.g., Suess and Urey, 1956), Burbidge *et al.* (1957) and Cameron (1957) identified the nuclear processes by which element formation occurs in stellar and supernova environments: (i) *hydrogen burning*, which powers stars for ~90% of their lifetimes; (ii) *helium burning*, which is responsible for the production of ^{12}C and ^{16}O, the two most abundant elements heavier than helium; (iii) the *α-process*, which we now understand as a combination of

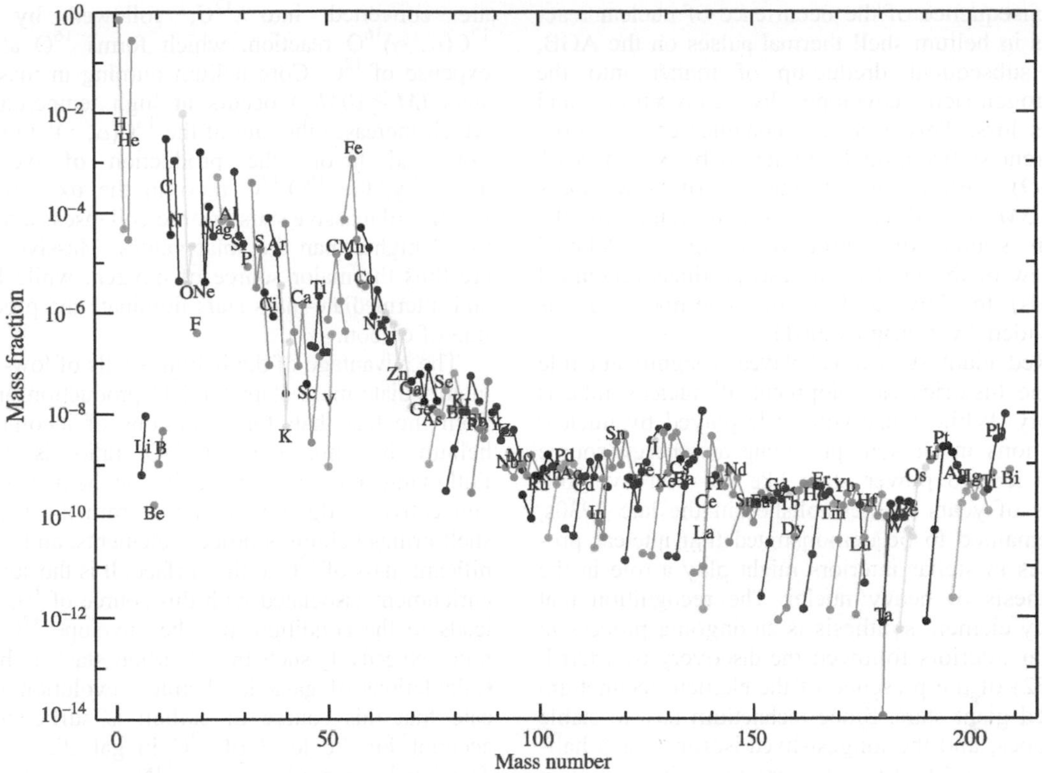

Figure 1 The abundances of the isotopes present in solar system matter are plotted as a function of mass number A (the solar system abundances for the heavy elements are those compiled by Palme and Jones.

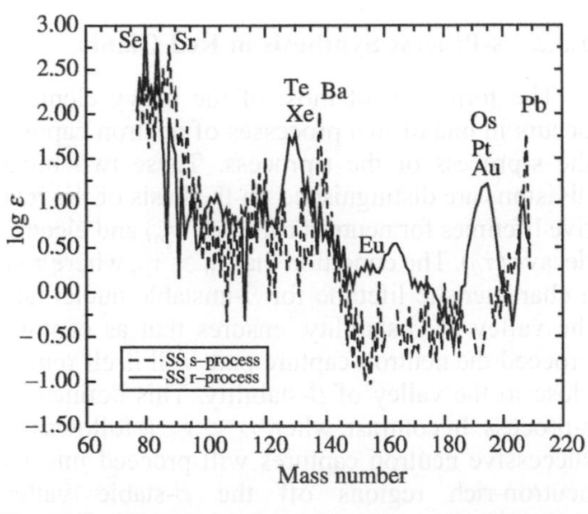

Figure 2 The s-process and r-process abundances in solar system matter (based upon the work by Käppeler *et al.*, 1989). Note the distinctive s-process signature at masses $A \sim 88$, 138, and 208 and the corresponding r-process signatures at $A \sim 130$ and 195, all attributable to closed-shell effects on neutron capture cross-sections. It is the r-process pattern thus extracted from solar system abundances that can be compared with the observed heavy element patterns in extremely metal-deficient stars (the total solar system abundances for the heavy elements are those compiled by Anders and Grevesse, 1989), which are very similar to those from the compilation of Palme and Jones.

carbon, neon, and oxygen burning; (iv) the *equilibrium process*, by which silicon burning proceeds to the formation of a nuclear statistical equilibrium abundance peak centered on mass $A = 56$; (v) the slow (*s-process*) and rapid (*r-process*) mechanisms of neutron capture synthesis of the heaviest elements ($A \gtrsim 60$–70); and (vi) the *p-process*, a combination of the γ*-process* and the ν*-process*, which we understand to be responsible for the synthesis of a number of stable isotopes of nuclei on the proton-rich side of the valley of beta stability. Our subsequent discussions will identify the astrophysical environments in which these diverse processes are now understood to occur.

1.3 INTERMEDIATE MASS STARS: EVOLUTION AND NUCLEOSYNTHESIS

Intermediate-mass red giant stars are understood to be the primary source both of ^{12}C and of the heavy s-process (slow neutron capture) elements, as well as a significant source of ^{14}N and other less abundant CNO isotopes. Their contributions to galactic nucleosynthesis are a consequence of the occurrence of nuclear reactions in helium shell thermal pulses on the AGB, the subsequent dredge-up of matter into the hydrogen-rich envelope by convection, and mass loss. This is a very complicated evolution. Current stellar models, reviewed by Busso *et al.* (1999), allow the formation of low mass ($\sim 1.5 M_{\odot}$) carbon stars, which represent the main source of s-process nuclei. A detailed review of the nucleosynthesis products (chemical yields) for low- and intermediate-mass stars is provided by Marigo (2001).

Red giant stars have played a significant role in the historical development of nucleosynthesis theory. While the pivotal role played by nuclear reactions in stars in providing an energy source sufficient to power stars like the Sun over billions of years was established in the late 1930s, it remained to be demonstrated that nuclear processes in stellar interiors might play a role in the synthesis of heavy nuclei. The recognition that heavy-element synthesis is an ongoing process in stellar interiors followed the discovery by Merrill (1952) of the presence of the element technetium in red giant stars. Since technetium has no stable isotopes, and the longest-lived isotope has a half-life $\tau_{1/2} \sim 4.6$ Myr, its presence in red-giant atmospheres indicates its formation in these stars. This confirmed that the products of nuclear reactions operating at high temperatures and densities in the deep interior can be transported by convection to the outermost regions of the stellar envelope.

The role of such convective "dredge-up" of matter in the red-giant phase of evolution of $1-10 M_{\odot}$ stars is now understood to be an extremely complex process (Busso *et al.*, 1999). On the first ascent of the giant branch (prior to helium ignition), convection can bring the products of CNO cycle burning (e.g., ^{13}C, ^{17}O, and ^{14}N) to the surface. A second dredge-up phase occurs following the termination of core helium burning. The critical third dredge up, occurring in the aftermath of thermal pulses in the helium shells of these AGB stars, is responsible for the transport of both ^{12}C and s-process nuclei (e.g., technetium) to the surface. The subsequent loss of this enriched envelope matter by winds and planetary nebula formation serves to enrich the interstellar media of galaxies, from which new stars are born. A brief review of the mechanisms of production of ^{12}C and the "main" component of the s-process of neutron capture nucleosynthesis is presented in the following sections.

1.3.1 Shell Helium Burning and ^{12}C Production

Stellar helium burning proceeds by means of the "triple-alpha" reaction in which three ^{4}He nuclei are converted into ^{12}C, followed by the ^{12}C(α, γ)^{16}O reaction, which forms ^{16}O at the expense of ^{12}C. Core helium burning in massive stars ($M \gtrsim 10 M_{\odot}$) occurs at high temperatures, which increases the rate of the ^{12}C(α, γ)^{16}O reaction and favors the production of oxygen. Typically, the ^{16}O/^{12}C ratio in the oxygen-rich mantles of massive stars prior to collapse is a factor ~2–3 higher than the solar values. Massive stars are thus the major source of oxygen, while low- and intermediate-mass stars dominate the production of carbon.

The advantage of the helium shells of low- and intermediate-mass stars for ^{12}C production arises from the fact that, for conditions of incomplete helium burning, the ^{12}C/^{16}O ratio is high. Following a thermal pulse in the helium shell, convective dredge-up of matter from the helium shell brings helium, s-process elements, and a significant mass of ^{12}C to the surface. It is the surface enrichment associated with this source of ^{12}C that leads to the condition that the envelope ^{12}C/^{16}O ratio exceeds 1, such that "carbon star" is born. Calculations of galactic chemical evolution indicate that this source of carbon is sufficient to account for the level of ^{12}C in galactic matter. The levels of production of ^{14}N, ^{13}C, and other CNO isotopes in this environment are significantly more difficult to estimate, and thus the corresponding contributions of AGB stars to the galactic abundances of these isotopes remain uncertain.

1.3.2 s-Process Synthesis in Red Giants

The formation of most of the heavy elements occurs in one of two processes of neutron capture: the s-process or the r-process. These two broad divisions are distinguished on the basis of the relative lifetimes for neutron captures (τ_n) and electron decays (τ_β). The condition that $\tau_n > \tau_\beta$, where τ_β is a characteristic lifetime for β-unstable nuclei near the valley of β-stability, ensures that as captures proceed the neutron-capture path will itself remain close to the valley of β-stability. This defines the s-process. In contrast, when $\tau_n < \tau_\beta$, it follows that successive neutron captures will proceed into the neutron-rich regions off the β-stable valley. Following the exhaustion of the neutron flux, the capture products approach the position of the valley of β-stability by β-decay, forming the r-process nuclei. The s-process and r-process patterns in solar system matter are those shown in Figure 2.

The environment provided by thermal pulses in the helium shells of intermediate-mass stars on the AGB provides conditions consistent with the synthesis of the bulk of the heavy s-process isotopes through bismuth. Neutron captures in AGB stars are driven by a combination of neutron sources: the ^{13}C(α, n)^{16}O reaction provides the bulk of the neutron budget at low-neutron densities, while the $^{22}Ne(\alpha, n)^{25}Mg$ operating at high temperatures helps to set the timescale for critical reaction branches. This s-process site (the main s-process component) is understood to operate in low-mass AGB stars ($M \sim 1-3M_\odot$) and to be responsible for the synthesis of the s-process nuclei in the mass range $A \gtrsim 90$. Calculations reviewed by Busso *et al.* (1999) indicate a great sensitivity both to the characteristics of the ^{13}C "pocket" in which neutron production occurs and to the initial metallicity of the star. In their view, this implies that the solar system abundances are not the result of a unique s-process but rather the consequence of a complicated galactic chemical evolutionary history which witnessed mixing of the products of s-processing in stars of different metallicity and a range of ^{13}C pockets. We can hope that observations of the s-process abundance patterns in stars as a function of metallicity will ultimately be better able to guide and to constrain such theoretical models.

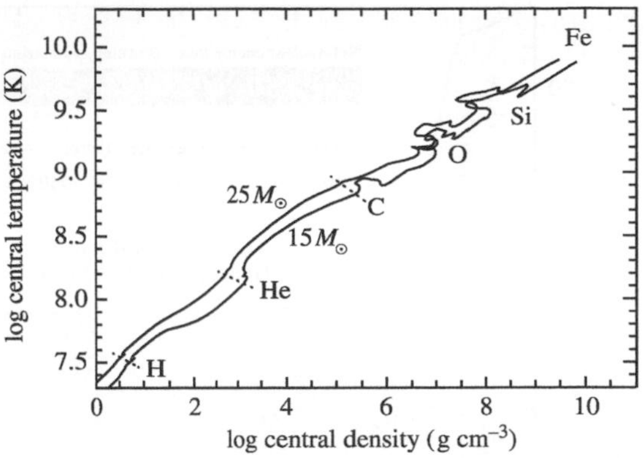

Figure 3 Evolution of the central temperature and density in stars of $15M_\odot$ and $25M_\odot$ from birth as hydrogen burning stars until iron core collapse (Table 1). In general, the trajectories follow a line of $\rho \propto T^3$, but with some deviation downward (towards higher ρ at a given T) due to the decreasing entropy of the core. Nonmonotonic behavior is observed when nuclear fuels are ignited and this is exacerbated in the $15M_\odot$ model by partial degeneracy of the gas (source Woosley *et al.*, 2002).

1.4 MASSIVE-STAR EVOLUTION AND NUCLEOSYNTHESIS

Generally speaking, the evolution of a massive star follows a well-understood path of contraction to increasing central density and temperature. The contraction is caused by the energy loss of the star, due to light radiated from the surface and neutrino losses (see below). The released potential energy is in part converted into internal energy of the gas (Virial theorem). This path of contraction is interrupted by nuclear fusion—first hydrogen is burned to helium, then helium to carbon and oxygen. This is followed by stages of carbon, neon, oxygen, and silicon burning, until finally a core of iron is produced, from which no more energy can be extracted by nuclear burning. The onsets of these burning phases as the star evolves through the temperature–density plane are shown in Figure 3, for stars of masses $15M_\odot$ and $25M_\odot$. Each fuel burns first in the center of the star, then in one or more shells (Figure 4). Most burning stages proceed convectively: i.e., the energy production rate by the burning is so large and centrally concentrated that the energy cannot be transported by radiation (heat diffusion) alone, and convective motions dominate the heat transport. The reason for this is the high-temperature sensitivity of nuclear reaction rates: for hydrogen burning in massive stars, nuclear energy generation has a $\propto T^{18}$ dependence, and the dependence is even stronger for later burning stages. The important consequence is that, due to the efficient mixing caused by the convection, the entire unstable region evolves essentially chemically homogeneously—replenishing the fuel at the bottom of burning region (central or shell burning) and depleting the fuel elsewhere in that region at the same time. As a result, shell burning of this fuel then commences outside that region.

Table 1 summarizes the burning stages and their durations for a $20M_\odot$ star. The timescale for helium burning is $\sim$10 times shorter than that of hydrogen burning, mostly because of the lower energy release per unit mass. The timescale of the burning stages and contraction beyond central helium burning is greatly reduced by thermal neutrino losses that carry away energy *in situ*, instead of requiring that it be transported to the stellar surface by diffusion or convection. These losses increase with temperature (as $\propto T^9$). When the star has built up a large-enough iron core, exceeding its effective Chandrasekhar mass (the maximum mass for which such a core can be stable), the core collapses to form a neutron star or a black hole (see Woosley *et al.* (2002) for a more extended review). A supernova explosion may result (e.g., Colgate and White, 1966) that ejects most of the layers outside the iron core, including many of the ashes from the preceding burning phase. However, when the supernova shock front travels outward, for a brief time peak temperatures are reached, that exceed the maximum temperatures that have been reached in each region in the preceding hydrostatic burning stages (Table 2). This defines the transient stage of "explosive nucleosynthesis" that is critical to the formation of an equilibrium peak dominated by ^{56}Ni (Truran *et al.*, 1967).

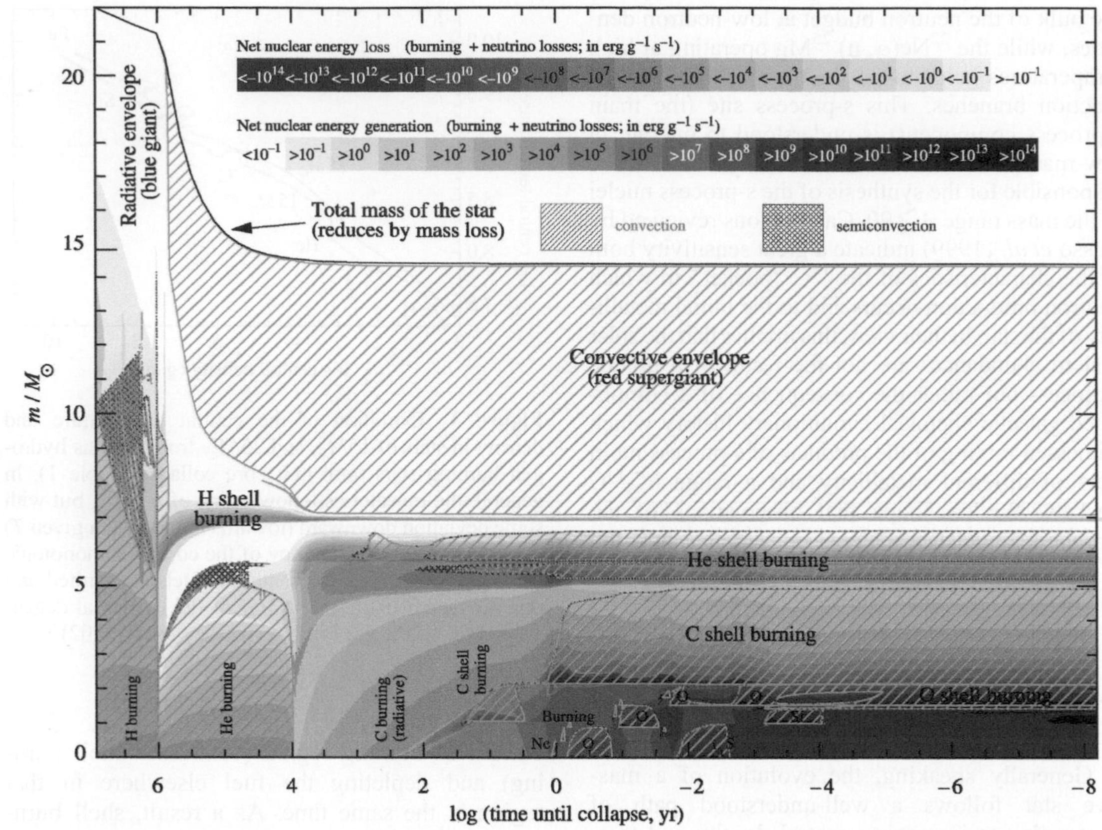

Figure 4 Interior structure of a $22M_\odot$ star of solar composition as a function of time (logarithm of time till core collapse) and enclosed mass. *Green hatching* and *red cross hatching* indicate convective and semiconvective regions. Convective regions are typically well mixed and evolve chemically homogeneously. *Blue shading* indicates energy generation and *pink shading* energy loss. Both take into account the sum of nuclear and neutrino loss contributions. The *thick black* line at the top indicates the total mass of the star, being reduced by mass loss due to stellar winds. Note that the mass loss rate actually increases at late times of the stellar evolution. The decreasing slope of the total mass of the star in the figure is due to the logarithmic scale chosen for the time axis.

Table 1 Hydrostatic nuclear burning stages in massive stars. The table gives burning stages, main and secondary products (ashes), typical temperatures and burning timescales for a $20M_\odot$ star, and the main nuclear reactions. An ellipsis (···) indicates more than one product of the double carbon and double oxygen reactions, and a chain of reactions leading to the buildup of iron group elements for silicon burning.

Fuel	*Main products*	*Secondary products*	*T* (10^9 K)	*Duration* (yr)	*Main reaction*
H	He	^{14}N	0.037	8.1×10^6	$4H \rightarrow {}^4He$ (CNO cycle)
He	O, C	^{18}O, ^{22}Ne s-Process	0.19	1.2×10^6	$3\,{}^4He \rightarrow {}^{12}C$ ${}^{12}C + {}^4He \rightarrow {}^{16}O$
C	Ne, Mg	Na	0.87	9.8×10^2	${}^{12}C + {}^{12}C \rightarrow \cdots$
Ne	O, Mg	Al, P	1.6	0.60	${}^{20}Ne \rightarrow {}^{16}O + {}^4He$ ${}^{20}Ne + {}^4He \rightarrow {}^{24}Mg$
O	Si, S	Cl, Ar, K, Ca	2.0	1.3	${}^{16}O + {}^{16}O \rightarrow \cdots$
Si	Fe	Ti, V, Cr, Mn, Co, Ni	3.3	0.031	${}^{28}Si \rightarrow {}^{24}Mg + {}^4He \cdots$ ${}^{28}Si + {}^4He \rightarrow {}^{24}Mg \cdots$

Massive stars build up most of the heavy elements from oxygen through the iron group from the initial hydrogen and helium of which they are formed. They also make most of the s-process heavy elements up to atomic mass numbers 80–90 from initial iron, converting initial carbon, oxygen, and nitrogen into ^{22}Ne, thus providing a neutron source for the s-process. Massive stars are probably

Table 2 Explosive nucleosynthesis in supernovae. Similar to Table 1, but for the explosion of the star. The "(A, B)" notation means "A" is on the *ingoing* channel and "B" is on the *outgoing* channel. An "α" is same as ^{4}He, "γ" denotes a photon, i.e., a photodisintegration reaction when on the ingoing channel, "n" is a neutron, β^- shows β-decay, and ν indicates neutrino-induced reactions.

Fuel	*Main products*	*Secondary products*	*T* (10^9 K)	*Duration* (s)	*Main reaction*
"*Innermost ejecta*"	r-Process (low Y_e)		>10	1	(n, γ), β^-
Si, O	^{56}Ni	Iron group	>4	0.1	(α, γ)
O	Si, S	Cl, Ar, K, Ca	3–4	1	^{16}O + ^{16}O
Ne, O	O, Mg, Ne	Na, Al, P	2–3	5	(γ, α)
"*Heavy elements*"		p-Process, ν-process:	2–3	5	(γ, n)
		^{11}B, ^{19}F, ^{138}La, ^{180}Ta		5	(ν, ν'), (ν_e, e$^-$)

also the source of most of the proton-rich isotopes of atomic mass number greater than 100 (p-process) and the site of the r-process that is responsible for many of the neutron-rich isotopes from barium through uranium and thorium.

1.4.1 Nucleosynthesis in Massive Stars

The most abundant product of the evolution of massive stars is oxygen, ^{16}O in particular—the third most abundant isotope in the Universe and the most abundant "metal." Massive stars are also the main source of most heavy elements up to atomic mass number $A \sim 80$, of some of the rare proton-rich nuclei, and of the r-process nuclei from barium to uranium. In the following, we will briefly review the burning stages and nuclear processes that characterize the evolution of massive stars and the resulting core collapse supernovae.

1.4.1.1 Hydrogen burning

The first stage of stellar burning and energy generation is hydrogen burning, during which hydrogen is transformed to helium by the so-called CNO cycle. An initial abundance of carbon, nitrogen, and oxygen (of which ^{16}O is the most abundant) is collected into ^{14}N—the proton capture on this isotope is the slowest reaction in this cycle. The total number of CNO isotopes remains unchanged—they operate only as nuclear catalysts in the conversion of hydrogen into helium.

1.4.1.2 Helium burning and the s-process

After hydrogen is depleted, the star contracts towards helium burning. Before helium burning is ignited, the ^{14}N is converted into ^{18}O by ^{14}N(α, γ)^{18}F(β^+)^{18}O and then burned to ^{22}Ne by ^{18}O(α, γ)^{22}Ne, where α stands for a ^{4}He nucleus and the bracket notation means that the first particle is on the ingoing channel and the second on the outgoing channel. The first stage of helium burning involves the interaction of three α-particles to form carbon (^{4}He(α, γ)^{8}Be*(α, γ)^{12}C; "3α reaction"). As helium becomes depleted, the α-capture on ^{12}C takes over and produces ^{16}O. Indeed, as helium is entirely depleted in the central regions of the star, most of the ashes are ^{16}O, while ^{12}C only comprises 10–20%, decreasing with increasing stellar mass.

Towards the end of the central helium burning phase, the temperature becomes high enough for the ^{22}Ne(α, n)^{25}Mg reaction to proceed, which provides a neutron source for the so-called "weak" component of the s-process. In massive stars, this process builds up elements of atomic mass number of up to 80–90, starting with neutron capture on original ^{56}Fe present in the interstellar medium from which the star was born. In this environment, the timescale for neutron capture is slow compared to the weak (β^-) decay of radioactive nuclei produced by the capture. The s-process path thus proceeds along the neutron-rich side of the valley of stable nuclei. The main competitor reaction for the neutron source is the destruction channel for ^{22}Ne that does not produce neutrons, ^{22}Ne(α,γ)^{26}Mg. The main neutron sink or "poison" for the s-process in this environment is neutron captures on the progeny of the ^{22}Ne(α, n) reaction, among which ^{25}Mg is itself a major neutron poison.

1.4.1.3 Hydrogen and helium shell burning

During central helium burning, hydrogen continues to burn in a shell at the outer edge of the helium core, leaving the ashes (helium) at the base of the shell and increasing the size of the helium core. This growth of the helium core by the continued addition of fresh helium fuel contributes to the destruction of carbon at the end of helium burning.

At the completion of core helium burning, helium continues to burn in a shell overlying the helium-free core. It first burns radiatively, but then becomes convective, especially in more massive stars. Since helium is not completely consumed until the death of the star, some s-processing may start here as well, as a consequence of the higher temperatures characterizing the shell burning phase. By the time the stable burning phases of evolution are completed and the iron core is on the verge of collapse, this shell consists of a mixture of helium, carbon, and oxygen, each of which comprises several tens of percent of the matter.

1.4.1.4 Carbon burning

After a contraction phase of several 10^4 yr carbon burning starts in the center. This produces mostly ^{20}Ne(via ^{12}C(^{12}C, α)^{20}Ne), but also some ^{24}Mg(via ^{12}C(^{12}C, γ)^{24}Mg) and ^{23}Na(via ^{12}C(^{12}C, p)^{23}Na). There is also some continuation of the s-process from "unburned" ^{22}Ne after the end of central helium burning, operating with neutrons released during carbon burning via the reaction ^{12}C(^{12}C, n)^{23}Mg.

1.4.1.5 Neon and oxygen burning

Neon burning is induced by photodisintegration of ^{20}Ne into an α-particle and ^{16}O. The α-particle is then captured by another ^{20}Ne nucleus to make ^{24}Mg or by ^{24}Mg to make ^{28}Si. Secondary products include ^{27}Al, ^{31}P, and ^{32}S.

This is closely followed by oxygen burning, for which the dominant reaction is ^{16}O + ^{16}O. The significant products are the α-particle nuclei ^{28}Si, ^{32}S, ^{36}Ar, and ^{40}Ca, together with such isotopes of odd-Z nuclei such as ^{35}Cl, ^{37}Cl, and ^{39}Cl.

1.4.1.6 Silicon burning

Finally, ^{28}Si (and ^{32}S) is burned in a similar way as ^{20}Ne: photodisintegration of ^{28}Si and resulting lighter isotopes accompanies the gradual buildup of iron peak nuclei with higher binding energies per nucleon. At the same time weak processes—positron decaying and electron capture—also become important, producing an neutron excess and allowing the formation of nuclei along the valley of stability. The ultimate result is the formation of an iron "equilibrium" peak of nuclei of maximal nuclear binding energies, from which no further energy can be gained by means of nuclear fusion. The extent in mass of convective central silicon burning is typically $\sim 1.05 M_\odot c$. Several brief stages of shell silicon burning follow until the star has built a core of iron group elements ("iron core") large enough to collapse, typically $1.3 M_\odot$ to $\gtrsim 2 M_\odot$. The core had been held up, in part, by electron degeneracy pressure, but when their Fermi energy becomes sufficiently large they can be captured on protons. This serves both to neutronize the core and to reduce pressure support, ultimately leading to the collapse of the core to a neutron star or black hole. It follows that virtually all of the iron-peak products of this phase of quasi-hydrostatic silicon burning do not escape from the star and thus do not contribute to galactic nucleosynthesis. It is the overlying regions of the core that will generally be ejected.

1.4.1.7 Explosive nucleosynthesis

Following the collapse of the core a supernova shock front, driven by energy deposition by neutrinos from the hot protoneutron star, travels outwards through the star, setting the stage for a brief phase of "explosive nucleosynthesis." In the inner regions, in silicon- and oxygen-rich layers, the photodisintegration of silicon into α-particles and free nucleons is accompanied by the capture of these particles onto silicon and heavier nuclei, leading to the formation of a nuclear statistical equilibrium peak centered on mass $A = 56$. Since the material here is characterized by a small excess of neutrons over protons, it follows that the final products of these explosive burning episodes must lie along or very near to the $Z = N$ line. The dominant species *in situ* therefore include the nuclei ^{44}Ti, ^{48}Cr, ^{52}Fe, ^{56}Ni, and ^{60}Zn. Following decay, these contribute to the production of the most proton-rich stable isotopes at these mass numbers: ^{44}Ca, ^{48}Ti, ^{52}Cr, ^{56}Fe, and ^{60}Ni. The neutron enrichments characteristic of matter of solar composition also allow the production of the odd-Z species ^{51}V (formed as ^{51}Mn), ^{55}Mn (formed as ^{55}Co), and ^{59}Co (formed as ^{59}Cu).

Further out, in the oxygen, neon, and carbon shells, explosive burning can act to modify somewhat the abundance patterns resulting from earlier hydrostatic burning phases. The final nucleosynthesis products of the evolution of massive stars and associated type II core collapse supernovae have been discussed by a number of authors (Woosley and Weaver, 1995; Thielemann *et al.*, 1996; Nomoto *et al.*, 1997; Limongi *et al.*, 2000; Woosley *et al.*, 2002). Most of the elements and isotopes in the region from ^{16}O to ^{40}Ca are formed in relative proportions consistent with solar system matter. The same may be said of the nuclei in the iron-peak region from approximately ^{48}Ti to ^{64}Zn. A unique signature of such massive star/type II supernova nucleosynthesis, however, is the fact that nuclei in the ^{16}O to ^{40}Ca region are overproduced by a factor $\sim$2–3 with respect to nuclei in the ^{48}Ti to ^{64}Zn region—relative to solar system abundances. We will see how this signature, reflected in the compositions of the oldest stars in our galaxy, provides important constraints on models of galactic chemical evolution.

1.4.1.8 The p-process

The "p-process" generally serves to include all possible mechanisms that can contribute to the formation of proton-rich isotopes of heavy nuclei. It was previously believed that these nuclei might be formed by proton captures. We now understand that, for the conditions that obtain in massive stars and accompanying supernovae, proton captures are not sufficient. The two dominant processes making proton-rich nuclei in massive stars are the γ-process and the ν-process.

The γ-process involves the photodisintegration of heavy elements. Obviously, this process is not very efficient, as the effective "seed" nuclei are the primordial heavy-element constituents of the star. Most of the heavy p-elements with atomic mass number greater than 100 that are produced in massive stars (see Figure 5) are formed by this mechanism. It is primarily for this reason that these proton-rich heavy isotopes are rare in nature.

The rarest of these proton-rich isotopes have a different and more exotic sources: the immense neutrino flux from the forming hot neutron star can either convert a neutron into a proton (making, e.g., ^{138}La, ^{180}Ta) or knock out a nucleon (e.g., ^{11}B, ^{19}F). Since the neutrino cross-sections are quite small, it is clear that the "parent" nucleus must have an abundance that is at least several thousand times greater than that of the neutrino-produced "daughter" if this process is to contribute significantly to nucleosynthesis. This is indeed true for the cases of ^{11}B (parent ^{12}C) and ^{19}F (parent ^{20}Ne). Production of ^{138}La and ^{180}Ta, the two rarest stable isotopes in the Universe, is provided by neutrino interactions with ^{138}Ba and ^{180}Hf, respectively.

1.4.1.9 The r-process

It is generally accepted that the r-process synthesis of the heavy neutron capture elements in the mass regime $A \gtrsim 130-140$ occurs in an environment associated with massive stars. This results from two factors: (i) the two most promising mechanisms for r-process synthesis—a neutrino heated "hot bubble" and neutron star mergers—are both tied to environments associated with core collapse supernovae; and (ii) observations of old stars (discussed in Section 1.6) confirm the early entry of r-process isotopes into galactic matter.

- The r-process model that has received the greatest study in recent years involves a high-entropy (neutrino-driven) wind from a core collapse supernova (Woosley *et al.*, 1994; Takahashi *et al.*, 1994). The attractive features of this model include the facts that it may be a natural consequence of the neutrino emission that must accompany core collapse in collapse events, that it would appear to be quite robust, and that it is indeed associated with massive stars of short lifetime. Recent calculations have, however, called attention to a significant problem associated with this mechanism: the entropy values predicted by current type II supernova models are too low to yield the correct levels of production of both the lighter and heavier r-process nuclei.
- The conditions estimated to characterize the decompressed ejecta from neutron star mergers (Lattimer *et al.*, 1977; Rosswog *et al.*, 1999) may also be compatible with the production of an r-process abundance pattern generally consistent with solar system matter. The most recent numerical study of r-process nucleosynthesis in matter ejected in such mergers (Freiburghaus *et al.*, 1999) show specifically that the r-process heavy nuclei in the mass range $A \gtrsim 130-140$ are produced in solar proportions. Here again, the association with a massive star/core collapse super-nova environment is consistent with the early appearance of r-process nuclei in the galaxy and the mechanism seems quite robust.

Massive stars may also contribute to the abundances of the lighter r-process isotopes ($A \lesssim 130-140$). The helium and carbon shells of massive stars undergoing supernova explosions can give rise to neutron production via such reactions as $^{13}C(\alpha,n)^{16}O$, $^{18}O(\alpha,n)^{21}Ne$, and $^{22}Ne(\alpha,n)^{25}Mg$, involving residues of hydrostatic burning phases. Early studies of this problem (Hillebrandt *et al.*, 1981; Truran *et al.*, 1978) indicated that these conditions might allow the production of at least the light ($A \lesssim 130-140$) r-process isotopes (see also Truran and Cowan (2000)). A recent analysis of nucleosynthesis in massive stars by Rauscher *et al.* (2002) has, however, found that this yields only a slight redistribution of heavy-mass nuclei at the base of the helium shell and no significant production of r-process nuclei above mass $A = 100$. This environment may, nevertheless, provide a source of r-process-like anomalies in grains.

1.5 TYPE Ia SUPERNOVAE: PROGENITORS AND NUCLEOSYNTHESIS

Two broad classes of supernovae are observed to occur in the Universe: type I and type II. We have learned from our discussion in the previous section that type II (core collapse) supernovae are products of the evolution of massive

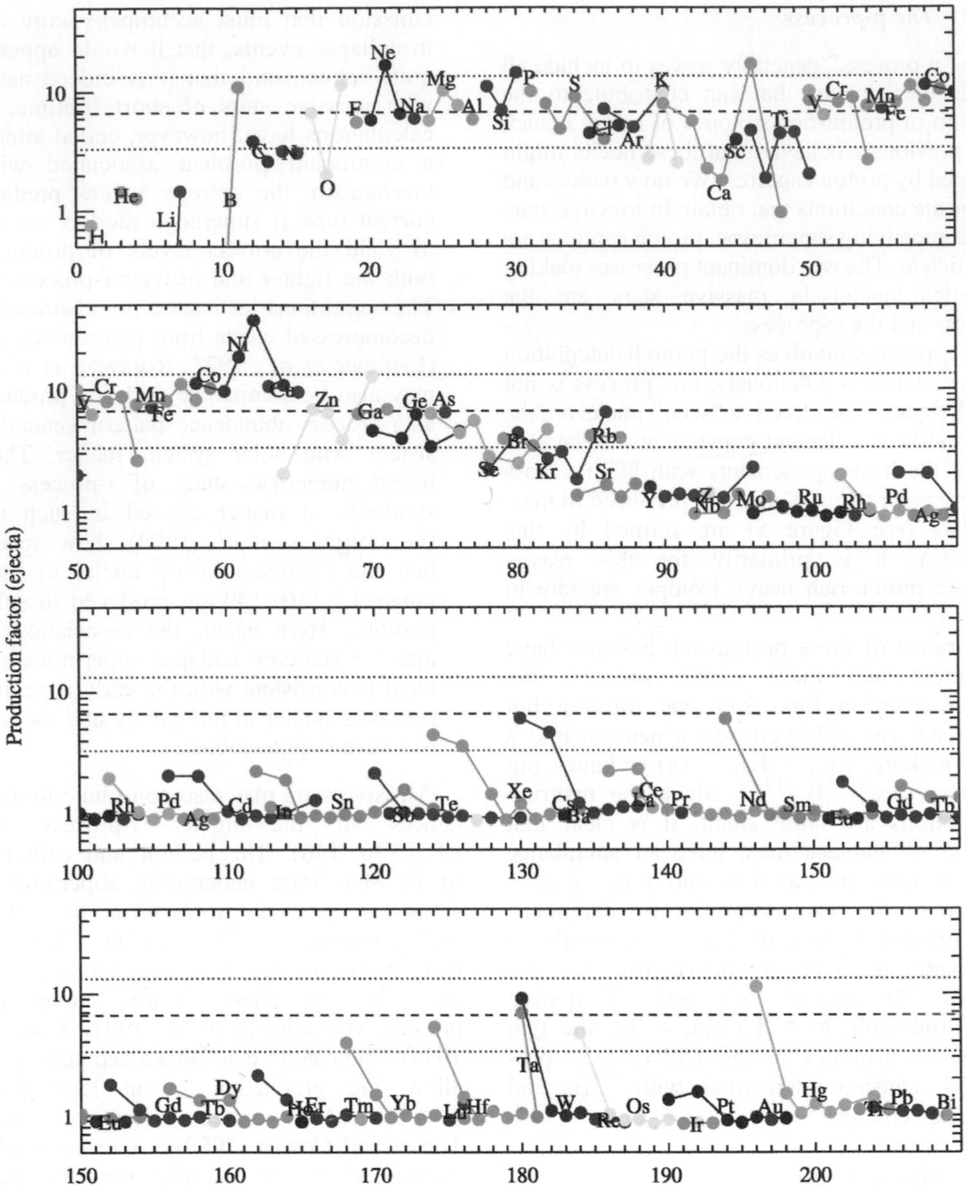

Figure 5 Production factor in the ejecta of a $15M_{\odot}$ star from Rauscher *et al.* (2002). The production factor is defined as the average abundance of the isotope in the stellar ejecta divided by its solar abundance. To guide the eye, we mark the production factor of ^{16}O by a *dashed* line and a range of "acceptable co-production," i.e., within a factor 2 of ^{16}O, by *dotted* lines. The absolute value of the production factors does not matter so much, as the ejecta will be diluted with the interstellar material after the supernova explosion. However, to reproduce a solar *abundance pattern* for certain isotopes in massive stars, they should be closely co-produced with the most abundant product of massive stars, ^{16}O.

stars ($M \gtrsim 10M_{\odot}$). Observationally, the critical distinguishing feature of type I supernovae is the absence of hydrogen features in their spectra at maximum light. Theoretical studies focus attention on models for type I events involving either exploding white dwarfs (type Ia) or the explosions of massive stars (similar to those discussed previously) which have, via wind-driven mass loss or binary effects, shed virtually their entire hydrogen envelopes prior to (iron) core collapse (e.g., types Ib and Ic). We are concerned here with supernovae of type Ia, a subclass of the type Is, which are understood to be associated with the explosion of a white dwarf in a close binary system. The standard model for SNe Ia involves specifically the growth of a carbon–oxygen white dwarf to the Chandrasekhar-limit mass in a close binary system and its subsequent incineration (see, e.g., the review of type Ia progenitor models by Livio, 2000).

The iron-peak nuclei observed in nature had their origin in supernova explosions. Type II and type Ia supernovae provide the dominant sites in which this "explosive nucleosynthesis" mechanism is known to operate. These two supernova sites operate on distinctively different timescales and eject different amounts of iron. As we shall see, an understanding of the detailed nucleosynthesis patterns emerging from these two classes of events provides important insights into the star formation histories of galaxies. Type II (core collapse) supernovae produce both the intermediate-mass nuclei from oxygen to calcium- and iron-peak nuclei. Calculations of charged-particle nucleosynthesis in massive stars and type II supernovae (Woosley and Weaver, 1995; Thielemann *et al.*, 1996; Nomoto *et al.*, 1997; Limongi *et al.*, 2000) reveal one particularly significant distinguishing feature of the emerging abundance patterns: the elements from oxygen through calcium (and titanium) are overproduced relative to iron (peak nuclei) by a factor ~2–3. This means that SNe II produce only ~1/3–1/2 of the iron in galactic matter. (We note that, while all such models necessarily make use of an artificially induced shock wave via thermal energy deposition (Thielemann *et al.*, 1996) or a piston (Woosley and Weaver, 1995), the general trends obtained from such nucleosynthesis studies are expected to be valid.) As we shall see in our discussion of chemical evolution, these trends are reflected in the abundance patterns of metal deficient stars in the halo of our galaxy (Wheeler *et al.*, 1989; McWilliam, 1997). This leaves to SNe Ia the need to produce the ~1/3–1/2 of the iron-peak nuclei from titanium to zinc (Ti–V–Cr–Mn–Fe–Co–Ni–Cu–Zn).

Calculations of explosive nucleosynthesis associated with carbon deflagration models for type Ia events (Thielemann *et al.*, 1986; Iwamoto *et al.*, 1999; Dominguez *et al.*, 2001) predict that sufficient iron-peak nuclei ~(0.6–0.8$M_\odot$ of ^{56}Fe in the form of ^{56}Ni) are synthesized to explain both the powering of the light curves due to the decays of ^{56}Ni and ^{56}Co and the observed mass fraction of ^{56}Fe in galactic matter. Estimates of the timescale for first entry of the ejecta of SNe Ia into the interstellar medium of our galaxy yield ~2×10^9 yr. at a metallicity [Fe/H] ~ −1 (Kobayashi *et al.*, 2000; Goswami and Prantzos, 2000).

A representative nucleosynthesis calculation for such a type Ia supernova event is shown in Figure 6. Note particularly the region of mass fraction between ~0.2$M_\odot$ and 0.8$M_\odot$, which is dominated by the presence of ^{56}Ni, is in nuclear statistical equilibrium. It is this nickel mass that is responsible—as a consequence of the decay of ^{56}Ni through ^{56}Co to ^{56}Fe—for the bulk of the luminosity of type Ia supernovae at maximum

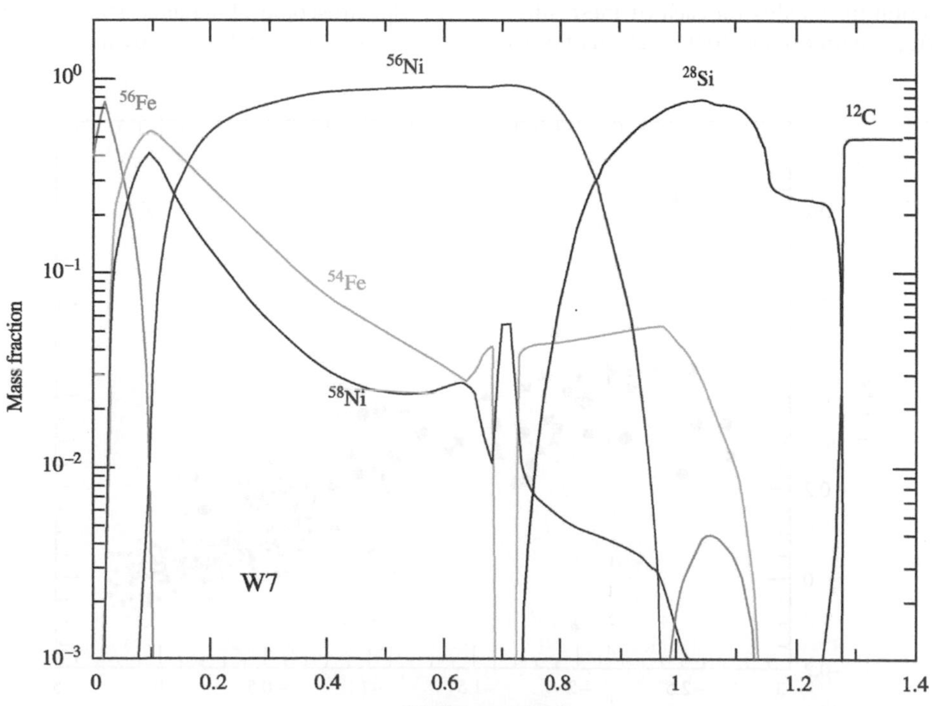

Figure 6 The composition of the core of a type Ia supernova as a function of interior mass. Note the region of ~0.6M$_\odot$ within which the dominant product is ^{56}Ni (source Timmes *et al.*, 2003).

light. This is a critical factor in making SNe Ia the tool of choice for the determination of the cosmological distance scale—as they are the brightest stellar objects known. From the point of view of nucleosynthesis, we then understand that it is the $\sim 0.6 M_\odot$ of iron-peak nuclei ejected per event by type Ia supernovae that represents the bulk of ^{56}Fe in galactic matter. The remaining mass, which is converted into nuclei from ^{16}O to ^{40}Ca, does not make a significant contribution to the synthesis of these elements.

1.6 NUCLEOSYNTHESIS AND GALACTIC CHEMICAL EVOLUTION

We have concentrated in this review on three broad categories of stellar and supernova nucleosynthesis sites: (i) the mass range $1 \lesssim M/M_\odot \lesssim 10$ of "intermediate"-mass stars, for which substantial element production occurs during the AGB phase of their evolution; (ii) the mass range $M \gtrsim 10 M_\odot$, corresponding to the massive star progenitors of type II ("core collapse") supernovae; and (iii) type Ia supernovae, which are understood to arise as a consequence of the evolution of intermediate mass stars in close binary systems.

In the context of models of galactic chemical evolution, it is extremely important to know as well the production timescales for each of these sites—i.e., the effective timescales for the return of a star's nucleosynthesis yields to the interstellar gas. The lifetime of a $10 M_\odot$ star is $\sim 5 \times 10^7$ yr. We can thus expect all massive stars $M \gtrsim 10 M_\odot$ to evolve on timescales $\tau_{SNII} < 10^8$ yr, and to represent the first sources of heavy-element enrichment of stellar populations. In contrast, intermediate-mass stars evolve on timescales $\tau_{IMS} \gtrsim 10^8 - 10^9$ yr. Finally, the timescale for SNe Ia product enrichment is a complicated function of the binary history of type Ia progenitors (see, e.g., Livio, 2000). Observations and theory suggest a timescale $\tau_{SNIa} \gtrsim (1.5-2) \times 10^9$ yr.

Many significant features of the evolution of our galaxy follow from a knowledge of the nuclear ashes and evolutionary timescales for the three sites we have surveyed. The primordial composition of the galaxy was that which it inherited from cosmological nucleosynthesis—primarily hydrogen and helium. The characteristics of the first stellar contributions to nucleosynthesis, whether associated with population II or population III, reflect (as might be expected) the nucleosynthesis products of the evolution of massive stars ($M \gtrsim 10 M_\odot$) of short lifetimes ($\tau \lesssim 10^8$ yr). The significant trends in galactic chemical evolution of concern here are those involving the timing of first entry of the products of the other two broad classifications of nucleosynthesis contributors that we have identified: low-mass stars (s-process) and type Ia supernovae (iron-peak nuclei).

Spectroscopic abundance studies of the oldest stars in our galaxy, down to metallicities

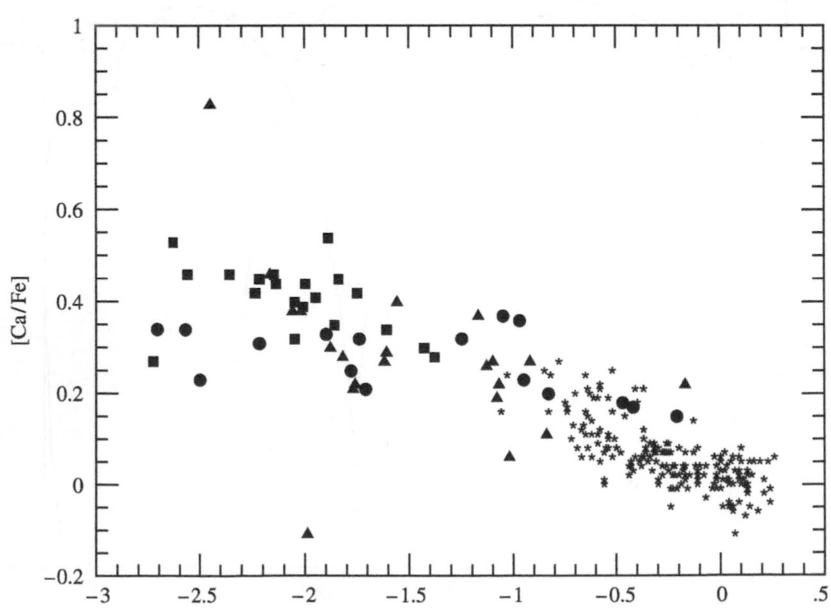

Figure 7 Observed evolution of the calcium to iron abundance ratio with metallicity (▲: Hartmann and Gehren (1998); ■: Zhao and Magain (1990); ●: Gratton and Sneden (1991); ★: Edvardsson *et al.*, (1993)).

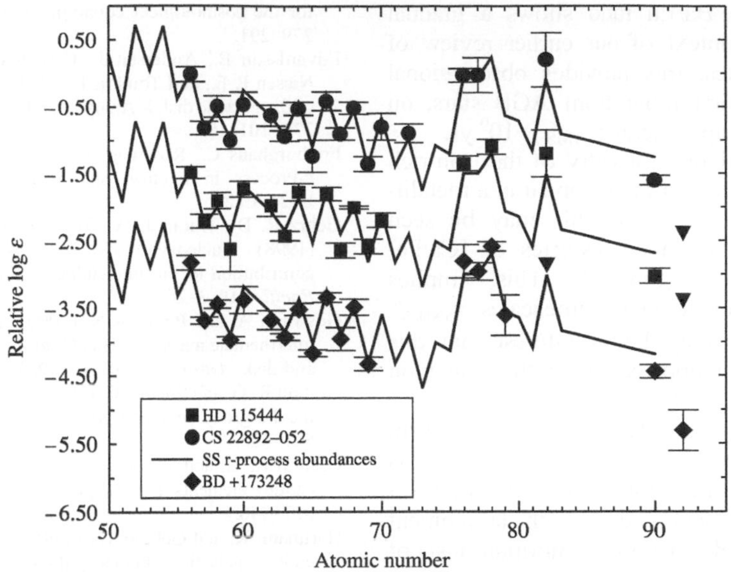

Figure 8 The heavy-element abundance patterns for the three stars CS 22892-052, HD 155444, and BD+17°3248 are compared with the scaled solar system r-process abundance distribution (solid line) (Sneden *et al.*, 2003; Westin *et al.*, 2000; Cowan *et al.*, 2002). Upper limits are indicated by inverted triangles (source Truran *et al.*, 2003).

[Fe/H] ~ -4 to -3, have been reviewed by Wheeler *et al.* (1989). Such studies typically reveal abundance trends which can best be understood as reflecting the nucleosynthesis products of the massive stars and associated type II supernovae that can be expected to evolve and to enrich the interstellar media of galaxies on rapid timescales ($\lesssim 10^8$ yr). Metal deficient stars ($-1.5 \lesssim$ [Fe/H] $\lesssim -3$) in our own galaxy's halo show two significant variations with respect to solar abundances: the elements in the mass range from oxygen to calcium are overabundant—relative to iron-peak nuclei—by a factor ~2–3, and the abundance pattern in the heavy element region $A \gtrsim 60-70$ closely reflects the r-process abundance distribution that is characteristic of solar system matter, with no evidence for an s-process nucleosynthesis contribution. For purposes of illustration, the trends in [Ca/Fe] as a function of metallicity [Fe/H] are shown in Figure 7. Note the factor 2–3 overabundance of calcium with respect to iron at metallicities below [Fe/H] ~ -1. We also display in Figure 8 the detailed agreement of metal-poor star heavy-element abundance pattern with that of solar system r-process abundances, for three metal-poor but r-process-rich stars: CS 22892-052 (Sneden *et al.*, 2003), HD 115444 (Westin *et al.*, 2000), and BD+17°3248 (Cowan *et al.*, 2002). Both of these features are entirely consistent with nucleosynthesis expectations for massive stars and associated type II supernovae.

The signatures of an increasing s-process contamination first appears at an [Fe/H] ~ -2.5 to -2.0.

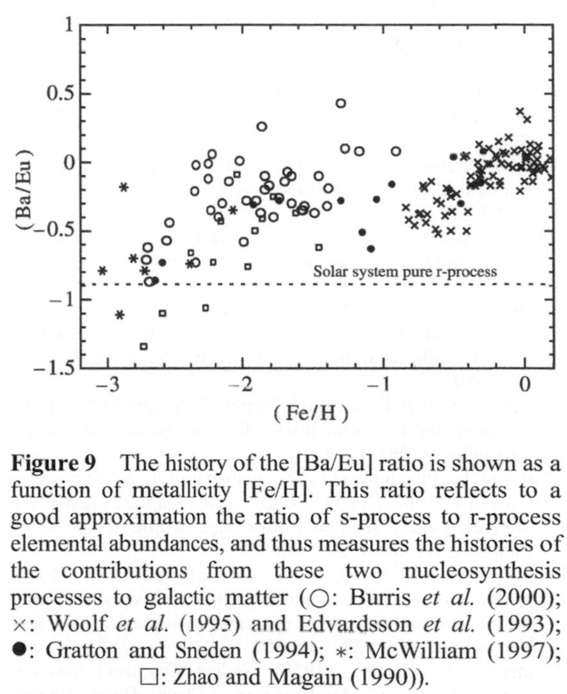

Figure 9 The history of the [Ba/Eu] ratio is shown as a function of metallicity [Fe/H]. This ratio reflects to a good approximation the ratio of s-process to r-process elemental abundances, and thus measures the histories of the contributions from these two nucleosynthesis processes to galactic matter (○: Burris *et al.* (2000); ×: Woolf *et al.* (1995) and Edvardsson *et al.* (1993); ●: Gratton and Sneden (1994); ∗: McWilliam (1997); □: Zhao and Magain (1990)).

The evidence for this is provided by an increase in the ratio of barium (the abundance of which in galactic matter is dominated by s-process contributions) to europium (almost exclusively an r-process product). The ratio [Ba/Eu] is shown in Figure 9 as a function of [Fe/H] for a large sample of halo and disk stars. Note that at the lowest metallicities the [Ba/Eu] ratio clusters around the "pure" r-process value ($[\mathrm{Ba/Eu}]_{\text{r-process}} \sim -0.9$); at a metallicity

[Fe/H] ~ -2.5, the Ba/Eu ratio shows a gradual increase. In the context of our earlier review of nucleosynthesis sites, this provides observational evidence for the first input from AGB stars, on timescales perhaps approaching $\tau_{\rm IMS} \gtrsim 10^9$ yr.

In contrast, evidence for entry of the iron-rich ejecta of SNe Ia is seen first to appear at a metallicity [Fe/H] ~ -1.5 to -1.0. This may be seen reflected in the abundance histories of [Ca/Fe] with [Fe/H], in Figure 7. This implies input from supernovae Ia on timescales $\tau_{\rm SNIa} \gtrsim (1-2) \times 10^9$ yr. It may be of interest that this seems to appear at approximately the transition from halo to (thick) disk stars.

These observed abundance histories confirm theoretical expectations for the nucleosynthesis contributions from different stellar and supernova sites and the timescales on which enrichment occurs. They provide extremely important tests of numerical simulations and illustrate how the interplay of theory and observation can provide important constraints on the star formation and nucleosynthesis history of our galaxy.

REFERENCES

Allende Prieto C., Lambert D. L., and Asplund M. (2001) The forbidden abundance of oxygen in the Sun. *Astrophys. J.* **556**, L63–L66.

Allende Prieto C., Lambert D. L., and Asplund M. (2002) A reappraisal of the solar photospheric C/O ratio. *Astrophys. J.* **573**, L137–L140.

Anders E. and Grevesse N. (1989) Abundances of the elements: meteoritic and solar. *Geochim. Cosmochim. Acta* **53**, 197–214.

Arnett W. D. (1995) Explosive nucleosynthesis revisited: yields. *Ann. Rev. Astron. Astrophys.* **33**, 115–132.

Burbidge E. M., Burbidge G. R., Fowler W. A., and Hoyle F. (1957) Synthesis of the elements in stars. *Rev. Mod. Phys.* **29**, 547–650.

Burles S., Nollett K. M., and Turner M. S. (2001) Big Bang nucleosynthesis predictions for precision cosmology. *Astrophys. J.* **552**, L1–L6.

Burris D. L., Pilachowski C. A., Armandroff T. A., Sneden C., Cowan J. J., and Roe H. (2000) Neutron-capture elements in the early galaxy: insights from a large sample of metal-poor giants. *Astrophys. J.* **544**, 302–319.

Busso M., Gallino R., and Wasserburg G. J. (1999) Nucleosynthesis in asymptotic giant branch stars: relevance for galactic enrichment and solar system formation. *Ann. Rev. Astron. Astrophys.* **37**, 239–309.

Cameron A. G. W. (1957) *Stellar Evolution Nuclear Astrophysics and Nucleogenesis.* Chalk River Report, AELC (Atomic Energy Canada), CRL-41.

Colgate S. A. and White R. H. (1966) The hydrodynamic behavior of supernovae explosions. *Astrophys. J.* **143**, 626–681.

Cowan J. J., Sneden C., Burles S., Ivans I. I., Beers T. C., Truran J. W., Lawler J. E., Primas F., Fuller G. M., Pfeiffer B., and Kratz K.-L. (2002) The chemical composition and age of the metal-poor halo star BD + 17°3248. *Astrophys. J.* **572**, 861–879.

Cyburt R. H., Fields B. D., and Olive K. A. (2002) Primordial nucleosynthesis with CMB inputs: probing the early universe and light element astrophysics. *Astropart. Phys.* **17**, 87–100.

Domínguez I., Höflich P., and Straniero O. (2001) Constraints on the progenitors of Type Ia supernovae and implications for the cosmological equation of state. *Astrophys. J.* **557**, 279–291.

Edvardsson B., Andersen J., Gustafsson B., Lambert D. L., Nissen P. E., and Tomkin J. (1993) The chemical evolution of the galactic disk I. Analysis and results. *Astron. Astrophys.* **275**, 101–152.

Freiburghaus C., Rosswog S., and Thielemann F.-K. (1999) *r*-Process in neutron star mergers. *Astrophys. J.* **525**, L121–L124.

Gehrz R. D., Truran J. W., Williams R. E., and Starrfield S. (1998) Nucleosynthesis in classical novae and its contribution to the interstellar medium. *Proc. Astron. Soc. Pacific* **110**, 3–26.

Goswami A. and Prantzos N. (2000) Abundance evolution of intermediate mass elements (C to Zn) in the Milky Way halo and disk. *Astron. Astrophys.* **359**, 191–212.

Gratton R. G. and Sneden C. (1991) Abundances of elements of the Fe-group in metal-poor stars. *Astron. Astrophys.* **241**, 501–525.

Gratton R. G. and Sneden C. (1994) Abundances of neutron-capture elements in metal-poor stars. *Astron. Astrophys.* **287**, 927–946.

Hartmann K. and Gehren T. (1989) Metal-poor subdwarfs and early galactic nucleosynthesis. *Astron. Astrophys.* **199**, 269–270.

Heger A., Fryer C. L., Woosley S. E., Langer N., and Hartmann D. H. (2003) How massive single stars end their life. *Astrophys. J.* **591**, 288–300.

Hillebrandt W., Klapdor H. V., Oda T., and Thielemann F.-K. (1981) The r-process during explosive helium burning in supernovae. *Astron. Astrophys.* **99**, 195–198.

Iwamoto K., Brachwitz F., Nomoto K., Kishimoto N., Hix R., and Thielemann F.-K. (1999) Nucleosynthesis in Chandrasekhar mass models for Type Ia supernovae and constraints on progenitor systems and burning-front propagation. *Astrophy. J. Suppl.* **125**, 439–462.

Käppeler F., Beer H., and Wisshak K. (1989) s-Process nucleosynthesis–nuclear physics and the classical model. *Rev. Prog. Phys.* **52**, 945–1013.

Kobayashi C., Tsujimoto T., and Nomoto K. (2000) The history of the cosmic supernova rate derived from the evolution of the host galaxies. *Astrophys. J.* **539**, 26–38.

Lattimer J. M., Mackie F., Ravenhall D. G., and Schramm D. N. (1977) The decompression of cold neutron star matter. *Astrophys. J.* **213**, 225–233.

Limongi M., Straniero O., and Chieffi A. (2000) Massive stars in the range 13–25$M_\odot$: evolution and nucleosynthesis: II. The solar metallicity models. *Astrophys. J. Suppl.* **125**, 625–644.

Livio M. (2000) The progenitors of Type Ia supernovae. In *Type Ia Supernovae: Theory and Cosmology* (eds. J. C. Niemeyer and J. W. Truran). Cambridge University Press, Cambridge, pp. 33.

Marigo P. (2001) Chemical yields from low- and intermediate-mass stars: model predictions and basic observational constraints. *Astron. Astrophys.* **370**, 194–217.

McWilliam A. (1997) Abundance ratios and galactic chemical evolution. *Ann. Rev. Astron. Astrophys.* **35**, 503–556.

Merrill P. (1952) Technetium in Red Giant Stars. *Science* **115**, 484–485.

Nomoto K., Hashimoto M., Tsujimoto T., Thielemann F.-K., Kishimoto N., Kubo Y., and Nakasato N. (1997) Nucleosynthesis in type II supernovae. *Nucl. Phys.* **616A**, 79c–91c.

Rauscher T., Heger A., Hoffman R. D., and Woosley S. E. (2002) Nucleosynthesis in massive stars with improved nuclear and stellar physics. *Astrophys. J.* **576**, 323–348.

Rosswog S., Liebendörfer M., Thielemann F.-K., Davies M. B., Benz W., and Piran T. (1999) Mass ejection in neutron star mergers. *Astron. Astrophys.* **341**, 499–526.

Sneden C., Cowan J. J., Lawler J. E., Ivans I. I., Burles S., Beers T. C., Primas F., Hill V., Truran J. W., Fuller G. M., Pfeiffer B., and Kratz K.-L. (2003) The extremely metal-poor, neutron-capture-rich star CS 22892-052: a comprehensive abundance analysis. *Astrophys. J.* (in press).

Suess H. E. and Urey H. C. (1956) Abundances of the elements. *Rev. Mod. Phys.* **28**, 53–74.

Takahashi K., Witti J., and Janka H.-T. (1994) Nucleosynthesis in neutrino-driven winds from protoneutron stars: II. The *r*-process. *Astron. Astrophys.* **286**, 857–869.

Thielemann F.-K., Nomoto K., and Yokoi K. (1986) Explosive nucleosynthesis in carbon deflagration models of Type I supernovae. *Astron. Astrophys.* **158**, 17–33.

Thielemann F.-K., Nomoto K., and Hashimoto M. (1996) Core-collapse supernovae and their ejecta. *Astrophys. J.* **460**, 408–436.

Timmes F. X., Brown E. F., and Truran J. W. (2003) On variations in the peak luminosity of Type Ia supernovae. *Astrophys. J.* (in press).

Tinsley B. (1980) Evolution of the stars and gas in galaxies. *Fundament. Cosmic Phys.* **5**, 287–388.

Trimble V. (1975) The origin and abundances of the chemical elements. *Rev. Mod. Phys.* **47**, 877–976.

Truran J. W. (1984) Nucleosynthesis. *Ann. Rev. Nucl. Part. Sci.* **34,** 53–97.

Truran J. W. and Cowan J. J. (2000) On the site of the weak *r*-process component. In *Proceedings Ringberg Workshop on Nuclear Astrophysics.* Max Planck Publication, MPA-P12, 64p.

Truran J. W., Arnett W. D., and Cameron A. G. W. (1967) Nucleosynthesis in supernova shock waves. *Can. J. Phys.* **45**, 2315–2332.

Truran J. W., Cowan J. J., and Cameron A. G. W. (1978) The helium-driven r-process in supernovae. *Astrophys. J.* **222**, L63–L67.

Truran J. W., Cowan J. J., Pilachowski C. A., and Sneden C. (2003) Probing the neutron-capture nucleosynthesis history of galactic matter. *Proc. Astron. Soc. Pacific* **114**, 1293–1308.

Wallerstein G., Iben I., Jr., Parker P., Boesgaard A. M., Hale G. M., Champagne A. E., Barnes C. A., Käppeler F., Smith V. V., Hoffman R. D., Timmes F. X., Sneden C., Boyd R. N., Meyer B. S., and Lambert D. L. (1997) Synthesis of the elements in stars: forty years of progress. *Rev. Mod. Phys.* **69**, 995–1084.

Westin J., Sneden C., Gustafsson B., and Cowan J. J. (2000) The r-process-enriched low-metallicity giant HD 115444. *Astrophys. J.* **530**, 783–799.

Wheeler J. C., Sneden C., and Truran J. W. (1989) Abundance ratios as a function of metallicity. *Ann. Rev. Astron. Astrophys.* **27**, 279–349.

Woolf V. M., Tomkin J., and Lambert D. L. (1995) The r-process element europium in galactic disk F and G dwarf stars. *Astrophys. J.* **453**, 660–672.

Woosley S. A. and Weaver T. A. (1995) The evolution and explosion of massive stars: II. Explosive hydrodynamics and nucleosynthesis. *Astrophys. J. Suppl.* **101**, 181–235.

Woosley S. E., Wilson J. R., Mathews G. J., Hoffman R. D., and Meyer B. S. (1994) The r-process and neutrino-heated supernova ejecta. *Astrophys. J.* **433**, 229–246.

Woosley S. E., Heger A., and Weaver T. A. (2002) The evolution and explosion of massive stars. *Rev. Mod. Phys.* **74**, 1015–1071.

Zhao G. and Magain P. (1990) The chemical composition of the extreme halo stars: II. Green spectra of 20 dwarfs. *Astron. Astrophys.* **238**, 242–248.

Readings from the Treatise on Geochemistry
ISBN: 978-0-12-381391-6

pp. 1–16

2

The Origin and Earliest History of the Earth

A. N. Halliday

University of Oxford, UK

2.1 INTRODUCTION

The purpose of this chapter is to explain the various lines of geochemical evidence relating to the origin and earliest development of the Earth, while at the same time clarifying current limitations on these constraints. The Earth's origins are to some extent shrouded in greater uncertainty than those of Mars or the Moon because, while vastly more accessible and extensively studied, the geological record of the first 500 Myr is almost entirely missing. This means that we have to rely heavily on theoretical modeling and geochemistry to determine the mechanisms and timescales involved. Both of these approaches have yielded a series of, sometimes strikingly different, views about Earth's origin and early evolution that have seen significant change every few years. There has been a great deal of discussion and debate in the past few years in particular, fueled by new kinds of data and more powerful computational codes.

The major issues to address in discussing the origin and early development of the Earth are as follows:

(i) What is the theoretical basis for our understanding of the mechanisms by which the Earth accreted?

(ii) What do the isotopic and bulk chemical compositions of the Earth tell us about the Earth's accretion?

(iii) How are the chemical compositions of the early Earth and the Moon linked? Did the formation of the Moon affect the Earth's composition?

(iv) Did magma oceans exist on Earth and how can we constrain this from geochemistry?

(v) How did the Earth's core form?

(vi) How did the Earth acquire its atmosphere and hydrosphere and how have these changed?

(vii) What kind of crust might have formed in the earliest stages of the Earth's development?

(viii) How do we think life first developed and how might geochemical signatures be used in the future to identify early biological processes?

Although these issues could, in principle, all be covered in this chapter, some are dealt with in more detail in other chapters and, therefore, are given only cursory treatment here. Furthermore, there are major gaps in our knowledge that render a comprehensive overview unworkable. The nature of the early crust (item (vii)) is poorly constrained, although some lines of evidence will be mentioned. The nature of the earliest life forms (item (viii)) is so loaded with projections into underconstrained hypothetical environments that not a great deal can be described as providing a factual basis suitable for inclusion in a reference volume at this time. Even in those areas in which geochemical constraints are more plentiful, it is essential to integrate them with astronomical observations and dynamic (physical) models of planetary growth and primary differentiation. In some cases, the various theoretical dynamic models can be tested with isotopic and geochemical methods. In other cases, it is the Earth's composition itself that has been used to erect specific accretion paradigms. Therefore, much of this background is provided in this chapter.

All these models and interpretations of geochemical data involve some level of assumption in scaling the results to the big picture of the Earth. Without this, one cannot erect useful concepts that address the above issues. It is one of the main goals of this chapter to explain what these underlying assumptions are. As a consequence, this chapter focuses on the range of interpretations and uncertainties, leaving many issues "open." The chapter finishes by indicating where the main sources of uncertainty remain and what might be done about these in the future.

2.2 OBSERVATIONAL EVIDENCE AND THEORETICAL CONSTRAINTS PERTAINING TO THE NEBULAR ENVIRONMENT FROM WHICH EARTH ORIGINATED

2.2.1 Introduction

The starting place for all accretion modeling is the circumstellar disk of gas and dust that formed during the collapse of the solar nebula. It has been theorized for a long time that a disk of rotating circumstellar material will form as a normal consequence of transferring angular momentum during cloud collapse and star formation. Such disks now are plainly visible around young stars in the Orion nebula, thanks to the Hubble Space Telescope (McCaughrean and O'Dell, 1996). However, circumstellar disks became clearly detectable before this by using ground-based interferometry. If the light of the star is canceled out, excess infrared can be seen being emitted from the dust around the disk. This probably is caused by radiation from the star itself heating the disk.

Most astronomers consider nebular timescales to be of the order of a few million years (Podosek and Cassen, 1994). However, this is poorly constrained because unlike dust, gas is very difficult to detect around other stars. It may be acceptable to assume that gas and dust stay together for a portion of nebular history. However, the dust in some of these disks is assumed to be the secondary product of planetary accretion. Colliding planetesimals and planets are predicted to form at an early stage, embedded in the midplane of such optically thick disks (Wetherill and Stewart, 1993; Weidenschilling, 2000). The age of Beta Pictoris (Artymowicz, 1997; Vidal-Madjar *et al.*, 1998) is rather unclear but it is probably more than ~20 Myr old (Hartmann, 2000) and the dust in this case probably is secondary, produced as a consequence of collisions. Some disks around younger (<10 Myr) stars like HR 4796A appear to show evidence of large inner regions entirely swept clear of dust. It has been proposed that in these regions the dust already may be incorporated into planetary objects (Schneider *et al.*, 1999). The Earth probably formed by aggregating planetesimals and small planets that had formed in the midplane within such a dusty disk.

2.2.2 Nebular Gases and Earth-like versus Jupiter-like Planets

What features of Earth's composition provide information on this early circumstellar disk of dust that formed after the collapse of the solar nebula? The first and foremost feature of the Earth that relates to its composition and accretion is its size and density. Without any other information, this immediately raises questions about how Earth could have formed from the same disk as Jupiter and Saturn. The uncompressed density of the terrestrial planets is far higher than that of the outer gas and ice giant planets. The four most abundant elements making up ~90% of the Earth are oxygen, magnesium, silicon, and iron. Any model of the Earth's accretion has to account for this. The general explanation is that most of the growth of terrestrial planets postdated the loss of nebular gases from the disk. However, this is far from certain. Some solar-like noble gases were trapped in the Earth and although other explanations are considered (Trieloff *et al.*, 2000; Podosek *et al.*, 2003), the one that is most widely accepted is that the nebula was still present at the time of Earth's accretion (Harper and Jacobsen, 1996a). How much nebular gas was originally present is unclear. There is xenon-isotopic evidence that the vast majority (>99%) of Earth's noble gases were lost subsequently (Ozima and Podosekm, 1999; Porcelli and Pepin, 2000). A detailed discussion of this is provided in Chapter 6. The dynamics and timescales for accretion will be very different in the presence or absence of nebular gas. In fact, one needs to consider the possibility that even Jupiter-sized gas giant planets may have formed in the terrestrial planet-forming region and were subsequently lost by being ejected from the solar system or by migrating into the Sun (Lin *et al.*, 1996). More than half the extrasolar planets detected are within the terrestrial planet-forming region of their stars, and these all are, broadly speaking, Jupiter-sized objects (Mayor and Queloz, 1995; Boss, 1998; Lissauer, 1999). There is, of course, a strong observational bias: we are unable to detect Earth-sized planets, which are not massive enough to induce a periodicity in the observed Doppler movement of the associated star or large enough to significantly occult the associated star (Boss, 1998; Seager, 2003).

For many years, it had been assumed that gas dissipation is a predictable response to radiative effects from an energetic young Sun. For example, it was theorized that the solar wind would have been ~100 times stronger than today and this, together with powerful ultraviolet radiation and magnetic fields, would have driven gases away from the disk (e.g., Hayashi *et al.*, 1985). However, we now view disks more as dynamic "conveyor belts" that transport mass *into* the star. The radiative effects on the materials that form the terrestrial planets may in fact be smaller than previously considered. Far from being "blown off" or "dissipated," the gas may well have been lost largely by being swept into the Sun or incorporated into planetary objects—some of which were themselves consumed by the Sun or ejected (Murray *et al.*, 1998; Murray and Chaboyer, 2001). Regardless of how the solar nebula was lost, its former presence, its mass, and the timing of Earth's accretion relative to that of gas loss from the disk will have a profound effect on the rate of accretion, as well as the composition and physical environment of the early Earth.

2.2.3 Depletion in Moderately Volatile Elements

Not only is there a shortage of nebular gas in the Earth and terrestrial planets today but the moderately volatile elements also are depleted (Figure 1) (Gast, 1960; Wasserburg *et al.*, 1964; Cassen, 1996). As can be seen from Figure 2, the depletion in the moderately volatile alkali elements, potassium and rubidium in particular, is far greater than that found in any class of chondritic meteorites (Taylor and Norman, 1990; Humayun and Clayton, 1995; Halliday and Porcelli, 2001; Drake and Righter,

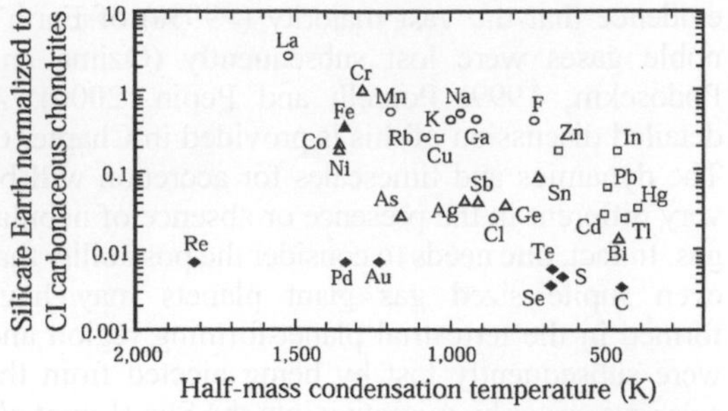

Figure 1 The estimated composition of the silicate portion of the Earth as a function of condensation temperature normalized to CI values in Anders and Grevesse (1989). Open circles: lithophile elements; shaded squares: chalcophile elements; shaded triangles: moderately siderophile elements; solid diamonds: highly siderophile elements. The spread in concentration for a given temperature is thought to be due to core formation. The highly siderophile element abundances may reflect a volatile depleted late veneer. Condensation temperatures are from Newsom (1995).

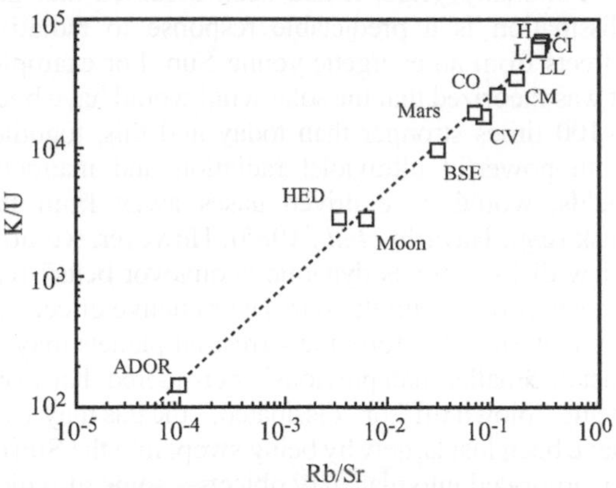

Figure 2 Comparison between the K/U and Rb/Sr ratios of the silicate Earth compared with other solar system objects. ADOR: Angra dos Reis; HED: howardite–eucrite–diogenite parent body; BSE: bulk silicate Earth; CI, CM, CV, CO, H, L, and LL are all classes of chondrites (source Halliday and Porcelli, 2001).

2002). The traditional explanation is that the inner "terrestrial" planets accreted where it was hotter, within the so-called "ice line" (Cassen, 1996; Humayun and Cassen, 2000). For several reasons, it has long been assumed that the solar nebula in the terrestrial planet-forming region started as a very hot, well-mixed gas from which all of the solid and liquid Earth materials condensed. The geochemistry literature contains many references to this hot nebula, as well to major *T-Tauri* heating events that may have further depleted the inner solar system in moderately volatile elements (e.g., Lugmair and Galer, 1992). Some nebula models predict early temperatures that were sufficiently high to prevent condensation of moderately volatile elements (Humayun and Cassen, 2000), which somehow were lost subsequently. To what extent these volatile elements condensed on grains that are now in the outer solar system and may be represented by interplanetary dust particles (Jessberger *et al.*, 1992) is unclear.

Nowadays, inner solar system models are undergoing major rethinking because of new observations of stars, theoretical models, and data from meteorites. It is important to keep track of the models and observational evidence on stars and disks as this continuously changes with greater resolution and detectability. The new data provide important insights into how our solar system may have developed. As of early 2000s, the linkage between temperature in the disk and accretion dynamics is anything but clear. There is no question that transient heating was important on some scale. But a large-scale hot nebula now is more difficult to accommodate. The depletion in volatile elements in the Earth is probably the result of several different processes and the latest astronomical evidence for these is summarized below. To understand these processes one has to have some idea of how solar mass stars and their disks are thought to "work."

2.2.4 Solar Mass Stars and Heating of the Inner Disk

Solar mass stars are thought to accrete rapidly. The pre-main-sequence solar mass *protostar* probably forms from collapse of a portion of a molecular cloud onto a "cloud core" in something like 10^5 yr (Hartmann, 2000). Strong outflows and jets are sometimes observable. Within a few hundred thousand years such protostars have developed into class I young stellar objects, as can be seen in the Orion nebula. These objects already have disks and are called *proplyds*. Remaining material from the cloud will accrete onto both the disk and onto the star itself. The disk also accretes onto the star and, as it does so, astronomers can track the accretion rate from the radiation produced at the innermost margin of the disk. In general terms, the accretion rate shows a very rough decrease with age of the star. From this, it can be shown that the mass of material being accreted from the disk onto the star is about the same as the minimum mass solar nebula estimated for our solar system (Hartmann, 2000).

This "minimum mass solar nebula" is defined to be the minimum amount of hydrogen–helium gas with dust, in bulk solar system proportions, that is needed in order to form our solar system's planets (Hoyle, 1960; Weidenschilling, 1977a). It is calculated by summing the assumed amount of metal (in the astronomical sense, i.e., elements heavier than hydrogen and helium) in all planets and adding enough hydrogen and helium to bring it up to solar composition. Usually, a value of

0.01 solar masses is taken to be the "minimum mass" (Boss, 1990). The strongest constraint on the value is the abundance of heavy elements in Jupiter and Saturn. This is at least partially independent of the uncertainty of whether these elements are hosted in planetary cores. Such estimates for the minimum mass solar nebula indicate that the disk was at least a factor of 10 more massive than the total mass of the current planets. However, the mass may have been much higher and because of this the loss of metals during the planet-forming process sometimes is factored in. There certainly is no doubt that some solids were consumed by the protosun or ejected into interstellar space. This may well have included entire planets. Therefore, a range of estimates for the minimum mass of 0.01–0.1 solar masses can be found. The range reflects uncertainties that can include the bulk compositions of the gas and ice giant planets (Boss, 2002), and the amount of mass loss from, for example, the asteroid belt (Chambers and Wetherill, 2001).

Some very young stars show enormous rapid changes in luminosity with time. These are called *FU Orionis* objects. They are young Sun-like stars that probably are temporarily accreting material at rapid rates from their surrounding disks of gas and dust; they might be consuming planets, for example (Murray and Chaboyer, 2001). Over a year to a decade, they brighten by a hundred times, then stay bright for a century or so before fading again (Hartmann and Kenyon, 1996). A protostar may go through this sequence many times before the accretion disk and surrounding cloud are dispersed. Radiation from the star on to the disk during this intense stage of activity could be partially responsible for volatile depletions in the inner solar system (Bell *et al.*, 2000), but the relative importance of this versus other heating processes has not been evaluated. Nor is it known if our Sun experienced such dramatic behavior.

T-Tauri stars also are pre-main-sequence stars. They are a few times 10^5 yr to a few million years in age and the *T-Tauri* effect appears to develop after the stages described above. They have many of the characteristics of our Sun but are much brighter. Some have outflows and produce strong stellar winds. Many have disks. The *T-Tauri* effect itself is poorly understood. It has long been argued that this is an early phase of heating of the inner portions of the disk. However, such disks are generally thought to have inclined surfaces that dip in toward the star (Chiang and Goldreich, 1997, 1999). It is these surfaces that receive direct radiation from the star and produce the infrared excess observed from the dust. The *T-Tauri* stage may last a few million years. Because it heats the disk surface it may not have any great effect on the composition of the gas and dust in the accretionary midplane of the disk, where planetesimal accretion is dominant.

Heating of inner solar system material in the midplane of the disk *will* be produced from compressional effects. The thermal effects can be calculated for material in the disk being swept into an increasingly dense region during migration toward the Sun during the early stages of disk development. Boss (1990) included compressional heating and grain opacity in his modeling and showed that temperatures in excess of 1,500 K could be expected in the terrestrial planet-forming region. The main heating takes place at the midplane, because that is where most of the mass is concentrated. The surface of the disk is much cooler. More recent modeling includes the detailed studies by Nelson *et al.* (1998, 2000), which provide a very similar overall picture. Of course, if the material is being swept into the Sun, one has to ask how much of the gas and dust would be retained from this portion of the disk. This process would certainly be very early. The timescales for subsequent cooling at 1 AU would have been very short (10^5 yr). Boss (1990), Cassen (2001), and Chiang *et al.* (2001) have independently modeled the thermal evolution of such a disk and conclude that in the midplane, where planetesimals are likely to accrete, temperatures will drop rapidly. Even at 1 AU, temperatures will be ~300 K after only 10^5 yr (Chiang *et al.*, 2001). Most of the dust settles to the midplane and accretes to form planetesimals over these same short timescales (Hayashi *et al.*, 1985; Lissauer, 1987; Weidenschilling, 2000); the major portion of the solid material may not be heated externally strongly after 10^5 yr.

Pre-main-sequence solar mass stars can be vastly (10^4 times) more energetic in terms of X-ray emissions from solar flare activity in their earliest stages compared with the most energetic flare activity of the present Sun (Feigelson *et al.*, 2002a). With careful sampling of large populations of young solar mass stars in the Orion nebula it appears that this is the normal behavior of stars like our Sun. This energetic solar flare activity is very important in the first million years or so, then decreases (Feigelson *et al.*, 2002a). From this it has been concluded that the early Sun had a 10^5-fold enhancement in energetic protons which may have contributed to short-lived nuclides (Lee *et al.*, 1998; McKeegan *et al.*, 2000; Gounelle *et al.*, 2001; Feigelson *et al.*, 2002b; Leya *et al.*, 2003).

Outflows, jets, and X-winds may produce a flux of material that is scattered across the disk from the star itself or the inner regions of the disk (Shu *et al.*, 1997). The region between the outflows and jets and the disk may be subject to strong magnetic fields that focus the flow of incoming material from the disk as it is being accreted onto the star and then project it back across the disk. These "X-winds" then produce a conveyor belt that cycle material through a zone where it is vaporized before being condensed and dispersed as grains of high-temperature condensates across the disk. If material from areas close to the Sun is scattered across the disk as proposed by Shu *et al.* (1997) it could provide a source for early heated and volatile depleted objects such as calcium-, aluminum-rich refractory inclusions (CAIs) and chondrules, as well as short-lived nuclides, regardless of any direct heating of the disk at 1 AU.

Therefore, from all of the recent examples of modeling and observations of circumstellar disks a number of mechanisms can be considered that might contribute to very early heating and depletion of moderately volatile elements at 1 AU. However, some of these are localized processes and the timescales for heating are expected to be short in the midplane.

It is unclear to what extent one can relate the geochemical evidence of extreme volatile depletion in the inner solar system (Figure 2) to these observations of processes active in other disks. It has been argued that the condensation of iron grains would act as a thermostat, controlling temperatures and evening out gradients within the inner regions of the solar nebula (Wood and Morfill, 1988; Boss, 1990; Wood, 2000). Yet the depletion in moderately volatile elements between different planetary objects is highly variable and does not even vary systematically with heliocentric distance (Palme, 2000). The most striking example of this is the Earth and Moon, which have very different budgets of moderately volatile elements. Yet they are at the same heliocentric distance and appear to have originated from an identical mix of solar system materials as judged from their oxygen isotopic composition (Figure 3) (Clayton and Mayeda, 1975; Wiechert *et al.*, 2001).

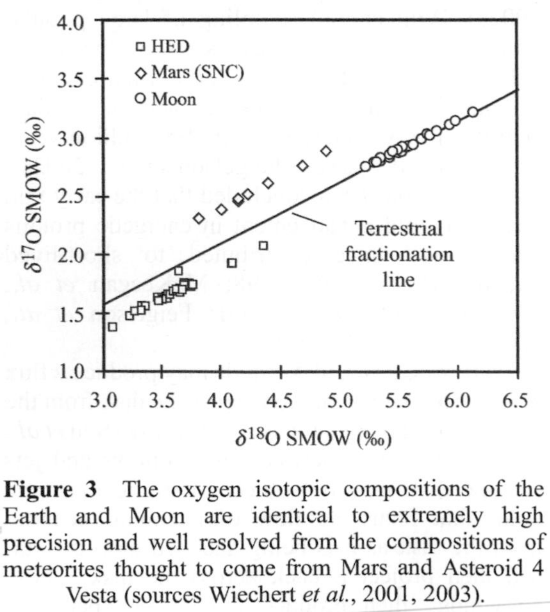

Figure 3 The oxygen isotopic compositions of the Earth and Moon are identical to extremely high precision and well resolved from the compositions of meteorites thought to come from Mars and Asteroid 4 Vesta (sources Wiechert *et al.*, 2001, 2003).

Oxygen isotopic compositions are highly heterogeneous among inner solar system objects (Clayton *et al.*, 1973; Clayton, 1986, 1993). Therefore, the close agreement in oxygen isotopic composition between the Earth and Moon (Clayton and Mayeda, 1975), recently demonstrated to persist to extremely high precision (Figure 3), is a striking finding that provides good evidence that the Earth and Moon were formed from material of similar origin and presumably similar composition (Wiechert *et al.*, 2001). The very fact that chondritic materials are not as heavily depleted in moderately volatile elements as the Earth and Moon provides evidence that other mechanisms of volatile loss must exist. Even the enstatite chondrites, with exactly the same oxygen isotopic composition as the Earth and Moon, are not as depleted in moderately volatile alkali elements (Newsom, 1995). The geochemical constraints on the origins of the components that formed the Earth are discussed below. But first it is necessary to review some of the history of the theories about how the Earth's chemical constituents were first incorporated into planetary building material.

2.2.5 The "Hot Nebula" Model

The current picture of the early solar system outlined above, with a dynamic dusty disk, enormous gradients in temperature, and a rapidly cooling midplane, is different from that prevalent in geochemistry literature 30 yr ago. The chemical condensation sequences modeled thermodynamically for a nebular gas cooling slowly and perhaps statically from 2,000 K were long considered a starting point for understanding the basic chemistry of the material accreting in the inner solar system (Grossman and Larimer, 1974). These traditional standard hot solar nebula models assumed that practically speaking *all* of the material in the terrestrial planet-forming region resulted from gradual condensation of such a nebular gas. Because so many of the concepts in the cosmochemistry literature relate to this hot nebula model, it is important to go through the implications of the newer ways of thinking about accretion of chemical components in order to better understand how the Earth was built.

Here are some of the lines of evidence previously used to support the theory of a large-scale hot nebula that now are being reconsidered.

(i) The isotopic compositions of a wide range of elements have long been known to be broadly similar in meteorites thought to come from Mars and the asteroid belt on the one hand and the Earth and Moon on the other. Given that stars produce huge degrees of isotopic heterogeneity it was assumed that the best way to achieve this homogenization was via a well-mixed gas from which all solids and liquids condensed (Suess, 1965; Reynolds, 1967). However, we know now that chondrites contain presolar grains that cannot have undergone the heating experienced by some of the other components in these meteorites, namely CAIs and chondrules. Presolar grains are unstable in a silicate matrix above a few hundred degrees Celsius (e.g., Mendybaev *et al.*, 2002). The ubiquitous former presence of presolar grains (Huss, 1997; Huss and Lewis, 1995; Nittler, 2003; Nittler *et al.*, 1994) provides unequivocal evidence of dust that has been physically admixed after the formation of the other components (CAIs and chondrules). It is this well-mixed cold dust that forms the starting point for the accretion of chondrite parent bodies, and probably the planets.

(ii) The models of Cameron (1978) using a 1 solar mass disk produced extremely high temperatures ($T > 2{,}000$ K) throughout most of the nebular disk. Such models fueled the hot nebula model but have been abandoned in favor of minimum mass nebula models. Some such viscous accretion disk models produced very low temperatures at 1 AU because these did not include compressional heating. However, Boss (1990) provided the first comprehensive thermal model including compressional heating and grain opacity, and this model does produce temperatures in excess of 1,500 K in the terrestrial planet region.

(iii) CAIs were found to have the composition of objects that condensed at high temperatures from a gas of solar composition (Grossman, 1972; Grossman and Larimer, 1974). Their old age confirmed that they were the earliest objects to form in the solar system (Göpel *et al.*, 1991, 1994; Amelin *et al.*, 2002). Although most CAIs have bulk compositions broadly consistent with high-temperature condensation (Wänke *et al.*, 1974), nearly all of them have been melted and recrystallized, destroying any textural record of condensation. It now appears that they condensed and then were reheated and possibly partially evaporated, all within a short time. It is suspected by some that these objects condensed at very high temperatures close to the Sun and that they were scattered across the disk to be admixed with other components (Gounelle *et al.*, 2001; Shu *et al.*, 1997). This is far from certain and some "FUN" CAIs have isotopic compositions that cannot be easily reconciled with such a model (MacPherson *et al.*, 1988). However, the important point is that their old age and refractory nature can be explained in ways other than just with a large-scale hot solar nebula.

(iv) The overall composition of the Earth is volatile-element depleted and this depletion is broadly consistent with that predicted from condensation theory (Cassen, 1996; Humayun and Cassen, 2000; Allègre *et al.*, 2001). However, this agreement has rather little genetic significance. Why should chondrites be less depleted in volatiles like potassium and rubidium than the terrestrial planets and asteroids (Figure 2) if this is a nebular phenomenon? One explanation is that the chondrites accreted at 2–3 AU, where Boss (1990) shows that the nebula was cooler (<1,000 K). However, this provides no explanation for the extreme depletion in alkalis in eucrites and the Moon. The latter could be related to impact-induced losses (Halliday and Porcelli, 2001) but then the question arises as to whether the Earth's depletion in alkalis also relates to this in part. There is as yet no basis for distinguishing the volatile depletion that might be produced in planetary collisions (O'Neill, 1991a,b; Halliday and Porcelli, 2001) from that predicted to occur as a result of incomplete condensation of nebular gas.

(v) Strontium isotope differences between early very rubidium-depleted objects and planetesimals such as CAIs, eucrites and angrites have long been thought to provide evidence that they must have been created within a high-Rb/Sr environment such as the solar nebula but at a temperature above the condensation of rubidium (Gray *et al.*, 1973; Wasserburg *et al.*, 1977b; Lugmair and Galer, 1992; Podosek *et al.*, 1991). The timescales over which the solar nebula has to be maintained above the condensation temperature of rubidium for this to work are a few million years. However, there is growing evidence that both cooling of the inner nebula and planetesimal growth may be very fast. Excluding the thermal effects from dense planetary atmospheres and the effects of planetary collisions, the timescale for major direct heating of the inner disk itself may be rather short (10^5 yr), but this view could change again with new observational data.

2.2.6 The "Hot Nebula" Model and Heterogeneous Accretion

It was at one time thought that even the terrestrial planets themselves formed directly by condensation from a hot solar nebula. This led to a class of models called heterogeneous accretion models, in which the composition of the material accreting to form the Earth changed with time as the nebula cooled. Eucken (1944) proposed such a heterogeneous accretion model in which early condensed metal formed a core to the Earth around which silicate accreted after condensation at lower temperatures. In this context the silicate-depleted, iron-enriched nature of Mercury makes sense as a body that accreted in an area of the solar nebula that was kept too hot to condense the same proportion of silicate as is found in the Earth (Lewis, 1972; Grossman and Larimer, 1974). Conversely, the lower density of Mars could partly reflect collection of an excess of silicate in cooler reaches of the inner solar nebula. So the concept of heliocentric "feeding zones" for accretion fitted this nicely. The discovery that iron metal condenses at a lower temperature than some refractory silicates made these models harder to sustain (Levin 1972). Nevertheless, a series of models involving progressive heterogeneous accretion at successively lower condensation temperatures were developed for the Earth (e.g., Turekian and Clark, 1969; Smith, 1977, 1980).

These models "produced" a zoned Earth with an early metallic core surrounded by silicate, without the need for a separate later stage of core formation. The application of condensation theory to the striking variations in the densities and compositions of the terrestrial planets, and how metal and silicate form in distinct reservoirs has been seen as problematic for some time. Heterogeneous accretion models require fast accretion and core formation if these processes reflect condensation in the nebula and such timescales can be tested with isotopic systems. The timescales for planetary accretion now are known to be far too long for an origin by partial condensation from a hot nebular gas. Nevertheless, heterogeneous accretion models have become embedded in the textbooks in Earth sciences (e.g., Brown and Mussett, 1981) and astronomy (e.g., Seeds, 1996).

An important development stemming from heterogeneous accretion models is that they introduced the concept that the Earth was built from more than one component and that these may have been accreted in separate stages. This provided an apparent answer to the problem of how to build a planet with a reduced metallic core and an oxidized silicate mantle. However, heterogeneous accretion is hard to reconcile with modern models for the protracted dynamics of terrestrial planet accretion compared with the shortness of nebular timescales. Therefore, they have been abandoned by most scientists and are barely mentioned in modern geochemistry literature any more.

2.3 THE DYNAMICS OF ACCRETION OF THE EARTH

2.3.1 Introduction

Qualitatively speaking, all accretion involves several stages, although the relative importance must differ between planets and some mechanisms are only likely to work under certain conditions that currently are underconstrained. Although the exact mechanisms of accretion of the gas and ice giant planets are poorly understood (Boss, 2002), all such objects need to accrete very rapidly in order to trap large volumes of gas before dissipation of the solar nebula. Probably this requires timescales of $<10^7$ yr (Podosek and Cassen, 1994). In contrast, the most widely accepted dynamic models advocated for the formation of the terrestrial planets (Wetherill, 1986), involve protracted timescales $\sim10^7$–10^8 yr. Application of these same models to the outer planets would mean even longer timescales. In fact, some of the outermost planets would not have yet formed. Therefore, the bimodal distribution of planetary density and its striking spatial distribution appear to require different accretion mechanisms in these two portions of the solar system. However, one simply cannot divide the accretion dynamics into two zones. A range of rate-limiting processes probably controlled accretion of both the terrestrial and Jovian planets and the debates about which of these processes may have been common to both is far from resolved. There almost certainly was some level of commonality.

2.3.2 Starting Accretion: Settling and Sticking of Dust at 1 AU

In most models of accretion at 1 AU, the primary process being studied is the advanced stage of gravitationally driven accretion. However, one first has to consider how accretion got started and in many respects this is far more problematic. Having established that the disk was originally dominated by gas and dust, it must be possible to get these materials to combine and form larger objects on a scale where gravity can play a major role. The starting point is gravitational settling toward the midplane. The dust and grains literally will "rain" into the midplane. The timescales proposed for achieving an elevated concentration of dust in the midplane of the disk are rapid, $\sim10^3$ yr (Hayashi *et al.*, 1985; Weidenschilling, 2000). Therefore, within a very short time the disk will form a concentrated midplane from which the growth of the planets ultimately must be fed.

Laboratory experiments on sticking of dust have been reviewed by Blum (2000), who concluded that sticking microscopic grains together with static and Van der Waals forces to build millimeter-sized compact objects was entirely feasible. However, building larger objects (fist- to football-pitch-sized) is vastly more problematic. Yet it is only when the objects are roughly kilometer-sized that gravity plays a major role. Benz (2000) has reviewed the dynamics of accretion of the larger of such intermediate-sized objects. The accretion of smaller objects is unresolved.

One possibility is that there was a "glue" that made objects stick together. Beyond the ice line, this may indeed have been relatively easy. But in the terrestrial planet-forming region in which early nebular temperatures were >1,000 K such a cement would have been lacking in the earliest stages. Of course, it already has been pointed out that cooling probably was fast at 1 AU. However, even this may not help. The baseline temperature in the solar system was then, and is now, above 160 K (the condensation temperature of water ice), so that no matter how rapid the cooling rate, the temperature would not have fallen sufficiently. The "stickiness" required rather may have been provided by carbonaceous coatings on silicate grains which might be stable at temperatures of >500 K (Weidenschilling, 2000). Waiting for the inner solar nebula to cool before accretion proceeds may not provide an explanation, anyway, because dynamic simulations provide evidence that these processes must be completed extremely quickly. The early Sun was fed with material from the disk and Weidenschilling (1977b, 2000) has argued that unless the dust and small debris are incorporated into much larger objects very quickly (in periods of less than $\sim 10^5$ yr), they will be swept into the Sun. Using a relatively large disk, Cuzzi *et al.* (2003) propose a mechanism for keeping a small fraction of smaller CAIs and fine debris in the terrestrial planet-forming region for a few million years. Most of the dust is lost. Another way of keeping the solids dust from migrating would have been the formation of gaps in the disk, preventing transfer to the Sun. The most obvious way of making gaps in the disk is by planet formation. So there is a "chicken and egg problem." Planets cannot form without gaps. Gaps cannot form without planets. This is a fundamental unsolved problem of terrestrial planetary accretion dynamics that probably deserves far more attention than has been given so far. Some, as yet uncertain, mechanism must exist for sticking small bodies together at 1 AU.

2.3.3 Starting Accretion: Migration

One mechanism to consider might be planetary migration (Lin *et al.*, 1996; Murray *et al.*, 1998). Observations of extrasolar planets provide strong evidence that planets migrate after their formation (Lin *et al.*, 1996). Resonances are observed in extrasolar planetary systems possessing multiple Jupiter-like planets. These resonances can only be explained if the planets migrated after their formation (Murray *et al.*, 1998). Two kinds of models can be considered.

(i) If accretion could not have started in the inner solar system, it might be that early icy and gas rich planets formed in the outer solar system and then migrated in toward the Sun where they opened up gaps in the disk prior to being lost into the Sun. They then left isolated zones of material that had time to accrete into planetesimals and planets.

(ii) Another model to consider is that the terrestrial planets themselves first started forming early in the icy outer solar system and migrated in toward the Sun, where gaps opened in the disk and prevented further migration. There certainly is evidence from noble gases that Earth acquired volatile components from the solar nebula and this might be a good way to accomplish this.

Both of these models have difficulties, because of the evidence against migration in the inner solar system. First, it is hard to see why the migrating planets in model (i) would not accrete most of the material in the terrestrial planet-forming region, leaving nothing for subsequent formation of the terrestrial planets themselves. Therefore, the very existence of the terrestrial planets would imply that such migration did not happen. Furthermore, there is evidence against migration in general in the inner solar system, as follows. We know that Jupiter had to form fast (<10 Myr) in order to accrete sufficient nebular gas. Formation of Jupiter is thought to have had a big effect (Wetherill, 1992) causing the loss of >99% of the material from the asteroid belt (Chambers and Wetherill, 2001). Therefore, there are good reasons for believing that the relative positions of Jupiter and the asteroid belt have been maintained in some approximate sense at least since the earliest history of the solar system. Strong supporting evidence against inner solar system migration comes from the fact that the asteroid belt is zoned today (Gaffey, 1990; Taylor, 1992). ^{26}Al heating is a likely cause of this (Grimm and McSween, 1993; Ghosh and McSween, 1999). However, whatever the reason it must be an early feature, which cannot have been preserved if migration were important.

Therefore, large-scale migration from the outer solar system is not a good mechanism for initiating accretion in the terrestrial planet-forming region unless it predates formation of asteroid belt objects or the entire solar system has migrated relative to the Sun. The outer solar system provides some evidence of ejection of material and migration but the inner solar system appears to retain much of its original "structure."

2.3.4 Starting Accretion: Gravitational Instabilities

Sticking together of dust and small grains might be aided by differences between gas and dust velocities in the circumstellar disk (Weidenschilling, 2000). However, the differential velocities of the grains are calculated to be huge and nobody has been able to simulate this adequately. An early solution that was proposed by Goldreich and Ward

(1973) is that gravitational instabilities built up in the disk. This means that sections of the swirling disk built up sufficient mass to establish an overall gravitational field that prevented the dust and gas in that region from moving away. With less internal differential movement there would have been more chance for clumping together and sticking. A similar kind of model has been advocated on a much larger scale for the rapid growth of Jupiter (Boss, 1997). Perhaps these earlier models need to be looked at again because they might provide the most likely explanation for the onset of terrestrial planet accretion. This mechanism has recently been reviewed by Ward (2000).

2.3.5 Runaway Growth

Whichever way the first stage of planetary accretion is accomplished, it should have been followed by *runaway gravitational growth* of these kilometer-scale planetesimals, leading to the formation of numerous Mercury- to Mars-sized planetary embryos. The end of this stage also should be reached very quickly according to dynamic simulations. Several important papers study this phase of planetary growth in detail (e.g., Lin and Papaloizou, 1985; Lissauer, 1987; Wetherill and Stewart, 1993; Weidenschilling, 2000; Kortenkamp *et al.*, 2000). With runaway growth, it is thought that Moon-sized "planetary-embryos" are built over timescales $\sim 10^5$ yr (Wetherill, 1986; Lissauer, 1993; Wetherill and Stewart, 1993). Exhausting the supply of material in the immediate vicinity prevents further runaway growth. However, there are trade-offs between the catastrophic and constructive effects of planetesimal collisions. Benz and Asphaug (1999) calculate a range of "weakness" of objects with the weakest in the solar system being ~300 km in size. Runaway growth predicts that accretion will be completed faster, closer to the Sun where the "feeding zone" of material will be more confined. On this basis material in the vicinity of the Earth would accrete into Moon-sized objects more quickly than material in the neighborhood of Mars, for example.

2.3.6 Larger Collisions

Additional growth to form Earth-sized planets is thought to require collisions between these "planetary embryos." This is a stochastic process such that one cannot predict in any exact way the detailed growth histories for the terrestrial planets. However, with Monte Carlo simulations and more powerful computational codes the models have become quite sophisticated and yield similar and apparently robust results in terms of the kinds of timescales that must be involved. The mechanisms and timescales are strongly dependent on the amount of nebular gas. The presence of nebular gas has two important effects on accretion mechanisms. First, it provides added friction and pressure that speeds up accretion dramatically. Second, it can have the effect of reducing eccentricities in the orbits of the planets. Therefore, to a first approximation one can divide the models for the overall process of accretion into three possible types that have been proposed, each with vastly differing amounts of nebular gas and therefore accretion rates:

(i) *Very rapid accretion in the presence of a huge nebula*. Cameron (1978) argued that the Earth formed with a solar mass of nebular gas in the disk. This results in very short timescales of $<10^6$ yr for Earth's accretion.

(ii) *Protracted accretion in the presence of a minimum mass solar nebula*. This is known as the Kyoto model and is summarized nicely in the paper by Hayashi *et al.* (1985). The timescales are 10^6–10^7 yr for accretion of all the terrestrial planets. The timescales increase with heliocentric distance. The Earth was calculated to form in ~5 Myr.

(iii) *Protracted accretion in the absence of a gaseous disk*. This model simulates the effects of accretion via planetesimal collisions assuming that all of the nebular gas has been lost. Safronov (1954) first proposed this model. He argued that the timescales for accretion of all of the terrestrial planets then would be very long, in the range of 10^7–10^8 yr.

Safronov's model was confirmed with the Monte Carlo simulations of Wetherill (1980), who showed that the provenance of material would be very broad and only slightly different for each of the terrestrial planets (Wetherill, 1994). The timescales for accretion of each planet also would be very similar. By focusing on the solutions that result in terrestrial planets with the correct (broadly speaking) size and distribution and tracking the growth of these objects, Wetherill (1986) noted that the terrestrial planets would accrete at something approaching exponentially decreasing rates. The half-mass accretion time (time for half of the present mass to accumulate) was comparable (~5–7 Myr), and in reality indistinguishable, for Mercury, Venus, Earth, and Mars using such simulations. Of course these objects, being of different size, would have had different absolute growth rates.

These models did not consider the effects of the growth of gas giant planets on the terrestrial planet-forming region. However, the growth of Jupiter is unlikely to have slowed down accretion at 1 AU (Kortenkamp and Wetherill, 2000). Furthermore, if there were former gas giant planets in the terrestrial planet-forming region they probably would have caused the terrestrial planets to be ejected from their orbits and lost.

In order to distinguish between these models one has to know the amount of nebular gas that was present at the time of accretion. For the terrestrial planets this is relatively difficult to estimate. Although the Safronov–Wetherill model, which specifically assumes no nebular gas, has become the main textbook paradigm for Earth accretion, the discovery that gas giant planets are found in the terrestrial planet-forming regions of other stars (Mayor and Queloz, 1995; Boss, 1998; Lissauer, 1999; Seager, 2003) has fueled re-examination of this issue. Furthermore, recent attempts of accretion modeling have revealed that terrestrial planets can indeed be formed in the manner predicted by Wetherill but that they have high eccentricities (Canup and Agnor, 1998). Thus, they depart strongly from circular orbits. The presence of even a small amount of nebular gas during accretion has the effect of reducing this eccentricity (Agnor and Ward, 2002). This, in turn, would have sped up accretion. As explained below, geochemical data provide strong support for a component of nebular-like gases during earth accretion.

The above models differ with respect to timing and therefore can be tested with isotopic techniques. However, not only are the models very different in terms of timescales, they also differ with respect to the environment that would be created on Earth. In the first two cases the Earth would form with a hot dense atmosphere of nebular gas that would provide a ready source of solar noble gases in the Earth. This atmosphere would have blanketed the Earth and could have caused a dramatic buildup of heat leading to magma oceans (Sasaki, 1990). Therefore, the evidence from dynamic models can also be tested with compositional data for the Earth, which provide information on the nature of early atmospheres and melting.

2.4 CONSTRAINTS FROM LEAD AND TUNGSTEN ISOTOPES ON THE OVERALL TIMING, RATES, AND MECHANISMS OF TERRESTRIAL ACCRETION

2.4.1 Introduction: Uses and Abuses of Isotopic Models

Radiogenic isotope geochemistry can help with the evaluation of the above models for accretion by determining the rates of growth of the silicate reservoirs that are residual from core formation. By far the most useful systems in this regard have been the $^{235}U/^{238}U$–$^{207}Pb/^{206}Pb$ and ^{182}Hf–^{182}W systems. These are discussed in detail below. Other long-lived systems, such as ^{87}Rb–^{87}Sr, ^{147}Sm–^{143}Nd, ^{176}Lu–^{176}Hf, and ^{187}Re–^{187}Os, have provided more limited constraints (Tilton, 1988; Carlson and Lugmair, 2000), although in a fascinating piece of work, McCulloch (1994) did attempt to place model age constraints on the age of the earth using strontium isotope data for Archean rocks (Jahn and Shih, 1974; McCulloch, 1994). The short-lived systems ^{129}I–^{129}Xe and ^{244}Pu–^{136}Xe have provided additional constraints (Wetherill, 1975a; Allègre *et al.*, 1995a; Ozima and Podosek, 1999; Pepin and Porcelli, 2002). Other short-lived systems that have been used to address the timescales of terrestrial accretion and differentiation are ^{53}Mn–^{53}Cr (Birck *et al.*, 1999), ^{92}Nb–^{92}Zr (Münker *et al.*, 2000; Jacobsen and Yin, 2001), ^{97}Tc–^{97}Mo (Yin and Jacobsen, 1998), and ^{107}Pd–^{107}Ag (Carlson and Hauri, 2001). None of these now appear to provide useful constraints. Either the model deployed currently is underconstrained (as with Mn–Cr) or the isotopic effects subsequently have been shown to be incorrect or better explained in other ways.

Hf–W and U–Pb methods both work well because the mechanisms and rates of accretion are intimately associated with the timing of core formation and this fractionates the parent/daughter ratio strongly. For a long while, however, it was assumed that accretion and core formation were completely distinct events. It was thought that the Earth formed as a cold object in less than a million years (e.g., Hanks and Anderson, 1969) but that it then heated up as a result of radioactive decay and later energetic impacts. On this basis, it was calculated that the Earth's core formed rather gradually after tens or even hundreds of millions of years following this buildup of heat and the onset of melting (Hsui and Toksöz, 1977; Solomon, 1979).

In a similar manner isotope geochemists have at various times treated core formation as a process that was distinctly later than accretion and erected relatively simple lead, tungsten and, most recently, zirconium isotopic model ages that "date" this event (e.g., Oversby and Ringwood, 1971; Allègre *et al.*, 1995a; Lee and Halliday, 1995; Galer and Goldstein, 1996; Harper and Jacobsen, 1996b; Jacobsen and Yin, 2001; Dauphas *et al.*, 2002; Kleine *et al.*, 2002; Schöenberg *et al.*, 2002). A more complex model was presented by Kramers (1998). Detailed discussions of U–Pb, Hf–W, and Nb–Zr systems are presented later in this chapter. However, some generalities should be mentioned first.

In looking at these models the following "rules" apply:

(i) Both U–Pb and Hf–W chronometry are unable to distinguish between early accretion with late core formation, and late accretion with concurrent late core formation because it is dominantly core formation that fractionates the parent/daughter ratio.

(ii) If accretion or core formation or both are protracted, the isotopic model age does not define any particular event. In the case of U–Pb it could define a kind of weighted average. In the case of short-lived nuclides, such as the ^{182}Hf–^{182}W system with a half-life of ~9 Myr, it cannot even provide this. Clearly, if a portion of the core formation were delayed until after ^{182}Hf had become effectively extinct, the tungsten isotopic composition of the residual silicate Earth would not be changed. Even if >50% of the mass of the core formed yesterday it would not change the tungsten isotopic composition of the silicate portion of the Earth! Therefore, the issue of how long core formation persisted is completely underconstrained by Hf–W but is constrained by U–Pb data. It also is constrained by trace element data (Newsom *et al.*, 1986).

(iii) Isotopic approaches such as those using Hf–W can only provide an indication of how quickly core formation may have started if accretion was early and very rapid. Clearly this is not a safe assumption for the Earth. If accretion were protracted, tungsten isotopes would provide only minimal constraints on when core formation started.

Tungsten and lead isotopic data can, however, be used to define the timescales for accretion, simply by assuming that core formation, the primary process that fractionates the parent/daughter ratio, started very early and that the core grew in constant proportion to the Earth (Halliday *et al.*, 1996, 2000; Harper and Jacobsen, 1996b; Jacobsen and Harper, 1996; Halliday and Lee, 1999; Halliday, 2000). There is a sound basis for the validity of this assumption, as follows.

(i) The rapid conversion of kinetic energy to heat in a planet growing by accretion of planetesimals and other planets means that it is inescapable that silicate and metal melting temperatures are achieved (Sasaki and Nakazawa, 1986; Benz and Cameron, 1990; Melosh, 1990). This energy of accretion would be sufficient to melt the entire Earth such that in all likelihood one would have magma oceans permitting rapid core formation.

(ii) There is strong observational support for this view that core formation was quasicontinuous during accretion. Iron meteorites and basaltic achondrites represent samples of small planetesimals that underwent core formation early. A strong theoretical basis for this was recently provided by Yoshino *et al.* (2003). Similarly, Mars only reached one-eighth of the mass of the Earth but clearly its size did not limit the opportunity for core formation. Also, most of the Moon is thought to come from the silicate-rich portion of a Mars-sized impacting planet, known as “Theia” (Cameron and Benz, 1991; Canup and Asphaug, 2001; Halliday, 2000), which also was already differentiated into core and silicate. The amount of depletion in iron in eucrites, martian meteorites, and lunar samples provides support for the view that the cores of all the planetesimals and planets represented were broadly similar in their proportions to Earth’s, regardless of absolute size. The slightly more extensive depletion of iron in the silicate Earth provides evidence that core formation was more efficient or protracted, but not that it was delayed.

There is no evidence that planetary objects have to achieve an Earth-sized mass or evolve to a particular state (other than melting), before core formation will commence. It is more reasonable to assume that the core grew with the accretion of the Earth in roughly the same proportion as today. If accretion were protracted, the rate-limiting parameter affecting the isotopic composition of lead and tungsten in the silicate Earth would be the timescale for accretion. As such, the “age of the core” is an average time of formation of the Earth itself. Therefore, simple tungsten and lead isotopic model ages do not define an event as such. The isotopic data instead need to be integrated with models for the growth of the planet itself to place modeled limits for the rate of growth.

The first papers exploring this approach were by Harper and Jacobsen (1996b) and Jacobsen and Harper (1996). They pointed out that the Monte Carlo simulations produced by Wetherill (1986) showed a trend of exponentially decreasing planetary growth with time. They emulated this with a simple expression for the accretionary mean life of the Earth, where the mean life is used in the same way as in nuclear literature as the inverse of a time constant. This model is an extension of the earlier model of Jacobsen and Wasserburg (1979) evaluating the mean age of the continents using Sm–Nd. Jacobsen and Harper applied the model to the determination of the age of the Earth based on (then very limited) tungsten isotope data. Subsequent studies (Halliday *et al.*, 1996, 2000; Halliday and Lee, 1999; Halliday, 2000; Yin *et al.*, 2002), including more exhaustive tungsten as well as lead isotope modeling, are all based on this same concept. However, the data and our understanding of the critical parameters have undergone major development.

Nearly all of these models assume that:

(i) accretion proceeded at an exponentially decreasing rate from the start of the solar system;

(ii) core formation and its associated fractionation of radioactive parent/radiogenic daughter ratios was coeval with accretion;

(iii) the core has always existed in its present proportion relative to the total Earth;

(iv) the composition of the accreting material did not change with time;

(v) the accreting material equilibrated fully with the silicate portion of the Earth just prior to fractionation during core formation; and

(vi) the partitioning of the parent and daughter elements between mantle and core remained constant.

The relative importance of these assumptions and the effects of introducing changes during accretion have been partially explored in several studies (Halliday *et al.*, 1996, 2000; Halliday and Lee, 1999; Halliday, 2000). The issue of metal–silicate equilibration has been investigated recently by Yoshino *et al.* (2003). However, the data upon which many of the fundamental isotopic and chemical parameters are based are in a state of considerable uncertainty.

2.4.2 Lead Isotopes

Until recently, the most widely utilized approach for determining the rate of formation of the Earth was U–Pb geochronology. The beauty of using this system is that one can deploy the combined constraints from both ^{238}U–^{206}Pb ($T_{1/2}$ = 4,468 Myr) and ^{235}U–^{207}Pb ($T_{1/2}$ = 704 Myr) decay. Although the atomic abundance of both of the daughter isotopes is a function of the U/Pb ratio and age, combining the age equations allows one to cancel out the U/Pb ratio. The relative abundance of ^{207}Pb and ^{206}Pb indicates when the fractionation took place. Patterson (1956) adopted this approach in his classic experiment to determine the age of the Earth. Prior to his work, there were a number of estimates of the age of the Earth based on lead isotopic data for terrestrial galenas. However, Patterson was the first to obtain lead isotopic data for early low-U/Pb objects (iron meteorites) and this defined the initial lead isotopic composition of the solar system. From this, it was clear that the silicate Earth's lead isotopic composition required between 4.5 Gyr and 4.6 Gyr of evolution as a high-U/Pb reservoir. Measurements of the lead isotopic compositions of other high-U/Pb objects such as basaltic achondrites and lunar samples confirmed this age for the solar system.

In detail it is now clear that most U–Pb model ages of the Earth (Allègre *et al.*, 1995a) are significantly younger than the age of early solar system materials such as chondrites (Göpel *et al.*, 1991) and angrites (Wasserburg *et al.*, 1977b; Lugmair and Galer, 1992). Such a conclusion has been reached repeatedly from consideration of the lead isotope compositions of early Archean rocks (Gancarz and Wasserburg, 1977; Vervoort *et al.*, 1994), conformable ore deposits (Doe and Stacey, 1974; Manhès *et al.*, 1979; Tera, 1980; Albarède and Juteau, 1984), average bulk silicate Earth (BSE) (Galer and Goldstein, 1996) and mid-ocean ridge basalts (MORBs) (Allègre *et al.*, 1995a), all of which usually yield model ages of $<$4.5 Ga. Tera (1980) obtained an age of 4.53 Ga using Pb–Pb data for old rocks but even this postdates the canonical start of the solar system by over 30 Myr.

The reason why nearly all such approaches yield similar apparent ages that postdate the start of the solar system by a few tens of millions of years is that there was a very strong U/Pb fractionation that took place during the protracted history of accretion. The U–Pb model age of the Earth can only be young if U/Pb is fractionated at a late stage. This fractionation was of far greater magnitude than that associated with any later processes. Thus it has left a clear and irreversible imprint on the $^{207}Pb/^{206}Pb$ and $^{207}Pb/^{204}Pb$ isotope ratios of the silicate portion of the Earth. Uranium, being lithophile, is largely confined to the silicate portion of the Earth. Lead is partly siderophile and chalcophile such that $>$90% of it is thought to be in the core (Allègre *et al.*, 1995a). Therefore, it was long considered that the lead isotopic "age of the Earth" dates core formation (Oversby and Ringwood, 1971; Allègre *et al.*, 1982).

The exact value of this fractionation is poorly constrained, because lead also is moderately volatile, so that the U/Pb ratio of the total Earth (mantle, crust, and core combined) also is higher than chondritic. In fact, some authors even have argued that the dominant fractionation in U/Pb in the BSE was caused by volatile loss (Jacobsen and Harper, 1996b; Harper and Jacobsen, 1996b; Azbel *et al.*, 1993). This is consistent with some compilations of data for the Earth, which show that lead is barely more depleted in the bulk silicate portion of the Earth than lithophile elements of similar volatility (McDonough and Sun, 1995). It is, therefore, important to know how much of the lead depletion is caused by accretion of material that was depleted in volatile elements at an early (nebular) stage. Galer and Goldstein (1996) and Allègre *et al.* (1995a) have compellingly argued that the major late-stage U/Pb fractionation was the result of core formation. However, the uncertainty over the U/Pb of the total Earth and whether it changed with accretion time remains a primary issue limiting precise application of lead isotopes.

Using exponentially decreasing growth rates and continuous core formation one can deduce an accretionary mean life assuming a $^{238}U/^{204}Pb$ for the total Earth of 0.7 (Halliday, 2000). This value is based on the degree of depletion of moderately volatile lithophile elements, as judged from the K/U ratio of the BSE (Allègre *et al.*, 1995a). Application of this approach to the lead isotopic compositions of the Earth (Halliday, 2000) provides evidence that the Earth accreted with an accretionary mean life of between 15 Myr and

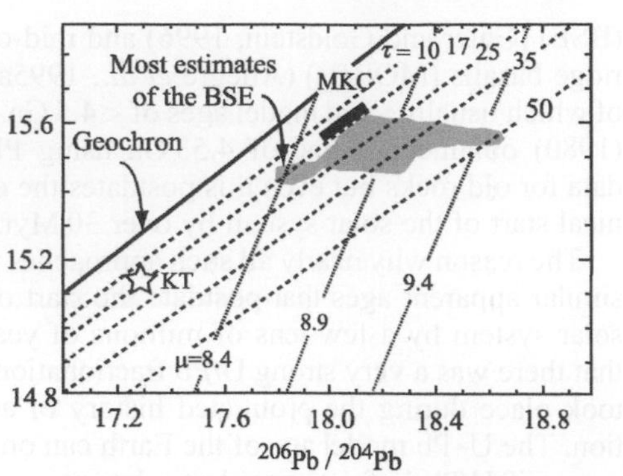

Figure 4 Lead isotopic modeling of the composition of the silicate Earth using continuous core formation. The principles behind the modeling are as in Halliday (2000). See text for explanation. The field for the BSE encompasses all of the estimates in Galer and Goldstein (1996). The values suggested by Kramers and Tolstikhin (1997) and Murphy *et al.* (2003) also are shown. The mean life (τ) is the time required to achieve 63% of the growth of the Earth with exponentially decreasing rates of accretion. The μ values are the $^{238}U/^{204}Pb$ of the BSE. It is assumed that the μ of the total Earth is 0.7 (Allègre *et al.*, 1995a). It can be seen that the lead isotopic composition of the BSE is consistent with protracted accretion over périods of 10^7–10^8 yr.

50 Myr, depending on which composition of the BSE is deployed (Figure 4). A similar figure to this in Halliday (2000) is slightly different (i.e., incorrect) because of a scaling error. The shaded region covers the field defined by eight estimates for the composition of the BSE as summarized by Galer and Goldstein (1996). The star shows the estimate of Kramers and Tolstikhin (1997). The thick bar shows the recent estimate provided by Murphy *et al.* (2003). Regardless of which of these 10 estimates of the lead isotopic composition is used, the mean life for accretion is at least 15 Myr. Therefore, there is no question that the lead isotopic data for the Earth provide evidence of a protracted history of accretion and concomitant core formation as envisaged by Wetherill (1986).

2.4.3 Tungsten Isotopes

While lead isotopes have been useful, the ^{182}Hf–^{182}W chronometer ($T_{1/2}$=9 Myr) has been at least as effective for defining rates of accretion (Halliday, 2000; Halliday and Lee, 1999; Harper and Jacobsen, 1996b; Jacobsen and Harper, 1996; Lee and Halliday, 1996, 1997; Yin *et al.*, 2002). Like U–Pb, the Hf–W system has been used more for defining a model age of core formation (Kramers, 1998; Horan *et al.*, 1998; Kleine *et al.*, 2002; Lee and Halliday, 1995, 1996, 1997; Quitté *et al.*, 2000; Dauphas *et al.*, 2002; Schöenberg *et al.*, 2002). As explained above this is not useful for an object like the Earth.

The half-life renders ^{182}Hf as ideal among the various short-lived chronometers for studying accretionary timescales. Moreover, there are two other major advantages of this method.

(i) Both parent and daughter elements (hafmium and tungsten) are refractory and, therefore, are in chondritic proportions in most accreting objects. Therefore, unlike U–Pb, we think we know the isotopic composition and parent/daughter ratio of the entire Earth relatively well.

(ii) Core formation, which fractionates hafmium from tungsten, is thought to be a very early process as discussed above. Therefore, the rate-limiting process is simply the accretion of the Earth.

There are several recent reviews of Hf–W (e.g., Halliday and Lee, 1999; Halliday *et al.*, 2000), to which the reader can refer for a comprehensive overview of the data and systematics. However, since these were written it has been shown that chondrites, and by inference the average solar system, have tungsten isotopic compositions that are resolvable from that of the silicate Earth (Kleine *et al.*, 2002; Lee and Halliday, 2000a; Schocnberg *et al.*, 2002; Yin *et al.*, 2002). Although the systematics, equations, and arguments have not changed greatly, this has led to considerable uncertainty over the exact initial ^{182}Hf abundance in the early solar system. Because this is of such central importance to our understanding of the timescales of accretion that follow from the data, it is discussed in detail below. Similarly, some of the tungsten isotopic effects that were once considered to reflect radioactive decay within the Moon (Lee *et al.*, 1997; Halliday and Lee, 1999) are now thought to *partly* be caused by production of cosmogenic ^{182}Ta (Leya *et al.*, 2000; Lee *et al.*, 2002).

The differences in tungsten isotopic composition are most conveniently expressed as deviations in parts per 10,000, as follows:

$$\varepsilon_W = \left[\frac{(^{182}W/^{184}W)_{sample}}{(^{182}W/^{184}W)_{BSE}} - 1\right] \times 10^4$$

where the BSE value $(^{182}W/^{184}W)_{BSE}$ is the measured value for an NIST tungsten standard. This should be representative of the BSE as found by comparison with the values for terrestrial standard rocks (Lee and Halliday, 1996; Kleine *et al.*, 2002; Schoenberg *et al.*, 2002). If ^{182}Hf was sufficiently abundant at the time of formation (i.e., at an early age), then minerals, rocks, and reservoirs with higher Hf/W ratios will produce tungsten that is significantly more radiogenic (higher $^{182}W/^{184}W$ or ε_W) compared with the initial tungsten isotopic composition of the solar system. Conversely, metals with low Hf/W that segregate at an early stage from bodies with chondritic Hf/W (as expected for most early planets and planetesimals) will sample unradiogenic tungsten.

Harper *et al.* (1991a) were the first to provide a hint of a tungsten isotopic difference between the iron meteorite Toluca and the silicate Earth. It is now clear that there exists a ubiquitous clearly resolvable deficit in ^{182}W in iron meteorites and the metals of ordinary chondrites, relative to the atomic abundance found in the silicate Earth (Lee and Halliday, 1995, 1996; Harper and Jacobsen, 1996b; Jacobsen and Harper, 1996; Horan *et al.*, 1998). A summary of most of the published data for iron meteorites is given in Figure 5. It can be seen that most early segregated metals are deficient by ~(3–4)ε_W units (300–400 ppm) relative to the silicate Earth. Some appear to be even more negative, but the results are not well resolved. The simplest explanation for this difference is that the metals, or the silicate Earth, or both, sampled early solar system tungsten before live ^{182}Hf had decayed.

The tungsten isotopic difference between early metals and the silicate Earth reflects the time integrated Hf/W of the material that formed the

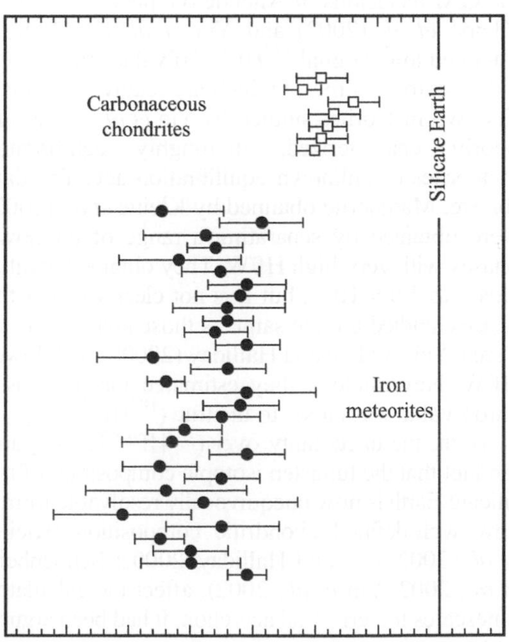

Figure 5 Well-defined deficiency in ^{182}W in early metals and carbonaceous chondrites relative to the silicate Earth (source Lee and Halliday, 1996; Horan *et al.*, 1998; Kleine *et al.*, 2002).

Earth and its reservoirs, during the lifetime of ^{182}Hf. The Hf/W ratio of the silicate Earth is considered to be in the range of 10–40 as a result of an intensive study by Newsom *et al.* (1996). This is an order of magnitude higher than in carbonaceous and ordinary chondrites and a consequence of terrestrial core formation. More recent estimates are provided in Walter *et al.* (2000).

If accretion and core formation were early, an excess of ^{182}W would be found in the silicate Earth, relative to average solar system (chondrites). However, the tungsten isotopic difference between early metals and the silicate Earth on its own does not provide constraints on timing. One needs to know the atomic abundance of ^{182}Hf at the start of the solar system (or the $(^{182}Hf/^{180}Hf)_{BSSI}$, the "bulk solar system initial") and the composition of the chondritic reservoirs from which most metal and silicate reservoirs were segregated. In other words, it is essential to know to what extent the "extra" ^{182}W in the silicate Earth relative to iron meteorites accumulated in the accreted chondritic precursor materials or proto-Earth with an Hf/W ~1 prior to core formation, and to what extent it reflects an accelerated change in isotopic composition because of the high Hf/W (~15) in the silicate Earth.

For this reason some of the first attempts to use Hf–W (Harper and Jacobsen, 1996b; Jacobsen and Harper, 1996) gave interpretations that are now known to be incorrect because the $(^{182}Hf/^{180}Hf)_{BSSI}$ was underconstrained. This is a central concern in Hf–W chronometry that does not apply to U–Pb; for the latter system, parent abundances can still be measured today. In order to determine the $(^{182}Hf/^{180}Hf)_{BSSI}$ correctly one can use several approaches with varying degrees of reliability:

(i) The first approach is to model the expected $(^{182}Hf/^{180}Hf)_{BSSI}$ in terms of nucleosynthetic processes. Wasserburg *et al.* (1994) successfully predicted the initial abundances of many of the short-lived nuclides using a model of nucleosynthesis in AGB stars. Extrapolation of their model predicted a low $(^{182}Hf/^{180}Hf)_{BSSI}$ of $<10^{-5}$, assuming that ^{182}Hf was indeed produced in this manner. Subsequent to the discovery that the $(^{182}Hf/^{180}Hf)_{BSSI}$ was $>10^{-4}$ (Lee and Halliday, 1995, 1996), a number of new models were developed based on the assumption that ^{182}Hf is produced in the same kind of r-process site as the actinides (Wasserburg *et al.*, 1996; Qian *et al.*, 1998; Qian and Wasserburg, 2000).

(ii) The second approach is to measure the tungsten isotopic composition of an early high-Hf/W phase. Ireland (1991) attempted to measure the amount of ^{182}W in zircons (with very high hafnium content) from the mesosiderite Vaca Muerta, using an ion probe, and from this deduced that the $(^{182}Hf/^{180}Hf)_{BSSI}$ was $<10^{-4}$. Unfortunately, these zircons are not dated with sufficient precision (Ireland and Wlotzka, 1992) to be very certain about the time extrapolation of the exact hafnium abundances. Nevertheless, on the basis of this work and the model of Wasserburg *et al.* (1994), Jacobsen and Harper (1996) assumed that the $(^{182}Hf/^{180}Hf)_{BSSI}$ was indeed low ($\sim 10^{-5}$). It was concluded that the difference in tungsten isotopic composition between the iron meteorite Toluca and the terrestrial value could only have been produced by radioactive decay within the silicate Earth with high Hf/W. Therefore, the fractionation of Hf/W produced by terrestrial core formation had to be early. They predicted that the Earth accreted very rapidly with a two-stage model age of core formation of <15 Myr after the start of the solar system.

(iii) The third approach is to simply assume that the consistent unradiogenic tungsten found in iron meteorites and metals from ordinary chondrites represents the initial tungsten isotopic composition of the solar system. This is analogous to the way in which the lead isotopic composition of iron meteorites has been used for decades. The difference between this and the present-day value of carbonaceous chondrites represents the effects of radiogenic ^{182}W growth *in situ* with chondritic Hf/W ratios. This in turn will indicate the $(^{182}Hf/^{180}Hf)_{BSSI}$. The difficulty with this approach has been to correctly determine the tungsten isotopic composition of chondrites. Using multiple collector ICPMS, Lee and Halliday (1995) were the first to publish such results and they reported that the tungsten isotopic compositions of the carbonaceous chondrites Allende and Murchison could not be resolved from that of the silicate Earth. The $(^{182}Hf/^{180}Hf)_{BSSI}$, far from being low, appeared to be surprisingly high at $\sim 2\times10^{-4}$ (Lee and Halliday, 1995, 1996). Subsequently, it has been shown that these data must be incorrect (Figure 5). Three groups have independently shown (Kleine *et al.*, 2002; Schoenberg *et al.*, 2002; Yin *et al.*, 2002) that there is a small but clear deficiency in ^{182}W ($\varepsilon_W = -1.5$ to -2.0) in carbonaceous chondrites similar to that found in enstatite chondrites (Lee and Halliday, 2000a) relative to the BSE. Kleine *et al.* (2002), Schoenberg *et al.* (2002), and Yin *et al.* (2002) all proposed a somewhat lower $(^{182}Hf/^{180}Hf)_{BSSI}$ of $\sim 1.0\times10^{-4}$. However, the same authors based this result on a solar system initial tungsten isotopic composition of $\varepsilon_W = -3.5$, which was derived from their own, rather limited, measurements. Schoenberg *et al.* (2002) pointed out that if instead one uses the full range of tungsten isotopic composition previously reported for iron meteorites (Horan *et al.*, 1998; Jacobsen and Harper, 1996; Lee and Halliday, 1995, 1996) one obtains a $(^{182}Hf/^{180}Hf)_{BSSI}$ of $>1.3\times10^{-4}$ (Table 1).

Table 1 Selected W-isotope data for iron meteorites.

Iron meteorite	$\varepsilon^{182}W$	$^{182}Hf/^{180}Hf$
Bennett Co.	-4.6 ± 0.9	$(1.74\pm0.57)\times10^{-4}$
Lombard	-4.3 ± 0.3	$(1.55\pm0.37)\times10^{-4}$
Mt. Edith	-4.5 ± 0.6	$(1.68\pm0.46)\times10^{-4}$
Duel Hill-1854	-5.1 ± 1.1	$(2.07\pm0.68)\times10^{-4}$
Tlacotopec	-4.4 ± 0.4	$(1.68\pm0.41)\times10^{-4}$

Source: Horan *et al.* (1998). The calculated difference between the initial and present-day W-isotopic composition of the solar system is equal to the $^{180}Hf/^{184}W$ of the solar system multiplied by the $(^{182}Hf/^{180}Hf)_{BSSI}$. The W-isotopic compositions of iron meteorites are maxima for the $(\varepsilon^{182}W)_{BSSI}$, and therefore provide a limit on the minimum $(^{182}Hf/^{180}Hf)_{BSSI}$. The $^{182}Hf/^{180}Hf$ at the time of formation of these early metals shown here is calculated from the W-isotopic composition of the metal, the W-isotopic composition of carbonaceous chondrites (Kleine *et al.*, 2002), and the average $^{180}Hf/^{184}W$ for carbonaceous chondrites of 1.34 (Newsom *et al.*, 1996).

(iv) The fourth approach is to determine an internal isochron for an early solar system object with a well-defined absolute age (Swindle, 1993). The first such isochron was for the H4 ordinary chondrite Forest Vale (Lee and Halliday, 2000a). The best-fit line regressed through these data corresponds to a slope ($=^{182}Hf/^{180}Hf$) of $(1.87\pm0.16)\times10^{-4}$. The absolute age of tungsten equilibration in Forest Vale is unknown but may be 5 Myr younger than the CAI inclusions of Allende (Göpel *et al.*, 1994). Kleine *et al.* (2002) and Yin *et al.* (2002) both obtained lower initial $^{182}Hf/^{180}Hf$ values from internal isochrons, some of which are relatively precise. The two meteorites studied by Yin *et al.* (2002) are poorly characterized, thoroughly equilibrated meteorites of unknown equilibration age. The data for Ste. Marguerite obtained by Kleine *et al.* (2002) were obtained by separating a range of unknown phases with very high Hf/W. They obtained a value closer to 1.0×10^{-4}, but it is not clear whether the phases studied are the same as those analyzed from Forest Vale by Lee and Halliday (2000b) with lower Hf/W. Nevertheless, they estimated that their isochron value was closer to the true $(^{182}Hf/^{180}Hf)_{BSSI}$.

Both, the uncertainty over $(^{182}Hf/^{180}Hf)_{BSSI}$ and the fact that the tungsten isotopic composition of the silicate Earth is now unequivocally resolvable from a now well-defined chondritic composition (Kleine *et al.*, 2002; Lee and Halliday, 2000a; Schoenberg *et al.*, 2002; Yin *et al.*, 2002), affect the calculated timescales for terrestrial accretion. It had been argued that accretion and core formation were fairly protracted and characterized by equilibration between accreting materials and the silicate Earth (Halliday, 2000; Halliday *et al.*, 1996, 2000; Halliday and Lee, 1999). In other words, the tungsten isotope data provide very strong confirmation of the models of Safronov (1954) and Wetherill (1986). This general scenario remains the same with the new data but in detail there are changes to the exact timescales.

Previously, Halliday (2000) estimated that the mean life, the time required to accumulate 63% of the Earth's mass with exponentially decreasing accretion rates, must lie in the range of 25–40 Myr based on the combined constraints imposed by the tungsten and lead isotope data for the Earth. Yin *et al.* (2002) have argued that the mean life for Earth accretion is more like ~11 Myr based on their new data for chondrites. The lead isotope data for the Earth are hard to reconcile with such rapid accretion rates as already discussed (Figure 4). Therefore, at present there is an unresolved apparent discrepancy between the models based on tungsten and those based on lead isotope data. Resolving this discrepancy highlights the limitations in both the tungsten and the lead isotope modeling. Here are some of the most important weaknesses to be aware of:

(i) The U/Pb ratio of the total Earth is poorly known.

(ii) In all of these models it is assumed that the Earth accretes at exponentially decreasing rates. Although the exponentially decreasing rate of growth of the Earth is based on Monte Carlo simulations and makes intuitive sense given the ever decreasing probability of collisions, the reality cannot be this simple. As planets get bigger, the average size of the objects with which they collide also must increase. As such, the later stages of planetary accretion are thought to involve major collisions. This is a stochastic process that is hard to predict and model. It means that the current modeling can only provide, at best, a rough description of the accretion history.

(iii) The Moon is thought to be the product of such a collision. The Earth's U/Pb ratio conceivably might have increased during accretion if a fraction of the moderately volatile elements were lost during very energetic events like the Moon-forming giant impact (Figure 6).

(iv) Similarly, as the objects get larger, the chances for equilibration of metal and silicate would seem to be less likely. This being the case, the tungsten and lead isotopic composition of the silicate Earth could reflect only partial equilibration with incoming material such that the tungsten and lead isotopic composition is partly inherited. This has been modeled in detail by Halliday (2000) in the context of the giant impact and more recently has been studied by Vityazev *et al.* (2003) and Yoshino *et al.* (2003) in the context of equilibration of asteroidal-sized objects. If correct, it would mean accretion was even slower than can be deduced from tungsten or lead isotopes. If lead equilibrated more readily than tungsten did, for whatever reason, it might help explain some of the discrepancy. One possible way to decouple lead from tungsten would be by their relative volatility. Lead could

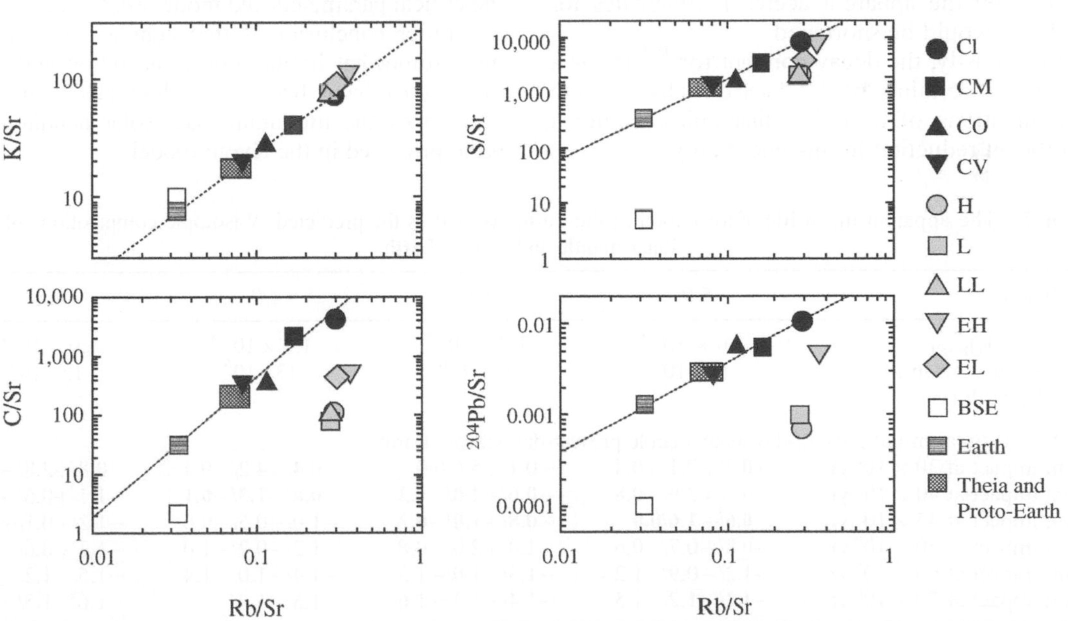

Figure 6 Volatile/refractory element ratio–ratio plots for chondrites and the silicate Earth. The correlations for carbonaceous chondrites can be used to define the composition of the Earth, the Rb/Sr ratio of which is well known, because the strontium isotopic composition of the BSE represents the time-integrated Rb/Sr. The BSE inventories of volatile siderophile elements carbon, sulfur, and lead are depleted by more than one order of magnitude because of core formation. The values for Theia are time-integrated compositions, assuming time-integrated Rb/Sr deduced from the strontium isotopic composition of the Moon (Figure 8) can be used to calculate other chemical compositions from the correlations in carbonaceous chondrites (Halliday and Porcelli, 2001). Other data are from Newsom (1995).

have been equilibrated by vapor-phase exchange, while tungsten would not have been able to do this and would require intimate physical mixing and reduction to achieve equilibration.

(v) $(^{182}Hf/^{180}Hf)_{BSSI}$ is, at present, very poorly defined and could be significantly higher (Table 1). This would result in more protracted accretion timescales deduced from Hf–W (Table 2), which would be more consistent with the results obtained from U–Pb. An improved and more reliable Hf–W chronometry will depend on the degree to which the initial hafnium and tungsten isotopic compositions at the start of the solar system can be accurately defined (Tables 1 and 2). Techniques must be developed for studying very early objects like CAIs.

(vi) There is also a huge range of uncertainty in the Hf/W ratio of the silicate Earth (Newsom *et al.*, 1996), with values ranging between 10 and 40. The tungsten isotope age calculations presented in the literature tend to assume a value at the lower end of this range. Adoption of higher values also would result in more protracted accretion timescales based on Hf–W. The most recent independent estimates (Walter *et al.*, 2000) are significantly higher than those used by Yin *et al.* (2002) and Kleine *et al.* (2002) in support of their proposed timescales.

(vii) The lead isotopic composition of the BSE is not well defined (Figure 4). If the correct value lies closer to the Geochron than previously recognized, then the apparent accretion timescales for the Earth would be shortened.

(viii) Lastly, the decay constant for ^{182}Hf has a reported uncertainty of ±22% and really accurate determinations of accretion timescales require a significant reduction in this uncertainty.

Having made all of these cautionary statements, one still can state something useful about the overall accretion timescales. All recent combined accretion/continuous core formation models (Halliday, 2000; Halliday *et al.*, 2000; Yin *et al.*, 2002) are in agreement that the timescales are in the range 10^7–10^8 yr, as predicted by Wetherill (1986). Therefore, we can specifically evaluate the models of planetary accretion proposed earlier as follows.

If the Earth accreted very fast, in $<10^6$ yr, as proposed by Cameron (1978), the silicate Earth would have a tungsten isotopic composition that is vastly more radiogenic than that observed today (Figure 5). Such objects would have $\varepsilon_W > +10$, rather than 0 (just two ε units above average solar system). Therefore, we can say with some confidence that this model does not describe the accretion of the Earth. Protracted accretion in the absence of nebular gas, as proposed by Safronov and Wetherill, is very consistent with the close agreement between chondrites and the silicate Earth (Figure 4). To what extent the Kyoto model, which involves a significant amount of nebular gas (Hayashi *et al.*, 1985), can be confirmed or discounted is unclear at present. However, even the timescales presented by Yin *et al.* (2002) are long compared with the 5 Myr for accretion of the Earth predicted by the Kyoto model. This could change somewhat with further tungsten and lead isotope data and the definition of the critical parameters and modeling. However, the first-order conclusion is that nebular gas was at most somewhat limited during accretion and that there must have been much less gas than that implied by the minimum mass solar nebula scenario proposed in the Kyoto model.

Table 2 The apparent mean life of formation of the Earth, as well as the predicted W-isotopic compositions of the lunar mantle and silicate Earth.

$\varepsilon^{182}W_{BSSI}$	−5.0	−4.5	−4.0	−3.5
$(^{182}Hf/^{180}Hf)_{BSSI}$	2.0×10^{-4}	1.7×10^{-4}	1.4×10^{-4}	1.0×10^{-4}
Accretionary mean life of the earth (yr)	15×10^6	14×10^6	13×10^6	12×10^6
ε^{182}W for: lunar initial/present-day lunar mantle/present-day silicate Earth				
Giant impact at 30×10^6 yr	+0.3/+7.1/+0.4	−0.1/+5.6/0.0	−0.4/+4.2/−0.4	−0.8/+2.8/−0.7
Giant impact at 40×10^6 yr	−0.3/+2.9/+0.8	**−0.6/+2.0/+0.3**	**−0.8/+1.3/−0.1**	−1.1/+0.6/−0.5
Giant impact at 45×10^6 yr	**−0.6/+1.6/0.0**	**−0.8/+1.0/−0.3**	−1.0/+0.5/−0.6	−1.2/−0.1/−0.9
Giant impact at 50×10^6 yr	−0.8/+0.7/−0.6	−1.0/+2.0/−0.8	−1.2/−0.2/−1.0	−1.3/−0.6/−1.2
Giant impact at 60×10^6 yr	−1.2/−0.9/−1.2	−1.3/−1.0/−1.3	−1.4/−1.0/−1.4	−1.5/−1.2/−1.6
Giant impact at 70×10^6 yr	−1.3/−1.2/−1.5	−1.4/−1.3/−1.6	−1.5/−1.4/−1.7	−1.6/−1.5/−1.7

All parameters are critically dependent on the initial W- and Hf- isotopic composition of the solar system, which at present are poorly known (see text). All calculated values assume that the W depletion in the silicate Earth and the lunar mantle are as given in Walter *et al.* (2000). The $(^{182}Hf/^{180}Hf)_{BSSI}$, model ages, and Earth and Moon compositions are all calculated using the W-isotopic composition of carbonaceous chondrites (Kleine *et al.*, 2002) and the average $^{180}Hf/^{184}W$ for carbonaceous chondrites of 1.34 (Newsom, 1990). The principles behind the modeling are as in Halliday (2000). The giant impact is assumed to have occurred when the Earth had reached 90% of its current mass, the impactor adding a further 9%. Exponentially decreasing accretion rates are assumed before and after (Halliday, 2000). The solutions in bold type provide the best match with the current data for the W-isotopic compositions of the initial Moon (~0), and the present-day lunar mantle (<2) and silicate Earth (0).

2.5 CHEMICAL AND ISOTOPIC CONSTRAINTS ON THE NATURE OF THE COMPONENTS THAT ACCRETED TO FORM THE EARTH

2.5.1 Chondrites and the Composition of the Disk from Which Earth Accreted

A widespread current view is that chondrites represent primitive undifferentiated material from which the Earth accreted. In reality, chondrites are anything but simple and, although they contain early and presolar objects, how primitive and how representative of the range of very early planetesimals they are is completely unclear. In all probability the Earth was built largely from more extensively differentiated materials (Taylor and Norman, 1990). Nevertheless, chondrites do represent a useful set of reference reservoirs in chemical and isotopic terms from which one can draw some conclusions about the "average stuff" from which the Earth may have been built (e.g., Ganapathy and Anders, 1974; Anders, 1977; Wolf *et al.*, 1980). Indeed, if geochemists did not have chondrite samples to provide a reference, many geochemical arguments about Earth's origin, not to say present-day interior structure, would be far weaker.

The degree to which different kinds of chondrites reflect bulk Earth composition has been extensively debated, since none of them provide a good match. CI carbonaceous chondrites define a reference reservoir for *undepleted* solar system compositions because they are strikingly similar in composition to the Sun when normalized to an element such as silicon (Grevesse and Sauval, 1998). However, they are not at all similar to the Earth's volatile-depleted composition. Nor are they much like the vast majority of other chondrites! Confirmation of the primitive nature of CIs is most readily demonstrated by comparing the relative concentrations of a well-determined moderately volatile major element to a well-determined refractory major element with the values in the Sun's photosphere. For example, the concentrations of sodium and calcium are both known to better than a percent in the photosphere (Grevesse and Sauval, 1998) and are identical to within a percent with the values found in CIs despite a big difference in volatility between the two elements (Anders and Grevesse, 1989; Newsom, 1995). It is tempting to ascribe CIs to a complete condensation sequence. However, they also contain dispersed isotopic anomalies in Cr, for example, that indicate a lack of homogenization on a fine scale (Podosek *et al.*, 1997). This cannot be reconciled with nebular condensation from high temperatures.

Volatile-element depletion patterns in other (e.g., CV, CM, or CO) carbonaceous chondrites (Larimer and Anders, 1970; Palme *et al.*, 1988; Takahashi *et al.*, 1978; Wolf *et al.*, 1980) and ordinary chondrites (Wasson and Chou, 1974) have long been considered to partly reflect incomplete condensation (Wasson, 1985), with the more volatile elements removed from the meteorite formation regions before cooling and the termination of condensation. However, this depletion must at least partially reflect the incorporation of volatile-depleted CAIs (Guan *et al.*, 2000) and chondrules (Grossman, 1996; Meibom *et al.*, 2000). The origins of CAIs have long been enigmatic, but recently it has been proposed that CAIs may be the product of rather localized heating close to the young Sun (Shu *et al.*, 1997) following which they were scattered across the disk. Regardless of whether this model is exactly correct, the incorporation of volatile-depleted CAIs must result in some degree of volatile-depleted composition for chondritic meteorites. Chondrules are the product of rapid melting of chondritic materials (Connolly *et al.*, 1998; Connolly and Love, 1998; Desch and Cuzzi, 2000; Jones *et al.*, 2000) and such events also may have been responsible for some degree of volatile-element depletion (Vogel *et al.*, 2002).

The discovery that pristine presolar grains are preserved in chondrites indicates that they were admixed at a (relatively) late stage. Many of the presolar grain types (Mendybaev *et al.*, 2002; Nittler, 2003) could not have survived the high temperatures associated with CAI and chondrule formation. Therefore, chondrites must include a widely dispersed presolar component (Huss, 1988, 1997; Huss and Lewis, 1995) that either settled or was swept into the chondrule-forming and/or chondrule-accumulating region and would have brought with it nebular material that was not depleted in volatiles. Therefore, volatile depletion in chondrites must reflect, to some degree at least, the mixing of more refractory material (Guan *et al.*, 2000; Kornacki and Fegley, 1986) with CI-like matrix of finer grained volatile-rich material (Grossman, 1996). CIs almost certainly represent accumulations of mixed dust inherited from the portion of the protostellar molecular cloud that collapsed to form the Sun within a region of the nebula that was never greatly heated and did not accumulate chondrules and CAIs (for whatever reason). Though undifferentiated, chondrites, with the possible exception of CIs, are not primitive and certainly do not represent the first stages of accretion of the Earth.

Perhaps the best current way to view the disk of debris from which the Earth accreted was as an environment of vigorous mixing. Volatile-depleted material that had witnessed very high temperatures at an early stage (CAIs) mixed with material that had been flash melted (chondrules) a few million years later. Then presolar grains that had escaped these processes rained into the midplane or more likely were swept in from the outer regions of the solar system. It is within these environments that dust, chondrules and CAIs mixed and accumulated into planetesimals. There also would have been earlier formed primary planetesimals (planetary embryos) that had accreted rapidly by runaway growth but were subsequently consumed. The earliest most primitive objects probably differentiated extremely quickly aided by the heat from runaway growth and live ^{26}Al.

It is not clear whether we have *any* meteorites that are samples of these very earliest planetesimals. Some iron meteorites with very unradiogenic tungsten (Horan *et al.*, 1998) might be candidates (Table 1). Nearly all silicate-rich basaltic achondrites that are precisely dated appear to have formed later (more than a million years after the start of the solar system). Some of the disk debris will have been the secondary product of collisions between these planetesimals, rather like the dust in Beta Pictoris. Some have proposed that chondrules formed in this way (Sanders, 1996; Lugmair and Shukolyukov, 2001), but although the timescales appear right this is not generally accepted (Jones *et al.*, 2000). A working model of the disk is of a conveyor belt of material rapidly accreting into early planetesimals, spiraling in toward the Sun and being fed by heated volatile-depleted material (Krot *et al.*, 2001) scattered outwards and pristine volatile-rich material being dragged in from the far reaches of the disk (Shu *et al.*, 1996). This mixture of materials provided the raw ingredients for early planetesimals and planets that have largely been destroyed or ejected. One of the paths of destruction was mutual collisions, and it was these events that ultimately led to the growth of terrestrial planets like the Earth.

2.5.2 Chondritic Component Models

Although it is now understood quite well that chondrites are complicated objects with a significant formation history, many have proposed that they should be used to define Earth's bulk composition. More detailed discussion of the Earth's composition is provided elsewhere in this treatise and only the essentials for this chapter are covered here. A recent overview of some of the issues is provided by Drake and Righter (2002). The carbonaceous chondrites often are considered to provide the best estimates of the basic building blocks (Agee and Walker, 1988; Allègre *et al.*, 1995a,b, 2001; Anders, 1977; Ganapathy and Anders, 1974; Herzberg, 1984; Herzberg *et al.*, 1988; Jagoutz *et al.*, 1979; Jones and Palme, 2000; Kato *et al.*, 1988a,b; Newsom, 1990). Because the Earth is formed from collisions between differentiated material, it is clear that one is simply using these reference chondrites to provide little more than model estimates of the total Earth's composition. Allègre *et al.* (1995a,b, 2001), for example, have plotted the various compositions of chondrites using major and trace element ratios and have shown that the refractory lithophile elements are most closely approximated by certain specific kinds of chondrites. Relative abundances of the platinum group elements and osmium isotopic composition of the silicate Earth have been used extensively to evaluate which class of chondrites contributed the late veneer (Newsom, 1990; Rehkämper *et al.*, 1997; Meisel *et al.*, 2001; Drake and Righter, 2002), a late addition of volatile rich material. Javoy (1998, 1999) has developed another class of models altogether based on enstatite chondrites.

If it is assumed that one can use chondrites as a reference, one can calculate the composition of the total Earth and predict the concentrations of elements that are poorly known. For example, in theory one can predict the amount of silicon in the total Earth and determine from this how much must have gone into the core. The Mainz group produced many of the classic papers pursuing this approach (Jagoutz *et al.*, 1979). However, it was quickly realized that the Earth does not fit any class of chondrite. In particular, Jagoutz presented the idea of using element ratios to show that the Earth's upper mantle had Mg/Si that was nonchondritic. Jagoutz *et al.* (1979) used CI, ordinary and enstatite chondrites to define the bulk Earth. Anders always emphasized that all chondrites were fractionated relative to CIs in his series of papers on "Chemical fractionations in meteorites" (e.g., Wolf *et al.*, 1980). That is, CMs and CVs, etc., are also fractionated, as are planets. Jagoutz *et al.* (1979) recognized that the fractionations of magnesium, silicon, and aluminum were found in both the Earth and in CI, O, and E chondrites, but not in C2–C3 chondrites, which exhibit a fixed Mg/Si but variable Al/Si ratio, probably because of addition of CAIs.

One also can use chondrites to determine the abundances of moderately volatile elements such as potassium. Allègre *et al.* (2001) and Halliday and Porcelli (2001) pursued this approach to show that the Earth's K/U is $\sim 10^4$ (Figure 2). One can calculate how much of the volatile chalcophile and siderophile elements such as sulfur, cadmium, tellurium, and lead may have gone in to the Earth's core (Figure 6; Yi *et al.*, 2000; Allègre *et al.*, 1995b; Halliday and Porcelli, 2001). Allègre *et al.* (1995a) also have used this approach to estimate the total Earth's $^{238}U/^{204}Pb$.

From the budgets of potassium, silicon, carbon, and sulfur extrapolated from carbonaceous chondrite compositions, one can evaluate the amounts of various light elements possibly incorporated in the core (Allègre *et al.*, 1995b; Halliday and

Porcelli, 2001) and see if this explains the deficiency in density relative to pure iron (Ahrens and Jeanloz, 1987). Such approaches complement the results of experimental solubility measurements (e.g., Wood, 1993; Gessmann *et al.*, 2001) and more detailed comparisons with the geochemistry of meteorites (Dreibus and Palme, 1996).

It should be noted that the Jagoutz *et al.* (1979) approach and the later studies by Allègre and co-workers (Allègre *et al.*, 1995a,b, 2001) are mutually contradictory, a fact that is sometimes not recognized. Whereas Jagoutz *et al.* (1979) used CI, ordinary, and enstatite chondrites, Allègre and co-workers used CI, CM, CV, and CO chondrites (the carbonaceous chondrite mixing line) to define the bulk earth, ignoring the others.

These extrapolations and predictions are based on two assumptions that some would regard as rendering the approach flawed:

(i) The carbonaceous chondrites define very nice trends for many elements that can be used to define the Earth's composition, but the ordinary and enstatite chondrites often lie off these trends (Figure 6). There is no obvious basis for simply ignoring the chemical compositions of the ordinary and enstatite chondrites unless they have undergone some parent body process of loss or redistribution in their compositions. This is indeed likely, particularly for volatile chalcophile elements but is not well understood at present.

(ii) The Earth as well as the Moon, Mars, and basaltic achondrite parent bodies are depleted in moderately volatile elements, in particular the alkali elements potassium and rubidium, relative to *most* classes of chondrite (Figure 2). This needs to be explained. Humayun and Clayton (1995) performed a similar exercise, showing that since chondrites are all more volatile rich than terrestrial planets, the only way to build planets from chondrite precursors was to volatilize alkalis and other volatile elements. This was not considered feasible in view of the identical potassium isotope compositions of chondrites, the Earth and the Moon. Taylor and Norman (1990) made a similar observation that planets formed from volatile-depleted and differentiated precursors and not from chondrites. Clearly, there must have been other loss mechanisms (Halliday and Porcelli, 2001) beyond those responsible for the volatile depletion in chondrites unless the Earth, Moon, angrite parent body and Vesta simply were accreted from parts of the disk that were especially enriched in some highly refractory component, like CAIs (Longhi, 1999).

This latter point and other similar concerns represent such a major problem that some scientists abandoned the idea of just using chondrites as a starting point and instead invoked the former presence of completely hypothetical components in the inner solar system. This hypothesis has been explored in detail as described below.

2.5.3 Simple Theoretical Components

In addition to making comparisons with chondrites, the bulk composition of the Earth also has been defined in terms of a "model" mixture of highly reduced, refractory material combined with a much smaller proportion of a more oxidized volatile-rich component (Wänke, 1981). These models follow on from the ideas behind earlier heterogeneous accretion models. According to these models, the Earth was formed from two components. Component A was highly reduced and free of all elements with equal or higher volatility than sodium. All other elements were in CI relative abundance. The iron and siderophile elements were in metallic form, as was part of the silicon. Component B was oxidized and contained all elements, including those more volatile than sodium in CI relative abundance. Iron and all siderophile and lithophile elements were mainly in the form of oxides.

Ringwood (1979) first proposed these models but the concept was more fully developed by Wänke (1981). In Wänke's model, the Earth accretes by heterogeneous accretion with a mixing ratio A:B$\sim$85:15. Most of component B would be added after the Earth had reached about two thirds of its present mass. The oxidized volatile-rich component would be equivalent to CI carbonaceous chondrites. However, the reduced refractory rich component is hypothetical and never has been identified in terms of meteorite components.

Eventually, models that involved successive changes in accretion and core formation replaced these. How volatiles played into this was not explained except that changes in oxidation state were incorporated. An advanced example of such a model is that presented by Newsom (1990). He envisaged the history of accretion as involving stages that included concomitant core formation stages (discussed under core formation).

2.5.4 The Nonchondritic Mg/Si of the Earth's Primitive Upper Mantle

It has long been unclear why the primitive upper mantle of the Earth has a nonchondritic proportion of silicon to magnesium. Anderson (1979) proposed that the mantle was layered with the lower mantle having higher Si/Mg. Although a number of coeval papers presented a similar view (Herzberg, 1984; Jackson, 1983) it is thought by many geochemists these days that the major element composition of the mantle is, broadly speaking, well mixed and homogeneous as a result of 4.5 Gyr of mantle convection (e.g., Hofmann, 1988; Ringwood, 1990).

A notable exception is the enstatite chondrite model of Javoy (1999).

Wänke (1981) and Allègre *et al.* (1995b) have proposed that a significant fraction of the Earth's silicon is in the core. However, this is not well supported by experimental data. To explain the silicon deficiency this way conditions have to be so reducing that niobium would be siderophile and very little would be left in the Earth's mantle (Wade and Wood, 2001).

Ringwood (1989a) proposed that the nonchondritic Mg/Si reflected a radial zonation in the solar system caused by the *addition* of more volatile silicon to the outer portions of the solar system. That is, he viewed the Earth as possibly more representative of the solar system than chondrites.

Hewins and Herzberg (1996) proposed that the midplane of the disk from which the Earth accreted was dominated by chondrules. Chondrules have higher Mg/Si than chondrites and if the Earth accreted in a particular chondrule rich area (as it may well have done) the sorting effect could dominate planetary compositions.

2.5.5 Oxygen Isotopic Models and Volatile Losses

Some have proposed that one can use the oxygen isotopic composition of the Earth to identify the proportions of different kinds of chondritic components (e.g., Lodders, 2000). The isotopic composition of oxygen is variable, chiefly in a mass-dependent way, in terrestrial materials. However, meteorites show mass-independent variations as well (Clayton, 1986, 1993). Oxygen has three isotopes and a three-isotope system allows discrimination between mass-dependent planetary processes and mass-independent primordial nebular heterogeneities inherited during planet formation (Clayton and Mayeda, 1996; Franchi *et al.*, 1999; Wiechert *et al.*, 2001, 2003). Clayton pursued this approach and showed that the mix of meteorite types based on oxygen isotopes would not provide a chemical composition that was similar to that of the Earth. In particular, the alkalis would be too abundant. In fact, the implied relative proportions of volatile-rich and volatile-poor constituents are in the opposite sense of those derived in the Wänke–Ringwood mixing models mentioned above (Clayton and Mayeda, 1996). Even the differences in composition between the planets do not make sense. Earth would need to be less volatile-depleted than Mars, whereas the opposite is true (Clayton, 1993; Halliday *et al.*, 2001). The oxygen isotope compositions of Mars (based on analyses of SNC meteorites) show a *smaller* proportion of carbonaceous-chondrite-like material in Mars than in Earth (Clayton, 1993; Halliday *et al.*, 2001). It thus appears that, unless the Earth lost a significant proportion of its moderately volatile elements after it formed, the principal carrier of moderately volatile elements involved in the formation of the terrestrial planets was not of carbonaceous chondrite composition. This, of course, leads to the viewpoint that the Earth was built from volatile-depleted material, such as differentiated planetesimals (Taylor and Norman, 1990), but it still begs the question of how volatile depletion occurred (Halliday and Porcelli, 2001).

Whether it is plausible, given the dynamics of planetary accretion discussed above, that the Earth lost a major fraction of its moderately volatile elements during its accretion history is unclear. This has been advanced as an explanation by Lodders (2000) and is supported by strontium isotope data for early solar system objects discussed below (Halliday and Porcelli, 2001). The difficulty is to come up with a mechanism that does not fractionate K isotopes (Humayun and Clayton, 1995) and permits loss of heavy volatile elements. The degree to which lack of fractionation of potassium isotopes offers a real constraint depends on the mechanisms involved (Esat, 1996; Young, 2000). There is no question that as the Earth became larger, the accretion dynamics would have become more energetic and the temperatures associated with accretion would become greater (Melosh and Sonett, 1986; Melosh and Vickery, 1989; Ahrens, 1990; Benz and Cameron, 1990; Melosh, 1990; Melosh *et al.*, 1993). However, the gravitational pull of the Earth would have become so large that it would be difficult for the Earth to lose these elements even if they were degassed into a hot protoatmosphere. This is discussed further below.

2.6 EARTH'S EARLIEST ATMOSPHERES AND HYDROSPHERES

2.6.1 Introduction

The range of possibilities to be considered for the nature of the earliest atmosphere provides such a broad spectrum of consequences for thermal and magmatic evolution that it is better to consider the atmospheres first before discussing other aspects of Earth's evolution. Therefore, in this section a brief explanation of the different kinds of early atmospheres and their likely effects on the Earth are given in cursory terms. More comprehensive information on atmospheric components is found elsewhere in this treatise.

2.6.2 Did the Earth Have a Nebular Protoatmosphere?

Large nebular atmospheres have at various times been considered a fundamental feature of the early Earth by geochemists (e.g., Sasaki and Nakazawa, 1988; Pepin, 2000; Porcelli *et al.*, 1998). Large amounts of nebular gases readily explain why the Earth has primordial ^{3}He (Clarke *et al.*, 1969; Mamyrin *et al.*, 1969; Lupton and Craig, 1975; Craig and Lupton, 1976) and why a component of solar-like neon with a solar He/Ne ratio can be found in some plume basalts (Honda *et al.*, 1991; Dixon *et al.*, 2000; Moreira *et al.*, 2001). Evidence for a solar component among the heavier noble gases has been more scant (Moreira and Allègre, 1998; Moreira *et al.*, 1998). There is a hint of a component with different $^{38}Ar/^{36}Ar$ in some basalts (Niedermann *et al.*, 1997) and this may reflect a solar argon component (Pepin and Porcelli, 2002). Also, Caffee *et al.* (1988, 1999) have made the case that a solar component of xenon can be found in well gases.

The consequences for the early behavior of the Earth are anticipated to be considerable if there was a large nebular atmosphere. Huge reducing protoatmospheres, be they nebular or impact-induced can facilitate thermal blanketing, magma oceans and core formation. For example, based on the evidence from helium and neon, Harper and Jacobsen (1996b) suggested that iron was reduced to form the core during a stage with a massive early H_2–He atmosphere. A variety of authors had already proposed that the Earth accreted with a large solar nebular atmosphere. Harper and Jacobsen's model builds upon many earlier such ideas primarily put forward by the Kyoto school (Hayashi *et al.*, 1979, 1985; Mizuno *et al.*, 1980; Sasaki and Nakazawa, 1988; Sasaki, 1990). The thermal effects of a very large protoatmosphere have been modeled by Hayashi *et al.* (1979), who showed that surface temperatures might reach >4,000 K. Sasaki (1990) also showed that incredibly high temperatures might build up in the outer portions of the mantle, leading to widespread magma oceans. Dissolving volatiles like noble gases into early silicate and metal liquids may have been quite easy under these circumstances (Mizuno *et al.*, 1980; Porcelli *et al.*, 2001; Porcelli and Halliday, 2001).

Porcelli and Pepin (2000) and Porcelli *et al.* (2001) recently summarized the noble gas arguments pointing out that first-order calculations indicate that significant amounts of noble gases with a solar composition are left within the Earth's interior but orders of magnitude more have been lost, based on xenon isotopic evidence (see Chapter 6). Therefore, one requires a relatively large amount of nebular gas during Earth's accretion. To make the Earth this way, the timescales for accretion need to be characteristically short ($\sim 10^6$–10^7 yr) in order to trap such amounts of gas before the remains of the solar nebula are accreted into the Sun or other planets. Therefore, a problem may exist reconciling the apparent need to acquire a large nebular atmosphere with the longer timescales ($\sim 10^7$–10^8 yr) for accretion implied by Safronov–Wetherill models and by tungsten and lead isotopic data. It is hard to get around this problem because the nebular model is predicated on the assumption that the Earth grows extremely fast such that it can retain a large atmosphere (Pepin and Porcelli, 2002).

Therefore, other models for explaining the incorporation of solar-like noble gases should be considered. The most widely voiced alternative is that of accreting material that formed earlier elsewhere that already had acquired solar-like noble gases. For example, it has been argued that the neon is acquired as "Ne–B" from accretion of chondritic material (Trieloff *et al.*, 2000). With such a component in meteorites, one apparently could readily explain the Earth's noble gas composition, which may have a $^{20}Ne/^{22}Ne$ ratio that is lower than solar (Farley and Poreda, 1992). However, this component is no longer well defined in meteorites and the argument is less than certain because of this. In fact, Ne–B is probably just a fractionated version of solar (Ballentine *et al.*, 2001). This problem apart, the idea that the Earth acquired its solar-like noble gases from accreting earlier-formed objects is an alternative that is worth considering. Chondrites, however, mainly contain very different noble gases that are dominated by the so-called "Planetary" component (Ozima and Podosek, 2002). This component is nowadays more precisely identified as "Phase Q" (Wieler *et al.*, 1992; Busemann *et al.*, 2000). In detail, the components within chondrites show little sign of having incorporated large amounts of solar-like noble gases. Early melted objects like CAIs and chondrules are, generally speaking, strongly degassed (Vogel *et al.*, 2002, 2003).

Podosek *et al.* (2003) recently have proposed that the noble gases are incorporated into early formed planetesimals that are irradiated by intense solar-wind activity from the vigorous early Sun. With the new evidence from young solar mass stars of vastly greater flare activity (Feigelson *et al.*, 2002a,b), there is strong support for the notion that inner solar system objects would have incorporated a lot of solar-wind implanted noble gases. The beauty of this model is that small objects with larger surface-to-volume ratio trap more noble gases. Unlike the nebular model it works best if large objects take a long time to form. As such, the model is easier to reconcile with the kinds of long timescales for accretion implied by tungsten and lead isotopes. Podosek *et al.* (2003) present detailed calculations to illustrate the feasibility of such a scenario.

To summarize, although large nebular atmospheres have long been considered the most likely explanation for primordial solar-like noble gases in the Earth, the implications of such models appear hard to reconcile with the accretionary timescales determined from tungsten and lead isotopes. Irradiating planetesimals with solar wind currently appears to be the most promising alternative. If this is correct, the Earth may still at one time have had a relatively large atmosphere, but it would have formed by degassing of the Earth's interior.

2.6.3 Earth's Degassed Protoatmosphere

The discovery that primordial ^{3}He still is being released from the Earth's interior (Clarke *et al.*, 1969; Mamyrin *et al.*, 1969; Lupton and Craig, 1975; Craig and Lupton, 1976) is one of the greatest scientific contributions made by noble gas geochemistry. Far from being totally degassed, the Earth has deep reservoirs that must supply ^{3}He to the upper mantle and thence to the atmosphere. However, the idea that the majority of the components in the present-day atmosphere formed by degassing of the Earth's interior is much older than this. Brown (1949) and Rubey (1951) proposed this on the basis of the similarity in chemical composition between the atmosphere and hydrosphere on the one hand and the compositions of volcanic gases on the other. In more recent years, a variety of models for thc history of dcgassing of the Earth have been developed based on the idea that a relatively undegassed lower mantle supplied the upper mantle with volatiles, which then supplied the atmosphere (Allègre *et al.*, 1983, 1996; O'Nions and Tolstikhin, 1994; Porcelli and Wasserburg, 1995).

Some of the elements in the atmosphere provide specific "time information" on the degassing of the Earth. Of course, oxygen was added to the atmosphere gradually by photosynthesis with a major increase in the early Proterozoic (Kasting, 2001). However, the isotopic composition and concentration of argon provides powerful evidence that other gases have been added to the atmosphere from the mantle over geological time (O'Nions and Tolstikhin, 1994; Porcelli and Wasserburg, 1995). In the case of argon, a mass balance can be determined because it is dominantly (>99%) composed of ^{40}Ar, formed by radioactive decay of ^{40}K. From the Earth's potassium concentration and the atmosphere's argon concentration it can be shown that roughly half the ^{40}Ar is in the atmosphere, the remainder presumably being still stored in the Earth's mantle (Allègre *et al.*, 1996). Some have argued that this cannot be correct on geophysical grounds, leading to the proposal that the Earth's K/U ratio has been overestimated (Davies, 1999). However, the relationship with Rb/Sr (Figure 2) provides strong evidence that the Earth's K/U ratio is $\sim 10^{4}$ (Allègre *et al.*, 2001; Halliday and Porcelli, 2001). Furthermore, support for mantle reservoirs that are relatively undegassed can be found in both helium and neon isotopes (Allègre *et al.*, 1983, 1987; Moreira *et al.*, 1998, 2001; Moreira and Allègre, 1998; O'Nions and Tolstikhin, 1994; Niedermann *et al.*, 1997; Porcelli and Wasserburg, 1995). The average timing of this loss is poorly constrained from argon data. In principle, this amount of argon could have been supplied catastrophically in the recent past. However, it is far more reasonable to assume that because the Earth's radioactive heat production is decaying exponentially, the amount of degassing has been decreasing with time. The argon in the atmosphere is the time-integrated effect of this degassing.

Xenon isotopes provide strong evidence that the Earth's interior may have undergone an early and catastrophic degassing (Allègre *et al.*, 1983, 1987). Allègre *et al.* discovered that the MORB-source mantle had elevated ^{129}Xe relative to atmospheric xenon, indicating that the Earth and atmosphere separated from each other at an early stage. These models are hampered by a lack of constraints on the xenon budgets of the mantle and by atmospheric contamination that pervades many mantle-derived samples. Furthermore, the model assumes a closed system. If a portion of the xenon in the Earth's atmosphere was added after degassing (Javoy, 1998, 1999) by cometary or asteroidal impacts (Owen and Bar-Nun, 1995, 2000; Morbidelli *et al.*, 2000), the model becomes underconstrained. The differences that have been reported between the elemental and atomic abundances of the noble gases in the mantle relative to the atmosphere may indeed be explained by heterogeneous accretion of the atmosphere (Marty, 1989; Caffee *et al.*, 1988, 1999).

Isotopic evidence aside, many theoretical and experimental papers have focused on the production of a steam atmosphere by impact-induced degassing of the Earth's interior (Abe and Matsui, 1985, 1986, 1988; O'Keefe and Ahrens, 1977; Lange and Ahrens, 1982, 1984; Matsui and Abe, 1986a,b; Sasaki, 1990; Tyburczy *et al.*, 1986; Zahnle *et al.*, 1988). Water is highly soluble in silicate melts at high pressures (Righter and Drake, 1999; Abe *et al.*, 2000). As such, a large amount could have been stored in the Earth's mantle and then released during volcanic degassing.

Ahrens (1990) has modeled the effects of impact-induced degassing on the Earth. He considers that the Earth probably alternated between two extreme states as accretion proceeded. When the Earth was degassed it would accumulate a large reducing atmosphere. This would provide a blanket that also allowed enormous surface temperatures to be reached:

Ahrens estimates ~1,500 K. However, when an impact occurred this atmosphere would be blown off. The surface of the Earth would become cool and oxidizing again (Ahrens, 1990). There is strong isotopic evidence for such early losses of the early atmospheres as explained in the next section.

2.6.4 Loss of Earth's Earliest Atmosphere(s)

The xenon isotope data provide evidence that much (>99%) of the Earth's early atmosphere was lost within the first 100 Myr. Several papers on this can be found in the literature and the most recent ones by Ozima and Podosek (1999) and Porcelli *et al.* (2001) are particularly useful. The basic argument for the loss is fairly simple and is not so different from the original idea of using xenon isotopes to date the Earth (Wetherill, 1975a). We have a rough idea of how much iodine exists in the Earth's mantle. The Earth's current inventory is not well constrained but we know enough (Déruelle *et al.*, 1992) to estimate the approximate level of depletion of this volatile element. It is clear from the degassing of noble gases that the present ratio of I/Xe in the Earth is orders of magnitude higher than chondritic values. We know that at the start of the solar system ^{129}I was present with an atomic abundance of ~10^{-4} relative to stable ^{127}I (Swindle and Podosek, 1988). All of this ^{129}I formed ^{129}Xe, and should have produced xenon that was highly enriched in ^{129}Xe given the Earth's I/Xe ratio. Yet instead, the Earth has xenon that is only slightly more radiogenic than is found in meteorites rich in primordial noble gases; the ^{129}Xe excesses in the atmosphere and the mantle are both minute by comparison with that expected from the Earth's I/Xe. This provides evidence that the Earth had a low I/Xe ratio that kept its xenon isotopic compositions close to chondritic. At some point xenon was lost and by this time ^{129}I was close to being extinct such that the xenon did not become very radiogenic despite a very high I/Xe.

The xenon isotopic arguments can be extended to fissionogenic xenon. The use of the combined ^{129}I–^{129}Xe and ^{244}Pu–^{132,134,136}Xe (spontaneous fission half-life ~80 Myr) systems provides estimates of ~100 Myr for loss of xenon from the Earth (Ozima and Podosek, 1999). The fission-based models are hampered by the difficulties with resolving the heavy xenon that is formed from fission of ^{244}Pu as opposed to longer-lived uranium. The amount of ^{136}Xe that is expected to have formed from ^{244}Pu within the Earth should exceed that produced from uranium as found in well gases (Phinney *et al.*, 1978) and this has been confirmed with measurements of MORBs (Kunz *et al.*, 1998). However, the relative amount of plutonogenic Xe/uranogenic Xe will be a function of the history of degassing of the mantle. Estimating these amounts accurately is very hard.

This problem is exacerbated by the lack of constraint that exists on the initial xenon isotopic composition of the Earth. The xenon isotopic composition of the atmosphere is strikingly different from that found in meteorites (Wieler *et al.*, 1992; Busemann *et al.*, 2000) or the solar wind (Wieler *et al.*, 1996). It is fractionated relative to solar, the light isotopes being strongly depleted (Figure 7). One can estimate the initial xenon isotopic composition of the Earth by assuming it was strongly fractionated from something more like the composition found in meteorites and the solar wind. By using meteorite data to determine a best fit to atmospheric Xe one obtains a composition called "U–Xe" (Pepin, 1997, 2000). However, this is based on finding a composition that is consistent with the present-day atmosphere. Therefore caution is needed when using the fissionogenic xenon components to estimate an accretion age for the Earth because the arguments become circular (Zhang, 1998).

The strong mass-dependent fractionation of xenon has long been thought to be caused by hydrodynamic escape (Hunten *et al.*, 1987; Walker, 1986) of the atmosphere. Xenon probably was entrained in a massive atmosphere of light gases presumably dominated by hydrogen and helium that was lost (Sasaki and Nakazawa, 1988). This is consistent with the view based on radiogenic and fissionogenic xenon that a large fraction of the Earth's atmosphere was lost during the lifetime of ^{129}I.

Support for loss of light gases from the atmosphere via hydrodynamic escape can be found in other "atmophile" isotopic systems. The ^{20}Ne/^{22}Ne ratio in the mantle is elevated relative to the atmosphere (Trieloff *et al.*, 2000). The mantle

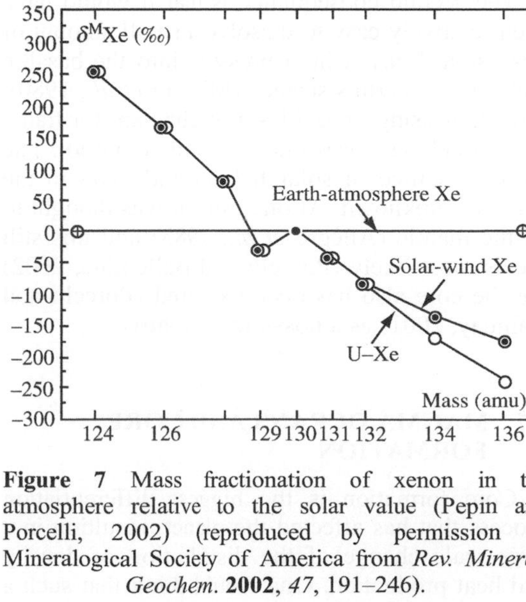

Figure 7 Mass fractionation of xenon in the atmosphere relative to the solar value (Pepin and Porcelli, 2002) (reproduced by permission of Mineralogical Society of America from *Rev. Mineral. Geochem.* **2002**, *47*, 191–246).

value is close to solar (Farley and Poreda, 1992), but the atmosphere plots almost exactly according to a heavy mass-dependent fractionated composition. Similarly, the hydrosphere has a D/H ratio that is heavy relative to the mantle. However, argon and krypton isotopes provide no support for the theory (Pepin and Porcelli, 2002). Therefore, this model cannot be applied in any simple way to all of the atmophile elements.

It is possible that the atmosphere was blown off by a major impact like the Moon-forming giant impact, but this is far from clear at this stage. Another mechanism that often is considered is the effect of strong ultraviolet wavelength radiation from the early Sun (Zahnle and Walker, 1982). This might affect Xe preferentially because of the lower ionization potential. It is of course possible that the Earth simply acquired an atmosphere, with xenon, like today's (Marty, 1989; Caffee *et al.*, 1999). However, then it is not clear how to explain the strong isotopic fractionation relative to solar and meteorite compositions.

Taken together, the noble gas data provide evidence that Earth once had an atmosphere that was far more massive than today's. If true, this would have had two important geochemical consequences. First, there would have been a blanketing effect from such an atmosphere. This being the case, the temperature at the surface of the Earth would have been very high. There may well have been magma oceans, rock vapor in the atmosphere, and extreme degassing of moderately volatile as well as volatile elements. Blow-off of this atmosphere may have been related to the apparent loss of moderately volatile elements (Halliday and Porcelli, 2001). With magma oceans there would be, at best, a weak crust, mantle mixing would have been very efficient, and core formation would have proceeded quickly.

The second consequence is that it would have been relatively easy to dissolve a small amount of this "solar" gas at high pressure into the basaltic melts at the Earth's surface (Mizuno *et al.*, 1980). This "ingassing" provides a mechanism for transporting nebular gases into the Earth's interior. The ultimate source of solar helium and neon in the mantle is unknown. At one time, it was thought to be the mantle (Allègre *et al.*, 1983) and this still seems most likely (Porcelli and Ballentine, 2002) but the core also has been explored (Porcelli and Halliday, 2001) as a possible alternative.

2.7 MAGMA OCEANS AND CORE FORMATION

Core formation is the biggest differentiation process that has affected the planet, resulting in a large-scale change of the distribution of density and heat production. One would think that such a basic feature would be well understood. However, the very existence of large amounts of iron metal at the center of an Earth with an oxidized mantle is problematic (Ringwood, 1977). Large reducing atmospheres and magma oceans together provide a nice explanation. For example, Ringwood (1966) considered that the iron metal in the Earth's core formed by reduction of iron in silicates and oxides and thereby suggested a huge CO atmosphere. Clearly, if the Earth's core formed by reduction of iron in a large atmosphere, the process of core formation would occur early and easily.

However, there is very little independent evidence from mantle or crustal geochemistry to substantiate the former presence of magma oceans. There has been no shortage of proposals and arguments for and against on the basis of petrological data for the Earth's upper mantle. The key problem is the uncertainty that exists regarding the relationship between the present day upper mantle and that that may have existed in the early Earth. Some have argued that the present day lower mantle is compositionally distinct in terms of major elements (e.g., Herzberg, 1984; Jackson, 1983), whereas others (e.g., Hofmann, 1988; Hofmann *et al.*, 1986; Davies, 1999) have presented strong evidence in favor of large-scale overall convective interchange. Although several papers have used the present-day major, trace, and isotopic compositions of the upper mantle to provide constraints on the earliest history of the Earth (Allègre *et al.*, 1983; Agee and Walker, 1988; Kato *et al.*, 1988a,b; McFarlane and Drake, 1990; Jones *et al.*, 1992; Drake and McFarlane, 1993; Porcelli and Wasserburg, 1995; O'Nions and Tolstikhin, 1994; Righter and Drake, 1996), it is unclear whether or not this is valid, given 4.5 Gyr of mantle convection.

For example, the trace element and hafnium isotopic compositions of the upper mantle provide no sign of perovskite fractionation in a magma ocean (Ringwood, 1990; Halliday *et al.*, 1995) and even the heavy REE pattern of the upper mantle appears to be essentially flat (e.g., Lee *et al.*, 1996). This is not to be expected if majorite garnet was a liquidus phase (Herzberg *et al.*, 1988). One explanation for the lack of such evidence is that the magma ocean was itself an efficiently mixed system with little if any crystal settling (Tonks and Melosh, 1990). Another possibility is that the entire mantle has been rehomogenized since this time. Ringwood (1990) suggested this possibility, which leaves some arguments regarding the relevance of the composition of the upper mantle in doubt. Of course, the subsequent introduction of heterogeneities by entrainment and radioactive decay and the development of an isotopically stratified mantle (Hofmann *et al.*, 1986, Moreira and Allègre, 1998) are not inconsistent with this.

Whether or not magma oceans are a necessary prerequisite for core formation is unclear. It is necessary to understand how it is possible for metallic iron to migrate through the silicate mantle (Shaw, 1978). Many have assumed that a part of the mantle at least was solid during core formation. A variety of mechanisms have been studied including grain boundary percolation (Minarik *et al.*, 1996; Rushmer *et al.*, 2000; Yoshino *et al.*, 2003) and the formation of large-scale metal structures in the upper mantle that sink like diapirs into the center of the Earth (Stevenson, 1981, 1990). Under some circumstances these may break up into small droplets of metal (Rubie *et al.*, 2003). To evaluate these models, it is essential to have some idea of the physical state of the early Earth at the time of core formation. All of these issues are addressed by modern geochemistry but are not yet well constrained. Most effort has been focused on using the composition of the silicate Earth itself to provide constraints on models of core formation.

The major problem presented by the Earth's chemical composition and core formation models is providing mechanisms that predict correctly the siderophile element abundances in the Earth's upper mantle. It long has been recognized that siderophile elements are more abundant in the mantle than expected if the silicate Earth and the core were segregated under low-pressure and moderate-temperature equilibrium conditions (Chou, 1978; Jagoutz *et al.*, 1979). Several explanations for this siderophile "excess" have been proposed, including:

(i) partitioning into liquid metal alloy at high pressure (Ringwood, 1979);

(ii) equilibrium partitioning between sulfur-rich liquid metal and silicate (Brett, 1984);

(iii) inefficient core formation (Arculus and Delano, 1981; Jones and Drake, 1986);

(iv) heterogeneous accretion and late veneers (Eucken, 1944; Turekian and Clark, 1969; Clark *et al.*, 1972; Smith, 1977, 1980; Wänke *et al.*, 1984; Newsom, 1990);

(v) addition of material to the silicate Earth from the core of a Moon-forming impactor (Newsom and Taylor, 1986);

(vi) very high temperature equilibration (Murthy, 1991); and

(vii) high-temperature equilibrium partitioning in a magma ocean at the upper/lower mantle boundary (Li and Agee, 1996; Righter *et al.*, 1997).

Although the abundances and partition coefficients of some of the elements used to test these models are not well established, sufficient knowledge exists to render all of them problematic. Model (vii) appears to work well for some moderately siderophile elements. Righter and Drake (1999) make the case that the fit of the siderophile element metal/silicate partion coefficient data is best achieved with a high water content (per cent level) in the mantle. This would have assisted the formation of a magma ocean and provided a ready source of volatiles in the Earth. Walter *et al.* (2000) reviewed the state of the art in this area. However, the number of elements with well-established high-pressure partition coefficients for testing this model is still extremely small.

To complicate chemical models further, there is some osmium isotopic evidence that a small flux of highly siderophile elements from the core could be affecting the abundances in the mantle (Walker *et al.*, 1995; Brandon *et al.*, 1998). This model has been extended to the interpretation of PGE abundances in abyssal peridotites (Snow and Schmidt, 1998). The inventories of many of these highly siderophile elements are not that well established and may be extremely variable (Rehkämper *et al.*, 1997, 1999b). In particular, the use of abyssal peridotites to assess siderophile element abundances in the upper mantle appears to be problematic (Rehkämper *et al.*, 1999a). Puchtel and Humayun (2000) argued that if PGEs are being fluxed from the core to the mantle then this is not via the Walker *et al.* (1995) mechanism of physical admixture by entrainment of outer core, but must proceed via an osmium isotopic exchange, since the excess siderophiles were not found in komatiite source regions with radiogenic ^{187}Os. Taken together, the status of core to mantle fluxes is very vague at the present time.

2.8 THE FORMATION OF THE MOON

The origin of the Moon has been the subject of intense scientific interest for over a century but particularly since the Apollo missions provided samples to study. The most widely accepted current theory is the giant impact theory but this idea has evolved from others and alternative hypotheses have been variously considered. Wood (1984) provides a very useful review. The main theories that have been considered are as follows:

Co-accretion. This theory proposes that the Earth and Moon simply accreted side by side. The difficulty with this model is that it does not explain the angular momentum of the Earth–Moon system, nor the difference in density, nor the difference in volatile depletion (Taylor, 1992).

Capture. This theory (Urey, 1966) proposes that the Moon was a body captured into Earth's orbit. It is dynamically difficult to do this without the Moon spiraling into the Earth and colliding. Also the Earth and Moon have indistinguishable oxygen isotope compositions (Wiechert *et al.*, 2001) in a solar system that appears to be highly heterogeneous in this respect (Clayton, 1986).

Fission. This theory proposes that the Moon split off as a blob during rapid rotation of a molten Earth. George Howard Darwin, the son of

Charles Darwin originally championed this idea (Darwin, 1878, 1879). At one time (before the young age of the oceanfloor was known) it was thought by some that the Pacific Ocean might have been the residual space vacated by the loss of material. This theory is also dynamically difficult. Detailed discussions of the mechanisms can be found in Binder (1986). However, this model does have certain features that are attractive. It explains why Earth and Moon have identical oxygen isotope compositions. It explains why the Moon has a lower density because the outer part of the Earth would be deficient in iron due to core formation. It explains why so much of the angular momentum of the Earth–Moon system is in the Moon's motion. These are key features of any successful explanation for the origin of the Moon.

Impact models. Mainly because of the difficulties with the above models, alternatives were considered following the Apollo missions. Hartmann and Davis (1975) made the proposal that the Moon formed as a result of major impacts that propelled sufficient debris into orbit to produce the Moon. However, an important new facet that came from sample return was the discovery that the Moon had an anorthositic crust implying a very hot magma ocean. Also it was necessary to link the dynamics of the Moon with that of the Earth's spin. If an impact produced the Moon it would be easier to explain these features if it was highly energetic. This led to a series of single giant impact models in which the Moon was the product of a glancing blow collision with another differentiated planet (Cameron and Benz, 1991). A ring of debris would have been produced from the outer silicate portions of the Earth and the impactor planet (named "Theia," the mother of "Selene," the goddess of the Moon). Wetherill (1986) calculated that there was a realistic chance of such a collision. This model explains the angular momentum, the "fiery start," the isotopic similarities and the density difference.

The giant-impact theory has been confirmed by a number of important observations. Perhaps most importantly, we know now that the Moon must have formed tens of millions of years after the start of the solar system (Lee *et al.*, 1997; Halliday, 2000). This is consistent with a collision between already formed planets. The masses of the Earth and the impactor at the time of the giant impact have been the subject of major uncertainty. Two main classes of models are usually considered. In the first, the Earth was largely (90%) formed at the time of the impact and the impacting planet Theia was roughly Mars-sized (Cameron and Benz, 1991). A recent class of models considers the Earth to be only half-formed at the time of the impact, and the mass ratio Theia/proto-Earth to be 3:7 (Cameron, 2000). The latter model is no longer considered likely; the most recent simulations have reverted to a Mars-sized impactor at the end of Earth accretion (Canup and Asphaug, 2001). The tungsten isotope data for the Earth and Moon do not provide a unique test (Halliday *et al.*, 2000).

The giant-impact model, though widely accepted, has not been without its critics. Geochemical arguments have been particularly important in this regard. The biggest concern has been the similarities between chemical and isotopic features of the Earth and Moon. Most of the dynamic simulations (Cameron and Benz, 1991; Cameron, 2000; Canup and Asphaug, 2001) predict that the material that forms the Moon is derived from Theia, rather than the Earth. Yet it became very clear at an early stage of study that samples from the Moon and Earth shared many common features that would be most readily explained if the Moon was formed from material derived from the Earth (Wänke *et al.*, 1983; Wänke and Dreibus, 1986; Ringwood, 1989b, 1992). These include the striking similarity in tungsten depletion despite a strong sensitivity to the oxidation state of the mantle (Rammensee and Wänke, 1977; Schmitt *et al.*, 1989). Other basaltic objects such as eucrites and martian meteorites exhibit very different siderophile element depletion (e.g., Treiman *et al.*, 1986, 1987; Wänke and Dreibus, 1988, 1994). Therefore, why should the Earth and Moon be identical if the Moon came from Theia (Ringwood, 1989b, 1992)? In a similar manner, the striking similarity in oxygen isotopic composition (Clayton and Mayeda, 1975), still unresolvable to extremely high precision (Wiechert *et al.*, 2001), despite enormous heterogeneity in the solar system (Clayton *et al.*, 1973; Clayton, 1986, 1993; Clayton and Mayeda, 1996), provides support for the view that the Moon was derived from the Earth (Figure 3).

One can turn these arguments around, however, and use the compositions of lunar samples to define the composition of Theia, assuming the impactor produced most of the material in the Moon (MacFarlane, 1989). Accordingly, the similarity in oxygen isotopes and trace siderophile abundances between the Earth and Moon provides evidence that Earth and Theia were neighboring planets made of an identical mix of materials with similar differentiation histories (Halliday and Porcelli, 2001). Their similarities could relate to proximity in the early solar system, increasing the probability of collision.

Certain features of the Moon may be a consequence of the giant impact itself. The volatile-depleted composition of the Moon, in particular, has been explained as a consequence of the giant impact (O'Neill, 1991a; Jones and Palme, 2000). It has been argued (Kreutzberger *et al.*, 1986; Jones and Drake, 1993) that the

Moon could not have formed as a volatile depleted residue of material from the Earth because it has Rb/Cs that is lower than that of the Earth and caesium supposedly is more volatile. However, the assumptions regarding the Earth's Rb/Cs upon which this is based are rather weak (McDonough *et al.*, 1992). Furthermore, the exact relative volatilities of the alkalis are poorly known. Using the canonical numbers, the Earth, Moon, and Mars are all more depleted in less volatile rubidium than more volatile potassium (Figure 2). From the slope of the correlation, it can be seen that the terrestrial depletion in rubidium (50% condensation temperature ~1,080 K) is ~80% greater than that in potassium (50% condensation temperature ~1,000 K) (Wasson, 1985). Similar problems are found if one compares sodium depletion, or alkali concentrations more generally for other early objects, including chondrites.

Attempts to date the Moon were initially focused on determining the ages of the oldest rocks and therefore providing a lower limit. These studies emphasized precise strontium, neodymium, and lead isotopic constraints (Tera *et al.*, 1973; Wasserburg *et al.*, 1977a; Hanan and Tilton, 1987; Carlson and Lugmair, 1988; Shih *et al.*, 1993; Alibert *et al.*, 1994). At the end of the Apollo era, Wasserburg *et al.* (1977a) wrote "The actual time of aggregation of the Moon is not precisely known, but the Moon existed as a planetary body at 4.45 Ga, based on mutually consistent Rb–Sr and U–Pb data. This is remarkably close to the ^{207}Pb–^{206}Pb age of the Earth and suggests that the Moon and the Earth were formed or differentiated at the same time." Although these collective efforts made a monumental contribution, such constraints on the age of the Moon still leave considerable scope (>100 Myr) for an exact age.

Some of the most precise and reliable early ages for lunar rocks are given in Table 3. They provide considerable support for an age of >4.42 Ga. Probably the most compelling evidence comes from the early ferroan anorthosite 60025, which defines a relatively low first-stage μ (or $^{238}U/^{204}Pb$) and an age of ~4.5 Ga. Of course, the ages of the oldest lunar rocks only date igneous events. Carlson and Lugmair (1988) reviewed all of the most precise and concordant data and concluded that the Moon had to have formed in the time interval 4.44–4.51 Ga. This is consistent with the estimate of 4.47 ± 0.02 Ga of Tera *et al.* (1973).

Model ages can provide upper and lower limits on the age of the Moon. Halliday and Porcelli (2001) reviewed the strontium isotope data for early solar system objects and showed that the initial strontium isotopic compositions of early lunar highlands samples (Papanastassiou and Wasserburg, 1976; Carlson and Lugmair, 1988) are all slightly higher than the best estimates of the solar system initial ratio (Figure 8). The conservative estimates of the strontium isotope data indicate that the difference between the $^{87}Sr/^{86}Sr$ of the bulk solar system initial at 4.566 Ga is 0.69891 ± 2 and the Moon at ~4.515 Ga is 0.69906 ± 2 is fully resolvable. An Rb–Sr model age for the Moon can be calculated by assuming that objects formed from material that separated from a solar nebula reservoir with the Moon's current Rb/Sr ratio. Because the Rb/Sr ratios of the lunar samples are extremely low, the uncertainty in formation age does not affect the calculated initial strontium isotopic composition, hence the model age, significantly. The CI chondritic Rb/Sr ratio ($^{87}Rb/^{86}Sr = 0.92$) is assumed to represent the solar nebula. This model provides an upper limit on the formation age of the object, because the solar nebula is thought to represent the most extreme Rb/Sr reservoir in which the increase in strontium isotopic composition could have been accomplished. In reality, the strontium isotopic composition probably evolved in a more complex manner over a longer time. The calculated time required to generate the difference in strontium isotopic composition in a primitive solar nebula environment is 11 ± 3 Myr. This is, therefore, the earliest point in time at which the Moon could have formed (Halliday and Porcelli, 2001).

A similar model-age approach can be used with the Hf–W system. In fact, Hf–W data provide the most powerful current constraints on the exact age of the Moon. The tungsten isotopic compositions of bulk rock lunar samples range from $\varepsilon_W \sim 0$ like the silicate Earth to $\varepsilon_W > 10$ (Lee *et al.*, 1997, 2002). This was originally interpreted as the result of radioactive decay of formerly live ^{182}Hf within the Moon, which has a variable but generally high Hf/W ratio in its mantle (Lee *et al.*, 1997). Now we know that a major portion of the ^{182}W excess in lunar samples is cosmogenic and the result of the reaction $^{181}Ta(n,\gamma)^{182}Ta(\beta^-)^{182}W$ while these rocks were exposed on the surface of the Moon (Leya *et al.*, 2000; Lee *et al.*, 2002). This can be corrected using (i) estimates of the cosmic ray flux from samarium and gadolinium compositions, (ii) the exposure age and Ta/W ratio, or (iii) internal isochrons of tungsten isotopic composition against Ta/W (Lee *et al.*, 2002). The best current estimates for the corrected compositions are shown in Figure 9. The spread in the data is reduced and the stated uncertainties are greater relative to the raw tungsten isotopic compositions (Lee *et al.*, 1997). Most data are within error of the Earth. A small excess ^{182}W is still resolvable for some samples, but these should be treated with caution.

Table 3 Recent estimates of the ages of early solar system objects and the age of the Moon.

Object	*Sample(s)*	*Method*	*References*	*Age (Ga)*
Earliest solar system	Allende CAIs	U–Pb	Göpel *et al.* (1991)	4.566 ± 0.002
Earliest solar system	Efremovka CAIs	U–Pb	Amelin *et al.* (2002)	4.5672 ± 0.0006
Chondrule formation	Acfer chondrules	U–Pb	Amelin *et al.* (2002)	4.5647 ± 0.0006
Angrites	Angra dos Reis and LEW 86010	U–Pb	Lugmair and Galer (1992)	4.5578 ± 0.0005
Early eucrites	Chervony Kut	Mn–Cr	Lugmair and Shukolyukov (1998)	4.563 ± 0.001
Earth accretion	Mean age	U–Pb	Halliday (2000)	≤ 4.55
Earth accretion	Mean age	U–Pb	Halliday (2000)	≥ 4.49
Earth accretion	Mean age	Hf–W	Yin *et al.* (2002)	≥ 4.55
Lunar highlands	Ferroan anorthosite 60025	U–Pb	Hanan and Tilton (1987)	4.50 ± 0.01
Lunar highlands	Ferroan anorthosite 60025	Sm–Nd	Carlson and Lugmair (1988)	4.44 ± 0.02
Lunar highlands	Norite from breccia 15445	Sm–Nd	Shih *et al.* (1993)	4.46 ± 0.07
Lunar highlands	Ferroan noritic anorthosite in breccia 67016	Sm–Nd	Alibert *et al.* (1994)	4.56 ± 0.07
Moon	Best estimate of age	U–Pb	Tera *et al.* (1973)	4.47 ± 0.02
Moon	Best estimate of age	U–Pb, Sm–Nd	Carlson and Lugmair (1988)	4.44–4.51
Moon	Best estimate of age	Hf–W	Halliday *et al.* (1996)	4.47 ± 0.04
Moon	Best estimate of age	Hf–W	Lee *et al.* (1997)	4.51 ± 0.01
Moon	Maximum age	Hf–W	Halliday (2000)	≤ 4.52
Moon	Maximum age	Rb–Sr	Halliday and Porcelli (2001)	≤ 4.55
Moon	Best estimate of age	Hf–W	Lee *et al.* (2002)	4.51 ± 0.01
Moon	Best estimate of age	Hf–W	Kleine *et al.* (2002)	4.54 ± 0.01

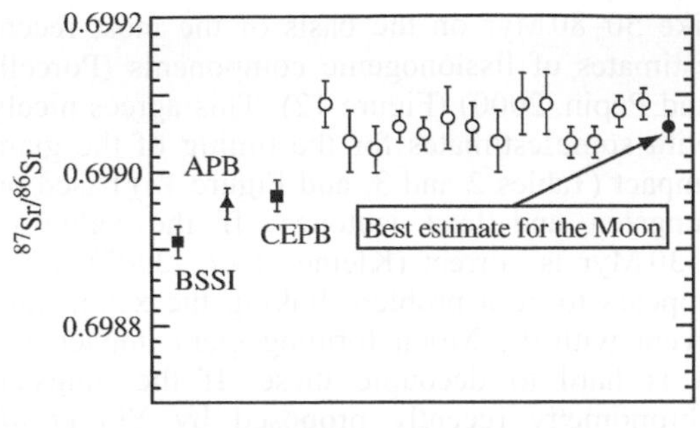

Figure 8 Initial strontium isotope composition of early lunar highland rocks relative to other early solar system objects. APB: Angrite Parent Body; CEPB: Cumulate Eucrite Parent Body; BSSI: Bulk Solar System Initial (source Halliday and Porcelli, 2001).

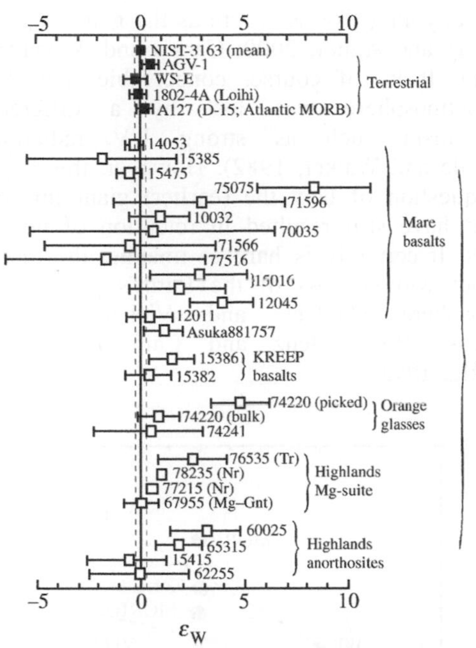

Figure 9 The tungsten isotopic compositions of lunar samples after calculated corrections for cosmogenic contributions (source Leya *et al.*, 2000).

The most obvious and clear implication of these data is that the Moon, a high-Hf/W object, must have formed late (Lee *et al.*, 1997). Halliday (2000) argued that the tungsten isotopic composition was hard to explain if the Moon formed before ~50 Myr after the start of the solar system. The revised parameters for the average solar system (Kleine *et al.*, 2002) mean that the Moon, like the Earth, has a well-defined excess of ^{182}W. This may have been inherited from the protolith silicate reservoirs from which the Moon formed. Alternatively, a portion might reflect ^{182}Hf decay within the Moon itself. Assuming that the Moon started as an isotopically homogeneous reservoir the most likely explanation for the small but well-defined excess ^{182}W found in Apollo basalts such as 15555 (Figure 10) relative to some of the other lunar rocks is that the Moon formed at a time when there was still a small amount of live ^{182}Hf in the lunar interior. This means that the Moon had to have formed within the first 60 Myr of the solar system (Halliday, 2000; Lee *et al.*, 2002). The Moon must also have formed by the time defined by the earliest lunar rocks. The earliest most precisely determined crystallization age of a lunar rock is that of 60025 which has a Pb–Pb age of close to 4.50 Ga (Hanan and Tilton, 1987). Therefore, the Moon appears to have formed before ~4.50 Ga.

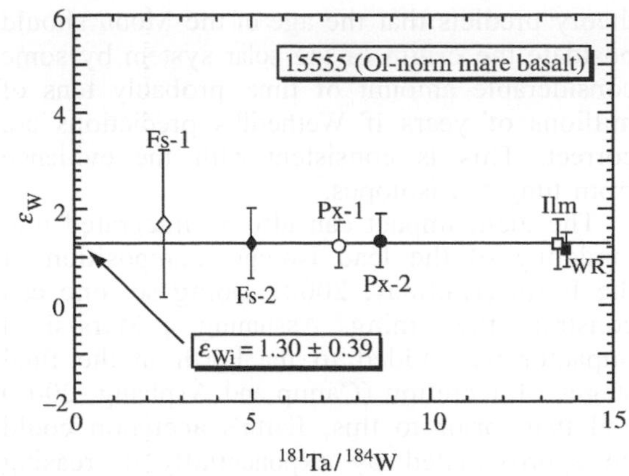

Figure 10 The tungsten isotopic composition of Apollo 15 basalt 15555 shows no internal variation as a function of Ta/W, consistent with its low exposure age (source Lee *et al.*, 2002).

Defining the age more precisely is proving difficult at this stage. First, more precise estimates of the Hf–W systematics of lunar rocks are needed. The amount of data for which the cosmogenically produced ^{182}W effects are well resolved is very limited (Lee *et al.*, 2002) and analysis is time consuming and difficult. Second, the $(^{182}Hf/^{180}Hf)_{BSSI}$ is poorly defined, as described above. If one uses a value of 1.0×10^{-4} (Kleine *et al.*, 2002; Schöenberg *et al.*, 2002; Yin *et al.*, 2002), the small tungsten isotopic effects of the Moon probably were produced $\geqslant$30 Myr after the start of the solar system (Kleine *et al.*, 2002). If, however, the $(^{182}Hf/^{180}Hf)_{BSSI}$ is slightly higher, as discussed above, the model age would be closer to 40–45 Myr (Table 3). The uncertainty in the ^{182}Hf decay constant (~±22%) also limits more precise constraints.

Either way, these Hf–W data provide very strong support for the giant-impact theory of lunar origin (Cameron and Benz, 1991). It is hard to explain how the Moon could have formed at such a late stage unless it was the result of a planetary collision. The giant-impact theory predicts that the age of the Moon should postdate the origin of the solar system by some considerable amount of time, probably tens of millions of years if Wetherill's predictions are correct. This is consistent with the evidence from tungsten isotopes.

The giant impact can also be integrated into modeling of the lead isotopic composition of the Earth (Halliday, 2000). Doing so, one can constrain the timing. Assuming a Mars-sized impactor was added to the Earth in the final stages of accretion (Canup and Asphaug, 2001) and that, prior to this, Earth's accretion could be approximated by exponentially decreasing rates (Halliday, 2000), one can calibrate the predicted lead isotopic composition of the Earth in terms of the time of the giant impact (Figure 11). It can be seen that all of the many estimates for the lead isotopic composition of the BSE compiled by Galer and Goldstein (1996) plus the more recent estimates of Kramers and Tolstikhin (1997) and Murphy *et al.* (2003) appear inconsistent with a giant impact that is earlier than ~45 Myr after the start of the solar system.

Xenon isotope data have also been used to argue specifically that the Earth lost its inventory of noble gases as a consequence of the giant impact (Pepin, 1997; Porcelli and Pepin, 2000). The timing of the "xenon loss event" looks more like 50–80 Myr on the basis of the most recent estimates of fissionogenic components (Porcelli and Pepin, 2000) (Figure 12). This agrees nicely with some estimates for the timing of the giant impact (Tables 2 and 3, and Figure 11) based on tungsten and lead isotopes. If the value of ~30 Myr is correct (Kleine *et al.*, 2002), there appears to be a problem linking the xenon loss event with the Moon forming giant impact, yet it is hard to decouple these. If the tungsten chronometry recently proposed by Yin *et al.* (2002) and Kleine *et al.* (2002) is correct it would seem that the giant impact cannot have been the last big event that blew off a substantial fraction of the Earth's atmosphere. On the basis of dynamic simulations, however, it is thought that subsequent events cannot have been anything like as severe as the giant impact (Canup and Agnor, 2000; Canup and Asphaug, 2001). It is, of course, conceivable that the protoatmosphere was lost by a different mechanism such as strong UV radiation (Zahnle and Walker, 1982). However, this begs the question of how the (earlier) giant impact could have still resulted in retention of noble gases. It certainly is hard to imagine the giant impact without loss of the Earth's primordial atmosphere (Melosh and Vickery, 1989; Ahrens, 1990; Benz and Cameron, 1990; Zahnle, 1993).

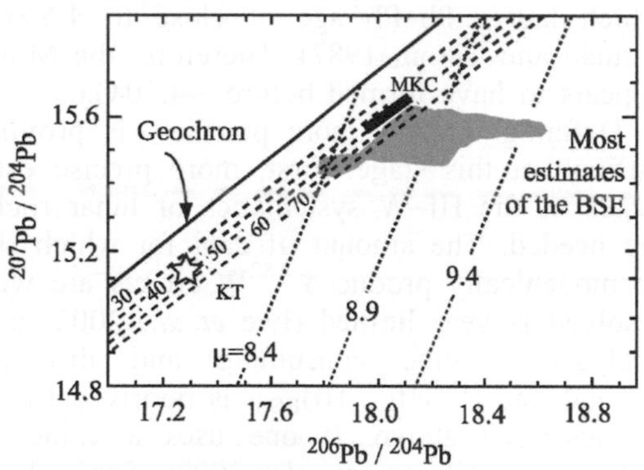

Figure 11 Lead isotopic modeling of the composition of the silicate Earth using continuous core formation and a sudden giant impact when the Earth is 90% formed. The impactor adds a further 9% to the mass of the Earth. The principles behind the modeling are as in Halliday (2000). See text for explanation. The field for the BSE encompasses all of the estimates in Galer and Goldstein (1996). The values suggested by Kramers and Tolstikhin (1997) and Murphy *et al.* (2002) are also shown. The figure is calibrated with the time of the giant impact (Myr). The μ values are the $^{238}U/^{204}Pb$ of the BSE. It is assumed that the μ of the total Earth is 0.7 (Allègre *et al.*, 1995a). It can be seen that the lead isotopic composition of the BSE is hard to reconcile with formation of the Moon before ~45 Myr after the start of the solar system.

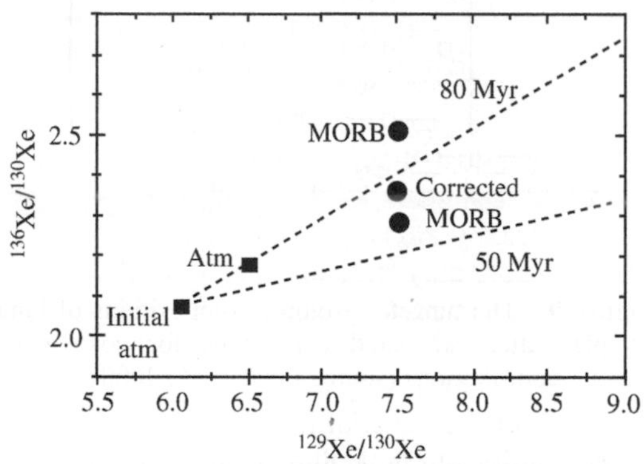

Figure 12 The relationship between the amounts of radiogenic ^{129}Xe inferred to come from ^{129}I decay within the Earth and fissionogenic ^{136}Xe thought to be dominated by decay of ^{244}Pu within the Earth. The differences in composition between the atmosphere and upper mantle relate to the timing of atmosphere formation. The compositions of both reservoirs are not very different from solar or initial U–Xe. This provides evidence that the strong depletion of xenon, leading to very high I/Xe, for example, was late. The data are shown modeled as a major loss event 50–80 Myr after the start of the solar system (Porcelli and Ballentine, 2002). The exact correction for uranium-derived fission ^{136}Xe in the MORB-mantle is unclear. Two values are shown from Phinney *et al.* (1978) and Kunz *et al.* (1998).

2.9 MASS LOSS AND COMPOSITIONAL CHANGES DURING ACCRETION

Collisions like the giant impact may well represent an important growth mechanism for terrestrial planets in general (Cameron and Benz, 1991; Canup and Agnor, 2000). These collisions are extraordinarily energetic and the question arises as to whether it is to be expected that accretion itself will lead to losses ("erosion") of material from the combined planetary masses. If this is the case it is only to be expected that the composition of the Earth does not add up to what one might expect from "chondrite building blocks" and, for example, the oxygen isotope composition (Clayton and Mayeda, 1996). There are several indications that the Earth may have lost a significant fraction of some elements during accretion and earliest development; all are circumstantial lines of evidence:

(i) As discussed above it has been argued that impact processes in particular are responsible for eroding protoatmospheres (Melosh and Vickery, 1989; Ahrens, 1990; Benz and Cameron, 1990; Zahnle, 1993). If these atmospheres were very dense and hot they may have contained a significant fraction of Earth's moderately volatile elements.

(ii) "Glancing blow" collisions between already differentiated planets, as during the giant impact, might be expected to preferentially remove major portions of the outer silicate portions of the planet as it grows. The Fe/Mg ratio of a planet is the simplest chemical parameter relating to planetary density and indicates the approximate size of the core relative to the silicate mantle. Mercury, with its high density, is a prime candidate for a body that lost a great deal of its outer silicate material by giant impacts (Benz *et al.*, 1987). Therefore, by analogy the proportional size of the Earth's core may have increased as a consequence of such impact erosion. Conversely, Mars (Halliday *et al.*, 2001) with a density lower than that of the Earth, may actually be a closer approximation of the material from which Earth accreted than Earth itself is (Halliday *et al.*, 2001; Halliday and Porcelli, 2001).

(iii) Such late collisional loss of Earth's silicate is clearly evident from the low density of the Moon. The disk of material from which the Moon accreted during the giant impact was silicate-rich. Most simulations predict that Theia provided the major source of lunar material. The density difference between silicate and metal leads to a loss of silicate from the combined Theia–proto-Earth system, even when the loss primarily is from the impactor. Note, however, that the giant-impact simulations retain most of the mass overall (Benz and Cameron, 1990). Very little is lost to space from a body as large as the Earth.

(iv) If there was early basaltic crust on the Earth (Chase and Patchett, 1988; Galer and Goldstein, 1991) or Theia or other impacting planets, repeated impact erosion could have had an effect on the Si/Mg ratio of the primitive mantle. However, the maximum effect will be very small because silicon is a major element in the mantle. Earth's Si/Mg ratio is indeed low, as discussed above, but this may instead represent other loss processes. Nevertheless, the erosion effects could have been accentuated by the fact that silicon is relatively volatile and there probably was a magma ocean. With extremely high temperatures and a heavy protoatmosphere (Ahrens, 1990; Sasaki, 1990; Abe *et al.*, 2000), it seems possible that one could form a "rock atmosphere" by boiling the surface of the magma ocean. This atmosphere would in turn be very vulnerable to impact-induced blow-off.

(v) The budgets of other elements that are more heavily concentrated in the outer portions of the Earth, such as the highly incompatible lithophile elements caesium, barium, rubidium, thorium, uranium, niobium, and potassium and the light rare earths, were possibly also depleted by impact erosion. Indeed some people have argued that the primitive mantle is slightly higher than chondritic in Sm/Nd (Nägler and Kramers, 1998). However, it is worth noting that there is no evidence for europium anomalies in the BSE as might be expected from impact loss of feldspathic flotation cumulates. These issues are further complicated by the fact that the early Earth (or impacting planets) may have had magma oceans with liquidus phases such as majorite or calcium- or magnesium-perovskite (Kato *et al.*, 1988a,b). This in turn could have led to a very variable element distribution in a stratified magma ocean. There is no compelling evidence in hafnium isotopes, yet, for loss of a portion of the partially molten mantle fractionated in the presence of perovskite (Ringwood, 1990).

(vi) The oxygen isotopic compositions of lunar samples have been measured repeatedly to extremely high precision (Wiechert *et al.*, 2001) using new laser fluorination techniques. The $\Delta^{17}O$ of the Moon relative to the terrestrial fractionation line can be shown to be zero (Figure 3). Based on a 99.7% confidence interval (triple standard error of the mean) the lunar fractionation line is, within $\pm 0.005‰$, identical to that of the Earth. There now is no doubt that the mix of material that accreted to form the Earth and the Moon was effectively identical in its provenance. Yet there is a big difference in the budgets of moderately volatile elements (e.g., K/U or Rb/Sr). One explanation is that there were major losses of moderately volatile elements during the giant impact (O'Neill, 1991a,b; Halliday and Porcelli, 2001).

(vii) The strontium isotopic compositions of early lunar highland rocks provide powerful support for late losses of alkalis (Figure 8). Theia had a time integrated Rb/Sr that was more than an order of magnitude higher than the actual Rb/Sr of the Moon, providing evidence that the processes of accretion resulted in substantial loss of alkalis (Halliday and Porcelli, 2001). Some of the calculated time-integrated compositions of the precursors to the present Earth and Moon are shown in Figure 6.

(viii) The abundances of volatile highly siderophile elements in the Earth are slightly depleted relative to refractory highly siderophile elements (Yi *et al.*, 2000). This appears to reflect the composition of a late veneer. If this is representative of the composition of material that accreted to form the Earth as a whole it implies that there were substantial losses of volatile elements from the protoplanets that built the Earth (Yi *et al.*, 2000).

Therefore, there exist several of lines of evidence to support the view that impact erosion may have had a significant effect on Earth's composition. However, in most cases the evidence is suggestive rather than strongly compelling. Furthermore, we have a very poor idea of how this is possible without fractionating potassium isotopes (Humayun and Clayton, 1995), unless the entire inventory of potassium is vaporized (O'Neill, 1991a,b; Halliday *et al.*, 1996; Halliday and Porcelli, 2001). We also do not understand how to lose heavy elements except via hydrodynamic escape of a large protoatmosphere (Hunten *et al.*, 1987; Walker, 1986). Some of the loss may have been from the proto-planets that built the Earth.

2.10 EVIDENCE FOR LATE ACCRETION, CORE FORMATION, AND CHANGES IN VOLATILES AFTER THE GIANT IMPACT

There are a number of lines of evidence that the Earth may have been affected by additions of further material subsequent to the giant impact. Similarly, there is limited evidence that there was additional core formation. Alternatively, there also are geochemical and dynamic constraints that strongly limit the amount of core formation and accretion since the giant impact. This is a very interesting area of research that is ripe for further development. Here are some of the key observations:

(i) It has long been recognized that there is an apparent excess of highly siderophile elements in the silicate Earth (Chou, 1978; Jagoutz *et al.*, 1979). These excess siderophiles already have been discussed above in the context of core formation. However, until Murthy's (1991) paper, the most widely accepted explanation was that there was a "late veneer" of material accreted after core formation and corresponding to the final one percent or less of the Earth's mass. Nowadays the effects of high temperatures and pressures on partitioning can be investigated and it seems clear that some of the excess siderophile signature reflects silicate-metal equilibration at depth. The volatile highly siderophile elements carbon, sulfur, selenium, and tellurium are more depleted in the silicate Earth than the refractory siderophiles (Figure 1; Yi *et al.*, 2000). Therefore, if there was a late veneer it probably was material that was on average depleted in volatiles, but not as depleted as would be deduced from the lithophile volatile elements (Yi *et al.*, 2000).

(ii) The light xenon isotopic compositions of well gases may be slightly different from those of the atmosphere in a manner that cannot be easily related to mass-dependent fractionation (Caffee *et al.*, 1988). These isotopes are not affected by radiogenic or fissionogenic additions. The effect is small and currently is one of the most important measurements that need to be made at higher precision. Any resolvable differences need to then be found in other reservoirs (e.g., MORBs) in order to establish that this is a fundamental difference indicating that a fraction of atmospheric xenon is not acquired by outgassing from the interior of the Earth. It could be that the Earth's atmospheric xenon was simply added later and that the isotopic compositions have no genetic link with those found in the Earth's interior.

(iii) Support for this possibility has come from the commonly held view that the Earth's water was added after the giant impact. Having lost so much of its volatiles by early degassing, hydrodynamic escape and impacts the Earth still has a substantial amount of water. Some have proposed that comets may have added a component of water to the Earth, but the D/H ratio would appear to be incorrect for this unless the component represented a minor fraction (Owen and Bar-Nun, 1995, 2000). An alternative set of proposals has been built around volatile-rich chondritic planetary embryos (Morbidelli *et al.*, 2000).

(iv) Further support for this latter "asteroidal" solution comes from the conclusion that the asteroid belt was at one time relatively massive. More than 99% of the mass of the asteroid belt has been ejected or added to other objects (Chambers and Wetherill, 2001). The Earth, being *en route* as some of the material preferentially travels toward the Sun may well have picked up a fraction of its volatiles in this way.

(v) The Moon itself provides a useful monitor of the amount of late material that could have been added to the Earth (Ryder *et al.*, 2000). The Moon provides an impact history (Hartmann *et al.*, 2000) that can be scaled to the Earth (Ryder *et al.*, 2000). In particular, there is evidence of widespread and intense bombardment of the Moon during the Hadean and this can be scaled up, largely in terms of relative cross-sectional area, to yield an impact curve for the Earth (Sleep *et al.*, 1989).

(vi) However, the Moon also is highly depleted in volatiles and its surface is very depleted in highly siderophile elements. Therefore, the Moon also provides a limit on how much can be added to the Earth. This is one reason why the more recent models of Cameron (2000) involving a giant impact that left an additional third are hard to accommodate. However, the database for this currently is poor (Righter *et al.*, 2000).

(vii) It has been argued that the greater depletion in iron and in tellurium in the silicate Earth relative to the Moon reflects an additional small amount of terrestrial core formation following the giant impact (Halliday *et al.*, 1996; Yi *et al.*, 2000). It could also simply reflect differences between Theia and the Earth. If there was further post-giant-impact core formation on Earth, it must have occurred prior to the addition of the late veneer.

2.11 THE HADEAN

2.11.1 Early Mantle Depletion

2.11.1.1 Introduction

Just as the I–Pu–Xe system is useful for studying the rate of formation of the atmosphere and U–Pb and Hf–W are ideal for studying the rates of accretion and core formation, lithophile element isotopic systems are useful for studying the history of melting of the silicate Earth. Two in particular, ^{92}Nb ($T_{1/2}=36$ Myr) and ^{146}Sm ($T_{1/2}=106$ Myr), have sufficiently long half-lives to be viable but have been explored with only limited success. Another chronometer of use is the long-lived chronometer ^{176}Lu ($T_{1/2}=34$–38 Gyr).

2.11.1.2 ^{92}Nb–^{92}Zr

^{92}Nb decays by electron capture to ^{92}Zr with a half-life of 36 ± 3 Myr. At one time it was thought to offer the potential to obtain an age for the Moon by dating early lunar ilmenites and the formation of ilmenite-rich layers in the lunar mantle. Others proposed that it provided constraints on the timescales for the earliest formation of continents on Earth (Münker *et al.*, 2000). In addition, it was argued that it would date terrestrial core formation (Jacobsen and Yin, 2001). There have been many attempts to utilize this isotopic system over the past few years. To do so, it is necessary to first determine the initial ^{92}Nb abundance in early solar system objects accurately and various authors have made claims that differ by two orders of magnitude.

Harper *et al.* (1991b) analyzed a single niobium-rutile found in the Toluca iron meteorite and presented the first evidence for the former existence of ^{92}Nb from which an initial ^{92}Nb/^{93}Nb of $(1.6\pm0.3)\times10^{-5}$ was inferred. However, the blank correction was very large. Subsequently, three studies using multiple collector inductively coupled plasma mass spectrometry proposed that the initial ^{92}Nb/^{93}Nb ratio of the solar system was more than two orders of magnitude higher (Münker *et al.*, 2000; Sanloup *et al.*, 2000; Yin *et al.*, 2000). Early processes that should fractionate Nb/Zr include silicate partial melting because niobium is more incompatible than zirconium (Hofmann *et al.*, 1986). Other processes relate to formation of titanium-rich (hence niobium-rich) and zirconium-rich minerals, the production of continental crust, terrestrial core formation (Wade and Wood, 2001) and the differentiation of the Moon. Therefore, on the basis of the very high ^{92}Nb abundance proposed, it was argued that, because there was no difference between the zirconium isotopic compositions of early terrestrial zircons and chondrites, the Earth's crust must have formed relatively late (Münker *et al.*, 2000). Similarly, because it is likely that a considerable amount of the Earth's niobium went into the core it was argued that core formation must have been protracted or delayed (Jacobsen and Yin, 2001). We now know that these arguments are incorrect. Precise internal isochrons have provided evidence that the initial abundance of ^{92}Nb in the early solar system is indeed low and close to 10^{-5} (Schönbächler *et al.*, 2002).

Therefore, rather than proving useful, the ^{92}Nb–^{92}Zr has no prospect of being able to provide constraints on these issues because the initial ^{92}Nb abundance is too low.

2.11.1.3 ^{146}Sm–^{142}Nd

High-quality terrestrial data now have been generated for the ^{146}Sm–^{142}Nd (half-life=106 Myr) chronometer (Goldstein and Galer, 1992; Harper and Jacobsen, 1992; McCulloch and Bennett, 1993; Sharma *et al.*, 1996). Differences in ^{142}Nd/^{144}Nd in early Archean rocks would indicate that the development of a crust on Earth was an early process and that subsequent recycling had failed to eradicate these effects. For many years, only one sample provided a hint of such an effect (Harper and Jacobsen, 1992) although these data have been questioned (Sharma *et al.*, 1996). Recently very high precision measurements of Isua sediments have resolved a 15 ± 4ppm effect (Caro *et al.*, 2003).

Any such anomalies are clearly small and far less than might be expected from extensive, repeated depletion of the mantle by partial melting in the Hadean. It seems inescapable that there was melting on the early Earth. Therefore, the interesting and important result of these studies is that such isotopic effects must largely have been eliminated. The most likely mechanism is very efficient mantle convection. In the earliest Earth convection may have been much more vigorous (Chase and Patchett, 1988; Galer and Goldstein, 1991) because of the large amount of heat left from the secular cooling and the greater radioactive heat production.

2.11.1.4 ^{176}Lu–^{176}Hf

A similar view is obtained from hafnium isotopic analyses of very early zircons. The ^{176}Lu–^{176}Hf isotopic system ($T_{1/2}$ = 34–38 Gyr) is ideally suited for studying early crustal evolution, because hafnium behaves in an almost identical fashion to zirconium. As a result, the highly resistant and easily dated mineral zircon typically contains ~1 wt% Hf, sufficient to render hafnium isotopic analyses of single zircons feasible using modern methods (Amelin *et al.*, 1999). The concentration of lutetium in zircon is almost negligible by comparison. As a result, the initial hafnium isotopic composition is relatively insensitive to the exact age of the grain and there is no error magnification involved in extrapolating back to the early Archean. Furthermore, one can determine the age of the single zircon grain very precisely using modern U–Pb methods. One can obtain an extremely precise initial hafnium isotopic composition for a particular point in time on a single grain, thereby avoiding the problems of mixed populations. The U–Pb age and hafnium isotopic compositions of zircons also are extremely resistant to resetting and define a reliable composition at a well-defined time in the early Earth. The hafnium isotopic composition that zircon had when it grew depends on whether the magma formed from a reservoir with a time-integrated history of melt depletion or enrichment. Therefore, one can use these early zircons to search for traces of early mantle depletion.

Note that this is similar to the approach adopted earlier with ^{147}Sm–^{143}Nd upon which many ideas of Hadean mantle depletion, melting processes and early crust were based (Chase and Patchett, 1988; Galer and Goldstein, 1991). However, the difficulty with insuring closed-system behavior with bulk rock Sm–Nd in metamorphic rocks and achieving a robust age correction of long-lived ^{147}Sm over four billion years has meant that this approach is now viewed as suspect (Nägler and Kramers, 1998). The ^{176}Lu–^{176}Hf isotopic system and use of low-Lu/Hf zircons is far more reliable in this respect (Amelin *et al.*, 1999, 2000). In practice, however, the interpretation is not that simple, for two reasons:

(i) The hafnium isotopic composition and Lu/Hf ratio of the Earth's primitive mantle is poorly known. It is assumed that it is broadly chondritic (Blichert-Toft and Albarède, 1997), but which exact kind of chondrite class best defines the isotopic composition of the primitive mantle is unclear. Without this information, one cannot extrapolate back in time to the early Earth and state with certainty what the composition of the primitive mantle reservoir was. Therefore, one cannot be sure what a certain isotopic composition means in terms of the level of time-integrated depletion.

(ii) The half-life of ^{176}Lu is *not* well established and is the subject of current debate and research (Scherer *et al.*, 2001). Although the determination of the initial hafnium isotopic composition of zircon is not greatly affected by this, because the Lu/Hf ratio is so low that the age correction is tiny, the correction to the value for the primitive mantle is very sensitive to this uncertainty.

With these caveats, one can deduce the following. Early single grains appear to have recorded hafnium isotopic compositions that provide evidence for chondritic or enriched reservoirs. There is no evidence of depleted reservoirs in the earliest (Hadean) zircons dated thus far (Amelin *et al.*, 1999). Use of alternative values for the decay constants or values for the primitive mantle parameters increases the proportion of hafnium with an enriched signature (Amelin *et al.*, 2000), but does not provide evidence for early mantle depletion events. Therefore, there is little doubt that the Hadean mantle was extremely well mixed. Why this should be is unclear, but it probably relates in some way to the lack of preserved continental material from prior to 4.0 Ga.

2.11.2 Hadean Continents

Except for the small amount of evidence for early mantle melting we are in the dark about how and when Earth's continents first formed (Figure 13). We already have pointed out that in its early stages Earth may have had a magma ocean, sustained by heat from accretion and the blanketing effects of a dense early atmosphere. With the loss of the early atmosphere during planetary collisions, the Earth would have cooled quickly, the outer portions would have solidified and it would thereby have developed its first primitive crust.

We have little evidence of what such a crust might have looked like. Unlike on the Moon and Mars, Earth appears to have no rock preserved that is more than 4.0 Gyr old. There was intense bombardment of the Moon until ~3.9 Gyr ago (Wetherill, 1975b; Hartmann *et al.*, 2000; Ryder *et al.*, 2000). Earth's earlier crust may therefore have been decimated by concomitant impacts. It may also be that a hotter Earth had a surface that

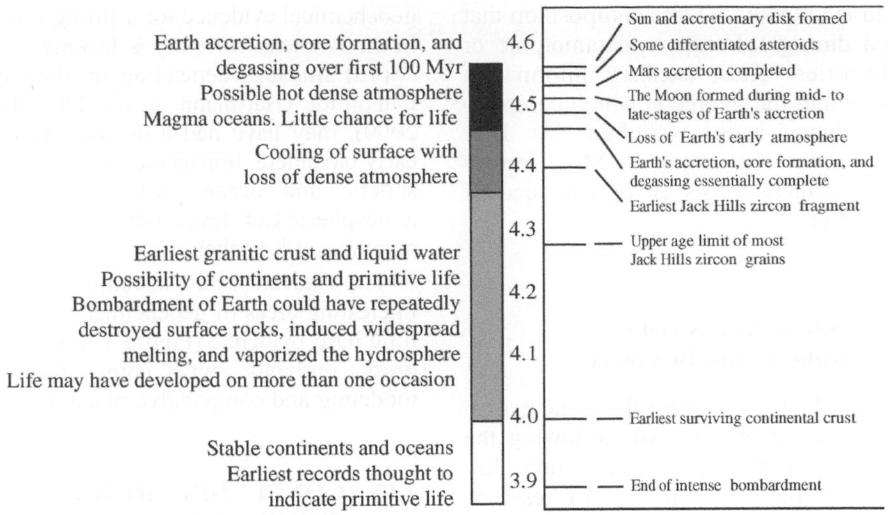

Figure 13 Schematic showing the timescales for various events through the "Dark Ages" of the Hadean.

was inherently unstable. Some argued that the earliest crust was like the lunar highlands—made from a welded mush of crystals that had previously floated on the magma ocean. Others have suggested that it was made of denser rocks more like those of the Earth's present oceanfloor (Galer and Goldstein, 1991). But firm evidence has so far been sparse.

Froude *et al.* (1983) reported the exciting discovery of pre-4.0 Ga zircon grains formed on Earth. The host rock from which these grains were recovered is not so old. The pre-4.0 Gyr rocks were largely destroyed but the zircons survived by becoming incorporated in sands that formed a sedimentary rock that now is exposed in Australia as the Jack Hills Metaconglomerate. By measuring uranium-lead ages a sizeable population of grains between 4.1 Gyr and 4.2 Gyr was discovered (Froude *et al.*, 1983). Subsequently, Wilde *et al.* (2001) and Mojzsis *et al.* (2001) reported uranium–lead ages and oxygen isotopic compositions of further old zircon grains. A portion of one grain appears to have formed 4.40 Gyr ago and this is the oldest terrestrial solid yet identified. More recent work has been published by Peck *et al.* (2001) and Valley *et al.* (2002).

These zircons provide powerful evidence for the former existence of some unknown amount of continental crust in the Hadean. Nearly all zircons grew from granite magmas, not similar at all to those forming the oceanfloor or the lunar highlands. Granite magmas usually form at >700 °C and >20 km depth, mainly by melting preexisting crust above subduction zones. Being buoyant they are typical of continental mountain regions such as the Andes—sites of very active erosion. The existence of such zircons would be consistent with continental crust as far back as 4.40 Gyr–~100 Myr after the formation of the Moon and the earliest atmosphere.

It would be nice to have more data and one is extrapolating through many orders of magnitude in mass when inferring extensive continents from single zircons. Large ocean islands like Iceland have small volumes of granitic magma and so one can conceive of protocontinents that started from the accumulation of such basaltic nuclei, the overall mass of silicic crustal material increasing gradually.

Oxygen isotopic measurements can be used to infer the presence of liquid water on the Earth's early surface. The oxygen isotopic composition reflects that of the magma from which the zircon crystallized, which in turn reflects that of the rocks that were melted to form the magma. Heavy oxygen (with a high proportion of ^{18}O) is produced by low-temperature interactions between a rock and liquid water such as those that form clay by weathering. The somewhat heavy oxygen of these zircons provides evidence that the rocks that were melted to form the magma included components that had earlier been at the surface in the presence of liquid water. Early on, when Earth was hotter, they might also have formed by melting of wet oceanfloor basalt that was taken back into the mantle by a process potentially comparable to modern subduction. Either way, the data indicate that surface rocks affected by low-temperature fluids were probably being transported to significant depths and melted, as occurs today.

These grains represent a unique archive for information on the early Earth. The potential is considerable. For example, Wilde *et al.* (2001) and Peck *et al.* (2001) also used trace elements and tiny inclusions to reconstruct the composition of the parent magma. In all of this work there is a need to be very aware that the grains may have been disturbed after they formed. Wet diffusion of oxygen could lead to an ^{18}O-rich composition that was acquired during subsequent metamorphic or magmatic histories. Thus, ancillary information on diffusivities and the degree to which compositions might have been perturbed by later metamorphism must be acquired. Most importantly, however, there exists an urgent need of many more grains.

2.11.3 The Hadean Atmosphere, Hydrosphere, and Biosphere

The nature of the Earth's atmosphere and hydrosphere after the Earth had cooled, following the cessation of the main stages of accretion, has been the subject of a fair amount of research (Wiechert, 2002), particularly given the *lack* of data upon which to base firm conclusions. An interesting and important question is to what extent it may have been possible for life to develop during this period (Mojzsis *et al.*, 1996; Sleep *et al.*, 2001; Zahnle and Sleep, 1996). A great deal has been written on this and because it is covered elsewhere in this treatise only cursory background is provided here.

The first-order constraints that exist on the nature of Earth's early exosphere (Sleep *et al.*, 1989; Sleep and Zahnle, 2001) are as follows:

(i) The Sun was fainter and cooler than today because of the natural start-up of fusion reactions that set it on the Main Sequence (Kasting and Grinspoon, 1991; Sagan and Chyba, 1997; Pavlov *et al.*, 2000). Therefore, the level of radiation will have been less.

(ii) Earth's interior was a few 100 K hotter because of secular cooling from accretion and far greater radiogenic heat production. The Earth's heat flow was 2–3 times higher. Therefore, one can assume that more heat was escaping via mantle melting and production of oceanic crust.

(iii) This, in turn, means more mantle-derived volatiles such as CO_2 were being released.

(iv) It also means that there was more hydrothermal alteration of the oceanfloor. Therefore, CO_2 was converted to carbonate in altered basalt and returned to the mantle at subduction zones (if those really existed already).

(v) There may have been far less marine carbonate. We can infer this from the geological record for the Archean. Therefore, it appears that atmospheric CO_2 levels were low—most of the CO_2 was being recycled to the mantle.

(vi) Because atmospheric CO_2 exerts a profound effect on temperature as a greenhouse gas, low concentrations of CO_2 imply that atmospheric temperatures were cold, unless another greenhouse gas such as methane (CH_4) was very abundant (Pavlov *et al.*, 2000). However, clear geochemical evidence for a strong role of methane in the Archean currently is lacking.

(vii) Impacts, depending on their number and magnitude (Hartmann *et al.*, 2000; Ryder *et al.*, 2000), may have had a devastating effect on the early biosphere. Impact ejecta will react with atmospheric and oceanic CO_2 and thereby lower atmospheric CO_2 levels reducing atmospheric temperatures still further.

The Hadean is fast becoming one of the most interesting areas of geochemical research. With so little hard evidence (Figure 13), much of the progress probably will come from theoretical modeling and comparative planetology.

2.12 CONCLUDING REMARKS—THE PROGNOSIS

Since the early 1990s, there has been great progress in understanding the origins and early development of the Earth. In some cases, this has been a function of improved modeling. This is particularly true for noble gases. However, in most cases it has been the acquisition of new kinds of data that has proven invaluable. The most obvious examples are in isotope geochemistry and cosmochemistry.

It is perhaps worth finishing by pointing out the kinds of developments that can be expected to have an impact on our understanding of the early Earth. One can ask a question like "what if we could measure. . .?" Here are some things that would be very interesting and useful to explore:

(i) If planets such as the Earth formed by very energetic collisions that were sufficient to cause vaporization of elements and compounds that normally are solid, it may be possible to find evidence for the kinetic effects of boiling in isotopic fractionations. Humayun and Clayton (1995) explored potassium at per mil levels of precision and found no significant evidence of fractionation. Poitrasson *et al.* (2003) have found evidence that the iron on the Moon may be very slightly heavy relative to other planetary objects. Note that this has nothing to do with the fractionations produced in the lunar surface during irradiation and implantation (Wiesli *et al.*, 2003). Perhaps boiling during the giant impact caused this. This is just a preliminary inference at this stage. But if so, small fractionations also should be found in other elements of similar volatility such as lithium, magnesium, silicon, and nickel. There is much to be done to explore these effects at very high precision.

(ii) A vast amount of work still is needed on core formation, understanding the depletion of siderophile elements in the Earth's mantle (Walter *et al.*, 2000), and determining the abundances of light elements in the core (Gessmann *et al.*, 2001). Much of this depends on using proxy elements that are very sensitive to pressure, temperature, water content or f_{O_2} (e.g., Wade and Wood, 2001). A problem at present is that too many of the elements of interest are sensitive to more than one parameter and, therefore, the solutions are under constrained. Also, new experiments are needed at very high pressures—close to those of the core–mantle boundary.

(iii) The origins of Earth's water and volatiles more generally are the subject of considerable debate (e.g., Anders and Owen, 1977; Carr and Wänke, 1992; Owen and Bar-Nun, 1995, 2000; Javoy, 1998; Caffee *et al.*, 1999; Righter and Drake, 1999; Abe *et al.*, 2000; Morbidelli *et al.*, 2000). A great deal probably will be learned from further modeling. For example, it is very important to understand what kinds of processes in the early Earth could have caused loss of early atmospheres. We do not understand how volatiles are retained during a Moon-forming giant impact (Melosh and Sonett, 1986; Melosh, 1990; Melosh *et al.*, 1993) or what the early Sun might have done to the atmosphere. We need to acquire more reliable data on the isotopic compositions of volatiles in the deep Earth. We might be able to learn much more about Earth's volatile history from more precise measurements of the volatile components in other planets. In this respect more detailed studies of Mars and how closely it resembles the Earth (Carr and Wänke, 1992) could prove critical.

(iv) Mojzsis *et al.* (2001) and Wilde *et al.* (2001) have made major advances in studying the Hadean using single zircons. Apart from needing many more such samples, one has to ask what other kinds of information might be extractable from such zircons. The oxygen isotopic composition provides evidence for early low-temperature water. Exploring the melt inclusions and the trace-element concentrations also has been shown to have potential (Peck *et al.*, 2001). Zircons are also iron-rich and conceivably could eventually be used to provide evidence for biological processes in the Hadean. However, the sensitivities of the techniques need to be improved vastly for this to be achieved with single grains. Furthermore, the current status of the rapidly expanding field of iron isotope geochemistry provides no clear basis for assuming a distinctive signal of biotic effects will be realizable. Also, the really interesting zircons are so precious that one should use minimally destructive techniques like SIMS on single grains. However, the required precision for measuring isotopic ratios in a useful manner is not available for trace elements in minerals using this method at present. Similarly, hafnium isotopes on single zircon grains provide the most reliable and powerful constraints on the extent of mantle depletion (Amelin *et al.*, 1999) but require destruction of a part of the grain. Developing improved methods that achieve far higher overall sensitivity is critical.

(v) Determining the rates of accretion of the Earth and Moon more reliably will be critically dependent on the correct and precise determination of the initial tungsten and hafnium isotopic compositions of the solar system, and the ^{182}Hf decay constant. The initial isotopic compositions really require the more widespread application of negative ion thermal ionization mass spectrometry (N-TIMS) (Quitté *et al.*, 2000). The decay constant work is going to require the acquisition of ^{182}Hf, probably from neutron-irradiated ^{180}Hf.

(vi) Similarly, progress in using hafnium isotopes to study the degree of early mantle depletion is being thwarted by the uncertainties associated with the ^{176}Lu decay constant (Scherer *et al.*, 2001), and some new experimental work is needed in this area.

(vii) A major need is for a closer integration of the modeling of different isotopic systems. As of early 2000s, this has really only been attempted for tungsten and lead (Halliday, 2000). In the future it will be essential to integrate xenon isotopes in with these and other accretion models.

(viii) Finally, another major area of modeling has to occur in the area of "early Earth system science." There needs to be integrated modeling of the evolution of the atmosphere, oceans, surface temperature, mantle convection, and magma oceans. This is now being attempted. For example, the studies by Sleep and Zahnle (2001) and Sleep *et al.* (1989, 2001) are paving the way for more comprehensive models that might involve the fluid dynamics of mantle convection.

ACKNOWLEDGMENTS

This chapter benefited enormously from discussion with, and comments and criticism received from Tom Ahrens, Alan Boss, Pat Cassen, Andy Davis, Martin Frank, Tim Grove, Munir Humayun, Don Porcelli, Norm Sleep, Mike Walter, Uwe Wiechert, Rainer Wieler, Kevin Zahnle, and two anonymous reviewers.

REFERENCES

Abe Y. and Matsui T. (1985) The formation of an impact-generated H_2O atmosphere and its implications for the early thermal history of the Earth. *Proc. 15th Lunar Planet. Sci. Conf.: J. Geophys. Res.* **90**(suppl.), C545–C559.

Abe Y. and Matsui T. (1986) Early evolution of the earth: accretion, atmosphere formation, and thermal history. *Proc. 17th Lunar Planet. Sci. Conf.: J. Geophys. Res.* **91**(no. B13), E291–E302.

Abe Y. and Matsui T. (1988) Evolution of an impact-generated H_2O–CO_2 atmosphere and formation of a hot proto-ocean on Earth. *Am. Meteorol. Soc.* **45**, 3081–3101.

Abe Y., Ohtani E., and Okuchi T. (2000) Water in the early earth. In *Origin of the Earth and Moon* (eds. R. M. Canup and K. Righter). University of Arizona Press, Tucson, pp. 413–433.

Agee C. B. and Walker D. (1988) Mass balance and phase density constraints on early differentiation of chondritic mantle. *Earth Planet. Sci. Lett.* **90**, 144–156.

Agnor C. B. and Ward W. R. (2002) Damping of terrestrial-planet eccentricities by density-wave interactions with a remnant gas disk. *Astrophys. J.* **567**, 579–586.

Ahrens T. J. (1990) Earth accretion. In *Origin of the Earth* (eds. H. E. Newsom and J. H. Jones). Oxford University Press, Oxford, pp. 211–217.

Ahrens T. J. and Jeanloz R. (1987) Pyrite shock compression, isentropic release and composition of the Earth's core. *J. Geophys. Res.* **92**, 10363–10375.

Albarède F. and Juteau M. (1984) Unscrambling the lead model ages. *Geochim. Cosmochim. Acta* **48**, 207–212.

Alibert C., Norman M. D., and McCulloch M. T. (1994) An ancient Sm–Nd age for a ferroan noritic anorthosite clast from lunar breccia 67016. *Geochim. Cosmochim. Acta* **58**, 2921–2926.

Allègre C. J., Dupré B., and Brévart O. (1982) Chemical aspects of formation of the core. *Phil. Trans. Roy. Soc. London* **A306**, 49–59.

Allègre C. J., Staudacher T., Sarda P., and Kurz M. (1983) Constraints on evolution of Earth's mantle from rare gas systematics. *Nature* **303**, 762–766.

Allègre C. J., Staudacher T., and Sarda P. (1987) Rare gas systematics: formation of the atmosphere, evolution, and structure of the Earth's mantle. *Earth Planet. Sci. Lett.* **81**, 127–150.

Allègre C. J., Manhès G., and Göpel C. (1995a) The age of the Earth. *Geochim. Cosmochim. Acta* **59**, 1445–1456.

Allègre C. J., Poirier J.-P., Humler E., and Hofmann A. W. (1995b) The chemical composition of the Earth. *Earth Planet. Sci. Lett.* **134**, 515–526.

Allègre C. J., Hofmann A. W., and O'Nions R. K. (1996) The argon constraints on mantle structure. *Geophys. Res. Lett.* **23**, 3555–3557.

Allègre C. J., Manhès G., and Lewin E. (2001) Chemical composition of the Earth and the volatility control on planetary genetics. *Earth Planet. Sci. Lett.* **185**, 49–69.

Amelin Y., Lee D.-C., Halliday A. N., and Pidgeon R. T. (1999) Nature of the Earth's earliest crust from hafnium isotopes in single detrital zircons. *Nature* **399**, 252–255.

Amelin Y., Lee D.-C., and Halliday A. N. (2000) Early-middle Archean crustal evolution deduced from Lu–Hf and U–Pb isotopic studies of single zircon grains. *Geochim. Cosmochim. Acta* **64**, 4205–4225.

Amelin Y., Krot A. N., Hutcheon I. D., and Ulyanov A. A. (2002) Lead isotopic ages of chondrules and calcium-aluminum-rich inclusions. *Science* **297**, 1678–1683.

Anders E. (1977) Chemical compositions of the Moon, Earth, and eucrite parent body. *Phil. Trans. Roy. Soc. London* **A295**, 23–40.

Anders E. and Grevesse N. (1989) Abundances of the elements: meteoritic and solar. *Geochim. Cosmochim. Acta* **53**, 197–214.

Anders E. and Owen T. (1977) Mars and Earth: origin and abundance of volatiles. *Science* **198**, 453–465.

Anderson D. L. (1979) Chemical stratification of the mantle. *J. Geophys. Res.* **84**, 6297–6298.

Arculus R. J. and Delano J. W. (1981) Siderophile element abundances in the upper mantle: evidence for a sulfide signature and equilibrium with the core. *Geochim. Cosmochim. Acta* **45**, 1331–1344.

Artymowicz P. (1997) Beta Pictoris: an early solar system? *Ann. Rev. Earth Planet. Sci.* **25**, 175–219.

Azbel I. Y., Tolstikhin I. N., Kramers J. D., Pechernikova G. V., and Vityazev A. V. (1993) Core growth and siderophile element depletion of the mantle during homogeneous Earth accretion. *Geochim. Cosmochim. Acta* **57**, 2889–2898.

Ballentine C. J., Porcelli D., and Wieler R. (2001) A critical comment on Trieloff *et al.*. *Science* **291**, 2269 (online).

Bell K. R., Cassen P. M., Wasson J. T., and Woolum D. S. (2000) The FU Orionis phenomenon and solar nebular material. In *Protostars and Planets IV* (eds. V. Mannings, A. P. Boss, and S. S. Russell). University of Arizona Press, Tucson, pp. 897–926.

Benz W. (2000) Low velocity collisions and the growth of planetesimals. *Space Sci. Rev.* **92**, 279–294.

Benz W. and Asphaug E. (1999) Catastrophic disruptions revisited. *Icarus* **142**, 5–20.

Benz W. and Cameron A. G. W. (1990) Terrestrial effects of the giant impact. In *Origin of the Earth* (eds. H. E. Newsom and J. H. Jones). Oxford University Press, Oxford, pp. 61–67.

Benz W., Cameron A. G. W., and Slattery W. L. (1987) Collisional stripping of Mercury's mantle. *Icarus* **74**, 516–528.

Binder A. B. (1986) The binary fission origin of the Moon. In *Origin of the Moon* (eds. W. K. Hartmann, R. J. Phillips, and G. J. Taylor). Lunar and Planetary Institute, Houston, pp. 499–516.

Birck J.-L., Rotaru M., and Allègre C. J. (1999) ^{53}Mn–^{53}Cr evolution of the early solar system. *Geochim. Cosmochim. Acta* **63**, 4111–4117.

Blichert-Toft J. and Albarède F. (1997) The Lu–Hf isotope geochemistry of chondrites and the evolution of the mantle–crust system. *Earth Planet. Sci. Lett.* **148**, 243–258.

Blum J. (2000) Laboratory experiments on preplanetary dust aggregation. *Space Sci. Rev.* **92**, 265–278.

Boss A. P. (1990) 3D Solar nebula models: implications for Earth origin. In *Origin of the Earth* (eds. H. E. Newsom and J. H. Jones). Oxford University Press, Oxford, pp. 3–15.

Boss A. P. (1997) Giant planet formation by gravitational instability. *Science* **276**, 1836–1839.

Boss A. P. (1998) *Looking for Earths.* Wiley, New York, 240pp.

Boss A. P. (2002) The formation of gas and ice giant planets. *Earth Planet. Sci. Lett.* **202**, 513–523.

Brandon A. D., Walker R. J., Morgan J. W., Norman M. D., and Prichard H. M. (1998) Coupled ^{186}Os and ^{187}Os evidence for core–mantle interaction. *Science* **280**, 1570–1573.

Brett R. (1984) Chemical equilibrium between Earth's core and upper mantle. *Geochim. Cosmochim. Acta* **48**, 1183–1188.

Brown G. C. and Mussett A. E. (1981) *The Inaccessible Earth.* Allen and Unwin, London.

Brown H. (1949) Rare gases and the formation of the Earth's atmosphere. In *The Atmospheres of the Earth and Planets* (ed. G. P. Kuiper). University of Chicago Press, Chicago, pp. 258–266.

Busemann H., Baur H., and Wieler R. (2000) Primordial noble gases in "phase Q" in carbonaceous and ordinary chondrites studied by closed-system stepped etching. *Meteorit. Planet. Sci.* **35**, 949–973.

Caffee M. W., Hudson G. B., Velsko C., Alexander E. C., Jr., Huss G. R., and Chivas A. R. (1988) Non-atmospheric noble gases from CO_2 well gases. In *Lunar Planet. Sci.* **XIX**. The Lunar and Planetary Institute, Houston, pp. 154–155.

Caffee M. W., Hudson G. B., Velsko C., Huss G. R., Alexander E. C., Jr., and Chivas A. R. (1999) Primordial noble gases from Earth's mantle: identification of a primitive volatile component. *Science* **285**, 2115–2118.

Cameron A. G. W. (1978) Physics of the primitive solar accretion disk. *Moons and Planets* **18**, 5–40.

Cameron A. G. W. (2000) Higher-resolution simulations of the Giant Impact. In *Origin of the Earth and Moon* (eds. K. Righter and R. Canup). University of Arizona Press, Tucson, pp. 133–144.

Cameron A. G. W. and Benz W. (1991) Origin of the Moon and the single impact hypothesis: IV. *Icarus* **92**, 204–216.

Canup R. M. and Agnor C. (1998) Accretion of terrestrial planets and the earth–moon system. In *Origin of the Earth and Moon,* LPI Contribution No. **597** Lunar and Planetary Institute, Houston, pp. 4–7.

Canup R. M. and Asphaug E. (2001) Origin of the Moon in a giant impact near the end of the Earth's formation. *Nature* **412**, 708–712.

Carlson R. W. and Hauri E. H. (2001) Extending the ^{107}Pd–^{107}Ag chronometer to Low Pd/Ag meteorites with the MC-ICPMS. *Geochim. Cosmochim. Acta* **65**, 1839–1848.

Carlson R. W. and Lugmair G. W. (1988) The age of ferroan anorthosite 60025: oldest crust on a young Moon? *Earth Planet. Sci. Lett.* **90**, 119–130.

Carlson R. W. and Lugmair G. W. (2000) Timescales of planetesimal formation and differentiation based on extinct and extant radioisotopes. In *Origin of the Earth and Moon* (eds. K. Righter and R. Canup). University of Arizona Press, Tucson, pp. 25–44.

Caro G., Bourdon B., Birck J.-L., and Moorbath S. (2003) ^{146}Sm–^{142}Nd evidence from Isua metamorphosed sediments for early differentiation of the Earth's mantle. *Nature* **423**, 428–431.

Carr M. H. and Wänke H. (1992) Earth and Mars: water inventories as clues to accretional histories. *Icarus* **98**, 61–71.

Cassen P. (1996) Models for the fractionation of moderately volatile elements in the solar nebula. *Meteorit. Planet. Sci.* **31**, 793 806.

Cassen P. (2001) Nebular thermal evolution and the properties of primitive planetary materials. *Meteorit. Planet. Sci.* **36**, 671–700.

Chambers J. E. and Wetherill G. W. (2001) Planets in the asteroid belt. *Meteorit. Planet. Sci.* **36**, 381–399.

Chase C. G. and Patchett P. J. (1988) Stored mafic/ultramafic crust and Early Archean mantle depletion. *Earth Planet. Sci. Lett.* **91**, 66–72.

Chiang E. I. and Goldreich P. (1997) Spectral energy distributions of T Tauri stars with passive circumstellar disks. *Astrophys. J.* **490**, 368.

Chiang E. I. and Goldreich P. (1999) Spectral energy distributions of passive T Tauri disks: inclination. *Astrophys. J.* **519**, 279.

Chiang E. I., Joung M. K., Creech-Eakman M. J., Qi C., Kessler J. E., Blake G. A., and van Dishoeck E. F. (2001) Spectral energy distributions of passive T Tauri and Herbig Ae disks: grain mineralogy, parameter dependences, and comparison with ISO LWS observations. *Astrophys. J.* **547**, 1077.

Chou C. L. (1978) Fractionation of siderophile elements in the Earth's upper mantle. *Proc. 9th Lunar Sci. Conf.* 219–230.

Clark S. P., Jr., Turekian K. K., and Grossman L. (1972) Model for the early history of the Earth. In *The Nature of the Solid Earth* (ed. E. C. Robertson). McGraw-Hill, New York, pp. 3–18.

Clarke W. B., Beg M. A., and Craig H. (1969) Excess ^{3}He in the sea: evidence for terrestrial primordial helium. *Earth Planet. Sci. Lett.* 6, 213–220.

Clayton R. N. (1986) High temperature isotope effects in the early solar system. In *Stable Isotopes in High Temperature Geological Processes* (eds. J. W. Valley, H. P. Taylor, and J. R. O'Neil). Mineralogical Society of America, Washington, DC, pp. 129–140.

Clayton R. N. (1993) Oxygen isotopes in meteorites. *Ann. Rev. Earth Planet. Sci.* 21, 115–149.

Clayton R. N. and Mayeda T. K. (1975) Genetic relations between the moon and meteorites. *Proc. 11th Lunar Sci. Conf.* 1761–1769.

Clayton R. N. and Mayeda T. K. (1996) Oxygen isotope studies of achondrites. *Geochim. Cosmochim. Acta* **60**, 1999–2017.

Clayton R. N., Grossman L., and Mayeda T. K. (1973) A component of primitive nuclear composition in carbonaceous meteorites. *Science* **182**, 485–487.

Connolly H. C., Jr. and Love S. G. (1998) The formation of chondrules: petrologic tests of the shock wave model. *Science* **280**, 62–67.

Connolly H. C., Jr., Jones B. D., and Hewins R. G. (1998) The flash melting of chondrules: an experimental investigation into the melting history and physical nature of chondrules. *Geochim. Cosmochim. Acta* **62**, 2725–2735.

Craig H. and Lupton J. E. (1976) Primordial neon, helium, and hydrogen in oceanic basalts. *Earth Planet. Sci. Lett.* **31**, 369–385.

Cuzzi J. N., Davis S. S., and Dobrovolskis A. R. (2003) Creation and distribution of CAIs in the protoplanetry nebula. In *Lunar Planet. Sci.* XXXIV, #1749. The Lunar and Planetary Institute, Houston (CD-ROM).

Darwin G. H. (1878) On the precession of a viscous spheroid. *Nature* **18**, 580–582.

Darwin G. H. (1879) On the precession of a viscous spheroid and on the remote history of the Earth. *Phil Trans. Roy. Soc. London* **170**, 447–538.

Dauphas N., Marty B., and Reisberg L. (2002) Inference on terrestrial genesis from molybdenum isotope systematics. *Geophys. Res. Lett.* **29**, 1084 doi:10.1029/2001GL014237.

Davies G. F. (1999) Geophysically constrained mass flows and the ^{40}Ar budget: a degassed lower mantle? *Earth Planet. Sci. Lett.* **166**, 149–162.

Déruelle B., Dreibus G., and Jambon A. (1992) Iodine abundances in oceanic basalts: implications for Earth dynamics. *Earth Planet. Sci. Lett.* **108**, 217–227.

Desch S. J. and Cuzzi J. N. (2000) The generation of lightning in the solar nebula. *Icarus* **143**, 87–105.

Dixon E. T., Honda M., McDougall I., Campbell I. H., and Sigurdsson I. (2000) Preservation of near-solar neon isotopic ratios in Icelandic basalts. *Earth Planet. Sci. Lett.* **180**, 309–324.

Doe B. R. and Stacey J. S. (1974) The application of lead isotopes to the problems of ore genesis and ore prospect evaluation: a review. *Econ. Geol.* **69**, 755–776.

Drake M. J. and McFarlane E. A. (1993) Mg-perovskite/silicate melt and majorite garnet/silicate melt partition coefficients in the system CaO–MgO–SiO_2 at high temperatures and pressures. *J. Geophys. Res.* **98**, 5427–5431.

Drake M. J. and Righter K. (2002) Determining the composition of the Earth. *Nature* **416**, 39–44.

Dreibus G. and Palme H. (1996) Cosmochemical constraints on the sulfur content in the Earth's core. *Geochim. Cosmochim. Acta* **60**, 1125–1130.

Esat T. M. (1996) Comment on Humayun and Clayton (1995). *Geochim. Cosmochim. Acta* **60**, 2755–2758.

Eucken A. (1944) Physikalisch-Chemische Betrachtungen über die früheste Entwicklungsgeschichte der Erde. *Nachr. Akad. Wiss. Göttingen, Math-Phys. Kl.* **Heft 1**, 1–25.

Farley K. A. and Poreda R. (1992) Mantle neon and atmospheric contamination. *Earth Planet. Sci. Lett.* **114**, 325–339.

Feigelson E. D., Broos P., Gaffney J. A., III, Garmire G., Hillenbrand L. A., Pravdo S. H., Townsley L., and Tsuboi Y. (2002a) X-ray-emitting young stars in the Orion nebula. *Astrophys. J.* **574**, 258–292.

Feigelson E. D., Garmire G. P., and Pravdo S. H. (2002b) Magnetic flaring in the pre-main-sequence Sun and implications for the early solar system. *Astrophys. J.* **572**, 335–349.

Franchi I. A., Wright I. P., Sexton A. S., and Pillinger C. T. (1999) The oxygen isotopic composition of Earth and Mars. *Meteorit. Planet. Sci.* **34**, 657–661.

Froude D. O., Ireland T. R., Kinny P. D., Williams I. S., Compston W., Williams I. R., and Myers J. S. (1983) Ion microprobe identification of 4,100–4,200 Myr-old terrestrial zircons. *Nature* **304**, 616–618.

Gaffey M. J. (1990) Thermal history of the asteroid belt: implications for accretion of the terrestrial planets. In *Origin of the Earth* (eds. H. E. Newsom and J. H. Jones). Oxford University Press, Oxford, pp. 17–28.

Galer S. J. G. and Goldstein S. L. (1991) Early mantle differentiation and its thermal consequences. *Geochim. Cosmochim. Acta* **55**, 227–239.

Galer S. J. G. and Goldstein S. L. (1996) Influence of accretion on lead in the Earth. In *Isotopic Studies of Crust–Mantle*

Evolution (eds. A. R. Basu and S. R. Hart). American Geophysical Union, Washington, DC, pp. 75–98.

Ganapathy R. and Anders E. (1974) Bulk compositions of the Moon and Earth, estimated from meteorites. *Proc. 5th Lunar Conf.* 1181–1206.

Gancarz A. J. and Wasserburg G. J. (1977) Initial Pb of the Amîtsoq Gneiss, West Greenland, and implications for the age of the Earth. *Geochim. Cosmochim. Acta* **41**, 1283–1301.

Gast P. W. (1960) Limitations on the composition of the upper mantle. *J. Geophys. Res.* **65**, 1287–1297.

Gessmann C. K., Wood B. J., Rubie D. C., and Kilburn M. R. (2001) Solubility of silicon in liquid metal at high pressure: implications for the composition of the Earth's core. *Earth Planet. Sci. Lett.* **184**, 367–376.

Ghosh A. and McSween H. Y., Jr. (1999) Temperature dependence of specific heat capacity and its effect on asteroid thermal models. *Meteorit. Planet. Sci.* **34**, 121–127.

Goldreich P. and Ward W. R. (1973) The formation of planetesimals. *Astrophys. J.* **183**, 1051–1060.

Goldstein S. L. and Galer S. J. G. (1992) On the trail of early mantle differentiation: ^{142}Nd/^{144}Nd ratios of early Archean rocks. *Trans. Am. Geophys. Union,* 323.

Göpel C., Manhès G., and Allègre C. J. (1991) Constraints on the time of accretion and thermal evolution of chondrite parent bodies by precise U–Pb dating of phosphates. *Meteoritics* **26**, 73.

Göpel C., Manhès G., and Allègre C. J. (1994) U–Pb systematics of phosphates from equilibrated ordinary chondrites. *Earth Planet. Sci. Lett.* **121**, 153–171.

Gounelle M., Shu F. H., Shang H., Glassgold A. E., Rehm K. E., and Lee T. (2001) Extinct radioactivities and protosolar cosmic-rays: self-shielding and light elements. *Astrophys. J.* **548**, 1051–1070.

Gray C. M., Papanastassiou D. A., and Wasserburg G. J. (1973) The identification of early condensates from the solar nebula. *Icarus* **20**, 213–239.

Grevesse N. and Sauval A. J. (1998) Standard solar composition. *Space Sci. Rev.* **85**, 161–174.

Grimm R. E. and McSween H. Y., Jr. (1993) Heliocentric zoning of the asteroid belt by aluminum-26 heating. *Science* **259**, 653–655.

Grossman J. N. (1996) Chemical fractionations of chondrites: signatures of events before chondrule formation. In *Chondrules and the Protoplanetary Disk* (eds. R. H. Hewins, R. H. Jones, and E. R. D. Scott). Cambridge University of Press, Cambridge, pp. 243–253.

Grossman L. (1972) Condensation in the primitive solar nebula. *Geochim. Cosmochim. Acta* **36**, 597–619.

Grossman L. and Larimer J. W. (1974) Early chemical history of the solar system. *Rev. Geophys. Space Phys.* **12**, 71–101.

Guan Y., McKeegan K. D., and MacPherson G. J. (2000) Oxygen isotopes in calcium–aluminum-rich inclusions from enstatite chondrites: new evidence for a single CAI source in the solar nebula. *Earth Planet. Sci. Lett.* **183**, 557–558.

Halliday A. N. (2000) Terrestrial accretion rates and the origin of the Moon. *Earth Planet. Sci. Lett.* **176**, 17–30.

Halliday A. N. and Lee D.-C. (1999) Tungsten isotopes and the early development of the Earth and Moon. *Geochim. Cosmochim. Acta* **63**, 4157–4179.

Halliday A. N. and Porcelli D. (2001) In search of lost planets—the paleocosmochemistry of the inner solar system. *Earth Planet. Sci. Lett.* **192**, 545–559.

Halliday A. N., Lee D.-C., Tomassini S., Davies G. R., Paslick C. R., Fitton J. G., and James D. (1995) Incompatible trace elements in OIB and MORB and source enrichment in the sub-oceanic mantle. *Earth Planet. Sci. Lett.* **133**, 379–395.

Halliday A. N., Rehkämper M., Lee D.-C., and Yi W. (1996) Early evolution of the Earth and Moon: new constraints from Hf–W isotope geochemistry. *Earth Planet. Sci. Lett.* **142**, 75–89.

Halliday A. N., Lee D.-C., and Jacobsen S. B. (2000) Tungsten isotopes, the timing of metal-silicate fractionation and the origin of the Earth and Moon. In *Origin of the Earth and Moon* (eds. R. M. Canup and K. Righter). University of Arizona Press, Tucson, pp. 45–62.

Halliday A. N., Wänke H., Birck J.-L., and Clayton R. N. (2001) The accretion, bulk composition and early differentiation of Mars. *Space Sci. Rev.* **96**, 197–230.

Hanan B. B. and Tilton G. R. (1987) 60025: Relict of primitive lunar crust? *Earth Planet. Sci. Lett.* **84**, 15–21.

Hanks T. C. and Anderson D. L. (1969) The early thermal history of the earth. *Phys. Earth Planet. Int.* **2**, 19–29.

Harper C. L. and Jacobsen S. B. (1992) Evidence from coupled ^{147}Sm–^{143}Nd and ^{146}Sm–^{142}Nd systematics for very early (4.5-Gyr) differentiation of the Earth's mantle. *Nature* **360**, 728–732.

Harper C. L. and Jacobsen S. B. (1996a) Noble gases and Earth's accretion. *Science* **273**, 1814–1818.

Harper C. L. and Jacobsen S. B. (1996b) Evidence for ^{182}Hf in the early solar system and constraints on the timescale of terrestrial core formation. *Geochim. Cosmochim. Acta* **60**, 1131–1153.

Harper C. L., Völkening J., Heumann K. G., Shih C.-Y., and Wiesmann H. (1991a) ^{182}Hf–^{182}W: new cosmochronometric constraints on terrestrial accretion, core formation, the astrophysical site of the r-process, and the origin of the solar system. In *Lunar Planet. Sci.* **XXII.** The Lunar and Planetary Science, Houston, pp. 515–516.

Harper C. L., Wiesmann H., Nyquist L. E., Howard W. M., Meyer B., Yokoyama Y., Rayet M., Arnould M., Palme H., Spettel B., and Jochum K. P. (1991b) ^{92}Nb/^{93}Nb and ^{92}Nb/^{146}Sm ratios of the early solar system: observations and comparison of p-process and spallation models. In *Lunar Planet. Sci.* **XXII.** The Lunar and Planetary Institute, Houston, pp. 519–520.

Hartmann L. (2000) Observational constraints on transport (and mixing) in pre-main sequence disks. *Space Sci. Rev.* **92**, 55–68.

Hartmann L. and Kenyon S. J. (1996) The FU Orionis phenomena. *Ann. Rev. Astron. Astrophys.* **34**, 207–240.

Hartmann W. K. and Davis D. R. (1975) Satellite-sized planetesimals and lunar origin. *Icarus* **24**, 505–515.

Hartmann W. K., Ryder G., Dones L., and Grinspoon D. (2000) The time-dependent intense bombardment of the primordial Earth/Moon system. In *Origin of the Earth and Moon* (eds. R. Canup and K. Righter). University of Arizona Press, Tucson, pp. 493–512.

Hayashi C., Nakazawa K., and Mizuno H. (1979) Earth's melting due to the blanketing effect of the primordial dense atmosphere. *Earth Planet. Sci. Lett.* **43**, 22–28.

Hayashi C., Nakazawa K., and Nakagawa Y. (1985) Formation of the solar system. In *Protostars and Planets II* (eds. D. C. Black and D. S. Matthews). University of Arizona Press, Tucson, pp. 1100–1153.

Herzberg C. (1984) Chemical stratification in the silicate Earth. *Earth Planet. Sci. Lett.* **67**, 249–260.

Herzberg C., Jeigenson M., Skuba C., and Ohtani E. (1988) Majorite fractionation recorded in the geochemistry of peridotites from South Africa. *Nature* **332**, 823–826.

Hewins R. H. and Herzberg C. (1996) Nebular turbulence, chondrule formation, and the composition of the Earth. *Earth Planet. Sci. Lett.* **144**, 1–7.

Hofmann A. W. (1988) Chemical differentiation of the Earth: the relationship between mantle, continental crust and oceanic crust. *Earth Planet. Sci. Lett.* **90**, 297–314.

Hofmann A. W., Jochum K. P., Seufert M., and White W. M. (1986) Nb and Pb in oceanic basalts: new constraints on mantle evolution. *Earth Planet. Sci. Lett.* **79**, 33–45.

Honda M., McDougall I., Patterson D. B., Doulgeris A., and Clague D. A. (1991) Possible solar noble-gas component in Hawaiian basalts. *Nature* **349**, 149–151.

Horan M. F., Smoliar M. I., and Walker R. J. (1998) ^{182}W and ^{187}Re–^{187}Os systematics of iron meteorites: chronology for melting, differentiation, and crystallization in asteroids.*Geochim. Cosmochim. Acta* **62**, 545–554.

Hoyle F. (1960) On the origin of the solar nebula. *Quart. J. Roy.Astron. Soc.* **1**, 28–55.

Hsui A. T. and Toksöz M. N. (1977) Thermal evolution of planetary size bodies. *Proc. 8th Lunar Sci. Conf.* 447–461.

Humayun M. and Cassen P. (2000) Processes determining the volatile abundances of the meteorites and terrestrial planets.In *Origin of the Earth and Moon* (eds. R. M. Canup and K. Righter). University of Arizona Press, Tucson, pp. 3–23.

Humayun M. and Clayton R. N. (1995) Potassium isotope cosmochemistry: genetic implications of volatile element depletion. *Geochim. Cosmochim. Acta* **59**, 2131–2151.

Hunten D. M., Pepin R. O., and Walker J. C. G. (1987) Mass fractionation in hydrodynamic escape. *Icarus* **69**, 532–549.

Huss G. R. (1988) The role of presolar dust in the formation of the solar system. *Earth Moon Planet.* **40**, 165–211.

Huss G. R. (1997) The survival of presolar grains during the formation of the solar system. In *Astrophysical Implications of the Laboratory Study of Presolar Materials,* American Institute of Physics Conf. Proc. 402, Woodbury, New York (eds. T. J. Bernatowicz and E. Zinner), pp. 721–748.

Huss G. R. and Lewis R. S. (1995) Presolar diamond, SiC, and graphite in primitive chondrites: abundances as a function of meteorite class and petrologic type. *Geochim. Cosmochim. Acta* **59**, 115–160.

Ireland T. R. (1991) The abundance of ^{182}Hf in the early solar system. In *Lunar Planet. Sci.* **XXII.** The Lunar and Planetary Institute, Houston, pp. 609–610.

Ireland T. R. and Wlotzka F. (1992) The oldest zircons in the solar system. *Earth Planet. Sci. Lett.* **109**, 1–10.

Jackson I. (1983) Some geophysical constraints on the chemical composition of the Earth's lower mantle. *Earth Planet. Sci. Lett.* **62**, 91–103.

Jacobsen S. B. and Harper C. L., Jr. (1996) Accretion and early differentiation history of the Earth based on extinct radionuclides. In *Earth Processes: Reading the Isotope Code* (eds. E. Basu and S. Hart). American Geophysical Union, Washington, DC, pp. 47–74.

Jacobsen S. B. and Wasserburg G. J. (1979) The mean age of mantle and crust reservoirs. *J. Geophys. Res.* **84**, 7411–7427.

Jacobsen S. B. and Yin Q. Z. (2001) Core formation models and extinct nuclides. In *Lunar Planet. Sci.* **XXXII**, #1961. The Lunar and Planetary Institute, Houston (CD-ROM).

Jagoutz E., Palme J., Baddenhausen H., Blum K., Cendales M., Drebus G., Spettel B., Lorenz V., and Wänke H. (1979) The abundances of major, minor, and trace elements in the Earth's mantle as derived from primitive ultramafic nodules. *Proc. 10th Lunar Sci. Conf.* 2031–2050.

Jahn B.-M. and Shih C. (1974) On the age of the Onverwacht group, Swaziland sequence, South Africa. *Geochim. Cosmochim. Acta* **38**, 873–885.

Javoy M. (1998) The birth of the Earth's atmosphere: the behaviour and fate of its major elements. *Chem. Geol.* **147**, 11–25.

Javoy M. (1999) Chemical earth models. *C. R. Acad. Sci. Ed. Sci. Méd. Elseviers SAS* **329**, 537–555.

Jessberger E. K., Bohsung J., Chakaveh S., and Traxel K. (1992) The volatile enrichment of chondritic interplanetary dust particles. *Earth Planet. Sci. Lett.* **112**, 91–99.

Jones J. H. and Drake M. J. (1986) Geochemical constraints on core formation in the Earth. *Nature* **322**, 221–228.

Jones J. H. and Drake M. J. (1993) Rubidium and cesium in the Earth and Moon. *Geochim. Cosmochim. Acta* **57**, 3785–3792.

Jones J. H. and Palme H. (2000) Geochemical constraints on the origin of the Earth and Moon. In *Origin of the Earth and Moon* (eds. R. M. Canup and K. Righter). University of Arizona Press, Tucson, pp. 197–216.

Jones J. H., Capobianco C. J., and Drake M. J. (1992) Siderophile elements and the Earth's formation. *Science* **257**, 1281–1282.

Jones R. H., Lee T., Connolly H. C., Jr., Love S. G., and Shang H. (2000) Formation of chondrules and CAIs: theory vs.o bservation. In *Protostars and Planets IV* (eds. V. Mannings, A. P. Boss, and S. S. Russell). University of Arizona Press, Tucson, pp. 927–962.

Kasting J. F. (2001) The rise of atmospheric oxygen. *Science* **293**, 819–820.

Kasting J. F. and Grinspoon D. H. (1991) The faint young Sun problem. In *The Sun in Time* (eds. C. P. Sonett, M. S. Gimpapa, and M. S. Matthews). University of Arizona Press, Tucson, pp. 447–462.

Kato T., Ringwood A. E., and Irifune T. (1988a) Experimental determination of element partitioning between silicate perovskites, garnets and liquids: constraints on early differentiation of the mantle. *Earth Planet. Sci. Lett.* **89**, 123–145.

Kato T., Ringwood A. E., and Irifune T. (1988b) Constraints on element partition coefficients between $MgSiO_3$ perovskite and liquid determined by direct measurements. *Earth Planet. Sci. Lett.* **90**, 65–68.

Kleine T., Münker C., Mezger K., and Palme H. (2002) Rapid accretion and early core formation on asteroids and the terrestrial planets from Hf–W chronometry. *Nature* **418**, 952–955.

Kornacki A. S. and Fegley B., Jr. (1986) The abundance and relative volatility of refractory trace elements in Allende Ca, Al-rich inclusions: implications for chemical and physical processes in the solar nebula. *Earth Planet. Sci. Lett.* **79**, 217–234.

Kortenkamp S. J., Kokubo E., and Weidenschilling S. J. (2000) Formation of planetary embryos. In *Origin of the Earth and Moon* (eds. R. M. Canup and K. Righter). University of Arizona Press, Tucson, pp. 85–100.

Kortenkamp S. J. and Wetherill G. W. (2000) Terrestrial planet and asteroid formation in the presence of giant planets. *Icarus* **143**, 60–73.

Kramers J. D. (1998) Reconciling siderophile element data in the Earth and Moon, W isotopes and the upper lunar age limit in a simple model of homogeneous accretion. *Chem. Geol.* **145**, 461–478.

Kramers J. D. and Tolstikhin I. N. (1997) Two terrestrial lead isotope paradoxes, forward transport modeling, core formation and the history of the continental crust. *Chem. Geol.* **139**, 75–110.

Kreutzberger M. E., Drake M. J., and Jones J. H. (1986) Origin of the Earth's Moon: constraints from alkali volatile trace elements. *Geochim. Cosmochim. Acta* **50**, 91–98.

Krot A. N., Meibom A., Russell S. S., Alexander C. M. O'D., Jeffries T. E., and Keil K. (2001) A new astrophysical setting for chondrule formation. *Science* **291**, 1776–1779.

Kunz J., Staudacher T., and Allègre C. J. (1998) Plutonium-fision xenon found in Earth's mantle. *Science* **280**, 877–880.

Lange M. A. and Ahrens T. J. (1982) The evolution of an impact-generated atmosphere. *Icarus* **51**, 96–120.

Lange M. A. and Ahrens T. J. (1984) FeO and H_2O and the homogeneous accretion of the earth. *Earth Planet. Sci. Lett.* **71**, 111–119.

Larimer J. W. and Anders E. (1970) Chemical fractionations in meteorites: III. Major element fractionation in chondrites. *Geochim. Cosmochim. Acta* **34**, 367–387.

Lee D.-C. and Halliday A. N. (1995) Hafnium–tungsten chronometry and the timing of terrestrial core formation. *Nature* **378**, 771–774.

Lee D.-C. and Halliday A. N. (1996) Hf–W isotopic evidence for rapid accretion and differentiation in the early solar system. *Science* **274**, 1876–1879.

Lee D.-C. and Halliday A. N. (1997) Core formation on Mars and differentiated asteroids. *Nature* **388**, 854–857.

Lee D-C. and Halliday A. N. (2000a) Accretion of primitive planetesimals: Hf–W isotopic evidence from enstatite chondrites. *Science* **288**, 1629–1631.

Lee D.-C. and Halliday A. N. (2000b) Hf–W isotopic systematics of ordinary chondrites and the initial $^{182}Hf/^{180}Hf$ of the solar system. *Chem. Geol.* **169**, 35–43.

Lee D.-C., Halliday A. N., Davies G. R., Essene E. J., Fitton J. G., and Temdjim R. (1996) Melt enrichment of shallow depleted mantle: a detailed petrological, trace element and isotopic study of mantle derived xenoliths and megacrysts from the Cameroon line. *J. Petrol.* 37, 415–441.

Lee D.-C., Halliday A. N., Snyder G. A., and Taylor L. A. (1997) Age and origin of the Moon. *Science* **278**, 1098–1103.

Lee D.-C., Halliday A. N., Leya I., Wieler R., and Wiechert U. (2002) Cosmogenic tungsten and the origin and earliest differentiation of the Moon. *Earth Planet. Sci. Lett.* **198**, 267–274.

Lee T., Shu F. H., Shang H., Glassgold A. E., and Rehm K. E. (1998) Protostellar cosmic rays and extinct radioactivities in meteorites. *Astrophys. J.* **506**, 898–912.

Levin B. J. (1972) Origin of the earth. *Tectonophysics* **13**, 7–29.

Lewis J. S. (1972) Metal/silicate fractionation in the solar system. *Earth Planet. Sci. Lett.* **15**, 286–290.

Leya I., Wieler R., and Halliday A. N. (2000) Cosmic-ray production of tungsten isotopes in lunar samples and meteorites and its implications for Hf–W cosmochemistry. *Earth Planet. Sci. Lett.* **175**, 1–12.

Leya I., Wieler R., and Halliday A. N. (2003) The influence of cosmic-ray production on extinct nuclide systems. *Geochim.Cosmochim. Acta* **67**, 527–541.

Li J. and Agee C. B. (1996) Geochemistry of mantle–core differentiation at high pressure. *Nature* **381**, 686–689.

Lin D. N. C. and Papaloizou J. (1985) On the dynamical origin of the solar system. In *Protostars and Planets II* (eds. D. C. Black and M. S. Matthews). University of Arizona Press, Tucson, pp. 981–1072.

Lin D. N. C., Bodenheimer P., and Richardson D. C. (1996) Orbital migration of the planetary companion of 51 Pegasi to its present location. *Nature* **380**, 606–607.

Lissauer J. J. (1987) Time-scales for planetary accretion and the structure of the protoplanetry disk. *Icarus* **69**, 249–265.

Lissauer J. J. (1993) Planet formation. *Ann. Rev. Astron.Astrophys.* **31**, 129–174.

Lissauer J. J. (1999) How common are habitable planets?*Nature* 402(suppl.2), C11–C14.

Lodders K. (2000) An oxygen isotope mixing model for the accretion and composition of rocky planets. *Space Sci. Rev.* **92**, 341–354.

Longhi J. (1999) Phase equilibrium constraints on angrite petrogenesis. *Geochim. Cosmochim. Acta* **63**, 573–585.

Lugmair G. W. and Galer S. J. G. (1992) Age and isotopic relationships between the angrites Lewis Cliff 86010 and Angra dos Reis. *Geochim. Cosmochim. Acta* 56, 1673–1694.

Lugmair G. W. and Shukolyukov A. (1998) Early solar system timescales according to ^{53}Mn– ^{53}Cr systematics. *Geochim. Cosmochim. Acta* **62**, 2863–2886.

Lugmair G. W. and Shukolyukov A. (2001) Early solar system events and timescales. *Meteorit. Planet. Sci.* **36**, C17–C26.

Lupton J. E. and Craig H. (1975) Excess ^3He in oceanic basalts: evidence for terrestrial primordial helium, *Earth Planet. Sci. Lett.* **26,** 133–139.

MacFarlane E. A. (1989) Formation of the Moon in a giant impact: composition of impactor. *Proc. 19th Lunar Planet. Sci. Conf.* 593–605.

MacPherson G. J., Wark D. A., and Armstrong J. T. (1988) Primitive material surviving in chondrites: refractory inclusions. In *Meteorites and the Early Solar System* (eds. J. F. Kerridge and M. S. Matthews). University of Arizona Press, Tucson, pp. 746–807.

Mamyrin B. A., Tolstikhin I. N., Anufriev G. S., and Kamensky I. L. (1969) Anomalous isotopic composition of helium in volcanic gases. *Dokl. Akad. Nauk. SSSR* **184**, 1197–1199 (in Russian).

Manhès G., Allègre C. J., Dupré B., and Hamelin B. (1979) Lead–lead systematics, the "age of the Earth" and the chemical evolution of our planet in a new representation space. *Earth Planet. Sci. Lett.* **44**, 91–104.

Marty B. (1989) Neon and xenon isotopes in MORB: implications for the earth-atmosphere evolution. *Earth Planet. Sci. Lett.* **94**, 45–56.

Matsui T. and Abe Y. (1986a) Evolution of an impact-induced atmosphere and magma ocean on the accreting Earth. *Nature* **319**, 303–305.

Matsui T. and Abe Y. (1986b) Impact-induced atmospheres and oceans on Earth and Venus. *Nature* **322,** 526–528.

Mayor M. and Queloz D. (1995) A Jupiter-mass companion to a solar-type star. *Nature* **378**, 355–359.

McCaughrean M. J. and O'Dell C. R. (1996) Direct imaging of circumstellar disks in the Orion nebula. *Astronom. J.* **111**, 1977–1986.

McCulloch M. T. (1994) Primitive $^{87}Sr/^{86}Sr$ from an Archean barite and conjecture on the Earth's age and origin, *Earth Planet. Sci. Lett.* **126**, 1–13.

McCulloch M. T. and Bennett V. C. (1993) Evolution of the early Earth: constraints from ^{143}Nd–^{142}Nd isotopic systematics. *Lithos* **30**, 237–255.

McDonough W. F. and Sun S.-S. (1995) The composition of the Earth. *Chem. Geol.* **120**, 223–253.

McDonough W. F., Sun S.-S., Ringwood A. E., Jagoutz E., and Hofmann A. W. (1992) Potassium, rubidium, and cesium in the Earth and Moon and the evolution of the mantle of the Earth, *Geochim. Cosmochim. Acta* **53**, 1001–1012.

McFarlane E. A. and Drake M. J. (1990) Element partitioning and the early thermal history of the Earth. In *Origin of the Earth* (eds. H. E. Newsom and J. H. Jones). Lunar and Planetary Institute, Houston, pp. 135–150.

McKeegan K. D., Chaussidon M., and Robert F. (2000) Incorporation of short-lived Be-10 in a calcium–aluminum-rich inclusion from the Allende meteorite. *Science* **289**, 1334–1337.

Meibom A., Desch S. J., Krot A. N., Cuzzi J. N., Petaev M. I., Wilson L., and Keil K. (2000) Large-scale thermal events in the solar nebula: evidence from Fe, Ni metal grains in primitive meteorites. *Science* **288**, 839–841.

Meisel T., Walker R. J., Irving A. J., and Lorand J.-P. (2001) Osmium isotopic compositions of mantle xenoliths: a global perspective. *Geochim. Cosmochim. Acta* **65**, 1311–1323.

Melosh H. J. (1990) Giant impacts and the thermal state of the early Earth. In *Origin of the Earth* (eds. H. E. Newsom and J. H. Jones). Oxford University Press, Oxford, pp. 69–83.

Melosh H. J. and Sonett C. P. (1986) When worlds collide: jetted vapor plumes and the Moon's origin. In *Origin of the Moon* (eds. W. K. Hartmann, R. J. Phillips, and G. J. Taylor). Lunar and Planetary Institute, Houston, pp. 621–642.

Melosh H. J. and Vickery A. M. (1989) Impact erosion of the primordial atmosphere of Mars. *Nature* **338**, 487–489.

Melosh H. J., Vickery A. M., and Tonks W. B. (1993) Impacts and the early environment and evolution of the terrestrial planets. In *Protostars and Planets III* (eds. E. H. Levy and J. I. Lunine). University of Arizona Press, Tucson, pp. 1339–1370.

Mendybaev R. A., Beckett J. R., Grossman L., Stolper E., Cooper R. F., and Bradley J. P. (2002) Volatilization kinetics of silicon carbide in reducing gases: an experimental study with applications to the survival of presolar grains in the solar nebula. *Geochim. Cosmochim. Acta* **66**, 661–682.

Minarik W. G., Ryerson F. J., and Watson E. B. (1996) Textural entrapment of core-forming melts. *Science* **272**, 530–533.

Mizuno H., Nakazawa K., and Hayashi C. (1980) Dissolution of the primordial rare gases into the molten Earth's material. *Earth Planet. Sci. Lett.* **50**, 202–210.

Mojzsis S. J., Arrhenius G., McKeegan K. D., Harrison T. M., Nutman A. P., and Friend C. R. L. (1996) Evidence for life on Earth before 3,800 million years ago. *Nature* **384**, 55–59.

Mojzsis S. J., Harrison T. M., and Pidgeon R. T. (2001) Oxygen isotope evidence from ancient zircons for liquid water at the Earth's surface 4300 Myr ago Jack Hills, evidence for more very old detrital zircons in Western Australia. *Nature* **409**, 178–181.

Morbidelli A., Chambers J., Lunine J. I., Petit J. M., Robert F., Valsecchi G. B., and Cyr K. E. (2000) Source regions and time-scales for the delivery of water to the Earth. *Meteorit. Planet. Sci.* **35**, 1309–1320.

Moreira M. and Allègre C. J. (1998) Helium–neon systematics and the structure of the mantle. *Chem. Geol* **147**, 53–59.

Moreira M., Kunz J., and Allègre C. J. (1998) Rare gas systematics in Popping Rock: isotopic and elemental compositions in the upper mantle. *Science* **279**, 1178–1181.

Moreira M., Breddam K., Curtice J., and Kurz M. D. (2001) Solar neon in the Icelandic mantle: new evidence for an undegassed lower mantle. *Earth Planet. Sci. Lett.* **185**, 15–23.

Münker C., Weyer S., Mezger K., Rehkämper M., Wombacher F., and Bischoff A. (2000) ^{92}Nb–^{92}Zr and the early differentiation history of planetary bodies. *Science* **289**, 1538–1542.

Murphy D. T., Kamber B. S., and Collerson K. D. (2003) A refined solution to the first terrestrial Pb-isotope paradox. *J. Petrol.* **44**, 39–53.

Murray N. and Chaboyer B. (2001) Are stars with planets polluted? *Ap. J.* **566**, 442–451.

Murray N., Hansen B., Holman M., and Tremaine S. (1998) Migrating planets. *Science* **279**, 69–72.

Murthy V. R. (1991) Early differentiation of the Earth and the problem of mantle siderophile elements: a new approach. *Science* **253**, 303–306.

Nägler Th. F. and Kramers J. D. (1998) Nd isotopic evolution of the upper mantle during the Precambrian: models, data and the uncertainty of both. *Precambrian Res.* **91**, 233–252.

Nelson A. F., Benz W., Adams F. C., and Arnett D. (1998) Dynamics of circumstellar disks. *Ap. J.* **502**, 342–371.

Nelson A. F., Benz W., and Ruzmaikina T. V. (2000) Dynamics of circumstellar disks: II. Heating and cooling. *Ap. J.* **529**, 357–390.

Newsom H. E. (1990) Accretion and core formation in the Earth: evidence from siderophile elements. In *Origin of the Earth* (eds. H. E. Newsom and J. H. Jones). Oxford University Press, Oxford, pp. 273–288.

Newsom H. E. (1995) Composition of the solar system, planets, meteorites, and major terrestrial reservoirs. In *Global Earth Physics: A Handbook of Physical Constants,* AGU Reference Shelf 1 (ed. T. J. Ahrens). American Geophysical Union, Washington, DC.

Newsom H. E. and Taylor S. R. (1986) The single impact origin of the Moon. *Nature* **338**, 29–34.

Newsom H. E., White W. M., Jochum K. P., and Hofmann A. W. (1986) Siderophile and chalcophile element abundances in oceanic basalts, Pb isotope evolution and growth of the Earth's core. *Earth Planet. Sci. Lett.* **80**, 299–313.

Newsom H. E., Sims K. W. W., Noll P. D., Jr., Jaeger W. L., Maehr S. A., and Bessera T. B. (1996) The depletion of W in the bulk silicate Earth. *Geochim. Cosmochim. Acta* **60**, 1155–1169.

Niedermann S., Bach W., and Erzinger J. (1997) Noble gas evidence for a lower mantle component in MORBs from the southern East Pacific Rise: decoupling of helium and neon isotope systematics. *Geochim. Cosmochim. Acta* **61**, 2697–2715.

Nittler L. R. (2003) Presolar stardust in meteorites: recent advances and scientific frontiers. *Earth Planet. Sci. Lett.* **209**, 259–273.

Nittler L. R., Alexander C. M. O'D., Gao X., Walker R. M., and Zinner E. K. (1994) Interstellar oxide grains from the Tieschitz ordinary chondrite. *Nature* **370**, 443–446.

O'Keefe J. D. and Ahrens T. J. (1977) Impact-induced energy partitioning, melting, and vaporization on terrestrial planets. *Proc. 8th Lunar Sci. Conf.* 3357–3374.

O'Neill H. St. C. (1991a) The origin of the Moon and the early history of the Earth—a chemical model: Part I. The Moon. *Geochim. Cosmochim. Acta* **55**, 1135–1158.

O'Neill H. St. C. (1991b) The origin of the Moon and the early history of the Earth—a chemical model: Part II. The Earth. *Geochim. Cosmochim. Acta* **55**, 1159–1172.

O'Nions R. K. and Tolstikhin I. N. (1994) Behaviour and residence times of lithophile and rare gas tracers in the upper mantle. *Earth Planet. Sci. Lett.* **124**, 131–138.

Oversby V. M. and Ringwood A. E. (1971) Time of formation of the Earth's core. *Nature* **234**, 463–465.

Owen T. and Bar-Nun A. (1995) Comets, impacts and atmosphere. *Icarus* **116**, 215–226.

Owen T. and Bar-Nun A. (2000) Volatile contributions from icy planetesimals. In *Origin of the Earth and Moon* (eds. R. M. Canup and K. Righter). University of Arizona Press,Tucson, pp. 459–471.

Ozima M. and Podosek F. A. (1999) Formation age of Earth from ^{129}I/^{127}I and ^{224}Pu/^{238}U systematics and the missing Xe. *J. Geophys. Res.* **104**, 25493–25499.

Ozima M. and Podosek F. A. (2002) *Noble Gas Geochemistry,* 2nd edn. Cambridge University Press, Cambridge, 286p.

Palme H. (2000) Are there chemical gradients in the inner solar system? *Space Sci. Rev.* **92**, 237–262.

Palme H., Larimer J. W., and Lipschultz M. E. (1988) Moderately volatile elements. In *Meteorites and the Early Solar System* (eds. J. F. Kerridge and M. S. Matthews).University of Arizona Press, Tucson, pp. 436–461.

Papanastassiou D. A. and Wasserburg G. J. (1976) Early lunar differentiates and lunar initial ^{87}Sr/^{86}Sr. In *Lunar Sci.* **VII.** The Lunar Science Institute, Houston, pp. 665–667.

Patterson C. C. (1956) Age of meteorites and the Earth.*Geochim. Cosmochim. Acta* **10**, 230–237.

Pavlov A. A., Kasting J. F., Brown L. L., Rages K. A., and Freedman R. (2000) Greenhouse warming by CH4 in the atmosphere of early Earth. *J. Geophys. Res.* **105**, 11981–11990.

Peck W. H., Valley J. W., Wilde S. A., and Graham C. M. (2001) Oxygen isotope ratios and rare earth elements in 3.3. to 4.4 Ga zircons: ion microprobe evidence for high δ^{18}O continental crust and oceans in the Early Archean. *Geochim Cosmochim. Acta* **65**, 4215–4229.

Pepin R. O. (1997) Evolution of Earth's noble gases: consequences of assuming hydrodynamic loss driven by giant impact. *Icarus* **126**, 148–156.

Pepin R. O. (2000) On the isotopic composition of primordial xenon in terrestrial planet atmospheres. *Space Sci. Rev.* **92**, 371–395.

Pepin R. O. and Porcelli D. (2002) Origin of noble gases in the terrestrial planets. In *Noble Gases in Geochemistry and Cosmochemistry,* Rev. Mineral. Geochem. 47 (eds. D. Porcelli, C. J. Ballentine, and R. Wieler). Mineralogical Society of America, Washington, DC, pp. 191–246.

Phinney D., Tennyson J., and Frick U. (1978) Xenon in CO_2 well gas revisited. *J. Geophys. Res.* **83**, 2313–2319.

Podosek F. A. and Cassen P. (1994) Theoretical, observational, and isotopic estimates of the lifetime of the solar nebula. *Meteoritics* **29**, 6–25.

Podosek F. A., Zinner E. K., MacPherson G. J., Lundberg L. L., Brannon J. C., and Fahey A. J. (1991) Correlated study of initial Sr-87/Sr-86 and Al–Mg isotopic systematics and petrologic properties in a suite of refractory inclusions from the Allende meteorite. *Geochim. Cosmochim. Acta* **55**, 1083–1110.

Podosek F. A., Ott U., Brannon J. C., Neal C. R., Bernatowicz T. J., Swan P., and Mahan S. E. (1997) Thoroughly anomalous chromium in Orgueil. *Meteorit. Planet. Sci.* **32**, 617–627.

Podosek F. A., Woolum D. S., Cassen P., Nicholls R. H., Jr., and Weidenschilling S. J. (2003) Solar wind as a source of terrestrial light noble gases. *Geochim. Cosmochim. Acta* (in press).

Poitrasson F., Halliday A. N., Lee D.-C., Levasseur S., and Teutsch N. (2003) Iron isotope evidence for formation of the Moon through partial vaporisation. In *Lunar Planet. Sci.* **XXXIV**, #1433. The Lunar and Planetary Institute, Houston (CD-ROM).

Porcelli D. and Ballentine C. J. (2002) Models for the distribution of terrestrial noble gases and evolution of the atmosphere. In *Noble Gases in Geochemistry and Cosmochemistry,* Rev. Mineral. Geochem. 47 (eds. D. Porcelli, C. J. Ballentine, and R. Wieler). Mineralogical Society of America, Washington, DC. pp. 411–480.

Porcelli D. and Halliday A. N. (2001) The possibility of the core as a source of mantle helium. *Earth Planet. Sci. Lett.* **192**, 45–56.

Porcelli D. and Pepin R. O. (2000) Rare gas constraints on early earth history. In *Origin of the Earth and Moon* (eds. R. M. Canup and K. Righter). University of Arizona Press, Tucson, pp. 435–458.

Porcelli D. and Wasserburg G. J. (1995) Mass transfer of helium, neon, argon and xenon through a steady-state upper mantle. *Geochim. Cosmochim. Acta* **59**, 4921–4937.

Porcelli D., Cassen P., Woolum D., and Wasserburg G. J. (1998) Acquisition and early losses of rare gases from the deep Earth. In *Origin of the Earth and Moon,* LPI Contribution No. 597. Lunar and Planetary Institute, Houston, pp. 35–36.

Porcelli D., Cassen P., and Woolum D. (2001) Deep Earth rare gases: initial inventories, capture from the solar nebula and losses during Moon formation. *Earth Planet. Sci. Lett* **193**, 237–251.

Puchtel I. and Humayun M. (2000) Platinum group elements in Kostomuksha komatiites and basalts: implications for oceanic crust recycling and core–mantle interaction. *Geochim. Cosmochim. Acta* **64**, 4227–4242.

Qian Y. Z. and Wasserburg G. J. (2000) Stellar abundances in the early galaxy and two r-process components. *Phys. Rep.* **333–334**, 77–108.

Qian Y. Z., Vogel P., and Wasserburg G. J. (1998) Diverse supernova sources for the r-process. *Astrophys. J.* **494**, 285–296.

Quitté G., Birck J.-L., and Allègre C. J. (2000) ^{182}Hf–^{182}W systematics in eucrites: the puzzle of iron segregation in the early solar system. *Earth Planet. Sci. Lett.* **184**, 83–94.

Rammensee W. and Wänke H. (1977) On the partition coefficient of tungsten between metal and silicate and its bearing on the origin of the Moon. *Proc. 8th Lunar Sci. Conf.* 399–409.

Rehkämper M., Halliday A. N., Barfod D., Fitton J. G., and Dawson J. B. (1997) Platinum group element abundance patterns in different mantle environments. *Science* **278**, 1595–1598.

Rehkämper M., Halliday A. N., Alt J., Fitton J. G., Zipfel J., and Takazawa E. (1999a) Non-chondritic platinum group element ratios in abyssal peridotites: petrogenetic signature of melt percolation? *Earth Planet. Sci. Lett.* **172**, 65–81.

Rehkämper M., Halliday A. N., Fitton J. G., Lee D.-C., and Wieneke M. (1999b) Ir, Ru, Pt, and Pd in basalts and komatiites: new constraints for the geochemical behavior of the platinum-group elements in the mantle. *Geochim. Cosmochim. Acta* **63**, 3915–3934.

Reynolds J. H. (1967) Isotopic abundance anomalies in the solar system. *Ann. Rev. Nuclear Sci.* **17**, 253–316.

Righter K. and Drake M. J. (1996) Core formation in Earth's Moon, Mars, and Vesta. *Icarus* **124**, 513–529.

Righter K. and Drake M. J. (1999) Effect of water on metal-silicate partitioning of siderophile elements: a high pressure and temperature terrestrial magma ocean and core formation. *Earth Planet. Sci. Lett.* **171**, 383–399.

Righter K., Drake M. J., and Yaxley G. (1997) Prediction of siderophile element metal/silicate partition coefficients to 20 GPa and 2800 °C: the effects of pressure, temperature, oxygen fugacity, and silicate and metallic melt compositions. *Phys. Earth Planet. Int.* **100**, 115–142.

Righter K., Walker R. J., and Warren P. W. (2000) The origin and significance of highly siderophile elements in the lunar and terrestrial mantles. In *Origin of the Earth and Moon* (eds. R. M. Canup and K. Righter). University of Arizona Press, Tucson, pp. 291–322.

Ringwood A. E. (1966) The chemical composition and origin of the Earth. In *Advances in Earth Sciences* (ed. P. M. Hurley). MIT Press, Cambridge, MA, pp. 287–356.

Ringwood A. E. (1977) Composition of the core and implications for origin of the Earth. *Geochem. J.* **11**, 111–135.

Ringwood A. E. (1979) *Origin of the Earth and Moon.* Springer, New York, 295p.

Ringwood A. E. (1989a) Significance of the terrestrial Mg/Si ratio. *Earth Planet. Sci. Lett.* **95**, 1–7.

Ringwood A. E. (1989b) Flaws in the giant impact hypothesis of lunar origin. *Earth Planet. Sci. Lett.* **95**, 208–214.

Ringwood A. E. (1990) Earliest history of the Earth–Moon system. In *Origin of the Earth* (eds. A. E. Newsom and J. H. Jones). Oxford University Press, Oxford, pp. 101–134.

Ringwood A. E. (1992) Volatile and siderophile element geochemistry of the Moon: a reappraisal. *Earth Planet. Sci. Lett.* **111**, 537–555.

Rubey W. W. (1951) Geological history of seawater. *Bull. Geol. Soc.Am.* **62**, 1111–1148.

Rubie D. C., Melosh H. J., Reid J. E., Liebske C., and Righter K. (2003) Mechanisms of metal-silicate equilibration in the terrestrial magma ocean. *Earth Planet. Sci. Lett.* **205**, 239–255.

Rushmer T., Minarik W. G., and Taylor G. J. (2000) Physical processes of core formation. In *Origin of the Earth and Moon* (eds. R. M. Canup and K. Righter). University of Arizona Press, Tucson, pp. 227–243.

Ryder G., Koeberl C., and Mojzsis S. J. (2000) Heavy bombardment of the Earth at ~3.85 Ga: the search for petrographic and geochemical evidence. In *Origin of the Earth and Moon* (eds. R. M. Canup and K. Righter). University of Arizona Press, Tucson, pp. 475–492.

Safronov V. S. (1954) On the growth of planets in the protoplanetary cloud. *Astron. Zh.* **31**, 499–510.

Sagan C. and Chyba C. (1997) The early faint Sun paradox: organic shielding of ultraviolet-labile greenhouse gases. *Science* **276**, 1217–1221.

Sanders I. S. (1996) A chondrule-forming scenario involving molten planetesimals. In *Chondrules and the Protoplanetary Disk* (eds. R. H. Hewins, R. H. Jones, and E. R. D. Scott). Cambridge University Press, Cambridge, UK, pp. 327–334.

Sanloup C., Blichert-Toft J., Télouk P., Gillet P., and Albarède F. (2000) Zr isotope anomalies in chondrites and the presence of live ^{92}Nb in the early solar system. *Earth Planet. Sci. Lett.* **184**, 75–81.

Sasaki S. (1990) The primary solar-type atmosphere surrounding the accreting Earth: H_2O-induced high surface temperature. In *Origin of the Earth* (eds. H. E. Newsom and J. H. Jones). Oxford University Press, Oxford, pp. 195–209.

Sasaki S. and Nakazawa K. (1986) Metal-silicate fractionation in the growing Earth: energy source for the terrestrial magma ocean. *J. Geophys. Res.* **91**, B9231–B9238.

Sasaki S. and Nakazawa K. (1988) Origin of isotopic fractionation of terrestrial Xe: hydrodynamic fractionation during escape of the primordial H_2–He atmosphere. *Earth Planet. Sci. Lett.* **89**, 323–334.

Scherer E. E., Münker C., and Mezger K. (2001) Calibration of the lutetium–hafnium clock. *Science* **293**, 683–687.

Schmitt W., Palme H., and Wänke H. (1989) Experimental determination of metal/silicate partition coefficients for P, Co, Ni, Cu, Ga, Ge, Mo, and W and some implications for the early evolution of the Earth. *Geochim. Cosmochim. Acta* **53**, 173–185.

Schneider G., Smith B. A., Becklin E. E., Koerner D. W., Meier R., Hines D. C., Lowrance P. J., Terrile R. I., and Rieke M.

(1999) Nicmos imaging of the HR 4796A circumstellar disk. *Astrophys. J.* **513**, L127–L130.

Schönberg R., Kamber B. S., Collerson K. D., and Eugster O. (2002) New W isotope evidence for rapid terrestrial accretion and very early core formation. *Geochim. Cosmochim. Acta* **66**, 3151–3160.

Schönbächler M., Rehkämper M., Halliday A. N., Lee D. C., Bourot-Denise M., Zanda B., Hattendorf B., and Günther D. (2002) Niobium–zirconium chronometry and early solar system development. *Science* **295**, 1705–1708.

Seager S. (2003) The search for Earth-like extrasolar planets.*Earth Planet. Sci. Lett.* **208**, 113–124.

Seeds M. A. (1996) *Foundations of Astronomy.* Wadsworth Publishing Company, Belmont, California, USA.

Sharma M., Papanastassiou D. A., and Wasserburg G. J. (1996) The issue of the terrestrial record of ^{146}Sm. *Geochim. Cosmochim. Acta* **60**, 2037–2047.

Shaw G. H. (1978) Effects of core formation. *Phys. Earth Planet. Int.* **16**, 361–369.

Shih C.-Y., Nyquist L. E., Dasch E. J., Bogard D. D., Bansal B. M., and Wiesmann H. (1993) Age of pristine noritic clasts from lunar breccias 15445 and 15455. *Geochim. Cosmochim. Acta* **57**, 915–931.

Shu F. H., Shang H., and Lee T. (1996) Toward an astrophysical theory of chondrites. *Science* **271**, 1545–1552.

Shu F. H., Shang H., Glassgold A. E., and Lee T. (1997) X-rays and fluctuating x-winds from protostars. *Science* **277**, 1475–1479.

Sleep N. H. and Zahnle K. (2001) Carbon dioxide cycling and implications for climate on ancient Earth. *J. Geophys. Res.* **106**, 1373–1399.

Sleep N. H., Zahnle K. J., Kasting J. F., and Morowitz H. J. (1989) Annihilation of ecosystems by large asteroid impacts on the early Earth. *Nature* **342**, 139–142.

Sleep N. H., Zahnle K., and Neuhoff P. S. (2001) Initiation of clement surface conditions on the earliest Earth. *Proc. Natl. Acad. Sci.* **98**, 3666–3672.

Smith J. V. (1977) Possible controls on the bulk composition of the Earth: implications for the origin of the earth and moon.*Proc. 8th Lunar Sci. Conf.* 333–369.

Smith J. V. (1980) The relation of mantle heterogeneity to the bulk composition and origin of the Earth. *Phil. Trans. Roy.Soc. London A* **297**, 139–146.

Snow J. E. and Schmidt G. (1998) Constraints on Earth accretion deduced from noble metals in the oceanic mantle. *Nature* **391**, 166–169.

Solomon S. C. (1979) Formation, history and energetics of cores in the terrestrial planets. *Earth Planet. Sci. Lett.* **19**, 168–182.

Stevenson D. J. (1981) Models of the Earth's core. *Science* **214**, 611–619.

Stevenson D. J. (1990) Fluid dynamics of core formation. In *Origin of the Earth* (eds. H. E. Newsom and J. H. Jones). Oxford University Press, Oxford, pp. 231–249.

Suess (1965) Chemical evidence bearing on the origin of the solar system. *Rev. Astron. Astrophys.* **3**, 217–234.

Swindle T. D. (1993) Extinct radionuclides and evolutionary timescales. In *Protostars and Planets III* (eds. E. H. Levy and J. I. Lunine). University of Arirona Press, Tucson, pp. 867–881.

Swindle T. D. and Podosek (1988) Iodine–xenon dating. In *Meteorites and the Early Solar System* (eds. J. F. Kerridge and M. S. Matthews). University of Arizona Press, Tucson, pp. 1127–1146.

Takahashi H., Janssens M.-J., Morgan J. W., and Anders E. (1978) Further studies of trace elements in C3 chondrites. *Geochim. Cosmochim. Acta* **42**, 97–106.

Taylor S. R. (1992) *Solar System Evolution: A New Perspective.* Cambridge University Press, New York.

Taylor S. R. and Norman M. D. (1990) Accretion of differentiated planetesimals to the Earth. In *Origin of the Earth* (eds. H. E. Newsom and J. H. Jones). Oxford University Press, Oxford, pp. 29–43.

Tera F. (1980) Reassessment of the "age of the Earth." *Carnegie Inst. Wash. Yearbook* **79**, 524–531.

Tera F., Papanastassiou D. A., and Wasserburg G. J. (1973) A lunar cataclysm at ~3.95 AE and the structure of the lunar crust. In *Lunar Sci.* **IV.** The Lunar Science Institute, Houston, pp. 723–725.

Tilton G. R. (1988) Age of the solar system. In *Meteorites and the Early Solar System* (eds. J. F. Kerridge and M. S. Matthews). University of Arizona Press, Tucson, pp. 259–275.

Tonks W. B. and Melosh H. J. (1990) The physics of crystal settling and suspension in a turbulent magma ocean. In *Origin of the Earth* (eds. H. E. Newsom and J. H. Jones). Oxford University Press, Oxford, pp. 17–174.

Treiman A. H., Drake M. J., Janssens M.-J., Wolf R., and Ebihara M. (1986) Core formation in the Earth and shergottite parent body (SPB): chemical evidence from basalts. *Geochim. Cosmochim. Acta* **50**, 1071–1091.

Treiman A. H., Jones J. H., and Drake M. J. (1987) Core formation in the shergottite parent body and comparison with the Earth. *J. Geophys. Res.* **92**, E627–E632.

Trieloff M., Kunz J., Clague D. A., Harrison D., and Allègre C. J. (2000) The nature of pristine noble gases in mantle plumes. *Science* **288**, 1036–1038.

Turekian K. K. and Clark S. P., Jr. (1969) Inhomogeneous accumulation of the earth from the primitive solar nebula. *Earth Planet. Sci. Lett.* **6**, 346–348.

Tyburczy J. A., Frisch B., and Ahrens T. J. (1986) Shock-induced volatile loss from a carbonaceous chondrite: implications for planetary accretion. *Earth Planet. Sci. Lett.* **80**, 201–207.

Urey H. C. (1966) The capture hypothesis of the origin of the Moon. In *The Earth–Moon System* (eds. B. G. Marsden and A. G. W. Cameron). Plenum, New York, pp. 210–212.

Valley J. W., Peck W. H., King E. M., and Wilde S. A. (2002) A cool early Earth. *Geology* **30**, 351–354.

Vervoort J. D., White W. M., and Thorpe R. I. (1994) Nd and Pb isotope ratios of the Abitibi greenstone belt: new evidence for very early differentiation of the Earth. *Earth Planet. Sci. Lett.* **128**, 215–229.

Vidal-Madjar A., Lecavelier des Etangs A., and Ferlet R. (1998) β Pictoris, a young planetary system? A review. *Planet. Space Sci.* **46**, 629–648.

Vityazev A. V., Pechernikova A. G., Bashkirov A. G. (2003) Accretion and differentiation of terrestrial protoplanetary bodies and Hf–W chronometry. In *Lunar Planet. Sci.* **XXXIV**, #1656. The Lunar and Planetary Institute, Houston (CD-ROM).

Vogel N., Baur H., Bischoff A., and Wieler R. (2002) Noble gases in chondrules and metal-sulfide rims of primitive chondrites—clues on chondrule formation. *Geochim. Cosmochim. Acta* **66**, A809.

Vogel N., Baur H., Leya I., and Wieler R. (2003) No evidence for primordial noble gases in CAIs. *Meteorit. Planet. Sci.* **38**(suppl.), A75.

Wade J. and Wood B. J. (2001) The Earth's "missing" niobium may be in the core. *Nature* **409**, 75–78.

Walker J. C. G. (1986) Impact erosion of planetary atmospheres. *Icarus* **68**, 87–98.

Walker R. J., Morgan J. W., and Horan M. F. (1995) Osmium-187 enrichment in some plumes: evidence for core–mantle interaction? *Science* **269**, 819–822.

Walter M. J., Newsom H. E., Ertel W., and Holzheid A. (2000) Siderophile elements in the Earth and Moon: Metal/silicate partitioning and implications for core formation. In *Origin of the Earth and Moon* (eds. R. M. Canup and K. Righter). University of Arizona Press, Tucson, pp. 265–289.

Wänke H. (1981) Constitution of terrestrial planets. *Phil. Trans. Roy. Soc. London* **A303**, 287–302.

Wänke H. and Dreibus G. (1986) Geochemical evidence for formation of the Moon by impact induced fission of the proto-Earth. In *Origin of the Moon* (eds. W. K. Hartmann,

R. J. Phillips, and G. J. Taylor). Lunar and Planetary Institute, Houston, pp. 649–672.

Wänke H. and Dreibus G. (1988) Chemical composition and accretion history of terrestrial planets. *Phil. Trans. Roy. Soc. London A* **325**, 545–557.

Wänke H. and Dreibus G. (1994) Chemistry and accretion of Mars. *Phil. Trans. Roy. Soc. London A* **349**, 285–293.

Wänke H., Baddenhausen H., Palme H., and Spettel B. (1974) On the chemistry of the Allende inclusions and their origin as high temperature condensates. *Earth Planet. Sci. Lett.* **23**, 1–7.

Wänke H., Dreibus G., Palme H., Rammensee W., and Weckwerth G. (1983) Geochemical evidence for the formation of the Moon from material of the Earth's mantle. In *Lunar Planet. Sci.* **XIV.** The Lunar and Planetary Institute, Houston, pp. 818–819.

Wänke H., Dreibus G., and Jagoutz E. (1984) Mantle chemistry and accretion history of the Earth. In *Archaean Geochemistry* (eds. A. Kroner, G. N. Hanson, and A. M. Goodwin).Springer, New York, pp. 1–24.

Ward W. R. (2000) On planetesimal formation: the role of collective particle behavior. In *Origin of the Earth and Moon* (eds. R. M. Canup and K. Righter). University of Arizona Press, Tucson, pp. 75–84.

Wasserburg G. J., MacDonald F., Hoyle F., and Fowler W. A. (1964) Relative contributions of uranium, thorium, and potassium to heat production in the Earth. *Science* **143**, 465–467.

Wasserburg G. J., Papanastassiou D. A., Tera F., and Huneke J. C. (1977a) Outline of a lunar chronology. *Phil. Trans. Roy. Soc. London A* **285**, 7–22.

Wasserburg G. J., Tera F., Papanastassiou D. A., and Huneke J. C. (1977b) Isotopic and chemical investigations on Angra dos Reis. *Earth Planet. Sci. Lett.* **35**, 294–316.

Wasserburg G. J., Busso M., Gallino R., and Raiteri C. M. (1994) Asymptotic giant branch stars as a source of shortlived radioactive nuclei in the solar nebula. *Astrophys. J.* **424**, 412–428.

Wasserburg G. J., Busso M., and Gallino R. (1996) Abundances of actinides and short-lived nonactinides in the interstellar medium: diverse supernova sources for the r-processes. *Astrophys. J.* **466**, L109–L113.

Wasson J. T. (1985) *Meteorites: Their Record of Early Solar-system History.* W. H. Freeman, New York, 251p.

Wasson J. T. and Chou C.-L. (1974) Fractionation of moderately volatile elements in ordinary chondrites. *Meteoritics* **9**, 69–84.

Weidenschilling S. J. (1977a) The distribution of mass in the planetary system and solar nebula. *Astrophys. Space Sci.* **51**, 153–158.

Weidenschilling S. J. (1977b) Aerodynamics of solid bodies in the solar nebula. *Mon. Not. Roy. Astron. Soc.***180**, 57–70.

Weidenschilling S. J. (2000) Formation of planetesimals and accretion of the terrestrial planets. *Space Sci. Rev.* **92**, 295–310.

Wetherill G. W. (1975a) Radiometric chronology of the early solar system. *Ann. Rev. Nuclear Sci.* **25**, 283–328.

Wetherill G. W. (1975b) Late heavy bombardment of the moon and terrestrial planets. *Proc. 6th Lunar Sci. Conf.* 1539–1561.

Wetherill G. W. (1980) Formation of the terrestrial planets. *Ann. Rev. Astron. Astrophys.* **18**, 77–113.

Wetherill G. W. (1986) Accumulation of the terrestrial planets and implications concerning lunar origin. In *Origin of the Moon* (eds. W. K. Hartmann, R. J. Phillips, and G. J. Taylor). Lunar and Planetary Institute, Houston, pp. 519–551.

Wetherill G. W. (1992) An alternative model for the formation of the Asteroids. *Icarus* **100**, 307–325.

Wetherill G. W. (1994) Provenance of the terrestrial planets. *Geochim. Cosmochim. Acta* **58**, 4513–4520.

Wetherill G. W. and Stewart G. R. (1993) Formation of planetary embryos: effects of fragmentation, low relative velocity, and independent variation of eccentricity and inclination. *Icarus* **106**, 190–209.

Wiechert U. (2002) Earth's early atmosphere. *Science* **298**, 2341–2342.

Wiechert U., Halliday A. N., Lee D.-C., Snyder G. A., Taylor L. A., and Rumble D. A. (2001) Oxygen isotopes and the Moon-forming giant impact. *Science* **294**, 345–348.

Wiechert U., Halliday A. N., Palme H., and Rumble D. (2003) Oxygen isotopes in HED meteorites and evidence for rapid mixing in planetary embryos. *Earth Planet. Sci. Lett.* (in press).

Wieler R., Anders E., Baur H., Lewis R. S., and Signer P. (1992) Characterisation of Q-gases and other noble gas components in the Murchison meteorite. *Geochim. Cosmochim. Acta* **56**, 2907–2921.

Wieler R., Kehm K., Meshik A. P., and Hohenberg C. M. (1996) Secular changes in the xenon and krypton abundances in the solar wind recorded in single lunar grains. *Nature* **384**, 46–49.

Wiesli R. A., Beard B. L., Taylor L. A., Welch S. A., and Johnson C. M. (2003) Iron isotope composition of the lunar mare regolith: implications for isotopic fractionation during production of single domain iron metal. In *Lunar Planet Sci.* **XXXIV**, #1500. The Lunar and Planetary Institute, Houston (CD-ROM).

Wilde S. A., Valley J. W., Peck W. H., and Graham C. M. (2001) Evidence from detrital zircons for the existence of continental crust and oceans on the Earth 4.4 Gyr ago. *Nature* **409**, 175–178.

Wolf R., Richter G. R., Woodrow A. B., and Anders E. (1980) Chemical fractionations in meteorites: XI. C2 chondrites. *Geochim. Cosmochim. Acta* **44**, 711–717.

Wood B. J. (1993) Carbon in the core. *Earth Planet. Sci. Lett.* **117**, 593–607.

Wood J. A. (1984) Moon over Mauna Loa: a review of hypotheses of formation of Earth's moon. In *Origin of the Moon* (eds. W. K. Hartmann, R. J. Phillips, and G. J. Taylor). Lunar and Planetary Institute, Houston, pp. 17–55.

Wood J. A. (2000) Pressure ands temperature profiles in the solar nebula. *Space Sci. Rev.* **92**, 87–93.

Wood J. A. and Morfill G. E. (1988) A review of solar nebula models. In *Meteorites and the Early Solar System* (eds. J. F. Kerridge and M. S. Matthews). University of Arizona Press, Tucson, pp. 329–347.

Yi W., Halliday A. N., Alt J. C., Lee D.-C., Rehkämper M., Garcia M., Langmuir C., and Su Y. (2000) Cadmium, indium, tin, tellurium and sulfur in oceanic basalts: implications for chalcophile element fractionation in the earth. *J. Geophys. Res.* **105**, 18927–18948.

Yin Q. Z. and Jacobsen S. B. (1998) The ^{97}Tc–^{97}Mo chronometer and its implications for timing of terrestrial accretion and core formation. In *Lunar Planet Sci.* **XXIX**, #1802. The Lunar and Planetary Institute, Houston (CD-ROM).

Yin Q. Z., Jacobsen S. B., McDonough W. F., Horn I., Petaev M. I., and Zipfel J. (2000) Supernova sources and the ^{92}Nb–^{92}Zr p-process chronometer: *Astrophys. J.* **535**, L49–L53.

Yin Q. Z., Jacobsen S. B., Yamashita K., Blicher-Toft J., Télouk P., and Albarède F. (2002) A short timescale for terrestrial planet formation from Hf–W chronometry of meteorites. *Nature* **418**, 949–952.

Yoshino T., Walter M. J., and Katsura T. (2003) Core formation in planetesimals triggered by permeable flow. *Nature* **422**, 154–157.

Young E. D. (2000) Assessing the implications of K isotope cosmochemistry for evaporation in the preplanetary solar nebula. *Earth Planet. Sci. Lett.* **183**, 321–333.

Zahnle K. J. (1993) Xenological constraints on the impact erosion of the early Martian atmosphere. *J. Geophys. Res.* **98**, 10899–10913.

Zahnle K. J. and Sleep N. H. (1996) Impacts and the early evolution of life. In *Comets and the Origin and Evolution of Life* (eds. P. J. Thomas, C. F. Chyba, and C. P. McKay). Springer, Heidelberg, pp. 175–208.

Zahnle K. J. and Walker J. C. G. (1982) The evolution of solar ultraviolet luminosity. *Rev. Geophys.* **20**, 280.

Zahnle K. J., Kasting J. F., and Pollack J. B. (1988) Evolution of a steam atmosphere during Earth's accretion. *Icarus* **74**, 62–97.

Zhang Y. (1998) The young age of the Earth. *Geochim. Cosmochim. Acta* **62**, 3185–3189.

Readings from the Treatise on Geochemistry
ISBN: 978-0-12-381391-6

pp. 17–66

3
Sampling Mantle Heterogeneity through Oceanic Basalts: Isotopes and Trace Elements

A. W. Hofmann
Max-Plank-Institut für Chemie, Mainz, Germany and Lamont-Doherty Earth Observatory, Columbia University, Palisades, NY, USA

3.1 INTRODUCTION

3.1.1 Early History of Mantle Geochemistry

Until the arrival of the theories of plate tectonics and seafloor spreading in the 1960s, the Earth's mantle was generally believed to consist of peridotites of uniform composition. This view was shared by geophysicists, petrologists, and geochemists alike, and it served to characterize the compositions and physical properties of mantle and crust as "Sial" (silica-alumina) of low density and "Sima" (silica-magnesia) of greater density.

Thus, Hurley and his collaborators were able to distinguish crustal magma sources from those located in the mantle on the basis of their initial strontium-isotopic compositions (Hurley *et al.*, 1962; and Hurley's lectures and popular articles not recorded in the formal scientific literature). In a general way, as of early 2000s, this view is still considered valid, but literally thousands of papers have since been published on the isotopic and trace-elemental composition of oceanic basalts because they come from the mantle and they are rich sources of information about the composition of the mantle, its differentiation history and its internal structure. Through the study of oceanic basalts, it was found that the mantle is compositionally just as heterogeneous as the crust. Thus, geochemistry became a major tool to decipher the geology of the mantle, a term that seems more appropriate than the more popular "chemical geodynamics."

The pioneers of this effort were Gast, Tilton, Hedge, Tatsumoto, and Hart (Hedge and Walthall, 1963; Gast *et al.*, 1964; Tatsumoto, *et al.*, 1965; Hart, 1971). They discovered from isotope analyses of strontium and lead in young (effectively zero age) ocean island basalts (OIBs) and mid-ocean ridge basalts (MORBs) that these basalts are isotopically not uniform. The isotope ratios $^{87}Sr/^{86}Sr$, $^{206}Pb/^{204}Pb$, $^{207}Pb/^{204}Pb$, and $^{208}Pb/^{204}Pb$ increase as functions of time and the respective radioactive-parent/nonradiogenic daughter ratios, $^{87}Rb/^{86}Sr$, $^{238}U/^{204}Pb$, $^{235}U/^{204}Pb$, and $^{232}Th/^{204}Pb$, in the sources of the magmas. This means that the mantle must contain geologically old reservoirs with different Rb/Sr, U/Pb, and Th/Pb ratios. The isotope story was complemented by trace-element geochemists, led primarily by Schilling and Winchester (1967, 1969) and Gast (1968) on chemical trace-element fractionation during igneous processes, and by Tatsumoto *et al.* (1965) and Hart (1971). From the trace-element abundances, particularly rare-earth element (REE) abundances, it became clear that not only some particular parent–daughter element abundance ratios but also the light-to-heavy REE ratios of the Earth's mantle are quite heterogeneous. The interpretation of these heterogeneities has occupied mantle geochemists since the 1960s.

This chapter is in part an update of a previous, more abbreviated review (Hofmann, 1997). It covers the subject in greater depth, and it reflects some significant changes in the author's views since the writing of the earlier paper. In particular, the spatial range of equilibrium attained during partial melting may be much smaller than previously thought, because of new experimental diffusion data and new results from natural settings. Also, the question of "layered" versus "whole-mantle" convection, including the depth of subduction and of the origin of plumes, has to be reassessed in light of the recent breakthroughs achieved by seismic mantle tomography. As the spatial resolution of seismic tomography and the pressure range, accuracy, and precision of experimental data on melting relations, phase transformations, and kinetics continue to improve, the interaction between these disciplines and geochemistry *sensu stricto* will continue to improve our understanding of what is actually going on in the mantle. The established views of the mantle being engaged in simple two-layer, or simple single-layer, convection are becoming obsolete. In many ways, we are just at the beginning of this new phase of mantle geology, geophysics, and geochemistry.

3.1.2 The Basics

3.1.2.1 Major and trace elements: incompatible and compatible behavior

Mantle geochemists distinguish between major and trace elements. At first sight, this nomenclature seems rather trivial, because which particular elements should be called "major" and which "trace" depends on the composition of the system. However, this distinction actually has a deeper meaning, because it signifies fundamental differences in geochemical behavior. We define elements as "major" if they are essential constituents of the minerals making up a rock, namely, in the sense of the phase rule. Thus, on the one hand, silicon, aluminum, chromium, magnesium, iron, calcium, sodium, and oxygen are major elements because they are essential constituents of the upper-mantle minerals—olivine, pyroxene, garnet, spinel, and plagioclase. Adding or subtracting such elements can change the phase assemblage. Trace elements, on the other hand just replace a few atoms of the major elements in the crystal structures without affecting the phase assemblage significantly. They are essentially blind passengers in many mantle processes, and they are therefore immensely useful as tracers of such processes. During solid-phase transformations, they will redistribute themselves locally between the newly formed mineral phases but, during melting, they are partitioned to a greater or lesser degree into the melt. When such a melt is transported to the Earth's surface, where it can be sampled, its trace elements carry a wealth of information about the composition of the source rock and the nature of the melting processes at depth.

For convenience, the partitioning of trace elements between crystalline and liquid phases is usually described by a coefficient D, which is just a simple ratio of two concentrations at chemical equilibrium:

$$D^i = \frac{C_s^i}{C_l^i} \quad (1)$$

where D^i is the called the partition coefficient of trace element i, C_s^i and C_l^i are the concentrations (by weight) of this element in the solid and liquid phases, respectively.

Goldschmidt (1937, 1954) first recognized that the distribution of trace elements in minerals is strongly controlled by ionic radius and charge. The partition coefficient of a given trace element between solid and melt can be quantitatively described by the elastic strain this element causes by its presence in the crystal lattice. When this strain is large because of the magnitude of the misfit, the partition coefficient becomes small, and the element is partitioned into the liquid.

Most trace elements have values of $D \ll 1$, simply because they differ substantially either in ionic radius or ionic charge, or both, from the atoms of the major elements they replace in the crystal lattice. Because of this, they are called *incompatible*. Exceptions are trace elements such as strontium in plagioclase, ytterbium, lutetium, and scandium in garnet, nickel in olivine, and scandium in clinopyroxene. These latter elements actually fit into their host crystal structures slightly better than the major elements they replace, and they are therefore called *compatible*. Thus, most chemical elements of the periodic table are trace elements, and most of them are incompatible; only a handful are compatible.

Major elements in melts formed from mantle rocks are by definition compatible, and most of them are well buffered by the residual minerals, so that their concentrations usually vary by factors of less than two in the melts. In contrast, trace elements, particularly those having very low partition coefficients, may vary by as many as three orders of magnitude in the melt, depending on the degree of melting. This is easily seen from the mass-balance-derived equation for the equilibrium concentration of a trace element in the melt, C_l, given by (Shaw, 1970)

$$C_l = \frac{C_0}{F + D(1-F)} \quad (2)$$

where the superscript i has been dropped for clarity, C_0 is the concentration in the bulk system, and F is the melt fraction by mass. For highly incompatible elements, which are characterized by very low partition coefficients, so that $D \ll F$, this equation reduces to

$$C_l \approx \frac{C_0}{F} \quad (3)$$

This means that the trace-element concentration is then inversely proportional to the melt fraction F, because the melt contains essentially all of the budget of this trace element. An additional consequence of highly incompatible behavior of trace elements is that their concentration ratios in the melt become constant, independent of melt fraction, and identical to the respective ratio in the mantle source. This follows directly when Equation (3) is written for two highly incompatible elements:

$$\frac{C_l^1}{C_l^2} \approx \frac{C_0^1}{F} \times \frac{F}{C_0^2} = \frac{C_0^1}{C_0^2} \quad (4)$$

In this respect, incompatible trace-element ratios resemble isotope ratios. They are therefore very useful in complementing the information obtained from isotopes.

3.1.2.2 Radiogenic isotopes

The decay of long-lived radioactive isotopes was initially used by geochemists exclusively for the measurement of geologic time. As noted in the introduction, their use as tracers of mantle processes was pioneered by Hurley and co-workers in the early 1960s. The decay

$$^{87}\text{Rb} \rightarrow {}^{87}\text{Sr} \qquad (\lambda = 1.42 \times 10^{-11}\ \text{yr}) \quad (5)$$

serves as example. The solution of the decay equation is

$$^{87}\text{Sr} = {}^{87}\text{Rb} \times (e^{\lambda t} - 1) \quad (6)$$

Dividing both sides by one of the nonradiogenic isotopes, by convention ^{86}Sr, we obtain

$$\frac{^{87}\text{Sr}}{^{86}\text{Sr}} = \frac{^{87}\text{Rb}}{^{86}\text{Sr}} \times (e^{\lambda t} - 1) \approx \frac{^{87}\text{Rb}}{^{86}\text{Sr}} \times \lambda t \quad (7)$$

The approximation in Equation (7) holds only for decay systems with sufficiently long half-lives, such as the Rb–Sr and the Sm–Nd systems, so that $\lambda t \ll 1$ and $e^{\lambda t} - 1 \approx \lambda t$. Therefore, the isotope ratio ^{87}Sr/^{86}Sr in a system, such as some volume of mantle rock, is a linear function of the parent/daughter chemical ratio Rb/Sr and a nearly linear function of time or geological age of the system. When this mantle volume undergoes equilibrium partial melting, the melt inherits the ^{87}Sr/^{86}Sr ratio of the entire system. Consequently, radiogenic isotope ratios such as ^{87}Sr/^{86}Sr are powerful tracers of the parent–daughter ratios of mantle sources of igneous rocks. If isotope data from several decay systems are combined, a correspondingly richer picture of the source chemistry can be constructed.

Table 1 shows a list of long-lived radionuclides, their half-lives, daughter isotopes, and radiogenic-to-nonradiogenic isotope rates commonly used as tracers in mantle geochemistry. Noble-gas isotopes are not included here, because a separate chapter of this Treatise is devoted to them. Taken

Table 1 Long-lived radionuclides.

Parent nuclide	*Daughter nuclide*	*Half life (yr)*	*Tracer ratio (radiogenic/nonradiogenic)*
^{147}Sm	^{143}Nd	106×10^9	$^{143}Nd/^{144}Nd$
^{87}Rb	^{87}Sr	48.8×10^9	$^{87}Sr/^{86}Sr$
^{176}Lu	^{176}Hf	35.7×10^9	$^{176}Hf/^{177}Hf$
^{187}Re	^{187}Os	45.6×10^9	$^{187}Os/^{188}Os$
^{40}K	^{40}Ar	1.25×10^9	$^{40}Ar/^{36}Ar$
^{232}Th	^{208}Pb	14.01×10^9	$^{208}Pb/^{204}Pb$
^{238}U	^{206}Pb	4.468×10^9	$^{206}Pb/^{204}Pb$
^{235}U	^{207}Pb	0.738×10^9	$^{207}Pb/^{204}Pb$

together, they cover a wide range of geochemical properties including incompatible and compatible behavior. These ratios will be used, together with some incompatible trace-element ratios, as tracers of mantle reservoirs, crust–mantle differentiation processes, and mantle melting processes in later sections of this chapter.

3.2 LOCAL AND REGIONAL EQUILIBRIUM REVISITED

How do we translate geochemical data from basalts into a geological model of the present-day mantle and its evolution? The question of chemical and isotopic equilibrium, and particularly its spatial dimension, has always played a fundamental role in this effort of interpretation. The basic, simple tenet of isotope geochemists and petrologists alike has generally been that partial melting at mantle temperatures, pressures, and timescales achieves essentially complete chemical equilibrium between melt and solid residue. For isotope data in particular, this means that at magmatic temperatures, the isotope ratio of the melt is identical to that of the source, and this is what made isotope ratios of volcanic rocks apparently ideal tracers of mantle composition. The question of spatial scale seemed less important, because heterogeneities in the mantle were thought to be important primarily on the 10^2–10^4 km scale (Hart *et al.*, 1973; Schilling, 1973; White and Schilling, 1978; Dupré and Allègre, 1983). To be sure, this simple view was never universal. Some authors invoked special isotopic effects during melting, so that the isotopic composition of the melt could in some way be "fractionated" during melting, in spite of the high temperatures prevailing, so that the isotope ratios observed in the melts would *not* reflect those of the melt sources (e.g., O'Hara, 1973; Morse, 1983). These opinions were invariably raised by authors not directly familiar with the analytical methods of isotope geochemistry, so that they did not realize that isotopic fractionation occurs in every mass spectrometer and is routinely corrected in the reported results.

3.2.1 Mineral Grain Scale

Some authors invoked mineral-scale isotopic (and therefore also chemical) disequilibrium and preferential melting of phases, such as phlogopite, which have higher Rb/Sr, and therefore also higher $^{87}Sr/^{86}Sr$ ratios than the bulk rock, in order to explain unusually high $^{87}Sr/^{86}Sr$ ratios in OIBs (e.g., O'Nions and Pankhurst, 1974; Vollmer, 1976). Hofmann and Hart (1978) reviewed this subject in light of the available diffusion data in solid and molten silicates. They concluded that mineral-scale isotopic and chemical disequilibrium is exceedingly unlikely, if melting timescales are on the order of thousands of years or more. More recently, Van Orman *et al.* (2001) have measured REE diffusion coefficients in clinopyroxene and found that REE mobility in this mineral is so low at magmatic temperatures that chemical disequilibrium between grain centers and margins will persist during melting. Consequently, the melt will not be in equilibrium with the bulk residue for geologically reasonable melting times, if the equilibration occurs by volume diffusion alone. This means that the conclusions of Hofmann and Hart (1978) must be revised significantly: The slowest possible path of chemical reaction no longer guarantees attainment of equilibrium. However, it is not known whether other mechanisms such as recrystallization during partial melting might not lead to much more rapid equilibration. One possible test of this would be the examination of mantle clinopyroxenes from oceanic and ophiolitic peridotites. These rocks have undergone various extents of partial melting (Johnson *et al.*, 1990; Hellebrand *et al.*, 2001), and the residual clinopyroxenes should show compositional zoning if they had not reached equilibrium with the melt via volume diffusion. Although the above-cited studies were not specifically conducted to test this question, the clinopyroxenes were analyzed by ion microprobe, and these analyses showed no significant signs of internal compositional gradients. It is, of course, possible in principle that the internal equilibration occurred after extraction of the melt, so this evidence is not conclusive at present. Nevertheless, these results certainly leave open the possibility that the crystals re-equilibrated continuously with the melt during melt production and extraction. There is at present no definitive case from "natural laboratories" deciding the case one way or the other, at least with respect to incompatible lithophile elements such as the REE.

Osmium isotopes currently provide the strongest case for mineral-to-mineral disequilibrium, and for mineral–melt disequilibrium available from observations on natural rocks. Thus, both osmium alloys and sulfides from ophiolites and mantle xenoliths have yielded strongly heterogeneous osmium isotope ratios (Alard *et al.*, 2002; Meibom *et al.*, 2002). The most remarkable aspect of these results is that these ophiolites were emplaced in Phanerozoic times, yet they contain osmium-bearing phases that have retained model ages in excess of 2 Ga in some cases. The melts that were extracted from these ophiolitic peridotites contained almost certainly much more radiogenic osmium and could, in any case, not have been in osmium-isotopic equilibrium with all of these isotopically diverse residual phases.

Another strong indication that melts extracted from the mantle are not in osmium-isotopic equilibrium with their source is given by the fact that osmium isotopes in MORBs are, on average, significantly more radiogenic than osmium isotopes from oceanic peridotites (see also Figure 9 further below). Although it may be argued that there is no one-to-one correspondence between basalts and source peridotites, and further, that the total number of worldwide MORB and peridotite samples analyzed is still small, these results strongly suggest that, at least with regard to osmium, MORBs are generally not in isotopic equilibrium with their sources or residues. However, osmium-isotopic disequilibrium does not automatically mean strontium, neodymium, lead, or oxygen-isotopic disequilibrium or incompatible-trace-element disequilibrium. This is because osmium is probably incompatible in all silicate phases (Snow and Reisberg, 1995; Schiano *et al.*, 1997; Burton *et al.*, 2000) but very highly compatible with nonsilicate phases such as sulfides and, possibly, metal alloys such as osmiridium "nuggets," which may form inclusions within silicate minerals and might therefore be protected from reaction with a partial silicate melt. At the time of writing, no clear-cut answers are available, and for the time being, we will simply note that the geochemistry of osmium and rhenium is considerably less well understood than that of silicate-hosted major and trace elements such as strontium, neodymium, lead, and their isotopic abundances.

3.2.2 Mesoscale Heterogeneities

By "mesoscale" I mean scales larger than about a centimeter but less than a kilometer. This intermediate scale was addressed only briefly by Hofmann and Hart, who called it a "lumpy mantle" structure. It was specifically invoked by Hanson (1977) and Wood (1979), and others subsequently, who invoked veining in the mantle to provide sources for chemically and isotopically heterogeneous melts. Other versions of mesoscale heterogeneities were invoked by Sleep (1984), who suggested that preferential melting of ubiquitous heterogeneities may explain ocean island-type volcanism, and by Allègre and Turcotte (1986), who discussed a "marble cake" structure of the mantle generated by incomplete homogenization of subducted heterogeneous lithosphere. These ideas have recently been revived in several publications discussing a mantle containing pyroxenite or eclogite layers, which may melt preferentially (Phipps Morgan *et al.*, 1995; Hirschmann and Stolper, 1996; Phipps Morgan and Morgan, 1998; Yaxley and Green, 1998; Phipps Morgan, 1999).

One of the main difficulties with these mesoscale models is that they have been difficult to test by direct geochemical and petrological field observations. Recently, however, several studies have been published which appear to support the idea of selective melting of mesoscale heterogeneities. Most important of these are probably the studies of melt inclusions showing that single basalt samples, and even single olivine grains, contain chemically and isotopically extremely heterogeneous melt inclusions. Extreme heterogeneities in REE abundances from melt inclusions had previously been explained by progressive fractional melting processes of uniform sources (Sobolev and Shimizu, 1993; Gurenko and Chaussidon, 1995). In contrast, the more recent studies have demonstrated that source heterogeneities must (also) be involved to explain the extreme variations in isotopic and chemical compositions observed (Saal *et al.*, 1998; Sobolev *et al.*, 2000). While the spatial scale of these source heterogeneities cannot be directly inferred from these melt inclusion data, it seems highly plausible that it is in the range of what is here called "mesoscale."

Other, more circumstantial, evidence for preferential melting of mesoscale heterogeneities has been described by Regelous and Hofmann (2002/3), who found that the Hawaiian plume delivered MORB-like magmas $\sim$80 Ma ago, when the plume was located close to the Pacific spreading ridge.

Unless this is a fortuitous coincidence, this implies that the same plume produces "typical" OIB-like, incompatible-element-enriched melts with elevated $^{87}Sr/^{86}Sr$ and low $^{143}Nd/^{144}Nd$ ratios when the degree of melting is relatively low under a thick lithosphere, and typically MORB-like, incompatible-element-depleted melts when the degree of melting is high because of the shallow melting level near a spreading ridge. Such a dependence on the extent of melting is consistent with a marble-cake mantle containing incompatible-element-rich pyroxenite or eclogite layers having a lower melting temperature than the surrounding peridotite matrix. This melting model is further corroborated by the observation that at least three other plumes located at or near spreading ridges have produced MORB-like lavas, namely, the Iceland, the Galapagos, and the Kerguelen plume. The overall evidence is far from clear-cut, however, because the Iceland and Galapagos plumes have also delivered OIB-like tholeiites and alkali basalts more or less in parallel with the depleted MORB-like tholeiites or picrites.

To sum up, the question of grain-scale equilibration with partial melts, which had apparently been settled definitively by Hofmann and Hart (1978), has been reopened by the experimental work of Van Orman *et al.* (2001) and by recent osmium isotope data. The mesoscale equilibrium involving a veined or marble-cake mantle consisting of a mixture of lherzolite (or harzburgite) and pyroxenite (or eclogite) has also received substantial support in the recent literature. In either case, the isotopic composition of the melt is likely to change as a function of the bulk extent of melting, and the melts do not provide quantitative estimates of the isotopic composition of the bulk sources at scales of kilometers or more. It will be seen in subsequent sections that this has ramifications particularly with respect to quantitative estimates of the sizes and spatial distributions of the reservoirs hosting the geochemical mantle heterogeneities observed in basalts. While this defeats one of the important goals of mantle geochemistry, it will be seen in the course of this chapter that the geochemical data can still be used to map large-scale geochemical provinces of the mantle and to reveal much about the smaller-scale structure of the mantle heterogeneities. In addition, they remain powerful tracers of recycling and mixing processes and their history in the mantle.

3.3 CRUST–MANTLE DIFFERENTIATION

Before discussing the internal chemical structure of the mantle, it is necessary to have a general understanding of crust–mantle differentiation, because this has affected the incompatible trace-element and isotope budget of the mantle rather drastically. This topic has been covered by Hofmann (1988), but the most important points will be summarized here again. The treatment here differs in detail because more recent estimates have been used for the bulk composition of the continental crust and of the bulk silicate earth (BSE), also called "primitive mantle."

3.3.1 Enrichment and Depletion Patterns

The growth of the continental crust has removed major proportions of the highly incompatible elements from the mantle, and this depletion is the chief (but not the sole) cause of the specific isotope and trace-element characteristics of MORBs. The effects of ionic radius and charge, described in Section 3.1.2.1, on this enrichment–depletion process can be readily seen in a diagram (Figure 1) introduced by Taylor and McLennan (1985). It is obvious from this that those trace elements that have ionic properties similar to the major silicate-structure-forming elements, namely, nickel, cobalt, manganese, scandium, and chromium are not enriched in the continental crust but remain in the mantle. In contrast, elements with deviating ionic properties are more or less strongly enriched in the crust, depending on the magnitude of the deviation. Two main transfer mechanisms are available for this differentiation, both of them are ultimately driven by mantle convection. The first is partial melting and ascent of the melt to the surface or into the already existing crust. The second involves dehydration (and decarbonation) reactions during subduction, metasomatic transfer of soluble elements via hydrothermal fluid from the subducted crust-plus-sediment into the overlying

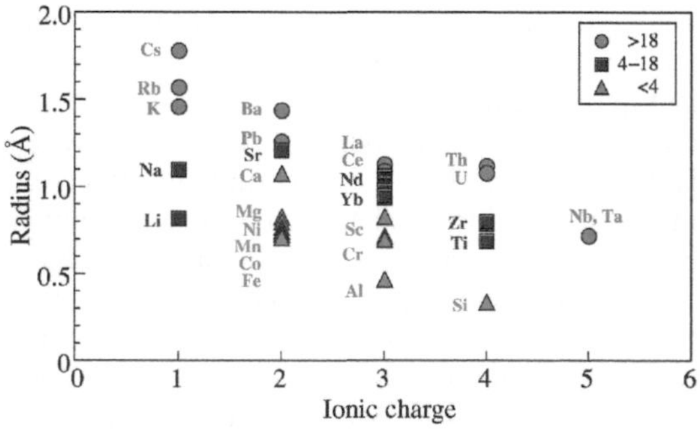

Figure 1 Ionic radius (in angstrom) versus ionic charge for lithophile major and trace elements in mantle silicates. Also shown are ranges of enrichment factors in average continental crust, using the estimate of (Rudnick and Fountain, 1995), relative to the concentrations in the primitive mantle (or "bulk silicate Earth") (source McDonough and Sun, 1995).

mantle "wedge," and partial melting of the metasomatized (or "fertilized") region. This partial melt ascends and is added to the crust, carrying the geochemical signature caused by mantle metasomatic transfer into the crust. Both mechanisms may operate during subduction, and a large body of geochemical literature has been devoted to the distinction between the two (Elliott *et al.*, 1997; Class *et al.*, 2000; Johnson and Plank, 2000). Continental crust may also be formed by mantle plume heads, which are thought to produce large volumes of basaltic oceanic plateaus. These may be accreted to existing continental crust, or continental flood basalts, which similarly add to the total volume of crust (e.g., Abouchami *et al.*, 1990; Stein and Hofmann, 1994; Puchtel *et al.*, 1998). The quantitative importance of this latter mechanism remains a matter of some debate (Kimura *et al.*, 1993; Calvert and Ludden, 1999).

Hofmann (1988) showed that crust formation by extraction of partial melt from the mantle could well explain much of the trace-element chemistry of crust–mantle differentiation. However, a few elements, notably niobium, tantalum, and lead, do not fit into the simple pattern of enrichment and depletion due to simple partial melting (Hofmann *et al.*, 1986). The fundamentally different behavior of these elements in the MORB–OIB environment on the one hand, and in the subduction environment on the other, requires the second, more complex, transfer mechanism via fluids (Miller *et al.*, 1994; Peucker-Ehrenbrink *et al.*, 1994; Chauvel *et al.*, 1995). Thus, local fluid transport is essential in preparing the mantle sources for production of continental crust, but the gross transport of incompatible elements from mantle to crust is still carried overwhelmingly by melting and melt ascent.

The simplest case discussed above, namely, crust–mantle differentiation by partial melting alone is illustrated in Figure 2. This shows the abundances of a large number of chemical elements in the continental crust, as estimated by (Rudnick and Fountain, 1995) and divided by their respective abundances in the primitive mantle or bulk silicate earth as estimated by (McDonough and Sun, 1995). Each element is assigned a nominal partition coefficient D as defined in Equation (1), calculated by rearranging Equation (2) and using a nominal "melt fraction" $F=0.01$. In this highly simplified view, the continental crust is assumed to have the composition of an equilibrium partial melt derived from primitive mantle material. Also shown is the hypothetical solid mantle residue of such a partial melt and a second-stage partial melt of this depleted residue. This second-stage melt curve may then be compared with the actual element abundances of "average" ocean crust. Although this "model" of the overall crust–mantle differentiation is grossly oversimplified, it can account for the salient features of the relationship

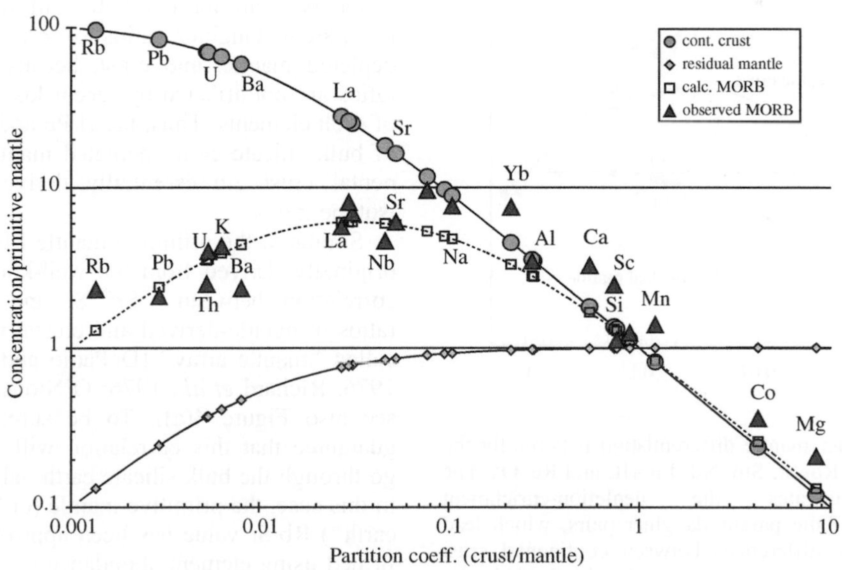

Figure 2 Comparison of the abundances of trace and (some) major elements in average continental crust and average MORB. Abundances are normalized to the primitive-mantle values (McDonough and Sun, 1995). The "partition coefficient" of each element is calculated by solving Equation (2) for D, using a melt fraction $F=0.009$ and its abundance value in the continental crust (Rudnick and Fountain, 1995). The respective abundances in average MORB are plotted using the same value of D and using the average ("normal") MORB values of (Su, 2002), where "normal" refers to ridge segments distant from obvious hotspots.

between primitive mantle, continental crust, depleted mantle, and oceanic crust quite well. This representation is remarkably successful because Equation (2) is essentially a mass-balance relationship and because these major reservoirs are in fact genetically related by enrichment and depletion processes in which partial melting plays a dominant role.

The above model of extracting continental crust and remelting the depleted residue also accounts approximately for the isotopic relationships between continental crust and residual mantle, where the isotopic composition is directly represented by MORB, using the assumption of complete local and mesoscale equilibrium discussed in Section 3.2. This is illustrated by Figure 3, which is analogous to Figure 2, but shows only the commonly used radioactive decay systems Rb–Sr, Sm–Nd, Lu–Hf, and Re–Os. Thus, the continental crust has a high parent–daughter ratios Rb/Sr and Re/Os, but low Sm/Nd and Lu/Hf, whereas the mantle residue has complementary opposite ratios. With time, these parent–daughter ratios will generate higher than primitive $^{87}Sr/^{86}Sr$ and $^{187}Os/^{188}Os$, and lower than primitive $^{143}Nd/^{144}Nd$ and $^{176}Hf/^{177}Hf$, ratios in the crust and complementary, opposite ratios in the mantle, and this is indeed observed for strontium,

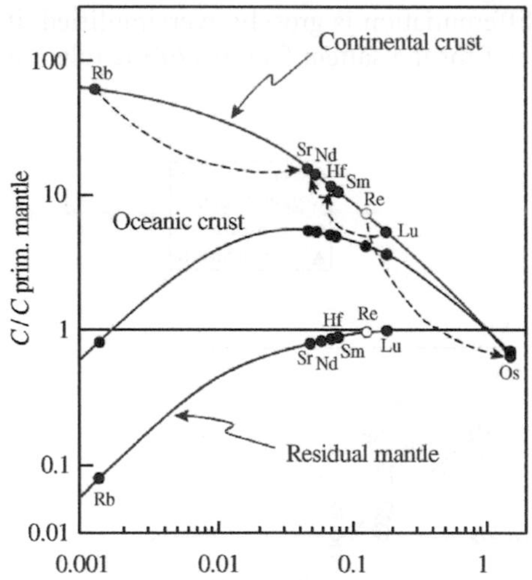

Figure 3 Crust–mantle differentiation patterns for the decay systems Rb–Sr, Sm–Nd, Lu–Hf, and Re–Os. The diagram illustrates the depletion-enrichment relationships of the parent–daughter pairs, which lead to the isotopic differences between continental crust and the residual mantle. For example, the Sm/Nd ratio is increased, whereas the Rb/Sr ratio is decreased in the residual mantle. This leads to the isotopic correlation in mantle-derived rocks plotted in Figure 4(a). The construction is similar to that used in Figure 2, but *D* values have been adjusted slightly for greater clarity.

neodymium, and hafnium, as will be seen in the review of the isotope data.

The case of lead isotopes is more complicated, because the estimates for the mean parent–daughter ratios of mantle and crust are similar. This similarity is not consistent with purely magmatic production of the crust, because the bulk partition coefficient of lead during partial mantle melting is expected to be only slightly lower than that of strontium (see Figure 1), but significantly higher than the coefficients for the highly incompatible elements uranium and thorium. In reality, however, the enrichment of lead in the continental crust shown in Figure 2 is slightly higher than the enrichments for thorium and uranium, and the $^{206}Pb/^{204}Pb$ and $^{208}Pb/^{204}Pb$ ratios of continental rocks are similar to those of MORB. This famous–infamous "lead paradox," first pointed out by Allègre (1969), will be discussed in a separate section below.

How do we know the parent–daughter ratios in crust and mantle? When both parent and daughter nuclides have refractory lithophile character, and are reasonably resistant to weathering and other forms of low-temperature alteration, as is the case for the pairs Sm–Nd and Lu–Hf, we can obtain reasonable estimates from measuring and averaging the element ratios in representative rocks of crustal or mantle heritage. But when one of the elements was volatile during terrestrial accretion, and/or is easily mobilized by low-temperature or hydrothermal processes, such as rubidium, uranium, or lead, the isotopes of the daughter elements yield more reliable information about the parent–daughter ratios of primitive mantle, depleted mantle, and crust, because the isotope ratios are not affected by recent loss (or addition) of such elements. Thus, the U/Pb and Th/Pb ratios of bulk silicate earth, depleted mantle and continental crust are essentially derived from lead isotope ratios.

Similarly, the primitive mantle Rb/Sr ratio was originally derived from the well-known negative correlation between $^{87}Sr/^{86}Sr$ and $^{143}Nd/^{144}Nd$ ratios in mantle-derived and crustal rocks, the so-called "mantle array" (DePaolo and Wasserburg, 1976; Richard *et al.*, 1976; O'Nions *et al.*, 1977; see also Figure 4(a)). To be sure, there is no guarantee that this correlation will automatically go through the bulk silicate earth value. However, in this case, the primitive mantle (or "bulk silicate earth") Rb/Sr value has been approximately confirmed using element abundance ratios of barium, rubidium, and strontium. Hofmann and White (1983) found that Ba/Rb ratios in mantle-derived basalts and continental crust are sufficiently similar, so that the terrestrial value of Ba/Rb can be estimated within narrow limits. The terrestrial Ba/Sr ratio (comprising two refractory, lithophile elements) can be assumed to be identical to the ratio in chondritic meteorites, so that the terrestrial Rb/Sr ratio can be estimated as

$$\left(\frac{\mathrm{Rb}}{\mathrm{Sr}}\right)_{\mathrm{terr.}} = \left(\frac{\mathrm{Rb}}{\mathrm{Ba}}\right)_{\mathrm{terr.}} \times \left(\frac{\mathrm{Ba}}{\mathrm{Sr}}\right)_{\mathrm{chondr.}} \quad (8)$$

The terrestrial Rb/Sr ratio estimated in this way turned out to be indistinguishable from the ratio estimated by isotope correlations, and therefore the consistency between isotope and element abundance data is not circular. This example of internal consistency has been disturbed by the more recent crustal estimate of Ba/Rb = 6.7 (Rudnick and Fountain, 1995), which is significantly lower than the above mantle estimate Ba/Rb = 11.0 (Hofmann and White, 1983), based mostly on MORB and OIB data. This shows that, for many elements, there are greater uncertainties about the composition of the continental crust than about the mantle. The reason for this is that the continental crust has become much more heterogeneous than the mantle because of internal differentiation processes including intracrustal melting, transport of metamorphic fluids, hydrothermal transport, weathering, erosion, and sedimentation. In any case, assuming that Rudnick's crustal estimate is correct, the primitive mantle Ba/Rb should lie somewhere between 7 and 11. The lesson from this is that we must be careful when using "canonical" element ratios to make mass-balance estimates for the sizes of different mantle reservoirs.

3.3.2 Mass Fractions of Depleted and Primitive Mantle Reservoirs

The simple crust–mantle differentiation model shown in Figure 2 contains three "reservoirs:" continental crust, depleted residue, and oceanic crust. However, the depleted reservoir may well be smaller than the entire mantle, in which case another, possibly primitive reservoir would be needed. Thus, if one assumes that the mantle consists of two reservoirs only, one depleted and one remaining primitive, and if one neglects the oceanic crust because it is thin and relatively depleted in highly incompatible elements, one can calculate the mass fractions of these two reservoirs from their respective isotopic and/or trace-element compositions (Jacobsen and Wasserburg, 1979; O'Nions *et al.*, 1979; DePaolo, 1980; Davies, 1981; Allègre *et al.*, 1983, 1996; Hofmann *et al.*, 1986; Hofmann, 1989a). The results of these estimates have yielded mass fractions of the depleted reservoir ranging from ~30–80%. Originally, the 30% estimate was particularly popular because it matches the mass fraction of the upper mantle above the 660 km seismic discontinuity. It was also attractive because at least some of the mineral physics data indicated that the lower mantle has a different, intrinsically denser, major-element composition. However, more recent data and their evaluations indicate that they do not require such compositional layering (Jackson and Rigden, 1998). Nevertheless, many authors argue that the 660 km boundary can isolate upper-mantle- from lower-mantle convection, either because of the endothermic nature of the phase changes at this boundary, or possibly because of extreme viscosity differences between upper and lower mantle. Although this entire subject has been debated in the literature for many years, there appeared to be good reasons to think that the 660 km seismic discontinuity is the fundamental boundary between an upper, highly depleted mantle and a lower, less depleted or nearly primitive mantle.

The most straightforward mass balance, assuming that we know the composition of the continental crust sufficiently well, can be calculated from the abundances of the most highly incompatible elements, because their abundances in the depleted mantle are so low that even relatively large relative errors do not affect the mass balance very seriously. The most highly enriched elements in the continental crust have estimated crustal abundances (normalized to the primitive mantle abundances given by McDonough and Sun (1995)) of Cs = 123, Rb = 97, and Th = 70 (Rudnick and Fountain, 1995). The estimate for Cs is rather uncertain because its distribution within the crust is particularly heterogeneous, and its primitive-mantle abundance is afflicted by special uncertainties (Hofmann and White, 1983; McDonough *et al.*, 1992). Therefore, a more conservative enrichment factor of 100 (close to the value of 97 for Rb) is chosen for elements most highly enriched in the continental crust. The simple three-reservoir mass balance then becomes

$$X_{\mathrm{lm}} = \frac{1 - C_{\mathrm{cc}} \times X_{\mathrm{cc}} - C_{\mathrm{um}} \times X_{\mathrm{um}}}{C_{\mathrm{lm}}} \quad (9)$$

where C refers to primitive-mantle normalized concentrations (also called "enrichment factors"), X refers to the mass fraction of a given reservoir, and the subscripts cc, lm, and um refer to continental crust, lower mantle, and upper mantle, respectively.

If the lower mantle is still primitive, so that $C_{\mathrm{lm}} = 1$, the upper mantle is extremely depleted, so that $C_{\mathrm{um}} = 0$, and $X_{\mathrm{cc}} = 0.005$, the mass balance yields

$$X_{\mathrm{lm}} = \frac{1 - 100 \times 0.005 - 0 \times X_{\mathrm{um}}}{1} = 0.5 \quad (10)$$

Remarkably, this estimate is identical to that obtained using the amounts of radiogenic argon in the atmosphere, the continental crust and the depleted, upper mantle (Allègre *et al.*, 1996). There are reasons to think that the abundances of potassium and rubidium in BSE used in these calculations have been overestimated, perhaps by as much as 30% (Lassiter, 2002), and this would of course decrease the remaining mass fraction of primitive-mantle material. Thus, we can conclude that at least half, and perhaps 80%, of the most highly incompatible element budget now resides either in the continental crust or in the atmosphere (in the case of argon).

Can we account for the entire silicate earth budget by just these three reservoirs (crust plus atmosphere, depleted mantle, primitive mantle), as has been assumed in all of the above estimates? Saunders *et al.* (1988) and Sun and McDonough (1989) (among others) have shown that this cannot be the case, using global systematics of a single trace-element ratio, Nb/La. Using updated, primitive-mantle normalized estimates for this ratio, namely, $(Nb/La)_n = 0.66$ for the continental crust (Rudnick and Fountain, 1995), and $(Nb/La)_n = 0.81$ for so-called N-type ("normal") MORB (Su, 2002), we see that both reservoirs have lower than primitive Nb/La ratios. Using the additional constraint that niobium is slightly more incompatible than lanthanum during partial melting, we find that the sources of all these mantle-derived basalts must have sources with Nb/La ratios equal to or lower than those of the basalts themselves. This means that all the major mantle sources as well as the continental crust have $(Nb/La)_n \leq 1$. By definition, the entire silicate earth has $(Nb/La)_n = 1$, so there should be an additional, hidden reservoir containing the "missing" niobium. A similar case has more recently been made using Nb/Ta, rather than Nb/La. Current hypotheses to explain these observations invoke either a refractory eclogitic reservoir containing high-niobium rutiles (Rudnick *et al.*, 2000), or partitioning of niobium into the metallic core (Wade and Wood, 2001). Beyond these complications involving special elements with unexpected geochemical "behavior," there remains the question whether ~50% portion of the mantle not needed to produce the continental crust has remained primitive, or whether it is also differentiated into depleted, MORB-source-like, and enriched, OIB-source like subreservoirs. In the past, the occurrence of noble gases with primordial isotope ratios have been used to argue that the lower part of the mantle must still be nearly primitive. However, it will be seen below that this inference is no longer as compelling as it once seemed to be.

3.4 MID-OCEAN RIDGE BASALTS: SAMPLES OF THE DEPLETED MANTLE

3.4.1 Isotope Ratios of Strontium, Neodymium, Hafnium, and Lead

The long-lived radioactive decay systems commonly used to characterize mantle compositions, their half-lives, and the isotope ratios of the respective radiogenic daughter elements are given in Table 1. The half-lives of ^{147}Sm, ^{87}Sr, ^{186}Hf, ^{187}Re, and ^{232}Th are several times greater than the age of the Earth, so that the accumulation of the radiogenic daughter nuclide is nearly linear with time. This is not the case for the shorter-lived ^{238}U and ^{235}U, and this is in part responsible for the more complex isotopic relationships displayed by lead isotopes in comparison with the systematics of strontium, neodymium, hafnium, and osmium isotopes. The mantle geochemistry of noble gases, although of course an integral part of mantle geochemistry.

Figures 4–6 show the isotopic compositions of MORBs from spreading ridges in the three major ocean basins. Figures 4(b) and 5(a) also show isotope data for marine sediments, because these are derived from the upper continental crust and should roughly represent the isotopic composition of this crust. In general, the isotopic relationships between the continental and oceanic crust are just what is expected from the elemental parent–daughter relationships seen in Figure 3. The high Rb/Sr and low Sm/Nd and Lu/Hf ratios of continental materials relative to the residual mantle are reflected by high ^{87}Sr/^{86}Sr and low ^{143}Nd/^{144}Nd and ^{176}Hf/^{177}Hf ratios (not shown).

In lead isotope diagrams, the differences are not nearly as clear, and they are expressed primarily by slightly elevated ^{207}Pb/^{204}Pb ratios for given values of ^{206}Pb/^{204}Pb (Figure 5(a)). This topology in lead-isotope space requires a comparatively complex evolution of the terrestrial U–Pb system. It involves an ancient period of high U/Pb ratios in continental history (with complementary, low ratios in the residual mantle). The higher ^{235}U/^{238}U ratios prevailing during that time led to elevated ^{207}Pb/^{206}Pb ratios in the crust. This subject is treated more fully in the section on the lead paradox (Section 3.6) further below.

Another important observation is that, while strontium, neodymium, and hafnium isotopes all correlate with each other, they form poorer, but still significant, correlations with ^{206}Pb/^{204}Pb (or ^{208}Pb/^{204}Pb, not shown) ratios in the Pacific and Atlantic, but no discernible correlation in the Indian Ocean MORB (Figure 6(a)). Nevertheless, if instead of ^{208}Pb/^{204}Pb or ^{206}Pb/^{204}Pb ratios one plots the so-called "radiogenic" ^{208}Pb*/^{206}Pb* ratio, the lead data do correlate with neodymium

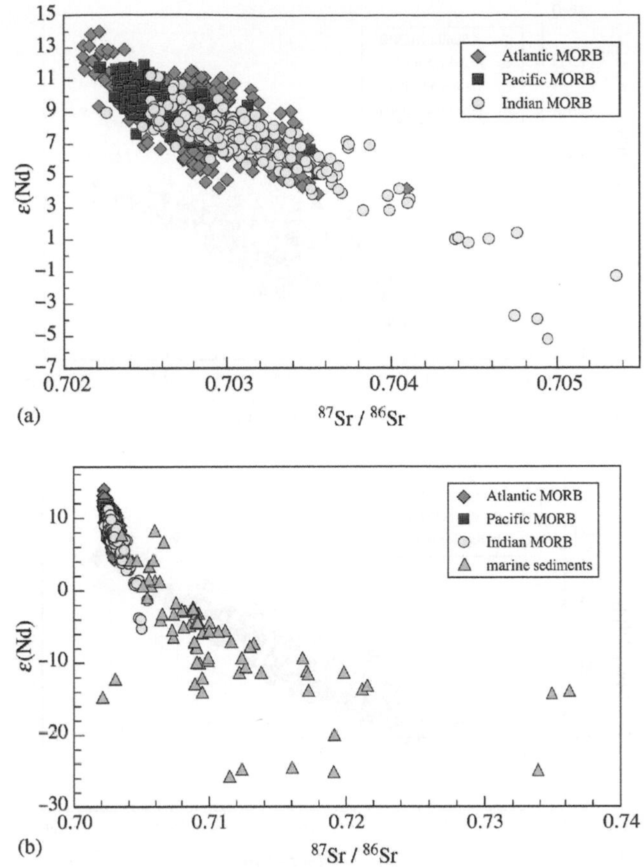

Figure 4 (a) $^{87}Sr/^{86}Sr$ versus $\varepsilon(Nd)$ for MORBs from the three major ocean basins. $\varepsilon(Nd)$ is a measure of the deviation of the $^{143}Nd/^{144}Nd$ ratio from the chondritic value, assumed to be identical to the present-day value in the bulk silicate earth. It is defined as $\varepsilon(Nd) = 10^4 \times (^{143}Nd/^{144}Nd_{measured} - ^{143}Nd/^{144}Nd_{Chondrite})/^{143}Nd/^{144}Nd_{Chondrite}$. The chondritic value used is $^{143}Nd/^{144}Nd_{Chondrite} = 0.512638$. The data are compiled from the PETDB database (http://petdb.ldeo.columbia.edu). (b) $^{87}Sr/^{86}Sr$ versus $\varepsilon(Nd)$ for MORBs compared with data for turbidites and other marine sediments (Ben Othman *et al.*, 1989; Hemming and McLennan, 2001). This illustrates the complementary nature of continent-derived sediments and MORB expected from the relationships shown in Figure 3.

isotopes in all three ocean basins (Figure 6(b)). This parameter is a measure of the radiogenic additions to $^{208}Pb/^{204}Pb$ and $^{206}Pb/^{204}Pb$ ratios during Earth's history; it is calculated by subtracting the primordial (initial) isotope ratios from the measured values. The primordial ratios are those found in the Th–U-free sulfide phase (troilite) of iron meteorites. Thus, the radiogenic $^{208}Pb^*/^{206}Pb^*$ ratio is defined as

$$\frac{^{208}Pb^*}{^{206}Pb^*} = \frac{^{208}Pb/^{204}Pb\text{-}(^{208}Pb/^{204}Pb)_{init}}{^{206}Pb/^{204}Pb\text{-}(^{206}Pb/^{204}Pb)_{init}} \qquad (11)$$

Unlike $^{208}Pb/^{204}Pb$ or $^{206}Pb/^{204}Pb$, which depend on Th/Pb and U/Pb, respectively, $^{208}Pb^*/^{206}Pb^*$ reflects the Th/U ratio integrated over the history of the Earth. The existence of global correlations between neodymium, strontium, and hafnium isotope ratios and $^{208}Pb^*/^{206}Pb^*$ and the absence of such global correlations with $^{208}Pb/^{204}Pb$ or $^{206}Pb/^{204}Pb$, shows that the elements neodymium, strontium, hafnium, thorium, and uranium behave in a globally coherent fashion during crust–mantle differentiation, whereas lead deviates from this cohesion.

Figures 4–6 show systematic isotopic differences between MORB from different ocean basins, reflecting some very large scale isotopic heterogeneities in the source mantle of these basalts. Also, the ranges of $\varepsilon(Nd)$ values present in a single ocean basin are quite large. For example, the range of neodymium isotope ratios in Atlantic MORB ($\sim 10\,\varepsilon(Nd)$ units) is somewhat smaller than the respective range of Atlantic OIB values of about $^{14}\varepsilon(Nd)$ units (see Section 3.5 on OIB), but this difference does not justify calling Atlantic MORB "isotopically homogeneous." This heterogeneity contradicts the widespread notion that the MORB-source mantle reservoir is isotopically nearly uniform, a myth that has persisted through many repetitions in the literature. One can just as easily argue that there is no such thing as a typical

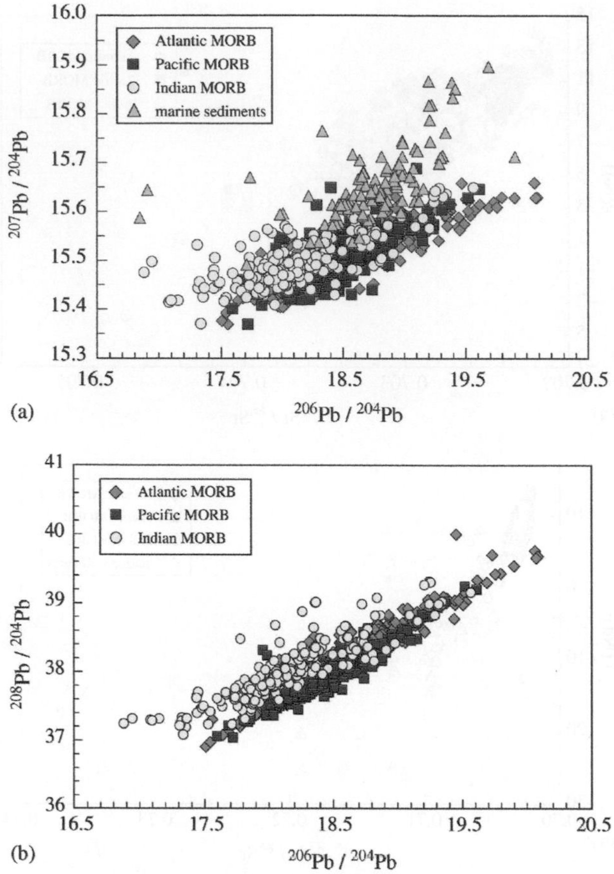

Figure 5 (a) $^{207}Pb/^{204}Pb$ versus $^{206}Pb/^{204}Pb$ for MORB from three major ocean basins and marine sediments. (b) $^{208}Pb/^{204}Pb$ versus $^{206}Pb/^{204}Pb$ for MORB from three major ocean basins. Sediments are not plotted because of strong overlap with the basalt data. For data sources see Figure 4.

"normal" (usually called N-type) MORB composition. In particular, the $^{208}Pb/^{204}Pb$ and $^{208}Pb^*/^{206}Pb^*$ ratios of Indian Ocean MORB show very little overlap with Pacific MORB (but both populations overlap strongly with Atlantic MORB). These very large scale regional "domains" were first recognized by Dupré and Allègre (1983) and named DUPAL anomaly by Hart (1984).

The boundary between the Indian Ocean and Pacific Ocean geochemical domains coincides with the Australian–Antarctic Discordance (AAD) located between Australia and Antarctica, an unusually deep ridge segment with several unusual physical and physiographic characteristics. The geochemical transition in Sr–Nd–Hf–Pb isotope space across the AAD is remarkably sharp (Klein *et al.*, 1988; Pyle *et al.*, 1992; Kempton *et al.*, 2002), and it is evident that very little mixing has occurred between these domains. The isotopic differences observed cannot be generated overnight. Rehkämper and Hofmann (1997), using lead isotopes, have estimated that the specific isotopic characteristics of the Indian Ocean MORB must be at least 1.5 Ga old. An important conclusion from this is that convective stirring of the mantle can be remarkably *ineffective* in mixing very large scale domains in the upper mantle.

When we further consider the fact that the present-day ocean ridge system, though globe encircling, samples only a geographically limited portion of the total, present-day mantle, it is clear that we must abandon the notion that we can characterize the isotopic composition of the depleted mantle reservoir by a single value of any isotopic parameter. What remains is a much broader, nevertheless limited, range of compositions, which, on average, differ from other types of oceanic basalts to be discussed further below. The lessons drawn from Section 3.2 (local and regional equilibrium) merely add an additional cautionary note: although it is possible to map the world's ocean ridge system using isotopic compositions of MORB, we cannot be sure how accurately these MORB compositions represent the underlying mantle. The differences between ocean basins are particularly obvious in the $^{208}Pb/^{204}Pb$ versus $^{206}Pb/^{204}Pb$ diagrams, where Indian Ocean MORBs have consistently higher $^{208}Pb/^{204}Pb$ ratios than Pacific Ocean

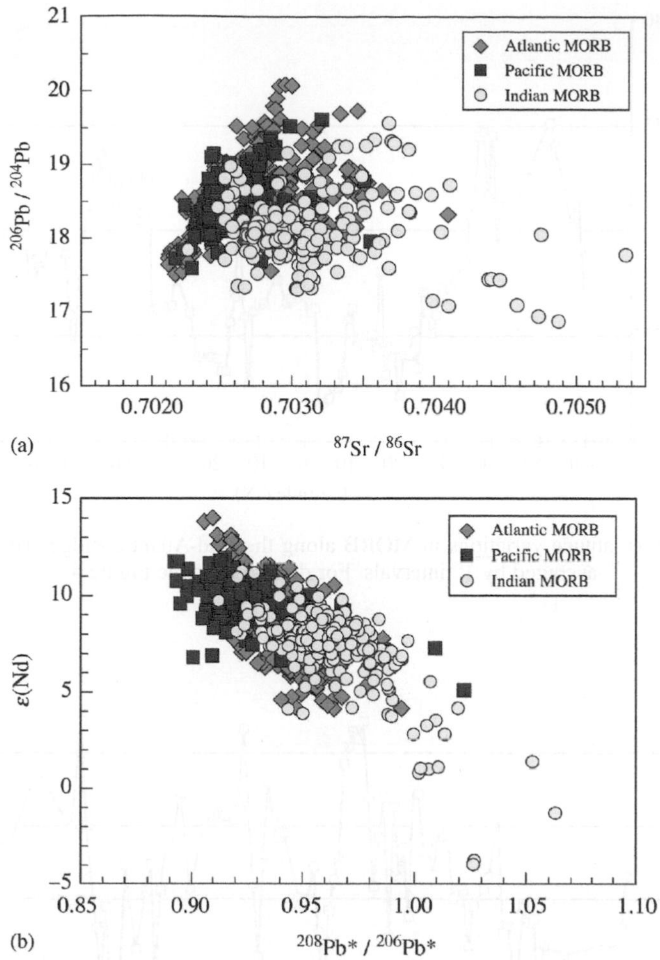

Figure 6 (a) $^{206}Pb/^{204}Pb$ versus $^{87}Sr/^{86}Sr$ for MORB from three major ocean basins. In contrast with the Sr–Nd and the Pb–Pb diagrams (Figures 4 and 5), the $^{206}Pb/^{204}Pb$–$^{87}Sr/^{86}Sr$ data correlate well only for Pacific MORB and not at all for Indian MORB. This indicates some anomalous behavior of the U–Pb decay system during global differentiation. For data sources see Figure 4. (b) $^{208}Pb^{*}/^{206}Pb^{*}$ versus ε(Nd) for MORB from three major ocean basins. The overall correlation is similar to the Sr–Nd correlation shown in Figure 4(a). This indicates that the Th/U ratios, which control $^{208}Pb^{*}/^{206}Pb^{*}$, do correlate with Sm/Nd (and Rb/Sr) ratios during mantle differentiation. Taken together, (a) and (b) identify Pb as the element displaying anomalous behavior.

MORBs. Diagrams involving neodymium isotopes show more overlap, but Indian Ocean MORBs have many ε(Nd) values lower than any sample from the Pacific Ocean.

An intermediate scale of isotopic variations is shown in Figures 7 and 8, using basalts from the Mid-Atlantic Ridge (MAR). The isotope ratios of strontium and neodymium (averaged over 1° intervals for clarity) vary along the ridge with obvious maxima and minima near the oceanic islands of Iceland, the Azores, and the Bouvet triple junction, and with large scale, relatively smooth gradients in the isotope ratios, e.g., between 20°S and 38°S. In general, the equatorial region between 30°S and 30°N is characterized by much lower strontium isotope ratios than the ridge segments to the north and the south. Some of this variation may be related to the vicinity of the mantle hotspots of Iceland, the Azores, or Bouvet, and the literature contains a continuing debate over the subject of "plume–asthenosphere interaction." Some authors argue the case where excess, enriched plume material spreads into the asthenosphere and mixes with depleted asthenospheric material to produce the geochemical gradients observed (e.g., Hart *et al.*, 1973; Schilling, 1973). Others argue that the hotspot or plume is internally heterogeneous and contains depleted, MORB-like material, which remains behind during normal plume-generated volcanism, spreads out in the asthenosphere, and becomes part of the asthenospheric mantle (Phipps Morgan and Morgan, 1998; Phipps Morgan, 1999). Irrespective of the specific process of plume–ridge interaction, the existence of compositional gradients up to ~2,000 km long implies some rather large-scale mixing processes,

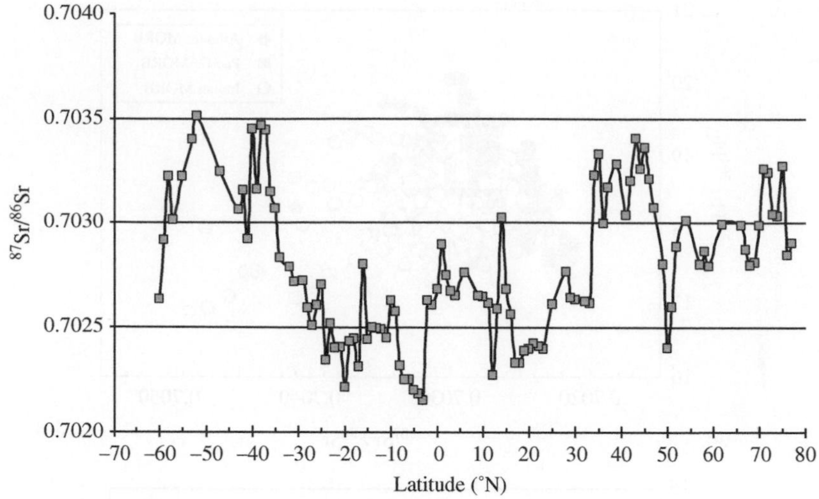

Figure 7 $^{87}Sr/^{86}Sr$ versus latitude variations in MORB along the Mid-Atlantic Ridge. The isotope data have been averaged by 1° intervals. For data sources see Figure 4.

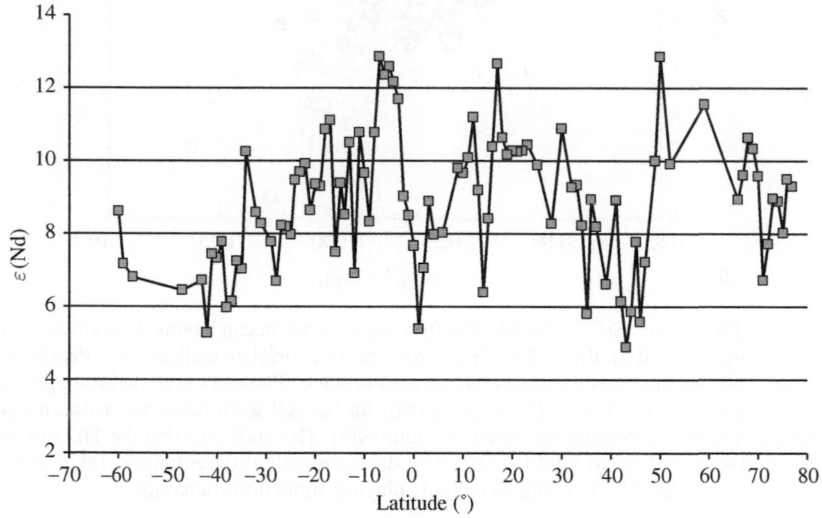

Figure 8 ε(Nd) versus latitude variations in MORB along the MAR. The isotope data have been averaged by 1° intervals. For data sources see Figure 4.

quite distinct from the sharpness of the compositional boundary seen at the AAD.

Isotopic heterogeneities are also observed on much smaller scales than so far discussed. For example, the region around 14 °N on the MAR shows a sharp "spike" in neodymium and strontium isotope ratios (see Figures 7 and 8) with an amplitude in ε(Nd) nearly as large as that of the entire Atlantic MORB variation, even though there is no obvious depth anomaly or other physiographic evidence for the possible presence of a mantle plume.

Finally, work on melt inclusions from phenocrysts has recently shown that sometimes extreme isotopic and trace-element heterogeneities exist within single hand specimens from mid-ocean ridges. Initially, the extreme chemical heterogeneities found in such samples were ascribed to the effects of progressive fractional melting of initially uniform source rocks (Sobolev and Shimizu, 1993), but recent Pb-isotope analyses of melt inclusions from a single MORB sample have shown a large range of isotopic compositions that require a locally heterogeneous source (Shimizu *et al.*, 2003). This phenomenon had previously also been observed in some OIBs (Saal *et al.*, 1998; Sobolev *et al.*, 2000), and future work must determine whether this is the exception rather than the rule. In any case, the normally observed local homogeneity of bulk basalt samples may turn out to be the result of homogenization in magma chambers rather than melting of homogeneous sources.

In summary, it should be clear that the mantle that produces MORB is isotopically heterogeneous on all spatial scales ranging from the size of ocean basins down to kilometers or possibly meters. The often-invoked homogeneity of MORB and MORB sources is largely a myth, and the definition of "normal" or N-type MORB actually applies to the depleted end of the spectrum rather than average MORB. It would be best to abandon these obsolete concepts, but they are likely to persist for years to come. Any unbiased evaluation of the actual MORB isotope data shows unambiguously, e.g., that Indian Ocean MORBs differ substantially from Pacific Ocean MORBs, and that only about half of Atlantic MORB conform to what is commonly referred to as "N-type" with $^{87}Sr/^{86}Sr$ ratios lower than 0.7028 and ε(Nd) values higher than +9. Many geochemists think that MORB samples with higher $^{87}Sr/^{86}Sr$ ratios and lower ε(Nd) values do not represent normal upper mantle but are generated by contamination of the normally very depleted, and isotopically extreme, upper mantle with plume material derived from the deeper mantle. Perhaps this interpretation is correct, especially in the Atlantic Ocean, where there are several hotspots/plumes occurring near the MAR, but its automatic application to all enriched samples (such as those near 14° N on the MAR) invites circular reasoning and it can get in the way of an unbiased consideration of the actual data.

3.4.2 Osmium Isotopes

The Re–Os decay system is discussed separately, in part because there are far fewer osmium isotope data than Sr–Nd–Pb data. This is true because, until ~10 years ago, osmium isotopes in silicate rocks were extraordinarily difficult to measure. The advent of negative-ion thermal ionization mass spectrometry has decisively changed this (Creaser *et al.*, 1991; Völkening *et al.*, 1991), and subsequently the number of publications providing osmium isotope data has increased dramatically.

Osmium is of great interest to mantle geochemists because, in contrast with the geochemical properties of strontium, neodymium, hafnium, and lead, all of which are incompatible elements, osmium is a compatible element in most mantle melting processes, so that it generally remains in the mantle, whereas the much more incompatible rhenium is extracted and enriched in the melt and ultimately in the crust. This system therefore provides information that is different from, and complementary to, what we can learn from strontium, neodymium, hafnium, and lead isotopes. However, at present there are still significant obstacles to the full use and understanding of osmium geochemistry. There are primarily three reasons for this:

(i) Osmium is present in oceanic basalts usually at sub-ppb concentration levels. Especially, in low-magnesium basalts, the concentrations can approach low ppt levels. The problem posed by this is that crustal rocks and seawater can have $^{187}Os/^{188}Os$ ratios 10 times higher than the (initial) ratios in mantle-derived melts. Thus, incorporation of small amounts of seawater-altered material in a submarine magma chamber may significantly increase the $^{187}Os/^{188}Os$ ratio of the magma. Indeed, many low-magnesium, moderately to highly differentiated oceanic basalts have highly radiogenic osmium, and it is not easy to know which basalts are unaffected by this contamination.

(ii) The geochemistry of osmium is less well understood than other decay systems, because much of the osmium resides in non-silicate phases such as sulfides, chromite and (possibly) metallic phases, and these phases can be very heterogeneously distributed in mantle rocks. This frequently leads to a "nugget" effect, meaning that a given sample powder is not necessarily representative of the system. Quite often, the reproducibility of concentration measurements of osmium and rhenium is quite poor by normal geochemical standards, with differences of several percent between duplicate analyses, and this may be caused either by intrinsic sample heterogeneity ("nugget effect") or by incomplete equilibration of sample and spike during dissolution and osmium separation.

(iii) There are legitimate doubts whether the osmium-isotopic composition of oceanic basalts are ever identical to those of their mantle source rocks (Section 3.2.1).

The Point (iii) above is illustrated in Figure 9, which shows osmium isotope ratios and osmium concentrations in abyssal peridotites and in MORB. This diagram shows two remarkable features: (i) The osmium isotope ratios of MORB and abyssal peridotites have very little overlap, the peridotites being systematically lower than those of seafloor basalts. (ii) The MORB data show a strong negative correlation between isotope ratios and osmium concentrations. These results suggest that the basalts may not be in isotopic equilibrium with their source rocks, but we have no proof of this, because we have no samples of specific source rocks for specific basalt samples. Also, the total number of samples represented in Figure 9 is rather small. Nevertheless, the apparently systematically higher $^{187}Os/^{188}Os$ ratios of the basalts compared with the peridotites seem to indicate that unradiogenic portions of the source peridotites

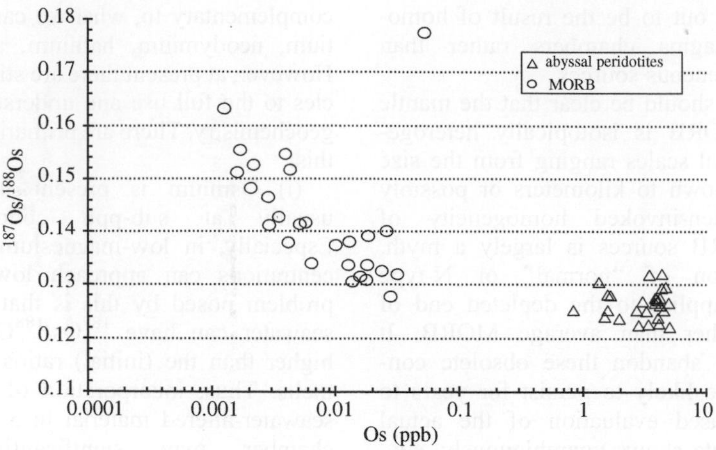

Figure 9 Osmium isotope ratios in MORB and abyssal peridotites. This diagram shows that osmium is generally compatible in peridotites during MORB melting. The systematic differences in $^{187}Os/^{188}Os$ ratios between MORB and peridotites suggest that the melts may not be in isotopic equilibrium with their residual peridotite (sources Martin, 1991; Roy-Barman and Allègre, 1994; Snow and Reisberg, 1995; Schiano *et al.*, 1997; Brandon *et al.*, 2000).

did not contribute to, or react with, the melt. The negative correlation displayed by the MORB data may mean that essentially *all* the melts are contaminated by seawater-derived osmium and that the relative contribution of the contaminating osmium to the measured isotopic compositions is inversely correlated with the osmium concentration of the sample. However, the MORB samples also show a strong positive correlation between $^{187}Os/^{188}Os$ and Re/Os (not shown). Therefore, it is also possible that MORB osmium is derived from heterogeneous sources in such a way that low-osmium, high-Re/Os samples are derived from high-Re/Os portions of the sources (such as pyroxenitic veins), whereas high-osmium, low-Re/Os samples are derived from the peridotitic or even harzburgitic matrix.

To avoid the risk of contamination by seawater, either through direct contamination of the samples or contamination of the magma by assimilation of contaminated material, many authors disregard samples with very low osmium concentrations. Unfortunately, this approach does not remove the inherent ambiguity of interpretation, and it may simply bias the sampling. What is clearly needed are independent measures of very low levels of magma chamber and sample contamination.

3.4.3 Trace Elements

The general model of crust–mantle differentiation predicts that, after crust formation, the residual mantle should be depleted in incompatible elements. Melts from this depleted mantle may be absolutely enriched but should still show a relative depletion of highly incompatible elements relative to moderately incompatible elements, as illustrated in Figure 2. Here I examine actual trace-element data of real MORB and their variability. An inherent difficulty is that trace-element abundances in a basalt depend on several factors, namely, the source composition, the degree, and mechanism of melting and melt extraction, the subsequent degree of magmatic fractionation by crystallization, and finally, on possible contamination of the magma during this fractionation process by a process called AFC (assimilation with fractional crystallization). This inherent ambiguity resulted in a long-standing debate about the relative importance of these two aspects. O'Hara, in particular, championed the case of fractional crystallization and AFC processes in producing enrichment and variability of oceanic basalts (e.g., O'Hara, 1977; O'Hara and Mathews, 1981). In contrast, Schilling and coworkers argued that variations in trace-element abundances, and in particular ratios of such abundances, are strongly controlled by source compositions. They documented several cases where REE patterns vary systematically along mid-ocean ridge segments, and they mapped such variations specifically in the vicinity of hotspots, which they interpreted as the products of mantle plumes relatively enriched in incompatible elements. As was the case for the isotopic variations, they interpreted the trace-element variations in terms of mixing of relatively enriched plume material with relatively depleted upper mantle, the asthenospheric mantle (Schilling, 1973; White and Schilling, 1978). Figure 10 shows a compilation of La/Sm ratios (normalized to primitive mantle values) of basalts dredged

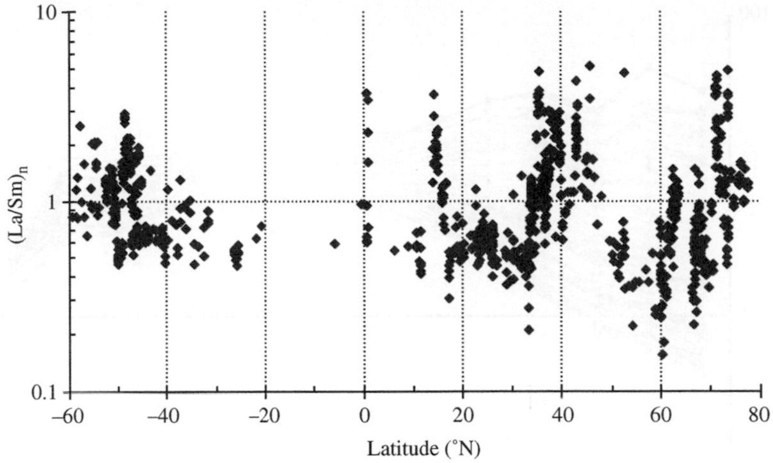

Figure 10 La/Sm ratios in MORB (not smoothed), normalized to primitive-mantle values, as a function of latitude along the MAR.

from the MAR. This parameter has been used extensively by Schilling and co-workers as a measure of source depletion or enrichment, where they considered samples with $(La/Sm)_n < 1.0$ as normal or "N-type" MORB derived from depleted sources with similar or even lower La/Sm ratios. As was found for the isotope ratios, only two-thirds of the MAR shows "typical" or "normal" La/Sm ratios lower than unity. In general, the pattern resembles that of the isotope variations, especially in the North Atlantic, where the coverage for both parameters is extensive. Thus, high La/Sm and $^{87}Sr/^{86}Sr$ values are found near the hotspots of Iceland and the Azores, between 45° S and 50° S, 14° N, and 43° N. Because of these correlations, the interpretation that the trace-element variations are primarily caused by source variations has been widely accepted. Important confirmation for this has come from the study of Johnson *et al.* (1990). They showed that peridotites dredged from near-hotspot locations along the ridge are more depleted in incompatible elements than peridotites from normal ridge segments. This implies that they have been subjected to higher degrees of melting (and loss of that melt). In spite of this higher degree of melting, the near-hotspot lavas are more enriched in incompatible elements, and therefore their initial sources must also have been more enriched.

Trace-element abundance patterns, often called "spidergrams," of MORB are shown in Figure 11 ("spidergram" is a somewhat inappropriate but a convenient term coined by R. N. Thompson (Thompson *et al.*, 1984), presumably because of a perceived resemblance of these patterns to spider webs, although the resemblance is tenuous at best). The data chosen for this plot are taken from le Roux *et al.* (2002) for MORB glasses from the MAR (40–55° S), which encompasses both depleted regions and enriched regions resulting from ridge–hotspot interactions. The patterns are highly divergent for the most incompatible elements, but they converge and become more parallel for the more compatible elements. This phenomenon is caused by the fact that variations in melt fractions produce the largest concentration variations in the most highly incompatible elements in both melts and their residues. This is a simple consequence of Equation (3), which states that for elements with very small values of D the concentration in the melt is inversely proportional to the melt fraction. At the other end of the spectrum, compatible elements, those with D values close to unity or greater, become effectively buffered by the melting assemblage. For an element having $D \gg F$, Equation (2) reduces to

$$C_l \approx \frac{C_0}{D} \qquad (12)$$

and for $D = 1$, it reduces to

$$C_l = C_0 \qquad (13)$$

In both cases, the concentration in the melt becomes effectively buffered by the residual mineral assemblage until the degree of melting is large enough, so that the specific mineral responsible for the high value of D is exhausted. This buffering effect is displayed by the relatively low and uniform concentrations of scandium (Figure 11). It is caused by the persistence of residual clinopyroxene during MORB melting.

These relationships lead to the simple consequence that the variability of element concentrations in large data sets of basalt analyses are related to the bulk partition coefficients of these elements (Hofmann, 1988; Dupré *et al.*, 1994). This can be verified by considering a set of trace

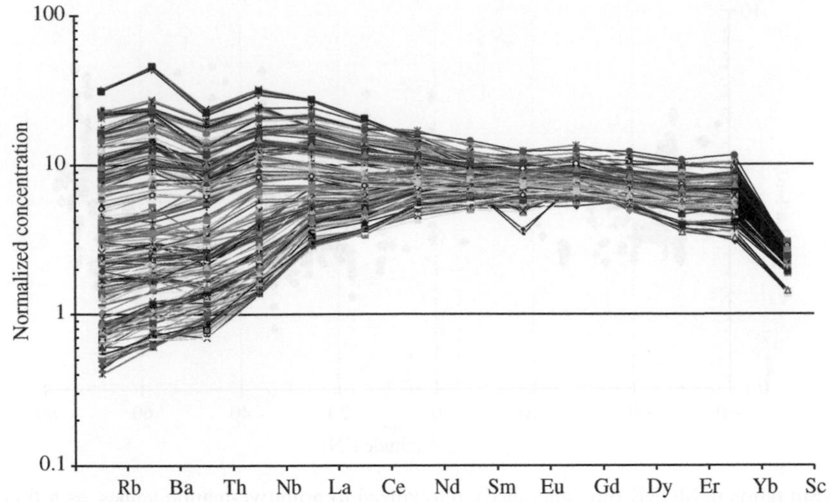

Figure 11 Trace element abundances of 250 MORB between 40°S and 55°S along the Mid-Atlantic Ridge. Each sample is represented by one line. The data are normalized to primitive-mantle abundances of (McDonough and Sun, 1995) and shown in the order of mantle compatibility. This type of diagram is popularly known as "spidergram". The data have been filtered to remove the most highly fractionated samples containing less than 5% MgO (source le Roux *et al.*, 2002).

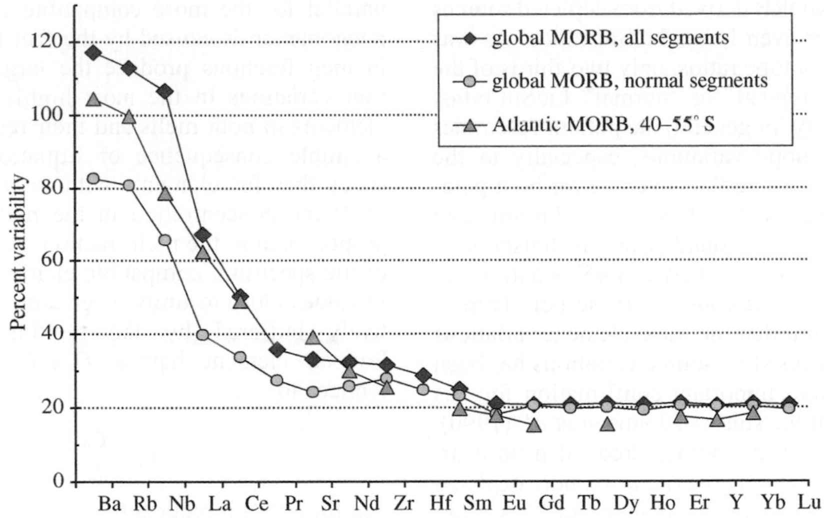

Figure 12 Variability of trace element concentrations in MORB, expressed as 100 * standard deviation/mean concentration. The data for "Global MORB" are from the PETDB compilation of (Su, 2002). "All segments" refers to ~250 ridge segments from all oceans. "Normal segments" refers to ~62 ridge segments that are considered not to represent any sort of "anomalous" ridges, because those might be affected by such factors as vicinity to mantle plumes or subduction of sediments (e.g., back-arc basins and the Southern Chile Ridge). The Atlantic MORB, 40–55°S, from which samples with less than 5% MgO have been removed (source le Roux *et al.*, 2002).

elements for which enough experimental data are available to be confident of the relative solid–melt partition coefficients, namely, the REE. These coefficients decrease monotonically from the heavy to light REEs, essentially because the ionic radii increase monotonically from heavy to light REEs (with the possible exception of europium, which has special properties because of its variable valence). Figure 12 shows three plots of variability of REEs and other trace elements in MORB as a function of mantle compatibility of the elements listed. Variability is defined as the standard deviation of the measured concentrations divided by the respective mean value. Compatibilities of elements other than the REEs are estimated from global correlations of trace-element ratios with absolute abundances as derived from simple partial melting theory

(Hofmann *et al.*, 1986; Hofmann, 1988) (see also Figures 17 and 18). Two sets of data are from a new, ridge segment-by-segment compilation made by Su (2002) using the MORB database http://petdb.ldeo.columbia.edu. The third represents data for 270 MORB glasses from the South Atlantic ridge (40–55° S) (le Roux *et al.*, 2002). The qualitative similarity of the three plots is striking. It indicates that the order of variabilities is a robust feature. For example, the variabilities of the heavy REEs (europium to lutetium) are all essentially identical at 20% in all three plots. For the light REEs, the variability increases monotonically from europium to lanthanum, consistent with their decreasing partition coefficients in all mantle minerals. As expected, variabilities are greatest for the very highly incompatible elements (VICEs) niobium, rubidium, and barium. The same is true for thorium and uranium in South Atlantic MORBs (le Roux *et al.*, 2002), which are not shown here because their averages and standard deviations were not compiled by Su (2002). All of this is consistent with the enrichment pattern in the continental crust (Figure 2), which shows the greatest enrichments for barium, rubidium, thorium, and uranium in the continental crust, as well as monotonically decreasing crustal abundances for the REEs from lanthanum to lutetium, with a characteristic flattening from europium to lutetium. Obvious exceptions to this general consistency are the elements niobium and lead. These will be discussed separately below.

Some additional lessons can be learned from Figure 12:

(i) The flattening of the heavy-REE variabilities in MORB are consistent with the flat heavy-REE patterns almost universally observed in MORB, and these are consistent with the flat pattern of HREE partition coefficients in clinopyroxene. This does not rule out some role of garnet during MORB melting, but it does probably rule out a major role of garnet.

(ii) Strontium, zirconium, and hafnium have very similar variabilities as the REEs neodymium and samarium. Again, this is consistent with the abundance patterns of MORB and with experimental data. Overall, the somewhat tentative suggestion made by Hofmann (1988) regarding the relationship between concentration variability and degree of incompatibility, based on a very small set of MORB data, is strongly confirmed by the very large data sets now available. A note of caution is in order for strontium, which has a high partition coefficient in plagioclase. Thus, when oceanic basalts crystallize plagioclase, the REEs tend to increase in the residual melt, but strontium is removed from the melt by the plagioclase. The net effect of this is that the overall variability of strontium is reduced in data sets incorporating plagioclase-fractionated samples. Such samples have been partly filtered out from the South Atlantic data set. This is the likely reason why strontium shows the greatest inconsistencies between the three plots shown in Figure 12.

3.4.4 N-MORB, E-MORB, T-MORB, and MORB Normalizations

It has become a widely used practice to define standard or average compositions of N-MORB, E-MORB, and T-MORB (for normal, enriched, and transitional MORB), and to use these as standards of comparison for ancient rocks found on land. In addition, many authors use "N-MORB" compositions as a normalization standard in trace-element abundance plots ("spidergrams") instead of chondritic or primitive mantle. This practice should be discouraged, because trace-element abundances in MORB form a complete continuum of compositions ranging from very depleted to quite enriched and OIB-like. A plot of global La/Sm ratios (Figure 13) demonstrates this: there is no obvious typical value but a range of lanthanum concentrations covering two orders of magnitude and a range of La/Sm ratios covering about one-and-a-half orders of magnitude. Although the term N-MORB was intended to describe "normal" MORB, it actually refers to depleted MORB, often defined by $(La/Sm)_n < 1$. Thus, while these terms do serve some purpose for characterizing MORB compositions, there is no sound basis for using any of them as normalizing values to compare other rocks with "typical" MORB.

The strong positive correlation seen in Figure 13 is primarily the result of the fact that lanthanum is much more variable than samarium. Still, the overall coherence of this relationship is remarkable. It demonstrates that the variations of the REE abundances are not strongly controlled by variations in the degree of crystal fractionation of MORB magmas, because these would cause similar variability of lanthanum and samarium. Although this reasoning is partly circular because highly fractionated samples containing less than 6% MgO have been eliminated, the total number of such samples in this population of $\sim$2,000 is less than 100. Thus, it is clear that the relationship is primarily controlled either by source or by partial melting effects. Figure 14 shows that the La/Sm ratios are also negatively correlated with $^{143}Nd/^{144}Nd$. Because this isotope ratio is a function of source Sm/Nd (and time), and neodymium is intermediate in bulk partition coefficient between lanthanum and samarium, such a negative correlation is expected if the variability of the

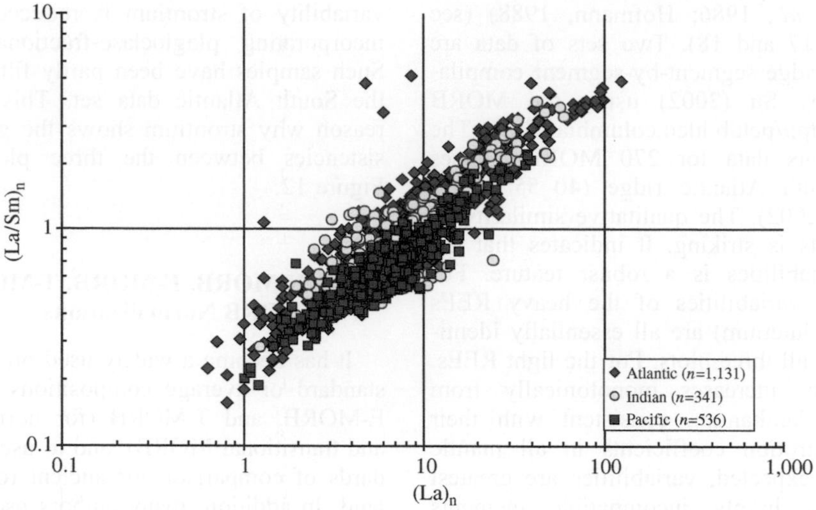

Figure 13 Primitive-mantle normalized La/Sm versus La for MORB from three Ocean basins. Numbers in parentheses refer to the number of samples from each ocean basin. Lanthanum concentrations vary by about two orders of magnitude; La/Sm varies by more than one order of magnitude. Data were extracted from PETDB.

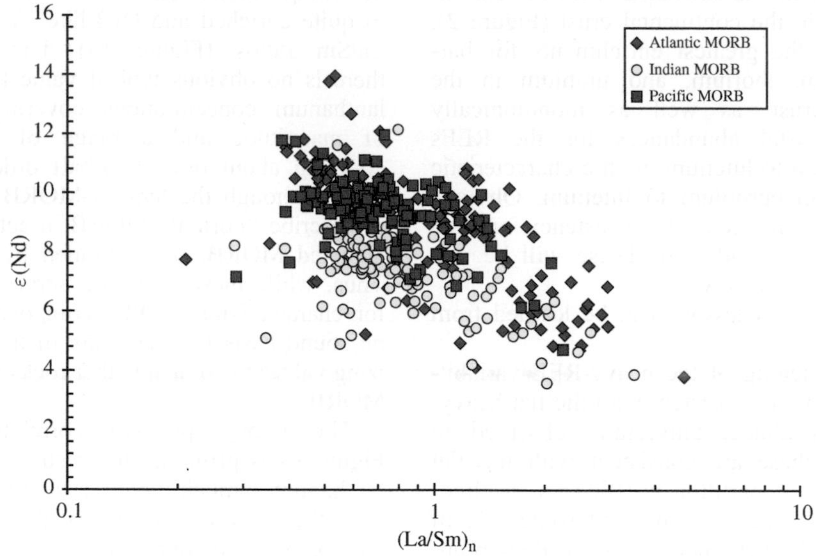

Figure 14 Primitive-mantle normalized La/Sm versus $^{143}Nd/^{144}Nd$ for MORB from three ocean basins. The (weak but significant) negative correlation is consistent with the inference that the variations in La/Sm in the basalts are to a significant part inherited from their mantle sources.

REE abundances is (at least in part) inherited from the source. Thus, while it would be perfectly possible to generate the relationship seen in Figure 13 purely by variations in partial melting, we can be confident that La/Sm ratios (and other highly incompatible element ratios) do track mantle source variations, as was shown by Schilling many years ago. It is important to realize, however, that such differences in source compositions were originally also produced by melting. These sources are simply the residues of earlier melting events during previous episodes of (continental or oceanic) crust formation.

3.4.5 Summary of MORB and MORB-Source Compositions

Klein and Langmuir (1987), in a classic paper, have shown that the element sodium is almost uniquely suited for estimating the degree of melting required to produce MORBs from their respective sources. This element is only slightly incompatible at low melt fractions produced at relatively high pressures. As a result, the extraction of the continental crust has reduced the sodium concentration of the residual mantle by no more than ~10% relative to the primitive-mantle value. We therefore know the approximate sodium concentration of all MORB sources. In contrast, this element behaves much more incompatibly during production of oceanic crust, where relatively high melt fractions at relatively low pressures produce ocean floor basalts. This allowed Klein and Langmuir to estimate effective melt fractions ranging from 8% to 20%, with an average of ~10%, from sodium concentrations in MORB. Once the melt fraction is known, the more highly incompatible-element concentrations of the MORB-source mantle can be estimated from the measured concentrations in MORB. For the highly incompatible elements, the source concentrations are therefore estimated at ~10% of their respective values in the basalts. This constitutes a significant revision of earlier thinking, which was derived from melting experiments and the assumption that essentially clinopyroxene-free harzburgites represent the typical MORB residue, and which led to melt fraction estimates of 20% and higher.

The compilation of extensive MORB data from all major ocean basins has shown that they comprise wide variations of trace-element and isotopic compositions and the widespread notion of great compositional uniformity of MORB is largely a myth. An exception to this may exist in helium-isotopic compositions. From the state of heterogeneity of the more refractory elements it is clear, however, that the apparently greater uniformity of helium compositions is not the result of mechanical mixing and stirring, because this process should homogenize all elements to a similar extent. Moreover, the isotope data of MORB from different ocean basins show that different regions of the upper mantle have not been effectively mixed in the recent geological past, where "recent" probably means approximately the last 10^9 yr.

The general, incompatible-element depleted nature of the majority of MORBs and their sources is well explained by the extraction of the continental crust. Nevertheless, the bulk continental crust and the bulk of the MORB sources are not *exact* chemical complements. Rather, the residual mantle has undergone additional differentiation, most likely involving the generation of OIBs and their subducted equivalents. In addition, there may be more subtle differentiation processes involving smaller-scale melt migration occurring in the upper mantle (Donnelly *et al.*, 2003). It is these additional differentiation processes that have generated much of the heterogeneity observed in MORBs and their sources.

3.5 OCEAN ISLAND, PLATEAU, AND SEAMOUNT BASALTS

These basalts represent the oceanic subclass of so-called intraplate basalts, which also include continental varieties of flood and rift basalts. They will be collectively referred to as "OIB," even though many of them are not found on actual oceanic islands either because they never rose above sea level or because they were formed on islands that have sunk below sea level. Continental and island arc basalts will not be discussed here, because at least some of them have clearly been contaminated by continental crust. Others may or may not originate in, or have been "contaminated" by, the subcontinental lithosphere. For this reason, they are not considered in the present chapter, which is concerned primarily with the chemistry of the sublithospheric mantle.

Geochemists have been particularly interested in OIB because their isotopic compositions tend to be systematically different from MORB, and this suggests that they come from systematically different places in the mantle (e.g., Hofmann *et al.*, 1978; Hofmann and Hart, 1978). Morgan's mantle plume theory (Morgan, 1971) thus provided an attractive framework for interpreting these differences, though not quite in the manner originally envisioned by Morgan. He viewed the entire mantle as a single reservoir, in which plumes rise from a lower boundary layer that is not fundamentally different in composition from the upper mantle. Geochemists, on the other hand, saw plumes being formed in a fundamentally different, more primitive, less depleted, or enriched, deeper part of the mantle than MORB sources (e.g., Wasserburg and Depaolo, 1979). The debate about these issues continues to the present day, and some of the mantle models based on isotopic and trace-element characteristics will be discussed below.

3.5.1 Isotope Ratios of Strontium, Neodymium, Hafnium, and Lead and the Species of the Mantle Zoo

Radiogenic isotope ratios of OIB are shown on Figure 15. These diagrams display remarkably similar topologies as the respective MORB data shown in Figures 4–6. Strontium isotope ratios are negatively correlated with neodymium and hafnium isotopes, and correlations between strontium, neodymium, and hafnium isotopes and lead isotopes are confined to $^{208}Pb^{*}/^{206}Pb^{*}$, and the ranges of isotope ratios are even greater (although not dramatically so) for OIBs than MORBs. There is one important difference, however, namely, a significant shift in all of these ratios between MORBs and OIBs. To be

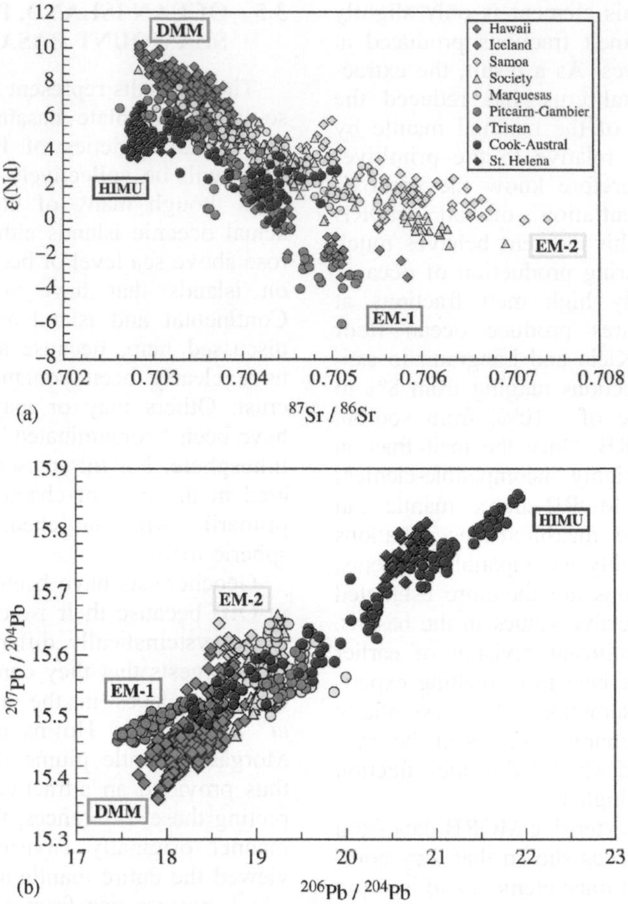

Figure 15 (a) $^{87}Sr/^{86}Sr$ versus $\varepsilon(Nd)$ for OIB (excluding island arcs). The islands or island groups selected are chosen to represent extreme isotopic compositions in isotope diagrams. They are the "type localities" for HIMU (Cook-Austral Islands and St. Helena), EM-1 (Pitcairn-Gambier and Tristan), EM-2 (Society Islands, Samoa, Marquesas), and PREMA (Hawaiian Islands and Iceland). See text for explanations of the acronyms. (b) $^{207}Pb/^{204}Pb$ versus $^{206}Pb/^{204}Pb$ for the same OIB as plotted in (a). Note that the $^{207}Pb/^{204}Pb$ ratios of St. Helena and Cook-Australs are similar but not identical, whereas they overlap completely in the other isotope diagrams. (c) $^{208}Pb/^{204}Pb$ versus $^{206}Pb/^{204}Pb$ for the same OIB as plotted in (a). (d) $^{206}Pb/^{204}Pb$ versus $^{87}Sr/^{86}Sr$ for the same OIB as plotted in (a). Note that correlations are either absent (e.g., for the EM-2 basalts from Samoa, the Society Islands and Marquesas) or point in rather different directions, a situation that is similar to the MORB data (Figure 6(a)). (e) $^{208}Pb^{*}/^{206}Pb^{*}$ versus $\varepsilon(Nd)$ for the same OIB as plotted in (a). Essentially all island groups display significant negative correlations, again roughly analogous to the MORB data. Data were assembled from the GEOROC database (htpp://georoc.mpch-mainz.gwdg.de).

sure, there is extensive overlap between the two populations, but OIBs are systematically more radiogenic in strontium and less radiogenic in neodymium and hafnium isotopes. In lead isotopes, OIBs overlap the MORB field completely but extend to more extreme values in $^{206}Pb/^{204}Pb$, $^{207}Pb/^{204}Pb$, and $^{208}Pb/^{204}Pb$. As was true for MORBs, OIB isotopic composition can be "mapped," and certain oceanic islands or island groups can be characterized by specific isotopic characteristics. Recognition of this feature has led to the well-known concept of end-member compositions or "mantle components" initially identified by White (1985) and subsequently labeled HIMU, PREMA, EM-1, and EM-2 by Zindler and Hart (1986). These acronyms refer to mantle sources characterized by high μ values (HIMU; $\mu = (^{238}U/^{204}Pb)_{t=0}$), "prevalent mantle" (PREMA), "enriched mantle-1" (EM-1), and "enriched mantle-2" (EM-2). "PREMA" has, in recent years, fallen into disuse. It has been replaced by three new terms, namely, "FOZO" (for "focal zone," Hart *et al.*, (1992)), "C" (for "common" component, Hanan and Graham (1996)), or "PHEM" (for "primitive helium mantle," Farley *et al.* (1992)), which differ from each other only in detail, if at all. In contrast with the illustration chosen by Hofmann (1997), which used color coding to illustrate how the isotopic characteristics of, e.g., extreme HIMU samples appear in different isotope diagrams,

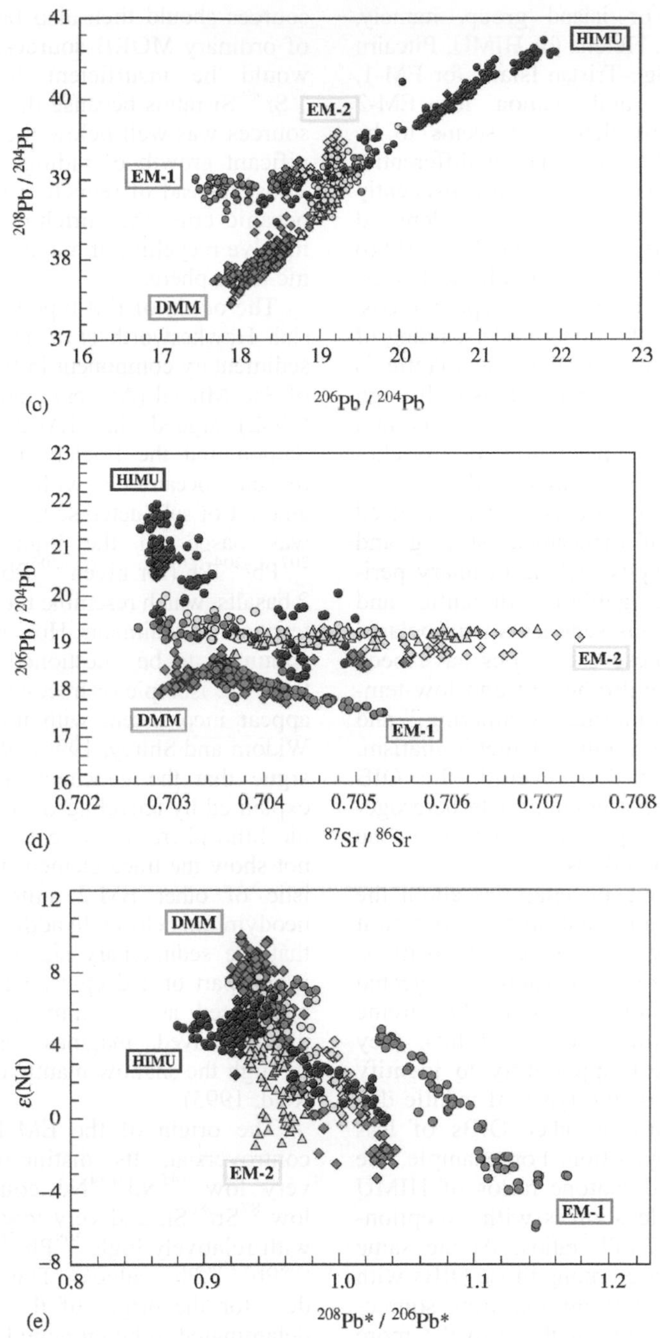

Figure 15 (continued).

irrespective of their geographic location, here I use the more conventional representation of identifying "type localities" of the various "species" of this mantle isotope zoo.

Two extreme notions about the meaning of these components or end members (sometimes also called "flavors") can be found in the literature. One holds that the extreme isotopic end members of these exist as identifiable "species," which may occupy separate volumes or "reservoirs" in the mantle. In this view, the intermediate compositions found in most oceanic basalts are generated by instantaneous mixing of these species during melting and emplacement of OIBs. The other notion considers them to be merely extremes of a continuum of isotopic compositions existing in mantle rocks.

Apparent support for the "species" hypothesis is provided by the observation that the isotopically extreme compositions can be found in more than a single ocean island or island group, namely, Austral Islands and St. Helena for HIMU, Pitcairn Island and Walvis-Ridge-Tristan Island for EM-1, and Society Islands and Samoa for EM-2 (Hofmann, 1997). Nevertheless, it seems to be geologically implausible that mantle differentiation, by whatever mechanism, would consistently produce just four (or five, when the "depleted MORB mantle" DMM (Zindler and Hart, 1986) is included) species of essentially identical ages, which would then be remixed in variable proportions. It is more consistent with current understanding of mantle dynamics to assume that the mantle is differentiated and remixed continuously through time. Moreover, we can be reasonably certain that a great many rock types with differing chemistries are continuously introduced into the mantle by subduction and are thereafter subjected to variable degrees of mechanical stirring and mixing. These rock types include ordinary peridotites, harzburgites, gabbros, tholeiitic and alkali basalts, terrigenous sediments, and pelagic sediments. Most of these rock types have been affected by seafloor hydrothermal and low-temperature alteration, submarine "weathering," and subduction-related alteration and metasomatism. Finally, it is obvious that, overall, the OIB isotopic data constitute a continuously heterogeneous spectrum of compositions, just as is the case for MORB compositions.

In spite of the above uncertainties about the meaning of mantle components and reservoirs, it is clear that the extreme isotopic compositions represent melting products of sources subjected to some sort of ancient and comparatively extreme chemical differentiation. Because of this, they probably offer the best opportunity to identify the specific character of the types of mantle differentiation also found in other OIBs of less extreme isotopic composition. For example, the highly radiogenic lead isotope ratios of HIMU samples require mantle sources with exceptionally high U/Pb and Th/Pb ratios. At the same time, HIMU samples are among those OIBs with the least radiogenic strontium, requiring source-Rb/Sr ratios nearly as low as those of the more depleted MORBs. Following the currently popular hypothesis of Hofmann and White (1980, 1982); Chase, 1981, it is widely thought that such rocks are examples of recycled oceanic crust, which has lost alkalis and lead during alteration and subduction (Chauvel *et al.*, 1992). However, there are other possibilities. For example, the characteristics of HIMU sources might also be explained by enriching oceanic lithosphere "metasomatically" by the infiltration of low-degree partial melts, which have high U/Pb and Th/Pb ratios because of magmatic enrichment of uranium and thorium over lead (Sun and McDonough, 1989). The Rb/Sr ratios of these sources should then also be elevated over those of ordinary MORB sources, but this enrichment would be insufficient to significantly raise $^{87}Sr/^{86}Sr$ ratios because the initial Rb/Sr of these sources was well below the level where any significant growth of radiogenic ^{87}Sr could occur. Thus, instead of recycling more or less ordinary oceanic crust the enrichment mechanism would involve recycling of magmatically enriched oceanic lithosphere.

The origin of EM-type OIBs is also controversial. Hawkesworth *et al.* (1979) had postulated a sedimentary component in the source of the island of Sao Miguel (Azores), and White and Hofmann (1982) argued that EM-2 basalt sources from Samoa and the Society Islands are formed by recycled ocean crust with an addition of the small amount of subducted sediment. This interpretation was based on the high $^{87}Sr/^{86}Sr$ and high $^{207}Pb/^{204}Pb$ (for given $^{206}Pb/^{204}Pb$) ratios of EM-2 basalts, which resemble the isotopic signatures of terrigenous sediments. However, this interpretation continues to be questioned on the grounds that there are isotopic or trace-element parameters that appear inconsistent with this interpretation (e.g., Widom and Shirey, 1996). Workman *et al.* (2003) argue that the geochemistry of Samoa is best explained by recycling of melt-impregnated oceanic lithosphere, because their Samoa samples do not show the trace-element fingerprints characteristic of other EM-2 suites (see discussion of neodymium below). In addition, it has been argued that the sedimentary signature is present, but it is not part of a deep-seated mantle plume but is introduced as a sedimentary contaminant into plume-derived magmas during their passage through the shallow mantle or crust (Bohrson and Reid, 1995).

The origin of the EM-1 flavor is similarly controversial. Its distinctive features include very low $^{143}Nd/^{144}Nd$ coupled with relatively low $^{87}Sr/^{86}Sr$, and very low $^{206}Pb/^{204}Pb$ coupled with relatively high $^{208}Pb/^{204}Pb$ (leading to high $^{208}Pb^*/^{206}Pb^*$ values). The two leading contenders for the origin of this are (i) recycling of delaminated subcontinental lithosphere and (ii) recycling of subducted ancient pelagic sediment. The first hypothesis follows a model originally proposed by McKenzie and O'Nions (1983) to explain the origin of OIBs in general. The more specific model for deriving EM-1 type basalts from such a source was developed by Hawkesworth *et al.* (1986), Mahoney *et al.* (1991), and Milner and le Roex (1996). It is based on the observation that mantle xenoliths from Precambrian shields display similar isotopic characteristics. The second hypothesis is based on the observation that many pelagic sediments are characterized by high Th/U and low (U,Th)/Pb ratios (Ben Othman *et al.*, 1989; Plank and

Langmuir, 1998), and this will lead to relatively unradiogenic lead with high $^{208}Pb^*/^{206}Pb^*$ ratios after passage of 1–2 Ga (Weaver, 1991; Chauvel *et al.*, 1992; Rehkämper and Hofmann, 1997; Eisele *et al.*, 2002). Additional support for this hypothesis has come from hafnium isotopes. Many (though not all) pelagic sediments have high Lu/Hf ratios (along with low Sm/Nd ratios), because they are depleted in detrital zircons, the major carrier of hafnium in sediments (Patchett *et al.*, 1984; Plank and Langmuir, 1998). This is expected to lead to relatively high $^{176}Hf/^{177}Hf$ ratios combined with low $^{143}Nd/^{144}Nd$ values, and these relationships have indeed been observed in lavas from Koolau volcano, Oahu (Hawaiian Islands) (Blichert-Toft *et al.*, 1999) and from Pitcairn (Eisele *et al.*, 2002). Gasperini *et al.* (2000) have proposed still another origin for EM-1 basalts from Sardinia, namely, recycling of gabbros derived from a subducted, ancient plume head.

Recycling of subducted ocean islands and oceanic plateaus was suggested by Hofmann (1989b) to explain not the extreme end-member compositions of the OIB source zoo, but the enrichments seen in the basalts forming the main "mantle array" of negatively correlated $^{143}Nd/^{144}Nd$ and $^{87}Sr/^{86}Sr$ ratios. The $^{143}Nd/^{144}Nd$ values of many of these basalts (e.g., many Hawaiian basalts) are too low, and their $^{87}Sr/^{86}Sr$ values too high, for these OIBs to be explained by recycling of depleted oceanic crust. However, if the recycled material consists of either enriched MORB, tholeiitic or alkaline OIB, or basaltic oceanic plateau material, such a source will have the pre-enriched Rb/Sr and Nd/Sm ratios capable of producing the observed range of strontium- and neodymium-isotopic compositions of the main OIB isotope array.

Melt inclusions in olivine phenocrysts have been shown to preserve primary melt compositions, and these have revealed a startling degree of chemical and isotopic heterogeneity occurring in single-hand specimens and even in single olivine crystals (Sobolev and Shimizu, 1993; Sobolev, 1996; Saal *et al.*, 1998; Sobolev *et al.*, 2000; Hauri, 2002). These studies have demonstrated that rather extreme isotopic and chemical heterogeneities exist in the mantle on scales considerably smaller than the melting region of a single volcano, as discussed in Section 3.2.2 on "mesoscale" heterogeneities. One of these studies, in particular, demonstrated the geochemical fingerprint of recycled oceanic gabbros in melt inclusions from Mauna Loa Volcano, Hawaii (Sobolev *et al.*, 2000). These rare melt inclusions have trace-element patterns that are very similar to those of oceanic and ophiolitic gabbros. They are characterized by very high Sr/Nd and low Th/Ba ratios that can be ascribed to cumulus plagioclase, which dominates the modes of many of these gabbros. Chemical and isotopic studies of melt inclusions therefore have great potential for unraveling the specific source materials found in oceanic basalts. These inclusions can preserve primary heterogeneities of the melts much better than the bulk melts do, because the latter go through magma chamber mixing processes that attenuate most of the primary melt features.

The origin of FOZO-C-PHEM-PREMA, simply referred to as "FOZO" hereafter, may be of farther-reaching consequence than any of the other isotope flavors, if the inference of (Hart *et al.*, 1992) is correct, namely, that it represents material from the lower mantle that is present as a mixing component in all deep-mantle plumes. The evidence for this is that samples from many individual OIB associations appear to form binary mixing arrays that radiate from this "focal zone" composition in various directions toward HIMU, EM-1, or EM-2. These relationships are shown in Figure 16. The FOZO composition is similar, but not identical, to DMM represented by MORB. It is only moderately more radiogenic in strontium, less radiogenic in neodymium and hafnium, but significantly more radiogenic in lead isotopes than DMM. If plumes originate in the very deep mantle and rise from the core–mantle boundary, rather than from the 660 km

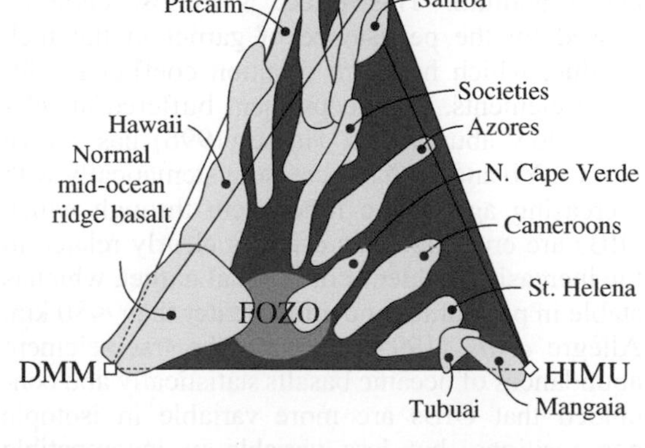

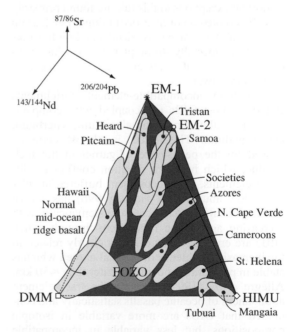

Figure 16 Three-dimensional projection of $^{87}Sr/^{86}Sr$, $^{143}Nd/^{144}Nd$, $^{206}Pb/^{204}Pb$ isotope arrays of a large number of OIB groups (after Hart *et al.*, 1992). Most of the individual arrays appear to radiate from a common region labeled "FOZO" (for focal zone) thought to represent the composition of the deep mantle. The diagram was kindly made available by S. R. Hart.

seismic discontinuity, they are likely to entrain far more deep-mantle material than upper-mantle material (Griffiths and Campbell, 1990; Hart *et al.*, 1992). It should be noted, however, that the amount of entrained material in plumes is controversial, with some authors insisting that plumes contain very little entrained material (e.g., Farnetani *et al.*, 2002).

3.5.2 Trace Elements in OIB

Most OIBs are much more enriched in incompatible trace elements than most MORB, and there are two possible reasons for this: (i) their sources may be more enriched than MORB sources, and (ii) OIBs may be produced by generally lower degrees of partial melting than MORBs. Most likely, both factors contribute to this enrichment. Source enrichment (relative to MORB sources) is required because the isotopic compositions require relative enrichment of the more incompatible of the parent–daughter element ratios. Low degrees of melting are caused by the circumstance that most OIB are also within-plate basalts, so a rising mantle diapir undergoing partial melting encounters a relatively cold lithospheric lid, and melting is confined to low degrees and relatively deep levels. This is also the reason why most OIBs are alkali basalts rather than tholeiites, the predominant rock type in MORBs. Important exceptions to this rule are found primarily in OIBs erupted on or near-ocean ridges (such as on Iceland and the Galapagos Islands) and on hotspots created by especially strong plume flux, which generates tholeiites at relatively high melt fractions, such as in Hawaii.

The high incompatible-element enrichments found in most OIBs are coupled with comparatively low abundances of aluminum, ytterbium, and scandium. This effect is almost certainly caused by the persistence of garnet in the melt residue, which has high partition coefficients for these elements, and keeps them buffered at relatively low abundances. Haase (1996) has shown that Ce/Yb and Tb/Yb increase systematically with increasing age of the lithosphere through which OIBs are erupting. This effect is clearly related to the increasing influence of residual garnet, which is stable in peridotites at depths greater than ~80 km. Allègre *et al.* (1995) analyzed the trace-element abundances of oceanic basalts statistically and concluded that OIBs are more variable in isotopic compositions, but less variable in incompatible element abundances than MORBs. However, their sampling was almost certainly too limited to properly evaluate the effect of lithospheric thickness on the abundances and their variability. In particular, their near-ridge sampling was confined to 11 samples from Iceland and 4 samples from Bouvet. The actual range of abundances of highly incompatible elements, such as thorium and uranium, from Iceland spans nearly three orders of magnitude. In contrast, ytterbium varies by only a factor of 10. This means that either the source of Iceland basalts is internally extremely heterogeneous, or the melt fractions are highly variable, or both. Because of this ambiguity, the REE abundance patterns and most of the moderately incompatible elements, sometimes called "MICE," are actually not very useful to unravel the relative effects of partial melting and source heterogeneity.

The very highly incompatible elements, sometimes called "VICE," are much less fractionated from each other in melts (through normal petrogenetic processes), but they are more severely fractionated in melt residues. This is the reason why their relative abundances vary in the mantle and why these variations can be traced by VICE ratios in basalts, which are in this respect similar to, though not as precise as, isotope ratios. VICE ratios thus enlarge the geochemical arsenal for determining mantle chemical heterogeneities and their origins. These differences are conventionally illustrated either by the "spidergrams" (primitive-mantle normalized element abundance diagrams), or by plotting trace-element abundance ratios.

Spidergrams have the advantage of representing a large number of trace element abundances of a given sample by a single line. However, they can be confusing because there are no standard rules about the specific sequence in which the elements are shown or about the normalizing abundances used. The (mis)use of N-MORB or E-MORBs as reference values for normalizations has already been discussed and discouraged in Section 3.4.4. However, there are other pitfalls to be aware of: One of the most widely used normalizations is that given by Sun and McDonough (1989), which uses primitive-mantle estimates for all elements *except* lead, the abundance of which is adjusted by a factor of 2.5, presumably in order to generate smoother abundance patterns in oceanic basalts. The great majority of authors using this normalization simply call this "primitive mantle" without any awareness of the fudge factor applied. Such *ad hoc* adjustments for esthatic reasons should be strongly discouraged. Spidergrams communicate their message most effectively if they are standardized as much as possible, i.e., if they use only one standard for normalization, namely primitive-mantle abundances, and if the sequence of elements used is in the order of increasing compatibility (see also Hofmann, 1988). The methods for determining this order of incompatibility are addressed in the subsequent Section 3.5.2.1.

Spidergrams tend to carry a significant amount of redundant information, most of which is useful for determining the general level of incompatible-element enrichment, rather than specific information about the sources. Therefore, diagrams of critical trace-element (abundance) ratios can be very effective in focusing on specific source differences. Care should be taken to use ratios of elements with similar bulk partition coefficients during partial melting (or, more loosely speaking, similar incompatibilities). Otherwise, it may be difficult or impossible to separate source effects from melting effects. Some rather popular element pairs of mixed incompatibility, such as Zr/Nb, which are almost certainly fractionated at the relatively low melt fractions prevailing during intraplate melting, are often used in a particularly confusing manner. For example, in the popular plot of Zr/Nb versus La/Sm, the more incompatible element is placed in the numerator of one ratio (La/Sm) and in the denominator of the other (Zr/Nb). The result is a hyperbolic relationship that looks impressive, but carries little if any useful information other than showing that the more enriched rocks have high La/Sm and low Zr/Nb, and the more depleted rocks have low La/Sm and high Zr/Nb.

Trace-element ratios of similarly incompatible pairs, such as Th/U, Nb/U, Nb/La, Ba/Th, Sr/Nd, or Pb/Nd, tend to be more useful in identifying source differences, because they are fractionated relatively little during partial melting. Elements that appear to be diagnostic of distinctive source types in the mantle are niobium, tantalum, lead, and to a lesser extent strontium, barium, potassium, and rubidium. These will be discussed in connection with the presentation of specific "spidergrams" in Section 3.5.2.2.

3.5.2.1 "Uniform" trace-element ratios

In order to use geochemical anomalies for tracing particular source compositions, it is necessary to establish "normal" behavior first. Throughout the 1980s, Hofmann, Jochum, and co-workers noticed a series of trace-element ratios that are globally more or less uniform in both MORBs and OIBs. For example, the elements barium, rubidium, and caesium, which vary by about three orders of magnitude in absolute abundances, have remarkably uniform relative abundances in many MORBs and OIBs (Hofmann and White, 1983). This became clear only when sufficiently high analytical precision (isotope dilution at the time) was applied to fresh glassy samples. Hofmann and White (1983) argued that this uniformity must mean that the Ba/Rb and Rb/Cs ratios found in the basalts reflect the respective ratios in the source rocks. And because these ratios were so similar in highly depleted MORBs and in enriched OIBs, these authors concluded that these element ratios have not been affected by processes of global differentiation, and they therefore also reflect the composition of the primitive mantle. Similarly, Jochum *et al.* (1983) estimated the K/U ratio of the primitive mantle to be 1.27×10^4, a value that became virtually canonical for 20 years, even though it was based on remarkably few measurements. Other such apparently uniform ratios were Sn/Sm and Sb/Pr (Jochum *et al.*, 1993; Jochum and Hofmann, 1994) and Sr/Nd (Sun and McDonough, 1989). Zr/Hf and Nb/Ta were also thought to be uniform (Jochum *et al.*, 1986), but recent analyses carried out at higher precision and on a greater variety of rock types have shown systematic variations of these ratios.

The above approach of determining primitive-mantle abundances from apparently globally unfractionated trace-element ratios was up-ended by the discovery that Nb/U and Ce/Pb are also rather uniform in MORBs and OIBs the world over, but these ratios are higher by factors of $\sim$5–10 than the respective ratios in the continental crust (Hofmann *et al.*, 1986). This invalidated the assumption that primitive-mantle abundances could be obtained simply from MORB and OIB relations, because the continental crust contains such a large portion of the total terrestrial budget of highly incompatible elements. However, these new observations meant that niobium and lead could potentially be used as tracers for recycled continental material in oceanic basalts. In other words, while ratios such as Nb/U show only limited variation when comparing oceanic basalts as a function of enrichment or depletion on a global or local scale, this uniformity can be interpreted to mean that such a specific ratio is not significantly fractionated during partial melting. If this is true, then the variations that do exist may be used to identify differences in source composition.

Figure 17 shows updated versions of the Nb/U variation diagram introduced by Hofmann *et al.* (1986). It represents an attempt to determine which other highly incompatible trace element is globally most similar to niobium in terms bulk partition coefficient during partial melting. The form of the diagram was chosen because an element ratio will systematically increase as the melt fractions decrease and the absolute concentration of the elements increase. It is obvious that Nb/Th and Nb/La ratios vary systematically with niobium concentration, but Nb/U does not. Extending the comparison to other elements, such as the heavier REEs (not shown), simply increases the slopes of such plots. Thus, while Nb/U is certainly *not constant* in oceanic basalts, its variations are the lowest and the least systematic. To be sure, there is possible circularity in this argument, because it is possible, in principle, that enriched sources have systematically lower Nb/U ratios, which are systematically (and relatively precisely) compensated by partition coefficients that are lower for niobium

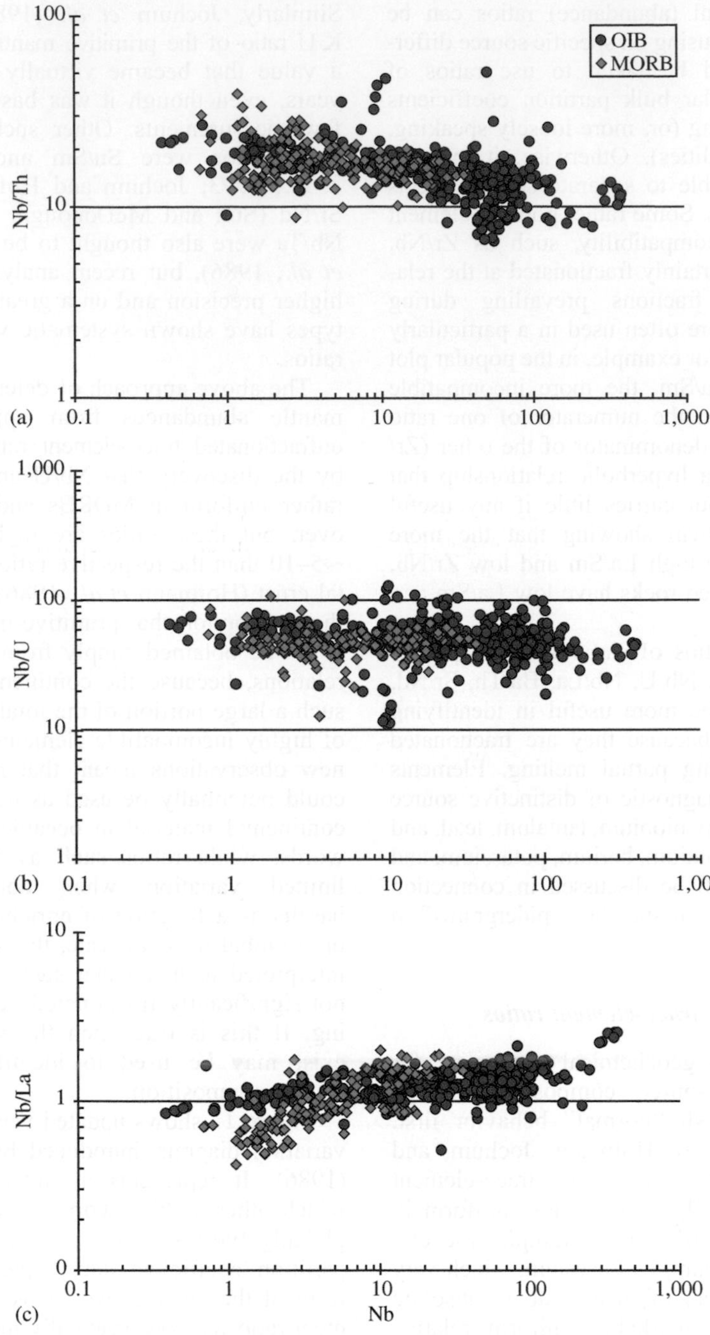

Figure 17 Nb/Th, Nb/U, and Nb/La ratios versus Nb concentrations of global MORB and (non-EM-2-type) OIBs (Hawaiian Isl., Iceland, Australs, Pitcairn, St. Helena, Cnary, Bouvet, Gough Tristan, Ascension, Madeira, Fernando de Noronha, Cameroon Line Isl., Comores, Cape Verdes, Azores, Galapagos, Easter, Juan Fernandez, San Felix). The diagram shows a systematic increase of Nb/Th, approximately constant Nb/U, and systematic decrease of La/Nb as Nb concentrations increase over three orders of magnitude.

than uranium, thus systematically compensating the lower source ratio during partial melting. Such a compensating mechanism has been advocated by Sims and DePaolo (1997), who criticized the entire approach of Hofmann *et al.* (1986) on this basis. Such fortuitously compensating circumstances, as postulated in the model of Sims and DePaolo (1997), may be *ad hoc* assumptions, but they are not *a priori* impossible.

The model of Hofmann *et al.* (1986) can be tested by examining more local associations of oceanic basalts characterized by large variations in melt fraction. Figure 18 shows Nb–Th–U–La–Nd relationships on Iceland, for volcanic

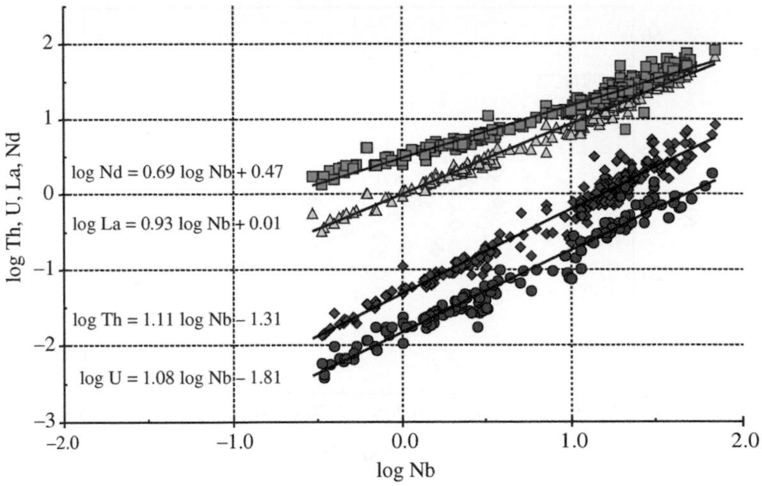

Figure 18 Concentrations of Th, U, La, and Nd versus Nb for basalts and picrites from Iceland. On this logarithmic plot, a regression line of slope 1 represents a constant element concentration ratio (corresponding to a horizontal line in Figure 17). Slopes greater than 1 correspond to positive slopes, and slopes lower than 1 correspond to negative slopes in Figure 17. The correlations of U and La versus Nb yields the slope closest to unity (= 1.08 and 0.93).

rocks ranging from picrites to alkali basalts, as compiled from the recent literature. The representation differs from Figure 17, which has the advantage of showing the element ratios directly, but the disadvantage pointed out by Sims and DePaolo (1997) that the two variables used are not independent. The simple log–log plot of Figure 18 is less intuitively obvious but in this sense more rigorous. In this plot, a constant concentration ratio yields a slope of unity. The Th–Nb, U–Nb, La–Nb, and Nd–Nb plots show progressively increasing slopes, with the log U–log Nb and the log La–log Nb plots being closest to unity. The data from Iceland shown in Figure 18 are therefore consistent with the global data set shown in Figure 17. This confirms that uranium and niobium have nearly identical bulk partition coefficients during mantle melting in most oceanic environments.

The point of these arguments is not that an element ratio such as Nb/U in a melt will *always* reflect the source ratio very precisely. Rather, because of varying melting conditions, the specific partition coefficients of two such chemically different elements must vary *in detail*, as expected from the partitioning theory of Blundy and Wood (1994). The nephelinites and nepheline melilitites of the Honolulu Volcanic Series, which represent the post-erosional, highly alkalic phase of Koolau Volcano, Oahu, Hawaii, may be an example where the partition coefficients of niobium and uranium are significantly different. These melts are highly enriched in trace elements and must have been formed by very small melt fractions from relatively depleted sources, as indicated by their nearly MORB-like strontium and neodymium isotopic compositions. Their Nb/U ratios average 27, whereas the alkali basalts average Nb/U = 44 (Yang *et al.*, 2003). This may indicate that under melting conditions of very low melt fractions, Nb is significantly more compatible than uranium, and the relationships that are valid for basalts cannot necessarily be extended to more exotic rock types such as nephelinite.

In general, the contrast between Nb/U in most OIBs and MORBs and those in sediments, island arcs and continental rocks is so large that it appears to provide an excellent tracer of recycled continental material in oceanic basalts. A significant obstacle in applying this tracer is the lack of high-quality Nb–U data, partly because of analytical limitations and partly because of sample alteration. The latter can, however, often be overcome by "interpolating" the uranium concentration between thorium and lanthanium (the nearest neighbors in terms of compatibility) and replacing Nb/U by the primitive-mantle normalized Nb/(Th + La) ratio (e.g., Weaver, 1991; Eisele *et al.*, 2002).

Having established that Nb/U or Nb/(Th + La) ratios can be used to trace-mantle source compositions of basalts, this parameter can be turned into a tool to trace recycled continental material in the mantle. The mean Nb/U of 166 MORBs is Nb/U = 47 ± 11, and mean of nearly 500 "non-EM-type" OIBs is Nb/U = 52 ± 15. This contrasts with a mean value of the continental crust of Nb/U = 8 (Rudnick and Fountain, 1995). As is evident from Figure 4(b), continent-derived sediments also have consistently higher $^{87}Sr/^{86}Sr$ ratios than ordinary mantle rocks; therefore, any OIB containing significant amounts of recycled sediments should be distinguished by high $^{87}Sr/^{86}Sr$ and low Nb/U

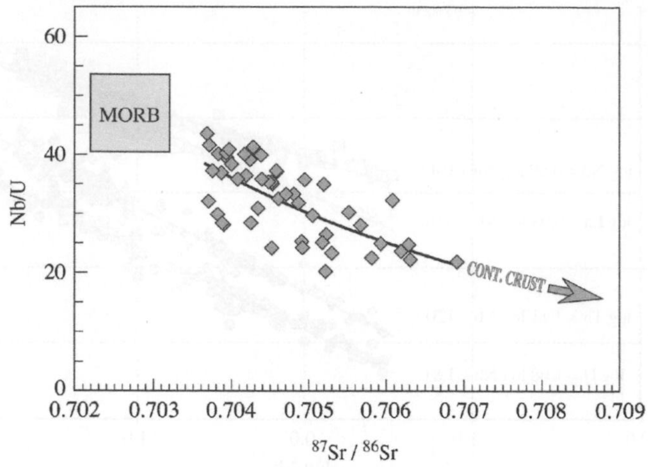

Figure 19 Nb/U versus $^{87}Sr/^{86}Sr$ for basalts from the Society Islands using data of White and Duncan (1996). Two samples with Th/U > 6.0 have been removed because they form outliers on an Nb/Th versus Nb/U correlation and are therefore suspected of alteration or analytical effects on the U concentration. One strongly fractionated trachyte sample has also been removed. This correlation and a similar one of Nd/Pb versus $^{87}Sr/^{86}Sr$ (not shown) is consistent with the addition of a sedimentary or other continental component to the source of the Society Island (EM-2) basalts.

ratios. Figure 19 shows that this is indeed observed for EM-2 type OIBs and, to a lesser extent, for EM-1 OIBs as well. Of course, this does not "prove" that EM-type OIBs contain recycled sediments. However, there is little doubt that sediments have been subducted in geological history. Much of their trace-element budget is likely to have been short-circuited back into island arcs during subduction. But if any of this material has entered the general mantle circulation and is recycled at all, then EM-type OIBs are the best candidates to show it. Perhaps the greater surprise is that there are so few EM-type ocean islands.

Finding the "constant-ratio partner" for lead has proved to be more difficult. Originally, Hofmann *et al.* (1986) chose cerium because, on average, the Ce/Pb ratio of their MORB data was most similar to their OIB average. However, Sims and DePaolo (1997) pointed out one rather problematic aspect, namely, that even in the original, very limited data set, each separate population showed a distinctly positive slope. In addition, they showed that Ce/Pb ratios appear to correlate with europium anomalies in the MORB population, and this is a strong indication that both parameters are affected by plagioclase fractionation. Recognizing these problems, Rehkämper and Hofmann (1997) argued on the basis of more extensive and more recent data that Nd/Pb is a better indicator of source chemistry than Ce/Pb. Unfortunately, lead concentrations are not often analyzed in oceanic basalts, partly because lead is subject to alteration and partly because it is difficult to analyze, so a literature search tends to yield highly scattered data. Nevertheless, the average MORB value of Pb/Nd = 0.04 is lower than the average continental value of 0.63 by a factor of 15. Because of this great contrast, this ratio is potentially an even more sensitive tracer of continental contamination or continental recycling in oceanic basalts.

But why are Pb/Nd ratios so different in continental and oceanic crust in the first place? An answer to that question will be attempted in the following section.

3.5.2.2 Normalized abundance diagrams ("Spidergrams")

The techniques illustrated in Figures 17 and 18 can be used to establish an approximate compatibility sequence of trace elements for mantle-derived melts. In general, this sequence corresponds to the sequence of decreasing (normalized) abundances in the continental crust shown in Figure 2, but this does not apply to niobium, tantalum, and lead for which the results discussed in the previous section demand rather different positions (see also Hofmann, 1988). Here I adopt a sequence similar to that used by Hofmann (1997), but with slightly modified positions for lead and strontium.

Figure 20 shows examples of "spidergrams" for representative samples of HIMU, EM-1, EM-2, and Hawaiian basalts, in addition to average MORB and average "normal" MORB, average subducting sediment, and average continental crust. Prominent features of these plots are negative spikes for niobium in average continental crust and in sediment, and corresponding positive anomalies in most oceanic basalts except EM-type basalts.

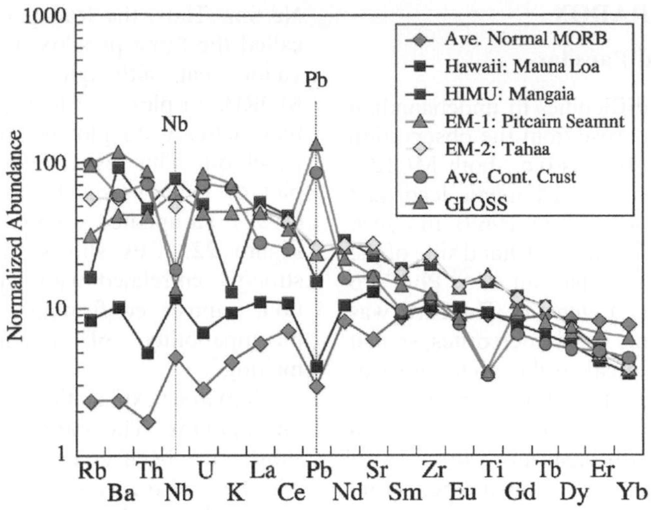

Figure 20 Examples of primitive-mantle normalized trace element abundance diagrams ("spidergrams") for representative samples of HIMU (Mangaia, Austral Islands, sample M-11; Woodhead (1996)), EM-1 (Pitcairn Seamount sample 49DS1; Eisele *et al.* (2002)), EM-2 (Tahaa, Society Islands, sample 73-190; White and Duncan (1996)); Average Mauna Loa (Hawaii) tholeiite (Hofmann, unpublished data), average continental crust (Rudnick and Fountain, 1995), average subducting sediment, GLOSS (Plank and Langmuir, 1998), and Average Normal MORB (Su, 2002). Th, U, and Pb values for MORB were calculated from average Nb/U = 47, and Nd/Pb = 26. All abundances are normalized to primitive/mantle values of McDonough and Sun (1995).

Similarly, the positive spikes for lead in average continental crust and in sediments is roughly balanced by negative anomalies in most oceanic basalts. More subtle features distinguishing the isotopically different OIB types are relative deficits for potassium and rubidium in HIMU basalts and high Ba/Th ratios coupled with elevated Sr/Nd ratios in Mauna Loa basalts.

The prominent niobium and lead spikes of continental materials are not matched by any of the OIBs and MORBs reviewed here. They are, however, common features of subduction-related volcanic rocks found on island arcs and continental margins. It is therefore likely that the distinctive geochemical features of the continental crust are produced during subduction, where volatiles can play a major role in the element transfer from mantle to crust. The net effect of these processes is to transfer large amounts of lead (in addition to mobile elements like potassium and rubidium) into the crust. At the same time, niobium and tantalum are retained in the mantle, either because of their low solubility in hydrothermal solutions, or because they are partitioned into residual mineral phases such as Ti-minerals or certain amphiboles. These processes are the subject of much ongoing research, but are beyond the scope of this chapter.

For the study of mantle circulation, these chemical anomalies can help trace the origin of different types of OIBs and some MORBs. Niobium and lead anomalies, coupled with high $^{87}Sr/^{86}Sr$ ratios, seem to be the best tracers for material of continental origin circulating in the mantle. They have been found not only in EM-2-type OIBs but also in some MORBs found on the Chile Ridge (Klein and Karsten, 1995). Other trace-element studies, such as the study of the chemistry of melt inclusions, have already identified a specific recycled rock type, namely, a gabbro, which could be recognized by its highly specific trace-element "fingerprint" (Sobolev *et al.*, 2000).

Until quite recently, the scarceness of high-quality data for the diagnostic elements has been a serious impediment to progress in gaining a full interpretation of the origins of oceanic basalts using complete trace-element data together with complete isotope data. All data compilations aimed at detecting global geochemical patterns are currently seriously hampered by spotty literature data of uncertain quality on samples of unknown freshness. This is now changing, because of the advent of new instrumentation capable of producing large quantities of high-quality data of trace elements at low abundances. The greater ease of obtaining large quantities of data also poses significant risks from lack of quality control. Nevertheless, we are currently experiencing a dramatic improvement in the general quantity and quality of geochemical data, and we can expect significant further improvements in the very near future. These developments offer a bright outlook for the future of deciphering the chemistry and history of mantle differentiation processes.

3.6 THE LEAD PARADOX

3.6.1 The First Lead Paradox

One of the earliest difficulties in understanding terrestrial lead isotopes arose from the observation that almost all oceanic basalts (i.e., both MORBs and OIBs) have more highly radiogenic lead than does the primitive mantle (Allègre, 1969). In effect, most of these basalts lie to the right-hand side of the so-called "geochron" on a diagram of $^{207}Pb/^{204}Pb$ versus $^{206}Pb/^{204}Pb$ ratios (Figure 21). The Earth was assumed to have the same age as meteorites, so that the "geochron" is identical to the meteorite isochron of 4.56Ga. If the total silicate portion of the Earth remained a closed system involved only in internal (crust–mantle) differentiation, the sum of the parts of this system must lie on the geochron. The reader is referred to textbooks (e.g., Faure, 1986) on isotope geology for fuller explanations of the construction and meaning of the geochron and the construction of common-lead isochrons.

The radiogenic nature of MORB lead was surprising because uranium is expected to be considerably more incompatible than lead during mantle melting. The MORB source, being depleted in highly incompatible elements, is therefore expected to have had a long-term history of U/Pb ratios lower than primitive ones, just as was found to be the case for Rb/Sr and Nd/Sm. Thus, the lead paradox (sometimes also called the "first paradox") is given by the observation that, although one would expect most MORBs to plot well to the left of the geochron, they actually do plot mostly to the right of the geochron. This expectation is reinforced by a plot of U/Pb versus U, as first used by White (1993), an updated version of which is shown in Figure 22. This shows that the U/Pb ratio is strongly correlated with the uranium concentration, thus confirming the much greater incompatibility of uranium during mantle melting.

Numerous explanations have been advanced for this paradox. The most recent treatment of the subject has been given by Murphy *et al.* (2003), who have also reviewed the most important solutions to the paradox. These include delayed uptake of lead by the core ("core pumping," Allègre *et al.* (1982)) and storage of unradiogenic lead (to balance the excess radiogenic lead seen in MORBs, OIBs, and upper crustal rocks) in the lower continental crust or the subcontinental lithosphere (e.g., Zartman and Haines, 1988; Kramers and Tolstikhin, 1997).

An important aspect not addressed by Murphy *et al.* is the actual position of the geochron. This is the locus of any isotopic mass balance of a closed-system silicate earth. This is

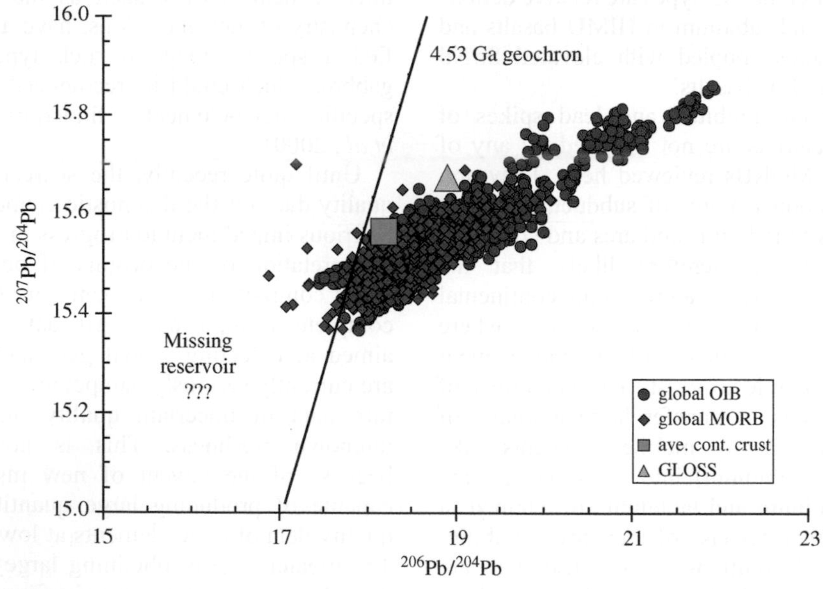

Figure 21 Illustration of the first lead paradox. Estimates of the average composition of the continental crust (Rudnick and Goldstein, 1990), of average "global subducted sediments" (GLOSS, Plank and Langmuir, 1998) mostly derived from the upper continental crust, and a global compilation of MORB and OIB data (from GEOROC and PETDB databases) lie overwhelmingly on the right-hand side of the 4.53Ga geochron. One part of the paradox is that these data require a hidden reservoir with lead isotopes to the left of the geochron in order to balance the reservoirs represented by the data from the continental and oceanic crust. The other part of the paradox is that both continental and oceanic crustal rocks lie rather close to the geochron, implying that there is surprisingly little net fractionation of the U/Pb ratio during crust–mantle differentiation, even though U is significantly more incompatible than Pb (see Figure 22).

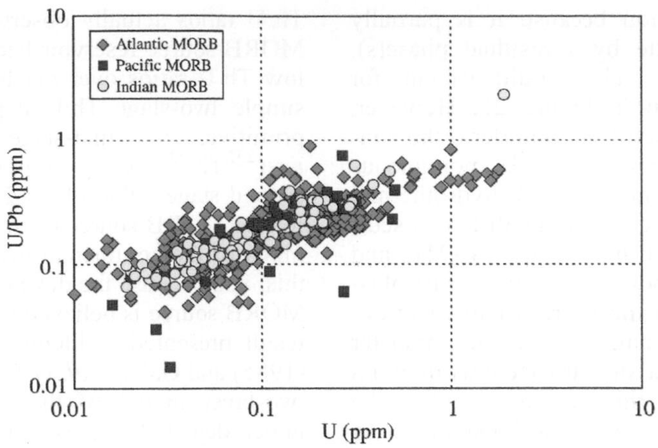

Figure 22 U/Pb versus U concentrations for MORB from three ocean basins. The positive slope of the correlation indicates that U is much more incompatible than Pb during mantle melting. This means that the similarity of Pb isotopes in continental and oceanic crust (see Figure 21) is probably caused by a nonmagmatic transport of lead from mantle to the continental crust.

not the meteorite isochron of 4.56 Ga, because later core formation and giant impact(s) are likely to have prevented closure of the bulk silicate earth with regard to uranium and lead until lead loss by volatilization and/or loss to the core effectively ended. Therefore, the reference line (geochron) that is relevant to balancing the lead isotopes from the various silicate reservoirs is younger than, and it lies to the right of, the meteorite isochron. The analysis of the effect of slow accretion on the systematics of terrestrial lead isotopes (Galer and Goldstein, 1996) left reasonably wide latitude as to where the relevant geochron should actually be located. This was further reinforced by publication of early tungsten isotope data (daughter product of the short-lived ^{182}Hf), which appeared to require terrestrial core formation to have been delayed by at least 50 Myr (Lee and Halliday, 1995). This would have moved the possible locus of bulk silicate lead compositions closer to the actual positions of oceanic basalts, thus diminishing the magnitude of the paradox, or possibly eliminating it altogether. However, the most recent redeterminations of tungsten isotopes in chondrites by Kleine *et al.* (2002) and Yin *et al.* (2002) have shown the early tungsten data to be in error, so that core and moon formation now appear to be constrained at ~4.53 Ga. This value is not sufficiently lower than the meteorite isochron of 4.56 Ga to resolve the problem.

This means that the lead paradox is alive and well, and the search for the unradiogenic, hidden reservoir continues. The lower continental crust remains (in the author's opinion) a viable candidate, even though crustal xenolith data appear to be, on the whole, not sufficiently unradiogenic (see review of these data by Murphy *et al.* (2003)). It is not clear how representative the xenoliths are, particularly of the least radiogenic, Precambrian lower crust. Another hypothetical candidate is a garnetite reservoir proposed by Murphy *et al.* (2003).

The above discussion, and most of the relevant literature, does not address the perhaps geochemically larger and more interesting question, namely, why are the continental crust and the oceanic basalts so similar in lead isotopes in the first place? It is quite remarkable that most MORBs and most continent-derived sediments cover the same range of $^{206}Pb/^{204}Pb$ ratios, namely, ~17.5–19.5. This means that MORB sources and upper continental crust, from which these sediments are derived, have very similar U/Pb ratios, when integrated over the entire Earth history. The main offset between lead isotope data for oceanic sediments and MORBs is in terms of $^{207}Pb/^{204}Pb$, and even this offset is only marginally outside the statistical scatter of the data. In the previous section we have seen that lead behaves as a moderately incompatible element such as neodymium or cerium, but both uranium and thorium are highly incompatible elements. So the important question remains why lead and uranium are nearly equally enriched in the continental crust, whereas they are very significantly fractionated during the formation of the oceanic crust and ocean islands. Therefore, the unradiogenic reservoir (lower crust or hidden mantle reservoir) needed to balance the existing, slightly radiogenic reservoirs in order to obtain full, bulk-silicate-earth lead isotope values represents only a relatively minor aspect of additional adjustment to this major discrepancy. The answer is, in my opinion, that lead behaves relatively compatibly during

MORB–OIB production because it is partially retained in the mantle by a residual phase(s), most likely sulfide(s). This would account for the relationships seen in Figure 22. However, when lead is transferred from mantle to the continental crust, two predominantly nonigneous processes become important: (i) hydrothermal transfer from oceanic crust to metalliferous sediment (Peucker-Ehrenbrink *et al.*, 1994), and (ii) transfer from subducted oceanic crust-plus-sediment into arc magma sources (Miller *et al.*, 1994). This additional, nonigneous transfer enriches the crust to a similar extent as uranium and thorium, and this explains why the $^{206}Pb/^{204}Pb$ ratios of crustal and mantle rocks are so similar to each other and so close to the geochron. Thus, the anomalous geochemical behavior of lead is the main cause of the "lead paradox," the high Pd/Nd ratios of island arc and continental rocks, and the lead "spikes" seen in Figure 20.

The above explanation does not account for the elevated $^{207}Pb/^{204}Pb$ ratios of continental rocks and their sedimentary derivatives relative to mantle-derived basalts (Figures 5(a) and 21). This special feature can be explained by a more complex evolution of continents subsequent to their formation. New continental crust formed during Archean time by subduction and accretion processes must have initially possessed a U/Pb ratio slightly higher than that of the mantle. At that time, the terrestrial $^{235}U/^{238}U$ ratio was significantly higher than today, and this produced elevated $^{207}Pb/^{204}Pb$ relative to $^{206}Pb/^{204}Pb$. Some of this crust was later subjected to high-grade metamorphism, causing loss of uranium relative to lead in the lower crust, and transporting the excess uranium into the upper crust. From there, uranium was lost by the combined action of oxidation, weathering, dissolution, and transport into the oceans. This uranium loss retarded the growth of $^{206}Pb/^{204}Pb$ while preserving the relatively elevated $^{207}Pb/^{204}Pb$ of the upper crust. The net result of this two-stage process is the present position of sediments directly above the mantle-derived basalts in $^{207}Pb/^{204}Pb$–$^{207}Pb/^{204}Pb$ space. Another consequence of this complex behavior of uranium will be further discussed in the following section. Here it is important to reiterate that lead, not uranium or thorium, is the *major* player in generating the main part of the lead paradox.

3.6.2 The Second Lead Paradox

Galer and O'Nions (1985) made the important observation that measured $^{208}Pb^*/^{206}Pb^*$ ratios (see Equation (11)) in most MORBs are higher than can be accounted for by the relatively low Th/U ratios actually observed in MORBs, if the MORB source reservoir had maintained similarly low Th/U ratios over much of Earth's history. A simple two-stage Th/U depletion model with a primitive, first-stage value of $\kappa=(^{232}Th/^{238}U)_{today}=3.9$ changing abruptly to a second stage value of $\kappa=2.5$ yields a model age for the MORB source of only ~600 Ma. Because much of the continental mass is much older than this, and because the development of the depleted MORB source is believed to be linked to this, this result presented a dilemma. Galer and O'Nions (1985) and Galer *et al.* (1989) resolved this with a two-layer model of the mantle, in which the upper, depleted layer is in a steady state of incompatible-element depletion by production of continental crust and replenishment by leakage of less depleted material from the lower layer. This keeps the $^{208}Pb^*/^{206}Pb^*$ ratio of the upper mantle at a relatively high value in spite of the low chemical Th/U ratio. However, such two-layer convection models have fallen from grace in recent years, mostly because of the results of seismic mantle tomography (see below). Therefore, other solutions to this second lead paradox have been sought.

A fundamentally different mechanism for lowering the Th/U ratio of the mantle has been suggested by Hofmann and White (1982), namely, preferential recycling of uranium through dissolution of oxidized (hexavalent) uranium at the continental surface, riverine transport into the oceans, and fixation by ridge-crest hydrothermal circulation and reduction to the tetravalent state. The same mechanism was invoked in the quantitative "plumbotectonic" model of Zartman and Haines (1988). Staudigel *et al.* (1995) introduced the idea that this preferential recycling of uranium into the mantle may be connected to a change toward oxidizing conditions at the Earth's surface relatively late in Earth's history. Geological evidence for a rapid atmospheric change toward oxidizing conditions during Early Proterozoic time (i.e., relatively late in Earth's history) has been presented by Holland (1994), and this has been confirmed by new geochemical evidence showing that sulfides and sulfates older than ~2.4 Ga contain non-mass-dependent sulfur isotope fractionations, which can be explained by high-intensity UV radiation in an oxygen-absent atmosphere (Farquhar *et al.*, 2000). Kramers and Tolstikhin (1997) and Elliott *et al.* (1999) have developed quantitative models to resolve the second lead paradox by starting to recycle uranium into the mantle ~2.5 Ga ago. This uranium cycle is an excellent example how mantle geochemistry, surface and atmospheric chemistry, and the evolution of life are intimately interconnected.

3.7 GEOCHEMICAL MANTLE MODELS

The major aim of mantle geochemistry has been, from the beginning, to elucidate the structure and evolution of the Earth's interior, and it was clear that this can only be done in concert with observations and ideas derived from conventional field geology and from geophysics. The discussion here will concentrate on the chemical structure of the recent mantle, because the early mantle evolution and dynamics and history of convective mixing.

The isotopic and chemical heterogeneities found in mantle-derived basalts, reviewed on the previous pages, mandate the existence of similar or even greater heterogeneities in the mantle. The questions are: how are they spatially arranged in the mantle? When and how did they originate? Because these heterogeneities have their primary expression in trace elements, they cannot be translated into physical parameters such as density differences. Rather, they must be viewed as passive tracers of mantle processes. These tracers are separated by melting and melt migration, as well as fluid transport, and they are stirred and remixed by convection. Complete homogenization appears to be increasingly unlikely. Diffusion distances in the solid state have ranges of centimeters at best (Hofmann and Hart, 1978), and possibly very much less (Van Orman *et al.*, 2001), and homogenization of a melt source region, even via the movement and diffusion through a partial melt, becomes increasingly unlikely. This is attested by the remarkable chemical and isotopic heterogeneity observed in melt inclusions preserved in magmatic crystals from single basalt samples (Saal *et al.*, 1998; Sobolev *et al.*, 2000). The models developed for interpreting mantle heterogeneities have, with some exceptions, largely ignored the possibly extremely small scale of these heterogeneities. Instead, they have usually relied on the assumption that mantle-derived basalts are, on the whole, representative of some chemical and isotopic average for a given volcanic province or source volume.

In the early days of mantle geochemistry, the composition of the bulk silicate earth, also called "primitive mantle" (i.e., mantle prior to the formation of any crust) was not known for strontium isotopes because of the obvious depletion of rubidium of the Earth relative to chondritic meteorites (Gast, 1960). The locus of primitive-mantle lead was assumed to be on the meteorite isochron (which was thought to be identical to the geochron until the more recent realization of delayed accretion and core formation; see above), but the interpretation of the lead data was confounded by the lead paradox discussed above. This situation changed in the 1970s with the first measurements of neodymium isotopes in oceanic basalts (DePaolo and Wasserburg, 1976; Richard *et al.*, 1976; O'Nions *et al.*, 1977) and the discovery that $^{143}Nd/^{144}Nd$ was negatively correlated with $^{87}Sr/^{86}Sr$ (see Figures 6 and 15). Because both samarium and neodymium are refractory lithophile elements, the Sm/Nd and $^{143}Nd/^{144}Nd$ ratios of the primitive mantle can be safely assumed to be chondritic. With this information, a primitive-mantle value for the Rb/Sr and $^{87}Sr/^{86}Sr$ was inferred, and it became possible to estimate the size of the MORB source reservoir primarily from isotopic abundances. The evolution of a silicate Earth consisting of three boxes—primitive mantle, depleted mantle, and continental crust—were subsequently modeled by Jacobsen and Wasserburg (1979), O'Nions *et al.* (1979), DePaolo (1980), Allègre *et al.* (1983), and Davies (1981) cast this in terms of a simple mass balance similar to that given above in Equation (11) but using isotope ratios:

$$X_{\rm dm} = \frac{X_{\rm cc}C_{\rm cc}(R_{\rm dm}-R_{\rm cc})}{C_{\rm pm}(R_{\rm dm}-R_{\rm pm})} \qquad (13)$$

where X are mass fractions, R are isotope or trace-element ratios, C are the concentrations of the denominator element of R, and the subscripts cc, dm, and pm refer to continental crust, depleted mantle, and primitive mantle reservoirs, respectively.

With the exception of Davies, who favored whole-mantle convection all along, the above authors concluded that it was only the upper mantle above the 660 km seismic discontinuity that was needed to balance the continental crust. The corollary conclusion was that the deeper mantle must be in an essentially primitive, nearly undepleted state, and consequently convection in the mantle had to occur in two layers with only little exchange between these layers. These conclusions were strongly reinforced by noble gas data, especially $^3He/^4He$ ratios and, more recently, neon isotope data. These indicated that hotspots such as Hawaii are derived from a deep-mantle source with a more primordial, high $^3He/^4He$ ratio, whereas MORBs are derived from a more degassed, upper-mantle reservoir with lower $^3He/^4He$ ratios. In the present context, two points must be mentioned. Essentially all quantitative evolution models dealing with the noble gas evidence concluded that, although plumes carry the primordial gas signature from the deep mantle to the surface, the plumes themselves do not originate in the deep mantle. Instead they rise from the base of the upper mantle, where they entrain very small quantities of lower-mantle, noble-gas-rich material. However, all these models have been constrained by the present-day very low flux of helium from the mantle into the oceans. This flux does not allow the lower mantle to be significantly degassed over the Earth's history. Other authors, who do not consider this constraint on the evolution models to be binding, have interpreted the noble gas data quite differently: they argue that the entire plume comes from a nearly primitive deep-mantle source and rises through the upper, depleted mantle. The former view is consistent with the geochemistry of the refractory elements, which strongly favors some type of recycled, not primitive mantle material to supply the bulk of the plume source. The latter interpretation can be reconciled with the refractory-element geochemistry, if the deep-mantle reservoir is not actually primitive (or close to primitive), but consists of significantly processed mantle with the geochemical characteristics of the FOZO (C, PHEM, etc.) composition, which is characterized by low $^{87}Sr/^{86}Sr$ but relatively high $^{206}Pb/^{204}Pb$ ratios, together with high $^{3}He/^{4}He$ and solar-like neon isotope ratios. The processed nature of this hypothetical deep-mantle reservoir is also evident from its trace-element chemistry, which shows the same nonprimitive (high) Nb/U and (low) Pb/Nd ratios as MORBs and other OIBs. It is not clear how and why the near-primordial noble gas compositions would have survived this processing. So far, except for the two-layer models, no internally consistent mantle evolution model has been published that accounts for all of these observations.

The two-layer models have been dealt a rather decisive blow by recent results of seismic mantle tomography. The images of the mantle produced by this discipline appear to show clear evidence for subduction reaching far into the lower mantle (Grand, 1994; van der Hilst *et al.*, 1997). If this is correct, then there must be a counterflow from the lower mantle across the 660 km boundary, which in the long run would surely destroy the chemical isolation between upper- and lower-mantle reservoirs. Most recently, mantle tomography appears to be able to track some of the major mantle plumes (Hawaii, Easter Island, Cape Verdes, Reunion) into the lowermost mantle (Montelli *et al.*, 2003). If these results are confirmed, there is, at least in recent mantle history, no convective isolation, and mantle evolution models reconciling all the geochemical aspects with the geophysical evidence clearly require new ideas.

As of early 2000s, the existing literature and scientific conferences all show clear signs of a period preceding a significant or even major paradigm shift as described by Thomas Kuhn in his classic work *The Structure of Scientific Revolutions* (Kuhn, 1996): The established paradigm is severely challenged by new observations. While some scientists attempt to reconcile the observations with the paradigm by increasingly complex adjustments of the paradigm, others throw the established conventions overboard and engage in increasingly free speculation. This process continues until a new paradigm evolves or is discovered, which is consistent with all the observations. Examples of the effort of reconciliation are the papers by Stein and Hofmann (1994) and Allègre (1997). These authors point out that the current state of whole-mantle circulation may be episodic or recent, so that whole-mantle chemical mixing has not been achieved.

Examples of the more speculative thinking are the papers by Albarède and van der Hilst (1999) and Kellogg *et al.* (1999), who essentially invent new primitive reservoirs within the deep mantle, which are stabilized by higher chemical density, and may be very irregularly shaped. In the same vein, Porcelli and Halliday (2001) have proposed that the storage reservoir of primordial noble gases may be the core, whereas Tolstikhin and Hofmann (2002) speculate that the lowermost layer of the mantle, called D″ by seismologists, contains the "missing" budget of heat production and primordial noble gases. Finally, an increasing number of convection and mantle evolution modelers are throwing the entire concept of geochemical reservoirs overboard.For example, Phipps Morgan and Morgan (1998) and Phipps Morgan (1999) suggest that the specific geochemical characteristics of plume-type mantle are randomly distributed in the deeper mantle. Plumes rising from the core–mantle boundary constitute the main upward flux balancing the subduction flux. They preferentially lose their enriched components during partial melting, leaving a depleted residue that replenishes the depleted upper mantle.

Starting with the contribution of Christensen and Hofmann (1994), a steadily increasing number of models have recently been published, in which geochemical heterogeneities are specifically incorporated in mantle convection models (e.g., van Keken and Ballentine, 1998; Tackley, 2000; van Keken *et al.*, 2001; Davies, 2002; Farnetani *et al.*, 2002; Tackley, 2002). Thus, while the current state of understanding of the geochemical heterogeneity of the mantle is unsatisfactory, to say the least, the formerly quite separate disciplines of geophysics and geochemistry have begun to interact intensely. This process surely offers the best approach to reach a new paradigm and an understanding of how the mantle really works.

ACKNOWLEDGMENTS

I am most grateful for the hospitality and support offered by the Lamont-Doherty Earth Observatory where much of this article was written. Steve Goldstein made it all possible, while being under duress with the similar project of his own. Cony Class and the other members of the

Petrology/Geochemistry group at Lamont provided intellectual stimulation and good spirit, and the BodyQuest Gym and the running trails of Tallman Park kept my body from falling apart.

REFERENCES

Abouchami W., Boher M., Michard A., and Albarède F. (1990) A major 2.1 Ga event of mafic magmatism in West-Africa: an early stage of crustal accretion. *J. Geophys. Res.* **95**, 17605–17629.

Alard O., Griffin W. L., Pearson N. J., Lorand J. P., and O'Reilly S. Y. (2002) New insights into the Re–Os systematics of sub-continental lithospheric mantle from *in situ* analysis of sulphides. *Earth Planet. Sci. Lett.* **203**(2), 651–663.

Albarède F. (1998) Time-dependent models of U–Th–He and K–Ar evolution and the layering of mantle convection. *Chem. Geol.* **145**, 413–429.

Albarède F. and Van der Hilst R. D. (1999) New mantle convection model may reconcile conflicting evidence. *EOS. Trans. Am. Geophys. Union* **80**(45), 535–539.

Allègre C. J. (1969) Comportement des systèmes U–Th–Pb dans le manteau supérieur et modèle d'évolution de ce dernier au cours de temps géologiques. *Earth Planet. Sci. Lett.* **5**, 261–269.

Allègre C. J. (1997) Limitation on the mass exchange between the upper and lower mantle: the evolving convection regime of the Earth. *Earth Planet. Sci. Lett.* **150**, 1–6.

Allègre C. J. and Turcotte D. L. (1986) Implications of a two-component marble-cake mantle. *Nature* **323**, 123–127.

Allègre C. J., Dupré B., and Brévart O. (1982) Chemical aspects of the formation of the core. *Phil Trans. Roy. Soc. London A* **306**, 49–59.

Allègre C. J., Hart S. R., and Minster J.-F. (1983) Chemical structure and evolution of the mantle and continents determined by inversion of Nd and Sr isotopic data: **II**. Numerical experiments and discussion. *Earth Planet. Sci. Lett.* **66**, 191–213.

Allègre C. J., Schiano P., and Lewin E. (1995) Differences between oceanic basalts by multitrace element ratio topology. *Earth Planet. Sci. Lett.* **129**, 1–12.

Allègre C. J., Hofmann A. W., and O'Nions K. (1996) The argon constraints on mantle structure. *Geophys. Res. Lett.* **23**, 3555–3557.

Ben Othman D., White W. M., and Patchett J. (1989) The geochemistry of marine sediments, island arc magma genesis, and crust–mantle recycling. *Earth Planet. Sci. Lett.* **94**, 1–21.

Ben-Avraham Z., Nur A., Jones D., and Cox A. (1981) Continental accretion: from oceanic plateaus to allochthonous terranes. *Science* **213**, 47–54.

Bird J. M., Meibom A., Frei R., and Nagler T. F. (1999) Osmium and lead isotopes of rare OsIrRu minerals: derivation from the core–mantle boundary region? *Earth Planet. Sci. Lett.* **170**(1–2), 83–92.

Blichert-Toft J., Frey F. A., and Albarède F. (1999) Hf isotope evidence for pelagic sediments in the source of Hawaiian basalts. *Science* **285**, 879–882.

Blundy J. D. and Wood B. J. (1994) Prediction of crystal–melt partition coefficients from elastic moduli. *Nature* **372**, 452–454.

Bohrson W. A. and Reid M. R. (1995) Petrogenesis of alkaline basalts from Socorro Island, Mexico: trace element evidence for contamination of ocean island basalt in the shallow ocean crust. *J. Geophys. Res.* **100**(B12), 24555–24576.

Brandon A. D., Snow J. E., Walker R. J., Morgan J. W., and Mock T. D. (2000) Pt-190–Os-186 and Re-187–Os-187 systematics of abyssal peridotites. *Earth Planet. Sci. Lett.* **177**(3–4), 319–335.

Burton K. W., Schiano P., Birck J.-L., Allègre C. J., and Rehkämper M. (2000) The distribution and behavior of Re and Os amongst mantle minerals and the consequences of metasomatism and melting on mantle lithologies. *Earth Planet. Sci. Lett.* **183**, 93–106.

Calvert A. J. and Ludden J. N. (1999) Archean continental assembly in the southeastern Superior Province of Canada. *Tectonics* **18**(3), 412–429.

Campbell I. H. and Griffiths R. W. (1990) Implications of mantle plume structure for the evolution of flood basalts. *Earth Planet. Sci. Lett.* **99**, 79–93.

Carlson R. W. and Hauri E. (2003) Mantle–crust mass balance and the extent of Earth differentiation. *Earth Planet. Sci. Lett.* (submitted 2002).

Chamorro E. M., Brooker R. A., Wartho J. A., Wood B. J., Kelley S. P., and Blundy J. D. (2002) Ar and K partitioning between clinopyroxene and silicate melts to 8 GPa. *Geochim. Cosmochim. Acta* **66**, 507–519.

Chase C. G. (1981) Oceanic island Pb: two-state histories and mantle evolution. *Earth Planet. Sci. Lett.* **52**, 277–284.

Chauvel C., Hofmann A. W., and Vidal P. (1992) HIMU-EM:the French Polynesian connection. *Earth Planet. Sci. Lett.* **110**, 99–119.

Chauvel C., Goldstein S. L., and Hofmann A. W. (1995) Hydration and dehydration of oceanic crust controls Pb evolution in the mantle. *Chem. Geol.* ***126***, 65–75.

Christensen U. R. and Hofmann A. W. (1994) Segregation of subducted oceanic crust in the convecting mantle. *J. Geophys. Res.* **99**, 19867–19884.

Class C., Miller D. M., Goldstein S. L., and Langmuir C. H. (2000) Distinguishing melt and fluid subduction components in Umnak volcanics, Aleutian Arc. *Geochem. Geophys. Geosyst.* **1** #1999GC000010.

Coltice N. and Ricard Y. (1999) Geochemical observations and one layer mantle convection. *Earth Planet. Sci. Lett.* **174**(1–2), 125–137.

Creaser R. A., Papanastasiou D. A., and Wasserburg G. J. (1991) Negative thermal ion mass spectrometry of osmium, rhenium, and iridium. *Geochim. Cosmochim. Acta* **55**, 397–401.

Davidson J. P. and Bohrson W. A. (1998) Shallow-level processes in ocean-island magmatism. *J. Petrol.* **39**, 799–801.

Davies G. F. (1981) Earth's neodymium budget and structure and evolution of the mantle. *Nature* **290**, 208–213.

Davies G. F. (1998) Plates, plumes, mantle convection and mantle evolution. In *The Earth's Mantle-composition, Structure, and Evolution* (ed. I. Jackson). Cambridge University Press, Cambridge, pp. 228–258.

Davies G. F. (2002) Stirring geochemistry in mantle convection models with stiff plates and slabs. *Geochim. Cosmochim.Acta* **66,** 3125–3142.

DePaolo D. J. (1980) Crustal growth and mantle evolution: inferences from models of element transport and Nd and Sr isotopes. *Geochim. Cosmochim. Acta* **44**, 1185–1196.

DePaolo D. J. and Wasserburg G. J. (1976) Nd isotopic variations and petrogenetic models. *Geophys. Res. Lett.* **3**, 249–252.

Donnelly K. E., Goldstein S. L., Langmuir C. H., and Spiegelmann M. (2003) Origin of enriched ocean ridge basalts and implications for mantle dynamics. *Nature* (ms submitted April 2003).

Dupré B. and Allègre C. J. (1983) Pb–Sr isotope variation in Indian Ocean basalts and mixing phenomena. *Nature* **303**, 142–146.

Dupré B., Schiano P., Polvé M., and Joron J.-L. (1994) Variability: a new parameter which emphasizes the limits of extended rare earth diagrams. *Bull. Soc. Geol. Francais* **165**(1), 3–13.

Eisele J., Sharma M., Galer S. J. G., Blichert-toft J., Devey C. W., and Hofmann A. W. (2002) The role of sediment recycling in EM-1 inferred from Os, Pb, Hf, Nd, Sr isotope and trace element systematics of the Pitcairn hotspot. *Earth Planet. Sci. Lett.* **196**, 197–212.

Elliott T., Plank T., Zindler A., White W., and Bourdon B. (1997) Element transport from slab to volcanic front at the Mariana Arc. *J. Geophys. Res.* **102**, 14991–15019.

Elliott T., Zindler A., and Bourdon B. (1999) Exploring the kappa conundrum: the role of recycling on the lead isotope evolution of the mantle. *Earth Planet. Sci. Lett.* **169**, 129–145.

Farley K. A., Natland J. H., and Craig H. (1992) Binary mixing of enriched and undegassed (primitive?) mantle components (He, Sr, Nd, Pb) in Samoan lavas. *Earth Planet. Sci. Lett.* **111**, 183–199.

Farnetani C. G., Legrasb B., and Tackley P. J. (2002) Mixing and deformations in mantle plumes. *Earth Planet. Sci. Lett.* **196**, 1–15.

Farquhar J., Bao H., and Thiemens M. H. (2000) Atmospheric influence of Earth's earliest sulfur cycle. *Science* **290**, 756–758.

Faure G. (1986) *Principles of Isotope Geology.* Wiley, New York.

Foulger G. R. and Natland J. H. (2003) Is "Hotspot" volcanism a consequence of plate tectonics? *Science* **300**, 921–922.

Galer S. J. G. (1999) Optimal triple spiking for high precision lead isotopic measurement. *Chem. Geol.* **157**, 255–274.

Galer S. J. G. and Goldstein S. L. (1996) Influence of accretion on lead in the Earth. In *Earth Processes: Reading the Isotopic Code,* Geophys. Monograph 95 (eds. A. Basu and S. R. Hart). Am. Geophys. Union, Washington, pp. 75–98.

Galer S. J. G. and O'Nions R. K. (1985) Residence time of thorium, uranium and lead in the mantle with implications for mantle convection. *Nature* **316**, 778–782.

Galer S. J. G., Goldstein S. L., and O'Nions R. K. (1989) Limits on chemical and convective isolation in the Earth's interior. *Chem. Geol.* **75**, 252–290.

Gasperini G., Blichert-Toft J., Bosch D., Del Moro A., Macera P., Télouk P., and Albarède F. (2000) Evidence from Sardinian basalt geochemistry for recycling of plume heads into the Earth's mantle. *Nature* **408**, 701–704.

Gast P. W. (1960) Limitations on the composition of the upper mantle. *J. Geophys. Res.* **65**, 1287.

Gast P. W. (1968) Trace element fractionation and the origin of tholeiitic and alkaline magma types. *Geochim. Cosmochim. Acta* **32**, 1057–1086.

Gast P. W., Tilton G. R., and Hedge C. (1964) Isotopic composition of lead and strontium from Ascension and Gough Islands. *Science* **145**, 1181–1185.

Goldschmidt V. M. (1937) Geochemische Verteilungsgesetze der Elemente: IX. Die Mengenverhältnisse der Elemente and Atomarten. *Skr. Nor. vidensk.-Akad. Oslo* **1**, 148.

Goldschmidt V. M. (1954) *Geochemistry.* Clarendon Press, zOxford.

Grand S. P. (1994) Mantle shear structure beneath the Americas and surrounding oceans. *J. Geophys. Res.* **99**, 11591–11621.

Griffiths R. W. and Campbell I. H. (1990) Stirring and structure in mantle starting plumes. *Earth Planet. Sci. Lett.* **99**, 66–78.

Gurenko A. A. and Chaussidon M. (1995) Enriched and depleted primitive melts included in olivine from Icelandic tholeiites: origin by continuous melting of a single mantle column. *Geochim. Cosmochim. Acta* **59**, 2905–2917.

Haase K. M. (1996) The relationship between the age of the lithosphere and the composition of oceanic magmas: constraints on partial melting, mantle sources and the thermal structure of the plates. *Earth Planet. Sci. Lett.* **144**(1–2), 75–92.

Hanan B. and Graham D. (1996) Lead and helium isotope evidence from oceanic basalts for a common deep source of mantle plumes. *Science* **272**, 991–995.

Hanson G. N. (1977) Geochemical evolution of the suboceanic mantle. *J. Geol. Soc. London* **134**, 235–253.

Hart S. R. (1971) K, Rb, Cs, Sr, Ba contents and Sr isotope ratios of ocean floor basalts. *Phil. Trans. Roy. Soc. London Ser. A* **268**, 573–587.

Hart S. R. (1984) A large-scale isotope anomaly in the southern hemisphere mantle. *Nature* **309**, 753–757.

Hart S. R., Schilling J.-G., and Powell J. L. (1973) Basalts from Iceland and along the Reykjanes Ridge: Sr isotope geochemistry. *Nature* **246**, 104–107.

Hart S. R., Hauri E. H., Oschmann L. A., and Whitehead J. A. (1992) Mantle plumes and entrainment: isotopic evidence. *Science* **256**, 517–520.

Hauri E. (2002) SIMS analysis of volatiles in silicate glasses: 2. Isotopes and abundances in Hawaiian melt inclusions. *Chem.Geol.* **183**, 115–141.

Hawkesworth C. J., Norry M. J., Roddick J. C., and Vollmer R. (1979) $^{143}Nd/^{144}Nd$ and $^{87}Sr/^{86}Sr$ ratios from the Azores and their significance in LIL-element enriched mantle. *Nature* **280**, 28–31.

Hawkesworth C. J., Mantovani M. S. M., Taylor P. N., and Palacz Z. (1986) Evidence from the Parana of South Brazil for a continental contribution to Dupal basalts. *Nature* **322**, 356–359.

Hedge C. E. and Walthall F. G. (1963) Radiogenic strontium 87 as an index of geological processes. *Science* **140**, 1214–1217.

Hellebrand E., Snow J. E., Dick H. J. B., and Hofmann A. W. (2001) Coupled major and trace elements as indicators of the extent of melting in mid-ocean-ridge peridotites. *Nature* **410**, 677–681.

Hemming S. R. and McLennan S. M. (2001) Pb isotope compositions of modern deep sea turbidites. *Earth Planet. Sci. Lett.* **184**, 489–503.

Hémond C. H., Arndt N. T., Lichtenstein U., Hofmann A.W., Oskarsson N., and Steinthorsson S. (1993) The heterogeneous Iceland plume: Nd–Sr–O isotopes and trace element constraints. *J. Geophys. Res.* **98**, 15833–15850.

Hirschmann M. M. and Stopler E. (1996) A possible role for garnet pyroxenite in the origin of the "garnet signature" in MORB. *Contrib. Mineral. Petrol.* **124**, 185–208.

Hofmann A. W. (1988) Chemical differentiation of the Earth: the relationship between mantle, continental crust, and oceanic crust. *Earth Planet. Sci. Lett.* **90**, 297–314.

Hofmann A. W. (1989a) Geochemistry and models of mantle circulation. *Phil. Trans. Roy. Soc. London A* **328**, 425–439.

Hofmann A. W. (1989b) A unified model for mantle plume sources. *EOS* **70**, 503.

Hofmann A. W. (1997) Mantle geochemistry: the message from oceanic volcanism. *Nature* **385**, 219–229.

Hofmann A. W. and Hart S. R. (1978) An assessment of local and regional isotopic equilibrium in the mantle. *Earth Planet. Sci. Lett.* **39**, 44–62.

Hofmann A. W. and White W. M. (1980) The role of subducted oceanic crust in mantle evolution. *Carnegie Inst. Wash. Year Book* **79**, 477–483.

Hofmann A. W. and White W. M. (1982) Mantle plumes from ancient oceanic crust. *Earth Planet. Sci. Lett.* **57**, 421–436.

Hofmann A. W. and White W. M. (1983) Ba, Rb, and Cs in the Earth's mantle. Z. *Naturforsch.* **38**, 256–266.

Hofmann A. W., White W. M., and Whitford D. J. (1978) Geochemical constraints on mantle models: the case for a layered mantle. *Carnegie Inst. Wash. Year Book* **77**, 548–562.

Hofmann A. W., Jochum K.-P., Seufert M., and White W. M. (1986) Nb and Pb in oceanic basalts: new constraints on mantle evolution. *Earth Planet. Sci. Lett.* **79**, 33–45.

Holland H. D. (1994) Early Proterozoic atmospheric change. In *Early Life on Earth* (ed. S. Bengtson). Columbia University Press, New York, pp. 237–244.

Hurley P. M., Hughes H., Faure G., Fairbairn H., and Pinson W. (1962) Radiogenic strontium-87 model for continent formation. *J. Geophys. Res.* **67**, 5315–5334.

Jackson I. N. S. and Rigden S. M. (1998) Composition and temperature of the mantle: seismological models interpreted through experimental studies of mantle minerals. In *The Earth's Mantle: Composition, Structure and Evolution* (ed. I. N. S. Jackson). Cambridge University Press, Cambridge, pp. 405–460.

Jacobsen S. B. and Wasserburg G. J. (1979) The mean age of mantle and crustal reservoirs. *J. Geophys. Res.* **84**, 7411–7427.

Jochum K. P. and Hofmann A. W. (1994) Antimony in mantle-derived rocks: constraints on Earth evolution from moderately siderophile elements. *Min. Mag.* **58A**, 452–453.

Jochum K.-P., Hofmann A. W., Ito E., Seufert H. M., and White W. M. (1983) K, U, and Th in mid-ocean ridge basalt glasses and heat production, K/U, and K/Rb in the mantle. *Nature* **306**, 431–436.

Jochum K. P., Seufert H. M., Spettel B., and Palme H. (1986) The solar system abundances of Nb, Ta, and Y. and the relative abundances of refractory lithophile elements in differentiated planetary bodies. *Geochim. Cosmochim. Acta* **50**, 1173–1183.

Jochum K. P., Hofmann A. W., and Seufert H. M. (1993) Tin in mantle-derived rocks: constraints on Earth evolution. *Geochim. Cosmochim. Acta* **57**, 3585–3595.

Johnson K. T. M., Dick H. J. B., and Shimizu N. (1990) Melting in the oceanic upper mantle: an ion microprobe study of diopsides in abyssal peridotites. *J. Geophys. Res.* **95**, 2661–2678.

Johnson M. C. and Plank T. (2000) Dehydration and melting experiments constrain the fate of subducted sediments. *Geochem. Geophys. Geosys.* **1**, 1999GC000014.

Kellogg L. H., Hager B. H., and van der Hilst R. D. (1999) Compositional stratification in the deep mantle. *Science* **283**, 1881–1884.

Kempton P. D., Pearce J. A., Barry T. L., Fitton J. G., Langmuir C. H., and Christie D. M. (2002) Sr–Nd–Pb–Hf isotope results from ODP Leg 187: evidence for mantle dynamics of the Australian-Antarctic Discordance and origin of the Indian MORB source. *Geochem. Geophys. Geosys.* **3**, 10.29/ 2002GC000320.

Kimura G., Ludden J. N., Desrochers J. P., and Hori R. (1993) A model of ocean-crust accretion for the superior province, Canada. *Lithos* **30**(3–4), 337–355.

Klein E. M. and Karsten J. L. (1995) Ocean-ridge basalts with convergent-margin geochemical affinities from the Chile Ridge. *Nature* **374**, 52–57.

Klein E. M. and Langmuir C. H. (1987) Global correlations of ocean ridge basalt chemistry with axial depth and crustal thickness. *J Geophys. Res.* **92**, 8089–8115.

Klein E. M., Langmuir C. H., Zindler A., Staudigel H., and Hamelin B. (1988) Isotope evidence of a mantle convection boundary at the Australian–Antarctic discordance. *Nature* **333**, 623–629.

Kleine T., Munker C., Mezger K., and Palme H. (2002) Rapid accretion and early core formation on asteroids and the terrestrial planets from Hf–W chronometry. *Nature* 418, 952–955.

Kramers J. D. and Tolstikhin I. N. (1997) Two terrestrial lead isotope paradoxes, forward transport modelling, core formation and the history of the continental crust. *Chem. Geol.* **139**(1–4), 75–110.

Krot A. N., Meibom A., Weisberg M. K., and Keil K. (2002) The CR chondrite clan: implications for early solar system processes. *Meteorit. Planet. Sci.* 37(11), 1451–1490.

Kuhn T. S. (1996) *The Structure of Scientific Revolutions*. University of Chicago Press, Chicago.

Lassiter J. C. (2002) The influence of recycled oceanic crust on the potassium and argon budget of the Earth. *Geochim. Cosmochim. Acta* **66**(15A), A433–A433(suppl.).

Lassiter J. C. and Hauri E. H. (1998) Osmium-isotope variations in Hawaiian lavas: evidence for recycled oceanic lithosphere in the Hawaiian plume. *Earth Planet. Sci. Lett.* **164**, 483–496.

le Roux P. J., le Roex A. P., and Schilling J.-G. (2002) MORB melting processes beneath the southern Mid-Atlantic Ridge (40–55 degrees S): a role for mantle plume-derived pyroxenite. *Contrib. Min. Pet.* **144**, 206–229.

Lee D. C. and Halliday A. N. (1995) Hafnium–tungsten chronometry and the timing of terrestrial core formation. *Nature* **378**, 771–774.

Mahoney J., Nicollet C., and Dupuy C. (1991) Madagascar basalts: tracking oceanic and continental sources. *Earth Planet. Sci. Lett.* **104**, 350–363.

Martin C. E. (1991) Osmium isotopic characteristics of mantle-derived rocks. *Geochim. Cosmochim. Acta* **55**, 1421–1434.

McDonough W. F. (1991) Partial melting of subducted oceanic crust and isolation of its residual eclogitic lithology. *Phil. Trans. Roy. Soc. London A* **335**, 407–418.

McDonough W. F. and Sun S.-S. (1995) The composition of the Earth. *Chem. Geol.* **120**, 223–253.

McDonough W. F., Sun S.-S., Ringwood A. E., Jagoutz E., and Hofmann A. W. (1992) Potassium, rubidium, and cesium in the Earth and Moon and the evolution of the mantle of the Earth. *Geochim. Cosmochim. Acta* **56**, 1001–1012.

McKenzie D. and O'Nions R. K. (1983) Mantle reservoirs and ocean island basalts. *Nature* **301**, 229–231.

Meibom A. and Frei R. (2002a) Evidence for an ancient osmium isotopic reservoir in Earth (vol. 298, p. 516, 2002). Science **297**(5584), 1120–1120.

Meibom A. and Frei R. (2002b) Evidence for an ancient osmium isotopic reservoir in Earth. *Science* **296**(5567), 516–518.

Meibom A., Frei R., Chamberlain C. P., Coleman R. G., Hren M., Sleep N. H., and Wooden J. L. (2002a) OS isotopes, deep-rooted mantle plumes and the timing of inner core formation. *Meteorit. Planet. Sci.* **37**(7), A98.

Meibom A., Frei R., Chamberlain C. P., Coleman R. G., Hren M. T., Sleep N. H., and Wooden J. L. (2002b) Os isotopes, deep-rooted mantle plumes and the timing of inner core formation. *Geochim. Cosmochim. Acta* **66**(15A), A504–A504.

Meibom A., Sleep N. H., Chamberlain C. P., Coleman R. G., Frei R., Hren M. T., and Wooden J. L. (2002c) Re–Os isotopic evidence for long-lived heterogeneity and equilibration processes in the Earth's upper mantle. *Nature* **419**(6908), 705–708.

Meibom A., Anderson D. L., Sleep N. H., Frei R., Chamberlain C. P., Hren M. T., and Wooden J. L. (2003) Are high He-3/He-4 ratios in oceanic basalts an indicator of deep-mantle plume components? *Earth Planet. Sci. Lett.* **208**(3–4), 197–204.

Miller D. M., Goldstein S. L., and Langmuir C. H. (1994) Cerium/lead and lead isotope ratios in arc magmas and the enrichment of lead in the continents. *Nature* **368**, 514–520.

Milner S. C. and le Roex A. P. (1996) Isotope characteristics of the Okenyena igneous complex, northwestern Namibia: constraints on the composition of the early Tristan plume and the origin of the EM 1 mantle component. *Earth Planet. Sci. Lett.* **141**, 277–291.

Montelli R., Nolet G., Masters G., Dahlen F. A., and Hung S.-H. (2003) Global P and PP traveltime tomography: rays versus waves. *Geophys. J. Int.* **142** (in press).

Morgan W. J. (1971) Convection plumes in the lower mantle. *Nature* **230**, 42–43.

Morse S. A. (1983) Strontium isotope fractionation in the Kiglapait intrusion. *Science* **220**, 193–195.

Murphy D. T., Kamber B. S., and Collerson K. D. (2003) A refined solution to the first terrestrial Pb-isotope paradox. *J. Petrol.* **44**, 39–53.

O'HaraM. J. (1973) Non-primary magmas and dubious mantle plume beneath Iceland. *Nature* **243**(5409), 507–508.

O'Hara M. J. (1975) Is there an icelandic mantle plume. *Nature* **253**(5494), 708–710.

O'Hara M. J. (1977) Open system crystal fractionation and incompatible element variation in basalts. *Nature* **268**, 36–38.

O'Hara M. J. and Mathews R. E. (1981) Geochemical evolution in an advancing, periodically replenished, periodically tapped, continuously fractionated magma chamber. *J. Geol. Soc. London* **138**, 237–277.

O'Nions R. K. and Pankhurst R. J. (1974) Petrogenetic significance of isotope and trace element variation in volcanic rocks from the Mid-Atlantic. *J. Petrol.* **15**, 603–634.

O'Nions R. K., Evensen N. M., and Hamilton P. J. (1977) Variations in $^{143}Nd/^{144}Nd$ and $^{87}Sr/^{86}/Sr$ ratios in oceanic basalts. *Earth Planet. Sci. Lett.* **34**, 13–22.

O'Nions R. K., Evensen N. M., and Hamilton P. J. (1979) Geochemical modeling of mantle differentiation and crustal growth. *J. Geophys. Res.* **84**, 6091–6101.

Patchett P. J., White W. M., Feldmann H., Kielinczuk S., and Hofmann A. W. (1984) Hafnium/rare earth fractionation in the sedimentary system and crust–mantle recycling. *Earth Planet. Sci. Lett.* **69**, 365–378.

Patterson C. C. (1956) Age of meteorites and the Earth. *Geochim. Cosmochim. Acta* **10**, 230–237.

Peucker-Ehrenbrink B., Hofmann A. W., and Hart S. R. (1994) Hydrothermal lead transfer from mantle to continental crust: the role of metalliferous sediments. *Earth Planet. Sci. Lett.* **125**, 129–142.

Phipps Morgan J. (1999) Isotope topology of individual hotspot basalt arrays: mixing curves of melt extraction trajectories? *Geochem. Geophys. Geosystems* **1**, 1999GC000004.

Phipps Morgan J., Morgan W. J., and Zhang Y.-S. (1995) Observational hints for a plume-fed, suboceanic asthenosphere and its role in mantle convection. *J. Geophys. Res.* **100**, 12753–12767.

Phipps Morgan J. and Morgan W. J. (1998) Two-stage melting and the geochemical evolution of the mantle: a recipe for mantle plum-pudding. *Earth Planet. Sci. Lett.* **170**, 215–239.

Plank T. and Langmuir C. H. (1993) Tracing trace elements from sediment input to volcanic output at subduction zones. *Nature* **362**, 739–742.

Plank T. and Langmuir C. H. (1998) The chemical composition of subducting sediment and its consequences for the crust and mantle. *Chem. Geol.* **145**, 325–394.

Porcelli D. and Halliday A. N. (2001) The core as a possible source of mantle helium. *Earth Planet. Sci. Lett.* **192**(1), 45–56.

Puchtel I. S., Hofmann A. W., Mezger K., Jochum K. P., Shchipansky A. A., and Samsonov A. V. (1998) Oceanic plateau model for continental crustal growth in the Archaean: a case study from the Kostomuksha greenstone belt, NW Baltic Shield. *Earth Planet. Sci. Lett.* **155**, 57–74.

Pyle D. G., Christie D. M., and Mahoney J. J. (1992) Resolving an isotopic boundary within the Australian–Antarctic Discordance. *Earth Planet. Sci. Lett.* **112**, 161–178.

Regelous M., Hofmann A. W., Abouchami W., and Galer J. S. G. (2002/3) Geochemistry of lavas from the Emperor Seamounts, and the geochemical evolution of Hawaiian magmatism 85–42 Ma. *J. Petrol.* **44**, 113–140.

Rehkämper M. and Hofmann A. W. (1997) Recycled ocean crust and sediment in Indian Ocean MORB. *Earth Planet. Sci. Lett.* **147**, 93–106.

Rehkämper M. and Halliday A. N. (1998) Accuracy and long-term reproducibility of Pb isotopic measurements by MC-ICPMS using an external method for the correction of mass discrimination. *Int. J. Mass Spectrom. Ion Processes* **181**, 123–133.

Richard P., Shimizu N., and Allègre C. J. (1976) $^{143}Nd/^{146}Nd$—a natural tracer: an application to oceanic basalt. *Earth Planet. Sci. Lett.* **31**, 269–278.

Roy-Barman M. and Allègre C. J. (1994) $^{187}Os/^{186}Os$ ratios of mid-ocean ridge basalts and abyssal peridotites. *Geochim. Cosmochim. Acta* **58**, 5053–5054.

Rudnick R. L. and Fountain D. M. (1995) Nature and composition of the continental crust: a lower crustal perspective. *Rev. Geophys.* **33**, 267–309.

Rudnick R. L. and Goldstein S. L. (1990) The Pb isotopic compositions of lower crustal xenoliths and the evolution of lower crustal Pb. *Earth Planet. Sci. Lett.* **98**, 192–207.

Rudnick R. L., McDonough W. F., and O'Connell R. J. (1998) Thermal structure, thickness and composition of continental lithosphere. *Chem. Geol.* **145**, 395–411.

Rudnick R. L., Barth M., Horn I., and McDonough W. F. (2000) Rutile-bearing refractory eclogites: missing link between continents and depleted mantle. *Science* **287**, 278–281.

Saal A. E., Hart S. R., Shimizu N., Hauri E. H., and Layne G. D. (1998) Pb isotopic variability in melt inclusions from oceanic island basalts, Polynesia. *Science* **282**, 1481–1484.

Saunders A. D., Norry M. J., and Tarney J. (1988) Origin of MORB and chemically-depleted mantle reservoirs: trace element constraints. *J. Petrol.* (Special Lithosphere Issue), 415–445.

Schiano P., Birck J.-L., and Allègre C. J. (1997) Osmium-strontium–neodymium–lead isotopic covariations in mid-ocean ridge basalt glasses and the heterogeneity of the upper mantle. *Earth Planet. Sci. Lett.* **150**, 363–379.

Schilling J.-G. (1973) Iceland mantle plume: geochemical evidence along Reykjanes Ridge. *Nature* **242**, 565–571.

Schilling J. G. and Winchester J. W. (1967) Rare-earth fractionation and magmatic processes. In *Mantles of Earth and Terrestrial Planets* (ed. S. K. Runcorn). Interscience Publ., London, pp. 267–283.

Schilling J. G. and Winschester J. W. (1969) Rare earth contribution to the origin of Hawaiian lavas. *Contrib. Mineral. Petrol.* **40**, 231.

Shaw D. M. (1970) Trace element fractionation during anatexis. *Geochim. Cosmochim. Acta* **34**, 237–242.

Shimizu N., Sobolev A. V., Layne G. D., and Tsameryan O. P. (2003) Large Pb isotope variations in olivine-hosted melt inclusions in a basalt fromt the Mid-Atlantic Ridge. *Science* (ms. submitted May 2003).

Sims K. W. W. and DePaolo D. J. (1997) Inferences about mantle magma sources from incompatible element concentration ratios in oceanic basalts. *Geochim. Cosmochim. Acta* **61**, 765–784.

Sleep N. H. (1984) Tapping of magmas from ubiquitous mantle heterogeneities: an alternative to mantle plumes? *J. Geophys. Res.* **89**, 10029–10041.

Snow J. E. and Reisberg L. (1995) Os isotopic systematics of the MORB mantle: results from altered abyssal peridotites. *Earth Planet. Sci. Lett.* **133**, 411–421.

Sobolev A. V. (1996) Melt inclusions in minerals as a source of principal petrological information. *Petrology* **4**, 209–220.

Sobolev A. V. and Shimizu N. (1993) Ultra-depleted primary melt included in an olivine from the Mid-Atlantic Ridge. *Nature* **363**, 151–154.

Sobolev A. V., Hofmann A. W., and Nikogosian I. K. (2000) Recycled oceanic crust observed in "ghost plagioclase" within the source of Mauna Loa lavas. *Nature* **404**, 986–990.

Staudigel H., Davies G. R., Hart S. R., Marchant K. M., and Smith B. M. (1995) Large scale isotopic Sr, Nd and O isotopic anatomy of altered oceanic crust: DSDP/ODP sites 417/418. *Earth Planet. Sci. Lett.* **130**, 169–185.

Stein M. and Hofmann A. W. (1994) Mantle plumes and episodic crustal growth. *Nature* **372**, 63–68.

Su Y. J. (2002) Mid-ocean ridge basalt trace element systematics: constraints from database management, ICP-MS analyses, global data compilation and petrologic modeling. PhD Thesis, Columbia University, 2002, 472pp.

Sun S.-S. and McDonough W. F. (1989) Chemical and isotopic systematics of oceanic basalts: implications for mantle composition and processes. In *Magmatism in the Ocean Basins,* Geological Society Spec. Publ. 42 (eds. A. D. Saunders and M. J. Norry). Oxford, pp. 313–345.

Tackley P. J. (2000) Mantle convection and plate tectonics: toward an integrated physical and chemical theory. *Science* **288**, 2002–2007.

Tackley P. J. (2002) Strong heterogeneity caused by deep mantle layering. *Geochem. Geophys. Geosystems* **3**(4), 101029/2001GC000167.

Tatsumoto M., Hedge C. E., and Engel A. E. J. (1965) Potassium, rubidium, strontium, thorium, uranium, and the ratio of strontium-87 to strontium-86 in oceanic tholeiitic basalt. *Science* **150**, 886–888.

Taylor S. R. and McLennan S. M. (1985) *The Continental Crust: its Composition and Evolution.* Blackwell, Oxford.

Taylor S. R. and McLennan S. M. (1995) The geochemical evolution of the continental crust. *Rev. Geophys.* **33**, 241–265.

Taylor S. R. and McLennan S. M. (2001) Chemical composition and element distribution in the Earth's crust. In *Encyclopedia of Physical Sciences and Technology,* Academic Press, vol. 2, pp. 697–719.

Thompson R. N., Morrison M. A., Hendry G. L., and Parry S. J. (1984) An assessment of the relative roles of crust and mantle in magma genesis: an elemental approach. *Phil. Trans. Roy. Soc. London* **A310**, 549–590.

Todt W., Cliff R. A., Hanser A., and Hofmann A. W. (1996) Evaluation of a ^{202}Pb–^{205} Pb double spike for high-precision lead isotope analysis. In *Earth Processes: Reading the Isotopic Code* (eds. A. Basu and S. R. Hart). Am. Geophys. Union, Geophys Monograph 95, Washington, pp. 429–437.

Tolstikhin I. N. and Hofmann A. W. (2002) Generation of a long-lived primitive mantle reservoir during late stages of Earth accretion. *Geochim. Cosmochim. Acta* **66**(15A), A779–A779.

van der Hilst R. D., Widiyantoro S., and Engdahl E. R. (1997) Evidence for deep mantle circulation from global tomography. *Nature* **386**, 578–584.

van Keken P. E. and Ballentine C. J. (1998) Whole-mantle versus layered mantle convection and the role of a high-viscosity lower mantle in terrestrial volatile evolution. *Earth Planet. Sci. Lett.* **156**, 19–32.

van Keken P. E., Ballentine C. J., and Porcelli D. (2001) A dynamical investigation of the heat and helium imbalance. *Earth Planet. Sci. Lett.* **188**(3–4), 421–434.

Van Orman J. A., Grove T. L., and Shimizu N. (2001) Rare earth element diffusion in diopsde: influence of temperature, pressure and ionic radius, and an elastic model for diffusion in silicates. *Contrib. Mineral. Petrol.* **141**, 687–703.

Völkening J., Walczyk T., and Heumann K. G. (1991) Osmium isotope ratio determinations by negative thermal ionization mass spectrometry. *Int. J. Mass Spectrom. Ion Process.* **105**, 147–159.

Vollmer R. (1976) Rb–Sr and U–Th–Pb systematics of alkaline rocks: the alkaline rocks from Italy. *Geochim. Cosmochim. Acta* **40**, 283–295.

Wade J. and Wood B. J. (2001) The Earth's "missing" niobium may be in the core. *Nature* **409**, 75–78.

Wasserburg G. J. and Depaolo D. J. (1979) Models of Earth structure inferred from neodymium and strontium isotopic abundances. *Proc. Natl. Acad. Sci. USA* **76**(8), 3594–3598.

Weaver B. L. (1991) The origin of ocean island basalt end-member compositions: trace element and isotopic constraints. *Earth Planet. Sci. Lett.* **104**, 381–397.

Wedepohl K. H. (1995) The composition of the continental crust. *Geochim. Cosmochim. Acta* **59**(7), 1217–1232.

White W. M. (1985) Sources of oceanic basalts: radiogenic isotope evidence. *Geology* **13**, 115–118.

White W. M. (1993) ^{238}U/^{204}Pb in MORB and open system evolution of the depleted mantle. *Earth Planet. Sci. Lett.* **115**, 211–226.

White W. M. and Duncan R. A. (1996) Geochemistry and geochronology of the Society Islands: new evidence for deep mantle recycling. In *Earth Processes: Reading the Isotopic Code.* (eds. A. Basu and S. R. Hart). Am. Geophys. Union, Geophys. Monograph 95, Washington, pp. 183–206.

White W. M. and Hofmann A. W. (1982) Sr and Nd isotope geochemistry of oceanic basalts and mantle evolution. *Nature* **296**, 821–825.

White W. M. and Schilling J.-G. (1978) The nature and origin of geochemical variation in Mid-Atlantic Ridge basalts from the central North Atlantic. *Geochim. Cosmochim. Acta* **42**, 1501–1516.

Widom E. and Shirey S. B. (1996) Os isotope systematics in the Azores: implications for mantle plume sources. *Earth Planet. Sci. Lett.* **142**, 451–465.

Wood B. J. and Blundy J. D. (1997) A predictive model for rare earth element partitioning between clinopyroxene and anhydrous silicate melt. *Contrib. Mineral. Petrol.* **129**, 166–181.

Wood D. A. (1979) A variably veined suboceanic mantle—genetic significance for mid-ocean ridge basalts from geochemical evidence. *Geology* **7**, 499–503.

Woodhead J. (1996) Extreme HIMU in an oceanic setting: the geochemistry of Mangaia Island (Polynesia), and temporal evolution of the Cook-Austral hotspot. *J. Volcanol. Geotherm. Res.* **72**, 1–19.

Woodhead J. D. and Devey C. W. (1993) Geochemistry of the Pitcairn seamounts: I. Source character and temporal trends. *Earth Planet. Sci. Lett.* **116**, 81–99.

Workman R. K., Hart S. R., Blusztajn J., Jackson M., Kurz M., and Staudigel H. (2003) Enriched mantle: II. A new view from the Samoan hotspot. *Geochem. Geophys. Geosys.* (submitted).

Yang H. J., Frey F. A., and Clague D. A. (2003) Constraints on the source components of lavas forming the Hawaiian north arch and honolulu volcanics. *J. Petrol.* **44**, 603–627.

Yaxley G. M. and Green D. H. (1998) Reactions between eclogite and peridotite: mantle refertilisation by subduction of oceanic crust. *Schweiz. Mineral. Petrogr. Mitt.* **78**, 243–255.

Yin Q., Jacobsen S. B., Yamashita K., Blichert-Toft J., Télouk P., and Albarède F. (2002) A short timescale for terrestrial planet formation from Hf–W chronometry of meteorites. *Nature* **418**, 949–952.

Zartman R. E. and Haines S. M. (1988) The plumbotectonic model for Pb isotopic systematics among major terrestrial reservoirs—a case for bi-directional transport. *Geochim. Cosmochim. Acta* **52**, 1327–1339.

Zindler A. and Hart S. (1986) Chemical geodynamics. *Ann. Rev. Earth Planet. Sci.* **14**, 493–571.

Readings from the Treatise on Geochemistry
ISBN: 978-0-12-381391-6
pp. 67–108

4

Compositional Model for the Earth's Core

W. F. McDonough

University of Maryland, College Park, USA

4.1 INTRODUCTION

The remote setting of the Earth's core tests our ability to assess its physical and chemical characteristics. Extending out to half an Earth radii, the metallic core constitutes a sixth of the planet's volume and a third of its mass (see Table 1 for physical properties of the Earth's core). The boundary between the silicate mantle and the core (CMB) is remarkable in that it is a zone of greatest contrast in Earth properties. The density increase across this boundary represents a greater contrast than across the crust-ocean surface. The Earth's gravitational acceleration reaches a maximum (10.7 m s^{-2}) at the CMB and this boundary is also the site of the greatest temperature gradient in the Earth. (The temperature at the base of the mantle (~2,900 °C) is not well established, and that at the top of the inner core is even less securely known (~3,500–4,500 °C).) The pressure range throughout the core (i.e., 136 GPa to >360 GPa) makes recreating environmental conditions in most experimental labs impossible, excepting a few diamond anvil facilities or those with high-powered, shock-melting guns. Thus, our understanding of the core is based on very few pieces of direct evidence and many fragments of indirect observations. Direct evidence comes from seismology, geodesy, geo- and paleomagnetism, and, relatively recently isotope geochemistry (see Section 4.6). Indirect evidence comes from geochemistry, cosmochemistry, and meteoritics; further constraints on the core

Table 1 Physical properties of the Earth's core.

		Units	*Refs.*
Mass			
Earth	5.9736E + 24	kg	1
Inner core	9.675E + 22	kg	1
Outer core	1.835E + 24	kg	1
Core	1.932E + 24	kg	1
Mantle	4.043E + 24	kg	1
Inner core to core (%)	5.0%		
Core to Earth (%)	32.3%		
Depth			
Core–mantle boundary	3,483 ± 5	km	2
Inner–outer core boundary	1,220 ± 10	km	2
Mean radius of the Earth	6,371.01 ± 0.02	km	1
Volume relative to planet			
Inner core	7.606E + 09 (0.7%)	km^3	
Inner core relative to the bulk core	4.3%		
Outer core	1.694E + 11 (15.6%)	km^3	
Bulk core	1.770E + 11 (16.3%)	km^3	
Silicate earth	9.138E + 11 (84%)	km^3	
Earth	1.083E + 12	km^3	
Moment of inertia constants			
Earth mean moment of inertia (I)	0.3299765	Ma^2	1
Earth mean moment of inertia (I)	0.3307144	MR_0^2	1
Mantle: I_m/Ma^2	0.29215	Ma^2	1
Fluid core: I_f/Ma^2	0.03757	Ma^2	1
Inner core: I_{ic}/Ma^2	2.35E−4	Ma^2	1
Core: $I_{f+ic}/M_{f+ic}a_f^2$	0.392	Ma^2	1

1—Yoder (1995), 2—Masters and Shearer (1995).
M is the Earth's mass, a is the Earth's equatorial radius, R_0 is the radius for an oblate spheroidal Earth, I_m is the moment of inertia for the mantle, I_f is the moment of inertia for the outer (fluid) core, I_{ic} is the moment of inertia for the inner core, and $I_{f+ic}/M_{f+ic}a_f^2$ is the mean moment of inertia for the core.

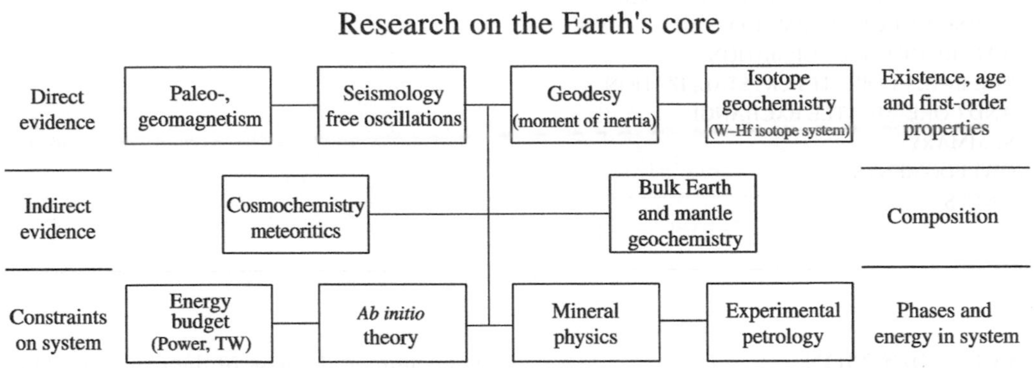

Figure 1 The relative relationship between disciplines involved in research on the Earth's core and the nature of data and information that come from these various investigations. Studies listed in the upper row yield direct evidence on properties of the core. Those in the middle row yield indirect evidence on the composition of the Earth's core, whereas findings from disciplines listed on the bottom row provide descriptions of the state conditions for the core and its formation.

system are gained from studies in experimental petrology, mineral physics, *ab initio* calculations, and evaluations of the Earth's energy budget (e.g., geodynamo calculations, core crystallization, heat flow across the core–mantle boundary). Figure 1 provides a synopsis of research on the Earth's core, and the relative relationship between disciplines. Feedback loops between all of these disciplines refine other's understanding of the Earth's core.

4.2 FIRST-ORDER GEOPHYSICS

The Earth's three-layer structure (the core, the silicate shell (mantle and crust), and the atmosphere–hydrosphere system) is the product of planetary differentiation and is identified as the most significant geological process to have occurred since the formation of the Earth. Each layer is distinctive in its chemical composition, the nature of its phase (i.e., solid, liquid, and gas), and physical properties. Evidence for the existence and nature of the Earth's core comes from laboratory studies coupled with studies that directly measure physical properties of the Earth's interior including its magnetic field, seismological profile, and orbital behavior, with the latter providing a coefficient of the moment of inertia and a model for the density distribution in the Earth.

There is a long history of knowing indirectly or directly of the existence of Earth's core. Our earliest thoughts about the core, albeit indirect and unwittingly, may have its roots in our understanding of the Earth's magnetic field. The magnetic compass and its antecedents appear to be ~2,000 yr old. F. Gies and J. Gies (1994) report that Chinese scholars make reference to a south-pointing spoon, and claim its invention to ca. AD 83 (Han dynasty). A more familiar form of the magnetic compass was known by the twelfth century in Europe. With the discovery of iron meteorites followed by the suggestion that these extraterrestrial specimens came from the interior of fragmented planets in the late-nineteenth century came the earliest models for planetary interiors. Thus, the stage was set for developing Earth models with a magnetic and metallic core. Later development of geophysical tools for peering into the deep Earth showed that with increasing depth the proportion of metal to rock increases with a significant central region envisaged to be wholly made up of iron.

A wonderful discussion of the history of the discovery of the Earth's core is given in the Brush (1980) paper. The concept of a core perhaps begins with understanding the Earth's magnetic field. Measurements of the Earth's magnetic field have been made since the early 1500s. By 1600 the English physician and physicist, William Gilbert, studied extensively the properties of magnets and found that their magnetic field could be removed by heating; he concluded that the Earth behaved as a large bar-magnet. In 1832, Johann Carl Friedrich Gauss, together with Wilhelm Weber, began a series of studies on the nature of Earth's magnetism, resulting in the 1839 publication of *Allgemeine Theorie des Erdmagnetismus* (*General Theory of the Earth's Magnetism*), demonstrating that the Earth's magnetic field was internally generated.

With the nineteenth-century development of the seismograph, studies of the Earth's interior and core accelerated rapidly. In 1897 Emil Wiechert subdivided the Earth's interior into two main layers: a silicate shell surrounding a metallic core, with the core beginning at ~1,400 km depth. This was the first modern model of the Earth's internal structure, which is now confirmed widely by many lines of evidence. Wiechert was a very interesting scientist; he invented a seismograph that saw widespread use in the early twentieth century, was one of the founders of the Institute of Geophysics at Göttingen, and was the PhD supervisor of Beno Gutenberg. The discoverer of the Earth's core is considered to be Richard Dixon Oldham, a British seismologist, who first distinguished *P* (compressional) and *S* (shear) waves following his studies of the Assam earthquake of 1897. In 1906 Oldham observed that *P* waves arrived later than expected at the surface antipodes of epicenters and recognized this as evidence for a dense and layered interior. Oldham placed the depth to the core–mantle boundary at 3,900 km. Later, Gutenberg (1914) established the core–mantle boundary at 2,900 km depth (cf. the modern estimate of 2,891 ± 5 km depth; Masters and Shearer, 1995) and suggested that the core was at least partly liquid (Gutenberg, 1914). Subsequently, Jeffreys (1926) established that the outer core is liquid, and Lehmann (1936) identified the existence of a solid inner core using seismographic records of large earthquakes, which was later confirmed by Anderson *et al.* (1971) and Dziewonski and Gilbert (1972) using Earth's free-oscillation frequencies. Finally, Washington (1925) and contemporaries reported that an iron core would have a significant nickel content, based on analogies with iron meteorites and the cosmochemical abundances of these elements.

The seismological profile of the Earth's core (Figure 2) combined with the first-order relationship between density and seismic wave speed velocity (i.e., $V_p = ((K + 4/3\mu)/\rho)^{0.5}$, $V_s = (\mu/\rho)^{0.5}$, $d\rho/dr = -GM_r\rho(r)/r^2\Phi$ (the latter being the Adams–Williamson equation),

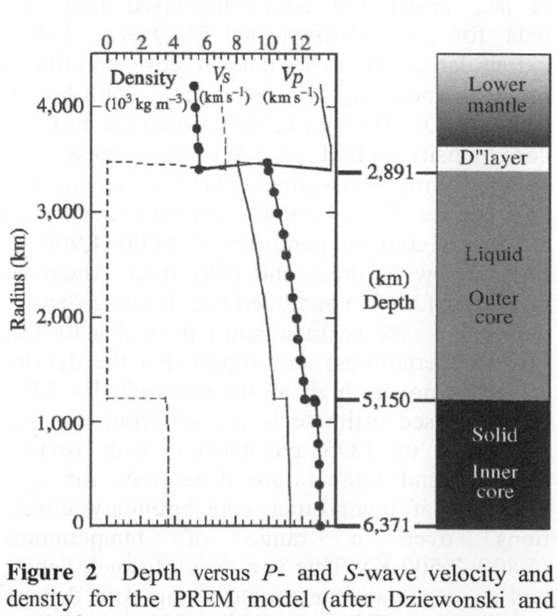

Figure 2 Depth versus *P*- and *S*-wave velocity and density for the PREM model (after Dziewonski and Anderson, 1981 and Masters and Shearer, 1995).

where V_p is the *P*-wave velocity, V_s is the *S*-wave velocity, *K* is the bulk modulus, μ is the shear modulus, ρ is the density, $\rho(r)$ is the density of the shell within radius *r, G* is gravitational constant, M_r is the mass of the Earth within radius *r*, and $\Phi = V_p^2 - (4/3)V_s^2$) provides a density profile for the core that, in turn, is perturbed to be consistent with free oscillation frequencies (Dziewonski and Anderson, 1981). Combining seismological data with mineral physics data (e.g., equation of state (EOS) data for materials at core appropriate conditions) from laboratory studies gives us the necessary constraints for identifying the mineralogical and chemical constituents of the core and mantle.

Birch (1952) compared seismically determined density estimates for the mantle and core with the available EOS data for candidate materials. He argued that the inner core was "a crystalline phase, mainly iron" and the liquid outer core is perhaps some 10–20% less dense than that expected for iron or iron–nickel at core conditions. Later, Birch (1964) showed that the Earth's outer core is ~10% less dense than that expected for iron at the appropriate pressures and temperatures and proposed that it contained (in addition to liquid iron and nickel) a lighter alloying element or elements such as carbon, or hydrogen (Birch, 1952) or sulfur, silicon, or oxygen (Birch, 1964).

Uncertainties in estimates of the composition of the Earth's core derive from uncertainties in the core density (or bulk modulus, or bulk sound velocity) and in that of candidate materials (including pure liquid iron) when calculated for the temperatures and pressures of the outer core. Although there is an excellent agreement between the static compression data for ε-Fe (and Fe–Ni mixtures) at core pressures (Mao *et al.*, 1990) and isothermal-based Hugoniot data for ε-Fe (Brown and McQueen, 1986), extrapolation of these data to core conditions requires knowledge of the thermal contribution to their EOS. Boehler (2000) calculated an outer core density deficit of ~9% using these data coupled with an assumed value for the pressure dependence of α (thermal expansion coefficient) and outer core temperatures of 4,000–4,900 K. In a review of these and other data, Anderson and Isaak (2002) concluded that the core density deficit is ~5% (with a range from 3% to 7%, given uncertainties) and argued that the density deficit is not as high as the often-cited ~10%. Their revised estimate is derived from a re-examination of EOS calculations with revised pressure and temperature derivatives for core materials at inner–outer core boundary conditions over a range of temperatures (4,800–7,500 K). This is a topic of much debate and a conservative estimate of the core density deficit is ~5–10%.

The solid inner core, which has a radius of 1,220 ± 10 km (Masters and Shearer, 1995), represents 5% of the core's mass and <5% of its volume. It is estimated to have a slightly lower density than solid iron and, thus, it too would have a small amount of a light element component (Jephcoat and Olson, 1987). Birch (1952) may have recognized this when he said that it is "a crystalline phase, mainly iron." Like the outer core, uncertainties in the amount of this light element component is a function of seismically derived density models for the inner core and identifying the appropriate temperature and pressure derivatives for the EOS of candidate materials. Hemley and Mao (2001) have provided an estimate of the density deficit of the inner core of 4–5%.

The presence of an iron core in the Earth is also reflected in the Earth's shape. The shape of the Earth is a function of its spin, mass distribution, and rotational flattening such that there is an equatorial bulge and flattening at the poles. The coefficient of the moment of inertia for the Earth is an expression that describes the distribution of mass within the planet with respect to its rotational axis. If the Earth was a compositionally homogenous planet having no density stratification, its coefficient of the moment of inertia would be $0.4Ma^2$, with *M* as the mass of the Earth and *a* as the equatorial radius. The equatorial bulge, combined with the precession of the equinoxes, fixes the coefficient of the moment of inertia for the Earth at $0.330Ma^2$ (Yoder, 1995) reflecting a marked concentration of mass at its center (see also Table 1).

Finally, studies of planets and their satellites show that internally generated magnetic fields do not require the existence of a metallic core, particularly given the diverse nature of planetary magnetic fields in the solar system (Stevenson, 2003). Alternatively, the 500+ years of global mapping of the Earth's magnetic field in time and space demonstrates the existence of the Earth's central magnetic core (Bloxham, 1995; Merrill *et al.*, 1996). The generation of this field in the core also requires the convection of a significant volume of iron (or similar electrically conducting material) as it creates a self-exciting dynamo (Buffett, 2000). In the Earth, as with the other terrestrial planets, iron is the most abundant element, by mass (Wänke and Dreibus, 1988). Its high solar abundance is the result of a highly stable nuclear configuration and processes of nucleosynthesis in stars.

4.3 CONSTRAINING THE COMPOSITION OF THE EARTH'S CORE

The major "core" issues in geochemistry include: (i) its composition (both inner and outer core), (ii) the nature and distribution of the light element, (iii) whether there are radioactive elements in the core, (iv) timing of core formation, and (v) what evidence exists for core–mantle exchange. The answers to some or all of these questions provide constraints on the conditions (e.g., P, T, f_{O_2}) under which the core formed.

4.3.1 Observations from Meteorites and Cosmochemistry

That the core is not solely an Fe–Ni alloy, but contains ~5–10% of a light mass element alloy, is about the extent of the compositional guidance that comes from geophysics. Less direct information on the makeup of the Earth is provided by studies of meteorites and samples of the silicate Earth. It is from these investigations that we develop models for the composition of the bulk Earth and primitive mantle (or the silicate Earth) and from these deduce the composition of the core.

The compositions of the planets in the solar system and those of chondritic meteorites provide a guide to the bulk Earth composition. However, the rich compositional diversity of these bodies presents a problem insofar as there is no single meteorite composition that can be used to characterize the Earth. The solar system is compositionally zoned; planets with lesser concentrations of volatile elements are closer to the Sun. Thus, as compared to Mercury and Jupiter, the Earth has an intermediate uncompressed density (roughly a proportional measure of metal to rock) and volatile element inventory, and is more depleted in volatile elements than CI-chondrites, the most primitive of all of the meteorites.

There is a wide range of meteorite types, which are readily divided into three main groups: the irons, the stony irons and the stones. With this simple classification, we obtain our first insights into planetary differentiation. All stony irons and irons are differentiated meteorites. Most stony meteorites are chondrites, undifferentiated meteorites, although lesser amounts are achondrites, differentiated stony meteorites. The achondrites make up ~4% of all meteorites, and <5% of the stony meteorites. A planetary bulk composition is analogous to that of a chondrite, and the differentiated portions of a planet—the core, mantle, and crust—have compositional analogues in the irons, stony irons (for core–mantle boundary regions), and achondrites (for mantle and crust).

Among the chondrites there are three main classes: the carbonaceous, enstatite, and ordinary chondrites. One simple way of thinking about these three classes is in terms of their relative redox characteristics. First, the carbonaceous chondrites, some of which are rich in organic carbon, have more matrix and Ca–Al inclusions and are the most oxidized of the chondrites, with iron existing as an FeO component in silicates. Second, the enstatite chondrites are the most reduced, with most varieties containing native metals, especially iron. Finally, the ordinary chondrites, the most abundant meteorite type, have an intermediate oxidation state (see review chapters in Volume 1 and Palme (2001)). Due to chemical and isotopic similarities, some researchers have argued that the bulk Earth is analogous to enstatite chondrites (Javoy, 1995). In contrast, others believe that the formation of the Earth initially began from materials such as the enstatite chondrites with the later 20–40% of the planet's mass forming from more oxidized accreting materials like the carbonaceous chondrites (Wänke, 1987; Wänke and Dreibus, 1988). As of early 2000s, we do not have sufficient data to resolve this issue and at best we should treat the chondrites and all meteoritic materials as only a guide to understanding the Earth's composition.

A subclass of the carbonaceous chondrites that uniquely stands out among all others is the CI (or C1) carbonaceous chondrite. These chondrites possess the highest proportional abundances of the highly volatile and moderately volatile elements, are chondrule free, and they possess compositions that match that of the solar photosphere when compared on a silicon-based scale. The photosphere is the top of the Sun's outer convection zone, which can be thought of as an analogue to the Sun's surface. The Sun's photospheric layer emits visible light and hence its composition can be measured spectroscopically. This, plus the fact that the Sun contains >99.9% of the solar system's mass, makes the compositional match with CI carbonaceous chondrites seem all that more significant.

For this review the Earth's composition will be considered to be more similar to carbonaceous chondrites and somewhat less like the high-iron end-members of the ordinary or enstatite chondrites, especially with regard to the most abundant elements (iron, oxygen, silicon, and magnesium) and their ratios. However, before reaching any firm conclusions about this assumption, we need to develop a compositional model for the Earth that can be compared with different chondritic compositions. To do this we need to: (i) classify the elements in terms of their properties in the nebula and the Earth and (2) establish the absolute abundances of the refractory and volatile elements in the mantle and bulk Earth.

4.3.2 Classification of the Elements

Elements can be classified according to their volatility in the solar nebular at a specific partial pressure (Larimer, 1988). This classification scheme identifies the major components (e.g., magnesium, iron, silicon, and nickel), which are intermediate between refractory and volatile, and then assigns the other, less abundant elements to groups based on volatility distinguishing refractory (condensation temperatures >1,250 K), moderately volatile (condensation temperatures <1,250 K and >600 K), and highly volatile (condensation temperatures <600 K) elements, depending on their sequence of condensation into mineral phases (metals, oxides, and silicates) from a cooling gas of solar composition (Larimer, 1988). In terms of accretionary models for chondrites and planetary bodies, it is often observed that a model assuming a 10^{-4} atm partial pressure best fits the available data (Larimer, 1988). Those with the highest condensation temperatures (>1,400 K) are the refractory elements (e.g., calcium, aluminum, titanium, zirconium, REE, molybdenum, and tungsten), which occur in all chondrites with similar relative abundances (i.e., chondritic ratios of Ca/Al, Al/Ti, Ti/Zr). Major component elements (aside from oxygen and the gases) are the most abundant elements in the solar system, including silicon, magnesium, and iron (as well as cobalt and nickel); these elements have condensation temperatures of ~1,250 K. Moderately volatile elements (e.g., chromium, lithium, sodium, potassium, rubidium, manganese, phosphorus, iron, tin, and zinc) have condensation temperatures of ~1,250–600 K (Palme *et al.*, 1988), whereas highly volatile elements (e.g., thallium, cadmium, bismuth, and lead) have condensation temperatures ~600–400 K. Below this temperature the gas-phase elements (carbon, hydrogen, and nitrogen) condense. The relative abundance ratios of the major components, moderately volatile and highly volatile elements all vary considerably between the different types of chondritic meteorites. Figure 3 illustrates the differing proportions of the major component elements in chondrite groups and the Earth, which together with oxygen make up some 90% of the material in the Earth and other terrestrial planets.

Elements can also be classified according to their chemical behavior based on empirical observations from meteorites and systems in the Earth; this leads to the following groups: lithophile, siderophile, chalcophile, or atmophile. The lithophile elements are ones that bond readily with oxygen and are concentrated in the silicate shell (crust and mantle) of the Earth. The siderophile elements readily bond with iron and are concentrated in the core. The chalcophile elements bond readily with sulfur and are distributed between the core and mantle, with a greater percentage of them likely to be in the core. Finally, the atmophile elements (e.g., hydrogen, carbon, nitrogen, oxygen, and noble gases) are gaseous and are concentrated in the atmosphere–hydrosphere system surrounding the planet. A combination of these two different classification schemes provides a better understanding of the relative behavior of the elements, particularly during accretion and large-scale planetary differentiation.

Developing a model for the composition of the Earth and its major reservoirs can be established in a four-step process. The first involves estimating the composition of the silicate Earth (or primitive mantle, which includes the crust plus mantle after core formation). The second step involves defining a volatility curve for the planet, based on the abundances of the moderately volatile and highly volatile lithophile elements in the silicate Earth, assuming that none have been sequestered into the core (i.e., they are truly lithophile). The third step entails calculating a bulk Earth composition using the planetary volatility curve established in step two, chemical data for chondrites, and the first-order

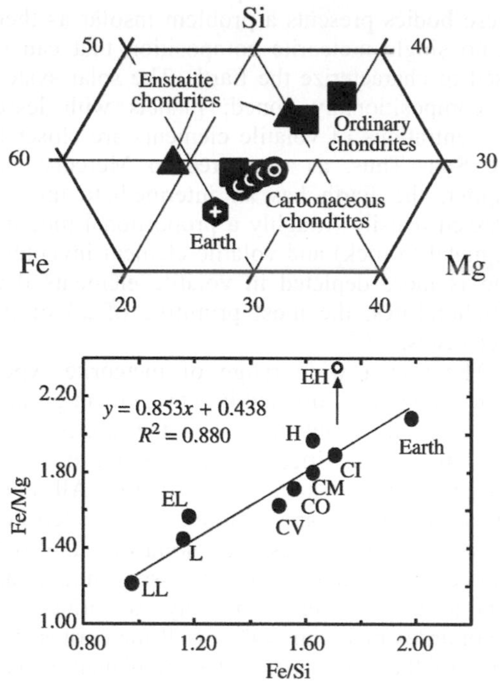

Figure 3 A ternary plot (upper) and binary ratio plot (lower) of the differing proportions (in wt.%) of Si, Fe, and Mg (three out of the four major elements) in chondrites and the Earth. These elements, together with oxygen, constitute >90% by mass of chondrites, the Earth, and other terrestrial planets. Data for the chondrites are from Wasson and Kellemeyn (1988) and for the Earth are from Table 2. The regression line is derived using only chondrites and does not include the EH data.

features of the planets in the solar system. Finally, a core composition is extracted by subtracting the mantle composition from the bulk planetary composition, revealing the abundances of the siderophile and chalcophile elements in the core. Steps three and four are transposable with different assumptions, with the base-level constraints being the compositions of meteorites and the silicate Earth and the solar system's overall trend in the volatile element abundances of planets outward from the Sun.

4.3.3 Compositional Model of the Primitive Mantle and the Bulk Earth

The silicate Earth describes the solid Earth minus the core. There is considerable agreement about the major, minor, and trace element abundances in the primitive mantle (Allegre *et al.*, 1995; McDonough and Sun, 1995). The relative abundances of the lithophile elements (e.g., calcium, aluminum, titanium, REE, lithium, sodium, rubidium, boron, fluorine, zinc, etc.) in the primitive mantle establish both the absolute abundances of refractory elements in the Earth and the planetary signature of the volatile element depletion pattern (Figure 4). The details of how these compositional models are developed can be found in Allegre *et al.* (1995); McDonough and Sun (1995), and Palme and O'Neill. A model composition for the silicate Earth is given in Table 2, which is adapted from McDonough (2001); Palme and O'Neill present a similar model.

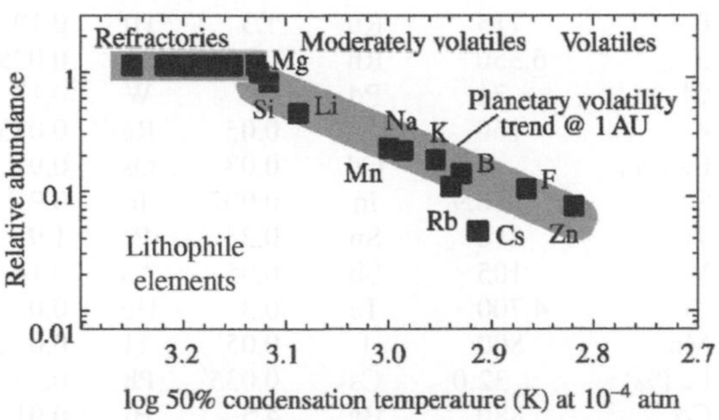

Figure 4 The relative abundances of the lithophile elements in the primitive mantle (or silicate Earth) plotted versus the log of the 50% condensation temperature (K) at 10^{-4} atm pressure. The relative abundances of the lithophile elements are reported as normalized to CI carbonaceous chondrite on an equal basis of Mg content. The planetary volatility trend (negative sloping shaded region enclosing the lower temperature elements) establishes integrated flux of volatile elements at 1 AU. Data for condensation temperatures are from Wasson (1985); chemical data for the chondrites are from Wasson and Kellemeyn (1988) and for the Earth are from Table 2.

A first-order assumption is that lithophile elements, inclusive of the refractory, moderately volatile, and highly volatile ones, are excluded from the core. The moderately volatile and highly volatile lithophiles are depleted relative to those in CI-chondrites. Together, the lithophiles describe a coherent depletion or volatility pattern. This negative correlation (Figure 4) thus establishes the planetary volatile curve at ~1 AU, which is an integrated signature of accreted nebular material in the coalescing region of the proto-Earth. By comparison, Mars has a less depleted abundance pattern (Wänke, 1981), whereas Mercury has a more depleted abundance pattern (BVSP, 1981). The most significant feature of this pattern is that potassium follows all of the other moderately volatile and highly volatile lithophiles. This observation demonstrates that the potassium budget of the silicate Earth is sufficient to describe that in the planet and argue against any sequestration of potassium into the core.

Data for the content of lithophile elements in the Earth plus knowledge of the iron content of the mantle and core together establish a bulk Earth compositional model (McDonough, 2001). This model assumes chondritic proportions of Fe/Ni in the Earth, given limited Fe/Ni variation in chondritic meteorites (see below). This approach yields an Fe/Al of 20 ± 2 for

Table 2 The composition of the silicate Earth.

H	100	Zn	55	Pr	0.25
Li	1.6	Ga	4	Nd	1.25
Be	0.07	Ge	1.1	Sm	0.41
B	0.3	As	0.05	Eu	0.15
C	120	Se	0.075	Gd	0.54
N	2	Br	0.05	Tb	0.10
O (%)	44	Rb	0.6	Dy	0.67
F	15	Sr	20	Ho	0.15
Na (%)	0.27	Y	4.3	Er	0.44
Mg (%)	22.8	Zr	10.5	Tm	0.068
Al (%)	2.35	Nb	0.66	Yb	0.44
Si (%)	21	Mo	0.05	Lu	0.068
P	90	Ru	0.005	Hf	0.28
S	250	Rh	0.001	Ta	0.037
Cl	17	Pd	0.004	W	0.029
K	240	Ag	0.008	Re	0.0003
Ca (%)	2.53	Cd	0.04	Os	0.003
Sc	16	In	0.01	Ir	0.003
Ti	1,200	Sn	0.13	Pt	0.007
V	82	Sb	0.006	Au	0.001
Cr	2,625	Te	0.012	Hg	0.01
Mn	1,045	I	0.01	Tl	0.004
Fe (%)	6.26	Cs	0.021	Pb	0.15
Co	105	Ba	6.6	Bi	0.003
Ni	1,960	La	0.65	Th	0.08
Cu	30	Ce	1.68	U	0.02

Concentrations are given in $\mu g\, g^{-1}$ (ppm), unless stated as "%," which are given in wt.%.

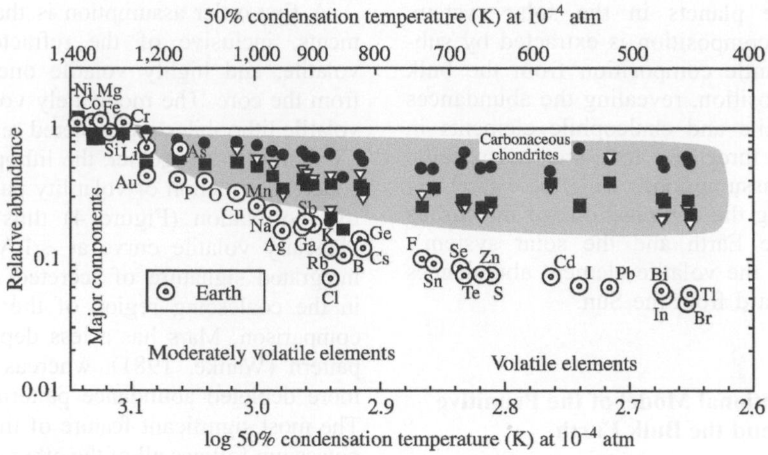

Figure 5 The relative abundances of the elements in the Earth and various carbonaceous chondrites plotted versus the log of the 50% condensation temperature (K) at 10^{-4} atm pressure. Data are normalized to CI carbonaceous chondrite on an equal basis of Mg content. The overall volatility trend for the Earth is comparable to that seen in these chondrites. The carbonaceous chondrites include CM (filled circles), CV (filled squares), and CO (open triangles) and define the shaded region. Data for condensation temperatures are from Wasson (1985); chemical data for the chondrites are from Wasson and Kellemeyn (1988) and for the Earth are from Table 3.

the Earth. Aluminum, a refractory lithophile element, is considered the least likely of the lithophile elements (e.g., silicon, magnesium, and calcium) to be incorporated in the core. Thus, an aluminum content for the mantle translates directly into the aluminum content for the bulk Earth. This tightly constrained Fe/Al value also provides a first-order compositional estimate of the planet that requires no knowledge of light elements in the core.

Chondritic meteorites display a range of Fe/Al ratios, with many having a value close to 20 (Allegre *et al.*, 1995), although high Fe/Al values (35) are found in the iron-rich (EH) enstatite chondrites (Wasson and Kallemeyn, 1988). Combining these data and extending the depletion pattern for the abundances of nonrefractory, nonlithophile elements provides a model composition for the bulk Earth (Figure 5). A model composition for the bulk Earth is given in Table 3, which is adapted from McDonough (2001); Palme and O'Neill present a similar model. In terms of major elements this Earth model is iron and magnesium rich and coincident with the Fe/Mg–Fe/Si compositional trend established by chondrites (Figure 3). The Earth's volatility trend is comparable, albeit more depleted, than that of other carbonaceous chondrites (data in gray field in Figure 5).

Table 3 The composition of the bulk Earth.

H	260	Zn	40	Pr	0.17
Li	1.1	Ga	3	Nd	0.84
Be	0.05	Ge	7	Sm	0.27
B	0.2	As	1.7	Eu	0.10
C	730	Se	2.7	Gd	0.37
N	25	Br	0.3	Tb	0.067
O (%)	29.7	Rb	0.4	Dy	0.46
F	10	Sr	13	Ho	0.10
Na (%)	0.18	Y	2.9	Er	0.30
Mg (%)	15.4	Zr	7.1	Tm	0.046
Al (%)	1.59	Nb	0.44	Yb	0.30
Si (%)	16.1	Mo	1.7	Lu	0.046
P	715	Ru	1.3	Hf	0.19
S	6,350	Rh	0.24	Ta	0.025
Cl	76	Pd	1	W	0.17
K	160	Ag	0.05	Re	0.075
Ca (%)	1.71	Cd	0.08	Os	0.9
Sc	10.9	In	0.007	Ir	0.9
Ti	810	Sn	0.25	Pt	1.9
V	105	Sb	0.05	Au	0.16
Cr	4,700	Te	0.3	Hg	0.02
Mn	800	I	0.05	Tl	0.012
Fe (%)	32.0	Cs	0.035	Pb	0.23
Co	880	Ba	4.5	Bi	0.01
Ni	18,200	La	0.44	Th	0.055
Cu	60	Ce	1.13	U	0.015

Concentrations are given in $\mu g\, g^{-1}$ (ppm), unless stated as "%," which are given in wt.%.

4.4 A COMPOSITIONAL MODEL FOR THE CORE

As stated earlier, the Earth's core is dominantly composed of a metallic Fe–Ni mixture. This fact is well established by seismic data (*P*-wave velocity, bulk modulus, and density), geodynamo observations (the need for it to be reasonably good electrical conductor), and cosmochemical constraints. This then requires that the core, an iron- and nickel-rich reservoir, chemically balances the silicate Earth to make up a primitive, chondritic planet. Many iron meteorites, which are mixtures of iron and nickel in various proportions, are pieces of former asteroidal cores. These meteorites provide insights into the compositions of smaller body cores, given they are products of low-pressure differentiation, whereas the Earth's core likely formed under markedly different conditions. Thus, the Earth's core superficially resembles an iron meteorite; however, such comparisons are only first-order matches and in detail we should anticipate significant differences given contrasting processes involved in their formation.

4.4.1 Major and Minor Elements

A compositional model for the primitive mantle and bulk Earth is described above, which indirectly prescribes a core composition, although it does not identify the proportion of siderophile and chalcophile elements in the core and mantle. The mantle abundance pattern for the lithophile elements shown in Figure 4 provides a reference state for reviewing the abundances of the siderophile and chalcophile elements in the silicate Earth, which are shown in Figure 6. All of the siderophile (except gallium) and chalcophile elements plot below the shaded band that defines the abundance pattern for the lithophile elements. That these nonlithophile elements fall below this band (i.e., the planetary volatility trend) indicates that they are depleted in the mantle, and therefore the remaining planetary complement of these elements are in the core. The relative effects of core subtraction are illustrated in both panels with light-gray arrows, extending downward from the planetary volatility trend. The displacement length below the volatility trend (or length of the downward-pointing arrow) reflects the element's bulk distribution coefficient between core and mantle (e.g., bulk $D^{\mathrm{metal/silicate}}$ for $\mathrm{Mo} > \mathrm{P} \approx \mathrm{Sb}$).

By combining the information derived from Figures 4–6, one can construct a compositional model for the Earth's core (Table 4), which is adapted from McDonough (1999). A first-order comparison of the composition of the bulk Earth, silicate Earth, and core in terms of weight percent and atomic proportion is presented in Table 5. The

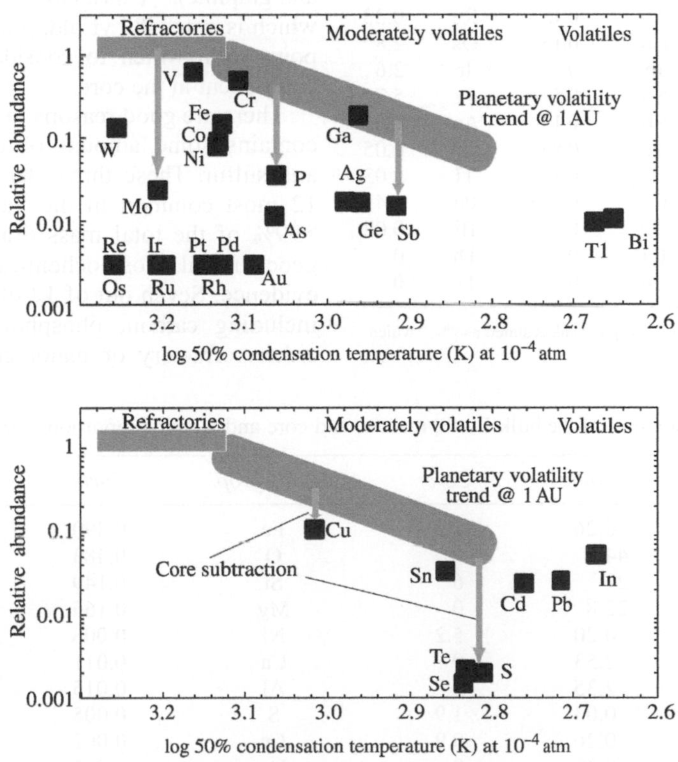

Figure 6 The relative abundances of the siderophile elements (upper panel) and chalcophile elements (lower panel) in the primitive mantle (or silicate Earth) plotted versus the log of the 50% condensation temperature (K) at 10^{-4} atm pressure. Data are normalized to CI carbonaceous chondrite on an equal basis of Mg content. The gray shaded region illustrates the relative abundances of the lithophile elements as reported in Figure 4. The light gray, downward pointing arrows reflect the element's bulk distribution coefficient between core and mantle during core formation; the longer the length of the arrow, the greater the bulk D (data sources are as in Figure 4).

compositional model for the core has a light element composition that seeks to fit the density requirements for the outer core and is consistent with cosmochemical constraints. Significantly, along with iron and nickel the core contains most of the planet's sulfur, phosphorus, and carbon budget. Finally, this model composition is notable in that it is devoid of radioactive elements. The discussion that follows reviews the issues associated with compositional models for the core.

Table 4 The composition of the Earth's core.

H	600	Zn	0	Pr	0
Li	0	Ga	0	Nd	0
Be	0	Ge	20	Sm	0
B	0	As	5	Eu	0
C (%)	0.20	Se	8	Gd	0
N	75	Br	0.7	Tb	0
O (%)	0	Rb	0	Dy	0
F	0	Sr	0	Ho	0
Na (%)	0	Y	0	Er	0
Mg (%)	0	Zr	0	Tm	0
Al (%)	0	Nb	0	Yb	0
Si (%)	6.0	Mo	5	Lu	0
P (%)	0.20	Ru	4	Hf	0
S (%)	1.90	Rh	0.74	Ta	0
Cl	200	Pd	3.1	W	0.47
K	0	Ag	0.15	Re	0.23
Ca (%)	0	Cd	0.15	Os	2.8
Sc	0	In	0	Ir	2.6
Ti	0	Sn	0.5	Pt	5.7
V	150	Sb	0.13	Au	0.5
Cr (%)	0.90	Te	0.85	Hg	0.05
Mn	300	I	0.13	Tl	0.03
Fe (%)	85.5	Cs	0.065	Pb	0.4
Co	0.25	Ba	0	Bi	0.03
Ni (%)	5.20	La	0	Th	0
Cu	125	Ce	0	U	0

Concentrations are given in $\mu g\, g^{-1}$ (ppm), unless stated as "%," which are given in wt.%.

4.4.2 The Light Element in the Core

Given constraints of an outer core density deficit of 5–10% and a host of candidate elements (e.g., hydrogen, carbon, oxygen, silicon, and sulfur), we need to evaluate the relative potential of these elements to explain core density deficit. Uniformly, the bolstering of one's view for these components in the core involve metallurgical or cosmochemical arguments, coupled with the identification of candidate minerals found in meteorites, particularly iron meteorites and reduced chondrites (the classic example being the high-iron (EH) enstatite chondrite).

Washington (1925), of the Carnegie Institution of Washington, developed a model for the chemical composition of the Earth based on the Wiechert structural model, the Oldham–Gutenburg revised core radius, and the newly derived Adams–Williamson relationship (Williamson and Adams, 1923) for determining the density profile of the planet. Washington's model for the core assumed an average density for the core of $\sim$10 g cm^{-3} (cf. $\sim$11.5 g cm^{-3} for today's models), and a "considerable amount, up to $\sim$5% or so, of phosphides (schreibersite, $(Fe,Ni)_3P$), carbides (cohenite, Fe_3C), sulfides (troilite, FeS) and carbon (diamond and graphite)." This amazing and insightful model, which is now $\sim$80 yr old, provides us with a good point from which to consider the light element component in the core.

There are good reasons to assume that the core contains some amount of carbon, phosphorus, and sulfur. These three elements are among the 12 most common in the Earth that account for >99% of the total mass (Table 5), as based on geochemical, cosmochemical, and meteoritical evidence. Seven out of 12 of these elements (not including carbon, phosphorus, and sulfur) are either refractory or major component elements,

Table 5 The composition of the bulk Earth, mantle, and core and atomic proportions for abundant elements.

wt.%	*Earth*	*Mantle*	*Core*	*Atomic prop.*	*Earth*	*Mantle*	*Core*
Fe	32.0	6.26	85.5	Fe	0.490	0.024	0.768
O	29.7	44	0	O	0.483	0.581	0.000
Si	16.1	21	6	Si	0.149	0.158	0.107
Mg	15.4	22.8	0	Mg	0.165	0.198	0.000
Ni	1.82	0.20	5.2	Ni	0.008	0.001	0.044
Ca	1.71	2.53	0	Ca	0.011	0.013	0.000
Al	1.59	2.35	0	Al	0.015	0.018	0.000
S	0.64	0.03	1.9	S	0.005	0.000	0.030
Cr	0.47	0.26	0.9	Cr	0.002	0.001	0.009
Na	0.18	0.27	0	Na	0.002	0.002	0.000
P	0.07	0.009	0.20	P	0.001	0.000	0.003
Mn	0.08	0.10	0.03	Mn	0.000	0.000	0.000
C	0.07	0.01	0.20	C	0.002	0.000	0.008
H	0.03	0.01	0.06	H	0.007	0.002	0.030
Total	99.88	99.83	99.97	Total	1.000	1.000	1.000

and so their abundances in the Earth are relatively fixed for all planetary models (see also Figure 2). The remaining five elements are sodium, chromium, carbon, phosphorus, and sulfur (Table 5); all of these are highly volatile to moderately volatile and estimates of their abundances in the bulk Earth and core are established from cosmochemical constraints. A significant question concerning the abundance of carbon, phosphorus, and sulfur in the core, however, is whether their incorporation into the core can account for the density discrepancy?

The planetary volatility trend illustrated in Figures 4 and 5 does not extend out to the lowest temperature components, including the ices and gases (e.g., hydrogen, carbon, nitrogen, oxygen, and the noble gases). Estimates for the Earth's content of these components (Figure 7) are from McDonough and Sun (1995) and McDonough (1999, 2001) and are based on data for the Earth's mantle and a comparison of carbonaceous chondrite data. Figure 7 provides a comparison of the Earth's estimate of these elements relative to the data for chondrites; the estimate for the Earth comes from an extrapolation of the trend shown in Figure 5. Although these extrapolations can only provide an approximate estimate, the abundance of carbon in the Earth is suggested to be of the order <0.1 wt.%. This estimate translates to a core having only ~0.2 wt.% carbon (Tables 3 and 4). By comparison Wood (1993) estimated a factor of 10–20 times more carbon in the core. Wood's estimate seems most unlikely insofar as it is inconsistent with data for meteorites, which are not markedly enriched in highly volatile elements (Figure 7). This view is untenable when compared with data

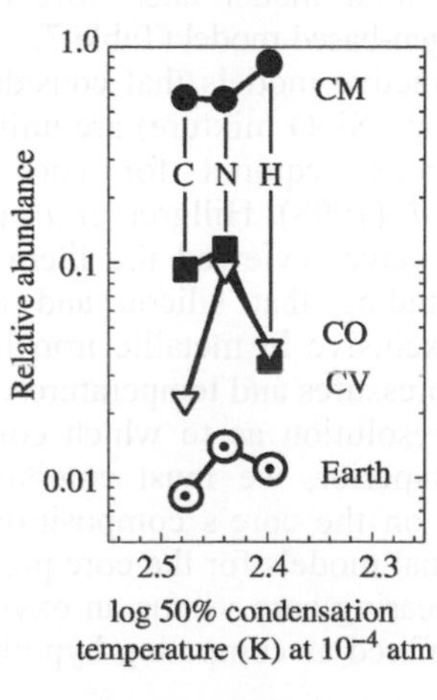

Figure 7 The relative abundances of C, N, and H in the Earth plotted versus the log of the 50% condensation temperature (K) at 10^{-4} atm pressure. The Earth's estimate is based on compositional estimates of these gases in the mantle and the Earth's surface, as well as by comparison with data for carbonaceous chondrites (data sources are as in Figure 5).

trends in Figure 5 for the Earth and the carbonaceous chondrites. It is also noted that the Earth's budget for hydrogen and nitrogen are such that the core would likely contain a minor amount of these elements. The consequences of having hydrogen in the core are significant and have been reviewed by Williams and Hemley (2001).

There is ~90 ppm of phosphorus in the silicate Earth (McDonough *et al.*, 1985), and the bulk Earth is estimated to have ~0.1 wt.% phosphorus. Using the relationships in Figure 6 the core is thus estimated to have ~0.20 wt.% phosphorus (Table 4). Thus, 90% of the planet's inventory of phosphorus is in the core (Table 6) and the core's metal/silicate phosphorus enrichment factor is ~22. Similarly, the core hosts ~90% of the planet's carbon budget, and has a metal/silicate enrichment factor only slightly lower at ~17.

The sulfur content of the core is said to be ~1.5–2 wt.% (McDonough and Sun, 1995; Dreibus and Palme, 1996). This number is based on calculating the degree of sulfur depletion in the silicate Earth relative to the volatility trend (Figure 6). Figure 8 illustrates the problem with suggesting that the core contains 10% sulfur, which is commonly invoked as the light element required to compensate for the density deficit in the outer core. Accordingly, the total sulfur, carbon, and phosphorus content of the core constitute only a minor fraction (~2.5 wt%) and this mixture of light elements cannot account for the core's density discrepancy. Thus, it is likely that there is another, more abundant, light element in the core in addition to these other components.

A model core composition has been constructed using silicon as the other light element in the outer core, which is also consistent with

Table 6 The metal/silicate enrichment factor and the proportion of element in the core relative to the planet.

Elements	*Metal/silicate enrichment factor*	*% of planetary inventory in the core*
Re, PGE	>800	98
Au	~500	98
S, Se, Te, Mo, As	~100	96
N	~40	97
Ni, Co, Sb, P	~25	93
Ag, Ge, C, W	~17	91
Fe (%)	~14	87
Cl, Br, and I	10–15	85
Bi and Tl	~10	80
H and Hg	~6	70
Cu, Sn, Cd, Cr	3–4	60–65
Cs and Pb	~3	55–60
V	~2	50
Si and Mn	0.3	~10

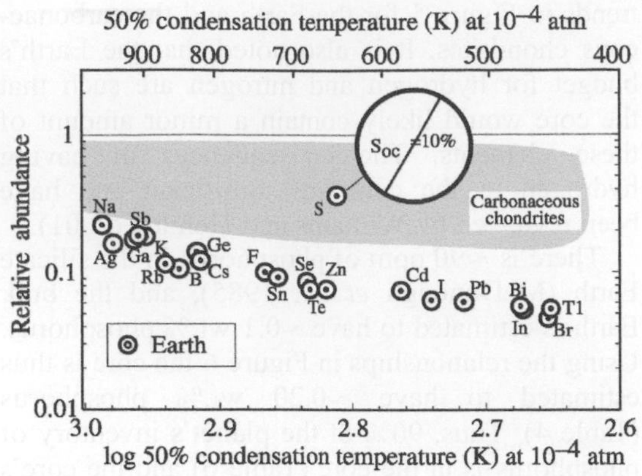

Figure 8 An illustration showing where S would plot if the core contained 10 wt.% sulfur so to account for the core's density discrepancy (see text for further discussion). The relative abundances of the elements in the Earth are plotted versus the log of the 50% condensation temperature (K) at 10^{-4} atm pressure. Data are normalized to CI carbonaceous chondrite on an equal basis of Mg content. The overall volatility trend for the Earth is comparable to that seen in these chondrites. The shaded region for the carbonaceous chondrites is the same as in Figure 5 (data sources are as in Figure 5).

evidence for core formation at high pressures (e.g., 20–30 GPa). This model is at best tentative, although comparisons of Mg/Si and Fe/Si in the Earth and chondrites (Figure 2) show that it is permissible. Silicon is known to have siderophilic behavior under highly reducing conditions and is found as a metal in some enstatite chondrites. A number of earlier models have suggested silicon as the dominant light element in core (Macdonald and Knopoff, 1958; Ringwood, 1959; Wänke, 1987; O'Neill, 1991b; Allegre *et al.*, 1995; O'Neill and Palme, 1997). The estimate for silicon in the core is based on the volatility curve for lithophile elements in the Earth (Figure 4).

An alternative case can be made for oxygen as the predominant light element in the core. On the grounds of availability, oxygen is a good candidate; it is the second most abundant element in the Earth and only a few percent might be needed to account for the core's density discrepancy. However, O'Neill *et al.* (1998) point out that oxygen solubility in iron liquids increases with temperature but decreases with pressure and thus showed that only ~2% or less oxygen could be dissolved into a core forming melt. The planetary volatility trend provides no guidance to the core's oxygen abundance. A plot of the log 50% condensation temperature versus element abundance (i.e., Figure 5) does not consider oxygen, because its 50% condensation temperature is not considered in systems where it is the dominant element in rocks and water ice. The core and Earth

Table 7 Compositional comparison of two models for the Earth and core.

wt.%	*Si-bearing*		*O-bearing*	
	Earth	*Core*	*Earth*	*Core*
Fe	32.0	85.5	32.9	88.3
O	29.7	0	30.7	3
Si	16.1	6	14.2	0
Ni	1.82	5.2	1.87	5.4
S	0.64	1.9	0.64	1.9
Cr	0.47	0.9	0.47	0.9
P	0.07	0.20	0.07	0.20
C	0.07	0.20	0.07	0.20
H	0.03	0.06	0.03	0.06
Mean atomic #		23.5		23.2
Atomic proportions				
Fe		0.768		0.783
O		0.000		0.093
Si		0.107		0.000
Ni		0.044		0.045
S		0.030		0.029
Cr		0.009		0.009
P		0.003		0.003
C		0.008		0.008
H		0.030		0.029
Total		1.000		1.000

model composition, assuming oxygen as the light element in the core, is presented in Table 7, along with that for the silicon-based model. Both model compositions attempt to fit the density requirements for the outer core by assuming a mean atomic number of ~23, following Birch (1966). In terms of the light-element-alloy component in the core, this results in ~9% (by weight) for the silicon-based model and ~6% (by weight) for the oxygen-based model (Table 7).

Less attractive models that consider complex mixtures (e.g., Si–O mixture) are unlikely, given the conditions required for core formation. O'Neill *et al.* (1998), Hillgren *et al.* (2000), and Li and Fei have reviewed the literature on the topic concluding that silicon and oxygen are mutually exclusive in metallic iron liquids over a range of pressures and temperatures. Until there is a clear resolution as to which compositional model is superior, we must entertain multiple hypotheses on the core's composition. The two compositional models for the core presented here (a silicon-bearing core versus an oxygen-bearing core) are offered as competing hypotheses.

4.4.3 Trace Elements in the Core

The abundance of trace siderophile elements in the bulk Earth (and that for the core) may be constrained by examining their abundance ratios

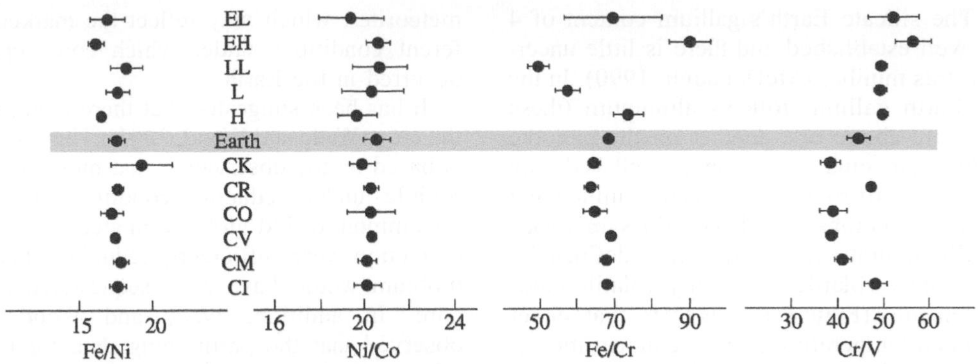

Figure 9 A plot of the variation in Fe/Ni, Ni/Co, Fe/Cr, and Cr/V values in various chondrites and the Earth. The different groups of carbonaceous chondrites include CI, CM, CV, CO, CR, and CK; the ordinary chondrites include H-, L-, and LL-types, and the enstatite chondrites include EH- and EL-types. The error bars represent the 1 SD of the data population. The Earth's composition is shown in the shaded bar and data are from Table 3 (including data are from various papers of Wasson and Kellemeyn cited in Wasson and Kellemeyn (1988)).

in chondrites. Figure 9 presents data for various groups of chondrites, which show limited variation for Fe/Ni (17.5) and Ni/Co (20), and slightly more variation for Fe/Cr (67) and Cr/V (45). Using the iron content for the core and the silicate Earth abundances for iron, nickel, cobalt, chromium, and vanadium, the Earth's core composition is established by assuming chondritic ratios of the elements for the planet. Following similar lines of reasoning for siderophile and chalcophile elements, the trace element composition of the core is also determined (Tables 4 and 6).

Based on these results, the core appears to be rich in chromium and vanadium (i.e., 50–60% of the planet's budget for these elements, Table 6), with a minor amount of manganese in the core (~10% of the planet's budget). Discussions relating to incorporation of chromium into the core usually also involve that for manganese and vanadium, because the partitioning behavior of these three elements during core formation may have been similar (Ringwood, 1966; Dreibus and Wänke, 1979; Drake *et al.*, 1989; Ringwood *et al.*, 1990; O'Neill, 1991a; Gessmann and Rubie, 2000). However the model presented here does not take into account element partitioning behavior during core formation, it is solely based on the planetary volatility trend and a model composition for the silicate Earth.

The minor amount of manganese in the core reflects the volatility model assumed for this element. O'Neill and Palme (1997) argue that manganese and sodium have similar volatilities based on the limited variation in Mn/Na ratios in chondrites. However, a plot of Na/Ti versus Mn/Na in chondrites (Figure 10) shows that indeed Mn/Na varies as a function of volatility; this illustration monitors volatility by comparing titanium, a refractory lithophile element, with sodium, a moderately

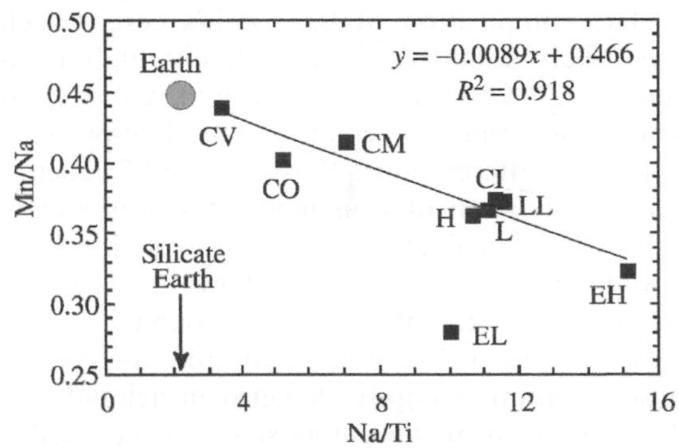

Figure 10 A plot of the variation in Na/Ti versus Mn/Na ratios in chondritic meterorites and the Earth. Data for chondrites are from Wasson and Kellemeyn (1988). The value for the Na/Ti ratio of the silicate Earth is indicated with an arrow (data from McDonough and Sun, 1995). The regression line, R^2 value, and the coefficients for the line equation are derived from the data for chondrites, not including the low-Fe enstatite chondrite. This regression and the Na/Ti ratio of the silicate Earth together provide a method to estimate the Mn/Na ratio for the Earth and indicate that the core is likely to contain a small fraction of the Earth's Mn budget.

lithophile element. Therefore, given the planetary volatility trend (Figure 4) and a reasonably well-constrained value for Na/Ti in the silicate Earth (McDonough and Sun, 1995), one estimates a planetary Mn/Na value of 0.45 for the Earth, implying that the core hosts ~10% of the planet's manganese budget.

The behavior of gallium, a widely recognized siderophile element, during core formation appears to be the most anomalous; this is most clearly illustrated by noting that gallium plots directly on the planetary volatility trend (Figure 6, top panel), indicating its undepleted character in the mantle. This result implies that there is little to no gallium in the Earth's core, which is a most unexpected result. The silicate Earth's gallium content of 4 ppm is well established and there is little uncertainty to this number (McDonough, 1990). In the silicate Earth gallium follows aluminum (these elements are above one another on the periodic table) during magma generation, as well as during the weathering of rocks, with overall limited and systematic variations in Al/Ga values in rocks. That gallium plots within the field defined by the moderately volatile and highly volatile lithophile elements (Figure 6) suggests that either the assumed temperature at 50% condensation is incorrect (unlikely given a wide spectrum of supporting meteorite data), or gallium behaves solely as a lithophile element during core formation. If the latter is true, then determining under what conditions gallium becomes wholly lithophile provides an important constraint on core formation.

The composition of the Earth's core, which was likely established at relatively high pressures (~20 GPa), can be compared with that of iron meteorites, which are low-pressure (<1 GPa) differentiates. Wasson's (1985) chemical classification of iron meteorites uses nickel, gallium, germanium, and iridium to divide them into 13 different groups. He shows that gallium is clearly a siderophile element found in abundance in the metal phases of iron meteorites. Also, gallium is highly depleted in achondrites. A comparison of the composition of the Earth's core with that of different iron meteorites is given in Figure 11. The Earth's core and some iron meteorites have comparable nickel, germanium, and iridium contents, albeit on the low end of the nickel spectrum. In contrast, the gallium content of the Earth's core (Figure 11) is substantially lower than that found in all iron meteorites, which may reflect the markedly different conditions under which core separation occurred in the Earth.

It has been suggested that there is niobium in the core (Wade and Wood, 2001). This suggestion is based on the observation that niobium is siderophile under reducing conditions (it is not uncommon to find niobium in steels) and if core extraction were sufficiently reducing, then some niobium would have been sequestered into the core. In addition, Wade and Wood (2001) observed that the partitioning data for niobium mimicked that for chromium and vanadium. Given the distribution of chromium and vanadium between the core and mantle, it is expected that a considerable portion of the Earth's niobium budget is hosted in the core. The Wade and Wood model was, in part, developed in response to the observations of McDonough (1991) and Rudnick *et al.* (2000), who reported that niobium and tantalum are depleted in the upper mantle and crust and that both reservoirs have low Nb/Ta values relative to chondrites. These observations lead to the suggestion that refractory components of subducting oceanic crust would contain the complementary niobium- and tantalum-enriched reservoir of the silicate Earth (McDonough, 1991; Rudnick *et al.*, 2000). However, Wade and Wood (2001) proposed an alternative model in which niobium, but not tantalum, is extracted into the core. To address this issue it is useful to examine the relative abundances of Nb–Ta–La in the crust–mantle system, because this triplet may characterize silicate Earth processes and reservoirs.

The range of Nb/Ta and La/Ta values in the continental crust and depleted mantle (MORB source) are given in Figure 12. This illustration

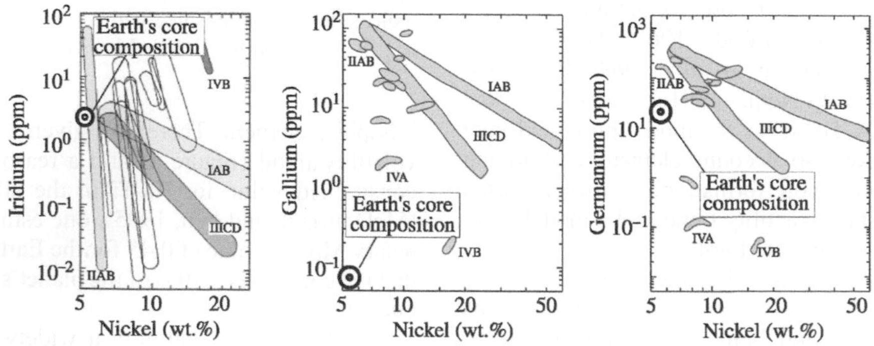

Figure 11 A plot of the variation in Ni versus Ir, Ni versus Ga, and Ni versus Ge in iron meteorites and the Earth. Data for the iron meteorites are adapted from the work of Wasson (1985). Data for the Earth's core are from Table 4. The plot of Ni–Ir shows that the composition of the Earth's core is comparable to that of various iron meteorites, whereas the Earth's core appears to have a slightly lower Ge content and a markedly lower Ga; the latter being unlike anything seen in iron meteorites. These four elements are the ones that are used to define the chemical classification of iron meteorites (reproduced by permission of W. H. Freeman from *Meteorites, Their Record of Early Solar-system History*, **1985**, p. 41, 42).

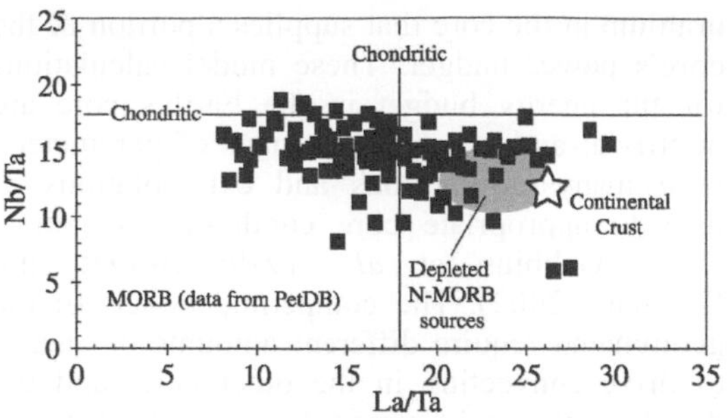

Figure 12 A plot of the Nb/Ta and La/Ta variation in MORB and the continental crust. The continental crust, MORBs, and their source regions all plot below the chondritic Nb/Ta value. Likewise, the continental crust plots and depleted MORB source regions are strongly depleted in Ta relative to La. See text for further details (data for MORB are from the PetDB resource on the web (http://petdb.ldeo.columbia.edu/petdb/); the estimate of the continental crust from Rudnick and Gao (Chapter 5).

shows that both of these major silicate reservoirs are clearly depleted in Nb/Ta relative to chondrites. In addition, both the continental crust and the depleted source regions of MORBs plot in the field that characterizes depletions of both niobium and tantalum relative to lanthanum, and tantalum relative to niobium. Niu and Batiza (1997) showed that during melting $D_{Nb} < D_{Ta} < D_{La}$ such that increasing melt extraction depletes the MORB source regions respectively and progressively in these elements so that they would plot in the same field as continental crust (Figure 12). This demonstrates that both the production of continental and oceanic crusts result in the production of a crustal component that, when processed through a subduction zone filter, generates residues with high Nb/Ta and low La/Ta that remain in the mantle. Is the core another niobium-enriched reservoir in the Earth? As of early 2000s, this is an unresolved issue, but crucial tests of this hypothesis will be gained by further examination of silicate Earth samples, iron meteorites, and further tests from experimental petrology.

Finally, Table 4 lists the halides—chlorine, bromium, and iodine—in the core. McDonough and Sun (1995) noted the marked depletion of these elements in the silicate Earth and suggested that this effect is due possibly to their incorporation into the Earth's core, or that the region of the nebula at 1 AU was anomalously depleted in the halides. There are iron halides, some of which are found in chondrites. However, such halides in chondrites are believed to be decompositional products created during terrestrial weathering (Rubin, 1997).

4.5 RADIOACTIVE ELEMENTS IN THE CORE

Those that have suggested the presence of radioactive elements in the Earth's core have usually done so in order to offer an alterative explanation for the energy needed to run the geodynamo, and/or as a way to explain Earth's volatile elements inventory. Potassium is commonly invoked as being sequestered into the Earth's core due to: (i) potassium sulfide found in some meteorites; (ii) effects of high-pressure *s–d* electronic transitions; and/or (iii) solubility of potassium in Fe–S (and Fe–S–O) liquids at high pressure. Each of these is considered below and rejected.

The cosmochemical argument for potassium in the core is based on the presence of a potassium iron sulfide (djerfisherite) and sodium chromium sulfide (carswellsilverite) in enstatite chondrites, and the plausibility of these phases in core-forming liquids (Lodders, 1995). However, this hypothesis does not consider that enstatite chondrites also contain a myriad of other (and more abundant) sulfides, including niningerite ((Mg,Fe)S), titanium-bearing troilite (FeS), ferroan alabandite ((Mn,Fe)S), and oldhamite (CaS). These common, higher-temperature sulfide phases contain substantial concentrations of REE and other refractory lithophile elements (see review in Brearley and Jones (1998)). If these were incorporated into the Earth's core, the composition of the silicate Earth would be grossly changed on both an elemental and isotopic level. However, there is no evidence, even at the isotopic level (e.g., Sm/Nd and Lu/Hf systems), for REE depletion in the silicate Earth and, thus, it is unlikely that such sulfides were incorporated into the core. The mere identification of a potassium-bearing sulfide does not demonstrate the existence of potassium in the core; it simply allows for the possibility. Plausibility arguments need to be coupled with corroborating paragenetic evidence that is also free of negating geochemical consequences.

The *s–d* electronic transitions occur at higher pressures, particularly for larger alkali metal ions (e.g., caesium, rubidium, and potassium). Under high confining pressures the outer most *s*-orbital electron transforms to a d^1-orbital configuration, resulting in transition metal-like ions. This electronic transition changes the chemical characteristics of the ion making it more siderophilic and potentially allowing it to be sequestered into the core. It has been suggested that some amount of caesium (see Figure 4) may have been sequestered into the Earth's core via this mechanism (McDonough and Sun, 1995). However, data for rubidium and potassium show that this effect is unlikely to have taken place based on the depletion pattern for the moderately volatile lithophile elements.

The third argument for the presence of radioactivity in the core usually involves finding a condition (P–T–X–f_{O_2}) under which potassium is soluble (Hall and Rama Murthy, 1971; Lewis, 1971; Chabot and Drake, 1999). Gessmann and Wood (2002) demonstrated that potassium is soluble in Fe–S and Fe–S–O liquids at high pressure; these authors argued that potassium was sequestered into the core. However, these experimental studies suffer from either examining only simple liquid systems (e.g., synthetic Fe–Ni–S), or overlooking the consequences of other minor and trace elements (e.g., Th, U, and REE). In the case of the Gessmann and Wood (2002) study, calcium is also incorporated into the metallic liquid and the consequences of this are that it creates even more problems. For example, the silicate Earth has a calcium content that is in chondritic proportions to other refractory lithophile elements (e.g., Ca/Al, Ca/Sc, Ca/Yb), demonstrating that there was neither calcium nor potassium incorporated into the core. Similar arguments can also be made for uranium and thorium, which are also based on ratios with other refractory lithophile elements. No experimental evidence exists that shows similar solubility for uranium and thorium (i.e., two elements with significantly different siderophilic behavior) in Fe–S and Fe–S–O liquids at high pressure that does not incorporate other refractory lithophile elements.

An Earth's core containing a significant amount of radioactive elements has been proposed by Herndon (1996). This model envisages a highly reduced composition for the whole Earth and, in particular, for the core. Unfortunately, Herndon has developed a core compositional model that is inconsistent with chemical and isotopic observations of the Earth's mantle and a chondritic planetary composition. Herndon's core contains significant quantities of calcium, magnesium, uranium, and other elements typically considered lithophile. Drawing upon analogies with enstatite chondrites (highly reduced meteorites), Herndon has suggested that these elements were extracted into a metal phase as sulfides (e.g., oldhamite and niningerite). However, these phases are known to grossly fractionate many lithophile elements from one another (Crozaz and Lundberg, 1995), which would lead to a mantle with significantly nonchondritic ratios of Sm/Nd, Lu/Hf, and Th/U (element pairs constrained by isotopic evidence), as well Ca/Ti, Ca/Al, Ti/Sc, and others (element pairs whose bulk mantle properties are well constrained to be chondritic in the mantle), and this is not observed (McDonough and Sun, 1995).

Finally, there is a question of the need for radioactive heating of the core to support the necessary energy budget. Some geophysicists have speculated that there is either potassium or uranium in the core that supplies a portion of the core's power budget. These model calculations for the energy budget in the Earth's core are nontrivial and involve a number of parameters, with many assumptions and extrapolations of data to appropriate core conditions (Gubbins, 1977; Gubbins *et al.*, 1979; Buffett and Bloxham, 2002). The competing models of the geodynamo require different amounts of energy to drive convection in the outer core, and the details of the various models are vastly different (Glatzmaier, 2002). Labrosse *et al.* (2001) proposed a model for the timing and rate of inner core solidification that requires radioactive heating. Likewise, Anderson (2002) examined the energy balance at the CMB and concluded that there is a need for some amount of radioactive heating in the core. In contrast, no radioactive heating is required in other models of the core's energy budget (Stacey, 1992), which is consistent with geochemical evidence for its general absence in the core.

4.6 TIMING OF CORE FORMATION

Defining the age and duration of core formation depends on having an isotope system in which the parent–daughter isotope pairs are fractionated by core subtraction over a time interval within the functional period of the system's half-life. Fortunately, analytical advances in the W–Hf isotope system provide us with a tool to gauge the timing of core formation (Kleine *et al.*, 2002; Yin *et al.*, 2002).

The W–Hf isotope system involves the decay of ^{182}Hf to ^{182}W with a half-life of 9 Myr (thus the system became extinct within the first 100 Myr of Earth history). Both hafnium and tungsten are refractory elements (lithophile and siderophile, respectively) and thus their relative concentration in the Earth is set at chondritic. Some 90% of the Earth's budget of tungsten is hosted in the core (Table 6), whereas all of the planet's hafnium is hosted in the silicate Earth. Early studies found that iron meteorites have lower $^{182}W/^{184}W$ isotopic compositions (by about some 4$\varepsilon_{^{182}W}$ units, where ε units express difference in parts per 10,000) than the Earth (Lee and Halliday, 1995; Harper and Jacobsen, 1996; Horan *et al.* (1998)); the Lee and Halliday (1995) study found no difference between the Earth and chondrites for their tungsten isotopic compositions. The findings of Lee and Halliday (1995), however, have been challenged by Yin *et al.* (2002), Kleine *et al.* (2002), and Schoenberg *et al.* (2002), who found that the Earth's tungsten isotopic composition is some 2$\varepsilon_{^{182}W}$ units higher than that of chondrites. (These studies measured some of the same chondrites as reported in the Lee and Halliday (1995) study and were able to resolve the compositional differences between the Earth and chondrites.) This difference means that core separation was very early, and happened prior to the effective decay of the ^{182}Hf system such that the tungsten remaining in the silicate Earth became enriched in ^{182}W relative to that in the core. These studies demonstrate that much of the core's separation must have been completed by $\sim$30 Ma after t_0 (4.56 Ga) in order to explain the Earth's higher ε_{182W} signature (Figure 13). There are possible scenarios in which one could argue for significantly shorter, but not longer time interval for core formation (Kleine *et al.*, 2002; Yin *et al.*, 2002). By implication the core must have an ε_{182W} of about -2.2 compared to the zero value for the silicate Earth.

The U–Pb, Tc–Ru (^{98}Tc has a half-life between 4 Ma and 10 Ma), and Pd–Ag (^{107}Pd has a half-life of 9.4 Ma) isotope systems have also been examined in terms of providing further insights into the timing of core separation. Overall, the results from these systems are definitive, but not very instructive. The extinct systems of Tc–Ru and Ag–Pd have parent and daughter isotopes that are siderophilic and so were strongly partitioned into the core during its formation. The absence of isotopic anomalies in these systems indicates that core separation left no signature on the silicate Earth. The extant U–Pb system (^{235}U with a half-life of 0.7 Ga) has also been examined with respect to the incorporation of lead into the core, with the result being that core separation must have happened within the first 100 Myr of the Earth's formation in order to reconcile the lead isotopic evolution of the silicate Earth (see review of Galer and Goldstein (1996)).

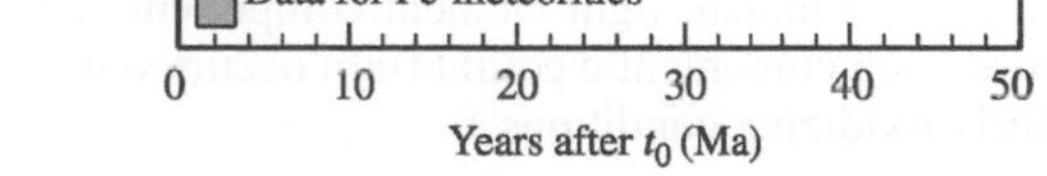

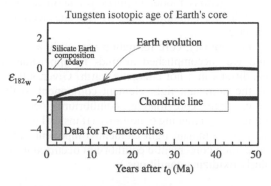

Figure 13 An evolutionary model of time versus the ε_{182W} composition of the silicate Earth for the first 50 of Earth's history. The higher ε_{182W} composition of the Earth relative to chondrites can only be balanced by a complementary lower than chondrites reservoir in the core. Extraction age models for the core are a function of the decay constant, the difference between the silicate Earth and chondrites, the proportion of W and Hf in the mantle and core and the rate of mass extraction to the core. Details of these models are given in the above citations, with the upper limit of the age curves shown here (sources Yin *et al.*, 2002; Kleine *et al.*, 2002; Schoenberg *et al.*, 2002).

4.7 NATURE OF CORE FORMATION

Core formation is not a well-understood process. Constraints for this process come from pinning down the timing of the event, characterizing its bulk chemical properties, and establishing a bulk Earth compositional model. The W–Hf isotope studies dictate that core formation happened early and was virtually completed within 30 Ma of solar system formation. The findings of Li and Agee (1996) and related studies demonstrate that the integrated pressure and temperature of core formation was accomplished at mid- to upper-mantle conditions, not in predifferentiated planetismals. This finding, however, does not preclude the accretion of predifferentiated planetismals; it simply requires that these additions were rehomogenized back into the larger and still evolving Earth system. Finally, the nickel content of the silicate Earth places significant restrictions on oxidation potential of the mantle during core formation. These findings have led to the competing hypotheses of homogeneous and heterogeneous planetary accretion (Wänke, 1981; Jones and Drake, 1986). The former envisages the composition of accreting materials to remain constant throughout Earth's growth history, whereas heterogeneous accretion models postulate that there was a significant compositional shift during the latter stages of the Earth's growth history. These models were developed in order to account for the observed chemical features of the mantle.

The homogeneous accretion model requires a fairly restricted set of conditions to attain the silicate Earth composition observed today (Jones and Drake, 1986). Continued support for this model is wanning given its failure to reconcile a number of rigorous chemical and isotopic constraints (see reviews of O'Neill, 1991b; O'Neill and Palme, 1997; and Palme and O'Neil). For example, it is well established from osmium isotope studies (Meisel *et al.*, 1996; Walker *et al.*, 1997, 2002) that the mantle abundances of rhenium, osmium, and platinum are in chondritic proportions (to within 3% and 10% uncertainty, respectively, for Re/Os and Pt/Os) and as of 2003 no model of homogeneous accretion has been successful in generating such a result. In order to address these and other issues (e.g., the high nickel, sulfur, and selenium content of the mantle) many have appealed to models of heterogeneous accretion (Morgan *et al.*, 1981; Wänke, 1981; Wänke and Dreibus, 1982; Ringwood, 1984; O'Neill, 1991b; O'Neill and Palme, 1997).

Heterogeneous accretion models for the formation of the Earth advocate the initial accretion of refractory, less-oxidized components that make up the bulk of the planet (some 50–80%), followed by the accretion of a lower-temperature, more oxidized component (e.g., perhaps comparable to carbonaceous chondrites). The overall nature of the initially refractory material is not well characterized, but it could have affinities to ordinary or enstatite chondrites. These two-component mixing models seek to reconcile the observational constraints from chemical and isotopic studies of the silicate Earth. As of early 2000s, we do not have sufficient data to identify in detail the nature of these two components of accretion if they existed.

Data for sulfur, selenium, and tellurium (the sulfonic elements, the latter two are also strongly chalcophile and sit below sulfur on the periodic table) show that these elements were sequestered into the core in equal proportions (Figure 6). The upper and lower panels of Figure 6 also show that the mantle content of the highly siderophile elements (HSEs)—rhenium and gold—and the platinum group elements (PGEs)—ruthenium, rhodium, palladium, osmium, iridium, and platinum—are depleted to approximately the same level as that for the sulfonic elements. This is consistent with all of these elements being delivered to the mantle by either the entrainment of small amounts of a core material in plumes coming off of the early CMB (McDonough, 1995) or a model that invokes the addition of a late stage veneer component added to the Earth (Kimura *et al.*, 1974; Chou, 1978; Morgan *et al.*, 1981). Quantitatively, it has been demonstrated that either an endogenous delivery mechanism (former model) or an exogenous delivery model (the latter model) is consistent with the early (ca. pre-4.0 Ga) addition of a sulfonic-HSE component to the mantle (McDonough, 1995).

The endogenous delivery mechanism is inefficient in that it requires core entrainment by plumes that arise off of a newly formed CMB. Kellogg and King (1993) and Kellogg (1997) have shown that such plumes can incorporate $\sim<1\%$ of core material and that this material can be re-entrained back into the mantle. However, such plumes would have been considerably more vigorous during the Hadean, assuming a significant temperature contrast across the core–mantle boundary (i.e., established some 10^5–10^8 yr following accretion and core formation) and a higher-temperature state of the planet resulting from accretion, Moon formation, and core separation. Therefore, it is likely that some degree of entrainment of core material into the mantle occurs in the aftermath of establishing a core–mantle boundary.

Walker *et al.* (2002) demonstrated that the primitive upper mantle has $^{187}Os/^{188}Os$ values similar to ordinary and enstatite chondrites, and that this mantle signature is distinct from that of carbonaceous chondrites. These observations translate to different Re/Os values in different chondrite classes (ordinary, enstatite, and carbonaceous), with the mantle having Re/Os a value unlike that of carbonaceous chondrites. This raises the importance of this late addition (i.e., the sulfonic-HSE signature material), given its distinctive composition. Therefore, the diagnostic sulfonic-HSE signature of the mantle reflects either the nature of the very earliest accreted material delivered to the forming Earth (the endogenous model) or that delivered at the final stages of accretion (the exogenous model).

If the sulfonic-HSE signature derives from material extracted from the core, we can use its HSE signature to characterize the nature of material delivered to the early accreting Earth. Standard heterogeneous accretion models argue that this early accreting material is reduced, with compositional characteristics comparable to ordinary chondrites. Thus, the observations of Walker *et al.* (2002) on Re/Os values of the silicate Earth are consistent with the early accretion of a reduced component. Alternatively, the exogenous delivery model (i.e., late veneer component) would contradict standard heterogeneous accretion models, which envisage accretion of an initial reduced component followed by the subsequent accretion of a more oxidized component. Thus, the exogenous model requires yet another, final shift in the oxidation state of the late accretion material.

In summary, core formation was early and fast and was accomplished at mid- to upper-mantle conditions in a hot energetic Earth. Given silicon as the dominant light element component in the core, then core–mantle equilibrium occurred under fairly reducing conditions. (If instead oxygen was the dominant light element component in the core, then core–mantle equilibrium occurred under fairly oxidizing conditions.)

4.8 THE INNER CORE, ITS CRYSTALLIZATION, AND CORE–MANTLE EXCHANGE

The solid inner core represents only $\sim$5% of the core's mass and $\sim$4% of its volume. Geophysical models of the inner core have identified its structure, elastic properties, and modeled its crystallization history. This, without question, is the most remote region of the planet and little is known of its properties and origins. There are no direct insights to be gained from either compositional studies of the mantle or those of meteorites. The high-pressure conditions of the inner core limit the applicability of any insights drawn from analogies with iron meteorites, which were formed at <1 GPa conditions.

Labrosse *et al.* (1997, 2001) examined the power budget for the core and its implications for inner core crystallization. These calculations generally find that inner core crystallization began in the latter half of Earth's history (ca. 1–2 Ga) and that some amount of radioactive heating is necessary to extend the age of crystallization back in time (Labrosse *et al.*, 2001). Similarly, Brodholt and Nimmo (2002) concluded that models for inner core crystallization could perhaps be developed with long lifetimes (~2.5 Ga) for inner core crystallization with some potassium in the core producing radioactive heating. A long protracted history for inner core crystallization, however, would satisfy those who view the paleomagnetic record in 3.5 Ga old rocks as evidence for an inner core, which gave rise to the early Earth's geomagnetic field. The fundamental problem with developing an early inner core (i.e., older than 2.5 Ga) is with Earth's cooling rates, and the power needed to drive the geodynamo.

Isotopic studies have also considered ways in which to constrain the timing of inner core crystallization. Walker *et al.* (1995) argued that enrichment of $^{187}Os/^{188}Os$ in some plume-derived systems relative to the ambient mantle was a signature from the outer core delivered by CMB-originating plumes. The origin of this isotopic signature would be due to inner core crystallization, which produces an outer core relatively enriched in ^{187}Re (the parent isotope) but still overall depleted in rhenium and osmium. Following this, Brandon *et al.* (1998) found coupled enrichments in $^{187}Os/^{188}Os$ and $^{186}Os/^{188}Os$ similar to those predicted by Walker *et al.* (1995), which provided further support for Walker's model of inner core crystallization that left behind an outer core relatively enriched in ^{187}Re and ^{190}Pt (the parent isotopes) with respect to osmium. These constraints, however, argued for significant element fractionation due to inner core crystallization that was relatively early and rapid in the Earth's history (<1 Ga) in order to obtain the observed elevated isotopic compositions. Brandon *et al.* (2003) have extended these observations to include the Cretaceous komatiite suite from Gorgona Island and again re-enforced a model of early and rapid growth of the inner core, conclusions that are not mutually exclusive with findings from Labrosse *et al.* (2001) but are more difficult to reconcile.

In summary, there are two models of inner core crystallization: one involving early and rapid growth (osmium isotopic model) and one involving later, slow growth (energy balance model). These models address very different problems and concerns, are mutually independent, and reach somewhat divergent conclusions. The inner core exists and began forming after core formation (i.e., after the first 30 Ma of Earth's history). In addition, the generation and maintenance of a planetary dynamo does not require inner core growth (Stevenson, 2003). Thus, as of 2003, we are unable to resolve the issue of when inner core crystallization began.

A final observation on the amount of core–mantle exchange, albeit on a less sensitive scale, can be gained from studies of peridotites. It is recognized that by ~3.8 Ga, when we begin to have a substantial suite of crust and mantle samples, the mantle's composition is relatively fixed as far as key ratios of lithophile-to-siderophile elements in mantle samples. McDonough and Sun (1995) showed that ratios of Mg/Ni and Fe/Mn in the mantle have been fixed (total of ±15% SD for both ratios) for mantle peridotites spanning the age range 3.8 Ga to present (their Figure 7), which is inconsistent with continued core–mantle exchange. At a finer scale, there is ~20% variation in P/Nd values of Archean to modern basalts and komatiites, which because of the core's high P/Nd value (virtually infinity) and mantle's low value ($\sim 70 \pm 15$) restricts mass exchange between the core and mantle to <1%. Collectively, these and other ratios of lithophile (mantle)-to-siderophile (core) elements bound the potential core–mantle exchange to <1% by mass since core formation. The suggested mass fraction of core–mantle exchange based on Re–Os and Pt–Os isotopic studies is at a much smaller scale by at least two orders of magnitude, the scaling is only relative to the mass of the upwelling plume.

4.9 SUMMARY

An estimate of the density deficit in the core is ~5–10% (Boehler, 2000; Anderson and Isaak, 2002); the uncertainty in this estimate is dominantly a function of uncertainties in the pressure and temperature derivatives of EOS data for candidate core materials and knowledge of the temperatures conditions in the core. A tighter constraint on this number will greatly help to refine chemical and petrological models of the core. A density deficit estimate for the inner core is 4–5% (Hemley and Mao, 2001).

The Fe/Ni value of the core (16.5) is well constrained by the limited variation in chondritic meteorites (17.5 ± 0.5) and the mantle ratio (32), as well as the mass fraction of these elements in the two terrestrial reservoirs. The total content of sulfur, carbon, and phosphorus in the core represents only a minor fraction (~2.5 wt.%) of the light element component and this mixture is insufficient to account for the core's density discrepancy. A model composition of the core using silicon as the additional light element in the outer core is preferred over an alternative composition using oxygen as the other light element. Within the limits of our resolving power, either model is tenable.

The trace element content of the core can be determined by using constraints derived from the composition of the mantle and that of chondritic meteorites. This approach demonstrates that there is no geochemical evidence for including any radioactive elements in the core. Relative to the bulk Earth, the core contains about half of the Earth's vanadium and chromium budget and it is equivocal as to whether the core hosts any niobium and tantalum. As compared to iron meteorites, the core is depleted in germanium and strongly depleted in (or void of) gallium. Collectively, the core's chemical signature provides a robust set of restrictions on core formation conditions (i.e., pressure, temperature, and gas fugacity).

The W–Hf isotope system constrains the age of core–mantle differentiation to within the first 30 Ma years of Earth's history (Kleine *et al.*, 2002; Yin *et al.*, 2002). However, the age of inner core crystallization is not resolved.

ACKNOWLEDGMENTS

I thank Rick Carlson for the invitation to contribute to this volume. Also, I am very grateful to Rick Carlson, Rus Hemley, Guy Masters, Hugh O'Neill, Herbert Palme, Bill Minarik and others for review comments on this manuscript and for the many discussions relating to core and mantle phenomena that we have had over the years.

REFERENCES

Allegre C. J., Poirier J.-P., Humler E., and Hofmann A. W. (1995) The chemical composition of the Earth. *Earth Planet. Sci. Lett.* **134**, 515–526.

Anderson D. L., Sammis C., and Jordan T. (1971) Composition and evolution of mantle and core. *Science* **171**, 1103.

Anderson O. L. (2002) The power balance at the core–mantle boundary. *Phys. Earth Planet. Inter.* **131**, 1–17.

Anderson O. L. and Isaak D. G. (2002) Another look at the core density deficit of Earth's outer core. *Phys. Earth Planet. Inter.* **131**, 19–27.

Birch F. (1952) Elasticity and constitution of the Earth's interior. *J. Geophys. Res.* **57**, 227–286.

Birch F. (1964) Density and composition of mantle and core.*J. Geophys. Res.* **69**, 4377–4388.

Birch F. (1966) Evidence from high-pressure experiments bearing on density and composition of Earth. *Geophys. J. Roy. Astro. Soc.* **11**, 256.

Bloxham J. (1995) Global magnetic field. In *Global Earth Physics: A Handbook of Physical Constants*, Vol. AGU Reference Shelf (ed. T. J. Ahrens). American Geophysical Union, Washington, DC, pp. 47–65.

Boehler R. (2000) High-pressure experiments and the phase diagram of lower mantle and core materials. *Rev. Geophys.* **38**, 221–245.

Brandon A., Walker R. J., Morgan J. W., Norman M. D., and Prichard H. (1998) Coupled ^{186}Os and ^{187}Os evidence for core–mantle interaction. *Science* **280**, 1570–1573.

Brandon A. D., Walker R. J., Puchtel I. S., Becker H., Humayun M., and Revillon S. (2003) ^{186}Os/^{187}Os systematics of Gorgona Island komatiites: implications for early growth of the inner core. *Earth Planet. Sci. Lett.* **206**, 411–426.

Brearley A. J. and Jones R. H. (1998) Chondritic meteorites. In *Planetary Materials* (ed. J. J. Papike). Mineralogical Soc. Amer., Washington, DC, vol. 36, pp. 3-01–3-398.

Brodholt J. and Nimmo F. (2002) Earth science—core values. *Nature* **418**(6897), 489–491.

Brown J. M. and McQueen R. G. (1986) Phase transitions Güneisen parameter and elasticity for shocked iron between 77 GPa and 400 GPa. *J. Geophys. Res.* **91**, 7485–7494.

Brush S. G. (1980) Discovery of the Earth's core. *Am. J. Phys.* **48**, 705–724.

Buffett B. A. (2000) Earth's core and the geodynamo. *Science* **288**(5473), 2007–2012.

Buffett B. A. and Bloxham J. (2002) Energetics of numerical geodynamo models. *Geophys. J. Int.* **149**, 211–224.

BVSP (1981) *Basaltic Volcanism on the Terrestrial Planets.* Pergamon, New York.

Chabot N. L. and Drake M. J. (1999) Potassium solubility in metal: the effects of composition at 15 kbar and 1,900 °C on partitioning between iron alloys and silicate melts. *Earth Planet. Sci. Lett.* **172**, 323–335.

Chou C. L. (1978) Fractionation of siderophile elements in the Earth's upper mantle. *Proc. Lunar Sci. Conf.* **9**, 219–230.

Crozaz G. and Lundberg L. (1995) The origin of oldhamite in unequilibrated enstatite chondrites. *Geochim. Cosmochim. Acta* **59**, 3817–3831.

Drake M. J., Newsom H. E., and Capobianco C. J. (1989) V, Cr, and Mn in the Earth, Moon, EPB and SPB and the origin of the Moon: experimental studies. *Geochim. Cosmochim. Acta* **53**, 2101–2111.

Dreibus G. and Palme H. (1996) Cosmochemical constraints on the sulfur content in the Earth's core. *Geochim. Cosmochim. Acta* **60**(7), 1125–1130.

Dreibus G. and Wänke H. (1979) On the chemical composition of the moon and the eucrite parent body and comparison with composition of the Earth: the case of Mn, Cr and V. *Lunar Planet Sci. (Abstr.)* **10**, 315–317.

Dziewonski A. and Anderson D. L. (1981) Preliminary reference Earth model. *Phys. Earth Planet. Inter.* **25**, 297–356.

Dziewonski A. and Gilbert F. (1972) Observations of normal modes from 84 recordings of the Alsakan earthquake of 1964 March 28. *Geophys. J. Roy. Astron. Soc.* **27**, 393–446.

Galer S. J. G. and Goldstein S. L. (1996) Influence of accretion on lead in the Earth. In *Earth Processes: Reading the Isotopic Code* (eds. A. Basu and S. R. Hart). American Geophysical Union, Washington, DC, pp. 75–98.

Gessmann C. K. and Rubie D. C. (2000) The origin of the depletion of V, Cr, and Mn in the mantles of the Earth and Moon. *Earth Planet. Sci. Lett.* **184**, 95–107.

Gessmann C. K. and Wood B. J. (2002) Potassium in the Earth's core? *Earth Planet. Sci. Lett.* **200**(1–2), 63–78.

Gies F. and Gies J. (1994) *Cathedral, Forge, and Waterwheel: Technology and Invention in the Middle Ages.* Harper Collins, New York.

Glatzmaier G. A. (2002) Geodynamo simulations: How realistic are they? *Ann. Rev. Earth Planet. Sci.* **30**, 237–257.

Gubbins D. (1977) Energetics of the Earth's core. *J. Geophys.* **47**, 453–464.

Gubbins D., Masters T. G., and Jacobs J. A. (1979) Thermal evolution of the Earth's core. *Geophys. J. Roy. Astron. Soc.* **59**(1), 57–99.

Gutenberg B. (1914) Über Erdbebenwellen VIIA. Nachr. Ges.Wiss. Göttingen Math. Physik. Kl, 166. *Nachr. Ges. Wiss.Göttingen Math. Physik.* **125**, 116.

Hall H. T. and Rama Murthy V. (1971) The early chemical history of the Earth: some critical elemental fractionations. *Earth Planet. Sci. Lett.* **11**, 239–244.

Harper C. L. and Jacobsen S. B. (1996) Evidence for Hf-182 in the early solar system and constraints on the timescale of terrestrial accretion and core formation. *Geochim. Cosmo-chim. Acta* **60**(7), 1131–1153.

Hemley R. J. and Mao H. K. (2001) *In situ* studies of iron under pressure: new windows on the Earth's core. *Int. Geol. Rev.* **43**(1), 1–30.

Herndon J. M. (1996) Substructure of the inner core of the Earth. *Proc. Natl. Acad. Sci. USA* **93**(2), 646–648.

Hillgren V. J., Gessmann C. K., and Li J. (2000) An experimental perspective on the light element in the earth's core. In *Origin of the Earth and Moon* (eds. R. M. Canup and K. Righter). University of Arizona Press, Tucson, pp. 245–263.

Horan M. F., Smoliar M. I., and Walker R. J. (1998) ^{182}W and ^{187}Re–^{187}Os systematics of iron meteorites: chronology for melting, differentiation and crystallization in asteroids. *Geochim. Cosmochim. Acta* 62, 545–554.

Javoy M. (1995) The integral enstatite chondrite model of the Earth. *Geophys. Res. Lett.* **22**, 2219–2222.

Jeffreys H. (1926) The rigidity of the Earth's central core. *Mon. Not. Roy. Astron. Soc. Geophys. Suppl.* **1**(7), 371–383.

Jephcoat A. and Olson P. (1987) Is the inner core of the Earth pure iron. *Nature* **325**(6102), 332–335.

Jones J. H. and Drake M. J. (1986) Geochemical constraints on core formation in the Earth. *Nature* **322**, 221–228.

Kellogg L. H. (1997) Growing the Earth's D″ layer: effect of density variations at the core–mantle boundary. *Geophys. Res. Lett.* **24**, 2749–2752.

Kellogg L. H. and King S. D. (1993) Effect of mantle plumes on the growth of D″ by reaction between the core and mantle. *Geophys. Res. Lett.* **20**, 379–382.

Kimura K., Lewis R. S., and Anders E. (1974) Distribution of gold and rhenium between nickel–iron and silicate melts:implications for the abundances of siderophile elements on the Earth and Moon. *Geochim. Cosmochim. Acta* **38**, 683–701.

Kleine T., Münker C., Mezger K., and Palme H. (2002) Rapid accretion and early core formation on asteroids and the terrestrial planets from Hf–W chronometry. *Nature* **418**, 952–955.

Labrosse S., Poirier J. P., and LeMouel J. L. (1997) On cooling of the Earth's core. *Phys. Earth Planet. Inter.* **99**(1–2),1–17.

Labrosse S., Poirier J. P., and LeMouel J. L. (2001) The age of the inner core. *Earth Planet. Sci. Lett.* **190**, 111–123.

Larimer J. W. (1988) The cosmochemical classification of the elements. In *Meteorites and the Early Solar System* (eds. J. F.Kerridge and M. S. Matthews). University of Arizona Press, Tucson, pp. 375–389.

Lee D.-C. and Halliday A. N. (1995) Hafnium–tungsten chronometry and the timing of terrestrial core formation.*Nature* **378**, 771–774.

Lehmann I. (1936) *P′ Publ. Bur. Cent. Seism. Int. Ser. A* **14**, 87–115.

Lewis J. S. (1971) Consequences of presence of sulfur in core of Earth. *Earth Planet. Sci. Lett.* **11**, 130.

Li J. and Agee C. B. (1996) Geochemistry of mantle–core differentiation at high pressure. *Nature* **381**(6584), 686–689.

Lodders K. (1995) Alkali elements in the Earths core—evidence from enstatite meteorites. *Meteoritics* **30**(1), 93–101.

Macdonald G. J. F. and Knopoff L. (1958) On the chemical composition of the outer core. *Geophys. J. Roy. Astron. Soc.* **1**(4), 284–297.

Mao H. K., Wu Y., Chen L. C., Shu J. F., and Jephcoat A. P. (1990) Static compression of iron to 300 GPa and $Fe_{0.8}Ni_{0.2}$ to 260 GPa: implications for composition of the core. *J. Geophys. Res.* **95**, 21737–21742.

Masters T. G. and Shearer P. M. (1995) Seismic models of the Earth: elastic and anelastic. In *Global Earth Physics: A Handbook of Physical Constants.* Vol. AGU Reference Shelf (ed. T. J. Ahrens). American Geophysical Union, Washington, DC, pp. 88–103.

McDonough W. F. (1990) Comment on "Abundance and distribution of gallium in some spinel and garnet lherzolites" by D. B. McKay and R. H. Mitchell. *Geochim. Cosmochim. Acta* **54**, 471–473.

McDonough W. F. (1991) Partial melting of subducted oceanic crust and isolation of its residual eclogitic lithology. *Phil. Trans. Roy. Soc. London A* **335**, 407–418.

McDonough W. F. (1995) An explanation for the abundance enigma of the highly siderophile elements in the earth's mantle. *Lunar Planet. Sci. (Abstr.)* 26, 927–928.

McDonough W. F. (1999) Earth's Core. In *Encyclopedia of Geochemistry* (eds. C. P. Marshall and R. W. Fairbridge). Kluwer Academic, Dordrecht, pp. 151–156.

McDonough W. F. (2001) The composition of the Earth. In *Earthquake Thermodynamics and Phase Transformations in the Earth's Interior* (eds. R. Teissseyre and E. Majewski). Academic Press, San Diego, vol. 76, pp. 3–23.

McDonough W. F. and Sun S.-S. (1995) The composition of the Earth. *Chem. Geol.* **120**, 223–253.

McDonough W. F., McCulloch M. T., and Sun S.-S. (1985) Isotopic and geochemical systematics in Tertiary–Recent basalts from southeastern Australia and implications for the evolution of the sub-continental lithosphere. *Geochim. Cosmochim. Acta* **49**, 2051–2067.

Meisel T., Walker R. J., and Morgan J. W. (1996) The osmium isotopic composition of the Earth's primitive upper mantle. *Nature* **383**, 517–520.

Merrill R. T., McElhinney M. W., and McFadden P. L. (1996) *The Magnetic Field of the Earth.* Academic Press, San Diego.

Morgan J. W., Wandless G. A., Petrie R. K., and Irving A. J. (1981) Composition of the Earth's upper mantle: 1. Siderophile trace-elements in Ultramafic Nodules. *Tectonophysics* **75**(1–2), 47–67.

Niu Y. and Batiza R. (1997) Trace element evidence from seamounts for recycled oceanic crust in eastern Pacific mantle. *Earth Planet. Sci. Lett.* **148**, 471–783.

O'Neill H. S. C. (1991a) The origin of the Moon and the early history of the Earth—a chemical model: Part 1.The Moon. *Geochim. Cosmochim. Acta* **55**, 1135–1157.

O'Neill H. S. C. (1991b) The origin of the Moon and the early history of the Earth—a chemical model: Part 2.The Earth. *Geochim. Cosmochim. Acta* **55**, 1159–1172.

O'Neill H. S. C. and Palme H. (1997) Composition of the silicate Earth: implications for accretion and core formation. In *The Earth's Mantle: Structure, Composition, and Evolution—The Ringwood Volume* (ed. I. Jackson). Cambridge University Press, Cambridge, pp. 1–127.

O'Neill H. S., Canil D., and Rubie D. C. (1998) Oxide–metal equilibria to 2,500 degrees C and 25 GPa: implications for core formation and the light component in the Earth's core. *J. Geophys. Res.-Solid Earth* **103**(B6), 12239–12260.

Palme H. (2001) Chemical and isotopic heterogeneity in protosolar matter. *Phil. Trans. Roy. Soc. London* **359**, 2061–2075.

Palme H., Larimer J. W., and Lipchutz M. E. (1988) Moderately volatile elements. In *Meteorites and the Early Solar System* (eds. J. F. Kerridge and M. S. Matthews). University of Arizona Press, Tucson, pp. 436–461.

Ringwood A. E. (1959) On the chemical evolution and density of the planets. *Geochim. Cosmochim. Acta* **15**, 257–283.

Ringwood A. E. (1966) The chemical composition and origin of the Earth. In *Advances in Earth Sciences* (ed. P. M. Hurley). MIT Press, Cambridge, pp. 287–356.

Ringwood A. E. (1984) The Bakerian Lecture, 1983—the Earth's core—its composition, formation and bearing upon the origin of the Earth. *Proc. Roy. Soc. London Ser. A: Math. Phys. Eng. Sci.* **395**(1808), 1–46.

Ringwood A. E., Kato T., Hibberson W., and Ware N. (1990) High-pressure geochemistry of Cr, V, and Mn and implications for the origin of the Moon. *Nature* **347**, 174–176.

Rubin A. E. (1997) Mineralogy of meteorite groups. *Meteorit. Planet. Sci.* **32**, 231–247.

Rudnick R. L., Barth M., Horn I., and McDonough W. F. (2000) Rutile-bearing refractory eclogites: missing link between continents and depleted mantle. *Science* **287**, 278–281.

Schoenberg R., Kamber B. S., Collerson K. D., and Eugster O. (2002) New W-isotope evidence for rapid terrestrial accretion and very early core formation. *Geochim. Cosmo-chim. Acta* **66**, 3151–3160.

Stacey F. D. (1992) *Physics of the Earth.* Brookfield Press, Brisbane.

Stevenson D. J. (2003) Planetary magnetic fields. *Earth Planet. Sci. Lett.* **208**, 1–11.

Wade J. and Wood B. J. (2001) The Earth's "missing" niobium may be in the core. *Nature* **409**(6816), 75–78.

Walker R. J., Morgan J. W., and Horan M. F. (1995) Osmium-187 enrichments in some plumes: evidence for core–mantle interactions. *Science* **269**, 819–822.

Walker R. J., Morgan J. W., Beary E., Smoliar M. I., Czamanske G. K., and Horan M. F. (1997) Applications of the ^{90}P– ^{186}Os isotope system to geochemistry and cosmochemistry. *Geochim. Cosmochhim. Acta* **61**, 4799–4808.

Walker R. J., Horan M. F., Morgan J. W., Becker H., Grossman J. N., and Rubin A. E. (2002) Comparative ^{187}Re–^{187} Os systematics of chondrites: implications regarding early solar system processes. *Geochim. Cosmochim. Acta* **66**, 4187–4201.

Wänke H. (1981) Constitution of terrestrial planets. *Phil. Trans. Roy. Soc. London A* **303**, 287–302.

Wänke H. (1987) Chemistry and accretion of Earth and Mars. *Bull. Soc. Geol. Fr.* **3**(1), 13–19.

Wänke H. and Dreibus G. (1982) Chemical and isotopic evidence for the early history of the Earth–Moon system. In *Tidal Friction and the Earth's Rotation* (eds. P. Brosche and J. Sundermann). Springer, Berlin, pp. 322–344.

Wänke H. and Dreibus G. (1988) Chemical composition and accretion history of terrestrial planets. *Phil. Trans. Roy. Soc.London A* **325**, 545–557.

Washington H. S. (1925) The chemical composition of the Earth. *Am. J. Sci.* **9**, 351–378.

Wasson J. T. (1985) *Meteorites, Their Record of Early Solar-system History.* W. H. Freeman, New York, 267pp.

Wasson J. T. and Kallemeyn G. W. (1988) Compositions of chondrites. *Phil. Trans. Roy. Soc. London A* **325**, 535–544.

Williams Q. and Hemley R. J. (2001) Hydrogen in the deep Earth. *Ann. Rev. Earth Planet. Sci.* **29**, 365–418.

Williamson E. D. and Adams L. H. (1923) Density distribution in the Earth. *J. Washington Acad. Sci.* **13**, 413–428.

Wood B. J. (1993) Carbon in the Core. *Earth Planet. Sci. Lett.* **117**(3–4), 593–607.

Yin Q., Jacobsen S. B., Yamashita K., Blichert-Toft J., Télouk P., and Albarede F. (2002) A short timescale for terrestrial planet formation from Hf–W chronometry of meteorites. *Nature* **418**, 949–952.

Yoder C. F. (1995) Astrometric and geodetic properties of the Earth and the solar system. In *Global Earth Physics: A Handbook of Physical Constants.* Vol. AGU Reference Shelf (ed. T. J. Ahrens). American Geophysical Union, Washington, DC, pp. 1–31.

Readings from the Treatise on Geochemistry
ISBN: 978-0-12-381391-6

pp. 109–130

5
Composition of the Continental Crust

R. L. Rudnick

University of Maryland, College Park, MD, USA

and

S. Gao

China University of Geosciences, Wuhan, People's Republic of China and Northwest University, Xi'an, People's Republic of China

5.1 INTRODUCTION

The Earth is an unusual planet in our solar system in having a bimodal topography that reflects the two distinct types of crust found on our planet. The low-lying oceanic crust is thin (~7 km on average), composed of relatively dense rock types such as basalt and is young (≤200 Ma old). In contrast, the high-standing continental crust is thick (~40 km on average), is composed of highly diverse lithologies (virtually every rock type known on Earth) that yield an average intermediate or "andesitic" bulk composition (Taylor and McLennan (1985) and references therein), and contains the oldest rocks and minerals yet observed on Earth (currently the 4.0 Ga Acasta gneisses (Bowring and Williams, 1999) and 4.4 Ga detrital zircons from the Yilgarn Block, Western Australia (Wilde *et al.*, 2001)), respectively. Thus, the continents preserve a rich geological history of our planet's evolution and understanding their origin is critical for understanding the origin and differentiation of the Earth.

The origin of the continents has received wide attention within the geological community, with hundreds of papers and several books devoted to the topic (the reader is referred to the following general references for further reading: Taylor and McLennan (1985), Windley (1995), and Condie (1997). Knowledge of the age and composition of the continental crust is essential for understanding its origin. Patchett and Samson review the present-day age distribution of the continental crust and Kemp and Hawkesworth review secular evolution of crust composition. Moreover, to understand fully the origin and evolution of continents requires an understanding of not only the crust, but also the mantle lithosphere that formed more-or-less contemporaneously with the crust and translates with it as the continents move across the Earth's surface.

This chapter reviews the present-day composition of the continental crust, the methods employed to derive these estimates, and the implications of the continental crust composition for the formation of the continents, Earth differentiation, and its geochemical inventories.

5.1.1 What is the Continental Crust?

In a review of the composition of the continental crust, it is useful to begin by defining the region under consideration and to provide some generalities regarding its structure. The continental crust, as considered here, extends vertically from the Earth's surface to the Mohorovicic discontinuity, a jump in compressional wave speeds from $\sim$7 km s^{-1} to $\sim$8 km s^{-1} that is interpreted to mark the crust–mantle boundary. In some regions the Moho is transitional rather than discontinuous and there may be some debate as to where the crust–mantle boundary lies (cf. Griffin and O'Reilly, 1987; McDonough *et al.*, 1991). The lateral extent of the continents is marked by the break in slope on the continental shelf. Using this definition, $\sim$31% of continental area is submerged beneath the oceans (Figure 1; Cogley, 1984), and is thus less accessible to geological sampling. For this reason, most estimates of continental crust composition derive from exposed regions of the continents. In some cases the limited geophysical data for submerged continental shelves reveal no systematic difference in bulk properties between the shelves and exposed continents; the shelves simply appear to be thinned regions of the crust. In other cases, such as volcanic rifted margins, the submerged continent is characterized by high-velocity layers interpreted to represent massive basaltic intrusions associated with continental breakup (Holbrook and Kelemen, 1993). Depending on the extent of the latter type of continental margin (which is yet to be quantified), crust compositional estimates derived from exposed regions may not be wholly representative of the total continental mass.

The structure of the continental crust is defined seismically to consist of upper-, middle-, and lower crustal layers (Christensen and Mooney, 1995;

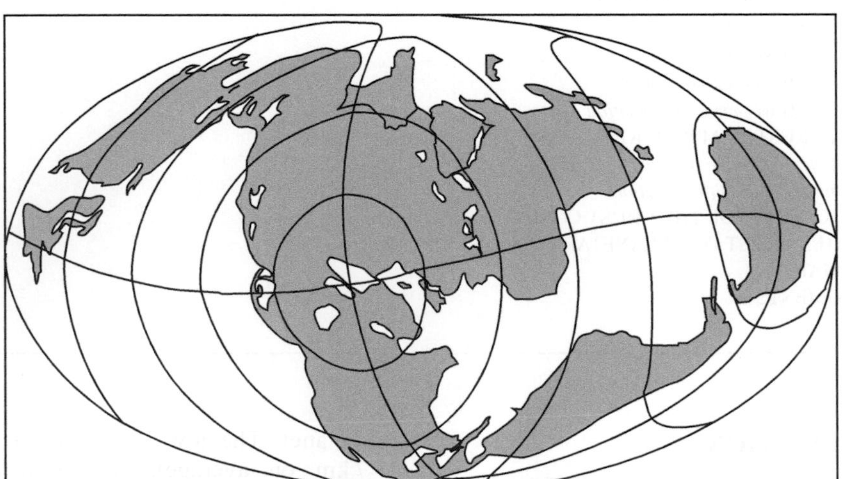

Figure 1 Map of continental regions of the Earth, including submerged continents (Cogley (1984); reproduced by permission of American Geophysical Union from *Rev. Geophys. Space Phys.*, **1984**, *22*, 101–122).

Holbrook *et al.*, 1992; Rudnick and Fountain, 1995). The upper crust is readily accessible to sampling and robust estimates of its composition are available for most elements (Section 5.2). These show the upper crust to have a granodioritic bulk composition, to be rich in incompatible elements, and generally depleted in compatible elements. The deeper reaches of the crust are more difficult to study. In general, three probes of the deep crust are employed to unravel its composition: (i) studies of high-grade metamorphic rocks (amphibolite or granulite facies) exposed in surface outcrops (Bohlen and Mezger, 1989) and, in some cases, in uplifted cross-sections of the crust reaching to depths of 20 km or more (Fountain *et al.*, 1990a; Hart *et al.*, 1990; Ketcham, 1996; Miller and Christensen, 1994); (ii) studies of granulite-facies xenoliths (foreign rock fragments) that are carried from great depths to the Earth's surface by fast-rising magmas (see Rudnick (1992) and references therein); and (iii) remote sensing of lower crustal lithologies through seismic investigations (Christensen and Mooney, 1995; Holbrook *et al.*, 1992; Rudnick and Fountain, 1995; Smithson, 1978) and surface heat-flow studies. Collectively, the observations from these probes show that the crust becomes more mafic with depth (Section 5.3). In addition, the concentration of heat-producing elements drops off rapidly from the surface downwards. This is due, in part, to an increase in metamorphic grade but is also due to increasing proportions of mafic lithologies. Thus, the crust is vertically stratified in terms of its chemical composition.

In addition to this stratification, the above studies also show that the crust is heterogeneous from place to place, with few systematics available for making generalizations about crustal structure and composition for different tectonic settings. For example, the crust of Archean cratons in some regions is relatively thin and has low seismic velocities, suggesting an evolved composition (e.g., Yilgarn craton (Drummond, 1988); Kaapvaal craton (Durrheim and Green, 1992; Niu and James, 2002); and North China craton (Gao *et al.*, 1998a,b)). However, in other cratons, the crust is thick (40–50 km) and the deep crust is characterized by high velocities, which imply mafic-bulk compositions (Wyoming craton (Gorman *et al.*, 2002) and Baltic shield (Luosto and Korhonen, 1986; Luosto *et al.*, 1990)). The reasons for these heterogeneities are not fully understood and we return to this topic in Section 5.3. Similar heterogeneities are observed for Proterozoic and Paleozoic regions (see Rudnick and Fountain (1995) and references therein). Determining an average composition of such a heterogeneous mass is difficult and, at first glance, may seem like a futile endeavor. Yet it is just such averages that allow insights into the relative contribution of the crust to the whole Earth-chemical budget and the origin of the continents. Thus, deriving average compositions is critical to studies of the continents and the whole Earth.

5.1.2 The Importance of Determining Crust Composition

Although the continental crust constitutes only ~0.6% by mass of the silicate Earth, it contains a very large proportion of incompatible elements (20–70%, depending on element and model considered; Rudnick and Fountain (1995)), which include the heat-producing elements and members of a number of radiogenic-isotope systems (Rb–Sr, U–Pb, Sm–Nd, Lu–Hf). Thus the continental crust factors prominently in any mass-balance calculation for the Earth as a whole and in estimates of the thermal structure of the Earth (Sclater *et al.*, 1980).

In addition, knowledge of the bulk composition of the crust and determining whether this composition has changed through time is important for: (i) understanding the processes by which the crust is generated and modified and (ii) determining whether there is any secular evolution in crust generation and modification processes. The latter has important implications for the evolution of our planet as a whole.

In this chapter we review the composition of the upper, middle, and lower continental crust (Sections 5.2 and 5.3). We then examine the bulk crust composition and the implications of this composition for crust generation and modification processes (Sections 5.4 and 5.5). Finally, we compare the Earth's crust with those of the other terrestrial planets in our solar system (Section 5.6) and speculate about what unique processes on Earth have given rise to this unusual crustal distribution.

5.2 THE UPPER CONTINENTAL CRUST

The upper continental crust, being the most accessible part of our planet, has long been the target of geochemical investigations (Clarke, 1889). There are two basic methods employed to determine the composition of the upper crust: (i) establishing weighted averages of the compositions of rocks exposed at the surface and (ii) determining averages of the composition of insoluble elements in fine-grained clastic sedimentary rocks or glacial deposits and using these to infer upper-crust composition.

The first method was utilized by F. W. Clarke and colleagues over a century ago (Clarke, 1889; Clarke and Washington, 1924) and entails

Table 1 Major element composition[a] (in weight percent oxide) of the upper continental crust. Columns 1–9 represent averages of surface exposures and glacial clays. Columns 10–11 are derivative compositions from these data. Column 12 shows our recommended values.

Element	1 Clarke (1889)	2 Clarke and Washington (1924)	3 Goldschmidt (1933)	4 Shaw et al. (1967)	5 Fahrig and Eade (1968)	6 Ronov and Yaroshevskiy (1976)	7 Condie (1993)	8 Gao et al. (1998a)	9 Borodin (1998)	10 Taylor and McLennan (1985)	11 Wedepohl (1995)	12 This Study[b]
SiO_2	60.2	60.30	62.22	66.8	66.2	64.8	67.0	67.97	67.12	65.89	66.8	66.62
TiO_2	0.57	1.07	0.83	0.54	0.54	0.55	0.56	0.67	0.60	0.50	0.54	0.64
Al_2O_3	15.27	15.65	16.63	15.05	16.10	15.84	15.14	14.17	15.53	15.17	15.05	15.40
FeO_T[c]	7.26	6.70	6.99	4.09	4.40	5.78	4.76	5.33	4.94	4.49	4.09	5.04
MnO	0.10	0.12	0.12	0.07	0.08	0.10		0.10	0.00	0.07	0.07	0.10
MgO	4.59	3.56	3.47	2.30	2.20	3.01	2.45	2.62	2.10	2.20	2.30	2.48
CaO	5.45	5.18	3.23	4.24	3.40	3.91	3.64	3.44	3.51	4.19	4.24	3.59
Na_2O	3.29	3.92	2.15	3.56	3.90	2.81	3.55	2.86	3.21	3.89	3.56	3.27
K_2O	2.99	3.19	4.13	3.19	2.91	3.01	2.76	2.68	3.01	3.39	3.19	2.80
P_2O_5	0.23	0.31	0.23	0.15	0.16	0.16	0.12	0.16	0.00	0.20	0.15	0.15
Mg#	53.0	48.7	46.9	50.1	47.4	48.1	47.9	46.7	43.2	46.6	50.1	46.7

Mg# = molar $100 \times Mg/(Mg+Fe_{tot})$.
[a] Major elements recast to 100% anhydrous. [b] See Table 3 for derivation of this estimate. [c] Total Fe as FeO.

large-scale sampling and weighted averaging of the wide variety of rocks that crop out at the Earth's surface. All major-element (and a number of soluble trace elements) determinations of upper-crust composition rely upon this method.

The latter method is based on the concept that the process of sedimentation averages wide areas of exposed crust. This method was originally employed by Goldschmidt (1933) and his Norwegian colleagues in their analyses of glacial sediments to derive average composition of the crystalline rocks of the Baltic shield and has subsequently been applied by a number of investigators, including the widely cited work by Taylor and McLennan (1985) to derive upper-crust composition for insoluble trace elements. In the following sections we review the upper-crust composition determined from each of these methods, then provide an updated estimate of the composition of the upper crust.

5.2.1 Surface Averages

In every model for the composition of the upper-continental crust, major-element data are derived from averages of the composition of surface exposures (Table 1). Several surface-exposure studies have also provided estimates of the average composition of a number of trace elements (Table 2). For soluble elements that are fractionated during the weathering process (e.g., sodium, calcium, strontium, barium, etc.), this is the only way in which a reliable estimate of their abundances can be obtained.

The earliest of such studies was the pioneering work of Clarke (1889), who, averaging hundreds of analyses of exposed rocks, determined an average composition for the crust that is markedly similar to present-day averages of the bulk crust (cf. Tables 1 and 9). Although Clarke's intention was to derive the average crust composition, his samples are limited to the upper crust; there was little knowledge of the structure of the Earth when these studies were undertaken; oceanic crust was not distinguished as different from continental and the crust was assumed to be only 16 km thick. Clarke's values are, therefore, most appropriately compared to upper crustal estimates. Later, Clarke, joined by H. S. Washington, used a larger data set to determine an average composition of the upper-crust that is only slightly different from his original 1889 average (Clarke and Washington, 1924; Table 1). Compared to more recent estimates of upper-crust composition, these earliest estimates are less evolved (lower silicon, higher iron, magnesium, and calcium), but contain similar amount of the alkali elements, potassium and sodium.

The next major undertakings in determining upper-crust composition from large-scale surface

Table 2 Estimates of the trace-element composition of the upper continental crust. Columns 1–4 represent averages of surface exposures. Columns 5–8 are estimates derived from sedimentary and loess data. Column 9 is a previous estimate, where bracketed data are values derived from surface exposure studies. Column 10 is our recommended value (see Table 3).

Element	*Units*	*1* *Shaw et al. (1967, 1976)*	*2* *Eade and Fahrig (1973)*	*3* *Condie (1993)*	*4* *Gao et al. (1998a)*	*5* *Sims et al. (1990)*	*6* *Plank and Langmuir (1998)*	*7* *Peucker-Eherenbrink and Jahn (2001)*	*8* *Taylor and McLennan (1985, 1995)*	*9* *Wedepohl (1995)*[a]	*10* *This study*[b]
Li	$\mu g\ g^{-1}$	22			20				20	[22]	21
Be	"	1.3			1.95				3	3.1	2.1
B	"	9.2			28				15	17	17
N	"									83	83
F	"	500			561					611	557
S	"	600			309					953	621
Cl	"	100			142					640	370
Sc	"	7	12	13.4	15				13.6[c]	[7]	14.0
V	"	53	59	86	98				107[c]	[53]	97
Cr	"	35	76	112	80				85[c]	[35]	92
Co	"	12		18	17				17[c]	[12]	17.3
Ni	"	19	19	60	38				44[c]	[19]	47
Cu	"	14	26		32				25	[14]	28
Zn	"	52	60		70				71	[52]	67
Ga	"	14			18				17	[14]	17.5
Ge	"				1.34				1.6	1.4	1.4
As	"				4.4	5.1			1.5	2	4.8
Se	"				0.15				0.05	0.083	0.09
Br	"									1.6	1.6
Rb	"	110	85	83	82				112	110	84
Sr	"	316	380	289	266				350	[316]	320
Y	"	21	21	24	17.4				22	[21]	21
Zr	"	237	190	160	188				190	[237]	193
Nb	"	26		9.8	12		13.7		12[c]	[26]	12
Mo	"				0.78	1.2			1.5	1.4	1.1
Ru	$ng\ g^{-1}$							0.34			0.34
Pd	"				1.46			0.52	0.5		0.52
Ag	"				55				50	55	53
Cd	$\mu g\ g^{-1}$	0.075			0.079				0.098	0.102	0.09
In	"								0.05	0.061	0.056
Sn	"				1.73				5.5	2.5	2.1
Sb	"				0.3	0.45			0.2	0.31	0.4

(continued)

Table 2 (continued).

Element	Units	1 Shaw et al. (1967, 1976)	2 Eade and Fahrig (1973)	3 Condie (1993)	4 Gao et al. (1998a)	5 Sims et al. (1990)	6 Plank and Langmuir (1998)	7 Peucker-Eherenbrink and Jahn (2001)	8 Taylor and McLennan (1985, 1995)	9 Wedepohl (1995)[a]	10 This study[b]
I	"									1.4	1.4
Cs	"				3.55		7.3		4.6[c]	5.8	4.9
Ba	"	1070	730	633	678				550	668	624
La	"	32.3	71	28.4	34.8				30	[32.3]	31
Ce	"	65.6		57.5	66.4				64	[65.7]	63
Pr	"								7.1	6.3	7.1
Nd	"	25.9		25.6	30.4				26		27
Sm	"	4.61		4.59	5.09				4.5	4.7	4.7
Eu	"	0.937		1.05	1.21				0.88	0.95	1.0
Gd	"			4.21					3.8	2.8	4.0
Tb	"	0.481		0.66	0.82				0.64	[0.5]	0.7
Dy	"	2.9							3.5	[2.9]	3.9
Ho	"	0.62							0.8	[0.62]	0.83
Er	"								2.3		2.3
Tm	"								0.33		0.30
Yb	"	1.47		1.91	2.26				2.2	[1.5]	2.0
Lu	"	0.233		0.32	0.35				0.32	[0.27]	0.31
Hf	"	5.8		4.3	5.12				5.8	[5.8]	5.3
Ta	"	5.7		0.79	0.74		0.96		1.0[c]	1.5	0.9
W	"				0.91	3.3			2	1.4	1.9
Re	$ng\ g^{-1}$							0.198	0.4		0.198
Os	"							0.031	0.05		0.031
Ir	"	0.02						0.022	[0.02]		0.022
Pt	"							0.51			0.5
Au	"	1.81			1.24				[1.8]		1.5
Hg	$\mu g\ g^{-1}$	0.096			0.0123					0.056	0.05
Tl	"	0.524			1.55				0.75	0.75	0.9
Pb	"	17	18	17	18				17[c]	17	17
Bi	"	0.035			0.23				0.13	0.123	0.16
Th	"	10.3	10.8	8.6	8.95				10.7	[10.3]	10.5
U	"	2.45	1.5	2.2	1.55				2.8	[2.5]	2.7

[a] Wedepohl's upper crust is largely derived from the Canadian Shield composites of Shaw *et al.* (1967, 1976). Values taken directly from Shaw *et al.* are shown in brackets. [b] See Table 3 for derviation of this estimate. [c] Updated in McLennan (2001b).

sampling campaigns did not appear until twenty years later in studies centered on the Canadian, Baltic, and Ukranian Shields. It is these studies that form the foundation on which many of the more recent estimates of upper-crust composition are constructed (e.g., Taylor and McLennan, 1985; Wedepohl, 1995).

Shaw *et al.* (1967, 1976, 1986) and Eade and Fahrig (1971, 1973) independently derived estimates for the average composition of the Canadian Precambrian shield. Both studies created composites from representative samples taken over large areas that were weighted to reflect their surface outcrop area. The estimates of Shaw *et al.* are based on a significantly smaller number of samples than that of Eade and Fahrig's (i.e., ~430 versus ~14,000) and cover different regions of the shield, but the results are remarkably similar (Figure 2). All major elements agree to within ~10% except for CaO, which is ~20% higher, and MnO, which is 15% lower in the estimates of Shaw *et al.* estimates.

Shaw *et al.* (1967, 1976, 1986) also measured a number of trace elements in their shield composites and these are compared to the smaller number of trace elements determined by Eade and Fahrig in Table 2 and Figure 3. As might be expected, considering the generally greater variability in trace-element concentrations and the greater analytical challenge, larger discrepancies exist between the two averages. For example, scandium, chromium, copper, lanthanum, and uranium values vary by

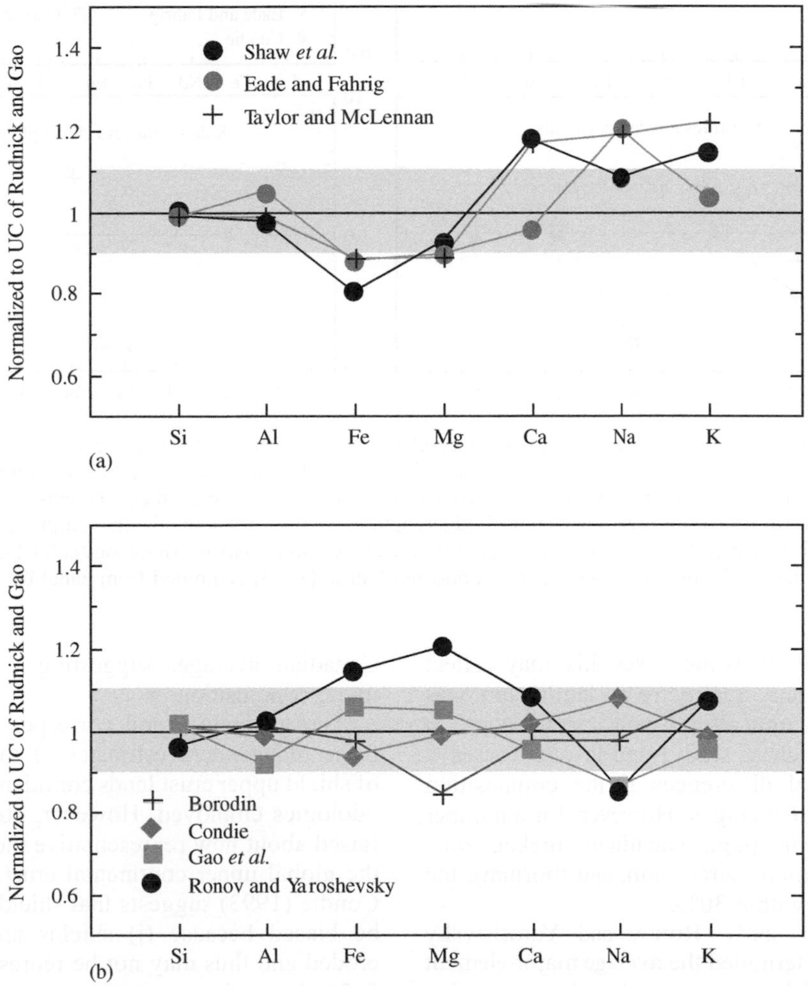

Figure 2 Comparison of different models for the major-element composition of the upper continental crust. All values normalized to the new composition provided in Table 3. Gray shaded field represents ±10% variation from this value. (a) Compositions derived from Canadian Shield samples (Shaw *et al.*, 1967, 1976, 1986; Fahrig and Eade, 1968; Eade and Fahrig, 1971, 1973) and the Taylor and McLennan model (1985, 1995, as modified by McLennan, 2001b). (b) Compositions derived from surface sampling of the former Soviet Union (Ronov and Yaroshevsky, 1967, 1976; Borodin, 1998) and China (Gao *et al.*, 1998a) and a global compilation of upper crustal rock types weighted in proportion to their areal distribution (Condie, 1993). The Canadian shield averages appear to be more evolved (having lower Mg, Fe, and higher Na and K) than other estimates.

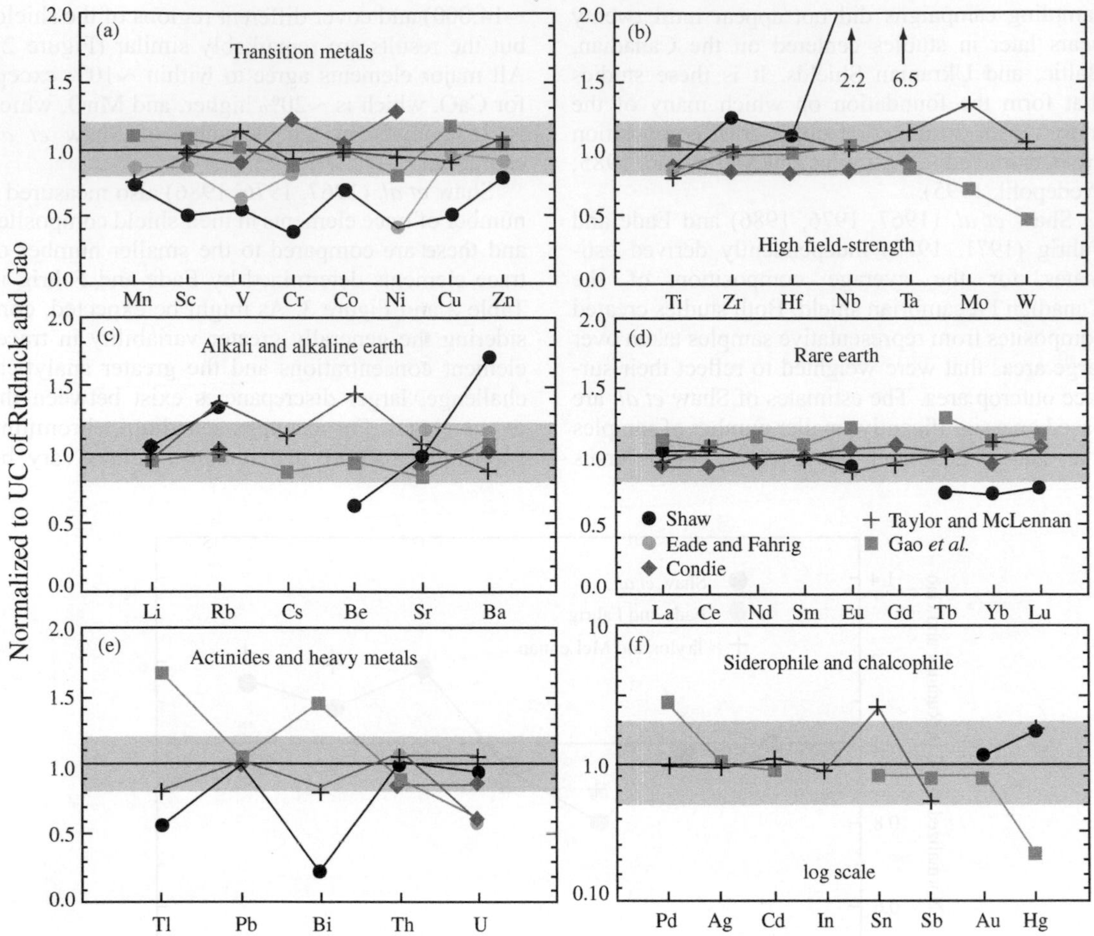

Figure 3 Comparison of different models for the trace-element composition of the upper-continental crust. All values normalized to the new composition provided in Table 3. Gray shaded field represents ±20% variation from this value for all panels except (f), in which gray field represents a factor of two variation. Trace elements are divided into the following groups: (a) transition metals, (b) high-field strength elements, (c) alkali, alkaline-earth elements, (d) REEs, (e) actinides and heavy metals, and (f) highly siderophile and chalcophile elements (note log scale). Data from Tables 1 and 2; lanthanum estimate from Eade and Fahrig (1973) is omitted from panel D.

~50% or more. In some cases this may reflect compromised data quality (e.g., lanthanum was determined by optical-emission spectroscopy in the Eade and Fahrig study) and in other cases it may reflect real differences in the composition between the two averages. However, for a number of trace elements (e.g., vanadium, nickel, zinc, rubidium, strontium, zirconium, and thorium), the averages agree within 30%.

In a similar study, Ronov and Yaroshevsky (1967, 1976) determined the average major-element composition of the upper crust based on extensive sampling of rocks from the Baltic and Ukranian shields and the basement of the Russian platform (Table 1). While the SiO_2, Al_2O_3, and K_2O values fall within 5% of those of the average Canadian Shield, as determined by Eade and Fahrig (1971, 1973), FeO_T, MgO, and CaO are ~10–30% higher, and Na_2O is ~30% lower than the Canadian average, suggesting a slightly more mafic composition.

The generally good correspondence between these independent estimates of the composition of shield upper crust lends confidence in the methodologies employed. However, questions can be raised about how representative the shields are of the global upper continental crust. For example, Condie (1993) suggests that shield averages may be biased because (i) shields are significantly eroded and thus may not be representative of the 5–20 km of uppermost crust that has been removed from them and (ii) they include only Precambrian upper crust and largely ignore any Phanerozoic contribution to upper crust. Condie (1993) derived an upper-crust composition based on over 3,000 analyses of upper crustal rock types weighted according to their distributions on geologic maps and stratigraphic sections, mainly covering regions of North America, Europe, and Australia. He utilized two methods in calculating an average upper-crust composition: (i) using the map distributions, irrespective of level of erosion and (ii) for areas that have been significantly eroded, restoring the eroded upper crust, assuming it has a ratio of supracrustal rocks to plutonic rocks similar to that seen in uneroded upper crustal regions. The latter approach was particularly important for his study, as one of his primary objectives was to evaluate whether there has been any secular change in upper crust composition. However, in this review, we are interested in the present-day composition of the upper crust (eroded or not), so it seems most appropriate to consider his "map model" for comparisons with other models (for a discussion of the secular evolution of the continents.

Condie's "map model" is compared with other estimates of the upper crust in Tables 1 and 2 and Figures 2 and 3. For major elements, his upper crust composition is within 10% of the Canadian Shield values of Eade and Fahrig. It is also within 10% of some of the major elements estimated by Shaw, but has generally higher magnesium and iron, and lower calcium and potassium compared to Shaw's estimate (Figure 2). Many trace elements in Condie's upper-crust composition are similar (i.e., within 20%) to those of Shaw's Canadian Shield composites (Figure 3), including the light rare-earth elements (LREEs), strontium, yttrium, thorium, and uranium. However, several trace elements in Condie's average vary by $\geqslant 50\%$ from those of Shaw *et al.* (1967, 1976, 1986) as can be seen in the figure. These include transition metals (scandium, vanadium, chromium, and nickel), which are considerably higher in Condie's upper crust, and niobium, barium and tantalum, which are significantly lower in Condie's upper crust compared to Shaw's. These differences may reflect regional variations in upper crust composition (i.e., the Canadian Shield is not representative of the worldwide upper crust) or inaccuracies in either of the estimates due to data quality or insufficiency. As will be discussed below, it is likely that Condie's values for transition metals, niobium, tantalum, and barium are the more robust estimates of the average upper crust composition.

A recent paper by Borodin (1998) provides an average composition of the upper crust that includes much Soviet shield and granite data not included in most other worldwide averages. For this reason, it makes an interesting comparison with other data sets. Like other upper crustal estimates, major elements in the Borodin average upper crust (Table 1 and Figure 2) fall within 10% of the Eade and Fahrig average for the Canadian Shield, except for TiO_2 and FeO, which are ~13% higher, and Na_2O, which is ~20% lower than the Canadian average. Borodin's limited trace element averages (for chromium, nickel, rubidium, strontium, zirconium, niobium, barium, lanthanum, thorium, and uranium—not given in table or figures) fall within 50% of Shaw's Canadian Shield values except for niobium, which, like other upper crustal estimates, is about a factor of 2 lower than the Canadian average.

The more recent and comprehensive study of upper-crust composition derived from surface exposures was carried out by Gao *et al.* (1998a). Nine hundred and five composite samples were produced from over 11,000 individual rock samples covering an area of $9.5 \times 10^5\ km^2$ in eastern China, which includes samples from Precambrian cratons as well as Phanerozoic fold belts. The samples comprised both crystalline basement rocks and sedimentary cover, the thickness of which was determined from seismic and aeromagnetic data. Averages were derived by combining compositions of individual map units weighted according to their thicknesses (in the cases of sedimentary cover) and areal exposure, for shields. The upper crust is estimated to be ~15 km thick based on seismic studies (Gao *et al.*, 1998a) and the crystalline rocks exposed at the surface area assumed to maintain their relative abundance through this depth interval. Average upper crust was calculated both as a grand average and on a carbonate-free basis; carbonates comprise a significant rock type (7–22%) in many of the areas sampled (e.g., Yangtze craton). The grand average (including carbonate) has a significantly different bulk composition than other estimates of the upper crust (Gao *et al.*, 1998a; Table 2). Most of the latter are derived from crystalline shields and so a difference is expected. However, Condie's map model incorporates sedimentary cover as well as crystalline basement. The differences between Condie's map model and Gao *et al.* grand-total upper crust suggest that the carbonate cover in eastern China is thicker than most other areas. For this reason, we use Gao *et al.* (1998a) carbonate-free compositions in further discussions, but with the caveat that carbonates may be an overlooked upper crustal component in many upper crustal estimates.

The Gao *et al.* (1998a) major- and trace-element results are presented in Tables 1 and 2 and plotted in Figures 2 and 3, respectively. Unlike the model of Condie (1993), several of the major elements fall beyond 10% of Eade and Fahrig's Canadian Shield data (Figures 2 and 3). These include TiO_2, FeO, MnO, and MgO, which are higher, and Na_2O, which is lower in the eastern China upper crust compared to the Canadian Shield. Gao *et al.* (1998a) attribute these differences to erosional differences between the two areas. Whereas the Canadian Shield composites comprised mainly metamorphic rocks of the amphibolite facies, the eastern China composites contain large proportions of unmetamorphosed supracrustal units that are considered to have, on average, higher proportions of mafic volcanics. In this respect, the Gao *et al.* model composition compares favorably to Condie's map model and the Russian estimates for all major elements. However, the Na_2O content of the eastern China upper crust is one of lowest of all (~20% lower than Condie's average and 10% lower than Borodin's values, but similar to Ronov and Yaroshevsky's average) (Figure 2).

The trace-element composition estimated by Gao *et al.* (1998a) for the Chinese upper crust is very similar to that of Condie (1993). Like the latter model, many lithophile trace elements in the Gao *et al.* model are within 50% of the Canadian Shield averages of Shaw *et al.* (e.g., LREEs, yttrium, rubidium, strontium, zirconium, hafnium, thorium, and uranium), and the Chinese average has significantly higher transition metals and lower niobium, barium, and tantalum than the Canadian Shield average. In addition, Gao *et al.* (1998a) provide values for some of the less well-constrained element concentrations. Of these, averages for lithium, beryllium, zinc, gallium, cadmium, and gold fall within 40% of the Shaw *et al.* averages, but boron, thallium, and bismuth are significantly higher, and mercury is significantly lower in Gao's average than in Shaw's. There is too little information for these elements in general to fully evaluate the significance of these differences.

Several generalizations can be made from the above studies of surface composites.

(i) Major element data are very consistent from study to study, with most major-element averages falling within 10% of Eade and Fahrig's Canadian Shield average. When differences do occur, they appear to reflect a lower percentage of mafic lithologies in the Canadian averages: all other estimates (including the Russian shield data) have higher FeO and TiO_2 than the Canadian averages and most also have higher CaO and MgO (Figures 2 and 3). The Eade and Fahrig average also has higher Na_2O than all other estimates (including Shaw's estimate for the Canadian Shield).

(ii) Trace elements show more variation than major elements from study to study, but some lithophile trace elements are relatively constant: rare earth elements (REEs), yttrium, lithium, rubidium, caesium, strontium, zirconium, hafnium, lead, thorium, and uranium do not vary beyond 50% between studies. Transition metals (scandium, cobalt, nickel, chromium, and vanadium) are consistently lower in the Canadian Shield estimates than in other studies, which may also be attributed to a lower percentage of mafic lithologies in the Canadian Shield (a conclusion supported by studies of sediment composition, as discussed in the next section). Barium is ~40% higher in the Shaw *et al.* average than in all other averages, including that of Eade and Fahrig, suggesting that this value is too high. Finally, niobium and tantalum are both about a factor of 2 higher in the Shaw *et al.* average than in any other average, suggesting that the former is not representative of the upper continental crust, a conclusion reached independently by Plank and Langmuir (1998) and Gallet *et al.* (1998) based on the composition of marine sediments and loess (see next section).

5.2.2 Sedimentary Rocks and Glacial Deposit Averages

While the large-scale sampling campaigns outlined above are the primary means by which the major-element composition of the upper continental crust has been determined, many estimates of the trace-element composition of the upper crust rely on the natural wide-scale sampling processes of sedimentation and glaciation. These methods are used primarily for elements that are insoluble during weathering and are, therefore, transported quantitatively from the site of weathering/glacial erosion to deposition. This methodology has been especially useful for determining the REE composition of the upper crust (see Taylor and McLennan (1985) and references therein). The averages derived from each of these natural large-scale samples are discussed in turn. When the upper crustal concentration of elements is discussed, the element name is printed in italic text so that the reader can quickly scan the text to the element of interest.

5.2.2.1 Sedimentary rocks

Processes that produce sedimentary rocks include weathering, erosion, transportation, deposition, and diagenesis. Elemental fractionation during weathering is discussed in detail by Taylor and McLennan (1985) (see also Chapter 7) and the interested reader is referred to these works for more extensive information. Briefly, elements with high solubilities in natural waters (Figure 4) have greater potential for being fractionated during sedimentary processing; thus, their concentration in fine-grained sedimentary rocks may not be representative of their source region. These elements include the alkali and alkaline-earth elements as well as boron, rhenium, molybdenum, gold, and uranium.

In contrast, a number of elements have very low solubilities in waters. Their concentrations in sedimentary rocks may, therefore, provide robust estimates of the average composition of their source regions (i.e., average upper-continental crust). Taylor and McLennan (1985) identified that REEs, yttrium, scandium, thorium, and possibly cobalt as being suitably insoluble and thus

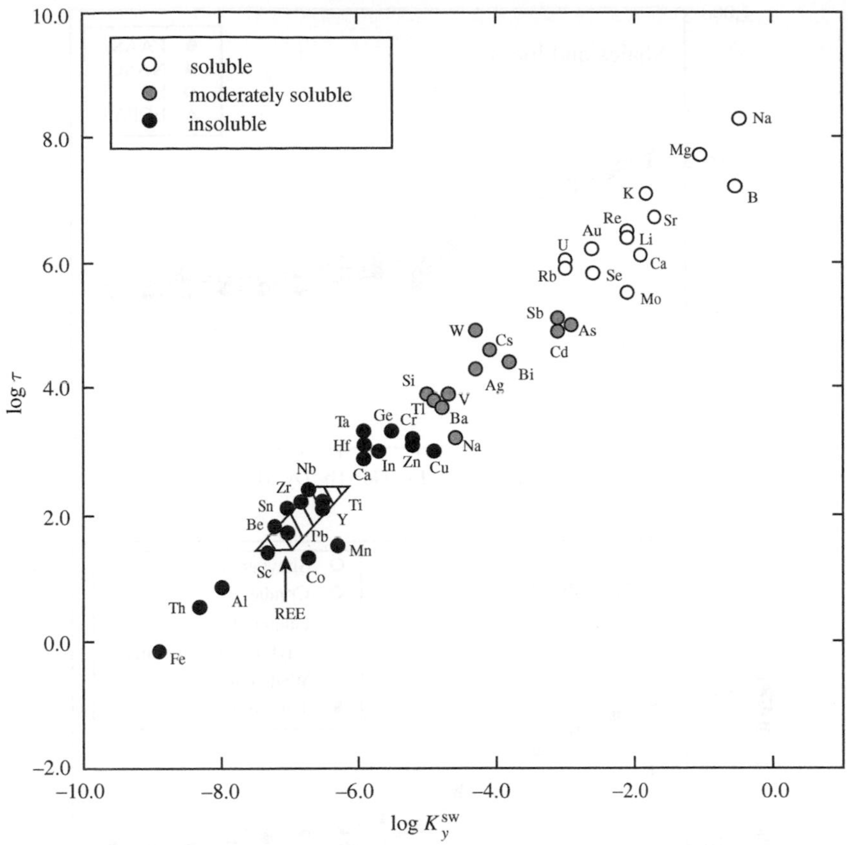

Figure 4 Plot of residence time (expressed as log τ) against seawater upper crust partition coefficient (expressed as K_y^{sw}) (source Taylor and McLennan, 1985).

providing useful information on upper crust composition.

The *REE* patterns for post-Archean shales show striking similarity worldwide (Figure 5): they are light REE enriched, with a negative europium anomaly and relatively flat heavy REEs. This remarkable consistency has led to the suggestion that the REE patterns of shales reflect that of the average upper-continental crust (Taylor and McLennan (1985) and references therein). Thus, Taylor and McLennan's (1985) upper crustal REE pattern is parallel to average shale, but lower in absolute abundances due to the presence of sediments with lower REE abundances such as sandstones, carbonates, and evaporites. Using a mass balance based on the proportions of different types of sedimentary rocks, they derive an upper crustal REE content that is 80% that of post-Archean average shale.

Comparison of various upper crustal REE patterns is provided in Figure 5. All estimates, whether from shales, marine sediments, or surface sampling, agree to within 20% for the LREEs and ~50% for the heavy rare-earth elements (HREEs). The estimate of Shaw *et al.* (1976) has the lowest HREEs and if these data are excluded, the HREEs agree to within 15% between the models of Condie (1993), Gao *et al.* (1998a), and Taylor and McLennan (1985). Thus, the REE content of the upper continental crust is established to within 10–25%, similar to the uncertainties associated with its major-element composition.

Once the REE concentration of the upper crust has been established, values for other insoluble elements can be determined from their ratios with an REE. Using the constant ratios of La/Th and La/Sc observed in shales, McLennan *et al.* (1980) and Taylor and McLennan (1985) estimated the upper crustal *thorium* and *scandium* contents at 11 ppm and 10.7 ppm, respectively. The scandium value increased slightly (to 13.7 ppm) and the thorium value remained unchanged when a more comprehensive sediment data set was employed by McLennan (2001b). The sediment-derived scandium and thorium averages agree to within 20% of the surface-sample averages (Table 2 and Figure 3).

Other insoluble elements include the high-field strength elements (HFSEs—titanium, zirconium, hafnium, niobium, tantalum, molybdenum,

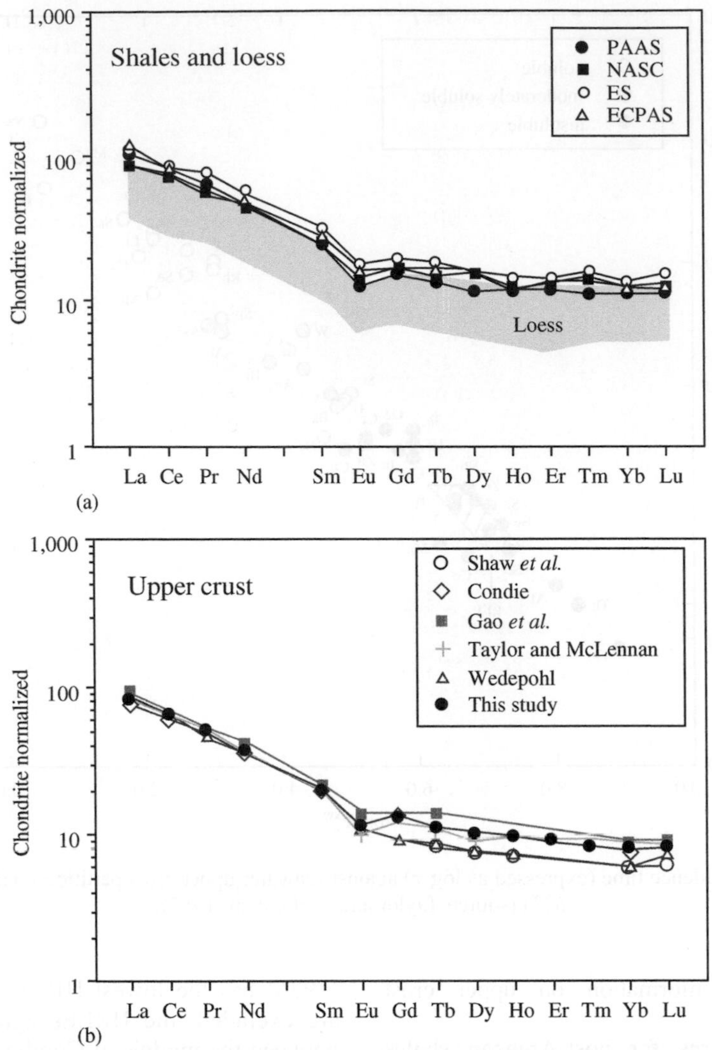

Figure 5 Comparison of REE patterns between (a) average post-Archean shales and loess and (b) various estimates of the upper continental crust composition. PAAS = post-Archean Australian Shale (Taylor and McLennan, 1985); NASC = North American shale composite (Haskin *et al.*, 1966); ES = European shale composite (Haskin and Haskin, 1966); ECPAS = Eastern China post-Archean shale (Gao *et al.*, 1998a). The loess range includes samples from China, Spitsbergen, Argentina, and France (Gallet *et al.*, 1998; Jahn *et al.*, 2001). Chondrite values are from Taylor and McLennan (1985).

tungsten), beryllium, aluminum, gallium, germanium, indium, tin, lead, and a number of transition metals (chromium, cobalt, nickel, copper, and zinc). Taylor and McLennan (1985) noted that some of these insoluble elements (e.g., HFSEs) may be fractionated during sedimentary processing if they reside primarily in heavy minerals. More recent evaluations have suggested that this effect is probably not significant for niobium and tantalum, and fractionations of zirconium and hafnium due to heavy mineral sorting are only really apparent in loess (Barth *et al.*, 2000; McLennan, 2001b; Plank and Langmuir, 1998). Plank and Langmuir (1998) noted that the *niobium, tantalum*, and *titanium* concentrations derived for the upper crust using marine sediments are considerably different from those of the Taylor and McLennan's upper continental crust composition. As oceanic processes are unlikely to fractionate these elements, Plank and Langmuir (1998) suggested that marine sediments provide a reliable estimate of the average composition of the upper continental crust. Using correlations between Al_2O_3 and niobium, they derived a niobium concentration for the upper crust of 13.7 ppm, and tantalum of 0.96 ppm (assuming Nb/Ta = 14); these values are about a factor of 2 lower than Taylor and McLennan's (1985) upper crustal estimates. Taylor and McLennan (1985) adopted their niobium value from Shaw *et al.* (1976), and their tantalum value was derived by assuming a Nb/Ta ratio of ~12 for the upper crust. The Plank and Langmuir niobium and tantalum values are similar to those derived from the surface-sampling studies of Condie (1993), Gao *et al.* (1998a), and Borodin (1998) and from more recent evaluations of these elements in shales, loess and other terrigenous sedimentary rocks (Barth *et al.*, 2000; Gallet *et al.*, 1998; McLennan, 2001b). All of these estimates range between 10 ppm and 14 ppm niobium and 0.74 ppm to 1.0 ppm tantalum, an overall variation of ~30%. Thus, niobium and tantalum concentrations now appear to be nearly as well constrained as the REE in the upper continental crust.

Plank and Langmuir (1998) also suggested, from their analyses of marine sediments, increasing the upper crustal TiO_2 values by ~40% (from 0.5 wt. % to 0.76 wt. %). Thus the TiO_2 content of the upper-continental crust probably lies between 0.55 wt.% and 0.76 wt.%, a difference of ~30%.

Of the remaining insoluble elements, recent evaluation of *zirconium* and *hafnium* concentrations derived from terrigenous sediment (McLennan, 2001b) show no significant differences with Taylor and McLennan's estimates, whose upper crustal zirconium value derives from the *Handbook of Geochemistry* (Wedepohl, 1969–1978), with hafnium determined from an assumed Zr/Hf ratio of 33. These values lie within ~20% of the surface-exposure averages (Table 2, Figure 3).

For the insoluble transition metals *chromium, cobalt*, and *nickel*, McLennan's (2001b) recent evaluation suggests approximate factor of 2 increases in average upper crustal values over those of Taylor and McLennan (1985). Taylor and McLennan's (1985) values were taken from a variety of sources (see Table 1 of Taylor and McLennan, 1981) and are similar to the Canadian Shield averages, which appear to represent a more felsic upper-crust composition, as discussed above. Even after eliminating these lower values, 30–40% variation exists for chromium, cobalt, and nickel between different estimates (Table 2 and Figure 3), and the upper crustal concentrations of these elements remains poorly constrained relative to REE.

McLennan (2001b) evaluated the upper crustal *lead* concentration from sediment averages and suggested a slight (~15%) downward revision (17 ppm) from the value of Taylor and McLennan (1985), whose value derives from a study by Heinrichs *et al.* (1980). McLennan's value is identical to that of surface averages (Table 2) and collectively these should be considered as a robust estimate for the lead content of the upper crust. For the remaining insoluble elements—*beryllium, copper, zinc, gallium, germanium, indium*, and *tin*—no newer data are available for terrigenous sediment averages. Estimates for some elements (e.g., zinc, gallium, germanium, and indium) vary by only ~20–30% between different studies, but others (beryllium, copper, and tin) vary by a factor of 2 or more (Table 2 and Figure 3).

It may also be possible to derive average upper crustal abundances of elements that have intermediate solubilities (e.g., vanadium, arsenic, silver, cadmium, antimony, caesium, barium, tungsten, and bismuth) using their concentrations in fine-grained sedimentary rocks, if they show significant correlations with lanthanum. Using this method McLennan (2001b) derived estimates of the upper crustal composition for *barium* (550 ppm) and *vanadium* (107 ppm). McLennan's barium value does not differ from that of Taylor and McLennan (1985), which derives from the *Handbook of Geochemistry* (Wedepohl, 1969–1978). This value is ~10% to a factor of 2 lower than the shield estimates; 630–700 ppm seems to be the most common estimate for barium from surface exposures. McLennan's vanadium estimate is ~50% higher than that of Taylor and McLennan (1985), which was derived from a 50 : 50 mixture of basalt:tonalite (Taylor and McLennan, 1981) and is similar to the Canadian Shield averages (Table 2). The revised vanadium value is similar to the surface-exposure averages from eastern China (Gao *et al.*, 1998a) and Condie's (1993) global average.

Several studies have used data for sedimentary rocks to derive the concentration of *caesium* in the upper crust. McDonough *et al.* (1992) found that a variety of sediments and sedimentary rocks (including loess) have an Rb/Cs ratio of 19 (± 11, 1σ), which is lower than the value of 30 in Taylor and McLennan's (1985) upper crust. Using this ratio and assuming a rubidium content of 110 ppm (from Shaw *et al.*, 1986; Shaw *et al.*, 1976; Taylor and McLennan, 1981), led them to an upper crustal caesium concentration of ~6 ppm. Data for marine sediments compiled by Plank and Langmuir (1998) also support a lower Rb/Cs ratio of the upper crust. Using the observed Rb/Cs ratio of 15 and a rubidium concentration of 112 ppm, they derived an upper crustal caesium concentration of 7.8 ppm. Although caesium data show only a poor correlation with lanthanum, the apparent La/Cs ratio of sediments led McLennan (2001b) to a revised caesium estimate of 4.6 ppm, which yields an Rb/Cs ratio of 24. Very few data exist for caesium from shield composites. Gao *et al.* determined a value of 3.6 ppm caesium, which is very similar to the estimate of Taylor and McLennan (1985). However, the Gao *et al.* rubidium estimate (83 ppm) is lower than Taylor and McLennan's (112 ppm), leading to an Rb/Cs ratio of 23 in the upper crust of eastern China. Caesium concentrations in all estimates vary by up to 70% and there thus appears to be substantial uncertainty in the upper crust's caesium concentration. Further evidence for the caesium content of the upper continental crust is derived from loess (see next section).

The upper crustal abundances of *arsenic, antimony,* and *tungsten* were determined by Sims *et al.* (1990), based on measurements of these elements in loess and shales. They find As/Ce to be rather constant at 0.08, leading to an arsenic content of 5.1 ± 1 ppm. In a similar fashion they estimate the upper crustal antimony content to be 0.45 ± 0.08 ppm and tungsten to be 3.3 ± 1.1 ppm. The antimony and arsenic values are factors of 2 and 3 higher, respectively, than the values given by Taylor and McLennan (1985), and the tungsten contents are a factor of 2 lower than Taylor and McLennan's (1985), which were adopted from the *Handbook of Geochemistry* (Wedepohl, 1969–1978). For all three elements, the Sims *et al.* estimates lie within uncertainty of the values given by Gao *et al.* (1998a) for the upper crust of eastern China, and these new estimates can thus be considered as representative of the upper crust to within ~30% uncertainty.

For the remaining moderately soluble elements *silver, cadmium,* and *bismuth,* there are no data for sedimentary composites. Taylor and McLennan (1985) adopted values from Heinrichs *et al.* (1980) for cadmium and bismuth and from the *Handbook of Geochemistry* (Wedepohl, 1969–1978) for silver. The only other data come from the study of Gao *et al.* (1998a). So essentially there are only two studies that address the concentrations of these elements in the upper crust: Gao *et al.* (1998a) and Wedepohl (1995) (which incorporates data from the *Handbook of Geochemistry* and Heinrichs *et al.* (1980)). For silver and cadmium, the two estimates converge: silver is identical and cadmium varies by 25% between Gao *et al.* and Wedepohl *et al.* estimates. In contrast, bismuth shows a factor of 2 of variation, with the Gao *et al.* estimates being higher.

5.2.2.2 Glacial deposits and loess

The concept of analyzing glacial deposits in order to determine average upper crustal composition originated with Goldschmidt (1933, 1958). The main attraction of this approach is that glaciers mechanically erode the rock types that they traverse, giving rise to finely comminuted sediments that represent averages of the bedrock lithologies. Because the timescale between erosion and sedimentation is short, glacial sediments experience little chemical weathering associated with their transport and deposition. In support of this methodology for determining upper crust composition, Goldschmidt noted that the major-element composition of composite glacial loams from Norway (analysed by Hougen *et al.*, 1925, as cited in Goldschmidt, 1933, 1958), which sample $\sim 2 \times 10^5$ km^2 of Norwegian upper crust, compares favorably with the average igneous-rock composition determined by Clarke and Washington (1924) (Table 1). It would take another fifty years before geochemists returned to this method of determining upper crustal composition.

More recent studies using glacial deposits to derive average upper-crust composition have focused on the chemical composition of loess—fine-grained eolian sediment derived from glacial outwash plains (Taylor *et al.*, 1983; Gallet *et al.*, 1998; Peucker-Ehrenbrink and Jahn, 2001; Hattori *et al.*, 2003). This can be accomplished in two ways: either using the average composition of loess as representative of the upper continental crust or, if an element correlates with an insoluble element such as lanthanum whose upper concentration is well established, using the average X/La ratio of loess (where "X" is the element of interest), and assuming an upper crustal lanthanum value to determine the concentration of "X" (e.g., McLennan, 2001b). In this and subsequent discussion of loess, we derive upper crustal concentrations for particular elements using this method and assuming an upper crustal lanthanum value of 31 ppm, and compare these to previous estimates for these elements. The quoted uncertainty reflects 1σ on that ratio.

Loess is rich in SiO_2 (most carbonate-free loess has 73 wt.% to 80 wt.% SiO_2 (Taylor *et al.*, 1983; Gallet *et al.*, 1998), which probably reflects both the preferential eolian transport of quartz into loess and sedimentary recycling processes. This enrichment causes other elemental concentrations to be diluted. In addition, some other elements may be similarly fractionated during eolian processing. For example, loess shows anomalously high concentrations of zirconium and hafnium (Taylor *et al.*, 1983; Barth *et al.*, 2000), which, like the SiO_2 excess, have been attributed to size sorting through eolian concentration of zircon (Taylor *et al.*, 1983). Thus, loess Zr/La and Hf/La are enriched relative to the upper continental crust and cannot be used to derive upper crustal zirconium and hafnium concentrations. In addition, a recent study of rhenium and osmium in loess suggests that osmium contents are enhanced in loess compared to its source regions (Hattori *et al.*, 2003). This is explained by Hattori *et al.* (2003) as being due to preferential sampling of the fine sediment fraction by the wind, which may be enriched in mafic minerals that are soft and hence more easily ground to finer, transportable particle sizes. Mafic-mineral enhancement could give rise to similar fractionations between lanthanum and elements that are found primarily in mafic minerals (e.g., nickel, vanadium, scandium, chromium, cobalt, manganese, etc.). In such cases neither averages nor La/X ratios can be used to determine a reliable estimate of upper crustal composition. However, it is not apparent that eolian processing has significantly fractionated incompatible elements from lanthanum (e.g., barium, strontium, potassium, rubidium, niobium, thorium, etc.) that are not hosted primarily in mafic minerals. Indeed, the close correspondence of the thorium content of the upper crust derived from loess La–Th correlations (10.5 ± 1 ppm; Figure 6) to that deduced from shales (10.7 ppm, Taylor and McLennan, 1985) suggests that upper crustal concentrations of these elements derived from loess La–X correlations are not significantly affected by eolian processing.

Taylor *et al.* (1983), and later Gallet *et al.* (1998), determined the trace-element composition of a variety of loess samples from around the world and found that their REE patterns are remarkably constant and similar to that of average shales (see previous section and Figure 5). Likewise, niobium, tantalum, and thorium show strong positive correlations with the REE (Figure 6; Barth *et al.*, 2000; Gallet *et al.*, 1998). Thus, it appears that loess provides a robust estimate of average upper crustal composition for insoluble, incompatible trace elements.

Because loess is glacially derived, weathering effects are significantly reduced compared to shales (Taylor *et al.*, 1983), raising the possibility that loess may provide robust upper crustal estimates for the more soluble trace elements. However, examination of the major-element compositions of loess shows that all bear the signature of chemical weathering (Gallet *et al.*, 1998). Gallet *et al.* attributed this to derivation of loess particles from rocks that had previously

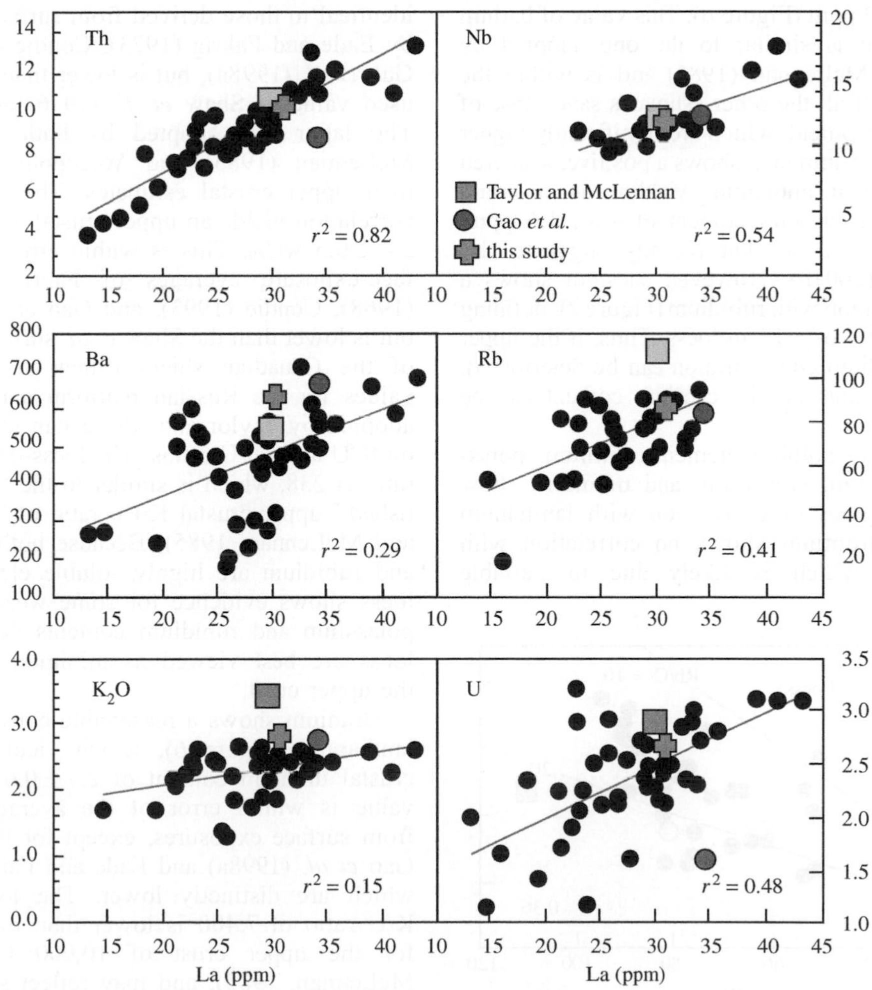

Figure 6 Lanthanum versus moderately to highly soluble incompatible trace elements in loess. Although loess is derived in part from weathered source regions, the positive correlations suggest that weathering has not completely obliterated the original, upper crustal mixing trends. Lines represent linear fit to data. Various models for the average upper crustal composition are superimposed (Taylor and McLennan, 1995, as modified by McLennan, 2001b; Gao *et al.*, 1998a and this study—Table 3) (sources Taylor *et al.*, 1983; Gallet *et al.*, 1998; Barth *et al.*, 2000; Jahn *et al.*, 2001; Peucker-Ehernbrink and Jahn, 2001).

experienced sedimentary differentiation. Likewise, Peucker-Ehernbrink and Jahn (2001) noted a positive correlation between $^{87}Rb/^{86}Sr$ and $^{87}Sr/^{86}Sr$ in loess, indicating that the weathering-induced fractionation is an ancient feature, and therefore inherited from the glacially eroded bedrocks. Even so, the degree of weathering in loess, as measured by the "chemical index of alteration" (CIA = molar $Al_2O_3/(Al_2O_3 + CaO + Na_2O + K_2O)$ Nesbit and Young (1984)), is small relative to that seen in shales (Gallet *et al.*, 1998), and it is likely that loess would provide a better average upper crustal estimate for moderately soluble trace elements (e.g., arsenic, silver, cadmium, antimony, caesium, barium, tungsten, and bismuth) than shales. Unfortunately, few measurements of these elements in loess are available (Barth *et al.*, 2000; Gallet *et al.*, 1998; Jahn *et al.*, 2001; Taylor *et al.*, 1983). Barium data show a scattered, positive correlation with lanthanum, yielding an upper crustal average of 510 ± 139 ppm (Figure 6). This value of barium concentration is similar to the one adopted by Taylor and McLennan (1985) and is within the uncertainty of all the other estimates save those of the Canadian Shield, which are significantly higher (Table 2). Caesium also shows a positive, scattered correlation with lanthanum, yielding an uncertain upper crustal caesium content of 4.8 ± 1.6 ppm, which is similar to that recently suggested by McLennan (2001b). However, caesium shows a better correlation with rubidium (Figure 7), defining an Rb/Cs ratio of ~17 in loess. Thus, if the upper crustal rubidium concentration can be determined, better constraints on the caesium content can be derived.

The highly soluble elements (lithium, potassium, rubidium, strontium, and uranium) show variable degrees of correlation with lanthanum in loess. Strontium shows no correlation with lanthanum, which is likely due to variable

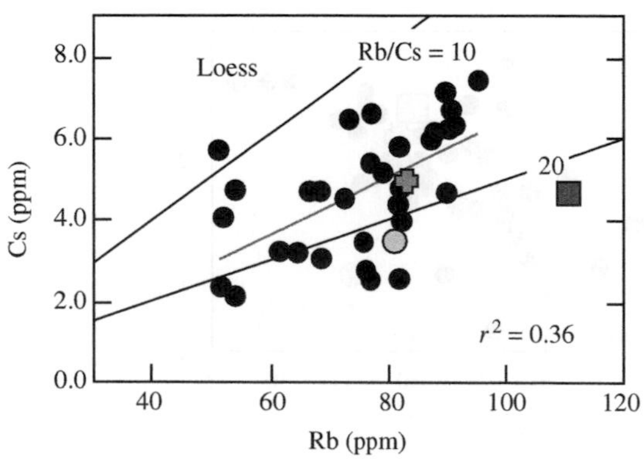

Figure 7 Rubidium versus caesium concentrations in loess samples. Short line is linear regression of data, thin, labeled lines represent constant Rb/Cs ratios. Symbols for crustal models and data sources as in Figure 6.

amounts of carbonate in the loess samples (Taylor *et al.*, 1983). Teng *et al.* (2003) recently reported lithium contents and isotopic compositions of shales and loess. Lithium contents of loess show no correlation with lanthanum, but fall within a limited range of compositions (17–41 ppm), yielding an average of 29 ± 10 ppm ($n = 14$). A similar value is derived using the correlation observed between lithium and niobium in shales. Thus, Teng *et al.* (2002) estimated the upper crustal lithium content at 31 ± 10 ppm, which is within error of previous estimates (Shaw *et al.*, 1976; Taylor and McLennan, 1985; Gao *et al.*, 1998a).

Potassium and rubidium show scattered, positive correlations with lanthanum (the Rb–La correlation is better, and the K–La is worse than the Ba–La correlation) (Figure 6). These correlations yield an upper crustal rubidium concentration of 84 ± 17 ppm. This rubidium value is identical to those derived from surface sampling by Eade and Fahrig (1973), Condie (1993), and Gao *et al.* (1998a), but is lower than the widely used value of Shaw *et al.* (1976) at 110 ppm. The latter was adopted by both Taylor and McLennan (1985) and Wedepohl (1995) for their upper crustal estimates. The weak K–La correlation yields an upper crustal K_2O value of 2.4 ± 0.5 wt.%. This is within error of the surface-exposure averages of Fahrig and Eade (1968), Condie (1993), and Gao *et al.* (1998a), but is lower than the Shaw *et al.* surface averages of the Canadian shield (Shaw *et al.*, 1967), values for the Russian platform and the value adopted by Taylor and McLennan (1985) based on K/U and Th/U ratios. The loess-derived K/Rb ratio is 238, which is similar to the "well established" upper crustal K/Rb ratio of 250 (Taylor and McLennan, 1985). Because both potassium and rubidium are highly soluble elements, and loess shows evidence for some weathering, the potassium and rubidium contents derived from loess are best viewed as minimum values for the upper crust.

Uranium shows a reasonable correlation with lanthanum (Figure 6), which yields an upper crustal uranium content of 2.7 ± 0.6 ppm. This value is within error of the averages derived from surface exposures, except for the value of Gao *et al.* (1998a) and Eade and Fahrig (1973), which are distinctly lower. The loess-derived K/U ratio of 7,400 is lower than that assumed for the upper crust of 10,000 (Taylor and McLennan, 1985), and may reflect some potassium loss due to weathering, as discussed above.

Peucker-Ehrenbrink and Jahn (2001) analyzed loess in order to determine the concentrations of the platinum-group element (PGE) in the upper continental crust. To do this they examined

PGE-major-element trends and used previously determined major-element compositions of the upper continental crust to infer the PGE concentrations (Table 2). Of the elements analyzed (ruthenium, palladium, osmium, iridium, and platinum), they found positive correlations for ruthenium, palladium, osmium, and iridium with major and trace elements for which upper crustal values had previously been established, leading to suggested upper crustal abundances of 340 ppt, 520 ppt, 31 ppt, and 22 ppt, respectively. They found no correlation between platinum contents and other elements, and so they simply used the average loess platinum content (510 ppt) as representative of the upper continental crust. We have estimated uncertainty on these values (shown in Table 2) by using the 95% confidence limit on the correlations published by Peucker-Ehrenbrink and Jahn (2001) and, for platinum, the standard deviation of the mean (Table 3). Recently, Hattori *et al.* (2003) suggested that preferential sampling of mafic minerals in loess may lead to enhancement of PGE and thus, loess-derived estimates may represent maximum concentrations for the upper crust. Based on samples of glacially derived desert sands and glacial moraines, Hattori *et al.* (2003) estimated an upper crustal osmium abundance of $\sim$10 ppt.

Prior to these studies, few estimates were available for the PGE content of the upper continental crust. Peucker-Ehrenbrink and Jahn's loess-derived palladium value is similar to the value published by Taylor and McLennan (1985), which derives from the *Handbook of Geochemistry* (S. R. Taylor, personal communication), but is a factor of 3 smaller than that determined by Gao *et al.* (1998a) for the upper crust of eastern China. Peucker-Ehrenbrink and Jahn's (2001) loess-derived osmium abundance is $\sim$65% lower than the estimate of Esser and Turekian (1993), which Peucker-Ehrenbrink and Jahn attribute to the hydrogenous uptake of osmium by the riverine sediments used in that study. Furthermore, the desert-sand and glacial moraine-derived osmium value of Hattori *et al.* (2003) is a factor of 3 lower than the estimate of Peucker-Ehrenbrink and Jahn (2001). Peucker-Ehrenbrink and Jahn's (2001) loess-derived iridium content is the same as that published for the Canadian Shield by Shaw *et al.* (1976). Thus, the upper crustal concentration of some PGE may be reasonably well constrained (e.g., palladium and iridium), while considerable uncertainty remains for others (e.g., platinum and osmium).

Rhenium is a highly soluble element that is easily leached during weathering, so the rhenium abundances of loess cannot be used directly to infer its upper crustal abundance. Following Esser and Turekian (1993), Peucker-Ehrenbrink and Jahn (2001) used the average $^{187}Os/^{188}Os$ ratio, osmium concentration, and average neodymium-model age of the crust to calculate the rhenium content of the upper continental crust. Their value (198 ppt) is about half that reported in Taylor and McLennan (1985) and calculated by Esser and Turekian (1993), who used the higher osmium abundance in their calculation. Using a similar methodology and osmium-isotopic composition, and the lower osmium abundance determined for glacially derived desert sands, Hattori *et al.* (2003) determined an upper crustal $^{187}Re/^{188}Os$ ratio of 35, which (assuming an average neodymium model age of 2.2 Ga for the crust) corresponds to a rhenium content of 74 ppt, about a third of the concentration determined by Peucker-Ehrenbrink and Jahn from loess data. Sun *et al.* (2003) used the rhenium contents of undegassed arc lavas to estimate the rhenium content of the bulk continental crust, assuming that the crust grows primarily by arc accretion. Their value of 2.0 ± 0.1 ppb is over an order of magnitude higher than that estimated by Peucker-Ehrenbrink and Jahn (2001) and Hattori *et al.* (2003) and is $\sim$5 times higher than the Esser and Turekian (1993) and Taylor and McLennan (1985) values. Because rhenium is a moderately incompatible element, the rhenium concentration of the upper crust should be comparable to or higher than the bulk crust value (similar to ytterbium). However, this extreme rhenium concentration would require an order of magnitude higher osmium concentration in the crust or an extremely radiogenic crust composition, neither of which are consistent with any current estimates. Sun *et al.* (2003) suggest that rhenium may be lost from the continents by either rhenium degassing during arc volcanism or continental rhenium deposition into anoxoic sediments that are recycled into the mantle. Thus the value of 2 ppb rhenium is a maximum value for the upper continental crust and our knowledge of the rhenium content of the upper crust remains uncertain.

5.2.3 An Average Upper-crustal Composition

In Table 3 we present our best estimate for the chemical composition of the upper continental crust. The footnote provides detailed information on how the value for each element was derived. In general, major-element values represent averages of the different surface-exposure studies, and errors represent one standard deviation of the mean. Because two independent studies are available for the Canadian Shield, and because it appears the Canadian Shield has lower abundances of mafic lithologies and higher abundances of sodium-rich tonalitic–trondhjemitic granitic gneisses compared to other areas (see Section 5.2.1), we include only values from one of these studies—the Fahrig and Eade (1968) study, which encompasses a greater number of samples compared to that of Shaw *et al.*

Table 3 Recommended composition of the upper continental crust. Major elements in weight percent.

Element	Units	Upper crust	1 Sigma	%	Source[a]	Element	Units	Upper crust	1 Sigma	%	Source[a]
SiO_2	wt.%	66.6	1.18	2	1	Ag	$ng\,g^{-1}$	53	3	5	4
TiO_2	"	0.64	0.08	13	2	Cd	$\mu g\,g^{-1}$	0.09	0.01	15	4
Al_2O_3	"	15.4	0.75	5	1	In	"	0.056	0.008	14	4
FeO_T	"	5.04	0.53	10	1	Sn	"	2.1	0.5	26	14
MnO	"	0.10	0.01	13	1	Sb	"	0.4	0.1	28	12
MgO	"	2.48	0.35	14	1	I	"	1.4		50	5
CaO	"	3.59	0.20	6	1	Cs	"	4.9	1.5	31	15
Na_2O	"	3.27	0.48	15	1	Ba	"	628	83	13	16
K_2O	"	2.80	0.23	8	3	La	"	31	3	9	4
P_2O_5	"	0.15	0.02	15	1	Ce	"	63	4	6	4
Li	$\mu g\,g^{-1}$	24	5	21	11	Pr	"	7.1			4
Be	"	2.1	0.9	41	4	Nd	"	27	2	8	4
B	"	17	8	50	4	Sm	"	4.7	0.3	6	4
N	"	83			5	Eu	"	1.0	0.1	14	4
F	"	557	56	10	4	Gd	"	4.0	0.3	7	4
S	"	62	33	53	4	Tb	"	0.7	0.1	21	4
Cl	"	370	382	103	4	Dy	"	3.9			17
Sc	"	14.0	0.9	6	6	Ho	"	0.83			17
V	"	97	11	11	6	Er	"	2.3			4
Cr	"	92	17	19	6	Tm	"	0.30			17
Co	"	17.3	0.6	3	6	Yb	"	1.96	0.4	18	4
Ni	"	47	11	24	6	Lu	"	0.31	0.05	17	4
Cu	"	28	4	14	7	Hf	"	5.3	0.7	14	4
Zn	"	67	6	9	7	Ta	"	0.9	0.1	13	11
Ga	"	17.5	0.7	4	8	W	"	1.9	1	54	18
Ge	"	1.4	0.1	9	4	Re	$ng\,g^{-1}$	0.198			13
As	"	4.8	0.5	10	9	Os	"	0.031	0.009	29	13
Se	"	0.09	0.05	54	4	Ir	"	0.022	0.007	32	13
Br	"	1.6			5	Pt	"	0.5	0.5	95	13
Rb	"	84	17	20	10	Au	"	1.5	0.4	26	4
Sr	"	320	46	14	4	Hg	$\mu g\,g^{-1}$	0.05	0.04	76	4
Y	"	21	2	11	4	Tl	"	0.9	0.5	57	4
Zr	"	193	28	14	4	Pb	"	17	0.5	3	4
Nb	"	12	1	12	11	Bi	"	0.16	0.06	38	19
Mo	"	1.1	0.3	28	12	Th	"	10.5	1.0	10	20
Ru	$ng\,g^{-1}$	0.34	0.02	6	13	U	"	2.7	0.6	21	20
Pd	"	0.52	0.02	3	13						

[a] Sources: (1) Average of all surface exposure data from Table 1, excluding Shaw *et al.* (1967), which is replicated by Fahrig and Eade (1968). (2) As (1) above, but including sediment-derived data from Plank and Langmuir (1998) and McLennan (2001b). (3) As (1) above, but also including K_2O value derived from loess (see text). (4) Average of all values in Table 2, excluding Wedepohl (1995) value or Taylor and McLennan (1985) value for Au, if it is derivitive from Shaw *et al.* (1976) and Taylor and McLennan (1985). (5) Wedepohl (1995). (6) Average of all surface composite data in Table 2, excluding Shaw *et al.* (1976), and including additional data from sediments (McLennan, 2001b). (7) Average of all surface composite data in Table 2, excluding Shaw *et al.* (1976), and including Taylor and McLennan (1985) values. (8) Average of all surface composite data in Table 2, excluding Shaw *et al.* (1976) due to their fractionated Ga/Al ratio. (9) Average of sedimentary data from Table 2 (Sims *et al.*, 1990) and Gao *et al.* (1998a) surface averages. (10) Dervied from La/Rb correlation in loess (see text). Value is identical to surface exposure data except for the Shaw *et al.* (1976) values. Data from Handbook of Geochemistry are about a factor of 2 lower than the latter and are not included in the average. (11) Average of all surface exposure data in Table 2 (minus Shaw *et al.* (1976), values) plus data from sediments and loess (Plank and Langmuir, 1998; Barth *et al.*, 2000; McLennan, 2001b; Teng *et al.*, 2003). (12) Average of all data in Table 2, excluding Taylor and McLennan (1985), which derive from same source as Wedephol's. (13) From Peucker-Ehrenbrink and Jahn (2001); see text for origin of error estimates. (14) Average of all data in Table 2, excluding Taylor and McLennan (1985), which is a factor of two higher than all other estimates. (15) Derived from Rb/Cs = 17 and upper crustal Rb value (see text). (16) Average of all data in Table 2, excluding the Shaw *et al.* (1976) and including additional data from loess (see text). (17) Value interpolated from REE pattern. (18) Average of all values in Table 2, plus correlation from Newsom *et al.* (1996), assuming W/Th = 0.2 (19) Average of all values in Table 2, excluding the Shaw *et al.* (1976) value, which is a factor of 5 lower than the others. (20) From loess correlations with La (see text). Both values are within error of the average of all surface exposure data and other sedimentary data (Taylor and McLennan, 1985; McLennan, 2001b).

(1967). We also incorporate TiO_2 values derived from recent sedimentary studies (McLennan, 2001b; Plank and Langmuir, 1998) and the K_2O value from loess (Section 5.2.2) into the upper crustal averages (note that including the latter value in the average does not significantly change it). The standard deviation for most major-element averages is 10% or less. Only the ferromagnesian elements (iron, manganese, magnesium, and titanium), Na_2O, and P_2O_5 vary by up to 15%.

The trace-element abundances shown in Table 3 derive from different methods depending on their solubility. For most insoluble elements (see Section 5.2.2 and Figure 4 for definitions of solubility), we average the surface composites in addition to sediment or loess-derived estimates to derive the upper crustal composition. The uncertainty reported represents 1σ from the mean of all estimates. For moderately and highly soluble elements, we use the data derived from loess, if the elements show correlations with lanthanum ($r^2 > 0.4$), to infer their concentrations. In this case, the error represents the SD of the X/La ratio (where X is the element of interest). For elements that show no or only a poor correlation with lanthanum in loess and sediments (e.g., K_2O, Li, Ba, and Sr), we use the average of surface composites and sedimentary data (if some correlations exist with lanthanum) to derive an average. In most cases, the loess or sediment-derived values are within error of the surface-composite averages and these are noted in the footnote. Caesium is a special case. The loess caesium data show a poor correlation with lanthanum, but good correlation with rubidium. We thus use the observed Rb/Cs ratio of 17 (which is similar to the previous determination of this ratio in sedimentary rocks (McDonough *et al.*, 1992)) and the upper crustal rubidium concentration of 84 ppm to derive the caesium concentration of 4.9 ppm in the upper crust. The error on this estimate derives from the standard deviation of the Rb/Cs ratio. For some elements, only single estimates are available (e.g., bromine, nitrogen, iodine), and these are adopted as reported. The uncertainty of these estimates is likely to be very high, but there is no way to estimate uncertainty quantitatively with such few data. Remarkably, the SD on a large number of trace elements is below 20%, and the concentrations of a few (flourine, scandium, vanadium, cobalt, zinc, gallium, germanium, arsenic, yttrium, niobium, LREEs, tantalum, lead, and thorium) would appear to be known within ~10% (Table 3). However, in a number of these cases (e.g., flourine, cobalt, gallium, germanium, and arsenic), the small uncertainties undoubtedly reflect the fact that there have been few independent estimates made of the upper crust composition for these elements. It is likely that the true uncertainty for these elements is considerably greater than expressed in Table 3.

The upper crustal composition in Table 3 has many similarities to the widely used estimate of Taylor and McLennan (1985, with recent revision by McLennan (2001b)), but also some notable differences (Figure 8). Most of thc elements that vary by more than 20% from the estimate of Taylor and McLennan are elements for which new data are recently available and few data exist overall (i.e., beryllium, arsenic, selenium, molybdenum, tin, antimony, rhenium, osmium, iridium, thallium, and bismuth). However, a number of estimates exist for K_2O, P_2O_5, and rubidium contents of the upper crust and our estimates are significantly lower (by 20–40%) than Taylor and McLennan's upper crust. The difference in P_2O_5 may simply be due to rounding errors. Taylor and McLennan (1985) report P_2O_5 of 0.2 wt.% versus 0.15 wt.% in our and other estimates of the upper crust—Tables 2 and 3). Taylor and McLennan (1985) derived their upper crustal K_2O indirectly from thorium abundances by assuming Th/U = 3.8 and K/U = 10,000. The resulting K_2O value is the

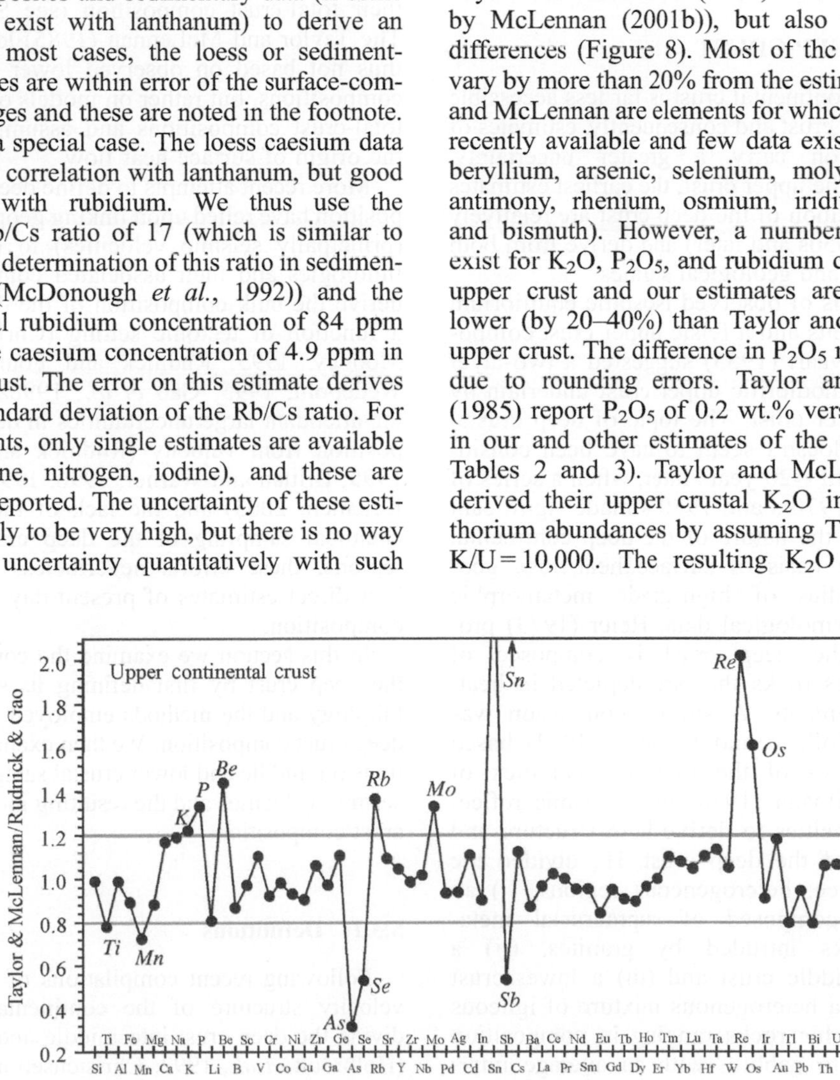

Figure 8 Plot of upper crustal compositional estimate of Taylor and McLennan (1995) (updated with values from McLennan, 2001b), divided by recommended values from this study. Horizontal lines mark 20% variation. Most elements fall within the ±20% bounds; elements falling beyond these bounds are labeled. Of the elements that differ by over 20%, potassium and rubidium are probably the most significant, since these elements are commonly analyzed to high precision in crustal rocks.

highest of any of the estimates (Table 2 and Figure 2). Likewise, Taylor and McLennan's rubidium value was determined from their K_2O content, assuming a K/Rb ratio of the upper crust of 250. Their rubidium concentration matches the Canadian Shield value of Shaw *et al.* (1976), but is higher than all other surface-exposure studies, including Fahrig and Eades Canadian Shield estimate (Table 2 and Figure 3). In contrast, the remaining surface-exposure studies match the rubidium value we derived from the loess Rb–La correlation (Figure 6). We conclude that the upper crust may have lower potassium and rubidium contents than estimated by Taylor and McLennan (1985). This finding has implications for total crustal heat production (see Section 5.4).

5.3 THE DEEP CRUST

The deep continental crust is far less accessible than the upper crust and consequently, estimates of its composition carry a greater uncertainty. Compared to the upper crust, the earliest estimates of the composition of the deep crust are relatively recent (i.e., 1950s and later) and derive from both seismological and geological studies.

On the basis of observed isostatic equilibrium of the continents and a felsic upper crust composition, Poldervaart (1955) suggested a two-layer crust with granodioritic upper crust underlain by a basaltic lower crust. The topic of deep crustal composition doesn't seem to have been considered again until ~20 years later, when a series of works in the 1970 s and 1980 s made significant headway into the nature of the deep continental crust. On the basis of surface heat-flow, geochemical studies of high-grade metamorphic rocks and seismological data, Heier (1973) proposed that the deep crust is composed of granulite-facies rocks that are depleted in heat-producing elements. A similar conclusion was reached by Holland and Lambert (1972) based on their studies of the Lewisian complex of Scotland. Smithson (1978) used seismic reflections and velocities to derive both structure and composition of the deep crust. He divided the crust into three, heterogeneous regions: (i) an upper crust composed of supracrustal metamorphic rocks intruded by granites, (ii) a migmatitic middle crust and (iii) a lower crust composed of a heterogenous mixture of igneous and metamorphic rocks ranging in composition from granite to gabbro, with an average intermediate (dioritic) composition. This three-layer model of the crust survives today in most seismologically based studies. Weaver and Tarney (1980, 1981, 1984) derived a felsic and intermediate composition for the Archean middle and lower crust, respectively, based on studies of amphibolite to granulite-facies rocks exposed in the Lewisian complex, Scotland. R. W. Kay and S. M. Kay (1981) were one of the first to stress the importance of xenolith studies to unravelling deep-crustal composition. They highlighted the heterogeneous nature of the deep crust and suggested its composition should vary depending on tectonic setting, cautioning against the use of singular cross sections or deep-crustal exposures to derive global models. Taylor and McLennan (1985) considered the lower crust to be the portion of the crust from 10 km depth to the Moho. Their "lower crust" thus includes both middle and lower crust, as used here (see Section 5.3.1). Taylor and McLennan's (1985) lower-crust composition was derived by subtracting the upper crust from their total-crust composition (see Section 5.4). The Taylor and McLennan (1985) lower crust is thus not based on observed lower crustal rock compositions, but rather on models of upper- and total-crust compositions and assumptions about the origin of surface heat flow.

More recent attempts to define deep crust composition have relied upon linking geophysical data (principally seismic velocities) to deep crustal lithologies and their associated compositions to derive the bulk composition of the deep crust as a function of tectonic setting (Christensen and Mooney, 1995; Rudnick and Fountain, 1995; Wedepohl, 1995; Gao *et al.*, 1998a,b). Despite the attendant large uncertainties in deriving composition from velocity (Rudnick and Fountain, 1995; Brittan and Warner, 1996, 1997; Behn and Kelemen, 2003) and the lack of thorough geochemical sampling of the deep crust in many regions, these efforts nevertheless provide the best direct estimates of present-day deep crustal composition.

In this section we examine the composition of the deep crust by first defining its structure and lithology and the methods employed to determine deep crust composition. We then examine observations on middle and lower crustal samples, average seismic velocities and the resulting models of deep crust composition.

5.3.1 Definitions

Following recent compilations of the seismic-velocity structure of the continental crust, we divide the deep crust into middle and lower crust (Holbrook *et al.*, 1992; Christensen and Mooney, 1995; Rudnick and Fountain, 1995). Holbrook *et al.* (1992) defined the middle crust as: (i) the middle-third, where the velocity structure suggests a natural division of the crust into thirds; (ii) the region beneath the upper crust and above a Conrad discontinuity, if there is a layer beneath the Conrad;

and (iii) the region immediately beneath the Conrad if there are two distinct velocity layers beneath a Conrad discontinuity. The lower crust is thus the layer beneath the middle crust and above the Moho.

For a ~40 km thick average global continental crust (Christensen and Mooney, 1995; Rudnick and Fountain, 1995), the middle crust is 11 km thick and ranges in depth from 12 km, at the top, to 23 km at the bottom (Gao *et al.* (1998b) based on the compilations of data for crustal structure in various tectonic settings by Rudnick and Fountain (1995)). The average lower crust thus begins at 23 km depth and is 17 km thick. However, the depth and thickness of both middle and lower crust vary from setting to setting. In fore-arcs, active rifts, and rifted margins, the crust is generally thinner: middle crust extends from 8 km to 17 km depth and lower crust from 17 km to 27 km depth. In Mesozoic–Cenozoic orogenic belts the crust is thicker and middle crust extends from 16 km to 27 km depth and the lower crust from 27 km to 51 km depth (Rudnick and Fountain, 1995).

5.3.2 Metamorphism and Lithologies

Studies of exposed crustal cross-sections and xenoliths indicate that the middle crust is dominated by rocks metamorphosed at amphibolite facies to lower granulite facies, while the lower crust consists mainly of granulite facies rocks (Fountain *et al.*, 1990a; Fountain and Salisbury, 1981; Mengel *et al.*, 1991; Weber *et al.*, 2002). However, exceptions to these generalities do occur. For thin crust in rifted areas, greenschist-facies and amphibolite-facies rocks may predominate in the middle and lower crust, respectively. In overthickened Mesozic and Cenozoic orogenic belts (e.g., Alps, Andes, Tibet, and Himalyas), and paleo-orogenic belts that now have normal crustal thicknesses (e.g., Appalachains, Adirondacks, Variscan belt), granulite-facies and eclogite-facies rocks may be important constituents of the middle and lower crust (Leech, 2001; LePichon *et al.*, 1997; Lombardo and Rolfo, 2000). In contrast, amphibolite-facies lithologies may be present in the deep crust of continental arcs (Aoki, 1971; Miller and Christensen, 1994; Weber *et al.*, 2002), where hydrous fluids are fluxed from the subducting slab and the water contents of underplating magmas are high.

Lithologically, both middle and lower crust are highly heterogeneous, as seen in surface exposures of high-grade metamorphic rocks, crustal cross-sections, and deep-crustal xenolith suites. However, there is a general tendency for the middle crust to have a higher proportion of evolved rock compositions (as observed in cross-sections and granulite-facies terranes) while the lower crust has a higher proportion of mafic rock types (as observed in xenolith suites (Bohlen and Mezger, 1989)). Metasedimentary lithologies are often present, albeit in small proportions. The exact proportions of felsic to mafic lithologies in the deep crust varies from place to place and can only be established through the study of crustal cross-sections or inferred from seismic velocity profiles of the crust (Christensen and Mooney, 1995; Rudnick and Fountain, 1995; Wedepohl, 1995; Gao *et al.*, 1998b).

5.3.3 Methodology

There are three approaches to derive the composition of the deep crust (see Rudnick and Fountain (1995) for a review).

(i) *By studying samples derived from the deep crust.* These occur as surface outcrops of high-grade metamorphic terranes (e.g., Bohlen and Mezger, 1989; Harley, 1989), tectonically uplifted crustal cross-sections (e.g., Fountain and Salisbury, 1981; Percival *et al.*, 1992), and as deep-crustal xenoliths carried in volcanic pipes (Rudnick, 1992; Downes, 1993).

(ii) *By correlating seismic velocities with rock lithologies* (Christensen and Mooney, 1995; Rudnick and Fountain, 1995; Wedepohl, 1995, Gao *et al.*, 1998a,b).

(iii) From surface heat-flow measurements.

As pointed out by Jaupart and Mareschal, surface heat flow is the only geophysical parameter that is a direct function of crustal composition. In general, however, heat flow provides only very broad constraints on deep-crust composition due to the ambiguity involved in distinguishing the amount of surface heat flow arising from crustal radioactivity versus the Moho heat flux; Rudnick *et al.*, 1998). Most models of the deep-crust composition fall within these broad constraints. The exception is the global model of Wedepohl, 1995, which produces more heat than the average surface heat flow in the continents, thereby allowing no mantle heat flux into the base of the crust (Rudnick *et al.*, 1998). In addition, the regional model of Gao *et al.* (1998a) for eastern China produces too much heat to be globally representative of the continental crust composition (see discussion in Rudnick *et al.* (1998) and Jaupart and Mareschal). However, the Gao *et al.* composition may be representative of the continental crust of eastern China, where the crust is relatively thin (30–35 km) and the heat flow is high (>60 mW m^{-2}). In the remaining discussion of deep-crust composition, we rely most heavily on methods (i)–(ii), above, but return to the question of heat flow when considering the bulk crust composition in Section 5.4.

In addition to mineralogy, which is in turn a function of bulk composition and metamorphic grade, factors affecting the seismic velocities of the continental crust include temperature, pressure, and the presence or absence of volatiles, fractures, and mineralogical anisotropy. It is generally assumed that cracks and fractures are closed under the ambient confining pressures of the middle to lower crust (0.4–1.2 GPa). In addition, although evidence for volatile transport is present in many rocks derived from the deep crust, the low density of these fluids allows for their escape to the upper crust shortly after their formation. Hence, most studies assume the deep crust does not, in general, contain an ambient, free volatile phase (Yardley, 1986).

Some minerals are particularly anisotropic with respect to seismic-wave speeds (e.g., olivine, sillimanite, mica (Christensen, 1982)), which can lead to pronounced seismic anisotropy in rocks if these minerals are crystallographically aligned through deformational processes (Meltzer and Christensen, 2001). This, in turn, could lead to over- or underestimation of representative seismic velocities of the deep crust if deformed rocks with such anisotropic minerals occur there. Olivine is not commonly stable in the deep crust, but other strongly anisotropic minerals are (e.g., mica, which is predominantly stable in the middle crust, and sillimanite, which is found in metapelitic rocks in the middle-to-lower crust). Some of the largest seismic anisotropies have been recorded in mica schists and gneisses, which can have average anisotropies over 10% (Christensen and Mooney, 1995; Meltzer and Christensen, 2001). Amphibole is also anisotropic and the average anisotropies for amphibolite are also ~10% (Christensen and Mooney, 1995; Kern *et al.*, 1996). In general, anisotropy is expected to be highest in metapelitic rocks and amphibolites, which contain the highest proportions of anisotropic minerals. These lithologies appear to be subordinate in middle-crustal sections and outcrops (described in the next section) compared to felsic gneisses, which typically have low anisotropies (<5%). In contrast, studies of xenoliths show metapelite to be a common lithology in the lower crust, albeit proportionally minor, and amphibolite may be important in some regions (Section 5.3.5.1). Thus, seismic anisotropy could be especially important in regions having large amounts of metasedimentary rocks (e.g., accretionary wedges) and amphibolite (arc crust?) in the deep crust, but is less likely to be important in crust dominated by felsic metaigneous rocks or mafic granulites.

Changes in *P*-wave velocity of a rock as a function of temperature and pressure are generally assumed to be on the order of $-4\times10^{-4}\,\mathrm{km\,s^{-1}\,°C^{-1}}$ and $2\times10^{-4}\,\mathrm{km\,s^{-1}MPa^{-1}}$ (see Rudnick and Fountain (1995 and references7 therein). Because most laboratory measurements of ultrasonic velocities are carried out at confining pressures of 0.6–1.0 GPa, no pressure correction needs to be made in order to compare field and laboratory-based velocity measurements. However, temperature influence on seismic-wave speeds can be significant, especially when comparing laboratory data collected at room temperature to field-based measurements in areas of high heat flow (e.g., rifts, arcs, extentional settings). The decrease in compressional wave velocities in the deep crust under these high geotherms can be as much as 0.3 km s^{-1} (see Rudnick and Fountain, 1995, figure 1). For these reasons, Rudnick and Fountain (1995) used regional surface heat flow and assumed a conductive geothermal gradient, to correct the field-based velocities to room-temperature conditions. In this way, direct comparisons can be made between velocity profiles and ultrasonic velocities of lower-crustal rock types measured in the laboratory. Another benefit of this correction is that deep-crustal velocities from areas with grossly different geotherms can be considered directly in light of possible lithologic variations. In subsequent sections we quote deep-crustal velocities corrected to room-temperature conditions as "temperature-corrected velocities."

5.3.4 The Middle Crust

5.3.4.1 Samples

The best evidence for the compositional makeup of the middle crust comes from studies of high-grade metamorphic terranes and crustal cross-sections. There are far fewer studies of amphibolite-facies xenoliths derived from mid-crustal depths (Grapes, 1986; Leeman *et al.*, 1985; Mattie *et al.*, 1997; Mengel *et al.*, 1991; Weber *et al.*, 2002) compared to their granulite-facies counterparts. This may be due to the fact that it can be difficult to distinguish such xenoliths from the exposed or near-surface amphibolite-facies country rocks through which the xenolith-bearing volcanic rocks erupted. For this reason, xenolith studies have not been employed to any large extent in understanding the composition of the middle crust, and most information about the middle crust comes from studies of high-grade terranes, crustal cross-sections, and seismic profiles.

Interpreting the origin of granulite-facies terranes and hence their significance towards determining deep-crustal composition depends on unraveling their pressure–temperature–time history. Those showing evidence for a "clockwise" *P–T* path (i.e., heating during decompression) are often interpreted as having been only transiently in the lower crust; they represent upper crustal assemblages that passed through high *P–T* conditions on their way back to the surface during continent-scale collisional orogeny. In contrast, granulite terrains showing evidence for isobaric cooling can have extended lower-crustal histories, and thus may shed light on deep-crustal composition (see discussion in Rudnick and Fountain (1995)). Bohlen and Mezger (1989) pointed out that isobarically cooled granulite-facies terranes show evidence of equilibration at relatively low pressures (i.e., $\leq$0.6–0.8 GPa), corresponding to mid-crustal depths ($\leq$25 km). Although a number of high-pressure and even ultra-high-pressure metamorphic belts have been recognized since their study, it remains true that the majority of isobarically cooled granulite-facies terranes show only moderate equilibration depths and, therefore, may provide evidence regarding the composition of the middle crust.

Although lithologically diverse, the average composition of rocks analyzed from granulite terrains is evolved (Rudnick and Presper, 1990), with median compositions corresponding to granodiorite/dacite (64–66 wt.% SiO_2, 4.1–5.2 wt.% $Na_2O + K_2O$, based on classification of Le Bas and Streckeisen (1991)). Rudnick and Fountain (1995) suggested that isobarically cooled granulite terrains have a higher proportion of mafic lithologies than granulites having clockwise *P–T* paths. However, the median composition of rocks analyzed from isobarically cooled terranes is indistinguishable (62 wt.% SiO_2, 4.6 wt.% $Na_2O + K_2O$) from the median composition of the entire granulite-terrane population given in Rudnick and Presper (1990). Collectively, these data point to a chemically evolved mid-crustal composition.

Observations from crustal cross-sections also point to an evolved mid-crust composition (Table 4). Most of these cross-sections have been exposed by compressional uplift due to thrust faulting (e.g., Kapuskasing, Ivrea, Kohistan, and Musgrave). Other proposed origins for the uplift include wide, oblique transitions (Pikwitonei), impactogenesis (Vredefort), and transpression (Sierra Nevada) (Percival *et al.*, 1992). In nearly all these sections, sampling depth ranges from upper to middle crust; only a few (e.g., Vredefort, Ivrea, Kohistan) appear to penetrate into the lower crust. In the following paragraphs we review the insights into middle (and lower) crust lithologies gained from the studies of these crustal cross-sections.

The Vredefort dome represents a unique, upturned section through ~36 km of crust of the Kaapvaal craton, possibly exposing a paleo-Moho at its base (Hart *et al.*, 1981, 1990; Tredoux *et al.*, 1999; Moser *et al.*, 2001). The origin of this structure is debated, but one likely scenario is that it was produced by crustal rebound following meteorite impact. The shallowest section of basement (corresponding to original depths of 10–18 km depth) is composed of amphibolite-facies rocks consisting of granitic gneiss (the outer granite gneiss). The underlying granulite-facies rocks (original depths of 18–36 km) are composed of charnockites and leucogranofels with ~10% mafic and ultramafic granulites (the Inlandsee Leucogranofels terrain). The mid-crust, as defined here, is thus composed of amphibolite-facies felsic gneisses in fault contact with underlying charnockites and mixed felsic granulites and mafic/ultramafic granulites (Hart *et al.*, 1990). The lower crust, which is only partially exposed, consists of mixed felsic and mafic/ultramafic granulites, with the proportion of mafic rocks increasing with depth. The mantle beneath the proposed paleo-Moho, as revealed by borehole drilling, is dominated by 3.3–3.5 Ga serpentinized amphibole-bearing harzburgite (Tredoux *et al.*, 1999).

The Kapuskasing Structural Zone represents an exposed middle-to-lower crustal section through a greenstone belt of the Archean Canadian Shield, where the middle crust is represented by the amphibolite-facies Wawa gneiss dome and lower granulite-facies litihologies along the Kapuskasing uplift. Altogether, ~25 km of crust are exposed out of a total crustal thickness of 43 km (Fountain *et al.*, 1990b; Percival and Card, 1983). The Wawa gneiss dome is dominated by tonalite–granodiorite gneisses and their igneous equivalents (87%), but also contains small amounts of paragneiss (5%) and mafic gneiss and intrusives (8%) (Burke and Fountain, 1990; Fountain *et al.*, 1990b; Shaw *et al.*, 1994). The slightly deeper-level Kapuskasing Structural Zone has a greater proportion of paragneisses and mafic lithologies. It contains 35% mafic or anorthositic gneisses, 25% dioritic gneisses, 20% paragneiss, and only 20% tonalite gneisses.

Like the high-grade rocks of the Kapusksasing Structural Zone, those in the Pikwitonei crustal cross-section represent high-grade equivalents of granite–gneiss–greenstone successions (Fountain and Salisbury, 1981; Percival *et al.*, 1992). Approximately 25 km of upper-to-middle crust is exposed in this section out of a total-crustal thickness of 37 km (Fountain *et al.*, 1990b). Both amphibolite- and granulite-facies rocks are dominated by tonalitic gneiss with minor mafic gneiss, and metasedimentary rocks.

The Wutai–Jining Zone is suggested to be an exposed cross-section through the Archean North China craton (Kern *et al.*, 1996). Rocks from this exposure equilibrated at depths of up to ~30 km, thus sampling middle and uppermost lower crust, but leaving the lowermost 10 km of crust unexposed (Kern *et al.*, 1996). Like the previously described cross-sections, felsic gneisses dominate the middle crust; tonalitic–trondhjemitic–granodioritc, and granitic gneisses comprise 89% of the dominant amphibolite to granulite-facies

Table 4 Chemical and petrological composition of crustal cross-sections.

	Reference[a]	*Age*	*Setting, (uplift origin)*[b]	*Current crustal thickness (km)*	*Maximum depth (km)*	*Middle crust lithologies*	*Lower crust lithologies*
Archean							
Vredefort Dome	1–3	2.6–3.6 Ga	Kaapvaal craton, (3)	36	36 (w/paleo Moho)	Amphibolite-faciesgranitic gneiss	Granulite-facies charnockites, leucogranofels, mafic, and ultramafic granulites
KapuskasingUplift	4–5	2.5–2.7 Ga	Superior craton, (1)	43	25	*Amphibolite-facies87% felsic* 8% mafic-intermediate 5% metasediment	*Lower granulite-facies35% mafic/anorthositic* 25% diorite 20% metasediments 20% felsic
Pikwitonei granulite domain	6–9	2.5–3.1 Ga	Superior craton, (2)	37	25	*Amphibolite-lower granulite facies*. Dominately tonalite gneiss, minor mafic gneisses, quartzites, anorthosites.	*Granulite-facies*. Predominantly silicic to intermediate gneiss, with minor paragneiss, mafic-ultramafic bodies and anorthosites
Wutai-Jining zone	10	2.5–2.8 Ga	North China Craton, (2)	40	30	*Amphibolite-lower granuilte facies.89% tonalitic-trondhjemitic-granodioritic-granitic gneiss* 8% amphibolite and mafic granulite 3% metapelite	*Granulite-facies.54% tonalitic-trondhjemitic-granodioritic-granitic gneiss* 32% mafic granulite 6% metapelite 8% metasandstone
Proterozoic							
Musgrave ranges	6, 8, 11	1.1–2.0 Ga	Central Australia, (1)	40	Unknown	Quartzofeldspathic gneiss, amphibolite, metapelite, marble, calc-silicate gneiss	Silicic to intermediate gneiss, mafic granulite, layered mafic-ultramafic intrusions
S. Norway	12–13	1.5–2.0 Ga	Baltic Shield, (2)	35	Unknown	Quartzofeldspathic gneiss, amphibolite, metasediments	Felsic granulite, mafic granulite, metasediments
Phanerozoic							
Ivrea-Verbano zone	14–17	Permian	Alps, (1)	35	30	*Amphibolite-facies*. felsic gneiss, amphibolite, metapelite (kinzigite), marble	*Granulite facies*. mafic intrusivesand ultramafic cumulates, resistic metapelite (stronalite), diorite
Sierra Nevada, California	8, 18–20	Cretaceous	Continental arc, (4)	27–43	30	Mafic to felsic gneiss, amphibolite, diorite–tonalite	Granofels, mafic granulite, graphite-bearing metasediments
Kohistan, Pakistan	8, 21	Late Jurassic-Eocene	Oceanic arc, (1)	Unknown	45	Diroite, metadiorite, gabbronorite	Amphibolite, metagabbro, gabbronorite, garnet gabbro, garnet hornblendite, websterite
Talkeetna, Alaska	22–23	Jurassic	Oceanic arc, (1)	25–35	13	Gabbro, tonalite, diorite	Garnet gabbro, amphibole gabbro, dunite, wehrlite, pyroxenite

[a] References: 1. Hart *et al.* 1990, 2. Tredoux *et al.*, 1999, 3. Moser *et al.*, 2001, 4. Fountain *et al.*, 1990a, 5. Shaw *et al.*, 1994, 6. Fountain and Salisbury, 1981, 7. Fountain *et al.*, 1987, 8. Percival *et al.*, 1992, 9. Fountain and Salisbury 1995, 10. Kern *et al.*, 1996, 11. Clitheroe *et al.*, 2000, 12. Pinet and Jaupart, 1987, 13. Alirezaei and Cameron, 2002, 14. Mehnert, 1975, 15. Fountain, 1976, 16. Voshage *et al.*, 1990, 17. Mayer *et al.*, 2000, 18. Ross, 1985, 19. Saleeby, 1990, 20. Ducea, 2001, 21. Miller and Christensen, 1994, 22. Pearcy *et al.*, 1990, 23. [b] The different mechanisms responsible for uplift of these crustal cross sections include (1) compressional uplifts along thrust faults, (2) wide, oblique transitions, which are also compressional in origin, but over wide transitions, with no one thrust fault obviously responsible for their uplift, (3) meteorite impact, and (4) transpressional uplifts, which are vertical uplifts along a transcurrent faults (Percival *et al.*, 1992).

Henshan-Fuping terrains, the remaining lithologies are amphibolite-mafic granulite (8%) and metapelite (3%). Tonalitic–trondhjemitic–granodioritc and granitic gneiss (54%) are less significant but still dominant in the lower-crustal Jining terrain.

The Musgrave Range (Fountain and Salisbury, 1981; Percival *et al.*, 1992) and the Bamble Sector of southern Norway (Pinet and Jaupart, 1987; Alirezaei and Cameron, 2002)) represent two crustal sections through Proterozoic crust of central Australia and the Baltic Shield, respectively. In both sections, the middle crust is dominated by quartzofeldspathic gneiss. The lower crust consists of silicic to intermediate gneiss, felsic granulite, and mafic granulite with layered mafic and ultramafic intrusions being important lithological components in the Musgrave Range and metasediments being important in the lower crust of southern Norway.

The Ivrea–Verbano Zone in the southern Alps of Italy was the first to be proposed as an exposed deep-crustal section by Berckhemer (1969) and has subsequently been the focus of extensive geological, geochemical, and geophysical studies (e.g., Mehnert, 1975; Fountain, 1976; Dostal and Capedri, 1979; Voshage *et al.*, 1990; Quick *et al.*, 1995). The Paleozoic rocks of the Ivrea zone are unusual when compared with Precambrian granulite outcrops because they contain a large proportion of mafic lithologies and, as such, closely resemble granulite xenoliths in composition (Rudnick, 1990b). Amphibolite-facies rocks of the middle crust consist of felsic gneiss, amphibolite, metapelite (kinzigite), and marble, whereas the lower crustal section comprises mafic granulite and diorite, which formed by intrusion and subsequent fractionation of basaltic melts that partially melted the surrounding metasediments (now resistic stronalite) (Mehnert, 1975; Dostal and Capedri, 1979; Fountain *et al.*, 1976; Voshage *et al.*, 1990). Detailed mapping by Quick *et al.* (1995) demonstrated that mantle peridotites in the southern Ivrea Zone are lenses that were tectonically interfingered with metasedimentary rocks prior to intrusion of the gabbroic complex and the present exposures reside an unknown distance above the pre-Alpine contiguous mantle. Thus reference to the section as a complete crust–mantle transition could be misleading. Altogether, the exposed rocks represent ~30 km of crust with ~5 km lowermost crust remaining unexposed (Fountain *et al.*, 1990a). The similarity in isotope composition and age between the Ivrea zone cumulates and Hercynian granites in the upper crust led Voshage *et al.* (1990) to speculate that these granites were derived from lower-crustal magma chambers similar to those in the Ivrea Zone, suggesting that basaltic underplating may be important in the formation and modification of the lower continental crust (Rudnick, 1990a).

Three sections through Mesozoic arcs show contrasting bulk compositions, depending on their settings (continental versus oceanic). In the southern Sierra Nevada, a tilted section exposes the deeper reaches of the Sierra Nevada batholith, which is part of a continental arc formed during the Mesozoic. This section is dominated by arc-related granitoids to depths of ~30 km, which have a tonalitic bulk chemistry (Ducea and Saleeby, 1996; Ducea, 2001). At the deepest structural levels, the mafic Tehachapi Complex comprises mafic and felsic gneiss, amphibolite, diorite, tonalite, granulite, and rare metasediments (Percival *et al.*, 1992; Ross, 1985). In contrast, two sections through accreted intraoceanic arcs have considerably more mafic middle-crust compositions. In the Jurassic Talkeetna section of southeastern Alaska, the middle crust comprises gabbro and tonalite (4.5 km), which is underlain by variably deformed garnet gabbro and gabbro with cumulate dunite, wehrlite, and pyroxenite (2.2 km) in the lower crust (Pearcy *et al.*, 1990). The upper, middle, and lower crustal units are estimated to have an average SiO_2 of 57%, 52%, and 44–45%, respectively. The Late Jurassic–Eocene Kohistan arc of Pakistan represents a 45 km thick reconstructed crustal column through a deformed, intruded intraoceanic arc sequence exposed in the Himalayan collision zone (Miller and Christensen, 1994). The depth interval from 10 km to 18 km is dominated by diorite and metadiorite. Rocks below this level, from ~18 km to the Moho, are dominated by metamorphosed mafic to ultramafic rocks from a series of layered mafic intrusions.

In summary, exposed amphibolite- to granulite-facies terranes and middle crustal cross-sections contain a wide variety of lithologies, including metasedimentary rocks, but they are dominated by igneous and metamorphic rocks of the diorite–tonalite–trondhjemite–granodiorite (DTTG), and granite suites. This is true not only for Precambrian shields but also for Phanerozoic crust and continental arcs, as documented in the crustal cross-sections described above. However, intra-oceanic arcs may contain substantially greater proportions of mafic rocks in the middle and lower crust, as illustrated by the Kohistan and Talkeetna arc sections (Pearcy *et al.*, 1990; Miller and Christensen, 1994).

5.3.4.2 Seismological evidence

The samples described above provide evidence of the lithologies likely to be present in the middle crust. By definition, however, these samples no longer reside in the middle crust and additional information is required in order to determine the composition of the present-day middle crust. For this, we turn to seismological data for continental crust from a variety of tectonic settings.

Except for active rifts and some intra-oceanic island arcs, which exhibit the highest middle-crust *P*-wave velocities (6.7 ± 0.3 km s^{-1} (Rudnick and Fountain, 1995) and 6.8 ± 0.2 km s^{-1} (data from Holbrook *et al.*, 1992) corrected to room temperature), other continental tectonic units have room-temperature middle-crustal *P*-wave velocities between 6.4 km s^{-1} and 6.6 km s^{-1} (Rudnick and Fountain, 1995). This range overlaps the average velocity of *in situ* middle crust, which was determined by Christensen and Mooney (1995) to be from 6.3 km s^{-1} to 6.6 ± 0.3 km s^{-1}, with an average of 6.5 ± 0.2 km s^{-1} over the depth range of 15–25 km. When corrected for temperature (an increase of 0.1–0.2 km s^{-1}, depending on the regional geotherm), these average middle-crustal velocities are similar to the room-temperature velocities considered by Rudnick and Fountain (1995). Thus, the middle crust has average, room-temperature-corrected velocity between 6.4 km s^{-1} and 6.7 km s^{-1}.

Amphibolite-facies felsic gneisses have room temperature *P*-wave velocities of 6.4 ± 0.1 km s^{-1} (Rudnick and Fountain, 1995). This compares well with the room-temperature velocity of average biotite (tonalite) gneiss at 6.32 ± 0.17 km s^{-1} (20 km depth; Christensen and Mooney, 1995). Granitic gneiss has a slightly lower velocity (6.25 ± 0.11 km s^{-1}; Christensen and Mooney (1995)), but is within uncertainty of the tonalite. A mixture of such gneisses with 0–30% amphibolite or mafic gneiss of the same metamorphic grade ($Vp = 7.0$ km s^{-1}; Rudnick and Fountain, 1995, Christensen and Mooney, 1995) yields *P*-wave velocities in the range observed for most middle crust. The above seismic data are thus consistent with the observations from granulite terranes and crustal cross-sections, and suggest that the middle crust is dominated by felsic gneisses.

5.3.4.3 Middle-crust composition

Compared to other regions of the crust (upper, lower, and bulk), few estimates have been made of the composition of the middle crust (Table 5, and Figures 9 and 10). Moreover, these estimates provide data for a far more limited number of elements, and large differences exist between different estimates. The estimates of Weaver and Tarney (1984), Shaw *et al.* (1994) and Gao *et al.* (1998a) are based on surface sampling of amphibolite-facies rocks in the Lewisian Complex, the Canadian Shield, and Eastern China, respectively. Rudnick and Fountain (1995) modeled the middle crust as 45% intermediate amphibolite-facies gneisses, 45% mixed amphibolite and felsic amphibolite-facies gneisses, and 10% metapelite. This mixture is very similar to that of Christensen and Mooney (1995), who proposed a middle crust of 50% tonalitic gneiss, 35% amphibolite, and 15% granitic gneiss. Unfortunately, compositional data are not available for Christensen and Mooney's samples and so the chemical composition of their middle crust cannot be calculated.

The estimates of Rudnick and Fountain (1995) and Gao *et al.* (1998a) show a broad similarity, although the latter is more evolved, having higher SiO_2, K_2O, barium, lithium, zirconium, and LREEs and La_N/Yb_N and lower total FeO, scandium, vanadium, chromium, and cobalt with a significant negative europium anomaly (Figures 9 and 10). These differences are expected, based on the slightly higher compressional velocity of Rudnick and Fountain's global middle crust compared to that of Eastern China (6.6 km s^{-1} versus 6.4 km s^{-1}; Gao *et al.*, 1998b). The consistency is surprising considering that the two estimates are based on different sample bases and different approaches, one global and the other regional.

The middle-crustal compositions of Weaver and Tarney (1984) and Shaw *et al.* (1994) deviate from the above estimates by being markedly higher in SiO_2 and lower in TiO_2, FeO, MgO, and CaO. Moreover, these middle-crust compositions are more felsic (based on the above elements) than all estimates of the upper-continental crust composition given in Table 1. Thus, it is unlikely that the Weaver and Tarney (1984) and Shaw *et al.* (1994) compositions are representative of the global average middle crust, as both heat flow and seismic observations require that the crust becomes more mafic with depth. It should be noted, however, that heat production for Shaw's middle-crust composition is indistinguishable from those of Rudnick and Fountain (1995) and Gao *et al.* (1998a) at ~ 1.0 μW m^{-3}, due largely to the very high K/Th and K/U of Shaw *et al.* estimate. The middle crust of Weaver and Tarney (1984) has significantly higher heat production, at 1.4 μW m^{-3}.

Generally speaking, it would be best to derive the middle-crust composition from observed seismic-wave speeds and chemical analyses of amphibolite-facies rocks. However, few such data sets exist. Only two studies attempt to define the global average seismic-wave speeds for the middle crust (Christensen and Mooney, 1995; Rudnick and Fountain, 1995) and neither provides chemical data for amphibolite facies samples. Rudnick and Fountain used compiled chemical data for granulite-facies rocks and inferred the concentrations of fluid-mobile elements (e.g., rubidium, uranium) of their amphibolite facies counterparts, while Christensen and Mooney

Table 5 Compositional estimates of the middle continental crust. Major elements in weight percent. Trace element concentration units the same as in Table 2.

	1 *Weaver and Tarney (1984)*	*2* *Shaw et al. (1994)*	*3* *Rudnick and Fountain (1995)*	*4* *Gao et al. (1998a)*	*5* *This study*[a]	*1* *Sigma*[a]	%
SiO_2	68.1	69.4	62.4	64.6	63.5	2	2
TiO_2	0.31	0.33	0.72	0.67	0.69	0.04	6
Al_2O_3	16.33	16.21	15.96	14.08	15.0	1	9
FeO_T[b]	3.27	2.72	6.59	5.45	6.02	0.8	13
MnO	0.04	0.03	0.10	0.11	0.10	0.00	2
MgO	1.43	1.27	3.50	3.67	3.59	0.1	3
CaO	3.27	2.96	5.25	5.24	5.25	0.01	0
Na_2O	5.00	3.55	3.30	3.48	3.39	0.1	4
K_2O	2.14	3.36	2.07	2.52	2.30	0.3	14
P_2O_5	0.14	0.15	0.10	0.19	0.15	0.06	43
Mg#	43.8	45.5	48.6	54.5	51.5		
Li		20.5	7	16	12	6	55
Be				2.29	2.29		
B		3.2		17	17		
N							
F				524	524		
S				20	20		
Cl				182	182		
Sc		5.4	22	15	19	5	27
V		46	118	95	107	16	15
Cr	32	43	83	69	76	10	13
Co		30	25	18	22	5	23
Ni	20	18	33	34	33.5	0.7	2
Cu		8	20	32	26	8	33
Zn		50	70	69	69.5	0.7	1
Ga			17	18	17.5	0.7	4
Ge				1.13	1.13		
As				3.1	3.1		
Se				0.064	0.064		
Br							
Rb	74	92	62	67	65	4	5
Sr	580	465[c]	281	283	282	1	1
Y	9	16	22	17.0	20	4	18
Zr	193	129	125	173	149	34	23
Nb	6	8.7	8	11	10	2	22
Mo		0.3		0.60	0.60		
Ru							
Pd				0.76	0.76		
Ag				48	48		
Cd				0.061	0.061		
In							
Sn				1.30	1.30		
Sb				0.28	0.28		
I							
Cs		0.98	2.4	1.96	2.2	0.3	14
Ba	713	1376	402	661	532	183	34
La	36	22.9	17	30.8	24	10	41
Ce	69	42.1	45	60.3	53	11	21
Pr			5.8		5.8		
Nd	30	18.3	24	26.2	25	2	6
Sm	4.4	2.8	4.4	4.74	4.6	0.2	5
Eu	1.09	0.78	1.5	1.20	1.4	0.2	16
Gd		2.11	4.0		4.0		
Tb	0.41	0.28	0.58	0.76	0.7	0.1	19
Dy		1.54	3.8		3.8		
Ho			0.82		0.82		
Er			2.3		2.3		
Tm	0.14				0.32		

(continued)

Table 5 (continued).

	1 Weaver and Tarney (1984)	*2* Shaw et al. (1994)	*3* Rudnick and Fountain (1995)	*4* Gao et al. (1998a)	*5* This study[a]	*1* Sigma[a]	%
Yb	0.76	0.63	2.3	2.17	2.2	0.09	4
Lu	0.1	0.12	0.41	0.32	0.4	0.06	17
Hf	3.8	3.3	4.0	4.79	4.4	0.6	13
Ta		1.8	0.6	0.55	0.6	0.04	6
W				0.60	0.60		
Re							
Os							
Ir							
Pt				0.85	0.85		
Au				0.66	0.66		
Hg				0.0079	0.0079		
Tl				0.27	0.27		
Pb	22	9.0	15.3	15	15.2	0.2	1
Bi				0.17	0.17		
Th	8.4	6.4	6.1	6.84	6.5	0.5	8
U	2.2	0.9	1.6	1.02	1.3	0.4	31

Units for trace elements are the same as in Table 2. Major elements recast to 100% anhydrous.
[a] Averages and standard deviations of middle crustal composition by Rudnick and Fountain (1995) and Gao *et al.* (1998a), or from either of these two studies if data from the other one are unavailable. [b] Total Fe as FeO. [c] Recalculated from original data given by Shaw *et al.* (1994; Table 4), due to a typographical error in the published table. Mg# = molar 100 × Mg/(Mg + Fe_{tot}).

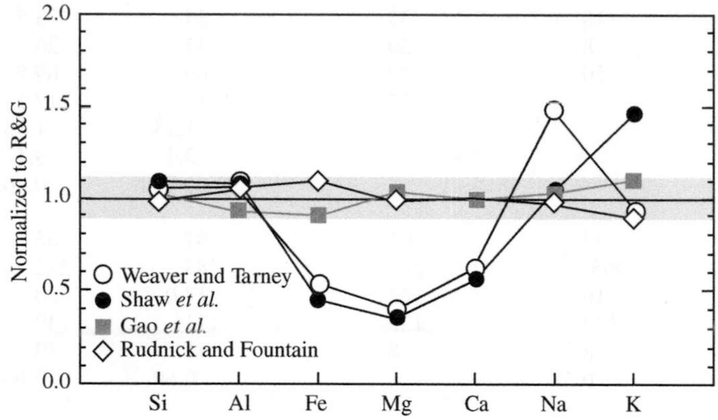

Figure 9 Comparison of the major-element composition of the middle continental crust as determined by sampling of surface exposures (Shaw *et al.*, 1994; Weaver and Tarney, 1984) and inferred from middle-crustal seismic velocities combined with surface and xenolith samples (Rudnick and Fountain, 1995; Gao *et al.*, 1998a). All values normalized to the new composition provided in Table 5 ("R&G"), which is an average between the values of Gao *et al.* (1998a) and Rudnick and Fountain (1995). Gray shaded field represents ±10% variation from this value.

(1995) did not publish their chemical data for the amphibolite-facies rocks they studied. For this reason, we have chosen to estimate the middle-crust composition by averaging the estimates of Rudnick and Fountain (1995) and Gao *et al.* (1998a) (Table 5), where corresponding data are available. Although the latter study is regional in nature, its similarity to the global model of Rudnick and Fountain (1995) suggests that it is not anomalous from a global perspective (unlike the lower crust of Eastern China as described in Section 5.3.5) and it provides additional estimates for little-measured trace elements.

This middle crust has an intermediate composition with lower SiO_2 and K_2O concentrations and higher FeO, MgO, and CaO concentrations than average upper crust (Table 1), consistent with the geophysical evidence (cited above) of a chemically stratified crust. Differences in trace-element concentrations between these two estimates are generally less than 30%, with the exceptions of P_2O_5, lithium, copper, barium, lanthanum, and uranium (Figure 10). The concentrations of these elements are considered to be less constrained. The middle crust is LREE enriched and exhibits the characteristic depletion of

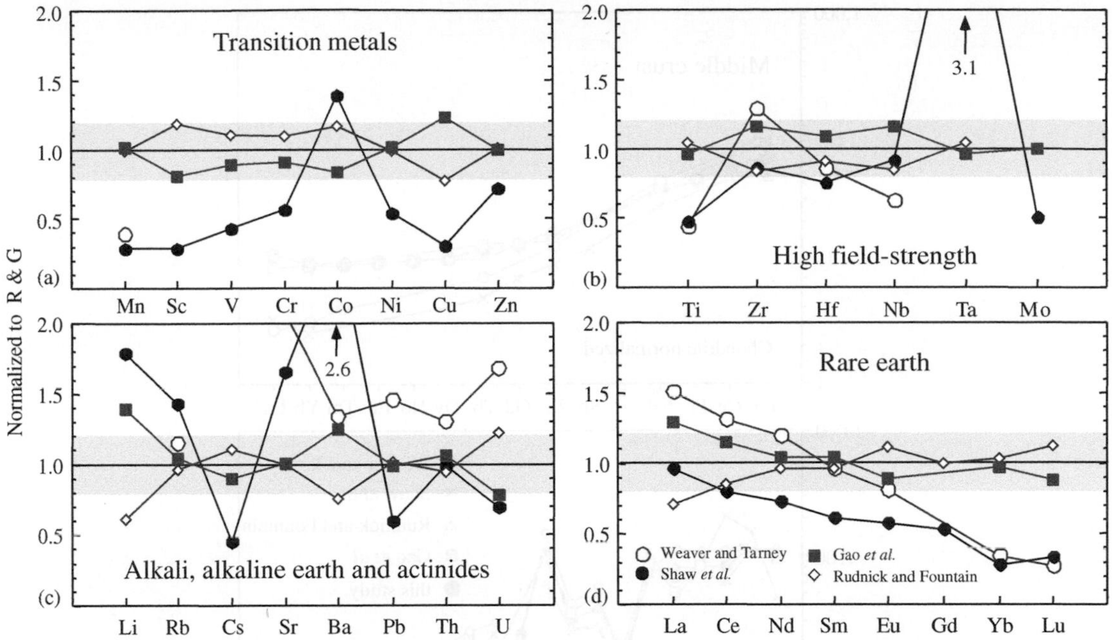

Figure 10 Comparison of the trace-element composition of the middle continental crust as determined by sampling of surface exposures (Shaw *et al.*, 1994; Weaver and Tarney, 1984) and inferred from middle-crustal seismic velocities combined with surface and xenolith samples (Rudnick and Fountain, 1995; Gao *et al.*, 1998a). All values normalized to the new composition provided in Table 5 ("R&G"), which is an average of the values of Gao *et al.* (1998a) and Rudnick and Fountain (1995). Gray shaded field represents ±20% variation from this value. (a) transition metals, (b) high-field strength elements, (c) alkali, alkaline earth and actinides, and (d) REEs.

niobium relative to lanthanum and enrichment of lead relative to cerium seen in all other parts of the crust (Figure 11).

In summary, our knowledge of middle-crustal composition is limited by the small number of studies that have focused on the middle crust and the ambiguity in deriving chemical compositions from seismic velocities. Thus, the average composition given in Table 5 is poorly constrained for a large number of elements. Seismological and heat-flow data suggest an increase in seismic-wave speeds and a decrease in heat production with depth in the crust. Studies of crustal cross-sections show the middle crust to be dominated by felsic gneisses of tonalitic bulk composition. The average middle-crust composition given in Table 5 is consistent with these broad constraints and furthermore suggests that the middle crust contains significant concentrations of incompatible trace elements. However, the uncertainty on the middle-crust composition, particularly the trace elements, remains large.

5.3.5 The Lower Crust

5.3.5.1 Samples

Like the middle crust, the lower crust also contains a wide variety of lithologies, as revealed by granulite xenoliths, exposed high-pressure granulite terranes and crustal cross-sections. Metaigneous lithologies range from granite to gabbro, with a predominance of the latter in most lower crustal xenolith suites. Exceptions include xenolith suites from Argentina (Lucassen *et al.*, 1999) and central Spain (Villaseca *et al.*, 1999), where the xenoliths are dominated by intermediate to felsic granulites and the Massif Central (Leyreloup *et al.*, 1977; Downes *et al.*, 1990) and Hannuoba, China (Liu *et al.*, 2001), where intermediate to felsic granulites comprise nearly half the population. Metapelites occur commonly in both terranes and xenoliths, but only rarely do other metasedimentary lithologies occur in xenolith suites; unique xenolith localities have been documented with meta-arenites (Upton *et al.*, 1998) and quartzites (Hanchar *et al.*, 1994), but so far marbles occur only in terranes. The reason for their absence in lower crustal xenolith suites is uncertain—they may be absent in the lower crust sampled by volcanoes, they may not survive transport in the hot magma, or they may simply have been overlooked by xenolith investigators. These issues related to the representativeness of xenolith sampling are the reason why robust estimates of lower-crustal composition must rely on a grand averaging technique, such as using seismic velocities to infer composition.

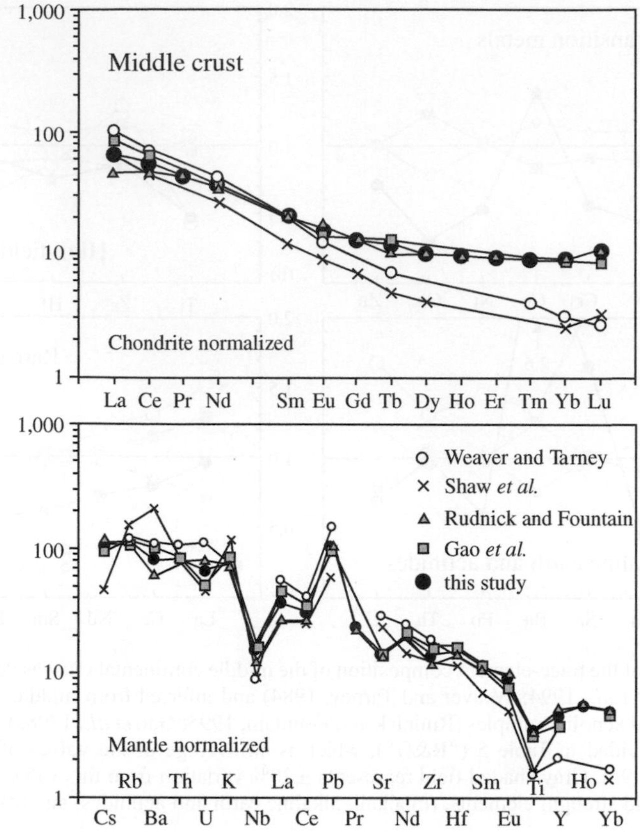

Figure 11 REE (upper) and multi-element plot (lower) of the compositions of the middle crust given in Table 5. Chondrite values from Taylor and McLennan (1985) and primitive mantle values from McDonough and Sun (1995).

Information on the lower crust derived from crustal cross-sections has been given in Section 5.3.4.1 and only the main points are summarized here. All crustal cross-sections show an increase in mafic lithologies with depth and most of those in which possible crust–mantle boundaries are exposed reveal a lower crust that is dominated by mafic compositions. For example, in the Ivrea Zone, Italy, the lower crust is dominated by mafic granulite formed from basaltic underplating of country rock metapelite (Voshage *et al.*, 1990). The same is true for the Kohistan sequence, Pakistan (Miller and Christensen, 1994), although here metapelites are lacking. Although the crust–mantle boundary is not exposed in the Wutai-Jining terrain, the granulite-facies crust exposed in this cross-section has a more mafic composition than the rocks of the middle-crust section. Even in the Vredefort and Sierra Nevada cross-sections, which are dominated by granitic rocks throughout most of the crustal sections (Ducea, 2001; Hart *et al.*, 1990), the deepest reaches of exposed crust are characterized by more mafic lithologies (Ross, 1985; Hart *et al.*, 1990; Table 4).

There have been a number of studies of granulite-facies xenoliths since the reviews of Rudnick (1992) and Downes (1993) and a current tabulation of xenolith studies is provided in Table 6, which provides a summary of most lower crustal xenolith studies published through 2002. Perhaps most significant are the studies of lower-crustal xenoliths from Archean cratons, which had been largely lacking prior to 1992 (Kempton *et al.*, 1995, 2001; Davis, 1997; Markwick and Downes, 2000; Schmitz and Bowring, 2000, 2003a,b; Downes *et al.*, 2002). These studies reveal a great diversity in lower-crustal lithologies beneath Archean cratons, which appear to correlate with seismic structure of the crust.

Lower-crustal xenoliths from the Archean part of the Baltic (or Fennoscandian) Shield, like their post-Archean counterparts, are dominated by mafic lithologies (Kempton *et al.*, 1995, 2001; Markwick and Downes, 2000; Hölttä *et al.*, 2000). Most equilibrated at depths of 22–50 km and contain hydrous phases (amphibole ± biotite). Partial melting and restite development is evident in some migmatitic xenoliths, but cumulates are absent (Kempton *et al.*, 1995, 2001; Hölttä *et al.*, 2000). A curious feature of these samples is the common occurrence of potassic phases (e.g., potassium feldspar,

Table 6 Geochemical and mineral chemical studies of lower crustal xenoliths.

Locality	*Host*	*Xenolith types*	*Types of analyses*	*Pipe age*	*Crust age*	*Age*	*References*
North America							
Nunivak Island, Alaska	AB	MG	ME, Min.	<5 Ma	Phanerozoic		Francis (1976)
Central Slave Province, Canada	K	MGG, FG, MG	U–Pb	50–70 Ma	Archean	2.5 Ga, 1.3 Ga (meta)	Davis (1997)
Kirkland Lake, Ontario, Superior craton	K	AN, MG	U–Pb	160 Ma	Archean	2.6–2.8 Ga, 2.4–2.5 Ga (meta)	Moser and Heaman (1997)
Ayer's cliff, Quebec	K	MG, PG	Min.	~100 Ma	Proterozoic		Trzcienski and Marchildon (1989)
Popes Harbour[a], Nova Scotia	K	MG, PG, FG	ME, TE, Min.	<400 Ma	Phanerozoic		Owen (1988), Eberz (1991)
Snake River Plains[a], Idaho	Evol. B	PG	Sr, Nd, Pb	<2 Ma	Archean	~2.8 Ga	Leeman *et al.* (1985, 1992)
Bearpaw Mts., Montana	K	MGG, MG, FG	ME, TE, O, U–Pb	45 Ma	Archean	~2.6 Ga	Collerson *et al.* (1989), Kempton and Harmon (1992), Moecher *et al.* (1994), Rudnick *et al.* (1999)
Simcoe Volcanic Field, Washington Cascades	AB	MG	FI	<1 Ma	Phanerozoic		Ertan and Leeman (1999)
Riley County, Kansas	K	MGG, EC	Min.	<230 Ma	Proterozoic		Meyer and Brookins (1976)
Central Sierra Nevada, California	AB	MGG, EC, FG, MP	ME, TE, Min., Sr, Nd, O, U–Pb	8–11 Ma	Proterozoic	180 Ma	Dodge *et al.* (1986, 1988), Domenick *et al.* (1983), Ducea and Saleeby (1996, 1998)
Colorado/Wyoming	K	MGG, MG	ME, Min.		Proterozoic		Bradley and McCallum (1984)
Mojave Desert, California	AB	MP, FG, IG, QZ, MGG	ME, TE, Min., Sr, Nd, Pb, U–Pb		Archean	~1,7 Ga	Hanchar *et al.* (1994)
Navajo Volcanic Field, Colorado Plateau	K	MGG, FG, AM, EC, MP	ME, TE, Min., Sr, Nd	25–30 Ma	Proterozoic	~1.8 Ga	Ehrenberg and Griffin (1979), Broadhurst (1986), Wendlandt *et al.* (1993, 1996), Mattie *et al.* (1997), Condie *et al.* (1999)
Camp Creek, Arizona	L	MGG, AM, EC	ME, TE, Min., Sr, Nd, Pb	23–27 Ma	Proterozoic	1.2–1.9 Ga	Esperanca *et al.* (1988)

(continued)

Table 6 (continued).

Locality	*Host*	*Xenolith types*	*Types of analyses*	*Pipe age*	*Crust age*	*Age*	*References*
Chino Valley, Arizona	L	MGG, AM	ME, TE, Min.	25 Ma	Proterozoic		Arculus and Smith (1979), Schulze and Helmstadt (1979), Arculus *et al.* (1988)
San Francisco Volcanic Field	AB	MG, IG	ME, TE, Min., Sr, Nd	<20 Ma	Proterozoic	~1.9 Ga	Chen and Arculus (1995)
Geronimo Volcanic Field, New Mexico	AB	MG, IG	ME, TE, Min., Sr, Nd, Pb, O	<3 Ma	Proterozoic	1.1–1.4 Ga	Kempton *et al.* (1990), Kempton and Harmon (1992)
Kilbourne Hole, New Mexico	AB	MG, AN, PG, FG	Min., ME, Sr, Nd, Pb, O, U–Pb	<1 Ma	Proterozoic	1.5 Ga	Padovani and Carter (1977), Davis and Grew (1977), James *et al.* (1980), Padovani *et al.* (1982), Reid *et al.* (1989), Leeman *et al.* (1992), Scherer *et al.* (1997).
Elephant Butte, New Mexico	AB	MG	Min., ME	<3 Ma	Proterozoic		Baldridge (1979)
Engle Basin, New Mexico	AB	MG	Min.	<3 Ma	Proterozoic		Warren *et al.* (1979)
West Texas	AB	MGG, IG, FG	ME, Sr, Nd, Pb, U–Pb	<40 Ma	Proterozoic	1.1 Ga	Cameron and Ward (1998)
Northern Mexico	AB	MG, PG, FG	ME, TE, Min., Sr, Nd, C, U–Pb	<25 Ma	Proterozoic	From 1 to 1400 Ma	Nimz *et al.* (1986), Ruiz *et al.* (1988a,b), Roberts and Ruiz (1989), Hayob *et al.* (1989), Rudnick and Cameron (1991), Cameron *et al.* (1992), Moecher *et al.* (1994), Smith *et al.* (1996), Scherer *et al.* (1997)
Central Mexico	AB	MG	ME, Min, Nd		Phanerozoic	1.5 Ga	Urrutia-Fucagauchi and Uribe-Cifuentes (1999)
San Luis Potosi, Central Mexico	AB	MG, MGG, IG	ME, TE, Min., Sr, Nd	<1 Ma	Proterozoic	~1.2 Ga	Schaaf *et al.* (1994)
South America							
Mercaderes, SW Columbia	AB	MG, MGG, IG, HB,	ME, TE, Min., Sr, Nd, Pb	<10 Ma	Phanerozoic		Weber *et al.* (2002)
Salta Rift, NW Argentina	AB	FG, MG	Min., ME, TE, Sr, Nd, Pb	Mesozoic	Proterozoic	~ 1.8 Ga	Lucassen *et al.* (1999)

Calbuco Volcano, Chile	AND	MG	ME, TE, Min., Sr, Nd	<1,000 yr	Paleozoic		Hickey-Vargas *et al.* (1995)
Pali Aike, Southern Chile	AB	MG	ME, Min., FI	<3 Ma	Phanerozoic		Selverstone and Stern (1983)
Europe							
Scotland, Northern Uplands	AB	MG, AN, IG, MP, MGG, HB	ME, TE, Min., Nd, Sr, U–Pb	~300 Ma	Archean/ Proterozoic	360 Ma, 1.8 Ga	van Breeman and Hawkesworth (1980), Upton *et al.* (1983), Halliday *et al.* (1984), Hunder *et al.* (1984), Upton *et al.* (1998, 2001)
Eastern Finland	K	MG, MGG, AM	ME, TE, Min., Nd, U–Pb	525 Ma	Archean	1.7–2.6 Ga	Hölttä *et al.* (2000)
Arkhangelsk Kimberlite, Baltic shield, Russia	K	MGG	ME, TE, Min., Sr, Nd	360 Ma	Archean	1.7–1.9 Ga	Markwick and Downes (2000)
Elovy island, Baltic shield, Russia	K	MGG, EC, FG, AM	ME, TE, Min., Sr, Nd, Pb, U–Pb	360–380 Ma	Archean	~1.8 Ga, 2.4–2.5 Ga	Kempton *et al.* (1995, 2001), Downes *et al.* (2002)
Belarus, Russia	K	MGG, EC, HB	ME, TE, Min., Sr, Nd	370 Ma			Markwick *et al.* (2001)
Pannonian basin, W. Hungary	AB	MG, MGG	ME, TE, Min., Sr, Nd, O	2–5 Ma	Phanerozoic		Embey-Isztin *et al.* (1990), Kempton *et al.* (1997), Embey-Isztin *et al.* (2003), Dobosi *et al.* (2003)
Kampernich, E. Eifel, Germany	AB	MGG, AM	ME, TE, Min., Sr, Nd, Hf, Pb, O	<1 Ma	Phanerozoic	1.5 Ga or ~450 Ma?	Okrusch *et al.* (1979), Stosch and Lugmair (1984), Rudnick and Goldstein (1990), Loock *et al.* (1990), Kempton and Harmon (1992), Sachs and Hansteen (2000)
Wehr Volcano[a], E. Eifel, Germany	AB	AM	ME, TE, Min.	<1 Ma	Phanerozoic		Worner *et al.* (1982), Grapes (1986)
N. Hessian Depression, Germany	AB	MGG, MG, PG, FG	ME, TE, Min., O	<50 Ma	Phanerozoic		Mengel and Wedepohl (1983), Mengel (1990)
Massif Central, France	vAB	MGG, MG, PG, FG	ME, TE, Nd, Sr, Pb, O, C	<5 Ma	Phanerozoic	~350 Ma	Leyreloup *et al.* (1977), Dostal *et al.* (1980), Vidal and Postaire (1985), Downes and Leyreloup (1986), Kempton and Harmon (1992), Downes *et al.* (1991), Moecher *et al.* (1994)

(continued)

Table 6 (continued).

Locality	*Host*	*Xenolith types*	*Types of analyses*	*Pipe age*	*Crust age*	*Age*	*References*
Central Queensland	AB	MG, MGG	ME, TE, Min., Sr, Nd	≤50 Ma	Phanerozoic		Griffin *et al.* (1987), O'Reilly *et al.* (1988)
Gloucester, NSW	AB	MGG, MG	ME, TE, Sr, Nd	≤50 Ma	Phanerozoic		Griffin *et al.* (1986), O'Reilly *et al.* (1988)
Sydney Basin	AB	MG	ME, TE, Sr, Nd	≤50 Ma	Phanerozoic		Griffin *et al.* (1986), O'Reilly *et al.* (1988)
Boomi Creek, NSW	AB	MG	ME, TE, Min.	≤50 Ma	Phanerozoic		Wilkinson (1975), Wilkinson and Taylor (1980)
Delegate, NSW	AB	MG, MGG, FG, EC	ME, TE, Min., Sr, Nd, U–Pb	∼140 Ma	Phanerozoic	400 Ma	Lovering and White, (1964, 1969), Griffin and O'Reilly (1986), O'Reilly *et al.* (1988), Arculus *et al.* (1988), Chen *et al.* (1998)
Jugiong, NSW	K	MG, MGG	Min.	<17 Ma	Phanerozoic		Arculus *et al.* (1988)
White Cliffs, NSW	K	MGG	Min.	∼260 Ma	Proterozoic		Arculus *et al.* (1988)
Anakies, Victoria	AB	MG, MGG	ME, TE, Min., Sr, Nd	<2 Ma	Phanerozoic		Sutherland and Hollis (1982), Wass and Hollis (1983), O'Reilly *et al.* (1988)
El Alamein, South Australia	K	MGG, EC	ME, TE, Min.	∼170 Ma	Proterozoic		Edwards *et al.* (1979), Arculus *et al.* (1988)
Calcutteroo, South Australia	K	MGG, FG, EC	ME, TE, Min., Sr, Nd, U–Pb	∼170 Ma	Proterozoic	1.6–1.5 Ga, 780 Ma, 620 Ma, 330 Ma	McCulloch *et al.* (1982), Arculus *et al.* (1988), Chen *et al.* (1994)
Banks Penninsula, New Zealand	AB	MG	ME, TE, Min				Sewell *et al.* (1993)
Mt. Erebus Volcanic Field,	AB	MG, MGG	ME, TE, Min., Sr, O	<5 Ma	Phanerozoic/ Proterozoic		Kyle *et al.* (1987), Kalamarides *et al.* (1987), Berg *et al.* (1989)
Indian Ocean							
Kergulen Archipelago	AB	MG, SG	ME, Min.		Phanerozoic/ Proterozoic		McBirney and Aoki (1973), Gregoire *et al.* (1998, 1994)

Only papers in which data are reported are listed here. *Abbreviations*

Host types: AB = alkali basaltic association; AND = andesite; K = kimberlitic association (including lamproites, minettes, kimberlites), Evol. B = evolved basalt; L = latite.Xenolith types: AM = amphibolite; AN = anorthosite; EC = elcogite; FG = felsic granuilte; HB = hornblendite; IG = intermediate granulite; MG = mafic granulite; MGG = mafic garnet granulite; MP = metapelite, PG = paragneiss, QZ = quartzite, SG = saphirine granulites. Types of analyses: FI = fluid inclusions, ME = major element analyses; Min = mineral analyses; TE = trace element analyses; Sr = Sr isotope analyses; Nd = Nd isotope analyses; Pb = Pb isotope analyses, O = oxygen isotope analyses, C = carbon isotope analyses, U–Pb = U–Pb geochronology on accessory phases (zircon, rutile, titanite, etc.).

[a] Xenoliths from these localities are probably derived from mid-crustal levels based on either: equilibration pressures, lack of mantle derived xenoliths in the same hosts and/or chemically evolved character of the host.

hornblende, biotite) in otherwise mafic granulites. These mafic xenoliths have been interpreted to represent gabbroic intrusions that underplated the Baltic Shield during the Paleoproterozoic flood-basalt event (2.4–2.5 Ga) and later experienced potassium-metasomatism coincident with partial melting at ~1.8 Ga, a major period of granitic magmatism in this region (Kempton *et al.*, 2001; Downes *et al.*, 2002). The dominately mafic compositions of these xenoliths is consistent with the thick layer of high-velocity ($\geq$7 km s^{-1}) material imaged beneath the Archean crust of the Baltic Shield (Luosto *et al.*, 1989, 1990). The xenolith studies suggest that this layer formed during Paleoproterozoic basaltic underplating and is not part of the original Archean architecture of this Shield.

In contrast to the Baltic Shield, mafic granulites appear to be absent in lower-crustal xenolith suites from the Archean Kaapvaal craton, which are dominated by metapelite and unique ultra-high-temperature granulites of uncertain petrogenesis (Dawson *et al.*, 1997; Dawson and Smith, 1987; Schmitz and Bowring, 2003a,b). These xenoliths derive from depths of >30 km and show evidence for multiple thermal metamorphic overprints starting with ultrahigh temperature metamorphism at ~2.7 Ga, which is associated with Ventersdorp magmatism (Schmitz and Bowring, 2003a,b). The absence of mafic granulites is consistent with the relatively low *P*-wave velocities in the lower crust of the Kaapvaal craton (Durrheim and Green, 1992; Nguuri *et al.*, 2001; Niu and James, 2002), but it is not clear whether the lack of a mafic lower crust reflects the original crustal structure of this Archean craton (Nguuri *et al.*, 2001) or reflects loss of a mafic complement some time after crust formation in the Archean (Niu and James, 2002).

Lower crustal xenoliths from the Hannuoba basalts, situated in the central zone of the North China Craton, show a diversity of compositions ranging from felsic to mafic metaigneous granulites and metapelites (Gao *et al.*, 2000; Chen *et al.*, 2001; Liu *et al.*, 2001; Zhou *et al.*, 2002); approximately half the xenoliths have evolved compositions (Liu *et al.*, 2001). All granulite xenoliths equilibrated under high temperatures (700–1,000 °C), corresponding to depths of 25–40 km (Chen *et al.*, 2001), but mafic granulites yield higher temperatures than metapelitic xenoliths, suggesting their derivation from deeper crustal levels (Liu *et al.*, 2001). Liu *et al.* used regional seismic refraction data and the lithologies observed in the Hannuoba xenoliths to infer the lower-crust composition in this part of the North China craton. They describe a layered lower crust in which the upper portion (from 24 km to ~38 km, $V_p \sim 6.5\,\mathrm{km\,s^{-1}}$), consists largely of felsic granulites and metasediments, and is underlain by a "lowermost" crust (38–42 km, $V_p \sim 7.0\,\mathrm{km\,s^{-1}}$) composed of intermediate granulites, mafic granulites, pyroxenite, and peridotite. Thus, the bulk lower crust in this region is intermediate in composition, consistent with the relatively large proportion of evolved granulites at Hannuoba. Zircon geochronology shows that mafic granulites and some intermediate granulites were formed by basaltic underplating in the Cretaceous. This mafic magmatism intruded pre-existing Precambrian crust consisting of metapelites that had experienced high-grade metamorphism at 1.9 Ga (Liu *et al.* (2001) and references therein).

Fragmentary xenolithic evidence for the composition of the lower crust is available for three other Archean cratons. Two mafic garnet granulites from the Udachnaya kimberlite in the Siberian craton yield Archean lead–lead and Proterozoic samarium–neodymium mineral isochrons (Shatsky, Rudnick and Jagoutz, unpublished data). It is likely that that lead–lead isochrons are frozen isochrons yielding anomalously old ages due to ancient uranium loss; the best estimate of the true age of these mafic granulites is Proterozoic. Moser and Heaman (1997) report Archean uranium–lead ages for zircons derived from mafic lower-crustal xenoliths from the Superior Province, Canada. They suggest these samples represent the mafic lower crust presently imaged seismically beneath the Abitibi greenstone belt, but which is not exposed in the Kapuskasing uplift. These granulites experienced an episode of high-grade metamorphism at 2.4 Ga, which Moser and Heaman (1997) attribute to underplating of basaltic magmas associated with the opening of the Matachewan Ocean. Davis (1997) reports mafic to felsic granulite xenoliths from the Slave craton, Canada, that have Archean to Proterozoic uranium–lead zircon ages. The mafic granulites appear to derive from basaltic magmas that underplated the felsic-Archean crust during the intrusion of the 1.3 Ga McKenzie dike swarm.

The above case studies illustrate the utility of lower crustal xenolith studies in defining the age, lithology, and composition of the lower crust beneath Archean cratons. When viewed collectively, an interesting generality emerges: when mafic granulites occur within the lower crust of Archean cratons they are generally inferred to have formed from basaltic underplating related to post-Archean magmatic events (In addition to the studies mentioned above is the case of the thick, high-velocity lower crust beneath the Archean Wyoming Province and Medicine Hat Block, western North America, which is also inferred to have formed by Proterozoic underplating based on uranium–lead zircon ages from lower crustal xenoliths (Gorman *et al.*, 2002)). Only granulites from the Superior Province appear to represent Archean mafic lower crust (Moser and Heaman, 1997). This generality is based on still only a handful of studies of xenoliths from Archean cratons and more such studies are clearly needed. However, if this generality proves robust, it implies that the processes responsible for generation of crust in most Archean cratons did not leave behind a mafic lower crust, the latter of which is commonly observed in post-Archean regions (Rudnick (1992) and references therein). It may be that this mafic lower crust was never produced, or that it formed but was removed from the crust, perhaps via density foundering (R. W. Kay and S. M. Kay, 1991; Gao *et al.*, 1998b; Jull and Kelemen, 2001). In either case, the apparent contrast in lower-crustal composition between Archean and post-Archean regions, originally pointed out by Durrheim and Mooney (1994), suggests different processes may have been operative in the formation of Archean crust. We return to the issue of what crust composition tells us about crustal generation processes in Section 5.5.

In summary, despite the uncertainties regarding the representativeness of any given lower-crustal xenolith suite (Rudnick, 1992), the above studies show that an accurate picture of the deep crust can be derived from such studies, especially when xenolith studies are combined with seismological observations of lower-crust velocities, to which we now turn.

5.3.5.2 Seismological evidence

The *P*-wave velocity of the lower crust varies from region to region, but average, temperature-corrected velocities for lower crust from a variety of different tectonic settings are high (6.9–7.2 $km\,s^{-1}$; Rudnick and Fountain, 1995; Christensen and Mooney, 1995). Such velocities are consistent with the dominance of mafic lithologies (mafic granulite and/or amphibolite) in these lower-crustal sections. High-grade metapelite, in which much of the quartz and feldspars have been removed by partial melting, is also characterized by high seismic velocities and thus may also be present (Rudnick and Fountain, 1995). Although seismically indistinct, some limit on the amount of metapelite in these high-velocity layers can be made on the basis of heat-flow and xenolith studies; these suggest that metapelite is probably a minor constituent of the lower crust (i.e., <10%; Rudnick and Fountain, 1995). In addition, average *P*-wave velocities for mafic granulite or amphibolite are higher than those observed in many lower-crustal sections (corrected to room-temperature velocities). Average room-temperature *P*-wave velocities for a variety of mafic lower crustal rock types are generally equal to or higher than 7 $km\,s^{-1}$: $7.0 \pm 0.2\,km\,s^{-1}$ for amphibolite, 7.0 to $7.2 \pm 0.2\,km\,s^{-1}$ for garnet-free mafic granulites, and 7.2 to 7.3 ± 0.2 km s^{-1} for garnet-bearing mafic granulites at 600 MPa (Rudnick and Fountain, 1995; Christensen and Mooney, 1995). Lower-crustal sections having temperature-corrected *P*-wave velocities of 6.9–7.0 $km\,s^{-1}$ (e.g., Paleozoic orogens and Mesozoic/Cenozoic extensional and contractional terranes), are thus likely to have lower-velocity rock types present (up to 30% intermediate to felsic granulites), in addition to mafic granulites or amphibolites (Rudnick and Fountain, 1995).

Although the average lower-crustal seismic sections discussed above show high velocities, some sections are characterized by much lower velocities, indicating a significantly more evolved lower-crust composition. For example, the crust of a number of Archean cratons is relatively thin (~35 km) with low seismic velocities in the lower crust (6.5–6.7 km s^{-1}), suggesting an evolved composition (e.g., Yilgarn craton (Drummond, 1988), Kaapvaal craton (Durrheim and Green, 1992; Niu and James, 2002), and North China craton (Gao *et al.*, 1998a,b)). As discussed above, it is not clear whether these thin and relatively evolved regions of Archean crust represent the original crustal architecture, formed by processes distinct from those responsible for thicker and more mafic crustal regions (e.g., Nguuri *et al.*, 2001), or reflect loss of a mafic layer from the base of the original crust (Gao *et al.*, 1998b; Niu and James, 2002). In addition, some Cenozoic–Mesozoic extensional and contractional regions, Paleozoic orogens, and active rifts show relatively slow lower-crustal velocities of 6.7–6.8 km s^{-1} and may contain >40% felsic and intermediate granulites (Rudnick and Fountain, 1995). Two extreme examples are the southern Sierra Nevada and Central Andean back-arc. In both cases, the entire crustal columns are characterized by *P*-wave velocities of less than 6.4 km s^{-1} (Beck and Zandt, 2002; Wernicke *et al.*, 1996). A relatively high-velocity ($Vp = 6.4 - 6.8\,km\,s^{-1}$) layer of <5 km in thickness occurs only at the base of the Central Andean backarc at ~60 km depth.

In summary, the seismic velocity of the lower crust is variable from region to region, but is generally high, suggesting a dominance of mafic lithologies. However, most seismic sections require the presence of evolved compositions in addition to mafic lithologies in the lower crust (up to 30% for average velocity of 6.9 km s^{-1}) and a few regions (e.g., continental arcs and some Archean cratons) are characterized by slow lower crust, indicating a highly evolved average composition. This diversity of lithologies is consistent with that seen in both crustal cross-sections and lower-crustal xenolith suites and also provides a mechanism (lithological layering) to explain the common occurrence of seismic reflections observed in many seismic reflection profiles (Mooney and Meissner, 1992).

5.3.5.3 Lower-crust composition

Table 7 lists previous estimates of the composition of the lower crust. These estimates include averages of exposed granulites (columns 1 and 2; Weaver and Tarney, 1984; Shaw *et al.*, 1994), averages of individual lower-crustal xenolith suites (columns 3–6, Condie and Selverstone, 1999; Liu *et al.*, 2001; Rudnick and Taylor, 1987; Villaseca *et al.*, 1999), the median composition of lower-crustal xenoliths (column 7; updated from Rudnick and Presper (1990), with data from papers cited in Table 6 (the complete geochemical database for lower crustal xenoliths is available on the GERM web site http://earth-ref.org/cgi-bin/erda.cgi?*n*=1,2,3,8 and also on the Treatise web site), averages derived from linking seismic velocity data for the lower crust with the compositions of lower-crustal rock types (columns 8–10; Rudnick and Fountain, 1995; Wedepohl, 1995; Gao *et al.*, 1998a), and Taylor and McLennan's model lower crust (column 11). It is readily apparent from this table and Figures 12 and 13 that, compared to estimates of the upper-crust composition (Table 1), there is much greater variability in estimates of the lower-crust composition. For example, TiO_2, MgO, FeO_T, and Na_2O all vary by over a factor of 2, CaO varies by almost a factor of 7, and K_2O varies by over an order of magnitude between the different estimates (Figures 12 and 13). Trace elements show correspondingly large variations (Figure 13). In contrast, modern estimates of major elements in the upper crust generally fall within 20% of each other (Table 1 and Figure 2—gray shading) and most trace elements fall with 50%. We now explore the possible reasons for these variations,

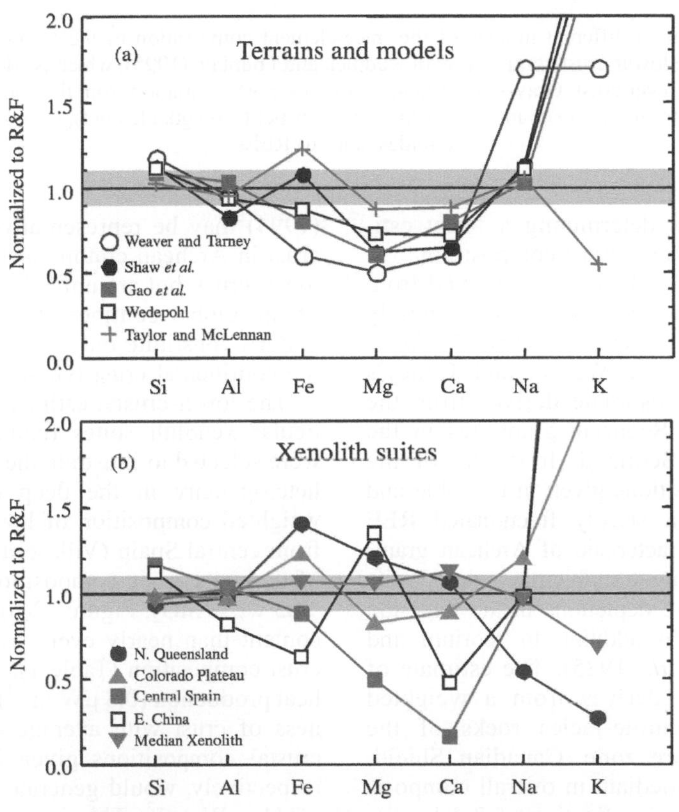

Figure 12 Comparison of different major-element estimates of the composition of the lower continental crust. All data normalized to the lower-crust composition of Rudnick and Fountain (1995), which is adopted here. Gray shaded field represents ±10% variation from the model of Rudnick and Fountain (1995). (a) Models based on granulite terrains (Scourian granulites: Weaver and Tarney, 1984; Kapuskasing Structure Zone: Shaw *et al.*, 1986), seismological models (Eastern China: Gao *et al.*, 1998a,b; western Europe: Wedepohl, 1995) and Taylor and McLennan (1985, 1995; modified by McLennan, 2001b) model lower crust. (b) Models based on weighted averages of lower crustal xenoliths. These include: Northern Queensland, Australia (Rudnick and Taylor, 1987); Colorado Plateau, USA (Condie and Selverstone, 1999); Central Spain (Villaseca *et al.*, 1999); eastern China (Liu *et al.*, 2001) and the median global lower crustal xenolith composition, updated from Rudnick and Presper (1990). Note that K data for eastern China and Central Spain are co-incident on this plot, making it hard to distinguish the separate lines.

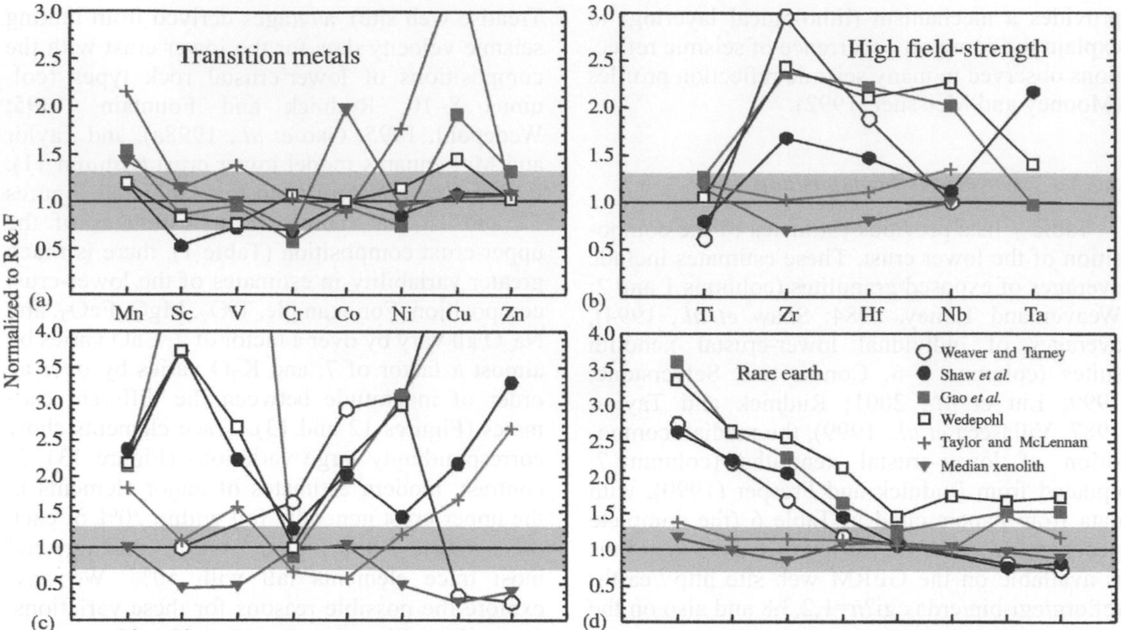

Figure 13 Comparison of different models of the trace-element composition of the lower continental crust. All values normalized to the lower-crust composition of Rudnick and Fountain (1995), which is adopted here as the "best estimate" of the global lower crust. Gray-shaded field represents ±30% variation from this value. Trace elements are divided into the following groups: (a) transition metals, (b) high-field strength elements, (c) alkali, alkaline earth, and actinides, and (d) REEs.

with an eye towards determining a "best estimate" of the global lower-crust composition.

The two lower-crustal estimates derived from averages of surface granulites are generally more evolved than other estimates (Table 7, and Figures 12 and 13). Weaver and Tarney's (1984) lower-crustal estimate derives from the average of Archean Scourian granulites in the Lewisian complex, Scotland. It is one of the most evolved compositions given in the table and is characterized by a steeply fractionated REE pattern, which is characteristic of Archean granitoids of the tonalite–trondhjemite–granodiorite assemblage and severe depletions in the large-ion lithophile elements, in addition to thorium and uranium (Rudnick *et al.*, 1985). The estimate of Shaw *et al.* (1994) derives from a weighted average of the granulite-facies rocks of the Kapuskasing Structure zone, Canadian Shield. The average is intermediate in overall composition. As discussed in Section 5.3.4.1, the Kapuskasing cross-section provides samples down to depths of ~25 km, leaving the lower 20 km of lower crust unexposed. Seismic-velocity data show this unexposed deepest crust to be mafic in bulk composition, consistent with the limited data for lower-crustal xenoliths from the Superior province (Moser and Heaman, 1997). Thus, the lower-crustal estimates of Weaver and Tarney (1984) and Shaw *et al.* (1994) may be representative of evolved lower crust in Archean cratons lacking a high-velocity lower crust, but are unlikely to be representative of the global continental lower crust (Archean cratons constitute only ~7% of the total area of the continental crust (Goodwin, 1991)).

The lower-crustal estimates derived from particular xenolith suites (columns 3–6, Table 7) were selected to illustrate the great compositional heterogeneity in the deep crust. The average weighted composition of lower-crustal xenoliths from central Spain (Villaseca *et al.*, 1999) is one of the most felsic compositions in Table 7 (with ~63 wt.% SiO_2, Figure 12(b)). It has higher K_2O content than nearly every estimate of the upper crust composition (Table 1), and has such a high heat production (0.8 $\mu W\ m^{-3}$), that a 40 km thickness of crust with average upper- and middle-crustal compositions given in Tables 3 and 5, respectively, would generate a surface heat flow of 41 $mW\ m^{-2}$. This is equivalent to 100% of surface heat flow through Archean crust, 85% of surface heat flow through Proterozoic crust, and 71% of the surface heat flow through Paleozoic crust. Assuming the heat flux through the Moho is ~17 $mW\ m^{-2}$ this lower-crust composition could thus be representative of the lower crust in Phanerozoic regions with high surface heat flow, but clearly cannot be representative of the global average lower crust. Likewise, the "evolved"

Table 7 Compositional estimates of the lower continental crust. Major elements in weight percent.

	1 Weaver and Tamey (1984)	2 Shaw et al. (1994)	3 Rudnick and Taylor (1987)	4 Condie and Selverstone (1999)	5 Villaseca et al. (1999)	6 Liu et al. (2001)	7 Updated from Rudnick and Presper (1990)	8 Rudnick and Fountain (1995)	9 Wedepohl (1995)	10 Gao et al. (1998a)	11 Taylor and McLennan (1985, 1995)
SiO_2	62.9	58.3	49.6	52.6	62.7	59.6	52.0	53.4	59.0	59.8	54.3
TiO_2	0.5	0.65	1.33	0.95	1.04	0.60	1.13	0.82	0.85	1.04	0.97[b]
Al_2O_3	16.0	17.4	16.4	16.4	17.4	13.9	17.0	16.9	15.8	14.0	16.1
FeO_T[a]	5.4	7.09	12.0	10.5	7.52	5.44	9.08	8.57	7.47	9.30	10.6
MnO	0.08	0.12	0.22	0.16	0.10	0.08	0.15	0.10	0.12	0.16	0.22
MgO	3.5	4.36	8.72	6.04	3.53	9.79	7.21	7.24	5.32	4.46	6.28
CaO	5.8	7.68	10.1	8.50	1.58	4.64	10.28	9.59	6.92	6.20	8.48
Na_2O	4.5	2.70	1.43	3.19	2.58	2.60	2.61	2.65	2.91	3.00	2.79
K_2O	1.0	1.47	0.17	1.37	3.41	3.30	0.54	0.61	1.61	1.75	0.64[b]
P_2O_5	0.19	0.24		0.21	0.16	0.13	0.13	0.10		0.21	
Mg#	53.4	52.3	56.5	50.5	45.6	76.2	58.6	60.1	55.9	46.1	51.4
Li		14				3.3	5	6	13	13	11
Be									1.7	1.1	1.0
B		3.2							5	7.6	8.3
N									34		
F									429	703	
S									408	231	
Cl									278	216	
Sc		16	33	28	17	20	29	31	25	26	35[b]
V		140	217		139	100	189	196	149	185	271[b]
Cr	88	168	276	133	178	490	145	215	228	123	219[b]
Co		38	31	20	22	31	41	38	38	36	33[b]
Ni	58	75	141	73	65	347	80	88	99	64	156[b]
Cu		28	29		40		32	26	37	50	90
Zn		83			83	89	85	78	79	102	83
Ga						15	17	13	17	19	18
Ge									1.4	1.24	1.6
As									1.3	1.6	0.8
Se									0.17	0.17	0.05
Br									0.28		
Rb	11	41	12	37	90	51	7	11	41	56	12[b]

(continued)

Table 7 (continued).

	1 *Weaver and Tamey (1984)*	*2* *Shaw et al. (1994)*	*3* *Rudnick and Taylor (1987)*	*4* *Condie and Selverstone (1999)*	*5* *Villaseca et al. (1999)*	*6* *Liu et al. (2001)*	*7* *Updated from Rudnick and Presper (1990)*	*8* *Rudnick and Fountain (1995)*	*9* *Wedepohl (1995)*	*10* *Gao et al. (1998a)*	*11* *Taylor and McLennan (1985, 1995)*
Sr	569	447	196	518	286	712	354	348	352	308	230
Y	7	16	28		40	8	20	16	27	18	19
Zr	202	114	127	86	206	180	68	68	165	162	70
Nb	5	5.6	13	7.75	15	6.4	5.6	5.0	11	10	6.7[b]
Mo			0.8				0.8		0.6	0.54	0.8
Ru											
Pd										2.78	1
Ag									80	51	90
Cd									0.101	0.097	0.098
In									0.052		0.050
Sn							1.3		2.1	1.34	1.5
Sb									0.30	0.09	0.2
I									0.14		
Cs		0.67	0.07			0.15	0.19	0.3	0.8	2.6	0.47[b]
Ba	757	523	212	564	994	1434	305	259	568	509	150
La	22	21	12	22	38	18	9.5	8	27	29	11
Ce	44	45	28	46	73	36	21	20	53	53	23
Pr			3.6				[2.1]		7.4		2.8
Nd	19	23	16	24	30	14	13.3	11	28	25	13
Sm	3.3	4.1	4.1	5.17	6.6	2.59	3.40	2.8	6.0	4.65	3.17
Eu	1.18	1.18	1.36	1.30	1.8	0.97	1.20	1.1	1.6	1.39	1.17
Gd			4.31	4.67	6.8		3.6	3.1	5.4		3.13
Tb	0.43	0.28	0.79	0.72		0.33	0.50	0.48	0.81	0.86	0.59
Dy			5.05		6.7		3.9	3.1	4.7		3.6
Ho			1.12				0.6	0.68	0.99		0.77
Er			3.25				2.0	1.9			2.2
Tm	0.19										0.32
Yb	1.2	1.13	3.19	2.09	4.0	0.79	1.70	1.5	2.5	2.29	2.2
Lu	0.18	0.2		0.37	0.65	0.12	0.30	0.25	0.43	0.38	0.29
Hf	3.6	2.8	3.3	1.9		4.6	1.9	1.9	4.0	4.2	2.1
Ta		1.3		0.5	2.1	0.3	0.5	0.6	0.8	0.6	0.7[b]
W			0.5				0.5		0.6	0.51	0.6[b]

	1	2	3	4	5	6	7	8	9	10	11
Re											0.4
Os											0.05
Ir											0.13
Pt										2.87	
Au										1.58	3.4
Hg									0.021	0.0063	
Tl									0.26	0.38	0.23
Pb	13	6	3.3	9.8		12.9	4.1	4	12.5	13	5.0[b]
Bi									0.037	0.38	0.038
Th	0.42	2.6	0.54	1.64	5.74	0.49	0.50	1.2	6.6	5.23	2.0[b]
U	0.05	0.66	0.21	1.38	0.47	0.18	0.18	0.2	0.93	0.86	0.53[b]

[a] Total Fe as FeO. [b] Value from McLennan (2001b).1.
Weighted average of Scourian granulites, Scotland, from Weaver and Tarney (1984). 2. Weighted average of Kapuskasing Structural Zone granulites, from Shaw *et al.* (1994). 3. Average lower crustal xenoliths from the McBride Province, Queensland, Australia from Rudnick and Taylor (1987). 4. Average lower crustal xenoliths from the four corners region, Colorado Plateau, USA from Condie and Selverstone (1999). 5. Weighted mean composition calculated from lithologic proportions of lower crustal xenoliths from Central Spain from Villaseca *et al.*, (1999). 6. Weighted average of lower crustal xenoliths from Hannuoba according to seismic crustal model of North China Craton from Liu *et al.* (2001). 7. Median worldwide lower crustal xenoliths from Rudnick and Presper (1990), updated with data from more recent publications. Complete database available at http://earthref.org/cgi-bin/erda.cgi?*n* = 1, 2, 3, 8, and on Treatise website. 8. Average lower crust derived from global average seismic velocities and granulites from Rudnick and Fountain (1995). 9. Average lower crust in western Europe derived from seismic data and granulite xenolith compositions from Wedepohl (1995). 10. Average lower crust derived from seismic velocities and granulite data from the North China craton from Gao *et al.* (1998a). 11. Average lower crust from Taylor and McLennan (1985, 1995), updated by McLennan and Taylor (1996) and McLennan (2001b). Mg# = molar $100 \times Mg/(Mg+Fe_{tot})$.

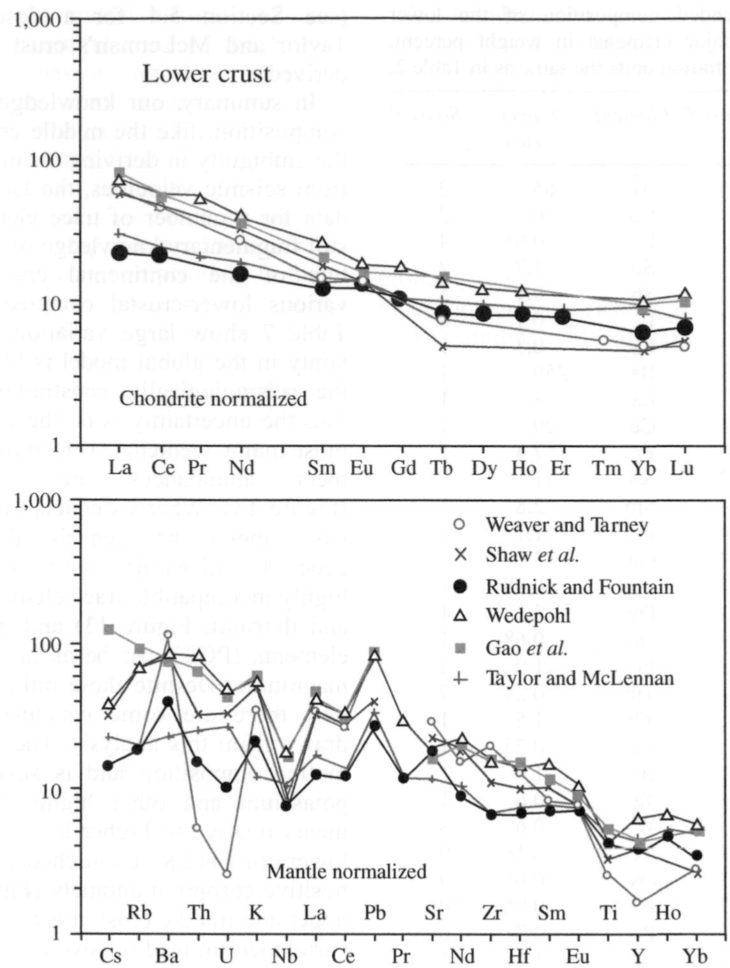

Figure 14 REE (upper) and multi-element plot (lower) of the compositions of the lower crust given in Table 7. Compositions derived from individual xenolith suites not shown. Chondrite values from Taylor and McLennan (1985) and primitive mantle values from McDonough and Sun (1995).

approximately equal proportions. His composition (Table 9) is also remarkably similar to more modern estimates for a large number of elements, but like the earliest estimates, is more appropriately considered an upper crustal estimate since sediments derive strictly from upper crustal sources. Following the plate-tectonic revolution, Taylor (1967) modified his crust-composition model. Recognizing that the present site of continental growth is at convergent-plate margins, he developed the "island arc" or "andesite" model for crustal growth and hence crust composition. In this model (Taylor, 1967, 1977), the crust is assumed to have a composition equal to average convergent-margin andesite. Taylor and McLennan (1985) discussed the difficulties with this approach. Moreover, it is now recognized that basalts dominate present intra-oceanic arcs.

Crust-composition estimates made since the 1970s derive from a variety of approaches. Smithson (1978) was the first to use seismic velocities to determine the lithological makeup of the deep crust. His crust composition is similar to other estimates, save for the very high alkali element contents (4 wt.% Na_2O and 2.7 wt.% K_2O), which presumably reflects the choice of granitic rocks used in his calculations. Holland and Lambert (1972), Weaver and Tarney (1984), and Shaw *et al.* (1986) recognized the importance of granulite-facies rocks in the deep crust and based their crustal models on the composition of rocks from high-grade terranes exposed at the Earth's surface and previous estimates of upper crustal composition.

Taylor and McLennan (1985, 1995), like Taylor's previous estimates (Taylor, 1967, 1977), derived their crust composition using an approach based on assumptions about its formation processes. They assumed that 75% of the crust grew during the Archean from bimodal volcanism and the remaining 25% originated from post-Archean accretion of island arcs

Table 9 Compositional estimates of the bulk continental crust. Major elements in weight percent. Trace element concentration units the same as in Table 2.

	1 *Taylor (1964)*	2 *Ronov and Yaroshevsky (1967)*	3 *Holland and Lambert (1972)*	4 *Smithson (1978)*	5 *Weaver and Tarney (1984)*	6 *Shaw et al. (1986)*	7 *Christensen and Mooney (1995)*	8 *Rudnick and Fountain (1995)*	9 *Wedepohl (1995)*	10 *Gao et al. (1998a)*	11 *Taylor and McLennan (1985, 1995)*	12 *This study*[a]
SiO_2	60.4	62.2	62.8	63.7	63.9	64.5	62.4	60.1	62.8	64.2	57.1	60.6
TiO_2	1.0	0.8	0.7	0.7	0.6	0.7	0.9	0.7	0.7	0.8	0.9	0.72
Al_2O_3	15.6	15.7	15.7	16.0	16.3	15.1	14.9	16.1	15.4	14.1	15.9	15.9
FeO_T[b]	7.3	6.3	5.5	5.3	5.0	5.7	6.9	6.7	5.7	6.8	9.1	6.71
MnO	0.12	0.10	0.10	0.10	0.08	0.09	0.10	0.11	0.10	0.12	0.18	0.10
MgO	3.9	3.1	3.2	2.8	2.8	3.2	3.1	4.5	3.8	3.5	5.3	4.66
CaO	5.8	5.7	6.0	4.7	4.8	4.8	5.8	6.5	5.6	4.9	7.4	6.41
Na_2O	3.2	3.1	3.4	4.0	4.2	3.4	3.6	3.3	3.3	3.1	3.1	3.07
K_2O	2.5	2.9	2.3	2.7	2.1	2.4	2.1	1.9	2.7	2.3	1.3	1.81
P_2O_5	0.24		0.20		0.19	0.14	0.20	0.20		0.18		0.13
Mg#	48.7	47.0	50.9	49.0	50.5	50.1	44.8	54.3	54.3	48.3	50.9	55.3
Li	20							11	18	17	13	17
Be	2.8								2.4	1.7	1.5	1.9
B	10					9.3			11	18	10	11
N	20								60			56
F	625								525	602		553
S	260								697	283		404
Cl	130								472	179		244
Sc	22					13		22	16	19	30	21.9
V	135					96		131	98	128	230	138
Cr	100				56	90		119	126	92	185	135
Co	25					26		25	24	24	29	26.6
Ni	75				35	54		51	56	46	105	59
Cu	55					26		24	25	38	75	27
Zn	70					71		73	65	81	80	72
Ga	15							16	15	18	18	16
Ge	1.5								1.4	1.25	1.6	1.3
As	1.8								1.7	3.1	1.0	2.5
Se	0.05								0.12	0.13	0.05	0.13
Br	2.5								1.0			0.88
Rb	90				61	76		58	78	69	37[c]	49
Sr	375				503	317		325	333	285	260	320
Y	33				14	26		20	24	17.5	20	19
Zr	165				210	203		123	203	175	100	132
Nb	20				13	20		8[d]	19	11	8[d]	8
Mo	1.5								1.1	0.65	1.0	0.8
Ru									0.1			0.57
Pd									0.4	1.74	1	1.5

(continued)

Table 9 (continued).

	1 *Taylor (1964)*	2 *Ronov and Yaroshevsky (1967)*	3 *Holland and Lambert (1972)*	4 *Smithson (1978)*	5 *Weaver and Tarney (1984)*	6 *Shaw et al.* (1986)	7 *Christensen and Mooney (1995)*	8 *Rudnick and Fountain (1995)*	9 *Wedepohl (1995)*	10 *Gao et al.* (1998a)	11 *Taylor and McLennan (1985, 1995)*	12 *This study*[a]
Ag	70								70	52	80	56
Cd	0.20								0.10	0.08	0.10	0.08
In	0.1								0.05		0.05	0.05
Sn	2.0								2.3	1.5	2.5	1.7
Sb	0.2								0.3	0.2	0.2	0.2
I	0.5								0.8			0.7
Cs	3.0							2.6	3.4	2.8	1.5[c]	2
Ba	425				707	764		390	584	614	250	456
La	30				28			18	30	31.6	16	20
Ce	60				57			42	60	60.0	33	43
Pr	8.2								6.7		3.9	4.9
Nd	28				23			20	27	27.4	16	20
Sm	6				4.1			3.9	5.3	4.84	3.5	3.9
Eu	1.2				1.09			1.2	1.3	1.27	1.1	1.1
Gd	5.4								4.0		3.3	3.7
Tb	0.9				0.53			0.56	0.65	0.82	0.60	0.6
Dy	3								3.8		3.7	3.6
Ho	1.2								0.80		0.78	0.77
Er	2.8								2.1		2.2	2.1
Tm	0.48				0.24				0.30		0.32	0.28
Yb	3.0				1.5			2.0	2.0	2.2	2.2	*1.9*
Lu	0.50				0.23			0.33	0.35	0.35	0.30	0.30
Hf	3				4.7	5		3.7	4.9	4.71	3.0	3.7
Ta	2					4		0.7[d]	1.1	0.6	0.8[c]	0.7
W	1.5								1.0	0.7	1.0	1
Re									0.4		0.4	0.19
Os									0.05		0.05	0.041
Ir									0.05		0.10	0.037
Pt									0.4	1.81		0.5
Au	40								2.5	1.21	3.0	1.3
Hg	0.08								0.040	0.009		0.03
Tl	0.45								0.52	0.39	0.36	0.50
Pb	12.5				15	20		12.6	14.8	15	8.0	11
Bi	0.17								0.085	0.27	0.06	0.18
Th	9.6				5.7	9		5.6	8.5	7.1	4.2	5.6
U	2.7				1.3	1.8		1.4	1.7	1.2	1.1	1.3

Major elements recast to 100% anhydrous.
[a] See Table 10 for derviation of this estimate. [b] Total Fe as FeO. Mg# = molar 100 × Mg/(Mg + Fe_{tot}). [c] Updated by McLennan (2001b). [d] Updated by Barth *et al.* (2000).

having an average andesite composition (from Taylor, 1977). To constrain the proportions of mafic- to felsic-Archean volcanics, they used heat-flow data from Archean cratons, which is relatively low and uniform at ~40 mW m^{-2}. Assuming that half the surface heat flow derives from the mantle yielded a mafic to felsic proportion of 2 : 1. This dominantly mafic Archean-crustal component is reflected in the major- and trace-element composition of their crust. Their crust has low SiO_2 and K_2O, high MgO and CaO, and very high FeO content. It also has the highest transition-metal concentrations and lowest incompatible element concentrations of all the models presented in Table 9.

Refinements of the Taylor and McLennan (1985) model are provided by McLennan and Taylor (1996) and McLennan (2001b). The latter is a modification of several trace-element abundances in the upper crust and as such, should not affect their compositional model for the bulk crust, which does not rely on their upper crustal composition. Nevertheless, McLennan (2001b) does provide modified bulk-crust estimates for niobium, rubidium, caesium, and tantalum (and these are dealt with in the footnotes of Table 9). McLennan and Taylor (1996) revisited the heat-flow constraints on the proportions of mafic and felsic rocks in the Archean crust and revised the proportion of Archean-aged crust to propose a more evolved bulk crust composition. This revised composition is derived from a mixture of 60% Archean crust (which is a 50 : 50 mixture of mafic and felsic end-member lithologies), and 40% average-andesite crust of Taylor (1977). McLennan and Taylor (1996) focused on potassium, thorium, and uranium, and did not provide amended values for other elements, although other incompatible elements will be higher (e.g., rubidium, barium, LREEs) and compatible elements lower in a crust composition so revised.

More recently, a number of studies have estimated the bulk-crust composition by deriving lithological proportions for the deep crust from seismic velocities (as discussed in Section 5.3) with upper crustal contributions based on data for surface rocks or previous estimates of the upper crust (Christensen and Mooney, 1995; Rudnick and Fountain, 1995; Wedepohl, 1995; Gao *et al.*, 1998a) (Table 9). Like previous estimates of the crust, all of these show intermediate bulk compositions with very similar major-element contents. The greatest differences in major elements between these recent seismologically based estimates are for MgO, CaO, and K_2O, which show ~30% variation, with the Rudnick and Fountain (1995) estimate having the highest MgO and CaO, and lowest K_2O (Table 9, Figure 15). Most trace elements from these estimates fall within 30% total variation as well

Table 10 Recommended composition of the bulk continental crust.

Element	*Units*		*Element*	*Units*	
SiO_2	wt.%	60.6	Ag	ng g^{-1}	56
TiO_2	"	0.7	Cd	μg g^{-1}	0.08
Al_2O_3	"	15.9	In	"	0.052
FeOT	"	6.7	Sn	"	1.7
MnO	"	0.10	Sb	"	0.2
MgO	"	4.7	I[a]	"	0.7
CaO	"	6.4	Cs	"	2
Na_2O	"	3.1	Ba	"	456
K_2O	"	1.8	La	"	20
P_2O_5	"	0.1	Ce	"	43
Li	μg g^{-1}	16	Pr	"	4.9
Be	"	1.9	Nd	"	20
B	"	11	Sm	"	3.9
N[a]	"	56	Eu	"	1.1
F	"	553	Gd	"	3.7
S	"	404	Tb	"	0.6
Cl	"	244	Dy	"	3.6
Sc	"	21.9	Ho	"	0.77
V	"	138	Er	"	2.1
Cr	"	135	Tm	"	0.28
Co	"	26.6	Yb	"	1.9
Ni	"	59	Lu	"	0.30
Cu	"	27	Hf	"	3.7
Zn	"	72	Ta	"	0.7
Ga	"	16	W	"	1
Ge	"	1.3	Re[a]	ng g^{-1}	0.188
As	"	2.5	Os[a]	"	0.041
Se	"	0.13	Ir[a]	"	0.037
Br[a]	"	0.88	Pt	"	1.5
Rb	"	49	Au	"	1.3
Sr	"	320	Hg	μg g^{-1}	0.03
Y	"	19	Tl	"	0.50
Zr	"	132	Pb	"	11
Nb	"	8	Bi	"	0.18
Mo	"	0.8	Th	"	5.6
Ru[a]	ng g^{-1}	0.6	U	"	1.3
Pd	"	1.5			

The total-crust composition is calculated according to the upper, middle and lower-crust compositions obtained in this study and corresponding weighing factors of 0.317, 0.296 and 0.388. The weighing factors are based on the layer thickness of the global continental crust, recalculated from crustal structure and areal proportion of various tectonic units given by Rudnick and Fountain (1995).
[a] Middle crust is not considered due to lack of data.

(Figure 16); the exceptions are trace elements for which very limited data exist (i.e., sulfur, chlorine, arsenic, tin, mercury, bismuth, and the PGEs). Niobium and tantalum also show >30% total variation, but this is due to the very high niobium and tantalum contents of Wedepohl's estimate, which reflects his reliance on the old Canadian Shield data, for which niobium content is anomalously high (see discussion in Section 5.2.1). These elements were also compromised in the Rudnick and Fountain (1995) crust composition, which relied (indirectly) on the Canadian Shield data for the upper crust by adopting the Taylor and McLennan upper-crust composition (see discussions in Plank range of trace elements than given in that model. The heat production of this new estimate is 0.89 μW m^{-2}, which falls in the middle of the range estimated for average-crustal heat production by Jaupart and Mareschal.

Figures 15–17 show how this composition compares to other estimates of crust composition. Figure 15 shows that our new composition has generally higher MgO, CaO, and FeO, and lower Na_2O and K_2O than most other seismically based models. The differences between our model and that of Wedepohl (1995) and Gao *et al.* (1998a) likely reflect the regional character of these latter models (western Europe, eastern China), where the crust is thinner and more evolved than the global averages (Chirstensen and Mooney, 1995, and Rudnick and Fountain, 1995). The lower MgO and higher alkali elements in Christensen and Mooney's model compared to ours must stem from the differences in the chemical databases used to construct these two models, as the lithological proportions of the deep crust are very similar (Section 5.3). Rudnick and Fountain (1995) (and hence our current composition) used the compositions of lower-crustal xenoliths to constrain the mafic end-member of the deep crust. These xenoliths have high Mg# and low alkalis (Table 7), and thus may be chemically distinct from mafic rocks exposed on the Earth's surface (the chemical data used by Christensen and Mooney, 1995).

The variations between the different seismological-based crust compositions can be considered representative of the uncertainties that exist in our understanding of the bulk crust composition. Some elemental concentrations (e.g., silicon, aluminum, sodium) are known to within 20% uncertainty. The remaining major-element and many trace-element (transition metals, high-field strength elements, most REE) concentrations are known to within 30% uncertainty. Still some trace element concentrations are yet poorly constrained in the crust,

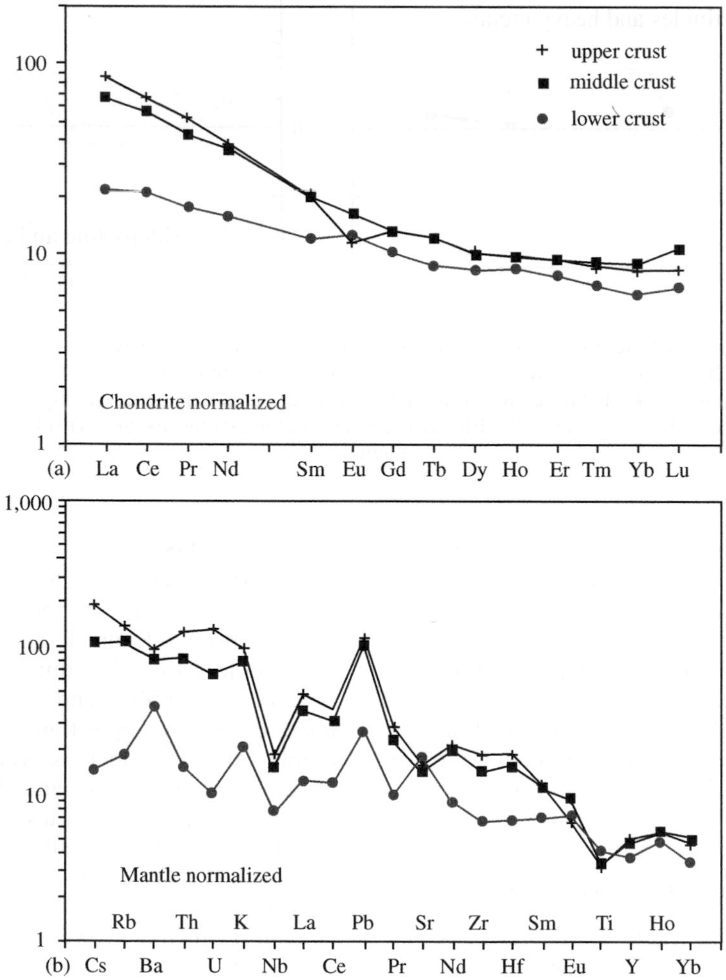

Figure 17 Comparison of (a) rare-earth and (b) additional trace-element compositions of the upper, middle, and lower crust recommended here. Chondrite values from Taylor and McLennan (1985), mantle-normalizing values from McDonough and Sun (1995).

including many of the highly siderophile elements (Figure 16).

5.4.2 Intracrustal Differentiation

Table 11 provides the composition of the upper, middle, lower, and bulk crust for comparison purposes and Figure 17 compares their respective REE and extended trace-element patterns. The upper crust has a large negative europium anomaly (Eu/Eu*, Table 11) that is largely complemented by the positive europium anomaly of the lower crust; the middle crust has essentially no europium anomaly. Similar complementary anomalies exist for strontium. These features, in addition to the greater LREE enrichment of the upper crust relative to the lower crust, suggests that the upper crust is largely the product of intracrustal magmatic differentiation in the presence of plagioclase (see Taylor and McLennan, 1985). That is, the upper crust is dominated by granite that differentiated from the lower crust through partial melting, crystal fractionation and mixing processes. The middle crust has an overall trace-element pattern that is very similar to the upper crust, indicating that it too is dominated by the products of intracrustal differentiation. All segments of the crust are characterized by an overall enrichment of the most incompatible elements, as well as high La/Nb and low Ce/Pb ratios. These are characteristics of convergent margin magmas and thus have implications for the processes responsible for generation of the continental crust as discussed in the next section.

5.5 IMPLICATIONS OF THE CRUST COMPOSITION

Despite the uncertainties in estimating crust composition discussed in the previous section, there are a number of similarities that all crust-compositional models share and these may be important for understanding the origin of the crust. The crust is characterized by an overall *intermediate* igneous-rock composition, with relatively high Mg#. It is enriched in incompatible elements (Figure 18), and contains up to 50% of the silicate Earth's budget of these elements (Rudnick and Fountain, 1995). It is also well established that the crust is depleted in niobium relative to lanthanum, and has a subchondritic Nb/Ta ratio. These features are not consistent with formation of the crust by single-stage melting of peridotitic mantle, as discussed in Rudnick (1995), Kelemen (1995) and Rudnick *et al.* (2000).

If one assumes that the crust grows ultimately by igneous processes (i.e., magmatic transport of mass from the mantle into the crust), then the disparity between crust composition and the composition of primary mantle melts requires the operation of additional processes to produce the present crust composition. As reviewed in Rudnick (1995) and Kelemen (1995), these additional processes could include (but are not limited to):

(i) *Recycling of mafic/ultramafic lower crust and upper mantle via density foundering (often referred to as delamination within the geochemical literature).* In this process, lithologically stratified continental crust is thickened during an orogenic event, causing the mafic lower crust to transform to eclogite, which has a higher density than the underlying mantle peridotite. Provided the right temperatures and viscosities exist (i.e., hot and goey), the base of the lithosphere will sink into the underlying asthenosphere. Numerical simulations of this process show that it is very likely to occur at the time of arc–continent collision (and in fact, may be impossible to avoid)(Jull and Kelemen, 2001).

(ii) *Production of crust from a mixture of silicic melts derived from subducted oceanic crust, and basaltic melts from peridotite.* This process is likely to have been more prevalent in a hotter, Archean Earth and would have involved extensive silicic melt–peridotite reaction as the slab melts traverse the mantle wedge (Kelemen, 1995). The abundance of Archean-aged granitoids of the so-called "TTG" suite (trondhjemite, tonalite, granodiorite) are often cited as the surface manifestations of these processes (Drummond and Defant, 1990; Martin, 1994.

(iii) *Weathering of the crust, with preferential recycling of Mg ± Ca into the mantle via hydrothermally altered mid-ocean ridge basalt (Albarede, 1998; Anderson, 1982).* This hypothesis states that during continental weathering, soluble cations such as Ca^{2+}, Mg^{2+}, and Na^{+} are carried to the oceans while silicon and aluminum remain behind in the continental regolith. Whereas other elements (e.g., sodium) may be returned to the continents via arc magmatism, magnesium may be preferentially sequestered into altered seafloor basalts and returned to the mantle via subduction, producing a net change in the crust composition over time. However, one potential problem with this hypothesis is that examination of altered ocean-floor rocks suggests that magnesium may not be significantly sequestered there.

A fourth possibility, that ultramafic cumulates representing the chemical complement to the andesitic crust are present in the uppermost mantle, is not supported by studies of peridotite xenoliths, which show a predominance of restitic peridotite

Table 11 Comparison of the upper, middle, lower and total continental crust compositions recommended here.

Element	*Upper crust*	*Middle crust*	*Lower crust*	*Total crust*
SiO_2	66.6	63.5	53.4	60.6
TiO_2	0.64	0.69	0.82	0.72
Al_2O_3	15.4	15.0	16.9	15.9
FeO_T	5.04	6.02	8.57	6.71
MnO	0.10	0.10	0.10	0.10
MgO	2.48	3.59	7.24	4.66
CaO	3.59	5.25	9.59	6.41
Na_2O	3.27	3.39	2.65	3.07
K_2O	2.80	2.30	0.61	1.81
P_2O_5	0.15	0.15	0.10	0.13
Total	100.05	100.00	100.00	100.12
Mg#	46.7	51.5	60.1	55.3
Li	24	12	13	16
Be	2.1	2.3	1.4	1.9
B	17	17	2	11
N	83		34	56
F	557	524	570	553
S	621	249	345	404
Cl	294	182	250	244
Sc	14.0	19	31	21.9
V	97	107	196	138
Cr	92	76	215	135
Co	17.3	22	38	26.6
Ni	47	33.5	88	59
Cu	28	26	26	27
Zn	67	69.5	78	72
Ga	17.5	17.5	13	16
Ge	1.4	1.1	1.3	1.3
As	4.8	3.1	0.2	2.5
Se	0.09	0.064	0.2	0.13
Br	1.6		0.3	0.88
Rb	82	65	11	49
Sr	320	282	348	320
Y	21	20	16	19
Zr	193	149	68	132
Nb	12	10	5	8
Mo	1.1	0.60	0.6	0.8
Ru	0.34		0.75	0.57
Pd	0.52	0.76	2.8	1.5
Ag	53	48	65	56
Cd	0.09	0.061	0.10	0.08
In	0.056		0.05	0.052
Sn	2.1	1.30	1.7	1.7
Sb	0.4	0.28	0.10	0.2
I	1.4		0.14	0.71
Cs	4.9	2.2	0.3	2
Ba	628	532	259	456
La	31	24	8	20
Ce	63	53	20	43
Pr	7.1	5.8	2.4	4.9
Nd	27	25	11	20
Sm	4.7	4.6	2.8	3.9
Eu	1.0	1.4	1.1	1.1
Gd	4.0	4.0	3.1	3.7
Tb	0.7	0.7	0.48	0.6
Dy	3.9	3.8	3.1	3.6
Ho	0.83	0.82	0.68	0.77
Er	2.3	2.3	1.9	2.1
Tm	0.30	0.32	0.24	0.28
Yb	2.0	2.2	1.5	1.9

(continued)

Table 11 (continued).

Element	*Upper crust*	*Middle crust*	*Lower crust*	*Total crust*
Lu	0.31	0.4	0.25	0.30
Hf	5.3	4.4	1.9	3.7
Ta	0.9	0.6	0.6	0.7
W	1.9	0.60	0.60	1
Re	0.198		0.18	0.188
Os	0.031		0.05	0.041
Ir	0.022		0.05	0.037
Pt	0.5	0.85	2.7	1.5
Au	1.5	0.66	1.6	1.3
Hg	0.05	0.0079	0.014	0.03
Tl	0.9	0.27	0.32	0.5
Pb	17	15.2	4	11
Bi	0.16	0.17	0.2	0.18
Th	10.5	6.5	1.2	5.6
U	2.7	1.3	0.2	1.3
Eu/Eu^*	0.72	0.96	1.14	0.93
Heat production ($\mu W\ m^{-3}$)	1.65	1.00	0.19	0.89
Nb/Ta	13.4	16.5	8.3	12.4
Zr/Hf	36.7	33.9	35.8	35.5
Th/U	3.8	4.9	6.0	4.3
K/U	9475	15607	27245	12367
La/Yb	15.4	10.7	5.3	10.6
Rb/Cs	20	30	37	24
K/Rb	283	296	462	304
La/Ta	36	42	13	29

over cumulates (e.g., Wilshire *et al.*, 1988). If such cumulates were originally there, they must have been subsequently removed via a process such as density foundering.

All of the above processes require return of mafic to ultramafic lithologies to the convecting mantle. These lithologies are the chemical complement of the present-day andesitic crust. Thus crustal recycling, in various forms, must have been important throughout Earth history.

Another implication of the distinctive trace element composition of the continental crust is that the primary setting of crust generation is most likely to be that of a convergent margin. The characteristic depletion of niobium relative to lanthanum seen in the crust (Figure 18) is a ubiquitous feature of convergent margin magmas (see review of Kelemen *et al.*) and is virtually absent in intraplate magmas. Simple mixing calculations indicate that the degree of niobium depletion seen in the crust suggests that at least 80% of the crust was generated in a convergent margin (Barth *et al.*, 2000; Plank and Langmuir, 1998).

5.6 EARTH'S CRUST IN A PLANETARY PERSPECTIVE

The other terrestrial planets show a variety of crustal types, but none that are similar to that of the Earth. Mercury has an ancient, heavily cratered crust with a high albedo (see review of Taylor and Scott). Its brightness plus the detection of sodium, and more recently the refractory element calcium, in the Mercurian atmosphere (Bida *et al.*, 2000) has led to the speculation that Mercury's crust may be anorthositic, like the lunar highlands (see Taylor, 1992 and references therein). The MESSENGER mission (http://messenger.jhuapl.edu/), currently planned to rendezvous with Mercury in 2007, should considerably illuminate the nature of the crust on Mercury.

In contrast to Mercury's ancient crust, high-resolution radar mapping of Venus' cloaked surface has revealed an active planet, both tectonically and volcanically (see review of Fegley and references therein). Crater densities are relatively constant, suggesting a relatively young surface (~300–500 Ma, Phillips *et al.*, 1992; Schaber *et al.*, 1992; Strom *et al.*, 1994). It has been suggested that this statistically random crater distribution may reflect episodes of mantle overturn followed by periods of quiescence (Schaber *et al.*, 1992; Strom *et al.*, 1994). Most Venusian volcanoes appear to erupt basaltic magmas, but a few are pancake-shaped, which may signify the eruption of a highly viscous lava such as rhyolite (e.g., Ivanov and Head, 1999). The unimodal topography of Venus is distinct from that of the Earth and there appear to be no equivalents to Earth's oceanic and continental dichotomy. It is possible

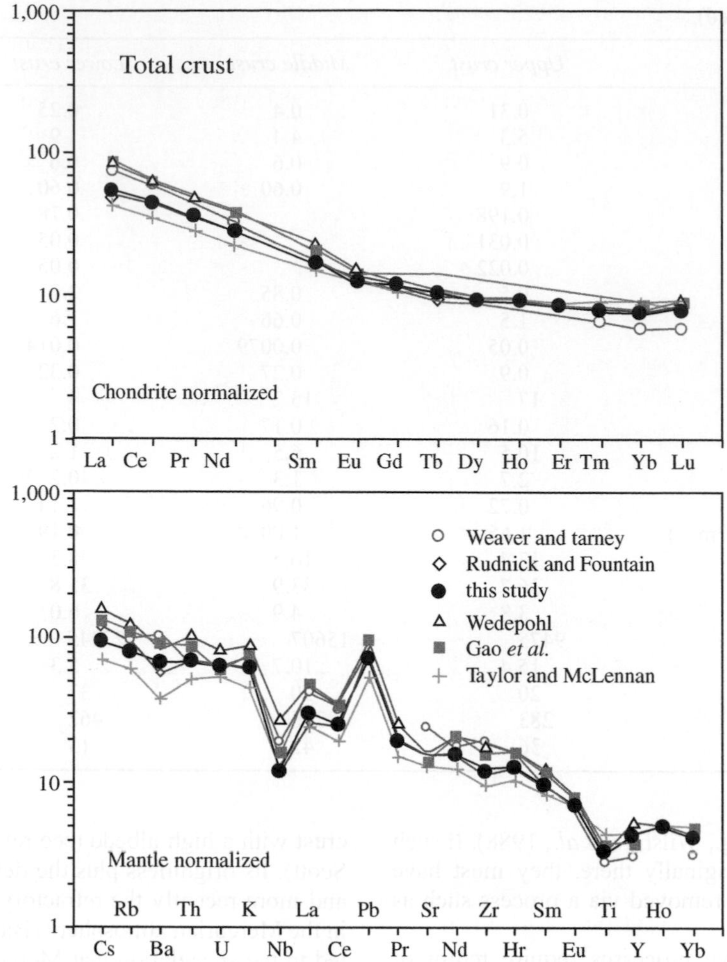

Figure 18 REE (upper) and multi-element plot (lower) of the compositions of the continental crust given in Table 9. Chondrite values from Taylor and McLennan (1985), mantle-normalizing values from McDonough and Sun (1995).

that the high elevations on Venus were produced tectonically by compression of basaltic rocks made rigid by the virtual absence of water (Mackwell *et al.*, 1998).

Of the terrestrial planets, only Mars has the bimodal topographic distribution seen on the Earth (Smith *et al.*, 1999). In addition, evolved igneous rocks, similar to the andesites found in the continents on Earth, have also been observed on the Martian surface, although their significance and relative abundance is a matter of contention (see review by McSween). However, the bimodal topography of Mars appears to be an ancient feature (Frey *et al.*, 2002), unlike the Earth's, which is a product of active plate tectonics. It remains to be seen whether the rocks that compose the high-standing southern highlands of Mars bear any resemblance to those of Earth's continental crust (McLennan, 2001a; Wanke *et al.*, 2001).

5.7 SUMMARY

The crust is the Earth's major repository of incompatible elements and thus factors prominently into geochemical mass-balance calculations for the whole Earth. For this reason, and to understand the processes by which it formed, determining the composition of the continental crust has been a popular pursuit of geochemists from the time the first rocks were analyzed.

It has been known for over a century that the continental crust has an average composition approximating to andesite (when cast as an igneous rock type) (Clarke, 1889, Clarke and Washington, 1924). The myriad studies on continental crust composition carried out in the intervening years have refined our picture of the crust's composition, particularly for trace elements.

Based on seismic investigations the crust can be divided into three regions: upper, middle, and lower continental crust. The upper crust is the most accessible region of the solid earth and its composition is estimated from both weighted averages of surface samples and studies of shales and loess. The latter is a particularly powerful means of estimating the average upper crustal concentrations of insoluble to moderately soluble trace elements. Most estimates of the major-element composition of the upper continental crust fall within 20% standard deviation of the mean and thus the composition of this important reservoir appears to be reasonably well known. The concentrations of some trace elements also appear to be known to within 20% (most of the transition metals, rubidium, strontium, yttrium, zirconium, niobium, barium, REE, hafnium, tantalum, lead, thorium, and uranium), whereas others are less-precisely known. In particular, very few estimates have been made of the upper crust's halogen, sulfur, germanium, arsenic, selenium, indium, and platinum-group element concentration.

Lacking the access and widescale natural sampling by sediments afforded the upper crust, the composition of the deep crust must be inferred from more indirect means. Both heat flow and seismic velocities have been employed towards this end. Heat flow provides bounds on the potassium, thorium, and uranium content of the crust, and seismic-wave speeds can be interpreted, with some caveats, in terms of rock types, whose compositions are derived from averages of appropriate deep-crustal lithologies. The middle crust is perhaps the least well characterized of the three crustal regions. This is due to the lack of systematic geochemical studies of amphibolite-facies crustal lithologies. In contrast, the lower crust has been the target of a number of geochemical investigations, yet there is wide variation in different estimates of lower-crust composition. This reflects, in part, the highly heterogeneous character of this part of the Earth. However, some generalities can be made. Heat production must decrease and seismic velocities are observed to increase with depth in the crust. Thus the lower crust is, on an average, mafic in composition and depleted in heat-producing elements. Curiously, the lower crust of many Archean cratons, where heat flow is lowest, has relatively slow *P*-wave velocities. Such low velocities imply the dominance of evolved rock types and thus these rocks must be highly depleted in potassium, thorium, and uranium compared to their upper crustal counterparts.

The andesitic continental crust composition is difficult to explain if the crust is generated by single-stage melting of peridotitic mantle, and additional processes must therefore be involved in its generation. All of these processes entail return of mafic or ultramafic crustal material (which is complementary to the present continental crust) to the convecting mantle. Thus crustal recycling, in various forms, must have been important throughout Earth history and is undoubtedly related to the plate-tectonic cycle on our planet. Crustal recycling, along with the presence of abundant water to facilitate melting (Campbell and Taylor, 1985), may be the major factor responsible for our planet's unique crustal dichotomy.

ACKNOWLEDGMENTS

We thank Sandy Romeo and Yongshen Liu for their able assistance in the preparation of this manuscript and updating of the lower-crustal xenolith database. Reviews by Kent Condie and Scott McLennan and comments from Herbert Palme and Rich Walker improved the presentation. This work was supported by NSF grant EAR 99031591 to R.L.R., a National Nature Science Foundation of China grant (40133020) and a Chinese Ministry of Science and Technology grant (G1999043202) to S.G.

REFERENCES

Albarede F. (1998) The growth of continental crust. In *Continents and their Mantle Roots,* Tectonophysics (eds. A. Vauchez and R. O. Meissner). Elsevier, Amsterdam, vol. 296 (1–2), pp. 1–14.

Alirezaei S. and Cameron E. M. (2002) Mass balance during gabbro-amphibolite transition, Bamble sector, Norway: implications for petrogenesis and tectonic setting of the gabbros. *Lithos* **60**(1–2), 21–45.

Anderson A. T. (1982) Parental basalts in subduction zones: implications for continental evolution. *J. Geophys. Res.* **87**, 7047–7060.

Aoki K.-I. (1971) Petrology of mafic inclusion from Itinome-gata, Japan. *Contrib. Mineral. Petrol.* **30**, 314–331.

Arculus R. J. and Smith D. (1979) Eclogite, pyroxenite, and amphibolite inclusions in the Sullivan Buttes latite, Chino Valley, Yavapai County, Arizona. In *The Mantle Sample: Inclusions in Kimberlites and Other Volcanics* (eds. F. R. Boyd and H. O. A. Meyer). American Geophysical Union, Washington, DC, pp. 309–317.

Arculus R. J., Ferguson J., Chappell B. W., Smith D., McCulloch M. T., Jackson I., Hensel H. D., Taylor S. R., Knutson J., and Gust D. A. (1988) Trace element and isotopic characteristics of eclogites and other xenoliths derived from the lower continental crust of southeastern Australia and southwestern Colorado Plateau USA. In *Eclogites and Eclogite-Facies Rocks* (ed. D. C. Smith). Elsevier, Amsterdam, pp. 335–386.

Baldridge W. S. (1979) Mafic and ultramafic inclusion suites from the Rio Grande Rift (New Mexico) and their bearing on the composition and thermal state of the lithosphere. *J. Volcanol. Geotherm. Res.* **6**, 319–351.

Barth M., McDonough W. F., and Rudnick R. L. (2000) Tracking the budget of Nb and Ta in the continental crust. *Chem. Geol.* **165**, 197–213.

Barth M., Rudnick R. L., Carlson R. W., Horn I., and McDonough W. F. (2002) Re–Os and U–Pb geochronological constraints on the eclogite-tonalite connection in the

Archean Man Shield, West Africa. *Precamb. Res.* **118**(3–4), 267–283.
Beck S. L. and Zandt G. (2002) The nature of orogenic crust in the central Andes. *J. Geophys. Res. Solid Earth* **107**(B10) (article no. 2230).
Behn M. D. and Kelemen P. B. (2003) The relationship between seismic P-wave velocity and the composition of anhydrous igneous and meta-igneous rocks. *Geochem. Geophys. Geosys.* **(4)**1041, doi: 10.1029/2002GC000393.
Berckhemer H. (1969) Direct evidence for the composition of the lower crust and Moho. *Tectonophysics* **8**, 97–105.
Berg J. H., Moscati R. J., and Herz D. L. (1989) A petrologic geotherm from a continental rift in Antarctica. *Earth Planet. Sci. Lett.* **93**, 98–108.
Bida T. A., Killen R. M., and Morgan T. H. (2000) Discovery of calcium in Mercury's atmosphere. *Nature* **404**, 159–161.
Bohlen S. R. and Mezger K. (1989) Origin of granulite terranes and the formation of the lowermost continental crust. *Science* **244**, 326–329.
Borodin L. S. (1998) Estimated chemical composition and petrochemical evolution of the upper continental crust. *Geochem. Int.* **37**(8), 723–734.
Bowring S. A. and Williams I. S. (1999) Priscoan (4.00–4.03 Ga) orthogneisses from northwestern Canada. *Contrib. Mineral. Petrol.* **134**(1), 3–16.
Bradley S. D. and McCallum M. E. (1984) Granulite facies and related xenoliths from Colorado-Wyoming kimberlite. In *Kimberlites: II. The Mantle and Crust–Mantle Relationships* (ed. J. Kornprobst). Elsevier, Amsterdam, vol. 11B, pp. 205–218.
Broadhurst J. R. (1986) Mineral reactions in xenoliths from the Colorado Plateau: implications for lower crustal conditions and fluid composition. In *The Nature of the Lower Continental Crust,* Geol. Soc. Spec. Publ. (eds. J. B. Dawson, D. A. Carswell, J. Hall, and K. H. Wedepohl). London, vol. 24, pp. 331 349.
Brittan J. and Warner M. (1996) Seismic velocity, heterogeneity, and the composition of the lower crust. *Tectonophysics* **264**, 249–259.
Brittan J. and Warner M. (1997) Wide-angle seismic velocities in heterogeneous crust. *Geophys. J. Int.* **129**, 269–280.
Burke M. M. and Fountain D. M. (1990) Seismic properties of rocks from an exposure of extended continental crust—new laboratory measurements from the Ivrea zone. *Tectonophysics* **182**, 119–146.
Cameron K. L. and Ward R. L. (1998) Xenoliths of Grenvillian granulite basement constrain models for the origin of voluminous Tertiary rhyolites, Davis Mountains, west Texas. *Geology* **26**(12), 1087–1090.
Cameron K. L., Robinson J. V., Niemeyer S., Nimz G. J., Kuentz D. C., Harmon R. S., Bohlen S. R., and Collerson K. D. (1992) Contrasting styles of pre-Cenozoic and mid-Tertiary crustal evolution in northern Mexico: evidence from deep crustal xenoliths from La Olivina. *J. Geophys. Res.* **97**, 17353–17376.
Campbell I. H. and Taylor S. R. (1985) No water, no granites—no oceans, no continents. *Geophys. Res. Lett.* **10**, 1061–1064.
Chen S., O'Reilly S. Y., Zhou X., Griffin W. L., Zhang G., Sun M., Feng J., and Zhang M. (2001) Thermal and petrological structure of the lithosphere beneath Hannuoba, Sino-Korean craton, China: evidence from xenoliths. *Lithos* **56**, 267–301.
Chen W. and Arculus R. J. (1995) Geochemical and isotopic characteristics of lower crustal xenoliths, San Francisco volcanic field, Arizona, USA. *Lithos* **36**(3–4), 203–225.
Chen Y. D., O'Reilly S. Y., Kinny P. D., and Griffin W. L. (1994) Dating lower crust and upper-mantle events—an ion microprobe study of xenoliths from kimberlitic pipes, South-Australia. *Lithos* **32**(1–2), 77–94.
Chen Y. D., O'Reilly S. Y., Griffin W. L., and Krogh T. E. (1998) Combined U–Pb dating and Sm–Nd studies on lower crustal and mantle xenoliths from the delegate basaltic pipes, southeastern Australia. *Contrib. Mineral. Petrol.* **130**(2), 154–161.
Christensen N. I. (1982) Seismic velocities. In *Handbook of Physical Properties of Rocks* (ed. R. S. Carmichael). CRC Press, Boca Raton, FL, vol. II, pp. 1–228.
Christensen N. I. and Mooney W. D. (1995) Seismic velocity structure and composition of the continental crust: a global view. *J. Geophys. Res.* **100**(B7), 9761–9788.
Clarke F. W. (1889) The relative abundance of the chemical elements. *Phil. Soc. Washington Bull.* XI, 131–142.
Clarke F. W. and Washington H. S. (1924) The composition of the Earth's crust. *USGS Professional Paper* **127**, 117pp.
Clitheroe G., Gudmundsson O., and Kennett B. L. N. (2000) The crustal thickness of Australia. *J. Geophys. Res. Solid Earth* **105**(B6), 13697–13713.
Cogley J. G. (1984) Continental margins and the extent and number of the continents. *Rev. Geophys. Space Phys.* **22**, 101–122.
Cohen R. S., O'Nions R. K., and Dawson J. B. (1984) Isotope geochemistry of xenoliths from East Africa: implications for development of mantle reservoirs and their interaction. *Earth Planet. Sci. Lett.* **68**, 209–220.
Collerson K. D., Hearn B. C., MacDonald R. A., Upton B. F., and Park J. G. (1988) Granulite xenoliths from the Bearpaw mountains, Montana: constraints on the character and evolution of lower continental crust. *Terra Cognita* **8**, 270.
Condie K. C. (1993) Chemical composition and evolution of the upper continental crust: contrasting results form surface samples and shales. *Chem. Geol.* **104**, 1–37.
Condie K. C. (1997) *Plate Tectonics and Crustal Evolution.* Butterworth-Heinemann, Oxford, UK.
Condie K. C. and Selverstone J. (1999) The crust of the Colorado plateau: new views of an old arc. *J. Geol.* **107**(4), 387–397.
Condie K. C., Latysh N., Van Schmus W. R., Kozuch M., and Selverstone J. (1999) Geochemistry, Nd and Sr isotopes, and U/Pb zircon ages of Navajo Volcanic Field, Four Corners area, southwestern United States. *Chem. Geol.* **156**(1–4), 95–133.
Davis G. L. (1977) The ages and uranium contents of zircons from kimberlites and associated rocks. *Carnegie Inst. Wash. Yearbook* **76**, 631–635.
Davis G. L. and Grew E. S. (1977) Age of zircon from a crustal xenolith, Kilbourne Hole, New Mexico. *Carnegie Inst. Wash. Yearbook* **77**, 897–898.
Davis W. J. (1997) U–Pb zircon and rutile ages from granulite xenoliths in the Slave province: evidence for mafic magmatism in the lower crust coincident with Proterozoic dike swarms. *Geology* **25**(4), 343–346.
Dawson J. B. (1977) Sub-cratonic crust and upper-mantle models based on xenolith suites in kimberlite and nephelinitic diatremes. *J. Geol. Soc.* **134**, 173–184.
Dawson J. B. and Smith J. V. (1987) Reduced sapphirine granulite xenoliths from the Lace Kimberlite, South Africa: implications for the deep structure of the Kaapvaal Craton. *Contrib. Mineral. Petrol.* **95**, 376–383.
Dawson J. B., Harley S. L., Rudnick R. L., and Ireland T. R. (1997) Equilibration and reaction in Archaean quartz-sapphirine granulite xenoliths from the Lace Kimberlite pipe, South Africa. *J. Metamorph. Geol.* **15**(2), 253–266.
Dessai A. G., and Vaselli O. (1999) Petrology and geochemistry of xenoliths in lamprophyres from the Deccan Traps: implications for the nature of the deep crust boundary in western India. *Min. Mag.* **63**(5), 703–722.
Dessai A. G., Knight K., and Vaselli O. (1999) Thermal structure of the lithosphere beneath the Deccan Trap along the western Indian continental margin: evidence from xenolith data. *J. Geol. Soc. India* **54**(6), 585–598.
Dobosi G., Kempton P. D., Downes H., Embey-Isztin A., Thirl wall M., and Greenwood P. (2003) Lower crustal granulite xenoliths from the Pannonian Basin, Hungary: Part 2. Sr–Nd–Pb–Hf and O isotope evidence for formation of

continental lower crust by tectonic emplacement of oceanic crust. *Contrib. Mineral. Petrol.* **144**, 671–683.

Dodge F. C. W., Lockwood J. P., and Calk L. C. (1988) Fragments of the mantle and crust from beneath the Sierra Nevada batholith: xenoliths in a volcanic pipe near Big Creek, California. *Geol. Soc. Am. Bull.* **100**, 938–947.

Dodge F. W., Calk L. C., and Kistler R. W. (1986) Lower crustal xenoliths, Chinese Peak lava flow, central Sierra Nevada. *J. Petrol.* **27**, 1277–1304.

Domenick M. A., Kistler R. W., Dodge F. W., and Tatsumoto M. (1983) Nd and Sr isotopic study of crustal and mantle inclusions from the Sierra Nevada and implications for batholith petrogenesis. *Geol. Soc. Am. Bull.* **94**, 713–719.

Dostal J. and Capedri S. (1979) Rare earth elements in high-grade metamorphic rocks from the western Alps. *Lithos* **12**, 41–49.

Dostal J., Dupuy C., and Leyreloup A. (1980) Geochemistry and petrology of meta-igneous granulitic xenoliths in Neogene volcanic rocks of the Massif Central, France—implications for the lower crust. *Earth Planet. Sci. Lett.* **50**, 31–40.

Downes H. (1993) The nature of the lower continental crust of Europe petrological and geochemical evidence from xenoliths. *Phys. Earth Planet. Inter.* **79**(1–2), 195–218.

Downes H., and Leyreloup A. F. (1986) Granulitic xenoliths from the French massif central—petrology, Sr and Nd isotope systematics and model age estimates. In *The Nature of the Lower Continental Crust* (eds. B. Dawson, D. A. Carswell, J. Hall, and K. H. Wedepohl). Geological Society of London, London, pp. 319–330.

Downes H., Dupuy C., and Leyreloup A. F. (1990) Crustal evolution of the Hercynian belt of western Europe: evidence from lower-crustal granulitic xenoliths (French Massif-Central). *Chem. Geol.* **83**(3–4), 209–231.

Downes H., Kempton P. D., Briot D., Harmon R. S., and Leyreloup A. F. (1991) Pb and O isotope systematics in granulite facies xenoliths, French Massif-Central—implications for crustal processes. *Earth Planet. Sci. Lett.* **102**(3–4), 342–357.

Downes H., Peltonen P., Manttari I., and Sharkov E. V. (2002) Proterozoic zircon ages from lower crustal granulite xenoliths, Kola Peninsula, Russia: evidence for crustal growth and reworking. *J. Geol. Soc.* **159**, 485–488.

Drummond B. J. (1988) A review of crust upper mantle structure in the Precambrian areas of Australia and implications for Precambrian crustal evolution. *Precamb. Res.* **40**(1), 101–116.

Drummond M. S. and Defant M. J. (1990) A model for trondhjemite-tonalite-dacite genesis and crustal growth via slab melting: archean to modern comparisons. *J. Geophys. Res.* **95**(B13), 21503–21521.

Ducea M. (2001) The California arc: thick granitic batholiths, eclogitic residues, lithospheric-scale thrusting, and magmatic flare-ups. *GSA Today* **11**(11), 4–10.

Ducea M. N. and Saleeby J. B. (1996) Buoyancy sources for a large, unrooted mountain range, the Sierra Nevada, California: evidence from xenolith thermobarometry. *J. Geophys. Res. Solid Earth* **101**(B4), 8229–8244.

Ducea M. N. and Saleeby J. B. (1998) The age and origin of a thick mafic-ultramafic keel from beneath the Sierra Nevada batholith. *Contrib. Mineral. Petrol.* **133**(1–2), 169–185.

Durrheim R. J. and Green R. W. E. (1992) Aseismic refraction investigation of the Archaean Kaapvaal craton, South Africa, using mine tremors as the energy source. *Geophys. J. Int.* **108**, 812–832.

Durrheim R. J. and Mooney W. D. (1994) The evolution of the Precambrian lithosphere: seismological and geochemical constraints. *J. Geophys. Res.* **99**, 15359–15374.

Eade K. E. and Fahrig W. F. (1971) Chemical Evolutionary Trends of Continental Plates—preliminary Study of the Canadian Shield. *Geol. Sur. Can. Bull.* **179**, 51pp.

Eade K. E. and Fahrig W. F. (1973) Regional, Lithological, and Temporal Variation in the Abundances of some Trace Elements in the Canadian Shield. Geol. Sur. Canada Paper 72–46, Ottawa, Ontario.

Eberz G. W., Clarke D. B., Chatterjee A. K., and Giles P. S. (1991) Chemical and isotopic composition of the lower crust beneath the Meguma Lithotectonic Zone, Nova Scotia: evidence from granulite facies xenoliths. *Contrib. Mineral. Petrol.* **109**, 69–88.

Edwards A. C., Lovering J. F., and Ferguson J. (1979) High pressure basic inclusions from the Kayrunnera kimberlitic breccia pipe in New South Wales, Australia. *Contrib. Mineral. Petrol.* **69**, 185–192.

Ehrenberg S. N. and Griffin W. L. (1979) Garnet granulite and associated xenoliths in minette and serpentinite diatremes of the Colorado Plateau. *Geology* **7**, 483–487.

Embey-Isztin A., Scharbert H. G., Deitrich H., and Poultidis H. (1990) Mafic granulite and clinopyroxenite xenoliths from the Transdanubian volcanic region (Hungary): implication for the deep structure of the Pannonian Basin. *Min. Mag.* **54**, 463–483.

Embey-Isztin A., Downes H., Kempton P. D., Dobosi G., and Thirlwall M. (2003) Lower crustal granulite xenoliths from the Pannonian Basin, Hungary: Part 1. Mineral chemistry, thermobarometry and petrology. *Contrib. Mineral. Petrol.* **144**, 652–670.

Ertan I. E. and Leeman W. P. (1999) Fluid inclusions in mantle and lower crustal xenoliths from the Simcoe volcanic field, Washington. *Chem. Geol.* **154**(1–4), 83–95.

Esperança S. and Garfunkel Z. (1986) Ultramafic xenoliths from the Mt. Carmel area (Karem Maharal Volcano), Israel. *Lithos* **19**, 43–49.

Esperança S., Carlson R. W., and Shirey S. B. (1988) Lower crustal evolution under central Arizona: Sr, Nd, and Pb isotopic and geochemical evidence from the mafic xenoliths of Camp Creek. *Earth Planet. Sci. Lett.* **90**, 26–40.

Esser B. K. and Turekian K. K. (1993) The osmium isotopic composition of the continental crust. *Geochim. Cosmochim. Acta* **57**, 3093–3104.

Fahrig W. F., and Eade K. E. (1968) The chemical evolution of the Canadian Shield. *Geochim. Cosmochim. Acta* **5**, 1247–1252.

Fountain D. M. (1976) The Ivrea-Verbano and Strona-Ceneri zones, northern Italy: a cross-section of the continental crust—new evidence from seismic velocities of rock samples. *Tectonophysics* **33**, 145–165.

Fountain D. M. and Salisbury M. H., (1981) Exposed cross-sections through the continental crust: implications for crustal structure, petrology, and evolution. *Earth Planet. Sci. Lett.* **56**, 263–277.

Fountain D. M., and Salisbury M. H. (1995) Seismic properties of rock samples from the Pikwitonei granulite belt—God's lake domain crustal cross section, Manitoba. *Can. J. Earth Sci.* **33**(5), 757–768.

Fountain D. M., Percival J., and Salisbury M. H. (1990a) Exposed cross sections of the continental crust-synopsis. In *Exposed Cross-sections of the Continental Crust* (eds. M. H. Salisbury and D. M. Fountain). Kluwer, Amsterdam, pp. 653–662.

Fountain D. M., Salisbury M. H. and Percival J. (1990b) Seismic structure of the continental crust based on rock velocity measurements from the Kapuskasing uplift. *J. Geophys. Res.* **95**, 1167–1186.

Fountain D. M., Salisbury M. H., and Furlong K. P. (1987) Heat production and thermal conductivity of rocks from the Pikwitonei-Sachigo continental cross section, central Manitoba: implications for the thermal structure of Archean crust. *Can. J. Earth Sci.* **24**, 1583–1594.

Francis D. M. (1976) Corona-bearing pyroxene granulite xenoliths and the lower crust beneath Nunivak Island, Alaska. *Can. Mineral.* **14**, 291–298.

Frey H. V., Roark J. H., Shockey K. M., Frey E. L., and Sakimoto S. E. H. (2002) Ancient lowlands on Mars. *Geophys. Res. Lett.* **29**(10), 22-1–22-4.

Gallet S., Jahn B.-M., van Vliet Lanoë B., Dia A., and Rossello E. (1998) Loess geochemistry and its implications for particle origin and composition of the upper continental crust. *Earth Planet. Sci. Lett.* **156**, 157–172.

Gao S., Luo T.-C., Zhang B.-R., Zhang H.-F., Han Y.-W., Hu Y.-K., and Zhao Z.-D. (1998a) Chemical composition of the continental crust as revealed by studies in east China. *Geochim. Cosmochim. Acta* **62**, 1959–1975.

Gao S., Zhang B.-R., Jin Z.-M., Kern H., Luo T.-C., and Zhao Z.-D. (1998b) How mafic is the lower continental crust? *Earth Planet. Sci. Lett.* **106**, 101–117.

Gao S., Kern H., Liu Y. S., Jin S. Y., Popp T., Jin Z. M., Feng J. L., Sun M., and Zhao Z. B. (2000) Measured and calculated seismic velocities and densities for granulites from xenolith occurrences and adjacent exposed lower crustal sections: a comparative study from the North China craton. *J. Geophys. Res. Solid Earth* **105**(B8), 18965–18976.

Goldschmidt V. M. (1933) Grundlagen der quantitativen Geochemie. *Fortschr. Mienral. Kirst. Petrogr.* **17**, 112.

Goldschmidt V. M. (1958) *Geochemistry.* Oxford University Press, Oxford.

Goodwin A. M. (1991) *Precambrian Geology.* Academic Press, London.

Gorman A. R., Clowes R. M., Ellis R. M., Henstock T. J., Spence G. D., Keller G. R., Levander A., Snelson C. M., Burianyk M. J. A., Kanasewich E. R., Asudeh I., Hajnal Z., and Miller K. C. (2002) Deep probe: imaging the roots of western North America. *Can. J. Earth Sci.* **39**(3), 375–398.

Grapes R. H. (1986) Melting and thermal reconstitution of pelitic xenoliths, Wehr volcano, East Eifel, West Germany. *J. Petrol.* **27**, 343–396.

Gregoire M., Mattielli N., Nicollet C., Cottin J. Y., Leyrit H., Weis D., Shimizu N., and Giret A. (1994) Oceanic mafic granulite xenoliths from the Kerguelen archipelago. *Nature* **367**, 360–363.

Gregoire M., Cottin J. Y., Giret A., Mattielli N., and Weis D. (1998) The meta-igneous granulite xenoliths from Kerguelen archipelago: evidence of a continent nucleation in an oceanic setting. *Contrib. Mineral. Petrol.* **133**(3), 259–283.

Griffin W. L. and O'Reilly S. Y. (1986) The lower crust in eastern Australia: xenolith evidence. In *The Nature of the Lower Continental Crust,* Geol. Soc. London Spec. Publ. (eds. B. Dawson, D. A. Carswell, J. Hall, and K. H. Wedepohl). London, vol. 25, pp. 363–374.

Griffin W. L. and O'Reilly S. Y. (1987) Is the continental Moho the crust–mantle boundary? *Geology* **15**, 241–244.

Griffin W. L., Carswell D. A., and Nixon P. H. (1979) Lower-crustal granulites and eclogites from Lesotho, southern Africa. In *The Mantle Sample: Inclusions in Kimberlites* (eds. F. R. Boyd and H. O. A. Meyer). American Geophysical Union, Washington, DC, pp. 59–86.

Griffin W. L., Sutherland F. L., and Hollis J. D. (1987) Geothermal profile and crust–mantle transition beneath east-central Queensland: volcanology, xenolith petrology and seismic data. *J. Volcanol. Geotherm. Res.* **31**, 177–203.

Griffin W. L., Jaques A. L., Sie S. H., Ryan C. G., Cousens D. R., and Suter G. F. (1988) Conditions of diamond growth: a proton microprobe study of inclusions in West Australian diamonds. *Contrib. Mineral. Petrol.* **99**, 143–158.

Hacker B. R., Gnos E., Ratschbacher L., Grove M., McWilliams M., Sobolev S. V., Wan J., and Wu Z. H. (2000) Hot and dry deep crustal xenoliths from Tibet. *Science* **287**(5462), 2463–2466.

Halliday A. N., Aftalion M., Upton B. G. J., Aspen P., and Jocelyn J. (1984) U–Pb isotopic ages from a granulite-facies xenolith from Partan Craig in the Midland Valley of Scotland. *Trans. Roy. Soc. Edinburgh: Earth Sci.* **75**, 71–74.

Hanchar J. M., Miller C. F., Wooden J. L., Bennett V. C., and Staude J.-M. G. (1994) Evidence from xenoliths for a dynamic lower crust, eastern Mojave desert, California. *J. Petrol.* **35**, 1377–1415.

Harley S. L. (1989) The origin of granulites: a metamorphic perspective. *Geol. Mag.* **126**, 215–247.

Hart R. J., Nicolaysen L. O., and Gale N. H. (1981) Radioelement concentrations in the deep profile through Precambrian basement of the Vredefort structure. *J. Geophys. Res.* **86**, 10639–10652.

Hart R. J., Andreoli M. A. G., Tredoux M., and Dewit M. J. (1990) Geochemistry across an exposed section of Archean crust at Vredefort, South Africa with implications for midcrustal discontinuities. *Chem. Geol.* **82**(1–2), 21–50.

Haskin M. A. and Haskin L. A. (1966) Rare earths in European shales: a redetermination. *Science* **154**, 507–509.

Haskin L. A., Wildeman T. R., Frey F. A., Collins K. A., Keedy C. R., and Haskin M. A. (1966) Rare earths in sediments. J. *Geophys. Res. B: Solid Earth* **71**(24), 6091–6105.

Hattori Y., Suzuki K., Honda M., and Shimizu H. (2003) Re–Os isotope systematics of the Taklimakan Desert sands, moraines and river sediments around the Taklimakan Desert, and of Tibetan soils. *Geochim. Cosmochim. Acta* **67**, 1195–1206.

Hayob J. L., Essene E. J., Ruiz J., Ortega-Gutiérrez F., and Aranda-Gómez J. J. (1989) Young high-temperature granulites form the base of the crust in central Mexico. *Nature* **342**, 265–268.

Heier K. S. (1973) Geochemistry of granulite facies rocks and problems of their origin. *Phil. Trans. Roy. Soc. London* **A273**, 429–442.

Heinrichs H., Schulz-Dobrick B., and Wedepohl K. H. (1980) Terrestrial geochemistry of Cd, Bi, Tl, Pb, Zn, and Rb. *Geochim. Cosmochim. Acta* **44**, 1519–1533.

Hickey-Vargas R., Abdollahi M. J., Parada M. A., Lopezescobar L., and Frey F. A. (1995) Crustal xenoliths from Calbuco volcano, Andean southern volcanic zone—implications for crustal composition and magma–crust interaction. *Contrib. Mineral. Petrol.* **119**(4), 331–344.

Holbrook W. S. and Kelemen P. B. (1993) Large igneous province on the US Atlantic margin and implications for magmatism during continental breakup. *Nature* **364**, 433–436.

Holbrook W. S., Mooney W. D., and Christensen N. I. (1992) The seismic velocity structure of the deep continental crust. In *Continental Lower Crust* (eds. D. M. Fountain, R. Arculus, and R. W. Kay). Elsevier, Amsterdam, pp. 1–44.

Holland J. G. and Lambert R. S. J. (1972) Major element chemical composition of shields and the continental crust. *Geochim. Cosmochim. Acta* **36**, 673–683.

Hölttä P., Huhma H., Manttari I., Peltonen P., and Juhanoja J. (2000) Petrology and geochemistry of mafic granulite xenoliths from the Lahtojoki kimberlite pipe, eastern Finland. *Lithos* **51**(1–2), 109–133.

Hu S., He L., and Wang J. (2000) Heat flow in the continental area of China: a new data set. *Earth Planet. Sci. Lett.* **179**, 407–419.

Huang Y. M., van Calsteren P., and Hawkesworth C. J. (1995) The evolution of the lithosphere in southern Africa—a perspective on the basic granulite xenoliths from kimberlites in South-Africa. *Geochim. Cosmochim. Acta* **59**(23), 4905–4920.

Hunter R. H., Upton B. G. J., and Aspen P. (1984) Meta-igneous granulite and ultramafic xenoliths from basalts of the Midland Valley of Scotland: petrology and mineralogy of the lower crust and upper mantle. *Trans. Roy. Soc. Edinburgh* **75**, 75–84.

Ivanov M. A. and Head J. W. (1999) Stratigraphic and geographic distribution of steep-sided domes on Venus: preliminary results from regional geological mapping and implications for their origin. *J. Geophys. Res. Planet.* **104**(E8), 18907–18924.

Jahn B. M., Gallet S., and Han J. M. (2001) Geochemistry of the Xining, Xifeng, and Jixian sections, Loess Plateau of China: eolian dust provenance and paleosol evolution during the last 140 ka. *Chem. Geol.* **178**(1–4), 71–94.

James D. E., Padovani E. R., and Hart S. R. (1980) Preliminary results on the oxygen isotopic composition of the lower crust,

Kilbourne Hole Maar, New Mexico. *Geophys. Res. Lett.* **7**, 321–324.

Jones A. P., Smith J. V., Dawson J. B., and Hansen E. C. (1983) Metamorphism, partial melting, and K-metasomatism of garnet-scapolite-kyanite granulite xenoliths from Lashaine, Tanzania. *J. Geol.* **91**, 143–166.

Jull M. and Kelemen P. B. (2001) On the conditions for lower crustal convective instability. *J. Geophys. Res. B: Solid Earth* **106**(4), 6423–6446.

Kalamarides R. I., Berg J. H., and Hank R. A. (1987) Lateral isotopic discontinuity in the lower crust: an example from Antarctica. *Science* **237**, 1192–1195.

Kay R. W. and Kay S. M. (1981) The nature of the lower continental crust: inferences from geophysics, surface geology, and crustal xenoliths. *Rev. Geophys. Space Phys.* **19**, 271–297.

Kay R. W. and Kay S. M. (1991) Creation and destruction of lower continental crust. *Geol. Rundsch.* **80**, 259–278.

Kay S. M. and Kay R. W. (1983) Thermal history of the deep crust inferred from granulite xenoliths, Queensland, Australia. *Am. J. Sci.* **283**, 486–513.

Kelemen P. B. (1995) Genesis of high Mg# andesites and the continental crust. *Contrib. Mineral. Petrol.* **120**, 1–19.

Kempton P. D. and Harmon R. S. (1992) Oxygen-isotope evidence for large-scale hybridization of the lower crust during magmatic underplating. *Geochim. Cosmochim. Acta* **55**, 971–986.

Kempton P. D., Harmon R. S., Hawkesworth C. J., and Moorbath S. (1990) Petrology and geochemistry of lower crustal granulites from the Geronimo volcanic field, southeastern Arizona. *Geochim. Cosmochim. Acta* **54**, 3401–3426.

Kempton P. D., Downes H., Sharkov E. V., Vetrin V. R., Ionov D. A., Carswell D. A., and Beard A. (1995) Petrology and geochemistry of xenoliths from the northern Baltic shield: evidence for partial melting and metasomatism in the lower crust beneath an Archaean terrane. *Lithos* **36**(3–4), 157–184.

Kempton P. D., Downes H., and Embey-Isztin A. (1997) Mafic granulite xenoliths in Neogene alkali basalts from the western Pannonian Basin: insights into the lower crust of a collapsed orogen. *J. Petrol.* **38**(7), 941–970.

Kempton P. D., Downes H., Neymark L. A., Wartho J. A., Zartman R. E., and Sharkov E. V. (2001) Garnet granulite xenoliths from the northern Baltic shield the underplated lower crust of a palaeoproterozoic large igneous province. *J. Petrol.* **42**(4), 731–763.

Kern H., Gao S., and Liu Q.-S. (1996) Seismic properties and densities of middle and lower crustal rocks exposed along the North China geoscience transect. *Earth Planet. Sci. Lett.* **139**, 439–455.

Ketcham R. A. (1996) Distribution of heat-producing elements in the upper and middle crust of southern and west central Arizona: evidence from core complexes. *J. Geophys. Res.* (B) **101**, 13611–13632.

Kopylova M. G., O'Reilly S. Y., and Genshaft Y. S. (1995) Thermal state of the lithosphere beneath Central Mongolia: evidence from deep-seated xenoliths from the Shavaryn-Saram volcanic centre in the Tariat depression, Hangai, Mongolia. *Lithos* **36**, 243–255.

Kuno H. (1967) Mafic and ultramafic nodules from Itinome-gata, Japan. In *Ultramafic and Related Rocks* (ed. P. J. Wiley). Wiley, New York, pp. 337–342.

Kyle P. R., Wright A., and Kirsch I. (1987) Ultramafic xenoliths in the late Cenozoic McMurdo volcanic group, western Ross Sea embayment, Antarctica. In *Mantle Xenoliths* (ed. P. H. Nixon). Wiley, New York, pp. 287–294.

Le Bas M. J. and Streckeisen A. L. (1991) The IUGS systematics of igneous rocks. *J. Geol. Soc. London* **148**, 825–833.

Lee C.-Y., Chung S. L., Chen C.-H., and Hsieh Y. L. (1993) Mafic granulite xenoliths from Penghu Islands: evidence for basic lower crust in SE China continental margin. *J. Geol. Soc. China* **36**(4), 351–379.

Leech M. L. (2001) Arrested orogenic development: eclogitization, delamination, and tectonic collapse. *Earth Planet. Sci. Lett.* **185**(1–2), 149–159.

Leeman W. P., Menzies M. A., Matty D. J., and Embree G. F. (1985) Strontium, neodymium, and lead isotopic compositions of deep crustal xenoliths from the Snake River Plain: evidence for Archean basement. *Earth Planet. Sci. Lett.* **75**, 354–368.

Leeman W. P., Sisson V. B., and Reid M. R. (1992) Boron geochemistry of the lower crust-evidence from granulite terranes and deep crustal xenoliths. *Geochim. Cosmochim. Acta* **56**(2), 775–788.

LePichon X., Henry P., and Goffe B. (1997) Uplift of Tibet: from eclogites to granulites-implications for the Andean Plateau and the Variscan belt. *Tectonophysics* **273**(1–2), 57–76.

Leyreloup A., Dupuy C., and Andriambololona R. (1977) Catazonal xenoliths in French Neogene volcanic rocks: constitution of the lower crust: 2. Chemical composition and consequences of the evolution of the French Massif Central Precambrian crust. *Contrib. Mineral. Petrol.* **62**, 283–300.

Leyreloup A., Bodinier J. L., Dupuy C., and Dostal J. (1982) Petrology and geochemistry of granulite xenoliths from Central Hoggar (Algeria)—implications for the lower crust. *Contrib. Mineral. Petrol.* **79**, 68–75.

Liu Y. S., Gao S., Jin S. Y., Hu S. H., Sun M., Zhao Z. B., and Feng J. L. (2001) Geochemistry of lower crustal xenoliths from Neogene Hannuoba basalt, North China craton: implications for petrogenesis and lower crustal composition. *Geochim. Cosmochim. Acta* **65**(15), 2589–2604.

Lombardo B. and Rolfo F. (2000) Two contrasting eclogite types in the Himalayas: implications for the Himalayan orogeny. *J. Geodynam.* **30**(1–2), 37–60.

Loock G., Seck H. A., and Stosch H.-G. (1990) Granulite facies lower crustal xenoliths from the Eifel, West Germany: petrological and geochemical aspects. *Contrib. Mineral. Petrol.* **105**, 25–41.

Lovering J. F. and White A. J. R. (1964) The significance of primary scapolite in granulitic inclusions from deep-seated pipes. *J. Petrol.* **5**, 195–218.

Lovering J. F. and White A. J. R. (1969) Granulitic and eclogitic inclusions from basic pipes at Delegate, Australia. *Contrib. Mineral. Petrol.* **21**, 9–52.

Lucassen F., Lewerenz S., Franz G., Viramonte J., and Mezger K. (1999) Metamorphism, isotopic ages and composition of lower crustal granulite xenoliths from the Cretaceous Salta Rift, Argentina. *Contrib. Mineral. Petrol.* **134**(4), 325–341.

Luosto U. and Korhonen H. (1986) Crustal structure of the baltic shield based on off-fennolora refraction data. *Tectonophysics* **128**, 183–208.

Luosto U., Flüh E. R., Lund C.-E., and Group W. (1989) The crustal structure along the POLAR profile from seismic refraction investigations. *Tectonophysics* **162**, 51–85.

Luosto U., Tiira T., Korhonen H., Azbel I., Burmin V., Buyanov A., Kosminskaya I., Ionkis V., and Sharov N. (1990) Crust and upper mantle structure along the DSS Baltic profile in SE Finland. *Geophys. J. Int.* **101**, 89–110.

Mackwell S. J., Zimmerman M. E., and Kohlstedt D. L. (1998) High-temperature deformation of dry diabase with application to tectonics on Venus. *J. Geophys. Res. B: Solid Earth* **103**, 975–984.

Markwick A. J. W. and Downes H. (2000) Lower crustal granulite xenoliths from the Arkhangelsk kimberlite pipes: petrological, geochemical and geophysical results. *Lithos* **51**(1–2), 135–151.

Markwick A. J. W., Downes H., and Veretennikov N. (2001) The lower crust of SE Belarus: petrological, geophysical, and geochemical constraints from xenoliths. *Tectonophysics* **339**(1–2), 215–237.

Martin H. (1994) The Archean grey gneisses and the genesis of continental crust. In *Archean Crustal Evolution* (ed. K. C. Condie). Elsevier, Amsterdam, pp. 205–259.

Mattie P. D., Condie K. C., Selverstone J., and Kyle P. R. (1997) Origin of the continental crust in the Colorado Plateau: geochemical evidence from mafic xenoliths from the Navajo volcanic field, southwestern USA. *Geochim. Cosmochim. Acta* **61**(10), 2007–2021.

Mayer A., Mezger K., and Sinigoi S. (2000) New Sm–Nd ages for the Ivrea-Verbano Zone, Sesia and Sessera valleys (Northern-Italy). *J. Geodynamics* **30**(1–2), 147–166.

McBirney A. R. and Aoki K.-I. (1973) Factors governing the stability of plagioclase at high pressures as shown by spinel-gabbro xenoliths from the Kerguelen archipelago. *Am. Mineral.* **58**, 271–276.

McCulloch M. T., Arculus R. J., Chappell B. W., and Ferguson J. (1982) Isotopic and geochemical studies of nodules in kimberlite have implications for the lower continental crust. *Nature* **300**, 166–169.

McDonough W. F. and Sun S.-S. (1995) Composition of the Earth. *Chem. Geol.* **120**, 223–253.

McDonough W. F., Rudnick R. L., and McCulloch M. T. (1991) The chemical and isotopic composition of the lower eastern Australian lithosphere: a review. In *The Nature of the Eastern Australian Lithosphere,* Geol. Soc. Austral. Spec. Publ. (ed. B. Drummond). Sydney, vol. 17, pp. 163–188.

McDonough W. F., Sun S.-S., Ringwood A. E., Jagoutz E., and Hofmann A. W. (1992) Potassium, rubidium, and cesium in the Earth and Moon and the evolution of the mantle of the Earth. *Geochim. Cosmochim. Acta* **56**, 1001–1012.

McLennan S. M. (2001a) Crustal heat production and the thermal evolution of Mars. *Geophys. Res. Lett.* **28**(21), 4019–4022.

McLennan S. M. (2001b) Relationships between the trace element composition of sedimentary rocks and upper continental crust. *Geochem. Geophys. Geosys.* 2 (article no. 2000GC000109).

McLennan S. M. and Taylor S. R. (1996) Heat flow and the chemical composition of continental crust. *J. Geol.* **104**, 396–377.

McLennan S. M., Nance W. B., and Taylor S. R. (1980) Rare earth element-thorium correlations in sedimentary rocks, and the composition of the continental crust. *Geochim. Cosmochim. Acta* **44**, 1833–1839.

Mehnert K. R. (1975) The Ivrea zone: a model of the deep crust. *Neus Jahrb. Mineral. Abh.* **125**, 156–199.

Meltzer A. and Christensen N. (2001) Nanga Parbat crustal anisotrophy: implication for interpretation of crustal velocity structure and shear-wave splitting. *Geophys. Res. Lett.* **28**(10), 2129–2132.

Mengel K. (1990) Crustal xenoliths from Tertiary volcanics of the northern Hessian depression: petrological and chemical evolution. *Contrib. Mineral. Petrol.* **104**, 8–26.

Mengel K. and Wedepohl K. H. (1983) Crustal xenoliths in Tertiary volcanics from the northern Hessian depression. In *Plateau Uplift* (eds. K. Fuchs, *et al.).* Springer, Berlin, pp. 332–335.

Mengel K., Sachs P. M., Stosch H. G., Worner G., and Loock G. (1991) Crustal xenoliths from Cenozoic volcanic fields of West Germany implications for structure and composition of the continental crust. *Tectonophysics* **195**(2–4), 271.

Meyer H. O. A. and Brookins D. G. (1976) Sapphirine, sillimanite, and garnet in granulite xenoliths from Stockdale kimberlite, Kansas. *Am. Mineral.* **61**, 1194–1202.

Miller J. D. and Christensen N. I. (1994) Seismic signature and geochemistry of an island arc: a multidisciplinary study of the Kohistan accreted terrane, northern Pakistan. *J. Geophys. Res.* (B)**99**, 11623–11642.

Mittlefehldt D. W. (1984) Genesis of clinopyroxene-amphibole xenoliths from Birket Ram: trace element and petrologic constraints. *Contrib. Mineral. Petrol.* **88**, 280–287.

Mittlefehldt D. W. (1986) Petrology of high pressure clinopyroxenite series xenoliths, Mount Carmel, Israel. *Contrib. Mineral. Petrol.* **94**, 245–252.

Moecher D. P., Valley J. W., and Essene E. J. (1994) Extraction and carbon isotope analysis of CO_2 from scapolite in deep crustal granulites and xenoliths. *Geochim. Cosmochim. Acta* **58**(2), 959–967.

Mooney W. D. and Meissner R. (1992) Multi-genetic origin of crustal reflectivity: a review of seismic reflection profiling of the continental lower crust and Moho. In *Continental Lower Crust* (eds. D. M. Fountain, R. Arculus, and R. W. Kay). Elsevier, pp. 45–80.

Moser D. E. and Heaman L. M. (1997) Proterozoic zircon growth in Archean lower crust xenoliths, southern Superior craton: a consequence of Matachewan ocean opening. *Contrib. Mineral. Petrol.* **128**, 164–175.

Moser D. E., Flowers R. M., and Hart R. J. (2001) Birth of the Kaapvaal tectosphere 3.08 billion years ago. *Science* **291**(5503), 465–468.

Nasir S. (1992) The lithosphere beneath the northwestern part of the Arabian plate (Jordan)—evidence from xenoliths and geophysics. *Tectonophysics* **201**(3–4), 357–370.

Nasir S. (1995) Mafic lower crustal xenoliths from the northwestern part of the Arabian plate. *Euro. J. Mineral.* **7**(1), 217–230.

Nasir S. and Safarjalani A. (2000) Lithospheric petrology beneath the northern part of the Arabian plate in Syria: evidence from xenoliths in alkali basalts. *J. African Earth Sci.* **30**(1), 149–168.

Nesbitt H. W. and Young G. M. (1984) Prediction of some weathering trends of plutonic and volcanic rocks based on thermodynamic and kinetic considerations. *Geochim. Cosmochim. Acta* **48**, 1523–1534.

Newsom H. E., Sims K. W. W., Noll P. D., Jr., Jaeger W. L., Maehr S. A., and Beserra T. B. (1996) The depletion of tungsten in the bulk silicate Earth: constraints on core formation. *Geochim. Cosmochim. Acta* **60**, 1155–1169.

Nguuri T. K., Gore J., James D. E., Webb S. J., Wright C., Zengeni T. G., Gwavava O., and Snoke J. A. (2001) Crustal structure beneath southern Africa and its implications for the formation and evolution of the Kaapvaal and Zimbabwe cratons. *Geophys. Res. Lett.* **28**(13), 2501–2504.

Nimz G. J., Cameron K. L., Cameron M., and Morris S. L. (1986) Petrology of the lower crust and upper mantle beneath southeastern Chihuahua, Mexico. *Geofisica Int.* **25**, 85–116.

Niu F. L. and James D. E. (2002) Fine structure of the lowermost crust beneath the Kaapvaal craton and its implications for crustal formation and evolution. *Earth Planet. Sci. Lett.* **200**(1–2), 121–130.

Nyblade A. A. and Pollack H. N. (1993) A global analysis of heat flow from Precambrian terrains: implications for the thermal structure of Archean and Proterozoic lithosphere. *J. Geophys. Res.* **98**, 12207–12218.

Okrusch M., Schröder B., and Schnütgen A. (1979) Granulite-facies metabasite ejecta in the Laacher Sea area, Eifel, West Germany. *Lithos* **12**, 251–270.

O'Reilly S. Y., Griffin W. L., and Stabel A. (1988) Evolution of Phanerozoic eastern Australian lithosphere: isotopic evidence for magmatic and tectonic underplating. In *Oceanic and Continental Lithosphere: Similarities and Differences,* J. Petrol. Spec. Vol. (eds. M. A. Menzies and K. G. Cox). Oxford University Press, Oxford, pp. 89–108.

Owen J. V., Greenough J. D., Hy C., and Ruffman A. (1988) Xenoliths in a mafic dyke at Popes Harbour, Nova Scotia: implications for the basement to the Meguma Group. *Can. J. Earth Sci.* **25**, 1464–1471.

Padovani E. R. and Carter J. L. (1977) Aspects of the deep crustal evolution beneath south central New Mexico. In *The Earth's Crust* (ed. J. G. Heacock). American Geophysical Union, Washington, DC, pp. 19–55.

Padovani E. R. and Hart S. R. (1981) Geochemical constraints on the evolution of the lower crust beneath the Rio Grande rift. In *Conference on the Processes of Planetary Rifting.* Lunar and Planetary Science Institute, pp. 149–152.

Padovani E. R., Hall J., and Simmons G. (1982) Constraints on crustal hydration below the Colorado Plateau from V_p measurements on crustal xenoliths. *Tectonophysics* **84**, 313–328.

Pearcy L. G., DeBari S. M., and Sleep N. H. (1990) Mass balance calculations for two sections of island arc crust and implications for the formation of continents. *Earth Planet. Sci. Lett.* **96**, 427–442.

Pearson N. J., O'Reilly S. Y., and Griffin W. L. (1995) The crust–mantle boundary beneath cratons and craton margins:a transect across the south-west margin of the Kaapvaal craton. *Lithos* **36**(3–4), 257–287.

Percival J. A. and Card K. D. (1983) Archean crust as revealed in the Kapuskasing uplift, Superior province, Canada. *Geology* **11**, 323–326.

Percival J. A., Fountain D. M., and Salisbury M. H. (1992)Exposed cross sections as windows on the lower crust. In *Continental Lower Crust* (eds. D. M. Fountain, R. Arculus, and R. W. Kay). Elsevier, Amsterdam, pp. 317–362.

Peucker-Ehrenbrink B. and Jahn B.-M. (2001) Rhenium-osmium isotope systematics and platinum group element concentations: loess and the upper continental crust. *Geochem. Geophys. Geosys.* **2**, 2001GC000172.

Phillips R. J., Raubertas R. F., Arvidson R. E., Sarkar I. C., Herrick R. R., Izenberg N., and Grimm R. E. (1992) Impact craters and Venus resurfacing history. *J. Geophys. Res. Planet.* **97**(E10), 15923–15948.

Pinet C. and Jaupart C. (1987) The vertical distribution of radiogenic heat production in the Precambrian crust of Norway and Sweden: geothermal implications. *Geophys. Res. Lett.* **14**, 260–263.

Plank T. and Langmuir C. H. (1998) The chemical composition of subducting sediment and its consequences for the crust and mantle. *Chem. Geol.* **145**, 325–394.

Poldervaart A. (1955) The chemistry of the Earth's crust. *Geol. Soc. Am. Spec. Pap.* **62**, 119–144.

Quick J. E., Sinigoi S., and Mayer A. (1995) Emplacement of mantle peridotite in the lower continental crust, Ivrea-Verbano zone, northwest Italy. *Geology* **23**(8), 739–742.

Reid M. R., Hart S. R., Padovani E. R., and Wandless G. A. (1989) Contribution of metapelitic sediments to the composition, heat production, and seismic velocity of the lower crust of southern New Mexico. *Earth Planet. Sci. Lett.* **95**, 367–381.

Roberts S. and Ruiz J. (1989) Geochemical zonation and evolution of the lower crust in Mexico. *J. Geophys. Res.* **94**, 7961–7974.

Rogers N. W. (1977) Granulite xenoliths from Lesotho kimberlites and the lower continental crust. *Nature* **270**, 681–684.

Rogers N. W. and Hawkesworth C. J. (1982) Proterozoic age and cumulate origin for granulite xenoliths, Lesotho. *Nature* **299**, 409–413.

Ronov A. B. and Yaroshevsky A. A. (1967) Chemical structure of the Earth's crust. *Geokhimiya* 11, 1285–1309.

Ronov A. B. and Yaroshevsky A. A. (1976) A new model for the chemical structure of the Earth's crust. *Geokhimiya* **12**, 1761–1795.

Ross D. C. (1985) Mafic gneissic complex (batholithic root?) in the southernmost Sierra Nevada, California. *Geology* **13**, 288–291.

Rudnick R. L. (1990a) Continental crust: growing from below. *Nature* **347**, 711–712.

Rudnick R. L. (1990b) Nd and Sr isotopic compositions of lower crustal xenoliths from North Queensland, Australia:implications for Nd model ages and crustal growth processes. *Chem. Geol.* **83**, 195–208.

Rudnick R. L. (1992) Xenoliths—samples of the lower continental crust. In *Continental Lower Crust* (eds. D. M.Fountain, R. Arculus, and R. W. Kay). Elsevier, Amsterdam, pp. 269–316.

Rudnick R. L. (1995) Making continental crust. *Nature* **378**, 571–578.

Rudnick R. L. and Cameron K. L. (1991) Age diversity of the deep crust in northern Mexico. *Geology* **19**, 1197–1200.

Rudnick R. L. and Goldstein S. L. (1990) The Pb isotopic compositions of lower crustal xenoliths and the evolution of lower crustal Pb. *Earth Planet. Sci. Lett.* **98**, 192–207.

Rudnick R. L. and Fountain D. M. (1995) Nature and composition of the continental crust: a lower crustal perspective. *Rev. Geophys.* **33**(3), 267–309.

Rudnick R. L. and Presper T. (1990) Geochemistry of intermediate to high-pressure granulites. In *Granulites and Crustal Evolution* (eds. D. Vielzeuf and P. Vidal). Kluwer, Amsterdam, pp. 523–550.

Rudnick R. L. and Taylor S. R. (1987) The composition and petrogenesis of the lower crust: a xenolith study. *J. Geophys. Res.* **92**(B13), 13981–14005.

Rudnick R. L. and Taylor S. R. (1991) Petrology and geochemistry of lower crustal xenoliths from northern Queensland and inferences on lower crustal composition. In *The Eastern Australian Lithosphere,* Geol. Soc. Austral. Spec. Publ. (ed. B. Drummond), 189–208.

Rudnick R. L. and Williams I. S. (1987) Dating the lower crust by ion microprobe. *Earth Planet. Sci. Lett.* **85**, 145–161.

Rudnick R. L., McLennan S. M., and Taylor S. R. (1985) Large ion lithophile elements in rocks from high-pressure granulite facies terrains. *Geochim. Cosmochim. Acta* **49**, 1645–1655.

Rudnick R. L., McDonough W. F., McCulloch M. T., and Taylor S. R. (1986) Lower crustal xenoliths from Queensland, Australia: evidence for deep crustal assimilation and fractionation of continental basalts. *Geochim. Cosmochim. Acta* **50**, 1099–1115.

Rudnick R. L., McDonough W. F., and O'Connell R. J. (1998) Thermal structure, thickness and composition of continental lithosphere. *Chem. Geol.* **145**, 399–415.

Rudnick R. L., Ireland T. R., Gehrels G., Irving A. J., Chesley J. T., and Hanchar J. M. (1999) Dating mantle metasomatism: U–Pb geochronology of zircons in cratonic mantle xenoliths from Montana and Tanzania. In *Proceedings of the VIIth International Kimberlite Conference* (eds. J. J. Gurney, J. L. Gurney, M. D. Pascoe, and S. R. Richardson). Red Roof Design, Cape Town, pp. 728–735.

Rudnick R. L., Barth M., Horn I., and McDonough W. F. (2000) Rutile-bearing refractory eclogites: missing link between continents and depleted mantle. *Science* **287**, 278–281.

Ruiz J., Patchett P. J., and Arculus R. J. (1988a) Nd–Sr isotope composition of lower crustal xenoliths—evidence for the origin of mid-Tertiary felsic volcanics in Mexico. *Contrib. Mineral. Petrol.* **99**, 36–43.

Ruiz J., Patchett P. J., and Ortega-Gutierrez F. (1988b) Proterozoic and Phanerozoic basement terranes of Mexico from Nd isotopic studies. *Geol. Soc. Am. Bull.* **100**, 274–281.

Rutter M. J. (1987) The nature of the lithosphere beneath the Sardinian continental block: mantle and deep crustal inclusions in mafic alkaline lavas. *Lithos* **20**, 225–234.

Saal A. E., Rudnick R. L., Ravizza G. E., and Hart S. R. (1998) Re–Os isotope evidence for the composition, formation and age of the lower continental crust. *Nature* **393**, 58–61.

Sachs P. M. and Hansteen T. H. (2000) Pleistocene under-plating and metasomatism of the lower continental crust: a xenolith study. *J. Petrol.* **41**(3), 331–356.

Saleeby J. B. (1990) Progress in tectonic and petrogenetic studies in an exposed cross-section of young (c. 100Ma) continental crust southern Sierra Nevada, California. In *Exposed Cross-sections of the Continental Crust* (eds. M. H. Salisbury, and D. M. Fountain). Kluwer Academic, Norwell, MA, pp. 137–159.

Schaaf P., Heinrich W., and Besch T. (1994) Composition and Sm–Nd isotopic data of the lower crust beneath San-Luis-Potosi, Central Mexico—evidence from a granulite-facies xenolith suite. *Chem. Geol.* **118**(1–4), 63–84.

Schaber G. G., Strom R. G., Moore H. J., Soderblom L. A., Kirk R. L., Chadwick D. J., Dawson D. D., Gaddis L. R., Boyce J. M., and Russell J. (1992) Geology and distribution of impact craters on Venus-what are they telling us. *J. Geophys. Res. Planet.* **97**(E8), 13257–13301.

Scherer E. K., Cameron K. L., Johnson C. M., Beard B. L., Barovich K. M., and Collerson K. D. (1997) Lu–Hf geochronology applied to dating Cenozoic events affecting lower crustal xenoliths from Kilbourne Hole, New Mexico. *Chem. Geol.* **142**, 63–78.

Schmitz M. D. and Bowring S. A. (2000) The significance of U–Pb zircon dates in lower crustal xenoliths from the southwestern margin of the Kaapvaal craton, southern Africa. *Chem. Geol.* **172,** 59–76.

Schmitz M. D. and Bowring S. A. (2003a) Constraints on the thermal evolution of continental lithosphere from U–Pb accessory mineral thermochronometry of lower crustal xenoliths, southern Africa. *Contrib. Mineral. Petrol.* **144**, 592–618.

Schmitz M. D. and Bowring S. A. (2003b) Ultrahigh-temperature metamorphism in the lower crust during Neoarchean Ventersdorp rifting and magmatism, Kaapvaal craton, southern Africa. *Geol. Soc. Am. Bull.* **115**, 533–548.

Schulze D. J. and Helmstaedt H. (1979) Garnet pyroxenite and eclogite xenoliths from the Sullivan Buttes latite, Chino valley, Arizona. In *The Mantle Sample: Inclusions in Kimberlites and Other Volcanics* (eds. F. R. Boyd and H. O. A. Meyer). American Geophysics Union, Washington, DC, pp. 318–329.

Sclater J. G., Jaupart C. J., and Galson D. (1980) The heat flow through oceanic and continental crust and the heat loss of the earth. *Rev. Geophys. Space Phys.* **18**, 269–311.

Selverstone J. and Stern C. R. (1983) Petrochemistry and recrystallization history of granulite xenoliths from the Pali-Aike volcanic field, Chile. *Am. Mineral.* **68**, 1102–1111.

Sewell R. J., Hobden B. J., and Weaver S. D. (1993) Mafic and ultramafic mantle and deep-crustal xenoliths from Banks Peninsula, South-Island, New-Zealand. *NZ J. Geol. Geophys.* **36**(2), 223–231.

Shatsky V., Rudnick R. L., and Jagoutz E. (1990) Mafic granulites from Udachnaya pipe, Yakutia: samples of Archean lower crust? *Deep Seated Magmatism and Evolution of Lithosphere of the Siberian Platform,* 23–24.

Shatsky V. S., Sobolev N. V., and Pavlyuchenko V. S. (1983) Fassa'ite-garnet-anorthite xenolith from the Udachnaya kimberlite pipe, Yakutia. *Dokl. Akad. Nauk. SSSR* **272**(1), 188–192.

Shaw D. M., Reilly G. A., Muysson J. R., Pattenden G. E., and Campbell F. E. (1967) An estimate of the chemical composition of the Canadian Precambrian shield. *Can. J. Earth Sci.* **4**, 829–853.

Shaw D. M., Dostal J., and Keays R. R. (1976) Additional estimates of continental surface Precambrian shield composition in Canada. *Geochim. Cosmochim. Acta* **40**, 73–83.

Shaw D. M., Cramer J. J., Higgins M. D., and Truscott M. G. (1986) Composition of the Canadian Precambrian shield and the continental crust of the Earth. In *The Nature of the Lower Continental Crust* (eds. J. B. Dawson, D. A. Carswell, J. Hall, and K. H. Wedepohl). Geol. Soc. London, London, vol. **24**, pp. 257–282.

Shaw D. M., Dickin A. P., Li H., McNutt R. H., Schwarcz H. P., and Truscott M. G. (1994) Crustal geochemistry in the Wawa-Foleyet region, Ontario. *Can. J. Earth Sci.* **31**(7), 1104–1121.

Sims K. W. W., Newsom H. E., and Gladney E. S. (1990) Chemical fractionation during formation of the Earth's core and continental crust: clues from As, Sb, W, and M., and In *Origin of the Earth* (eds. H. E. Newsom, J. H. Jones, and J. H. Newson). Oxford University Press, Oxford, pp. 291–317.

Smith D. E., Zuber M. T., Solomon S. C., Phillips R. J., Head J. W., Garvin J. B., Banerdt W. B., Muhleman D. O., Pettengill G. H., Neumann G. A., Lemoine F. G., Abshire J. B., Aharonson O., Brown C. D., Hauck S. A., Ivanov A. B., McGovern P. J., Zwally H. J., and Duxbury T. C. (1999) The global topography of Mars and implications for surface evolution. *Science* **284**(5419), 1495–1503.

Smith R. D., Cameron K. L., McDowell F. W., Niemeyer S., and Sampson D. E. (1996) Generation of voluminous silicic magmas and formation of mid-Cenozoic crust beneath north-central Mexico: evidence from ignimbrites, associated lavas, deep crustal granulites, and mantle pyroxenites.*Contrib. Mineral. Petrol.* **123**, 375–389.

Smithson S. B. (1978) Modeling continental crust-structural and chemical constraints. *Geophys. Res. Lett.* **5**(9),749–752.

Stolz A. J. (1987) Fluid activity in the lower crust and upper mantle: mineralogical evidence bearing on the origin of amphibole and scapolite in ultramafic and mafic granulite xenoliths. *Min. Mag.* **51**, 719–732.

Stolz A. J. and Davies G. R. (1989) Metasomatized lower crustal and upper mantle xenoliths from north Queensland:chemical and isotopic evidence bearing on the composition and source of the fluid phase. *Geochim. Cosmochim. Acta* **53**, 649–660.

Stosch H.-G. and Lugmair G. W. (1984) Evolution of the lower continental crust: granulite facies xenoliths from the Eifel, West Germany. *Nature* **311**, 368–370.

Stosch H.-G., Ionov D. A., Puchtel I. S., Galer S. J. G., and Sharpouri A. (1995) Lower crustal xenoliths from Mongolia and their bearing on the nature of the deep crust beneath cental Asia. *Lithos* **36**, 227–242.

Strom R. G., Schaber G. G., and Dawson D. D. (1994) The global resurfacing of Venus. *J. Geophys. Res. Planet.* **99**(E5), 10899–10926.

Sun W., Bennett V. C., Eggins S. M., Kamenetsky V. S., and Arculus R. J. (2003) Evidence for enhanced mantle to crust rhenium transfer from undegassed arc magmas. *Nature* **422**, 294–297.

Sutherland F. L. and Hollis J. D. (1982) Mantle–lower crust petrology from inclusions in basaltic rocks in eastern Australia—an outline. *J. Volcanol. Geotherm. Res.* **14**, 1–29.

Tanaka T. and Aoki K.-I. (1981) Petrogenetic implications of REE and Ba data on mafic and ultramafic inclusions from Itinome-gata, Japan. *J. Geol.* **89**, 369–390.

Taylor S. R. (1964) Abundance of chemical elements in the continental crust—a new table. *Geochim. Cosmochim. Acta* **28**, 1273–1285.

Taylor S. R. (1967) The origin and growth of continents. *Tectonophysics* **4**, 17–34.

Taylor S. R. (1977) Island arc models and the composition of the continental crust. In *Island Arcs, Deep Sea Trenches and Back-Arc Basins* (ed. M. Talwani). American Geophysical Union, Washington, DC, pp. 325–336.

Taylor S. R. (1992) *Solar System Evolution.* Cambridge University Press, Cambridge. Taylor S. R. and McLennan S. M. (1981) The composition and evolution of the continental crust: rare Earth element evidence from sedimentary rocks. *Phil. Trans. Roy. Soc. London* A **301**, 381–399.

Taylor S. R. and McLennan S. M. (1985) *The Continental Crust: Its Composition and Evolution.* Blackwell, Oxford. Taylor S. R., and McLennan S. M. (1995) The geochemical evolution of the continental crust. *Rev. Geophys.* **33**, 241–265.

Taylor S. R., McLennan S. M., and McCulloch M. T. (1983) Geochemistry of loess, continental crustal composition and crustal model ages. *Geochim. Cosmochim. Acta* **47**, 1897–1905.

Teng F., McDonough W. F., Rudnick R. L., Dalpé C.,Tomascak P. B., Chappell B. W., and Gao S. (2003) Lithium isotopic composition and concentration of the upper continental crust. *Geochim. Cosmochim. Acta* (submitted).

Thomas C. W. and Nixon P. H. (1987) Lower crustal granulite xenoliths in carbonatite volcanoes of the western rift of East Africa. *Min. Mag.* **51**, 621–633.

Toft P. B., Hills D. V., and Haggerty S. E. (1989) Crustal evolution and the granulite to eclogite transition in xenoliths from kimberlites in the West African craton. *Tectonophysics* **161**, 213–231.

Tredoux M., Hart R. J., Carlson R. W., and Shirey S. B. (1999) Ultramafic rocks at the center of the Vredefort structure:further evidence for the crust on edge model. *Geology* **27**(10), 923–926.

Trzcienski W. E. and Marchildon N. (1989) Kyanite-garnet-bearing Cambrian rocks and Grenville granulites from the Ayers Cliff, Quebec, Canada, Lamprophyre Dike

Suite—deep crustal fragments from the northern appalachians. *Geology* **17**(7), 637–640.

Upton B. G. J., Aspen P., and Chapman N. A. (1983) The upper mantle and deep crust beneath the British Isles: evidence from inclusion suites in volcanic rocks. *J. Geol. Soc. London* **140**, 105–122.

Upton B. G. J., Aspen P., Rex D. C., Melcher F., and Kinny P. (1998) Lower crustal and possible shallow mantle samples from beneath the Hebrides: evidence from a xenolithic dyke at Gribun, western Mull. *J. Geol. Soc.* **155**, 813–828.

Upton B. G. J., Aspen P., and Hinton R. W. (2001) Pyroxenite and granulite xenoliths from beneath the Scottish northern Highlands terrane: evidence for lower-crust/upper-mantle relationships. *Contrib. Mineral. Petrol.* **142**(2), 178–197.

Urrutia-Fucugauchi J. and Uribe-Cifuentes R. M. (1999) Lower-crustal xenoliths from the Valle de Santiago maar field, Michoacan-Guanajuato volcanic field, central Mexico. *Int. Geol. Rev.* **41**(12), 1067–1081.

van Breeman O. and Hawkesworth C. J. (1980) Sm–Nd isotopic study of garnets and their metamorphic host rocks. *Trans. Roy. Soc. Edinburgh* **71**, 97–102.

van Calsteren P. W. C., Harris N. B. W., Hawkesworth C. J., Menzies M. A., and Rogers N. W. (1986) Xenoliths from southern Africa: a perspective on the lower crust. In *The Nature of the Lower Continental Crust,* Geol. Soc. London Spec. Publ. (eds. J. B. Dawson, D. A. Carswell, J. Hall, and K. H. Wedepohl). London, vol. 25, pp. 351–362.

Vidal P. and Postaire B. (1985) Étude par la méthode Pb–Pb de roches de haut grade métamorphique impliquées dans la chaîne Hercynienne. *Chem. Geol.* **49**, 429–449.

Vielzeuf D. (1983) The spinel and quartz associations in high grade xenoliths from Tallante (S. E. Spain) and their potential use in geothermometry and barometry. *Contrib. Mineral. Petrol.* **82**, 301–311.

Villaseca C., Downes H., Pin C., and Barbero L. (1999) Nature and composition of the lower continental crust in central Spain and the granulite-granite linkage: inferences from granulitic xenoliths. *J. Petrol.* **40**(10), 1465–1496.

Voshage H., Hofmann A. W., Mazzucchelli M., Rivalenti G., Sinigoi S., Raczek I., and Demarchi G. (1990) Isotopic evidence from the Ivrea zone for hybrid lower crust formed by magmatic underplating. *Nature* **347**, 731–736.

Wanke H., Bruckner J., Dreibus G., Rieder R., and Ryabchikov I. (2001) Chemical composition of rocks and soils at the pathfinder site. *Space Sci. Rev.* **96**(1–4), 317–330.

Warren R. G., Kudo A. M., and Keil K. (1979) Geochemistry of lithic and single-crystal inclusions in basalts and a characterization of the upper mantle-lower crust in the Engle Basin, Rio Grande Rift, New Mexico. In *Rio Grande Rift: Tectonics and Magmatism* (ed. R. E. Riecker). American Geophysical Union, Washington, DC, pp. 393–415.

Wass S. Y. and Hollis J. D. (1983) Crustal growth in southeastern Australia—evidence from lower crustal eclogitic and granulitic xenoliths. *J. Metamorph. Geol.* **1**, 25–45.

Weaver B. L. and Tarney J. (1980) Continental crust composition and nature of the lower crust: constraints from mantle Nd–Sr isotope correlation. *Nature* **286**, 342–346.

Weaver B. L. and Tarney J. (1981) Lewisian gneiss geochemistry and Archaean crustal development models. *Earth Planet. Sci. Lett.* **55**, 171–180.

Weaver B. L. and Tarney J. (1984) Empirical approach to estimating the composition of the continental crust. *Nature* **310**, 575–577.

Weber M. B. I., Tarney J., Kempton P. D., and Kent R. W. (2002) Crustal make-up of the northern Andes: evidence based on deep crustal xenolith suites, Mercaderes, SW Colombia. *Tectonophysics* **345**(1–4), 49–82.

Wedepohl H. (1995) The composition of the continental crust. *Geochim. Cosmochim. Acta* **59**, 1217–1239.

Wedepohl K. H. (1969–1978) *Handbook of Geochemistry.* Springer, Berlin.

Wendlandt E., DePaolo D. J., and Baldridge W. S. (1993) Nd and Sr isotope chronostratigraphy of Colorado Plateau lithosphere: implications for magmatic and tectonic underplating of the continental crust. *Earth Planet. Sci. Lett.* **116**, 23–43.

Wendlandt E., DePaolo D. J., and Baldridge W. S. (1996) Thermal history of Colorado Plateau lithosphere from Sm–Nd mineral geochronology of xenoliths. *Geol. Soc. Am. Bull.* **108**(7), 757–767.

Wernicke B., Clayton R., Ducea M., Jones C. H., Park S., Ruppert S., Saleeby J., Snow J. K., Squires L., Fliedner M., Jiracek G., Keller R., Klemperer S., Luetgert J., Malin P., Miller K., Mooney W., Oliver H., and Phinney R. (1996) Origin of high mountains in the continents: the southern Sierra Nevada. *Science* **271**, 190–193.

Wilde S. A., Valley J. W., Peck W. H., and Graham C. M. (2001) Evidence from detrital zircons for the existence of continental crust and oceans on the Earth 4.4 Gyr ago. *Nature* **409**(6817), 175–178.

Wilkinson J. F. G. (1975) An Al-spinel ultramafic-mafic inclusion suite and high pressure megacrysts in an analcimite and their bearing on basaltic magma fractionation at elevated pressures. *Contrib. Mineral. Petrol.* **53**, 71–104.

Wilkinson J. F. G. and Taylor S. R. (1980) Trace element fractionation trends of thoeiiticmagma at moderate pressure: evidence from an Al-spinel ultramafic-mafic inclusion suite. *Contrib. Mineral. Petrol.* **75**, 225–233.

Wilshire H. W., Meyer C. E., Nakata J. K., Calk L. C., Shervais J. W., Nielson J. E., and Schwarzman E. C. (1988) *Mafic and Ultramafic Xenoliths from Volcanic Rocks of the Western United States.* Prof. Paper, USGS, Washington, DC, West Sussex, UK.

Windley B. F. (1995) *The Evolving Continents.* Wiley.Wörner G., Schmincke H.-U., and Schreyer W. (1982) Crustal xenoliths from the Quaternary Wehr volcano (East Eifel). *Neus. Jahrb. Mineral. Abh.* **144**(1), 29–55.

Yardley B. W. D. (1986) Is there water in the deep continental crust? *Nature* **323**, 111.

Yu. J. H., O'Reilly S. Y., Griffin W. L., Xu X. S., Zhang M., and Zhou X. M. (2003) The thermal state and composition of the lithospheric mantle beneath the Leizhou Peninsula South China. *J. Volcanol. Geotherm. Res.* **122**(3–4), 165–189.

Zashu S., Kaneoka I., and Aoki K.-I. (1980) Sr isotope study of mafic and ultramafic inclusions from Itinome-gata, Japan. *Geochem. J.* **14**, 123–128.

Zheng J. P., Sun M., Lu F. X., and Pearson N. (2003) Mesozoic lower crustal xenoliths and their significance in lithospheric evolution beneath the Sino-Korean craton. *Tectonophysics* **361**(1–2), 37–60.

Zhou X. H., Sun M., Zhang G. H., and Chen S. H. (2002) Continental crust and lithospheric mantle interaction beneath North China: isotopic evidence from granulite xenoliths in Hannuoba, Sino-Korean craton. *Lithos* **62**(3–4), 111–124.

Readings from the Treatise on Geochemistry
ISBN: 978-0-12-381391-6

pp. 131–196

6
The History of Planetary Degassing as Recorded by Noble Gases

D. Porcelli
University of Oxford, UK
and
K. K. Turekian
Yale University, New Haven, CT, USA

6.1 INTRODUCTION

Noble gases provide unique clues to the structure of the Earth and the degassing of volatiles into the atmosphere. Since the noble gases are highly depleted in the Earth, their isotopic compositions are prone to substantial changes due to radiogenic additions, even from scarce parent elements and low-yield nuclear processes. Therefore, noble gas isotopic signatures of major reservoirs reflect planetary differentiation processes that generate fractionations between these volatiles and parent elements. These signatures can be used to construct planetary degassing histories that have relevance to the degassing of a variety of chemical species as well.

It has long been recognized that the atmosphere is not simply a remnant of the volatiles that surrounded the forming Earth with the composition of the early solar nebula. It was also commonly thought that the atmosphere and oceans were derived from degassing of the solid Earth over time (Brown, 1949; Suess, 1949; Rubey, 1951). Subsequent improved understanding of the processes of planet formation, however, suggests that substantial volatile inventories could also have been added directly to the atmosphere. The characteristics of the atmosphere therefore reflect the acquisition of volatiles by the solid Earth during formation (see Pepin and Porcelli, 2002, as well as the history of degassing from the mantle. The precise connection between volatiles now emanating from the Earth and the long-term evolution of the atmosphere are key subjects of modeling efforts, and are discussed below.

Major advances in understanding the behavior of terrestrial volatiles have been made based upon observations on the characteristics of noble gases that remain within the Earth. Various models have been constructed that define different components and reservoirs in the planetary interior, how materials are exchanged between them, and how the noble gases are progressively transferred to the atmosphere. While there remain many uncertainties, an overall process of planetary degassing can be discerned. The present chapter discusses the constraints provided by the noble gases and how these relate to the degassing of the volatile molecules formed from nitrogen, carbon, and hydrogen. The evolution of particular atmospheric molecular species, such as CO_2, that are controlled by interaction with other crustal reservoirs and which reflect surface chemical conditions, are primarily discussed elsewhere.

Noble gases provide the most detailed constraints on planetary degassing. A description of the available noble gas data that must be incorporated into any Earth degassing history is provided first in Section 6.2, and the constraints on the total extent of degassing of the terrestrial interior are provided in Section 6.3. Noble gas degassing models that have been used to describe and calculate degassing histories of both the mantle (Section 6.4) and the crust (Section 6.5) are then presented. These discussions then provide the context for an evaluation of major volatile cycles in the Earth (Section 6.6), and speculations about the degassing of the other terrestrial planets (Section 6.7), Mars and Venus, that are obviously based on much more limited data. The processes controlling mantle degassing are clearly related to the structure of the mantle, as discussed in Section 6.4. An important aspect is the origin of planetary volatiles and whether initial incorporation was into the solid Earth or directly to the atmosphere. Basic noble gas elemental and isotopic characteristics are given in Ozima and Podosek (2001) and Porcelli *et al.* (2002). The major nuclear processes that produce noble gases within the solid Earth, and the half-lives of the major parental nuclides, are given in Table 1.

6.2 PRESENT-EARTH NOBLE GAS CHARACTERISTICS

There are various terrestrial reservoirs that have distinct volatile characteristics. Data from mid-ocean ridge basalts (MORBs) characterize the underlying convecting upper mantle, and are described here without any assumptions about the depth of this reservoir. Other mantle reservoirs are sampled by ocean island basalts (OIBs) and may represent a significant fraction of the mantle. Note that significant krypton isotopic variations due to radiogenic additions are neither expected nor observed, and there are no isotopic fractionation observed between any terrestrial noble gas reservoirs. Therefore, no constraints on mantle degassing can be obtained from krypton, and so krypton is not discussed further.

6.2.1 Surface Inventories

The atmosphere is the largest accessible terrestrial noble gas reservoir, and its composition serves as a reference for measurements of other materials. The major volatile molecules of carbon, nitrogen, and hydrogen, have considerable inventories in the crust that are part of the volatile budget that has been either degassed from the mantle or initially incorporated into the atmosphere. The total surface inventory is summarized in Table 2 and includes the atmosphere, hydrosphere, and continental crust.

Table 1 Major nuclear processes producing noble-gas isotopes in the solid earth.[a]

Daughter	*Nuclear process*	*Parent half-life*	*Yield (atoms/decay)*	*Comments*
^{3}He	$^{6}Li(n, \alpha)^{3}H(\beta^{-})^{3}He$			$^{3}He/^{4}He = 1\times10^{-8}$[b]
^{4}He	α-decay of ^{238}U decay series nuclides	4.468 Ga	8[c]	
^{4}He	α-decay of ^{235}U decay series nuclides	0.7038 Ga	7[c]	$^{238}U/^{235}U = 137.88$
^{4}He	α-decay of ^{232}Th decay series nuclides	14.01 Ga	6[c]	Th/U = 3.8 in bulk Earth
^{21}Ne	$^{18}O(\alpha, n)^{21}Ne$			$^{21}Ne/^{4}He = 4.5\times10^{-8}$[b]
^{21}Ne	$^{24}Mg(n, \alpha)^{21}Ne$			$^{21}Ne/^{4}He = 1\times10^{-10}$[b]
^{40}Ar	$^{40}K\ \beta^{-}$ decay	1.251 Ga	0.1048[b]	$^{40}K = 0.01167\%$ total K
^{129}Xe	$^{129}I\ \beta^{-}$ decay	15.7 Ma	1	$^{129}I/^{127}I = 1.1\times10^{-4}$ at 4.56 Ga[d]
^{136}Xe	^{238}U spontaneous fission		4×10^{-8}[e]	
^{136}Xe	^{244}Pu spontaneous fission	80.0 Ma	7.00×10^{-5}	$^{244}Pu/^{238}U = 6.8\times10^{-3}$ at 4.56 Ga[f]

[a] From data compilations of Blum (1995), Ozima and Podosek (2001), and Pfennig *et al.* (1998). [b] Production ratio for upper crust (Ballentine and Burnard, 2002). [c] Per decay of series parent, assuming secular equilibrium for entire decay series. [d] Hohenberg *et al.* (1967). [e] Eikenberg *et al.* (1993) and Ragettli *et al.* (1994). [f] Hudson *et al.* (1989).

Table 2 Volatile surface inventories.

Constituent	*Atmosphere* (mol)	*Crust* (mol)
N	2.760×10^{20}	4×10^{19}
O_2	3.702×10^{19}	
Ar	1.651×10^{18}	
C	5.568×10^{16}	8.3×10^{21}
Ne	3.213×10^{15}	
He	9.262×10^{14}	
Kr	2.015×10^{14}	
Xe	1.537×10^{13}	

Note: Based on dry tropospheric air. Water generally accounts for ≤4% of air. Other chemical constituents have mixing ratios less than Xe. Data from compilation by Ozima and Podosek (2001). C atmosphere data from Keeling and Whorf (2000). Crustal C from Hunt (1972) and Ronov and Yaroshevsky (1976). Crustal N from Marty and Dauphas (2003).

Since helium is lost from the atmosphere, the atmospheric abundance has no significance for determining long-term evolution. Atmospheric isotopic compositions, which are generally used as standards for comparison and measurement normalization, are provided in Table 3.

6.2.2 Helium Isotopes

There are two isotopes of helium. In addition to the cosmologically produced ^{4}He and ^{3}He, ^{4}He is produced as α-particles during radioactive decay of various parent radionuclides, and the much less abundant ^{3}He is produced from ^{6}Li (Tables 1 and 2). Overall, radiogenic helium is primarily ^{4}He, with a ratio of $^{3}He/^{4}He \sim 0.01R_A$ (Morrison and Pine, 1955), where R_A is the air value of 1.39×10^{-6}. The initial value for the Earth depends upon the origin of terrestrial noble gases, and is presumed to be that of the solar nebula of $^{3}He/^{4}He = 120R_A$ (Mahaffy *et al.*, 1998). The solar wind value of $330R_A$ (Benkert *et al.*, 1993) was established after deuterium burning in the Sun; if terrestrial helium

Table 3 Noble-gas and major volatile isotope composition of the atmosphere.

Isotope	*Relative abundances*	*Percent molar abundance*
^{3}He	$(1.399 \pm 0.013)\times10^{-6}$	0.000140
^{4}He	$\equiv 1$	100
^{20}Ne	9.80 ± 0.08	90.50
^{21}Ne	0.0290 ± 0.0003	0.268
^{22}Ne	$\equiv 1$	9.23
^{36}Ar	$\equiv 1$	0.3364
^{38}Ar	0.1880 ± 0.0004	0.0632
^{40}Ar	295.5 ± 0.5	99.60
^{78}Kr	0.6087 ± 0.0020	0.3469
^{80}Kr	3.9599 ± 0.0020	2.2571
^{82}Kr	20.217 ± 0.004	11.523
^{83}Kr	20.136 ± 0.021	11.477
^{84}Kr	$\equiv 100$	57.00
^{86}Kr	30.524 ± 0.025	17.398
^{124}Xe	2.337 ± 0.008	0.0951
^{126}Xe	2.180 ± 0.011	0.0887
^{128}Xe	47.15 ± 0.07	1.919
^{129}Xe	649.6 ± 0.9	26.44
^{130}Xe	$\equiv 100$	4.070
^{131}Xe	521.3 ± 0.8	21.22
^{132}Xe	660.7 ± 0.5	26.89
^{134}Xe	256.3 ± 0.4	10.430
^{136}Xe	217.6 ± 0.3	8.857
^{14}N	0.0037	0.37
^{15}N	$\equiv 1$	99.63
^{12}C	$\equiv 1$	98.63
^{13}C	0.0113	1.11
^{1}H	$\equiv 1$	99.985
^{2}H	0.00015	0.015

After Ozima and Podosek (2001) and Porcelli *et al.* (2002).

was captured after significant deuterium burning, then this higher value would be the composition of the initial helium. The first clear evidence for the degassing of primordial volatiles still remaining within the solid Earth came from helium isotopes (Figure 1). MORB has an average of $8R_A$ (Clarke *et al.*, 1969; Mamyrin *et al.*, 1969; see Graham, 2002), and so is a mixture of radiogenic helium (that accounts for most of the ^{4}He) with initially trapped "primordial" helium (that accounts for most of the ^{3}He).

OIB has more variable ^{3}He/^{4}He ratios. Some are below those of MORB, probably due to radiogenic recycled components (e.g., Kurz *et al.*, 1982; Hanyu and Kaneoka, 1998). Due to their limited occurrence, these values are likely to represent only a small fraction of the total mantle. ^{3}He/^{4}He ratios greater than $\sim 10R_A$ provide evidence for a long-term noble gas reservoir distinct from MORB that has a time-integrated ^{3}He/(U+Th) ratio greater than that of the upper mantle (Kurz *et al.*, 1982; Allègre *et al.*, 1983) and so leads to the highest ^{3}He/^{4}He ratios of 32–38R_A found in Loihi Seamount, the youngest Hawaiian volcano (Kurz *et al.*, 1982; Rison and Craig, 1983; Honda *et al.*, 1993; Valbracht *et al.*, 1997), and Iceland (Hilton *et al.*, 1998b). A major issue has been determining the nature of this reservoir, the abundances of noble gases it contains, and how it degasses (see Section 6.4).

Helium isotopes have a $\sim$1 Myr residence time in the atmosphere prior to loss to space; therefore, their atmospheric abundances do not contain information about the integrated degassing history of the Earth. However, the large variations in helium isotope compositions in the Earth constrain mantle degassing models by (i) providing clear fingerprints of mantle volatile fluxes into the crust and atmosphere (see Sections 6.2.7 and 6.2.8); (ii) requiring several mantle noble gas reservoirs (see Section 6.4.3); (iii) indicating that the upper mantle is relatively well-mixed with respect to noble gases; and (iv) relating the sources of noble gases with heat production, since uranium and thorium are the dominant sources of both ^{4}He and heat in the mantle (see Section 6.2.7).

6.2.3 Neon Isotopes

There are three neon isotopes. The more abundant ^{20}Ne and ^{22}Ne are both essentially all primordial, as there is no significant global production of these isotopes. In contrast, ^{21}Ne is produced by nuclear reactions. In mantle-derived materials, measured ^{20}Ne/^{22}Ne ratios are greater than that of the atmosphere of 9.8, and extend toward the values of the solar wind (13.8) or implanted solar wind (12.5) (Figure 2). Since these isotopes are not produced in significant quantities in the Earth, this is unequivocal evidence for storage in the Earth of at least one nonradiogenic mantle component that is

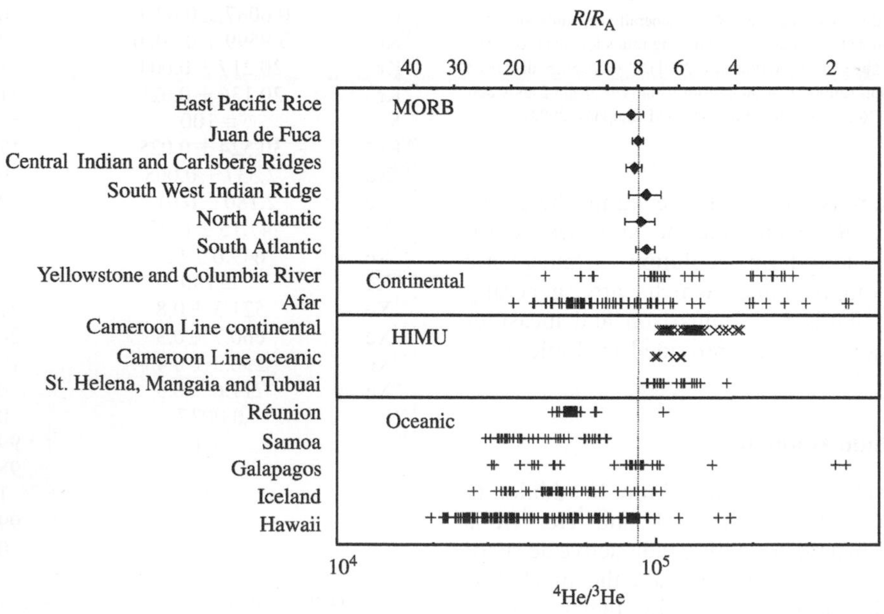

Figure 1 Helium isotope data from various mantle-derived volcanics. The upper axis is the ^{3}He/^{4}He ratio (R) normalized to the atmospheric ratio (R_A). As indicated by the data for selected segments of the MORB away from ocean islands falls almost entirely within the range of (7–9)R_A. While there are hotspot basalts that are characterized by high U/Pb ratios and low ^{3}He/^{4}He ratios (HIMU), many major oceanic hotspots, as well as continental hotspots, have high ^{3}He/^{4}He ratios (source Porcelli and Ballentine, 2002).

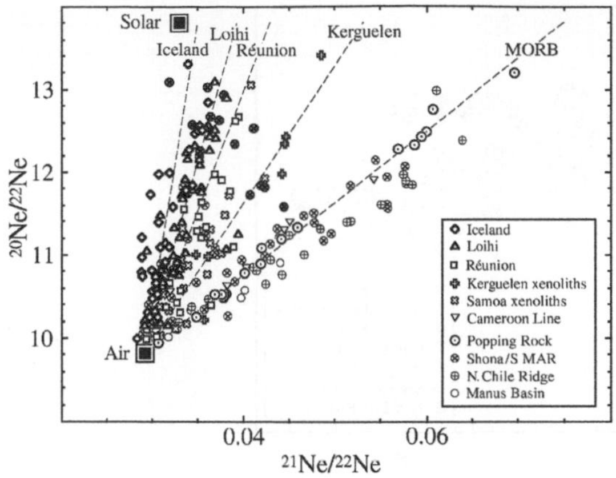

Figure 2 Neon isotope compositions of MORB and selected OIBs. The data generally fall on correlations that extend from air contamination to higher values that characterize the trapped mantle components. Islands with high $^{3}He/^{4}He$ ratios have lower-mantle $^{21}Ne/^{22}Ne$ ratios, reflecting high $^{3}He/(U+Th)$ and $^{22}Ne/(U+Th)$ ratios (source Graham, 2002).

distinctive from the atmosphere and has remained trapped separately since formation of the Earth. It is likely that the $^{20}Ne/^{22}Ne$ ratio of the atmosphere was originally similar to the higher values now found in the mantle, and was fractionated during losses to space. This could only have occurred early in Earth history. Since the neon remaining in the mantle preserves the original isotopic composition, the difference from that of the atmosphere also limits the amount of mantle neon that can have subsequently degassed. The exact proportion depends upon how much fractionation of atmospheric neon originally occurred. While this is unconstrained, the observed fractionation is already considered quite extreme, and so it is unlikely that much lower $^{20}Ne/^{22}Ne$ ratios had been generated, and so only a small proportion of mantle neon is likely to have been degassed subsequently.

MORB $^{20}Ne/^{22}Ne$ and $^{21}Ne/^{22}Ne$ ratios generally are correlated (Sarda *et al.*, 1988; Moreira *et al.*, 1998), and this is likely due to mixing of variable amounts of air contamination with uniform mantle neon. The upper-mantle $^{21}Ne/^{22}Ne$ ratio of $\sim$0.074 is higher than the solar value (0.033) due to additions of nucleogenic ^{21}Ne (Table 1). OIBs with high $^{3}He/^{4}He$ ratios span a similar range in $^{20}Ne/^{22}Ne$ ratios, but with lower corresponding $^{21}Ne/^{22}Ne$ ratios (Sarda *et al.*, 1988; Honda *et al.*, 1991, 1993). The OIB sources therefore have higher time-integrated He/(U+Th) and Ne/(U+Th) ratios (Honda and McDougall, 1993).

The MORB helium and neon isotopic compositions can be used to calculate the $^{3}He/^{22}Ne$ ratio of the source region prior to any recent fractionations created during transport and eruption. Since the production ratio of ^{4}He to ^{21}Ne is fixed (Table 1), the shifts in $^{3}He/^{4}He$ and $^{21}Ne/^{22}Ne$ isotope ratios from the initial, primordial values of the Earth due to radiogenic and nucleogenic additions can be used to calculate the reservoir $^{3}He/^{22}Ne$ ratio. Using an uncontaminated MORB value of $^{21}Ne/^{22}Ne=0.074$ and $^{21*}Ne/^{4*}He=4.5\times10^{-8}$, then $^{3}He/^{22}Ne=11$. Calculating a source value for each MORB and OIB sample individually, a mantle average of 7.7 was found (Honda and McDougall, 1998). For comparison, the solar nebula value is $^{3}He/^{22}Ne=1.9$ (see Porcelli and Pepin, 2000). This value can be used to relate degassing of helium with the other noble gases (see Section 6.2.7).

Overall, neon isotopes clearly identify the noble gases presently degassing from the mantle as solar in origin. Taking this composition as the original value of the neon in the atmosphere, most of the atmospheric neon was degassed very early in Earth history and suffered substantial fractionating losses. The atmospheric noble gas inventory was therefore highly modified from that which was originally degassed from the solid Earth or added to the surface. Since noble gases generally behave similarly within the mantle and during degassing (see Section 6.2.6), the constraints on neon degassing can be applied to the other noble gases.

6.2.4 Argon Isotopes

The two minor isotopes of argon, ^{36}Ar and ^{38}Ar, are essentially all primordial, with no significant radiogenic production on a global scale. The initial $^{40}Ar/^{36}Ar$ ratio of the solar system was $<10^{-3}$ (Begemann *et al.*, 1976) and orders of magnitude less than any planetary values, so essentially all ^{40}Ar is radiogenic. A large range in $^{40}Ar/^{36}Ar$ ratios has been measured in MORBs that is likely due to mixing of variable proportions of air argon (with $^{40}Ar/^{36}Ar=296$) with a single, more radiogenic, mantle composition (Figure 3). The minimum value for this mantle composition is represented by the highest measured values of 2.8×10^{4} (Staudacher *et al.*, 1989) to 4×10^{4} (Burnard *et al.*, 1997). From correlations between $^{20}Ne/^{22}Ne$ and $^{40}Ar/^{36}Ar$ during step heating of a gas-rich MORB that was designed to separate contaminant air noble gases from those trapped within the glass, a maximum value of $^{40}Ar/^{36}Ar=4.4\times10^{4}$ was obtained (Moreira *et al.*, 1998).

Like terrestrial $^{20}Ne/^{22}Ne$ ratios, it might be expected that $^{38}Ar/^{36}Ar$ ratios vary due to different initial sources building the Earth or to early fractionation events. Measurements of MORB and OIB $^{38}Ar/^{36}Ar$ ratios typically are atmospheric within error, but have been of low precision due to the low abundance of these isotopes. While some

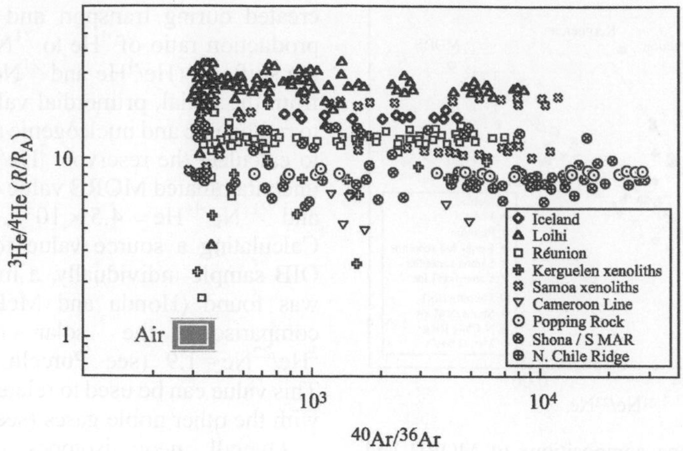

Figure 3 Helium and argon isotope compositions of MORB ("Popping Rock" from the Atlantic, Shona ridge section, and N. Chile ridge) and selected islands. $^{40}Ar/^{36}Ar$ ratios vary widely due to variable amounts of atmospheric contamination (source Graham, 2002).

high-precision analyses of MORB and OIB samples show $^{38}Ar/^{36}Ar$ ratios lower than that of the atmosphere and approaching solar values (see Pepin, 1998), but others do not (Kunz, 1999). While nonatmospheric ratios would limit the amount of argon transfer between the mantle and atmosphere, atmospheric ratios in the mantle could be explained either by early trapping of argon that had been fractionated or entrainment of atmospheric argon during subduction or melt formation (see Porcelli and Wasserburg, 1995b).

OIBs with high $^3He/^4He$ ratios reflecting high $^3He/(U+Th)$ ratios have been expected to have low $^{40}Ar/^{36}Ar$ ratios reflecting correspondingly high $^{36}Ar/K$ ratios. Measurements of $^{40}Ar/^{36}Ar$ in OIBs with $^3He/^4He > 10R_A$ are indeed consistently lower than MORB values. However, early values for Loihi glasses of $^{40}Ar/^{36}Ar < 10^3$ appear to reflect overwhelming contamination with air argon (Fisher, 1985; Patterson *et al.*, 1990). A study of basalts from Juan Fernandez (Farley *et al.*, 1993) found atmospheric contamination contained within phenocrysts introduced into the magma chamber, thus providing an explanation for the prevalence of air contamination of OIBs. Recent measurements of Loihi samples found higher $^{40}Ar/^{36}Ar$ values of 2,600–2,800 associated with high $^3He/^4He$ ratios (Hiyagon *et al.*, 1992; Valbracht *et al.*, 1997), while Trieloff *et al.* (2000) found values up to 8,000 on samples with $^3He/^4He = 24$ (and so with a helium composition midway between MORBs and the highest OIBs). Poreda and Farley (1992) found values of $^{40}Ar/^{36}Ar \leq 1.2 \times 10^4$ in Samoan xenoliths that have intermediate $^3He/^4He$ ratios (9–20R_A). Kola Peninsula carbonatites with high $^{20}Ne/^{22}Ne$ ratios were used to calculate a mantle $^{40}Ar/^{36}Ar$ value of 5,000 (Marty *et al.*, 1998). Other attempts to remove the effects of air contamination have used associated neon isotopes and the debatable assumption that the contaminant Ne/Ar ratio is constant, and have also found $^{40}Ar/^{36}Ar$ values substantially lower than in MORBs (Sarda *et al.*, 2000). Overall, it appears that $^{40}Ar/^{36}Ar$ ratios in the high $^3He/^4He$ OIB source are >3,000 but probably $<10^4$, and so lower than that of the MORB source (see also Matsuda and Marty, 1995).

The unambiguous measurement of past atmospheric $^{40}Ar/^{36}Ar$ ratios would provide an important constraint on the degassing history of the atmosphere. If ^{36}Ar had largely degassed early (see below), then the past atmospheric $^{40}Ar/^{36}Ar$ ratio would reflect the past atmospheric abundance of ^{40}Ar due to subsequent ^{40}Ar degassing. Unfortunately, it has been difficult to find samples that have captured and retained atmospheric argon but do not contain significant amounts of either inherited radiogenic ^{40}Ar or potassium. Nonetheless, various studies have sought to find $^{40}Ar/^{36}Ar$ ratios that are lower than the present atmosphere, and so clearly are not dominated by either present atmosphere contamination or radiogenic ^{40}Ar. Cadogan (1977) reported a value of $^{40}Ar/^{36}Ar = 291.0 \pm 1.5$ from the 380 Myr old Rhynie chert, while Hanes *et al.* (1985) reported 258 ± 3 for an old pyroxenite sill sample from the Abitibi Greenstone Belt that contains amphibole apparently produced during deuteric alteration at 2.70 Ga. Both samples are presumed to have atmospheric argon trapped at these times. Using these data, the maximum possible changes in the atmospheric ^{40}Ar abundance is obtained by assuming that the ^{36}Ar abundance has been constant since that time, so that the differences between these $^{40}Ar/^{36}Ar$ ratios and the present atmospheric value reflect only lower ^{40}Ar abundances. In this case, the Rhynie chert data suggest that the atmospheric ^{40}Ar abundance was 1.5% lower 380 Ma ago than today. At that time, there was 1.9% less ^{40}Ar in the Earth (assuming a bulk Earth potassium concentration of 270 ppm), and it thus appears that the same fraction of terrestrial ^{40}Ar as today was in the atmosphere at that time. At 2.7 Ga ago, the calculated atmospheric ^{40}Ar abundance is 12.7% lower than at present. However, there was 66% less ^{40}Ar 2.7 Ga ago, and assuming a BSE concentration of 270 ppm K, even complete degassing of the entire Earth at that time (compared to 40% today) would not provide sufficient ^{40}Ar for the atmosphere. The alternative that the atmospheric ^{36}Ar abundance has doubled since 2.7 Ga ago is also unlikely if current arguments for early degassing of nonradiogenic nuclides are valid (see Sections 6.2.3 and 6.4.1). However, before pursuing further speculation, the possibility that the Abitibi sample contains either excess ^{40}Ar trapped during formation or subsequently produced radiogenic ^{40}Ar, and so provides an unreasonably high $^{40}Ar/^{36}Ar$ atmospheric ratio for that time, must be discounted.

Overall, argon isotope compositions indicate that the mantle is much more radiogenic than the atmosphere, and this provides an important starting point for degassing history models. It appears that there are variations in $^{40}Ar/^{36}Ar$ associated with different $^{3}He/^{4}He$ ratios in the mantle, although atmospheric contamination of samples has made this difficult to quantify. It appears that the same proportion of ^{40}Ar has been in the atmosphere over the last 380 Ma, although reliable data are required from much earlier to effectively discriminate between different degassing histories (see Section 6.4).

6.2.5 Xenon Isotopes

There are nine isotopes of xenon (see Table 3). On a global scale, there have been additions to ^{129}Xe through decay of ^{129}I ($t_{1/2}$ = 15.7 Myr), which as a short-lived nuclide was only present in significant quantities early in Earth history. Additions to the heavy isotopes ^{131}Xe, ^{132}Xe, ^{134}Xe, and ^{136}Xe have also occurred by fission of ^{244}Pu ($t_{1/2}$ = 80 Myr), another short-lived nuclide, and ^{238}U ($t_{1/2}$ = 4.5 Gyr). The largest fission contributions are to ^{136}Xe, so this isotope is usually used as an index for fissiogenic contributions. Primordial components completely account for the other isotopes and even a dominant proportion of those with later additions.

The starting point for examining xenon isotope systematics is defining the nonradiogenic isotope composition; i.e., the proportion of primordial xenon underlying the radiogenic and fissiogenic contributions. The light isotopes of atmospheric xenon (^{124}Xe, ^{126}Xe, ^{128}Xe, and ^{130}Xe) are related to both bulk chondritic and solar xenon by very large fractionation of ~4.2% per amu (Krummenacher *et al.*, 1962), which demands a strongly fractionating planetary process. However, both chondritic and solar xenon, when fractionated to match the light isotopes of the atmosphere, have proportionately more ^{134}Xe and ^{136}Xe than presently in the atmosphere and so neither can serve as the primordial terrestrial composition (see Pepin, 2000). There is no other commonly observed solar system composition that can be used to account for atmospheric xenon by simple mass fractionation and the addition of radiogenic and fissiogenic xenon. Pepin (2000) has used isotopic correlations of chondrite data to infer the presence of a mixing component, U–Xe, that has solar light isotope ratios and when highly mass-fractionated yields the light-isotope ratios of terrestrial xenon and is relatively depleted in heavy xenon isotopes. The composition of the present atmosphere then can be obtained from fractionated U–Xe by the addition of a heavy isotope component that has the composition of ^{244}Pu-derived fission xenon (Pepin, 2000). Therefore, the proportions of atmospheric xenon isotopes that are fissiogenic can be calculated. The budgets of radiogenic and fissiogenic xenon in the atmosphere are discussed in detail in Section 6.3.2 below.

MORB $^{129}Xe/^{130}Xe$ and $^{136}Xe/^{130}Xe$ ratios lie on a correlation extending from atmospheric ratios to higher values (Staudacher and Allègre, 1982; Kunz *et al.*, 1998), and likely reflect mixing of variable proportions of contaminant air xenon with an upper-mantle component having more radiogenic $^{129}Xe/^{130}Xe$ and $^{136}Xe/^{130}Xe$ ratios (Figure 4). The highest measured values thus provide lower limits for the MORB source. The MORB data demonstrate that the xenon presently in the atmosphere was in an environment with a higher Xe/I ratio than that of the xenon in the mantle, at least during the lifetime of ^{129}I. As discussed further below, this difference can be generated either by degassing processes or by early differences in mantle reservoirs related to the formation processes of the planet.

Contributions to $^{136*}Xe$ enrichments in MORBs can be from either decay of ^{238}U over Earth history or early ^{244}Pu decay, which in theory can be distinguished based on the spectrum of contributions to other xenon isotopes, although analyses have typically not been sufficiently precise to do so. More precise measurements can be obtained from the abundant xenon in some CO_2 well gases that have ^{129}Xe and ^{136}Xe enrichments similar to those found in MORBs, and are likely to

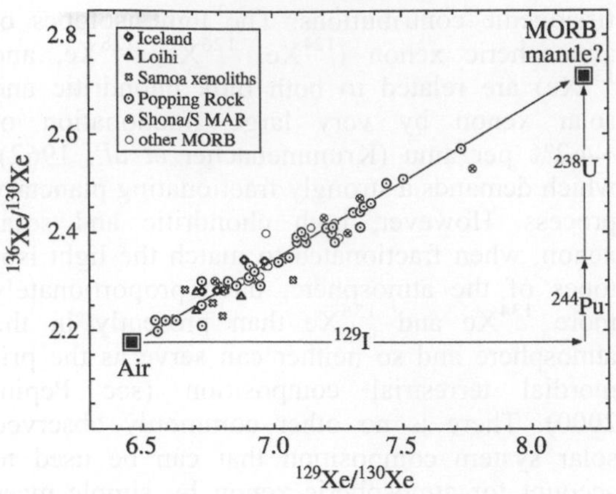

Figure 4 Xenon isotope compositions of MORB and selected ocean islands. The excesses in ^{129}Xe due to ^{129}I, and in ^{136}Xe due to ^{238}U and ^{244}Pu, are correlated due to mixing between mantle xenon and atmospheric contamination. The fraction of ^{136}Xe from ^{244}Pu calculated by Kunz *et al.* (1998) is shown for illustration (source Graham, 2002).

be from the upper mantle (Staudacher, 1987). Precise measurements indicate that ^{244}Pu has contributed <10–20% of the ^{136}Xe that is in excess of the atmospheric composition (Phinney *et al.*, 1978; Caffee *et al.*, 1999). An error-weighted best fit to recent precise MORB data (Kunz *et al.*, 1998) yielded a value of $32 \pm 10\%$ for the fraction of ^{136}Xe excesses relative to the atmospheric composition that are ^{244}Pu-derived, although with considerable uncertainties (Marti and Mathew, 1998). The atmosphere itself, therefore, contains ^{244}Pu-derived ^{136}Xe. Clearly, further work is warranted on the proportion of plutonium-derived heavy xenon in the mantle, although it appears that the fissiogenic xenon is dominantly derived from uranium.

It has proven to be more difficult to characterize the xenon in OIB source regions. Xenon with atmospheric isotopic ratios in high-^{3}He/^{4}He OIB samples (e.g., Allègre *et al.*, 1983) appears to be dominated by air contamination (Patterson *et al.*, 1990; Harrison *et al.*, 1999) rather than represent mantle xenon with an air composition. Although Samoan samples with intermediate (9–20R_A) helium isotope ratios have been found with xenon isotopic ratios distinct from those of the atmosphere (Poreda and Farley, 1992), the xenon in these samples may have been derived largely from the MORB source. Recently, Harrison *et al.* (1999) found slight ^{129}Xe excesses in Icelandic samples with ^{129*}Xe/^{3}He ratios that are compatible with the ratio in a gas-rich MORB, but due to the uncertainties in the data it cannot be determined whether there are indeed differences between the MORB and OIB sources. Trieloff *et al.* (2000) reported xenon isotope compositions in Loihi dunites and Icelandic glasses that were on the MORB correlation line and had values up to ^{129}Xe/^{130}Xe = 6.9. These were accompanied by ^{3}He/^{4}He ratios up to 24R_A, and so may contain noble gases from both MORB ($\sim$8R_A) and the highest ^{3}He/^{4}He ratio (37R_A) OIB source. From these data it appears that the OIB source may have xenon that is similar to that in MORB, although there may be some differences that have not been resolvable.

The relatively imprecise measured ratios of the nonradiogenic isotopes in MORB are indistinguishable from those in the atmosphere. However, more precise measurements of mantle-derived xenon in CO_2 well gases have been found to have higher $^{124-128}$Xe/^{130}Xe ratios (Phinney *et al.*, 1978; Caffee *et al.*, 1999) that can be explained by either: (i) a mixture of $\sim$10% xenon trapped within the Earth of solar isotopic composition and $\sim$90% atmospheric xenon (subducted or added in the crust); or (ii) a mantle xenon component that has not been isotopically fractionated relative to solar xenon to the same extent as air xenon.

In sum, nonradiogenic atmospheric xenon isotopes, like those of neon, require that early degassing of noble gases occurred when fractionating losses to space were still operating. Like argon, xenon in the upper mantle is more radiogenic than the atmosphere, and this must be a feature of any reasonable degassing model. The problems of atmospheric contamination are greatest for xenon, and so there is little definitive evidence regarding the isotopic variations in the mantle.

6.2.6 Noble Gas Abundance Patterns

Noble gas abundance patterns in MORBs and OIBs scatter greatly. This is due to sample alteration as well as fractionation during noble gas partitioning between basaltic melts and a vapor phase that may then be preferentially gained or lost by the sample. Nonetheless, MORB Ne/Ar and Xe/Ar ratios that are greater than the air values are common. An example of this pattern was found in a gas-rich, relatively uncontaminated MORB sample (Figure 5) with high ^{40}Ar/^{36}Ar and ^{129}Xe/^{130}Xe ratios and a ^{4}He/^{40}Ar ratio ($\sim$3) that is near that of production in the upper mantle (and so not fractionated) (Staudacher *et al.*, 1989; Moreira *et al.*, 1998). An upper-mantle pattern also can be calculated by assuming that the noble gases have been degassed from the mantle without substantial elemental fractionation, and the radiogenic nuclides are present in their production ratios. Using ratios of estimated upper-mantle production rates to determine the relative abundances

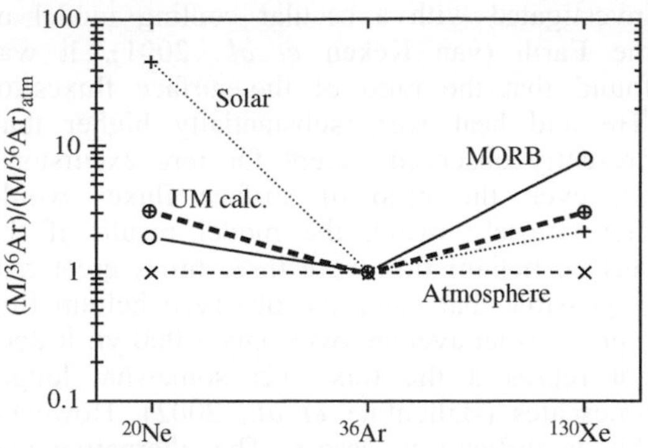

Figure 5 The relative abundances of noble gases in gas-rich MORB (Moreira *et al.*, 1998), as calculated for the upper mantle (see text), air (Tables 2 and 3), and in the solar composition (see Ozima and Podosek, 2001). The upper mantle is enriched in neon and xenon, relative to argon, compared to the air composition.

of radiogenic ^{4}He, ^{21}Ne, ^{40}Ar, and ^{136}Xe, measured isotopic compositions can be used to determine ratios of nonradiogenic isotopes. For example, using the production ratio of ^{21}Ne/^{40}Ar and the ratio of ^{21}Ne/^{22}Ne and ^{40}Ar/^{36}Ar (taking into account the amounts of non-nucleogenic ^{21}Ne) measured in basalts that are not altered by melting and transport processes, the ^{22}Ne/^{36}Ar ratio in the mantle source can be calculated. In this case, $^{22}\text{Ne}/^{36}\text{Ar}_{\text{MORB}} = 0.15 \times 10^{-4}$ and $^{130}\text{Xe}/^{36}\text{Ar}_{\text{MORB}} = 3.3 \times 10^{-4}$, significantly higher than the corresponding atmospheric values of 0.05×10^{-4} and 1.1×10^{-4}, respectively.

It is often assumed that all the noble gases are highly incompatible during basalt genesis and so are efficiently extracted from the mantle without elemental fractionation. Experimental data for partitioning between basaltic melts and olivine are consistent with this for helium but not for the heavier noble gases, which have been found in higher concentrations in olivine than expected (Hiyagon and Ozima, 1986; Broadhurst *et al.*, 1992). However, these results may be due to experimental difficulties. Recent data (Chamorro-Perez *et al.*, 2002) indicate that the argon clinopyroxene/silicate melt partition coefficient is relatively constant and equal to $\sim 4 \times 10^{-4}$ at pressures up to at least 80 kbar, and more recent data indicate that neon, krypton, and xenon are similarly incompatible (Brooker *et al.*, 2003). Therefore, it appears that the noble gases are all highly incompatible in the mantle, and there is no elemental fractionation between the melt and the mantle source region. However, while noble gases transported from the mantle may not be elementally fractionated, some fractionation may occur in the highly depleted melt residue due to small differences in partition coefficients. This remains a possibility since reliable partition coefficients for most of the noble gases are unavailable.

Any fractionations that might occur during degassing of the upper mantle would be expected to be monotonic across the noble gases. Therefore, degassing alone does not explain high values for both ^{22}Ne/^{36}Ar and ^{130}Xe/^{36}Ar with respect to atmospheric values. If the heavy noble gases were slightly more compatible (as might be plausibly assumed), then a higher ^{130}Xe/^{36}Ar value for the mantle would be generated, but accompanied by a lower ^{22}Ne/^{36}Ar ratio. It is possible that the atmospheric ^{22}Ne/^{36}Ar ratio was lowered instead during volatile losses to space. However, if the starting noble gas composition of the Earth was solar, with $^{22}\text{Ne}/^{36}\text{Ar} = 37$ (Geiss *et al.*, 1972), then the upper-mantle value, which is lower than the solar value, still requires explanation. The possibilities that remain are a mantle composition that was generated during Earth formation that is different from the solar value, and subduction of atmospheric argon (see Section 6.2.9).

6.2.7 MORB Fluxes and Upper-mantle Concentrations

Noble gas concentrations vary widely in mantle-derived volcanics due to degassing during ascent and eruption, as well as from volatile redistribution in vesicles. Therefore, these data cannot be readily used to constrain the upper-mantle concentrations. An alternative approach is to compare fluxes to the atmosphere from a region with a known rate of volcanism. The largest and most clearly defined mantle volatile flux, from mid-ocean ridges (Clarke *et al.*, 1969; Craig *et al.*, 1975), is $1,060 \pm 250\,\text{mol yr}^{-1}$ ^{3}He (Lupton and Craig, 1975; Farley *et al.*, 1995), and is obtained by combining seawater ^{3}He concentrations in excess of dissolved air helium with seawater advection models. This value represents an average over the last 1,000 years, and while this is generally assumed to be the long-term average, this remains to be confirmed. Using the ^{3}He/^{4}He in vented gas, the flux of ^{4}He can be determined once the ^{3}He flux is determined. The flux of ^{4}He is approximately equal to the production rate for the upper mantle above 670 km (O'Nions and Oxburgh, 1983). This fact has been incorporated in some mantle degassing models, and is discussed further below. The fluxes of other noble gases from the mantle can be obtained from the ^{3}He fluxes and the relative abundances in MORBs (see Section 6.2.6). Assuming that radiogenic ^{4}He and ^{40}Ar are degassing together, a ^{40}Ar flux of 1.9×10^{31} atoms a^{-1} is obtained by using a radiogenic ratio of $^4\text{He}/^{40}\text{Ar} = 3$, a MORB ^{3}He/^{4}He ratio of

$8R_A$, and noting that ~90% of mantle ^{4}He is radiogenic.

The concentration of ^{3}He in the mantle can be determined by dividing the flux of ^{3}He into the oceans by the rate of production of melt that is responsible for carrying this ^{3}He from the mantle, which is equivalent to the rate of ocean crust production of 20 $km^3\,yr^{-1}$ (Parsons, 1981). MORBs that degas quantitatively to produce a ^{3}He flux of 1,060 mol yr^{-1} must have an average ^{3}He content of 1.96×10^{-14} mol g^{-1} or 4.4×10^{-10} cm^3 ^{3}He(STP) g^{-1}. This ^{3}He concentration is within a factor of 2 of that obtained for the most gas-rich basalt glass of $\sim 10.0 \times 10^{-10}$ cm^3 ^{3}He(STP) g^{-1} (Sarda *et al.*, 1988; Moreira *et al.*, 1998). Assuming that MORB is generated by an average of 10% partial melting, the source region contains 2×10^{-15} mol g^{-1} (1.2×10^9 atoms ^{3}He g^{-1}). The mantle concentration of ^{3}He and ^{4}He cannot be usefully compared with the ^{3}He and ^{4}He inventories of the atmosphere since ^{3}He and ^{4}He are lost to space with a residence time of a few million years. A comparison, however, can be made with argon. Using the MORB ^{3}He/^{36}Ar ratio of 1.7, then the upper mantle has $(0.8–2.9) \times 10^9$ atoms ^{36}Ar g^{-1}. This is highly depleted compared to the benchmark value of 3×10^{12} atoms ^{36}Ar g^{-1} obtained by dividing the atmospheric inventory by the mass of the upper mantle. The abundances of other noble gas isotopes can be obtained similarly.

The flux of ^{4}He can be compared with that of heat, since ^{4}He is produced along with heat during radioactive decay. The Earth's global heat loss amounts to 44 TW (Pollack *et al.*, 1993). Subtracting the heat production from the continental crust (4.8–9.6 TW), and the core (3–7 TW; Buffett *et al.*, 1996) leaves 9.6–14.4 TW to be accounted for by present-day radiogenic heating and 17.8–21.8 TW as a result of secular cooling (largely from earlier heat production). O'Nions and Oxburgh (1983) pointed out that the present-day ^{4}He mantle flux was produced along with only 2.4 TW of heat, an order of magnitude less than that produced along with all the radiogenic heat presently reaching the surface, and suggested that this could be achieved by a boundary layer in the mantle through which heat could pass, but behind which helium was trapped. It has been suggested that heat and helium are transported to the surface at hotspots, where the bulk of the helium is lost, while the heat is lost subsequently at ridges (Morgan, 1998). However, evidence for such a hotspot helium flux at present or in the past, and formulation of a mantle noble gas model incorporating this suggestion, are unavailable. The possibility that the separation is due to the different mechanisms that extract heat and helium from the mantle was investigated with a secular cooling model of the Earth (van Keken *et al.*, 2001). It was found that the ratio of the surface fluxes of ^{4}He and heat were substantially higher than presently observed except for rare excursions. However, the ratio of surface fluxes would more closely match the model results if the surface helium flux were three times greater. It is possible that since the observed helium flux represents an average over only 1,000 yr, it does not represent the flux over somewhat longer timescales (Ballentine *et al.*, 2002). However, this is difficult to assess. The alternative that remains is that some boundary is needed for separating uranium and thorium, from the upper-mantle reservoir, and allowing heat to pass more efficiently than helium, although the configuration of this boundary is unconstrained by geochemistry. The implications of this are considered further in discussions of mantle models. This significance for mantle degassing is that there are considerable amounts of radiogenic noble gases maintained behind such a boundary. Bercovici and Karato (2003) have proposed a model of extraction of incompatible elements at depth in the upper mantle by melting and entrainment in the down going slab.

6.2.8 Other Mantle Fluxes

The flux of ^{3}He from intraplate volcanic systems is dominantly subaerial and so it is not possible to obtain directly time-integrated flux values for even a short geological period. While the Loihi hotspot in the Pacific is submarine, calculation of ^{3}He fluxes into the ocean using ocean circulation models have not required a large flux from this location that is comparable to the ^{3}He plumes seen over ridges (Gamo *et al.*, 1987; Farley *et al.*, 1995), although recent data has seen an extensive ^{3}He plume from Loihi (Lupton, 1996).

Helium fluxes from OIB could be calculated if the rates of magmatism are known, along with the helium concentrations of the source regions or concentrations of undegassed magmas (see Porcelli and Ballentine, 2002). Estimates of the rate of intraplate magma production vary from 1% to 12% of the MORB production rate (Reymer and Schubert, 1984; Schilling *et al.*, 1978; Batiza, 1982; Crisp, 1984). The total noble gas flux from ocean islands is related to the mantle source concentrations and whether some source domains are more gas-rich than the MORB source mantle. OIBs are typically more extensively degassed than MORBs. This is readily explained for those that are erupted at shallower depths or subaerially. Also, OIBs may be more volatile-rich than MORBs and, therefore, may degas more effectively (Dixon and Stolper, 1995). High water contents of basalts lower CO_2 solubility, and appear to have lowered helium contents in lavas with high $^3He/^4He$ ratios along the Reykjanes Ridge (Hilton *et al.*, 2000). Estimates for the mantle source with high $^3He/^4He$ ratios have ranged from much lower than that of the MORB source, perhaps due to prior melting (Hilton *et al.*, 1997), to up to 15 times higher (Moreira and Sarda, 2000; Hilton *et al.*, 2000). It should be emphasized that in many locations, the high $^3He/^4He$ source component makes up a small fraction of the sample source and, if gas-rich, may substantially affect the helium composition but not substantially increase the helium flux. Therefore, in many cases, source concentrations may be closer to that of MORBs. In this case, the overall ^3He flux relative to that from MORBs is proportional to the relative melt production rate, or 1–12% that of MORBs. It is not clear to what extent more gas-rich OIBs augment this flux.

Continental settings provide a small but significant flux from the mantle. Regional groundwater systems provide time-integrated records of this flux over large areas, but are based on short timescales and dependent on the hydrogeological model used. In the Pannonian basin (4,000 km^2 in Hungary), for example, the flux of ^3He has been estimated to be from $0.8–5 \times 10^4$ ^3He atoms $m^{-2}\,s^{-1}$ (Stute *et al.*, 1992) to 8×10^4 ^3He atoms $m^{-2}\,s^{-1}$ (Martel *et al.*, 1989). Taking an area of $2 \times 10^{14}\,m^2$ for continents and assuming 10% is under extension, this yields a total ^3He flux of 8.4–84 mol $^3He\,yr^{-1}$ (Porcelli and Ballentine, 2002). This value compares with a mantle flux of <3 mol $^3He\,yr^{-1}$ through the stable continental crust (O'Nions and Oxburgh, 1988).

The flux of mantle ^3He to the atmosphere at subduction zones from the upper mantle can be estimated from the volume of convergent zone volcanics. The estimate of Reymer and Schubert (1984) is 5% that at mid-ocean ridges. If this is largely generated by similar degrees of melting of a similar source as MORB, then the ^3He flux is only 5% that at the mid-ocean ridges, or 50 mol $^3He\,yr^{-1}$ (see also Hilton *et al.*, 2002). An estimated CO_2 flux at convergent margins of 3×10^{11} mol yr^{-1} (Sano and Williams, 1996) can also be used to estimate the ^3He flux. Using a mid-ocean ridge ratio of $CO_2/^3He = 2 \times 10^9$ (Marty and Jambon, 1987) gives 150 mol $^3He\,yr^{-1}$, although the convergent margin $CO_2/^3He$ ratio may be much greater due to the presence of recycled carbon.

In summary, the dominant noble gas flux from the mantle is from mid-ocean ridges. Other fluxes may be negligible or augment the ridge flux by up to an additional 25%.

6.2.9 Subduction Fluxes

It is possible that atmospheric noble gases that are incorporated in oceanic crust materials, including altered MORBs, hydrothermally altered crust, and sediments are subducted into the mantle and thus are mixed with primordial noble gas constituents. Available data indicate that holocrystalline MORBs and oceanic sediments contain atmospheric noble gases that are greatly enriched in the heavier noble gases. However, because concentrations vary by several orders of magnitude, an accurate average value cannot be easily determined. Measurements of pelagic sediments, with $3 \times 10^{15}\,g\,yr^{-1}$ subducted (Von Huene and Scholl, 1991), have $(0.05–7) \times 10^{10}$ atoms $^{130}Xe\,g^{-1}$ (Podosek *et al.*, 1980; Matsuda and Nagao, 1986; Staudacher and Allègre, 1988) with a mean of 6×10^9 atoms g^{-1}. Measurements of holocrystalline basalts found $(4–42) \times 10^7$ atoms $^{130}Xe\,g^{-1}$ (Dymond and Hogan, 1973; Staudacher and Allègre, 1988), with a mean of 2×10^8 atoms $^{130}Xe\,g^{-1}$. While the depth over which addition of atmospheric gases are added by alteration is unknown, it might be assumed that this occurs over the same depth as low temperature enrichment of alkaline elements of ~600 m (Hart and Staudigel, 1982). In this case, an estimated $7 \times 10^{15}\,g\,yr^{-1}$ of this material is subducted. In total, these numbers result in 2×10^{25} atoms $^{130}Xe\,yr^{-1}$ (33 mol yr^{-1}) reaching subduction zones in sediments and altered basalt (Porcelli and Wasserburg, 1995a). This can be compared to the estimate of Staudacher and Allègre (1988) determined by assuming that 40–80% of the subducting flux of oceanic crust ($6.3 \times 10^{16}\,g\,yr^{-1}$) is altered and contains atmosphere-derived noble gases, and that 18% of this mass is ocean sediment. In this case, a similar flux of 13.8–39 mol $^{130}Xe\,yr^{-1}$ was obtained, along with $1.9–2.5 \times 10^6$ mol $^4He\,yr^{-1}$, $1.9–5.9 \times 10^3$ mol $^{20}Ne\,yr^{-1}$, $5.8–21.9 \times 10^3$ mol $^{36}Ar\,yr^{-1}$, and $3.0–12.3 \times 10^3$ mol $^{84}Kr\,yr^{-1}$. Subduction of noble gases at these rates over 10^9 yr would have resulted in 1%, 90%, 110%, and 170% of the respective inventories of ^{20}Ne, ^{36}Ar, ^{84}Kr, and ^{130}Xe to be in the upper mantle (Allègre *et al.*, 1986); however, the fraction stripped from subducting materials during deformation and arc volcanism before entering the mantle is unknown but it is likely to be substantial.

Subduction zone processing and volcanism may return much of the noble gases in the slab to the atmosphere. However, it is possible that the total amounts of noble gases reaching subduction zones are sufficiently high that subduction into the deeper mantle (i.e., beyond the zone of magma generation) of only a small fraction may have a considerable impact upon the composition of argon and xenon in the upper mantle (Porcelli and Wasserburg, 1995a,b). Staudacher and Allègre (1988) argued that subducting argon and xenon must be almost completely lost to the atmosphere during subduction zone magmatism, or the high $^{129}Xe/^{130}Xe$ and $^{136}Xe/^{130}Xe$ in the upper mantle would not have been preserved throughout Earth history. However, this conclusion is dependent upon a model of unidirectional degassing to generate the upper-mantle xenon isotope composition (see Section 6.4.3), and as discussed in Section 6.4.5 the contrary view that subducted noble gases are mixed with nonrecycled, mantle-derived xenon to produce the upper-mantle composition (Porcelli and Wasserburg, 1995a,b) is compatible with the mantle data. Note that, as discussed above, the $^{128}Xe/^{130}Xe$ ratio measured in mantle-derived xenon trapped in CO_2 well gases may be interpreted as a mixture of ~90% subducted xenon with 10% trapped solar xenon. Further, direct input of the subducted slab into a gas-rich deeper reservoir that has $^{40}Ar/^{36}Ar$ values that are significantly lower than in the MORB source mantle (e.g., Trieloff *et al.*, 2000) is also possible.

6.3 BULK DEGASSING OF RADIOGENIC ISOTOPES

The total fraction of a species that has been degassed to the atmosphere can only be calculated for radiogenic nuclides, which have total planetary abundances that are constrained by the parent element abundances. Most attention has focused on ^{40}Ar, although as discussed below ^{136}Xe also provides valuable constraints. Similar calculations cannot be done for ^{4}He, which does not accumulate in the atmosphere, nor for neon isotopes, since the production rate of ^{21}Ne, as well as the amount of nonradiogenic ^{21}Ne in the atmosphere, are too uncertain. Also, the amount of radiogenic ^{129}Xe in the bulk Earth is not well constrained due to early losses from the planet.

6.3.1 The ^{40}K–^{40}Ar Budget

Potassium is a moderately volatile element and is depleted by a factor of ~8 in the bulk silicate Earth compared to CI chondrites, but a precise and unambiguous concentration is difficult to obtain. Estimates have been made by comparison with uranium, which like potassium is highly incompatible during melting and so is not readily fractionated between MORB and the upper mantle. There is little debate regarding the concentration of uranium, which is obtained from concentration in carbonaceous chondrites and, by assuming that refractory elements (e.g., calcium, uranium, thorium) are unfractionated from solar values in the bulk Earth (e.g., O'Nions *et al.*, 1981). In this case, a bulk silicate Earth concentration of 21 ppb uranium is obtained (Rocholl and Jochum, 1993). If it is assumed that the MORB source value of $K/U = 1.27 \times 10^4$ (Jochum *et al.*, 1983) is the same as that of the bulk silicate Earth, then there is 270 ppm potassium that has produced a total of 2.4×10^{42} atoms ^{40}Ar. This is a widely accepted value. Note that the core is not a significant repository of either potassium or ^{40}Ar (Chabot and Drake, 1999).

The ^{40}Ar in the atmosphere (9.94×10^{41} atoms) is essentially entirely radiogenic. Assuming the crust has 0.91–2.0 wt.% K (Taylor and McLennan, 1985; Rudnick and Fountain, 1995; Wedepohl, 1995), and noting that the continents have a mean K–Ar age of 1×10^9 a (Hurley and Rand, 1969), equivalent to a ratio of $^{40}Ar/K = 9.1 \times 10^{-6}$, yields an amount of crustal ^{40}Ar that is only 3.1–6.8% of that in the atmosphere. Therefore, in total, 41% of the ^{40}Ar that has been produced is now in the atmosphere (Allègre *et al.*, 1986, 1996; Turcotte and Schubert, 1988). Thus, a significant reservoir of ^{40}Ar remains in the Earth. How this relates to degassing of nonradiogenic noble gas isotopes is the subject of degassing models, but identifying the fraction of radiogenic isotopes in the atmosphere provides an important overall constraint. It has sometimes been assumed that the mantle reservoir that is rich in ^{40}Ar is the same as that with high $^{3}He/^{4}He$ and so is also rich in ^{3}He, although this is not the case of all mantle models (see Porcelli and Ballentine, 2002).

Uncertainty arises in the above calculation when considering that the depleted MORB-source mantle may not have a bulk silicate Earth K/U ratio, since a significant fraction of either element may have been preferentially added into the upper mantle by subduction. The bulk of the potassium and uranium originally in the upper mantle is now in the continental crust, which, therefore, may be expected to have a bulk silicate Earth K/U ratio. But this is not sufficiently well constrained due to wide variations in the potassium and uranium contents of the more differentiated continental crustal rocks. It has been suggested that the bulk Earth K/U ratio is much lower than the MORB value (Albarède, 1998; Davies, 1999). In this case, the potassium content of the Earth is much lower, and so a greater fraction of the total ^{40}Ar has degassed. This would make the ^{40}Ar budget compatible with geophysical models that have convection and mixing throughout the mantle, and so would imply that the depleted mantle that serves as the MORB source represents the bulk of the whole mantle. In such a case, there is no need to isolate and maintain a gas-rich mantle reservoir making up 60% of the mantle. However, it has been argued that the relative proportions of moderately volatile elements in the Earth lie on compositional trends defined by chondritic meteorite classes (Allègre *et al.*, 1995; Halliday and Porcelli, 2001), and trends in meteoritic Rb/Sr versus K/U are compatible with a terrestrial value of $K/U = 1.27 \times 10^4$. Only a modest reduction in the terrestrial potassium content would still be compatible with these relationships.

6.3.2 The ^{129}I–^{129}Xe and ^{244}Pu–^{136}Xe Budgets

The amount of short-lived isotope ^{129}I ($t_{1/2}$ = 15.7 Myr) that was in the Earth or Earth-forming materials and produced ^{129*}Xe can be estimated from the present bulk silicate Earth concentration of stable ^{127}I. Iodine is a volatile element and is highly depleted in the Earth relative to chondrites. Wänke *et al.* (1984) estimated a bulk silicate Earth stable ^{127}I concentration of 13 ppb based on the analysis of a fertile xenolith judged to represent the undepleted mantle, and this is the generally accepted estimate. McDonough and Sun (1995) arrived at a similar value of 11 ppb. Alternatively, using a bulk crust abundance of 8.6×10^{18} g I (Muramatsu and Wedepohl, 1998) and an upper-mantle concentration of 0.8 ppb (Déruelle *et al.*, 1992), and assuming that the crust was derived from 25% of the mantle, a bulk silicate Earth value of 9 ppb is obtained (Porcelli and Ballentine, 2002). However, based on unpublished data, Déruelle *et al.* (1992) quote a concentration for the crust that is 4.3 times higher, raising the possibility that the silicate Earth concentration of iodine is substantially higher than typically estimated, although such high values have not been adopted. At 4.57 Ga ago, $(^{129}\text{I}/^{127}\text{I})_0 = 1.1 \times 10^{-4}$ based on meteorite data (Hohenberg *et al.*, 1967; Brazzle *et al.*, 1999). For a silicate Earth value of 13 ppb ^{127}I, 2.7×10^{37} atoms of ^{129*}Xe were produced in the Earth or Earth-forming materials since formation of the solar system.

Plutonium is a highly refractory element represented naturally by a single isotope, ^{244}Pu ($t_{1/2}$ = 80 Myr), which produces heavy xenon isotopes by fission. It is assumed that plutonium was incorporated into Earth-forming materials unfractionated relative to other refractory elements such as uranium. Meteorite data suggests that at 4.56 Ga, $(^{244}\text{Pu}/^{238}\text{U})_0 = 6.8 \times 10^{-3}$ (Hudson *et al.*, 1989) and this is the commonly accepted value. In this case the silicate Earth, or Earth-forming materials, with a present-day bulk silicate Earth value of 21 ppb U (O'Nions *et al.*, 1981), initially had 0.29 ppb Pu. This produced 2.0×10^{35} atoms of ^{136*}Xe since the start of the solar system. Other work on meteorites (Hagee *et al.*, 1990) calculated values of $(4–7) \times 10^{-3}$, with the higher number considered more likely to represent the solar value, although a lower value remains a possibility. The amount produced by ^{238}U in the bulk silicate Earth (7.5×10^{33} atoms ^{136}Xe) is much less, and so bulk silicate Earth (and atmospheric) ^{136*}Xe is dominantly plutonium-derived; even if xenon was lost over the first 10^8 a of Earth history so that half of the plutonium-derived xenon was lost, plutonium-derived ^{136}Xe is still 13 times more abundant in the Earth.

The greatest difficulty in constraining the global xenon budget has been in calculating the abundances of radiogenic xenon in the atmosphere. The composition for nonradiogenic atmospheric xenon (Section 6.2.5) provides ratios of $^{129}\text{Xe}/^{130}\text{Xe} = 6.053$ and $^{136}\text{Xe}/^{130}\text{Xe} = 2.075$ as the present best estimates of the isotopic composition of nonradiogenic terrestrial xenon (Pepin, 2000). Therefore, $6.8 \pm 0.30\%$ of atmospheric ^{129}Xe ($^{129*}\text{Xe}_{atm} = 1.7 \times 10^{35}$ atoms) and $4.65 \pm 0.5\%$ of atmospheric ^{136}Xe ($^{136*}\text{Xe}_{atm} = 3.81 \times 10^{34}$ atoms) are radiogenic. The ^{136*}Xe in the atmosphere is 20% of the total ^{136}Xe produced by ^{244}Pu in the bulk silicate Earth. However, the $^{129*}\text{Xe}_{atm}$ is only 0.8% of the total ^{129}Xe produced since 4.57 Ga; such a low value cannot be accounted for by incomplete degassing of the mantle nor from any uncertainties in the estimated amount of ^{129*}Xe, and requires losses to space over an early period that is short relative to the longer time constant of ^{136}Xe production.

The depletion of radiogenic xenon in the atmosphere due to losses from the Earth to space must have occurred during early Earth history, when such heavy species could have been lost either from protoplanetary materials or from the growing Earth. Wetherill (1975) proposed that a "closure age" of the Earth could be calculated by assuming a two-stage history that involved essentially complete loss of ^{129*}Xe and ^{136*}Xe initially, followed by complete closure against further loss. The "closure age" also can be calculated by combining the ^{129}I–^{129}Xe and ^{244}Pu–^{136}Xe systems (Pepin and Phinney, 1976) to obtain a closure age of 82 Myr. If radiogenic ^{136}Xe was lost from the entire planet over about one half-life of ^{244}Pu (80 Myr), then ~40% of the ^{136}Xe remaining in the Earth is in the atmosphere, compatible with the fraction of ^{40}Ar in the atmosphere (using a silicate Earth value of 270 ppm K). Note that in the first 100 Ma, only 6% of the ^{40}Ar now present was produced, and losses over this time would not have significantly changed the ^{40}Ar budget (Davies, 1999).

The coincidence between the ^{40}K–^{40}Ar and ^{244}Pu–^{136}Xe budgets is remarkable considering that these values are the result of a series of independent, albeit somewhat uncertain, estimates. As mentioned above, in order to adjust these numbers to accommodate a greater fraction of degassing, it has been suggested that there is greater depletion of the moderately volatile potassium in the Earth

(Albarède, 1998); however, the coincidence with the Pu–^{136}Xe budget requires a similarly lowered estimate of the amount of short-lived, refractory ^{244}Pu in the solar nebula. Clearly, these two factors are unrelated. It might be assumed that the ^{40}Ar budget might reflect processing of a greater fraction of the mantle, but with subduction returning a considerable fraction of the potassium over geological time, thereby creating domains in the mantle that have been degassed early but now contain a considerable budget of ^{40}Ar by subsequent production. However, this is not possible for the ^{136}Xe budget; all plutonium-derived ^{136}Xe was produced early, and the present budget reflects the total processing of the mantle. Therefore, it appears that the noble gas budget requires that a considerable fraction of the mantle has not been degassed to the atmosphere.

6.4 DEGASSING OF THE MANTLE

The present value for the mid-ocean ridge flux of ^{3}He to the atmosphere, if constant over 4.5 Ga, would result in a total of 4.5×10^{12} mol ^{3}He degassed. Using the mantle value of $^{3}He/^{22}Ne = 11$ (see Section 6.2.3), this corresponds to a total of 5×10^{13} mol ^{22}Ne. This is only 2% of the ^{22}Ne presently in the atmosphere (Table 2). Therefore, present fluxes of noble gases from the mantle, applied over the history of the Earth, are insufficient to provide the inventories in the atmosphere. A degassing history that involves stronger degassing in the past is required, and a consideration of degassing models is needed to address the issue of the time dependence of the fluxes. An important factor in degassing models is the extent of noble gas recycling into the mantle by subduction. While subduction is unlikely to have a significant effect on the atmospheric inventory, it may impact the characteristics of noble gases in depleted mantle reservoirs may be significant. Therefore, models either assume that subduction of noble gases does not occur, or explicitly incorporates the effects of atmospheric inputs to the mantle.

6.4.1 Early Earth Degassing

Models for the evolution of terrestrial noble gases must necessarily consider appropriate starting conditions. The initial incorporation of noble gases and the establishment of terrestrial characteristics are discussed. Early degassing is likely to be very vigorous due to high accretional impact energies during Earth formation over 10^{8} yr. Loss of major volatiles from impacting materials, and noble gases, may occur after only 10% of the Earth has accreted (Ahrens *et al.*, 1989). In this case, much of the volatiles are added directly to the atmosphere rather than being degassed from the solid planet. Volatiles may also have been captured directly from the solar nebula (e.g., Porcelli *et al.*, 2001), followed by modifications due to losses to space. Various models for the origin of relatively volatile elements in the Earth have accounted for terrestrial volatiles by late infall of volatile-rich material (e.g., Turekian and Clark, 1969, 1975; Dreibus and Wänke, 1989; Owen *et al.*, 1992). The relative uniformity of lead and strontium isotopes in the mantle suggests that relatively volatile elements such as rubidium and lead that would also be supplied by late-accreting volatile-rich material were subsequently mixed into the deep mantle (Gast, 1960). However, loss of noble gases from impacting materials directly into the atmosphere likely inhibited their incorporation into the growing solid Earth. Therefore, noble gases supplied to the Earth in this way were unlikely to have been initially uniformly distributed in the solid Earth. It is also clear that very strong degassing of the Earth occurred during the extended period of planetary formation.

Atmospheric noble gases were also likely to have been lost to space during accretion by atmospheric erosion (Ahrens, 1993), when large impactors impart sufficient energy to the atmosphere for the constituents to reach escape velocity. In this way, each large impact during later accretion can drive away a substantial portion of the previously degassed atmosphere. Therefore, the present atmospheric abundances do not necessarily reflect the total amounts of nonradiogenic and early-produced nuclides that were degassed from accreting materials. Also, it has been suggested that the strong fractionation of neon and xenon isotopes in the atmosphere is due to hydrodynamic escape (Hunten *et al.*, 1987; Sasaki and Nakazawa, 1988; Pepin, 1991), where loss of hydrogen from the atmosphere entrains heavier species, leaving behind a fractionated residue. Such losses would not have affected gases within the Earth, and so would have generated isotopic contrasts between the atmosphere and internal terrestrial reservoirs. Such loss processes can account for the losses of xenon isotopes produced by short-lived ^{129}I and ^{244}Pu. In sum, degassing histories must include strong early degassing of the Earth as it accretes, and consider that the abundances degassed were not fully retained in the atmosphere.

6.4.2 Degassing from One Mantle Reservoir

The simplest case for atmosphere formation is unidirectional degassing from a single solid

Earth reservoir, which is represented by the MORB source region. Early models focused on argon isotopes. Generally, the key assumption is that the rate of degassing at any time is directly proportional to the total amount of argon present in the mantle at that time. Also, there is no return flux from the atmosphere by subduction. Then

$$\frac{d^{36}Ar_m}{dt} = -\alpha(t)^{36}Ar_m \quad (1)$$

and

$$\frac{d^{40}Ar_m}{dt} = -\alpha(t)^{40}Ar_m + \lambda_{40}y^{40}K_m \quad (2)$$

where $\alpha(t)$ is the time-dependent degassing proportionality constant, $\lambda_{40}=5.543\times10^{-10}\,a^{-1}$ is the total decay rate of ^{40}K, $y=0.1048$ is the fraction of decays of ^{40}K that yield ^{40}Ar (Table 1), and $^{36}Ar_m$, $^{40}Ar_m$, and $^{40}K_m$ are the total Earth abundances. In the simplest case, $\alpha(t)$ is a constant (Turekian, 1959; Ozima and Kudo, 1972; Fisher, 1978). This is reasonable if the mantle is well mixed and has been melted and degassed at a constant rate at mid-ocean ridges. Assuming there was no argon initially in the atmosphere, then the only free variable is α; in this case

$$\left(\frac{^{40}Ar}{^{36}Ar}\right)_{atm} = \left(\frac{\alpha}{\alpha-\lambda}(1-e^{-\lambda t}) - \frac{\lambda}{\alpha-\lambda}\times(1-e^{-\alpha t})\right)\left(\frac{y^{40}K_m e^{\lambda t}}{^{36}Ar_{atm}}\right) \quad (3)$$

Using a BSE value of 270 ppm K, a value of $\alpha = 1.82\times10^{-10}$ is obtained. From Equation (1), $^{36}Ar_{atm} = {}^{36}Ar_{m0}(1-e^{-\alpha t})$, so that the fraction of nonradiogenic ^{36}Ar that has degassed, $^{36}Ar_{atm}/^{36}Ar_{m0}$, is 0.62. The $^{40}Ar/^{36}Ar$ ratio of the mantle is then

$$\left(\frac{^{40}Ar}{^{36}Ar}\right)_m = \frac{\lambda_{40}}{\alpha-\lambda_{40}}\left(e^{(\alpha-\lambda_{40})t}-1\right)\frac{y^{40}K_m e^{\lambda_{40}t}}{^{36}Ar_{m0}} \quad (4)$$

A value of $(^{40}Ar/^{36}Ar)_{mn}=520$ is calculated from Equation (4) for the mantle. Once higher values were measured in MORB samples, such a simple formulation no longer appeared valid. Higher ratios can be obtained if an early catastrophic degassing event occurred, removing a fraction f of the ^{36}Ar from the mantle into the atmosphere (Ozima, 1973). In this case, the term $(1-f)^{36}Ar_{m0}$ can be substituted for $^{36}Ar_{m0}$ (see Ozima and Podosek, 1983). Then for a mantle with $^{40}Ar/^{36}Ar=4\times10^4$ (see Section 6.2.4), 98.6% of ^{36}Ar was degassed initially. Alternatively, a more complicated degassing function that is steeply diminishing with time (such as $\alpha(t) = \alpha e^{-\beta t}$) can be used to match the present isotope compositions (Sarda *et al.*, 1985; Turekian, 1990), and so also involves early degassing of the bulk of the atmospheric ^{36}Ar. Regardless of the formulation used, such early degassing is required by the high measured $^{40}Ar/^{36}Ar$ ratios and may reflect extensive devolatilization of impacting material during accretion or a greater rate of mantle melting very early in Earth history due to higher heat flow.

Another aspect of mantle degassing that can be represented in model calculations is the transfer of potassium from the upper mantle into the continental crust so that the mantle potassium content becomes time dependent. This requires including the crust as an additional model reservoir. The continents may be modeled as either attaining their complete mass very early or more gradually using some growth function (see, e.g., Ozima, 1975; Hamano and Ozima, 1978; Sarda *et al.*, 1985). However, all model formulations qualitatively agree that ^{36}Ar degassing dominantly occurred very early in Earth history.

A different perspective on ^{40}Ar degassing has been provided by Schwartzman (1973), who argued that potassium is likely to be transferred "coherently" out of the mantle with ^{40}Ar; i.e., any ^{40}Ar that has been produced by potassium will be degassed when the potassium is transferred to the crust. While the $K/^{40}Ar$ ratio of magmas leaving the mantle and that of the mantle source region are thus assumed to be approximately equal, the highly depleted residue could still be fractionated, so that very radiogenic-derived $^{40}Ar/^{36}Ar$ ratios could develop. It has been pointed out that, using the budgets of potassium and ^{40}Ar, this implies that the potassium in the crust should fully account for the ^{40}Ar in the atmosphere (Coltice *et al.*, 2000). This is only the case if the continental crust contains 2.0% K, which is higher than most estimates (see Section 6.5.1), and so additional ^{40}Ar was provided by additional potassium that has been recycled and which corresponds with up to ~30% of the potassium that is now in the continents (Coltice *et al.*, 2000).

Similar considerations used in the ^{40}Ar modeling have been applied to xenon (see Thomsen, 1980; Staudacher and Allègre, 1982; Turner, 1989). In these studies, there is greater resolution of the timing of early degassing due to the short half-lives of the parent nuclides ^{129}I and ^{244}Pu. The higher $^{129}Xe/^{130}Xe$ ratio of the upper mantle is interpreted as resulting from an increase in the I/Xe ratio during the lifetime of ^{129}I due to degassing of xenon to the atmosphere. Regardless of the exact degassing history used, this requires that strong degassing occur very early in Earth history, compatible with the results of the argon studies. Models of degassing from a single mantle reservoir have also been applied to helium (Turekian, 1959; Tolstikhin, 1975), although there are greater degrees of freedom since the atmosphere does not preserve a record

occurred before these parent elements became extinct, then the source would have an increased $^{129}I/^{130}Xe$ and $^{244}Pu/^{130}Xe$ ratio, and so would be expected to have higher $^{136*Pu}Xe/^{130}Xe$ and $^{129}Xe/^{130}Xe$ ratios now than the atmosphere. Evidence from MORBs indicates that the upper mantle indeed has such higher ratios. Another feature of any xenon that has remained within this source reservoir is that it will have a higher $^{136*Pu}Xe/^{129*}Xe$ ratio than that of air xenon; i.e., it will lie above the line that defining this ratio in Figure 4 (Ozima *et al.*, 1985). This can be most easily envisaged by considering that degassing occurred early and in a single event. At the time of xenon loss from the mantle to the atmosphere, the proportion of undecayed ^{244}Pu will be greater than that of ^{129}I (which has a much shorter half-life). Therefore, the remaining ^{244}Pu and ^{129}I will produce xenon with a higher $^{136*Pu}Xe/^{129*}Xe$ ratio than is in the atmosphere. This is true regardless of whether or not degassing actually occurred as a single event. Note that mantle xenon will also have fissiogenic ^{136}Xe from decay of ^{238}U produced during later mantle evolution (see Section 6.2.5), and so to compare the present upper-mantle $^{136*}Xe/^{129*}Xe$ ratio generated from short-lived parent nuclides to that of the atmosphere, the proportion of mantle $^{136*}Xe$ that is plutonium-derived must be determined. As seen in Figure 7, MORB data do fall above the line through the atmosphere. However, any correction for additions from ^{238}U will lower the $^{136}Xe/^{130}Xe$ ratio, and unless essentially all of $^{136*}Xe$ in MORB is ^{244}Pu-derived, the corrected value for upper-mantle xenon will fall below the atmosphere line and so cannot be the residue left after atmospheric xenon degassing. The data for CO_2 well gases and MORB indicate that a large fraction of the $^{136*}Xe$ is in fact from uranium (see Section 6.2.5), which is consistent with grow-in in the upper mantle due to the presently inferred $^{238}U/^{130}Xe$ ratio (see Porcelli and Wasserburg, 1995a). Therefore, it appears inescapable that the xenon presently found in the upper mantle is not the residue from degassing of the atmosphere. Since the daughter xenon isotopes how found in the atmosphere must have been derived from the ^{244}Pu and ^{129}I that were in the upper mantle (see Section 6.3.2), the xenon now found there must have been introduced from a deeper reservoir after the atmosphere was removed.

The main uncertainty in the evaluation of xenon isotopes is the composition of terrestrial nonradiogenic xenon (see Section 6.2.5). However, the arguments presented above appear to be robust when considering the possible compositions. As shown in Figure 8, solar wind xenon or U–Xe do not match the terrestrial light xenon isotope composition without extensive fractionation. Only U–Xe, when fractionated, provides a plausible

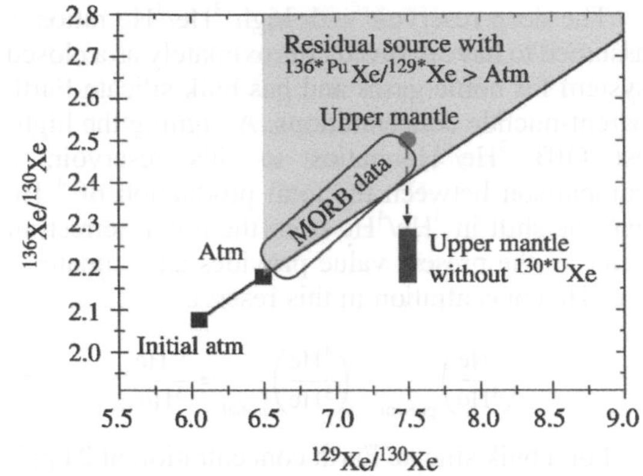

Figure 7 The relationship between the xenon isotope compositions of the atmosphere, initial atmosphere, and the upper mantle as sampled by MORB. The line connecting the initial atmosphere (fractionated U–Xe) and the present atmosphere has a slope equal to the ratio of plutonium-derived ^{136}Xe to radiogenic ^{129}Xe ($^{136*}Xe/^{129*}Xe$) in the atmosphere. Any xenon that remained in the solid reservoir from where this was degassed must have a greater value for this ratio and so lie in the shaded region (Ozima *et al.*, 1985). While measured MORBs do so, upper-mantle compositions that have been corrected for ^{238}U-derived ^{136}Xe (based on MORB data of Kunz *et al.*, 1998 and on CO_2 well gas data of Phinney *et al.*, 1978) do not, indicating that upper-mantle xenon is not the residual left from atmosphere degassing.

precursor for the atmosphere (Figure 8). While the errors on this composition are small (Pepin, 2000), it is worth considering whether another composition is plausible. Due to the magnitude of the excess of ^{129}Xe in the atmosphere, the amount of radiogenic ^{129}Xe present has not been debated, so that the $^{129}Xe/^{130}Xe$ ratio of nonradiogenic terrestrial xenon is relatively well fixed. As shown in Figure 8, the CO_2 well gas data can be residual of atmospheric degassing only if there is almost no plutonium-derived ^{136}Xe in the atmosphere (i.e., the $^{136}Xe/^{130}Xe$ ratios of the initial and present atmospheres are very similar). However, this would require a substantial downward revision of the amount of ^{244}Pu that was present in Earth-forming materials that has been estimated completely independently and is consistent with the ^{40}Ar budget of the Earth (Section 6.3.2). There appears to be no reason to make such a re-evaluation, and, as discussed in Section 6.4.5 below, there are mantle models that can satisfy these constraints.

Another constraint on the origin of the xenon now found in the upper mantle is obtained from the composition of fissiogenic ^{136}Xe. In a closed-system reservoir, plutonium-derived $^{136*}Xe$ will dominate over uranium-derived $^{136*}Xe$ (see Section 6.3.2). In a system that has been closed throughout solar system history and starting with a

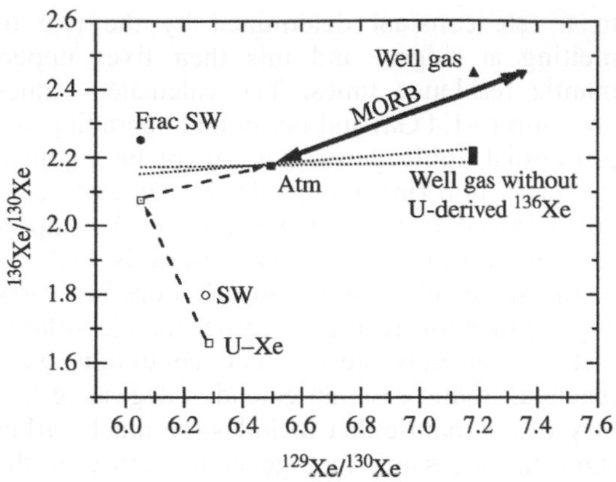

Figure 8 The xenon isotope compositions of U–Xe and solar wind xenon, both nonfractionated and fractionated to match the light xenon isotopes, are compared to the value of the atmosphere. It is clear that fractionated solar wind xenon cannot serve as the nonradiogenic composition of the atmosphere since it has a higher $^{136}Xe/^{130}Xe$ ratio. In order for the upper mantle to have the same $^{136*}Xe/^{129*}Xe$ ratio as the atmosphere, the nonradiogenic composition of the atmosphere must lie on the dotted lines, implying that there is very little plutonium-derived ^{136}Xe. This is contrary to the inferred ^{244}Pu budget of the Earth.

chondritic $^{244}Pu/^{238}U$ ratio of 0.0068, fission of ^{244}Pu will be 27 times that produced by ^{238}U (Porcelli and Wasserburg, 1995a). If xenon was lost over the first 100 Myr (about one half-life of ^{244}Pu), plutogenic ^{136}Xe will still dominate. Therefore, xenon from such a reservoir that has ^{129}Xe excesses will have accompanying $^{136*}Xe$ that is dominantly plutonium-derived. MORB and CO_2 well gas data indicate that uranium-derived $^{136*}Xe$ dominates upper-mantle $^{136*}Xe$. This requires that xenon in this mantle reservoir has been in an environment with a much higher $^{238}U/^{130}Xe$ ratio after extinction of ^{244}Pu, either due to a large decrease in Xe/U after the decay of ^{244}Pu, or has received xenon from another reservoir. As discussed above, based on xenon isotope systematics the xenon in the upper mantle has clearly been introduced from the deep mantle, and therefore this suggests that the $^{136*Pu}Xe$ and $^{129*}Xe$ was generated in the deep mantle, and the upper mantle has served as the locale for significant late $^{136*U}Xe$ additions.

Another constraint on the relationship between atmospheric and mantle xenon is obtained by considering the nonradiogenic xenon isotopes. As discussed, the nonradiogenic xenon isotopes presently in the atmosphere are highly fractionated with respect to the composition of xenon that is likely to have been initially trapped by the Earth. The fractionation most likely occurred during subsequent losses to space under conditions that were only present very early in Earth history. In this case, the degassing of nonradiogenic isotopes occurred early while such processes were still possible (see Pepin, 1991). In contrast, as discussed above, nonradiogenic xenon isotopes in the mantle are fractionated with respect to atmospheric xenon, with e.g., $^{128}Xe/^{130}Xe$ ratios that are intermediate between the solar and atmospheric ratio. This requires retention in the mantle of a component that was either only somewhat fractionated relative to solar xenon (due to fractionation processes before accretion by presently unknown processes), and therefore is less fractionated than atmospheric xenon, or more likely that is unfractionated solar xenon but has been mixed with subducted xenon. In the latter case, the xenon in the upper mantle is a mixture between xenon that has been stored in the deeper mantle since initial trapping with subducted xenon. Models of xenon evolution in the mantle therefore should have a return flux from the atmosphere, rather than simple unidirectional solid Earth degassing.

In sum, the xenon isotopes point to a more complicated mantle degassing history than simple unidirectional degassing into the atmosphere. It appears that first nonradiogenic noble gases were degassed to the atmosphere early. The solid reservoir volume from which these nonradiogenic gases were lost is unconstrained. This atmospheric xenon was fractionated by losses to space over 10^8 yr, so that the atmosphere became enriched in the heavy isotopes. Radiogenic xenon was then degassed quantitatively from at least 40% of the mantle (based on the ^{136}Xe–Pu budget). This occurred either by degassing of xenon directly from the mantle, or by transport of parent plutonium and iodine to the crust first, followed by crustal degassing. Xenon contained in another, deep-mantle reservoir and which is not residual from atmosphere removal was subsequently added to the upper mantle. The U/Xe ratio in the upper mantle was greater than that of the deeper reservoir, so that radiogenic $^{136*}Xe$ in the upper mantle became dominated by uranium-derived xenon.

An important implication of these considerations is that since the noble gases in the upper mantle are not the residual from atmosphere degassing, they provide no information regarding the degassing history of the atmosphere. Rather, the present composition of the upper mantle reflects interactions between different noble gas reservoirs. The disposition of these reservoirs, and the transfer of noble gases toward the surface, controls the overall degassing of the Earth, and these factors are considered further in the degassing models below. It should be emphasized that this is a crucial conclusion, since it renders invalid all calculations of degassing of the solid Earth based upon matching upper-mantle isotope compositions by simple degassing into the atmosphere. Rather, the relationship between mantle noble gas compositions and the degassing of the atmosphere is more complex.

6.4.5 Open-system Models

Another set of models still includes two mantle reservoirs, a MORB-source upper mantle and a deeper, gas-rich reservoir. However, these steady-state box models (Figure 6) are distinguished from the limited interaction box models discussed in Section 6.4.2 in being based upon the open interaction between the upper mantle and both the lower mantle (O'Nions and Oxburgh, 1983) and the atmosphere (Porcelli and Wasserburg, 1995a). Therefore, the constraints from xenon isotopes discussed above (see Section 6.4.4) are naturally accommodated. The steady-state model has been applied to helium isotope and heat fluxes (O'Nions and Oxburgh, 1983; Kellogg and Wasserburg, 1990), the U–Pb system (Galer and O'Nions, 1985), and the other noble gases (O'Nions and Tolstikhin, 1994; Porcelli and Wasserburg, 1995a,b). The central focus of the model is not degassing of the upper mantle to form the atmosphere, but rather mixing in the upper mantle. There are upper-mantle noble gas inputs by radiogenic production from decay of uranium, thorium, and potassium. In addition, atmospheric argon and xenon are subducted into the upper mantle, and lower-mantle noble gases are transported into the upper mantle within mass fluxes of upwelling material that carries all the deep-mantle noble gases together. The lower mantle is assumed to be an approximately closed system, in common with other mantle models (see Section 6.4.3). Noble gases from these sources comprise the outflows at mid-ocean ridges. The isotopic systematics of the different noble gases are linked by the assumption that transfer of noble gases from the upper mantle to the atmosphere by volcanism, as well as the transfer from the lower into the upper mantle by bulk mass flow, occurs without elemental fractionation. All model calculations assume that the upper mantle extends down to the 670 km discontinuity and so comprises 25% of the mantle. Plumes carrying lower-mantle material arise from the 670 km discontinuity.

It has been assumed that upper-mantle concentrations and isotopic compositions are in steady state, so that the inflows and outflows are equal. Since the main outflow is at mid-ocean ridges, this flux can be used to determine the upper-mantle residence times. The upper mantle is degassed to the atmosphere according to a rate constant determined by the rate of melting at ridges, and this then fixes upper-mantle residence times. The calculated values are short ($\sim$1.4 Ga) and imply that nonradiogenic primordial noble gases that cannot be supplied by subduction (including solar helium and neon) are provided from the lower mantle. An important implication of this conclusion is that the composition of the upper mantle does not carry any information regarding earlier mantle volatile history, nor how steady state conditions were reached. In this case, an extended degassing history of the mantle that includes the much earlier rigorous degassing that generated most of the atmosphere, cannot be inferred from noble gases presently found in the upper mantle.

In the open-system models, primordial nonradiogenic noble gases presently seen in the upper mantle originate in the lower mantle, and their fluxes to the atmosphere are moderated by the rates of transfer from the source region and through the upper mantle. As discussed above, the lower-mantle source region is assumed to be a closed system, and so has bulk Earth concentrations of radiogenic isotopes. However, since the lower-mantle argon and xenon isotope compositions, as represented by OIB noble gases, cannot be well constrained by available data, the nonradiogenic argon and xenon isotope concentrations are also not known. However, these can be calculated in the model from the balance of fluxes into the upper mantle. The MORB $^{3}He/^{4}He$ ratio is a result of mixing between lower-mantle helium and production of ^{4}He in the upper mantle (O'Nions and Oxburgh, 1983). Since the flux of the latter is fixed, the rate of helium transfer from the lower mantle can be calculated (Kellogg and Wasserburg, 1990), and this defines the degassing rate of the lower mantle. Neon, which follows helium, is transferred similarly, with decreases in $^{3}He/^{4}He$ ratios accompanied by increases in $^{21}Ne/^{22}Ne$ ratios due to coupled production of ^{4}He and ^{21}Ne. The MORB $^{40}Ar/^{36}Ar$ ratio (when corrected for atmospheric contamination of basalts) is a mixture of lower-mantle argon (obtained from the coupling with helium fluxes), radiogenic $^{40*}Ar$ produced in the upper mantle (calculated from the upper-mantle potassium concentration), and subducted air argon. When there is no subduction, the lower mantle is calculated to have $(^{40}Ar/^{36}Ar) = 9{,}700$. The MORB $^{136}Xe/^{130}Xe$ ratio (when corrected for air contamination) is the result of mixing between lower-mantle xenon, fissiogenic $^{136*U}Xe$ produced in the upper mantle, and subducted air xenon (Porcelli and Wasserburg, 1995a). The lower-mantle ratios are established early in Earth history by decay of ^{129}I and ^{244}Pu, and the $^{129}Xe/^{130}Xe$ ratio is constrained to be at least as great as that in the upper mantle. Once lower-mantle xenon is transported into the upper mantle, it is augmented by U-derived ^{136*U}Xe. Subduction of atmospheric xenon is also possible, lowering the ^{136}Xe/^{130}Xe and ^{129}Xe/^{130}Xe ratios. Therefore, the model is consistent with the isotopic evidence that upper-mantle xenon does not have a simple direct relationship to atmospheric xenon.

The lower mantle has a ^{40}Ar/^{36}Ar ratio that is much higher than the atmospheric value. This can be compared to the reservoirs that supplied the atmosphere, which includes all late-accreting sources and the uppermost fraction of the mantle that has degassed ^{36}Ar early and ^{40}Ar subsequently. The higher ^{40}Ar/^{36}Ar ratio of the lower mantle implies the ^{36}Ar/K ratio of the lower mantle is much lower than the bulk ratio of the reservoirs that supplied the atmosphere. Assuming potassium was initially uniformly distributed throughout the BSE, this implies very heterogeneous incorporation of ^{36}Ar. This contrasts with the typical *a priori* assumption of an initially uniform distribution of ^{36}Ar. The ^{244}Pu-derived ^{136}Xe and radiogenic ^{129}Xe in the upper mantle are derived from the lower mantle. This lower-mantle compositions, obtained by subtracting from MORB xenon the ^{238}U-derived ^{136}Xe produced in the upper mantle, corresponds to closure times that are similar to that of the atmosphere, indicating that early losses occurred from the deep mantle as well (Porcelli *et al.*, 2001). These losses must have been prior to the assumed closed-system evolution.

One feature of the model is that the atmosphere generally has no *a priori* connection with other reservoirs. It simply serves as a source for subducted gases, and no assumptions are made about its origin. Since no assumption is made about the initial distribution of noble gases in the Earth, the atmospheric abundances cannot be used to derive lower-mantle concentrations. The daughter isotopes that are presently found in the atmosphere clearly were originally degassed from the upper mantle, possibly along with nonradiogenic nuclides. However, this occurred before the present character of the upper mantle was established. Therefore, the upper mantle no longer contains information regarding atmosphere formation.

6.4.6 Boundaries within the Mantle

The principal objection has been based on geophysical arguments advanced for greater mass exchange with the lower mantle below 670 km, and so the difficulty of maintaining a distinctive deep-mantle reservoir. The clearest indication of the scale of mantle convection comes from seismic tomography imaging of subducting slabs and mantle plumes that cross the 670 km discontinuity (e.g., Creager and Jordan, 1986; Grand, 1987, 1994; van der Hilst *et al.*, 1997). Plumes may arise from boundary layer instabilities within the mantle or at the core–mantle boundary. There is growing evidence that plumes such as those that form the Hawaiian and Icelandic hotspots have their origin at the core–mantle boundary rather than at a depth of 670 km. Ultrahigh seismic velocity zones at the boundary beneath these islands have been described as plume-induced (Helmberger *et al.*, 1998; Russell *et al.*, 1998). Recent images of the Iceland plume are compatible with a deep origin for this plume (Shen *et al.*, 1998), and a mantle structure extending to the core (Bijwaard and Spakman, 1999). A plume extending to a depth of at least 2,000 km also has been imaged beneath central Europe (Goes *et al.*, 1999), and hotspots in Hawaii, Iceland, South Pacific, and East Africa have been shown to be located above slow anomalies in the lower mantle that extend to the core (Zhao, 2001).

Although geophysical observations point to the present-day mantle convecting as a single layer, it has been argued (Allègre, 1997) that models requiring long-term mantle layering can be reconciled with geophysical observations for present-day whole mantle convection if the mode of mantle convection changed less than 1 Ga ago from layered to whole mantle convection. In this case, the mode of volatile transfer from the deep mantle, and so the rate of deep-mantle degassing, has changed dramatically. However, this idea is at odds with the thermal history of the mantle.

Kellogg *et al.* (1999) developed a model in which mantle below ~1,700 km has a composition, and so density, that is sufficiently different from that of the shallower mantle to largely avoid being entrained and homogenized in the overlying convecting mantle. The boundary with the overlying mantle has a variable depth, and is much deeper where there is geophysical evidence for deeper slab penetration or plumes arising from the core–mantle boundary. This model preserves a region in the mantle behind which the radioelements and primitive noble gases can be preserved, while accommodating many geophysical observations. There is no geophysical evidence for such a boundary, although whether it could elude seismic detection is debated (Kellogg *et al.*, 1999; Vidale *et al.*, 2001). Also, if the overlying mantle has the composition of the MORB source, then the abyssal layer must contain a large proportion of the heat-producing elements, and it is not yet clear what this effect would have on the thermal stability of the layer or temperature contrast with the overlying mantle. Coltice and Ricard (1999) suggested an alternative model in which helium with high $^{3}He/^{4}He$ ratios are stored in the narrow zone around the core that is composed of subducted material. This would also be consistent with the geophysical evidence for whole mantle convection, and provide a source for all primordial noble gases. However, it does not provide a reservoir for the undegassed radionuclides (Section 6.3), and it is unclear how such a reservoir would have high initial ^{3}He concentrations (see Porcelli and Ballentine, 2002). Other reservoir configurations have also been suggested for reservoirs containing high $^{3}He/^{4}He$ ratios, such as in small domains embedded throughout the mantle, but have not been incorporated into comprehensive models that explain all of the noble gas and geophysical observations (Porcelli and Ballentine, 2002).

The core has also been considered as a source of noble gases. During formation of the Earth, the core may have incorporated sufficient quantities of solar helium to now supply ocean islands with high $^{3}He/^{4}He$ ratios (Porcelli and Halliday, 2001). In this case, such ratios trace the interaction between the core and mantle. However, the core cannot be used to explain the radiogenic isotope budgets (see Section 6.3) nor the heat–^{4}He budget, and so must be coupled with a mantle model for these features.

6.4.7 Summary

Early degassing models have described the formation of the atmosphere by progressive degassing of the upper mantle, leaving the presently observed noble gases behind to generate strongly radiogenic compositions. However, it is now clear that upper-mantle noble gases are not residual from atmosphere degassing, and have been introduced into the upper mantle from a deeper reservoir after quantitative removal of atmospheric constituents. Nonetheless, the conclusions regarding strong early degassing are compatible with the events of early Earth history. A complete model of noble gas evolution in the mantle and atmosphere remains to be constructed, although elements of earlier models may survive. It might be possible to reformulate the open-system models for a larger upper mantle, or greater mass fluxes between reservoirs, although in some cases non-steady-state upper-mantle concentrations may be required. Unfortunately, the particular configuration of the reservoirs will remain speculative until the geophysical constraints are fully clarified.

A general outline of the history of mantle degassing can be constructed from the observations and constraints discussed above (Porcelli *et al.*, 2003):

(i) Noble gases are acquired throughout the mantle during formation of the Earth, and a substantial inventory is retained within the growing planet, although concentrations in the deeper mantle are lower.

(ii) Radiogenic nuclides produced in the first 10^{8} a are degassed from throughout the entire mantle, while isotopic differences are generated due to lower ^{130}Xe concentrations in the deep mantle.

(iii) Strong fractionation of xenon and neon occurred during losses to space. The most plausible mechanism yet formulated involves hydrodynamic escape. Xenon is depleted by at least 10^{2}. This sets the currently observed fractionation of nonradiogenic xenon in the atmosphere. This may have overlapped with the degassing of the mantle. Nonfractionating losses by atmospheric erosion may have also occurred. All daughter noble gas isotopes that had been degassed to that time were lost to space.

(iv) After losses to space has terminated at $\sim$100 Ma, the radiogenic and fissiogenic xenon produced by the ^{244}Pu and ^{129}I remaining in the upper 40% of the Earth is degassed into the atmosphere, either directly from the mantle, or by transfer of the parent elements into any early crust followed by crustal degassing. This was not accompanied by a significant amount of nonradiogenic (and nonfractionated) xenon.

(v) 40% of the mantle loses ^{40}Ar to the atmosphere and potassium to the crust, from where any further ^{40}Ar produced is largely degassed.

(vi) Gases in 60% of the mantle are largely isolated from the atmosphere. However, a small amount of noble gases has leaked into the upper mantle, where it has mixed with radiogenic nuclides produced there, to generate the presently observed MORB compositions.

The composition of the upper mantle clearly does not provide information on the degassing history of the mantle, but rather reflects the small fluxes into this reservoir from the deep mantle, from production within the upper mantle, and possibly from subduction. The transfer of ^{40}Ar from the crust throughout much of Earth history has been dictated by the degassing of the crust, where the potassium has been stored.

6.5 DEGASSING OF THE CRUST

The processes involved in the formation of the crust are likely to release any volatiles into the atmosphere that were present in the mantle source region, and so the crust is not expected to be a significant reservoir of primordial volatiles. However, gaseous radiogenic nuclides that are produced within the crust do not readily escape. The most extensively studied has been ^{40}Ar.

6.5.1 Crustal Potassium and ^{40}Ar Budget

There have been a number of studies examining the total potassium content of the continental crust, which is a key parameter for the average petrologic composition of the crust. Estimates of K_2O in the crust, which has a total mass of 2.05×10^{25} g (Cogley, 1984) range from 0.96% to 2.4% (e.g., Weaver and Tarney, 1984; Taylor and McLennan, 1985; Rudnick and Fountain, 1995; Wedepohl, 1995). Taylor and McLennan (1985) and McLennan and Taylor (1996) have emphasized that the extensive data for heat flow can be used to constrain the total abundances of the heat-producing elements. For an average crustal heat flux away from recently disturbed regions of ~48 mW m^{-2} (Nyblade and Pollack, 1993) and a reduced heat flux (i.e., heat conducted through the crust from the mantle) of 27 mW m^{-2} (Morgan, 1984), and assuming that the heat-producing elements are in the ratios Th/U = 3.8×10^4 and K/U = 1.27×10^4, a value of 0.96% K_2O was obtained (McLennan and Taylor, 1996). The corresponding total crustal budget is therefore 1.6×10^{23} g K, and this produces 1.7×10^{31} atoms $^{40}Ar\,a^{-1}$. Using the values of 21 ppb U and K/U = 1.27×10^4 for the bulk silicate Earth, 15% of the total terrestrial potassium is in the crust. Further consideration of the mantle contribution to crustal heat flow has suggested that models with K_2O up to ~2% may be possible (Rudnick *et al.*, 1998), so that ~30% of the terrestrial potassium may be in the crust, and producing twice as much ^{40}Ar.

As discussed above, the K/U ratio was obtained for the MORB source, and is often taken to represent the bulk Earth. This can be applied to the crust only by assuming that there has been no fractionation of potassium and uranium during crust formation, and no preferential recycling of either element. It should be noted that using this ratio, ~85% of the heat currently produced in the crust is generated by uranium and thorium (see, e.g., Turcotte and Schubert, 1982), and so heat flow is a good constraint for the total crustal uranium and thorium budgets, but a poor constraint for the potassium budget, unless the K/U ratio is known very accurately. Therefore, it is possible that the K/U ratio of the crust is somewhat different from the MORB ratio largely due to a different potassium content. In this case, the composition of the deeper mantle (not represented by MORB) may be different.

The amount of ^{40}Ar that is in the crust can be calculated from the mean K–^{40}Ar age of the continental crust of 960 Ga (Hurley and Rand, 1969), which corresponds to a $^{40}Ar/K$ ratio of 1.6×10^{17} atoms $^{40}Ar\,g^{-1}$ K. This is 3–5% of the amount of ^{40}Ar in the atmosphere.

The composition of the crust provides some interesting constraints on the composition of the mantle. If the upper mantle that supplies MORB has ~50 ppm K (e.g., Hofmann, 1988), and extends down to a depth of 670 km (about one-quarter of the mantle), then it has only a small fraction of the total terrestrial potassium. Therefore, the 70–85% of the terrestrial potassium (for a bulk silicate Earth K/U = 1.27×10^4) is in the deeper mantle. Since ~40% of the terrestrial ^{40}Ar is in the atmosphere, some of this deep-mantle material that is not depleted in potassium must have degassed. It has been argued that the upper mantle has a higher potassium concentration of ~100 ppm (Korenaga and Kelemen, 2000), so that ~40% of the terrestrial potassium is in the crust and upper mantle and can supply the atmosphere. In this case, there has been net transfer of potassium to the upper 25% of the Earth. Clearly, how much of the Earth has been degassed and depleted requires further constraints on the potassium budget of both the upper mantle and crust. Coltice *et al.* (2000) pointed out that more potassium is needed to supply the atmosphere than is in the crust, and since potassium and argon are both highly incompatible and so extracted together from the mantle, the additional potassium needed has been subducted.

6.5.2 Formation Time of the Crust

There has been considerable discussion about the age of the crust. Various growth curves have been proposed, from slow steady growth based on K–Ar ages (Hurley and Rand, 1969) to rapid early growth (see Armstrong, 1991), and various intermediate histories (see discussion in Taylor and McLennan, 1985). The current rate of crustal formation is ~1 $km^3\,yr^{-1}$, and so is insufficient to generate the full mass of the crust over 4.5 Ga. Crustal formation was clearly much more rapid in the past, and the general consensus is that the average crustal formation age is substantially greater than the K–^{40}Ar age. Therefore, it appears that ^{40}Ar has been progressively degassed from the crust.

Considering that potassium now in the crust accounts for ~30–70% of the ^{40}Ar in atmosphere, crustal degassing may be an important component of the solid Earth degassing history (see Hamano and Ozima, 1978). If the crust formed very early, then release of ~30–70% of the ^{40}Ar in the atmosphere has been controlled by crustal processing. For slower crustal growth histories, the proportion degassed directly from the crust would depend upon the time constant for growth compared to the 1.4×10^9 a half-life of ^{40}K.

6.5.3 Present Degassing

The present rate of continental degassing of ^{40}Ar into the atmosphere cannot be readily measured, and so must be inferred by considering the mechanisms of degassing. Studies of radiogenic ^{4}He in groundwaters from the Great Artesian Basin in Australia (Torgersen and Clarke, 1985; Torgersen and Ivey, 1985) concluded that the higher ^{4}He concentrations in older waters could be explained by a steady influx from production in the underlying crust. The required calculated flux was found to equal the entire production rate in the crust, and so suggested that the continental crust degasses by continuous release of all radiogenic ^{4}He from uranium- and thorium-bearing host rocks. A similar conclusion was drawn from data from another basin as well (Heaton, 1984). Additional work on ^{40}Ar in the Great Artesian Basin extended this conclusion to ^{40}Ar (Torgersen *et al.*, 1989), implying that the continental flux to the atmosphere is $(1.7–3.3)\times 10^{31}$ atoms $^{40}Ar\,a^{-1}$. However, there have been doubts that helium and argon can be so readily released from minerals and transported effectively across the crust. These doubts have generated discussions regarding the interpretations of groundwater flow rates and so noble gas accumulation rates (e.g., Mazor, 1995; Bethke *et al.*, 1999).

An important factor is how radiogenic ^{40}Ar is released from host rocks. The release of ^{40}Ar from potassium-bearing minerals by diffusion has been extensively examined as part of thermal evolution studies, and the blocking temperatures of major minerals such as feldspars and biotites correspond to quantitative retention within the top ~10 km of the crust (McDougall and Harrison, 1999). The upper continental crust, down to ~10 km, is estimated to have 3.4% K_2O and 2.8 ppm U (Taylor and McLennan, 1985). This is consistent with the observation that the concentration of heat-producing elements decreases with depth (e.g., Jaupart *et al.*, 1981), although potassium may not decrease with depth as strongly as uranium and thorium (e.g., Rudnick and Fountain, 1995). Nonetheless, a dominant fraction of the potassium in the crust is below the blocking temperature of ^{40}Ar. Further, the ^{40}Ar produced in the deeper crust is more likely to be transported episodically to the surface during mobilization of fluids that may also be responsible for metamorphic processes (e.g., Etheridge *et al.*, 1984).

A process that clearly leads to widespread release of ^{40}Ar is weathering at the surface. In a comprehensive survey of river chemistry, Martin and Meybeck (1979) reported an average concentration of 1.4 ppm K in a total water discharge of $3.74\times 10^{16}\,L\,yr^{-1}$, leading to a total of 5.2×10^{13} g K. This potassium is likely to be released largely from the weathering of feldspars (e.g., Wollast and Mackenzie, 1983). Assuming that this K is derived from weathering of crystalline rocks with a mean age of 960 Ma (Hurley and Rand, 1969) with 1.6×10^{17} atoms $^{40}Ar\,g^{-1}$ of potassium, this leads to release of 0.83×10^{31} atoms a^{-1}. Mechanical weathering of the continental crust leads to discharge to the oceans of $1.6\times 10^{16}\,g\,a^{-1}$ (Milliman and Meade, 1983), which contains an average concentration of 2% K (Martin and Meybeck, 1979) or 3×10^{14} g K. If this material is derived from cannibalization of earlier sediments (see, e.g., Veizer and Jansen, 1985) with an average age of ~500 Ma (with 0.6×10^{17} atoms $^{40}Ar\,g^{-1}$ of potassium), then this material contained 1.8×10^{31} atoms ^{40}Ar when mobilized, which is twice that released during dissolution of potassium. Of course, the average K–Ar ages of materials that provide either detrital or dissolved constituents remains highly uncertain. The long-term rate of mechanical weathering may differ considerably from this because of recent changes in discharge due to human activities and due to the deposition within drainage basins. However, a much greater uncertainty lies in estimating how much of this material has lost ^{40}Ar. Limited studies have found that the K–Ar ages of surface sediments often reflect those of the source rock, even where significant clay mineral formation has occurred (e.g., Hurley *et al.*, 1961). Hurley (1966) summarized available data and suggested that detrital ages are gradually wiped out during burial. Later studies in the Gulf of Mexico confirmed that K–Ar ages decreased with depth at rates that depend upon mineral size fractions (Weaver and Wampler, 1970; Aronson and Hower, 1976), and likely reflect the redistribution of potassium, and loss of ^{40}Ar, during diagenetic illite formation. Overall, it is clear that some poorly constrained proportion of the ^{40}Ar contained in sediments is released. Perhaps as much ^{40}Ar is released by diagenesis of sediments as derived from primary chemical weathering.

Another mechanism for degassing ^{40}Ar is thermal processing due to tectonic thickening of the crust in orogenic belts (see, e.g., England and Thompson, 1984). Veizer and Jansen (1979, 1985) used the distribution of basement K–Ar ages to model the thermal cycling rate of the crust, and obtained a rate of $\sim 2\times 10^{16}\,g\,a^{-1}$. It is possible that during such events, ^{40}Ar in the lower crust is also transported to the surface, and so degassing of the entire crustal thickness may occur together. For an average crustal composition with 1% K and 1.6×10^{17} atoms $^{40}Ar\,g^{-1}$ of potassium (assuming an average K/Ar age of the crust of 960 Ma), 3.2×10^{31} atoms $^{40}Ar\,a^{-1}$ are released. Given the range of possible crustal potassium contents discussed above, this value may be as much as 2 times greater.

In summary, chemical weathering and orogenic processing of the crust generate a flux of $\sim(3.8\text{–}7)\times 10^{31}$ atoms $^{40}Ar\,a^{-1}$ to the atmosphere, and this may be augmented by a presently poorly constrained flux associated with potassium in riverine sediments. Nonetheless, thermal processing of the continental crust appears to dominate the ^{40}Ar flux. The total flux is more than the present-day crustal production rate of $1.7\text{–}3.4\times 10^{31}$ atoms $^{40}Ar\,a^{-1}$. If the flux does indeed exceed the present production rate, then the ^{40}Ar reservoir size of the continents is diminishing.

The present flux of ^{40}Ar from mid-ocean ridges is 1.9×10^{31} atoms a^{-1} (Section 6.2.7), and so substantially lower than the continental flux, indicating that at present continental processing is more important than mantle processing for ^{40}Ar release to the atmosphere.

6.6 MAJOR VOLATILE CYCLES

Many of the important characteristics of the atmosphere are related to the concentrations of particular molecular species, such as O_2. Here the absolute abundances of carbon, nitrogen, and hydrogen (as H_2O) as supplied from the mantle are considered in the context of models of noble gas degassing from the solid Earth. A convenient reference for comparing surface volatile inventories to mantle reservoirs is obtained by dividing the surface volatiles into the mass of the upper mantle, the minimum size of the source reservoir. However, this should not be taken to imply a particular model of degassing.

6.6.1 Carbon

Crustal carbon. In contrast to the noble gases, carbon near the planetary surface is concentrated in crustal rocks, and is largely divided between carbonates, with an isotopic composition of $\delta^{13}C=0‰$ and sedimentary organic carbon with $\delta^{13}C=-25‰$. While Hoefs (1969) estimated a total of 2.6×10^{22} g of C, with 66% in sedimentary rocks, Hunt (1972) reassessed the budget using sedimentary rock abundance data from Ronov and Yaroshevsky (1969) and carbon concentration data compilations to obtain a higher value of 9×10^{22} g. Ronov and Yaroshevsky (1976) updated their earlier work with new additional data to obtain a total budget of 1.2×10^{23} g. A similar value (6×10^{22} g C) was obtained by Li (1972) when requiring that the crust have an isotopic composition similar to that of MORB (see below) of approximately $\delta^{13}C=-5$; conversely, this agreement supports the notion that the crust and mantle are isotopically similar. Therefore, considering the uncertainties involved, a value of 1.0×10^{23} g C, or 8.3×10^{21} mol, with perhaps a 20% uncertainty, appears to be reasonable (Table 3).

Carbon in MORB and the upper mantle. MORB entering the ocean crust are generally oversaturated in carbon as CO_2, and exhibit exsolved CO_2-rich vesicles. Carbon contents of MORB samples vary widely due to degassing, but a ratio of $C/^3He=2\times 10^9$ appears to characterize the undegassed magmas (Marty and Jambon, 1987). Combined with the initial concentration of ^3He obtained from the total ^3He ridge flux (see Section 6.2.7), this provides a value of 510 ppm carbon in undegassed MORB, within the range of other estimates of undegassed MORB of 900–1,800 ppm carbon (Holloway, 1998) and 400 ppm (Pineau and Javoy, 1994). Assuming carbon is incompatible, this corresponds to an upper mantle (to a depth of 670 km) concentration of 50 ppm. Therefore, the upper mantle of 1×10^{27} g contains about half the carbon that is present in the crust. If this value applies to the entire mantle, then only about a third of the terrestrial carbon is at the surface. The carbon isotope composition of the upper mantle is $\delta^{13}C=-5‰$ based on MORB data (e.g., Javoy and Pineau, 1991), and is similar to that of the surface. The carbon brought to the surface by MORB is quantitatively degassed during eruption and subsequent alteration, and results in a flux of 2×10^{12} mol C yr^{-1}.

There may also be a significant amount of carbon in the core, with values of up to 4% possible, although this depends upon the amount available in the early Earth (Wood, 1993; Halliday and Porcelli, 2001). The amount of carbon that is supplied across the core–mantle boundary into the mantle is not known.

Carbon at hotspots. Ocean islands with high $^3He/^4He$ ratios clearly contain distinctive volatile components. However, there are also some data regarding the carbon flux of hotspot sources. The ratio of hotspot volcanics is an important parameter in further defining the carbon cycle. It has been argued that the $C/^3He$ ratio is equal to that of the upper mantle (Trull *et al.*, 1993). Alternatively, Poreda *et al.* (1992) argued that the ratio is higher and is 6×10^9 in Iceland, while Kingsley and Schilling (1995) argued from mixing relationships along the Mid-Atlantic Ridge that the Iceland source was $\sim 10^{10}$. Hilton *et al.* (1998a) found values of $(2\text{–}5)\times 10^9$ for Loihi, but also argued that degassing highly modified the ratio, and concluded that the source ratio was unknown. It should be noted that plumes appear to contain contributions not only from a high $^3He/^4He$ source region, but also subducted components (e.g., Sobolev *et al.*, 2000) that may contribute carbon that increases the $C/^3He$ ratio and mask the carbon accompanying helium with high $^3He/^4He$ ratios. Measured $\delta^{13}C$ values scatter around upper-mantle values (Trull *et al.*, 1993), and a consistent composition distinctive from that of MORB has not been unambiguously documented. The flux of carbon from ocean islands depends on the accepted $C/^3He$ as well as the flux of ^3He relative to MORB ^3He flux (see above).

Subduction. There is a significant amount of carbon on the ocean floor that is available for subduction. The amount of carbon released by arc volcanism is $(1.6–3.1)\times10^{12}$ mol a^{-1}. This can be compared with estimates of the amount of subducted carbon of $(0.35–2.3)\times10^{13}$ mol yr^{-1} (Bebout, 1995; Hilton *et al.*, 2002). Therefore, while it is possible that there is no carbon return flux into the upper mantle beyond island arcs, the ranges suggest that some carbon does survive in the downgoing slabs, and this flux may even be greater than the degassing flux observed at mid-ocean ridges of 0.2×10^{13} mol a^{-1}. Volcanic arcs have a range of $\sim\delta^{13}C=-2‰$ to $-12‰$ (Bebout, 1995), reflecting the variable proportions of subducted components.

The global carbon cycle. The flux of carbon brought to the surface by MORB of 2×10^{12} molC yr^{-1} can supply the entire crustal carbon inventory in 4.2 Ga; if there was stronger earlier degassing, there must have been a carbon return flux to the mantle. Much of the surface inventory indeed may have been degassed early along with noble gases (Section 6.4.1). It is possible that the surface inventory was initially greater than at present due to the difficulties of subducting carbon in hotter, younger slabs (Des Marais, 1985; McCulloch, 1993; Zhang and Zindler, 1993), although this effect may have been countered by increased incorporation of carbon into ocean crust by hydrothermal alteration (Sleep and Zahnle, 2001). If the carbon in the upper mantle is largely recycled, it may now be present in a steady-state abundance. This would naturally explain the high upper-mantle carbon concentration. However, since surface carbon is divided at the surface between carbonate carbon (with $\delta^{13}C=0‰$) and organic carbon (with $\delta^{13}C\sim-25‰$), the subducted carbon at any location need not have the average crustal $\delta^{13}C$. Therefore, in order to reconcile the isotopic similarity between the mantle and surface, carbonate and organic carbon must be subducted in the same proportions as at the surface. Otherwise, shifts in the composition away from that of the bulk Earth will occur in both the crust and mantle. There are no obvious controls that would demand this condition. The isotopic compositions of carbon in back arc basin basalts scatter, and can be $\sim5‰$ lighter than that of MORB carbon (Mattey *et al.*, 1984), reflecting mixing of organic carbon with MORB carbon.

An alternative view that naturally explains the isotopic similarity between the crust and upper mantle is that the carbon degassing from the mantle is primordial carbon that was trapped during solid Earth formation, representing the continuing unidirectional upward transfer of carbon. In this case, any carbon that is subducted does not constitute a dominant fraction of carbon in the upper mantle. However, carbon is not depleted in the upper mantle to the same extent as other highly incompatible elements such as the noble gases. An indication of this is the mantle $C/^{36}Ar$ ratio, which is 10^2 times greater than that of the surface (Marty and Jambon, 1987). This observation has been explained by having less net degassing of carbon early in Earth history, either due to higher recycling rates or more compatible behavior under more reducing conditions (Marty and Jambon, 1987). As discussed above (Section 6.4), the nonradiogenic noble gases may be supplied from a deeper mantle reservoir to the MORB source reservoir; if carbon is similarly supplied, then using the calculated ^3He content of a gas-rich reservoir (see Section 6.4.3) and MORB $^{36}Ar/^3He$ (see Section 6.2.6) and $^{36}Ar/C$ ratios as reflecting the deeper mantle, a carbon concentration of 3,000 ppm is obtained. This value is an order of magnitude less than values for carbonaceous chondrites (Kerridge, 1985), and would be even lower if some of the upper-mantle ^{36}Ar is from subduction.

6.6.2 Nitrogen

Nitrogen at the surface. Nitrogen is highly atmophile, although there is a significant fraction of surface nitrogen in the continental crust (see Table 3). Crustal nitrogen is isotopically heavier than atmospheric nitrogen, and the total nitrogen at the surface has $\delta^{15}N=2‰$ relative to the atmospheric value of 0‰ (Marty and Humbert, 1997). The total budget of 4.5×10^{21} g, when divided into the upper mantle, yields 4.5 ppm for the bulk silicate Earth.

Nitrogen in MORB and the upper mantle. The concentration of N_2 in undegassed MORB can be determined from the value of $N_2/^{40}Ar=120\pm20$ obtained from MORB data (Marty and Humbert, 1997; Marty and Zimmermann, 1999) to be 1.63 ppm. This corresponds to an upper-mantle concentration of 0.16 ppm, assuming nitrogen is incompatible during melting (Marty and Humbert, 1997). The amount of nitrogen in the upper mantle is therefore 1.6×10^{20} g, which is only $\sim3\%$ that in the crust and atmosphere. The MORB flux is equivalent to a flux of 5.0×10^{10}

mol yr^{-1}, or 9% of the surface N_2 over 4.5 Ga. Note that Javoy (1998) argued that nitrogen is relatively compatible, with an upper-mantle concentration of up to 40 ppm (Cartigny *et al.*, 2001). However, this is from mass balance calculations based upon model assumptions that volatiles in the upper mantle and on the surface are a mixture of enstatite chondrites and a late veneer of CI chondrites, and such a model has not been widely adopted.

The nitrogen isotopic composition of the upper mantle has been estimated from MORB (Marty and Humbert, 1997; Marty and Zimmermann, 1999) and diamond (Javoy *et al.*, 1984; Cartigney *et al.*, 1998; Boyd and Pillinger, 1994) data to be approximately $\delta^{15}N = -4‰$. Values as low as $\delta^{15}N = -25‰$ have been found in diamonds, indicating that there is another reservoir in the mantle.

Nitrogen at hotspots. $\delta^{15}N$ values from hotspots has been found to be largely positive and as high as $\delta^{15}N = +8‰$ as (Dauphas and Marty, 1999; Marty and Dauphas, 2003). It has been argued that these values are due to the presence of subducted components in mantle plumes (Marty and Dauphas, 2003), and is supported by the lack of correlation with helium isotope compositions.

Subduction. There are some data for the subduction of nitrogen. Metasedimentary rocks and organics in pelagic sediments are isotopically heavy, with organic nitrogen having $\delta^{15}N = +2‰$ to $+10‰$ in marine sediments (Peters *et al.*, 1978) and $\delta^{15}N = +2‰$ to $+15‰$ in metasediments (Haendel *et al.*, 1986; Bebout and Fogel, 1992). Metamorphosed complexes are also isotoipcally heavy (Bebout, 1995). Arc volcanics have an average of $\delta^{15}N = +7‰$ (Sano *et al.*, 1998). The subduction flux has been estimated from an average sediment subduction rate of 3.5×10^{15} g yr^{-1} (Ito *et al.*, 1983) and an average concentration of 125–600 ppm from the Catalina Schist to be $(0.44–2.1) \times 10^{12}$ g yr^{-1} (Bebout, 1995), although N in altered seafloor basalts may also carry nitrogen (Hall, 1989). The flux to the atmosphere due to arc volcanism is 4.5×10^{11} g a^{-1} (Hilton *et al.*, 2002). Therefore, it is possible that no nitrogen is returned to the mantle, although a flux equal to that at mid-ocean ridges (and so supporting steady state mantle and surface abundances), or higher (supporting arguments for the net inflow of nitrogen (Javoy, 1998) is also possible.

The global nitrogen cycle. A salient feature of the nitrogen cycle is that mantle nitrogen is isotopically lighter than both the surface and subducted nitrogen. The atmosphere is either derived from another component than the mantle or has suffered preferential loss of ^{14}N. The atmospheric composition therefore is inherited from the processes of Earth formation. Subsequently, the rates of mantle–crust exchange are constrained by these isotopic differences. There are limited possibilities for the source of isotopically light nitrogen in the mantle. Data for carbonaceous and ordinary chondrites typically show $\delta^{15}N > 0‰$ (Kerridge, 1985). However, E chondrites have been found to have consistently lower values, with an average of about $\delta^{15}N < -35‰$, and with concentrations on the order of 500 ppm (Grady *et al.*, 1986). Therefore, it has been proposed that E chondrites are the source of mantle nitrogen (Javoy, 1998). Since the meteorite concentrations are 4×10^3 times greater than presently found in the mantle, if a substantial portion of the Earth was derived from E chondrites, substantial nitrogen losses must have occurred during accretion (Javoy, 1998); conversely, only a small fraction of material needed to have been incorporated in an early protoplanet prior to sufficiently large impacts cause volatile loss (Ahrens, 1993). Alternatively, solar nitrogen may also be isotopically very light. Recent data for lunar samples indicate that solar nitrogen may be as light as $\delta^{15}N = -380‰$ (Hashizume *et al.*, 2000; Owen *et al.*, 2001). The consequences of incorporating solar nitrogen in the Earth has not been fully explored, although due to the high solar Ne/N ratio, incorporation of a significant amount of solar nitrogen will require strong elemental fractionation.

If nitrogen is highly incompatible along with the noble gases, then it likely is introduced into the upper mantle from another reservoir as well. Since this nitrogen may be isotopically very light, it appears that there is also a subducted, isotopically heavy, nitrogen component in the upper mantle, and this provides the dominant fraction of nitrogen. The specific proportions depend upon the isotopic compositions of each. In the case of solar nitrogen, only 3% of MORB nitrogen is solar (3 ppb), with the remainder provided by subduction. A much larger fraction of nitrogen derived from enstatite chondrites is required. If this nitrogen was introduced with ^3He from a deep-mantle source that also supplies plumes, then OIBs would be expected to have very light $\delta^{15}N$ values. This is not the case; rather, nitrogen appears to be dominated by subducted components. However, it is clear that both deep-mantle and subducted components are present in OIB source regions, and so primordial nitrogen may be masked. Alternatively, the composition of nitrogen subducted into the upper mantle prior to the Archean was isotopically light, and has persisted in the upper mantle over a long residence time (Marty and Dauphas, 2003).

6.6.3 Water

Crustal H_2O. The terrestrial oceans have $\delta D = 0‰$ with an inventory of 1.4×10^{24} g H_2O, equivalent to 120 ppm H when divided into the mass of the upper mantle. There is an additional 20% in other surface reservoirs and crustal rocks, and the total composition is $\sim\delta D = -10‰$ to $-18‰$ (Taylor and Sheppard, 1986; Lécuyer *et al.*, 1998).

H_2O in MORB and the upper mantle. Water is not so readily lost by vesiculation, with measurements of unaltered MORB glasses of 0.12% to 0.33 wt.% H_2O (Pineau and Javoy, 1994; Jambon, 1994; Sobolev and Chaussidon, 1996). Water behaves incompatibly during melting (see Jambon and Zimmermann, 1990), and so a concentration of 120–330 ppm is obtained for the mantle source region (see also Thompson, 1992). The upper mantle, therefore, contains at least as much water as the hydrosphere. The flux at ridges totals $(1–3) \times 10^{14}\,g\,a^{-1}$. The hydrogen isotopic composition is distinctive from that of the surface, with $\delta D = -71‰$ to $-91‰$ (Craig and Lupton, 1976; Kyser and O'Neil, 1984; Poreda *et al.*, 1986).

H_2O at hotspots. There are some hydrogen isotope data available for hotspots. Poreda *et al.* (1986) found a correlation between helium and hydrogen isotopes along the Reykjanes Ridge, with values of up to $\delta D = -50‰$, significantly higher than in MORB. These samples also had an increase in water content toward Iceland, with values of up to $\sim 0.35\%$. A higher water content was also found in samples from the Azores Platform, with 0.52 wt.% H_2O. Rison and Craig (1983) found glasses from Loihi had $\delta D = -69‰$ to $-74‰$, within the MORB range. Lighter compositions, down to $-125‰$, have been found in Hawaii and elsewhere (Deloule *et al.*, 1991; Hauri, 2002), which suggests that isotopically light, juvenile hydrogen remains in the mantle.

Subduction. The flux of H_2O entering subduction zones of $8.8 \times 10^{14}\,g\,yr^{-1}$ or greater (Ito *et al.*, 1983; Bebout, 1995) is substantially greater than the ridge flux, and has been interpreted as a net inflow of water (Ito *et al.*, 1983). However, a substantial fraction is returned during arc metamorphism, dewatering, and volcanism. Data for boron isotopes and B/H_2O relationships suggest that only $\sim 20\%$ of the water is recycled (Chaussidon and Jambon, 1994), compatible with a steady-state mantle concentration. Due to the difficulties of obtaining total long-term values for subduction and return of water, concrete constraints on the balance of H_2O will probably be impossible.

The global H_2O cycle. The main feature of the solid Earth hydrogen cycle is the large contrast between the isotopic composition of the surface of $\delta D = -10‰$ to $-20‰$ and the upper mantle of $\delta D = -80‰$. There have been two contrasting ideas. It has been argued that upper-mantle water is derived entirely from subduction. The low δD ratio of MORB could be controlled by those of dewatered metasedimetary and metamorphosed mafic subducted rocks that have compositions that are close to upper-mantle values, with $\delta D = -50‰$ to $-80‰$ (Magaritz and Taylor, 1976; Bebout, 1995). The average of these rocks may be somewhat heavier than MORB values, although Bell and Ihinger (2000) suggested that somewhat lighter hydrogen might be in nominally anhydrous minerals that may preferentially retain hydrogen during subduction. Alternatively, this isotopic signature has been ascribed to a primordial component established and isolated early in Earth history (e.g., Craig and Lupton, 1976; Lécuyer *et al.*, 1998; Dauphas *et al.*, 2000). It is possible that this is stored with ^3He in the deep mantle. If the H_2O and ^3He are both derived from the same reservoir, then using the MORB ratio of H_2O to ^3He, and an undegassed mantle ^3He concentration (Section 6.4.3), this reservoir contains 900–2,300 ppm H, which is much higher than the surface reservoir divided into the upper mantle, although somewhat below CI chondrites (Kerridge, 1985). Isotopically, CI chondrites are too heavy, and enstatite chondrites, with approximately $-460‰$, provide a better source, but contain only 50 ppm (Javoy, 1998). Overall, it appears likely that water in the upper mantle is dominated by subducted components, although hydrogen from the lower mantle may accompany ^3He upwards, and may also account for some isotopically light mantle values.

6.7 DEGASSING OF OTHER TERRESTRIAL PLANETS

Understanding of the degassing of Mars and Venus is incomplete due to the lack of data regarding interior volatile reservoirs, and uncertainties in whether differences between the surface inventories with those on Earth are due to different degassing histories or to initial differences generated during planet formation. There are manifestations of volcanic activity that can be used to deduce at least a generalized history of mantle melting, but of course the precise timing and relationships to mantle structure are not well constrained. In this review, some of the information regarding the extent of planetary degassing relative to that of the Earth is considered.

6.7.1 Mars

Mars is characterized by low noble gas abundances in the atmosphere, with equivalent total planet concentrations that are 10 times less than on Earth. The amount of martian carbon is also not well known; while the atmosphere has only

7×10^{18} g C, equivalent to 0.01 ppm for the bulk planet (Owen *et al.*, 1977), a large fraction may be stored in the polar regolith.

While the amount of water that has been observed on Mars is low, there is evidence that there was substantially more in the past. Measurements of atmospheric water vapor on Mars have found D/H values ~5 times that of the Earth and has been fractionated due to Jeans escape of hydrogen to space (Owen *et al.*, 1988). Models of hydrogen atmospheric losses (Donahue, 1995) and morphological data of features generated by surface water (Carr, 1986) both suggest that originally there may have been the equivalent of up to 500 m of water, or $\sim 7 \times 10^{22}$ g H_2O. The total mass of Mars is 0.11 that of the Earth, and so both planets originally may have had similar bulk water concentrations. This would also require degassing of water on both planets to similar extents.

The atmosphere of Mars has a high value of $\delta^{15}N = (+620 \pm 160)$‰ (Nier and McElroy, 1977) that may be due to fractionating losses, and $(4.7 \pm 1.2) \times 10^{17}$ g N, which is equivalent to 0.7 ppb when divided into the mass of the entire planet (Owen *et al.*, 1977). Therefore, Mars appears to have 10^{-4} times the nitrogen on the Earth. Mathew *et al.* (1998) reported evidence for a component in a martian meteorite with $\delta^{15}N < -22$‰, suggesting that, like the Earth, the solid planet may contain nitrogen that is isotopically lighter than the atmosphere, and consistent with the modification of atmospheric nitrogen isotopes by losses.

Radiogenic ^{40}Ar. The atmospheric $^{40}Ar/^{36}Ar$ ratio has been measured by Viking to be 3,000±400 (Pepin and Carr, 1992), although a lower value of ~1,800 has been deduced from meteorite data (Pepin and Carr, 1992; Bogard, 1997; Bogard *et al.*, 2001). Based on the Viking data, there are $(7.0 \pm 1.4) \times 10^{39}$ atoms ^{40}Ar in the atmosphere. The Mars mantle has been estimated to have 305 ppm K (Wänke and Dreibus, 1988), so that 3.3×10^{41} atoms ^{40}Ar have been produced in Mars. Therefore, only 2% of martian ^{40}Ar has degassed to the atmosphere, and most of the planet interior has retained the ^{40}Ar produced throughout its history. The history of degassing of ^{40}Ar from the interior has been discussed in several studies (Volkov and Frenkel, 1993; Sasaki and Tajika, 1995; Hutchins and Jakosky, 1996; Tajika and Sasaki, 1996). There are considerable uncertainties in the history of martian volcanism and the amounts of volatiles that have been lost to space that must be resolved before more definitive degassing histories can be constructed. Tajika and Sasaki (1996) argue that much of the ^{40}Ar has degassed from relatively recent volcanic regions, in contrast to early degassing of other volatiles, while Hutchins and Jakosky (1996) conclude that in order to account for volatiles that have been lost to space, much degassing must have occurred by processes other than volcanic outgassing.

The radiogenic ^{129}Xe budget. Martian atmospheric xenon clearly contains a considerable fraction of radiogenic ^{129}Xe. It has been estimated that the silicate portion of Mars contains 32 ppb I (Wänke and Dreibus, 1988). Assuming that $^{129}I/^{127}I = 1.1 \times 10^{-4}$ at 4.57 Ga (Hohenberg *et al.*, 1967), then 8.44×10^{36} atoms ^{129}Xe have been produced in Mars or precursor materials. Using fractionated CI chondrite xenon or solar xenon for the nonradiogenic light xenon isotope composition, the atmosphere is calculated to contain only 0.092% of what has been produced. Assuming there is none remaining in the planet, this corresponds to a closure age of 160 Myr. However, if only 2% has degassed (like ^{40}Ar) to the atmosphere, then a closure age of 70 Myr is obtained. This value is similar to that of the Earth; it suggests that there may also have been losses of volatiles from Mars over the same extended period of accretion.

The fissiogenic ^{136}Xe budget. The amount of ^{136}Xe produced in Mars or accreting materials, assuming that the silicate portion of Mars has 16 ppb of ^{238}U at present (Wänke and Dreibus, 1988) and initially had $^{244}Pu/^{238}U = 0.0068$ (Hudson *et al.*, 1989), is 1.9×10^{34} atoms ^{136}Xe from ^{244}Pu and 7.2×10^{32} atoms ^{136}Xe from ^{238}U. In contrast, there is a total of 2.8×10^{33} atoms ^{136}Xe in the atmosphere. Up to ~5% of the atmospheric ^{136}Xe may be plutonium-derived; if so, and the closure age for Mars is 70 Ma, then 1–2% of the ^{244}Pu produced in the solid planet has degassed. This is consistent with the ^{129}Xe and ^{40}Ar budgets. It has been argued that plutogenic $^{136*}Xe$ could make up much less that 5% of the total atmospheric inventory, requiring even less planetary degassing and greater very early isolation of interior volatiles from the atmosphere. Reports of significant abundances of ^{244}Pu fission xenon in several SNC meteorites do in fact point strongly to its efficient retention in the martian crust (Marty and Marti, 2002; Mathew and Marti, 2002). Further discussion of the abundances of daughter xenon isotopes in the atmosphere is provided by Swindle and Jones (1997).

Martian mantle noble gases. Martian meteorites contain components other than those derived directly from the atmosphere (see detailed discussion by Swindle (2002)). Information on the relative abundances of the heavier noble gases in the mantle (Ott, 1988; Mathew and Marti, 2001) suggests that the $^{84}Kr/^{132}Xe$ ratio is at least 10 times lower than both the martian atmosphere and the solar composition. If this is truly a source feature, it indicates that heavy noble gases trapped within the planet suffered substantially different elemental fractionation than the atmosphere and have not subsequently formed a dominant fraction of the atmosphere. However, it is not possible at present to conclusively determine whether the measured elemental abundance ratios reflect an interior reservoir that was initially different from atmospheric noble gases, rather than due either to planetary processing or transport and incorporation into the samples.

The noble gas isotopic composition of the martian interior is only available for xenon. Data for the martian meteorite Chassigny found xenon with little scope for radiogenic additions (Ott, 1988; Mathew and Marti, 2001), indicating that this reservoir had a high Xe/Pu ratio, at least during the lifetime of ^{244}Pu. Data from other meteorites indicate that there are other interior martian reservoirs that contain solar xenon but with resolvable fissiogenic contributions (see Mathew and Marti, 2002; Marty and Marti, 2002), and so have had lower Xe/Pu ratios. Data for ^{129}Xe and ^{136}Xe have been used to argue that there were substantial losses of xenon from the mantle within the first 35 Myr (Marty and Marti, 2002). As discussed, the martian mantle appears to have nonfractionated solar xenon (Ott, 1988). This contrasts with the fractionated character of atmospheric xenon, and is consistent with fractionation of the atmosphere after strong early degassing and minimal recycling of xenon into the mantle.

Martian degassing history. The budget for radiogenic argon and xenon indicate that only a small fraction of the planet has degassed since very early in planetary history. A consequence of this is that noble gases were retained within the planet during formation from a very early stage, as indicated by the ^{129}Xe budget. This is supported by evidence for xenon-rich mantle domains that for a body of at least this size, accretion does not necessarily lead to strong degassing. The low atmospheric abundances of nonradiogenic noble gases may be partly due to retention in the largely undegassed planet, but the much higher ^{40}Ar/^{36}Ar ratio relative to the terrestrial value indicates that the initial inventory of nonradiogenic species after planet formation was likely lower. The composition of nitrogen and water at the surface of Mars has been strongly affected by the history of losses to space, with no evidence for significant fluxes back into the planet.

The observation that the martian atmosphere has a higher ^{129}Xe/^{130}Xe ratio than the mantle, in contrast to the situation observed on Earth, has led to speculations about the mechanisms fractionating iodine from xenon. Both of these elements are highly incompatible, and so melting and crust formation are expected to transport them together out of the mantle. Available data suggest that iodine is more incompatible, so that the residue is expected to have a higher I/Xe ratio. In order to generate a higher atmospheric ^{129}Xe/^{130}Xe ratio, Musselwhite *et al.* (1991) suggested that the iodine was sequestered in the crust by hydrothermal alteration while xenon was lost to space. Alternatively, Musselwhite and Drake (2000) suggest that the iodine was preferentially retained within a magma ocean. These models are based upon the assumption that the mantle noble gases are the residue that is complementary to the atmosphere; this is not the case on the Earth and it is not clear whether this is true for Mars.

6.7.2 Venus

Venus is similar in size to the Earth and might be expected to have differentiated to a similar extent. However, while the early accretion history might have been similar (with the exception of the absence of a moon-forming event), silicate differentiation did not proceed according to the familiar plate tectonic mechanisms. There is, of course, no data on the interior of Venus, and so planetary degassing characteristics must be deduced from limited atmospheric data and observations of volcanic activity at the surface.

The measured ^{40}Ar/^{36}Ar ratio is 1.11 ± 0.02, substantially less radiogenic than the terrestrial atmosphere. For a mixing ratio of 21–48 ppm (Donahue and Pollack, 1983), the atmosphere contains $(1.8–4.4) \times 10^{41}$ atoms ^{40}Ar. Divided by the mass of the planet, this corresponds to $(3.6–9.0) \times 10^{13}$ atoms ^{40}Ar g^{-1}. This is 0.2–0.5 times the value for the Earth. However, Venus appears to be deficient in potassium. Data for the K/U ratio of the surface indicate that $K/U = 7{,}220 \pm 1{,}220$ (Kaula, 1999), or 0.57 ± 0.10 times that of the value of 1.27×10^4 commonly taken for the Earth. Assuming that Venus has the same uranium concentration as the total Earth (including the core) of 14 ppb, then $12 - 28\%$ of the ^{40}Ar produced in Venus is now in the atmosphere (see Kaula, 1999). This indicates that a substantial inventory of ^{40}Ar remains within the planet, possibly also accompanied by up to an equivalent fraction of nonradiogenic noble gases. In contrast, at least 40% of terrestrial ^{40}Ar is in the atmosphere.

Venus has about twice as much carbon (like nitrogen) at the surface than the Earth, equivalent to 26 ppm when divided into the bulk planet (von Zahn *et al.*, 1983); whether Venus is more rich in carbon therefore depends upon what volume of the mantle has degassed and how much remains in the mantle (Lécuyer *et al.*, 2000). From the radionuclide budget discussed above, it appears that Venus may have degassed to a similar extent as the Earth, but is unlikely to have been substantially more degassed. Therefore, it appears that the high abundances of carbon and nitrogen reflect greater total planetary abundances. The Venus atmosphere has

~200 ppm H_2O (Hoffman *et al.*, 1980) and a D/H ratio of $(1.6 \pm 0.2) \times 10^{-2}$, i.e., $\sim 10^2$ times that of the Earth (Donahue *et al.*, 1982). It has been suggested that Venus originally had the same D/H value as the Earth, but has lost at least one terrestrial ocean volume of water by hydrodynamic escape, thereby generating an enrichment in deuterium (Donahue *et al.*, 1982). The ratio of water to carbon and nitrogen therefore may have been similar to that of the Earth.

Venus is also rich in nonradiogenic noble gases, with the absolute abundance of ^{36}Ar on Venus exceeding that on Earth by a factor >70. This is clearly due to the amount of noble gases initially supplied and retained by the planet.

Venus is similar to the Earth in mass and composition, and so might be expected to evolve similarly. However, Venus has a hot, insulating atmosphere and does not have the features of plate tectonics. These two features appear to be related. Venus and Earth may have started with similar atmospheres, but the Earth suffered the consequences of a late moon-forming impact. The consequence is that the atmosphere of Venus remained sufficiently insulating to maintain temperatures that prevented the formation of liquid water at the surface, and so oceans. This may have led to less effective crustal recycling, and ultimately the inhibition of plate tectonics (see Kaula, 1990). However, the styles of early Venus evolution, and the transitions between different tectonic styles, are debated. Various models have been presented for the degassing of ^{40}Ar from the mantle and crust and into the Venus atmosphere, but these are dependent upon the history of tectonic activity and heat loss on the planet (Sasaki and Tajika, 1995; Turcotte and Schubert, 1988; Namiki and Solomon, 1998; Kaula, 1999). Indeed, the amount of ^{40}Ar in the atmosphere provides a constraint on the total amount of mantle melting and transfer of potassium to the crust (Kaula, 1999). The fraction of ^{40}Ar that has degassed, ~12–28%, is substantially lower than that of the Earth. Kaula (1999) has discussed various mechanisms that may account for decreased degassing of Venus. The possibility that this may reflect layering on Venus is difficult to assess, considering that relating the limited apparent degassing of the Earth to mantle structure has proven so controversial.

6.8 CONCLUSIONS

The degassing of the Earth is an integral part of the formation and thermal evolution of the planet. Degassing histories have often naturally been based on noble gas abundances and isotopic compositions. Radiogenic isotopes provide the strongest constraints on the total volume of the silicate Earth that has degassed to the atmosphere, and indicate that a large portion of the Earth remains undegassed. Nonradiogenic noble gases, incorporated during Earth formation, strongly degassed during accretion and the extreme thermal conditions on the forming Earth. The upper mantle, down to some as yet unresolved depth, is now highly degassed. Silicate differentiation producing the continental crust also likely promoted degassing, and continual processing of the continental crust has left it relatively degassed of even radiogenic daughters produced within the crust. Other volatiles have also been effectively removed from the upper mantle, although elements such as carbon may have substantial subduction fluxes that are the dominant inputs.

An outstanding question is how much of the mantle still maintains high volatile concentrations. This involves resolution of the nature of the high ^{3}He/^{4}He OIB-source region. Most models equate this with undepleted, undegassed mantle, although some models invoke depletion mechanisms. However, none of these has matched the end-member components seen in OIB lithophile isotope correlations. It remains to be demonstrated that a primitive component is present and so can dominate the helium and neon isotope signatures in OIB. The heavy-noble-gas characteristics in OIB must still be documented. It is not known to what extent major volatiles are stored in the deep Earth and associated with these noble gas components.

As understanding of terrestrial noble gas geochemistry has evolved, various earlier conclusions now appear to be incorrect. As discussed above, the very radiogenic argon and xenon isotope ratios of the upper mantle are not the result of early degassing of this mantle reservoir, since this is a model-dependent conclusion based on the assumption that upper-mantle noble gases are residual from atmosphere degassing. However, it is now clear that xenon isotope systematics precludes such a relationship (Ozima *et al.*, 1985). While a complete description of mantle noble gases remains to be formulated, it is clear that there are other mechanisms that can account for the observed xenon isotope variations. Nonetheless, early transfer of volatiles to the atmosphere probably did occur and was caused by impact degassing.

The interactions between the atmosphere and the mantle now appear to be more complex. The subduction of heavy noble gases may have a marked impact on mantle isotope compositions. The earlier conclusion that this was not possible was based on models of the isolation of the upper mantle or arguments about preservation of nonatmospheric ^{129}Xe/^{130}Xe ratios in the mantle. In fact, upper-mantle nonradiogenic xenon isotopes could be dominated by subducted xenon and admixed with very radiogenic xenon, and some models explicitly incorporate subducted xenon fluxes.

Until more is conclusively known about argon and xenon isotopic variations in the mantle, subduction must be considered a potentially important process.

In devising atmosphere formation histories, it is also now clear that the present is not the key to the past. Although primordial noble gases continue to degass, their isotopic compositions do not match those of the atmosphere and limit their contribution to a small fraction of the present atmospheric inventory. Volatile species continue to be added to the atmosphere, but the dominant inputs occurred during very early Earth history.

There are, of course, many questions regarding terrestrial noble gases that remain to be explored. Some of the issues that are critical to making advances in global models of noble gas behavior include the partitioning of noble gases into the core and between mantle minerals and silicate melts. While the common assumptions regarding general behavior may very well be correct, the effects of fractionation between noble gases cannot be clearly assessed. Further refinement of planetary degassing models will come from greater resolution of the nature of the reservoirs that remain undegassed within the planet.

ACKNOWLEDGMENT

A very useful review by N. Daup that was provided under very short notice is much appreciated.

REFERENCES

Ahrens T. J. (1993) Impact erosion of terrestrial planetary atmospheres. *Ann. Rev. Earth Planet. Sci.* **21**, 525–555.

Ahrens T. J., O'Keefe J. D., and Lange M. A. (1989) Formation of atmospheres during accretion of the terrestrial planets. In *Origin and Evolution of Planetary and Satellite Atmospheres* (eds. S. K. Atreya, J. B. Pollack, and M. S. Matthews). University of Arizona Press, Tucson, pp. 328–385.

Albarède F. (1998) Time-dependent models of U–Th–He and K–Ar evolution and the layering of mantle convection. *Chem. Geol.* **145**, 413–429.

Allègre C. J. (1997) Limitation on the mass exchange between the upper and lower mantle: the evolving convection regime of the Earth. *Earth Planet. Sci. Lett.* **150**, 1–6.

Allègre C. J., Staudacher T., Sarda P., and Kurz M. (1983) Constraints on evolution of Earth's mantle from rare gas systematics. *Nature* **303**, 762–766.

Allègre C. J., Staudacher T., and Sarda P. (1986) Rare gas systematics: *formation* of the atmosphere, evolution and structure of the Earth's mantle. *Earth Planet. Sci. Lett.* **87**, 127–150.

Allègre C. J., Poirier J.-P., Humler E., and Hofmann A. W. (1995) The chemical composition of the Earth. *Earth Planet. Sci. Lett.* **134**, 515–526.

Allègre C. J., Hofmann A. W., and O'Nions R. K. (1996) The argon constraints on mantle structures. *Geophys. Res. Lett.* **23**, 3555–3557.

Armstrong R. L. (1991) The persistent myth of crustal growth. *Austral. J. Earth Sci.* **38**, 613–630.

Aronson J. L. and Hower J. (1976) Mechanism of burial metmorphism of argillaceous sediment: 2. Radiogenic argon evidence. *Geol. Soc. Am. Bull.* **87**, 738–744.

Ballentine C. J. and Burnard P. G. (2002) Production, release, and transport of noble gases in the continental crust. *Rev. Mineral. Geochem.* **47**, 481–538.

Ballentine C. J., van Keken P. E., Porcelli D., and Hauri E. K. (2002) Numerical models, geochemistry and the zero-paradox noble-gas mantle. *Phil. Trans. Roy. Soc. London* A **360**, 2611–2631.

Batiza R. (1982) Abundances, distribution and sizes of volcanoes in the Pacific Ocean and implications for the origin of non-hotspot volcanoes. *Earth Planet. Sci. Lett.* **60**, 195–206.

Bebout G. E. (1995) The impact of subduction-zone metamorphism on mantle-ocean chemical cycling. *Chem. Geol.* **126**, 191–218.

Bebout G. E. and Fogel M. (1992) Nitrogen isotope composition of metasedimentary rocks in the Catalina Schist, California: implications for metorphic devolitization histor. *Geochim. Cosmochim. Acta* **56**, 2839–2849.

Begemann R., Weber H. W., and Hintenberger H. (1976) On the primordial abundance of argon-40. *Astrophys. J.* **203**, L155–L157.

Bell D. R. and Ihinger P. D. (2000) The isotopic composition of hydrogen in nominally anhydrous mantle minerals. *Geochim. Cosmochim. Acta* **64**, 2109–2118.

Benkert J.-P., Baur H., Signer P., and Wieler R. (1993) He, Ne, and Ar from solar wind and solar energetic particles in lunar ilmenites and pyroxenes. *J. Geophys. Res.* **98**, 13147–13162.

Bercovici D. and Karato S.-I. (2003) Whole-mantle convection and the transition-zone water filter. *Nature* **425**, 39–44.

Bethke C. M., Zhao X., and Torgersen T. (1999) Groundwater flow and the ^{4}He distribution in the Great Artesian Basin of Australia. *J. Geophys. Res.* **104**, 12999–13011.

Bijwaard H. and Spakman W. (1999) Tomographic evidence for a narrow whole mantle plume below Iceland. *Earth Planet. Sci. Lett.* **166**, 121–126.

Blum J. D. (1995) Isotopic decay data. In *Global Earth Physics: A Handbook of Physical Constants* (ed. T. J. Ahrens). American Geophysical Union, Washington, DC, pp. 271–282.

Bogard D. D. (1997) A reappraisal of the Martian Ar-36/Ar-38 ratio. *J. Geophys. Res.* **102**, 1653–1661.

Bogard D. D., Clayton R. N., Marti K., Owen T., and Turner G. (2001) Martian volatiles: isotopic composition, origin, and evolution. *Space Sci. Rev.* **96**, 425–458.

Boyd S. R. and Pillinger C. T. (1994) A preliminary study of ^{15}N/ ^{14}N in octahedral growth form diamonds. *Chem. Geol.* **116**, 43–59.

Brazzle R. H., Pravdivtseva O. V., Meshik A. P., and Hohenberg C. M. (1999) Verification and interpretation of the I–Xe chronometer. *Geochim. Cosmochim. Acta* **63**, 739–760.

Broadhurst C. L., Drake M. J., Hagee B. E., and Bernatowicz T. J. (1992) Solubility and partitioning of Ne, Ar, Kr, and Xe in minerals and synthetic basaltic melts. *Geochim. Cosmochim. Acta* **56**, 709–723.

Brooker R. A., Du Z., Blundy J. D., Kelley S. P., Allan N. L., Wood B. J., Chamorro E. M., Wartho J.-A., and Purton J. A. (2003) The "zero charge" partitioning behaviour of noble gases during mantle melting. *Nature* **423**, 738–741.

Brown H. (1949) Rare gases and the formation of the Earth's atmosphere. In *The Atmospheres of the Earth and Planets* (ed. G. P. Kuiper). University of Chicago Press, Chicago, pp. 258–266.

Buffett B. A., Huppert H. E., Lister J. R., and Woods A. W. (1996) On the thermal evolution of the Earth's core. *J. Geophys. Res.* **101**, 7989–8006.

Burnard P. G., Graham D., and Turner G. (1997) Vesicle specific noble gas analyses of Popping Rock: implications for primordial noble gases in Earth. *Science* **276**, 568–571.

Cadogan P. H. (1977) Paleaoatmospheric argon in Rhynie chert. *Nature* **268**, 38–41.

Caffee M. W., Hudson G. U., Velsko C., Huss G. R., Alexander E. C., Jr., and Chivas A. R. (1999) Primordial noble cases from Earth's mantle: identification of a primitive volatile component. *Science* **285**, 2115–2118.

Carr M. H. (1986) Mars: a water-rich planet? *Icarus* **68**, 187–216.

Cartigny P., Harris J. W., and Javoy M. (1998) Eclogitic diamond formation at Jwaneng: no room for a recycled component. *Science* **280**, 1421–1424.

Cartigny P., Jendrzejewski N., Pineau F., Petit E., and Javoy M. (2001) Volatile (C, N, Ar) variability in MORB and the respective roles of mantle source heterogeneity and degassing: the case of the Southwest Indian Ridge. *Earth Planet. Sci. Lett.* **194**, 241–257.

Chabot N. L. and Drake M. J. (1999) Potassium solubility in metal: the effects of composition at 15 kbar and 1900 degrees C on partitioning between iron alloys and silicate melts. *Earth Planet. Sci. Lett.* **172**, 323–335.

Chamorro-Perez E. M., Brooker R. A., Wartho J.-A., Wood B. J., Kelley S. P., and Blundy J. D. (2002) Ar and K partitioning between clinopyroxene and silicate melt to 8 GPa. *Geochim. Cosmochim. Acta* **66**, 507–519.

Chaussidon M. and Jambon A. (1994) Boron content and isotopic composition of oceanic basalts: geochemical and cosmochemical implications. *Earth Planet. Sci. Lett.* **121**, 277–291.

Clarke W. B., Beg M. A., and Craig H. (1969) Excess ^{3}He in the sea: evidence for terrestrial primordial helium. *Earth Planet. Sci. Lett.* **6**, 213–220.

Cogley J. C. (1984) Continental margins and the extent and number of the continents. *Rev. Geophys. Space Phys.* **22**, 101–122.

Coltice N. and Ricard Y. (1999) Geochemical observations and one layer mantle convection. *Earth Planet. Sci. Lett.* **174**, 125–137.

Coltice N., Albarède F., and Gillet P. (2000) ^{40}K–^{40}Ar constraints on recycling continental crust into the mantle. *Science* **288**, 845–847.

Craig H. and Lupton J. E. (1976) Primordial neon, helium, and hydrogen in oceanic basalts. *Earth Planet. Sci. Lett.* **31**, 369–385.

Craig H., Clarke W. B., and Beg M. A. (1975) Excess ^{3}He in deep water on the East Pacific Rise. *Earth Planet. Sci. Lett.* **2**, 125–132.

Creager K. C. and Jordan T. H. (1986) Slab penetration into the lower mantle beneath the Marianas and other island arcs of the northwest Pacific. *J. Geophys. Res.* **91**, 3573–3589.

Crisp J. A. (1984) Rates of magma emplacement and volcanic output. *J. Volcanol. Geotherm. Res.* **89**, 3031–3049.

Dauphas N. and Marty B. (1999) Heavy nitrogen in carbonatities of the Kola Peninsula: a possible signature of the deep mantle. *Science* **286**, 2488–2490.

Dauphas N., Robert F., and Marty B. (2000) The late asteroidal and cometary bombardment of Earth as recorded in water deuterium to protium ratio. *Icarus* **148**, 508–512.

Davies G. F. (1999) Geophysically constrained mantle mass flows and the Ar-40 budget: a degassed lower mantle? *Earth Planet. Sci. Lett.* **166**, 149–162.

Deloule E., Albarède F., and Sheppard S. M. F. (1991) Hydrogen isotope heterogeneities in the mantle from ion probe analysis of amphiboles from ultramafic rocks. *Earth Planet. Sci. Lett.* **105**, 543–553.

Déruelle B., Dreibus G., and Jambon A. (1992) Iodine abundances in oceanic basalts: implications for Earth dynamics. *Earth Planet. Sci. Lett.* **108**, 217–227.

Des Marais D. J. (1985) Carbon exchange between the mantle and crust and its effect upon the atmosphere: today compared to Archean time. In *The Carbon Cycle and Atmospheric CO_2: Natural Variations Archean to Present* (eds. E. T. Sundquist and W. S. Broecker). American Geophysical Union, Washington, DC, pp. 602–611.

Dixon J. E. and Stolper E. M. (1995) An experimental study of water and carbon dioxide solubilities in mid-ocean ridge basaltic liquids: 2. Applications to degassing. *J. Petrol.* **36**, 1633–1646.

Doe B. R. and Zartman R. E. (1979) Plumbotectonics: I. The Phanerozoic. In *Geochemistry of Hydrothermal Ore Deposits* (ed. H. L. Barnes). Wiley, New York, pp. 22–70.

Donahue T. M. (1995) Evolution of water reservoirs on Mars from D/H ratios in the atmosphere and crust. *Nature* **374**, 432–434.

Donahue T. M. and Pollack J. B. (1983) Origin and evolution of the atmosphere of Venus. In *Venus* (eds. D. Hunten, L. Colin, T. Donahue, and V. Moroz). University of Arizona Press, Tucson, pp. 1003–1036.

Donahue T. M., Hoffman J. H., Hodges R. R., Jr., and Watson A. J. (1982) Venus was wet: a measurement of the ratio of deuterium to hydrogen. *Science* **216**, 630–633.

Dreibus G. and Wänke H. (1989) Supply and loss of volatile constituents during accretion of terrestrial planets. In *Origin and Evolution of Planetary and Satellite Atmospheres* (eds. S. K. Atreya, J. B. Pollack, and M. S. Matthews). University of Arizona Press, Tucson, pp. 268–288.

Dymond J. and Hogan L. (1973) Noble gas abundance patterns in deep sea basalts—Primordial gases from the mantle. *Earth Planet. Sci. Lett.* **20**, 131–139.

Eikenberg J., Signer P., and Wieler R. (1993) U–Xe, U–Kr, and U–Pb systematics for dating uranium minerals and investigations of the production of nucleogenic neon and argon. *Geochim. Cosmochim. Acta* **57**, 1053–1069.

England P. C. and Thompson A. B. (1984) Pressure–temperature–time paths of regional metamorphism: I. Heat transfer during evolution of regions of thickened continental crust. *J. Petrol.* **25**, 894–928.

Etheridge M. A., Wall V. J., Cox S. F., and Vernon R. H. (1984) High fluid pressures during regional metamorphism and deformation: implications for mass transport and deformation mechanisms. *J. Geophys. Res.* **89**, 4344–4358.

Farley K. A., Basu A. R., and Craig H. (1993) He, Sr, and Nd isotopic variations in lavas from the Juan Fernandez archipelago. *Contrib. Mineral. Petrol.* **115**, 75–87.

Farley K. A., Maier-Reimer E., Schlosser P., and Broecker W. S. (1995) Constraints on mantle He-3 fluxes and deep-sea circulation from an oceanic general circulation model. *J. Geophys. Res.* **100**, 3829–3839.

Fisher D. E. (1978) Terrestrial potassium abundances as limits to models of atmospheric evolution. In *Terrestrial Rare Gases* (ed. M. Ozima). Japan Scientific Societies Press, Tokyo, pp. 173–183.

Fisher D. E. (1985) Noble gases from oceanic island basalts do not require an undepleted mantle source. *Nature* **316**, 716–718.

Galer S. J. G. and O'Nions R. K. (1985) Residence time of thorium, uranium, and lead in the mantle with implications for mantle convection. *Nature* **316**, 778–782.

Gamo T., Ishibashi J.-I., Sakai H., and Tilbrook B. (1987)Methane anomalies in seawater above the Loihi seamount summit area, Hawaii. *Geochim. Cosmochim. Acta* **51**, 2857–2864.

Gast P. W. (1960) Limitations on the composition of the upper mantle. *J. Geophys. Res.* **65**, 1287–1297.

Geiss J., Buehler F., Cerutti H., Eberhardt P., and Filleaux C. H. (1972) Solar wind composition experiments. *Apollo 15 Preliminary Sci. Report,* chap. 15.

Goes S., Spakman W., and Bijwaard H. (1999) A lower mantle source for central European volcanism. *Science* **286**, 1928–1931.

Grady M. M., Wright I. P., Carr L. P., and Pillinger C. T. (1986) Compositional differences in estatite chondrites based on carbon and nitrogen stable isotope measurements. *Geochim. Cosmochim. Acta* **50**, 2799–2813.

Graham D. W. (2002) Noble gases in MORB and OIB:observational constraints for the characterization of mantle source reservoirs. *Rev. Mineral. Geochem.* **47**, 247–318.

Graham D. W., Lupton F., Albarède F., and Condomines M. (1990) Extreme temporal homogeneity of helium isotopes at Piton de la Fournaise, Réunion Island. *Nature* **347**, 545–548.

Grand S. P. (1987) Tomographic inversion for shear velocity beneath the North American plate. *J. Geophys. Res.* **92**, 14065–14090.

Grand S. P. (1994) Mantle shear structure beneath the Americas and surrounding oceans. *J. Geophys. Res.* **99**, 66–78.

Haendel D., Mühle K., Nitzsche H., Stiehl G., and Wand U. (1986) Isotopic variations of the fixed nitrogen in metamorphic rocks. *Geochim. Cosmochim. Acta* **50**, 749–758.

Hagee B., Bernatowicz T. J., Podosek F. A., Johnson M. L., Burnett D. S., and Tatsumoto M. (1990) Actinide abundances in ordinary chondrites. *Geochim. Cosmochim. Acta* **54**, 2847–2858.

Hall A. (1989) Ammonium in spilitized basalts of southwest England and its implications for the recycling of nitrogen. *Geochem. J.* **23**, 19–23.

Halliday A. N. and Porcelli D. (2001) In search of lost planets—the paleocosmochemistry of the inner solar system. *Earth Planet. Sci. Lett.* **192**, 545–559.

Hamano Y. and Ozima M. (1978) Earth-atmosphere evolution model based on Ar isotopic data. In *Terrestrial Rare Gases* (ed. M. Ozima). Japan Scientific Societies Press, Tokyo, pp. 155–171.

Hanes J. A., York D., and Hall C. M. (1985) An $^{40}Ar/^{39}Ar$ geochronological and electron microprobe investigation of an Archaean pyroxenite and its bearing on ancient atmospheric compositions. *Can. J. Earth Sci.* **22**, 947–958.

Hanyu T. and Kaneoka I. (1998) Open system behavior of helium in case of the HIMU source area. *Geophys. Res. Lett.* **25**, 687–690.

Harrison D., Burnard P., and Turner G. (1999) Noble gas behaviour and composition in the mantle: constraints from the Iceland plume. *Earth Planet. Sci. Lett.* **171**, 199–207.

Hart R., Dymond J., and Hogan L. (1979) Preferential formation of the atmosphere-sialic crust system from the upper mantle. *Nature* **278**, 156–159.

Hart S. R. and Staudigel H. (1982) The control of alkalies and uranium in seawater by ocean crust alteration. *Earth Planet. Sci. Lett.* **58**, 202–212.

Hashizume K., Chaussidon M., Marty B., and Robert F. (2000) Solar wind record on the moon: deciphering presolar from planetary nitrogen. *Science* **290**, 1142–1145.

Hauri E. (2002) SIMS analysis of volatiles in silicate glasses: 2. Isotopes and abundances in Hawaiian melt inclusions. *Chem. Geol.* **183**, 115–141.

Heaton T. H. E. (1984) Rates and sources of 4He accumulation in groundwater. *Hydrol. Sci. J.* **29**, 29–47.

Helmberger D. V., Wen L., and Ding X. (1998) Seismic evidence that the source of the Iceland hotspot lies at the core–mantle boundary. *Nature* **396**, 251–255.

Hilton D. R., McMurty G. M., and Kreulen R. (1997) Evidence for extensive degassing of the Hawaiian mantle plume from helium–carbon relationships at Kilauea volcano. *Geophys. Res. Lett.* **24**, 3065–3068.

Hilton D. R., McMurtry G. M., and Goff F. (1998a) Large variations in vent fluid $CO_2/^3$ He ratios signal rapid changes in magma chemistry at Loihi seamount, Hawaii. *Nature* **396**, 359–362.

Hilton D. R., Grönvold K., Sveinbjornsdottir A. E., and Hammerschmidt K. (1998b) Helium isotope evidence for off-axis degassing of the Icelandic hotspot. *Chem. Geol.* **149**, 173–187.

Hilton D. R., Thirlwall M. F., Taylor R. N., Murton B. J., and Nichols A. (2000) Controls on magmatic degassing along the Reykjanes Ridge with implications for the helium paradox. *Earth Planet. Sci. Lett.* **183**, 43–50.

Hilton D. R., Fischer T. P., and Marty B. (2002) Noble gases in subduction zones and volatile recycling. *Rev. Mineral. Geochem.* **47**, 319–370.

Hiyagon H. and Ozima M. (1986) Partition of gases between olivine and basalt melt. *Geochim. Cosmochim. Acta* **50**, 2045–2057.

Hiyagon H., Ozima M., Marty B., Zashu S., and Sakai H. (1992) Noble gases in submarine glasses from mid-oceanic ridges and Loihi Seamount—Constraints on the early history of the Earth. *Geochim. Cosmochim. Acta* **56**, 1301–1316.

Hoefs J. (1969) Carbon. In *Handbook of Geochemistry* (ed.K. H. Wedepohl). Springer, Berlin. Hofmann A. W. (1988) Chemical differentiation of the Earth: the relationship between mantle, continental crust, and oceanic crust. *Earth Planet. Sci. Lett.* **90**, 297–314.

Hoffman J. H., Hodges R. R., Donahue T. M., and McElroy M. B. (1980) Composition of the Venus lower atmosphere from the Pioneer Venus mass spectrometer. *J. Geophys. Res.* **85**, 7882–7890.

Hohenberg C. M., Podosek F. A., and Reynolds J. H. (1967) Xenon–iodine dating: sharp isochronism in chondrites. *Science* **156**, 233–236.

Holloway J. R. (1998) Graphite-melt equilibria during mantle melting: constraints on CO_2 in MORB magmas and the carbon content of the mantle. *Chem. Geol.* **147**, 89–97.

Honda M. and McDougall I. (1993) Solar noble gases in the Earth—the systematics of helium–neon isotopes in mantle-derived samples. *Lithos* **30**, 257–265.

Honda M. and McDougall I. (1998) Primordial helium and neon in the Earth—a speculation on early degassing. *Geophys. Res. Lett.* **25**, 1951–1954.

Honda M., McDougall I., Patterson D. B., Doulgeris A., and Clague D. A. (1991) Possible solar noble-gas component in Hawaiian basalts. *Nature* **349**, 149–151.

Honda M., McDougall I., Patterson D. B., Doulgeris A., and Clague D. A. (1993) Noble gases in submarine pillow basalt glasses from Loihi and Kilauea, Hawaii—a solar component in the Earth. *Geochim. Cosmochim. Acta* **57**, 859–874.

Hudson G. B., Kennedy B. M., Podosek F. A., and Hohenberg C. M. (1989) The early solar system abundance of ^{244}Pu as inferred from the St. Severin chondrite. *Proc. 19th Lunar Planet. Sci. Conf.* 547–557.

Hunt J. M. (1972) Distribution of carbon in crust of Earth. *Bull. Am. Assoc. Petrol. Geol.* **56**, 2273–2277.

Hunten D. M., Pepin R. O., and Walker J. C. G. (1987) Mass fractionation in hydrodynamic escape. *Icarus* **69**, 532–549.

Hurley P. M. (1966) K–Ar dating of sediments. In *Potassium Argon Dating* (eds. O. A. Schaeffer and J. Zähringer). Springer, New York, pp. 134–151.

Hurley P. M. and Rand J. R. (1969) Pre-drift continental nuclei. *Science* **164**, 1229–1242.

Hurley P. M., Brookins D. G., Pinson W. H., Hart S. R., and Fairbairn H. W. (1961) K–Ar age studies of Mississippi and other river sediments. *Geol. Soc. Am. Bull.* **72**, 1807–1816.

Hutchins K. S. and Jakosky B. M. (1996) Evolution of martian atmospheric argon: implications for sources of volatiles. *J. Geophys. Res.* **101**, 14933–14949.

Ito E., Harris D. M., and Anderson A. T., Jr. (1983) Alteration of oceanic crust and geologic cycling of chlorine and water. *Geochim. Cosmochim. Acta* **47**, 1613–1624.

Jambon A. (1994) Earth degassing and large-scale geochemical cycling of volatile elements. *Rev. Mineral.* **30**, 479–517.

Jambon A. and Zimmermann J. L. (1990) Water in oceanic basalts—evidence for dehydration of recycled crust. *Earth Planet. Sci. Lett.* **101**, 323–331.

Jaupart C., Sclater J. G., and Simmons G. (1981) Heat flow studies: constraints on the distribution of uranium, thorium, and potassium in the continental crust. *Earth Planet. Sci. Lett.* **52**, 328–344.

Javoy M. (1998) The birth of the Earth's atmosphere: the behavior and fate of its major elements. *Chem. Geol.* **147**, 11–25.

Javoy M. and Pineau F. (1991) The volatiles record of a "Popping Rock" from the Mid-Atlantic Ridge at 14 °N: chemical and isotopic composition of gas trapped in the vesicles. *Earth Planet. Sci. Lett.* **107**, 598–611.

Javoy M., Pineau F., and Demaiffe Dl. (1984) Nitrogen and carbon isotopic composition in the diamonds of Mbuji Mayi (Zaire). *Earth Planet. Sci. Lett.* **68**, 399–412.

Jochum K. P., Hofmann A. W., Ito E., Seufert H. M., and White W. M. (1983) K, U, and Th in mid-ocean ridge basalt glasses and heat production, K/U and K/Rb in the mantle. *Nature* **306**, 431–436.

Kaula W. M. (1990) Venus: a contrast in evolution to Earth. *Science* **247**, 1191–1196.

Kaula W. M. (1999) Constraints on Venus evolution from radiogenic argon. *Icarus* **139**, 32–39.

Keeling C. D. and Whorf T. P. (2000) Atmospheric CO_2 records from sites in the SIO air sampling network. In *Trends: A Compendium of Data on Global Change.* Carbon Dioxide Information Analysis Center, Oak Ridge National Laboratory, Oak Ridge, TN.

Kellogg L. H. and Wasserburg G. J. (1990) The role of plumes in mantle helium fluxes. *Earth Planet. Sci. Lett.* **99**, 276–289.

Kellogg L. H., Hager B. H., and van der Hilst R. D. (1999) Compositional stratification in the deep mantle. *Science* **283**, 1881–1884.

Kerridge J. F. (1985) Carbon, hydrogen, and nitrogen in carbonaceous chondrites: abundances and isotopic compositions in bulk samples. *Geochim. Cosmochim Acta* **49**, 1707–1714.

Kingsley R. H. and Schilling J.-G. (1995) Carbon in Mid-Atlantic Ridge basalt glasses from 28-degrees-N to 63-degrees-N—evidence for a carbon-enriched azores mantle plume. *Earth Planet. Sci. Lett.* **129**, 31–53.

Korenaga J. and Kelemen P. B. (2000) Major element heterogeneity in the mantle source of the North Atlantic igneous province. *Earth Planet. Sci. Lett.* **184**, 251–268.

Krummenacher D., Merrihue C. M., Pepin R. O., and Reynolds J. H. (1962) Meteoritic krypton and barium versus the general isotopic anomalies in meteoritic xenon. *Geochim. Cosmochim. Acta* **26**, 231–249.

Kunz J. (1999) Is there solar argon in the Earth's mantle? *Nature* **399**, 649–650.

Kunz J., Staudacher T., and Allègre C. J. (1998) Plutonium-fission xenon found in Earth's mantle. *Science* **280**, 877–880.

Kurz M. D., Jenkins W. J., and Hart S. R. (1982) Helium isotopic systematics of oceanic islands and mantle heterogeneity. *Nature* **297**, 43–46.

Kyser T. K. and O'Neil J. R. (1984) Hydrogen isotope systematics of submarine basalts. *Geochim. Cosmochim. Acta* **48**, 2123–2133.

Lécuyer C., Gillet P., and Robert F. (1998) The hydrogen isotope composition of seawater and the global water cycle. *Chem. Geol.* **45**, 249–261.

Lécuyer C., Simon L., and Guyot F. (2000) Comparison of carbon, nitrogen, and water budgets on Venus and the Earth. *Earth Planet. Sci. Lett.* **181**, 33–40.

Li Y.-H. (1972) Geochemical mass balance among lithosphere, hydrosphere, and atmosphere. *Am. J. Sci.* **272**, 119–137.

Lupton J. E. (1996) A far-field hydrothermal plume from Loihi Seamount. *Science* **272**, 976–979.

Lupton J. E. and Craig H. (1975) Excess ^{3}He in oceanic basalts, evidence for terrestrial primordial helium. *Earth Planet. Sci. Lett.* **26**, 133–139.

Magaritz M. and Taylor H. P., Jr. (1976) Oxygen, hydrogen, and carbon isotope studies of the Franciscan Formation Coast Ranges, California. *Geochim. Cosmochim. Acta* **40**, 215–234.

Mahaffy P. R., Donahue T. M., Atreya S. K., Owen T. C., and Niemann H. B. (1998) Galileo probe measurements of D/H and ^{3}He/^{4}He in Jupiter's atmosphere. *Space Sci. Rev.* **84**, 251–263.

Mamyrin B. A., Tolstikhin I. N., Anufriev G. S., and Kamensky I. L. (1969) Anomalous isotopic composition of helium in volcanic gases. *Dokl. Akad. Nauk SSSR* **184**, 1197–1199 (in Russian).

Marti K. and Mathew K. J. (1998) Noble-gas components in planetary atmospheres and interiors in relation to solar wind and meteorites. *Proc. Indian Acad. Sci. (Earth Planet. Sci.)* **107**, 425–431.

Martel D. J., Deak J., Dovenyi P., Horvath F., O'Nions R. K., Oxburgh E. R., Stegna L., and Stute M. (1989) Leakage of helium from the Pannonian Basin. *Nature* **432**, 908–912.

Martin J.-M. and Meybeck M. (1979) Elemental mass-balance of material carried by major world rivers. *Mar. Chem. 7,* 173–206.

Marty B. and Dauphas N. (2003) The nitrogen record of crust–mantle interaction and mantle convection from Archean to present. *Earth Planet. Sci. Lett.* **206**, 397–410.

Marty B. and Humbert F. (1997) Nitrogen and argon isotopes in oceanic basalts. *Earth Planet. Sci. Lett.* **152**, 101–112.

Marty B. and Jambon A. (1987) C/ ^{3}He in volatile fluxes from the solid Earth—implications for carbon geodynamics. *Earth Planet. Sci. Lett.* **83**, 16–26.

Marty B. and Marti K. (2002) Signatures of early differentiation on Mars. *Earth Planet. Sci. Lett.* **196**, 251–263.

Marty B. and Zimmermann L. (1999) Volatiles (He, C, N, Ar) in mid-ocean ridge basalts: assessment of shallow-level fractionation and characterization of source composition. *Geochim. Cosmochim. Acta* **63**, 3619–3633.

Marty B., Tolstikhin I., Kamensky I. L., Nivin V., Balagans-kaya E., and Zimmermann J.-L. (1998) Plume-derived rare gases in 380 Ma carbonatities from the Kola region (Russia) and the argon isotopic composition of the deep mantle. *Earth Planet. Sci. Lett.* **164**, 179–192.

Mathew K. J. and Marti K. (2001) Early evolution of martian volatiles: nitrogen and noble gas components in ALH84001 and Chassigny. *J. Geophys. Res.* **106**, 1401–1422.

Mathew K. J. and Marti K. (2002) Martian atmospheric and interior volatiles in the meteorite Nakhla. *Earth Planet. Sci. Lett.* **199**, 7–20.

Mathew K. J., Kim J. S., and Marti K. (1998) martian atmospheric and indigenous components of xenon and nitrogen in the Shergotty, Nakhla, and Chassigny group meteorites. *Meteorit. Planet. Sci.* **33**, 655–664.

Matsuda J.-I. and Marty B. (1995) The $^{40}Ar/^{36}Ar$ ratio of the undepleted mantle: a re-evaluation. *Geophys. Res. Lett.* ***22***, 1937–1940.

Matsuda J.-I. and Nagao K. (1986) Noble gas abundances in a deep-sea core from eastern equatorial Pacific. *Geochem. J.* **20**, 71–80.

Mattey D. P., Carr R. H., Wright I. P., and Pillinger C. T. (1984) Carbon isotopes in submarine basalts. *Earth Planet. Sci. Lett.* **70**, 196–206.

Mazor E. (1995) Stagnant aquifer concept: 1. Large scale artesian systems—Great Artesian Basin, Australia. *J. Hydrol.* **173**, 219–240.

McCulloch M. T. (1993) The role of subducted slabs in an evolving Earth. *Earth Planet. Sci. Lett.* **115**, 89–100.

McDonough W. F. and Sun S. S. (1995) The composition of the Earth. *Chem. Geol.* **120**, 223–253.

McDougall I. and Harrison T. M. (1999) *Geochronology and Thermochronology by the $^{40}Ar/^{39}Ar$ Method.* Oxford University Press, Oxford. McLennan S. M. and Taylor S. R. (1996) Heat flow and the chemical composition of continental crust. *J. Geol.* **104**, 369–377.

Milliman J. D. and Meade R. H. (1983) World-wide delivery of river sediment to the oceans. *J. Geol.* **91**, 1–21.

Moreira M. and Sarda P. (2000) Noble gas constraints on degassing processes. *Earth Planet. Sci. Lett.* **176**, 375–386.

Moreira M., Kunz J., and Allègre C. J. (1998) Rare gas systematics in Popping Rock: isotopic and elemental compositions in the upper mantle. *Science* **279**, 1178–1181.

Morgan P. (1984) The thermal structure and thermal evolution of the continental lithosphere. *Phys. Chem. Earth* **15**, 107–193.

Morgan J. P. (1998) Thermal and rare gas evolution of the mantle. *Chem. Geol.* **145**, 431–445.

Morrison P. and Pine J. (1955) Radiogenic origin of the helium isotopes in rock. *Ann. NY Acad. Sci.* **62**, 69–92.

Muramatsu Y. W. and Wedepohl K. H. (1998) The distribution of iodine in the Earth's crust. *Chem. Geol.* **147**, 201–216.

Musselwhite D. S. and Drake M. J. (2000) Early outgassing of Mars: implications from experimentally determined solubility of iodine in silicate magmas. *Icarus* **148**, 160–175.

Musselwhite D. S., Drake M. J., and Swindle T. D. (1991) Early outgassing of Mars supported by differential water solubility of iodine and xenon. *Nature* **352**, 697–699.

Namiki N. and Solomon S. C. (1998) Volcanic degassing of argon and helium and the history of crustal production on Venus. *J. Geophys. Res.* **103**, 3655–3677.

Nier A. O. and McElroy M. B. (1977) Composition and structure of Mars' upper atmosphere: results from the neutral mass spectrometers on Viking 1 and 2. *J. Geophys. Res.* **82**, 4341–4349.

Nyblade A. A. and Pollack H. N. (1993) A global analysis of heat flow from Precambrian terrains: implications for the thermal structure of Archean and Proterozoic lithosphere. *J. Geophys. Res.* ***98***, 12207–12218.

O'Nions R. K. and Oxburgh E. R. (1983) Heat and helium in the Earth. *Nature* **306**, 429–431.

O'Nions R. K. and Oxburgh E. R. (1988) Helium, volatile fluxes and the development of continental crust. *Earth Planet. Sci. Lett.* **90**, 331–347.

O'Nions R. K. and Tolstikhin I. N. (1994) Behaviour and residence times of lithophile and rare gas tracers in the upper mantle. *Earth Planet. Sci. Lett.* **124**, 131–138.

O'Nions R. K., Carter S. R., Evensen N. M., and Hamilton P. J. (1981) Upper mantle geochemistry. In *The Sea* (ed. C. Emiliani). Wiley, vol. 7, pp. 49–71.

Ott U. (1988) Noble gases in SNC meteorites: Shergotty, Nakhla, Chassigny. *Geochim. Cosmochim. Acta* **52**, 1937–1948.

Owen T., Biemann K., Rushneck D. R., Biller J. E., Howarth D. W., and Lafleur A. L. (1977) The composition of the atmosphere at the surface of Mars. *J. Geophys. Res.* **82**, 4635–4639.

Owen T., Maillard J. P., Debergh C., and Lutz B. (1988) Deuterium on Mars: the abundance of HDO and the value of D/H. *Science* **240**, 1767–1770.

Owen T., Bar Nun A., and Kleinfeld I. (1992) Possible cometary origin of heavy noble gases in the atmospheres of Venus, Earth, and Mars. *Nature* **358**, 43–46.

Owen T., Mahaffay P. R., Niemann H. B., Atreya S., and Wong M. (2001) Protosolar nitrogen. *Astrophys. J.* **553**, L77–L79.

Ozima M. (1973) Was the evolution of the atmosphere continuous or catastrophic? *Nature Phys. Sci.* **246**, 41–42.

Ozima M. (1975) Ar isotopes and Earth-atmosphere evolution models. *Geochim. Cosmochim. Acta* **39**, 1127–1134.

Ozima M. and Kudo K. (1972) Excess argon in submarine basalts and an Earth-atmosphere evolution model. *Nature Phys. Sci.* **239**, 23–24.

Ozima M. and Podosek F. A. (1983) *Noble Gas Geochemistry.* Cambridge University Press, Cambridge.

Ozima M. and Podosek F. A. (2001) *Noble Gas Geochemistry,* 2nd edn. Cambridge University Press, Cambridge. Ozima M., Podozek F. A., and Igarashi G. (1985) Terrestrial xenon isotope constraints on the early history of the Earth. *Nature* **315**, 471–474.

Parsons B. (1981) The rates of plate creation and consumption. *Geophys. J. Roy. Astron. Soc.* **67**, 437–448.

Patterson D. B., Honda M., and McDougall I. (1990) Atmospheric contamination: a possible source for heavy noble gases basalts from Loihi Seamount, Hawaii. *Geophys. Res. Lett.* **17**, 705–708.

Pepin R. O. (1991) On the origin and early evolution of terrestrial planet atmospheres and meteoritic volatiles. *Icarus* **92**, 1–79.

Pepin R. O. (1998) Isotopic evidence for a solar argon component in the Earths mantle. *Nature* **394**, 664–667.

Pepin R. O. (2000) On the isotopic composition of primordial xenon in terrestrial planet atmospheres. *Space Sci. Rev.* **92**, 371–395.

Pepin R. O. and Carr M. (1992) Major issues and outstanding questions. In *Mars* (eds. H. H. Kieffer, B. M. Jakosky, C. W. Snyder, and M. S. Matthews). University of Arizona Press,Tucson, pp. 120–143.

Pepin R. O. and Phinney D. (1976) The formation interval of the Earth. *Lunar Sci.* **VII**, 682–684.

Pepin R. O. and Porcelli D. (2002) Origin of noble gases in the terrestrial planets. *Rev. Mineral. Geochem.* **47**, 191–246.

Peters K. E., Sweeney R. E., and Kaplan I. R. (1978) Correlation of carbon and nitrogen stable isotope ratios in sedimentary organic matter. *Limnol. Oceanogr.* **23**, 598–604.

Pfennig G., Klewe-Nebenius H., and Seelann-Eggebert W. (1998) *Karlsruhe Chart of the Nuclide,* 6th edn. (revised reprint). Institut für Instrumentelle Analytik, Karlsruhe. Phinney D., Tennyson J., and Frick U. (1978) Xenon in CO_2 well gas revisited. *J. Geophys. Res.* **83**, 2313–2319.

Pineau F. and Javoy M. (1994) Strong degassing at ridge crests: the behaviour of dissolved carbon and water in basalt glasses at 14 °N, Mid-Atlantic Ridge. *Earth Planet. Sci. Lett.* **123**, 179–198.

Podosek F. A., Honda M., and Ozima M. (1980) Sedimentary noble gases. *Geochim. Cosmochim. Acta* **44**, 1875–1884.

Pollack H. N., Hurter S. J., and Johnson J. R. (1993) Heat flow from the Earth's interior: analysis of the global data set. *Rev. Geophys.* **31**, 267–280.

Porcelli D. and Ballentine B. J. (2002) Models for the distribution of terrestrial noble gases and evolution of the atmosphere. *Rev. Mineral. Geochem.* **47**, 411–480.

Porcelli D. and Halliday A. N. (2001) The core as a possible source of mantle helium. *Earth Planet. Sci. Lett.* **192**, 45–56.

Porcelli D. and Pepin R. O. (2000) Rare gas constraints on early Earth history. In *Origin of the Earth and Moon* (eds. R. M. Canup and K. Righter). University of Arizona Press, Tucson, pp. 435–458.

Porcelli D. and Wasserburg G. J. (1995a) Mass transfer of xenon through a steady-state upper mantle. *Geochim. Cosmochim. Acta* **59**, 1991–2007.

Porcelli D. and Wasserburg G. J. (1995b) Mass transfer of helium, neon, argon, and xenon through a steady-state upper mantle. *Geochim. Cosmochim. Acta* **59**, 4921–4937.

Porcelli D., Woolum D., and Cassen P. (2001) Deep Earth rare gases: initial inventories, capture from the solar nebula, and losses during Moon formation. *Earth Planet. Sci. Lett.* **193**, 237–251.

Porcelli D., Ballentine C. J., and Wieler R. (2002) An introduction to noble gas geochemistry and cosmochemistry. *Rev. Mineral. Geochem.* **47**, 1–18.

Porcelli D., Pepin R. O., Ballentine C. J., and Halliday A. (2003) Xe and degassing of the Earth (in preparation). Poreda R. J. and Farley K. A. (1992) Rare gases in Samoan xenoliths. *Earth Planet. Sci. Lett.* **113**, 129–144.

Poreda R., Schilling J. G., and Craig H. (1986) Helium and hydrogen isotopes in ocean ridge basalts north and south of Iceland. *Earth Planet. Sci. Lett.* 78, 1–17.

Poreda R., Craig H., Arnorsson S., and Welhan J. A. (1992) Helium isotopes in Icelandic geothermal systems: 1. He, gas chemistry, and ^{13}C relations. *Geochim. Cosmochim.Acta* **56**, 4221–4228.

Ragettli R. A., Hebeda E. H., Signer P., and Wieler R. (1994) Uranium–xenon chronology: precise determiantion of $\alpha_{sf}{}^{*136}Y_{sf}$ for spontaneous fission of U. *Earth Planet. Sci. Lett.* **128**, 653–670.

Reymer A. and Schubert G. (1984) Phanerozoic addition rates to the continental crust. *Tectonics* **3**, 63–77.

Rison W. and Craig H. (1983) Helium isotopes and mantle evolatiles in Loihi Seamount and Hawaiian Island basalts and xenoliths. *Earth Planet. Sci. Lett.* **66**, 407–426.

Rocholl A. and Jochum K. P. (1993) Th, U, and other trace elements in carbonaceous chondrites—implications for the terrestrial and solar system Th/U ratios. *Earth Planet. Sci. Lett.* **117**, 265–278.

Ronov A. B. and Yaroshevsky A. A. (1969) Chemical composition of the Earth's crust. *Am. Geophys. Union Geophys. Mon. Ser.* **13**, 37–57.

Ronov A. B. and Yaroshevsky A. A. (1976) A new model for the chemical structure of the Earth's crust. *Geochem. Int.* **13**, 89–121.

Rubey W. W. (1951) Geological history of seawater. *Bull. Geol. Soc. Am.* **62**, 1111–1148.

Rudnick R. and Fountain D. M. (1995) Nature and composition of the continental crust: a lower crustal perspective. *Rev. Geophys.* **33**, 267–309.

Rudnick R. L., McDonough W. F., and O'Connell R. J. (1998) Thermal structure, thickness, and composition of continental lithosphere. *Chem. Geol.* **145**, 395–412.

Russell S. A., Lay T., and Garnero E. J. (1998) Seismic evidence for small-scale dynamics in the lowermost mantle at the root of the Hawaiian hotspot. *Nature* **369**, 225–258.

Sano Y. and Williams S. (1996) Fluxes of mantle and subducted carbon along convergent plate boundaries. *Geophys. Res. Lett.* **23**, 2746–2752.

Sano Y., Takahata N., Nishio Y., and Marty B. (1998) Nitrogen recycling in subduction zones. *Geophys. Res. Lett.* **25**, 2289–2292.

Sarda P., Staudacher T., and Allègre C. J. (1985) $^{40}Ar/^{36}Ar$ in MORB glasses: constraints on atmosphere and mantle evolution. *Earth Planet. Sci. Lett.* **72**, 357–375.

Sarda P., Staudacher T., and Allègre C. J. (1988) Neon isotopes in submarine basalts. *Earth Planet. Sci. Lett.* **91**, 73–88.

Sarda P., Moreira M., Staudacher T., Schilling J. G., and Allègre C. J. (2000) Rare gas systematics on the southernmost Mid-Atlantic Ridge: constraints on the lower mantle and the Dupal source. *J. Geophys. Res.* **105**, 5973–5996.

Sasaki S. and Nakazawa K. (1988) Origin and isotopic fractionation of terrestrial Xe: hydrodynamic fractionation during escape of the primordial H_2–He atmosphere. *Earth Planet. Sci. Lett.* **89**, 323–334.

Sasaki S. and Tajika E. (1995) Degassing history and evolution of volcanic activities of terrestrial planets based on radiogenic noble gas degassing models. In *Volatiles in the Earth and Solar System,* AIP Conf. Proc. 341 (ed. K. A. Farley). AIP Press, New York, pp. 186–199.

Schilling J.-G., Unni C. K., and Bender M. L. (1978) Origin of chlorine and bromine in the oceans. *Nature* **273**, 631–636.

Schwartzman D. W. (1973) Argon degassing and the origin of the sialic crust. *Geochim. Cosmochim. Acta* **37**, 2479–2495.

Shen Y., Solomon S. C., Bjarnason I. T., and Wolfe C. J. (1998) Seismic evidence for a lower-mantle origin of the Iceland plume. *Nature* **395**, 62–65.

Sleep N. H. and Zahnle K. (2001) Carbon dioxide cycling and implications for climate on ancient Earth. *J. Geophys. Res.* **106**, 1373–1399.

Sobolev A. V. and Chaussidon M. (1996) H_2O concentrations in primary melts from supra-subduction zones and mid-ocean ridges: implications for H_2O storage and recycling in the mantle. *Earth Planet. Sci. Lett.* **137**, 45–55.

Sobolev A. V., Hofmann A. W., and Nikogosian I. K. (2000) Recycled oceanic crust observed in "ghost plagioclase" within the source of Mauna Loa lavas. *Nature* **404**, 986–990.

Staudacher T. (1987) Upper mantle origin for Harding County well gases. *Nature* **325**, 605–607.

Staudacher T. and Allègre C. J. (1982) Terrestrial xenology. *Earth Planet. Sci. Lett.* **60**, 389–406.

Staudacher T. and Allègre C. J. (1988) Recycling of oceanic crust and sediments: the noble gas subduction barrier. *Earth Planet. Sci. Lett.* **89**, 173–183.

Staudacher T., Sarda P., and Allègre C. J. (1989) Noble gases in basalt glasses from a Mid-Atlantic Ridge topographic high at 14° N: geodynamic consequences. *Earth Planet. Sci. Lett.* **96**, 119–133.

Stute M., Sonntag C., Deak J., and Schlosser P. (1992) Helium in deep circulating groundwater in the Great Hungarian Plain—Glow dynamics and crustal and mantle helium fluxes. *Geochim. Cosmochim. Acta* **56**, 2051–2067.

Suess H. E. (1949) The abundance of noble gases in the Earth and the cosmos. *J. Geol.* **57**, 600–607 (in German).

Swindle T. D. (2002) Martian noble gases. *Rev. Mineral. Geochem.* **47**, 171–190.

Swindle T. D. and Jones J. H. (1997) The xenon isotopic composition of the primordial martian atmosphere: contributions from solar and fission components. *J. Geophys. Res.* **102**, 1671–1678.

Tajika E. and Sasaki S. (1996) Magma generation on Mars constrained from an ^{40}Ar degassing model. *J. Geophys. Res.* **101**, 7543–7554.

Taylor S. R. and McLennan S. M. (1985) *The Continental Crust: Its Composition and Evolution.* Blackwell, Oxford.

Taylor H. P., Jr. and Sheppard S. M. F. (1986) Igneous rocks: I. Processes of isotopic fractionation and isotope systematics. *Rev. Mineral.* **16**, 227–271.

Thompson A. B. (1992) Water in the Earth's upper mantle. *Nature* **358**, 295–302.

Thomsen L. (1980) ^{129}Xe on the out gassing of the atmosphere. *J. Geophys. Res.* **85**, 4374–4378.

Tolstikhin I. N. (1975) Helium isotopes in the Earth's interior and in the atmosphere: a degassing model of the Earth. *Earth Planet. Sci. Lett.* **26**, 88–96.

Torgersen T. and Clarke W. B. (1985) Helium accumulation in groundwater: I. An evaluation of sources and the continental flux of crustal ^{4}He in the Great Artesian Basin, Australia. *Geochim. Cosmochim. Acta* **49**, 1211–1218.

Torgersen T. and Ivey G. N. (1985) Helium accumulation in groundwater: II. A model for the accumulation of crustal ^{4}He degassing flux. *Geochim. Cosmochim. Acta* **49**, 2445–2452.

Torgersen T., Kennedy B. M., Hiyagon H., Chiou K. Y., Reynolds J. H., and Clarke W. B. (1989) Argon accumulation and the crustal degassing flux of ^{40}Ar in the Great Artesian Basin, Australia. *Earth Planet. Sci. Lett.* **92**, 43–56.

Trieloff M., Kunz J., Clague D. A., Harrison D., and Allègre C. J. (2000) The nature of pristine noble gases in mantle plumes. *Science* **288**, 1036–1038.

Trull T., Nadeau S., Pineau F., Polvé M., and Javoy M. (1993) C–He systematics in hotspot xenoliths: implications for mantle carbon contents and carbon recycling. *Earth Planet.Sci. Lett.* **118**, 43–64.

Turcotte D. L. and Schubert G. (1982) *Geodynamics.* Wiley, New York. Turcotte D. L. and Schubert G. (1988) Tectonic implications of radiogenic noble gases in planetary atmospheres. *Icarus* **74**, 36–46.

Turekian K. K. (1959) The terrestrial economy of helium and argon. *Geochim. Cosmochim. Acta* **17**, 37–43.

Turekian K. K. (1990) The parameters controlling planetary degassing based on ^{40}Ar systematics. In *From Mantle to Meteorites* (eds. K. Gopolan, V. K. Gaur, B. L. K.Somayajulu, and J. D. MacDougall). Indian Academy of Sciences, Bangalore, pp. 147–152.

Turekian K. K. and Clark S. P., Jr. (1969) Inhomogeneous accumulation of the Earth from the primitive solar nebula. *Earth Planet. Sci. Lett.* **6,** 346–348.

Turekian K. K. and Clark S. P., Jr. (1975) The non-homogeneous accumulation model for terrestrial planet formation and the consequences for the atmosphere of Venus. *J. Atmos. Sci.* **32,** 1257–1261.

Turner G. (1989) The outgassing history of the Earth's atmosphere. *J. Geol. Soc. London* **146**, 147–154.

Valbracht P. J., Staudacher T., Malahoff A., and Allègre C. J. (1997) Noble gas systematics of deep rift zone glasses fromLoihi Seamount, Hawaii. *Earth Planet. Sci. Lett.* **150**, 399–411.

van der Hilst R. D., Widiyantoro S., and Engdahl E. R. (1997) Evidence for deep mantle circulation from global tomography. *Nature* **386**, 578–584.

van Keken P. E., Ballentine C. J., and Porcelli D. (2001) A dynamical investigation of the heat and helium imbalance. *Earth Planet. Sci. Lett.* **188**, 421–443.

Veizer J. and Jansen S. L. (1979) Basement and sedimentary recycling and continental evolution. *Am. J. Sci.* **87**, 341–370.

Veizer J. and Jansen S. L. (1985) Basement and sedimentary recycling: 2. Time dimension to global tectonics. *J. Geol.* **93**, 625–643.

Vidale J. E., Schubert G., and Earle P. S. (2001) Unsuccessful initial search for a mid-mantle chemical boundary with seismic arrays. *Geophys. Res. Lett.* **28**, 859–862.

Volkov V. P. and Frenkel M. Y. (1993) The modelling of Venus degassing in terms of K–Ar system. *Earth Moon Planets* **62**, 117–129.

Von Huene R. and Scholl D. W. (1991) Observations at convergent margins concerning sediment subduction,

subduction erosion, and the growth of continental crust. *Rev. Geophys.* **29**, 279–316.

von Zahn U., Kumar S., Niemann H., and Prinn R. (1983) Composition of the Venus atmosphere. In *Venus* (eds. D. Hunten, L. Colin, T. Donahue, and V. Moroz). University of Arizona Press, Tucson, pp. 299–430.

Wänke H. and Dreibus G. (1988) Chemical composition and accretion history of terrestrial planets. *Phil. Trans. Roy. Soc. London* **A325**, 545–557.

Wänke H., Dreibus G., and Jagoutz E. (1984) Mantle chemistry and accretion history of the Earth. In *Archaean Geochemistry* (eds. A. Kröner, G. N. Hanson, and A. M. Goodwin). Springer, Berlin, pp. 1–24.

Weaver B. L. and Tarney J. (1984) Empirical approach to estimating the composition of the continental crust. *Nature* **310**, 575–577.

Weaver C. E. and Wampler J. M. (1970) K, Ar, Illite burial. *Geol. Soc. Am. Bull.* **81**, 3423–3430.

Wedepohl K. H. (1995) The composition of the continental crust. *Geochim. Cosmochim. Acta* **59**, 1217–1232.

Wetherill G. (1975) Radiometric chronology of the early solar system. *Ann. Rev. Nuclear Sci.* **25**, 283–328.

Wollast R. and Mackenzie F. T. (1983) The global cycle of silica. In *Silicon Geochemistry and Biogeochemistry* (ed. S. R. Aston). Academic Press, New York, pp. 39–76.

Wood B. J. (1993) Carbon in the core. *Earth Planet. Sci. Lett.* **117**, 593–607.

Zhao D. (2001) Seismic structure and origin of hotspots and mantle plumes. *Earth Planet. Sci. Lett.* **192**, 251–265.

Zhang Y. and Zindler A. (1993) Distribution and evolution of carbon and nitrogen in Earth. *Earth Planet. Sci. Lett.* **117**, 331–345.

Readings from the Treatise on Geochemistry
ISBN: 978-0-12-381391-6

pp. 197–234

7
Soil Formation

R. Amundson
University of California, Berkeley, CA, USA

. . .that the Earth has not always been here—that it came into being at a finite point in the past and that everything here, from the birds and fishes to the loamy soil underfoot, was once part of a star. I found this amazing, and still do.

Timothy Ferris (1998)

7.1 INTRODUCTION

Soil is the biogeochemically altered material that lies at the interface between the lithosphere and the atmosphere. *Pedology* is the branch of the natural sciences that concerns itself, in part, with the biogeochemical processes that form and distribute soil across the globe. Pedology originated during the scientific renaissance of the nineteenth century as a result of conceptual breakthroughs by the Russian scientist Vassali Dochuchaev (Krupenikov, 1992; Vil'yams, 1967) and conceptual and administrative efforts by the American scientist Eugene Hilgard (Jenny, 1961; Amundson and Yaalon, 1995).

Soil is the object of study in pedology, and while the science of pedology has a definition that commands some general agreement, there is no precise definition for soil, nor is there likely ever to be one. The reason for this paradox is that soil is a part of a continuum of materials at the Earth's surface (Jenny, 1941). At the soil's base, the exact line of demarcation between "soil" and "nonsoil" will forever elude general agreement, and horizontal changes in soil properties may occur so gradually that similar problems exist in delineating the boundary between one soil "type" and another. The scientific path out of this conundrum is to divide the soil continuum, albeit arbitrarily, into *systems* that suit the need of the investigator. Soil systems are necessarily open to their surroundings, and through them pass matter and energy which measurably alter the properties of the system over timescales from seconds to millennia. It was the recognition by Dokuchaev (1880), and later the American scientist Hans Jenny (1941), that the properties of the soil system are controlled by *state factors* that ultimately formed the framework of the fundamental paradigm of pedology.

The purpose of this chapter is to present an abridged overview of the factors and processes that control soil formation, and to provide, where possible, some general statements of soil formation processes that apply broadly and commonly.

7.2 FACTORS OF SOIL FORMATION

Jenny (1941) applied principles from the physical sciences to the study of soil formation. Briefly, Jenny recognized that soil systems (or if the aboveground flora and fauna are considered, *ecosystems*) exchange mass and energy with their surroundings and that their properties can be defined by a limited set of *independent variables*. From comparisons with other sciences, Jenny's *state factor model* of soil formation states that

$$\underbrace{\text{Soils/ecosystems}}_{\text{dependent variables}} = f\underbrace{\begin{pmatrix}\text{initial state of system,}\\ \text{surrounding environment,}\\ \text{elapsed time}\end{pmatrix}}_{\text{independent variables}} \quad (1)$$

From field observations and the conceptual work of Dokuchaev, a set of more specific environmental factors have been identified which encompass the controls listed above:

$$\underbrace{\text{Soils/ecosystems}}$$
$$= f(\underbrace{\text{climate, organisms,}}_{\text{surrounding environment}}$$
$$\underbrace{\text{topography, parent material,}}_{\text{initial state of system}} \text{time,}\ldots) \quad (2)$$

These so-called "state factors of soil formation" have the following important characteristics: (i) they are independent of the system being studied and (ii) in many parts of the Earth, the state factors vary independently of each other (though, of course, not always). As a result, through judicious site (system) selection, the influence of a single factor can be observed and quantified in nature.

Table 1 provides a brief definition of the state factors of soil formation. A field study designed to observe the influence of one state factor on soil properties or processes is referred to as a *sequence*, e.g., a series of sites which have similar state factor values except climate is referred to as a *climosequence*. Similar sequences can, and have been, established to examine the effect of other state factors on soils. An excellent review of soil state factor studies is presented by Birkeland (1999). An informative set of papers discussing the impact of Jenny's state factor model on advances in pedology, geology, ecology, and related sciences is presented in Amundson *et al.* (1994a,b).

The state factor approach to studying soil formation has been, and continues to be, a powerful quantitative means of linking soil properties to important variables (Amundson and Jenny, 1997). As an example, possibly the best characterized soil versus factor relationship is the relationship of soil organic carbon and nitrogen storage to climate (mean annual temperature and precipitation) (Figure 1). The pattern—increasing carbon storage with decreasing temperature and increasing precipitation—illustrated in Figure 1 is the result of nearly six decades of work, and is based on thousands of soil observations (Miller *et al.*, 2002). This climatic relationship is important in global change research and in predicting the response of soil carbon storage to climate change (Schlesinger and Andrews, 2000). However, the relationship, no matter how valid, provides no insight into the rates at which soils achieve their carbon storage, nor the mechanisms involved in the accumulation. Thus, other approaches, again amenable to systems studies, have been applied in pedology to quantify soil formation. These are discussed in later sections.

Table 1 The major state factors of soil and ecosystem formation, and a brief outline of their characteristics.

State factor	*Definition and characteristics*
Climate	Regional climate commonly characterized by mean annual temperature and precipitation
Organisms	Potential biotic flux into system (as opposed to what is present at any time)
Topography	Slope, aspect, and landscape configuration at time $t=0$
Parent material	Chemical and physical characteristics of soil system at $t=0$
Time	Elapsed time since system was formed or rejuvenated
Humans	A special biotic factor due to magnitude of human alteration of Earth's and humans' possession of variable cultural practices and attitudes that alter landscapes

Sources: Jenny (1941) and Amundson and Jenny (1997).

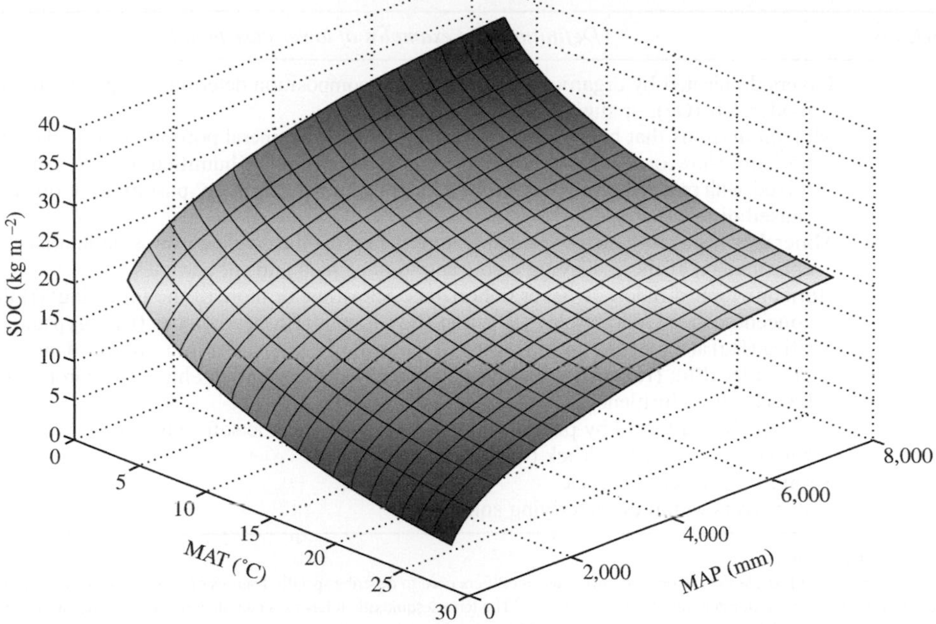

Figure 1 The distribution of global soil C in relation to variations in mean average temperature (MAT) and precipitaiton (MAP). The curve is derived from a multiple regression model of published soil C data versus climate (Amundson, 2001) (reproduced by permission of Annual Reviews from *Ann. Rev. Earth Planet. Sci.* **2001**, *29*, 535–562).

7.3 SOIL MORPHOLOGY

A trend in present-day pedology is to incorporate ever more sophisticated chemical and mathematical tools into our understanding of soil and their formation. Yet, an examination of soils *in situ* is required, in order to develop the appropriate models or to even logically collect samples for study.

Soil *profiles* are two-dimensional, vertical exposures of the layering of soils. The net result of the transport of matter and energy is a vertical differentiation of visible, distinctive layers called soil *horizons*. Soil horizons reflect the fact that soil formation is a depth-dependent process. They are layers that are readily identified by visual and tactile procedures (field based) that have been developed over the years (Soil Survey Staff, 1993). A nomenclature has developed over the past century, first started by the pioneering Russian scientists in the nineteenth century, that involves the "naming"of soil horizons on the basis of *how they differ from the starting parent material*. Therefore, horizon naming requires data acquisition and hypothesis development. Soil horizon names are commonly assigned from field-based data, and may ultimately be modified as a result of subsequent laboratory investigations.

The present US horizon nomenclature has two components: (i) an upper case, "master horizon" symbol and (ii) a lower case "modifier" that provides more information on the horizon characteristics or the processes that formed it. Tables 2(a) and (b) provide definitions of both the common master and modifier symbols. The detailed rules for their use can be found in the *Soil Survey Manual* (Soil Survey Staff, 1993).

Most soil process models are (roughly) continuous with depth. However, during the observation of many soil profiles, it is apparent that horizons do not always, or even commonly, grade gradually into one another. Sharp or abrupt horizon boundaries are common in soils around the world. This indicates that our concepts and models of soil formation capture only a part of the long-term trajectory of soil development. Some processes are not continuous with depth (the formation of carbonate horizons for example), while some may be continuous for some time period and then, due to feedbacks, change their character (the formation of clay-rich horizons which, if they reach a critical clay content, restrict further water and clay transport. This causes an abrupt buildup of additional clay at the top of the horizon). In the following sections, the author examines various approaches to understanding soil formation, and examines some of their attributes and limitations.

7.4 MASS BALANCE MODELS OF SOIL FORMATION

Detailed chemical analyses of soils and the interpretation of that data relative to the composition of the parent material have been performed

Table 2(a) A listing, and brief definitions, of the nomenclature used to identify master soil horizons.

Master horizons	*Definition and examples of lower case modifiers*
O	Layers dominated by organic matter. State of decomposition determines type: highly (Oa), moderately (Oe), or slightly (Oi)[a] decomposed
A	Mineral horizons that have formed at the surface of the mineral portion of the soil or below an O horizon. Show one of the following: (i) an accumulation of humified organic matter closely mixed with minerals or (ii) properties resulting from cultivation, pasturing, or other human-caused disturbance (Ap)
E	Mineral horizons in which the main feature is loss of silicate clay, iron, aluminum, or some combination of these, leaving a concentration of sand and silt particles
B	Horizons formed below A, E, or O horizons. Show one or more of the following: (i) illuvial[b] concentration of silicate clay (Bt), iron (Bs), humus (Bh), carbonates (Bk), gypsum (By), or silica (Bq) alone or in combination; (ii) removal of carbonates (Bw); (iii) residual concentration of oxides (Bo); (iv) coatings of sesquioxides[c] that make horizon higher in chroma or redder in hue (Bw); (v) brittleness (Bx); or (vi) gleying[d] (Bg).
C	Horizons little affected by pedogenic processes. May include soft sedimentary material (C) or partially weathered bedrock (Cr)
R	Strongly indurated[e] bedrock
W	Water layers within or underlying soil

Source: Soil Survey Staff (1999).

[a] The symbols in parentheses illustrate the appropriate lower case modifiers used to describe specific features of master horizons. [b] The term illuvial refers to material transported into a horizon from layers above it. [c] The term sesquioxide refers to accumulations of secondary iron and/or aluminum oxides. [d] Gleying is a process of reduction (caused by prolonged high water content and low oxygen concentrations) that results in soil colors characterized by low chromas and gray or blueish hues. [e] The term indurated means strongly consolidated and impenetrable to plant roots.

Table 2(b) Definitions used to identify the subordinate characteristics of soil horizons.

Lower case modifiers of master horizons	*Definitions (relative to soil parent material)*
a	Highly decomposed organic matter (O horizon)
b	Buried soil horizon
c	Concretions or nodules of iron, aluminum, manganese, or titanium
d	Noncemented, root restricting natural or human-made (plow layers, etc.) root restrictive layers
e	Intermediate decomposition of organic matter (O horizon)
f	Indication of presence of permafrost
g	Strong gleying present in the form of reduction or loss of Fe and resulting color changes
h	Accumulation of illuvial complexes of organic matter which coat sand and silt particles
i	Slightly decomposed organic matter (O horizon)
j	Presence of jarosite (iron sulfate mineral) due to oxidation of pyrite in previously reduced soils
k	Accumulation of calcium carbonate due to pedogenic processes
m	Nearly continuously cemented horizons (by various pedogenic minerals)
n	Accumulation of exchangeable sodium
o	Residual accumulation of oxides due to long-term chemical weathering
p	Horizon altered by human-related activities
q	Accumulation of silica (as opal)
r	Partially weathered bedrock
s	Illuvial accumulation of sesquioxides
ss	Presence of features (called slickensides) caused by expansion and contraction of high clay soils
t	Accumulation of silicate clay by weathering and/or illuviation
v	Presence of plinthite (iron rich, reddish soil material)
w	Indicates initial development of oxidized (or other) colors and/or soil structure
x	Indicates horizon of high firmness and brittleness
y	Accumulation of gypsum
z	Accumulation of salts more soluble than gypsum (e.g., Na_2CO_3)

Source: Soil Survey Staff (1999).

since nearly the origins of pedology (Hilgard, 1860). Yet, quantitative estimates of total chemical denudation, and associated physical changes that occur during soil formation, were not rigorously performed until the late 1980s when Brimhall and co-workers (Brimhall and Dietrich, 1987; Brimhall *et al.*, 1991) began applying a mass balance model originally derived for ore body studies to the soil environment. Here the author presents the key components of this model, and reports the results of its application to two issues: (i) the behavior of many of the chemical elements in soil formation and (ii) general trends of soil physical and chemical behavior as a function of time during soil formation.

A representation of a soil system during soil formation is shown in Figure 2. While the figure illustrates a loss of volume during weathering, volumetric increases can also occur, as will be shown later. The basic expression, describing mass gains or losses of a given chemical element (j), in the transition from parent material (p) to soil (s) in terms of volume (V), bulk density (ρ), and chemical composition (C) is

$$m_{j,\text{flux}} = m_{j,\text{s}} - m_{j,\text{p}} \tag{3}$$

where m is the mass of element j added/lost (flux) in the soil (s) or parent material (p). Incorporating volume, density, and concentration (in percent) into the model gives

$$\underbrace{m_{j,\text{flux}}}_{\text{mass of element}(j)\text{ into/out of parent material volume}} = \underbrace{\frac{V_\text{s}\rho_\text{s}C_{j,\text{s}}}{100}}_{\text{mass of element}(j)\text{ in soil volume of interest}} - \underbrace{\frac{V_\text{p}\rho_\text{p}C_{j,\text{p}}}{100}}_{\text{mass of element}(j)\text{ in parent material volume}} \tag{4}$$

Definitions of all terms used in these mass balance equations are given in Table 3. The 100 in the denominator is needed only if concentrations are in percent.

During soil development, volumetric collapse (ΔV, Figure 2) may occur through weathering losses while expansion may occur through biological or physical processes. Volumetric change is defined in terms of strain (ε):

$$\varepsilon_{i,\text{s}} = \frac{\Delta V}{V_\text{p}} = \left(\frac{V_\text{s}}{V_\text{p}} - 1\right) = \left(\frac{\rho_\text{p}C_{i,\text{p}}}{\rho_\text{s}C_{i,\text{s}}} - 1\right) \tag{5}$$

where the subscript i refers to an immobile, index element. Commonly zirconium, titanium, or other members of the titanium or rare earth groups of the periodic table are used as index elements. The

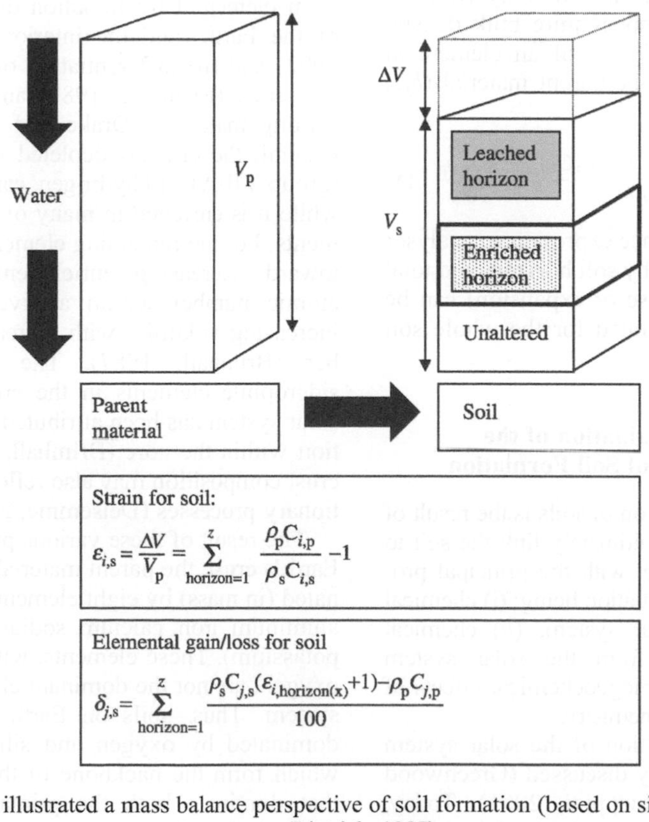

Figure 2 Diagram illustrated a mass balance perspective of soil formation (based on similar figures in Brimhall and Dietrich, 1987).

Table 3 Definition of parameters used in mass balance model.

Parameter	*Definition*
V_p (cm^3)	Volume of parent material
V_s (cm^3)	Volume of soil
ρ_p ($g\,cm^{-3}$)	Parent material bulk density
ρ_s ($g\,cm^{-3}$)	Soil bulk density
$C_{j,p}$ (%,ppm)	Concentration of mobile element *j* in parent material
$C_{j,s}$ (%,ppm)	Concentration of mobile element *j* in soil
$C_{i,p}$ (%,ppm)	Concentration of immobile element *i* in parent material
$C_{i,s}$ (%,ppm)	Concentration of immobile element *i* in soil
$m_{j,\text{flux}}$ ($g\,cm^{-3}$)	Mass of element *j* added or removed via soil formation
$\varepsilon_{i,s}$	Net strain determined using element *i*
τ	Fractional mass gain or loss of element *j* relative to immobile element *i*
$\delta_{j,s}$ ($g\,cm^{-3}$)	Mass gain or loss per unit volume of element *j* relative to immobile element *i*.

Sources: Brimhall and Dietrich (1987) and Brimhall *et al.* (1992).

fractional mass gain or loss of an element *j* relative to the mass in the parent material (τ) is defined by combining Equations (3)–(5):

$$\tau = \frac{m_{j,\text{flux}}}{m_{j,p}} = \left(\frac{\rho_s C_{j,s}}{\rho_p C_{j,p}}(\varepsilon_{i,s}+1)-1\right) \quad (6)$$

Through substitution, Equation (6) reduces to

$$\tau = \frac{R_s}{R_p} - 1 \quad (7)$$

where $R_s = C_{j,s}/C_{i,s}$ and $R_p = C_{j,p}/C_{i,p}$. Thus, τ can be calculated readily from commonly available chemical data and does not require bulk density data. Absolute gains or losses of an element in mass per unit volume of the parent material ($\delta_{j,s}$) can be expressed as

$$\delta_{j,s} = \frac{m_{j,\text{flux}}}{V_p} = \frac{\rho_s C_{j,s}(\varepsilon_{i,s}+1) - \rho_p C_{j,p}}{100} = \frac{\tau C_{j,p}\rho_p}{100} \quad (8)$$

In applying the mass balance expressions, analyses are commonly performed by soil horizon, and total gains or losses (or collapse or expansion) can be plotted by depth or integrated for the whole soil profile (Figure 2).

7.4.1 Mass Balance Evaluation of the Biogeochemistry of Soil Formation

The chemical composition of soils is the result of a series of processes that ultimately link the soil to the history of the universe, with the principal processes of chemical differentiation being: (i) chemical evolution of universe/solar system; (ii) chemical differentiation of Earth from the solar system components; and (iii) the biogeochemical effects of soil formation on crustal chemistry.

The chemical composition of the solar system (Figure 3) has been widely discussed (Greenwood and Earnshaw, 1997; Chiappini, 2001). Today, 99% of the universe is comprised of hydrogen and helium, which were formed during the first few minutes following the big bang. The production of elements of greater atomic number requires a series of nuclear processes that occur during star formation and destruction. Thus, the relative elemental abundance versus atomic number is a function of the age of the universe and the number of cycles of star formation/termination that have occurred (e.g., Allègre, 1992).

The chemical composition of average crustal rock (Taylor and McLennan, 1985; Bowen, 1979) relative to the solar system reveals systematic differences (Brimhall, 1987) (Figure 4) that result from elemental fractionation during: (i) accretion of the Earth (and the interior planets) (Allègre, 1992) and (ii) differentiation of the core, mantle, and crust (Brimhall, 1987) and possibly unique starting materials (Drake and Righter, 2002). In general, the crust is depleted in the noble gases (group VIIIA) and hydrogen, carbon, and nitrogen, while it is enriched in many of the remaining elements. For the remaining elements, there is a trend toward decreasing enrichment with increasing atomic number within a given period, due to increasing volatility with increasing atomic number (Brimhall, 1987). The depletion of the siderophile elements in the crust relative to the solar system has been attributed to their concentration within the core (Brimhall, 1987), though the crust composition may also reflect late-stage accretionary processes (Delsemme, 2001).

The result of these various processes is that the Earth's crust, the parent material for soils, is dominated (in mass) by eight elements (oxygen, silicon, aluminum, iron, calcium, sodium, magnesium, and potassium). These elements, with the exception of oxygen, are not the dominant elements of the solar system. Thus, soils on Earth form in a matrix dominated by oxygen and silicon, the elements which form the backbone of the silicate minerals that dominate both the primary and secondary minerals found in soils.

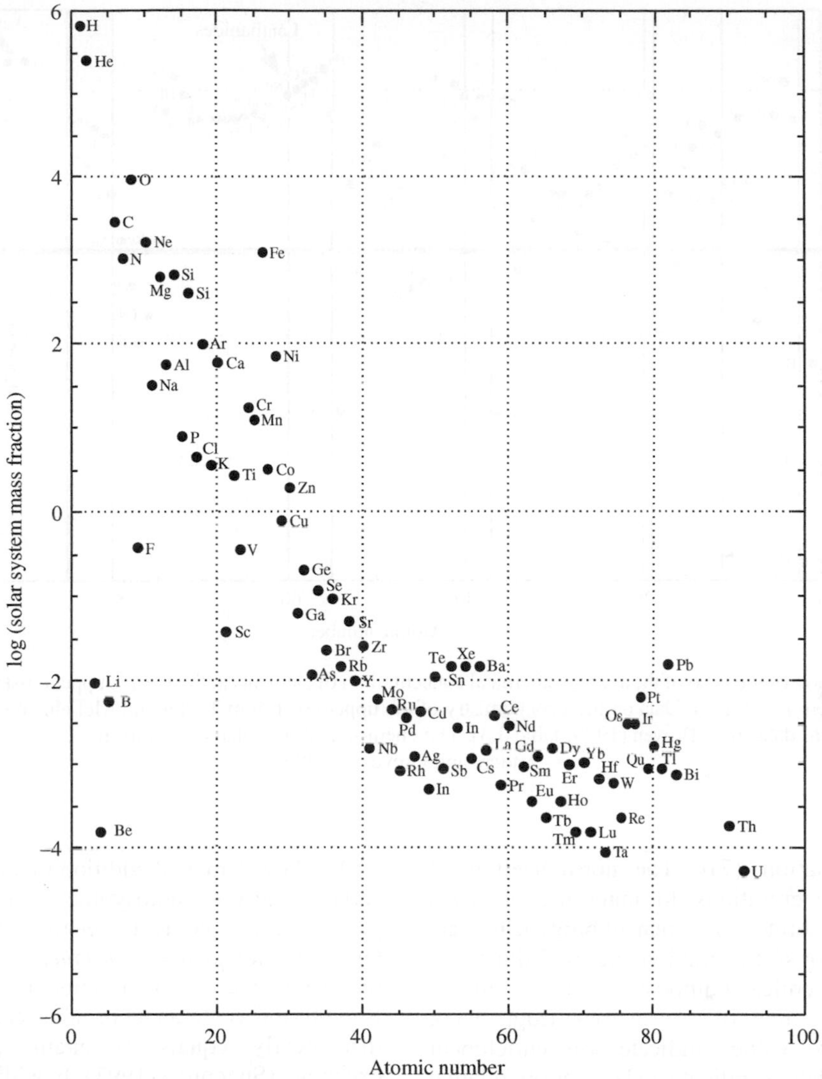

Figure 3 The log of the mass fraction (mg kg^{-1}) of elements in the solar system arranged by atomic number (source Anders and Grevesses, 1989, table 1).

There are a variety of compilations of the concentrations of many of the chemical elements for both crustal rocks (see above) and for soils (Bowen, 1979; Shacklette and Boerngen, 1984). In the case of soils, the samples analyzed are usually from a standard surface sampling depth, or from the uppermost horizon. Thus, these samples give a somewhat skewed view of the overall process of soil formation because, as will be discussed, soil formation is a depth-dependent process. Nonetheless, the data do provide a general overview of soil biogeochemistry that is applicable across broad geographical gradients.

When analyzing large chemical data sets, it is common to evaluate the behavior of elements in soils, and how they change during soil formation, by dividing the mass concentration of the elements in soils by that in crustal rocks, with the resulting ratio being termed the *enrichment factor*—values less than 1 indicating loss, more than 1 indicating gains. A disconcerting artifact of this analysis is that some mobile elements, particularly silicon, commonly show enrichment factors greater than 1. Silicon is one of the major elements lost via chemical weathering, having an annual flux to the ocean of 6.1×10^{12} mol Si yr^{-1} (Tréguer *et al.*, 1995), so that there is a large net loss of the element from landscapes. The reason for the apparent enrichment is that although silicon is lost via weathering, the concentration of chemically resistant silicates (e.g., quartz) leads to a relative retention of the element. These discrepancies can be avoided by relating soil and parent material concentrations to immobile index elements such as zirconium.

The present analysis uses τ, the fractional elemental enrichment factor relative to the parent

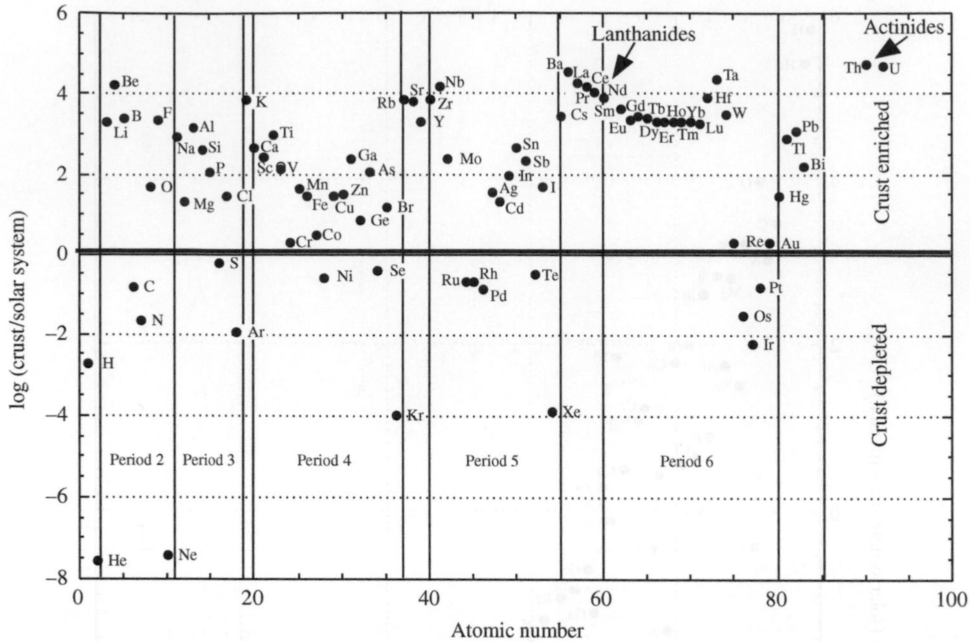

Figure 4 The log of the ratio of the average chemical concentration of elements in the Earth's upper crust (mg kg^{-1}) to that of the solar system (mg kg^{-1}). Data on the geochemistry of the upper crust from Taylor and McLennan (1985, table 2.15) with supplemental data from Bowen (1979, table 3.3). The chemistry of the solar system from Anders and Grevesse (1989).

material (Equation (7)). The normalization of elemental concentrations to immobile elements provides an accurate assessment of biogeochemical behavior during soil formation. Figure 5 illustrates the relative chemical composition of soil surface samples versus that of crustal rock ($\log(\tau+1)$), where positive values indicate soil enrichment and negative values indicate soil depletion, relative to the crust. Elemental losses are due to chemical weathering, and the ultimate removal of weathering products to oceans. Elemental gains are due primarily to biological processes—the addition of elements to soils, primarily by land plants.

The comparison (soils to average continental crust) indicates that soils are: (i) particularly depleted, due to aqueous weathering losses, in the alkali metals and alkaline earths (particularly magnesium, sodium, calcium, potassium, and beryllium) and some of the halides; (ii) depleted, to a lesser degree, in silicon, iron, and aluminum; and (iii) enriched in carbon, nitrogen, and sulfur. The losses are clearly due to chemical weathering, as the chemical composition of surface waters illustrates an enrichment of these same elements relative to that of the crust (Figure 6). Plants directly assimilate elements from soil water (though they exhibit elemental selectivity across the root interface (Clarkson, 1974)), and are therefore enriched, relative to the crust, in elements derived from chemical weathering.

The key elemental addition to soils by plants is carbon, because photosynthesis greatly increases plant carbon content relative to the crust. Globally, *net primary production* (NPP) (gross photosynthetic carbon fixation–plant respiration) is $\sim$60 Gt C yr^{-1}, an enormous carbon flux rate that nearly equals ocean/atmosphere carbon exchange (Sundquist, 1993). In addition to enriching the soil in carbon, the variety of organic molecules produced during the cycling of this organic material, coupled with the CO_2 generated in the soil by the decomposition of the organic compounds by heterotrophic microorganisms, greatly accelerate rates of chemical weathering. As a result, plants are responsible not only for enrichments of soil carbon, but also for enhanced rates of chemical weathering.

Second only to carbon inputs, nitrogen fixation by both symbiotic and nonsymbiotic organisms comprises an enormous biologically driven elemental influx to soils. Biological nitrogen fixation occurs via the following reaction (Allen *et al.*, 1994):

$$N_{2(\text{atmosphere})} + 10H^+ + n\text{MgATP} + 8e^- = 2NH_4^+ + H_2 + n\text{MgADP} + nP_i (n \geq 16)$$

where P_i is inorganic P. The breakage of the triple bonds in N_2 is a highly energy demanding process (thus the consumption of ATP), and in nature microorganisms have developed symbiotic relations with certain host plants (particularly

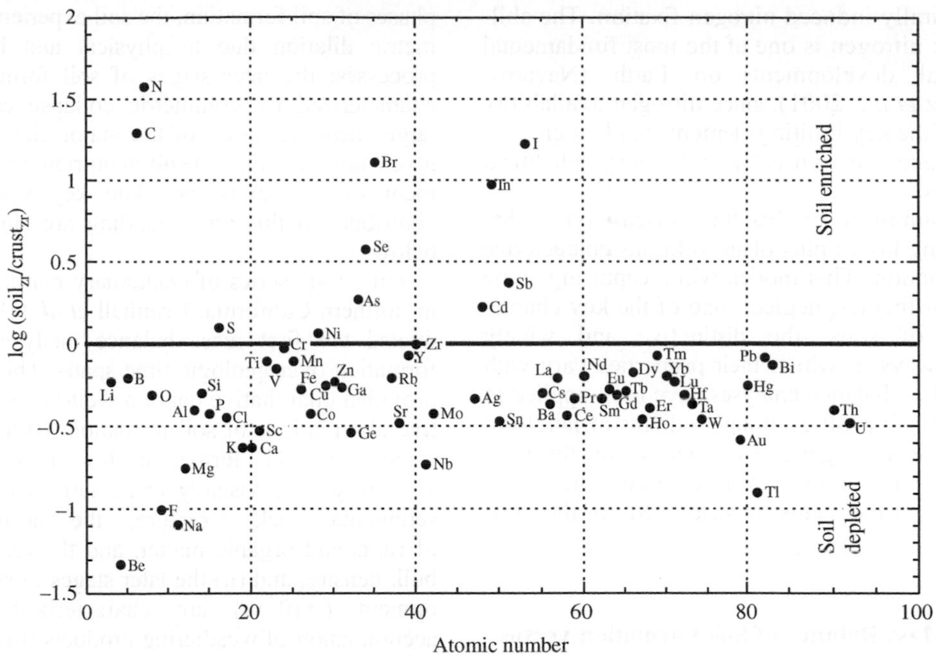

Figure 5 The log of the ratio of the mass fraction of an element in soil (relative to Zr) to that in the crust (relative to Zr). Soil data from Bowen (1979, table 4.4) and the crust data from Taylor and McLennan (1985, table 2.25) with supplemental data from Bowen (1979, table 3.3).

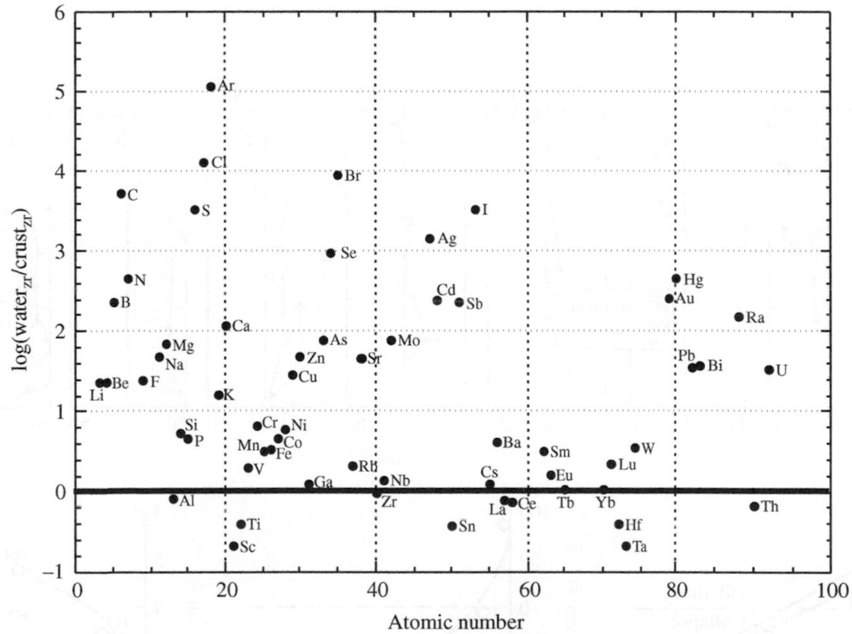

Figure 6 The log of the ratio of the mass fraction of an element in freshwater (rivers) (relative to Zr) to that of the crust (relative to Zr) as a function of atomic number. Water data from Bowen (1979, table 2.3) and crust data from Taylor and McLennan (1985, table 2.25) with supplemental data from Bowen (1979, table 3.3).

legumes), deriving carbon sources from the host plant, and in turn enzymatically reducing atmospheric N_2 to NH_4^+, a form which is plant available and becomes part of plant proteins.

Globally, it is estimated that, prior to extensive human activity, biological nitrogen fixation was ~(90–140)×10^{12} g N yr^{-1} (Vitousek *et al.*, 1997a). This rate is increasing because of the agriculturally induced nitrogen fixation. The ability to fix nitrogen is one of the most fundamental biological developments on Earth (Navarro-Gonzalez *et al.*, 2001), since nitrogen availability is one of the key limiting elements to plant growth, and hence to virtually all biogeochemical processes.

The summary of this brief discussion is that weathering losses plus plant additions characterize soil formation. This model, while capturing some important themes, neglects one of the key characteristics of soils—the distinctive and widely varying ways in which their properties vary with depth. Mass balance analyses have been applied to complete soil profiles along gradients of landform age, giving us a general perspective on the rates and directions of physical and chemical changes of soils with time. This is discussed in the next section.

7.4.2 Mass Balance of Soil Formation versus Time

7.4.2.1 Temperate climate

The main conclusions that can be summarized by mass balance analyses of soil formation over time in nonarid environments are that: in early phases of soil formation, the soil experiences volumetric dilation due to physical and biological processes; the later stages of soil formation are characterized by volumetric collapse caused by large chemical losses of the major elements that, given sufficient time, result in nutrient impoverishment of the landscape. The key studies that contribute to this understanding are summarized below.

On a time series of Quaternary marine terraces in northern California, Brimhall *et al.* (1992) conducted the first mass balance analysis of soil formation over geologic time spans. This analysis provided quantitative data on well-known qualitative observations of soil formation: (i) the earliest stages of soil formation (on timescales of 10^1–10^3 yr) are visually characterized by loss of sedimentary/rock structure, the accumulation of roots and organic matter, and the reduction of bulk density; and (ii) the later stages of soil development ($>10^3$ yr) are characterized by the accumulation of weathering products (iron oxides, silicate clays, and carbonates) and the loss of many products of weathering.

Figure 7 shows the trend in ε, volumetric strain (Equation (5)), over $\sim 2.40\times10^5$ yr. The data show the following physical changes: (i) large volumetric expansion ($\varepsilon>0$) occurred in the

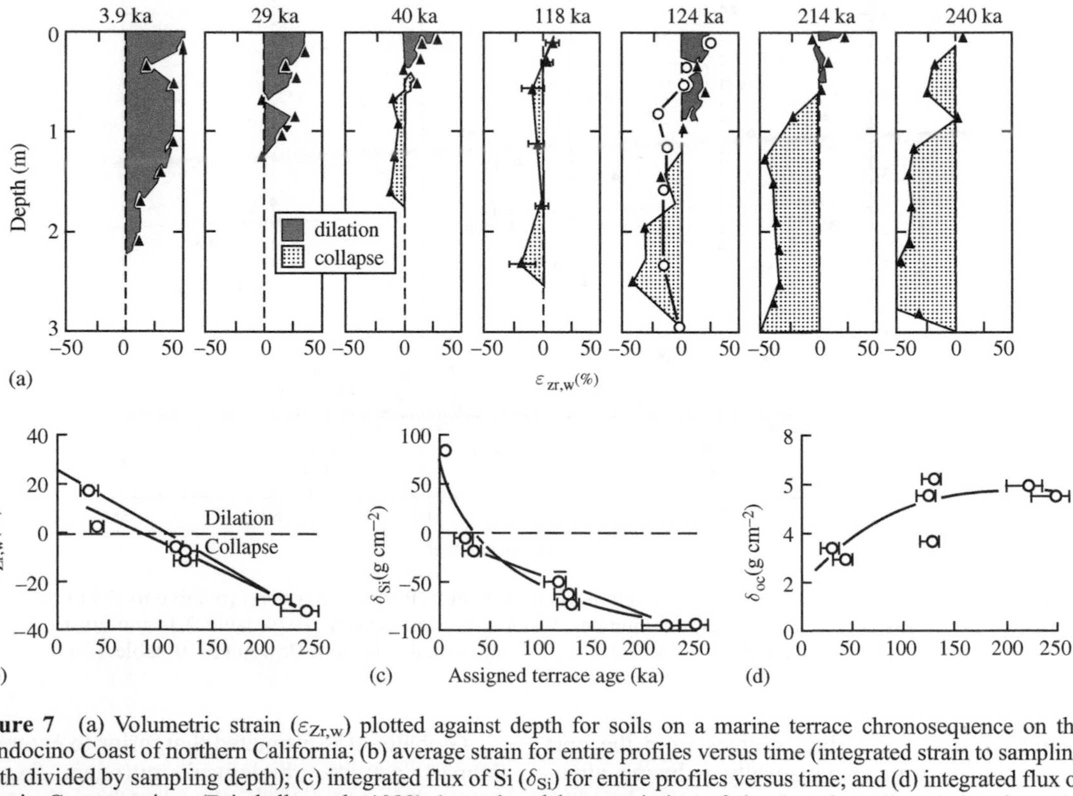

Figure 7 (a) Volumetric strain ($\varepsilon_{Zr,w}$) plotted against depth for soils on a marine terrace chronosequence on the Mendocino Coast of northern California; (b) average strain for entire profiles versus time (integrated strain to sampling depth divided by sampling depth); (c) integrated flux of Si (δ_{Si}) for entire profiles versus time; and (d) integrated flux of organic C versus time (Brimhall *et al.*, 1992) (reproduced by permission of the American Association from the Advancement of Science from *Science* **1992**, *255*, 695).

young soil (Figure 7(a)); (ii) integrated expansion for the whole soil declined with age (Figure 7(b)); and (iii) the cross-over point between expansion and collapse ($\varepsilon < 0$) moved progressively toward the soil surface with increasing age (Figure 7(a)).

Biological processes, along with abiotic mixing mechanisms, drive the distinctive first phases of soil formation. The large positive strain (expansion) measured in the young soil on the California coast was due to an influx of silicon-rich beach sand (Figure 7(c)) and the accumulation of organic matter from plants (Figure 7(d)). In many cases, there is a positive relationship between the mass influx of carbon to soil (δ_{oc}) and strain; Jersak *et al.* (1995)). Second, in addition to adding carbon mass relative to the parent material, the plants roots (and other subterranean organisms) expand the soil, create porosity, and generally assist in both mixing and expansion. Pressures created by growing roots can reach 15 bar (Russell, 1977), providing adequate forces to expand soil material. Brimhall *et al.* (1992) conducted an elegant lab experiment showing the rapid manner in which roots can effectively mix soil, and incorporate material derived from external sources. Over several hundred "root growth cycles" using an expandable/collapsible tube in a sand mixture (Figure 8(a)), they demonstrated considerable expansion and depth of mixing (Figure 8(b)), with an almost linear relation between expansion and depth of translocation of externally added materials (Figures 8(c) and (d)).

The rate of physical mixing and volumetric expansion caused by carbon additions declines quickly with time. Soil carbon accumulation with time (Figure 7(d)) can be described by the following first-order decay model (Jenny *et al.*, 1949):

$$\frac{dC}{dt} = I - kC \tag{9}$$

where I is plant carbon inputs (kg m^{-2} yr^{-1}), C the soil carbon storage (kg m^{-2}), and k the decay constant (yr^{-1}). Measured and modeled values of k for soil organic carbon (Jenkinson *et al.*, 1991; Raich and Schlesinger, 1992) indicate that steady state should be reached for most soils within $\sim 10^2$–10^3 yr. Thus, as rates of volumetric expansion decline, the integrated effects of mineral weathering and the leaching of silicon (Figure 7(c)), calcium, magnesium, sodium,

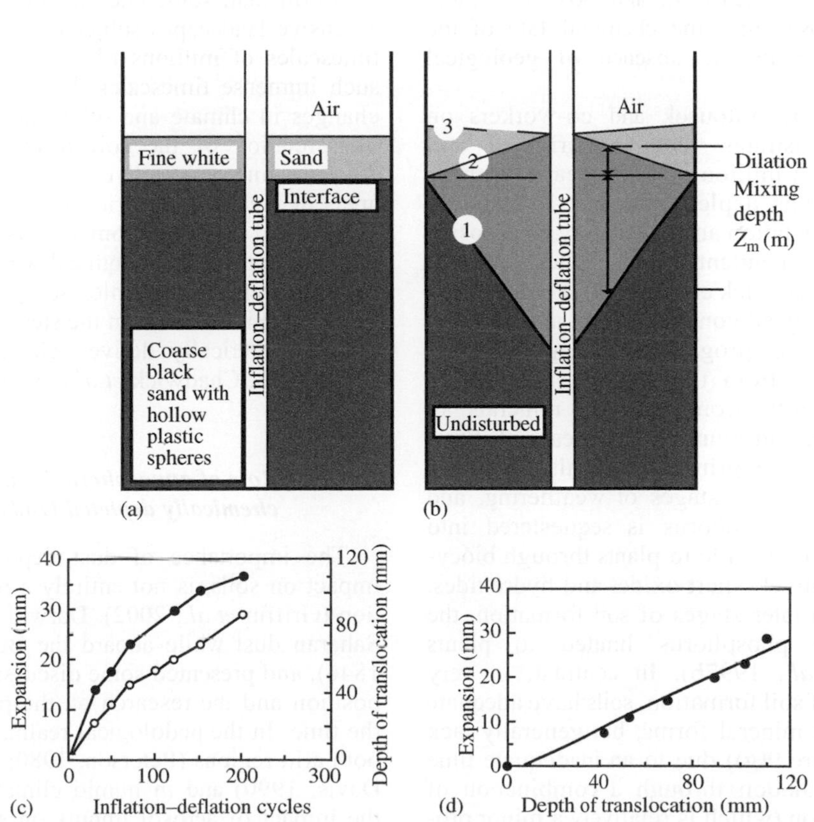

Figure 8 (a) Initial state of a cyclical dilation mixing experiment, with a surgical rubber tube embedded in a sandy matrix; (b) features after mixing: line 1 is depth of mixing after mixing, line 2 is the dilated surface, and line 3 is the top of the overlying fine sand lense; (c) expansion (o) and depth of mixing (•) as a function of mixing cycles; and (d) relationship of soil expansion to mixing depth (Brimhall *et al.*, 1992) (reproduced by permission of the American Association from the Advancement of Science from *Science* **1992**, *255*, 695).

potassium, and other elements begin to become measurable, and over time tend to eliminate the measured expansion not only near the surface (Figure 7(a)), but also for the whole profile (Figure 7(b)).

7.4.2.2 Cool tropical climate

The integrated mass losses of elements over time are affected by parent material mineralogy, climate, topography, etc. The mass balance analysis of soil formation of the temperate California coast (Brimhall *et al.*, 1991; Chadwick *et al.*, 1990; Merritts *et al.*, 1992) is complemented by an even longer time frame on the Hawaiian Islands (Vitousek *et al.*, 1997b; Chadwick *et al.*, 1999). The Hawaiian chronosequence encompasses ~4 Myr in a relatively cool, but wet, tropical setting. Because of both steady and cyclic processes of erosion and deposition, few geomorphic surfaces in temperate settings on Earth are older than Pleistocene age. Yet the exceptions to this rule: the Hawaiian Island chronosequence, river terrace/glacial outwash sequence in the San Joaquin Valley of California (e.g., Harden, 1987; White *et al.*, 1996), and possibly others provide glimpses into the chemical fate of the Earth's surface in the absence of geological rejuvenation.

The work by Vitousek and co-workers on Hawaii demonstrates that uninterrupted soil development on million-year timescales in those humid conditions depletes the soil in elements essential to vegetation and, ultimately, the ecosystem becomes dependent on atmospheric sources of nutrients (Chadwick *et al.*, 1999). Figures 9(a)–(g) illustrate: (i) silicon and alkali and alkaline earth metals are progressively depleted, and nearly removed from the upper 1 m; (ii) soil mineralogy shifts from primary minerals to secondary iron and aluminum oxides with time; (iii) phosphorus in primary minerals is rapidly depleted in the early stages of weathering, and the remaining phosphorus is sequestered into organic forms (available to plants through biocycling) and relatively inert oxides and hydroxides. As a result, in later stages of soil formation, the soils become phosphorus limited to plants (Vitousek *et al.*, 1997b). In contrast, in very early phases of soil formation, soils have adequate phosphorus in mineral forms, but generally lack nitrogen (Figure 9(g)) due to an inadequate time for its accumulation through a combination of nitrogen fixation (which is relatively a minor process on Hawaii; Vitousek *et al.* (1997b) and atmospheric deposition of NO_3^-, NH_4^+, and organic N (at rates of $\sim 5-50\,\mathrm{kg\,N\,yr^{-1}}$; Heath and Huebert (1999)). The rate of nitrogen accumulation in soil and the model describing it generally parallel the case of organic carbon, because carbon storage hinges on the availability of nitrogen (e.g., Parton *et al.*, 1987).

In summary, soils in the early stages of their development contain most of the essential elements for plant growth with the exception of nitrogen. Soil nitrogen, like carbon, reaches maximum steady-state values in periods on the order of thousands of years and, as a result, NPP of the ecosystems reach maximum values at this stage of soil development (Figure 10). Due to progressive removal of phosphorus and other plant essential elements (particularly calcium), plant productivity declines (Figure 10), carbon inputs to the soil decline, and both soil carbon and nitrogen storage begin a slow decline (Figure 9). This trend, because of erosive rejuvenating processes, is rarely observed in climatically and tectonically active parts of the Earth. Alternatively, low latitude, tectonically stable continental regions may reflect these long-term processes. Brimhall and Dietrich (1987) and Brimhall *et al.* (1991, 1992) discuss the pervasive weathering and elemental losses from cratonal regions such as Australia and West Africa. These regions, characterized by an absence of tectonic activity and glaciation, and by warm (and sometimes humid) climates, have extensive landscapes subjected to weathering on timescales of millions of years. However, over such immense timescales, known and unknown changes in climate and other factors complicate interpretation of the soil formation processes. An emerging perspective is that these and other areas of the Earth experience atmospheric inputs of dust and dissolved components that wholly or partially compensate chemical weathering losses, ultimately creating complex soil profiles and ecosystems which subsist on the steady but slow flux of atmospherically derived elements (Kennedy *et al.*, 1998; Chadwick *et al.*, 1999).

7.4.2.3 Role of atmospheric inputs on chemically depleted landscapes

The importance of dust deposition, and its impact on soils, is not entirely a recent observation (Griffin *et al.*, 2002). Darwin complained of Saharan dust while aboard the Beagle (Darwin, 1846), and presented some discussion of its composition and the research on the phenomenon at the time. In the pedological realm, researchers in both arid regions (Peterson, 1980; Chadwick and Davis, 1990) and in humid climates recognized the impact of aerosol inputs on soil properties. With respect to the Hawaiian Islands—a remarkably isolated volcanic archipelago—Jackson *et al.* (1971) recognized the presence of prodigous quantities of quartz in the basaltic soils of Hawaii. Oxygen-isotope analyses of these quartz

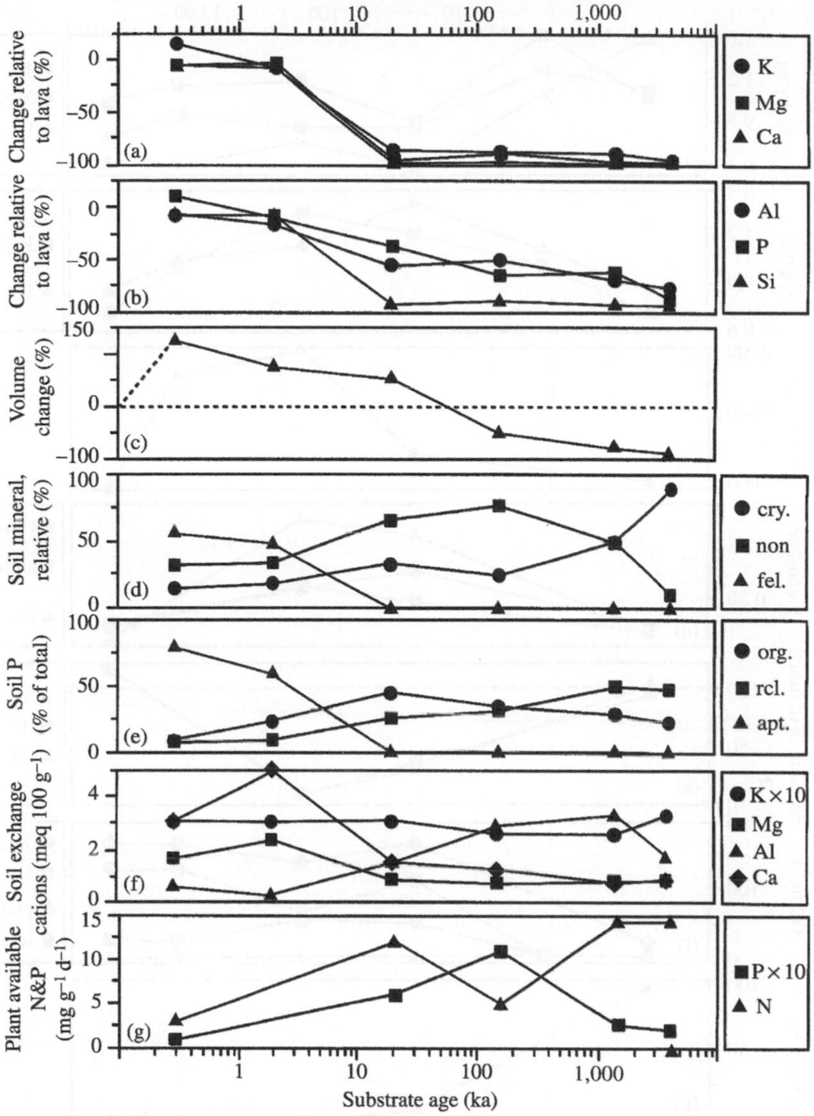

Figure 9 Weathering, mineralogical changes, and variations in plant-available elements in soils (to 1 m in depth) as a function of time in Hawaii: (a) total soil K, Mg, and Ca; (b) total soil Al, P, and Si; (c) volumetric change; (d) soil feldspar, crystalline and noncrystalline secondary minerals; (e) soil P pools (organic, recalcitrant, apatite); (f) exchangeable K, Mg, Al, and Ca; and (g) changes in resin-extractable (biologically available) inorganic N and P (Vitousek *et al.*, 1997b) (reproduced by permission of the Geological Society of America from *GSA Today*, **1997b**, *7*, 1–8).

grains showed that the quartz was derived from continental dust from the northern hemisphere, a source now well constrained by atmospheric observations (Nakai *et al.*, 1993) and the analysis of Pacific Ocean sediment cores (Rea, 1994).

The work by Chadwick *et al.* (1999) has demonstrated that the calcium and phosphorus nutrition of the older Hawaiian Island ecosystems depends almost entirely on atmospheric sources. With respect to calcium, strontium-isotope analyses of soils and plants indicates that atmospherically derived calcium (from marine sources) increases from less than 20% to more than 80% of total plant calcium with increasing soil age. With respect to phosphorus, the use of rare earth elements and isotopes of neodymium all indicated that from ~0.5 to more than 1.0 mg P m^{-2} yr^{-1} is delivered in the form of dust each year and, in the old soils, the atmospheric inputs approach 100% of the total available phosphorus at the sites. Brimhall *et al.* (1988, 1991) demonstrated that much of the zirconium in the upper part of the soils in Australia and presumably other chemical constituents are derived from atmospheric inputs. In summary, it is clear that the ultimate fate of soils in the absence of geological

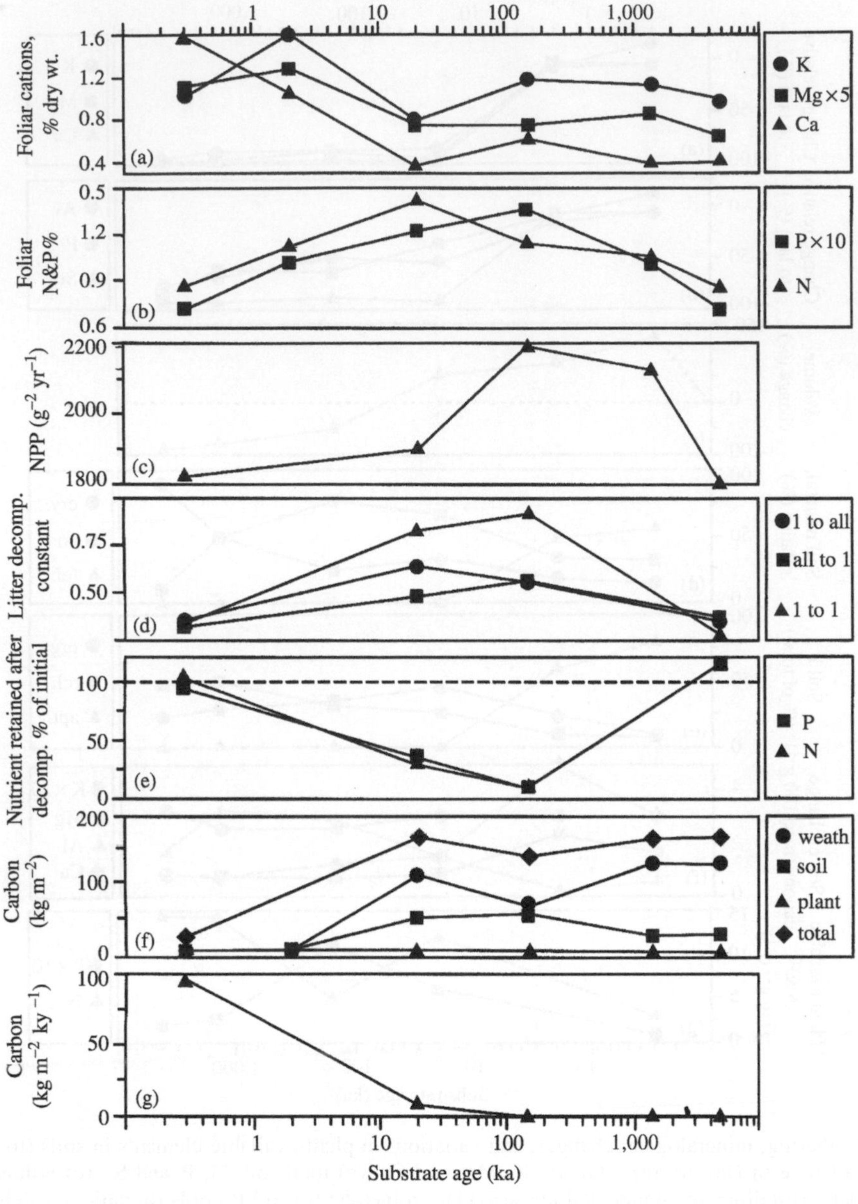

Figure 10 Plant nutrients, production and decomposition, and carbon sinks during soil development in Hawaii: (a) K, Mg, and Ca in canopy leaves of dominant tree (*Metrosideros*); (b) N and P in leaves of dominant tree; (c) changes in NPP of forests; (d) decomposition rate constant (*k*) of *metrosideros* leaves decomposed in the site collected (1-to-1), collected at each site and transferred to common site (all-to-1), and collected from one site and transferred to all sites (1-to-all); (e) fraction of N and P in original leaves remaining after 2 yr of decomposition; (f) carbon storage in the ecosystems in the form of plant biomass, soil organic matter, and as CO_2 consumed during silicate weathering; and (g) instantaneous rate of total ecosystem carbon storage (Vitousek *et al.*, 1997b) (reproduced by permission of the Geological Society of America from *GSA Today*, **1997b**, *7*, 1–8).

rejuvenation, *in humid climates*, is a subsistence on atmospheric elemental sources.

7.4.3 Mass Balance Evaluation of Soil Formation versus Climate

Hyperarid regions offer a unique view of the importance of atmospheric elemental inputs to soils, because in these areas, the inputs are not removed by leaching. The western coasts of southern Africa and America lie in hyperarid climates that have likely persisted since the Tertiary (Alpers and Brimhall, 1988). These regions, particularly the Atacama desert of Chile, are known for their commercial-grade deposits of sulfates, iodates, bromates, and particularly nitrates (Ericksen, 1981; Böhlke *et al.*, 1997). These deposits may form due to several processes and salt sources, including deflation around playas, spring deposits (Rech *et al.*, 2003), and other settings. However, many of the deposits are simply heavily concentrated soil horizons, formed by the long accumulation of soluble constituents over long time spans (Erickson, 1981). The ultimate origin of these salts (and their elements) is in some cases obscure, but it is clear that they arrive at the soils via the atmosphere. In Chile, sources of elements may be from fog, marine spray, reworking of playa crusts, and more general long-distance atmospheric sources (Ericksen, 1981; Böhlke *et al.*, 1997).

Recent novel research by Theimens and co-workers has convincingly demonstrated the atmospheric origin (as opposed to sea spray, playa reworking, etc.) of sulfates in Africa and Antarctica (Bao *et al.*, 2000a,b, 2001). Briefly, these researchers have shown that as sulfur undergoes chemical reactions in the stratosphere and troposphere, a mass-independent fractionation of oxygen isotopes in the sulfur oxides occurs. The mechanism for these fractionations is obscure (Thiemens, 1999), but the presence of mass-independent ratios of ^{17}O and ^{18}O in soil sulfate accumulations is a positive indicator of an atmospheric origin. In both the Nambian desert (Bao *et al.*, 2001) and Antarctica (Bao *et al.*, 2000a), there was an increase in the observed ^{17}O isotopic anomaly with distance from the ocean, possibly due to a decrease in the ratio of sea salt-derived sulfate in atmospheric sulfate.

There have been few, if any, systematic pedological studies of the soils of these hyperarid regions. Here, we present a preliminary mass balance analysis of soil formation along a precipitation gradient in the presently hyperarid region of the Atacama desert, northern Chile (Sutter *et al.*, 2002). Along a south-to-north gradient, precipitation decreases from $\sim$15 mm yr^{-1} to $\sim$2 mm yr^{-1} (http://www.worldclimate.com/worldclimate/index.htm). For the study, three sites were chosen $\sim$50 km inland on the oldest (probably Mid- to Early Pleistocene) observable fluvial landform (stream terrace or alluvial fan) in the region. Depths of observation were restricted by the presence of salt-cemented soil horizons.

Using the mass balance equations presented earlier and titanium as an immobile index element, the volumetric and major element changes with precipiation were calculated (Figure 11). The calculations for ε show progressive increases in volumetric expansion with decreasing precipitation (approaching 400% in some horizons at Yungay, the driest site) (Figure 11(a)). These measured expansions are due primarily to the accumulation and retention of NaCl and $CaSO_4$ minerals. For example, Yungay has large expansions near the surface and below 120 cm. Figure 11(c) shows that sulfur (in the form of $CaSO_4$) is responsible for the upper expansion, while chlorine (in the form of NaCl) is responsible for the lower horizon expansion. As the chemical data indicate, the type and depth of salt movement is climatically related: the mass of salt changes from (Cl, S) to (S, $CaCO_3$) to ($CaCO_3$) with increasing rainfall (north to south). Additionally, the depth of S and $CaCO_3$ accumulations increase with rainfall.

These data suggest that for the driest end-member of the transect, the virtual absence of chemical weathering (for possibly millions of years), and the pervasive input and retention of atmospherically derived chemical constituents (due to a lack of leaching), drives the long-term trajectory of these soils toward continued volumetric expansion (due to inputs) and the accretion, as opposed to the loss, of plant-essential elements. Therefore, it might be hypothesized that there is a critical water balance for soil formation (precipitation–evapotranspiration) at which the long-term accumulation of atmospherically derived elements exceeds weathering losses, and landscapes undergo continual dilation as opposed to collapse. The critical climatic cutoff point is likely to be quite arid. In the Atacama desert, the crossover point between the accretion versus the loss of soluble atmospheric inputs such as nitrate and sulfate is somewhere between 5 mm and 20 mm of precipitation per year. These Pleistocene (or older) landscapes have likely experienced changes in climate (Betancourt *et al.*, 2000; Latorre *et al.*, 2002), so the true climatic barrier to salt accumulations is unknown. Nonetheless, it is clear that the effect of climate drives strongly contrasting fates of soil formation (collapse and nutrient impoverishment versus dilation and nutrient accumulation) over geological time spans.

7.5 PROCESSES OF MATTER AND ENERGY TRANSFER IN SOILS

Chadwick *et al.* (1990) wrote that depth-oriented mass balance analyses change the study of soil formation from a "black to gray box." The "grayness" of the mass balance approach is due to the fact that it does not directly provide insight into mechanisms of mass transfer, and it does not address the transport of heat, water, and gases. The mechanistic modeling and quantification of these fluxes in field settings is truly in its infancy, but some general principles along with some notable success stories have emerged on the more mechanistic front of soil formation. In this section, the author discusses the general models that describe mass transfer in soils, and examines in some detail how these models have been

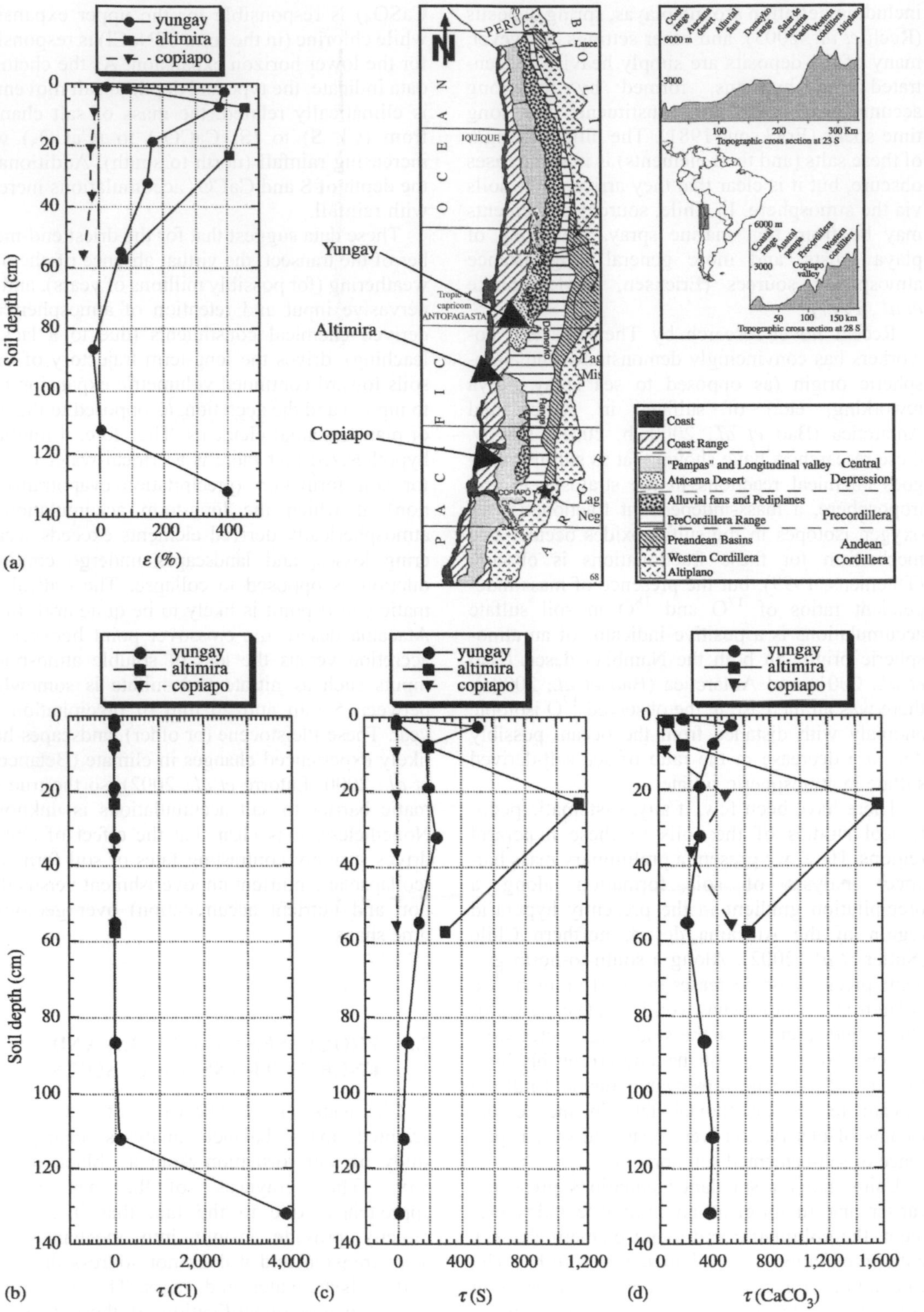

Figure 11 (a) Volumetric change along climate gradient in Atacama desert, northern Chile; (b) fractional mass gains of Cl; (c) fractional mass gains of S; and (d) fractional mass gains of $CaCO_3$ for three sites sampled along a precipitation gradient (Sutter *et al.*, 2002).

successfully used to describe observed patterns in soil gas and organic matter concentrations, and their isotopic composition.

The movement of most constituents in soils can ultimately be described as variants of diffusive (or in certain cases, advective) processes. The study of these processes, and their modeling, has long been the domain of *soil physics*, an experimental branch of the soil sciences. There are several good textbooks in soil physics that provide introductions to these processes (Jury *et al.*, 1991; Hillel, 1998, 1980a,b). However, it is fair to say that the application of this work to natural soil processes, and to natural soils, has been minimal given the focus on laboratory or highly controlled field experiments. However, notable exceptions to this trend exist, exceptions initiated by biogeochemists who adapted or modified these principles to illuminate the soil "black box."

There are few, if any, cases where these various models have been fully coupled to provide an integrated view of soil formation. However, one group of soil processes that has been extensively studied and modeled comprise the soil carbon cycle. We review the mechanisms of this cycle and the modeling approaches that have received reasonably wide acceptance in describing the processes.

7.5.1 Mechanistic Modeling of the Organic and Inorganic Carbon Cycle in Soils

The processing of carbon in soils has long received attention due to its importance to agriculture (in the form of organic matter) and the marked visual impact it imparts to soil profiles. A schematic perspective of the flow of carbon through soils is given in Figure 12. Carbon is fixed from atmospheric CO_2 by plants, enters soil in organic forms, undergoes decomposition, and is cycled back to the atmosphere as CO_2. In semi-arid to arid regions, a fraction of the CO_2 may ultimately become locked in pedogenic $CaCO_3$, whereas in humid regions a portion may be leached out as dissolved organic and inorganic carbon with groundwater. On hillslopes, a portion of the organic carbon may be removed by erosion (Stallard, 1998). Most studies of the organic part of the soil carbon assume the latter three mechanisms to be of minor importance (to be discussed more fully below) and consider the respiratory loss of CO_2 as the main avenue of soil carbon loss.

Jenny *et al.* (1949) were among the first to apply a mathematical framework to the soil carbon cycle. For the organic layer at the surface of forested soils, Jenny applied, solved, and evaluated the mass balance model given by Equation (9) discussed earlier. This approach can be applied to the soil organic carbon pool as a whole. This has proved useful in evaluating the response of soil carbon to climate and environmental change (Jenkinson *et al.*, 1991).

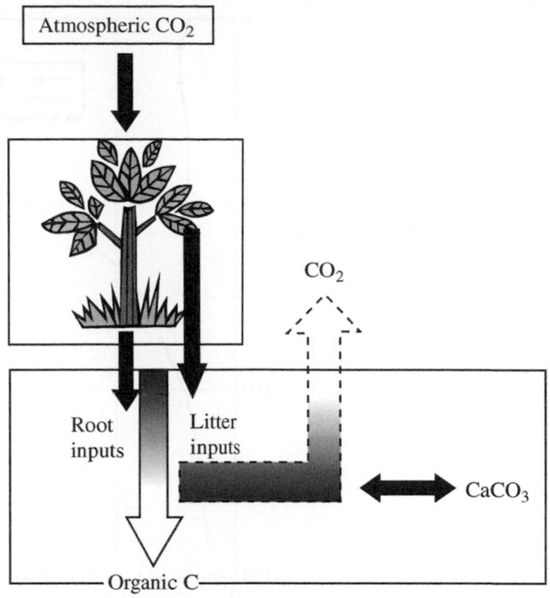

Figure 12 Schematic diagram illustrating C flow through terrestrial ecosystems.

The deficiency of the above model of soil carbon is that it ignores the interesting and important variations in soil carbon content with soil depth. The depth variations of organic carbon in soils vary widely, suggesting a complex set of processes that vary from one environment to another. Figure 13 illustrates just three commonly observed soil carbon (and nitrogen) trends with depth: (i) exponentially declining carbon with depth (common in grassland soils or Mollisols); (ii) randomly varying carbon with depth (common on young fluvial deposits or, as here, in highly stratified desert soils or Aridisols); and (iii) a subsurface accumulation of carbon below a thick plant litter layer on the soil surface (sandy, northern forest soils or Spodosols). The mechanisms controlling these distributions will be discussed, in reverse order, culminating with a discussion of transport models used to describe the distribution of carbon in grasslands.

Spodosols are one of the 12 soil orders in the USDA Soil Taxonomy. The soil orders, and their key properties, are listed in Table 4. The distribution of the soil orders in the world is illustrated in Figure 14. With respect to Spodosols, which are common to the NE USA, Canada, Scandinavia, and Russia, the key characteristics that lead to their formation are: sandy or coarse sediment

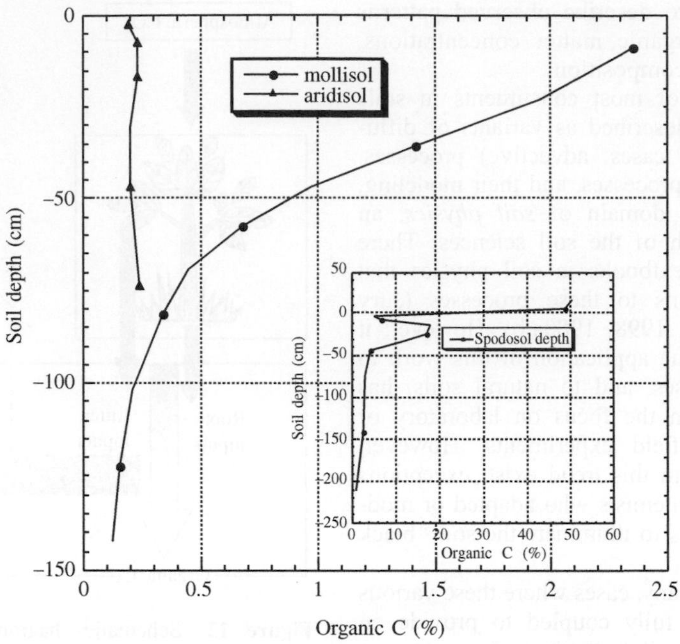

Figure 13 Soil organic C depth in three contrasting soil orders (source Soil Survey Staff, 1975).

Table 4 The soil orders of the US Soil Taxonomy and a brief definition of their characteristics.

Order	*Characteristics*
Alfisols	Soils possessing Bt horizons with >35% base saturation; commonly Pleistocene aged
Andisols	Soils possessing properties derived from weathered volcanic ash, such as low bulk density, high extractable Al, and high P retention
Aridisols	Soils of arid climates that possess some degree of subsurface horizon development; commonly Pleistocene aged
Entisols	Soils lacking subsurface horizon development due to young age, resistant parent materials, or high rates of erosion
Gelisols	Soils possessing permafrost and/or evidence of cryoturbation
Histosols	Soils dominated by organic matter in 50% or more of profile
Inceptisols	Soils exhibiting "incipient" stages of subsurface horizon development due to age, topographic position, etc.
Mollisols	Soils possessing a relatively dark and high C and base saturation surface horizon; commonly occur in grasslands
Oxisols	Soils possessing highly weathered profiles characterized by low-cation-exchange-capacity clays (kaolinite, gibbsite, etc.), few remaining weatherable minerals, and high clay; most common on stable, tropical landforms
Spodosols	Soils of northern temperate forests characterized by intense (but commonly shallow) biogeochemical downward transport of humic compounds, Fe, and Al; commonly Holocene aged
Ultisols	Soils possessing Bt horizons with <35% base saturation; commonly Pleistocene aged
Vertisols	Soils composed of >35% expandable clay, possessing evidence of shrink/swell in form of cracks, structure, or surface topography

After Soil Survey Staff (1999).

(commonly glacial outwash or till), deciduous or coniferous forest cover, and humid, cool to cold, climates. Organic matter added by leaf and branch litter at the surface accumulates to a steady-state thickness of organic horizons (Figure 13). During the decomposition of this surface material, soluble organic molecules are released which move downward with water. The organic molecules have reactive functional groups which complex with iron and aluminum, stripping

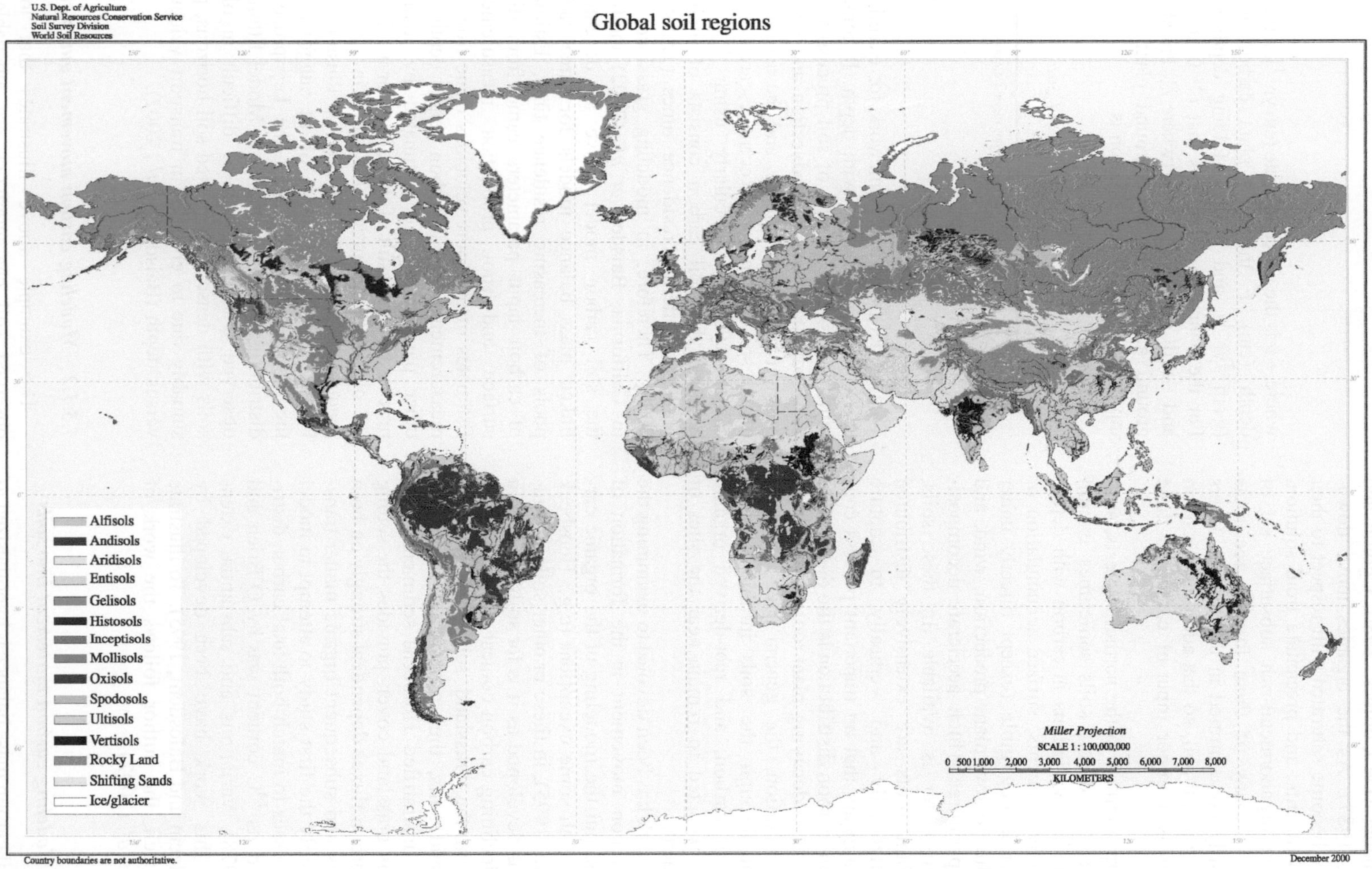

Figure 14 The global distribution of soil orders (source http://www.nrcs.usda.gov/technical/worldsoils/mapindx/).

the upper layers of the mineral soils that contain these elements and leaving a white, bleached layer devoid, or depleted in iron, aluminum, and organic matter (De Coninck, 1980; Ugolini and Dahlgren, 1987). As the organics move downward, they become saturated with respect to their metal constituents, and precipitate from solution, forming an organo/metal-rich subsurface set of horizons. In many of these forests, tree roots are primarily concentrated in the organic layers above the mineral soil, so that addition of carbon from roots is a minor input of carbon to these systems.

In sharp contrast to the northern forests, stratified, gravelly desert soils sometimes exhibit almost random variations in carbon with depth. In these environments, surface accumulation of plant litter is negligible (except directly under shrubs) due to low plant production, wind, and high temperatures which accelerate decomposition when water is available. In these soils, where plant roots are extensively distributed both horizontally and vertically to capture water, it appears that an important process controlling soil carbon distribution is the direct input of carbon from decaying plant roots or root exudates. In addition, the general lack of water movement through the soils inhibits vertical transport of carbon, and root-derived organic matter is expected to remain near the sites of emplacement.

Much work has been devoted to examining the role of carbon movement in the formation of Spodosols, but the modeling of the organic carbon flux, with some exceptions (e.g., Hoosbeek and Bryant, 1995), in these or other soils is arguably not as developed as it is for soils showing a steadily declining carbon content with depth that is found in the grassland soils of the world. In grassland soils, the common occurrence of relatively unstratified Holocene sediments, and continuous grassland cover, provides the setting for soil carbon fluxes dependent strongly on both root inputs and subsequent organic matter transport. Possibly the first study to attempt to model these processes to match both total carbon distribution and its ^{14}C content was by O'Brien and Stout (1978). Variations and substantial extensions of this work have been developed by others (Elzein and Balesdent, 1995). To illustrate the approach, the author follows the work of Baisden *et al.* (2002).

7.5.1.1 *Modeling carbon movement into soils*

The soil carbon mass balance is hypothesized to be, for grassland soils, a function of plant inputs (both surface and root), transport, and decomposition:

$$\frac{dC}{dt} = \underbrace{-v\frac{dC}{dz}}_{\substack{\text{downward}\\ \text{advective}\\ \text{transport}}} - \underbrace{kC}_{\text{decomposition}} + \underbrace{\frac{F}{L}e^{-z/L}}_{\substack{\text{plant inputs}\\ \text{distributed}\\ \text{exponentially}}} \qquad (10)$$

where $-\nu$ is the advection rate (cm yr^{-1}), z the soil depth (cm), F the total plant carbon inputs (g cm^{-2}yr^{-1}), and L the e-folding depth (cm). For the boundary conditions that $C=0$ at $z=\infty$ and $-\nu(dC/dz)=F_A$ at $z=0$ (where F_A are aboveground and F_B the belowground plant carbon inputs), the steady-state solution is

$$C(z) = \underbrace{\frac{F_A}{v}e^{-kz/v}}_{\substack{\text{above-ground}\\ \text{input/transport}}} + \underbrace{\frac{F_B}{kL-v}e^{-kz/v}\left(e^{z(kL-v)/vL}-1\right)}_{\text{root input/transport}} \qquad (11)$$

This model forms the framework for examining soil carbon distribution with depth. It contains numerous simplifications of soil processes such as steady state, constant advection and decomposition rates versus depth, and the assumption of one soil carbon pool. Recent research on soil carbon cycling, particularly using ^{14}C, has revealed that soil carbon consists of multiple pools of differing residence times (Trumbore, 2000). Therefore, in modeling grassland soils in California, Baisden *et al.* (2002) modified the soil carbon model above by developing linked mass balance models for three carbon pools of increasing residence time. Estimates of carbon input parameters came from direct surface and root production measurements. Estimates of transport velocities came from ^{14}C measurements of soil carbon versus depth, and other parameters were estimated by iterative processes. The result of this effort for a $\sim 2\times 10^5$ yr old soil (granitic alluvium) in the San Joaquin Valley of California is illustrated in Figure 15. The goodness of fit suggests that the model captures at least the key processes distributing carbon in this soil. Model fitting to observed data became more difficult in older soils with dense or cemented soil horizons, presumably due to changes in transport velocities versus depth (Baisden *et al.*, 2002).

7.5.1.2 *Modeling carbon movement out of soils*

The example above illustrates that long-observed soil characteristics are amenable to analytical or numerical modeling, and it illustrates the importance of transport in the vertical distribution of soil properties, in this case

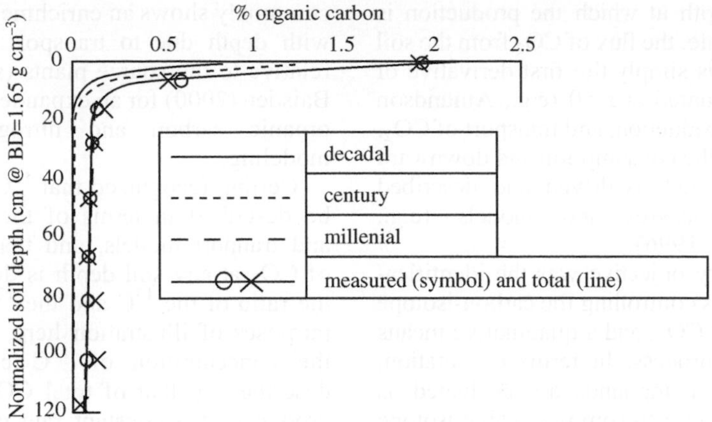

Figure 15 Measured total organic C versus depth and modeled amounts of three fractions of differing residence time (~10^0 yr, 10^2 yr, and 10^3 yr) for a ~600 ka soil formed on granitic alluvium in the San Jaoquin Valley of California (source Baisden, 2000).

organic carbon. As the model and observations indicate, the primary pathway for carbon loss from soil is the production of CO_2 from the decomposition of the organic carbon. It has long been recognized that CO_2 leaves the soil via diffusion, and various forms of Fick's law have been applied to describing the transport of CO_2 and other gases in soils (e.g., Jury *et al.*, 1991; Hillel, 1980b). However, the application of these models to natural processes and to the issue of stable isotopes in the soil gases is largely attributable to the work of Thorstenson *et al.* (1983) and, in particular, of Cerling (1984). Cerling's (1984) main interest was in describing the carbon-isotope composition of soil CO_2, which he recognized ultimately controls the isotope composition of pedogenic carbonate. We begin with the model describing total CO_2 diffusion, then follow that with the extension of the model to soil carbon isotopes.

Measurements of soil CO_2 concentrations versus depth commonly reveal an increase in CO_2 content with depth. The profiles and the maximum CO_2 levels found at a given depth are climatically controlled (Amundson and Davidson, 1990) due to rates of C inputs from plants, decomposition rates, etc. Given that most plant roots and soil C are concentrated near the surface, the production rates of CO_2 would be expected to decline with depth. Cerling developed a production/diffusion model to describe steady-state soil CO_2 concentrations:

$$\varepsilon\frac{dCO_2}{dt} = 0 = \underbrace{D\frac{d^2CO_2}{dz^2}}_{\text{net diffusion}} + \underbrace{\phi}_{\text{biological production}} \quad (12)$$

where ε is free air porosity in soil, CO_2 the concentration of CO_2 (mol cm^{-3}), and D the effective diffusion coefficient of CO_2 in soil (cm^2 s^{-1}), z the

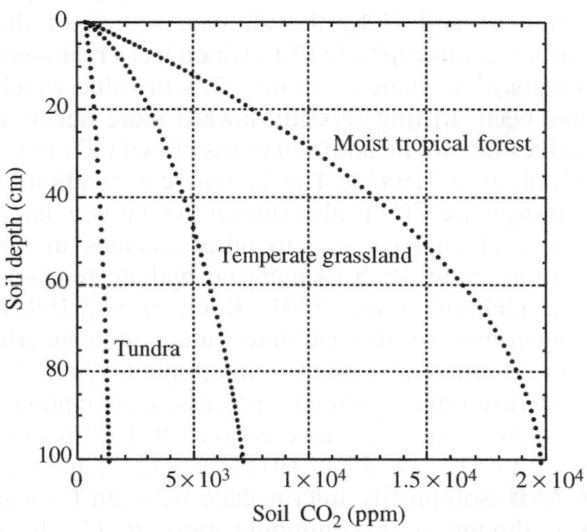

Figure 16 Calculated soil CO_2 concentrations versus depth for three contrasting ecosystems using Equation (13) and the following data: total soil respiration rates (tundra = 60 g C m^{-2}yr^{-1}, grassland = 442 g C m^{-2} yr^{-1}, and tropical forest = 1,260 g C m^{-2}yr^{-1}), D_s = 0.021 cm^{2}s^{-1}, atmospheric CO_2 = 350 ppm, and L = 100 cm (respiration data from Raich and Schlesinger, 1992).

soil depth (cm), and ϕ = CO_2 production (mol cm^{-3} s^{-1}). Using reported soil respiration rates, and reasonable parameter values for Equation (11), the CO_2 concentration profiles for three strongly contrasting ecosystems are illustrated in Figure 16.

For the boundary conditions of an impermeable lower layer and $CO_2(0) = CO_2(\text{atm})$, the solution to the model (with exponentially decreasing CO_2 production with depth) is

$$CO_2(z) = \frac{\phi_{z=0}z_0^2}{D}\left(1 - e^{-z/z_0}\right) + C_{\text{atm}} \quad (13)$$

where z_0 is the depth at which the production is $\phi_{z=0}/e$. At steady state, the flux of CO_2 from the soil to the atmosphere is simply the first derivative of Equation (13) evaluated at $z=0$ (e.g., Amundson *et al.*, 1998). The production, and transport of CO_2, is accompanied by the consumption and downward transport of O_2, which is driven and described by analogous processes and models (e.g., Severinghaus *et al.*, 1996).

Cerling's primary objective was the identification of the processes controlling the carbon-isotope composition of soil CO_2, and a quantitative means of describing the process. In terms of notation, carbon isotopes in compounds are evaluated as the ratio (R) of the rare to common stable isotope of carbon ($^{13}C/^{12}C$) and are reported in delta notation: $\delta^{13}C(‰) = (R_s/R_{std}-1)1{,}000$, where R_s and R_{std} refer to the carbon-isotope ratios of the sample and the international standard, respectively (Friedman and O'Neil, 1977).

In terms of the controls on the isotopic composition of soil CO_2, the ultimate source of the carbon is atmospheric CO_2, which has a relatively steady $\delta^{13}C$ value of about −7‰ (a value which has been drifting recently toward more negative values due to the addition of fossil fuel CO_2 (e.g., Mook *et al.*, 1983)). The isotopic composition of atmospheric CO_2 is also subject to relatively large temporal changes due to other changes in the carbon cycle, such as methane hydrate releases, etc. (Jahren *et al.*, 2001; Koch *et al.*, 1995). Regardless of the isotopic value, atmospheric CO_2 is utilized by plants through photosynthesis. As a result of evolutionary processes, three photosynthetic pathways have evolved in land plants: (i) C_3: $\delta^{13}C = \sim -27‰$, (ii) C_4: $\sim -12‰$, and (iii) CAM: isotopically intermediate between C_3 and C_4, though it is commonly close to C_3. It is believed that C_3 photosynthesis is an ancient mechanism, whereas C_4 photosynthesis (mainly restricted to tropical grasses) is a Cenozoic adaptation to decreasing CO_2 levels (Cerling *et al.*, 1997) or to strong seasonality and water stress (e.g., Farquhar *et al.*, 1988). Once atmospheric carbon is fixed by photosynthesis, it may eventually be added to soil as dead organic matter through surface litter, root detritus, or as soluble organics secreted by living roots. This material is then subjected to microbial decomposition, and partially converted to CO_2 which then diffuses back to the overlying atmosphere. An additional source of CO_2 production is the direct respiration of living roots, which is believed to account for ~50% of the total CO_2 flux out of soils (i.e., soil respiration; Hanson *et al.* (2000)). During decomposition of organic matter, there is a small (~2‰ or more) discrimination of carbon isotopes, whereby ^{12}C is preferentially lost as CO_2 and ^{13}C remains as humic substances (Nadelhoffer and Fry, 1988). Thus, soil organic matter commonly shows an enrichment (which increases with depth due to transport processes) of ^{13}C relative to the source plants (see Amundson and Baisden (2000) for an expanded discussion of soil organic carbon and nitrogen isotopes and modeling.

Cerling recognized that $^{12}CO_2$ and $^{13}CO_2$ can be described in terms of their own production and tranport models, and that the isotope ratio of CO_2 at any soil depth is described simply by the ratio of the ^{13}C and the ^{12}C models. For the purposes of illustration here, if we assume that the concentration of ^{12}C can be adequately described by that of total CO_2 and that CO_2 is produced at a constant rate over a given depth L, then the model describing the steady-state isotopic ratio of CO_2 at depth z is (see Cerling and Quade (1993) for the solution where the above assumptions are not applied)

$$R_s^{13} = \frac{(\phi R_p^{13}/D_s^{13})(Lz - z^2/2) + C_{atm}R_{atm}^{13}}{(\phi/D_s)(Lz - z^2/2) + C_{atm}} \quad (14)$$

where R_s, R_p, and R_{atm} refer to the isotopic ratios of soil CO_2, plant carbon, and atmospheric CO_2, respectively, and D^{13} is the diffusion coefficient of $^{13}CO_2$ which is $D_s/1.0044$. The $\delta^{13}C$ value of the CO_2 can be calculated by inserting R_s into the equation given above.

Quade *et al.* (1989a) examined the $\delta^{13}C$ value of soil CO_2 and pedogenic carbonate along elevation (climate) gradients in the Mojave Desert/Great Basin and utilized the models described above to analyze the data. Quade *et al.* found that there were systematic trends in soil CO_2 concentrations, and $\delta^{13}C$ values, due to changes in CO_2 production rates with increasing elevation, and soil depth (Figure 17). The primary achievement of this work was that the observations were fully explainable using the mechanistic model represented here by Equation (13), a result that opened the door for the use of pedogenic carbonates in paleoenvironmental (e.g., Quade *et al.*, 1989b) and atmospheric p_{CO_2} (Cerling, 1991) studies.

7.5.1.3 *Processes and isotope composition of pedogenic carbonate formation*

In arid and semi-arid regions of the world, where precipitation is exceeded by potential evapotranspiration, soils are incompletely leached and $CaCO_3$ accumulates in significant quantities. Figure 18 illustrates the global distribution of carbonate in the upper meter of soils. As the figure illustrates, there is a sharp boundary between calcareous and noncalcareous soil in the USA at about the 100th meridian. This long-recognized boundary reflects the soil water balance. Jenny and Leonard (1939)

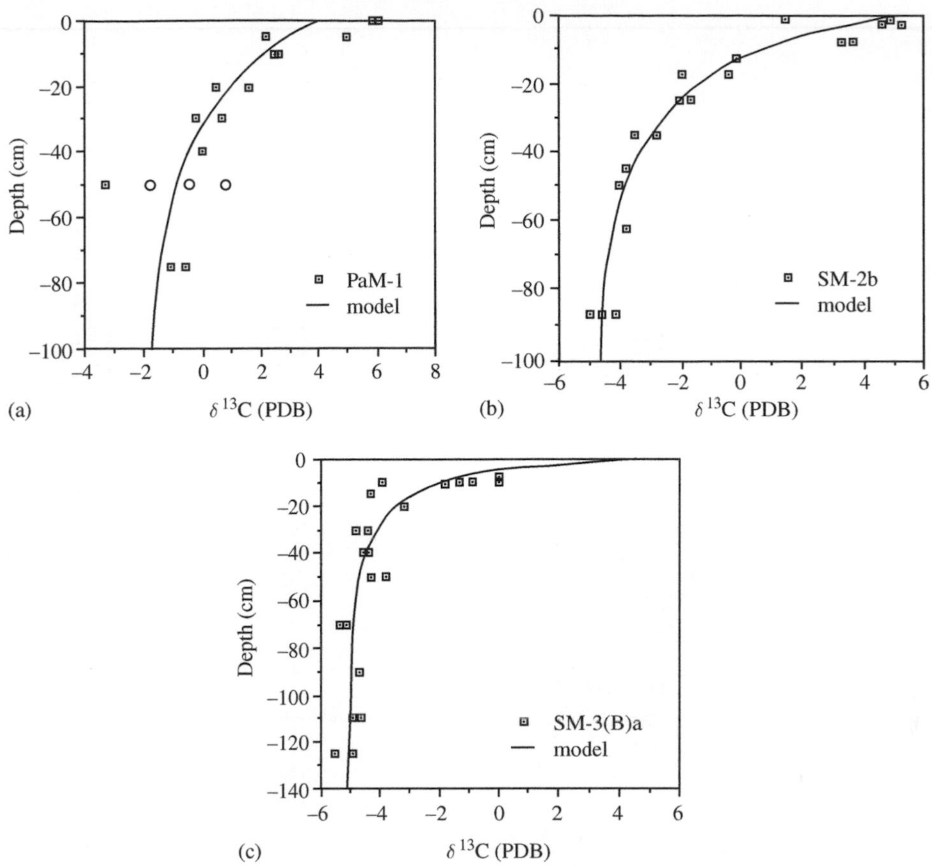

Figure 17 Measured (points) and modeled (curves) soil carbonate $\delta^{13}C$ values for three soils along an elevation (climate) gradient in the Mojave desert/Great Basin of California and Nevada. Modeled carbonate values based on Equation (14) plus the fractionation between CO_2 and carbonate (~10‰). The elevations (and modeled soil respiration rates that drive the curve fit) are: (a) 330 m (0.18 mmol $CO_2\,m^{-2}\,h^{-1}$); (b) 1,550 m (0.4 mmol $CO_2\,m^{-2}\,h^{-1}$); and (c) 1,900 m (1.3 mmol $CO_2\,m^{-2}\,h^{-1}$). The $\delta^{13}C$ value of soil organic matter (CO_2 source) was about −21‰ at all sites. Note that depth of atmospheric CO_2 isotopic signal decreases with increasing elevation and biological CO_2 production (Quade *et al.*, 1989a) (reproduced by permission of Geological Society of America from *Geol. Soc. Am. Bull.* **1989**, *101*, 464–475).

examined the depth to the top of the carbonate-bearing layer in soils by establishing a climosequence (precipitation gradient) along an east to west transect of the Great Plains (Figure 19). They observed that at constant mean average temperature (MAT) below 100 cm of mean average precipitation (MAP), carbonate appeared in the soils, and the depth to the top of the carbonate layer decreased with decreasing precipitation. An analysis has been made of the depth to carbonate versus precipitation relation for the entire USA (Royer, 1999). She found that, in general, the relation exists broadly but as the control on other variables between sites (temperature, soil texture, etc.) is relaxed, the strength of the relationship declines greatly.

In addition to the depth versus climate trend, there is a predictable and repeatable trend of carbonate amount and morphology with time (Gile *et al.* (1966); Figure 20) due to the progressive accumulation of carbonate over time, and the ultimate infilling of soil porosity with carbonate cement, which restricts further downward movement of water and carbonate.

The controls underlying the depth and amount of soil carbonate hinge on the water balance, Ca^{+2} availability, soil CO_2 partial pressures, etc. Arkley (1963) was the first to characterize these processes mathematically. His work has been greatly expanded by McFadden and Tinsley (1985), Marian *et al.* (1985), and others to include numerical models. Figure 21 illustrates the general concepts of McFadden and Tinsley's numerical model, and Figure 22 illustrates the results of model predictions for a hot, semi-arid soil (see the figure heading for model parameter values). These predictions generally mimic observations of carbonate distribution in desert soils, indicating that many of the key processes have been identified.

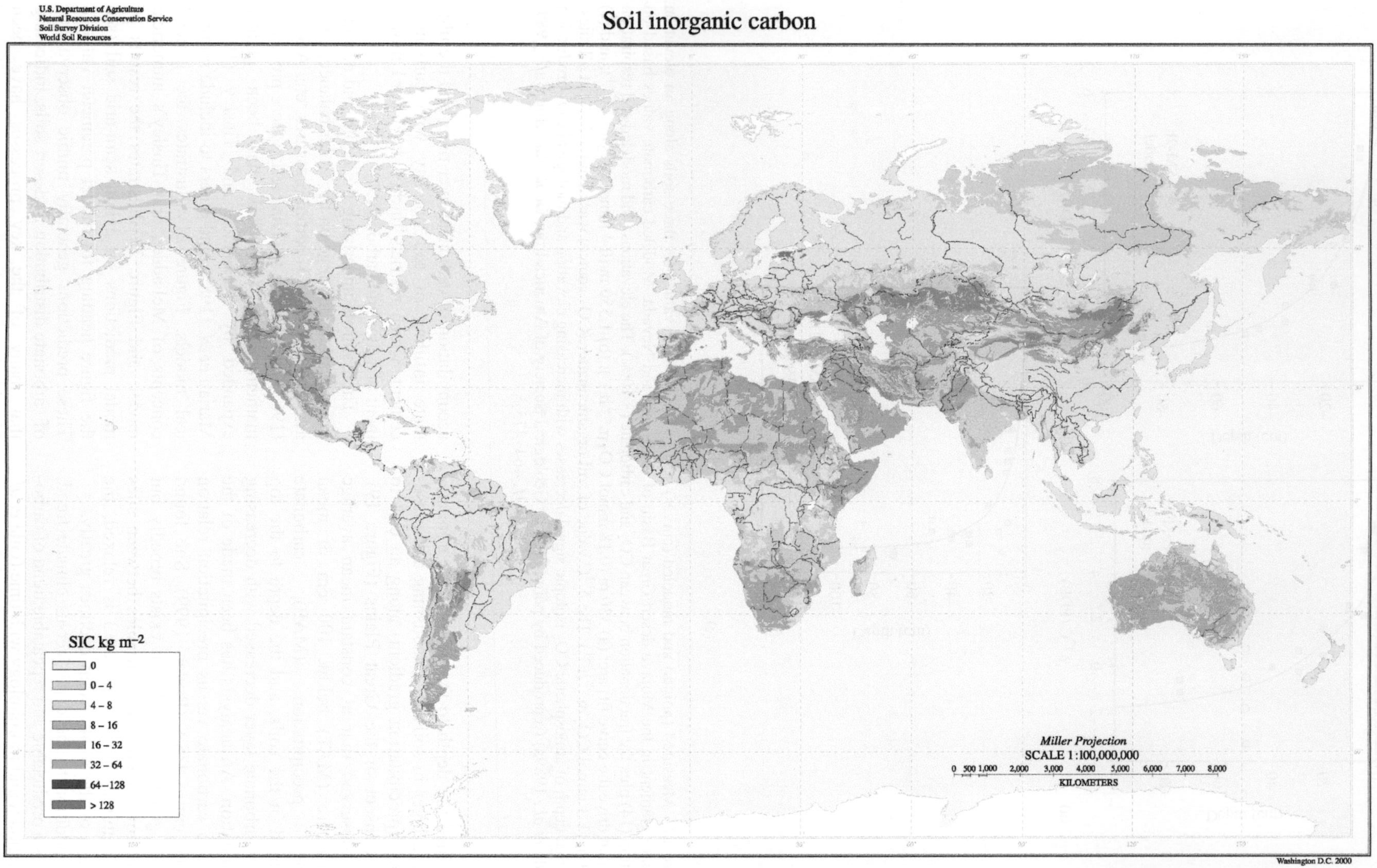

Figure 18 Global distribution of pedogenic carbonate (source http://www.nrcs.usda.gov/technical/worldsoils/mapindx/).

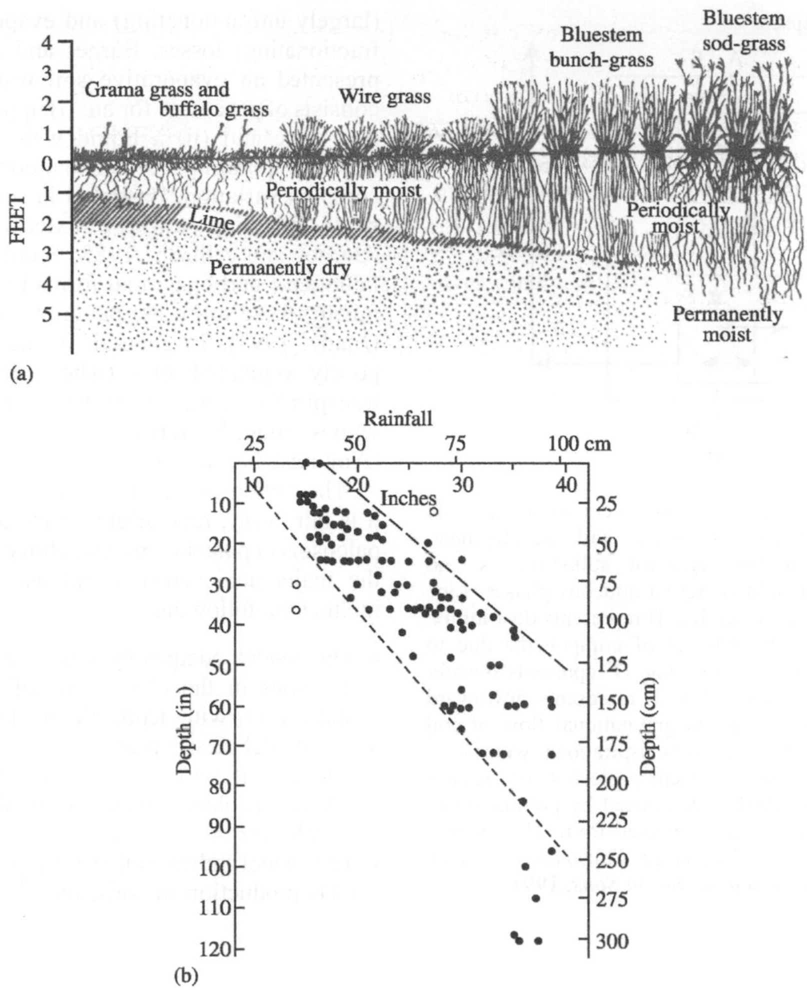

Figure 19 (a) Schematic view of plant and soil profile changes on an east (right) to west (left) transect of the Great Plains and (b) measured depth to top of pedogenic carbonate along the same gradient (source Jenny, 1941).

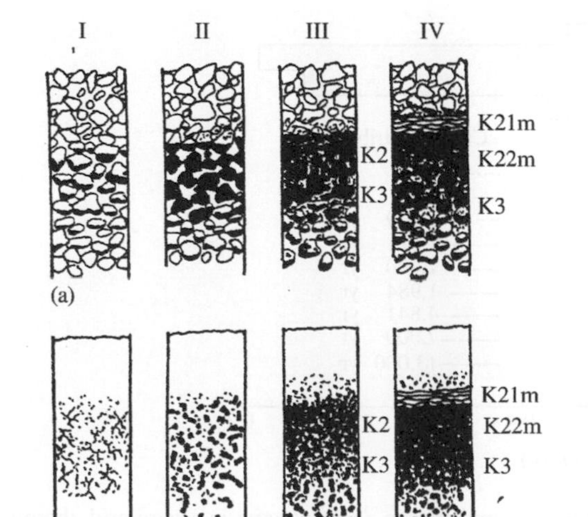

Figure 20 Soil carbonate morphology and amount versus time for: (a) gravelly and (b) fine-grained soils (Gile *et al.*, 1966) (reproduced by permission of Williams and Wilkins from *Soil Sci.* **1966**, *101*, 347–360).

The general equation describing the formation of carbonate in soils is illustrated by the reaction

$$CO_2 + H_2O + Ca^{+2} = CaCO_3 + 2H^+$$

From an isotopic perspective, in unsaturated soils, soil CO_2 represents an infinite reservoir of carbon and soil water an infinite reservoir of oxygen, and the $\delta^{13}C$ and $\delta^{18}O$ values of the pedogenic carbonate (regardless of whether its calcium is derived from silicate weathering, atmospheric sources, or limestone) are entirely set by the isotopic composition of soil CO_2 and H_2O. Here we focus mainly on the carbon isotopes. However, briefly for completeness, we outline the oxygen-isotope processes in soils. The source of soil H_2O is precipitation, whose oxygen-isotope composition is controlled by a complex set of physical processes (Hendricks *et al.*, 2000), but which commonly shows a positive correlation with MAT (Rozanski *et al.*, 1993). Once this water enters the soil, it is subject to transpirational

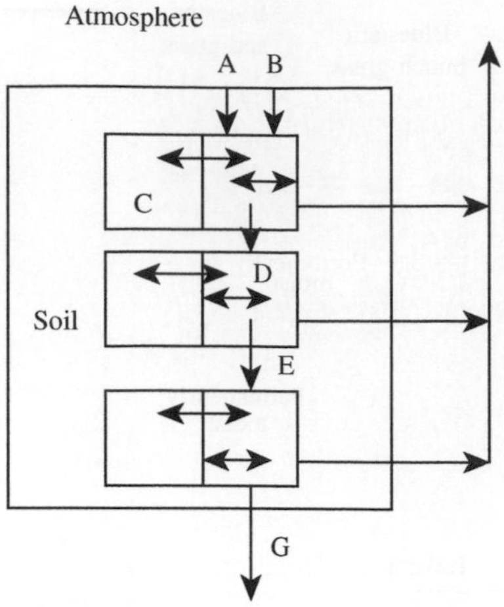

Figure 21 Diagram of compartment model for the numerical simulation of calcic soil development. Compartments on left represent solid phases and compartments on right represent aqueous phases. Line A represents precipitation, line B represents dust influx, line C represents the transfer of components due to dissolution and precipitation, line D represents transfer between aqueous phases, line E represents downward movement of solutes due to gravitational flow of soil water, line F represents evapotranspirational water loss, and line G represents leaching losses of solutes (McFadden *et al.*, 1991) (reproduced by permission of Soil Science Society of America from *Occurrence, Characteristics, and Genesis of Carbonate, Gypsum and Silica Accumulations in Soils*, **1991**).

(largely unfractionating) and evaporative (highly fractionating) losses. Barnes and Allison (1983) presented an evaporative soil water model that consists of processes for an: (i) upper, vapor transport zone and (ii) a liquid water zone with an upper evaporating front. The model describes the complex variations observed in soil water versus depth following periods of extensive evaporation (Barnes and Allison, 1983; Stern *et al.*, 1999). Pedogenic carbonate that forms in soils generally mirrors these soil water patterns (e.g., Cerling and Quade, 1993). In general, in all but hyperarid, poorly vegetated sites (where the evaporation/transpiration ratio is high), soil $CaCO_3$ $\delta^{18}O$ values roughly reflect those of precipitation (Amundson *et al.*, 1996).

The carbon-isotope model and its variants have, it is fair to say, revolutionized the use of soils and paleosols in paleobotany and climatology. Some of the major achievements and uses of the model include the following.

- The model adequately describes the observed increases in the $\delta^{13}C$ value of both soil CO_2 and $CaCO_3$ with depth (Figure 17).
- The model clearly provides a mechanistic understanding of why soil CO_2 is enriched in ^{13}C relative to plant inputs (steady-state diffusional enrichment of ^{13}C).
- The model indicates that for reasonable rates of CO_2 production in soils, the $\delta^{13}C$ value of soil

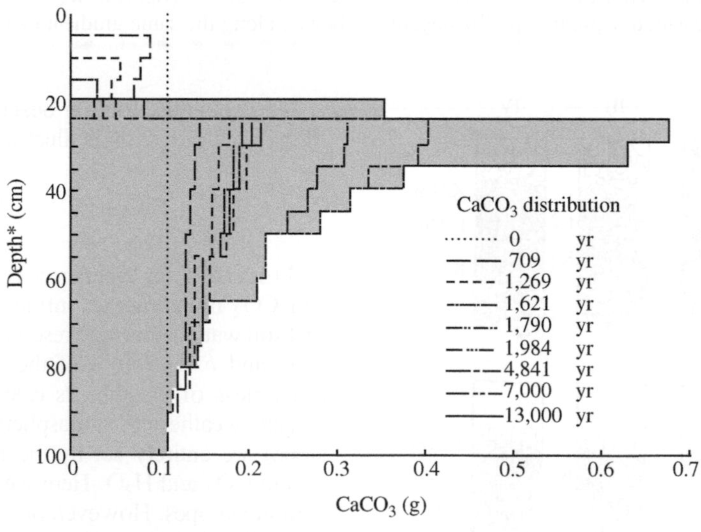

Figure 22 Predicted pattern of Holocene pedogenic carbonate accumulation in a cm^2 column in a semi-arid, thermic climate (leaching index = 3.5 cm). External carbonate flux rate = 1.5×10^{-4} g cm^{-2} yr^{-1}, $p_{CO_2} = 1.5\times10^{-3.3}$ atm in compartment 1 increasing to $10^{-2.5}$ atm in compartment 5 (20–25 cm). Below compartment 5, the p_{CO_2} decreases to a minimum value of 10^{-4} atm in compartment 20 (95–100 cm). Dotted line shows carbonate distribution at $t = 0$. Gray area indicates final simulated distribution. Depth* = absolute infiltration depth in <2 mm fraction (McFadden *et al.*, 1991) (reproduced by permission of Soil Science Society of America from *Occurrence, Characteristics, and Genesis of Carbonate, Gypsum and Silica Accumulations in Soils*, **1991**).

CO_2 should, at a depth within 100 cm of the surface, represent the $\delta^{13}C$ value of the standing biomass plus 4.4‰. The $\delta^{13}C$ of $CaCO_3$ will also reflect this value, plus an equilibrium fractionation of ~10‰ (depending on temperature). Therefore, if paleosols are sampled below the "atmospheric mixing zone," whose thickness depends on CO_2 production rates (Figure 17), the $\delta^{13}C$ value of the carbonate will provide a guide to the past vegetation (Cerling *et al.*, 1989).

Cerling (1991) recognized that Equation (14), if rearranged and solved for C_{atm}, could provide a means of utilizing paleosol carbonates for reconstructing past atmospheric CO_2 partial pressures. To do so, the values of the other variables (including the $\delta^{13}C$ value of the atmospheric CO_2—see Jahren *et al.* (2001)) must be known, which for soils of the distant past is not necessarily a trivial problem. Nonetheless, an active research field has developed using this method, and a compilation of calculated atmospheric CO_2 levels is emerging (Ekart *et al.*,1999; Mora *et al.*, 1996), with estimates that correlate well with model calculations by Berner (1992).

Cerling's (1984) approach to modeling stable carbon isotopes in soil CO_2 has been expanded and adapted to other isotopes in soil CO_2: (i) $^{14}CO_2$—for pedogenic carbonate dating (Wang *et al.*, 1994; Amundson *et al.*, 1994a,b) and soil carbon turnover studies (Wang *et al.*, 2000) and (ii) $C^{18}O^{16}O$—for hydrological tracer applications and, more importantly, as a means to constrain the controls on global atmospheric CO_2–^{18}O budgets (Hesterberg and Siegenthaler, 1991; Amundson *et al.*, 1998; Stern *et al.*, 1999, 2001; Tans, 1998). The processes controlling the isotopes, and the complexity of the models, greatly increase from ^{14}C to ^{18}O (see Amundson *et al.* (1998) for a detailed account of all soil CO_2 isotopic models).

7.5.2 Lateral Transport of Soil Material by Erosion

The previous section, devoted to the soil carbon cycle, emphasized the conceptual and mathematical focus on the vertical transport of materials. Models emphasizing vertical transport dominate present pedological modeling efforts. However, for soils on hillslopes or basins and floodplains, the lateral transport of soil material via erosive processes exerts an overwhelming control on a variety of soil properties. The study of the mechanisms by which soil is physically moved—even on level terrain—is an evolving aspect of pedology. Here we focus on new developments in the linkage between geomorphology and pedology in quantifying the effect of sediment transport on soil formation. The focus here is on natural, undisturbed landscapes as opposed to agricultural landscapes, where different erosion models may be equally or more important.

This section begins by considering soils on divergent, or convex, portions of the landscape. On divergent hillslopes, slope increases with distance downslope such that (for a two-dimensional view of a hillslope)

$$\frac{\Delta\text{slope}}{\Delta\text{distance}} = \left(\frac{\partial}{\partial x}\right)\left(\frac{\partial z}{\partial x}\right) = \text{negative quantity} \quad (15)$$

where z and x are distances in a vertical and horizontal direction, respectively. On convex slopes, there is an ongoing removal of soil material due to transport processes. In portions of the landscape where the derivative of slope versus distance is positive (i.e., convergent landscapes), there is a net accumulation of sediment, which will be discussed next.

In many divergent landscapes (those not subject to overland flow or landslides), the ongoing soil movement may be almost imperceptible on human timescales. Research by geomorphologists (and some early naturalists such as Darwin) has shed light on both the rate and mechanisms of this process. In general, a combination of physical and biological processes, aided by gravity, can drive the downslope movement of soil material. These processes are sometimes viewed as roughly diffusive in that soil is randomly moved in many directions, but if a slope gradient is present, the net transport is down the slope. Kirkby (1971), in laboratory experiments, showed that wetting and drying cycles caused soil sensors to move in all directions, but in a net downslope direction. Black and Montgomery (1991) examined pocket gopher activity on sloping landscapes in California, discussing their underground burrowing and transport mechanisms, and the rate at which upthrown material then moves downslope. Tree throw also contributes to diffusive-like movement (Johnson *et al.*, 1987). In a truly far-sighted study, Darwin (1881) quantified both the mass and the volume of soil thrown up by earthworms, and the net amount moved downslope by wind, water, and gravity.

The general transport model for diffusive-like soil transport is (Kirkby, 1971)

$$Q = -\rho_s K \frac{dz}{dx} \quad (16)$$

where Q is soil flux per unit length of a contour line $(\rho_s)_2$ $(g\,cm^{-1}yr^{-1})$ and K is transport coefficient $(cm^2\,yr^{-1})$. The concepts behind this expression are generally attributed to Davis (1892) and Gilbert (1909). However, Darwin (1881) clearly recognized this principle in relation to sediment transport by earthworms on slopes. For several sites in England (based on a few short-term measurements), Darwin reported Q and slope, allowing us to calculate a K based solely on earthworms. The values of $\sim 0.02\ \mathrm{cm}^2\,\mathrm{yr}^{-1}$, while significant, are about two orders of magnitude lower than overall K values calculated for parts of the western USA that capture the integrated biological and abiotic mixing/transport mechanisms (McKean *et al.*, 1993; Heimsath *et al.*, 1999).

At this point, it is worth commenting further on the effect and role of bioturbation in soil formation processes whether it results in net downslope movement or not. Beginning with Darwin, there has been a continued, but largely unappreciated, series of studies concerning the effects of biological mixing on soil profile features (Johnson *et al.*, 1987; Johnson, 1990). These studies concluded that most soils (those maintaining at least some plant biomass to support a food chain) have a "biomantle" (a highly mixed and sorted zone) that encompasses the upper portions of the soil profile (Johnson, 1990). The agents are varied, but include the commonly distributed earthworms, gophers, ground squirrels, ants, termites, wombats, etc. In many cases, the biomantle is equivalent to the soil's A horizons. The rapidity at which these agents can mix and sort the biomantle is impressive. The invasion of earthworms into the Canadian prairie resulted in complete homogenization of the upper 10 cm in 3 yr (Langmaid, 1964). Johnson (1990) reports that Borst (1968) estimated that the upper 75 cm of soils in the southern San Joaquin Valley of California are mixed in 360 yr by ground squirrels. Darwin reported that in England, $\sim 1.05\times 10^4$ kg of soil per hectare are passed through earthworms yearly, resulting in a turnover time for the upper 50 cm of the soil of just over 700 yr. The deep burial of Roman structures and artifacts in England by earthworm casts supports this estimated cycling rate.

Bioturbation on level ground may involve a large gross flux of materials (as indicated above), but no net movement in any direction due to the absence of a slope gradient. Nonetheless, the mixing has important physical and chemical implications for soil development: (i) rapid mixing of soil and loss of stratification at initial stages of soil formation; (ii) rapid incorporation of organic matter (and the subsequent slowed decomposition of this material below ground); (iii) periodic cycling of soil structure which prevents soil horizonation within the biomantle; and (iv) in certain locations, striking surficial expressions of biosediment movement caused by nonrandom sediment transport. The famed "mima mounds" of the Plio-Pleistocene fluvial terraces and fans of California's Great Valley (Figure 23(a)) are a series of well-drained sandy loam to clay loam materials that overlie a relatively level, impermeable layer (either a dense, clay-rich horizon or a silica cemented hardpan)(Figure 23(b)). The seasonally waterlogged conditions that develop over the impermeable layers have probably caused gophers to move soil preferentially into better drained landscape segments, thereby producing this unusual landscape. These landform features are now relatively rare due to the expansion of agriculture in the state, but the original extent of this surface feature is believed to have been more than 3×10^5 ha (Holland, 1996).

The conceptual model for diffusive soil transport down a hillslope is shown in Figure 24 (Heimsath *et al.*, 1997, 1999). In any given section of the landscape, the mass of soil present is the balance of transport in, transport out, and soil production (the conversion of rock or sediment to soil). If it is assumed that the processes have been operating for a sufficiently long period of time, then the soil thickness is at steady state. The model describing this condition is

$$\rho_s \frac{\partial h}{\partial t} = 0 = \underbrace{(\rho_r \phi)}_{\text{soil production}} + \underbrace{\left(K \rho_s \frac{\partial^2 z}{\partial x^2}\right)}_{\substack{\text{balance between} \\ \text{diffusive inputs losses}}} \tag{17}$$

where h is the soil thickness (cm), ρ_s and ρ_r the soil and rock bulk density ($\mathrm{g\,cm}^{-3}$), respectively, and ϕ the soil production rate ($\mathrm{cm\,yr}^{-1}$).

Soil production is a function of soil depth (Heimsath *et al.*, 1997), parent material, and environmental conditions (Heimsath *et al.*,1999). As soil thickens, the rate of the conversion of the underlying rock or sediment to soil decreases. This has been shown using field observations of the relation between soil thickness and the abundance of cosmogenic nuclides (^{10}Be and ^{26}Al) in the quartz grains at the rock–soil interface (Figure 24). From this work, soil production can be described by

$$\phi = \rho_r \phi_0 e^{-\alpha h} \tag{18}$$

where ϕ and ϕ_0 are soil production for a given soil thickness and no soil cover, respectively; α the constant (cm^{-1}), and h the soil thickness (cm). The soil production model described by Equation (18) is applicable where physical disruption of the bedrock is the major soil production mechanism. By inserting Equation (18) into Equation (17), and rearranging, one can solve for soil thickness at any position on a hillslope:

$$h = \frac{1}{\alpha}\left(-\ln\left(-\frac{K\rho_s \partial^2 z}{\phi_0 \rho_r \partial x^2}\right)\right) \tag{19}$$

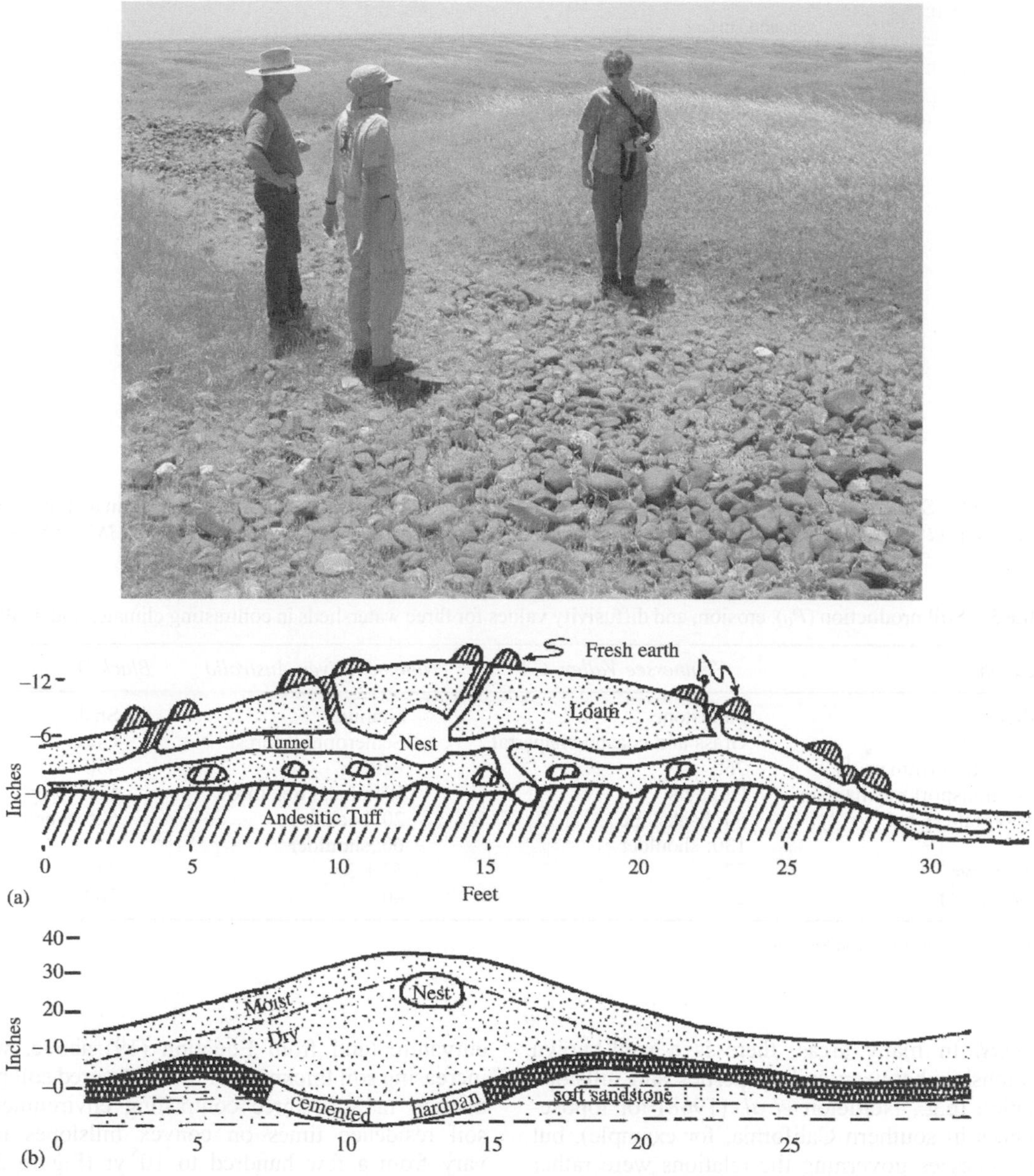

Figure 23 (a) Mima mounds (grass-covered areas) and vernal pools (gravel veneered low areas) on a Plio-Pleistocene aged fluvial terrace of the Merced river, California. Observations indicate that the landscape is underlain by a dense, clay pan (formed by long intervals of soil Bt horizon formation) capped by a gravel lense (the vernal pool gravels extend laterally under the Mima mounds). (b) Schematic diagram of "pocket gopher theory" of mima mound formation, illustrating preferential nesting and soil movement in/toward the well-drained mound areas and away from the seasonally wet vernal pools (Arkley and Brown, 1954) (reproduced by permission of Soil Science Society of America from *Soil Sci. Soc. Am. Proc.* **1954**, *18*, 195–199).

As this equation indicates, the key variables controlling soil thickness on hillslopes are slope curvature, the transport coefficient, and the soil production rate. The importance of curvature is intuitive in that changes in slope gradient drive the diffusive process. The value of the transport coefficient K varies greatly from one location to another (Fernandes and Dietrich, 1997; Heimsath *et al.*,1999), and seems to increase with increasing humidity and decreasing rock competence (Table 5). Production, like transport, is likely dependent on climate and rock composition (Heimsath *et al.*,1999).

The production/transport model of sediment transport provides a quantitative and mechanistic insight into soil profile thickness on hillslopes, an area of research that has received limited, but in some cases insightful, attention in the pedological literature. The study of the distribution of soil along hillslope gradients are called *toposequences* or

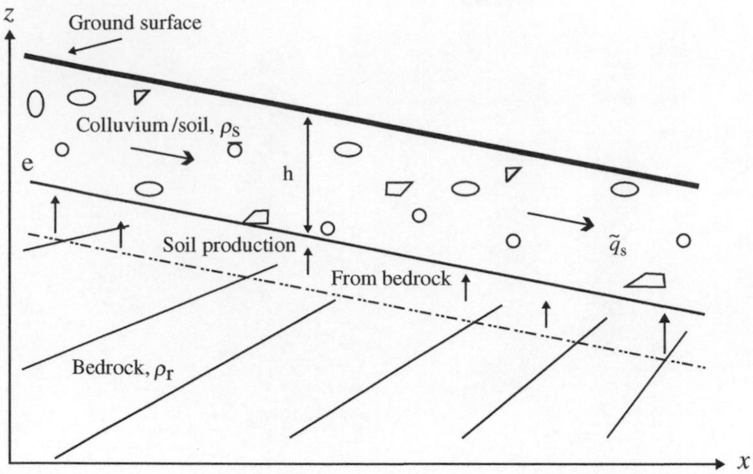

Figure 24 Schematic diagram of soil production and downslope transport on diffusion-dominated hillslopes (Heimsath *et al.*, 1997) (reproduced by permission of Nature Publishing Group from *Nature* **1997**, *388*, 358–361).

Table 5 Soil production (P_0), erosion, and diffusivity values for three watersheds in contrasting climates and geology.

Location	*Tennessee Valley, CA*	*Nunnock River, Australia*	*Black Diamond, CA*
Bedrock	Sandstone	Granitic	Shale
Vegetation	Grass and coastal chaparral	Schlerophyll forest	Grass
Precipitation (mm yr^{-1})	760	910	450
Major transport mechanism	Gophers	Wombats, tree throw	Soil creep
Erosion rate (g m^{-2} yr^{-1})	50: backslope	20: backslope	625 average
	130: shoulder	60: shoulder	
P_0 (mm kyr^{-1})	77 ± 9	53 ± 3	2,078
K (cm^2 yr^{-1})	25	40	360

Sources: Heimsath (1999) and McKean *et al.* (1993).

catenas. In many areas, there is a well-known relationship between soil properties and hillslope position (e.g., Nettleton *et al.* (1968) for toposequences in southern California, for example), but the processes governing the relations were rather poorly known—at least on a quantitative level.

Beyond the effect of slope position (and environmental conditions) on soil thickness, a key factor is the amount of time that a soil on a given hillslope has to form. Time is a key variable that determines the amount of weathering and horizon formation that has occurred (Equation (2)). Yet, constraining "soil age" on erosional slopes has historically been a challenging problem. Here the author takes a simple approach and views soil age on slopes in terms of residence time (τ), where

$$\tau = \frac{\text{soil mass/area}}{\rho_r \phi_0 e^{-\alpha h}} = \frac{\text{soil mass/area}}{\rho_s K(\partial^2 z/\partial x^2)} \qquad (20)$$

This is a pedologically meaningful measure that defines the rate at which soil is physically moved through a soil "box." As the expression indicates, soil residence time increases with decreasing curvature (and from Equation (19) above, with increasing soil thickness). From measured soil production rates in three contrasting environments, soil residence times on convex hillslopes may vary from a few hundred to 10^5 yr (Figure 25). These large time differences clearly help to explain many of the differences in soil profile development long observed on hillslope gradients.

On depositional landforms, sloping areas with concave slopes (convergent curvature) or level areas on floodplains, the concept of residence time can also be applied and quantitative models of soil formation can be derived. In these settings, residence time can be viewed as the amount of time required to fill a predetermined volume (or thickness) of soil with incoming sediment.

On active floodplains of major rivers, the soil residence time can be in most cases no longer than a few hundred years. Soil profiles here exhibit stratification, buried, weakly developed horizons, and little if any measurable chemical weathering.

In a landmark study of soils on depositional settings, Jenny (1962) examined soils on the

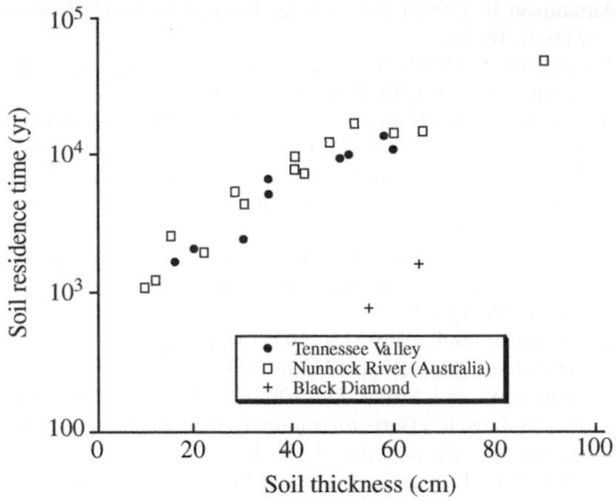

Figure 25 Calculated soil residence time versus soil thickness in three contrasting watersheds using Equation (20) and data from Table 5.

floodplain of the Nile River, Eygpt. Available data suggested that the sedimentation rate prior to the construction of the Aswan Dam was 1 mm yr^{-1}, giving a residence time of 1,500 yr for a soil 150 cm thick. Jenny observed that in these soils, used for agriculture for millennia, nitrogen (the most crop-limiting element in agriculture) decreased with depth. This trend was not due to plant/atmospheric nitrogen inputs and crop cycling which creates the standard decrease in carbon and nitrogen with depth (see previous section on soil organic carbon). Instead, the continual deposition of nitrogen-rich sediment derived from soils in the Ethopian highlands was the likely source.

Jenny demonstrated that the observed nitrogen decrease with depth was a result of microbial degradation of the nitrogen-bearing organic matter in the sediment. The mineralized nitrogen was used by the crops and was then removed from the site. Stated mathematically

$$N(z) = N(0)e^{-kz} \tag{21}$$

where $N(z)$ and $N(0)$ are the soil nitrogen at depth z and 0, respectively, and k the decomposition rate constant. Jenny's analysis showed that on the Nile floodplain, in the absence of nitrogen fertilization, long-term high rates of crop production could only be maintained by continued flood deposition of Nile sediments.

More generally, on concave (or convergent) slopes, where sedimentation is caused by long-term diffusional soil movement, one can calculate rates of sedimentation (or, of course, measure them directly) using a version of the geomorphic models discussed above. In deposition settings, soil production from bedrock or sediment can be viewed as being zero, and the change in soil height with time is (Fernandes and Dietrich, 1997)

$$\frac{\partial h}{\partial t} = K\frac{\partial^2 z}{\partial x^2} \tag{22}$$

Soils in these depositional settings are sometimes referred to as "cumulic" soils in the US Soil Taxonomy, and commonly exhibit high concentrations of organic carbon and nitrogen to depths of several meters. This organic matter is partially due to the burial of organic matter produced *in situ*, but also reflects the influx of organic-matter-rich soil from upslope positions. This process of organic carbon burial has attracted the attention of the global carbon cycle community, because geomorphic removal and burial of organic carbon may represent a large, neglected, global carbon flux (Stallard, 1998; Rosenblum *et al.*, 2001). Beyond the carbon burial, the relatively short residence time of any soil profile thickness inhibits the accumulation of chemical weathering products, vertical transport, and horizon development.

In summary, an under-appreciated fact is that all soils—regardless of landscape position—are subject to slow but measurable physical and (especially) biological turbation. On level terrain, the soil/sediment flux is relatively random. It results in extensive soil surface mixing and, only in certain circumstances, striking directional movement. The addition of a slope gradient, in combination with this ongoing soil turbation, results in a net downslope movement dependent on slope curvature and the characteristics of the geological and climatic settings. The thickness of a soil, and its residence time, is directly related to the slope curvature. These conditions also determine the expression of chemical weathering and soil horizonation that can occur on a given landscape position.

7.6 SOIL DATA COMPILATIONS

Soils, and their properties, are now being used in many regional to global biogeochemical analyses. The scientific and education community is fortunate to have a wealth of valuable, high-quality, data on soil properties available on the worldwide web and in other electronic avenues. Here the author identifies some of these data sources.

During soil survey operations, numerous soil profile descriptions, and large amounts of laboratory data, are generated for the soils that are being mapped. Generally, the soils being described, sampled, and analyzed are representatives of *soil series*, the most detailed (and restrictive) classification of soils in the USA. There are $\sim 2.1\times 10^4$ soil series that have been identified and mapped in the USA. The locations, soil descriptions, and lab data for many of these series are now available on the worldwide web via the USDA-NRCS (Natural Resource Conservation Service) National Soil Survey Center web site (http://www.statlab.iastate.edu/soils/ssl/natch_data.html). The complete data set is archived at the National Data Center, and is available to researchers.

The "characterization" data sheet is likely to be of interest to most investigators. In general, these data sheets are somewhat regionally oriented, reporting different chemical analyses for different climatic regions. For example, salt chemistry is reported for arid regions, whereas a variety of iron and aluminum oxide analyses are reported for humid, northern regions. Some major analyses of each soil horizon that are commonly reported include: particle size data, organic carbon and nitrogen, chemically extractable metals, cation exchange capacity and base saturation, bulk density, water retention, extractable bases and acidity, calcium carbonate, pH, chemistry of water extracted from saturated pastes, clay mineralogy, and sand and silt mineralogy (to list some of the major analyses). A complete and thorough discussion of the types of soil analyses, and the methods used, are presented by the Soil Survey Staff (1995), and on the web.

7.7 CONCLUDING COMMENTS

The scientific study of soils is just entering its second century. An impressive understanding of soil geography and, to a lesser degree, soil processes has evolved. There is growing interest in soils among scientists outside the agricultural sector, particularly geochemists and ecologists. Soils are central in the present attention to the global carbon cycle and its management, and as the human impact on the planet continues to increase, soils and their properties and services are being considered from a geobiodiversity perspective (Amundson, 1998; Amundson *et al.*, 2003). In many ways, the study of soils is at a critical and exciting point, and interdisciplinary cross-fertilization of the field is sure to lead to exciting new advances that bring pedology back to its multidisciplinary origins.

REFERENCES

Allègre C. (1992) *From Stone to Star: A View of Modern Geology.* Harvard University Press, Cambridge.

Allen R. M., Chatterjee R., Madden M. S., Ludden P. W., and Shar V. K. (1994) Biosynthesis of the iron–molybdenum co-factor of nitrogenase. *Crit. Rev. Biotech.* **14**, 225–249.

Alpers C. N. and Brimhall G. H. (1988) Middle Miocene climatic change in the Atacama desert, northern Chile: evidence for supergene mineralization at La Escondida. *Geol. Soc. Am. Bull.* **100**, 1640–1656.

Amundson R. (1998) Do soils need our protection? *Geotimes* March, 16–20.

Amundson R. (2001) The carbon budget in soils. *Ann. Rev. Earth Planet. Sci.* **29**, 535–562.

Amundson R. and Baisden W. T. (2000) Stable isotope tracers and mathematical models in soil organic matter studies. In *Methods in Ecosystem Science* (eds. O. E. Sala, R. B. Jackson, H. A. Mooney, and R. W. Howarth). Springer, New York, pp. 117–137.

Amundson R. and Davidson E. (1990) Carbon dioxide and nitrogenous gases in the soil atmosphere. *J. Geochem. Explor.* **38**, 13–41.

Amundson R. and Jenny H. (1997) On a state factor model of ecosystems. *Bioscience* **47**, 536–543.

Amundson R. and Yaalon D. (1995) E. W. Hilgard and John Wesley Powell: efforts for a joint agricultural and geological survey. *Soil Sci. Soc. Am. J.* **59**, 4–13.

Amundson R., Harden J., and Singer M. (1994a) *Factors of Soil Formation: A Fiftieth Anniversary Retrospective,* SSSA Special Publication No. 33. Soil Science Society of America, Madison, WI.

Amundson R., Wang Y., Chadwick O., Trumbore S., McFadden L., McDonald E., Wells S., and DeNiro M. (1994b) Factors and processes governing the ^{14}C content of carbonate in desert soils. *Earth Planet Sci. Lett.* **125**, 385–405.

Amundson R., Chadwick O. A., Kendall C., Wang Y., and DeNiro M. (1996) Isotopic evidence for shifts in atmospheric circulation patterns during the late Quaternary in mid-North America. *Geology* **24**, 23–26.

Amundson R., Stern L., Baisden T., and Wang Y. (1998) The isotopic composition of soil and soil-respired CO_2. *Geoderma* **82**, 83–114.

Amundson R., Guo Y., and Gong P. (2003) Soil diversity and landuse in the United States. *Ecosystems,* doi: 10.1007/s10021-002-0160-2.

Anders E. and Grevesse N. (1989) Abundances of the elements: meteoric and solar. *Geochim. Cosmochim. Acta* **53**, 197–214.

Arkley R. J. (1963) Calculations of carbonate and water movement in soil from climatic data. *Soil Sci.* **96**, 239–248.

Arkley R. J. and Brown H. C. (1954) The origin of mima mound (hog wallow) microrelief in the far western states. *Soil Sci. Soc. Am. Proc.* **18**, 195–199.

Baisden W. T. (2000) Soil organic matter turnover and storage in a California annual grassland chronosequence. PhD Dissertation, University of California, Berkeley.

Baisden W. T., Amundson R., Brenner D. L., Cook A. C., Kendall C., and Harden J. W. (2002) A multi-isotope C and N modeling analysis of soil organic matter turnover and transport as a function of soil depth in a California annual grassland chronosequence. *Glob. Biogeochem. Cycles,* 16(4), 1135, doi: 10.1029/2001/GB001823, 2002.

Bao H., Campbell D. A., Bockheim J. G., and Theimens M. H. (2000a) Origins of sulphate in Antarctic dry-valley soils as deduced fro anomalous ^{17}O compositions. *Nature* **407**, 499–502.

Bao H., Thiemens M. H., Farquhar J., Campbell D. A., Lee C. C.-W., Heine K., and Loope D. B. (2000b) Anomalous ^{17}O compositions in massive sulphate deposits on the Earth. *Nature* **406**, 176–178.

Bao H., Thiemens M. H., and Heine K. (2001) Oxygen-17 excesses of the Central Namib gypcretes: spatial distribution. *Earth Planet. Sci. Lett.* **192**, 125–135.

Barnes C. J. and Allison G. B. (1983) The distribution of deuterium and ^{18}O in dry soils: I. Theory. *J. Hydrol.* **60**, 141–156.

Betancourt J. L., Latorre C., Rech J. A., Quade J., and Rlander K. A. (2000) A 22,000-year record of monsoonal precipitation from Chile's Atacama desert. *Science* **289**, 1542–1546.

Black T. A. and Montgomery D. R. (1991) Sediment transport by burrowing animals, Marin county, California. *Earth Surf. Proc. Landforms* 16, 163–172.

Berner R. A. (1992) Palaeo-CO_2 and climate. *Nature* **358**, 114.

Birkeland P. W. (1999) *Soils and Geomorphology,* 3rd edn. Oxford University Press, New York.

Böhlke J. K., Erickson G. E., and Revesz K. (1997) Stable isotope evidence for an atmospheric origin of desert nitrate deposits in northern Chile and southern California, USA. *Chem. Geol.* **136**, 135–152.

Borst G. (1968) The occurrence of crotovinas in some southern California soils. *Trans. 9th Int. Congr. Soil Sci.,* Adelaide **2**, 19–22.

Bowen H. J. M. (1979) *Environmental Chemistry of the Elements.* Academic Press, London. Brimhall G. (1987) Preliminary fractionation patterns of ore metals through Earth history. *Chem. Geol.* **64**, 1–16.

Brimhall G. H. and Dietrich W. E. (1987) Constitutive mass balance relations between chemical composition, volume, density, porosity, and strain in metasomatic hydrochemical systems: results on weathering and pedogenesis. *Geochim. Cosmochim. Acta* **51**, 567–587.

Brimhall G. H., Lewis C. J., Ague J. J., Dietrich W. E., Hampel J., Teague T., and Rix P. (1988) Metal enrichment in bauxite by deposition of chemically-mature eolian dust. *Nature* **333**, 819–824.

Brimhall G. H., Lewis C. J., Ford C., Bratt J., Taylor G., and Warren O. (1991) Quantitative geochemical approach to pedogenesis: importance of parent material reduction, volumetric expansion, and eolian influx in laterization. *Geoderma* **51**, 51–91.

Brimhall G., Chadwick O. A., Lewis C. J., Compston W., Williams I. S., Danti K. J., Dietrich W. E., Power M. E., Hendricks D., and Bratt J. (1992) Deformational mass balance transport and invasive processes in soil evolution. *Science* **255**, 695–702.

Cerling T. E. (1984) The stable isotopic composition of modern soil carbonate and its relationship to climate. *Earth Planet. Sci. Lett.* **71**, 229–240.

Cerling T. E. (1991) Carbon dioxide in the atmosphere: evidence from Cenozoic and Mesozoic paleosols. *Am. J. Sci.* **291**, 377–400.

Cerling T. E. and Quade J. (1993) Stable carbon and oxygen isotopes in soil carbonates. In *Climate Change in Continental Isotopic Records* (eds. P. K. Swart, K. C. Lohmann, J. McKenzie, and S. Sabin). Geophysical Monograph 78. American Geophysical Union, Washington, DC, pp. 217–231.

Cerling T. E., Quade J., Wang Y., and Bowman J. R. (1989) Carbon isotopes in soils and palaeosols as paleoecologic indicators. *Nature* **341**, 138–139.

Cerling T. E., Harris J. M., MacFadden B. J., Leakey M. G., Quade J., Eisenmann V., and Ehleringer J. R. (1997) Global vegetation change through the Miocene/Pliocene boundary. *Nature* **389**, 153–158.

Chadwick O. A. and Davis J. O. (1990) Soil-forming intervals caused by eoloian sediment pulses in the Lahontan basin, northwestern Nevada. *Geology* **18**, 243–246.

Chadwick O. A., Brimhall G. H., and Hendricks D. M. (1990) From a black to a gray box—a mass balance interpretation of pedogenesis. *Geomorphology* **3**, 369–390.

Chadwick O. A., Derry L. A., Vitousek P. M., Huebert B. J., and Hedin L. O. (1999) Changing sources of nutrients during four million years of ecosystem development. *Nature* **397**, 491–497.

Chiappini C. (2001) The formation and evolution of the Milky Way. *Am. Sci.* **89**, 506–515.

Clarkson D. T. (1974) *Ion Transport and Cell Structure in Plants.* McGraw-Hill, New York, 350pp.

Darwin C. (1846) An account of the fine dust which often falls on vessels in the Atlantic ocean. *Quat. J. Geol. Soc. London* **2**, 26–30.

Darwin C. (1881) *The Formation of Vegetable Mould through the Action of Worms.* Appleton, New York.

Davis W. M. (1892) The convex profile of badland divides. *Science* **20**, 245.

De Coninck F. (1980) Major mechanisms in formation of spodic horizons. *Geoderma* **24**, 101–128.

Delsemme A. H. (2001) An argument for the cometary origin of the biosphere. *Am. Sci.* **89**, 432–442.

Dokuchaev V. V. (1880) Protocol of the meeting of the branch of geology and mineralogy of the St. Petersburg Society of Naturalists. *Trans. St. Petersburg Soc. Nat.* **XII**, 65–97. (Translated by the Department of Soils and Plant Nutrition, University of California, Berkeley).

Drake M. J. and Righter K. (2002) Determining the composition of the earth. *Nature* **416**, 39–51.

Ekart D. D., Cerling T. E., Montanañez I. P., and Tabor N. J. (1999) A 400 million year carbon-isotope record of pedogenic carbonate: implications for paleoatmospheric carbon dioxide. *Am. J. Sci.* **299**, 805–827.

Elzein A. and Balesdent J. (1995) Mechanistic simulation of vertical distribution of carbon concentrations and residence times in soils. *Soil Sci. Soc. Am. J.* **59**, 1328–1335.

Ericksen G. E. (1981) Geology and origin of the Chilean nitrate deposits. Geological Survey Professional Paper 1188, US Geological Survey, 37pp.

Farquhar G. D. Hubick K. T., Condon A. G., and Richards R. A. (1988) Carbon-isotope fractionation and plant water-use efficiency. In *Stable Isotopes in Ecological Research* (eds. P. W. Rundell, J. R. Ehleringer, and K. A. Nagy). Springer, New York, pp. 21–40.

Fernandes N. F. and Dietrich W. E. (1997) Hillslope evolution by diffusive processes: the timescale for equilibrium adjustments. *Water Resour. Res.* **33**, 1307–1318.

Ferris T. (1998) Seeing in the dark. *The New Yorker* August 10, 55–61.

Friedman I. and O'Neil J. R. (1977) Compilation of stable isotope fractionation factors of geochemical interest. In *Data of Geochemistry,* 6th edn., US Geological Survey Professional Paper 440-KK (ed. M. Fleischer) chap. KK.

Gilbert G. K. (1909) The convexity of hilltops. *J. Geol.* **17**, 344–350.

Gile L. H., Peterson F. F., and Grossman R. B. (1966) Morphological and genetic sequences of carbonate accumulation in desert soils. *Soil Sci.* **101**, 347–360.

Greenwood N. N. and Earnshaw A. (1997) *Chemistry of the Elements,* 2nd edn. Heinemann, Oxford.

Griffin D. W., Kellogg C. A., Garrison V. H., and Shinn E. A. (2002) The global transport of dust. *Am. Sci.* **90**, 228–235.

Hanson P. J., Edwards N. T., Garten C. T., and Andrews J. A. (2000) Separating root and soil microbial contributions to soil respiration: a review of methods and observations. *Biogeochemistry* **48**, 115–146.

Harden J. (1987) Soils developed in granitic alluvium near Merced, California. *US Geol. Surv. Bull.* **1590-A**, pp. A1–A65.

Heath J. A. and Huebert B. J. (1999) Cloudwater deposition as a source of fixed nitrogen in a Hawaiian montane forest. *Biogeochemistry* **44**, 119–134.

Hendricks M. B., DePaolo D. J., and Cohen R. C. (2000) Space and time variation of delta O-18 and delta D in precipitation: can paleotemperature be estimated from ice cores? *Global Biogeochem. Cycles* **14**, 851–861.

Heimsath A. M., Dietrich W. E., Nishiizumi K., and Finkel R. C. (1997) The soil production function and landscape equilibrium. *Nature* **388**, 358–361.

Heimsath A. M., Dietrich W. E., Nishiizumi K., and Finkel R. C. (1999) Cosmogenic nuclides, topography, and the spatial variation of soil depth. *Geomorphology* **27**, 151–172.

Hesterberg R. and Siegenthaler U. (1991) Production and stable isotopic composition of CO_2 in a soil near Bern, Switzerland. *Tellus, Ser. B* **43**, 197–205.

Hilgard E. W. (1860) Report on the geology and agriculture of the state of Mississippi, Jackson, MS.

Hillel D. (1980a) *Fundamentals of Soil Physics.* Academic Press, New York.

Hillel D. (1980b) *Applications of Soil Physics.* Academic Press, New York.

Hillel D. (1998) *Environmental Soil Physics.* Academic Press, San Diego.

Holland R. F. (1996) Great Valley vernal pool distribution, Photorevised 1996. In *Ecology, Conservation, and Management of Vernal Pool Ecosystems,* Proceedings of a 1996 Conference, California Native Plant Society, Sacramento, CA (eds. C. W. Witham, E. T. Bauder, D. Belk, W. R. Ferren, Jr., and R. Ornduff), pp. 71–75.

Hoosbeek M. R. and Bryant R. B. (1995) Modeling the dynamics of organic carbon in a Typic Haplorthod. In *Soils and Global Change* (eds. R. Lal, J. M. Kimble, E. R. Levine, and B. A. Stuart). CRC Lewis Publishers, Chelsea, MI, pp. 415–431.

Jackson M. L., Levelt T. W. M., Syers J. K., Rex R. W., Clayton R. N., Sherman G. D., and Uehara G. (1971) Geomorphological relationships of tropospherically derived quartz in the soils of the Hawaiian Islands. *Soil Sci. Soc. Am. Proc.* **35**, 515–525.

Jahren A. H., Arens N. C., Sarmiento G., Guerrero J., and Amundson R. (2001) Terrestrial record of methane hydrate dissociation in the early Cretaceous. *Geology* **29**, 159–162.

Jenkinson D. S., Adams D. E., and Wild A. (1991) Model estimates of CO_2 emissions from soil in response to global warming. *Nature* **351**, 304–306.

Jenny H. (1941) *Factors of Soil Formation: A System of Quantitative Pedology.* McGraw-Hill, New York.

Jenny H. (1961) *E. W. Hilgard and the Birth of Modern Soil Science.* Collan della Rivista "Agrochimica", Pisa, Italy.

Jenny H. (1962) Model of a rising nitrogen profile in Nile Valley alluvium, and its agronomic and pedogenic implications. *Soil Sci. Soc. Am. Proc.* **27**, 273–277.

Jenny H. and Leonard C. D. (1939) Functional relationships between soil properties and rainfall. *Soil Sci.* **38**, 363–381.

Jenny H., Gessel S. P., and Bingham F. T. (1949) Comparative study of decomposition rates of organic matter in temperate and tropical regions. *Soil Sci.* **68**, 419–432.

Jersak J., Amundson R., and Brimhall G. (1995) A mass balance analysis of podzolization: examples from the northeastern United States. *Geoderma* **66**, 15–42.

Johnson D. L. (1990) Biomantle evolution and the redistribution of earth materials and artifacts. *Soil Sci.* **149**, 84–102.

Johnson D. L., Watson-Stegner D., Johnson D. N., and Schaetzl R. J. (1987) Proisotopic and proanisotropic processes of pedoturbation. *Soil Sci.* **143**, 278–292.

Jury W. A., Gardner W. R., and Gardner W. H. (1991) *Soil Physics,* 5th edn. Wiley, New York.

Kirkby M. J. (1971) Hillslope process-response models based on the continuity equation. *Inst. Prof. Geogr. Spec. Publ.* **3**, 15–30.

Koch P. L., Zachos J. C., and Dettman D. L. (1995) Stable isotope stratigraphy and paleoclimatology of the Paleogene Bighorn Basin. *Palaeogeogr. Palaeoclimat. Palaeoecol.* **115**, 61–89.

Kennedy M. J., Chadwick O. A., Vitousek P. M., Derry L. A., and Hendricks D. M. (1998) Changing sources of base cations during ecosystem development, Hawaiian Islands. *Geology* **26**, 1015–1018.

Krupenikov I. A. (1992) *History of Soil Science: From its Inception to the Present.* Amerind Publishing, New Delhi.

Langmaid K. K. (1964) Some effects of earthworm inversion in virgin podzols. *Can. J. Soil Sci.* **44**, 34–37.

Latorre C., Betancourt J. L., Rylander K. A., and Quade J. (2002) Vegetation invasions into absolute desert: a 45,000 yr rodent midden record from the Calama-Salar de Atacama basins, northern Chile (lat. 22°–24° S). *Geol. Soc. Am. Bull.* **114**, 349–366.

Marian G. M., Schlesinger W. H., and Fonteyn P. J. (1985) Caldep: a regional model for soil $CaCO_3$ (calcite) deposition in southwestern deserts. *Soil Sci.* **139**, 468–481.

McFadden L. D. and Tinsley J. C. (1985) Rate and depth of pedogenic-carbonate accumulation in soils: formation and testing of a compartment model. In *Quaternary Soils and Geomorphology of the Southwestern United States.* (ed. D. L. Weide). Geological Society of America Special Paper 203. Geological Society of America, Boulder, pp. 23–42.

McFadden L. D., Amundson R. G., and Chadwick O. A. (1991) Numerical modeling, chemical, and isotopic studies of carbonate accumulation in soils of arid regions. In *Occurrence, Characteristics, and Genesis of Carboante, Gypsum, and Silica Accumulations in Soils* (ed. W. D. Nettleon). Soil Science Society of America, Madison, WI, pp. 17–35.

McKean J. A., Dietrich W. E., Finkel R. C., Southon J. R., and Caffee M. W. (1993) Quantification of soil production and downslope creep rates from cosmogenic ^{10}Be accumulations on a hillslope profile. *Geology* **21**, 343–346.

Merritts D. J., Chadwick O. A., Hendricks D. M., Brimhall G. H., and Lewis C. J. (1992) The mass balance of soil evolution on late Quaternary marine terraces, northern California. *Geol. Soc. Am. Bull.* **104**, 1456–1470.

Miller A. J., Amundson R., Burke I. C., and Yonker C. (2003) The effect of climate and cultivation on soil organic C and N. *Biogeochemistry* (in press).

Mook W. G., Koopmans M., Carter A. F., and Keeling C. D. (1983) Seasonal, latitudinal, and secular variations in the abundance and isotopic ratios of atmospheric carbon dioxide: 1. Results from land stations. *J. Geophys. Res.* 88(C15), 10915–10933.

Mora C. I., Driese S. G., and Colarusso L. A. (1996) Middle to late Paleozoic atmospheric CO_2 levels from soil carbonate and organic matter. *Science* **271**, 1105–1107.

Nadelhoffer K. and Fry B. (1988) Controls on the nitrogen-15 and carbon-13 abundances in forest soil organic matter. *Soil Sci. Soc. Am. J.* **52**, 1633–1640.

Nakai S., Halliday A. N., and Rea D. K. (1993) Provenance of dust in the Pacific Ocean. *Earth Planet. Sci. Lett.* **119**, 143–157.

Navarro-Gonzalez R., McKay C. P., and Mvondo D. N. (2001) A possible nitrogen crisis for Archaean life due to reduced nitrogen fixation by lightning. *Nature* **412**, 61–64.

Nettleton W. D., Flach K. W., and Borst G. (1968) *A Toposequence of Soils on Grus in the Southern California Peninsular Range.* US Dept. Agric. Soil Cons. Serv., Soil Surv. Invest. Rep. No. 21, 41p.

O'Brien J. B. and Stout J. D. (1978) Movement and turnover of soil organic matter as indicated by carbon isotope measurements. *Soil Biol. Biochem.* **10**, 309–317.

Parton W. J., Schimel D. S., Cole C. V., and Ojima D. S. (1987) Analysis of factors controlling soil organic matter levels in Great Plains grasslands. *Soil Sci. Soc. Am. J.* **5**, 1173–1179.

Peterson F. D. (1980) Holocene desert soil formation under sodium salt influence in a playa-margin environment. *Quat. Res.* **13**, 172–186.

Quade J., Cerling T. E., and Bowman J. R. (1989a) Systematic variations in the carbon and oxygen isotopic composition of pedogenic carbonate along elevation transects in the southern Great Basin, United States. *Geol. Soc. Am. Bull.* **101**, 464–475.

Quade J., Cerling T. E., and Bowman J. R. (1989b) Development of the Asian monsoon revealed by marked ecological shift during the latest Miocene in northern Pakistan. *Nature* **342**, 163–166.

Raich J. W. and Schlesinger W. H. (1992) The global carbon dioxide flux in soil respiration and its relationship to vegetation and climate. *Tellus* **B44**, 48–51.

Rea D. K. (1994) The paleoclimatic record provided by eolian deposition in the deep sea: the geological history of wind. *Rev. Geophys.* **5**, 193–259.

Rech J. A., Quade J., and Hart B. (2002) Isotopic evidence for the source of Ca and S in soil gypsum, anhyrite and calcite in the Atacama desert, Chile. *Geochim. Cosmochim. Acta* **67**(4), 575–586.

Rosenblum N. A., Doney S. C., and Schimel D. S. (2001) Geomorphic evolution of soil texture and organic matter in eroding landscapes. *Global Biogeochem. Cycles* **15**, 365–381.

Royer D. L. (1999) Depth to pedogenic carbonate horizon as a paleoprecipitation indicator? *Geology* **27**, 1123–1126.

Rozanski K., Araguás- Araguás L., and Gonfiantini R. (1993) Isotopic patterns in modern global precipitation.In *Climate Change in Continental Isotopic Records*. American Geophysical Union Monograph 78 (ed. P. K. Swart, *et al.).* American Geophysical Union, Washington, DC, pp. 1–36.

Russell R. S. (1977) *Plant Root Systems: Their Function and Interaction with the Soil.* McGraw-Hill, London.

Schlesinger W. H. and Andrews J. A. (2000) Soil respiration and the global carbon cycle. *Biogeochemistry* **48**, 7–20.

Severinghaus J. P., Bender M. L., Keeling R. F., and Broecker W. S. (1996) Fractionation of soil gases by diffusion of water vapor, gravitational settling, and thermal diffusion. *Geochim. Cosmochim. Acta* **60**, 1005–1018.

Shacklette H. T. and Boerngen J. G. (1984) Element concentrations in soils and other surficial materials of the conterminous United States. US Geological Survey Professional Paper 1270.

Soil Survey Staff (1975) *Soil Taxonomy: A Basic System of Soil Classification for Making and Interpreting Soil Surveys,* Agr. Handbook 426. US Government Printing Office, Washington, DC.

Soil Survey Staff (1993) *Soil Survey Manual.* USDA Handbook No. 18. US Government Printing Office, Washington, DC.

Soil Survey Staff (1995) *Soil Survey Laboratory Information Manual.* Soil Survey Investigations Report No. 45, Version 1.0, US Government Printing Office, Washington, DC.

Soil Survey Staff (1999) *Keys to Soil Taxonomy,* 8th edn. Pocahontas Press, Blacksburg, VA.

Stallard R. F. (1998) Terrestrial sedimentation and the carbon cycle: coupling weathering and erosion to the carbon cycle. *Global Biogeochem. Cycles* **12**, 231–257.

Stern L. A., Baisden W. T., and Amundson R. (1999) Processes controlling the oxygen-isotope ratio of soil CO_2: analytic and numerical modeling. *Geochim. Cosmochim. Acta* **63**, 799–814.

Stern L. A., Amundson R., and Baisden W. T. (2001) Influence of soils on oxygen-isotope ratio of atmospheric CO_2. *Global Biogeochem. Cycles* **15**, 753–759.

Sundquist E. T. (1993) The global carbon dioxide budget. *Science* **259**, 934–941.

Sutter B., Amundson R., Ewing S., Rhodes K. W., and McKay C. W. (2002) The chemistry and mineralogy of Atacama Desert soils: a possible analog for Mars soils. American Geophysical Union Fall Meeting, San Francisco, pp. 71A–0443.

Tans P. P. (1998) Oxygen isotopic equilibrium between carbon dioxide and water in soils. *Tellus* **B50**, 163–178.

Taylor S. R. and McLennan S. M. (1985) *The Continental Crust: Its Composition and Evolution.* Black-well, Oxford.

Thiemens M. H. (1999) Mass-independent isotope effects in planetary atmospheres and the early solar system. *Science* **283**, 341–345.

Thorstenson D. C., Weeks E. P., Haas H., and Fisher D. W. (1983) Distribution of gaseous $^{12}CO_2$, $^{13}CO_2$, and $^{14}CO_2$ in sub-soil unsaturated zone of the western US Great Plains. *Radiocarbon* **25**, 315–346.

Tréguer P., Nelson D. M., Van Bennekom A. J., DeMaster D. J., Leynaert A., and Quéguiner (1995) The silica balance in the world ocean: a reestimate. *Science* **268**, 375–379.

Trumbore S. E. (2000) Age of soil organic matter and soil respiration: radiocarbon constraints on belowground dynamics. *Ecol. Appl.* **10**, 399–411.

Ugolini F. C. and Dahlgren R. (1987) The mechanism of podzolization as revealed by soil solution studies. In *Podzols et Podolization* (eds. D. Fighiand and A. Chavell). Assoc. Fr. Estude Sol, Plassier, France, pp. 195–203.

Vil'yams V. R. (1967) V. V. Dokuchaev's role in the development of soil science. In *Russian Chernozem.* (translated by Israel Program for Scientific Translations, Jerusalem).

Vitousek P. M., Aber J. D., Howarth R. W., Likens G. E., Matson P. A., Schindler D. W., Schlesinger W. H., and Tilman D. G. (1997a) Human alterations of the global nitrogen cycle: sources and consequences. *Ecol. Appl.* **7**, 737–750.

Vitousek P. M., Chadwick O. A., Crews T. E., Fownes J. H., Hendricks D. M., and Herbert D. (1997b) Soil and ecosystem development across the Hawaiian Islands. *GSA Today* **7**, 1–8.

Wang Y., Amundson R., and Trumbore S. (1994) A model for soil $^{14}CO_2$ and its implications for using ^{14}C to date pedogenic carbonate. *Geochim. Cosmochim. Acta* **58**, 393–399.

Wang Y., Amundson R., and Niu X.-F. (2000) Seasonal and altitudinal variation in decomposition of soil organic matter inferred from radiocarbon measurements of soil CO_2 flux. *Global Biogeochem. Cycles* **14**, 199–211.

White A. F., Blum A. E., Schulz M. S., Bullen T. D., Harden J. W., and Peterson M. L. (1996) Chemical weathering rates of a soil chronosequence on granitic alluvium: I. Quantification of mineralogical and surface area changes and calculation of primary silicate reaction rates. *Geochim. Cosmochim. Acta* **60**, 2533–2550.

Readings from the Treatise on Geochemistry
ISBN: 978-0-12-381391-6
pp. 235–270

8
Global Occurrence of Major Elements in Rivers

M. Meybeck
University of Paris VI, CNRS, Paris, France

8.1 INTRODUCTION

Major dissolved ions (Ca^{2+}, Mg^{2+}, Na^{+}, K^{+}, Cl^{-}, SO_4^{2-}, HCO_3^{-}, and CO_3^{2-}) and dissolved silica (SiO_2) in rivers have been studied for more than a hundred years for multiple reasons: (i) geochemists focus on the origins of elements and control processes, and on the partitioning between dissolved and particulate forms; (ii) physical geographers use river chemistry to determine chemical denudation rates and their spatial distribution; (iii) biogeochemists are concerned with the use of carbon, nitrogen, phosphorus, silica species, and other nutrients by terrestrial and aquatic biota; (iv) oceanographers need to know the dissolved inputs to the coastal zones, for which rivers play the dominant role; (v) hydrobiologists and ecologists are interested in the temporal and spatial distribution of ions, nutrients, organic carbon, and pH in various water bodies; (vi) water users need to know if waters comply with their standards for potable water, irrigation, and industrial uses.

The concentrations of the major ions are commonly expressed in $mg\,L^{-1}$; they are also reported in $meq\,L^{-1}$ or $\mu eq\,L^{-1}$, which permits a check of the ionic balance of an analysis: the sum of cations (Σ^{+} in $eq\,L^{-1}$) should equal the sum of anions (Σ^{-} in $eq\,L^{-1}$). Dissolved silica is generally not ionized at pH values commonly found in rivers; its concentration is usually expressed in $mg\,L^{-1}$ or in $\mu mol\,L^{-1}$. Ionic contents can also be expressed as percent of Σ^{+} or Σ^{-} ($\%C_i$), which simplifies the determination of ionic types. Ionic ratios (C_i/C_j) in $eq\,eq^{-1}$ are also often tabulated (Na^{+}/Cl^{-}, Ca^{2+}/Mg^{2+}, Cl^{-}/SO_4^{2-}, etc.). As a significant fraction of sodium can be derived from atmospheric sea salt and from sedimentary halite, a chloride-corrected sodium concentration is commonly reported ($Na^{\#} = Na^{+}–Cl^{-}$ (in $meq\,L^{-1}$)). The export rate of ions and silica, or the yield ($Y_{Ca^{2+}}$, Y_{SiO_2}) at a given station is the average mass transported per year divided by the drainage area: it is expressed in units of $t\,km^{-2}\,yr^{-1}$ (equal to $g\,m^{-2}\,y^{-1}$) or in $eq\,m^{-2}\,yr^{-1}$.

This chapter covers the distribution of riverine major ions, carbon species, both organic and inorganic, and silica over the continents, including internal regions such as Central Asia, and also the major factors such as lithology and climate that control their distribution and yields.

Based on an unpublished compilation of water analyses in 1,200 pristine and subpristine basins, I am presenting here an idealized model of global river chemistry. It is somewhat different from the model proposed by Gibbs (1970), in that it includes a dozen major ionic types. I also illustrate the enormous range of the chemical composition of rivers—over three orders of magnitude for concentrations and yields—and provide two global average river compositions: the median composition and the discharge-weighted composition for both internally and externally draining regions of the world.

A final section draws attention to the human alteration of river chemistry during the past hundred years, particularly for Na^+, K^+, Cl^-, and SO_4^{2-}; it is important to differentiate anthropogenic from natural inputs. Trace element occurrence is covered by Gaillardet.

8.2 SOURCES OF DATA

Natural controls of riverine chemistry at the global scale have been studied by geochemists since Clarke (1924) and the Russian geochemists Alekin and Brazhnikova (1964). Regional studies performed prior to industrialization and/or in remote areas with very limited human impacts are rare (Kobayashi, 1959, 1960) and were collected by Livingstone (1963). Even some of his river water data are affected by mining, industries, and the effluents from large cities. These impacts are obvious when the evolution of rivers at different periods is compared; Meybeck and Ragu (1996, 1997) attempted this comparison for all the rivers flowing to the oceans from basins with areas exceeding 10^4 km^2. More recently interest in the carbon cycle and its riverine component has been the impetus for a new set of field studies (Stallard and Edmond, 1981, 1983; Degens *et al.*, 1991; Gaillardet *et al.*, 1997, 1999; Huh *et al.*, 1998/1999; Millot *et al.*, 2002; Guyot, 1993) that build on observations made since the 1970s (Reeder *et al.*, 1972; Stallard and Edmond, 1981, 1983). These data were used to construct the first global budgets of river dissolved loads (Meybeck, 1979) and their controls (Holland, 1978; Meybeck, 1987, 1994; Bluth and Kump, 1994; Berner, 1995; Stallard, 1995a,b).

Most of the annual means derived from these data have been collected into a global set of pristine or subpristine rivers and tributaries (PRISRI, $n = 1{,}200$) encompassing all the continents, including exorheic and endorheic (internal drainage) runoff. The largest basins, such as the Amazon, Mackenzie, Lena, Yenisei, and Mekong, have been subdivided into several smaller subbasins. In some regions (e.g., Indonesia, Japan) the size of the river basins included in these summaries may be less than 1,000 km^2. In the northern temperate regions only the most reliable historic analyses prior to 1920 have been included. Examples are analyses performed by the US Geological Survey (1909–1965) in the western and southwestern United States and in Alaska.

In order to study the influence of climate, total cationic contents (Σ^+) of PRISRI rivers have been split into classes based on annual runoff (q in mm yr^{-1}) and Σ^+ (meq L^{-1}). A medium-sized subset has also been used (basin area from 3,200 km^2 to 200,000 km^2, $n = 700$). The PRISRI data base covers the whole globe but has poor to very poor coverage of western Europe, Australia, South Africa, China, and India due to the lack of data for pre-impact river chemistry in these regions. PRISRI includes rivers that flow permanently ($q > 30$ mm yr^{-1}) and seasonally to occasionally ($3 < q < 30$ mm yr^{-1}). Lake outlets may be included in PRISRI.

8.3 GLOBAL RANGE OF PRISTINE RIVER CHEMISTRY

The most striking observation in global river chemistry is the enormous range in concentrations (C_i), ionic ratios (C_i/C_j), and the proportions of ions in cation and anion sums ($\%C_i$) as illustrated by the 1% and 99% quantiles (Q_1 and Q_{99}) of their distribution (Table 1). The Q_{99}/Q_1 ratio of solute concentrations is the lowest for potassium and silica, about two orders of magnitude, and it is very high for chloride and sulfate, exceeding three orders of magnitude. The concept of "global average river chemistry," calculated from the total input of rivers to oceans divided by the total volume of water, can only be applied for global ocean chemistry and elemental cycling; it is not a useful reference in either weathering studies, river ecology, or water quality.

Ionic ratios and ionic proportion distributions also show clearly that all major ions, except potassium, can dominate in multiple combinations: ionic ratios also range over two to three orders of magnitude, and they can be greater or less than unity for all except the Na^+/K^+ ratio, in which sodium generally dominates. River compositions found in less than 1% of analyses can be termed *rare*; for analyses from Q_1 to Q_{10} and Q_{90} to Q_{99}, I propose the term *uncommon*, from Q_{10} to Q_{25} and Q_{75} to Q_{90}, *common*, and between Q_{25} and Q_{75}, *very common*. An example of this terminology is shown in the next section (Figure 2) for dissolved inorganic carbon (DIC).

Table 1 Global range of pristine river chemistry (medium-sized basins).

	Ca^{2+}	Mg^{2+}	Na^{+}	K^{+}	Cl^{-}	SO_4^{2-}	HCO_3^{-}	SiO_2	Σ^{+}
Ionic contents (C_i)									
Q_{99}	9,300	5,900	14,500	505	17,000	14,500	5,950	680	32,000
Q_1	32	10	18	3.9	3.7	5	47	3.3	128
Q_{99}/Q_1	290	590	805	129	4,600	2,900	126	206	250
Ionic proportions (%C_i)									
Q_{99}	84	48	72	19.5	69	67	96		
Q_1	11	0.1	1	0.1	0.1	0.1	9		

	$\frac{Ca^{2+}}{Mg^{2+}}$	$\frac{Ca^{2+}}{Na^{+}}$	$\frac{Mg^{2+}}{Na^{+}}$	$\frac{Na^{+}}{K^{+}}$	$\frac{Na^{+}}{Cl^{-}}$	$\frac{Ca^{2+}}{SO_4^{2-}}$	$\frac{SO_4^{2-}}{HCO_3^{-}}$	$\frac{Cl^{-}}{SO_4^{2-}}$	$\frac{SiO_2}{\Sigma^{+}}$
Ionic ratios (C_i/C_j)									
Q_{99}	20.3	56	20	164	29	51	3.5	8.5	1.3
Q_1	0.01	0.14	0.01	0.95	0.33	0.19	0.01	0.01	0.0

C_i: ionic contents (μeq L^{-1} and μmol L^{-1} for silica); %C_i: proportion of ions in the sum of total cations or anions; C_i/C_j: ionic ratios; and Q_1 and Q_{99}: lowest and highest percentiles of distribution.

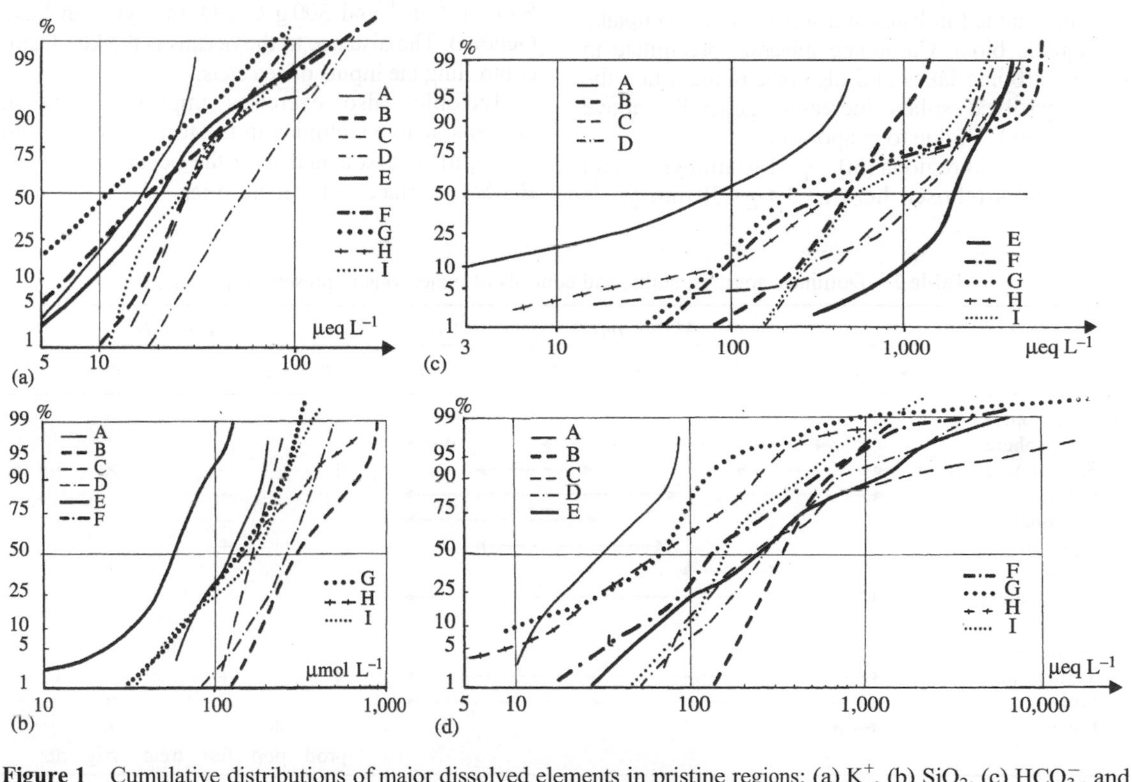

Figure 1 Cumulative distributions of major dissolved elements in pristine regions: (a) K^{+}, (b) SiO_2, (c) HCO_3^{-}, and (d) Na^{+}. A: Central and lower Amazon, B: Japan, C: Andean Amazon Basin, D: Thailand, E: Mackenzie Basin, F: French streams, G: temperate stream model, H: monolithologic miscellaneous French streams, I: major world rivers (source Meybeck, 1994).

When plotted on a log-probability scale (Henry's law diagram) the distributions of elemental concentrations show four patterns (Figure 1): (i) lognormal distribution as for potassium in Thailand (a, D), that can be interpreted as a single source of the element and limited control on its concentration; (ii) retention at lower concentrations (there is a significant break in the distribution as for silica in the Mackenzie Basin (b, E)); (iii) retention at higher concentrations (there is a significant break in the distribution as for bicarbonate (DIC) (c, C, E, F, G, I)); and (iv) additional source at higher concentrations (the break in the distribution is in the opposite direction, as for chloride, suggesting another source of material than the one observed between Q_{10} and Q_{90} (d, all except central Amazon)).

8.4 SOURCES, SINKS, AND CONTROLS OF RIVER-DISSOLVED MATERIAL

The observed distribution patterns have been interpreted by many authors (Likens *et al.*, 1977; Holland, 1978; Drever, 1988; Hem, 1989; E. K. Berner and R. A. Berner, 1996) in terms of multiple sources, sinks, and controls on riverine chemistry (Table 2). The sources of ions may be multiple: rainfall inputs, generally of oceanic origin, (rich in NaCl and also in $MgSO_4$), differential weathering of silicate minerals and carbonate minerals, dissolution of evaporitic minerals contained in some sedimentary rocks (gypsum and anhydrite, halite) or leached during rainstorms from surficial soils of semi-arid regions (Garrels and Mackenzie, 1971; Drever, 1988; Stallard, 1995a,b). Sinks are also multiple: silica may be retained in lakes and wetlands due to uptake by aquatic biota. Carbonate minerals precipitate in some eutrophic lakes and also precipitate when the total dissolved solids increases, generally above $\Sigma^+ = 6\,\text{meq L}^{-1}$, due to evaporation.

In semi-arid regions ($30 < q < 140\,\text{mm yr}^{-1}$) and arid regions (chosen here as $3 < q < 30\,\text{mm yr}^{-1}$) surface waters gradually evaporate and become concentrated, reaching saturation levels of calcium and magnesium carbonates first, and then of calcium sulfate (Holland, 1978): these minerals are deposited in soils and in river beds during their drying stage, a process termed evaporation/crystallization by Gibbs (1970).

There is also growing evidence of active recycling of most major elements, including silica, in terrestrial vegetation, particularly in forested areas (Likens *et al.*, 1977). Each element may, therefore, have multiple natural sources and sinks. The main controlling factors for each element are also multiple (Table 2). They can be climatic: higher temperature generally increases mineral dissolution; water runoff increases all weathering rates. Conversely, soil retention increases at lower runoff. The weathering of silicate and carbonate minerals is also facilitated by organic acids generated by terrestrial vegetation. Higher lake residence times ($\tau > 6$ months) favor biogenic silica retention and the precipitation of calcite (e.g., $30\,\text{g SiO}_2\,\text{m}^{-2}\,\text{yr}^{-1}$ and $300\,\text{g CaCO}_3\,\text{m}^{-2}\,\text{yr}^{-1}$ in Lake Geneva). The distance to the oceans is the key factor controlling the inputs of sea salts.

Tectonics also exercises important regional controls: active tectonics in the form of volcanism and uplift is associated with the occurrence at the Earth's surface of fresh rock that is more

Table 2 Dominant sources, sinks, and controls of major ions in present day rivers.

	Major ions								*Controls*						
	SiO_2	K^+	Na^+	Cl^-	SO_4^{2-}	Mg^{2+}	Ca^{2+}	DIC	$T°$	τ	q	d	QH	V	T
Natural sources															
Atmosphere		⇠⇢	⇠⇢	⇠⇢	⇠⇢	⇠⇢		↔				−			+
Silicate weathering	↔	↔	↔			↔	↔		+		+		×	+	+
Pyrite	↔	↔	↔		↔	↔	↔	↔			+				
Carbonate						↔	↔	↔			+			+	
Gypsum					↔	↔	↔				+				
Halite			↔	↔							+				
Deep waters	↔	↔	↔	↔	↔	↔	↔	↔							+
Natural sinks															
Terrestrial vegetation	↔	⇠⇢	⇠⇢	⇠⇢	⇠⇢	⇠⇢	⇠⇢	↔	+		+			+	−
Soils	↔	↔	↔	↔	↔	↔	↔	↔			−		×		−
Lakes	↔									+			×		
									prod	pop	fert	treat	irrig	nb	
Anthropogenic sources															
Mines		⇠⇢	⇠⇢	⇠⇢	⇠⇢	⇠⇢			+			+			
Industries		⇠⇢	⇠⇢	⇠⇢	⇠⇢	⇠⇢			+			+			
Cities			↔	↔	↔	↔				+					
Agriculture		↔	↔	↔	↔	↔					+				
Anthropogenic sinks															
Reservoirs	↔						↔	↔						+	
Irrigated soils	↔	↔	↔	↔	↔	↔	↔	↔					+		

+ , − : increase and decrease with related control; ×: complex relation; DIC: dissolved inorganic carbon.
$T°$: temperature; q: runoff; d: distance to ocean; QH: Quaternary history; V: terrestrial biomass; τ: water residence time; T: volcanism, tectonic uplift and rifting.
Prod: production; pop: urban population; fert: fertilization rate; treat: wastewater treatment and recycling; irrig: water loss through irrigation; nb: volume of reservoirs and eutrophication.

easily weathered than surficial rocks that have been exposed to weathering for many millions of years on a stable craton (Stallard, 1995b): there, the most soluble minerals have been dissolved and replaced by the least soluble ones such as quartz, aluminum and iron oxides, and clays. Active tectonics also limits the retention of elements by terrestrial vegetation, and retention in arid soils due to high mechanical erosion rates. Rifting is generally associated with inputs of saline deep waters that can have a major influence on surface water chemistry at the local scale. In formerly glaciated shields as in Canada and Scandinavia, glacial abrasion slows the development of a weathered soil layer. This limits silicate mineral weathering even under high runoff. These low relief glaciated areas are also characterized by a very high lake density, an order of magnitude higher than in most other regions of the world; lakes are sinks for silica and nutrients. The effects of Quaternary history in semi-arid and arid regions can be considerable, due to the inheritance of minerals precipitated under past climatic conditions. Due to the effect of local lithology, distance to the ocean, regional climate and tectonics, and the occurrence of lakes, river chemistry in a given area may be very variable.

8.4.1 Influence of Lithology on River Chemistry

Lithology is an essential factor in determining river chemistry (Garrels and Mackenzie, 1971; Drever, 1988, 1994), especially at the local scale (Strahler stream orders 1–3) (Miller, 1961; Meybeck, 1986). On more regional scales, there is generally a mixture of rock types, although some large river basins (area > 0.1 Mkm^2) may contain one major rock type such as a granitic shield or a sedimentary platform. When selecting nonimpacted monolithologic river basins in a given region such as France (Table 3, A–E), the influence of climate, tectonics, and distance from the ocean can be minimized, revealing the dominant control of lithology, which in turn depends on (i) the relative abundance of specific minerals and (ii) the sensitivity of each mineral to weathering. The weathering scale most commonly adopted is that of Stallard (1995b) for mineral stability in tropical soils: quartz ≫ K-feldspar, micas ≫ Na-feldspar > Ca-feldspar, amphiboles > pyroxenes > dolomite > calcite ≫ pyrite, gypsum, anhydrite ≫ halite (least stable).

Table 3F provides examples of stream and river chemistry under peculiar conditions: highly weathered quartz sand (Rio Negro), peridotite (Dumbea), hydrothermal inputs (Semliki and Tokaanu), evaporated (Saoura), black shales (Powder and Redwater), sedimentary salt deposits (Salt). They illustrate the enormous range of natural river chemistry (outlets of acidic volcanic lakes are omitted here).

Table 3 lists examples of more than a dozen different chemical types of river water. Although Ca^{2+} and HCO_3^- are generally dominant, Mg^{2+} dominance over Ca^{2+} can be found in rivers draining various lithologies such as basalt, peridotite, serpentinite, dolomite, coal, or where hydrothermal influence is important (Semliki). Sodium may dominate in sandstone basins, in black shales (Powder, Redwater in Montana), in evaporitic sedimentary basins (Salt), in evaporated basins (Saoura), and where hydrothermal and volcanic influence is important (Semliki, Tokaanu). K^+ rarely exceeds 4% of cations, except in some clayey sands, mica schists, and trachyandesite; it exceeds 15% in extremely dilute waters of Central Amazonia and in highly mineralized waters of rift lake outlets (Semliki, Ruzizi).

The occurrence of highly reactive minerals, such as evaporitic minerals, pyrite and even calcite, in low proportions—a percent or less—in a given rock, e.g., calcareous sandstone, pyritic shale, marl with traces of anhydrite, granite with traces of calcite, may determine the chemical character of stream water (Miller, 1961; Drever, 1988). In a study of 200 streams from monolithologic catchments underlain by various rock types under similar climatic conditions in France, the relative weathering rate based on the cation sum (Meybeck, 1986) ranges from 1 for quartz sandstone to 160 for gypsiferous marl.

When the lithology in a given region is fairly uniform, the distribution of major-element concentrations is relatively homogeneous with quantile ratios Q_{90}/Q_{10} well under 10 as, for example, in Japan (mostly volcanic), the Central Amazon Basin (detrital sand and shield) (Figure 1, distributions B and A); when a region is highly heterogeneous with regards to lithology, as in the Mackenzie Basin and in France, the river chemistry is much more heterogeneous (Figure 1, distributions E and F) and quantile ratios Q_{90}/Q_{10}, Q_{99}/Q_1 may reach those observed at the global scale (Figure 1, distributions G) (Meybeck, 1994). In the first set, regional geochemical background compositions can be easily defined, but not in the second set.

8.4.2 Carbon Species Carried by Rivers

The carbon cycle and its long-term influence on climate through the weathering of fresh silicate rocks (Berner *et al.*, 1983) has created a new interest in the river transfer of carbon. Bicarbonate (HCO_3^-) is the dominant form of DIC in the pH range of most world rivers

Table 3 A–E: composition of pristine waters draining single rock types in France (medians of analyses corrected for atmospheric inputs) (from 3 to 26 analyses in each class, except for estimates that are based on one analysis only). F: other river chemistry from various origins uncorrected for atmospheric inputs (Meybeck, 1986). Cation and anion proportions in percent of their respective sums (Σ^+ and Σ^-).

	SiO_2 ($\mu mol^{-1} L^{-1}$)	Σ^+ ($\mu eq^{-1} L^{-1}$)	Ca^{2+} (%)	Mg^{2+} (%)	Na^+ (%)	K^+ (%)	Cl^- (%)	SO_4^{2-} (%)	HCO_3^- (%)
A. *Noncarbonate detrital rocks*									
Quartz sand and sandstones	170	170	30	20	45	5	0	(40)	(60)
Clayey sands	135	300	53	17	13	17	0	(30)	(70)
Arkosic sands	200	400	48	35	12	5	0	(20)	(80)
Graywacke	90	350	58	22	18	2	0	(20)	(80)
Coal-bearing formations	150	5,000	30	55	14	1	0	20–90	80–10
B. *Carbonate-containing detrital rocks*									
Shales	90	500	60	30	8	1.5	0	25	75
Permian shales	175	2,200	53	35	8	4	(0–20)	10	90–70
Molasse	280	2,500	80	17	3	0.3	0	2	98
Flysch	50	2,200	79	19	1.5	0.5	0	2	98
Marl	90	3,000	83	14	2.5	0.5	0	2	98
C. *Limestones*									
Limestones	60	4,500	95	3	0.6	0.4	0	2.5	97.5
Dolomitic limestones	60	4,500	72	26	0.6	0.4	0	3.5	96.5
Chalk[a]	200	4,500	95	2.5	2.2	0.5	0	2.5	97.5
Dolomite[a]	67	5,900	54	46	0.1	0.1	0	8.5	91.5
D. *Evaporites*									
Gypsum marl[a]	160	22,000	77	22	0.2	0.2	0	83	17
Salt and gypsum marl[a]	133	27,500	34	32	33	(0.8)	36	44	20
E. *Plutonic, metamorphic, and volcanic rocks*									
Alkaline granite, gneiss, mica schists	140	130	15	15	65	5	0	(30)	(70)
Calc-alkaline granite, gneiss, mica schists	100	300	54	25	17	4	0	(15)	(85)
Serpentinite	225	1,500	38	60	7	1	0	(7)	93
Peridotite	180	600	5	93	1	1	0	(15)	(85)
Amphibolite	65	1,600	85	17	2	1	0	(7)	(93)
Marble[a]	150	3,400	86	11	1.8	0.6	0	(12)	88
Basalt	200	500	42	38	17	2.5	0	(2)	98
Trachyandesite	190	220	32	25	35	8	0	(2)	98
Rhyolite	190	550	53	15	25	2	0	(2)	98
Anorthosite	260	400	45	25	26	4	0	(2)	98
F. *Miscellaneous river waters (not rain corrected)*									
Rio Negro tributaries[b]	75	18.1	10.5	16.5	51.9	20.9			
Dumbea (New Caledonia)	232	1,175	3.9	84.7	10.9	0.3	14.2	7.1	78.5
Cusson (Landes, France)[b]	251	1,463	13.8	21.2	61.5	3.4	69	12.3	18.7
Semliki (Uganda)[c]	213	8,736	6.4	36.2	39.6	17.5	14.6	23.3	62.0
Powder (Montana)	148	20,200	27.2	20.3	51.5	0.9	15.0	62.8	22.2
Saoura (Marocco)[d]		26,150	23.3	16.8	59.2	0.7	62.8	27.7	9.6
Redwater (Montana)	116	40,700	10.7	24.5	64.1	0.6	1.0	72.2	26.7
Tokaanu (New Zeal.)[c]	4,760	41,600	14.4	3.0	79.5	3.1	90.6	5.8	3.5
Salt (NWT, Canada)	20	312,000	9.7	1.8	88.4	0.1	89.7	9.3	1.0

[a] Estimates. [b] Rain dominated. [c] Hydrothermal inputs. [d] Evaporated.

($6 < pH < 8.2$); carbonate (CO_3^{2-}) is significant only at higher pH, which occurs in a few eutrophic rivers such as the Loire River, where pH exceeds 9.2 during summer algal blooms, and in waters that have undergone evaporation. Undissociated dissolved CO_2 is significant only in very acidic waters rich in humic substances such as the Rio Negro (Amazonia), but this is unimportant at the global scale. In terms of fluxes, bicarbonate DIC dominates (Table 4). It has two different sources: (i) carbonic acid weathering of noncarbonate minerals, particularly of silicates such as feldspars, micas, and olivine, (ii) dissolution of carbonate minerals such as calcite and dolomite, in which half of the resulting DIC originates from soil and/or atmospheric CO_2, and

Table 4 Riverine carbon transfer and global change.

	Sources	*Age* (yr)	*Flux*[a] (10^{12} g C yr^{-1})	*Sensitivity to global change* A	B	C	D	E	F	
PIC	Geologic	10^4–10^8	170	•					•	
DIC	Geologic	10^4–10^8	140		•	•			•	
	Atmospheric	0–10^2	245		•	•			•	
DOC	Soils	10^0–10^3	200			•			•	
	Pollution	10^{-2}–10^{-1}	(15) ?					•		
CO_2	Atmospheric	0	(20–80)		•	•	•			TAC
POC	Soils	10^0–10^3	(100)	•					•	
	Algal	10^{-2}	(<10)				•		•	
	Pollution	10^{-2}–10^0	(15) ?					•		
	Geologic	10^4–10^8	(80)	•					•	

A = land erosion, B = chemical weathering, C = global warming and UV changes, D = eutrophication, E = organic pollution, F = basin management, TAC = Total atmospheric carbon.
[a] Present global flux to oceans mostly based on Meybeck (1993), 10^{12} g C yr^{-1}.

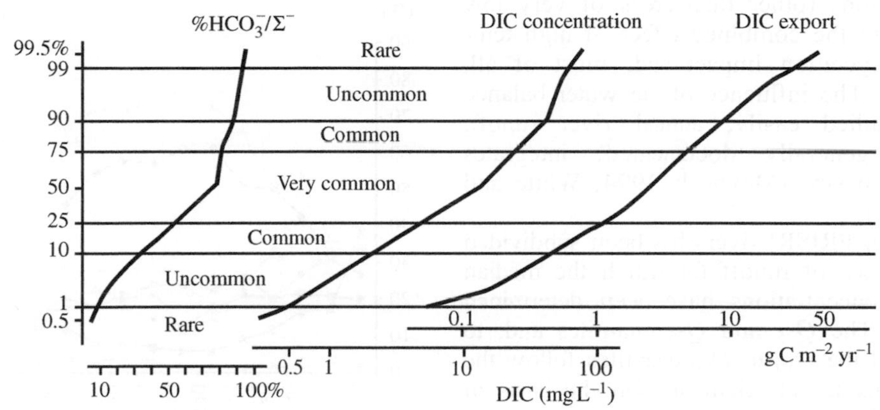

Figure 2 Global distribution of DIC concentration, ratio and export rate (yield) in medium-sized basins (3,500–200,000 km^2).

half from the weathered rock. Other forms of riverine carbon include particulate inorganic carbon (PIC) due to mechanical erosion in carbonate terrains, and dissolved and particulate organic carbon (DOC and POC) that are largely due to soil leaching and erosion; fossil POC in loess and shale may also contribute to river POC; organic pollution and algal growth in eutrophic lakes and rivers contribute minor fluxes (Meybeck, 1993). The "ages" of these carbon species, i.e., the time since their original carbon fixation fall into two categories: (i) those from 0 to ~1,000 yr, representing the fast cycling external part of the cycle are termed total atmospheric carbon (TAC); (ii) those from 50 kyr (Chinese Loess Plateau) to 100 Myr representing "old" carbon from sedimentary rocks. The sensitivity of these transfers to global change is complex (Table 4).

River carbon transfers are also very variable at the global scale: DIC concentrations range from 0.06 mg DIC L^{-1} (Q_1) to 71 mg DIC L^{-1} (Q_{99}); DIC export, or yield, ranges from 0.16 g C m^{-2} yr^{-1} (Q_1) to 33.8 g C m^{-2} yr^{-1} (Q_{99}). In 50% of river basins bicarbonate makes up more than 80% of the anionic charge. The distribution quantiles for these variables are shown in Figure 2.

Organic acids, particularly those present in wetland-rich basins, also play an important role in weathering (Drever, 1994; Viers *et al.*, 1997).

8.4.3 Influence of Climate on River Chemistry

The influence of air temperature is nearly impossible to measure for large basins due to its spatial heterogeneity and its absence in most databases. Some studies of mountain streams have considered the effect of air temperature (Drever and Zobrist, 1992). It is generally agreed that silicate weathering is more rapid in wet

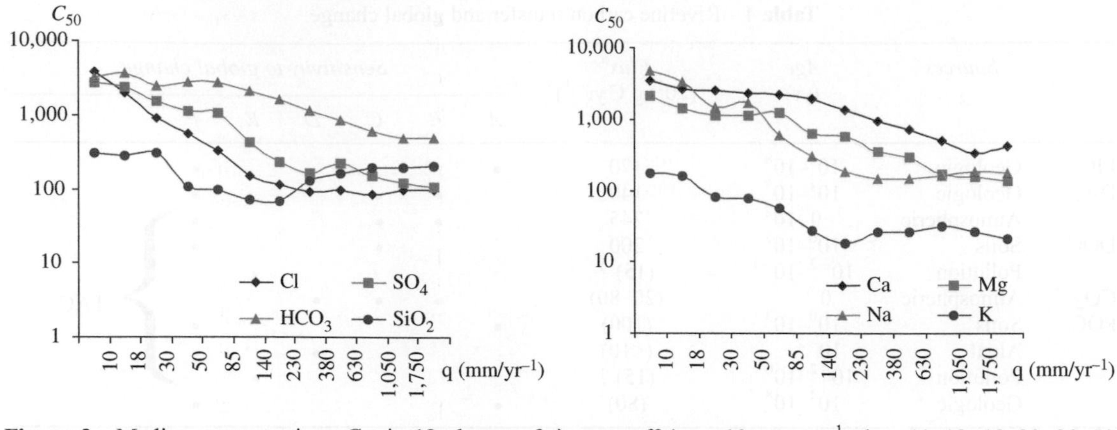

Figure 3 Median concentrations C_{50} in 12 classes of river runoff (q_1: <10 mm yr^{-1}, then 10–18, 18–30, 30–50, 50–85, 85–140, 140–230, 230–380, 380–630, 630–1,050, 1,050–1,750 and q_{12}: >1,750 mm yr^{-1}) (ions in µeq L^{-1}, silica in µmol L^{-1}) (total sample, $n = 1,091$).

tropical regions (other than areas of very low relief) due to the combined effect of high temperature, vegetation impact and, most of all, high runoff. The influence of the water balance can be studied easily; annual river runoff, which is generally documented, integrates entire river basins (Meybeck, 1994; White and Blum, 1995).

The set of PRISRI rivers has been subdivided into 12 classes of runoff for which the median elemental concentrations have been determined (Figure 3). The Q_{25} and Q_{75} quantiles and, to some extend the Q_{10} to Q_{90} quantiles follow the same patterns for all elements. On the basis of Na^+, K^+, Cl^-, and SO_4^{2-} two different clusters can be distinguished: (i) above 140 mm yr^{-1} the influence of runoff is not significant, (ii) below 140 mm yr^{-1} there is a gradual increase in elemental concentrations as runoff decreases. This pattern is interpreted as due to evaporation. For other elements there is a gradual increase of Ca^{2+}, Mg^{2+}, and HCO_3^- from the highest runoff to 85 mm yr^{-1} or 50 mm yr^{-1}, then a stabilization of these concentrations at lower runoff values ($q < 50$ mm yr^{-1}). The second part of this pattern is most probably due to the precipitation of carbonate minerals as a result of evaporative concentration. The decrease in Ca^{2+}, Mg^{2+}, and HCO_3^- concentrations at very high runoff, while the silicate-weathering related Na^+ and K^+ are stable, could be due to a lower occurrence of carbonate rocks in wetter regions. This does not imply a climate control of these rock types so much as the dominance of the Amazon and Congo basins, which are underlain by silicate rocks, and volcanic islands in the population of high-runoff rivers. The silica pattern is even more complex: the increase in silica concentration with runoff from 140 mm yr^{-1} to >1,750 mm yr^{-1} could be linked with a greater

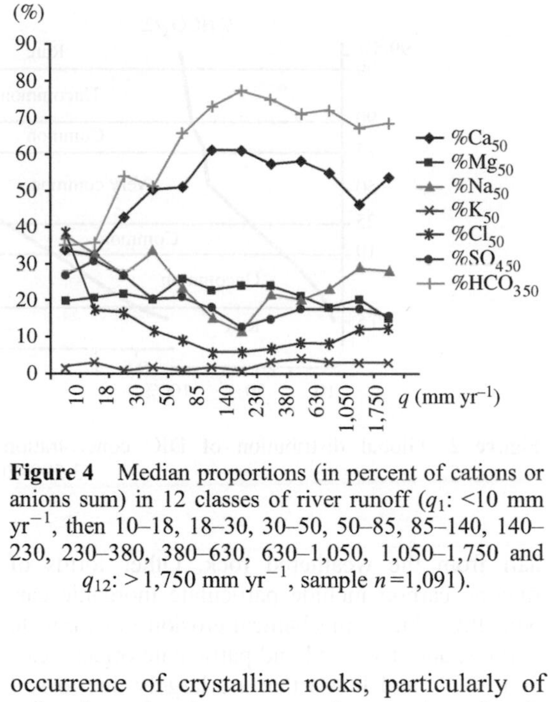

Figure 4 Median proportions (in percent of cations or anions sum) in 12 classes of river runoff (q_1: <10 mm yr^{-1}, then 10–18, 18–30, 30–50, 50–85, 85–140, 140–230, 230–380, 380–630, 630–1,050, 1,050–1,750 and q_{12}: > 1,750 mm yr^{-1}, sample $n = 1,091$).

occurrence of crystalline rocks, particularly of volcanic rocks, in wetter regions as in volcanic islands (Iceland, Japan, Indonesia) or coasts (Washington, Oregon), which are particularly abundant in the PRISRI database. These interactions between climate and lithology are possible but need to be verified. The lower silica value from 30 mm yr^{-1} to 140 mm yr^{-1} could also be attributed to the importance of the Canadian and Siberian rivers in these classes, where SiO_2 retention in lakes is important.

The relative median proportions of ions in the runoff classes of the PRISRI rivers confirm this pattern (Figure 4): below 140 mm yr^{-1} the $\%HCO_3^-$ and $\%Ca^{2+}$ drop from 75% and 60%, respectively, to ~33%, while those of

Cl^- and SO_4^{2-} increase from 6% to 39% and from 13% to 27%, respectively: the proportion of K^+ does not vary significantly. There is a significant increase in the median proportions of Na^+ and Cl^- above 1,050 mm yr^{-1} that could correspond with the increased marine influence in these categories. The median proportion of Mg^{2+} is the most stable of all ions: from 15% to 26%.

8.5 IDEALIZED MODEL OF RIVER CHEMISTRY

Gibbs (1970) defined three main categories of surface continental waters: rain dominated, with Na^+ and Cl^- as the major ions, weathering dominated, with Ca^{2+} and HCO_3^- as the major ions, and evaporation/crystallization dominated, with Na^+ and Cl^- as the major ions. This typology can still be used, but it is too simplified. According to the PRISRI database there are many more types of water and controlling mechanisms, although the ones described by Gibbs may account for ~80% of the observed chemical types. The sum of cations (Σ^+), which is a good indicator of weathering, ranges from 50 $\mu eq\,L^{-1}$ to 50,000 $\mu eq\,L^{-1}$, and the corresponding types of water, characterized by the dominant ions may reach as many as a dozen (Table 3). The true rain dominance type is found near the edges of continents facing oceanic aerosol inputs, as in many small islands and in Western Europe: it is effectively of the Na^+–Cl^- type and Σ^+ may be as high as 1 meq L^{-1}, for example, the Cusson R. (Landes, France) draining arkosic sands (Table 3, F). In continental interiors, in some rain forests, this water type occurs with much lower ionic concentrations ($\Sigma^+ < 0.1$ meq L^{-1}) in areas with highly weathered soils and very low mechanical erosion rates, such as the Central Amazon Basin (Rio Negro tributaries, Table 3F) or Cameroon: the very limited cationic inputs in rainfall are actively utilized and recycled by the forest and may even be stored (Likens *et al.*, 1977; Viers *et al.*, 1997). If the DOC level is high enough (~10 mg L^{-1}) the pH is often so low, that HCO_3^- is insignificant: the dominant anions are SO_4^{2-} and organic anions or Cl^-. This water chemistry is actually controlled by the terrestrial vegetation and can have a variety of ionic assemblages. The silica generated by chemical weathering is exported as dissolved SiO_2 and also as particulate biogenic SiO_2 from phytoliths or sponge spicules, as in the Rio Negro Basin.

Numerous water types reflect weathering control; these include Mg^{2+}–HCO_3^-, Ca^{2+}–HCO_3^-, Na^+–HCO_3^-, $Mg^{2+} - SO_4^{2-}$, $Ca^{2+} - SO_4^{2-}$, $Na^+ - SO_4^{2-}$, Na^+–Cl^- types (Table 3). The evaporation–crystallization control found in semi-arid and arid regions such as Central Asia (Alekin and Brazhnikova, 1964) also gives rise to multiple water types: $Mg^{2+} - SO_4^{2-}$, $Ca^{2+} - SO_4^{2-}$, $Na^+ - SO_4^{2-}$, Na^+–Cl^-, Mg^{2+}–HCO_3^-, and even Mg^{2+}–Cl^-. It is difficult to document the precipitation–redissolution processes (either during rare rain storms for occasional streams or in allochtonous rivers flowing from wetter headwaters such as the Saoura in Marocco, Table 3F), which are likely to have $\Sigma^+ > 6$ meq L^{-1} and/or runoff below 140–85 mm yr^{-1}. In some Canadian and Siberian basins, very low runoff (< 85 mm yr^{-1}) is not associated with very high evaporation but with very low precipitation: weathering processes still dominate.

In rift and/or volcanic regions and in recent mountain ranges such as the Caucasus, hydrothermal inputs may add significant quantities of dissolved material (Na^+, K^+, Cl^-, SO_4^{2-}, SiO_2) to surface waters. The Semliki River, outlet of Lake Edward, is particularly enriched in K^+ (Table 3, F); the Tokaanu River (New Zealand) drains a hydrothermal field with record values of silica, Na^+ and Cl^- (Table 3, F).

Overall, on the basis of ionic proportions and total concentrations, only 8.2% of the rivers (in number) in the PRISRI database can be described as evaporation controlled, 2.6% as rain dominated and vegetation controlled, and 89.2% as weathering dominated, including rivers affected by large water inputs.

A tentative reclassification of the major ionic types is presented on Figure 5 showing the occurrence of major ion sources for different

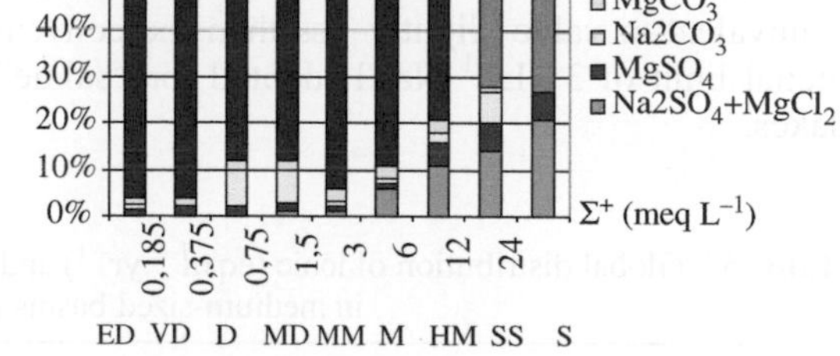

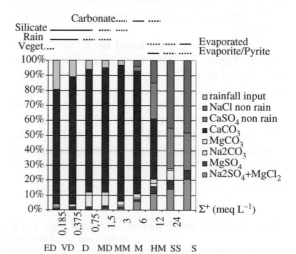

Figure 5 Idealized occurrence of water types (dominant ions and dominant control factors) per classes of increasing cationic content (Σ^+ in meq L^{-1}). ED: extremely dilute waters, VD: very dilute, D: dilute, MD: medium dilute, MM: medium mineralized, HM: highly mineralized, SS: subsaline, S: saline.

classes of Σ^+. The rainfall dominance types also include rivers with vegetation control and correspond to multiple ionic types ($Ca^{2+}-SO_4^{2-}$, Na^+-Cl^-, $Ca^{2+}-Cl^-$, $Na^+-HCO_3^-$). Other water types correspond to various rock weathering modes, including $Mg^{2+}-HCO_3^-$ waters observed in many volcanic regions. The most common type, $Ca^{2+}-HCO_3^-$, dominates from $\Sigma^+ < 0.185$ meq L^{-1} to 6 meq L^{-1}: in the most dilute waters it originates from silicate rock weathering, above 1.5 meq L^{-1} from weathering of carbonate minerals. This water type virtually disappears above 12 meq L^{-1}. "Non-rain" NaCl and $CaSO_4$ water types appear gradually above 6 meq L^{-1}. However, with these water types it is impossible to differenciate between evaporite rock dissolution, which can be observed even in the wet tropics (Stallard and Edmond, 1981) or in the Mackenzie (Salt R., Table 3F), from the leaching of salinized soils very common in Central Asia (Alekin and Brazhnikova, 1964). $Mg^{2+}-SO_4^{2-}$, $Na^+-SO_4^{2-}$, and $Mg^{2+}-Cl^-$ types are occasionally found in streams with Σ^+ values below 3 meq L^{-1}; they are commonly the result of pyrite weathering that can lead to very high Σ^+ as in the Powder and Redwater Rivers (Table 3, F).

The global proportions of the different water types depend on the global representativeness of the database. The exact occurrence of water types can only be estimated indirectly through modelling on the basis of lithologic maps, water balance, and oceanic fallout. Moreover, it will depend on spatial resolution: a very fine scale gives more importance to the smallest, rain-dominated coastal basins. In PRISRI, 70% of the basins have areas exceeding 3,200 km^2, thus limiting the appearance of oceanic influence.

A river salinity scale based on Σ^+ is also proposed (Figure 5). The least mineralized waters ($\Sigma^+ < 0.185$ meq L^{-1}), termed here "extremely dilute" correspond to a concentration of total dissolved solids of ~10mg L^{-1} in NaCl equivalent. The most mineralized waters ($\Sigma^+ > 24$ meq L^{-1}) are here termed "saline" up to 1.4 g L^{-1} NaCl equivalent, a value slightly less than the conventional limit of 3 g L^{-1} NaCl adopted for "saline" lakes.

8.6 DISTRIBUTION OF WEATHERING INTENSITIES AT THE GLOBAL SCALE

Present-day weathering intensities can theoretically be assessed by the export of dissolved material by rivers. Yet many assumptions and corrections have to be made: (i) human impacts (additional sources and sinks) should be negligible, (ii) atmospheric inputs should be subtracted, (iii) products of chemical weathering should not be carried as particulates (e.g., phytoliths), nor (iv) accumulated within river basins in lakes or soils. In addition, it must be remembered that 100% of the HCO_3^- may originate from the atmosphere in noncarbonate river basins (calcite is only found in trace amounts in granites) and ~50% in carbonate terrains. In basins of mixed lithologies the proportions range between 50% and 100%. The total cation export or yield Y^+ (which excludes silica) is used to express the weathering intensity. Y^+, expressed in eq m^{-2} yr^{-1}, the product of annual runoff (m yr^{-1}) and Σ^+ (meq L^{-1}), is extremely variable at the Earth's surface since it combines both runoff variability and river chemistry variability (Table 5).

The concentration of elements that are derived from rock weathering (Ca^{2+}, Mg^{2+}, Na^+, K^+), are less variable than runoff even in the driest conditions. The opposite is observed for Cl^- and SO_4^{2-}, which are characterized by very low Q_1 quantiles. The retention of silica, particularly under the driest conditions, makes its yield the most variable. The lowest yearly average runoff in this data set (3.1 mm yr^{-1} for Q_1) actually corresponds to the conventional limit for occasional river flow (3 mm yr^{-1}). Under such extremely arid conditions, flow may occur only few times per hundred years, as for some tributaries of Lake Eyre in Central Australia. The other runoff quantile Q_{99} corresponds to the wettest regions of the planet bordering the coastal zone.

Since the variability of runoff, represented by the percentile ratio Q_{99}/Q_1, is generally much greater than the variability of concentration (Tables 1 and 5), ionic yields primarily depend on runoff. With a given rock type, there is a strong correlation between yield and runoff, that corresponds to clusters of similar concentrations as for Ca^{2+} and silica (Figure 6). This influence was used

Table 5 Global distribution of ionic (eq m^{-2} yr^{-1}) and silica (mol m^{-2} yr^{-1}) yields and annual runoff (q in mm yr^{-1}) in medium-sized basins (3,200–200,000 km^2, n = 685).

Y_i	Ca^{2+}	Mg^{2+}	Na^+	K^+	Cl^-	SO_4^{2-}	HCO_3^-	Σ^+	SiO_2	q
Q_{99}	2.77	0.95	1.3	0.115	1.05	1.25	2.82	3.0	0.82	3,040
Q_1	0.0045	0.002	0.002	0.0002	0.0003	0.0004	0.0046	0.0005	0.0001	3.1
Q_{99}/Q_1	615	475	650	575	3500	3100	613	600	8200	980

Q_1 and Q_{99}: lowest and highest percentiles.

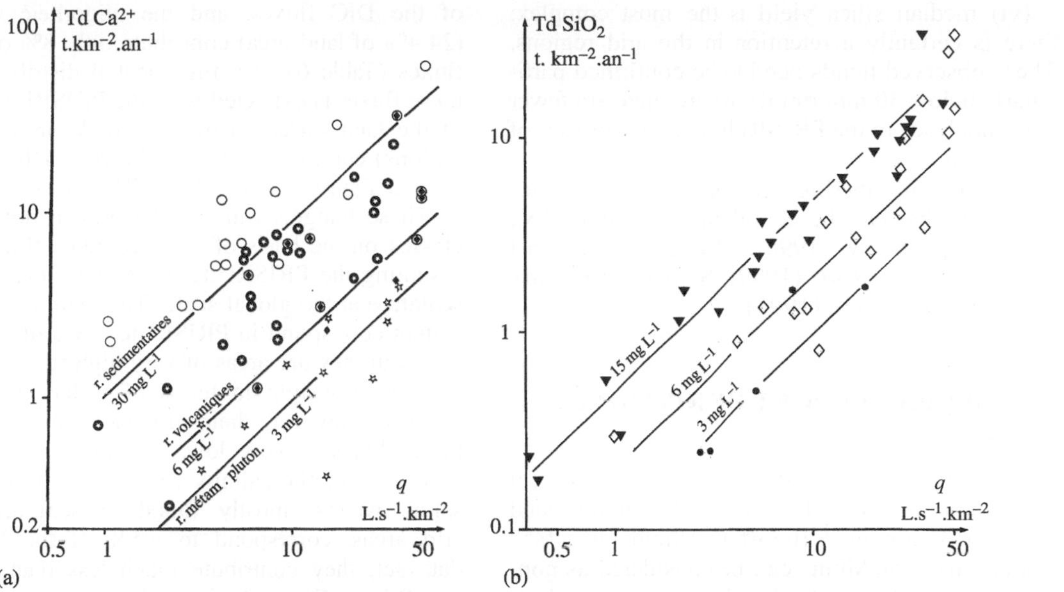

Figure 6 Relationship between calcium yield (a) and silica yield (b) versus river runoff for world rivers. Diagonals represent lines of equal concentrations. (a) Calcium yield for carbonated basins (average 30 mg $Ca^{2+}L^{-1}$), volcanic basins (6 mg L^{-1}), shield and plutonic (3 mg L^{-1}). (b) Silica yield for tropical regions (average 15 mg SiO_2 L^{-1}), temperate regions (6 mg L^{-1}), and cold regions (3 mg L^{-1}) (source Meybeck, 1994).

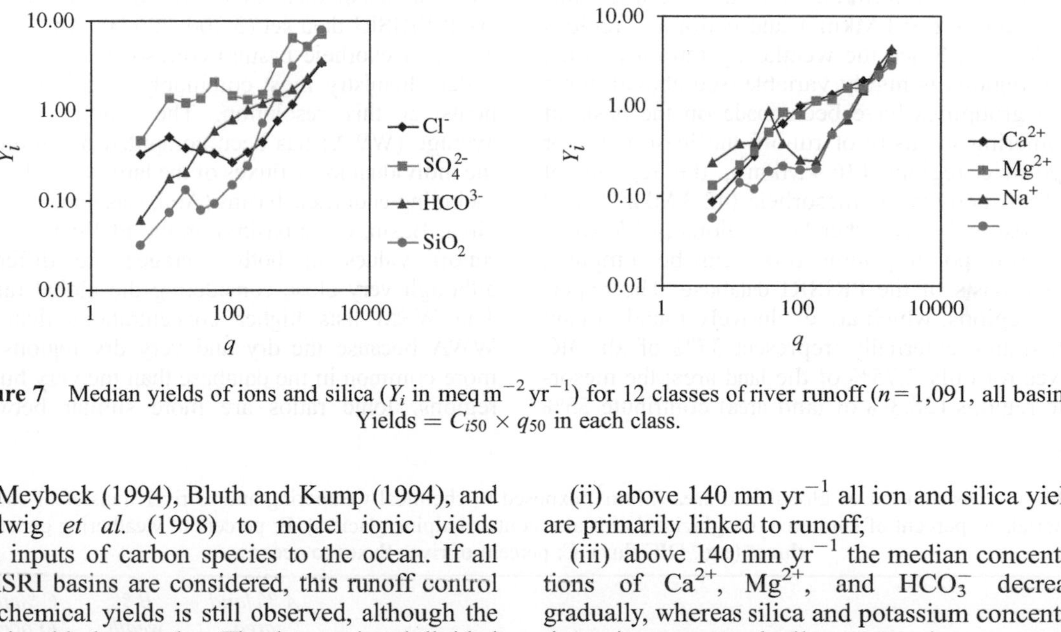

Figure 7 Median yields of ions and silica (Y_i in meq m^{-2} yr^{-1}) for 12 classes of river runoff ($n = 1{,}091$, all basins). Yields $= C_{i50} \times q_{50}$ in each class.

by Meybeck (1994), Bluth and Kump (1994), and Ludwig *et al.* (1998) to model ionic yields and inputs of carbon species to the ocean. If all PRISRI basins are considered, this runoff control on chemical yields is still observed, although the relationship is complex. The data set is subdivided into 12 classes of runoff from $q < 10$ mm yr^{-1} to $q > 1{,}750$ mm yr^{-1} (see Figure 3), for which the median yields of major ions and silica have been determined (Figure 7). In this log–log diagram, domains of equal concentration are parallel to the diagonal (1:1) line. Several types of evolution can be observed:

(i) sulfate and chloride yields are fairly constant below 140 mm yr^{-1}, suggesting that lower runoff is compensated by a concentration increase through evaporation (except for SO_4^{2-} below 10 mm yr^{-1});

(ii) above 140 mm yr^{-1} all ion and silica yields are primarily linked to runoff;

(iii) above 140 mm yr^{-1} the median concentrations of Ca^{2+}, Mg^{2+}, and HCO_3^- decrease gradually, whereas silica and potassium concentrations increase gradually, suggesting a greater influence of silicate weathering relative to carbonate weathering;

(iv) below 140 mm yr^{-1} median Ca^{2+} and Mg^{2+} concentrations are constant and median HCO_3^- concentration decreases slightly, which suggest a regulation mechanism through precipitation of carbonate minerals in semi-arid and arid regions;

(v) potassium yield is stable between 30 mm yr^{-1} and 140 mm yr^{-1}; below 30 mm yr^{-1} it decreases, suggesting retention; and

(vi) median silica yield is the most complex: there is certainly a retention in the arid regions. These observed trends need to be confirmed particularly below 30 mm runoff, where there are fewer silica analyses in the PRISRI base than analyses of major ions.

Detailed weathering controls, particularly for silicate rocks, can be found in Holland (1978), Drever and Clow (1995), Drever and Zobrist (1992), Gallard *et al.* (1999), Stallard (1995a,b), and White and Blum (1995).

8.7 GLOBAL BUDGET OF RIVERINE FLUXES

The unglaciated land surface of the present Earth amounts to ~133 Mkm^2. Excluding land below a 3 mm yr^{-1} runoff threshold (at a 0.5° resolution), ~50 Mkm^2 can be considered as non-exposed to surface water weathering (arheic, where $q<3$ mm yr^{-1}), whether draining to the ocean (exorheic) or draining internally such as the Caspian Sea basin (endorheic) (Table 6). The land area effectively exposed to weathering by meteoric water is estimated to be ~82.8 Mkm^2 (rheic regions); this, in turn, has to be divided into exorheic regions (76.1 Mkm^2) and endorheic regions (6.7 Mkm^2). Since the weathering intensity in the rheic regions is highly variable (see above), three main groupings have been made on the basis of weathering intensity or runoff: the least active or oligorheic regions (36.4 Mkm^2), the regions of medium activity or mesorheic (42.3 Mkm^2), and the most active or hyperrheic regions (4.1 Mkm^2). The corresponding ionic fluxes can be computed on the basis of the PRISRI database. The hyperrheic regions, which are exclusively found in land that drains externally, represent 37% of all DIC fluxes for only 2.75% of the land area; the mesorheic regions (28.3% of land area) contribute 55% of the DIC fluxes, and the oligorheic regions (24.4% of land area) contribute only 8% of these fluxes (Table 6). If a finer spatial distribution of these fluxes is extracted from the PRISRI data, 1% of the land surface (carbonate rocks in very wet regions) contributes 10% of the river DIC fluxes. Similar figures are found for all other ions.

These budgets can also be broken down into classes on the basis of ionic contents (Figure 8), assuming the PRISRI database to be fully representative at the global scale. The extremely dilute waters correspond in PRISRI to ~3% of the land area, but they are areas of very high runoff, therefore, their weight in terms of water volume is ~7.5%. However, their influence on the global ionic fluxes is very low, less than 1%. At the other end of the salinity scale the subsaline and saline waters, mostly found in semi-arid and arid areas, correspond to ~5% of the PRISRI data set; they contribute much less than 1% of runoff but ~7% of the ionic fluxes.

Although we have drawn attention to the extreme variability of ionic and silica concentrations, ionic proportions, ionic ratios, ionic and silica yields throughout this chapter, two sets of global averages are proposed here (Table 7). The world spatial median values (WSM) of the medium-sized PRISRI data set (3,200–200,000 km^2 endorheic and exorheic basins) correspond to the river water chemistry most commonly found on continents at this resolution. The world weighted average (WWA) has been computed by summing the individual ionic fluxes of the largest 680 basins, including endorheic basins (Aral, Caspian, Titicaca, Great Basin, Chad basins), in the PRISRI set. The runoff values in both averages are different, although very close considering the global range. The WSM lists higher concentrations than the WWA because the dry and very dry regions are more common in the database than the very humid regions. Ionic ratios are more similar between

Table 6 Distribution of global land area (Mkm^2) exposed to chemical weathering and to river transfer of soluble material. A: percent of land area (nonglaciated area also contains alpine glaciers). B: percent of weathering generated fluxes (e.g., DIC flux). C: percent of river fluxes to oceans.

					A % land area	*B % weath. flux*	*C % flux to ocean*
Total land 149 Mkm^2	Glaciated 16 Mkm^2				10.7	0.1?	0.1?
	Nonglaciated 133 Mkm^2	Arheic 50.2 Mkm^2		Endorheic[a] / Exorheic[a]	33.7	0	0
		Rheic 82.8 Mkm^2	Oligorheic	Endorheic	3.4	1	0
				Exorheic	21	7	7
			Mesorheic	Endorheic	1.1	2	0
				Exorheic	27.3	53	55
			Hyperrheic	Exorheic	2.75	37	38

[a] Potentially.

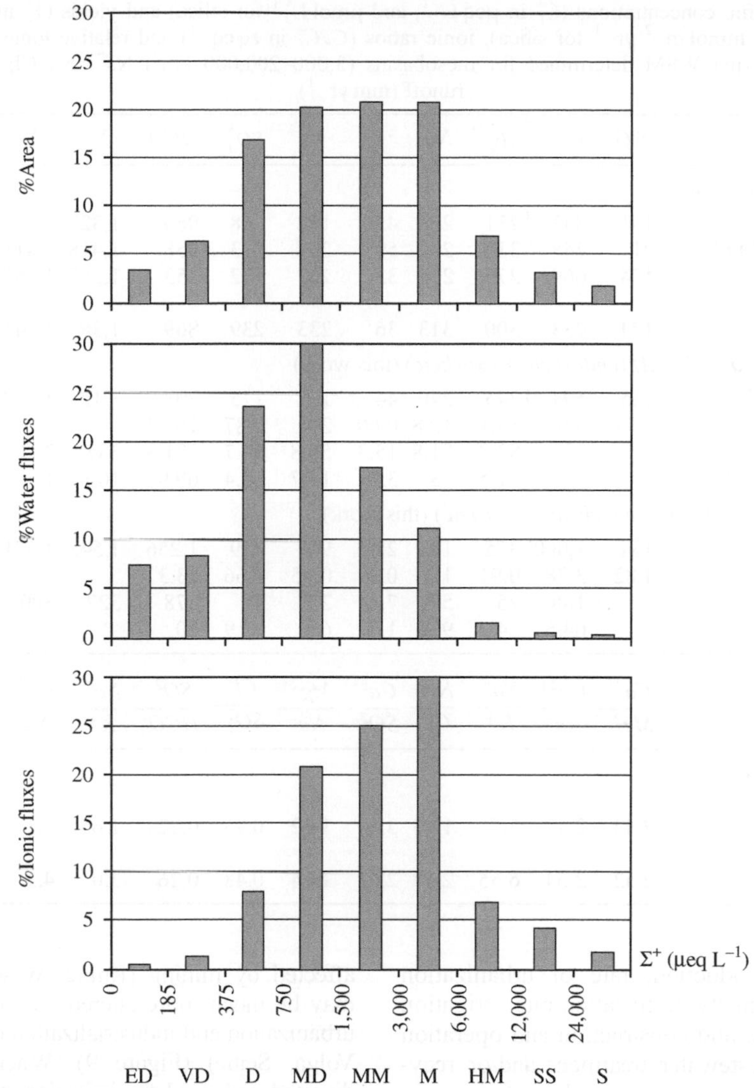

Figure 8 Distribution of global river attributes from the PRISRI database for classes of increasing total cations (Σ^+) based on PRISRI rivers (rheic regions). ED: extremely dilute waters, VD: very dilute, D: dilute, MD: medium dilute, MM: medium mineralized, HM: highly mineralized, SS: subsaline, S: saline.

the two averages. The composition of river inputs to the oceans is not stable: (i) it has varied since the Late Glacial Maximum and during the geological past (Kump and Arthur, 1997); (ii) present-day river chemistry is much altered by human activities. Four estimates of average exorheic rivers are presented in Table 7: considering the natural variations of river chemistry, these averages are relatively close to each other and to WWA and WSM, the differences are due to the nature of the data sets and to the inclusion or exclusion of presently altered rivers. A re-estimation of the pristine inputs (under present-day climatic conditions) should now be done with the new data, although human impacts are difficult to quantify.

8.8 HUMAN ALTERATION OF RIVER CHEMISTRY

River chemistry is very sensitive to alteration by many human activities, particularly mining and the chemical industries, but also to urbanization as urban wastewaters are much more concentrated than rural streams, and to agriculture through the use of fertilizers (Table 2) (Meybeck *et al.*, 1989; Meybeck, 1996; Flintrop *et al.*, 1996). New sinks are also created such as reservoirs (calcite and silica trapping; enhanced evaporation) and irrigated soils, which may retain soluble elements if they are poorly drained. New controls correspond to these anthropogenic influences, such as mining

Table 7 WWA ionic concentrations (C_i^* in μeq L^{-1}, and μmol L^{-1} for silica) and yields (Y_d^* in g m^{-2} yr^{-1}, Y_i^* in meq m^{-2} yr^{-1}, and mmol m^{-2} yr^{-1} for silica), ionic ratios (C_i/C_j^* in eq eq^{-1}) and relative ionic proportion (%C_i^*). Same variables for the WSM determined for mesobasins (3,000–200,000 km^2) $Na^{\#}$ = Na–Cl; q = global average runoff (mm yr^{-1}).

	SiO_2	Ca^{2+}	Mg^{2+}	Na^+	K^+	Cl^-	SO_4^{2-}	HCO_3^-	Σ^+	Σ^-	q
World river inputs to oceans											
Clarke (1924) C_i	161	843	231	208	45	132	208	969	1,327	1,309	(320)
Livingstone (1963) C_i	217	748	337	274	59	220	233	958	1,418	1,411	314
Meybeck (1979) pristine C_i	174	669	276	224	33	162	172	853	1,202	1,187	374
1970 C_i	174	733	300	313	36	233	239	869	1,382	1,341	374
World weighted average (WWA) (endorheic + exorheic) (this work)											340
C_i*	145	594	245	240	44	167	175	798	1,125	1,139	
Y_d*	2.98	4.04	1.01	1.88	0.60	2.0	2.87	16.54			
Y_i*		202	83.3	81.8	15.3	56.8	59.7	271.3	382	388	
%C_i*		52.2	21.5	23	3.3	14.7	15.4	69.9	100	100	
World spatial median (WSM) (endorheic + exorheic) (this work)											222
C_{i50}	134	1,000	375	148	25.5	96	219	1,256	1,548	1,571	
Y_{d50}	1.82	3.78	0.91	1.31	0.29	0.95	2.66	13.3			
Y_{i50}		188	75	57	7.4	22	55	278	327	300	
%C_{i50}		64.6	24.1	9.6	1.7	6.1	13.9	80			

	$\frac{Ca^{2+}}{Mg^{2+}}$	$\frac{Ca^{2+}}{Na^+}$	$\frac{Na^+}{K^+}$	$\frac{Na^+}{Cl^-}$	$\frac{Ca^{2+}}{SO_4^{2-}}$	$\frac{Mg^{2+}}{Na^+}$	$\frac{Cl^-}{SO_4^{2-}}$	$\frac{SO_4^{2-}}{HCO_3^-}$	$\frac{Na^{\#}}{K^+}$	$\frac{Ca^{2+}}{Na^{\#}}$	$\frac{Mg^{2+}}{Na^{\#}}$	$\frac{SiO_2}{\Sigma^+}$
WWA												
$(C_i/C_j)^*$	2.42	2.47	5.5	1.43	3.39	1.02	0.95	0.22	1.65	8.1	3.35	0.13
WSM												
$(C_i/C_j)_{50}$	2.32	2.54	6.55	2.27	2.97	1.14	0.48	0.26	3.0	4.3	1.9	0.12

and industrial production, rate of urbanization and population density, fertilization rate, irrigation rate and practices, and construction and operation of reservoirs. Wastewater treatment and/or recycling can be effective as a control on major ions originating from mines (petroleum and gas exploitation, coal and lignite, pyritic ores, potash and salt mines) and industries, yet their effects are seldom documented. Urban wastewater treatment does not generally affect the major ions. Human impacts on silica are still poorly studied apart from retention in reservoirs. At the pH values common in surface waters, silica concentration is limited and evidence of marked excess silica in rivers due to urban or industrial wastes has not been observed by this author. The gradual alteration of river chemistry was noted very early, for example, for SO_4^{2-} (Berner, 1971) and Cl^- (Weiler and Chawla, 1969) in the Mississippi and the Saint Lawrence systems. Regular surveys made since the 1960's and comparisons with river water analyses performed a hundred years ago reveal a worldwide increase in Na^+, Cl^-, and SO_4^{2-} concentrations, whereas Ca^{2+}, Mg^{2+}, and HCO_3^- concentrations are more stable (Meybeck *et al.*, 1989; Kimstach *et al.*, 1998). Some rivers affected by mining (Rhine, Weser, Vistula, Don) may be much more altered than rivers affected by urbanization and industrialization only (Mississippi, Volga, Seine) (Figure 9). When river water is diverted and used for irrigation, there is a gradual increase of ionic content, particularly Na^+, K^+, Cl^-, and SO_4^{2-} as for the Colorado, Murray, Amu Darya (Figure 9). Salinization may also result from agriculture as for the Neman River (Figure 9). Salinization is discussed in detail by Vengosh.

It is difficult to assess the different anthropogenic sources of major ions which depend on the factors mentioned above. These human factors also vary in time for a given society and reflect the different environmental concerns of these societies, resulting in multiple types of river–society relationships (Meybeck, 2002). In the developed regions of the northern temperate zone it is now difficult to find a medium-sized basin that is not significantly impacted by human activities. In industrialized countries, each person generates dissolved salt loadings that eventually reach river systems (Table 8).

These anthropogenic loads are higher for Na^+, Cl^-, and SO_4^{2-} relative to natural loads. This partially explains the higher sensitivity of river chemistry to human development. At a

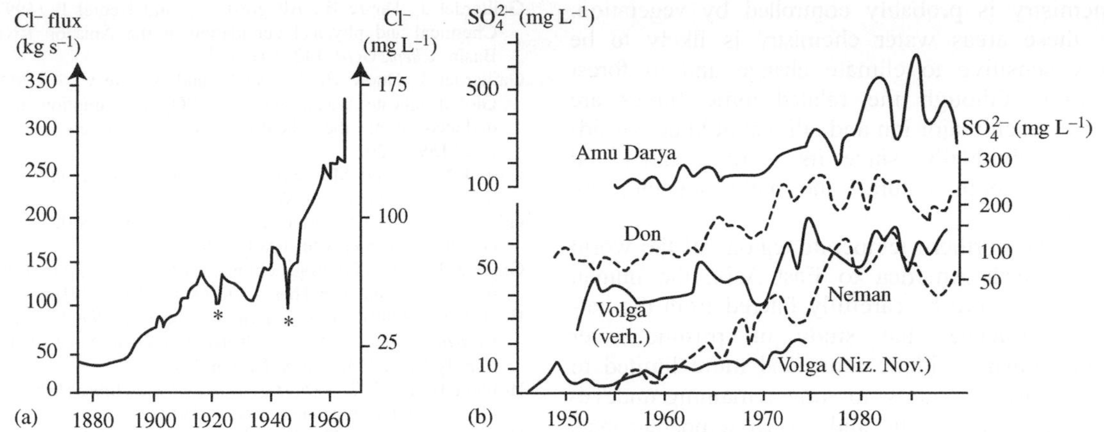

Figure 9 (a) Chloride evolution (fluxes, concentrations) in the Rhine River at mouth from 1,875 to 1,970 (ICPR, 1975). (b) Sulfate evolution in selected rivers of the former USSR, Volga at Nizhny Novgorod and Neman at Sovetsk (lower scale), Don at Aksai, and Volga at Verhnelebyazhye (medium scale), Amu Darya at Kzyl Djar (upper scale) (Tsikurnov in Kimstach *et al.*, 1998). * reduced mining activity in the Rhine Basin due to crisis and conflicts.

Table 8 Excess loads of major elements normalized to basin population in industrialized regions (kg cap^{-1} yr^{-1}).

	SiO_2	Ca^{2+}	Mg^{2+}	Na^+	K^+	Cl^-	SO_4^{2-}	HCO_3^-	*Period*
Per capita loads in residential urban sewage (kg $cap^{-1}yr^{-1}$)									
Montreal	ND	3.2	0.65	6.6	1.0	8.2	13.5	24	1970s
US sewer	2.4	3	1.5	14	2	15	6	20	1960s
Brussels	1.2	2.6	ND	9.5	1.6	8.4	5.8	14.7	1980s
Paris	0.2	1.2	0.7	6.4	1.5	6.3	11.0	14.5	1990s
Evolution of rivers[a] (kg $cap^{-1}yr^{-1}$)									
	ND	51	17	85	6	100	136	ND	1900–1970
Estimated anthropogenic input to ocean (10^6 t yr^{-1})									
	ND	47	10.5	78	5	93	124	100	1970

Sources: Meybeck (1979) and Meybeck *et al.* (1998).
[a]Mississippi, St. Lawrence, Seine, Rhine, Odra, Wisla.

certain population density in impacted river basins the anthropogenic loads equal (Na^+, K^+, Cl^-, SO_4^{2-}) or greatly exceed the natural ones (NO_3^-, PO_4^{2-}), defining a new era, the Anthropocene (Meybeck, 2002) when humans control geochemical cycles.

The silica trend in world rivers has recently attracted attention: dissolved silica is decreasing in impounded and/or eutrophied rivers, at the same time as nitrate is increasing in agricultural basins. As a result the Si : N ratio, which was generally well above 10 a hundred years ago, has dropped below 1.0 g g^{-1} in rivers such as the Mississippi, resulting in a major shift in the coastal algal assemblages (Rabalais and Turner, 2001), leading to dystrophic coastal areas. This trend, which is also observed for other river systems (Seine, Danube), could expand in the next decades due to increased fertilizer use and to increased reservoir construction.

8.9 CONCLUSIONS

Riverine chemistry is naturally highly variable at the global scale, which confirms the observation made by (the former) Soviet geochemists on a 20 Mkm^2 subset of land area (Alekin and Brazhnikova, 1964) and the conclusions of Clarke (1924), Livingstone (1963), Holland (1978), and Drever (1988). Ionic concentrations and yields (weathering rates) commonly vary over two to three orders of magnitude: only a fraction of the surface of the continents is actively exposed to weathering by meteoric water. The related river water types are multiple. There are at least a dozen types depending on surficial rock exposed to weathering, the water balance, and atmospheric inputs. Gibbs' (1970) global scheme for water chemistry holds for ~80% of river waters but is oversimplified for the remaining 20%. In very dilute waters (cation sum < 0.185 μeq L^{-1}) water chemistry is probably controlled by vegetation. In these areas water chemistry is likely to be very sensitive to climate change and to forest cutting, although the related ionic fluxes are small. Each major ion and silica should be considered individually, since its sources, sinks and controls, both natural and anthropogenic, are different.

In the northern temperate regions of the world from North America to East Asia, the human impacts should be carefully filtered from the natural influence: the study of pristine river geochemistry will be more and more limited to the subarctic regions, to some remaining undeveloped tropical regions, and to small temperate areas of the southern hemisphere such as South Chile and New Zealand. There is a global-scale increase of riverine Na^+, K^+, Cl^-, and SO_4^{2-}. HCO_3^- is still very stable. Impacts of riverine changes on the Earth System (Li, 1981) should be further addressed.

REFERENCES

Alekin O. A. and Brazhnikova L. V. (1964) *Runoff of Dissolved Substances from the USSR Territory* (in Russian). Nauka, Moscow.

Berner E. K. and Berner R. A. (1996) *Global Environment, Water, Air, and Geochemical Cycles.* Prentice Hall, Engle-woods Cliff, NJ.

Berner R. A. (1971) Worldwide sulfur pollution of rivers. *J. Geophys. Res.* **76**, 6597–6600.

Berner R. A. (1996) Chemical weathering and its effect on atmospheric CO_2 and climate. In *Chemical Weathering Rates of Silicate Minerals,* Reviews in Mineralogy (eds. A. F. White and S. L. Brantley). Mineralogical Society of America, Washington, DC, vol. 31, pp. 565–583.

Berner R. A., Lasaga A. C., and Garrels R. M. (1983) The carbonate–silicate geochemical cycle and its effect on atmospheric carbon dioxide over the past 100 millions years. *Am. J. Sci.* **301**, 182–204.

Bluth G. J. S. and Kump L. R. (1994) Lithologic and climatologic controls of river chemistry. *Geochim. Cosmochim. Acta* **58**, 2341–2359.

Clarke F. W. (1924) *The Data of Geochemistry,* 5th edn. US Geol. Surv. Bull. vol. 770, USGS, Reston, VA.

Degens E. T., Kempe S., and Richey J. E. (eds.) (1991) *Bio-Geochemistry of Major World Rivers.* Wiley, New York.

Drever J. I. (1988) *The Geochemistry of Natural Waters,* 2nd edn. Prentice Hall, Englewood Cliff, NJ, 437pp.

Drever J. I. (1994) The effect of land plants on weathering rates of silicate minerals. *Geochim. Cosmochim. Acta* **58**, 2325–2332.

Drever J. I. and Clow D. W. (1995) Weathering rates in catchments. In *Chemical Weathering Rates of Silicate Minerals,* Reviews in Mineralogy (eds. A. F. White and S. L. Brantley). Mineralogical Society of America, Washington, DC, vol. 31, pp. 463–483.

Drever J. I. and Zobrist J. (1992) Chemical weathering of silicate rocks as a function of elevation in the southern Swiss Alps. *Geochim. Cosmochim. Acta* **56**, 3209–3216.

Flintrop C., Hohlmann B., Jasper T., Korte C., Podlaha O. G., Sheele S., and Veizer J. (1996) Anatomy of pollution: rivers of North Rhine-Westphalia, Germany. *Am. J. Sci.* **296**, 58–98.

Gaillardet J., Dupré B., Allègre C. J., and Négrel P. (1997) Chemical and physical denudation in the Amazon River Basin. *Chem. Geol.* **142**, 141–173.

Gaillardet J., Dupré B., Louvat P., and Allègre C. J. (1999) Global silicate weathering and CO_2 consumption rates deduced from the chemistry of large rivers. *Chem. Geol.* **159**, 3–30.

Garrels R. M. and Mackenzie F. T. (1971) *The Evolution of Sedimentary Rocks.* W. W. Norton, New York.

Gibbs R. J. (1970) Mechanism controlling world water chemistry. *Science* **170**, 1088–1090.

Guyot J. L. (1993) Hydrogéochimie des fleuves de l'Amazonie bolivienne. Etudes et Thèses, ORSTOM, Paris, 261pp.

Hem J. D. (1989) *Study and Interpretation of the Chemical Characteristics of Natural Water.* US Geol. Surv., Water Supply Pap. 2254, USGS, Reston, VA.

Holland H. D. (1978) *The Chemistry of the Atmosphere and Oceans.* Wiley-Interscience, New York.

Huh Y. (1998/1999) The fluvial geochemistry of the rivers of Eastern Siberia: I to III. *Geochim. Cosmochim. Acta* **62**, 1657–1676; **62**, 2053–2075; **63**, 967–987.

ICPR (International Commission for the Protection of the Rhine) (1975) ICPR/IKSR Bundesh f. Gewässerkunde, Koblenz (Germany).

Kimstach V., Meybeck M., and Baroudy E. (eds.) (1998) *A Water Quality Assessment of the Former Soviet Union.* E and FN Spon, London.

Kobayashi J. (1959) Chemical investigation on river waters of southeastern Asiatic countries (report I). The quality of waters of Thailand. Bericht. Ohara Inst. für Landwirtschaft Biologie, vol. 11(2), pp. 167–233.

Kobayashi J. (1960) A chemical study of the average quality and characteristics of river waters of Japan. Bericht. Ohara Inst. für Landwirstschaft Biologie, vol. 11(3), pp. 313–357.

Kump L. R. and Arthur M. A. (1997) Global chemical erosion during the Cenozoic: weatherability balances the budget. In *Tectonic and Climate Change* (ed. W. F. Ruddiman). Plenum, New York, pp. 399–426.

Li Y. H. (1981) Geochemical cycles of elements and human perturbations. *Geochim. Cosmochim. Acta* **45**, 2073–2084.

Likens G. E., Bormann F. H., Pierce R. S., Eaton J. S., and Johnson N. M. (1977) *Biogeochemistry of a Forested Ecosystem.* Springer, Berlin.

Livingstone D.A. (1963) *Chemical Composition of Rivers and Lakes.* Data of Geochemistry. US Geol. Surv. Prof. Pap. 440 G, chap. G, G1–G64.

Ludwig W., Amiotte-Suchet P., Munhoven G., and Probst J. L. (1998) Atmospheric CO_2 consumption by continental erosion: present-day controls and implications for the last glacial maximum. *Global Planet. Change* **16–17**, 107–120.

Meybeck M. (1979) Concentration des eaux fluviales en éléments majeurs et apports en solution aux océans. *Rev. Géol. Dyn. Géogr. Phys.* **21**(3), 215–246.

Meybeck M. (1986) Composition chimique naturelle des ruisseaux non pollués en France. *Sci. Geol. Bull.* **39**, *3–77.*

Meybeck M. (1987) Global chemical weathering of surficial rocks estimated from river dissolved loads. *Am. J. Sci.* **287**, 401–428.

Meybeck M. (1993) Riverine transport of atmospheric carbon:sources, global typology and budget. *Water Air Soil Pollut.* **70**, 443–464.

Meybeck M. (1994) Origin and variable composition of present day riverborne material. In *Material Fluxes on the Surface of the Earth. Studies in Geophysics* (ed. Board on Earth Sciences and Resources–National Research Council). National Academic Press, Washington, DC, pp. 61–73.

Meybeck M. (1996) River water quality, global ranges time and space variabilities. *Vehr. Int. Verein. Limnol.* **26**, 81–96.

Meybeck M. (2002) Riverine quality at the Anthropocene: propositions for global space and time analysis, illustrated by the Seine River. *Aquat. Sci.* **64**, 376–393.

Meybeck M. and Ragu A. (1996) River Discharges to the Oceans. An assessment of suspended solids, major ions, and nutrients. Environment Information and Assessment Report. UNEP, Nairobi, 250p.

Meybeck M. and Ragu A. (1997) Presenting Gems Glori, a compendium of world river discharge to the oceans. *Int. Assoc. Hydrol. Sci. Publ.* **243**, 3–14.

Meybeck M., Chapman D., and Helmer R. (eds.) (1989) *Global Fresh Water Quality: A First Assessment.* Basil Blackwell, Oxford, 307pp.

Meybeck M., de Marsily G., and Fustec E. (eds.) (1998) *La Seine en Son Bassin.* Elsevier, Paris.

Miller J. P. (1961) *Solutes in Small Streams Draining Single Rock Types, Sangre de Cristo Range, New Mexico.* US Geol. Surv., Water Supply Pap., 1535-F.

Millot R., Gaillardet J., Dupré B., and Allègre C. J. (2002)The global control of silicate weathering rates and the coupling with physical erosion: new insights from rivers of the Canadian Shield. *Earth Planet. Sci. Lett.* **196**, 83–98.

Rabalais N. N. and Turner R. E. (eds.) (2001) *Coastal Hypoxia.* Coastal Estuar. Studies, **52**, American Geophysical Union.

Reeder S. W., Hitchon B., and Levinston A. A. (1972) Hydrogeochemistry of the surface waters of the Mackenzie river drainage basin, Canada: I. Factors controlling inorganic composition. *Geochim. Cosmochim. Acta* **36**, 825–865.

Stallard R. F. (1995a) Relating chemical and physical erosion. In *Chemical Weathering Rates of Silicate Minerals,* Reviews in Mineralogy (eds. A. F. White and S. L. Brantley). Mineralogical Society of America, Washington, DC, pp. 543–562.

Stallard R. F. (1995b) Tectonic, environmental and human aspects of weathering and erosion: a global review using a steady-state perspective. *Ann. Rev. Earth Planet. Sci.* **23**, 11–39.

Stallard R. F. and Edmond J. M. (1981) Geochemistry of the Amazon: 1. Precipitation chemistry and the marine contribution to the dissolved load at the time of peak discharge. *J. Geophys. Res.* **86**, 9844–9858.

Stallard R. F. and Edmond J. M. (1983) Geochemistry of the Amazon: 2. The influence of geology and weathering environment on the dissolved load at the time of peak discharge. *J. Geophys. Res.* **86**, 9844–9858.

USGS (1909–1965) *The Quality of Surface Waters of the United States.* US Geol. Surv., Water Supply Pap., 236, 237, 239, 339, 363, 364, 839, 970, 1953, Reston, VA.

Viers J., Dupré B., Polvé M., Schott J., Dandurand J. L., and Braun J. J. (1997) Chemical weathering in the drainage basin of a tropical watershed (Nsimi-Zoetele site, Cameroon): comparison between organic poor and organic rich waters. *Chem. Geol.* **140**, 181–206.

Weiler R. R. and Chawla V. K. (1969) Dissolved mineral quality of the Great Lakes waters. *Proc. 12th Conf. Great Lakes Res.: Int. Assoc. Great Lakes Res.,* 801–818.

White A. T. and Blum A. E. (1995) Effects of climate on chemical weathering in watersheds. *Geochim. Cosmochim. Acta* **59**, 1729–1747.

Readings from the Treatise on Geochemistry
ISBN: 978-0-12-381391-6

pp. 271–288

9 Geochemistry of Groundwater

F. H. Chapelle
US Geological Survey, Columbia, SC, USA

9.1 INTRODUCTION

Groundwater geochemistry is concerned with documenting the chemical composition of groundwater produced from different aquifer systems, and with understanding the processes that control this composition. This chapter provides a brief overview of the development of the science of groundwater geochemistry, and gives a series of examples of how aquifer lithology and hydrology affect groundwater composition. This, in turn, provides an introduction to the principles of mass balance, equilibrium chemistry, and microbiology that have proved useful in understanding the composition of groundwater in a variety of geologic settings. While the natural processes that affect groundwater geochemistry are complex, observational and quantitative methodologies have been developed that help to elucidate them. This chapter summarizes the more widely used methodologies in groundwater geochemistry, and shows how they have been applied to solve various hydrologic problems in both pristine and contaminated aquifer systems.

9.1.1 Historical Overview

Humans have used springs and wells as sources of water for thousands of years. Very early in human history, it was noted that the taste and chemical quality of spring and well waters differ from place to place (Back *et al.*, 1995). Some springs, for example, produced water that was cool, crisp, and excellent for drinking. Other springs produced hot, salty, or sulfurous waters that were undrinkable. Because springwaters were so important as sources of water, people paid careful attention to these chemical differences. Many cultures developed elaborate mythologies to explain why springwaters differed so much in their chemical quality. It was common, for example, to rationalize these differences in moral terms. Good-quality drinking water was associated with good deities, and poor-quality waters were associated with evil deities (Chapelle, 2000).

It was also noted that water from certain springs or wells appeared to have medicinal qualities (Burke, 1853). It was entirely logical to assume that these medicinal effects reflected the different composition of springwaters. One early motivation for studying groundwater geochemistry was to learn what dissolved constituents were present in springwaters, and to understand how these constituents imparted medicinal effects to the waters. Steel (1838), for instance, identified iodine in the springwaters of Saratoga Springs, NY, and associated this constituent with a variety of beneficial medicinal properties. It is not certain, however, whether Steel recognized that iodine in the springwaters could cure or prevent simple goiters, a medicinal use of iodine that later became widely known.

Similar practical concerns stimulated the earliest quantitative chemical analyses of groundwater. For example, Rogers (1917) noticed that oil-field brines associated with petroleum hydrocarbons lacked dissolved sulfate, whereas oil-field brines without associated petroleum hydrocarbons contained abundant sulfate. Rogers (1917) therefore raised the possibility that water chemistry could be useful in prospecting for oil. Other studies in the early twentieth century noted that water chemistry did not vary randomly. Rather, there were systematic changes in chemistry as water flowed along regional hydrologic gradients. For example, Renick (1924) noticed that shallow groundwater in the recharge areas of the Fort Union Formation of Montana were "hard," i.e., contained relatively high concentrations of calcium and magnesium. Downgradient, however, these waters became "soft," and contained relatively high concentrations of sodium. Renick (1924) attributed these changes to "base exchange" (i.e., cation exchange) processes that removed calcium and magnesium from solution and replaced it with sodium. Correlating water-chemistry changes with directions of groundwater flow, and attempting to identify specific chemical reactions occurring in aquifers, was an approach followed by several subsequent investigations (Cedarstrom, 1946; Chebotarev, 1955). This approach was formalized by Back (1966), who coined the term "hydrochemical facies" to refer to water chemistry changes commonly observed in the direction of groundwater flow.

The observation that water chemistry tended to change systematically and predictably suggested that the chemical and or biological processes involved in these changes could be identified. During the next thirty years, a number of approaches to studying these processes, and learning how to quantify their effects dominated groundwater geochemistry. These approaches assumed that waters entering aquifers are initially relatively unmineralized rainwater or snowmelt. Since these waters did contain dissolved carbon dioxide, the resulting carbonic acid attacks minerals present in aquifers and dissolves them. Thus, the chemistry of groundwaters could be explained in terms of initially nonequlibrium conditions moving toward chemical equilibrium by reacting with aquifer minerals. This equilibrium approach was formalized by Garrels and Christ (1965), and provided a quantitative framework within which the composition of groundwater could be understood. Application of the equilibrium approach soon showed that, once the minerals being dissolved were identified, it was possible to apply mass-balance constraints to calculate the amount of each mineral that was dissolved to produce groundwater of a particular composition (Garrels and Mackenzie, 1967). This mass-balance approach was soon expanded to include balances of electrons (accounting for redox reactions) and isotopes, and was formalized within a rigorous mathematical framework (Helgeson *et al.*, 1970; Plummer, 1977; Plummer and Back, 1980; Parkhurst *et al.*, 1982; Plummer *et al.*, 1983). This approach combined the concepts of chemical equilibrium and material balance, and came to be known as "geochemical modeling" (Plummer, 1984).

Applications of this quantitative approach soon demonstrated that there were many more mineral phases present in aquifers, including mineral phases with variable chemical compositions, than could be included in any one geochemical model. It was possible to construct a virtually limitless number of geochemical models that explained the observed water chemistry of even the simplest aquifer. The nonunique nature of these geochemical models is a reflection of the mineralogic complexity of aquifers and reflects the fact that aquifers are never at true thermodynamic equilibrium. In spite of this, much insight into the processes that control groundwater chemistry has been gained by applying the geochemical modeling approach.

9.1.2 Lithology, Hydrology, and Groundwater Geochemistry

In the early 1960s, Feth *et al.* (1964) published a compendium of analyses of springwaters in the

Sierra Nevada Mountains. This study suggested that the chemistry of the springwaters reflected the progressive alteration of plagioclase feldspar to kaolinite, which in turn produced the solutes present in groundwater. This hypothesis could readily be tested using the concepts of equilibrium chemistry and mass balance. In testing this hypothesis, Garrels and Mackenzie (1967) showed that the observed concentrations of sodium, silica, and the water pH were similar to those calculated by assuming that plagioclase was progressively dissolved by carbon dioxide-bearing waters. It was relatively easy to start with a given springwater and subtract quantities of calcium, sodium, and silica appropriate to the stoichiometry of feldspars and/or other reactive minerals. In this manner, it was possible to estimate the amount of minerals that had dissolved in order to generate the observed water composition.

The mass-balance approach was important because it provides a rational explanation for the chemical variability of groundwater. A general statement of the principle of conservation of mass for the chemical composition of groundwater flowing along a flow path is

$$\text{initial water} + \text{dissolving minerals} - \text{precipitating minerals} = \text{final water} \qquad (1)$$

The mass-balance approach clearly shows that groundwater composition is determined by the kinds and amounts of minerals that dissolve and/or precipitate in an aquifer.

9.1.3 Mineral Dissolution and Precipitation

The presence or absence of particular minerals in an aquifer places obvious constraints on which minerals may dissolve or precipitate (Equation (1)). A more subtle question has to do with mineral solubility. If a mineral is present but is insoluble, it cannot be the source of dissolved solutes. Mineral solubility constraints, in turn, can be addressed using the principles of equilibrium chemistry.

The *saturation index* (SI) of a particular mineral in the presence of water is defined as

$$SI = \log IAP/K \qquad (2)$$

where IAP is the ion activity product and K is the equilibrium constant for a particular mineral dissolution/precipitation reaction. Take, for example, the dissolution of calcite:

$$CaCO_3 = Ca^{2+} + CO_3^{2-} \qquad (3)$$

The IAP of this reaction is given by

$$IAP = a_{Ca^{2+}} a_{CO_3^{2-}} \qquad (4)$$

and the equilibrium constant is given by

$$a_{Ca^{2+}} a_{CO_3^{2-}} / a_{CaCO_3} = K \qquad (5)$$

At equilibrium, and assuming the solid phase ($CaCO_3$) has an activity of 1 (unit activity), it is clear that

$$IAP = K \qquad (6)$$

so that

$$IAP/K = 1 \qquad (7)$$

or

$$SI = \log IAP/K = 0 \qquad (8)$$

An SI index of zero indicates that a mineral is in thermodynamic equilibrium with the solution. If the SI is less than zero, the solution is undersaturated with respect to the mineral and that mineral may dissolve. Conversely, an SI greater than zero indicates the solution is oversaturated with respect to the mineral and that the mineral may precipitate. These considerations are useful in constraining the kinds of reactions that are possible in groundwater systems.

By taking into account aquifer lithology, direction of groundwater flow, and mineral solubility constraints, a great deal can be learned about groundwater geochemistry. These considerations can be taken into account in a variety of ways, ranging from entirely qualitative to highly constrained, rigorously mathematical approaches. The approach taken generally depends on the goals of the individual investigator. Regardless of the approach taken, however, the geochemistry of groundwater reflects the lithologic and hydrologic characteristics of different aquifers. The following sections, which describe the groundwater geochemistry of different aquifer systems in the United States, are designed to illustrate how lithology and hydrology affect the composition of groundwater, and how qualitative and quantitative approaches can be used to understand the groundwater geochemistry of different hydrologic systems.

9.2 GROUNDWATER GEOCHEMISTRY IN SOME COMMON GEOLOGIC SETTINGS

9.2.1 A Basalt Lithology: The Snake River Aquifer of Idaho

One of the principal difficulties in understanding groundwater geochemistry is due to the chemical heterogeneity of most aquifer systems. The rocks and sediments that form most aquifers were often deposited over long periods of time and under a variety of geologic conditions. Thus, the chemical composition of the rock or sediment matrix of aquifers is seldom constant. There are, however, examples of aquifer systems that are relatively homogeneous in mineral content and composition. The basalt aquifers of the Snake River Plain in Idaho are one of these relatively homogeneous systems. As such, it provides an excellent example of the effects of lithology on groundwater chemistry.

The Snake River Plain marks the track of the Yellowstone Hotspot, a deep mantle plume that has been in place and active since the Miocene (~18 Myr). As the North American Plate moved westward over the plume, at a rate 2.5–8.0 cm yr^{-1}, the surface manifestations of the associated volcanism moved eastward from what is now Nevada, along the Snake River Plain, and is now located underneath Yellowstone National Park in Wyoming (Armstrong, 1971; Embree *et al.*, 1982; Pierce and Morgan, 1992). The later stages of volcanism have been characterized by flood basalts, which have filled the valley created by the collapse of underlying magma chambers (Pierce and Morgan, 1992). As a result, the Snake River Plain is now underlain by several hundred meters of basalt flows, which are themselves underlain by rhyolitic volcanic rocks.

The Snake River Plain aquifer system, which has formed in the near-surface basalts, is largely unconfined. Groundwater moves from northeast to southwest until it discharges to the Snake River, mostly from a series of springs located between Twin Falls and King Hill (Lindholm *et al.*, 1983). The aquifer is recharged by atmospheric precipitation that falls on the topographically high areas to the east, north, and south of the Snake River Valley. This recharge pattern drives groundwater flow from east to west (Figure 1). The highly fractured and rubbly nature of the basalts reflects their rapid cooling at land surface. Consequently, the aquifer exhibits very high transmissivities of $\sim 10^5\ m^2\ d^{-1}$. This high transmissivity, combined with the relatively large amounts of recharge that enter the aquifer, leads to rates of groundwater flow that average 3 m d^{-1}. As a consequence of these rapid groundwater flow rates, the average residence time of water in the aquifer is relatively low, 200–250 yr.

The groundwater geochemistry of the Snake River Plain aquifer system has been described by Wood and Low (1988). It is an excellent example of how a knowledge of aquifer mineralogy can

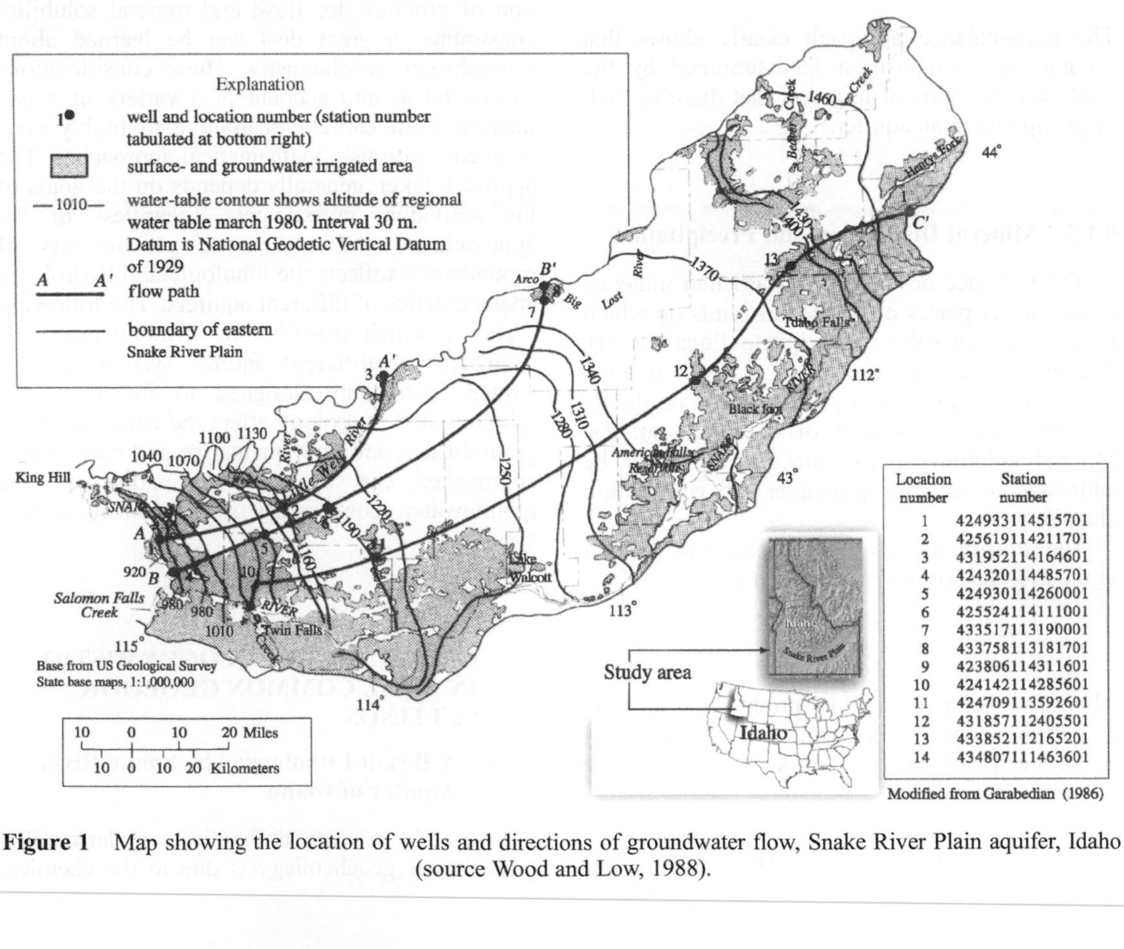

Figure 1 Map showing the location of wells and directions of groundwater flow, Snake River Plain aquifer, Idaho (source Wood and Low, 1988).

explain the observed composition of groundwaters. Some representative analyses of Snake River Plain groundwater are shown in Table 1. The water is characterized by moderate concentrations of calcium (36–69 mg L^{-1}), magnesium (9–28 mg L^{-1}), sodium (5–37 mg L^{-1}), bicarbonate (180–280 mg L^{-1}), chloride (1.5–55 mg L^{-1}), and sulfate (14–72 mg L^{-1}). The pH of the water is relatively high (7.5–8.0), and it contains relatively high concentrations of dissolved oxygen (7.8–9.6 mg L^{-1}), and relatively high concentrations of dissolved silica (16–45 mg L^{-1}).

Wood and Low (1988) began their investigation of Snake River Plain groundwater geochemistry by describing in detail the mineralogic composition of the basalt aquifer. Using bulk chemical analyses, mineralogic analyses, and scanning electron micrograph (SEM) images, they showed that the most reactive minerals present in the basalt were olivine, plagioclase feldspar, pyroxenes, and pyrite. The SEM images clearly showed that these minerals were actively dissolving in the aquifer. The SEM images also showed that secondary calcite, montmorillonite, and silica were being precipitated. It is a simple matter to estimate qualitatively the amounts of minerals that dissolved and precipitated in order to produce the observed groundwater geochemistry. Wood and Low (1988) normalized the amount of dissolution and precipitation to the annual water discharge from the aquifer, and reported the total amount of mineral dissolution/precipitation occurring throughout the aquifer in units of 10^9 mol yr^{-1}. In addition, they normalized the relative amounts of olivine and pyroxene dissolved based on petrographic observations. Specifically, they assumed that 76% of dissolved magnesium comes from olivine, and 24% of magnesium comes from pyroxene.

The procedure used by Wood and Low (1988) illustrates several important features of the mineralogic control on groundwater geochemistry in this aquifer system. First, while olivine comprises a relatively small percentage of the aquifer material, its dissolution has quite a large effect on the water chemistry, producing much of the observed magnesium and bicarbonate. Second, while calcite was not an important component of the original basalt, calcite precipitation is an important control on concentrations of calcium and bicarbonate, and on the pH of groundwater.

Equilibrium calculations based on mineral saturation indices also show how the concentration of solutes that are present in trace amounts are influenced by mineral dissolution/precipitation reactions. For example, the concentration of barium in groundwater appears to be buffered by the saturation index of barite ($BaSO_4$) (Figure 2).

Table 1 Representative analyses of groundwater from the Snake River Plain, Idaho.

Station no. (from Figure 2)	*Distance along flow path* (km)	*pH* (units)	*Dissolved oxygen* (mg L^{-1})	*Calcium* (mg L^{-1})	*Magnesium* (mg L^{-1})	*Sodium* (mg L^{-1})	*Potassium* (mg L^{-1})	*Bicarbonate* (mg L^{-1})	*Chloride* (mg L^{-1})	*Sulfate* (mg L^{-1})	*Silica* (mg L^{-1})
14	0	8.0	7.8	36	13	20	3.2	180	13	14	45
13	30	7.5	9.4	69	18	15	3.3	280	16	46	25
12	65	7.5	9.6	47	9.3	4.8	1.0	180	1.5	17	16
11	160	7.8	8.7	44	18	30	4.3	190	42	50	33
10	190	7.5	8.7	62	28	37	5.6	280	55	72	39

Source: Wood and Low (1988).

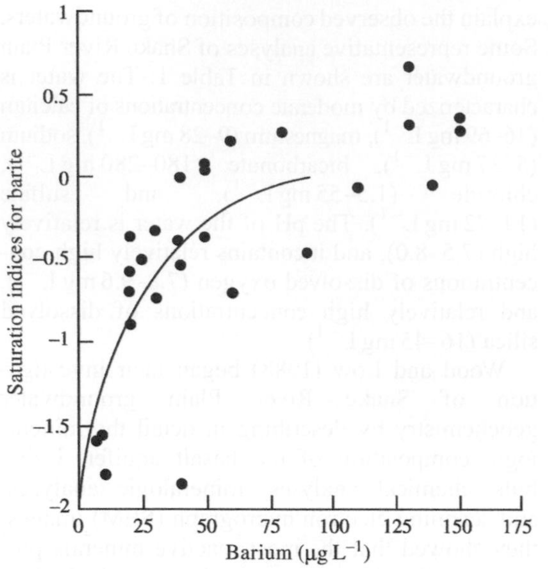

Figure 2 Graph showing barium concentrations plotted versus the saturation index of the mineral barite calculated for Snake River Plain groundwater (source Wood and Low, 1988).

The Snake River Plain aquifer is also an example of an aquifer in which groundwater chemistry is not strongly affected by directions and rates of groundwater flow. There are no well-defined changes in solute concentrations along the aquifer flow path (Table 1). As we shall see in subsequent examples, this insensitivity to directions of groundwater flow is unusual, and is due to the relatively rapid nature of ongoing dissolution and precipitation reactions in the Snake River Plain aquifer.

9.2.2 A Stratified Drift–Arkosic Sandstone Lithology: The Bedrock Aquifer of New England

The basaltic aquifers of the Snake River Plain are unusual in that the mineralogy of the aquifer matrix is (relatively) uniform. The basalts are composed of minerals (i.e., olivine, pyroxene, and plagioclase) that are relatively unstable in the presence of water at low temperature (~15 °C), and their rates of reaction with carbon dioxide-bearing waters are relatively rapid. In addition, the rate and direction of groundwater flow in the Snake River Plain aquifer can be determined with fair confidence. This combination of attributes makes it relatively easy to apply a simple mass-balance approach to document the geochemical processes that determine the groundwater geochemistry. But most aquifers are lithologically heterogeneous, are composed of minerals that react with water at slower rates, and are characterized by uncertain directions and rates of groundwater flow. It is worth asking, therefore, whether the simple mass-balance approach is applicable to more complex hydrologic systems. One such highly complex hydrologic system, characterized by glacial sediments overlying arkosic bedrock, underlies part of New England (Rogers, 1987). As such, it provides an opportunity to apply the mass-balance approach to a more complex hydrologic system.

The study area described by Rogers (1987) is in the central lowlands of Connecticut. The area is underlain by lithified sedimentary rocks of the Newark Supergroup (Mesozoic) that are locally referred to as "bedrock." The bedrock is mantled by "stratified drift" and till deposits of Pleistocene age. The glacial deposits are primarily medium- to coarse-grained sands interbedded with gravels and silts. In places, these glacial sands and gravels are productive aquifers which are locally tapped for groundwater. The bedrock underlying the glacial deposits is typically faulted and fractured, and is characterized by partings along bedding planes. Groundwater movement in bedrock is primarily along these secondary fractures and bedding planes, and individual wells produce ~0–750 gal (1gal = 3.785×10^{-3} m^3) of water per minute. Because this hydrologic system is highly complex and heterogeneous, it is difficult to delineate individual flow paths other than to say that flow is generally from areas of higher elevation to areas of lower elevation.

The mineralogy of the bedrock–glacial drift aquifers is similarly complex. Rocks of the Newark Supergroup consist primarily of quartz (44%), potassium feldspar (20%), plagioclase feldspar (10%), and clays and micas (18%). The bedrock is cemented with hematite (3.5%) and in places by calcite (2.5%). The glacial drift deposits are often derived from the bedrock and, other than having a higher proportion of quartz, have a similar mineralogy as the bedrock.

In spite of this mineralogic and hydrologic complexity, the groundwater geochemistry exhibits some uniformity. The mean concentration of the major ions present in stratified drift and bedrock aquifers is shown in Table 2. These data show that calcium, sodium, bicarbonate, and silica are the most abundant solutes, with lower concentrations of magnesium, sulfate, and chloride. The pH of the water varies from 6.4 to 8.1, and tends to be lower in groundwater containing low concentrations of dissolved solids (Table 2). SI calculated from the analytical data for chalcedony range from −0.3 to +0.4, and the SI for calcite range from −3.0 to +2.0 (Figure 3). Interestingly, the saturation indices for calcite are more tightly grouped around 0.0 (equilibrium with calcite) in bedrock groundwater than glacial drift groundwater. Because of the generally

Table 2 Mean concentrations of dissolved constituents in the stratified drift and bedrock aquifers of Connecticut.

Sample group number	*Calcium* (mmol L^{-1})	*Magnesium* (mmol L^{-1})	*Sodium* (mmol L^{-1})	*Potassium* (mmol L^{-1})	*Chloride* (mmol L^{-1})	*Sulfate* (mmol L^{-1})	*Bicarbonate* (mmol L^{-1})	*SiO_2* (mmol L^{-1})	*pH* (units)
Water from stratified drift aquifers									
G1	0.085	0.062	0.11	0.010	0.073	0.024	0.21	0.20	6.4
G2	0.24	0.063	0.13	0.013	0.093	0.17	0.40	0.16	6.9
G3	0.61	0.14	0.24	0.022	0.23	0.16	0.96	0.25	7.5
G4	0.81	0.084	0.24	0.016	0.16	0.33	1.20	0.23	8.1
Water from bedrock aquifers									
B1	0.36	0.064	0.16	0.012	0.13	0.16	0.55	0.19	7.4
B2	0.41	0.068	0.11	0.020	0.086	0.081	1.00	0.24	7.9
B3	0.66	0.12	0.29	0.013	0.14	0.090	1.50	0.27	8.1

After Rogers (1987).

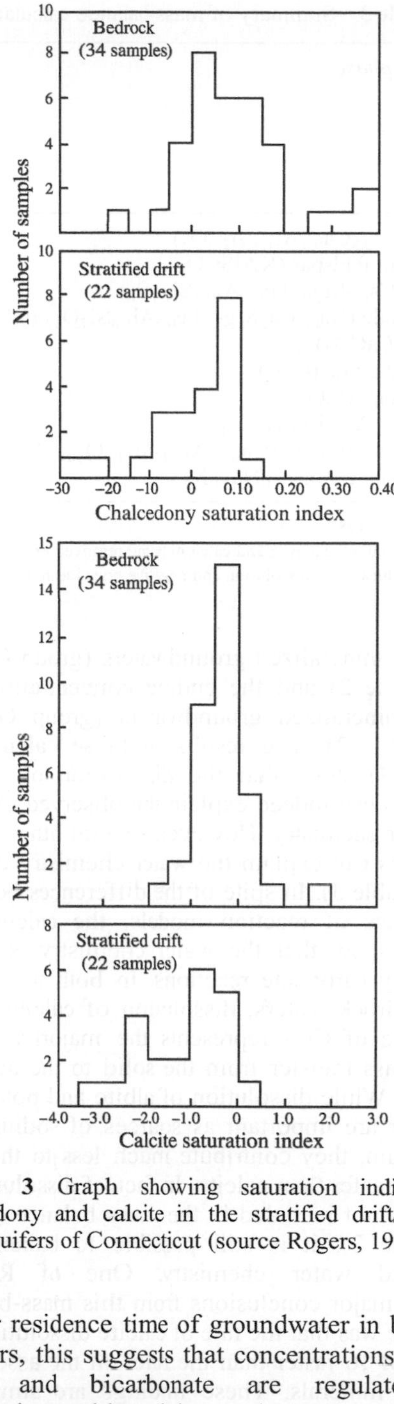

Figure 3 Graph showing saturation indices for chalcedony and calcite in the stratified drift/bedrock aquifers of Connecticut (source Rogers, 1987).

longer residence time of groundwater in bedrock aquifers, this suggests that concentrations of calcium and bicarbonate are regulated by equilibrium with calcite, and suggests that the solubility of chalcedony regulates the concentration of dissolved silica.

Dissolution of feldspars is a logical source of dissolved silica, calcium, sodium, and potassium in groundwater. Similarly, the reaction of carbon dioxide-charged water with silicate minerals is a logical source of bicarbonate. Rogers (1987) examined these and other hypotheses using a mass-balance approach. In these calculations, chloride and sulfate were not considered, and the beginning concentrations were considered to be

Table 3 Summary of mass-balance calculations for the stratified drift and bedrock aquifers of Connecticut.

Mineral phase	*The reaction model is identified by a number*				
	Stratified drift (mmol kg^{-1} H_2O)		*Bedrock* (mmol kg^{-1} H_2O)		
	1	*2*	*3*	*4*	*5*
Albite ($Na_{0.9}Ca_{0.1}Al_{1.1}Si_{2.9}O_8$)	0.048	0.044	0.133	0.124	
Potassium feldspar ($KAlSi_3O_8$)	0.006	0.006	0.001	0.001	0.001
Augite ($Ca_{0.7}Mg_{0.9}Fe_{0.3}Al_{0.1}Si_{1.9}O_6$)	0.004		0.062		0.062
Hornblende ($Na_{0.5}Ca_2Mg_{3.5}Fe_{0.5}Al_{1.8}Si_7O_{22})(OH_2)$		0.006		0.016	
Calcite ($CaCO_3$)	0.703	0.708	0.277	0.284	0.321
Carbon dioxide (CO_2)	0.287	0.282	0.693	0.686	0.649
Chalcedony (SiO_2)	−0.112	−0.094			
Kaolinite ($Al_2Si_2O_5(OH)_4$)	−0.031	−0.033	0.161	0.116	0.026
Ca-montmorillonite ($(Ca_{0.17}Al_{2.33}Si_{3.67}O_{10})(OH_2)$)			−0.025	−0.171	−0.025
Ion exchange ($Ca_{0-1.0}Na_{2.0}X$)					0.060[a]

After Rogers (1987).
[a] Indicates sodium entering and calcium being removed from solution.
Positive values mean dissolution and negative ones mean precipitation.

the less mineralized groundwaters (group G1 and B1, Table 2) and the ending concentrations the more mineralized groundwaters (group G4 and B4, Table 2). The results of these calculations (Table 3) show that the dissolution of aquifer minerals can indeed explain the observed changes in water chemistry. However, several other sets of reactions can explain the water chemistry equally well (Table 3). In spite of the differences between these sets of reaction models, the calculations clearly show that the water chemistry is dominated by carbonate reactions. In both the glacial and bedrock waters, dissolution of calcite in the presence of CO_2 represents the majority of the total mass transfer from the solid to the aqueous phases. While dissolution of albite and potassium feldspar are important as sources of sodium and potassium, they contribute much less to the total mass transfer than calcite. In fact, if dissolution of albite is not included in the mass balance (model 5, Table 3), it is still possible to balance the observed water chemistry. One of Rogers' (1987) major conclusions from this mass-balance exercise was that the rate of calcite dissolution is a factor of 16 faster than the rates of the associated silicate minerals. These findings are similar to those of Wood and Low (1988) in the Snake River Plain aquifer, which showed that the dissolution/precipitation of calcite was one of the most important controls on groundwater geochemistry.

The study of Rogers (1987) indicates that, in spite of the mineralogic and hydrologic complexity of this aquifer system, mass-balance calculations can be useful in understanding the important mineral dissolution/precipitation reactions controlling groundwater geochemistry. Thus, the utility of the mass-balance approach does not necessarily depend on the simplicity of aquifer hydrology and mineralogy. The case studies of the Snake River Plain basalt aquifers and the bedrock/stratified drift aquifers, however, show that mass-balance models are not unique. It can be argued, in fact, that the nonunique nature of mass-balance models injects more uncertainty into understanding groundwater geochemistry than lithologic and hydrologic uncertainties. This nonuniqueness of mass-balance models is an important characteristic of methods for studying groundwater geochemistry. This is discussed in detail by Bricker *et al.*, and will be examined further in the following example.

9.2.3 A Carbonate Lithology: The Floridan Aquifer of Florida

Aquifers that consist of carbonate rocks, which include limestones ($CaCO_3$) and dolomites ($CaMg(CO_3)_2$), are among the most productive sources of groundwater in the world. The Floridan aquifer, which underlies all of Florida, most of Georgia, and part of South Carolina, yields more than 3 billion gallons per day of freshwater. Because of its economic importance, and because the chemical quality of groundwater produced from the Floridan aquifer differs noticeably from place to place, it has been carefully studied by groundwater geochemists (Hanshaw *et al.*, 1965; Back and Hanshaw, 1970; Plummer, 1977; Plummer and Back, 1980; Plummer *et al.*, 1983; Sprinkle, 1989). These studies, in addition to elucidating the processes controlling the chemical quality of Floridan aquifer water, were also important in the development of methods for studying groundwater geochemistry.

The Floridan aquifer system consists of the Upper and Lower aquifers, which are separated by a less permeable confining unit (Miller, 1986). The Upper Floridan, which consists of the Tampa, Suwannee, and Ocala limestones as well as the Avon Park Formation, is more widely used for water supply than the Lower Floridan, and is the focus of this brief overview. The pre-pumping potentiometric surface of the Upper Floridan aquifer is shown in Figure 4. It indicates that the aquifer is recharged in the topographically high areas of central Florida, and that groundwater flows radially away from the recharge area. A cross-section from Polk City, which is located in the potentiometric high in the center of the Florida peninsula, directly south through Ft. Meade,

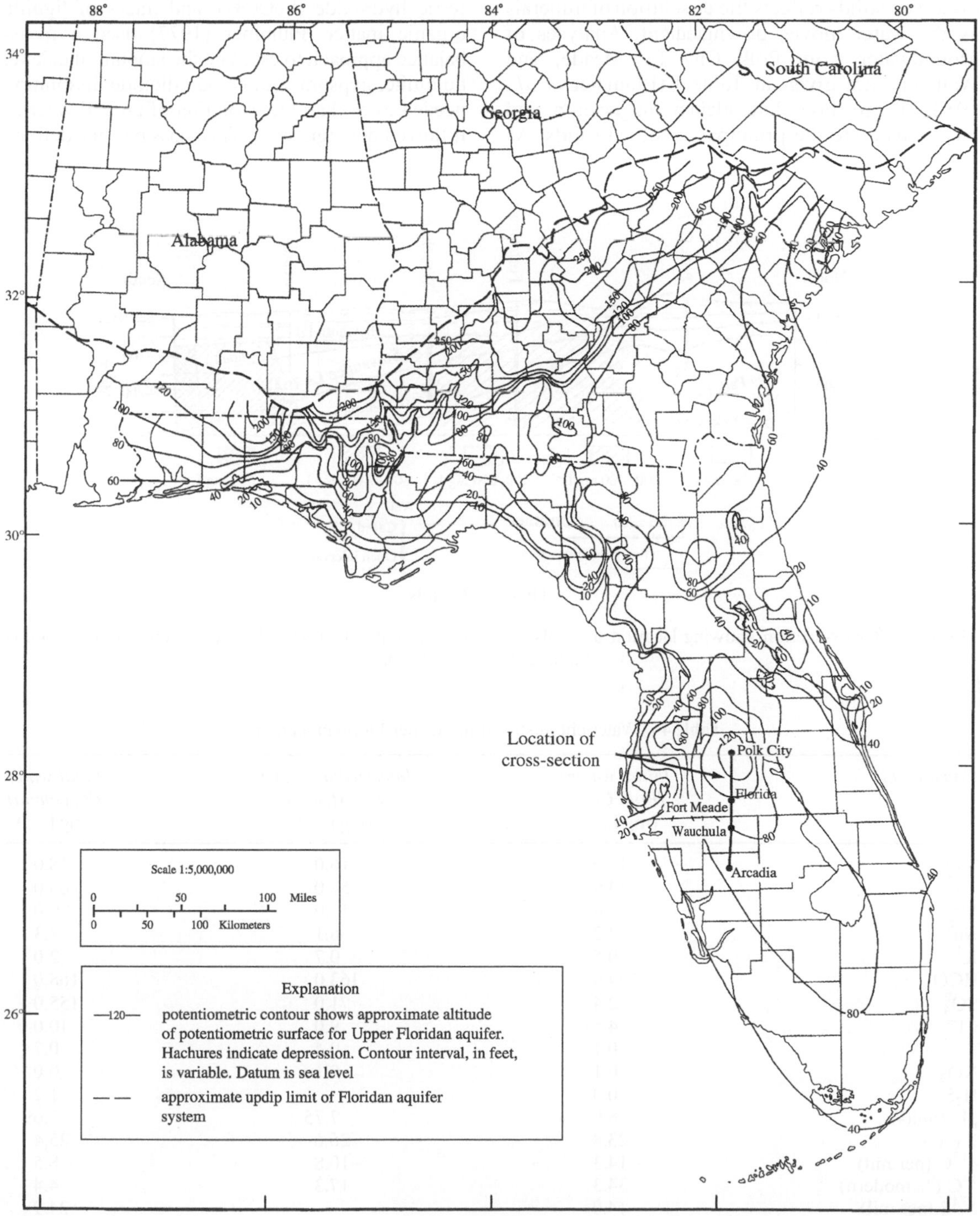

Figure 4 Map showing the prepumping potentiometric surface of the Floridan aquifer and locations of wells shown in Figure 5 (after Sprinkle, 1989).

Wauchula, and Arcadia is shown in Figure 5 (Plummer, 1977). This cross-section shows that in the aquifer recharge area near Polk City, the groundwater is characterized by relatively low concentrations of dissolved solids (<200 mg L^{-1}). As the water moves south along the regional hydrologic gradient, the concentration of dissolved solids progressively increases. In this freshwater portion of the aquifer this increase in dissolved solids reflects the dissolution of minerals as the water moves downgradient. Analyses of groundwater from Polk City, Ft. Meade, and Wauchula are given in Table 4 (Plummer *et al.*, 1983). These show that calcium, magnesium, and bicarbonate are the principal dissolved solids. As groundwater nears coastal areas, sodium and chloride concentrations increase due to the mixing of freshwater with seawater that is present in coastal and offshore portions of the Upper Floridan aquifer (Figure 5).

The lithology of the Floridan aquifer system consists predominantly of calcite and dolomite. However, the aquifer also contains nodular gypsum ($CaSO_4 \cdot 2H_2O$), anhydrite ($CaSO_4$), pyrite (FeS_2), ferric hydroxide (FeOOH), and traces of lignitic organic matter. Plummer (1977) used a mass-balance approach to estimate the amount of calcite, dolomite, gypsum, and carbon dioxide dissolution needed to explain the evolution of groundwater as it flowed downgradient. This mass balance defined

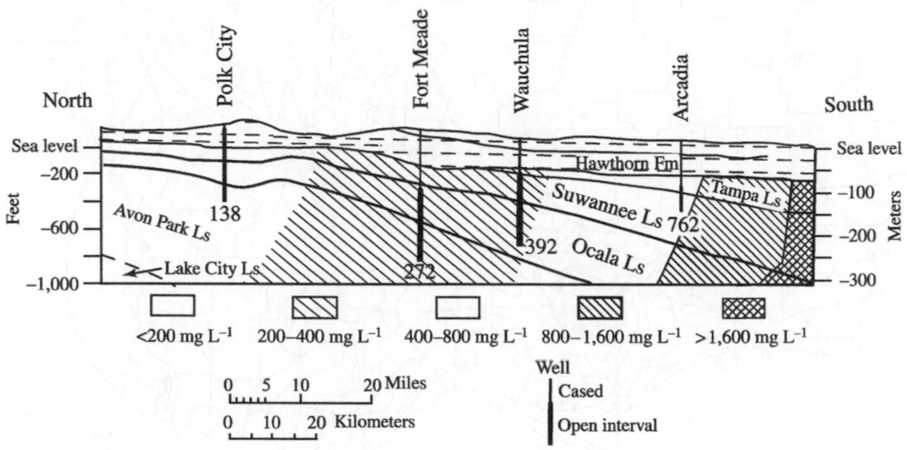

Figure 5 Cross-section showing location of wells and concentrations of dissolved solids in the Floridan aquifer (source Plummer, 1977).

Table 4 Water chemistry of the Upper Floridan aquifer.

Constituent	*Initial water (Polk City)* (mg L^{-1})	*Intermediate water (Ft. Meade)* (mg L^{-1})	*Final water (Wauchula)* (mg L^{-1})
SiO_2	12.0	16.0	18.0
Ca^{2+}	34.0	58.0	66.0
Mg^{2+}	5.6	17.0	29.0
Na^+	3.2	6.1	8.3
K^+	0.5	0.7	2.0
HCO_3^-	124.0	163.0	168.0
SO_4^{2-}	2.4	71.0	155.0
Cl^-	4.5	9.0	10.0
F^-	0.1	0.4	0.7
NO_3^-	0.1	0.1	0.0
H_2S	0.0		1.2
pH (units)	8.0	7.75	7.69
T (°C)	23.8	26.6	25.4
$\delta^{13}C$ (per mil)	−14.3	−10.8	−8.5
^{14}C (% modern)	34.3	17.3	4.4
$\delta^{34}S$ (per mil)	−14.0		+24.9
$\delta^{34}S_{(H_2S)}$ (per mil)			−32.9

Source: Plummer *et al.* (1983).

changes in solute concentrations and the dissolution of minerals as

$$\sum_{p=1}^{P} \alpha_p b_{p,k} = \Delta m_{\mathrm{tot},k}, \quad k = 1, \ldots, J \qquad (9)$$

which states that the molar change in any element k, $\Delta m_{\mathrm{tot},k}$, along a flow path is equal to the sum of all the sources and sinks for element k. Furthermore, this equation states that this relationship holds for 1 through J elements. The sources and sinks may include mineral dissolution, precipitation, microbial degradation, gas transfer and so forth, of P phases (minerals, gases) along a flow path where α_p is the number of moles reacting and b is the stoichiometric coefficient for the element in the pth mineral phase. Values for $\Delta m_{\mathrm{tot},k}$ are derived from groundwater chemistry data.

If it is assumed that calcite, dolomite, gypsum, and carbon dioxide are the phases to be considered, mass balance can be described by four linear equations of the form given by Equation (9). Simultaneous solution of these four equations for water chemistry changes between unmineralized rainwater to the water composition of Polk City yields the mass balance:

$$\sum_{p=1}^{P} \alpha_p = 0.482\mathrm{CaCO_3} + 0.230\mathrm{CaMg(CO_3)_2} + 0.025\mathrm{CaSO_4{\cdot}2H_2O} + 1.095\mathrm{CO_2} \qquad (10)$$

Similar solutions of these mass-balance equations can be used to explain water chemistry changes from Polk City and Ft. Meade, and from Ft. Meade to Wauchula (Plummer, 1977), as well as other parts of the Floridan aquifer (Sprinkle, 1989).

One reason why mass-balance models of groundwater geochemistry are useful, as illustrated by this example (Plummer, 1977), is that they can explain regional changes in water chemistry (Table 4). In this case, the net dissolution of carbonate minerals and gypsum increases concentrations of total dissolved solids as water flows along the hydrologic gradient (Figure 5). An inherent problem with the mass-balance approach, however, is that aquifers are much more lithologically complex then can be fully taken into account. The mass balance given above for the Floridan aquifer, for example, considers just four mineral phases and four dissolved constituents. Clearly, this is a simplification of the natural system. The question is, how much information is being obscured by making such simplifications?

This important question was addressed for the Floridan aquifer, using the same hydrologic data (Table 4) in a subsequent study by Plummer *et al.* (1983). In this study, the lithologic complexities of the aquifer were addressed by considering more mineral phases (pyrite, ferric hydroxide, organic carbon, and methane), by considering variations in the composition of calcite (which may contain up to 5 mol.% magnesium), dolomite (which may contain different proportions of Ca^{2+}, Mg^{2+}, and Fe^{2+}), and by considering the balance of sulfur and carbon isotopes. Because the possible number of minerals (termed "plausible phases") exceeded the number of chemical constituents in groundwater, it was necessary to construct multiple sets of mass-balance models. As the Floridan is a confined aquifer in this part of the hydrologic system, it was initially assumed that the system was closed to sources of carbon dioxide (models 1–6; Table 5). However, while each of the mass-balance models constructed was capable of explaining the observed major ion water chemistry (Table 5), they were not consistent with the observed isotopic ratios of carbon and sulfur. This led the investigators to reconsider

Table 5 Results of mass-balance calculations for the Floridan aquifer assuming a system closed to CO_2.

Plausible phases	*Reaction model number*					
	1	*2*	*3*	*4*	*5*	*6*
$CaCO_3$	−2.45	−19.79				
$CaMg(CO_3)_2$	0.96					
$CaSO_4{\cdot}2H_2O$	2.29	2.29	2.24	2.06	−7.60	6.13
CO_2						
CH_2O	1.32	1.32	1.21	1.03		
FeOOH	0.33	0.33	0.25	0.37	−4.66	4.44
FeS_2	−0.33	−0.33	−0.31		4.61	
$Ca_{0.95}Mg_{0.05}CO_3$		19.26				
$Ca_{0.98}Mg_{0.02}CO_3$			−2.68	−2.49	7.61	−6.74
$Ca_{1.05}Mg_{0.90}Fe_{0.05}(CO_3)_2$			1.13	1.13	0.90	1.22
FeS				−0.43		−4.50
CH_4					−8.62	5.09

Source: Plummer *et al.* (1983).
Values of α_p in mmol kg^{-1} H_2O. Positive values mean dissolution and negative ones mean precipitation.

Table 7 Chemical composition of Black Creek aquifer groundwater and confining bed pore waters.

	LE-37	*OLANTA*	*LM-IA*	*FLO-162*	*HO-338*	*Confining bed 38 m*	*Confining bed 67 m*
Distance down-gradient (km)	10.0	42.5	50.0	69.0	135.0		
pH (units)	4.5	7.6	8.4	8.6	8.4		
Temperature (°C)	19.5	19.5	18.0	20.2	23.0		
Ca (mmol L^{-1})	0.00998	0.274	0.324	0.0170	0.0599	0.157	0.177
Mg (mmol L^{-1})	0.00823	0.132	0.0905	0.00400	0.0370	0.0580	0.0410
Na (mmol L^{-1})	0.0478	0.135	0.826	1.48	12.6	2.26	2.83
K (mmol L^{-1})	0.0281	0.384	0.187	0.0480	0.128	0.169	0.307
DIC (mmol L^{-1})	0.675	1.16	0.170	1.24	10.4	2.57	2.70
Cl (mmol L^{-1})	0.0536	0.0536	0.0508	0.116	2.48	0.0880	0.105
SO_4 (mmol L^{-1})	0.0729	0.0802	0.0708	0.0820	0.00208	0.101	0.381
SiO_2 (mmol L^{-1})	0.216	0.649	0.350	0.328	0.266		
Al (mmol L^{-1})	0.0200	<0.0004	<0.0004	<0.0004	0.00185		
Fe (mmol L^{-1})	0.0134	0.00394	0.0005	0.0004	0.0004		
Sr (mmol L^{-1})			0.0025				
H_2S (as S) (mg L^{-1})	<0.03	<0.03	<0.03	<0.03	<0.03		
CH_4(μmol L^{-1})	0.05	0.05	0.08				
$\delta^{13}C$ of DIC (per mil)	−27.8	−18.1	−12.8				
Aragonite$_{si}$	−7.7	−1.1	−0.1	−1.6	−1.6		
Low-magnesium calcite$_{si}$	−7.5	−0.9	0.1	−1.4	−1.4		
Chalcedony$_{si}$	−0.1	0.4	0.1	0.1	0.1		

Source: McMahon and Chapelle (1991b).

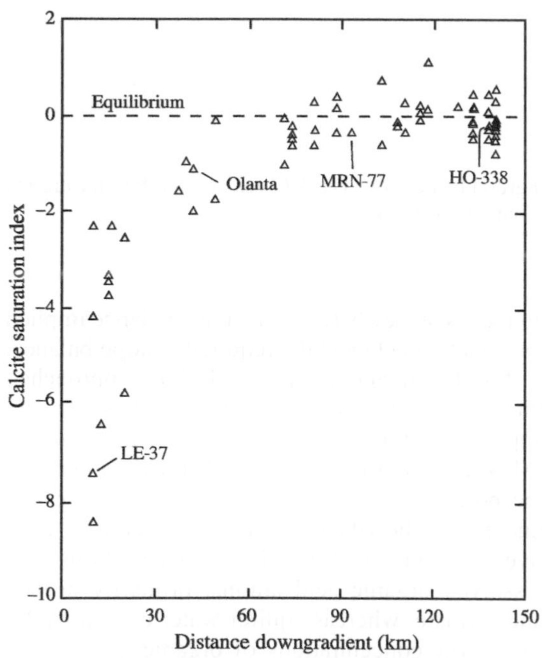

Figure 7 Calcite saturation indices plotted along the direction of groundwater flow in the Black Creek aquifer (source McMahon and Chapelle, 1991b).

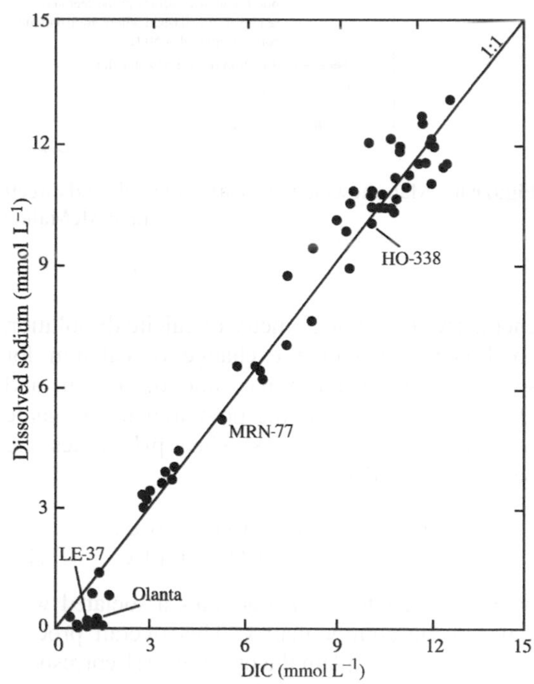

Figure 8 Observed 1:1 correspondence of DIC and sodium in the Black Creek aquifer of South Carolina (source McMahon and Chapelle, 1991b).

evolution of aquifer water chemistry in this system?

The locations of representative wells along the flow path are shown in Figure 7, and analyses of groundwater associated with these wells are shown in Table 7. In addition to well-water chemistry, Table 7 shows the composition of pore water in confining beds overlying the Black Creek aquifer at the Lake City Core Hole (Figure 7). These analytical data were used to

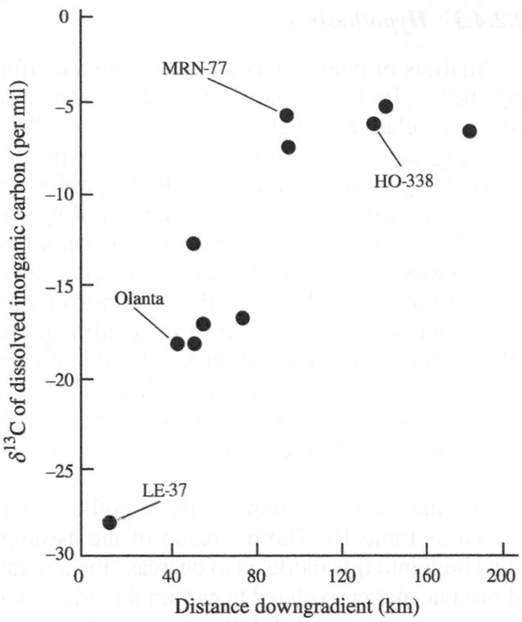

Figure 9 $\delta^{13}C$ composition of DIC in Black Creek aquifer water (source McMahon and Chapelle, 1991b).

construct material-balance models designed to test a series of hypotheses. These are:

Hypothesis 1. Mass balance given by the stoichiometry of Equation (11) can explain the major ion and carbon isotope composition of Black Creek aquifer water.

Hypothesis 2. Diffusion of DOC and sulfate from confining bed pore waters provides sources of electron donor (organic carbon) and electron acceptor (sulfate). Carbon dioxide produced by this reaction drives shell material dissolution/calcite cement precipitation which can explain the major ion and carbon isotope composition of Black Creek aquifer water.

Hypothesis 3. Diffusion of DIC together with DOC, sulfate, and cations from confining bed pore waters to the Black Creek aquifer provides sources of electron donor (organic carbon) and electron acceptor (sulfate) for microbial metabolism and additional inorganic carbon to drive low-magnesium calcite precipitation. The combination of magnesium-calcite dissolution from shell material driven by microbially produced carbon dioxide, and the precipitation of more thermodynamically stable low-magnesium calcite cement in the aquifer, can explain major ion and carbon isotope composition of Black Creek aquifer water.

Each of these hypotheses is testable using mass-balance and isotopic-balance considerations.

9.2.4.1 Hypothesis 1

The mass balance implied by hypothesis 1 (Table 8) shows several inconsistencies with what

Table 8 Calculated mass transfer for hypothesis 1.

Mineral phase	*Mass transfer*	
	Olanta to MRN-77	*MRN-77 to HO-338*
$CaCO_3$	2.297	2.531
NaX/CaX[a]	2.536	2.550
NaX/MgX[b]	0.0061	0.152
Illite[c]	−0.472	−0.186
Glauconite[d]	0.0020	0.0458
SiO_2	1.211	0.136
Pyrite[e]	−0.0080	−0.0914
CO_2	1.796	2.165
CH_2O[f]	0.0363	0.427
Seawater	0.0	0.0043

Source: McMahon and Chapelle (1991b).
[a] Ca-for-Na cation exchange. [b] Mg-for-Na cation exchange.
[c] $K_{0.6}Mg_{0.25}Al_{2.5}Si_{3.5}O_{10}$. [d] $K_2Fe_2Al_6(Si_4O_{10})3(OH)_{12}$. [e] FeS_2.
[f] Organic carbon, oxidation state +1.
Positive values mean dissolution and negative ones mean precipitation.

is known about this hydrologic system. First, this model requires a source of carbon dioxide (CO_2) independent of organic matter (CH_2O) oxidation to achieve mass balance. This requirement stems from the lack of electron acceptors (ferric iron, sulfate) needed to drive oxidation of organic matter, and the lack of significant methane production (in which DIC acts as an electron acceptor). Since carbon dioxide in low temperature groundwater systems is produced only by microbial metabolism, it would seem logical that there must be a source of electron acceptors to support this metabolism. In addition, isotope balance using the mass-balance coefficients calculated from hypothesis 1 predicts $\delta^{13}C$ of DIC values at wells MRN-77 and HO-338 of −12.9‰ and −12.5‰, respectively. However, the measured $\delta^{13}C$ values at these wells are −5.6‰ and −6.1‰, respectively. Thus, based on both mass-balance and isotopic-balance considerations, hypothesis 1 can be ruled out.

9.2.4.2 Hypothesis 2

Another possible explanation for the observed water chemistry is that confining beds provide sources of both organic carbon (electron donor) and sulfate (electron acceptor), which drives shell material dissolution in the Black Creek aquifer. The relatively large pool of solid organic carbon present in confining beds adjacent to the Black Creek aquifer is subject to fermentation, which in turn produces relatively high concentrations of organic acids in confining bed pore waters (McMahon and Chapelle, 1991b). Confining bed pore waters contain higher concentrations of sulfate than adjacent aquifer waters (Table 7). Oxidation of organic acids as they diffuse to the aquifer, coupled to reduction of sulfate as it diffuses to the aquifer, would produce carbon dioxide. This carbon dioxide, in turn, could drive dissolution/precipitation of carbonate shell material in the aquifer. Because carbon-12 tends to be used preferentially in carbonate dissolution/precipitation reactions, a process called Rayleigh distillation (Plummer *et al.*, 1983) causes the $\delta^{13}C$ of DIC to become progressively heavier. These coupled processes, therefore, might account for the observed mass and carbon isotope balance in this system.

The mass balance implied by hypothesis 2 is shown in Table 9. Again, however, while this model accounts for mass balance, predicted $\delta^{13}C$ of DIC values are not consistent with observed values. Isotope balance at wells MRN-77 and HO-338 using the mass balance of Table 9 predicts $\delta^{13}C$ of DIC values of −25.6‰ and −24.5‰, which are at variance with the observed values of −5.6‰ and −6.1‰, respectively. The relatively light predicted values reflect the light isotopic signature of organic matter in the confining beds (McMahon and Chapelle, 1991a), and the fact that the model of hypothesis 2 draws much of the inorganic carbon from oxidation of this organic matter. Rayleigh distillation accompanying shell material dissolution and calcite cement precipitation could account for the observed carbon isotope values (Plummer *et al.*, 1983), but it would require over 50 mM of shell material to dissolve per liter of water, and with an equal amount of calcite cement to precipitate in the downgradient part of the aquifer. This, in turn, would cause complete cementation of the Black Creek aquifer. While secondary calcite cement is present in the downgradient portion of the aquifer, it cements less than 10% of the aquifer by volume. Thus, hypothesis 2 can be ruled out by mass-balance and isotope-balance considerations.

9.2.4.3 *Hypothesis 3*

Analysis of pore waters associated with confining beds (Table 7) shows that the water also contains relatively high concentrations of DIC. This suggests that as solid organic matter present in confining beds is fermented with the production of organic acids, carbon dioxide produced by fermentation reacts with shell material in confining beds, producing relatively high DIC concentrations. Unlike hypothesis 2, the stoichiometry of dissolution suggests that half of the confining bed DIC would come from shell material and half from organic matter. Diffusion of sulfate, DIC, and organic acids to the Black Creek aquifer would significantly alter the isotope balance of the system.

The mass balance implied by hypothesis 3 is shown in Table 10. The net result of the assumptions built into this model is to decrease the amount of organic matter oxidized to carbon dioxide and to increase the amount of DIC from shell material dissolution. This, in turn, decreases the amount of shell material dissolution/calcite cement precipitation needed to achieve isotope balance. Between Olanta and MRN-77, the amount of dissolution/precipitation needed for isotope balance is 2.0 mmol $CaCO_3$ kg^{-1} of H_2O, and 25 mmol $CaCO_3$ from MRN-77 to HO-338. This, in turn, implies that 1–13 vol.% of the aquifer would be cemented by calcite, which is roughly in line with observed calcite cementation.

The overall carbon balance in the Black Creek aquifer implied by hypothesis 3 is shown in Table 11. According to this balance, shell material has contributed 83–87% of the carbon to DIC in the aquifer, compared to 13–17% from organic carbon. Of the total carbon added from shell material and organic carbon, ~74% was subsequently removed as calcite cement in the aquifer.

Table 9 Calculated mass transfer for hypothesis 2.

Mineral phase	*Mass transfer*	
	Olanta to MRN-77	*MRN-77 to HO-338*
$CaCO_3$	−7.196	−10.102
NaX/CaX[a]	−6.957	−10.084
NaX/MgX[b]	−1.0019	−1.170
Illite[c]	−4.504	−5.471
Glauconite[d]	1.212	1.631
SiO_2	0.808	0.393
Pyrite[e]	−2.427	−3.262
Pore water 1	11.326	15.225
Pore water 2	20.076	26.316
Seawater	0.0	0.0043

Source: McMahon and Chapelle (1991b).
Footnote symbols 'a' to 'e' are same as in Table 8.
Positive values mean dissolution and negative ones mean precipitation.

Table 10 Calculated mass transfer for hypothesis 3.

Mineral phase	*Mass transfer*	
	Olanta to MRN-77	*MRN-77 to HO-338*
$CaCO_3$	0.526	0.0193
NaX/CaX[a]	0.951	0.282
NaX/MgX[b]	−0.105	0.0062
Illite[c]	−1.138	−1.060
Glauconite[d]	0.0692	0.134
SiO_2	2.737	2.137
Pyrite[e]	−0.142	−0.268
Pore water 1	0.662	1.247
Pore water 2	1.116	1.462
Seawater	0.0	0.0043

Source: McMahon and Chapelle (1991b).
Footnote symbols 'a' to 'e' are same as in Table 8.
Positive values mean dissolution and negative ones mean precipitation.

Table 11 Sources and sinks of organic and inorganic carbon to Black Creek aquifer groundwater.

	Olanta to MRN-77		*MRN-77 to HO-338*	
	$mmol\,kg^{-1}\,H_2O$	*% of total*	$mmol\,kg^{-1}\,H_2O$	*% of total*
Carbon source				
Shell material aquifer	12.526	78	14.0193	73
Shell material confining bed	1.471	9	1.928	10
Organic carbon aquifer	0.0	0	0.0	0
Organic carbon confining bed	2.133	13	3.175	17
Seawater	0.0	0	0.00998	<1
Carbon sink				
Calcite cement	12.0	74	14.0	73

Source: McMahon and Chapelle (1991b).

This relatively large component of carbonate dissolution–precipitation is reflected in the relatively heavy $\delta^{13}C$ of DIC observed in the downgradient portion of the aquifer. Confining beds contributed 22–27% of the total carbon input, compared to 73–78% from the aquifer.

9.2.4.4 *Evaluation of hypotheses*

It is clear from this hypothesis testing approach that an infinite number of mass-balance scenarios can be constructed to explain the observed water-chemistry changes in the Black Creek aquifer. This nonuniqueness/uncertainty is a general characteristic, and is an unavoidable characteristic of material-balance calculations in systems that are not at full thermodynamic equilibrium. However, much can still be learned by carefully considering the implications of each material balance. Each of the hypotheses considered above produced testable predictions. Hypothesis 1 could not account for observed carbon isotope values, and did not predict the presence of calcite cement in the aquifer. Similarly, for hypothesis 2 to account for isotopic balance, it would require almost complete cementation of the aquifer by secondary calcite cement. Of the three hypotheses considered, hypothesis 3 gave the most plausible explanation of carbon isotope balance and observed calcite cementation of the aquifer. This does not, however, mean that hypothesis 3 is "correct." It simply provides an explanation for the observed geochemical features of the Black Creek aquifer, which includes aquifer water and confining bed pore-water composition, aquifer and confining bed mineralogy, and the observed presence and abundance of secondary calcite cements.

The chief utility of this hypothesis testing is that it indicates that microbial processes occurring in the aquifer (sulfate reduction) as well as microbial processes in confining beds (organic matter fermentation) have an important impact on the geochemistry of groundwater in adjacent aquifers. These nonintuitive results could not have been arrived at without undertaking a quantitative material balance.

9.3 REDUCTION/OXIDATION PROCESSES

Reduction–oxidation (redox) processes affect the chemical composition of groundwater in all aquifer systems. In particular, redox processes affect the mobility of organic chemicals and metals in both pristine and contaminated systems. Thus, methods for characterizing redox processes are an important part of groundwater geochemistry. The purpose of this section is to review equilibrium and kinetic frameworks for documenting the spatial and temporal distribution of redox processes in groundwater systems.

9.3.1 The Equilibrium Approach

The traditional approach for characterizing redox processes in groundwater is based on conventions and methods developed in classical physical chemistry (Sillén, 1952). Back and Barnes (1965) used platinum electrode measurements to determine the Eh of groundwater samples. This approach was systematized by Stumm and Morgan (1981), who suggested that the theoretical activity of electrons in aqueous solution (p_e), could be used by direct analogy to hydrogen ion activity (pH) as a "master variable" to describe redox processes. In this treatment, the p_e of a water sample is a linear function of Eh ($p_e = 16.9\mathrm{Eh}$ at 25 °C).

The definition of Eh, and thus p_e, is given by the Nernst equation, in which the Eh of a solution is related to concentrations of aqueous redox couples at chemical equilibrium and the voltage of a standard hydrogen electrode (E^0). For example, when concentrations of aqueous Fe^{3+} and Fe^{2+} are at equilibrium, Eh is defined as

$$\mathrm{Eh} = E^0 + \frac{2.303RT}{nF}\log\frac{a_{Fe^{3+}}}{a_{Fe^{2+}}} \qquad (12)$$

Equation (12) illustrates an important point. Eh *is uniquely defined only when a system is at thermodynamic equilibrium* (Drever, 1982, p. 257). If the activities of Fe^{3+} and Fe^{2+} ions are not at equilibrium, an infinite number of apparent Eh values can be calculated or measured with a platinum electrode, but will not be "Eh" as defined by Equation (12). Similarly, if the Eh defined by the Fe^{3+}/Fe^{2+} pair is not the same as that defined by the SO_4^{2-}/H_2S pair or the O_2/H_2O pair, it is impossible to define the Eh of the solution as a whole.

In the 1960s and 1970s, when groundwater systems were thought to be largely sterile environments devoid of microbial life, assuming equilibrium or near-equilibrium conditions seemed to be quite reasonable. However, in the early 1980s it became clear that groundwater systems contained active, respiring, reproducing microorganisms (Wilson *et al.*, 1983; Chapelle *et al.*, 1987). Furthermore, it gradually became clear that many of the important redox processes occurring in groundwater systems were catalyzed by microorganisms (Baedecker *et al.*, 1988; Lovley *et al.*, 1989; Chapelle and Lovley, 1992). This realization coincided with growing evidence that measuring the Eh of groundwaters was problematic. In particular, it was shown that Eh measurements with platinum electrodes were not consistent with Eh calculated from the Nernst equation, and that different redox couples gave widely different Eh values (Lindberg and Runnells, 1984). There are several reasons for these problems. These include:

(i) The aqueous activity of free electrons in water are so low ($\sim 10^{-55}$ M) that they are essentially zero (Thorstenson, 1984). Thus, while electron activity is a thermodynamically definable quantity, it is not measurable in the same way that hydrogen ion activity (pH) is.

(ii) The pH electrode responds to aqueous concentrations of hydrogen ions. In contrast, the Eh electrode responds to *electron transfers* between solutes (Thorstenson, 1984). A platinum Eh electrode, therefore, readily responds to concentrations of Fe^{2+} and Fe^{3+} because they react rapidly and reversibly with platinum. Because solutes such as oxygen, carbon dioxide, and methane react sluggishly on a platinum surface, the Eh electrode is relatively insensitive to the O_2/H_2O and CO_2/CH_4 redox couples.

(iii) Groundwater usually contains multiple redox couples such as O_2/H_2O, Fe^{3+}/Fe^{2+}, SO_4/H_2S, and CO_2/CH_4 that are not in mutual equilibrium.

(iv) Microorganisms cannot actively respire and reproduce unless there is available free energy to drive their metabolism. That is, microorganisms *require that their immediate environment* not *be at thermodynamic equilibrium*. Thus, using Eh to describe redox processes driven by microbial processes violates the underlying equilibrium assumption of Eh.

In light of these difficulties, it is not surprising that Eh measurements in groundwater systems are so often problematic.

9.3.2 The Kinetic Approach

Equilibrium considerations are not the only way to describe redox processes in groundwater systems. The metabolism of microorganisms is based on the cycling of electrons from electron donors (often organic carbon) to electron acceptors such as molecular oxygen, nitrate, ferric iron, sulfate, carbon dioxide, or other mineral electron acceptors. This flow of electrons is capable of doing work. Microorganisms capture this electrical energy, convert it to chemical energy, and use it to support their life functions. If it is assumed that redox processes in groundwater systems are driven predominantly by microbial metabolism, it becomes possible to describe these processes by the cycling of electron donors, electron acceptors, and intermediate products of microbial metabolism. Because this is an inherently nonequilibrium, kinetic description, it has been termed the "kinetic approach" (Lovley *et al.*, 1994).

A kinetic description of redox processes in groundwater systems includes two components. These are documenting (i) the source of electrons (electron donor) that supports microbial metabolism, and (ii) the final sink for electrons (electron acceptors) that supports microbial metabolism. In many groundwater systems, identifying electron donors is not a difficult problem since particulate or DOC is the most common source of electrons for subsurface microorganisms. Another common source of electrons is the aerobic oxidation of sulfide minerals, which facilitates the growth of sulfide-oxidizing microorganisms such as *Thiobacillus* sp. Alternatively, the aerobic oxidation of ferrous iron can facilitate the growth of iron-oxidizing bacteria such as *Gallionella* sp. A more difficult problem is determining the terminal electron accepting processes (TEAPs) that occur in a system. This problem is made even more difficult by the inherent heterogeneity of groundwater systems. This heterogeneity causes both spatial (Chapelle and Lovley, 1992) and temporal (Vroblesky and Chapelle, 1994) variations in TEAPs.

9.3.3 Identifying TEAPs

Studies in aquatic sediment microbiology have clearly demonstrated that microbially mediated redox processes tend to become segregated into discrete zones. When this happens, the observed sequence of redox zones follows a predicable pattern. At the sediment–water interface, oxic metabolism predominates. This oxic zone is underlain by zones dominated by nitrate reduction, manganese reduction, and ferric iron reduction (Froelich *et al.*, 1979). In organic rich marine sediments, a sulfate-reducing zone overlies a zone dominated by methanogenesis (Martens and Berner, 1977).

The mechanisms causing the observed segregation of redox zones remained unclear for a long time. However, studies with pure cultures of methanogens and sulfate reducers (Lovley and Klug, 1983), followed by studies with aquatic sediments (Lovley and Klug, 1986; Lovley and Goodwin, 1988), showed that redox zonation resulted from the ecology of aquatic sediments. In anoxic sediments, organic matter oxidation is carried out by food chains in which fermentative microorganisms partially oxidize organic matter with the production of fermentation products such as acetate and hydrogen. These fermentation products are then consumed by terminal electron-accepting microorganisms such as Fe(III) reducers or sulfate reducers. Because Fe(III) reduction produces more energy per mole of acetate or hydrogen oxidation, Fe(III) reducers are able to lower environmental concentrations of these fermentation products below levels required by less efficient sulfate reducers. Thus, when Fe(III) is available, Fe(III) reducers can outcompete sulfate reducers for available hydrogen, and sequester the majority (although generally not all) of the available electron flow. Similarly, when sulfate is available, sulfate reducers can outcompete methanogens for available hydrogen, and sequester the majority of electron flow. Thus, by considering concentrations of potential electron acceptors, and by considering concentrations of dissolved hydrogen (Chapelle *et al.*, 1995), it is often possible to deduce the distribution of TEAPs in groundwater systems (Figure 10).

9.3.4 Comparison of Equilibrium and Kinetic Approaches

A comparison of the equilibrium (Eh) and kinetic (TEAPs) approaches to describe redox processes in a petroleum hydrocarbon-contaminated aquifer was given by Chapelle *et al.* (1996). In this study, Eh measurements were made with a platinum electrode, and the results plotted on a standard Eh–pH diagram (Sillén, 1952). The results of this analysis are shown in Figure 11. Based on this analysis, it can be concluded that Fe(III) reduction is the predominant redox process, as none of the measured Eh values are sufficiently negative to indicate sulfate reduction or methanogenesis. In contrast, measured hydrogen concentrations (Figure 12) indicate the presence of methanogenesis near the water table surrounded by zones dominated by sulfate reduction and Fe(III) reduction. By considering other redox-sensitive solutes (Figure 10), including the depletion of potential electron acceptors such as oxygen ($<0.01\ mg\ L^{-1}$), sulfate ($<1.0\ mg\ L^{-1}$), and nitrate ($<0.02\ mg\ L^{-1}$) inside the plume, as well as the generation of sulfide ($\sim 1.0\ mg\ L^{-1}$), methane ($5–20\ mg\ L^{-1}$), and Fe^{2+} ($3–20\ mg\ L^{-1}$), it is clear that methanogenesis and sulfate reduction are occurring (Figure 12) in addition to Fe(III)-reduction indicated in Figure 11. In some groundwater systems, therefore, the kinetic approach appears to give a more complete evaluation of ambient redox processes than the more traditional equilibrium approach.

9.4 CONTAMINANT GEOCHEMISTRY

The principles and methods used to understand groundwater geochemistry were largely developed by considering pristine aquifer systems (Back, 1966; Plummer *et al.*, 1983). In the 1970s, however, it became clear that decades of unregulated disposal of industrial wastes had contaminated shallow groundwater underlying thousands of sites throughout the world. Many scientists involved with assessing and remediating this environmental contamination had been trained in the use of the mass-balance and equilibrium methods used in groundwater geochemistry. It was natural, therefore, to apply the mass-balance approach to problems of environmental contamination. This section gives a brief overview of the application of these principles to petroleum hydrocarbon and chlorinated solvent contamination of groundwater systems.

9.4.1 Petroleum Hydrocarbon Contamination

During the 1980s, enormous efforts were made in order to assess and monitor petroleum hydrocarbon contamination of groundwater in the United States and Europe. From this mass of information came several unanticipated and surprising results. It was widely observed, for example, that plumes of petroleum-hydrocarbon contaminated groundwater stopped expanding over time and assumed a dynamic steady-state configuration. Perhaps the best-documented example of this behavior was a crude oil spill in northern Minnesota near the town of Bemidji (Baedecker *et al.*, 1988; Cozzarelli *et al.*, 1999). In 1979 an oil pipeline ruptured and spilled 1,670 m^3 of crude oil on the land surface. During the

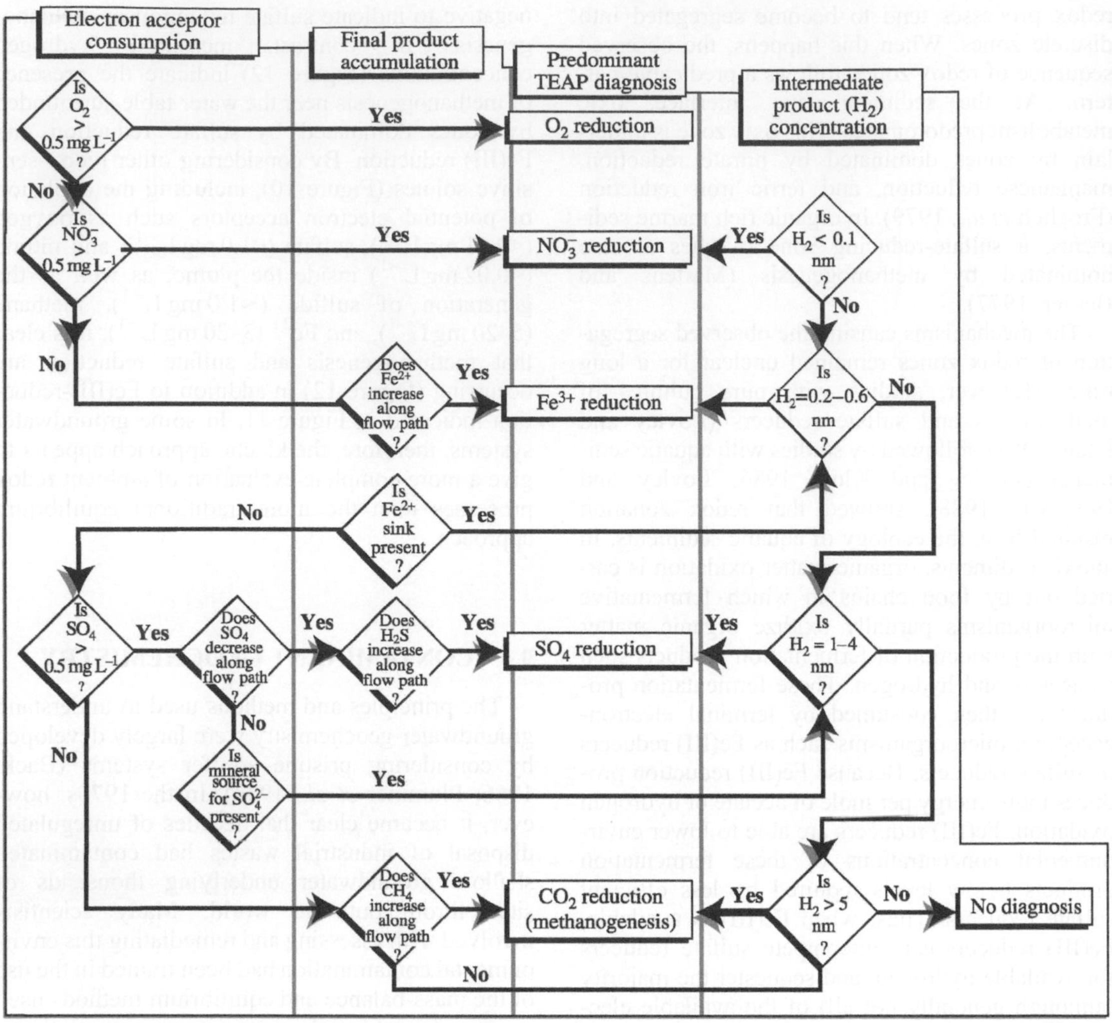

Figure 10 Flowchart for deducing operative TEAPs in groundwater systems (source Chapelle *et al.*, 1995).

following year, oil migrated downward and formed a lens floating on the water table. The site was instrumented with observation wells and monitored throughout the 1980s and 1990s. A plume of dissolved hydrocarbons, principally composed of soluble benzene, toluene, ethylbenzene, and xylene (BTEX) compounds, developed downgradient of the oil lens. However, by 1985, the BTEX plume appeared to stop spreading, extending only ~150 m downgradient of the oil lens. This result, was unanticipated by many scientists. Its explanation, however, came directly from mass-balance considerations (Baedecker *et al.*, 1993). The dynamic steady state of the plume reflected a balance between the rate at which soluble hydrocarbons were leached into the groundwater, and the rate at which biodegradation processes consumed the hydrocarbons (Lovley *et al.*, 1989; Baedecker *et al.*, 1993).

Similar observations were being made in other hydrologic systems. In a controlled experimental release to a shallow sandy aquifer, Barker *et al.* (1987) showed that naturally occurring biodegradation processes effectively removed benzene, toluene, and all three xylene isomers from solution in about one year. Barker *et al.* (1987) used the term *natural attenuation* to describe the combined dilution, dispersion, sorption, and biodegradation processes that caused contaminant concentrations to decrease. Similarly, Chiang *et al.* (1989) showed significant losses of benzene over a three-year period in a contaminated aquifer underlying a gas plant, and attributed this loss to natural attenuation processes.

On a much larger scale, a study of 7,167 municipal supply wells in California showed that, while leaking gasoline was by far the most common contaminant being released into groundwater systems, benzene was found in only 10 wells (Hadley and Armstrong, 1991). This study concluded that biodegradation processes were actively consuming BTEX

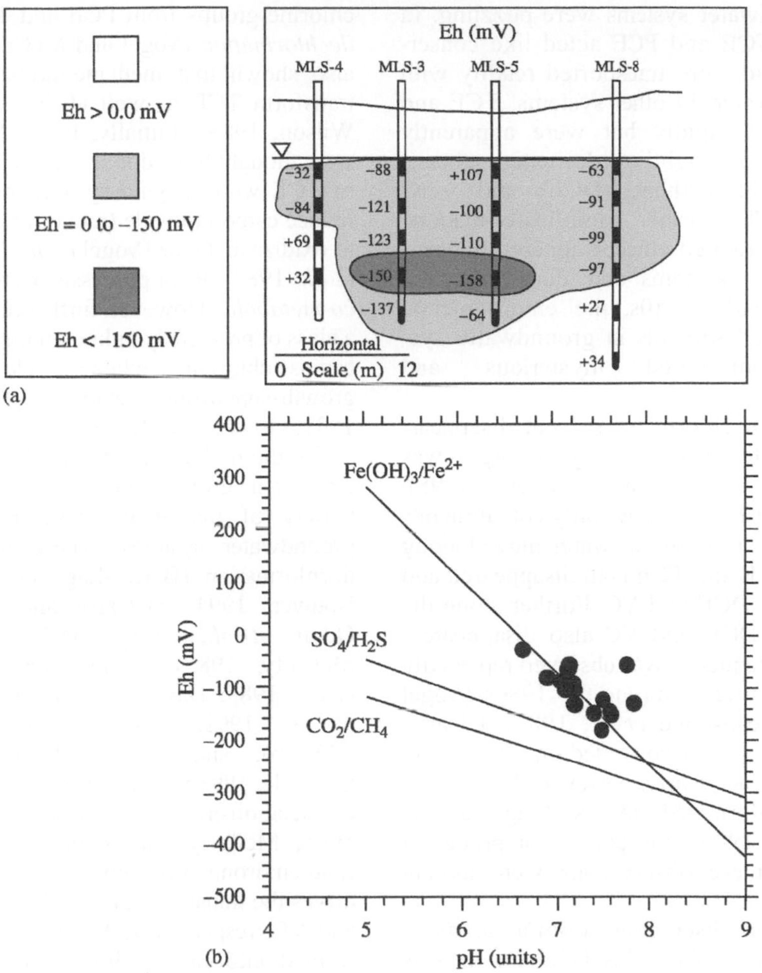

Figure 11 Measured Eh values in a cross-section of a contaminant plume, and the Eh values plotted on an Eh–pH diagram (source Chapelle *et al.*, 1996).

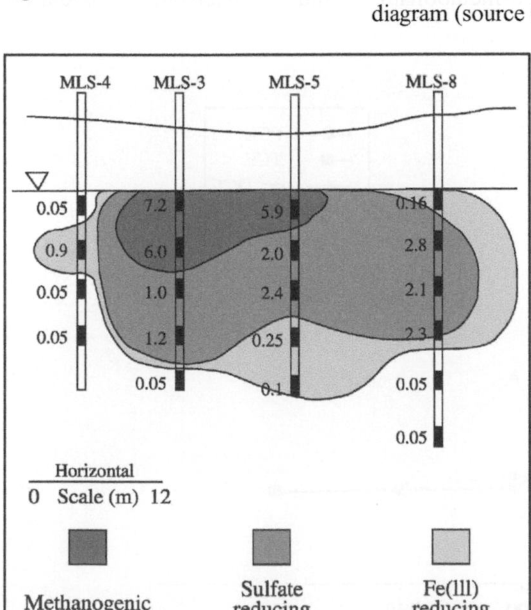

Figure 12 Concentrations of hydrogen in a cross-section of a contaminant plume, and the associated diagnosis of TEAPs (source Chapelle *et al.*, 1996).

compounds, and that these processes were protecting California groundwater supplies from widespread petroleum hydrocarbon contamination. These and other studies showed that natural biodegradation processes are far more important in affecting the migration of soluble petroleum hydrocarbons than had been foreseen in the early 1980s, and that they are a major factor in the remediation of environmental contaminants (US EPA, 1999).

9.4.2 Chlorinated Solvent Contamination

Mass-balance considerations, in particular the observed consumption of contaminants, were useful in showing the importance of biodegradation processes for limiting the mobility of petroleum hydrocarbons in groundwater systems. The mass-balance approach also contributed to our understanding of the environmental fate of chlorinated solvents in groundwater systems. In the 1980s, the observed behavior of chlorinated ethenes such as trichloroethene (TCE) and perchloroethene

(PCE) in groundwater systems were puzzling. In some systems, TCE and PCE acted like conservative solutes and were transported readily with flowing groundwater. In other systems, TCE and PCE disappeared rapidly but were apparently replaced with more lightly chlorinated ethenes such as *cis*-dichloroethene (DCE) and vinyl chloride (VC). To further complicate matters, these lightly chlorinated ethenes apparently accumulated in some systems and disappeared in others. In the early 1980s, the environmental fate of chlorinated solvents in groundwater systems was considered mysterious and unpredictable.

An example of chlorinated solvent disappearance, from a site in Kings Bay, Georgia was described by Chapelle and Bradley (1998). Initially, PCE and TCE were the only contaminants present (Figure 13). As groundwater moved along the flow path, PCE and TCE both disappeared and were replaced by DCE and VC. Further along the flow path, both DCE and VC also disappeared. This same basic sequence was observed repeatedly in groundwater systems during the 1980s (Vogel *et al.*, 1987; Barrio-Lage *et al.*, 1987). Clearly, transformation processes converted one compound into another. Just as clearly, however, there was a net loss of all chlorinated ethenes along the flow path. In the late 1980s, the process or processes contributing to these observations were largely unknown.

Soon thereafter, observations from a variety of sources began to explain this behavior. It was shown experimentally that pure cultures of methanogenic microorganisms were capable of stripping chlorine groups from PCE and TCE by *reductive dechlorination* (Vogel and McCarty, 1985). It was also shown that methane-oxidizing bacteria can transform TCE as well (J. T. Wilson and B. H. Wilson, 1985). Initially, both of these processes were thought to be due to the accidental interaction of TCE with enzyme systems designed either to reduce carbon dioxide (reductive dechlorination) or to oxidize methane (Vogel *et al.*, 1987; Barrio-Lage *et al.*, 1987). Both processes were initially termed *co-metabolic*. However, further research has shown a class of previously unknown microorganisms that can use chlorinated ethenes as electron acceptors in growth-supporting metabolism (DiStefano *et al.*, 1991; Gossett and Zinder, 1996).

By the mid-1990s, it was clear that chlorinated ethenes in groundwater systems are subject to a variety of microbial degradation processes in groundwater systems. These include reductive dechlorination (Barrio-Lage *et al.*, 1987, 1990; Bouwer, 1994; McCarty and Semprini, 1994; Odum *et al.*, 1995; Vogel, 1994; Vogel and McCarty, 1985), aerobic oxidation (Hartmans *et al.*, 1985; Davis and Carpenter, 1990; Phelps *et al.*, 1991; Bradley and Chapelle, 1996, 1998a,b), anaerobic oxidation (Bradley and Chapelle, 1998b; Bradley *et al.*, 1998), and aerobic co-metabolism (J. T. Wilson and B. H. Wilson, 1985; McCarty and Semprini, 1994). In many field environments, initial reductive dechlorination drives the transformation of PCE and TCE to DCE and VC, respectively. The latter in turn are transformed into carbon dioxide, chloride and water by the combined effects of methane-oxidizing co-metabolism and anaerobic oxidation.

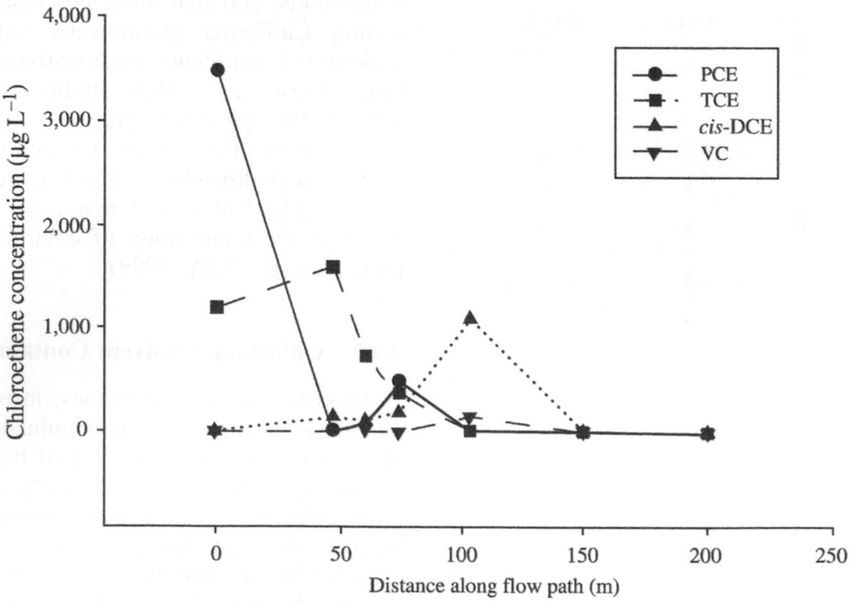

Figure 13 Observed transformation of chlorinated ethenes in a groundwater system (source Chapelle and Bradley, 1998).

Furthermore, because these biodegradation processes are all reduction/oxidation processes, the efficiency of biodegradation is very sensitive to ambient redox conditions (Chapelle, 1996).

A simple use of mass-balance considerations that shows this net transformation of chlorinated ethenes is the observed accumulation of chloride in contaminant plumes. For example, in a study of a large plume of TCE-contaminated groundwater at Dover Air Force Base, Delaware, Witt *et al.* (2002) showed that the concentration of chloride increased as TCE concentrations decreased, suggesting the net transformation of the chlorine in TCE to chloride. This chloride "tracer" of chlorinated ethene biodegradation is of use for mass-balance calculations to demonstrate biodegradation processes in the field. These field observations formed the basis of experimental studies under controlled laboratory conditions that documented the many and varied microbial processes that transform and destroy chlorinated ethenes in groundwater systems.

9.5 SUMMARY AND CONCLUSIONS

This brief overview shows that while the scope of groundwater geochemistry has changed significantly over time, the themes of practicality and the mass-balance approach have remained constant. The major questions of groundwater geochemistry have come directly from practical, human problems. Whether it was understanding the apparent medicinal properties of well and springwaters (Steel, 1838), helping to prospect for petroleum hydrocarbons (Rogers, 1917), explaining the distribution of "hard" and "soft" groundwater (Renick, 1924), investigating the origin of high concentrations of sodium and bicarbonate (Cedarstrom, 1946; Foster, 1950), showing how chemical changes of groundwater were related to directions of groundwater flow (Back, 1966), understanding the composition of springwaters (Garrels and Mackenzie, 1967), or documenting biodegradation of environmental contaminants (Baedecker *et al.*, 1993), all involved practical questions. The guiding principle used to address these questions has always been the notion of mass balance, that groundwater chemistry reflects the difference of what is put into solution minus what is taken out. Rigorous application of the mass-balance approach, and constraining material balances for thermodynamic consistency (Plummer *et al.*, 1983) has led to many unexpected and surprising discoveries. One of these is the realization that nonequilibrium, microbially mediated processes have an important effect on groundwater chemistry (Chapelle, 2001). In recent years, the mass-balance approach has helped identify microbial and nonmicrobial transformations that serve to detoxify environmental contaminants (US EPA, 1999), and has served to identify numerous remediation technologies for dealing with contamination problems. There is every reason to expect that the twin principles of practicality and mass balance will continue to be major themes in the development of groundwater geochemistry.

REFERENCES

Armstrong R. L. (1971) K–Ar chronology of Snake River Plain, Idaho (abs): Geological Society of America Abstracts with Programs, p. 366.

Back W. (1966) Hydrochemical facies and groundwater flow patterns in northern part of Atlantic Coastal Plain. *US Geol. Surv. Prof. Pap.* **498-A**, 42pp.

Back W. and Barnes I. (1965) Relation of electrochemical potentials and iron content to groundwater flow patterns. *US Geol. Surv. Prof. Pap.* **498-C, 16pp.**

Back W. and Hanshaw B. B. (1970) Comparison of chemical hydrogeology of the carbonate peninsulas of Florida and Yucatan. *J. Hydrol.* **10**, 330–368.

Back W., Landa E. R., and Meeks L. (1995) Bottled water, spas, and early years of water chemistry. *Ground Water* **33**, 605–613.

Baedecker M. J., Siegel D. I., Bennett P. C., and Cozzarelli I. M. (1988) The fate and effects of crude oil in a shallow aquifer: 1. The distribution of chemical species and geochemical facies. In *US Geological Survey Toxic Substances Hydrology Program*, Proceedings of the Technical Meeting, Phoenix, Arizona, September 26–30, 1988 (eds. G. E., Mallard and S. E. Ragone). US Geological Survey Water-Resources Investigations Report 88-4220, Reston, VA, pp. 13–20.

Baedecker M. J., Cozzarelli I. M., Eganhouse R. P., Siegel D. I., and Bennett P. C. (1993) Crude oil in a shallow sand and gravel aquifer: III. Biogeochemical reactions and mass balance modeling in anoxic ground water. *Appl. Geochem.* **8**, 569–586.

Barker J. F., Patrick G. C., and Major D. (1987) Natural attenuation of aromatic hydrocarbons in a shallow sand aquifer. *Ground Water Monitoring Rev.* **7**, 64–71.

Barrio-Lage G. A., Parsons F. Z., Nassar R. S., and Lorenzo P. A. (1987) Biotransformation of trichloroethene in a variety of subsurface materials. *Environ. Toxicol. Chem.* **6**, 571–578.

Barrio-Lage G. A., Parsons F. Z., Barbitz R. M., Lorenzo P. L., and Archer H. E. (1990) Enhanced anaerobic biodegradation of vinyl chloride in ground water. *Environ. Toxicol. Chem.* **9**, 403–415.

Bouwer E. J. (1994) Bioremediation of chlorinated solvents using alternate electron acceptors. *In Handbook of Bioremediation* (eds. R. D. Norris, R. E. Hinchee, R. Brown, P. L. McCarty, J. T. Wilson, M.. Reinhard, E. J. Bouwer, R. C. Bordon, T. M. Vogel, J. M. Thomas, and C. H. Ward). Lewis Publishers, Boca Raton, pp. 149–175.

Bradley P. M. and Chapelle F. H. (1996) Anaerobic mineralization of vinyl chloride in Fe(III)-reducing aquifer sediments. *Environ. Sci. Technol.* **30**, 2084–2086.

Bradley P. M. and Chapelle F. H. (1998a) Effect of contaminant concentration on aerobic microbial mineralization of DCE and VC in stream-bed sediments. *Environ. Sci. Technol.* **32**, 553–557.

Bradley P. M. and Chapelle F. H. (1998b) Microbial mineralization of VC and DCE under different terminal electron accepting conditions. *Anaerobe* **4**, 81–87.

Bradley P. M., Chapelle F. H., and Wilson J. T. (1998) Field and laboratory evidence for intrinsic biodegradation of vinyl

chloride contamination in a Fe(III)-reducing aquifer. *J. Contamin. Hydrol.* **31**, 111–127.
Burke W. (1853) *The Virginia Mineral Springs with Remarks on Their Use, The Diseases to Which They are Applicable, and in Which They are Contra-indicated.* Ritchies and Dunnavant, Richmond, VA, 376pp.
Cedarstrom D. J. (1946) Genesis of groundwaters in the coastal plain of Virginia. *Econ. Geol.* **41**(3), 218–245.
Chapelle F. H. (1996) Identifying redox conditions that favor the natural attenuation of chlorinated ethenes in contaminated groundwater systems. In *Symposium on Natural Attenuation of Chlorinated Organics in Ground Water*, EPA/540/R-96/509, pp. 17–20.
Chapelle F. H. (2000) *The Hidden Sea: Ground Water, Springs, and Wells*. The National Ground Water Association, Westerville, OH, 232pp.
Chapelle F. H. (2001) *Groundwater Microbiology and Geochemistry*, 2nd edn. Wiley, New York, 477pp.
Chapelle F. H. and Bradley P. M. (1998) Selecting remediation goals by assessing the natural attenuation capacity of groundwater systems. *Bioremed. J.* **2**, 227–238.
Chapelle F. H. and Lovley D. R. (1992) Competitive exclusion of sulfate-reduction by Fe(Ill)-reducing bacteria: a mechanism for producing discrete zones of high-iron ground water. *Ground Water* **30**, 29–36.
Chapelle F. H., Zelibor J. L., Grimes D. J., and Knobel L. L. (1987) Bacteria in deep coastal plain sediments of Maryland: a possible source of CO_2 to ground water. *Water Resour. Res.* **23**(8), 1625–1632.
Chapelle F. H., McMahon P. B., Dubrovsky N. M., Fujii R. F., Oaksford E. T., and Vroblesky D. A. (1995) Deducing the distribution of terminal electron-accepting processes in hydrologically diverse groundwater systems. *Water Resour. Res.* **31**, 359–371.
Chapelle F. H., Haack S. K., Adriaens P., Henry M. A., and Bradley P. M. (1996) Comparison of Eh and H_2 measurements for delineating redox processes in a contaminated aquifer. *Environ. Sci. Technol* **30**, 3565–3569.
Chebotarev I. I. (1955) Metamorphism of natural waters in the crust of weathering. *Geochim. Cosmochim. Acta* **8**(Part 1), 22–48; **8**(Part 2), 137–170; **8**(Part 3), 198–212.
Chiang C. Y., Salanitro J. P., Chai E. Y., Colthart J. D., and Klein C. L. (1989) Aerobic biodegradation of benzene, toluene, and xylene in a sandy aquifer: data analysis and computer modeling. *Ground Water* **27**, 823–834.
Cozzarelli I. M., Herman J. S., Baedecker M. J., and Fischer J. M. (1999) Geochemical heterogeneity of a gasoline-contaminated aquifer. *J. Contamin. Hydrol.* **40**, 261–284.
Davis J. W. and Carpenter C. L. (1990) Aerobic biodegradation of vinyl chloride in groundwater samples. *Appl. Environ. Microbiol.* **56**, 3870–3880.
DiStefano T. D., Gossett J. M., and Zinder S. H. (1991) Reductive dechlorination of high concentrations of tetrachloroethene to ethene by an anaerobic enrichment culture in the absence of methanogenesis. *Appl. Environ. Microbiol.* **57**, 2287–2292.
Drever J. I. (1982) *The Geochemistry of Natural Waters*. Prentice Hall, Englewood Cliffs, NJ, 388pp.
Embree G. F., McBroome L. A., and Soherty D. J. (1982) Preliminary stratigraphic framework of the Pliocene and Miocene rhyolite, Eastern Snake River Plain. *Idaho Bureau Mines Geol. Bull.* **26**, 333–346.
Feth J. H., Robertson C. E., and Polzer W. L. (1964) Sources of mineral constituents in water from granitic rocks in Sierra Nevada, California and Nevada. *US Geol. Surv. Water-Supply Pap.* **1535-I**, 70p.
Froelich P. N., Klinkhammer G. P., Bender M. L., Luedtke N. A., Heath G. R., Cullen D., and Dauphin P. (1979) Early oxidation of organic matter in pelagic sediments of the eastern equatorial Atlantic: suboxic diagenesis. *Geochim. Cosmochim. Acta* **43**, 1075–1090.
Foster M. D. (1950) The origin of high sodium bicarbonate waters in the Atlantic and Gulf Coast Plains. *Geochim. Cosmochim. Acta* **1**(1), 33–48.
Garrels R. M. and Christ C. L. (1965) *Solutions, Minerals, and Equilibria*. Freeman-Cooper, San Francisco, CA, 450pp.
Garrels R. M. and Mackenzie F. T. (1967) Origin of the chemical compositions of some springs and lakes. *In Equilibrium Concepts in Natural Water Chemistry,* Advances in Chemistry Series 67 (ed. R. F. Gould). American Chemical Society, Washington, DC, pp. 222–242.
Gossett J. M. and Zinder S. H. (1996) Microbiological aspects relevant to natural attenuation of chlorinated ethenes. In *Symposium on Natural Attenuation of Chlorinated Organics in Ground Water*, Denver, Co, EPA/540/R-96/509, pp. 10–13.
Hadley P. W. and Armstrong R. (1991) "Where's the benzene?"—Examining California groundwater quality surveys. *Ground Water* **29**, 35–40.
Hanshaw B. B., Back W., and Rubin M. (1965) Radiocarbon determinations for estimating groundwater flow velocities in central Florida. *Science* **148**, 494–495.
Hartmans S., deBont J. A. M., Tramper J., and Luyben K. C. A. M. (1985) Bacterial degradation of vinyl chloride. *Biotechnol. Lett.* **7**, 383–388.
Helgeson H. C., Brown T. H., Nigrini A., and Jones T. A. (1970) Calculation of mass transfer in geochemical processes involving aqueous solutions. *Geochim. Cosmochim. Acta* **34**, 569–592.
Lindberg R. D. and Runnells D. D. (1984) Groundwater redox reactions: an analysis of equilibrium state applied to Eh measurements and geochemical modeling. *Science* **225**, 925–927.
Lindholm G. F., Garabedian S. P., Newton G. D., and Whitehead R. L. (1983) *Configuration of the Water Table, March 1980 in the Snake River Plain Regional Aquifer System, Idaho and Eastern Oregon*. US Geological Survey Open-file Report 82-1022, scale 1:1,000,000.
Lovley D. R. and Goodwin S. (1988) Hydrogen concentrations as an indicator of the predominant terminal electron-accepting reactions in aquatic sediments. *Geochim. Cosmochim. Acta* **52**, 2993–3003.
Lovley D. R. and Klug M. J. (1983) Sulfate reducers can outcompete methanogens at freshwater sulfate concentrations. *Appl. Environ. Microbiol.* **45**, 187–192.
Lovley D. R. and Klug M. J. (1986) Model for the distribution of methane production and sulfate reduction in freshwater sediments. *Geochim. Cosmochim. Acta* **50**, 11–18.
Lovley D. R., Baedecker M. J., Lonergan D. J., Cozzarelli I. M., Phillips E. J. P., and Siegel D. I. (1989) Oxidation of aromatic contaminants coupled to microbial iron reduction. *Nature* **339**, 297–299.
Lovley D. R., Chapelle F. H., and Woodward J. C. (1994) Use of dissolved H_2 concentrations to determine distribution of microbially catalyzed redox reactions in anoxic groundwater. *Environ. Sci. Technol.* **28**, 1205–1210.
Martens C. S. and Berner R. A. (1977) Interstitial water chemistry of anoxic Long Island Sound sediments: I. Dissolved gases. *Limnol. Oceanogr.* **22**, 10–25.
McCarty P. L. and Semprini L. (1994) Groundwater treatment for chlorinated solvents. *In Handbook of Bioremediation* (eds. R. D. Norris, R. E. Hinchee, R. Brown, P. L. McCarty, J. T. Wilson, M.. Reinhard, E. J. Bouwer, R. C. Bordon, T. M. Vogel, J. M. Thomas, and C. H. Ward). Lewis Publishers, Boca Raton, pp. 87–116.
McMahon P. B. and Chapelle F. H. (1991a) Microbial production of organic acids in aquitard sediments and its role in aquifer geochemistry. *Nature* **349**, 233–235.
McMahon P. B. and Chapelle F. H. (1991b) Geochemistry of dissolved inorganic carbon in a Coastal Plain aquifer: 2. Modeling carbon sources, sinks, and $\delta^{13}C$ evolution. *J. Hydrol.* **127**, 109–135.
McMahon P. B., Chapelle F. H., Falls W. F., and Bradley P. M. (1992) Role of microbial processes in linking sandstone

diagenesis with organic-rick clays. *J. Sedimen. Petrol.* **62**, 1–10.

Miller J. A. (1986) Hydrogeologic framework of the Floridan aquifer system in Florida and in parts of Georgia, South Carolina, and Alabama. *US Geol. Surv. Prof. Pap. 1403-B,* 91pp.

Odum J. M., Tabinowski J., Lee M. D., and Fathepure B. Z. (1995) Anaerobic biodegradation of chlorinated solvents: comparative laboratory study of aquifer microcosms. In *Handbook of Bioremediation* (ed. R. D. Norris, R. E. Hinchee, R. Brown, P. L. McCarty, J. T. Wilson, M. Reinhard, E. J. Bouwer, R. C. Bordon, T. M. Vogel, J. M. Thomas, and C. H. Ward). Lewis Publishers, Boca Raton, pp. 17–24.

Parkhurst D. L., Plummer L. N., and Thorstenson D. C. (1982) BALANCE—a computer program for calculation of chemical mass balance. *US Geol. Surv. Water Res. Investigations Report* 82-14.

Phelps T. J., Malachowsky K., Schram R. M., and White D. C. (1991) Aerobic mineralization of vinyl chloride by a bacterium of the order *Actinomycetales*. *Appl. Environ. Microbiol.* **57**, 1252–1254.

Pierce K. L. and Morgan L. A. (1992) The track of the Yellowstone hot spot: volcanism, faulting, and uplift. *Geol. Soc. Am. Memoir 179.*

Plummer L. N. (1977) Defining reactions and mass transfer in part of the Floridan aquifer. *Water Resour. Res.* **13**, 801–812.

Plummer L. N. (1984) Geochemical modeling: a comparison of forward and reverse methods. In: B H. and E. I W. (eds.) *In First Canadian/American Conference on Hydrogeology. Practical Applications of Ground Water Geochemistry* (eds. B. Hitchon and E. I. Walick). National Water Well Association, Dublin, OH, pp. 149–177.

Plummer L. N. and Back W. (1980) The mass balance approach: application to interpreting the chemical evolution of hydrologic systems. *Am. J. Sci.* **280**, 130–142.

Plummer L. N., Parkhurst D. L., and Thorstenson D. C. (1983) Development of reaction models for groundwater systems. *Geochim. Cosmochim. Acta* **4**, 665–686.

Renick B. C. (1924) Base exchange in ground water by silicates as illustrated in Montana. *US Geol. Surv. Water Supply Pap.* **520-D**, 53–72.

Rogers G. S. (1917) Chemical relations of the oil-field waters in San Joaquin Valley, California. *US Geol. Surv. Bull.* **653**, 93–99.

Rogers R. J. (1987) Geochemical evolution of groundwater in stratified-drift and arkosic bedrock aquifers in north central Connecticut. *Water Resour. Res.* **23**, 1531–1545.

Sillén L. G. (1952) Redox diagrams. *J. Chem. Educat.* **29**, 600–608.

Sprinkle C. L. (1989) Geochemistry of the Floridan aquifer system in Florida and in parts of Georgia, South Carolina, and Alabama. *US Geol. Surv. Prof. Pap.* **1403-I**, 105pp.

Steel J. H. (1838) *An Analysis of the Mineral Waters of Saratoga and Ballston*. G. M. Davison, Saratoga Springs, NY, 203pp.

Stumm W. and Morgan J. J. (1981) *Aquatic Chemistry*. Wiley, New York, 780pp.

Thorstenson D. C. (1984) The concept of electron activity and its relation to redox potentials in aqueous geochemical systems. *US Geol. Surv. Open File Report* 84-072.

US Environmental Protection Agency (US EPA) (1999) Final OSWER Monitored Natural Attenuation Policy (Oswer Directive 9200.4-17P). *United States Environmental Protection Agency, Office of Solid Waste and Emergency Response.*

Vogel T. M. (1994) Natural bioremediation of chlorinated solvents. In *Handbook of Bioremediation* (eds. R. D. Norris, R. E. Hinchee, R. Brown, P. L. McCarty, J. T. Wilson, M. Reinhard, E. J. Bouwer, R. C. Bordon, T. M. Vogel, J. M. Thomas, and C. H. Ward). Lewis Publishers, Boca Raton, pp. 201–225.

Vogel T. M. and McCarty P. L. (1985) Biotransformation of tetrachloroethylene to trichloroethylene, dichloroethylene, vinyl chloride, and carbon dioxide under methanogenic conditions. *Appl. Environ. Microbiol.* **49**, 1080–1083.

Vogel T. M., Criddle C. S., and McCarty P. L. (1987) Transformation of halogenated aliphatic compounds. *Environ. Sci. Technol.* **21**, 722–736.

Vroblesky D. A. and Chapelle F. H. (1994) Temporal and spatial changes of terminal electron-accepting processes in a petroleum hydrocarbon-contaminated aquifer and the significance for contaminant biodegradation. *Water Resour. Res.* **30**, 1561–1570.

Wilson J. T. and Wilson B. H. (1985) Biotransformation of trichloroethylene in soil. *Appl. Environ. Microbiol.* **49**, 242–243.

Wilson J. T., McNabb J. F., Balkwill D. L., and Ghiorse W. C. (1983) Enumeration and characterization of bacteria indigenous to a shallow water-table aquifer. *Ground Water* **21**, 134–142.

Witt M. E., Klecka G. M., Lutz E. J., Ei T. A., Grosso N. R., and Chapelle F. H. (2002) Natural attenuation of chlorinated solvents at Area 6 Dover Air Force Base: groundwater biogeochemistry. *J. Contamin. Hydrol.* **57**, 61–80.

Wood W. W. and Low W. H. (1988) Solute geochemistry of the Snake River Plain Regional Aquifer System, Idaho and Eastern Oregon. *US Geol. Surv. Prof. Pap.* **1408-D**, 79pp.

Published by Elsevier Ltd.

Readings from the Treatise on Geochemistry
ISBN: 978-0-12-381391-6
pp. 289–314

10 The Oceanic $CaCO_3$ Cycle

W. S. Broecker

Lamont-Doherty Earth Observatory, Columbia University, Palisades, NY

10.1 INTRODUCTION

Along with the silicate debris carried to the sea by rivers and wind, the calcitic hard parts manufactured by marine organisms constitute the most prominent constituent of deep-sea sediments. On high-standing open-ocean ridges and plateaus, these calcitic remains dominate. Only in the deepest portions of the ocean floor, where dissolution takes its toll, are sediments calcite-free. The foraminifera shells preserved in marine sediments are the primary carriers of paleoceanographic information. Mg/Ca ratios in these shells record past surface water temperatures; temperature corrected $^{18}O/^{16}O$ ratios record the volume of continental ice; $^{13}C/^{12}C$ ratios yield information about the strength of the ocean's biological pump and the amount of carbon stored as terrestrial biomass; the cadmium and zinc concentrations serve, respectively, as proxies for the distribution of dissolved phosphate and dissolved silica in the sea. While these isotopic ratios and trace element concentrations constitute the workhorses of the field of paleoceanography, the state of preservation of the calcitic material itself has an important story to tell. It is this story with which this chapter is concerned.

In all regions of the ocean, plots of sediment composition against water depth have a characteristic shape. Sediments from mid-depth are rich in $CaCO_3$ and those from abyssal depths are devoid of $CaCO_3$. These two realms are separated by a transition zone spanning several hundreds of meters in water depth over which the $CaCO_3$ content drops toward zero from the 85–95% values which characterize mid-depth sediment. The upper bound of this transition zone has been termed the "lysocline" and signifies the depth at which dissolution impacts become noticeable. The lower bound is termed the "compensation depth" and signifies the depth at which the $CaCO_3$ content is reduced to 10%. While widely used (and misused), both of these terms suffer from ambiguities. My recommendation is that they be abandoned in favor of the term "transition zone." Where quantification is appropriate, the depth of the mid-point of $CaCO_3$ decline should be given. While the width of the zone is also of interest, its definition suffers from the same

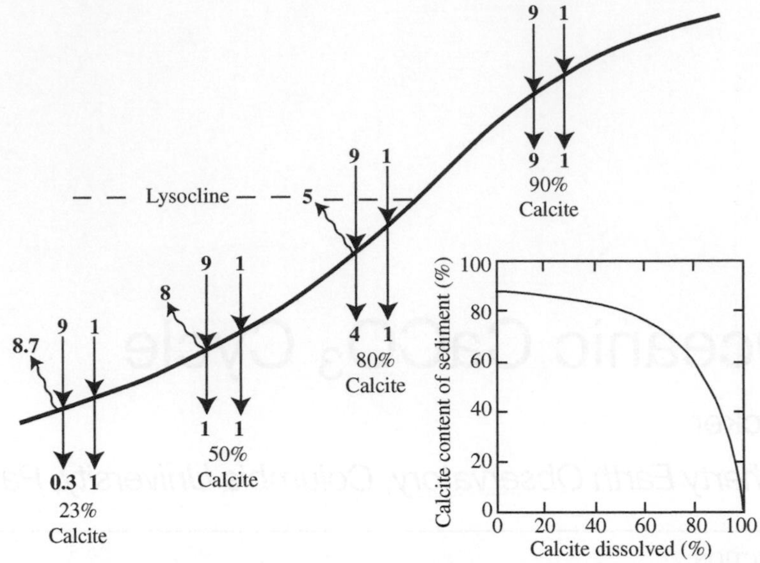

Figure 1 A diagrammatic view of how the extent of dissolution impacts the percent calcite in the sediment. In each example, the right-hand vertical arrows give the rain rate and accumulation rate of non-$CaCO_3$ debris and the left-hand vertical arrows the rain rate and accumulation rate of calcite. The wavy arrows represent the dissolution rates of calcite. As can be seen from the graph on the lower right, the percent of calcite in the sediment gives a misleading view of the fraction of the raining calcite which has dissolved, for large amounts of dissolution are required before the calcite content of the sediment drops significantly.

problems associated with the use of the terms "lysocline" and "compensation depth," namely, the boundaries are gradual rather than sharp.

While determinations of sediment $CaCO_3$ content as a function of water depth in today's ocean or at any specific time in the past constitute a potentially useful index of the extent of dissolution, it must be kept in mind that this relationship is highly nonlinear. Consider, for example, an area where the rain rate of calcite to the seafloor is 9 times that of noncarbonate material. In such a situation, were 50% of the calcite to be dissolved, the $CaCO_3$ content would drop only from 90% to only 82%, and were 75% dissolved away, it would drop only to 69% (see Figure 1). One might counter by saying that as the $CaCO_3$ content can be measured to an accuracy of ±0.5% or better, one could still use $CaCO_3$ content as a dissolution index. The problem is that in order to obtain a set of sediment samples covering an appreciable range of water depth, topographic gradients dictate that the cores would have to be collected over an area covering several degrees. It is unlikely that the ratio of the rain rate of calcite to that of noncalcite would be exactly the same at all the coring sites. Hence, higher accuracy is not the answer.

10.2 DEPTH OF TRANSITION ZONE

As in most parts of today's deep ocean the concentrations of Ca^{2+} and of CO_3^{2-} are nearly constant with water depth, profiles of $CaCO_3$ content with depth reflect mainly the increase in the solubility of the mineral calcite with pressure (see Figure 2). This increase occurs because the volume occupied by the Ca^{2+} and CO_3^{2-} ions dissolved in seawater is smaller than when they are combined in the mineral calcite. Unfortunately, a sizable uncertainty exists in the magnitude of this volume difference. The mid-depth waters in the ocean are everywhere supersaturated with respect to calcite. Because of the pressure dependence of solubility, the extent of supersaturation decreases with depth until the saturation horizon is reached. Below this depth, the waters are undersaturated with respect to calcite. While it is tempting to conclude that the saturation horizon corresponds to the top of the transition zone, as we shall see, respiration CO_2 released to the pore waters complicates the situation by inducing calcite dissolution above the saturation horizon.

One might ask what controls the depth of the transition zone. The answer lies in chemical economics. In today's ocean, marine organisms manufacture calcitic hard parts at a rate several times faster than CO_2 is being added to the ocean–atmosphere system (via planetary outgassing and weathering of continental rocks) (see Figure 3). While the state of saturation in the ocean is set by the product of the Ca^{2+} and CO_3^{2-} concentrations, calcium has such a long residence (10^6 yr) that, at least on the timescale of a single glacial cycle (~10^5 yr), its concentration can be assumed to have remained unchanged. Further, its

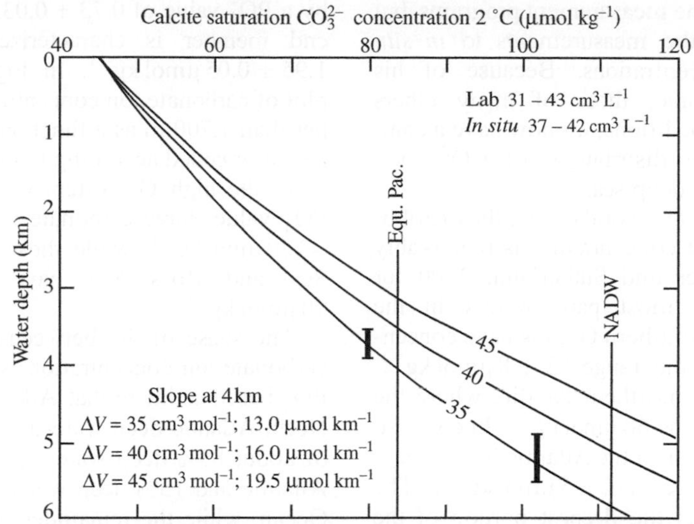

Figure 2 Saturation carbonate ion concentration for calcite at deep-water temperatures as a function of water depth (i.e., pressure). Curves are drawn for three choices of ΔV between Ca^{2+} and CO_3^{2-} ions when combined in calcite and when dissolved in seawater. The vertical black lines indicate the average CO_3^{2-} concentration in the deep equatorial Pacific and in NADW. The bold segments indicate the onset of dissolution as indicated by *in situ* experiments carried out in the deep North Pacific (Peterson, 1966) and in the deep North Atlantic (Honjo and Erez, 1978). As indicated in the upper right-hand corner, estimates of ΔV cover a wide range.

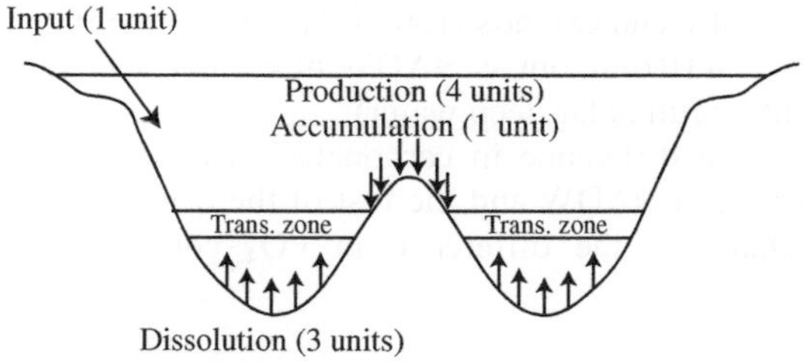

Figure 3 Marine organisms produce calcite at ~4 times the rate at which the ingredients for this mineral are supplied to the sea by continental weathering and planetary outgassing. A transition zone separates the mid-depth ocean floor where calcite is largely preserved from the abyssal ocean floor where calcite is largely dissolved.

concentration in seawater is so high that $CaCO_3$ cycling within the sea does not create significant gradients. In contrast, the dissolved inorganic carbon (i.e., CO_2, HCO_3^-, and CO_3^{2-}) in the ocean is replaced on a timescale roughly equal to that of the major glacial to interglacial cycle (10^5 yr). But, since in the deep sea CO_3^{2-} ion makes up only ~5% of the total dissolved inorganic carbon, its adjustment time turns out to be only about one-twentieth that for dissolved inorganic carbon or ~5,000 yr. Hence, the concentration of CO_3^{2-} has gradients within the sea and likely has undergone climate-induced changes.

Hence, at least on glacial to interglacial timescales, attention is focused on distribution of CO_3^{2-} concentration in the deep sea for it alone sets the depth of the transition zone. Thus, it is temporal changes in the concentration of carbonate ion which have captured the attention of those paleoceanographers interested in glacial to interglacial changes in ocean operation. These changes involve both the carbonate ion concentration averaged over the entire deep ocean and its distribution with respect to water depth and geographic location. Of course, it is the global average carbonate ion concentration in the deep sea that adjusts in order to assure that burial of $CaCO_3$ in the sediments matches the input of CO_2 to the ocean atmosphere system (or, more precisely, the input minus the fraction destined to be buried as organic residues). For example, were some anomaly to cause the burial of $CaCO_3$ in seafloor sediments to exceed supply, then the CO_3^{2-} concentration would be drawn down. This drawdown would continue until a balance between removal and supply was restored. As already mentioned, the time constant for this adjustment is on the order of 5,000 yr.

10.3 DISTRIBUTION OF CO_3^{2-} ION IN TODAY'S DEEP OCEAN

As part of the GEOSECS, TTO, SAVE and WOCE ocean surveys, $\sum CO_2$ and alkalinity measurements were made on water samples captured at various water depths in Niskin bottles. Given the depth, temperature and salinity for these samples, it is possible to compute the *in situ* carbonate ion concentrations. LDEO's Taro Takahashi played a key role not only in the measurement programs, but also in converting the measurements to *in situ* carbonate ion concentrations. Because of his efforts and, of course, those of many others involved in these expeditions, we now have a complete picture of the distribution of CO_3^{2-} ion concentrations in the deep sea.

Below 1,500 m in the world ocean, the distribution of carbonate ion concentration is remarkably simple (see Broecker and Sutherland, 2000 for summary). For the most part, waters in the Pacific, Indian, and Southern Oceans have concentrations confined to the range $83 \pm 8\ \mu mol\ kg^{-1}$. The exception is the northern Pacific, where the values drop to as low as $60\ \mu mol\ kg^{-1}$. In contrast, much of the deep water in the Atlantic has concentrations in the range $112 \pm 5\ \mu mol\ kg^{-1}$. The principal exception is the deepest portion of the western basin where Antarctic bottom water (AABW) intrudes.

As shown by Broecker *et al.*, the deep waters of the ocean can be characterized as a mixture of two end members, i.e., deep water formed in the northern Atlantic and deep water formed in the Southern Ocean. These end members are characterized by quite different values of a quasi-conservative property, PO_4^* (i.e., $PO_4 - 1.95 + O_2/175$). Although these two deep-water sources have similar initial O_2 contents, those formed in the northern Atlantic have only roughly half the PO_4 concentration of the deep waters descending in the Southern Ocean. Thus, the northern end member is characterized by a PO_4^* value of 0.73 ± 0.03, while the southern end member is characterized by a value of $1.95 \pm 0.05\ \mu mol\ kg^{-1}$. In Figure 4 is shown a plot of carbonate ion concentration for waters deeper than 1,700 m as a function of PO_4^*. The points are color coded according to O_2 content. As can be seen, the high O_2 waters with northern Atlantic PO_4^* values have carbonate ion concentrations of $\sim 120\ \mu mol\ kg^{-1}$, while those formed in Weddell Sea and Ross Sea have values closer to $90\ \mu mol\ kg^{-1}$.

The sense of the between-ocean difference in carbonate ion concentration is consistent with the PO_4^*-based estimate that Atlantic deep water (i.e., North Atlantic deep water (NADW)) is a mixture of about 85% deep water formed in the northern Atlantic and 15% deep water formed in Southern Ocean, while the remainder of the deep ocean is flooded with a roughly 50–50 mixture of these two source waters (Broecker, 1991). The interocean difference in carbonate ion concentration relates to the fact that deep water formed in the northern Atlantic has a higher CO_3^{2-} concentration than that produced in the Southern Ocean. The transition zone between NADW and the remainder of the deep ocean is centered in the western South Atlantic and extends around Africa into the Indian Ocean (fading out as NADW mixes into the ambient circumpolar deep water).

The difference in carbonate ion concentration between NADW and the rest of the deep ocean is related to the difference in PO_4 concentration.

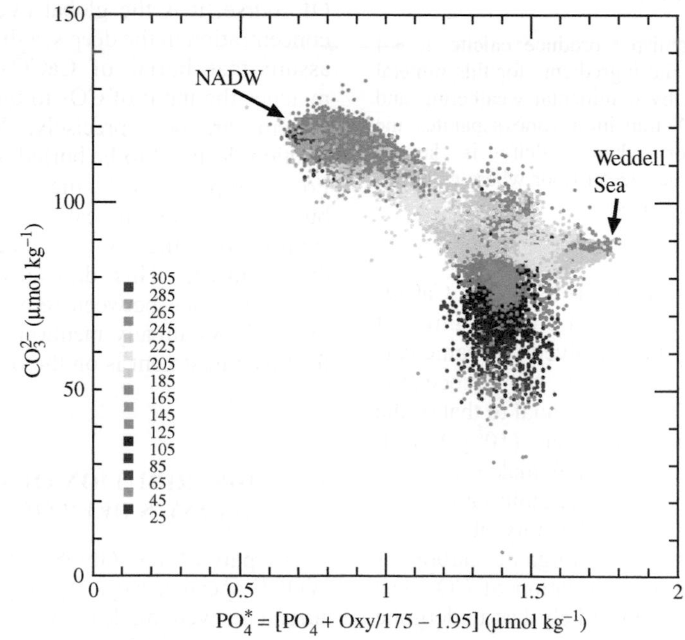

Figure 4 Plot of carbonate ion concentration as a function of PO_4^* for waters deeper than 1,700 m in the world ocean. The points are coded according to O_2 content (in $\mu mol\ kg^{-1}$). This plot was provided by LDEO'S Stew Sutherland and Taro Takahashi. It is based on measurements made as part of the GEOSECS, TTO, SAVE and WOCE surveys.

NADW has only about half the concentration of PO_4 as does, for example, deep water in equatorial Pacific. This is important because, for each mole of phosphorus released during respiration, ~120 mol of CO_2 are also produced. This excess CO_2 reacts with CO_3^{2-} ion to form two HCO_3^- ions. Were PO_4 content the only factor influencing the interocean difference in carbonate ion concentration, then it would be expected to be more like 90 μmol kg^{-1} rather than the observed 30 μmol kg^{-1}. So, something else must be involved.

This something is CO_2 transfer through the atmosphere (Broecker and Peng, 1993). The high-phosphate-content waters upwelling in the Southern Ocean lose part of their excess CO_2 to the atmosphere. This results in an increase in their CO_3^{2-} ion content. In contrast, the low-PO_4-content waters reaching in the northern Atlantic have CO_2 partial pressures well below that in the atmosphere and hence they absorb CO_2. This reduces their CO_3^{2-} concentration. Hence, it is the transfer of CO_2 from surface waters in the Southern Ocean to surface waters in the northern Atlantic reduces the contrast in carbonate ion concentration between deep waters in the deep Atlantic and those in the remainder of the deep ocean.

One other factor expected to have an impact on the carbonate ion concentration in deep Pacific Ocean and Indian Ocean turns out to be less important. Much of the floor of these two oceans lies below the transition zone. Hence, most of the $CaCO_3$ falling into the deep Pacific Ocean and Indian Ocean dissolves. One would expect then that the older the water (as indicated by lower $^{14}C/C$ ratios), the higher its CO_3^{2-} ion concentration would be. While to some extent this is true, the trend is much smaller than expected. The reason is that in the South Pacific Ocean and South Indian Ocean an almost perfect chemical titration is being conducted, i.e., for each mole of respiration CO_2 released to the deep ocean, roughly one mole of $CaCO_3$ dissolves (Broecker and Sutherland, 2000). So indeed, the older the water, the higher its $\sum CO_2$ content. But, due to $CaCO_3$ dissolution, there is a compensating increase in alkalinity such that the carbonate ion concentration remains largely unchanged. Only in the northern reaches of these oceans does the release of metabolic CO_2 overwhelm the supply of $CaCO_3$ allowing the CO_3 concentration to drop.

As in the depth range of transition zone, the solubility of $CaCO_3$ increases by ~14 μmol kg^{-1} km^{-1} increase in water depth, the 30 μmol kg^{-1} higher CO_3^{2-} concentration in NADW should (other things being equal) lead to a 2 km deeper transition zone in the Atlantic than in the Pacific Ocean and the Indian Ocean. In fact, this is more or less what is observed.

10.4 DEPTH OF SATURATION HORIZON

A number of attempts have been made to establish the exact depth of the calcite saturation horizon. The most direct way to do this is to suspend preweighed calcite entities at various water depths on a deep-sea mooring, then months later, recover the mooring and determine the extent of weight loss (see Figure 5). Peterson (1966)

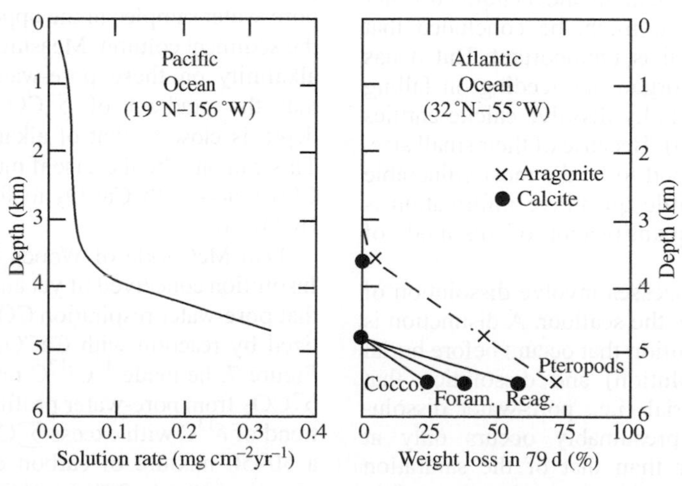

Figure 5 Results of *in situ* dissolution experiments. Peterson (1966) re-weighed polished calcite spheres after a 250 d deployment on a mooring in the North Pacific. Honjo and Erez (1978) observed the weight loss for calcitic samples (coccoliths, foraminifera and reagent calcite) and an aragonitic sample (pteropods) held at depth for a period of 79 d. While Peterson hung his spheres directly in seawater, the Honjo–Erez samples were held in containers through which water was pumped. The results suggest that the calcite saturation horizon lies at 4,800 ± 200 m in the North Atlantic and at about 3,800 ± 200 m in the North Pacific. For aragonite, which is 1.4 times more soluble than calcite, the saturation horizon in the North Atlantic is estimated to be in the range 3,400 ± 200 m.

performed such an experiment at 19 °N in the Pacific Ocean using polished calcite spheres and observed a pronounced depth-dependent increase in weight loss that commenced at ~3,900 m. Honjo and Erez (1978) performed a similar experiment at 32 °N in the Atlantic and found that coccoliths, foraminifera shells and reagent calcite experienced a 25–60% weight loss at 5,500 m but no measurable weight loss at 4,900 m. Thus the North Atlantic–North Pacific depth difference in the depth of the onset of dissolution is more or less consistent with expectation. Broecker and Takahashi (1978) used a combination of the depth of the onset of sedimentary $CaCO_3$ content decline and the results of a technique referred to as *in situ* saturometry (Ben-Yaakov and Kaplan, 1971) to define the depth dependence of solubility. While fraught with caveats, these results are broadly consistent with those from the mooring experiment. By measuring the composition of pore waters extracted *in situ* from sediments at various water depths, Sayles (1985) was able to calculate what he assumed to be saturation CO_3^{2-} concentrations. Finally, several investigators have performed laboratory equilibrations of calcite and seawater as a function of confining pressure. But, as each approach is subject to biases, more research is needed before the exact pressure dependence of the solubility of calcite can be pinned down.

10.5 DISSOLUTION MECHANISMS

Three possible dissolution processes come to mind. The first of these is termed water column dissolution. As foraminifera shells fall quite rapidly and as they encounter calcite undersaturated water only at great depth, it might be concluded that dissolution during fall is unimportant. But it has been suggested that organisms feeding on falling debris ingest and partially dissolve calcite entities (Milliman *et al.*, 1999). Because of their small size, coccoliths are presumed to be the most vulnerable in this regard. But little quantitive information is available to permit quantification of this mode of dissolution.

The other two processes involve dissolution of calcite after it reaches the seafloor. A distinction is made between dissolution that occurs before burial (i.e., interface dissolution) and dissolution that takes place after burial (i.e., pore-water dissolution). The former presumably occurs only at water depths greater than that of the saturation horizon. But the latter has been documented to occur above the calcite saturation horizon. It is driven by respiration CO_2 released to the pore waters.

Following the suggestion of Emerson and Bender (1981) that the release of respiration CO_2 in pore waters likely drives calcite dissolution above the saturation horizon, a number of investigators took the bait and set out to explore this possibility. David Archer, as part of his PhD thesis research with Emerson, developed pH microelectrodes that could be slowly ratcheted into the upper few centimeters of the sediment from a bottom lander. He deployed these pH microelectrodes along with the O_2 microelectrodes and was able to show that the release of respiration CO_2 (as indicated by a reduction in pore-water O_2) was accompanied by a drop in pH (and hence also of CO_3^{2-} ion concentration). Through modeling the combined results, Archer *et al.* (1989) showed that much of the CO_2 released by respiration reacted with $CaCO_3$ before it had a chance to escape (by molecular diffusion) into the overlying bottom water. As part of his PhD research, Burke Hales, a second Emerson student, improved Archer's electrode system and made measurements on the Ceara Rise in the western equatorial Atlantic (Hales and Emerson, 1997) and on the Ontong–Java Plateau in the western equatorial Pacific (Hales and Emerson, 1996) (see Figure 6). Taken together, these two studies strongly support the proposal that dissolution in pore waters of sediments leads to substantial dissolution of calcite. This approach has been improved upon by the addition of an LIX electrode to measure CO_2 itself and a micro-optode to measure Ca^{2+} (Wenzhöfer *et al.*, 2001).

In another study designed to confirm that most of the CO_2 released into the upper few centimeters of the sediments reacts with $CaCO_3$ before escaping to the overlying bottom water, Martin and Sayles (1996) deployed a very clever device that permitted the *in situ* collection of closely spaced pore-water samples in the upper few centimeters of the sediment column. Measurements of $\sum CO_2$ and alkalinity on these pore-water samples revealed that the gradient of $\sum CO_2$ (μmol km^{-1}) with depth is close to that of alkalinity (μequiv. kg^{-1}). This can only be the case if much of the respiration CO_2 reacts with $CaCO_3$ to form a Ca^{2+} and two HCO_3^- ions.

Dan McCorkle of Woods Hole Oceanographic Institution conceived of yet another way to confirm that pore-water respiration CO_2 was largely neutralized by reaction with $CaCO_3$. As summarized in Figure 7, he made $^{13}C/^{12}C$ ratio measurements on $\sum CO_2$ from pore-water profiles and found that the trend of $\delta^{13}C$ with excess $\sum CO_2$ is consistent with a 50–50 mixture of carbon derived from marine organic matter (−20‰) and that derived from marine calcite (+1‰) (Martin *et al.*, 2000). Again, these results require that a large fraction of the metabolic CO_2 reacts with $CaCO_3$.

There is, however, a fly in the ointment. Benthic flux measurements made by deploying chambers on the seafloor reveal a curious pattern

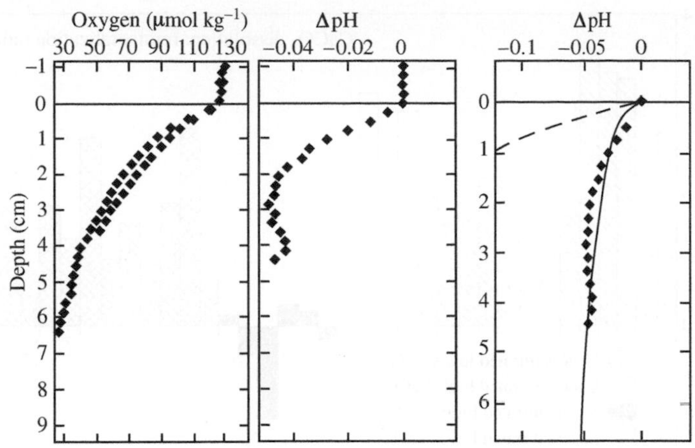

Figure 6 Microelectrode profiles of dissolved O_2 and ΔpH obtained by Hales and Emerson (1996) at 2.3 km depth on the Ontong–Java Plateau in the western equatorial Pacific. On the right are model curves showing the pH trend expected if none of the CO_2 released during the consumption of the O_2 was neutralized by reaction with sediment $CaCO_3$ (dashed curve) and a best model fit to the measured ΔpH trend (solid curve). The latter requires that much of the respiration CO_2 reacts with $CaCO_3$ before it escapes into the overlying bottom water.

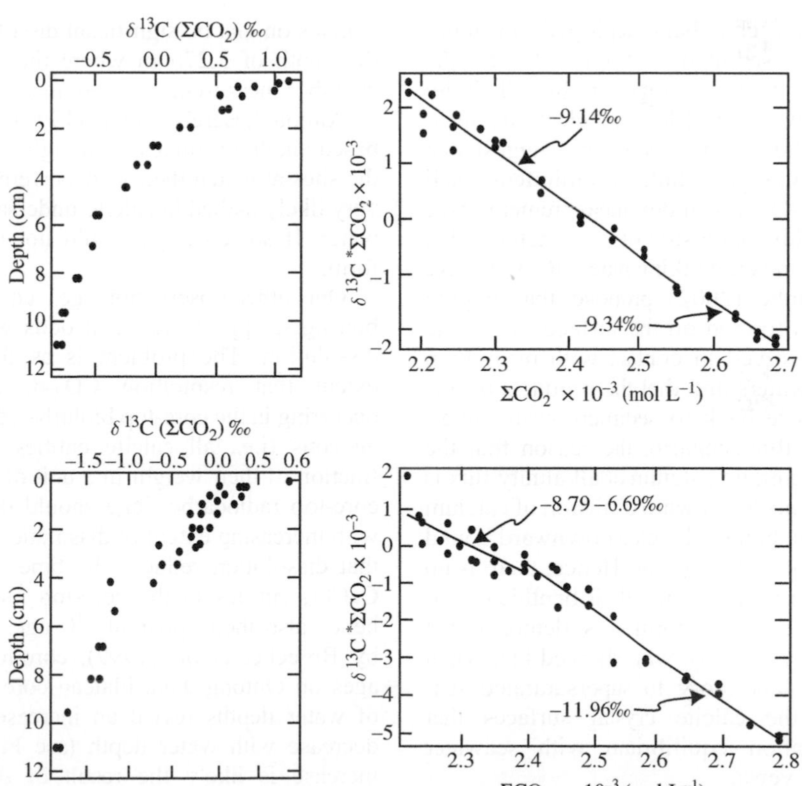

Figure 7 On the left are shown plots versus depth in the sediment of $\delta^{13}C$ in pore water total dissolved inorganic carbon (i.e., $\sum CO_2$) for two sites on the Ceara Rise (5°S in the western Atlantic) (Martin *et al.*, 2000). On the right are plots of $\delta^{13}C$ versus $\sum CO_2$. The slopes yield the isotopic composition of the excess CO_2. As can be seen, it requires that the respiration CO_2 be diluted with a comparable contribution from dissolved $CaCO_3$.

(see Figure 8). R. A. Jahnke and D. B. Jahnke (2002) found that alkalinity and calcium fluxes from sediments (both high and low in $CaCO_3$ content) below the calcite saturation horizon and on low-$CaCO_3$-content sediments from above the saturation horizon yield more or less the expected

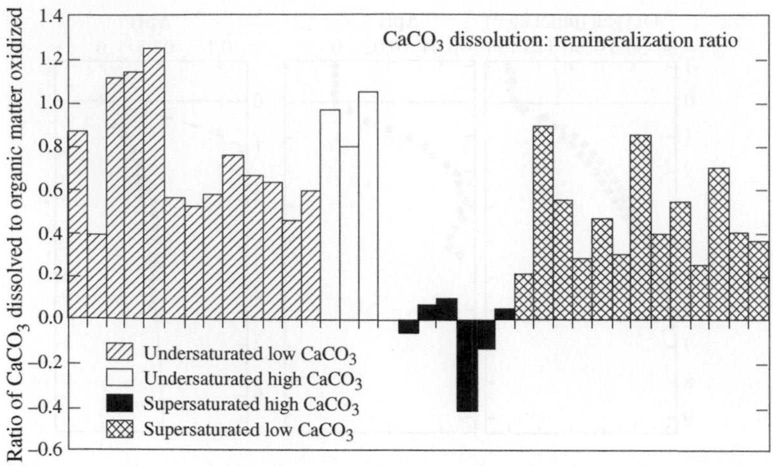

Figure 8 Summary of the ratio of $CaCO_3$ dissolved and organic material oxidized for bottom chamber deployments in the northeastern Pacific, Ontong–Java Plateau, Ceara Rise, Cape Verde Plateau, northwestern Atlantic continental rise and California borderland basins (R. A. Jahnke and D. B. Jahnke, 2002). The absence of measurable alkalinity fluxes from high-$CaCO_3$ sites bathed in supersaturated bottom water appears to be inconsistent with observations (see text).

fluxes. However, chambers deployed on high-$CaCO_3$-content sediments from above the saturation horizon yield no measurable alkalinity flux. Yet pore-water profiles and electrode measurements for these same sediments suggest that calcite is dissolving. Whole foraminifera shell weight and $CaCO_3$ size index measurements (see below) agree with conclusion of these authors that calcite dissolution is not taking place. R. A. Jahnke and D. B. Jahnke (2002) propose that impure $CaCO_3$ coatings formed on the surfaces of calcite grains are redissolved in contact with respiration CO_2-rich pore waters and that the products of this dissolution diffuse back to sediment–water interface. Based on this scenario, the reason that the these authors record no calcium or alkalinity flux is that the ingredients for upward diffusion of calcium and alkalinity are being advected downward bound to the surfaces of calcite grains. Hence, there is no net flux of either property into their benthic chamber. That such coatings form was demonstrated long ago by Weyl (1965), who showed that when exposed in the laboratory to supersaturated seawater it was the calcite crystal surfaces that achieved saturation equilibrium with seawater rather than vice versa.

Broecker and Clark (2003) fortify the mechanism proposed by R.A. Jahnke and D. B. Jahnke, providing additional evidence by proposing that it must be coatings rather than the biogenic calcite itself that dissolve. As shown in Figure 9, while on the Ontong–Java Plateau, there is a progressive decrease in shell weight and $CaCO_3$ size index with water depth; on the Ceara Rise, neither of these indices shows a significant decrease above a water depth of 4,100 m. This is consistent with the conclusion that no significant dissolution occurs at the depth of 3,270 m where the pore-water and chamber measurements were made.

Although Berelson *et al.* (1994) report chamber-based alkalinity fluxes from high-calcite sediment, the sites at which their studies were performed are very likely bathed in calcite-undersaturated bottom water. If so, coatings would not be expected to form.

One other observation, i.e., core-top radiocarbon ages, appears to be at odds with pore-water dissolution. The problem is as follows. To the extent that respiration CO_2-driven dissolution occurring in the core-top bioturbated zone is homogeneous (i.e., all calcite entities lose the same fraction of their weight in a unit of time), then the core-top radiocarbon age should decrease slowly with increasing extent of dissolution. The reason is that dissolution reduces the time of residence of $CaCO_3$ entities in the core-top mixed layer, and hence also their apparent ^{14}C age. But, as shown by Broecker *et al.* (1999), core-top radiocarbon ages on Ontong–Java Plateau cores from a range of water depths reveal an increase rather than a decrease with water depth (see Figure 10). This increase is likely the result of dissolution that occurs on the seafloor in calcite-undersaturated bottom waters before the calcite is incorporated into the core-top mixed layer. In this case, the reduction of $CaCO_3$ input to the sediment leads to an increase in the average residence time of calcite in the bioturbated layer. It may be that competition between pore-water dissolution and seafloor dissolution changes with depth. As shown in Figure 3, down to about 3 km pore-water dissolution appears to have the upper hand (and hence the ^{14}C ages

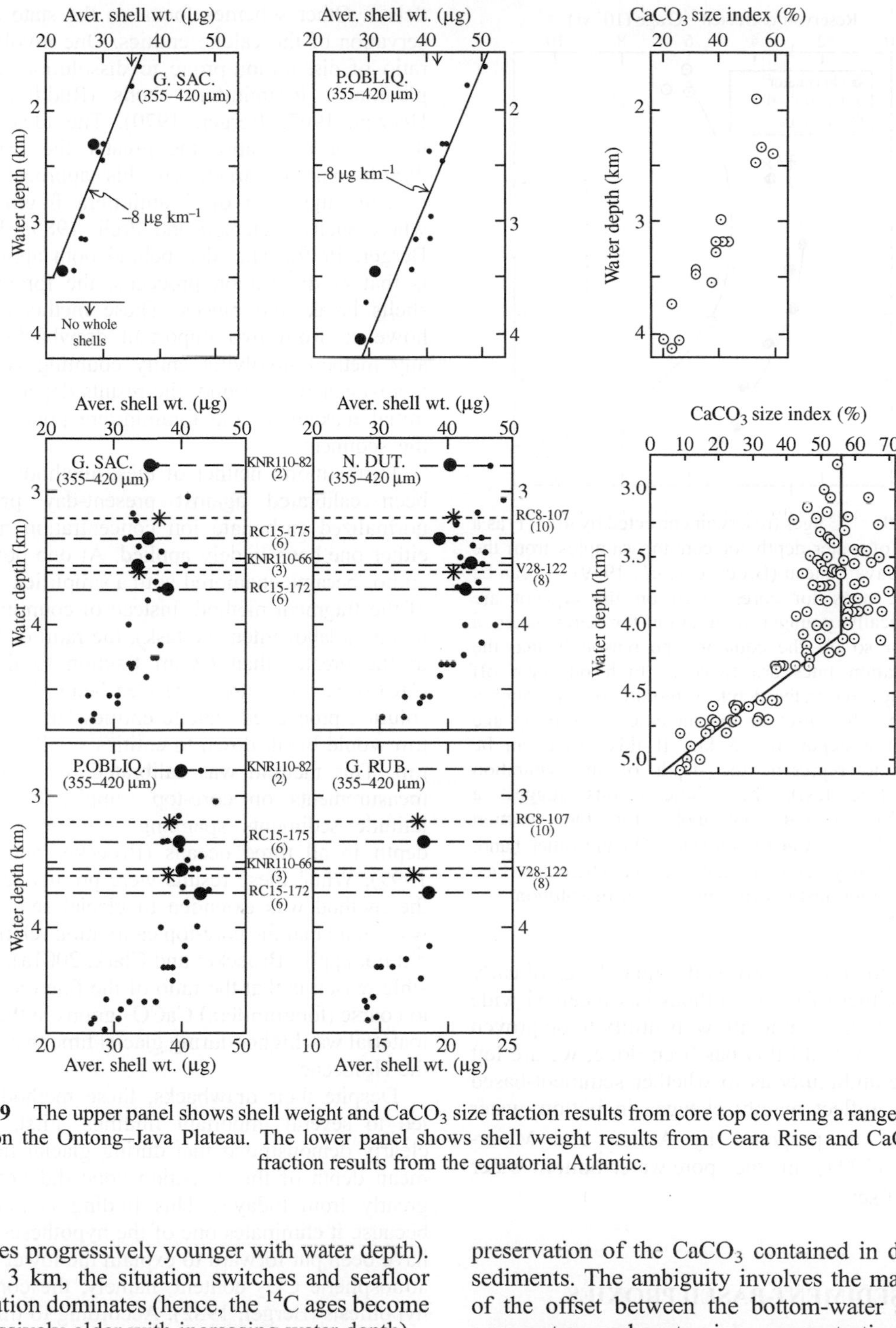

Figure 9 The upper panel shows shell weight and $CaCO_3$ size fraction results from core top covering a range of water depth on the Ontong–Java Plateau. The lower panel shows shell weight results from Ceara Rise and $CaCO_3$ size fraction results from the equatorial Atlantic.

becomes progressively younger with water depth). Below 3 km, the situation switches and seafloor dissolution dominates (hence, the ^{14}C ages become progressively older with increasing water depth).

10.6 DISSOLUTION IN THE PAST

One of the consequences of dissolution of $CaCO_3$ in pore waters is that it creates an ambiguity in all of the sediment-based methods for reconstructing past carbonate ion distributions in the deep sea. By "sediment-based" methods, one means methods involving some measure of the preservation of the $CaCO_3$ contained in deep-sea sediments. The ambiguity involves the magnitude of the offset between the bottom-water and the pore-water carbonate ion concentrations. The results obtained using any such methods can be applied to time trends in bottom-water carbonate ion concentration only if the pore-water–bottom-water offset is assumed to have remained nearly constant.

Fortunately, two methods have been proposed for which this ambiguity does not exist. One involves measurements of boron isotope ratios in benthic foraminifera (Sanyal *et al.*, 1995) and the other Zn/Cd ratios in benthic foraminifera

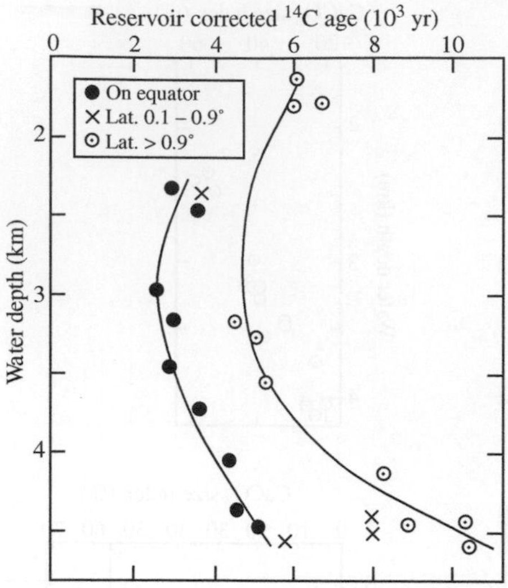

Figure 10 ^{14}C ages (reservoir corrected by 400 yr) as a function of water depth for core-top samples from the Ontong–Java Plateau (Broecker *et al.*, 1999). As can be seen, the ages for cores taken on the equator are systematically younger than those for cores taken a degree or so off the equator. The reason is that the sedimentation rates are twice as high on, than off the equator, while the depth of bioturbation is roughly the same. The onset of the increase in core-top age occurs at a depth of ~3 km. If this onset can be assumed to represent the depth of the saturation horizon (see text), then those results suggest a value ΔV of ~45 $cm^3\,mol^-$ for the reaction $CO_3^{2-} + Ca^{2+} \Leftrightarrow CaCO_3$ (calcite). On the other hand, this depth may represent the horizon where interface dissolution just matches pore-water dissolution.

(Marchitto *et al.*, 2000). Unfortunately, as of early 2003, neither of these methods has received wide enough application to allow its utility to be proven (see below). Until this has been done, we are left with the ambiguity as to whether sediment-based methods reflect mainly changes in bottom-water CO_3^{2-} or as proposed by Archer and Maier-Reimer (1994) in the pore-water-bottom-water CO_3^{2-} offset.

10.7 SEDIMENT-BASED PROXIES

A number of schemes have been proposed by which changes in the carbonate ion concentration in the deep sea might be reconstructed. The most obvious of these is the record of the $CaCO_3$ content of the sediment. Unfortunately, as already discussed, the $CaCO_3$ content depends on the ratio of the rain rate of $CaCO_3$ to that of silicate debris as well as on the extent of dissolution of the calcite. Unless quite large, changes in the extent of dissolution cannot be reliably isolated from changes in the composition of the raining debris. Other schemes focus on the state of preservation of the calcite entities. One involves the ratio of dissolution-prone to dissolution-resistant planktonic foraminifera shells (Ruddiman and Heezen, 1967; Berger, 1970). The idea is that the lower this ratio, the greater the extent of dissolution. A variant on this approach is to measure the ratio of foraminifera fragments to whole shells (Peterson and Prell, 1985; Wu and Berger, 1989). The idea behind both approaches is that as dissolution proceeds, the foraminifera shells break into pieces. These methods suffer, however, from two important drawbacks. First, any method involving entity counting is highly labor-intensive. Second, the results depend on the initial makeup of the foraminifera population in the sediment.

Furthermore, neither of these methods has yet been calibrated against present-day pressure-normalized carbonate ion concentration nor has either one been widely applied. At one point, the author became enamored with a simplified version of the fragment method. Instead of counting fragments (a labor-intensive task), the ratio of $CaCO_3$ in the greater than 63 μm fraction to the total $CaCO_3$ was measured, the idea being that as dissolution proceeded, calcite entities larger than 63 μm would break down to entities smaller than 63 μm. This method was calibrated by conducting measurements on core-top samples from low-latitude sediments spanning a range of water depth in all three oceans (Broecker and Clark, 1999). While these results were promising, when the method was extended to glacial sediment, it was found that the core-top calibration relationship did not apply (Broecker and Clark, 2001a). A possible reason is that the ratio of the fine (coccolith) to coarse (foraminifera) $CaCO_3$ grains in the initial material was higher during glacial time than during the Holocene.

Despite their drawbacks, these methods have led to several important findings. First, it was clearly demonstrated that during glacial time the mean depth of the transition zone did not differ greatly from today's. This finding is important because it eliminates one of the hypotheses which have been put forward to explain the lower glacial atmospheric CO_2 content, namely, the coral reef hypothesis (Berger, 1982). According to this idea, shallow-water carbonates (mainly coral and coralline algae) formed during the high-sea stands of periods of interglaciation would be eroded and subsequently dissolved during the low-sea stands of periods of glaciation, alternately reducing and increasing the sea's CO_3^{2-} concentration. But in order for this hypothesis to be viable, the transition zone would have to have been displaced downward by several kilometers during glacial time. Rather, the reconstructions suggest that the displacement was no more than a few hundred meters.

Two other findings stand out. First, as shown by Farrell and Prell (1989), at water depths in the 4 km range in the eastern equatorial Pacific, the impact of dissolution was greater during interglacials than during glacials (i.e., the transition zone was deeper during glacial time). Second, fragment-to-whole foraminifera ratios measured on a series of cores from various depths in the Caribbean Sea clearly demonstrate better preservation during glacials than interglacials (Imbrie, 1992). These findings have been confirmed by several investigators using a range of methods. Taken together, these findings gave rise to the conclusion that the difference between the depth of the transition zone in the Atlantic from that in the Pacific was somewhat smaller than now during glacial time. In addition, the existence of a pronounced dissolution event in the Atlantic Ocean at the onset of the last glacial cycle has been documented (Curry and Lohmann, 1986).

10.8 SHELL WEIGHTS

An ingenious approach to the reconstruction of the carbonate ion concentration in the deep sea was developed by WHOI's Pat Lohmann (1995). Instead of focusing on ratios of one entity to another, he developed a way to assess the extent of dissolution experienced by shells of a given species of planktonic foraminifera. He did this by carefully cleaning and sonification of the greater than 63 μm material sieved from a sediment sample. He then picked and weighed 75 whole shells of a given species isolated in a narrow size fraction range (usually 355–420 μm). In so doing, he obtained a measure of the average shell wall thickness. By obtaining shell weights for a given species from core-top samples spanning a range in water depth, Lohmann was able to show that the lower the pressure-normalized carbonate ion concentration, the smaller the whole shell weight (and hence the thinner the shell walls) (see Figure 11).

Lohmann's method seemingly has the advantage over those used previously in that no assumptions need to be made about the initial composition of the sediment. However, Barker and Elderfield (2002) make a strong case that the thickness of the foraminifera shell walls varies with growth conditions. They did so by weighing shells of temperate foraminifera from core tops from a number of locales in the North Atlantic. They found strong correlations between shell weight and both water temperature and carbonate ion concentration, the warmer the water and the higher its carbonate ion concentration, the thicker the shells. If, as Barker and Elderfield (2002) content, it is the carbonate ion concentration that drives the change in initial wall thickness, then glacial-age shells should have formed with thicker shells than do their Late Holocene counterparts. Fortunately, the ice-core-based atmospheric CO_2 record allows the carbonate ion concentration in the glacial surface waters to be reconstructed and hence presumably also the growth weight of glacial foraminifera. At this point, however, several questions remain unanswered. For example, does the dependence of shell weight on surface water carbonate ion concentration established for temperate species apply to tropical species? Perhaps the shell weight dependence flattens as the high carbonate ion concentrations characteristic of tropical surface waters are approached. Is carbonate ion concentration the only environmental parameter on which initial shell weights depends? As discussed below, there is reason to believe that the situation is perhaps more complicated.

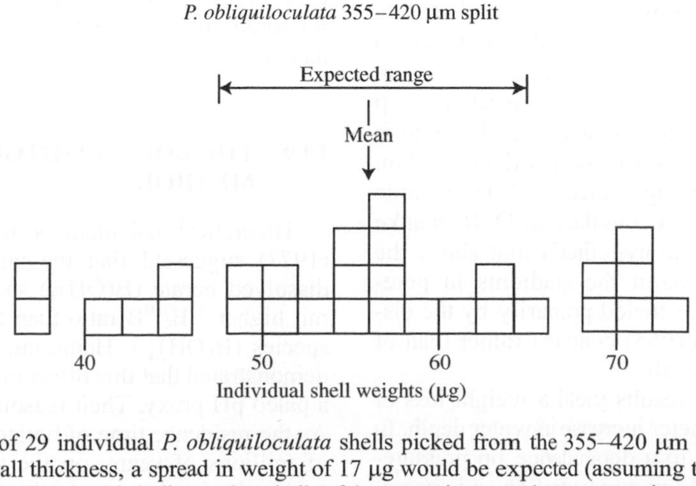

Figure 11 Weights of 29 individual *P. obliquiloculata* shells picked from the 355–420 μm size fraction. If all the shells had the same wall thickness, a spread in weight of 17 μg would be expected (assuming that shell weight varies with the square of size). Clearly, this indicates that shells of the same size must have a range in wall thickness. As can be seen, the observed range in weight is twice the expected range.

Table 1 The calcite-saturation carbonate ion concentration in cold seawater and the slope of this solubility as a function of water depth based on a 1 atm solubility of 45 μmol CO_3^{2-} kg^{-1} and a ΔV of 40 cm^3 mol^{-1}.

Water depth (km)	*Calcite sat.* (μmol CO_3^{2-} kg^{-1})	*Sol. slope* (μmol kg^{-1} km^{-1})	CO_3^{2-} *versus shell wt.* (μmol kg^{-1} μg^{-1})
1.5	58.8	10.3	1.4
2.0	64.2	11.4	1.5
2.5	70.2	12.5	1.6
3.0	76.7	13.6	1.8
3.5	83.8	14.9	2.0
4.0	91.6	16.3	2.1
4.5	100.1	17.9	2.3

Also shown is the slope of the shell-weight loss–carbonate ion concentration relationship for various water depths. The 0.7 μmol kg^{-1} km^{-1} increase in carbonate ion concentration in the Ontong–Java Plateau deep-water column is taken into account.

Lohmann's method has other drawbacks. Along with all sediment-based approaches, it suffers from an inability to distinguish changes in bottom-water carbonate ion concentration from changes in bottom-water to pore-water concentration offset. Shell thickness may also depend on growth rate and hence nutrient availability. Finally, a bias is likely introduced when dissolution becomes sufficiently intense to cause shell breakup, in which case the shells with the thickest walls are likely to be the last to break up. Nevertheless, Lohmann's method opens up a realm of new opportunities.

The sensitivity of shell-weight to pressure-normalized carbonate ion concentration (i.e., after correction for the increase in the solubility of calcite with water depth) was explored by determining the weight of Late Holocene shells from various water depths in the western equatorial Atlantic (Ceara Rise) and western equatorial Pacific (Ontong–Java Plateau). This strategy takes advantage of the contrast in carbonate ion concentration between the Atlantic and Pacific deep waters. As shown in Figure 9, shell weights for Ontong–Java Plateau samples do decrease with water depth and hence with decreasing pressure-normalized carbonate ion concentrations (Broecker and Clark, 2001a). However, the surprise is that there is no evidence of either weight loss or shell break at depths less than 4,200 m for Ceara Rise core-top samples. Rather, weight loss and shell breakup is evident only for samples from deeper than 4,200 m. This observation is in agreement with the benthic chamber results of R. A. Jahnke and D. B. Jahnke and hence supports the hypothesis that above the calcite saturation horizon the gradients in pore-water composition are fueled primarily by the dissolution of "Weyl" (1965) coatings rather than of the biogenic calcite itself.

The Ontong–Java results yield a weight loss of ~8 μg for each kilometer increase in water depth. In order to convert this to a dependence on pressure-normalized carbonate ion concentration, it is necessary to take into account the change in *in situ* carbonate ion concentration in the water column over the Ontong–Java Plateau water column (i.e., $CO_3^{2-} = 72 + 3(z-2)$ μmol kg^{-1}, where z is the water depth in km) and the pressure dependence of the saturation carbonate ion concentration. The latter depends on the difference in volume between Ca^{2+} and CO_3^{2-} ions when in solution and when they are bound into calcite. The relationship is as follows:

$$(CO_{3\,sat}^{2-})^z = (CO_{3\,sat}^{2-})^0 e^{PV/RT}$$

where the units of z are km, of ΔV, L mol^{-1}, of R, L atm, and T, K. If ΔV is re-expressed as cm^3 mol^{-1}, the relationship becomes

$$(CO_{3\,sat}^{2-})^z = (CO_{3\,sat}^{2-})^0 e^{z\Delta V/225}$$

where ΔV is the volume of the ions when bound into calcite minus that when they are dissolved in seawater. $(CO_3^{2-})^0$ is 45 mol kg^{-1} and while the exact value of ΔV remains uncertain, 40 cm^3 mol^{-1} fits most ocean observations (Peterson, 1966; Honjo and Erez, 1978; Ben-Yaakov and Kaplan, 1971; Ben-Yaakov *et al.*, 1974). Listed in Table 1 are the saturation concentrations based on this ΔV and the slope of the solubility as a function of water depth. Also given are estimates of the weight loss for foraminifera shells per unit decrease in carbonate ion concentration.

10.9 THE BORON ISOTOPE PALEO pH METHOD

Theoretical calculations by Kakihana *et al.* (1977) suggested that the uncharged species of dissolved borate ($B(OH)_3$) should have a 21 per mil higher $^{11}B/^{10}B$ ratio than that for the charged species ($B(OH)_4^-$). Hemming and Hanson (1992) demonstrated that this offset might be harnessed as a paleo pH proxy. Their reasoning was as follows. As the residence time of borate in seawater is tens of millions of years, on the timescale of glacial cycles the isotope composition of oceanic borate could not have changed. They further reasoned that

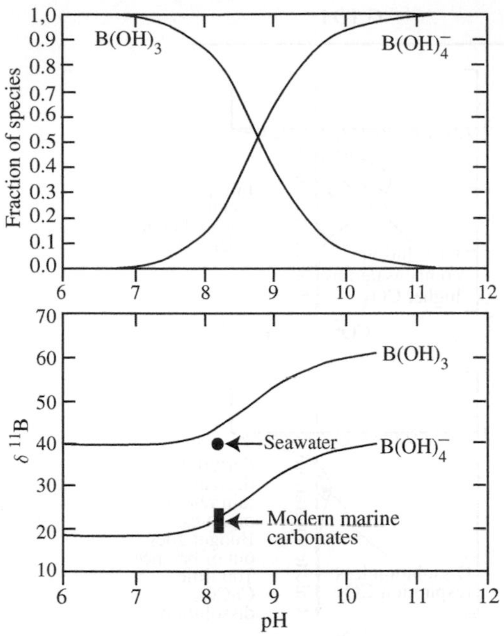

Figure 12 Speciation of borate in seawater as a function of pH (upper panel). Isotopic composition of the uncharged ($B(OH)_3$) and charged ($B(OH)_4^-$) species as a function of pH(lower panel) (Hemming and Hanson, 1992). As marine carbonates incorporate only the charged species, their isotopic composition is close to that of $B(OH)_4^-$.

it must be the charged borate species that is incorporated into marine $CaCO_3$ and hence marine calcite should have an isotope composition close to that of the charged species in seawater. This is important because as shown in Figure 12 the isotopic composition of the charged species must depend on the pH of the seawater. The higher the pH, the larger the fraction of the borate in the charged form and hence the closer its isotopic composition will be to that for bulk seawater borate. In contrast, for pH values the isotopic composition of the residual amount of charged borate must approach a value 21 per mil lower than that for bulk seawater borate. Working with a graduate student, Abhijit Sanyal, Hemming applied his method to foraminifera shells and demonstrated that indeed foraminifera shells record pH (Sanyal *et al.*, 1995). Benthic foraminifera had the expected offset from planktonics. Glacial age *G. sacculifer*, as dictated by the lower glacial atmospheric CO_2 content, recorded a pH about 0.15 units higher than that for Holocene shells. Sanyal went on to grow planktonic foraminifera shells at a range of pH values (Sanyal *et al.*, 1996, 2001). He also precipitated inorganic $CaCO_3$ at a range of pH values (Sanyal *et al.*, 2000). These results yielded the expected pH dependence of boron isotope composition. However, they also revealed sizable species-to-species offsets (as do the carbon and oxygen isotopic compositions).

The waterloo of this method came when glacial-age benthic foraminifera were analyzed. The results suggested that the pH of the glacial deep ocean was 0.3 units greater than today (Sanyal *et al.*, 1995). This corresponds to a whopping 90 $\mu mol\,kg^{-1}$ increase in carbonate ion concentration. The result was exciting because, if correct, the lowering of the CO_2 content of the glacial atmosphere would be explained by a whole ocean carbonate ion concentration change. But this result was clearly at odds with reconstructions of the depth of the glacial transition zone. Such a large increase in deep-water carbonate ion concentration would require that it deepened by several kilometers. Clearly, it did not. Archer and Maier-Reimer (1994) proposed a means by which this apparent disagreement might be explained. They postulated that if during glacial time the release of metabolic CO_2 to sediment pore waters (relative to the input of $CaCO_3$) was larger than today's, this would have caused a shoaling of the transition zone and thereby thrown the ocean's $CaCO_3$ budget out of kilter. Far too little $CaCO_3$ would have been buried relative to the ingredient input. The result would be a steady increase in the ocean's carbonate ion inventory (see Figure 13) and a consequent progressive deepening of the transition zone. This deepening would have continued until a balance between input and loss was once again achieved. In so doing, a several kilometer offset between the depth of the saturation horizon and the depth of the transition zone would have been created. However, this explanation raised three problems so serious that the boron isotope-based deep-water pH change has fallen into disrepute. First, it required that the change in glacial ecology responsible for the increase in the rain of organic matter be globally uniform. Otherwise, there would have been very large "wrinkles" in the depth of the glacial transition zone. No such wrinkles have been documented. Second, at the close of each glacial period when the flux of excess organic matter was shut down, there must have been a prominent global preservation event. In order to restore the saturation horizon to its interglacial position, an excess over ambient $CaCO_3$ accumulation of $\sim$3 g cm^{-2} would have to have occurred over the entire seafloor. It would be surprising if some residue from this layer were not to be found in sediments lining the abyssal plains. It has not. These sediments have no more than 0.2% by weight $CaCO_3$. In other words, of the 3 g cm^{-2} deposited during the course of the carbonate ion drawdown, almost nothing remains. Finally, based on model simulations, Sigman *et al.* (1998) have shown that it is not possible to maintain for tens of thousands of years a several-kilometer separation

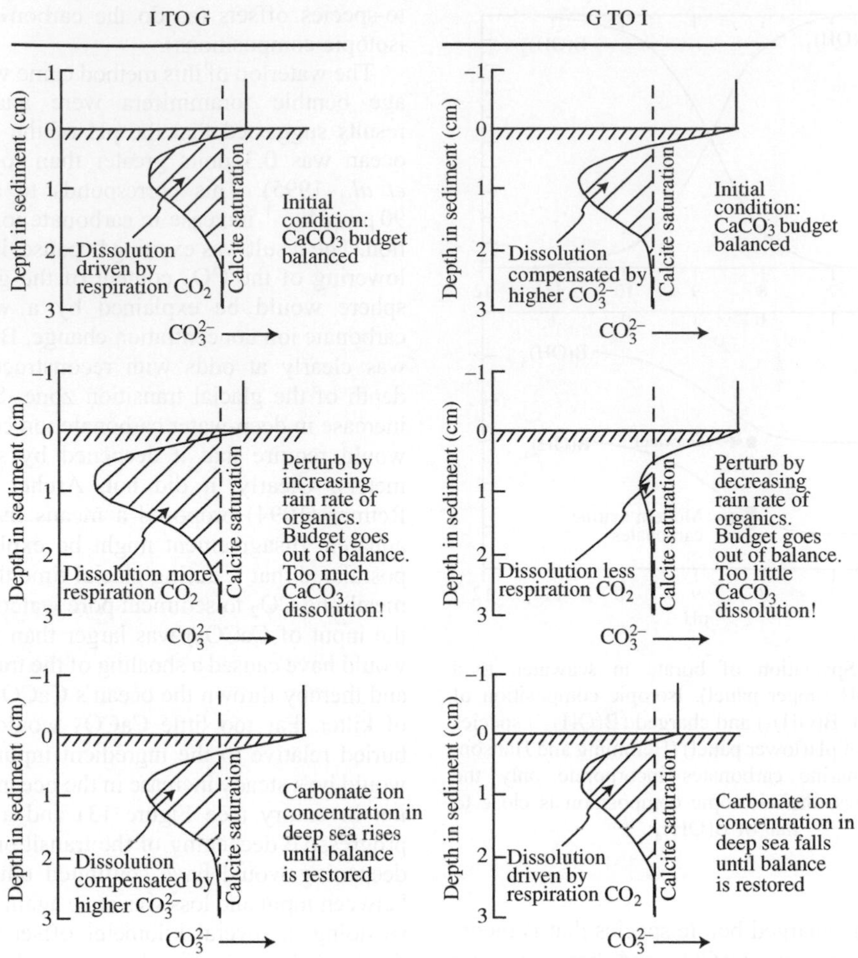

Figure 13 Shown on the left is the sequence of events envisioned by Archer and Maier-Reimer (1994) for the transition from interglacial (I) to glacial (G) conditions. An increase in respiration CO_2 release to the sediment pore waters enhances calcite dissolution, thereby unbalancing the $CaCO_3$ budget. This imbalance leads to a buildup in CO_3^{2-} ion concentration in the deep sea until it compensates for the extra respiration CO_2. On the right is the sequence of events envisioned for the transition from G to I conditions. The input of excess respiration CO_2 to the sediments ceases, thereby reducing the rate of calcite dissolution. This leads to an excess accumulation of $CaCO_3$ on the seafloor and hence to a reduction in carbonate ion concentration which continues until steady state is reestablished.

between the saturation horizon and the transition zone.

This "waterloo" was unfortunate for the author considers the boron method to be basically sound and potentially extremely powerful. The answer to the benthic enigma may lie in species-to-species differences in the boron isotope "vital" effect for benthic foraminifera. The measurement method use by Sanyal *et al.* (1996) required a large number of benthic shells in order to get enough boron to analyze. This created a problem because, as benthics are rare among foraminifera shells, mixed benthics rather than a single species were analyzed. If the boron isotope pH proxy is to become an aid to deep-ocean studies, then techniques requiring smaller amounts of boron will have to be created. There also appears to be a problem associated with variable isotopic fractionation of boron during thermal ionization. As this fractionation depends on the ribbon temperature and perhaps other factors, it may introduce biases in the results for any particular sample. Hopefully, a more reproducible means of ionizing boron will be found.

10.10 Zn/Cd RATIOS

The other bottom-water CO_3^{2-} ion concentration proxy is based on the Zn/Cd ratio in benthic foraminifera shells. As shown by Marchitto *et al.* (2000), the distribution coefficient of zinc between shell and seawater depends on CO_3^{2-} ion concentration, such that the lower the carbonate ion concentration, the large the Zn/Cd ratio in the foraminifera shell. Assuming that the Zn/Cd ratio in seawater was the same during the past as it is today, the ratio of these two trace elements should serve as a paleo carbonate ion proxy. However, there are problems to be overcome. For example, in today's ocean, zinc correlates with silica and cadmium with phosphorus. As silica is 10-fold enriched in deep Pacific water relative to deep Atlantic water while phosphorus is only twofold enriched, differential redistribution of silica and phosphorus in the glacial ocean poses a potential bias. However, as at high carbonate supersaturation the distribution coefficient for zinc flattens out, it may be possible to use measurements on benthic foraminifera from sediments bathed in highly supersaturated waters to sort this out. But, as is the case for the boron isotope proxy, much research will be required before reconstructions based on Zn/Cd ratios can be taken at face value.

10.11 DISSOLUTION AND PRESERVATION EVENTS

There are several mechanisms that might lead to carbonate ion concentration transients at the beginning and end of glacial periods. One such instigator is changes in terrestrial biomass. Shackleton (1997) was the first to suggest that the mass of carbon stored as terrestrial biomass was smaller during glacial than during interglacial periods. He reached this conclusion based on the fact that measurements on glacial-age benthic foraminifera yielded lower $\delta^{13}C$ values than those for their interglacial counterparts. Subsequent studies confirmed that this was indeed the case and when benthic foraminifera ^{13}C results were averaged over the entire deep sea, it was found that the ocean's dissolved inorganic carbon had a $^{13}C/^{12}C$ ratio 0.35 ± 0.10 per mil lower during glacial time than during the Holocene (Curry *et al.*, 1988). If this decrease is attributed to a lower inventory of wood and humus, then the magnitude of the glacial biomass decrease would have been 500 ± 150 Gt of carbon. The destruction of this amount of organic material at the onset of a glacial period would create a 20 μmol kg^{-1} drop in the ocean's CO_3^{2-} concentration and hence produce a calcite dissolution event. Correspondingly, the removal of this amount of CO_2 from the ocean–atmosphere reservoir at the onset of an interglacial period would raise the carbonate ion concentration by 20 μmol kg^{-1} and hence produce a calcite preservation event. This assumes that the time over which the biomass increase occurred was short compared to the CO_3^{2-} response (i.e., ~5,000 yr). If this is not the case, then the magnitude of the carbonate ion changes would be correspondingly smaller.

Another possible instigator of such transients was proposed by Archer and Maier-Reimer (1994). Their goal was to create a scenario by which the lower CO_2 content of the glacial atmosphere might be explained. As already mentioned, it involved a higher ratio of organic carbon to $CaCO_3$ carbon in the material raining to the deep-sea floor during glacial times than during interglacial times, and hence an intensification of pore-water dissolution. As in the case for the terrestrial biomass change, such an increase would have thrown the ocean's carbon budget temporarily out of kilter. The imbalance would have been remedied by a buildup of carbonate ion concentration at the onset of glacials and a drawdown of carbonate ion concentration at the onset of interglacials (see Figure 14). Hence, it would also lead to a dissolution event at the onset of glacial episodes and a preservation event at the onset of interglacial episodes. Were the changes in organic to $CaCO_3$ rain proposed by Archer and Maier-Reimer to have explained the entire glacial to interglacial CO_2 change, then the magnitude of the transients would have been ~4 times larger than that resulting from 500 Gt C changes in terrestrial biomass.

Regardless of their origin, these dissolution events and preservation events would be short-lived. As they would disrupt the balance between burial and supply, they would be compensated by either decreased or increased burial of $CaCO_3$ and

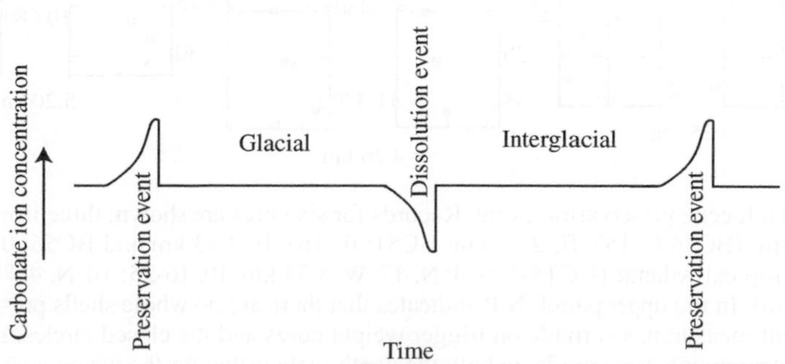

Figure 14 Idealized scenario for carbonate ion concentration changes associated with dissolution and preservation events.

the balance would be restored with a constant time of ~5,000 yr (see Figure 14).

Clear evidence for the compensation for an early Holocene preservation event is seen in shell weight results form a core from 4.04 km depth on the Ontong–Java Plateau in the western equatorial Pacific (see Figure 15). A drop in the weight of *P. obliquiloculata* shells of 11 μg between about 7,500 yr ago and the core-top bioturbated zone (average age 4,000 yr) requires a decrease in carbonate ion concentration between 7,500 y ago and today (see Table 1). This Late Holocene CO_3^{2-} ion concentration drop is characterized by an up-water column decrease in magnitude becoming imperceptible at 2.31 km. It is interesting to note that during the peak of the preservation event, the shell weights showed only a small decrease with water depth (see Figure 15), suggesting either that the pressure effect on calcite solubility was largely compensated by an increase with depth in the *in situ* carbonate ion concentration or that the entire water column was supersaturated with respect to calcite.

In the equatorial Atlantic only in the deepest core (i.e., that from 5.20 km) is the Late Holocene intensification of dissolution strongly imprinted. As in this core no whole shells are preserved, the evidence for an Early Holocene preservation event

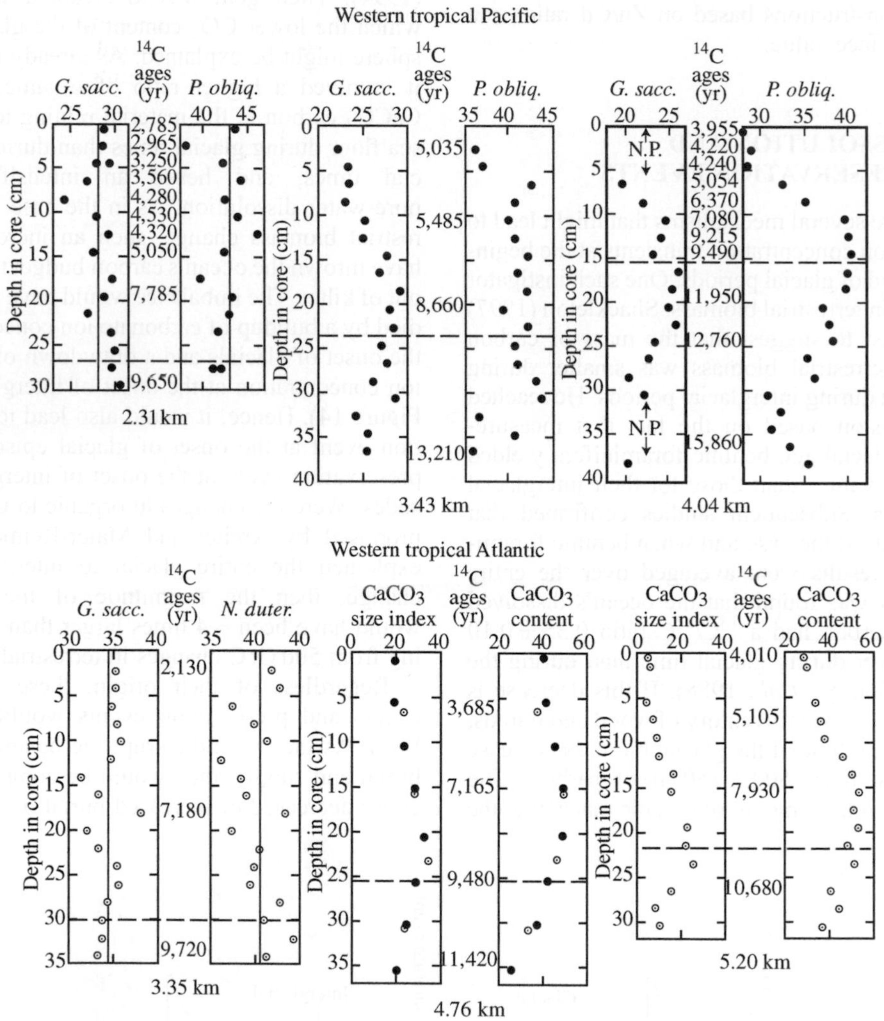

Figure 15 The Early Holocene preservation event. Records for six cores are shown: three from Ontong–Java Plateau in western tropical Pacific (BC36:0° 158°E, 2.31 km; BC51: 0° 161°E, 3.43 km and BC56: 0° 162°E, 4.04 km) and three from the western tropical Atlantic (RC15-175: 4°N, 47°W, 3.35 km; RC16-55: 10°N, 45°W, 4.76 km and RC17-30: 11°N, 41°W, 5.20 km). In the upper panel, N.P. indicates that there are no whole shells present. In the lower panel, the open circles represent measurements made on trigger weight cores and the closed circles, measurements made on piston cores. Where measurements were made on both the depth scale is that for the piston core, and the trigger weight sample depths have been multiplied by a factor of 1.5 to compensate for foreshortening (Broecker *et al.*, 1993). The shell weights are in μg, the size index is the percentage of the $CaCO_3$ contained in the >63 μm fraction and the calcium carbonate content is in percent. The dashed lines show the depth of the 9,500 B.P. radiocarbon-age horizon.

is based on $CaCO_3$ size-index and $CaCO_3$ content measurements. As can be seen in Figure 15, both show a dramatic decrease starting about 7,500 yr ago. As for the Pacific, the magnitude of the imprint decreases up-water column, becoming imperceptible at 3.35 km.

If either the biomass or the respiration CO_2 mechanisms are called upon, the magnitude of the Early Holocene CO_3^{2-} maximum must have been uniform throughout the deep ocean. The most straightforward explanation for the up-water column reduction in the magnitude of the preservation event is that at mid-depths; the sediment pore waters are presently close to saturation with respect to calcite. Hence, the Early Holocene maximum in deep-sea CO_3 ion concentration pushed them into the realm of supersaturation. If so, there is no need to call on a depth dependence for the magnitude of the preservation event.

The post-8,000-year-ago decrease in CO_3^{2-} ion concentration of 23 $\mu mol\,kg^{-1}$ required to explain the Late Holocene 11 μg decrease in *P. obliquiloculata* shell weights observed in the deepest Ontong–Java Plateau core is twice too large to be consistent with the 20 ppm increase in atmospheric CO_2 content over this time interval (Indermühle *et al.*, 1999). The significance of this remains unknown.

In the equatorial Atlantic $CaCO_3$ content, $CaCO_3$ size-index and shell-weight measurements reveal three major dissolution events, one during marine isotope stage 5d, one during 5b and one during stage 4 (see Figures 16(a) and (b)). As these events are only weakly imprinted on Pacific sediments, it appears that a major fraction of the carbonate ion reduction was the result of enhanced penetration into the deep Atlantic of low carbonate ion concentration Southern Ocean water. If this conclusion proves to be correct, then it suggests that the balance between the density of deep waters formed in the northern Atlantic and those formed in the Southern Ocean is modulated by the strength of northern hemisphere summer insolation (i.e., by Milankovitch cycles).

10.12 GLACIAL TO INTERGLACIAL CARBONATE ION CHANGE

In addition to the preservation and dissolution event transients, there were likely carbonate ion concentration changes that persisted during the entire glacial period. These changes could be placed in two categories. One involves a change in the average CO_3^{2-} concentration of the entire deep sea necessary to compensate for a change in the ratio of calcite production by marine organisms to ingredient supply. The other involves a redistribution of carbonate ion within the deep sea due to a redistribution of phosphate (and hence also of respiration CO_2) and/or to a change in the magnitude of the flux of CO_2 through the atmosphere from the Southern Ocean to the northern Atlantic.

Based on shell-weight measurements, Broecker and Clark (2001c) attempted to reconstruct the depth distribution of carbonate ion concentration during late glacial time for the deep equatorial Atlantic Ocean and Pacific Ocean. At the time their paper was published, these authors were unaware of the dependence of initial shell weight on carbonate ion concentration in surface water established by Barker and Elderfield (2002) for temperate species. Since during the peak glacial time the atmosphere's p_{CO_2} was ~80 ppm lower than during the Late Holocene, the carbonate ion concentration in tropical surface waters must have been 40–50 $\mu mol\,kg^{-1}$ higher at that time. Based on the Barker and Elderfield (2002) trend of ~1 μg increase in shell weight per 9 $\mu mol\,kg^{-1}$ increase in carbonate ion concentration, this translates to an 8 μg heavier initial shell weights during glacial time. Figure 17 shows, while this correction does not change the depth dependence or interocean concentration difference, it does greatly alter the magnitude of the change. In fact, were the correction made, it would require that the carbonate ion concentration in virtually the entire glacial deep ocean was lower during glacial time than during interglacial time. For the deep Pacific Ocean and the Indian Ocean, this flies in the face of all previous studies which conclude that dissolution was less intense during periods of glaciation than during periods of interglaciation. However, as the Broecker and Clark study concentrates on the Late Holocene while earlier studies concentrate on previous periods of interglaciation, it is possible that the full extent of the interglacial decrease in carbonate ion during the present interglacial has not yet been achieved. Of course, it is also possible that significant thickening of foraminifera shells during glacial time did not occur. Until this matter can be cleared up, reconstruction of glacial-age deep-sea carbonate ion concentrations must remain on hold.

10.13 NEUTRALIZATION OF FOSSIL FUEL CO_2

The ultimate fate of much of the CO_2 released to the atmosphere through the burning of coal, oil, and natural gas will be to react with the $CaCO_3$ stored in marine sediments (Broecker and Takahashi, 1977; Sundquist, 1990; Archer *et al.*, 1997). The amount of $CaCO_3$ available for dissolution at any given place on the seafloor depends on the calcite content in the sediment and the depth to which sediments are stirred by organisms. The former is now well mapped and the latter has been documented in many places by radiocarbon

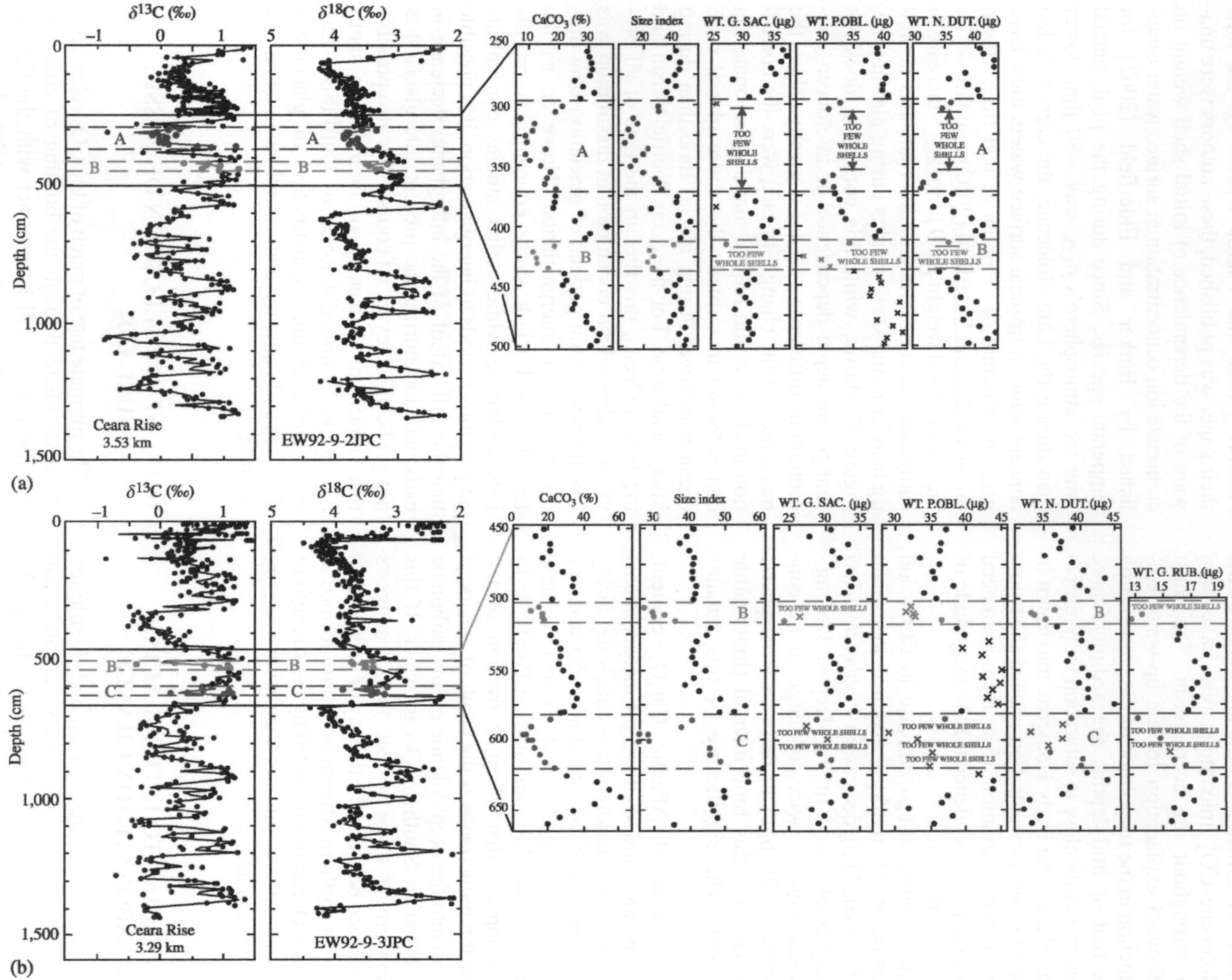

Figure 16 (A) $CaCO_3$, size index and shell weight results for a portion of jumbo pistoncore EW92-9-2 (a) and EW92-9-3 (b) from the northern flank of the Ceara Rise. The results in (a) document dissolution events A (stage 4) and B (stage 5b). (B) The results in (b) document dissolution events B (stage 5b) and C (stage 5d). The ^{18}O and ^{13}C records for benthic foraminifera are from Curry (1996). The *x*s represent samples in which only 15–30 whole shells were found.

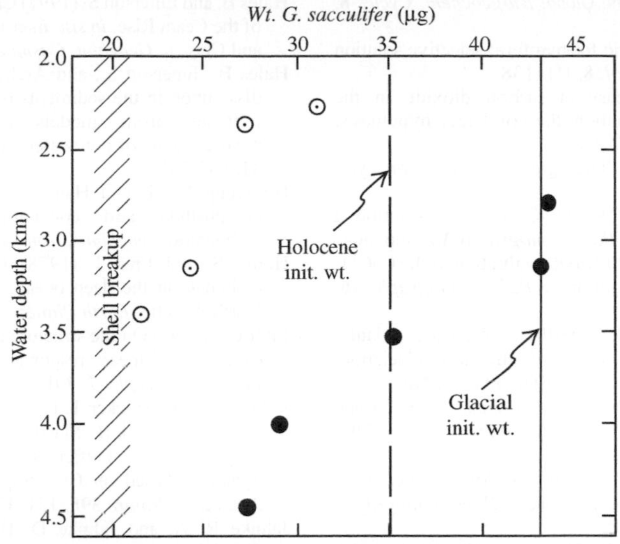

Figure 17 Average shell weights for whole *G. sacculifer* shells of glacial age as a function water depth. The open circles are for cores from the Ontong–Java Plateau, and the closed circles are for cores from the equatorial Atlantic. The vertical dashed line is the initial *G. sacculifer* weight obtained from measurements on Holocene samples from the Ceara Rise. The solid vertical line is the estimated initial weight for glacial-age *G. sacculifer* based on the assumption that the carbonate ion concentration dependence of initial shell weight established by Barker and Elderfield (2002) applies to tropical species.

measurements. The amount of $CaCO_3$ available for dissolution at any given site is given by

$$\sum CaCO_3 = \frac{h\rho f_c}{1-f_c}$$

where h is the depth of bioturbation, ρ is the water-free sediment density, and f_c is the weight-fraction calcite. The high-$CaCO_3$ sediments that drape the oceans' ridges and plateaus typically have ~90% $CaCO_3$ and a water-free density of 1 g cm^{-3}. The bioturbation depth in these sediments averages 8 cm. Hence, the upper limit on amount of $CaCO_3$ available for dissolution in such a sediment is 72 g cm^{-2}. As roughly one quarter of the seafloor is covered with calcite-rich sediments, this corresponds to $\sim 6.3 \times 10^{19}$ g $CaCO_3$ (i.e., 7,560 Gt C). This amount could neutralize 6.3×10^{17} mol of fossil fuel CO_2. This amount exceeds the combined oceanic inventory of dissolved CO_3^{2-} (1.6×10^{17} mol) and of dissolved HBO_3^- (0.8×10^{17} mol). It is comparable to the amount of recoverable fossil fuel carbon.

I say 'upper limit' because once this amount of $CaCO_3$ has been dissolved, the upper 8 cm of the sediment would consist entirely of a noncarbonate residue. As molecular diffusion through such a thick residue would be extremely slow, the rate of dissolution of $CaCO_3$ stored beneath this $CaCO_3$-free cap would be minuscule, and further neutralization would be confined to the fall to the seafloor of newly formed $CaCO_3$.

The rate of this dissolution of the $CaCO_3$ stored in the uppermost sediment will depend not only on the magnitude of the reduction of the deep ocean's CO_3^{2-} content, but also on the rate at which the insoluble residue is stirred into the sediment. This bioturbation not only homogenizes the mixed layer, but is also exhumes $CaCO_3$ from beneath the mixed layer.

REFERENCES

Archer D. and Maier-Reimer E. (1994) Effect of deep-sea sedimentary calcite preservation on atmospheric CO_2 concentration. *Nature* **367**, 260–263.

Archer D., Emerson S., and Reimers C. (1989) Dissolution of calcite in deep-sea sediments: pH and O_2 microelectrode results. *Geochim. Cosmochim. Acta* **53**, 2831–2845.

Archer D., Kheshgi H., and Maier-Reimer E. (1997) Multiple timescales for neutralization of fossil fuel CO_2. *Geophys.Res. Lett.* **24**, 405–408.

Archer D., Kheshgi H., and Maier-Riemer E. (1998) Dynamics of fossil fuel CO_2 neutralization by marine $CaCO_3$. *Global Biogeochem. Cycles* **12**, 259–276.

Barker S. and Elderfield H. (2002) Response of for aminiferal calcification to glacial–interglacial changes in atmospheric carbon dioxide. *Science* **297**, 833–836.

Ben-Yaakov S. and Kaplan I. R. (1971) Deep sea *in situ* calcium carbonate saturometry. *J. Geophys. Res.* **76**, 772–781.

Ben-Yaakov S., Ruth E., and Kaplan I. R. (1974) Carbonate compensation depth: relation to carbonate solubility in ocean waters. *Science* **184**, 982–984.

Berelson W. M., Hammond D. E., and Cutter G. A. (1990) *In situ* measurements of calcium carbonate dissolution rates in deep-sea sediments. *Geochim. Cosmochim. Acta* **54**, 3013–3020.

Berelson W. M., Hammond D. E., McManus J., and Kilgore T. E. (1994) Dissolution kinetics of calcium carbonate in

equatorial Pacific sediments. *Global Biogeochem. Cycles* **8**, 219–235.
Berger W. H. (1970) Planktonic foraminifera: selective solution and the lysocline. *Mar. Geol.* **8**, 111–138.
Berger W. H. (1982) Increase of carbon dioxide in the atmosphere during deglaciation: the coral reef hypothesis. *Naturwissenschaften* **69**, 87–88.
Broecker W. S. (1991) The great ocean conveyor. *Oceanography* **4**, 79–89.
Broecker W. S. and Clark E. (1999) $CaCO_3$ size distribution: a paleo carbonate ion proxy. *Paleoceanography* **14**, 596–604.
Broecker W. S. and Clark E. (2001a) Reevaluation of the $CaCO_3$ size index paleocarbonate ion proxy. *Paleoceanography* **16**, 669–771.
Broecker W. S. and Clark E. (2001b) A dramatic Atlantic dissolution event at the onset of the last glaciation. *Geochem. Geophys. Geosys.* **2**, 2001GC000185, Nov. 2.
Broecker W. S. and Clark E. (2001c) Glacial to Holocene redistribution of carbonate ion in the deep sea. *Science* **294**, 2152–2155.
Broecker W. S. and Clark E. (2001d) An evaluation of Lohmann's foraminifera-weight index. *Paleoceanography* **16**, 531–534.
Broecker W. S. and Clark E. (2002) A major dissolution event at the close of MIS 5e in the western equatorial Atlantic. *Geochem. Geophys. Geosys.* 3(2) 10.1029/2001GC000210.
Broecker W. S. and Clark E. (2003) Pseudo-dissolution of marine calcite. *Earth Planet. Sci. Lett.* **208**, 291–296.
Broecker W. S. and Peng T.-H. (1993) Interhemispheric transport of ΣCO_2 through the ocean. In *The Global Carbon Cycle.* NATO SI Series (ed. M. Heimann). Springer, vol. 115, pp. 551–570.
Broecker W. S. and Sutherland S. (2000) The distribution of carbonate ion in the deep ocean: support for a post-Little Ice Age change in Southern Ocean ventilation. *Geochem. Geophys. Geosys.* **1**, 2000GC000039, July 10.
Broecker W. S. and Takahashi T. (1977) Neutralization of fossil fuel CO_2 by marine calcium carbonate. In *The Fate of Fossil Fuel CO_2 in the Oceans* (eds. N. R. Andersen and A. Malahoff). Plenum, New York, pp. 213–248.
Broecker W. S. and Takahashi T. (1978) The relationship between lysocline depth and *in situ* carbonate ion concentration. *Deep-Sea Res.* **25**, 65–95.
Broecker W. S., Lao Y., Klas M., Clark E., Bonani G., Ivy S., and Chen C. (1993) A search for an early Holocene $CaCO_3$ preservation event. *Paleoceanography* **8**, 333–339.
Broecker W. S., Clark E., Hajdas I., Bonani G., and McCorkle D. (1999) Core-top ^{14}C ages as a function of water depth on the Ontong-Java Plateau. *Paleoceanography* **14**, 13–22.
Curry W. B. (1996) Late Quaternary deep circulation in the western equatorial Atlantic. In *The South Atlantic: Present and Past Circulation* (eds. G. Wefer, W. H. Berger, G. Siedler, and D. J. Webb). Springer, New York, pp. 577–598.
Curry W. B. and Lohmann G. P. (1986) Late Quaternary carbonate sedimentation at the Sierra Leone rise (eastern equatorial Atlantic Ocean). *Mar. Geol.* **70**, 223–250.
Curry W. B., Duplessy J. C., Labeyrie L. D., and Shackleton N. J. (1988) Changes in the distribution of $\delta^{13}C$ of deep water ΣCO_2 between the last glaciation and the Holocene. *Paleoceanography* **3**, 317–341.
Emerson S. and Bender M. (1981) Carbon fluxes at the sediment–water interface of the deep-sea: calcium carbonate preservation. *J. Mar. Res.* **39**, 139–162.
Farrell J. W. and Prell W. L. (1989) Climatic change and $CaCO_3$ preservation: an 800,000 year bathymetric reconstruction from the central equatorial Pacific Ocean. *Paleoceanography* **4**, 447–466.
Hales B. and Emerson S. (1996) Calcite dissolution in sediments of the Ontong-Java Plateau: *in situ* measurements of porewater O_2 and pH. *Global Biogeochem. Cycles* **5**, 529–543.
Hales B. and Emerson S. (1997) Calcite dissolution in sediments of the Ceara Rise: *in situ* measurements of porewater O_2, pH and $CO_{2(aq)}$. *Geochim. Cosmochim. Acta* **61**, 501–514.
Hales B., Emerson S., and Archer D. (1994) Respiration and dissolution in the sediments of the western North Atlantic: estimates from models of *in situ* microelectrode measurements of porewater oxygen and pH. *Deep-Sea Res.* **41**, 695–719.
Hemming N. G. and Hanson G. N. (1992) Boron isotopic composition and concentration in modern marine carbonates. *Geochim. Cosmochim. Acta* **56**, 537–543.
Honjo S. and Erez J. (1978) Dissolution rates of calcium carbonate in the deep ocean: an *in situ* experiment in the North Atlantic. *Earth Planet. Sci. Lett.* **40**, 226–234.
Imbrie J. (1992) On the structure and origin of major glaciation cycles: I. Linear responses to Milankovitch forcing. *Paleoceanography* **7**, 701–738.
Indermühle A., Stocker T. F., Joos F., Fischer H., Smith H. J., Wahlen M., Deck B., Mastroianni D., Techumi J., Blunier T., Meyer R., and Stauffer B. (1999) Holocene carbon-cycle dynamics based on CO_2 trapped in ice at Taylor Dome, Antarctica. *Nature* **398**, 121–126.
Jahnke R. A. and Jahnke D. B. (2002) Calcium carbonate dissolution in deep-sea sediments: implications of bottom water saturation state and sediment composition. *Geochim. Cosmochim. Acta* (submitted for publication, November, 2001).
Kakihana H., Kotaka M., Satoh S., Nomura M., and Okamoto M. (1977) Fundamental studies on the ion exchange separation of boron isotopes. *Bull. Chem. Soc. Japan* **50**, 158–163.
Lohmann G. P. (1995) A model for variation in the chemistry of planktonic foraminifera due to secondary calcification and selective dissolution. *Paleoceanography* **10**, 445–457.
Marchitto T. M., Jr., Curry W. B., and Oppo D. W. (2000) Zinc concentrations in benthic foraminifera reflect seawater chemistry. *Paleoceanography* **15**, 299–306.
Martin W. R. and Sayles F. L. (1996) $CaCO_3$ dissolution in sediments of the Ceara Rise, western equatorial Atlantic. *Geochim. Cosmochim. Acta* **60**, 243–263.
Martin W. R., McNichol A. P., and McCorkle D. C. (2000) The radiocarbon age of calcite dissolving at the sea floor: estimates from pore water data. *Geochim. Cosmochim. Acta* **64**, 1391–1404.
Milliman J. D., Troy P. J., Balch W. M., Adams A. K., Li Y.-H., and Mackenzie F. T. (1999) Biologically mediated dissolution of calcium carbonate above the chemical lysocline? *Deep-Sea Res.* **46**, 1653–1669.
Peterson M. N. A. (1966) Calcite: rates of dissolution in a vertical profile in the central Pacific. *Science* **154**, 1542–1544.
Peterson L. C. and Prell W. L. (1985) Carbonate preservation and rates of climatic change: an 800 kyr record from the Indian Ocean. In *The Carbon Cycle and Atmospheric CO_2: Natural Variations Archean to Present,* Geophys. Monogr. Ser. 32 (eds. E. T. Sundquist and W. S. Broecker) pp. 251–270.
Reimers C. E., Jahnke R. A., and Thomsen L. (2001) In situ sampling in the benthic boundary layer. In *The Benthic Boundary Layer: Transport Processes and Biogeochemistry* (eds. B. P. Boudreau and B. B. Jørgensen). Oxford University Press, Oxford, pp. 245–268.
Ruddiman W. F. and Heezen B. C. (1967) Differential solution of planktonic foraminifera. *Deep-Sea Res.* **14**, 801–808.
Sanyal A., Hemming N. G., Hanson G. N., and Broecker W. S. (1995) Evidence for a higher pH in the glacial ocean from boron isotopes in foraminifera. *Nature* **373**, 234–236.
Sanyal A., Hemming N. G., Broecker W. S., Lea D. W., Spero H. J., and Hanson G. N. (1996) Oceanic pH control on the boron isotopic composition of foraminifera: evidence from culture experiments. *Paleoceanography* **11**, 513–517.
Sanyal A., Nugent M., Reeder R. J., and Bijma J. (2000) Seawater pH control on the boron isotopic composition of

calcite: evidence from inorganic calcite precipitation experiments. *Geochim. Cosmochim. Acta* **64**, 1551–1555.

Sanyal A., Bijma J., Spero H., and Lea D. W. (2001) Empirical relationship between pH and the boron isotopic composition of *Globigerinoides sacculifer:* implications for the boron isotope paleo-pH proxy. *Paleoceanography* **16**, 515–519.

Sayles F. L. (1985) $CaCO_3$ solubility in marine sediments: evidence for equilibrium and non-equilibrium behavior. *Geochim. Cosmochim. Acta.* **49**, 877–888.

Shackleton N. J. (1977) Tropical rainforest history and the equatorial Pacific carbonate dissolution cycles. In *The Fate of Fossil Fuel CO_2 in the Oceans* (eds. N. R. Anderson and A. Malahoff). Plenum, New York, pp. 401–428.

Sigman D. M., McCorkle D. C., and Martin W. R. (1998) The calcite lysocline as a constraint on glacial/interglacial low-latitude production changes. *Global Biogeochem. Cycles* **12**, 409–427.

Sundquist E. T. (1990) Long-term aspects of future atmospheric CO_2 and sea-level changes. In *Sea-level Change.* National Research Council Studies in Geophysics (ed. R. Revelle). National Academy Press, Washington, pp. 193–207.

Wenzhöfer F., Adler M., Kohls O., Hensen C., Strotmann B., Boehme S., and Schulz H. D. (2001) Calcite dissolution driven by benthic mineralization in the deep-sea: *in situ* measurements. *Geochim. Cosmochim. Acta* **65**, 2677–2690.

Weyl P.K. (1965) The solution behavior of carbonate materials in seawater. In *Proc. Int. Conf. Tropical Oceanography, Miami Beach, Florida,* pp. 178–228.

Wu G. and Berger W. H. (1989) Planktonic foraminifera: differential dissolution and the quaternary stable isotope record in the west equatorial Pacific. *Paleoceanography* **4**, 181–198.

Readings from the Treatise on Geochemistry
ISBN: 978-0-12-381391-6

pp. 315–336

11 Hydrothermal Processes

C. R. German

Southampton Oceanography Centre, Southampton, UK

and

K. L. Von Damm†

University of New Hampshire, Durham, NH, USA

† Deceased.

11.1 INTRODUCTION

11.1.1 What is Hydrothermal Circulation?

Hydrothermal circulation occurs when seawater percolates downward through fractured ocean crust along the volcanic mid-ocean ridge (MOR) system. The seawater is first heated and then undergoes chemical modification through reaction with the host rock as it continues downward, reaching maximum temperatures that can exceed 400 °C. At these temperatures the fluids become extremely buoyant and rise rapidly back to the seafloor where they are expelled into the overlying water column. Seafloor hydrothermal circulation plays a significant role in the cycling of energy and mass between the solid earth and the oceans; the first identification of submarine hydrothermal venting and their accompanying chemosynthetically based communities in the late 1970s remains one of the most exciting discoveries in modern science. The existence of some form of hydrothermal circulation had been predicted almost as soon as the significance of ridges themselves was first recognized, with the emergence of plate tectonic theory. Magma wells up from the Earth's interior along "spreading centers" or "MORs" to produce fresh ocean crust at a rate of $\sim$20 $km^3\ yr^{-1}$, forming new seafloor at a rate of $\sim$3.3 $km^2\ yr^{-1}$ (Parsons, 1981; White *et al.*, 1992). The young oceanic lithosphere formed in this way cools as it moves away from the ridge crest. Although much of this cooling occurs by upward conduction of heat through the lithosphere, early heat-flow studies quickly established that a significant proportion of the total heat flux must also occur via some additional *convective* process (Figure 1), i.e., through circulation of cold seawater within the upper ocean crust (Anderson and Silbeck, 1981).

The first *geochemical* evidence for the existence of hydrothermal vents on the ocean floor came in the mid-1960s when investigations in the Red Sea revealed deep basins filled with hot, salty water (40–60 °C) and underlain by thick layers of metal-rich sediment (Degens and Ross, 1969). Because the Red Sea represents a young, rifting, ocean basin it was speculated that the phenomena observed there might also prevail along other young MOR spreading centers. An analysis of core-top sediments from throughout the world's oceans (Figure 2) revealed that such metalliferous sediments did, indeed, appear to be concentrated along the newly recognized global ridge crest (Boström *et al.*, 1969). Another early indication of hydrothermal activity came from the detection of plumes of excess ^3He in the Pacific

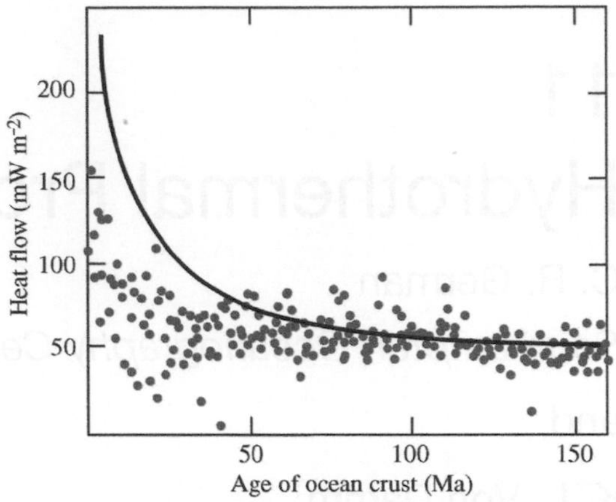

Figure 1 Oceanic heat flow versus age of ocean crust. Data from the Pacific, Atlantic, and Indian oceans, averaged over 2 Ma intervals (circles) depart from the theoretical cooling curve (solid line) indicating convective cooling of young ocean crust by circulating seawater (after C. A. Stein and S. Stein, 1994).

Ocean Basin (Clarke *et al.*, 1969)—notably the >2,000 km wide section in the South Pacific (Lupton and Craig, 1981)—because ^3He present in the deep ocean could only be sourced through some form of active degassing of the Earth's interior, at the seafloor.

One area where early heat-flow studies suggested hydrothermal activity was likely to occur was along the Galapagos Spreading Center in the eastern equatorial Pacific Ocean (Anderson and Hobart, 1976). In 1977, scientists diving at this location found hydrothermal fluids discharging chemically altered seawater from young volcanic seafloor at elevated temperatures up to 17 °C (Edmond *et al.*, 1979). Two years later, the first high-temperature (380 ± 30 °C) vent fluids were found at 21° N on the East Pacific Rise (EPR) (Spiess *et al.*, 1980)—with fluid compositions remarkably close to those predicted from the lower-temperature Galapagos findings (Edmond *et al.*, 1979). Since that time, hydrothermal activity has been found at more than 40 locations throughout the Pacific, North Atlantic, and Indian Oceans (e.g., Van Dover *et al.*, 2002) with further evidence—from characteristic chemical anomalies in the ocean water column—of its occurrence in even the most remote and slowly spreading ocean basins (Figure 3), from the polar seas of the Southern Ocean (German *et al.*, 2000; Klinkhammer *et al.*, 2001) to the extremes of the ice-covered Arctic (Edmonds *et al.*, 2003).

The most spectacular manifestation of seafloor hydrothermal circulation is, without doubt, the high-temperature (>400 °C) "black smokers" that expel

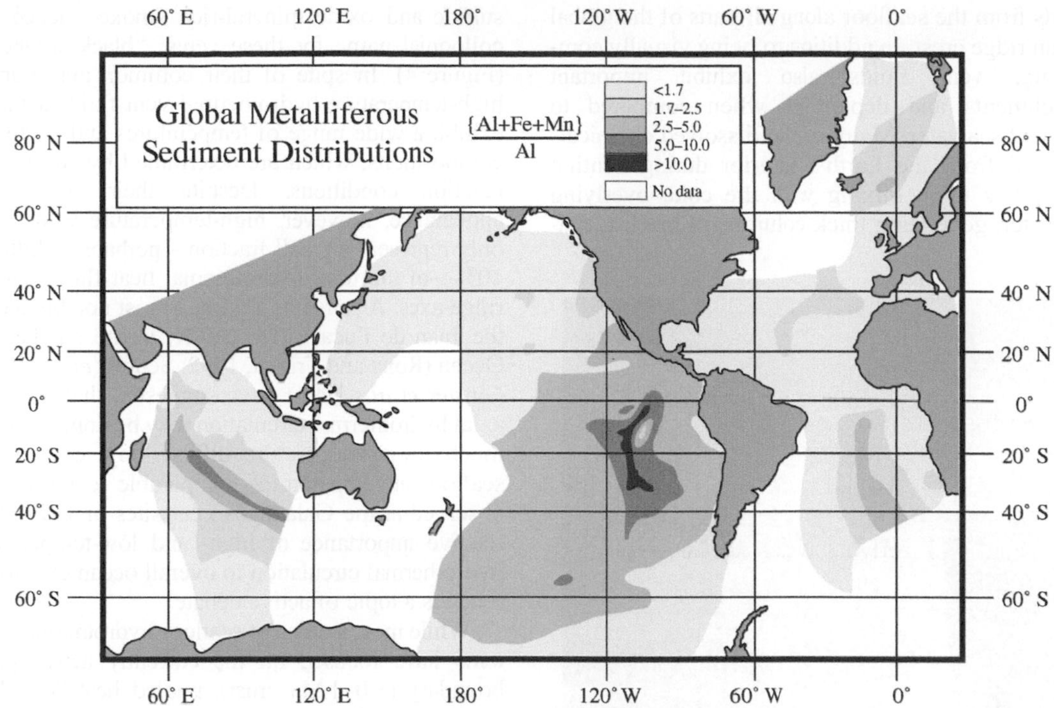

Figure 2 Global map of the (Al + Fe + Mn):Al ratio for surficial marine sediments. Highest ratios mimic the trend of the global MOR axis (after Boström *et al.*, 1969).

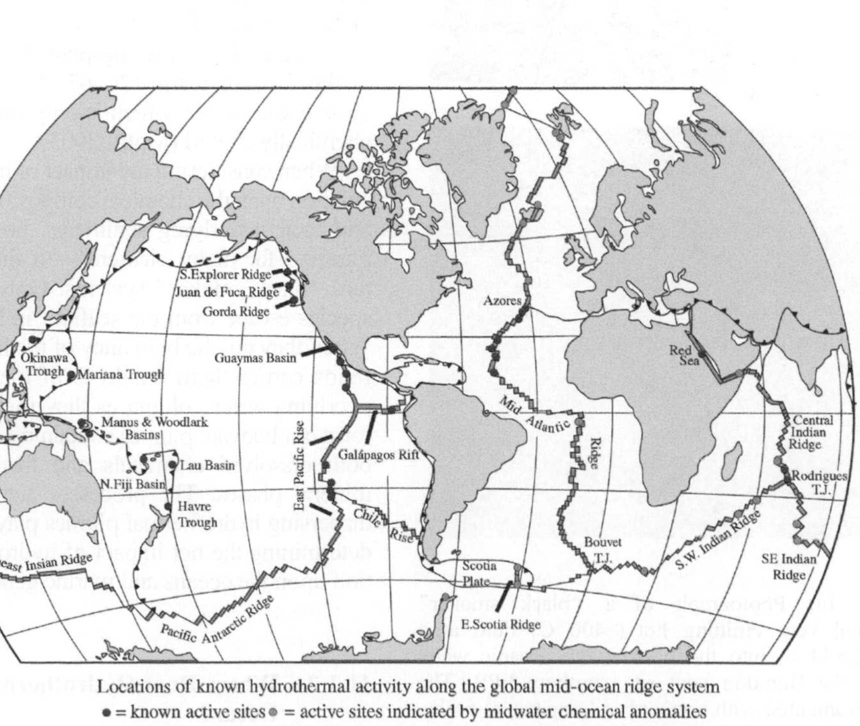

Figure 3 Schematic map of the global ridge crest showing the major ridge sections along which active hydrothermal vents have already been found (red circles) or are known to exist from the detection of characteristic chemical signals in the overlying water column (orange circles). Full details of all known hydrothermally active sites and plume signals are maintained at the InterRidge web-site: http://triton.ori.u-tokyo.ac.jp/~intridge/wg-gdha.htm

fluids from the seafloor along all parts of the global ocean ridge crest. In addition to being visually compelling, vent fluids also exhibit important enrichments and depletions when compared to ambient seawater. Many of the dissolved chemicals released from the Earth's interior during venting precipitate upon mixing with the cold, overlying seawater, generating thick columns of black metal-

(a)

(b)

Figure 4 (a) Photograph of a "black smoker" hydrothermal vent emitting hot (>400 °C) fluid at a depth of 2,834 m into the base of the oceanic water column at the Brandon vent site, southern EPR. The vent is instrumented with a recording temperature probe. (b) Diffuse flow hydrothermal fluids have temperatures that are generally <35 °C and, therefore, may host animal communities. This diffuse flow site at a depth of 2,500 m on the EPR at 9°50′ N is populated by *Riftia* tubeworms, mussels, crabs, and other organisms.

sulfide and oxide mineral-rich smoke—hence the colloquial name for these vents: "black smokers" (Figure 4). In spite of their common appearance, high-temperature hydrothermal vent fluids actually exhibit a wide range of temperatures and chemical compositions, which are determined by subsurface reaction conditions. Despite their spectacular appearance, however, high-temperature vents may only represent a small fraction—perhaps as little as 10%—of the total hydrothermal heat flux close to ridge axes. A range of studies—most notably along the Juan de Fuca Ridge (JdFR) in the NE Pacific Ocean (Rona and Trivett, 1992; Schultz *et al.*, 1992; Ginster *et al.*, 1994) have suggested that, instead, axial hydrothermal circulation may be dominated by much lower-temperature diffuse flow exiting the seafloor at temperatures comparable to those first observed at the Galapagos vent sites in 1977. The relative importance of high- and low-temperature hydrothermal circulation to overall ocean chemistry remains a topic of active debate.

While most studies of seafloor hydrothermal systems have focused on the currently active plate boundary (~0–1 Ma crust), pooled heat-flow data from throughout the world's ocean basins (Figure 1) indicate that convective heat loss from the oceanic lithosphere actually continues in crust from 0–65 Ma in age (Stein *et al.*, 1995). Indeed, most recent cstimates would indicate that hydrothermal circulation through this older (1–65 Ma) section, termed "flank fluxes," may be responsible for some 70% or more of the total hydrothermal heat loss associated with spreading-plate boundaries—either in the form of warm (20–65 °C) altered seawater, or as cooler water, which is only much more subtly chemically altered (Mottl, 2003).

When considering the impact of hydrothermal circulation upon the chemical composition of the oceans and their underlying sediments, however, attention returns—for many elements—to the high-temperature "black smoker" systems. Only here do many species escape from the seafloor in high abundance. When they do, the buoyancy of the high-temperature fluids carries them hundreds of meters up into the overlying water column as they mix and eventually form nonbuoyant plumes containing a wide variety of both dissolved chemicals and freshly precipitated mineral phases. The processes active within these dispersing hydrothermal plumes play a major role in determining the net impact of hydrothermal circulation upon the oceans and marine geochemistry.

11.1.2 Where Does Hydrothermal Circulation Occur?

Hydrothermal circulation occurs predominantly along the global MOR crest, a near-continuous volcanic chain that extends over ~6×10^4 km (Figure 3). Starting in the Arctic basin this ridge system extends south through the Norwegian-Greenland Sea as far as Iceland and then continues southward as the Mid-Atlantic Ridge (MAR), passing through the Azores and onward into the far South Atlantic, where it reaches the Bouvet Triple Junction, near 50° S. To the west, a major transform fault connects this triple junction to the Sandwich and Scotia plates that are separated by the East Scotia Ridge (an isolated back-arc spreading center). These plates are also bound to north and south by two major transform faults that extend further west between South America and the Antarctic Peninsula before connecting to the South Chile Trench. To the east of the Bouvet Triple Junction lies the SW Indian Ridge, which runs east and north as far as the Rodrigues Triple Junction (~25° S, 70° E), where the ridge crest splits in two. One branch, the Central Indian Ridge, extends north through the western Indian Ocean and Gulf of Aden ending as the incipient ocean basin that is the Red Sea (Section 11.1.1). The other branch of the global ridge crest branches south east from the Rodrigues Triple Junction to form the SE Indian and Pacific-Antarctic Ridges which extend across the entire southern Indian Ocean past Australasia and on across the southern Pacific Ocean as far as ~120° W, where the ridge again strikes north. The ridge here, the EPR, extends from ~55° S to ~30° N but is intersected near 30° S by the Chile Rise, which connects to the South Chile Trench. Further north, near the equator, the Galapagos Spreading Center meets the EPR at another triple junction. The EPR (and, hence, the truly continuous portion of the global ridge crest, extending back through the Indian and Atlantic Oceans) finally ends where it runs "on-land" at the northern end of the Gulf of California. There, the ridge crest is offset to the NW by a transform zone, more commonly known as the San Andreas Fault, which continues offshore once more, off northern California at ~40° N, to form the Gorda, Juan de Fuca, and Explorer Ridges—all of which hug the NE Pacific/N. American margin up to ~55° N. Submarine hydrothermal activity is also known to be associated with the back-arc spreading centers formed behind ocean–ocean subduction zones which occur predominantly around the northern and western margins of the Pacific Ocean, from the Aleutians via the Japanese archipelago and Indonesia all the way south to New Zealand. In addition to ridge-crest hydrothermal venting, similar circulation also occurs associated with hot-spot related intraplate volcanism—most prominently in the central and western Pacific Ocean (e.g., Hawaii, Samoa, Society Islands), but these sites are much less extensive, laterally, than ridge crests and back-arc spreading centers, combined. A continuously updated map of reported hydrothermal vent sites is maintained by the InterRidge community as a Vents Database (http://triton.ori.u-tokyo.ac.jp/~intridge/wg-gdha.htm).

As described earlier, the first sites of hydrothermal venting to be discovered were located along the intermediate to fast spreading Galapagos Spreading Center (6 cm yr^{-1}) and northern EPR (6–15 cm yr^{-1}). A hypothesis, not an unreasonable one, influenced heavily by these early observations but only formalized nearly 20 years later (Baker *et al.*, 1996) proposed that the incidence of hydrothermal venting along any unit length of ridge crest should correlate positively with spreading-rate because the latter is intrinsically linked to the magmatic heat flux at that location. Thus, the faster the spreading rate the more abundant the hydrothermal activity, with the most abundant venting expected (and found: Charlou *et al.*, 1996; Feely *et al.*, 1996; Ishibashi *et al.*, 1997) along the superfast spreading southern EPR (17–19° S), where ridge-spreading rate is among the fastest known (>14 cm yr^{-1}). Evidence for reasonably widespread venting has also been found most recently along some of the slowest-spreading sections of the global ridge crest, both in the SW Indian Ocean (German *et al.*, 1998a; Bach *et al.*, 2002) and in the Greenland/Arctic Basins (Connelly *et al.*, 2002; Edmonds *et al.*, 2003). Most explorations so far, however, have focused upon ridge crests closest to nations with major oceanographic research fleets and in the low- to mid-latitudes, where weather conditions are most favorable toward use of key research tools such as submersibles and deep-tow vehicles. Consequently, numerous active vent sites are known along the NE Pacific ridge crests, in the western Pacific back-arc basins and along the northern MAR (Figure 3), while other parts of the global MOR system remain largely unexplored (e.g., southern MAR, Central Indian Ridge, SE Indian Ridge, Pacific–Antarctic Ridge).

Reinforcing how little of the seafloor is well explored, as recently as December 2000 an entirely new form of seafloor hydrothermal activity, in a previously unexplored geologic setting was discovered (Kelley *et al.*, 2001). Geologists diving at the Atlantis fracture zone, which offsets part of the MAR near 30° N, found moderate-temperature fluids (40–75 °C) exiting from tall (up to 20 m) chimneys, formed predominantly from calcite [$CaCO_3$], aragonite [$CaCO_3$], and brucite [$Mg(OH)_2$]. These compositions are quite unlike previously documented hydrothermal vent fluids (Section 11.2), yet their geologic setting is one that may recur frequently along slow and very slow spreading ridges (e.g., Gracia *et al.*, 1999, 2000, Parson *et al.*, 2000; Sauter *et al.*, 2002). The Lost City vent site may, therefore, represent a new and important form of hydrothermal vent input to the oceans, which has hitherto been overlooked.

11.1.3 Why Should Hydrothermal Fluxes Be Considered Important?

Since hydrothermal systems were first discovered on the seafloor, determining the magnitude of their flux to the ocean and, hence, their importance in controlling ocean chemistry has been the overriding question that numerous authors have tried to assess (Edmond *et al.*, 1979, 1982; Staudigel and Hart, 1983; Von Damm *et al.*, 1985a; C. A. Stein and S. Stein, 1994; Elderfield and Schultz, 1996; Schultz and Elderfield, 1997; Mottl, 2003). Of the total heat flux from the interior of the Earth (~43 TW) ~32 TW is associated with cooling through oceanic crust and, of this, some 34% is estimated to occur in the form of hydrothermal circulation through ocean crust up to 65 Ma in age (C. A. Stein and S. Stein, 1994). The heat supply that drives this circulation is of two parts: magmatic heat, which is actively *emplaced* close to the ridge axis during crustal formation, and heat that is *conducted* into the crust from cooling lithospheric mantle, which extends out beneath the ridge flanks.

At the ridge axis, the magmatic heat available from crustal formation can be summarized as (i) heat released from the crystallization of basaltic magma at emplacement temperatures (latent heat), and (ii) heat mined from the solidified crust during cooling from emplacement temperatures to hydrothermal temperatures, assumed by Mottl (2003) to be 1175 ± 25 °C and 350 ± 25 °C, respectively. For an average crustal thickness of ~6 km (White *et al.*, 1992) the mass of magma emplaced per annum is estimated at 6×10^{16} g yr^{-1} and the maximum heat available from crystallization of this basaltic magma and cooling to hydrothermal temperatures is 2.8 ± 0.3 TW (Elderfield and Schultz, 1996; Mottl, 2003). If all this heat were transported as high-temperature hydrothermal fluids expelled from the seafloor at 350 °C and 350 bar, this heat flux would equate to a volume flux of $5–7 \times 10^{16}$ g yr^{-1}. It should be noted, however, that the heat capacity (c_p) of a 3.2% NaCl solution becomes extremely sensitive to increasing temperature under hydrothermal conditions of temperature and pressure, as the critical point is approached. Thus, for example, at 350 bar, a moderate increase in temperature near 400 °C could cause an increase in c_p approaching an order of magnitude resulting in a concomitant drop in the water flux required to transport this much heat (Bischoff and Rosenbauer, 1985; Elderfield and Schultz, 1996).

Of course, high-temperature hydrothermal fluids may not be entirely responsible for the transport of all the axial hydrothermal heat flux. Elderfield and Schultz (1996) considered a uniform distribution, on the global scale, in which only 10% of the total axial hydrothermal flux occurred as "focused" flow (heat FLUX = 0.2–0.4 TW; volume FLUX = $0.3–0.6 \times 10^{16}$ g yr^{-1}). In those calculations, the remainder of the axial heat flux was assumed to be transported by a much larger volume flux of lower-temperature fluid ($280–560 \times 10^{16}$ g yr^{-1} at ~5 °C). But how might such diffuse flow manifest itself? Should diffuse fluid be considered as diluted high-temperature vent fluid, conductively heated seawater, or some combination of the above? Where might such diffuse fluxes occur? Even if the axial hydrothermal heat flux were only restricted to 0–0.1 Ma crust, the associated fluid flow might still extend over the range of kilometers from the axis on medium-fast ridges—i.e., out onto young ridge flanks. For slow and ultraslow spreading ridges (e.g., the MAR) by contrast, all 0–0.1 Ma and, indeed 0–1 Ma crustal circulation would occur within the confines of the axial rift valley (order 10 km wide). The partitioning of "axial" and "near-axial" hydrothermal flow, on fast and slow ridges and between "focused" and "diffuse" flow, remains very poorly constrained in the majority of MOR settings and is an area of active debate.

On older oceanic crust (1–65 Ma) hydrothermal circulation is driven by upward conduction of heat from cooling of the underlying lithospheric mantle. Heat fluxes associated with this process are estimated at 7 ± 2 TW (Mottl, 2003). These values are significantly greater than the total heat fluxes associated with axial and near-axis circulation combined, and represent as much as 75–80% of Earth's total *hydrothermal* heat flux, >20% of the total *oceanic* heat flux and >15% of the Earth's *entire* heat flux. Mottl and Wheat (1994) chose to subdivide the fluid circulation associated with this heat into two components, warm (>20 °C) and cool (<20 °C) fluids, which exhibit large and small changes in the composition of the circulating seawater, respectively. Constraints from the magnesium mass balance of the oceans suggest that the cool (less altered) fluids carry some 88% of the total *flank* heat flux, representing a cool-fluid water flux (for 5–20 °C fluid temperatures) of $1–4 \times 10^{19}$ g yr^{-1} (Mottl, 2003).

To put these volume fluxes in context, the maximum flux of cool (<20 °C) hydrothermal fluids, calculated above is almost identical to the global riverine flux estimate of $3.7–4.2 \times 10^{19}$ g yr^{-1} (Palmer and Edmond, 1989). The flux of high-temperature fluids close to the ridge axis, by contrast, is ~1,000-fold lower. Nevertheless, for an ocean volume of ~1.4×10^{24} g, this still yields a (geologically short) oceanic residence time, with respect to high-temperature circulation, of ~20–30 Ma—and the hydrothermal fluxes will be important for those elements which exhibit high-temperature fluid concentrations more than 1,000-fold greater than river waters. Furthermore, high-temperature fluids emitted from "black smoker" hydrothermal systems typically entrain large volumes of ambient seawater during the formation of buoyant and neutrally buoyant plumes (Section 11.5) with typical dilution ratios of $\sim 10^4$:1 (e.g., Helfrich and Speer, 1995). If 50% of the fluids circulating at high temperature through young ocean crust are entrained into hydrothermal plumes then the total water flux through hydrothermal plumes would be approximately one order of magnitude greater than all other hydrothermal fluxes *and* the global riverine flux to the oceans (Table 1). The associated residence time of the global ocean, with respect to cycling through hydrothermal plume entrainment, would be just 4–8 kyr, i.e., directly comparable to the mixing time of the global deep-ocean conveyor (~1.5 kyr; Broecker and Peng, 1982). From that perspective, therefore, we can anticipate that hydrothermal circulation should play an important role in the marine geochemistry of any tracer which exhibits a residence time greater than ~1–10 kyr in the open ocean.

The rest of the chapter is organized as follows. In Section 11.2 we discuss the chemical composition of hydrothermal fluids, why they are important, what factors control their compositions, and how these compositions vary, both in space, from one location to another, and in time. Next (Section 11.3) we identify that the fluxes established thus far represent gross fluxes into and out of the ocean crust associated with high-temperature venting. We then examine the other source and sink terms associated with hydrothermal circulation, including alteration of the oceanic crust, formation of hydrothermal mineral deposits, interactions/uptake within hydrothermal plumes and settling into deep-sea sediments. Each of these "fates" for hydrothermal material is then considered in more detail. Section 11.4 provides a detailed discussion of near-vent deposits, including the formation of polymetallic sulfides and other minerals, as well as near-vent sediments. In Section 11.5 we present a detailed description of the processes associated with hydrothermal plumes, including a brief explanation of basic plume dynamics, a discussion of how plume processes modify the gross flux from high-temperature venting and further discussions of how plume chemistry can be both determined by, and influence, physical oceanographic, and biological interactions. Section 11.6 discusses the fate of hydrothermal products and concentrates on ridge-flank metalliferous sediments, including their potential for paleoceanographic investigations and role in "boundary scavenging" processes. We conclude (Section 11.7) by identifying some of the unresolved questions associated with hydrothermal circulation that are most in need of further investigation.

11.2 VENT-FLUID GEOCHEMISTRY

11.2.1 Why are Vent-fluid Compositions of Interest?

The compositions of vent fluids found on the global MOR system are of interest for several reasons; how and why those compositions vary has important implications. The overarching question, as mentioned in Section 11.1.3, is to determine how the fluids emitted from these systems influence and control ocean chemistry, on both short and long timescales. This question is very difficult to address in a quantitative manner because, in addition to all the heat flux and related water flux uncertainties discussed in Section 11.1, it also requires an understanding of the range of chemical variation in these systems and an understanding of the mechanisms and variables that control vent-fluid chemistries and temperatures. Essentially every hydrothermal vent that is discovered has a different composition (e.g., Von Damm, 1995) and we now

Table 1 Overview of hydrothermal fluxes: heat and water volume: data from Elderfield and Schultz (1996) and Mottl (2003).

(I) Summary of global heat fluxes		
Heat flux from the Earth's interior	43 TW	
Heat flux associated with ocean crust	32 TW	
Seafloor hydrothermal heat flux	11 TW	
(II) Global hydrothermal fluxes: heat and water		
	Heat flux (TW)	Water flux (10^{16}gyr^{-1})
Axial flow (0–1 Ma)		
All flow at 350 °C	2.8 ± 0.3	5.6 ± 0.6
10%@350 °C/90%@5 °C	2.8 ± 0.3	420 ± 140
Hydrothermal plumes (50%)		28,000 ± 3,000
Off-axis flow (1–65 Ma)	7 ± 2	1,000–4,000
Global riverine flux		3,700–4,200

know that these compositions often vary profoundly on short (minutes to years) timescales. Hence, the flux question remains a difficult one to answer. Vent fluid compositions also act as sensitive and unique indicators of processes occurring within young oceanic crust and at present, this same information cannot be obtained from any other source. The "window" that vent fluids provide into subsurface crustal processes is especially important because we can not yet drill young oceanic crust, due to its unconsolidated nature, unless it is sediment covered. The chemical compositions of the fluids exiting at the seafloor provide an integrated record of the reactions and the pressure and temperature (*P–T*) conditions these fluids have experienced during their transit through the crust. Vent fluids can provide information on the depth of fluid circulation (hence, information on the depth to the heat source), as well as information on the residence time of fluids within the oceanic crust at certain temperatures. Because the dissolved chemicals in hydrothermal fluids provide energy sources for microbial communities living within the oceanic crust, vent-fluid chemistries can also provide information on whether such communities are active at a given location. Vent fluids may also lead to the formation of metal-rich sulfide and sulfate deposits at the seafloor. Although the mineral deposits found are not economic themselves, they have provided important insights into how metals and sulfide can be transported in the same fluids and, thus, how economically viable mineral deposits are formed. Seafloor deposits also have the potential to provide an integrated history of hydrothermal activity at sites where actively venting fluids have ceased to flow.

11.2.2 Processes Affecting Vent-fluid Compositions

In all known cases the starting fluid for a submarine hydrothermal system is predominantly, if not entirely, seawater, which is then modified by processes occurring within the oceanic crust. Four factors have been identified: the two most important are (i) phase separation and (ii) water–rock interaction; the importance of (iii) biological processes and (iv) magmatic degassing has yet to be established.

Water–rock interaction and phase separation are processes that are inextricably linked. As water passes through the hydrothermal system it will react with the rock and/or sediment substrate that is present (Figure 5). These reactions begin in

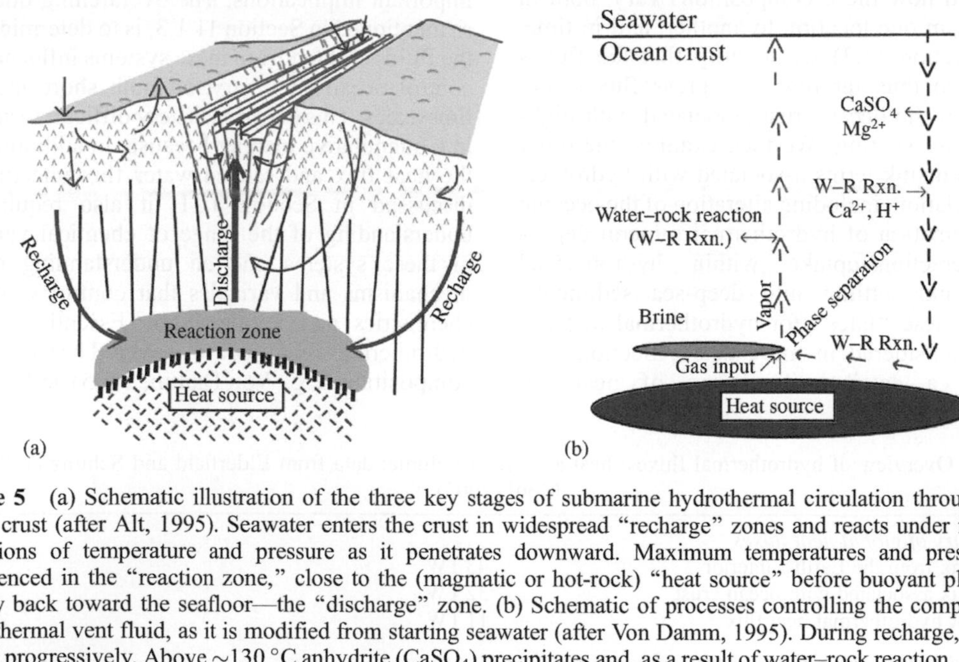

Figure 5 (a) Schematic illustration of the three key stages of submarine hydrothermal circulation through young ocean crust (after Alt, 1995). Seawater enters the crust in widespread "recharge" zones and reacts under increasing conditions of temperature and pressure as it penetrates downward. Maximum temperatures and pressures are experienced in the "reaction zone," close to the (magmatic or hot-rock) "heat source" before buoyant plumes rise rapidly back toward the seafloor—the "discharge" zone. (b) Schematic of processes controlling the composition of hydrothermal vent fluid, as it is modified from starting seawater (after Von Damm, 1995). During recharge, fluids are heated progressively. Above ~130 °C anhydrite ($CaSO_4$) precipitates and, as a result of water–rock reaction, additional calcium (Ca^{2+}) is leached from the rock in order to precipitate most of the sulfate (SO_4^{2-}) derived from seawater. Magnesium (Mg^{2+}) is also lost to the rock and protons (H^+) are added. As the fluid continues downward and up the temperature gradient, water–rock interactions continue and phase separation may occur. At at least two sites on the global MOR system, direct degassing of the magma must be occurring, because very high levels of gas (especially CO_2, and helium) are observed in the hydrothermal fluids. The buoyant fluids then rise to the seafloor. In most cases the fluids have undergone phase separation, and in at least some cases storage of the liquid or brine phase has occurred which has been observed to vent in later years from the same sulfide structure (Von Damm *et al.*, 1997). See Figure 6 for additional discussion of phase separation.

the downflow zone, and continue throughout. When vent fluids exit at the seafloor, what we observe represents the net result of all the reactions that have occurred along the entire hydrothermal flow path. Because the kinetics of most reactions are faster at higher temperatures, it is assumed that much of the reaction occurs in the "reaction zone." Phase separation may also occur at more than one location during the fluid's passage through the crust, and may continue as the *P–T* conditions acting on the fluid change as it rises through the oceanic crust, back toward the seafloor. However, without a direct view into any of the active seafloor hydrothermal systems, for simplicity of discussion, and because we lack better constraints, we often view the system as one of: (i) water–rock reaction on the downflow leg; (ii) phase separation and water–rock reaction in the "reaction zone"; (iii) additional water–rock reaction after the phase separation "event" (Figure 5). Unless confronted with clear inconsistencies in the (chemical) data that invalidate this approach, we usually employ this simple "flow-through" as our working conceptual model. Even though we are unable to rigorously constrain the complexities for any given system, it is always important to remember that the true system is likely far more complex than any model we employ.

In water–rock reactions, chemical species are both gained and lost from the fluids. In terms of differences from the major-element chemistry of seawater, magnesium and SO_4 are lost, and the pH is lowered so substantially that all the alkalinity is titrated. The large quantities of silicon, iron, and manganese that are frequently gained may be sufficient for these to become "major elements" in hydrothermal fluids. For example, silicon and iron can exceed the concentrations of calcium and potassium, two major elements in seawater. Much of the dissolved SO_4 in seawater is lost on the downflow leg of the hydrothermal system as $CaSO_4$ (anhydrite) precipitates at temperatures of ~130 °C—just by heating seawater. Because there is more dissolved SO_4 than calcium in seawater, on a molar basis, additional calcium would have to be leached from the host rock if more than ~33% of all the available seawater sulfate were to be precipitated in this way. In fact, it is now recognized that at least some dissolved SO_4 must persist down into the reaction zone, based on the inferred redox state at depth (see later discussion). Some seawater SO_4 is also reduced to H_2S, substantial quantities of which may be found in hydrothermal fluids at any temperature, based on information from sulfur isotopes (Shanks, 2001). The magnesium is lost by the formation of Mg–OH silicates. This results in the generation of H^+, which accounts for the low pH and titration of the alkalinity. Sodium can also be lost from the fluids due to Na–Ca replacement reactions in plagioclase feldspars, known as albitization. Potassium (and the other alkalis) are also involved in similar types of reaction that can also generate acidity. Large quantities of iron, manganese, and silicon are also leached out of the rocks and into the fluids.

An element that is relatively conservative through water–rock reaction is chlorine in the form of the anion chloride. Chloride is key in hydrothermal fluids, because with the precipitation and/or reduction of SO_4 and the titration of HCO_3^-/CO_3^{2-}, chloride becomes the overwhelming and almost only anion (Br is usually present in the seawater proportion to chloride). Chloride becomes a key component, therefore, because almost all of the cations in hydrothermal fluids are present as chloro-complexes; thus, the levels of chloride in a fluid effectively determine the total concentration of cationic species that can be present. A fundamental aspect of seawater is that the major ions are present in relatively constant ratios—this forms the basis of the definition of salinity (see Volume Editors Introduction). Because these constant proportions are not maintained in vent fluids and because chloride is the predominant anion, discussions of vent fluids are best discussed in terms of their *chlorinity*, not their *salinity*.

Although small variations in chloride may be caused by rock hydration/dehydration, there are almost no mineralogic sinks for chloride in these systems. Therefore, the main process that effects changes in the chloride concentrations in the vent fluids is phase separation (Figure 6). Phase separation is a ubiquitous process in seafloor hydrothermal systems. Essentially no hydrothermal fluids are found with chlorinities equal to the local ambient seawater value. To phase separate seawater at typical intermediate-to-fast spreading MOR depths of ~2,500 m requires temperatures $\gtrsim$389 °C (Bischoff, 1991). This sets a minimum temperature that fluids must have reached, therefore, during their transit through the oceanic crust. The greater the depth, the higher the temperature required for phase separation to occur. Known vent systems occur at depths of 800–3,600 m, requiring maximum temperatures in the range 297–433 °C to phase separate seawater. Seawater, being a two-component system, $H_2O+NaCl$ (to a first approximation) exhibits different phase separation behavior from pure water. The critical point for seawater is 407 °C and 298 bar (Bischoff and Rosenbauer, 1985) compared to 374 °C and 220 bar for pure water. For the salt solution, the two-phase curve does not stop at the critical point but, instead, continues beyond it. As a solution crosses the two-phase curve, it will separate into two phases, one with chlorinities greater than starting seawater, and the other with chlorinities less than starting seawater. If the fluid reaches the two-phase curve at temperature and pressure conditions *less* than the

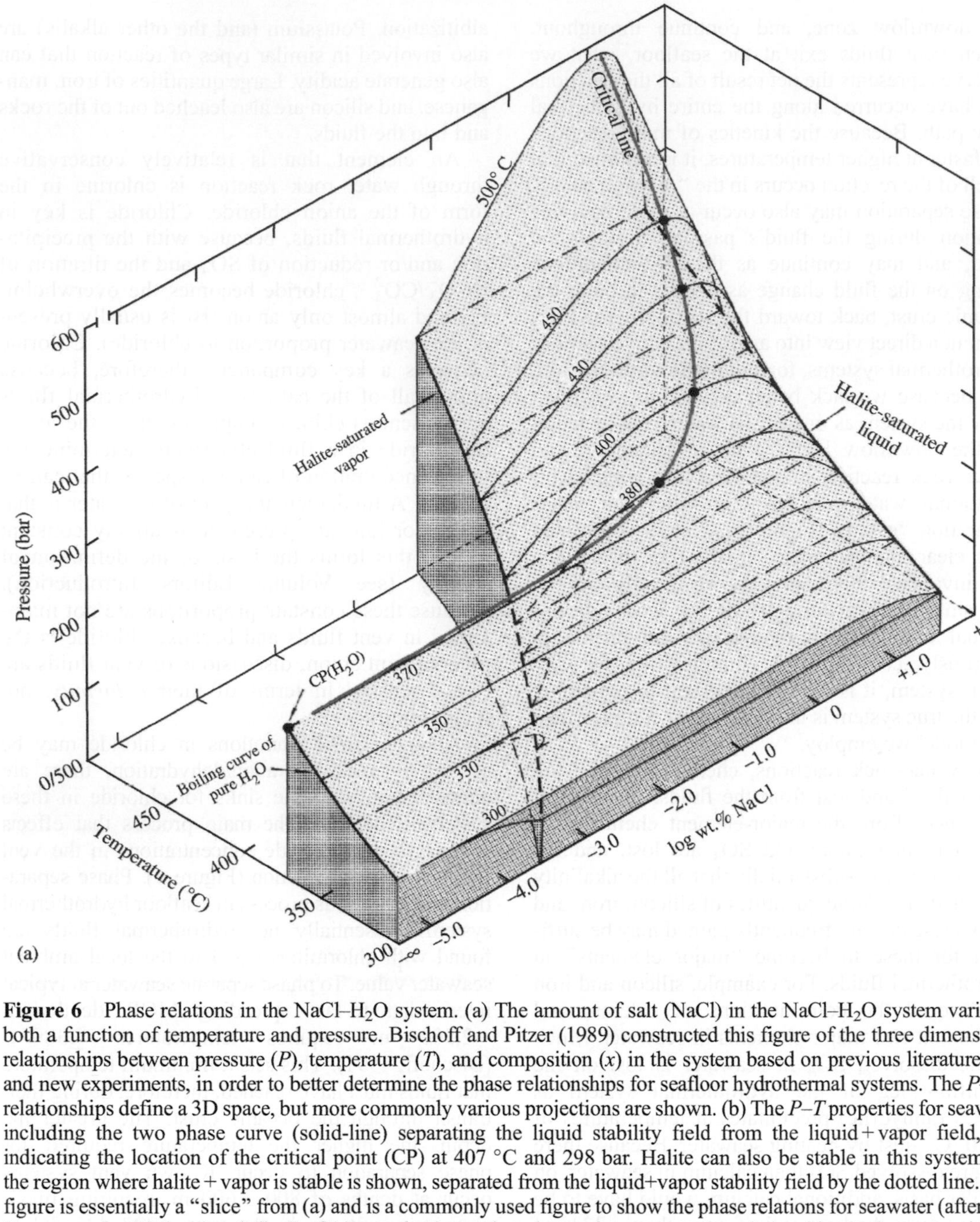

Figure 6 Phase relations in the NaCl–H_2O system. (a) The amount of salt (NaCl) in the NaCl–H_2O system varies as both a function of temperature and pressure. Bischoff and Pitzer (1989) constructed this figure of the three dimensional relationships between pressure (*P*), temperature (*T*), and composition (*x*) in the system based on previous literature data and new experiments, in order to better determine the phase relationships for seafloor hydrothermal systems. The *P–T–x* relationships define a 3D space, but more commonly various projections are shown. (b) The *P–T* properties for seawater including the two phase curve (solid-line) separating the liquid stability field from the liquid + vapor field, and indicating the location of the critical point (CP) at 407 °C and 298 bar. Halite can also be stable in this system and the region where halite + vapor is stable is shown, separated from the liquid+vapor stability field by the dotted line. This figure is essentially a "slice" from (a) and is a commonly used figure to show the phase relations for seawater (after Von Damm *et al.*, 1995). (c) A "slice" of (a) can also be made to better demonstrate the relationships in the system in *T–x* space. Here isobars show the composition of the conjugate vapor and brine (liquid) phases formed by the phase separation of seawater. This figure can be used to not only determine salt contents of the conjugate phases, but also their relative amounts. On this figure the compositions of the vapor and liquid phases sampled from "F" vent in 1991 and 1994, respectively, are shown (after Von Damm *et al.*, 1997).

critical point, subcritical phase separation (also called boiling) will occur, with the generation of a low chlorinity "vapor" phase. This phase contains salt, the amount of which will vary depending on where the two-phase curve was intersected (Bischoff and Rosenbauer, 1987). What is conceptually more difficult to grasp, is that when a fluid intersects the two-phase curve at *P–T* conditions *greater* than the critical point, the process is called supercritical phase separation (or condensation). In this case a small amount of a relatively high chlorinity liquid phase is condensed out of the fluid. Both sub- and super-critical phase separation occur in seafloor hydrothermal systems. To complete the phase relations in this system, halite may also precipitate (Figure 6). There is evidence that halite forms, and subsequently redissolves, in some seafloor hydrothermal systems (Oosting and Von Damm, 1996; Berndt and

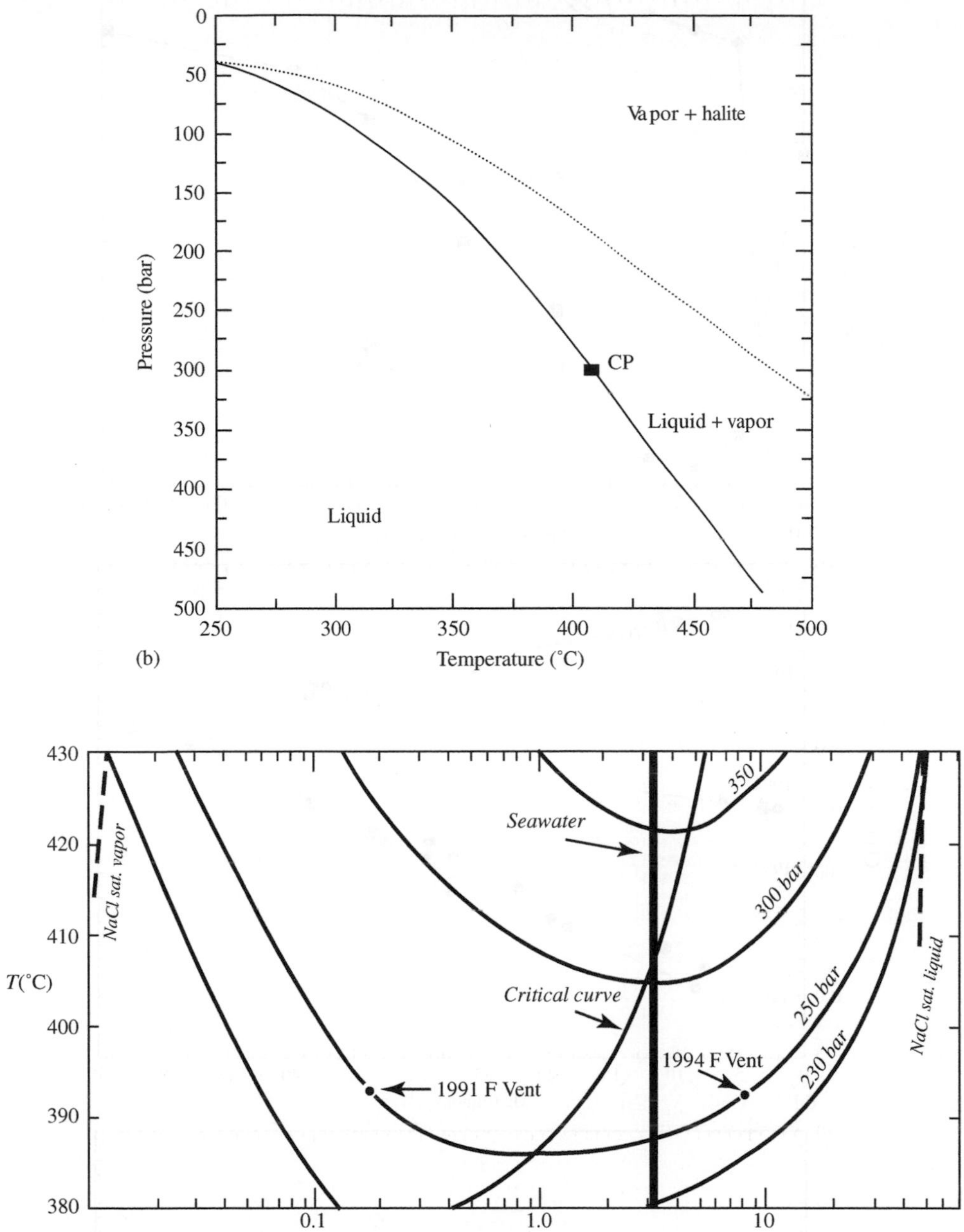

Figure 6 (continued).

Seyfried, 1997; Butterfield *et al.*, 1997; Von Damm, 2000). The *P–T* conditions at which the fluid intersects the two-phase curve, will determine the relative compositions of the two phases, as well as their relative amounts. Throughout this discussion, we have assumed the starting fluid undergoing phase separation is seawater (or, rather, an NaCl equivalent, because the initial magnesium and SO_4 will already be lost by this stage). If the NaCl content is different, the phase relations in this system change, forming a family of curves or surfaces that are a function of the NaCl content, as well as pressure and temperature. The critical point is also a function of the salt content, and hence is really a critical curve in *P–T–x* (*x* referring to composition) space.

As phase separation occurs, substantially changing the chloride content of vent fluids (values from $<6\%$ to $\sim 200\%$ of the seawater concentration have been observed), other chemical species will change in concert. It has been shown, both experimentally as well as in the field, that most of the cations (and usually bromide) maintain their element-to-Cl ratios during the phase separation process (Berndt and

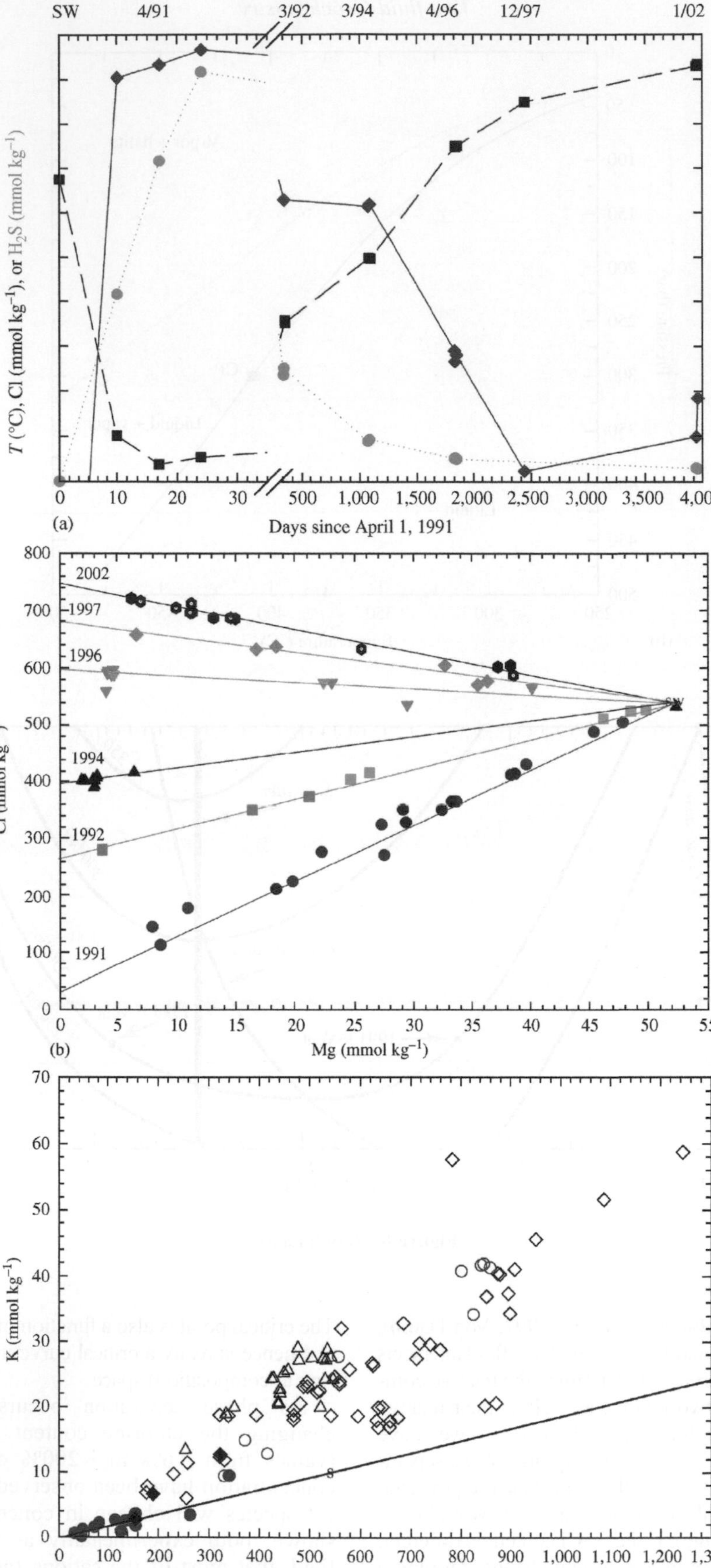
SW
4/91
3/92
3/94
4/96
12/97
1/02
T (°C), Cl (mmol kg−1), or H2S (mmol kg−1)
0 10 20 30 500 1,000 1,500 2,000 2,500 3,000 3,500 4,000
(a)
Days since April 1, 1991
2002
1997
1996
1994
1992
1991
SW
Cl (mmol kg−1)
800 700 600 500 400 300 200 100 0
0 5 10 15 20 25 30 35 40 45 50 55
(b)
Mg (mmol kg−1)
K (mmol kg−1)
70 60 50 40 30 20 10 0
0 100 200 300 400 500 600 700 800 900 1,000 1,100 1,200 1,300
s
(c)
Cl (mmol kg−1)

Seyfried, 1990; Von Damm, 2000; Von Damm *et al.*, 2003), i.e., most elements are conservative with respect to chloride. Exceptions do occur, however—primarily for those chemical species not present as chlorocomplexes. Dissolved gases (e.g., CO_2, CH_4, He, H_2, H_2S) are preferentially retained in the low chlorinity or vapor phase, and boron, which is present as a hydroxyl complex, is relatively unaffected by phase separation (Bray and Von Damm, 2003a). Bromide, as viewed through the Br/Cl ratio, is sometimes seen to be fractionated from chloride; this occurs whenever halite is formed or dissolves because bromide is preferentially excluded from the halite structure (Oosting and Von Damm, 1996; Von Damm, 2000). Fluids that have deposited halite, therefore, will have a high Br/Cl ratio, while fluids that have dissolved halite will have a low Br/Cl ratio, relative to seawater. It is because of the ubiquity of phase separation that vent-fluid compositions are often now viewed or expressed as ratios with respect to chloride, rather than as absolute concentrations. This normalization to chloride *must* be used when trying to evaluate net gains and losses of chemical species as seawater traverses the hydrothermal circulation cell, to correct for the fractionation caused by phase separation.

Aside from early eruptive fluids (discussed later), the chemical composition of most high-temperature fluids (Figure 7) appears to be controlled by equilibrium or steady state with the rock assemblage. (Equilibrium requires the assemblage to be at its lowest energy state, but the actual phases present may be metastable, in which case they are not at true thermodynamic equilibrium but, rather, have achieved a steady-state condition.) When vent-fluid data are modeled with geochemical modeling codes using modern thermodynamic databases, the results suggest equilibrium, or at least steady state, has been achieved. The models cannot be rigorously applied to many of the data, however, because the fluids are often close to the critical point and in that region the thermodynamic data are not as well constrained. Based on results from both geochemical modeling codes and elemental ratios, current data indicate that not only the major elements, but also many minor elements (e.g., rubidium, caesium, lithium, and strontium) are controlled by equilibrium, or steady-state, conditions between the fluids and their host-rocks (Bray and Von Damm, 2003b). The rare earth elements (REE) in vent fluids present one such example.

Figure 7 Compositional data for vent fluids. (a) Time series data from "A" vent for chloride and H_2S concentrations and measured temperature (T). When time-series data are available, this type of figure, demonstrating the change in fluid composition in a single vent over time, is becoming more common. The data plotted are referred to as the "end-member" data (data from Von Damm *et al.*, 1995 and unpublished). Points on the y-axis are values for ambient seawater. Note the low chlorinity (vapor phase) fluids venting initially from A vent; over time the chloride content has increased and the fluids sampled more recently, in 2002, are the liquid (brine) phase. As is expected, the concentration of H_2S, a gas that will be partitioned preferentially into the vapor phase, is anticorrelated with the chloride concentration. The vertical axis has 10 divisions with the following ranges: T (°C) 200–405 (ambient seawater is 2 °C); Cl (mmol kg^{-1}) 0–800 (ambient is 540 mmol kg^{-1}); H_2S (mmol kg^{-1}) 0–120 (ambient is 0 mmol kg^{-1}). (b) Whenever vent fluids are sampled, varying amounts of ambient seawater are entrained into the sampling devices. Vent fluids contain 0 mmol kg^{-1}, while ambient seawater contains 52.2 mmol kg^{-1}. Therefore, if actual sample data are plotted as properties versus magnesium, least squares linear regression fits can be made to the data. The calculated "end-member" concentration for a given species, which represents the "pure" hydrothermal fluid is then taken as the point where that line intercepts the y-axis (i.e., the calculated value at Mg = 0 mmol kg^{-1}). These plots versus magnesium are therefore mostly sampling artifacts and are referred to as "mixing" diagrams. While these types of plots were originally used to illustrate vent-fluid data, they have been largely superceded by figures such as (a) or (c). This figure (b) shows the data used to construct the time series represented in (a). Note the different lines for the different years. In some years samples were collected on more than one date. All samples for a given year are shown by the same shape, the different colors within a year grouping indicate different sample dates. In some years the chemical composition varied from day-to-day, but for simplicity a single line is shown for each year in which samples were collected (Von Damm *et al.*, 1995 and unpublished data). (c) As the chloride content of a vent fluid is a major control on the overall composition of the vent composition, most of the cations vary as a function of the chloride-content. Variations in the chloride content are a result of phase separation. This shows the relationship between the potassium (K) and chloride content in vent fluids in the global database, as of 2000. The line is the ratio of K/Cl in ambient seawater. Closed circles are from 9–10° N EPR following the 1991 eruption, open circles are other 9–10° N data not affected by the eruptive events; triangles are from sites where vents occur on enriched oceanic crust, diamonds are from bare-basalt (MORB) hosted sites, filled diamonds are other sites impacted by volcanic eruptions. Data sources and additional discussion in Von Damm (2000).

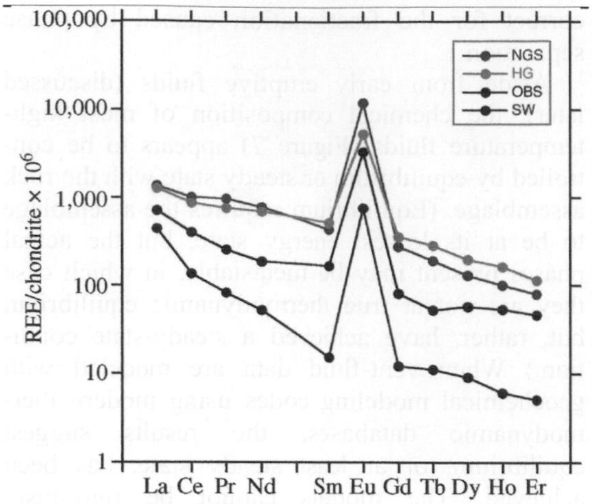

Figure 8 End-member REE concentrations in vent fluids from four different "black smokers" at the 21° N vent site, EPR, normalized to chondrite (REE data from Klinkhammer *et al.*, 1994). NGS=*National Geographic Smoker*; HG=*Hanging Gardens*; OBS=*Ocean Bottom Seismometer*; SW=*South West* vent.

REE distributions in hydrothermal fluids are light-REE enriched and exhibit strong positive europium anomalies, apparently quite unrelated to host-rock MORB compositions (Figure 8). However, Klinkhammer *et al.* (1994) have shown that when these same REE concentrations are plotted versus their ionic radii, the fluid trends not only become linear but also show the same fractionation trend exhibited by plagioclase during magma segregation, indicating that vent-fluid REE concentrations may be intrinsically linked to the high-temperature alteration of this particular mineral.

Two other processes are known to influence the chemistry of seafloor vent fluids: biological processes and magmatic degassing. Evidence for "magmatic degassing" has been identified at two sites along the global MOR system—at 9°50′ N and at 32° S on the EPR (M. D. Lilley, personal communication; Lupton *et al.*, 1999a). These sites have very high levels of CO_2, and very high He/heat ratios. The interpretation is that we are seeing areas with recent magma resupply within the crust and degassing of the lavas, resulting in very high gas levels in the hydrothermal fluids found at these sites. We do not know the spatial–temporal variation of this process, hence, we cannot yet evaluate its overall importance. Presumably, every site on the global MOR system undergoes similar processes episodically. What is not known, however, is the frequency of recurrence at any one site. Consequently, the importance of fluxes due to this degassing process, versus more "steady-state" venting, cannot currently be assessed. At 9°50′ N, high gas contents have now been observed for almost a decade; no signature of volatile-metal enrichment has been observed in conjunction with these high gas contents (Von Damm, 2003).

The fourth process influencing vent-fluid compositions is biological change, which can take the form of either consumption or production of various chemical species. As the current known limit to life on Earth is ~120 °C (e.g., Holland and Baross, 2003) this process can only affect fluids at temperatures lower than this threshold. This implies that high-temperature vents should not be subject to these effects whereas they may occur in both lower-temperature axial diffuse flow and beneath ridge flanks. From observations at the times of seafloor eruptions and/or diking events, it is known that there are microbial communities living within the oceanic crust (e.g., Haymon *et al.*, 1993). Their signatures can be seen clearly in at least some low-temperature fluids, as noted in particular by changes in the H_2, CH_4, and H_2S contents of those fluids (Von Damm and Lilley, 2003). Hence, biological influences have been observed; how widespread this is, which elements are affected, and what the overall impact on chemical fluxes may be all remain to be resolved.

11.2.3 Compositions of Hydrothermal Vent Fluids

11.2.3.1 Major-element chemistry

The known compositional ranges of vent fluids are summarized in Figure 9 and Table 2. Because no two vents yet discovered have exactly the same composition, these ranges often change with each new site. As discussed in Section 11.2.2, vent fluids are modified seawater characterized by the loss of magnesium, SO_4, and alkalinity and the gain of many metals, especially on a chloride normalized basis.

Vent fluids are acidic, but not as acid as may first appear from pH values measured at 25 °C and 1 atm (i.e., in shipboard laboratories). The cation H^+ in vent fluids is also present as a chloro-complex with the extent of complexation increasing as P and T increase. At the higher *in situ* conditions of P and T experienced at the seafloor, therefore, much of the H^+ is incorporated into the HCl–aqueous complex; hence, the activity of H^+ is reduced and the *in situ* pH is substantially higher than what is measured at laboratory temperatures. The K_w for water also changes as a function of P and T such that neutral pH is not necessarily 7 at other P–T conditions. For most vent fluids, the *in situ* pH is 1–2 pH units more acid than neutral, not the ~4 units of acidity that the measured (25 °C, 1 atm) data appear to imply. Most high-temperature vent fluids have (25 °C measured) pH values of 3.3 ± 0.5 but a few are more acidic whilst several are less acid. If fluids are more acid than pH 3.3 ± 0.5, it is often an indicator that metal sulfides have precipitated below the seafloor

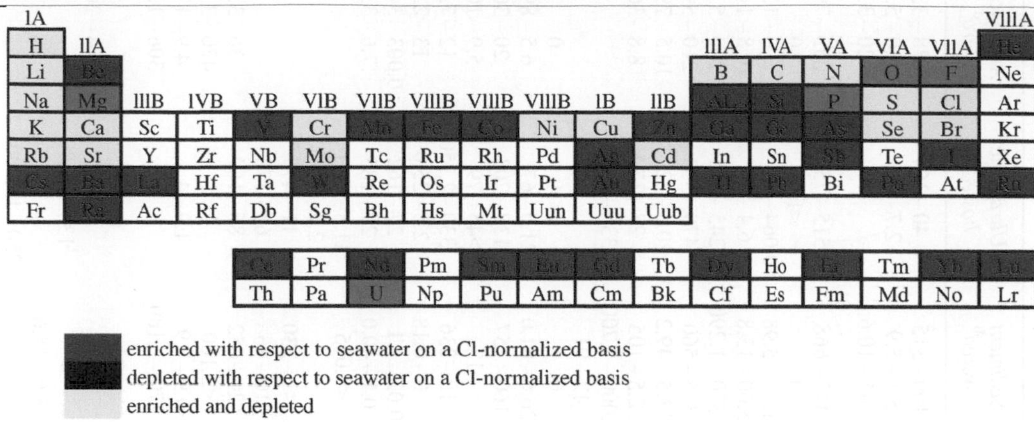

Figure 9 Periodic table of the elements showing the elements that are enriched in hydrothermal vent fluids relative to seawater (red), depleted (blue) and those which have been shown to exhibit both depletions and enrichments in different hydrothermal fluids (yellow) relative to seawater. All data are normalized to the chloride content of seawater in order to evaluate true gains and losses relative to the starting seawater concentrations.

because such reactions produce protons. Two mechanisms are known that can cause fluids to be less acidic than the norm: (i) cases where the rock substrate appears to be more highly altered—the rock cannot buffer the solutions to as low a pH; (ii) when organic matter is present, ammonium is often present and the NH_3/NH_4^+ couple serves to buffer the pH to a higher level.

Vent fluids are very reducing, as evidenced by the presence of H_2S rather than SO_4, as well as H_2, CH_4 and copious amounts of Fe^{2+} and Mn^{2+}. In rare cases, there can be more H_2S and/or H_2 than chloride on a molar basis and it is the prevailing high acidity that dictates that H_2S rather than HS– or S^{2-} is the predominant form in high-temperature vent fluids. Free H_2 is derived as a result of water–rock reaction and there is substantially more H_2 than O_2 in these fluids. Therefore, although redox calculations are typically given in terms of the $\log f_{O_2}$, the redox state is best calculated based on the H_2/H_2O couple for seafloor vent fluids. This can then be expressed in terms of the $\log f_{O_2}$. The K for the reaction:

$$H_2 + \frac{1}{2}O_2 = H_2O$$

also changes as a function of temperature and pressure. Another way to determine how reducing vent fluids are is by comparing them to various mineralogic buffers such as pyrite–pyrrhotite–magnetite (PPM) or hematite–magnetite–pyrite (HMP). Most vent fluids lie between these two extremes, but there is some systematic variation (Seyfried and Ding, 1995). The observation that vent fluids are more oxic than would be expected based on the PPM buffer, provides one line of evidence that the reaction zone is not as reducing as initially predicted, consistent with at least some dissolved seawater SO_4 penetrating into the deeper parts of the system rather than being quantitatively removed by anhydrite precipitation within shallower levels of the downflow limb.

Lower-temperature (< 100 °C) vent fluids found right at the axis are in most known cases a dilution of some amount of high-temperature fluids with seawater, or a low-temperature fluid with a composition close to seawater. There is some evidence for an "intermediate" fluid, perhaps most analogous to a crustal "ground water," with temperatures of ~150 °C within the oceanic crust. Evidence for the latter is found in some Ocean Drilling Program data (Magenheim *et al.*, 1992), some high-temperature vent fluids from 9°50′ N on the EPR (Ravizza *et al.*, 2001), and some very unusual ~90 °C fluids from the southern EPR (O'Grady, 2001). To conclude, the major-element composition of high-temperature vent fluids can be described as acidic, reducing, metal-rich NaCl solutions whilst lower-temperature fluids are typically a dilution of this same material with seawater. The few exceptions to this will be discussed below.

11.2.3.2 Trace-metal chemistry

Compared to the number of vent fluids sampled and analyzed for their major-element data, relatively little trace-metal data exist. This is because when hot, acidic vent fluids mix with seawater, or even just cool within submersible- or ROV-deployed sampling bottles, they become supersaturated with respect to many solid phases and, thus, precipitate. Once this occurs, everything in the sampling apparatus must be treated as one sample: a budget can only be constructed by integrating these different fractions back

Table 2 Ranges in chemical composition for all known vent fluids.

Chemical species	*Units*	*Seawater*	*Overall range*	*Slow*[a] *(0–2.5 cm yr^{-1})*	*Intermediate*[a] *>2.5–6*	*Fast*[a] *>6–12*	*Ultrafast*[a] *>12*	*Sediment covered*[b]	*Ultramafic hosted*[c]	*Arc, back-arc*[d]
T	°C	2	>2 – 405	40 – 369	13 – 382	8 – 403	16 – 405	100 – 315	40 – 364	278 – 334
pH	25 °C, 1 atm	7.8	2.0 – 9.8	2.5 – 4.85	2.8 – 4.5	2.45 – >6.63	2.96 – 5.53	5.1 – 5.9	2.7 – 9.8	2.0 – 5.0
Alkalinity	meq kg^{-1}	2.4	−3.75 – 10.6	−3.4 – 0.31	−3.75 – 0.66	−2.69 – <2.27	−1.36 – 0.915	1.45 – 10.6		−0.20 – 3.51
Cl	mmol kg^{-1}	540	30.5 – 1,245	357 – 675	176 – 1,245	30.5 – 902	113 – 1,090	412 – 668	515 – 756	255 – 790
SO_4	mmol kg^{-1}	28	<0 – <28	−3.5 – 1.9	−25 – 0.763	> – 8.76	−0.502 – 9.53	0	<12.9	0
H_2S	mmol kg^{-1}	0	0 – 110	0.5 – 5.9	0 – 19.5	0 – 110	0 – 35	1.10 – 5.98	0.064 – 1.0	2.0 – 13.1
Si	mmol kg^{-1}	0.03 – 0.18	<24	7.7 – 22	11 – 24	2.73 – 22.0	8.69 – 21.3	5.60 – 13.8	6.4 – 8.2	10.8 – 14.5
Li	umol kg^{-1}	26	4.04 – 5,800	238 – 1,035	160 – 2,350	4.04 – 1,620	248 – 1,200	370 – 1,290	245 – 345	200 – 5,800
Na	mmol kg^{-1}	464	10.6 – 983	312 – 584	148 – 924	10.6 – 983	109 – 886	315 – 560	479 – 553	210 – 590
K	mmol kg^{-1}	10.1	−1.17 – 79.0	17 – 28.8	6.98 – 58.7	−1.17 – 51	2.2 – 44.8	13.5 – 49.2	20.2 – 22	10.5 – 79.0
Rb	umol kg^{-1}	1.3	0.156 – 360	9.4 – 40.4	22.9 – 59	0.156 – 31.1	0.39 – 6.8	22.5 – 105	28 – 37.1	8.8 – 360
Cs	nmol kg^{-1}	2	2.3 – 7,700	100 – 285	168 – 364	2.3 – 264		1,000 – 7,700	331 – 385	
Be	nmol kg^{-1}	0	10 – 91		10 – 37			12 – 91		
Mg	mmol kg^{-1}	52.2	0	0	0	0	0	0	<19	0
Ca	mmol kg^{-1}	10.2	−1.31 – 109	9.9 – 43	9.75 – 109	−1.31 – 106	4.02 – 65.5	26.6 – 81.0	21.0 – 67	6.5 – 89.0
Sr	umol kg^{-1}	87	−29 – 387	42.9 – 133	0.0 – 348	−29 – 387	10.7 – 190	160 – 257	138 – 203	20 – 300
Ba	umol kg^{-1}	0.14	1.64 – 100	<52.2	>8 – >46		1.64 – 18.6	>12	>45 – 79	5.9 – 100
Mn	umol kg^{-1}	<0.001	10 – 7,100	59 – 1,000	140 – 4,480	62.7 – 3,300	20.6 – 2,750	10 – 236	330 – 2,350	12 – 7,100
Fe	mmol kg^{-1}	<0.001	0.007 – 25.0	0.0241 – 5.590	0.009 – 18.7	0.007 – 12.1	0.038 – 14.7	0 – 0.18	2.5 – 25.0	13 – 2,500
Cu	umol kg^{-1}	0.007	0 – 162	0 – 150	0.1 – 142	0.18 – 97.3	2.6 – 150	<0.02 – 1.1	27 – 162	0.003 – 34
Zn	umol kg^{-1}	0.012	0 – 3,000	0 – 400	2.2 – 600	13 – 411	1.9 – 740	0.1 – 40.0	29 – 185	7.6 – 3,000
Co	umol kg^{-1}	0.00003	<0.005 – 14.1	0.130 – 0.422	0.022 – 0.227			<0.005	11.8 – 14.1	
Ni	umol kg^{-1}	0.012							2.2 – 3.6	
Ag	nmol kg^{-1}	0.02	<1 – 230		<1 – 38			<1 – 230	11 – 47	
Cd	nmol kg^{-1}	1.0	0 – 180	75 – 146	0 – 180			<10 – 46	63 – 178	
Pb	nmol kg^{-1}	0.01	<20 – 3,900	221 – 376	183 – 360			<20 – 652	86 – 169	36 – 3,900
B	umol kg^{-1}	415	356 – 3,410	356 – 480	465 – 1,874	430 – 617	400 – 499	<2,160		470 – 3,410
Al	umol kg^{-1}	0.02	0.1 – 18.7	1.03 – 13.9	4.0 – 5.2	0.1 – 18	9.3 – 18.7	0.9 – 7.9	1.9 – 4	4.9 – 17.0
Br	umol kg^{-1}	840	29.0 – 1,910	666 – 1,066	250 – 1789	29.0 – 1370	216 – 1910	770 – 1,180		306 – 1,045
F	umol kg^{-1}	68	<38.8	16.1 – 38.8	"0"					
CO_2	mmol kg^{-1}			3.56 – 39.9	<5.7	<200	8.4 – 22			14.4 – 200
CH_4	umol kg^{-1}	0.0003		150 – 2,150	<52		7 – 133		130 – 2,200	
NH_4	mmol kg^{-1}	0	<15.6	<0.06	<0.65			5.6 – 15.6		
H_2	umol kg^{-1}	0.0003	<38,000	1.1 – 727	<0.45	<38,000	40 – 1300		250 – 13,000	

[a] These omit sedimented covered and um hosted. [b] Includes: Guaymas, Escanaba, Middle Valley. [c] Includes Rainbow, Lost City, kvd unpublished data for Logatchev. [d] Compilation from Ishibashi and Urabe (1995).

together. In the difficult sampling environment found at high-temperature vent sites, pieces of chimney structure are also sometimes entrained into the sampling apparatus. It is necessary, therefore, to be able to discriminate between particles that have precipitated from solution in the sampling bottle and "contaminating" particles that are extraneous to the sample. In addition, water samples are often subdivided into different fractions, aboard ship, making accurate budget reconstructions difficult if not impossible to complete. It is because of these difficulties that there are few robust analyses of many trace metals, especially those that precipitate as, or co-precipitate with, metal sulfide phases. Some general statements can, however, be made. In high-temperature vent fluids, most metals are enriched relative to seawater, sometimes by 7–8 orders of magnitude (as is sometimes true for iron). At least some data exist demonstrating the enrichment of all of vanadium, cobalt, nickel, copper, zinc, arsenic, selenium, aluminum, silver, cadmium, antimony, caesium, barium, tungsten, gold, thallium, lead, and REE relative to seawater. Data also exist showing that molybdenum and uranium are often lower than their seawater concentrations. These trace-metal data have been shown to vary with substrate and the relative enrichments of many of these trace metals varies significantly between MOR hydrothermal systems, those located in back arcs, and those with a significant sedimentary component. Even fewer trace-metal data exist for low-temperature "diffuse" fluids. The original work on the <20 °C GSC fluids (Edmond *et al.*, 1979) showed these fluids to be a mix of high-temperature fluids with seawater, with many of the transition metals present at *less than* their seawater concentrations due to precipitation and removal below the seafloor. Essentially the same results were obtained by James and Elderfield (1996) using the MEDUSA system to sample diffuse-flow fluids at TAG (26° N, MAR).

11.2.3.3 Gas chemistry of hydrothermal fluids

In general, concentrations of dissolved gases are highest in the lowest-chlorinity fluids, which represent the vapor phase. However, there are exceptions to this rule, and gas concentrations vary significantly between vents, even at a single location. In the lowest chlorinity and hottest fluids, H_2S may well be the dominant gas. However, because H_2S levels are controlled by metal-sulfide mineral solubility, this H_2S is often lost via precipitation. While the first vent sites discovered contained less than twice the CO_2 present in seawater (Welhan and Craig, 1983), more vents have higher levels of CO_2 than is commonly realized. Few MOR vent fluids have CO_2 levels less than or equal to the total CO_2 levels present in seawater ($\sim$2.5 mmol kg^{-1}). Instead, many fluids have concentrations approaching an order of magnitude more CO_2 than seawater; the highest approach two orders of magnitude more CO_2 than seawater, but these highest levels are uncommon (M. D. Lilley, personal communication). Back-arc systems more commonly have higher levels of CO_2 in their vent fluids, but concentrations two orders of magnitude greater than seawater are, again, close to the upper maximum of what has been sampled so far (Ishibashi and Urabe, 1995). CH_4 is much less abundant than CO_2 in most systems. Vent-fluid CH_4 concentrations are typically higher in sedimented systems and in systems hosted in ultramafic rocks, when compared to bare basaltic vent sites. CH_4 is also enriched in low-temperature vent fluids when compared to concentrations predicted from simple seawater/vent-fluid mixing (Von Damm and Lilley, 2003). Longer-chain organic molecules have also been reported from some sites, usually at even lower abundances than CO_2 and/or CH_4 (Evans *et al.*, 1988). The concentrations of H_2 gas in vent fluids vary substantially, over two orders of magnitude (M. D. Lilley, personal communication). Again the highest levels are usually observed in vapor phase fluids, especially those sampled immediately after volcanic eruptions or diking events. Relatively high values (several mmol kg^{-1}) have also been reported from sites hosted by ultramafic rocks. Of the noble gases, helium, especially ^{3}He, is most enriched in vent fluids. ^{3}He can be used as a conservative tracer in vent fluids, because its entire source in vent fluids is primordial, from within the Earth (see Section 11.5). Radon, a product of radioactive decay in the uranium series, is also greatly enriched in vent fluids (e.g., Kadko and Moore, 1988; Rudnicki and Elderfield, 1992). Less data are available for the other noble gases, at least some of which appear to be relatively conservative compared to their concentration in starting seawater (Kennedy, 1988).

11.2.3.4 Nutrient chemistry

The concentrations of nutrients available in seawater control biological productivity. Consequently, the dissolved nutrient concentrations in natural waters are always of great interest. Compared to deep-ocean seawater, the PO_4 contents of vent fluids are significantly lower, but are not zero. Much work remains to be done on the distribution of nitrogen species, and the nitrogen cycle in general, in vent fluids. Generally, in basalt-hosted systems, the nitrate + nitrite content is also lower than local deep-waters but, again, is not zero. Ammonium in these systems typically measures less than 10 μmol kg^{-1} but, in some systems in which no sediment cover is present, values of 10s to even 100s of μmol kg^{-1} are sometimes observed (cf. Von Damm, 1995). In Guaymas Basin, in contrast, ammonium concentrations as high as 15 mmol kg^{-1} have been measured (Von Damm *et al.*, 1985b). N_2 is also present. Silica concentrations are extremely high, due to interaction with host rocks at high temperature, at depth. Of course, in these systems, it could be debated what actually constitutes a "nutrient." For example, both dissolved H_2 and H_2S (as well as numerous other reduced species), represent important primary energy sources for the chemosynthetic communities invariably found at hydrothermal vent sites.

11.2.3.5 Organic geochemistry of hydrothermal vent fluids

Studies of the organic chemistry of vent fluids are truly in their infancy and little field data exist (Holm and Charlou, 2001). There are predictions of what should be present based on experimental work (e.g., Berndt *et al.*, 1996; Cruse and Seewald, 2001) and thermodynamic modeling (e.g., McCollom and Shock, 1997, 1998). These results await confirmation "in the field." Significant data on the organic geochemistry are uniquely available for the Guaymas Basin hydrothermal system (e.g., Simoneit, 1991), which underlies a very highly productive area of the ocean, and is hosted in an organic-rich sediment-filled basin.

11.2.4 Geographic Variations in Vent-fluid Compositions

11.2.4.1 The role of the substrate

There are systematic reasons for some of the variations observed in vent-fluid compositions. One of the most profound is the involvement of sedimentary material in the hydrothermal circulation cell, as seen at sediment-covered ridges (Von Damm *et al.*, 1985b; Campbell *et al.*, 1994; Butterfield *et al.*, 1994). The exact differences this imposes depend upon the nature of the sedimentary material involved: the source/nature of the aluminosilicate material, the proportions and type of organic matter it contains and the proportion and type of animal tests present, calcareous and/or siliceous. In the known sediment-hosted systems (Guaymas Basin, EPR; Escanaba Trough, Gorda Ridge; Middle Valley, JdFR; and perhaps the Red Sea) basalts are intercalated with the sediments or else underlie them. Hence, in these systems, reactions with basalt are overprinted by those with the sediments. In most cases, depending on the exact nature as well as thickness of the sedimentary cover, this causes a rise in the pH, which results in the precipitation of metal sulfides before the fluids reach the seafloor. The presence of carbonate and/or organic matter also buffers the pH to significantly higher levels (at least pH 5 at 25 °C and 1 atm).

The chemical composition of most seafloor vent fluids can be explained by reaction of unaltered basalt with seawater. However, in some cases the best explanation for the fluid chemistry is that the fluids have reacted with basalt that has already been highly altered (Von Damm *et al.*, 1998). Two indicators for this are higher pH values (pH $\sim$ 4 versus pH $\sim$ 3.3 at 25 °C), as well as lower K/Na molar ratios and lower concentrations of the REEs. In the last several years, several vent sites have been sampled that cannot be explained by these mechanisms. At some locations, vent fluids must be generated by reaction of seawater with ultramafic rocks (Douville *et al.*, 2002). These fluids can also have major variations from each other, depending on the temperature regime. In high-temperature fluids that have reacted with an ultramafic substrate, silicon contents are generally lower than in basalt-hosted fluids; H_2, calcium, and iron contents are generally higher, but these fluids remain acidic (Douville *et al.*, 2002). Not much is yet known about the seafloor fluids that are generated from lower-temperature ultramafic hydrothermal circulation. In the one example studied thus far (Lost City) the measured pH is greater than that in seawater, and fluid compositions are clearly controlled by a quite distinct set of serpentinization-related reactions (Kelley *et al.*, 2001). An illustration of this fundamental difference is given by magnesium which is quantitatively stripped from "black smoker" hydrothermal fluids but exhibits $\sim$20–40% of seawater concentrations (9–19 mM) in the Lost City vents, leading to the unusual magnesium-rich mineralization observed at this site (see later). The seafloor fluids from Lost City are remarkably similar to those found in continental hydrothermal systems hosted in ultramafic environments (Barnes *et al.*, 1972). In back-arc spreading centers such as those found in the western Pacific, andesitic rock types are common and profound differences in vent-fluid compositions arise (Fouquet *et al.*, 1991; Ishibashi and Urabe, 1995, Gamo *et al.*, 1997). These fluids can be both more acidic and more oxidizing than is typical and the relative enrichments of transition metals and volatile species in these fluids are quite distinct from what is observed in basalt-hosted systems.

Major differences in substrate are, therefore, reflected in the compositions of vent fluids. Insufficient trace-metal data for vent fluids exist, however, to discern more subtle substrate differences, e.g., between EMORB and NMORB on non-hotspot influenced ridges. Where the ridge axis is influenced by hot-spot volcanism, some differences may be seen in the fluid compositions, as, for example, the high barium in the Lucky Strike vent fluids, but in this case most of the fluid characteristics (e.g., potassium concentrations) do not show evidence for an enriched substrate (Langmuir *et al.*, 1997; Von Damm *et al.*, 1998).

11.2.4.2 The role of temperature and pressure

Temperature, of course, plays a major role in determining vent-fluid compositions. Pressure is often thought of as less important than temperature, but the relative importance of the two depends on the exact *P–T* conditions of the fluids. Because of the controls that pressure and temperature conditions exert on the thermodynamics as well as the physical properties of the fluids, the two effects cannot be discussed completely independently from each other. Not only do *P–T* conditions govern phase separation, as discussed above, they control transport in the fluids and mineral dissolution and precipitation reactions. Temperature, especially, plays a role in the quantities of elements that are leached from the host rocks. When temperature decreases, as it often does due to conductive cooling as fluids rise through the oceanic crust, most minerals become less soluble. Due to these decreasing mineral solubilities, transition metals and sulfide, in particular, may be lost from the ascending fluids. *P–T* conditions in the fluids also control the strength of the aqueous complexes. In general, as *P* and *T* rise, aqueous species become more associated. Because of the properties of water at the critical point (the dielectric constant goes to zero), all the species must become associated, as there can be no charged species in solution at the critical point. Therefore, transport of species can increase markedly as the critical point is approached because there will be smaller amounts of the (charged) species present in solution which are needed if mineral solubility products are to be exceeded (Von Damm *et al.*, 2003). It is in this critical point region that small changes in pressure can be particularly significant—for example, as a fluid is rising in the upflow ("discharge") limb of a hydrothermal cell (Figure 5). Because most vent-fluid compositions are controlled by equilibrium or steady state, and because the equilibrium constants for these reactions change as a function of pressure and temperature, *P–T* conditions will ultimately control all vent-fluid compositions. One problem associated with modeling vent fluids and trying to understand the controls on their compositions is that we really do not know the temperature in the "reaction zone." Basaltic lavas are emplaced at temperatures of 1,100–1,200 °C, but rocks must be brittle to retain fractures that allow fluid flow, and this brittle–ductile transition lies in the range 500–600 °C. A commonplace statement is that the reaction zone temperature is ~450 °C, but we do not really know this value with any accuracy, nor how variable it may be from one location to another. At the seafloor, we have sampled fluids with exit temperatures as high as ~405 °C. Hence, in addition to the constraints provided by the recognition that at least subcritical phase separation is pervasive (see above) we can further determine that (i) reaction zone temperatures must exceed 405 °C, at least in some cases, and (ii) that in cases where evidence for supercritical phase separation has been determined (e.g., Butterfield *et al.*, 1994; Von Damm *et al.*, 1998) temperatures must exceed 407 °C within the oceanic crust.

The pressure conditions at any hydrothermal field are largely controlled by the depth of the overlying water column. Pressure is most critical in terms of phase separation and vent fluids are particularly sensitive to small changes in pressure when close to the critical point. It is in this region, close to the critical point, when fluids are very expanded (i.e., at very low density) that small changes in pressure can cause significant changes in vent-fluid composition.

11.2.4.3 The role of spreading rate

When one considers tables of vent-fluid chemical data, one cannot separate vents from ultrafast- versus slow-spreading ridges (Table 2); the range of chemical compositions from each of these two end-member types of spreading regime overlap. There has been much debate in the marine geological literature whether rates of magma supply, rather than spreading rate, should more correctly be applied when defining ridge types (e.g., "magma-starved" versus "magma-rich" sections of ridge crest). While any one individual segment of ridge crest undoubtedly passes through different stages of a volcanic–tectonic cycle, regardless of spreading rate (e.g., Parson *et al.*, 1993; Haymon, 1996), it is generally the case that slow-spreading ridges are relatively magma-starved whilst fast spreading ridges are more typically magma-rich. Consequently, we continue to rely upon (readily quantified) spreading rate (De Mets *et al.*, 1994) as a proxy for magma supply. To a first approximation, ridge systems in the Atlantic Ocean are slow-spreading, while fast-spreading ridges are only found in the Pacific. The Pacific contains ridges that spread at rates from ~15 cm yr^{-1} full opening rate to a minimum of ~2.4 cm yr^{-1} on parts of the Gorda Ridge (comparable to the northern MAR). In the Indian Ocean, ridge spreading varies between intermediate (~6 cm yr^{-1}, CIR and SEIR) and very slow rates (<2.0 cm yr^{-1}, SWIR). In the Arctic Ocean the spreading rate is the slowest known, decreasing to <1.0 cm yr^{-1} from west to east as the Siberian shelf is approached. Discussion of vent-fluid compositions from different oceans, therefore, often approximates closely to variations in vent-fluid chemistries at different spreading rates. Although tables of the ranges of vent-fluid chemistries do not show distinct differences between ocean basins, some important differences do become apparent when those data are modeled. (NB: Although there is now firm evidence for hydrothermal activity in the Arctic, those systems have not yet been sampled for vent fluids; similarly, in the Indian Ocean, only two sites have recently been discovered). Consequently, meaningful comparisons can only readily be made, at present, between Atlantic and Pacific vent-fluid compositions. In systems on the slow spreading MAR, the calculated f_{O_2} of hydrothermal fluids is higher, suggesting that these systems are more oxidizing. Also, for example, both TAG and Lucky Strike vent fluids contain relatively little potassium compared to sodium (Edmond *et al.*, 1995; Von Damm *et al.*, 1998). Boron is also low in some of the Atlantic sites, especially at TAG and Logatchev (You *et al.*, 1994; Bray and Von Damm, 2003a). The explanation for these observations is that on the slow-spreading MAR, hydrothermal activity is active for a much longer period of time at any given site (also reflected in the relative sizes of the hydrothermal deposits formed: Humphris *et al.*, 1995; Fouquet *et al.*, submitted). Consequently, MAR vent fluids become more oxic because more dissolved seawater SO_4 has penetrated as deep as the reaction zone; the rocks within the hydrothermal flow cell have been more completely leached and altered. Because of the more pronounced tectonic (rather than volcanic) activity that is associated with slow spreading ridges, rock types that are normally found at greater depths within the oceanic crust can be exposed at or near the surface. Thus, hydrothermal sites have been located along the MARs that are hosted in ultramafic rocks: the Rainbow, Logatchev, and Lost City sites. No ultramafic-hosted systems are expected to occur, by contrast, along the fast-spreading ridges of the EPR.

11.2.4.4 The role of the plumbing system

Another fundamental difference observed in the nature of hydrothermal systems at fast- and slow-spreading ridges concerns intra-field differences in vent-fluid compositions. (Note that the terms "vent field" and "vent area" are often used interchangeably, have no specific size classifications, and may be used differently by different authors. In our usage, "vent area" is smaller, referring to a cluster of vents within 100 m or so, while a "vent field" may stretch for a kilometer or more along the ridge axis—but this is by no means a standard definition.) On a slow spreading ridge, such as the MAR, all of the fluids venting, for example, at the TAG site, can be shown to have a common source fluid that may have undergone some change in composition due to near surface processes such as mixing and/or conductive cooling. Many of the fluids can also be related to each other at the Lucky Strike site. In contrast, on fast spreading ridges, vents that may be within a few tens of meters of each other, clearly have different source fluids. A plausible explanation for these differences would be that vents on slow spreading ridges are fed from greater depths than those on fast-spreading ridges, with emitted fluids channeled upward from the subsurface along fault planes or other major tectonic fractures. Hence, in at least some cases, hydrothermal activity found on slow spreading ridges may be located wherever fluids have been preferentially channeled. Active vent sites on slow spreading ridges also appear to achieve greater longevity, based on the size of the sulfide structures and mounds they have produced. Fluids on fast spreading ridges, by contrast, are fed by much shallower heat sources and the conduits for these fluids appear to be much more localized, resulting in the very pronounced chemical differences often observed between immediately adjacent vent structures. Clearly, the plumbing systems at fast and slow spreading ridges must be characterized by significant, fundamental differences.

11.2.5 Temporal Variability in Vent-fluid Compositions

The MOR is, in effect, one extremely long, continuous submarine volcano. While volcanoes are commonly held to be very dynamic features, however, little temporal variability was observed for more than the first decade of work on hydrothermal systems. Indeed, a tendency arose not to view the MOR as an active volcano, at least on the timescales that were being worked on. This perspective was changed dramatically in the early 1990s. Together with evidence for recent volcanic eruptions at several sites, profound temporal variability in vent-fluid chemistries, temperatures, and styles of venting were also observed (Figure 7). In one case, the changes observed at a single vent almost span the full range of known compositions reported from throughout the globe. These temporal variations in hydrothermal venting reflect changes in the nature of the underlying heat source. The intrusion of a basaltic dike into the upper ocean crust, which may or may not be accompanied by volcanic extrusion at the seafloor, has been colloquially termed "the quantum unit of ocean accretion." These dikes are of the order 1 m wide, 10 km long, and can extend hundreds of meters upward through the upper crust toward the ocean floor. These shallow-emplaced and relatively small, transient, heat sources provide most, if not all of the heat that supports venting immediately following magmatic emplacement. Over timescales of as little as a year, however, an individual dike will have largely cooled and the heat source driving any continuing vent activity deepens. An immediate result is a decrease in measured exit temperatures for the vent fluids, because more heat is now lost, conductively, as the fluids rise from deeper within the oceanic crust. Vent-fluid compositions change, too, because the conditions of phase separation change; so, too, do the subsurface path length and residence time, such that the likelihood that circulating fluids reach equilibrium or steady state with the ocean crust also vary. Detailed time-series studies at sites perturbed by magmatic emplacements have shown that it is the vapor phase which vents first, in the earliest stages after a magmatic/volcanic event, while the high-chlorinity liquid phase is expelled somewhat later. In the best documented case study available, "brine" fluids were actively venting at a location some three years after the vapor phase had been expelled from the same individual hydrothermal chimney; at other event-affected vent sites, similar evolutions in vent-fluid composition have been observed over somewhat longer timescales.

The temporal variability that has now been observed has revolutionized our ideas about the functioning of hydrothermal systems and the timescales over which processes occur on the deep ocean floor. It is no exaggeration to state that processes we thought would take 100–1,000 yr, have been seen to occur in <10 yr. The majority of magmatic intrusions/eruptions that have been detected have been along the JdFR (Cleft Segment, Co-axial Segment, and Axial Volcano) where acoustic monitoring of the T-phase signal that accompanies magma migration in the upper crust has provided real-time data for events in progress and allowed "rapid response" cruises to be organized at these sites, within days to weeks. We also have good evidence for two volcanic events on the ultrafast spreading southern EPR, but the best-studied eruption site, to date, has been at 9°45–52′ N on the EPR. Serendipitously, submersible studies began at 9–10° N EPR less than one month after volcanic eruption at this site (Haymon *et al.*, 1993; Rubin *et al.*, 1994). Profound chemical changes (more than a factor of two in some cases) were noted at some of these vents during a period of less than a month (Von Damm *et al.*, 1995; Von Damm, 2000). Subsequently, it has become clear that very rapid changes occur within an initial one-year period which are related to changes in the conditions of phase separation and water rock reaction. These, in turn, are presumed to reflect responses to the mining of heat from the dike-intrusion, including lengthening of the reaction path and increases in the residence time of the fluids within the crust. At none of the other eruptive sites has it been possible to complete comparable direct sampling of vent fluids within this earliest "post-event" time period. It is now clear that the first fluid to be expelled is the vapor phase (whether formed as a result of sub- or supercritical phase separation), probably because of its lower density. What happens next, however, is less clear. In several cases, the "brine" (liquid) phase has been emitted next. In some vents this has occurred as a gradual progression to higher-chlorinity fluids; in other vents, the transition appears to occur more as a step function—although those observations may be aliased by the episodic nature of the sampling programs involved. What is most certainly the case at 9° N EPR, however, is that following initial vapor-phase expulsion, some vents have progressed to venting fluids with chlorinities greater than seawater much faster (≤ 3 yr), than others (~ 10 yr), and several have never made the transition. Further, in some parts of the eruptive area, vapor phase fluids are still the predominant type of fluid being emitted more than a decade after the eruption event. Fluids exiting from several of the vents have begun to decrease in chlorinity again, without ever having reached seawater concentrations. Conversely, other systems (most notably those from the Cleft segment) have been emitting vent fluids with chlorinities approximately twice that of seawater for more than a decade. If one wanted to sustain an argument that hydrothermal systems followed a vapor-to-brine phase evolution as a system ages (e.g., Butterfield *et al.*, 1997), it would be difficult to reconcile the observation that systems that are presumed to be relatively "older" (e.g., those on the northern Gorda Ridge) vent vapor-phase fluids, only (Von Damm *et al.*, 2001). Finally, one vent on the southern EPR, in an area with no evidence for a recent magmatic event, is emitting fluids which are phase separating "real time," with vapor exiting from the top of the structure, and brine from the bottom, simultaneously (Von Damm *et al.*, 2003). Fundamentally, there is a chloride-mass balance problem at many known hydrothermal sites. For example, at 21° N EPR, where high-temperature venting was first discovered and an active system is now known to have persisted for at least 23 years (based on sampling expeditions from 1979 and 2002) *only* low-chlorinity fluids are now being emitted (Von Damm *et al.*, 2002). Clearly there must be some additional storage and/or transport of higher-chlorinity fluids within the underlying crust. Our understanding of such systems is, at best, poor. What *is* clear is that pronounced temporal variability occurs at many vent sites, most notably at those that have been affected by magmatic events. There is also evidence for changes accompanying seismic events that are not related to magma-migration, but most likely related to cracking within the upper ocean crust (Sohn *et al.*, 1998; Von Damm and Lilley, 2003).

In marked contrast to those sites where volcanic eruptions and/or intrusions (diking events) have been detected, there are several other sites that have been sampled repeatedly over timescales of about two decades where no magmatic activity is known to have occurred. At some of these sites, chemical variations observed over the entire sampling period fall within the analytical error of our measurement techniques. The longest such time-series is for hydrothermal venting at 21° N on the EPR, where black smokers were first discovered in 1979, and where there has been remarkably little change in the composition of at least some of the vent fluids. Similarly, the Guaymas Basin vent site was first sampled in January 1982 (Von Damm *et al.*, 1985b), South Cleft on the JdF ridge in 1984 (Von Damm and Bischoff, 1987), and TAG on the MAR in 1986 (Campbell *et al.*, 1988). All these sites have exhibited very stable vent-fluid chemistries, although only TAG would be considered as a "slow spreading" vent site. Nevertheless, it is the TAG vent site that has shown perhaps the most remarkable stability in its vent-fluid compositions; these have remained invariant for more than a decade, even after perturbation from the drilling of a series of 5 ODP holes direct into the top of the active sulfide mound (Humphris *et al.*, 1995; Edmonds *et al.*, 1996).

Accounting for temporal variability (or lack thereof) when calculating hydrothermal fluxes is, clearly, problematic. It is very difficult to calculate the volume of fluid exiting from a hydrothermal system accurately. Many of the differences from seawater are most pronounced in the early eruptive period ($\sim$1 yr), which is also the time when fluid temperatures are hottest (Von Damm, 2000). Visual observations suggest that this is a time of voluminous fluid flow, which is not unexpected given that an enhanced heat source will have recently been emplaced directly at the seafloor in the case of an eruption, or, in the case of a dike intrusion, at shallow depths beneath. The upper oceanic crust exhibits high porosity, filled with ambient seawater. At eruption/intrusion, this seawater will be heated rapidly, its density will decrease, the resultant fluid will quickly rise, and large volumes of unreacted, cooler seawater will be drawn in and quickly expelled. It is not unreasonable to assume, therefore, that the water flux through a hydrothermal system may be at its largest during this initial period, just when chemical compositions are most extreme (Von Damm *et al.*, 1995; Von Damm, 2000). The key to the problem, therefore, probably lies in determining how much time a hydrothermal system spends in its "waxing" (immediate post-eruptive) stage when compared to the time spent at "steady state" (e.g., as observed at 21° N EPR) and in a "waning" period, together with an evaluation of the relative heat, water, and chemical fluxes associated with each of these different stages. If fluid fluxes and chemical anomalies are greatest in the immediate post-eruptive period, for example, the initial 12 months of any vent-field eruption may be geochemically more significant than a further 20 years of "steady-state" emission. At fast-spreading ridges, new eruptions might even occur faster than such a vent "lifecycle" can be completed. Alternately, the converse may hold true: early-stage eruptions may prove relatively insignificant over the full lifecycle of a prolonged, unperturbed hydrothermal site.

To advance our understanding of the chemical variability of vent fluids, it will be important to continue to find new sites that may be at evolutionary stages not previously observed. Equally, it will be important to continue studies of temporal variability at known sites: both those that have varied in the past and those that have appeared to be stable over the time intervals at which they have been sampled. Understanding the mechanisms and physical processes that control these vent-fluid compositions are key to calculating hydrothermal fluxes.

11.3 THE *NET* IMPACT OF HYDROTHERMAL ACTIVITY

It is important to remember that the gross chemical flux associated with expelled vent fluids (Section 11.2) is not identical to the net flux from hydrothermal systems. Subsurface hydrothermal circulation can have a net negative flux for some chemicals, the most obvious being magnesium which is almost quantitatively removed from the starting seawater and is added instead to the oceanic crust. Another example of such removal is uranium, which is also completely removed from seawater during hydrothermal circulation and then recycled into the upper mantle through subduction of altered oceanic crust. Even where, compared to starting seawater, there is no concentration gain or loss, an element may nevertheless undergo almost complete isotopic exchange within the oceanic crust indicating that none of the substance originally present in the starting seawater has passed conservatively through the hydrothermal system—the most obvious example being that of strontium. We present a brief summary of ocean crust mineralization in Section 11.4.1, but a more detailed treatment of ocean crust alteration is given elsewhere.

For the remainder of this chapter we concentrate, instead, upon the fate of hydrothermal discharge once it reaches the seafloor. Much of this material, transported in dissolved or gaseous form in warm or hot fluids, does not remain in solution but forms solid phases as fluids cool and/or mix with colder, more alkaline seawater. These products, whose formation may be mediated as well as modified by a range of biogeochemical processes, occur from the ridge axis out into the deep ocean basins. Massive sulfide as well as silicate, oxide, and sometimes carbonate deposits formed from high-temperature fluids are progressively altered by high-temperature metasomatism, as well as low-temperature oxidation and mass wasting—much of which may be biologically mediated. Various low-temperature deposits may also be formed, again often catalyzed by biological activity. In addition to these near-vent hydrothermal products, abundant fine-grained particles are formed in hydrothermal plumes, which subsequently settle to the seafloor to form metalliferous sediments, both close to vent sites and across ridge flanks into adjacent ocean basins. The post-depositional fates for these near- and far-field deposits remain poorly understood. Sulfide deposits, for example, may undergo extensive diagenesis and dissolution, leading to further release of dissolved chemicals into the deep ocean. Conversely, oxidized hydrothermal products may remain well-preserved in the sedimentary record and only be recycled via subduction back into the Earth's interior. On ridges where volcanic eruptions are frequent, both relatively fresh and more oxidized deposits may be covered over by subsequent lava flows (on the timescale of a decade) and, thus, become assimilated into the oceanic crust, isolated from the overlying water- and sediment-columns. In the following sections we discuss the fates of various hydrothermal "products" in order of their distance from the vent-source: near-vent deposits (Section 11.4); hydrothermal plumes (Section 11.5); and hydrothermal sediments (Section 11.6).

11.4 NEAR-VENT DEPOSITS

11.4.1 Alteration and Mineralization of the Upper Ocean Crust

Hydrothermal circulation causes extensive alteration of the upper ocean crust, reflected both in mineralization of the crust and in changes to physical properties of the basement (Alt, 1995). The direction and extent of chemical and isotopic exchange between seawater and oceanic crust depends on variations in temperature and fluid penetration and, thus, vary strongly as a function of depth. Extensive mineralization of the upper ocean crust can occur where metals leached from large volumes of altered crust become concentrated at, or close beneath, the sediment–water interface (Hannington *et al.*, 1995, Herzig and Hannington, 2000). As we have seen (Section 11.2), hydrothermal circulation within the ocean crust can be subdivided into three major components—the *recharge, reaction*, and *discharge* zones (Figure 5).

Recharge zones, which are broad and diffuse, represent areas where seawater is heated and undergoes reactions with the crust as it penetrates, generally downwards. (It is important to remember, however, that except where hydrothermal systems are sediment covered, the location of the recharge zone has never been established; debate continues, for example, whether recharge occurs *along* or *across* axis.) Important reactions in the recharge zone, at progressively increasing temperatures, include: low-temperature oxidation, whereby iron oxyhydroxides replace olivines and primary sulfides; fixation of alkalis (potassium, rubidium, and caesium) in celadonite and nontronite (ferric mica and smectite, respectively) and fixation of magnesium, as smectite (<200 °C) and chlorite (>200 °C). At temperatures exceeding ~130–150 °C, two other key reactions occur: formation of anhydrite ($CaSO_4$) and mobilization of the alkalis (potassium, rubidium, lithium) (Alt, 1995). Upon subduction, the altered mineralogy and composition of the ocean crust can lead to the development of chemical and isotopic heterogeneities, both in the mantle and in the composition of island arc volcanic rocks. This subject is discussed in greater detail by Staudigel.

The *reaction* zone represents the highest pressure and temperature (likely >400 °C) conditions reached by hydrothermal fluids during their subsurface circulation; it is here that hydrothermal vent fluids are believed to acquire much of their chemical signatures (Section 11.2). As they become buoyant, these fluids then rise rapidly back to the seafloor through *discharge* zones. The deep roots of hydrothermal discharge zones have only ever been observed at the base of ophiolite sequences (e.g., Nehlig *et al.*, 1994). Here, fluid inclusions and losses of metals and sulfur from the rocks indicate alteration temperatures of 350–440 °C (Alt and Bach, 2003) in reasonable agreement with vent-fluid observations (Section 11.2). Submersible investigations and towed camera surveys of the modern seafloor have allowed surficial

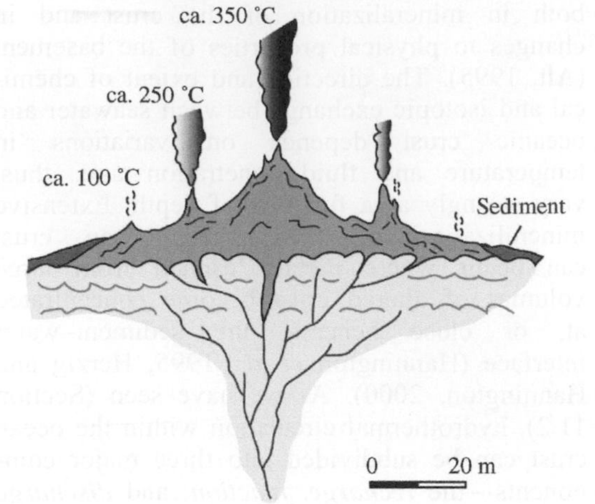

Figure 10 Schematic illustration of an idealized hydrothermal sulfide mound including: branching stockwork zone beneath the mound; emission of both high-temperature (350 °C) and lower-temperature (100–250 °C) fluids from the top of the mound together with more ephemeral diffuse flow; and deposition by mass-wasting of an apron of sulfidic sediments around the edges of the mound.

hydrothermal deposits to be observed in some detail (see next section). By contrast, the alteration pipes and "stockwork" zones that are believed to form the "roots" underlying all seafloor hydrothermal deposits (Figure 10) and which are considered to be quantitatively important in global geochemical cycles (e.g., Peucker-Ehrenbrink *et al.*, 1994) remain relatively inaccessible. Direct sampling of the stockwork beneath active seafloor hydrothermal systems has only been achieved on very rare occasions, e.g., through direct sampling from ODP drilling (Humphris *et al.*, 1995) or through fault-face exposure at the seabed (e.g., Fouquet *et al.*, 2003). Because of their relative inaccessibility, even compared to all other aspects of deep-sea hydrothermal circulation, study of sub-seafloor crustal mineralization remains best studied in ancient sulfide deposits preserved in the geologic record on-land (Hannington *et al.*, 1995).

11.4.2 Near-vent Hydrothermal Deposits

The first discoveries of hydrothermal vent fields (e.g., Galapagos; EPR, 21° N) revealed three distinctive types of mineralization: (i) massive sulfide mounds deposited from focused high-temperature fluid flow, (ii) accumulations of Fe–Mn oxyhydroxides and silicates from low-temperature diffuse discharge, and (iii) fine-grained particles precipitated from hydrothermal plumes. Subsequently, a wide range of mineral deposits have been identified that are the result of hydrothermal discharge, both along the global ridge crest and in other tectonic settings (Koski *et al.*, 2003). Of course, massive sulfides only contain a fraction of the total dissolved load released from the seafloor. Much of this flux is delivered to ridge flanks via dispersion in buoyant and nonbuoyant hydrothermal plumes (Section 11.5). In addition, discoveries such as the carbonate-rich "Lost City" deposits (Kelley *et al.*, 2001), silica-rich deposits formed in the Blanco Fracture Zone (Hein *et al.*, 1999), and metal-bearing fluids on the flanks of the JdFR (Mottl *et al.*, 1998) remind us that there is still much to learn about the formation of hydrothermal mineral deposits.

Haymon (1983) proposed the first model for how a black smoker chimney forms (Figure 11). The first step is the formation of an anhydrite ($CaSO_4$) framework due to the heating of seawater, and mixing of vent fluids with that seawater. The anhydrite walls then protect subsequent hydrothermal fluids from being mixed so extensively with seawater, as well as providing a template onto which sulfide minerals can precipitate as those fluids cool within the anhydrite structure. As the temperature and chemical compositions within the chimney walls evolve, a zonation of metal sulfide minerals develops, with more copper-rich phases being formed towards the interior, zinc-rich phases towards the exterior, and iron-rich phases ubiquitous. This model is directly analogous to the concept of an "intensifying hydrothermal system" developed by Eldridge *et al.* (1983), in which initial deposition of a fine-grained mineral carapace restricts mixing of hydrothermal fluid and seawater at the site of discharge. Subsequently, less-dilute, higher-temperature (copper-rich) fluids interact with the sulfides within this

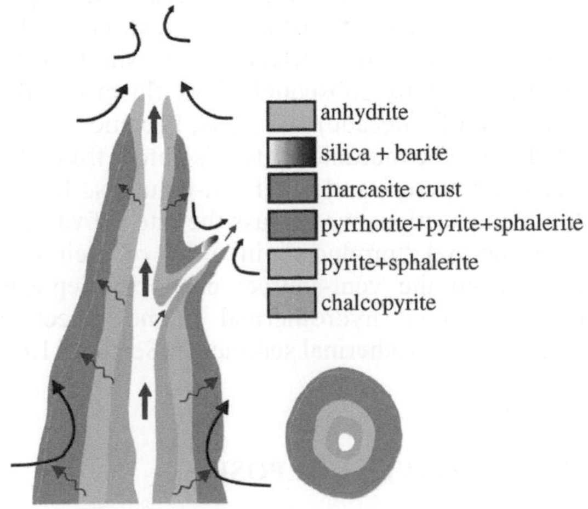

Figure 11 Schematic diagram showing mineral zonation in cross-section and in plan view for a typical black smoker chimney. Arrows indicate direction of inferred fluid flow (after Haymon, 1983).

carapace to precipitate chalcopyrite and mobilize more soluble, lower-temperature metals such as lead and zinc toward the outer, cooler parts of the deposit. Thus, it is the steep temperature and chemical gradients, caused by both mixing and diffusion, which account for the variations in wall mineralogy and Cu–Zn zonation observed in both chimneys and larger deposits. These processes, initially proposed as part of a conceptual model, have subsequently been demonstrated more rigorously by quantitative geochemical modeling of hydrothermal fluids and deposits (Tivey, 1995).

Drilling during Ocean Drilling Program Leg 158 revealed that similar internal variations can also occur on much larger scales—e.g., across the entire TAG mound (Humphris *et al.*, 1995; Petersen *et al.*, 2000). That work revealed the core of the mound to be dominated by chalcopyrite-bearing massive pyrite, pyrite–anhydrite and pyrite–silica breccias whilst the mound top and margins contained little or no chalcopyrite but more sphalerite and higher concentrations of metals soluble at lower temperatures (e.g., zinc, gold). The geochemical modeling results of Tivey (1995) point to a mechanism of entrainment of seawater into the focused upflow zone within the mound which would, almost simultaneously: (i) induce the precipitation of anhydrite, chalcopyrite, pyrite, and quartz; (ii) decrease the pH of the fluid; and (iii) mobilize zinc and other metals. When combined, the processes outlined above—zone refining and the entrainment of seawater into active sulfide deposits—appear to credibly explain mineralogical and chemical features observed both in modern hydrothermal systems (e.g., the TAG mound) and in "Cyprus-type" deposits found in many ophiolites of orogenic belts (Hannington *et al.*, 1998).

A quite different form of hydrothermal deposit has also been located, on the slow-spreading MAR. The Lost City vent site (Kelley *et al.*, 2001) occurs near 30° N on the MAR away from the more recently erupted volcanic ridge axis. Instead, it is situated high up on a tectonic massif where faulting has exposed variably altered peridotite and gabbro (Blackman *et al.*, 1998). The Lost City field hosts at least 30 active and inactive spires, extending up to 60 m in height, on a terrace that is underlain by diverse mafic and ultramafic lithologies. Cliffs adjacent to this terrace also host abundant white hydrothermal alteration both as flanges and peridotite mineralization, which is directly akin to deposits reported from Alpine ophiolites (Früh-Green *et al.*, 1990). The Lost City chimneys emit fluids up to 75 °C which have a very high pH (9.0–9.8) and compositions which are rich in H_2S, CH_4, and H_2—consistent with serpentinization reactions (Section 11.2)—but low in dissolved silica and metal contents (Kelley *et al.*, 2001). Consistent with this, the chimneys of the Lost City field are composed predominantly of magnesium and calcium-rich carbonate and hydroxide minerals, notably calcite, brucite, and aragonite.

In addition to the sulfide- and carbonate-dominated deposits described above, mounds and chimneys composed of Fe- and Mn-oxyhydroxides and silicate minerals also occur at tectonically diverse rift zones, from MORs such as the Galapagos Spreading Center to back-arc systems such as the Woodlark Basin (Corliss *et al.*, 1978; Binns *et al.*, 1993). Unlike polymetallic sulfides, Fe–Mn oxide-rich, low-temperature deposits should be chemically stable on the ocean floor. Certainly, metalliferous sediments in ophiolites—often referred to as "umbers"—have long been identified as submarine hydrothermal deposits formed in ancient ocean ridge settings. These types of ophiolite deposit may be intimately linked to the Fe–Mn–Si oxide "mound" deposits formed on pelagic ooze near the Galapagos Spreading Center (Maris and Bender, 1982). It has proven difficult, however, to determine the precise temporal and genetic relationship of umbers to massive sulfides, not least because no gradation of Fe–Mn–Si oxide to sulfide mineralization has yet been reported from ophiolitic terranes. The genetic relationship between sulfide and oxide facies deposits formed at modern hydrothermal sites also remains enigmatic. Fe–Mn–Si oxide deposits may simply represent "failed" massive sulfides. Alternately, it may well be that there are important aspects of, for example, axial versus off-axis plumbing systems (e.g., porosity, permeability, chemical variations caused by phase separation) or controls on the sulfur budget of hydrothermal systems that remain inadequately understood. What seems certain is that the three-dimensional problem of hydrothermal deposit formation (indeed, 4D if one includes temporal evolution) cannot be solved from seafloor observations alone. Instead, what is required is a continuing program of seafloor drilling coupled with analogue studies of hydrothermal deposits preserved on land.

11.5 HYDROTHERMAL PLUME PROCESSES

11.5.1 Dynamics of Hydrothermal Plumes

Hydrothermal plumes form wherever buoyant hydrothermal fluids enter the ocean. They represent an important dispersal mechanism for the thermal and chemical fluxes delivered to the oceans while the processes active within these plumes serve to modify the gross fluxes from venting, significantly. Plumes are of further interest to geochemists because they can be exploited in the detection and location of new hydrothermal fields *and* for the calculation of total integrated fluxes from any particular vent field. To biologists, hydrothermal plumes represent an effective transport mechanism for dispersing vent fauna, aiding gene-flow between adjacent vent sites along the global ridge crest (e.g., Mullineaux and France, 1995). In certain circumstances the heat and energy released into hydrothermal plumes could act as a driving force for mid-depth oceanic circulation (Helfrich and Speer, 1995).

Present day understanding of the dynamics of hydrothermal plumes is heavily influenced by the theoretical work of Morton *et al.* (1956) and Turner (1973). When high-temperature vent fluids are expelled into the base of the much colder, stratified, oceanic water column they are buoyant and begin to rise. Shear flow between the rising fluid and the ambient water column produces turbulence and vortices or eddies are formed which are readily visible in both still and video-imaging of active hydrothermal vents. These eddies or vortices act to entrain material from the ambient water column, resulting in a continuous dilution of the original vent fluid as the plume rises. Because the oceans exhibit stable density-stratification, this mixing causes progressive dilution of the buoyant plume with water which is denser than both the initial vent fluid and the overlying water column into which the plume is rising. Thus, the plume becomes progressively less buoyant as it rises and it eventually reaches some finite maximum height above the seafloor, beyond which it cannot rise (Figure 12). The first, rising stage of hydrothermal plume evolution is termed the *buoyant plume*. The later stage, where plume rise has ceased and hydrothermal effluent begins to be dispersed laterally, is termed the *nonbuoyant plume* also referred to in earlier literature as the *neutrally buoyant plume*.

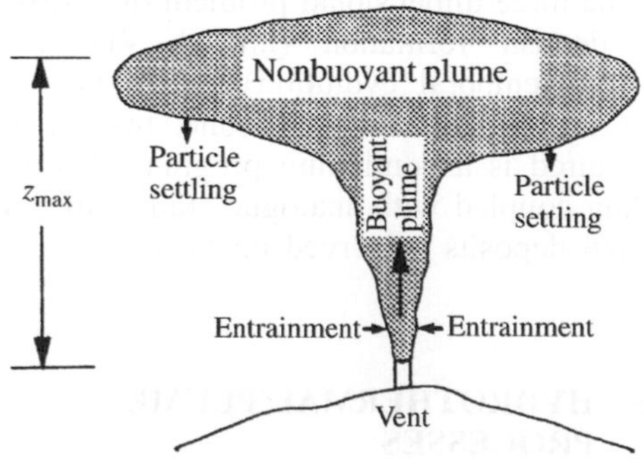

Figure 12 Sketch of the hydrothermal plume rising above an active hydrothermal vent, illustrating entrainment of ambient seawater into the buoyant hydrothermal plume, establishment of a nonbuoyant plume at height z_{max} (deeper than the maximum height of rise actually attained due to momentum overshoot) and particle settling from beneath the dispersing nonbuoyant plume (after Helfrich and Speer, 1995).

The exact height reached by any hydrothermal plume is a complex function involving key properties of both the source vent fluids and the water column into which they are injected—notably the initial buoyancy of the former and the degree of stratification of the latter. A theoretical approach to calculating the maximum height-of-rise that can be attained by any hydrothermal plume is given by Turner (1973) with the equation:

$$z_{max} = 3.76F_0^{1/4}N^{-3/4} \quad (1)$$

where F_0 and N represent parameters termed the buoyancy flux and the Brunt–Väisälä frequency, respectively. The concept of the buoyancy flux, F_0 (units $cm^4\,s^{-3}$) can best be understood from an explanation that the product $F_0\rho_0$ represents the total weight deficiency produced at the vent source per unit time (units $g\,cm\,s^{-3}$). The Brunt–Väisälä frequency, also termed the buoyancy frequency, N (units s^{-1}) is defined as:

$$N^2 = -(g/\rho_0)\mathrm{d}\rho/\mathrm{d}z \quad (2)$$

where g is the acceleration due to gravity, ρ_0 is the background density at the seafloor and $\mathrm{d}\rho/\mathrm{d}z$ is the ambient vertical density gradient. In practice, buoyant hydrothermal plumes always exceed this theoretical maximum because, as they reach the level z_{max} the plume retains some finite positive vertical velocity. This leads to "momentum overshoot" (Turner, 1973) and "doming" directly above the plume source before this (now negatively buoyant) dome collapses back to the level of zero buoyancy (Figure 12).

Note also the very weak dependence of emplacement height (z_{max}) upon the buoyancy flux or heat flux of any given vent source. A doubling of z_{max} for any plume, for example, could only be achieved by a 16-fold increase in the heat flux provided by its vent source. By contrast, the ambient water column with which the buoyant plume becomes progressively more diluted exerts a significant influence because the volumes entrained are nontrivial. For a typical plume with $F_0 = 10^{-2}\,m^4\,s^{-3}$, $N = 10^{-3}\,s^{-1}$ the entrainment flux is of the order $10^2\,m^3\,s^{-1}$ (e.g., Helfrich and Speer, 1995) resulting in very rapid dilution of the primary vent fluid (10^2–10^3:1) within the first 5–10 m of plume rise and even greater dilution (~10^4:1) by the time of emplacement within the nonbuoyant, spreading hydrothermal plume (Feely *et al.*, 1994). Similarly, the *time* of rise for a buoyant hydrothermal plume, is entirely dependent on the background buoyancy frequency, N (Middleton, 1979):

$$\tau = \pi N^{-1} \quad (3)$$

which, for a typical value of $N = 10^{-3}\,s^{-1}$, yields a plume rise-time of ≤ 1 h.

11.5.2 Modification of Gross Geochemical Fluxes

Hydrothermal plumes represent a significant dispersal mechanism for chemicals released from seafloor venting to the oceans. Consequently, it is important to understand the physical processes that control this dispersion (Section 11.5.1). It is also important to recognize that hydrothermal plumes represent non-steady-state fluids whose chemical compositions evolve with age (Figure 13). Processes active in hydrothermal plumes can lead to significant modification of *gross* hydrothermal fluxes (cf. Edmond *et al.*, 1979; German *et al.*, 1991b) and, in the extreme, can even reverse the sign of *net* flux to/from the ocean (e.g., German *et al.*, 1990, 1991a).

11.5.2.1 Dissolved noble gases

For one group of tracers, however, inert marine geochemical behavior dictates that they do undergo conservative dilution and dispersion within hydrothermal plumes. Perhaps the simplest example of such behavior is primordial

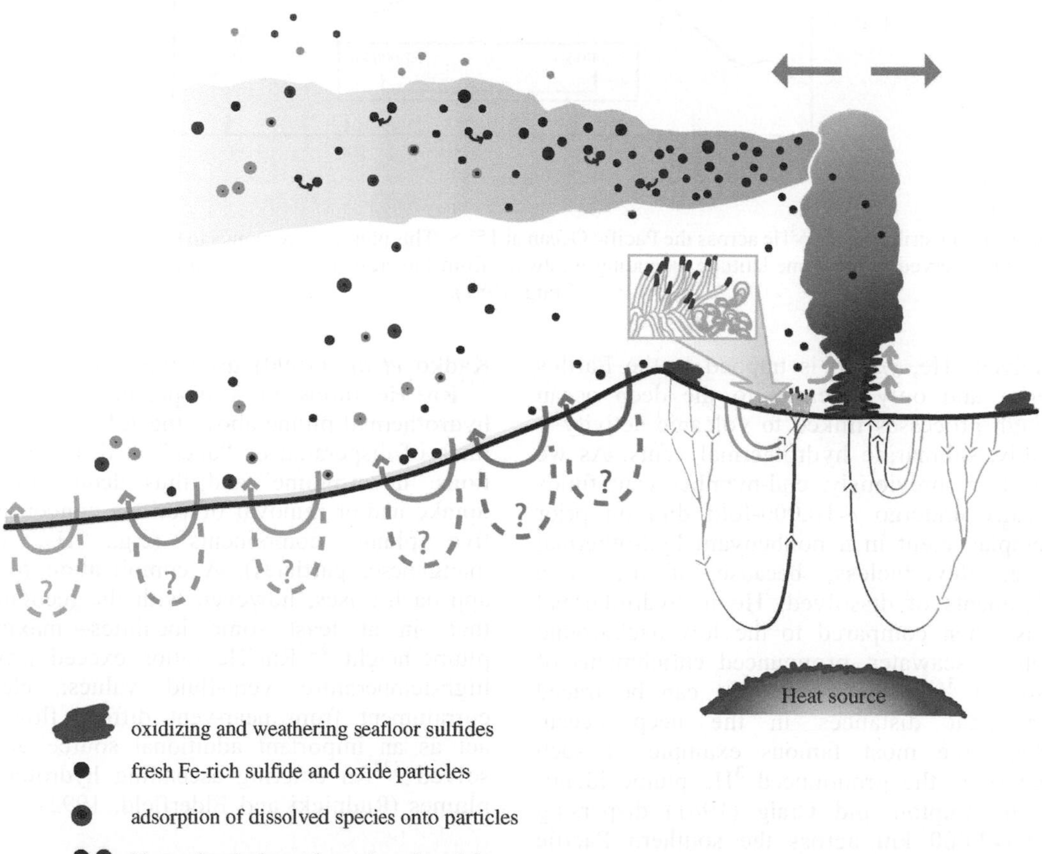

Figure 13 Schematic representation of an MOR hydrothermal system and its effects on the overlying water column. Circulation of seawater occurs within the oceanic crust, and so far three types of fluids have been identified and are illustrated here: high-temperature vent fluids that have likely reacted at >400 °C; high-temperature fluids that have then mixed with seawater close to the seafloor; fluids that have reacted at "intermediate" temperatures, perhaps ~150 °C. When the fluids exit the seafloor, either as diffuse flow (where animal communities may live) or as "black smokers," the water they emit rises and the hydrothermal plume then spreads out at its appropriate density level. Within the plume, sorption of aqueous oxyanions may occur onto the vent-derived particles (e.g., phosphate, vanadium, arsenic) making the plumes a sink for these elements; biogeochemical transformations also occur. These particles eventually rain-out, forming metalliferous sediments on the seafloor. While hydrothermal circulation is known to occur far out onto the flanks of the ridges, little is known about the depth to which it extends or its overall chemical composition because few sites of active ridge-flank venting have yet been identified and sampled (Von Damm, unpublished).

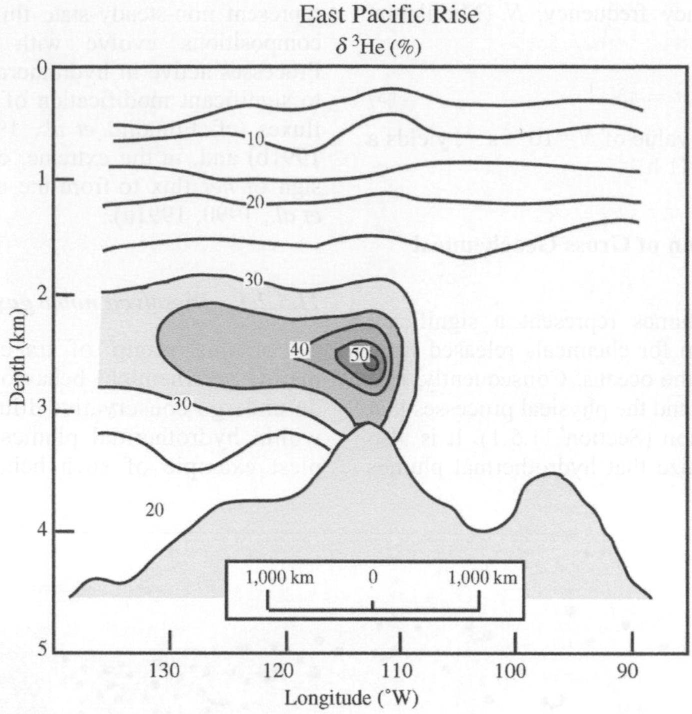

Figure 14 Distribution of δ^3He across the Pacific Ocean at 15° S. This plume corresponds to the lobe of metalliferous sediment observed at the same latitude extending westward from the crest of the EPR (Figure 2) (after Lupton and Craig, 1981).

dissolved ^{3}He, which is trapped in the Earth's interior and only released to the deep ocean through processes linked to volcanic activity—notably, submarine hydrothermal vents. As we have seen, previously, end-member vent fluids typically undergo ~10,000-fold dilution prior to emplacement in a nonbuoyant hydrothermal plume. Nevertheless, because of the large enrichments of dissolved ^{3}He in hydrothermal fluids when compared to the low background levels in seawater, pronounced enrichments of dissolved ^{3}He relative to ^{4}He can be traced over great distances in the deep ocean. Perhaps the most famous example of such behavior is the pronounced ^{3}He plume identified by Lupton and Craig (1981) dispersing over ~2,000 km across the southern Pacific Ocean, west of the southern EPR (Figure 14). A more recent example of the same phenomenon, however, is the large-scale ^{3}He anomaly reported by Rüth *et al.* (2000) providing the first firm evidence for high-temperature hydrothermal venting anywhere along the southern MAR. Rn-222, a radioactive isotope of the noble element radon, is also enriched in hydrothermal fluids; while it shares the advantages of being a conservative tracer with ^{3}He, it also provides the additional advantage of acting as a "clock" for hydrothermal plume processes because it decays with a half-life of 3.83 d. Kadko *et al.* (1990) used the fractionation of ^{222}Rn/^{3}He ratios in a dispersing nonbuoyant hydrothermal plume above the JdFR to estimate rates of dispersion or "ages" at different locations "down-plume" and, thus, deduce rates of uptake and/or removal of various nonconservative plume components (e.g., H_2, CH_4, manganese, particles). A complication to that approach arises, however, with the recognition that—in at least some localities—maximum plume-height ^{222}Rn/^{3}He ratios exceed pristine high-temperature vent-fluid values; clearly, entrainment from near-vent diffuse flow can act as an important additional source of dissolved ^{222}Rn entering ascending hydrothermal plumes (Rudnicki and Elderfield, 1992).

11.5.2.2 Dissolved reduced gases (H_2S, H_2, CH_4)

The next group of tracers that are important in hydrothermal plumes are the reduced dissolved gases, H_2S, H_2, and CH_4. As we have already seen, dissolved H_2S is commonly the most abundant dissolved reduced gas in high-temperature vent fluids (Section 11.2). Typically, however, any dissolved H_2S released to the oceans undergoes rapid reaction, either through precipitation of polymetallic sulfide phases close to the seafloor and in buoyant plumes, or through oxidation in the water column. As of the early 2000s, there has only been one report of measurable dissolved H_2S at nonbuoyant plume height anywhere in the deep ocean (Radford-Knöery *et al.*, 2001). That study revealed maximum dissolved H_2S concentrations of ≤ 2 nM, representing a 5×10^5-fold decrease from vent-fluid concentrations (Douville *et al.*, 2002) with complete oxidative removal occurring in just 4–5 h, within ~1–10 km of the vent site. Dissolved H_2, although not commonly abundant in such high concentrations in vent fluids (Section 11.2), exhibits similarly short-lived behavior within hydrothermal plumes. Kadko *et al.* (1990) and German *et al.* (1994) have reported maximum plume-height dissolved H_2 concentrations of 12 nM and 32 nM above the Main Endeavour vent site, JdFR and above the Steinahóll vent site, Reykjanes Ridge. From use of the $^{222}Rn/^{3}He$ "clock," Kadko *et al.* (1990) estimated an apparent "oxidative-removal" half-life for dissolved H_2 of ~10 h.

The most abundant and widely reported dissolved reduced gas in hydrothermal plumes is methane which is released into the oceans from both high- and low-temperature venting and the serpentinization of ultramafic rocks (e.g., Charlou *et al.*, 1998). Vent-fluid concentrations are significantly enriched over seawater values (10–2,000 μmol kg^{-1} versus <5 nmol kg^{-1}) but the behavior of dissolved methane, once released, appears variable from one location to another: e.g., rapid removal of dissolved CH_4 was observed in the Main Endeavour plume (half-life = 10 d; Kadko *et al.*, 1990) yet near-conservative behavior for the same tracer has been reported for the Rainbow hydrothermal plume, MAR, over distances up to 50 km from the vent site (Charlou *et al.*, 1998; German *et al.*, 1998b). Possible reasons for these significant variations are discussed later (Section 11.5.4).

11.5.2.3 Iron and manganese geochemistry in hydrothermal plumes

The two metals most enriched in hydrothermal vent fluids are iron and manganese. These elements are present in a reduced dissolved form (Fe^{2+}, Mn^{2+}) in end-member vent fluids yet are most stable as oxidized Fe(III) and Mn(IV) precipitates in the open ocean. Consequently, the dissolved concentrations of iron and manganese in vent fluids are enriched ~10^6:1 over open-ocean values (e.g., Landing and Bruland, 1987; Von Damm, 1995; Statham *et al.*, 1998). When these metal-laden fluids first enter the ocean, two important processes occur. First, the fluids are instantaneously cooled from >350 °C to ≤30 °C. This quenching of a hot saturated solution leads to precipitation of a range of metal sulfide phases that are rich in iron but not manganese because the latter does not readily form sulfide minerals. In addition, turbulent mixing between the sulfide-bearing vent fluid and the entrained, oxidizing, water column leads to a range of redox reactions resulting in the rapid precipitation of high concentrations of suspended iron oxyhydroxide particulate material. The dissolved manganese within the hydrothermal plume, by contrast, typically exhibits much more sluggish oxidation kinetics and remains predominantly in dissolved form at the time of emplacement in the nonbuoyant plume. Because iron and manganese are so enriched in primary vent fluids, nonbuoyant plumes typically exhibit total (dissolved and particulate) iron and manganese concentrations, which are ~100-fold higher than ambient water column values immediately following nonbuoyant plume emplacement. Consequently, iron and manganese, together with CH_4 and ^3He (above), act as extremely powerful tracers with which to identify the presence of hydrothermal activity from water-column investigations. The fate of iron in hydrothermal plumes is of particular interest because it is the geochemical cycling of this element, more than any other, which controls the fate of much of the hydrothermal flux from seafloor venting to the oceans (e.g., Lilley *et al.*, 1995).

Because of their turbulent nature, buoyant hydrothermal plumes have continued to elude detailed geochemical investigations. One approach has been to conduct direct sampling using manned submersibles or ROVs (e.g., Rudnicki and Elderfield, 1993; Feely *et al.*, 1994). An alternate, and indirect, method is to investigate the geochemistry of precipitates collected both rising in, and sinking from, buoyant hydrothermal plumes using near-vent sediment traps (e.g., Cowen *et al.*, 2001; German *et al.*, 2002). From direct observations it is apparent that up to 50% of the total dissolved iron emitted from hydrothermal vents is precipitated rapidly from buoyant hydrothermal plumes (e.g., Mottl and McConachy, 1990; Rudnicki and Elderfield, 1993) forming polymetallic sulfide phases which dominate (>90%) the iron flux to the near-vent seabed (Cowen *et al.*, 2001; German *et al.*, 2002). The remaining dissolved iron within the buoyant and nonbuoyant hydrothermal plume undergoes oxidative precipitation. In the well-ventilated N. Atlantic Ocean, very rapid Fe(II) oxidation is observed with a half-life for oxidative removal from solution of just 2–3 min (Rudnicki and Elderfield, 1993). In the NE Pacific, by contrast, corresponding half-times of up to 32 h have been reported from JdFR hydrothermal plumes (Chin *et al.*, 1994; Massoth *et al.*, 1994). Field and Sherrell (2000) have predicted that the oxidation rate for dissolved Fe^{2+} in hydrothermal plumes should decrease along the path of the thermohaline circulation, reflecting the progressively decreasing pH and dissolved oxygen content of the deep ocean (Millero *et al.*, 1987):

$$-d[Fe(II)]/dt = k[OH^-]^2[O_2][Fe(II)] \quad (4)$$

The first Fe(II) incubation experiments conducted within the Kairei hydrothermal plume, Central Indian Ridge, are consistent with that prediction, yielding an Fe(II) oxidation half-time of ~2 h (Statham *et al.*, 2003, submitted).

11.5.2.4 Co-precipitation and uptake with iron in buoyant and nonbuoyant plumes

There is significant co-precipitation of other metals enriched in hydrothermal fluids, along with iron, to form buoyant plume polymetallic sulfides. Notable among these are copper, zinc, and lead. Common accompanying phases, which also sink rapidly from buoyant plumes, are barite and anhydrite (barium and calcium sulfates) and amorphous silica (e.g., Lilley *et al.*, 1995). In nonbuoyant hydrothermal plumes, where Fe- and (to a lesser extent) Mn-oxyhydroxides predominate, even closer relationships are observed between particulate iron concentrations and numerous other tracers. To a first approximation, three differing iron-related behaviors have been identified (German *et al.*, 1991b; Rudnicki and Elderfield, 1993): (i) co-precipitation; (ii) fixed molar ratios to iron; and (iii) oxyhydroxide scavenging (Figure 15). The first is that already alluded to above and loosely termed "chalcophile" behavior—namely preferential co-precipitation with iron as sulfide phases followed by preferential settling from the nonbuoyant plume. Such elements exhibit a generally positive correlation with iron for plume particle concentrations but with highest *X*:Fe ratios closest to the vent site (*X* = Cu, Zn, Pb) and much lower values farther afield.

The second group are particularly interesting. These are elements that establish fixed *X*:Fe ratios in nonbuoyant hydrothermal plumes which do not vary with dilution or dispersal distance "down-plume" (Figure 15). Elements that have been shown to exhibit such "linear" behavior include potassium, vanadium, arsenic, chromium, and uranium (e.g., Trocine and Trefry, 1988; Feely *et al.*, 1990, 1991; German *et al.*, 1991a,b). Hydrothermal vent fluids are not particularly enriched in any of these elements, which typically occur as rather stable "oxyanion" dissolved species in seawater. Consequently, this uptake must represent a significant removal flux, for at least some of these elements, from the deep ocean. The P:Fe ratios observed throughout all Pacific hydrothermal plumes are rather similar (P:Fe = 0.17–0.21; Feely *et al.*, 1998) and distinctly higher than the value observed in Atlantic hydrothermal plumes (P:Fe=0.06–0.12). This has led to speculation that

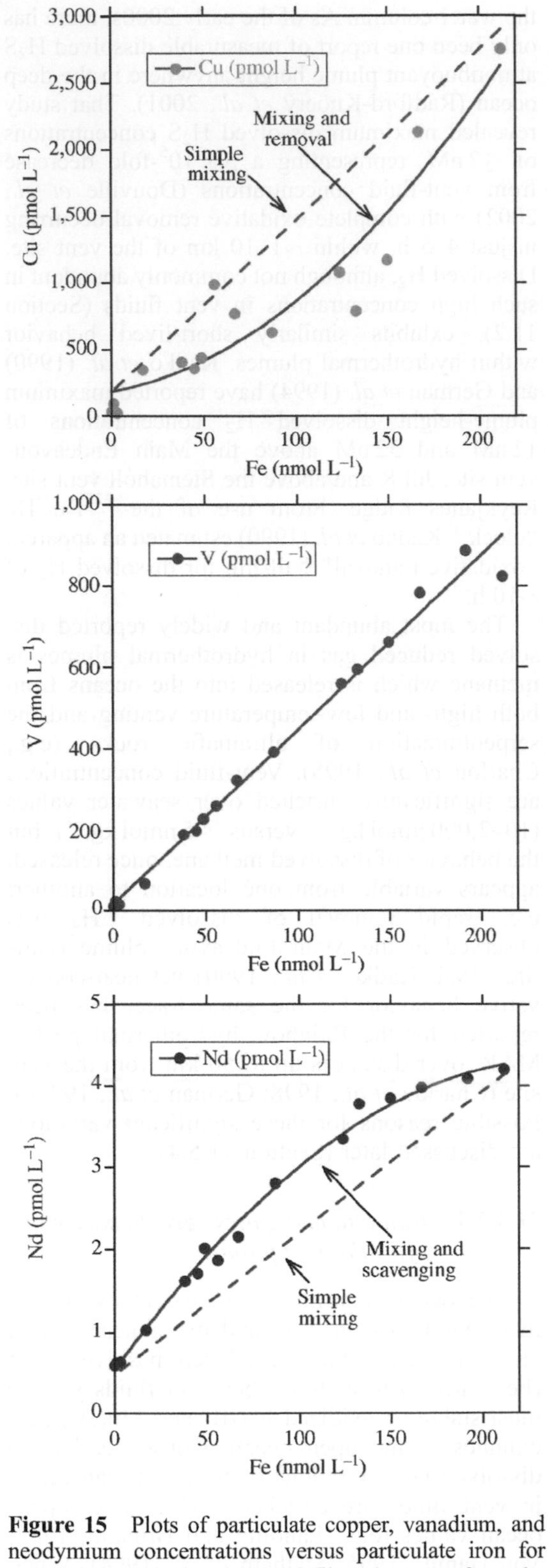

Figure 15 Plots of particulate copper, vanadium, and neodymium concentrations versus particulate iron for suspended particulate material filtered *in situ* from the TAG hydrothermal mound, MAR, 26° N (data from German *et al.*, 1990, 1991b). Note generally positive correlations with particulate Fe concentration for all three tracers but with additional negative (Cu) or positive (Nd) departure for sulfide-forming and scavenged elements, respectively.

plume P:Fe ratios may reflect the ambient dissolved PO_4^{3-} concentration of the host water column and, thus, may offer potential as a long-term tracer of past ocean circulation, if preserved faithfully in metalliferous marine sediments (Feely *et al.*, 1998).

11.5.2.5 Hydrothermal scavenging by Fe-oxyhydroxides

A final group of elements identified from hydrothermal plume particle investigations are of particular importance: "particle-reactive" tracers. Perhaps the best examples are the REE although other tracers that show similar behavior include beryllium, yttrium, thorium, and protactinium (German *et al.*, 1990, 1991a,b, 1993; Sherrell *et al.*, 1999). These tracers, like the two preceding groups, exhibit generally positive correlations with particulate iron concentrations in hydrothermal plumes (Figure 15). Unlike the "oxyanion" group, however, these tracers do not show constant *X*:Fe ratios within the nonbuoyant plume. Instead, highest values (e.g., for REE:Fe) are observed at increasing distances away from the vent site, rather than immediately above the active source (German *et al.*, 1990; Sherrell *et al.*, 1999). One possible interpretation of this phenomenon is that the Fe-oxyhydroxide particles within young nonbuoyant plumes are undersaturated with respect to surface adsorption and that continuous "scavenging" of dissolved, particle-reactive species occurs as the particles disperse through the water column (German *et al.*, 1990, 1991a,b). An alternate hypothesis (Sherrell *et al.*, 1999) argues, instead, for two-stage equilibration within a nonbuoyant hydrothermal plume: close to the vent source, a maximum in (e.g.,) particulate REE concentrations is reached, limited by equilibration at fixed distribution constants between the high particulate iron concentrations present and the finite dissolved tracer (e.g., REE) concentrations present in plume-water. As the plume disperses and undergoes dilution, however, particulate iron concentrations also decrease; re-equilibration between these particles and the diluting "pristine" ambient seawater, at the same fixed distribution constants, would then account for the higher REE/Fe ratios observed at lower particulate iron concentrations. More work is required to resolve which of these interpretations (kinetic versus equilibrium) more accurately reflects the processes active within hydrothermal plumes. What is beyond dispute concerning these particle-reactive tracers is that their uptake fluxes, in association with hydrothermal Fe-oxyhydroxide precipitates, far exceeds their dissolved fluxes entering the oceans from hydrothermal vents. Thus, hydrothermal plumes act as *sinks* rather than *sources* for these elements, even causing local depletions relative to the ambient water column concentrations (e.g., Klinkhammer *et al.*, 1983). Thus, even for those "particle-reactive" elements which are greatly enriched in vent fluids over seawater concentrations (e.g., REE), the processes active within hydrothermal plumes dictate that hydrothermal circulation causes a net *removal* of these tracers not just relative to the vent fluids themselves, but also from the oceanic water column (German *et al.*, 1990; Rudnicki and Elderfield, 1993). In the extreme, these processes can be sufficient to cause geochemical fractionations as pronounced as those caused by "boundary scavenging" in high-productivity ocean margin environments (cf.Anderson *et al.*, 1990; German *et al.*, 1997).

Thus far, we have treated the processes active in hydrothermal plumes as inorganic geochemical processes. However we know this is not strictly the case: microbial processes are well known to mediate key chemical reactions in hydrothermal plumes (Winn *et al.*, 1995; Cowen and German, 2003) and more recently the role of larger organisms such as zooplankton has also been noted (e.g., Burd *et al.*, 1992; Herring and Dixon, 1998). The biological modification of plume processes is discussed more fully below (Section 11.5.4).

11.5.3 Physical Controls on Dispersing Plumes

Physical processes associated with hydrothermal plumes may affect their impact upon ocean geochemistry; because of the fundamentally different hydrographic controls in the Pacific versus Atlantic Oceans, plume dispersion varies between these two oceans. In the Pacific Ocean, where deep waters are warmer and saltier than overlying water masses, nonbuoyant hydrothermal plumes which have entrained local deep water are typically warmer and more saline at the point of emplacement than that part of the water column into which they intrude (e.g., Lupton *et al.*, 1985). The opposite has been observed in the Atlantic Ocean where deep waters tend to be colder and less saline than the overlying water column. Consequently, for example, the TAG nonbuoyant plume is anomalously cold and fresh when compared to the background waters into which it intrudes, 300–400 m above the seafloor (Speer and Rona, 1989).

Of perhaps more significance, geochemically, are the physical processes which affect plume-dispersion *after* emplacement at nonbuoyant plume height. Here, topography at the ridge crest exerts particular influence. Along slow and ultra-slow spreading ridges (e.g., MAR, SW Indian Ridge, Central Indian Ridge) nonbuoyant plumes are typically emplaced within the confining bathymetry of the rift-valley walls (order 1,000 m).

Along faster spreading ridges such as the EPR, by contrast, buoyant plumes typically rise clear of the confining topography (order 100 m, only). Within rift-valley "corridors" plume dispersion is highly dependent upon along-axis current flow. At the TAG hydrothermal field (MAR 26° N), for example, residual currents are dominated by tidal excursions (Rudnicki *et al.*, 1994). A net effect of these relatively "stagnant" conditions is that plume material trapped within the vicinity of the vent site appears to be recycled multiple times through the TAG buoyant and nonbuoyant plume, enhancing the scavenging effect upon "particle-reactive" tracers (e.g., thorium isotopes) within the local water column (German and Sparks, 1993). At the Rainbow vent site (MAR 36° N) by contrast, much stronger prevailing currents ($\sim$10 cm s^{-1}) are observed and a more unidirectional, topographically controlled flow is observed (German *et al.*, 1998b). Failure to appreciate the potential complexities of such dispersion precludes the "informed" sampling required to resolve processes of geochemical evolution within a dispersing hydrothermal plume. Nor should it be assumed that such topographic steering is entirely a local effect confined to slow spreading ridges' rift-valleys. In recent work, Speer *et al.* (2003) used a numerical simulation of ocean circulation to estimate dispersion along and away from the global ridge crest. A series of starting points were considered along the entire ridge system, 200 m above the seafloor and at spacings of 30–100 km along-axis; trajectories were then calculated over a 10 yr integrated period. With few exceptions (e.g., major fracture zones) the net effect reported was that these dispersal trajectories tend to be constrained by the overall form of the ridge and flow parallel to the ridge axis over great distances (Speer *et al.*, 2003).

The processes described above are relevant to established "chronic" hydrothermal plumes. Important exceptions (only identified rarely–so far) are Event (or "*Mega-*") Plumes. One interpretation of these features is as follows: however a hydrothermal system may evolve, it must first displace a large volume of cold seawater from pore spaces within the upper ocean crust. Initial flushing of this system must be rapid, especially on fast ridges that are extending at rates in excess of ten centimeters per year. In circumstances where there is frequent recurrence, either of intrusions of magma close beneath the seafloor or dike-fed eruptions at the seabed, seafloor venting may commence with a rapid exhalation of a large volume of hydrothermal fluid to generate an "event" plume high up in the overlying water column (e.g., Baker *et al.*, 1995; Lupton *et al.*, 1999b). Alternately, Palmer and Ernst (1998, 2000) have argued that cooling of pillow basalts, erupted at $\sim$1,200 °C on the seafloor and the most common form of submarine volcanic extrusion, is responsible for the formation of these same "event" plume features. Whichever eruption-related process causes their formation, an important question that follows is: how much hydrothermal flux is overlooked if we fail to intercept "event" plumes at the onset of venting at any given location? To address this problem, Baker *et al.* (1998) estimated that the event plume triggered by dike-intrusion at the co-axial vent field (JdFR) contributed less than 10% of the total flux of heat and chemicals released during the $\sim$3 yr life span of that vent. If widely applicable, those deliberations suggest that event plumes can safely be ignored when calculating global geochemical fluxes (Hein *et al.*, 2003). They remain of great interest to microbiologists, however, as a potential "window" into the deep, hot biosphere (e.g., Summit and Baross, 1998).

11.5.4 Biogeochemical Interactions in Dispersing Hydrothermal Plumes

Recognition of the predominantly along-axis flow of water masses above MORs as a result of topographic steering has renewed speculation that hydrothermal plumes may represent important vectors for gene-flow along the global ridge crest, transporting both chemicals and vent larvae alike, from one adjacent vent site to another (e.g., Van Dover *et al.*, 2002). If such is the case, however, it is also to be expected that a range of biogeochemical processes should also be active within nonbuoyant hydrothermal plumes. One particularly good example of such a process is the microbial oxidation of dissolved manganese. In the restricted circulation regime of the Guaymas Basin, formation of particulate manganese is dominated by bacteria and the dissolved manganese residence time is less than a week (Campbell *et al.*, 1988). Similarly, uptake of dissolved manganese in the Cleft Segment plume (JdFR) is bacterially dominated, albeit with much longer residence times, estimated at 0.5 yr to >2 yr (Winn *et al.*, 1995). Distributions of dissolved CH_4 and H_2 in hydrothermal plumes have also been shown to be controlled by bacterially mediated oxidation (de Angelis *et al.*, 1993) with apparent residence times which vary widely for CH_4 (7–177 d) but are much shorter for dissolved H_2 (< 1 d).

The release of dissolved H_2 and CH_4 into hydrothermal plumes provides suitable substrates for both primary (chemolithoautotrophic) and secondary (heterotrophic) production within dispersing nonbuoyant hydrothermal plumes. Although the sinking organic carbon flux from hydrothermal plumes (Roth and Dymond, 1989; German *et al.*, 2002) may be less than 1% of the total oceanic photosynthetic production (Winn *et al.*, 1995)

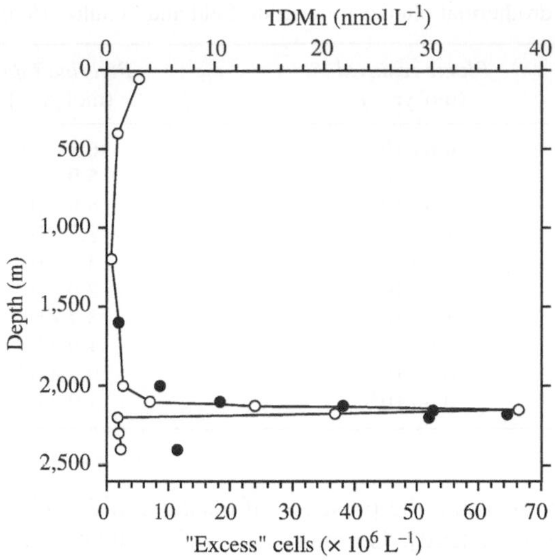

Figure 16 Plot showing co-registered enrichments of TDMn concentrations and "excess" cell counts (after subtraction of a "background" N. Atlantic cell-count profile) in the Rainbow nonbuoyant hydrothermal plume near 36° 14′ N, MAR (source German *et al.*, 2003).

hydrothermal production of organic carbon is probably restricted to a corridor extending no more than ~10 km to either side of the ridge axis. Consequently, microbial activity within hydrothermal plumes may have a pronounced local effect—perhaps 5–10-fold greater than the photosynthetic-flux driven from the overlying upper ocean (Cowen and German, 2003). Although the detailed microbiology of hydrothermal plumes remains poorly studied, as of the early 2000s, bacterial counts from the Rainbow plume have identified both an increase in total cell concentrations at plume-height compared to background (Figure 16) and, further, that 50–75% of the microbial cells identified in that work were particle-attached compared to typical values of just 15% for the open ocean. Detailed molecular biological analysis of those particle-associated microbes have also revealed that the majority (66%) are archeal in nature rather than bacterial (German *et al.*, 2003); in the open ocean, by contrast, it is the bacteria which typically dominate (Fuhrmann *et al.*, 1993; Mullins *et al.*, 1995). It is tempting to speculate that these preliminary data may provide testament to a long-established (even on geological timescales) chemical-microbial symbiosis at hydrothermal vents.

11.5.5 Impact of Hydrothermal Plumes Upon Ocean Geochemical Cycles

Hydrothermal plumes form by the entrainment of large volumes of ambient ocean water into rising buoyant plumes driven by the release of vent fluids at the seafloor. The effect of this dilution is such that the entire volume of the oceans is cycled through buoyant and nonbuoyant hydrothermal plumes, on average, every few thousand years—a timescale comparable to that for mixing of the entire deep ocean.

Close to the vent source, rapid precipitation of a range of polymetallic sulfide, sulfate, and oxide phases leads to a strong modification of the gross dissolved metal flux from the seafloor. Independent estimates by Mottl and McConachy (1990) and German *et al.* (1991b), from buoyant and nonbuoyant plume investigations in the Pacific and Atlantic Oceans respectively, concluded that perhaps only manganese and calcium achieved a significant dissolved metal flux from hydrothermal venting to the oceans. (For comparison, the following 27 elements exhibited no evidence for a significant dissolved hydrothermal flux: iron, beryllium, aluminum, magnesium, chromium, vanadium, cobalt, copper, zinc, arsenic, yttrium, molybdenum, silver, cadmium, tin, lanthanum, cerium, praseodymium, neodymium, samarium, europium, gadolinium, terbium, holmium, erbium, lead, and uranium.) In addition to rapid co-precipitation and deposition of vent-sourced metals to the local sediments (e.g., Dymond and Roth, 1988; German *et al.*, 2002), hydrothermal plumes are also the site of additional uptake of other dissolved tracers from the water column—most notably large dissolved "oxyanion" species (phosphorus, vanadium, chromium, arsenic, uranium) and particle-reactive species (beryllium, yttrium, REE, thorium, and protactinium). For many of these tracers, hydrothermal plume removal fluxes are as great as, or at least significant when compared to, riverine input fluxes to the oceans (Table 3). What remains less certain, however, is the extent to which these particle-associated species are subsequently retained within the hydrothermal sediment record.

11.6 HYDROTHERMAL SEDIMENTS

Deep-sea metalliferous sediments were first documented in the reports of the Challenger Expedition, 1873–1876 (Murray and Renard, 1891), but it took almost a century to recognize that such metalliferous material was concentrated along all the world's ridge crests (Figure 2). Boström *et al.* (1969) attributed these distributions to some form of "volcanic emanations"; the accuracy of those predictions was confirmed some ten years later with the discovery of ridge-crest venting (Corliss *et al.*, 1978; Spiess *et al.*, 1980) although metalliferous sediments had already been found in association with hot brine pools in the Red Sea (Degens and Ross, 1969). Following the discovery of active venting, it has become recognized that

Table 3 Global removal fluxes from the deep ocean into hydrothermal plumes (after Elderfield and Schultz, 1996)

Element	*Hydrothermal input* (mol yr^{-1})	*Plume removal* (mol yr^{-1})	*Riverine input* (mol yr^{-1})
Cr	0	4.8×10^{7}	6.3×10^{8}
V	0	4.3×10^{8}	5.9×10^{8}
As	$(0.01–3.6)\times10^{7}$	1.8×10^{8}	8.8×10^{8}
P	-4.5×10^{7}	1.1×10^{10}	3.3×10^{10}
U	-3.8×10^{5}	4.3×10^{7}	3.6×10^{7}
Mo	0	1.9×10^{6}	2.0×10^{8}
Be		1.7×10^{6}	3.7×10^{7}
Ce	9.1×10^{5}	1.0×10^{6}	1.9×10^{7}
Nd	5.3×10^{5}	6.3×10^{6}	9.2×10^{6}
Lu	2.1×10^{3}	0.6×10^{5}	1.9×10^{5}

hydrothermal sediments can be classified into two types: those derived from plume fall-out (including the majority of metalliferous sediments reported from ridge flanks) and those derived from mass wasting close to active vent sites (see, e.g., Mills and Elderfield, 1995).

11.6.1 Near-vent Sediments

Near-vent metalliferous sediments form from the physical degradation of hydrothermal deposits themselves, a process which begins as soon as deposition has occurred. Whilst there is ample evidence for extensive mass wasting in ancient volcanogenic sulfide deposits, only limited attention has been paid, to date, to this aspect of modern hydrothermal systems. Indeed, much of our understanding comes from a series of detailed investigations from a single site, the TAG hydrothermal field at 26° N on the MAR. It has been shown, for example, that at least some of the weathered sulfide debris at TAG is produced from collapse of the mound itself. This collapse is believed to arise from waxing and waning of hydrothermal circulation which, in turn, leads to episodic dissolution of large volumes of anhydrite within the mound (e.g., Humphris *et al.*, 1995; James and Elderfield, 1996). The mass-wasting process at TAG generates an apron of hydrothermal detritus with oxidized sulfides deposited up to 60 m out, away from the flanks of the hydrothermal mound.

Similar ponds of metalliferous sediment are observed close to other inactive sulfide structures throughout the TAG area (Rona *et al.*, 1993). Metz *et al.* (1988) characterized the metalliferous sediment in a core raised from a sediment pond close to one such deposit, ~2 km NNE of the active TAG mound. That core consisted of alternating dark red-brown layers of weathered sulfide debris and lighter calcareous ooze. Traces of pyrite, chalcopyrite, and sphalerite, together with elevated transition-metal concentrations were found in the dark red-brown layers, confirming the presence of clastic sulfide debris. Subsequently, German *et al.* (1993) investigated a short-core raised from the outer limit of the apron of "stained" hydrothermal sediment surrounding the TAG mound itself. That core penetrated through 7 cm of metal-rich degraded sulfide material into pelagic carbonate ooze. The upper "mass-wasting" layer was characterized by high transition-metal contents, just as observed by Metz *et al.* (1988), but also exhibited REE patterns similar to vent fluids (see earlier) and high uranium contents attributed to uptake from seawater during oxidation of sulfides (German *et al.*, 1993). Lead isotopic compositions in sulfidic sediments from both sites were indistinguishable from local MORB, vent fluids and chimneys (German *et al.*, 1993, Mills *et al.*, 1993). By contrast, the underlying/intercalated carbonate/calcareous ooze layers from each core exhibited lead isotope, REE, and U–Th compositions which much more closely reflected input of Fe-oxyhydroxide particulate material from nonbuoyant hydrothermal plumes (see below).

11.6.2 Deposition from Hydrothermal Plumes

Speer *et al.* (2003) have modeled deep-water circulation above the global ridge crest and concluded that this circulation is dominated by topographically steered flow along-axis. Escape of dispersed material into adjacent deep basins is predicted to be minimal, except in key areas where pronounced across-axis circulation occurs. If this model proves to be generally valid, the majority of hydrothermal material released into nonbuoyant hydrothermal plumes should not be dispersed more than ~100 km off-axis. Instead, most hydrothermal material should settle out in a near-continuous rain of metalliferous sediment along the length of the global ridge crest. Significant off-axis dispersion is only predicted (i) close to the equator (~5° N to 5° S), (ii) where the ridge intersects boundary currents or regions of deep-water formation, and (iii) in the Antarctic Circumpolar Current (Speer *et al.*, 2003). One good example of strong across-axis flow is at the equatorial MAR where pronounced eastward flow of both Antarctic Bottom Water and lower North Atlantic Deep Water has been reported, passing through the Romanche and Chain Fracture Zones (Mercier and Speer, 1998).

Another location where the large-scale off-axis dispersion modeled by Speer *et al.* (2003) has already been well documented is on the southern EPR (Figure 14). There, metalliferous sediment enrichments underlie the pronounced dissolved ^{3}He plume which extends westward across the southern Pacific Ocean at ~15° S (cf. Boström *et al.*, 1969; Lupton and Craig, 1981). Much of our understanding of ridge-flank metalliferous sediments comes from a large-scale study carried out at this latitude (19° S) by Leg 92 of the Deep Sea Drilling Project (DSDP). That work targeted sediments underlying the westward-trending plume to investigate both temporal and spatial variability in hydrothermal output at this latitude (Lyle *et al.*, 1987). A series of holes were drilled extending westward from the ridge axis into 5–28 Ma crust; the recovered cores comprised mixtures of biogenic carbonate and Fe–Mn oxyhydroxides. One important result of that work was the demonstration, based on lead isotopic analyses, that even the most distal sediments, collected at a range of >1,000 km from the ridge axis, contained 20–30% mantle-derived lead (Barrett *et al.*, 1987). In contrast, analysis of the same samples indicated that REE distributions in the metalliferous sediments were dominated by a seawater source (Ruhlin and Owen, 1986). This is entirely consistent with what has subsequently been demonstrated for hydrothermal plumes (see Section 11.5, above) with the caveat that REE/Fe ratios in DSDP Leg 92 sediments are everywhere higher than the highest REE/Fe ratios yet measured in modern nonbuoyant hydrothermal plume particles (German *et al.*, 1990; Sherrell *et al.*, 1999).

11.6.3 Hydrothermal Sediments in Paleoceanography

Phosphorus and vanadium, which are typically present in seawater as dissolved oxyanion species, have been shown to exhibit systematic plume-particle P:Fe and V:Fe variations which differ from one ocean basin to another (e.g., Trefry and Metz, 1989; Feely *et al.*, 1990). This has led to the hypothesis (Feely *et al.*, 1998) that (i) plume P:Fe and V:Fe ratios may be directly linked to local deep-ocean dissolved phosphate concentrations and (ii) ridge-flank metalliferous sediments, preserved under oxic diagenesis, might faithfully record temporal variations in plume-particle P:Fe and/or V:Fe ratios. Encouragingly, a study of slowly accumulating (~0.5 cm kyr^{-1}) sediments from the west flank of the JdFR has revealed that V:Fe ratios in the hydrothermal component from that core appear faithfully to record local plume-particle V:Fe ratios for the past ~200 kyr (German *et al.*, 1997; Feely *et al.*, 1998). More recently, however, Schaller *et al.* (2000) have shown that while cores from the flanks of the southern EPR (10° S) also exhibit V:Fe ratios that mimic modern plume-values, in sediments dating back to 60–70 kyr, the complementary P:Fe and As:Fe ratios in these samples are quite different from contemporaneous nonbuoyant plume values. These variations have been attributed to differences in the intensity of hydrothermal iron oxide formation between different hydrothermal plumes and/or significant uptake/release of phosphorus and arsenic, following deposition (Schaller *et al.*, 2000).

Unlike vanadium, REE/Fe ratios recorded in even the most recent metalliferous sediments are much higher than those in suspended hydrothermal plume particles (German *et al.*, 1990, 1997; Sherrell *et al.*, 1999). Further, hydrothermal sediments' REE/Fe ratios increase systematically with distance away from the paleo-ridge crest (Ruhlin and Owen, 1986; Olivarez and Owen, 1989). This indicates that the REE may continue to be taken up from seawater, at and near the sediment–water interface, long after the particles settle from the plume to the seabed. Because increased uptake of dissolved REE from seawater should also be accompanied by continuing fractionation across the REE series (e.g., Rudnicki and Elderfield, 1993) reconstruction of deep-water REE patterns from preserved metalliferous sediment records remain problematic. Much more tractable, however, is the exploitation of these same sample types for isotopic reconstructions.

Because seawater uptake dominates the REE content of metalliferous sediment, neodymium isotopic analysis of metalliferous carbonate can provide a reliable proxy for contemporaneous seawater, away from input of near-vent sulfide detritus (Mills *et al.*, 1993). Osmium also exhibits a similar behavior and seawater dominates the isotopic composition of metalliferous sediments even close to active vent sites (Ravizza *et al.*, 1996). Consequently, analysis of preserved metalliferous carbonate sediments has proven extremely useful in determining the past osmium isotopic composition of the oceans, both from modern marine sediments (e.g., Ravizza, 1993; Peucker-Ehrenbrink *et al.*, 1995) and those preserved in ophiolites (e.g., Ravizza *et al.*, 2001). Only in sediments close to an ultramafic-hosted hydrothermal system, have perturbations from a purely seawater osmium isotopic composition been observed (Cave *et al.*, 2003, in press).

11.6.4 Hydrothermal Sediments and Boundary Scavenging

It has been known for sometime that sediments underlying areas of high particle settling flux exhibit pronounced fractionations between particle-reactive tracers. Both ^{231}Pa and ^{10}Be, for example, exhibit pronounced enrichments relative to ^{230}Th, in ocean margin environments, when compared to sediments underlying mid-ocean gyres (e.g., Bacon, 1988; Anderson *et al.*, 1990; Lao *et al.*, 1992). Comparable fractionations between these three radiotracers (^{230}Th, ^{231}Pa, and ^{10}Be) have also been identified in sediments underlying hydrothermal plumes (German *et al.*, 1993; Bourlès *et al.*, 1994; German *et al.*, 1997). For example, a metalliferous sediment core raised from the flanks of the JdFR exhibited characteristic hydrothermal lead-isotopic and REE/Fe compositions, together with high $^{10}Be/^{230}Th$ ratios indicative of net focusing relative to the open ocean (German *et al.*, 1997). The degree of fractionation observed was high, even compared to high-productivity ocean-margin environments (Anderson *et al.*, 1990; Lao *et al.*, 1992), presumably due to intense scavenging onto hydrothermal Fe-oxyhydroxides. Of course, the observation that REE and thorium are scavenged into ridge-crest metalliferous sediments is not new; sediments from the EPR near 17° S, with mantle lead, excess ^{230}Th and seawater-derived REE compositions were reported more than thirty years ago by Bender *et al.* (1971). More recently, however, examination of ridge crest sediments and near-vent sediment-traps has revealed that the settling flux of scavenged tracers (e.g., ^{230}Th) from hydrothermal plumes is higher than can be sustained by *in situ* production in the overlying water column alone (German *et al.*, 2002). Thus, uptake onto Fe-oxyhydroxide material in hydrothermal plumes and sediments may act as a special form of deep-ocean "boundary scavenging" leading to the net focusing and deposition of these dissolved tracers in ridge-flank metalliferous sediments.

11.7 CONCLUSION

The field of deep-sea hydrothermal research is young; it was only in the mid-1970s when it was first discovered, anywhere in the oceans. To synthesize current understanding of its impact on marine geochemistry, therefore, could be considered akin to explaining the significance of rivers to ocean chemistry in the early part of the last century. This chapter has aimed to provide a brief synposis of the current state of the art, but much more surely remains to be learnt. There are three key questions that will continue to focus efforts within this vigorous research field:

(i) What are the geological processes that control submarine hydrothermal venting? How might these have varied during the course of Earth's history?

(ii) To what extent do geochemical and biological processes interact to regulate hydrothermal fluxes to the ocean? How might past-ocean processes have differed from the present-day ones?

(iii) What are the timescales relevant to hydrothermal processes? Whilst some long-term proxies do exist (sufide deposits, metalliferous sediments) for active processes, we do not have any time-series records longer than 25 years!

REFERENCES

Alt J. C. (1995) Subseafloor processes in mid-ocean ridge hydrothermal systems. *Geophys. Monogr. (AGU)* **91**, 85–114.

Alt J. C. and Bach W. (2003) Alteration of oceanic crust: subsurface rock–water interactions. In *Energy and Mass Transfer in Marine Hydrothermal Systems* (eds. P. Halbach, V. Tunnicliffe, and J. Hein), DUP, Berlin, pp. 7–28.

Anderson R. N. and Hobart M. A. (1976) The relationship between heat flow, sediment thickness and age in the eastern Pacific. *J. Geophys. Res.* **81**, 2968–2989.

Anderson R. N. and Silbeck J. N. (1981) Oceanic heat flow. In *The Oceanic Lithosphere, The Seas* (ed. C. Emiliani), Wiley, New York, vol 7, pp. 489–523.

Anderson R. F., Lao Y., Broecker W. S., Trumbore S. E., Hofmann H. J., and Wolfli W. (1990) Boundary scavenging in the Pacific Ocean: a comparison of ^{10}Be and ^{231}Pa. *Earth Planet. Sci. Lett.* **96**, 287–304.

Bach W., Banerjee N. R., Dick H. J. B., and Baker E. T. (2002) Discovery of ancient and active hydrothermal systems along the ultra-slow spreading southwest Indian ridge 10 degrees–16 degrees E. *Geophys. Geochem. Geosys.* **3**, paper 10.1029/2001GC000279.

Bacon M. P. (1988) Tracers of chemical scavenging in the ocean: boundary effects and largescale chemical fractionation. *Phil. Trans. Roy. Soc. London Ser. A* **325**, 147–160.

Baker E. T. (1998) Patterns of event and chronic hydrothermal venting following a magmatic intrusion: new perspectives from the 1996 Gorda ridge eruption. *Deep-Sea Res. II* **45**, 2599–2618.

Baker E. T., Massoth G. J., Feely R. A., Embley R. W., Thomson R. E., and Burd B. J. (1995) Hydrothermal event plumes from the CoAxial seafloor eruption site, Juan de Fuca ridge. *Geophys. Res. Lett.* **22**, 147–150.

Baker E. T., Chen Y. J., and Phipps Morgan J. (1996) The relationship between near-axis hydrothermal cooling and the spreading rate of mid-ocean ridges. *Earth Planet. Sci. Lett.* **142**, 137–145.

Baker E. T., Massoth G. J., Feely R. A., Cannon G. A., and Thomson R. E. (1998) The rise and fall of the CoAxial hydrothermal site, 1993–1996. *J. Geophys. Res.* **103**, 9791–9806.

Barnes I., Rapp J. T., O'Neill J. R., Sheppard R. A., and Gude A. J. (1972) Metamorphic assemblages and the flow of metamorphic fluid in four instances of serpentenization. *Contrib. Mineral. Petrol.* **35**, 263–276.

Barrett T. J., Taylor P. N., and Lugowski J. (1987) Metalliferous sediments from DSDP leg 92, the East Pacific rise transect. *Geochim. Cosmochim. Acta* **51**, 2241–2253.

Bender M., Broecker W., Gornitz V., Middel U., Kay R., Sun S., and Biscaye P. (1971) Geochemistry of three cores from the East Pacific Rise. *Earth Planet. Sci. Lett.* **12**, 425–433.

Berndt M. E. and Seyfried W. E. (1990) Boron, bromine, and other trace elements as clues to the fate of chlorine in mid-ocean ridge vent fluids. *Geochim. Cosmochim. Acta* **54**, 2235–2245.

Berndt M. E. and Seyfried W. E. (1997) Calibration of Br/Cl fractionation during sub-critical phase separation of seawater: possible halite at 9 to 10°N East Pacific Rise. *Geochim. Cosmochim. Acta* **61**, 2849–2854.

Berndt M. E., Allen D. E., and Seyfried W. E. (1996) Reduction of CO_2 during serpentinization of olivine at 300 degrees C and 500 bar. *Geology* **24**, 351–354.

Binns R. A., Scott S. D., Bogdanov Y. A., Lisitzin A. P., Gordeev V. V., Gurvcich E. G., Finlayson E. J., Boyd T., Dotter L. E., Wheller G. E., and Muravyev K. G. (1993) Hydrothermal oxide and gold-rich sulfate deposits of Franklin seamount, Western Woodlark Basin, Papua New Guinea. *Econ. Geol.* **88**, 2122–2153.

Bischoff J. L. (1991) Densities of liquids and vapors in boiling NaCl–H_2O solutions: a PVTX summary from 300° to 500 °C. *Am. J. Sci.* **291**, 309–338.

Bischoff J. L. and Pitzer K. S. (1989) Liquid-vapor relations for the system NaCl–H_2O: summary of the P–T–x surface from 300deg to 500deg C. *Am. J. Sci.* **289**, 217–248.

Bischoff J. L. and Rosenbauer R. J. (1985) An empirical equation of state for hydrothermal seawater (3.2% NaCl). *Am. J. Sci.* **285**, 725–763.

Bischoff J. L. and Rosenbauer R. J. (1987) Phase separation in seafloor systems—an experimental study of the effects on metal transport. *Am. J. Sci.* **287**, 953–978.

Blackman D. K., Cann J. R., Janssen B., and Smith D. K. (1998) Origin of extensional core complexes: evidence from the Mid-Atlantic Ridge at Atlantis fracture zone. *J. Geophys. Res.* **103**, 21315–21333.

Boström K., Peterson M. N. A., Joensuu O., and Fisher D. E. (1969) Aluminium-poor ferromanganoan sediments on active ocean ridges. *J. Geophys. Res.* **74**, 3261–3270.

Bourlès D. L., Brown E. T., German C. R., Measures C. I., Edmond J. M., Raïsbeck G. M., and Yiou F. (1994) Hydrothermal influence on oceanic beryllium. *Earth Planet. Sci. Lett.* **122**, 143–157.

Bray A. M. and Von Damm K. L. (2003a) The role of phase separation and water–rock reactions in controlling the boron content of mid-ocean ridge hydrothermal vent fluids. *Geochim. Cosmochim. Acta* (in revision).

Bray A. M. and Von Damm K. L. (2003b) Controls on the alkali metal composition of mid-ocean ridge hydrothermal fluids: constraints from the 9–10°N East Pacific Rise time series. *Geochim. Cosmochim. Acta* (in revision).

Broecker W. S. and Peng T. H. (1982) *Tracers in the Sea.* Eldigio Press, Columbia University, New York, 690pp.

Burd B. J., Thomson R. E., and Jamieson G. S. (1992) Composition of a deep scattering layer overlying a mid-ocean ridge hydrothermal plume. *Mar. Biol.* **113**, 517–526.

Butterfield D. A., McDuff R. E., Mottl M. J., Lilley M. D., Lupton J. E., and Massoth G. J. (1994) Gradients in the composition of hydrothermal fluids from the endeavour segment vent field: phase separation and brine loss. *J. Geophys. Res.* **99**, 9561–9583.

Butterfield D. A., Jonasson I. R., Massoth G. J., Feely R. A., Roe K. K., Embley R. E., Holden J. F., McDuff R. E., Lilley M. D., and Delaney J. R. (1997) Seafloor eruptions and evolution of hydrothermal fluid chemistry. *Phil. Trans. Roy. Soc. London* **A355**, 369–386.

Campbell A. C., Palmer M. R., Klinkhammer G. P., Bowers T. S., Edmond J. M., Lawrence J. R., Casey J. F., Thomson G., Humphris S., Rona P., and Karson J. A. (1988) Chemistry of hot springs on the Mid-Atlantic Ridge. *Nature* **335**, 514–519.

Campbell A. C., German C. R., Palmer M. R., Gamo T., and Edmond J. M. (1994) Chemistry of hydrothermal fluids from the Escanaba Trough, Gorda Ridge. In *Geologic, Hydrothermal and Biologic Studies at Escanaba Trough, Gorda Ridge*, US Geol. Surv. Bull. 2022 (eds. J. L. Morton, R. A. Zierenberg, and C. A. Reiss). Offshore Northern California, pp. 201–221.

Cave R. R., Ravizza G. E., German C. R., Thomson J., and Nesbitt R. W. (2003) Deposition of osmium and other platinum-group elements beneath the ultramafic-hosted rainbow hydrothermal plume. *Earth Planet. Sci. Lett.* **210**, 65–79.

Charlou J.-L., Fouquet Y., Donval J. P., Auzende J. M., Jean Baptiste P., Stievenard M., and Michel S. (1996) Mineral and gas chemistry of hydrothermal fluids on an ultrafast spreading ridge: East Pacific Rise, 17° to 19°S (NAUDUR cruise, 1993) phase separation processes controlled by volcanic and tectonic activity. *J. Geophys. Res.* **101**, 15899–15919.

Charlou J. L., Fouquet Y., Bougault H., Donval J. P., Etoubleau J., Jean-Baptiste P., Dapoigny A., Appriou P., and Rona P. A. (1998) Intense CH_4 plumes generated by serpentinization of ultramafic rocks at the intersection of the 15 degrees 20′ N. *Geochim. Cosmochim. Acta* **62**, 2323–2333.

Chin C. S., Coale K. H., Elrod V. A., Johnson K. S., Johnson G. J., and Baker E. T. (1994) *In situ* observations of dissolved iron and manganese in hydrothermal vent plumes, Juan de Fuca Ridge. *J. Geophys. Res.* **99**, 4969–4984.

Clarke W. B., Beg M. A., and Craig H. (1969) Excess ^3He in the sea: evidence for terrestrial primordial helium. *Earth Planet. Sci. Lett.* **6**, 213–220.

Connelly D. P., German C. R., Egorov A., Pimenov N. V., and Dohsik H. (2002) Hydrothermal plumes overlying the ultraslow spreading Knipovich Ridge, 72–78°N. *EOS Trans. AGU (abstr.)* **83**, T11A-1229.

Corliss J. B., Lyle M., and Dymond J. (1978) The chemistry of hydrothermal mounds near the Galapagos rift. *Earth Planet. Sci. Lett.* **40**, 12–24.

Cowen J. P. and German C. R. (2003) Biogeochemical cycling in hydrothermal plumes. In *Energy and Mass Transfer in Marine Hydrothermal Systems* (eds. P. Halbach, V. Tunnicliffe, and J. Hein), DUP, Berlin, pp. 303–316.

Cowen J. P., Bertram M. A., Wakeham S. G., Thomson R. E., Lavelle J. W., Baker E. T., and Feely R. A. (2001) Ascending and descending particle flux from hydrothermal plumes at endeavour segment, Juan de Fuca Ridge. *Deep-Sea Res. I* **48**, 1093–1120.

Cruse A. M. and Seewald J. S. (2001) Metal mobility in sediment-covered ridge-crest hydrothermal systems: experimental and theoretical constraints. *Geochim. Cosmochim. Acta* **65**, 3233–3247.

de Angelis M. A., Lilley M. D., Olson E. J., and Baross J. A. (1993) Methane oxidation in deep-sea hydrothermal plumes of the endeavour segment of the Juan de Fuca Ridge. *Deep-Sea Res.* **40**, 1169–1186.

Degens E. T. and Ross D. A. (eds.) (1969) *Hot Brines and Heavy Metal Deposits in the Red Sea.* Springer, New York.

De Mets C., Gordon R. G., Argus D. F., and Stein S. (1994) Effect of recent revisions to the geomagnetic reversal timescale on estimates of current plate motions. *Geophys. Res. Lett.* **21**, 2191–2194.

Douville E., Charlou J. L., Oelkers E. H., Bienvenu P., Colon C. F. J., Donval J. P., Fouquet Y., Prieur D., and Appriou P. (2002) The rainbow vent fluids (36 degrees 14′ N, MAR): the influence of ultramafic rocks and phase separation on trace metal content in Mid-Atlantic Ridge hydrothermal fluids. *Chem. Geol.* **184**, 37–48.

Dymond J. and Roth S. (1988) Plume dispersed hydrothermal particles—a time-series record of settling flux from the endeavour ridge using moored sensors. *Geochim. Cosmochim. Acta* **52**, 2525–2536.

Edmond J. M., Measures C. I., McDuff R. E., Chan L. H., Collier R., Grant B., Gordon L. I., and Corliss J. B. (1979) Ridge crest hydrothermal activity and the balances of the major and minor elements in the ocean: the Galapagos data. *Earth Planet. Sci. Lett.* **46**, 1–18.

Edmond J. M., Von Damm K. L., McDuff R. E., and Measures C. I. (1982) Chemistry of hot springs on the East Pacific Rise and their effluent dispersal. *Nature* **297**, 187–191.

Edmond J. M., Campbell A. C., Palmer M. R., Klinkhammer G. P., German C. R., Edmonds H. N., Elderfield H., Thompson G., and Rona P. (1995) Time-series studies of vent-fluids from the TAG and MARK sites (1986, 1990) Mid-Atlantic Ridge: a new solution chemistry model and a mechanism for Cu/Zn zonation in massive sulphide ore-bodies. In *Hydrothermal Vents and Processes*, Geol. Soc. Spec. Publ. 87 (eds. L.M. Parson, C.L. Walker, and D.R. Dixon), .The Geological Society Publishing House, Bath, UK, pp. 77–86.

Edmonds H. N., German C. R., Green D. R. H., Huh Y., Gamo T., and Edmond J. M. (1996) Continuation of the hydrothermal fluid chemistry time series at TAG and the effects of ODP drilling. *Geophys. Res. Lett.* **23**, 3487–3489.

Edmonds H. N., Michael P. J., Baker E. T., Connelly D. P., Snow J. E., Langmuir C. H., Dick H. J. B., German C. R., and Graham D. W. (2003) Discovery of abundant hydrothermal venting on the ultraslow-spreading Gakkel Ridge in the Arctic Ocean. *Nature* **421**, 252–256.

Elderfield H. and Schultz A. (1996) Mid-ocean ridge hydrothermal fluxes and the chemical composition of the ocean. *Ann. Rev. Earth Planet. Sci.* **24**, 191–224.

Eldridge C. S., Barton P. B., and Ohmoto H. (1983) Mineral textures and their bearing on formation of the Kuroko orebodies. *Econ. Monogr.* **5**, 241–281.

Evans W. C., White L. D., and Rapp J. B. (1988) Geochemistry of some gases in hydrothermal fluids from the southern Juan de Fuca Ridge. *J. Geophys. Res.* **93**, 15305–15313.

Feely R. A., Massoth G. J., Baker E. T., Cowen J. P., Lamb M. F., and Krogslund K. A. (1990) The effect of hydrothermal processes on mid-water phosphorous distributions in the northeast Pacific. *Earth Planet. Sci. Lett.* **96**, 305–318.

Feely R. A., Trefry J. H., Massoth G. J., and Metz S. (1991) A comparison of the scavenging of phosphorous and arsenic from seawater by hydrothermal iron hydroxides in the Atlantic and Pacific oceans. *Deep-Sea Res.* **38**, 617–623, 1991.

Feely R. A., Massoth G. J., Trefry J. H., Baker E. T., Paulson A. J., and Lebon G. T. (1994) Composition and sedimentation of hydrothermal plume particles from North Cleft Segment, Juan de Fuca Ridge. *J. Geophys. Res.* **99**, 4985–5006.

Feely R. A., Baker E. T., Marumo K., Urabe T., Ishibashi J., Gendron J., Lebon G. T., and Okamura K. (1996) Hydrothermal plume particles and dissolved phosphate over the superfast-spreading southern East Pacific Rise. *Geochim. Cosmochim. Acta* **60**, 2297–2323.

Feely R. A., Trefry J. H., Lebon G. T., and German C. R. (1998) The relationship between P/Fe and V/Fe in hydrothermal precipitates and dissolved phosphate in seawater. *Geophys. Res. Lett.* **25**, 2253–2256.

Field M. P. and Sherrell R. M. (2000) Dissolved and particulate Fe in a hydrothermal plume at 9°45' N, East Pacific Rise: slow Fe(II) oxidation kinetics in Pacific plumes. *Geochim. Cosmochim. Acta* **64**, 619–628.

Fouquet Y., Von Stackelberg U., Charlou J. L., Donval J. P., Erzinger J., Foucher J. P., Herzig P., Mühe R., Soakai S., Wiedicke M., and Whitechurch H. (1991) Hydrothermal activity and metallogenesis in the Lau back-arc basin. *Nature* **349**, 778–781.

Fouquet Y., Henry K., Bayon G., Cambon P., Barriga F., Costa I., Ondreas H., Parson L., Ribeiro A., and Relvas G. (2003) The Rainbow hydrothermal field: geological setting, mineralogical and chemical composition of sulfide deposits (MAR 36°14' N). *Earth Planet. Sci. Lett.* (submitted).

Früh-Green G. L., Weissert H., and Bernoulli D. (1990) A multiple fluid history recorded in alpine ophiolites. *J. Geol. Soc. London* **147**, 959–970.

Fuhrmann J. A., McCallum K., and Davis A. A. (1993) Phylogenetic diversity of subsurface marine microbial communities from the Atlantic and Pacific oceans. *Appl. Environ. Microbiol.* **59**, 1294–1302.

Gamo T., Okamura K., Charlou J. L., Urabe T., Auzende J. M., Ishibashi J., Shitashima K., and Chiba H. (1997) Acidic and sulfate-rich hydrothermal fluids from the Manus back-arc basin, Papua New Guinea. *Geology* **25**, 139–142.

German C. R. and Sparks R. S. J. (1993) Particle recycling in the TAG hydrothermal plume. *Earth Planet. Sci. Lett.* **116**, 129–134.

German C. R., Klinkhammer G. P., Edmond J. M., Mitra A., and Elderfield H. (1990) Hydrothermal scavenging of rare earth elements in the ocean. *Nature* **316**, 516–518.

German C. R., Fleer A. P., Bacon M. P., and Edmond J. M. (1991a) Hydrothermal scavenging at the Mid-Atlantic Ridge: radionuclide distributions. *Earth Planet. Sci. Lett.* **105**, 170–181.

German C. R., Campbell A. C., and Edmond J. M. (1991b) Hydrothermal scavenging at the Mid-Atlantic Ridge: modification of trace element dissolved fluxes. *Earth Planet. Sci. Lett.* **107**, 101–114.

German C. R., Higgs N. C., Thomson J., Mills R., Elderfield H., Blusztajn J., Fleer A. P., and Bacon M. P. (1993) A geochemical study of metalliferous sediment from the TAG hydrothermal mound, 26°08' N, Mid-Atlantic Ridge. *J. Geophys. Res.* **98**, 9683–9692.

German C. R., Briem J., Chin C., Danielsen M., Holland S., James R., Jónsdottir A., Ludford E., Moser C., Ólafsson J., Palmer M. R., and Rudnicki M. D. (1994) Hydrothermal activity on the Reykjanes ridge: the steinahóll vent-field at 63°06' N. *Earth Planet. Sci. Lett.* **121**, 647–654.

German C. R., Bourlès D. L., Brown E. T., Hergt J., Colley S., Higgs N. C., Ludford E. M., Nelsen T. A., Feely R. A., Raisbeck G., and Yiou F. (1997) Hydrothermal scavenging on the Juan de Fuca Ridge: Th-230(xs), Be-10 and REE in ridge-flank sediments. *Geochim. Cosmochim. Acta* **61**, 4067–4078.

German C. R., Baker E. T., Mevel C. A., Tamaki K., and the FUJI scientific team (1998a) Hydrothermal activity along the south-west Indian ridge. *Nature* **395**, 490–493.

German C. R., Richards K. J., Rudnicki M. D., Lam M. M., Charlou J. L., and FLAME scientific party (1998b) Topographic control of a dispersing hydrothermal plume. *Earth Planet. Sci. Lett.* **156**, 267–273.

German C. R., Livermore R. A., Baker E. T., Bruguier N. I., Connelly D. P., Cunningham A. P., Morris P., Rouse I. P., Statham P. J., and Tyler P. A. (2000) Hydrothermal plumes above the East Scotia Ridge: an isolated high-latitude back-arc spreading centre. *Earth Planet. Sci. Lett.* **184**, 241–250.

German C. R., Colley S., Palmer M. R., Khripounoff A., and Klinkhammer G. P. (2002) Hydrothermal sediment trap fluxes: 13°N, East Pacific Rise. *Deep Sea Res. I* **49**, 1921–1940.

German C. R., Thursherr A. M., Radford-Kröery J., Charlou J.-L., Jean-Baptiste P., Edmonds H. N., Patching J. W., and the FLAME 1 & II science teams (2003) Hydrothermal fluxes from the Rainbow vent-site, Mid-Atlantic Ridge: new constraints on global ocean vent-fluxes. *Nature* (submitted).

Ginster U., Mottl M. J., and VonHerzen R. P. (1994) Heat-flux from black smokers on the endeavor and Cleft segments, Juan de Fuca Ridge. *J. Geophys. Res.* **99**, 4937–4950.

Gracia E., Bideau D., Hekinian R., and Lagabrielle Y. (1999) Detailed geological mapping of two contrasting second-order segments of the Mid-Atlantic Ridge between oceanographer and Hayes fracture zones (33 degrees 30' N-35 degrees N). *J. Geophys. Res.* **104**, 22903–22921.

Gracia E., Charlou J. L., Radford-Knoery J., and Parson L. M. (2000) Non-transform offsets along the Mid-Atlantic Ridge south of the Azores (38 degrees N-34 degrees N): ultramafic exposures and hosting of hydrothermal vents. *Earth Planet. Sci. Lett.* **177**, 89–103.

Hannington M. D., Jonasson I., Herzig P., and Petersen S. (1995) Physical and chemical processes of seafloor mineralisation at mid-ocean ridges. *Geophys. Monogr. (AGU)* **91**, 115–157.

Hannington M. D., Galley A. G., Herzig P. M., and Petersen S. (1998) Comparison of the TAG mound and stock work complex with Cyprus-type massive sulfide deposits. In *Proceedings of the Ocean Drilling Program, Scientific Results*. **158**, pp. 389–415.

Haymon R. M. (1983) Growth history of hydrothermal black smoker chimneys. *Nature* **301**, 695–698.

Haymon R. M. (1996) The response of ridge-crest hydrothermal systems to segmented, episodic magma supply. In *Tectonic, Magmatic, Hydrothermal, and Biological Segmentation of Mid-ocean Ridges*, Geol. Soc. Spec. Publ. 118 (eds. C.J. McLeod, P.A. Tyler, and C.L. Walker), .Geological Society Publishing House, Bath, UK, pp. 157–168.

Haymon R., Fornari D., Von Damm K., Lilley M., Perfit M., Edmond J., Shanks W., Lutz R. A., Grebmeier J. M., Carbotte S., Wright D., McLaughlin E., Smith M., Beedle N., and Olson E. (1993) Volcanic eruption of the mid-ocean ridge along the EPR crest at 9°45–52' N: I. Direct submersible observations of seafloor phenomena associated with an eruption event in April 1991. *Earth Planet. Sci. Lett.* **119**, 85–101.

Hein J. R., Koski R. A., Embley R. W., Reid J., and Chang S.-W. (1999) Diffuse-flow hydrothermal field in an oceanic fracture zone setting, northeast Pacific: deposit composition. *Explor. Mining Geol.* **8**, 299–322.

Hein J. R., Baker E. T., Cowen J. P., German C. R., Holzbecher E., Koski R. A., Mottl M. J., Pimenov N. V., Scott S. D., and Thurnherr A. M. (2003) How important are material and chemical fluxes from hydrothermal circulation to the ocean? In *Energy and Mass Transfer in Marine Hydrothermal Systems* (eds. P. Halbach, V. Tunnicliffe, and J. Hein), DUP, Berlin, pp. 337–355.

Helfrich K. R. and Speer K. G. (1995) Ocean hydrothermal circulation: mesoscale and basin-scale flow. *Geophys. Monogr. (AGU)* **91**, 347–356.

Herring P. J. and Dixon D. R. (1998) Extensive deep-sea dispersal of postlarval shrimp from a hydrothermal vent. *Deep-Sea Res.* **45**, 2105–2118.

Herzig P. M. and Hannington M. D. (2000) Input from the deep: hot vents and cold seeps. In *Marine Geochemistry* (eds. H.D. Schultz and M. Zabel), Springer, Heidelberg, pp. 397–416.

Holland M. E. and Baross J. A. (2003) Limits to life in hydrothermal systems. In *Energy and Mass Transfer in Marine Hydrothermal Systems* (eds. P. Halbach, V. Tunnicliffe, and J. Hein), DUP, Berlin, pp. 235–248.

Holm N. G. and Charlou J. L. (2001) Initial indications of abiotic formation of hydrocarbons in the rainbow ultramafic hydrothermal system, Mid-Atlantic Ridge. *Earth Planet. Sci. Lett.* **191**, 1–8.

Humphris S. E., Herzig P. M., Miller D. J., Alt J. C., Becker K., Brown D., Brügmann G., Chiba H., Fouquet Y., Gemmel J. B., Guerin G., Hannington M. D., Holm N. G., Honnorez J. J., Iturrino G. J., Knott R., Ludwig R., Nakamura K., Petersen S., Reysenbach A.-L., Rona P. A., Smith S., Sturz A. A., Tivey M. K., and Zhao X. (1995) The internal structure of an active sea-floor massive sulphide deposit. *Nature* **377**, 713–716.

Ishibashi J.-I. and Urabe T. (1995) Hydrothermal activity related to arc-backarc magmatism in the western Pacific. In *Backarc Basins: Tectonics and Magmatism* (ed. B. Taylor), Plenum, NY, pp. 451–495.

Ishibashi J., Wakita H., Okamura K., Nakayama E., Feely R. A., Lebon G. T., Baker E. T., and Marumo K. (1997) Hydrothermal methane and manganese variation in the plume over the superfast-spreading southern East Pacific Rise. *Geochim. Cosmochim. Acta* **61**, 485–500.

James R. H. and Elderfield H. (1996) Chemistry of ore-forming fluids and mineral formation rates in an active hydrothermal sulfide deposit on the Mid-Atlantic Ridge. *Geology* **24**, 1147–1150.

Kadko D. and Moore W. (1988) Radiochemical constraints on the crustal residence time of submarine hydrothermal fluids—endeavour ridge. *Geochim. Cosmochim. Acta* **52**, 659–668.

Kadko D. C., Rosenberg N. D., Lupton J. E., Collier R. W., and Lilley M. D. (1990) Chemical reaction rates and entrainment within the endeavour ridge hydrothermal plume. *Earth Planet. Sci. Lett.* **99**, 315–335.

Kelley D. S., Karson J. A., Blackman D. K., Früh-Green G. L., Butterfield D. A., Lilley M. D., Olson E. J., Schrenk M. O., Roe K. K., Lebon G. T., Rivizzigno P., and the AT3-60 shipboard party, (2001) An off-axis hydrothermal vent field near the Mid-Atlantic Ridge at 30°N.. *Nature* **412**, 145–149.

Kennedy B. M. (1988) Noble gases in vent water from the Juan de Fuca Ridge. *Geochim. Cosmochim. Acta* **52**, 1929–1935.

Klinkhammer G., Elderfield H., and Hudson A. (1983) Rare earth elements in seawater near hydrothermal vents. *Nature* **305**, 185–188.

Klinkhammer G. P., Elderfield H., Edmond J. M., and Mitra A. (1994) Geochemical implications of rare earth element patterns in hydrothermal fluids from mid-ocean ridges. *Geochim. Cosmochim. Acta* **58**, 5105–5113.

Klinkhammer G. P., Chin C. S., Keller R. A., Dahlmann A., Sahling H., Sarthou G., Petersen S., and Smith F. (2001) Discovery of new hydrothermal vent sites in Bransfield strait, Antarctica. *Earth Planet. Sci. Lett.* **193**, 395–407.

Koski R. A., German C. R., and Hein J. R. (2003) Fate of hydrothermal products from mid-ocean ridge hydrothermal systems: near-field to global perspectives. In *Energy and Mass Transfer in Marine Hydrothermal Systems* (eds. P. Halbach, V. Tunnicliffe, and J. Hein), DUP, Berlin, pp. 317–335.

Landing W. M. and Bruland K. W. (1987) The contrasting biogeochemistry of iron and manganese in the Pacific Ocean. *Geochim. Cosmochim. Acta* **51**, 29–43.

Langmuir C., Humphris S., Fornari D., Van Dover C., Von Damm K., Tivey M. K., Colodner D., Charlou J.-L., Desonie D., Wilson C., Fouquet Y., Klinkhammer G., and Bougault H. (1997) Description and significance of hydrothermal vents near a mantle hot spot: the lucky strike vent field at 37°N on the Mid-Atlantic Ridge. *Earth Planet. Sci. Lett.* **148**, 69–92.

Lao Y., Anderson R. F., Broecker W. S., Trumbore S. E., Hoffman H. J., and Wölfli W. (1992) Transport and burial rates of ^{10}Be and ^{231}Pa in the Pacific ocean during the Holocene period. *Earth Planet. Sci. Lett.* **113**, 173–189.

Lilley M. D., Feely R. A., and Trefry J. H. (1995) Chemical and biochemical transformation in hydrothermal plumes. *Geophys. Monogr. (AGU)* **91**, 369–391.

Lupton J. E. and Craig H. (1981) A major ^{3}He source on the East Pacific Rise. *Science* **214**, 13–18.

Lupton J. E., Delaney J. R., Johnson H. P., and Tivey M. K. (1985) Entrainment and vertical transport of deep ocean water by buoyant hydrothermal plumes. *Nature* **316**, 621–623.

Lupton J. E., Butterfield D., Lilley M., Ishibashi J., Hey D., and Evans L. (1999a) Gas chemistry of hydrothermal fluids along the East Pacific Rise, 5°S to 32°S. *EOS Trans. AGU (abstr.)* **80**, F1099.

Lupton J. E., Baker E. T., and Massoth G. J. (1999b) Helium, heat and the generation of hydrothermal event plumes at mid-ocean ridges. *Earth Planet. Sci. Lett.* **171**, 343–350.

Lyle M., Leinen M., Owen R. M., and Rea D. K. (1987) Late tertiary history of hydrothermal deposition at the East Pacific Rise, 19°S—correlation to volcano-tectonic events. *Geophys. Res. Lett.* **14**, 595–598.

Magenheim A. J., Bayhurst G., Alt J. C., and Gieskes J. M. (1992) ODP leg 137, borehole fluid chemistry in hole 504B. *Geophys. Res. Lett.* **19**, 521–524.

Maris C. R. P. and Bender M. L. (1982) Upwelling of hydrothermal solutions through ridge flank sediments shown by pore-water profiles. *Science* **216**, 623–626.

Massoth G. J., Baker E. T., Lupton J. E., Feely R. A., Butterfield D. A., VonDamm K. L., Roe K. K., and LeBon G. T. (1994) Temporal and spatial variability of hydrothermal manganese and iron at Cleft segment, Juan de Fuca Ridge. *J. Geophys. Res.* **99**, 4905–4923.

McCollom T. M. and Shock E. L. (1997) Geochemical constraints on chemolithoautotrophic metabolism by microorganisms in seafloor hydrothermal systems. *Geochim. Cosmochim. Acta* **61**, 4375–4391.

McCollom T. M. and Shock E. L. (1998) Fluid-rock interactions in the lower oceanic crust: thermodynamic models of hydrothermal alteration. *J. Geophys. Res.* **103**, 547–575.

Mercier H. and Speer K. G. (1998) Transport of bottom water in the Romanche Fracture Zone and the Chain Fracture Zone. *J. Phys. Oc.* **28**, 779–790.

Metz S., Trefry J. H., and Nelsen T. A. (1988) History and geochemistry of a metalliferous sediment core from the Mid-Atlantic Ridge at 26°N. *Geochim. Cosmochim. Acta* **52**, 2369–2378.

Middleton J. H. (1979) Times of rise for turbulent forced plumes. *Tellus* **31**, 82–88.

Millero F. J., Sotolongo S., and Izaguirre M. (1987) The oxidation kinetics of Fe(II) in seawater. *Geochim. Cosmochim. Acta* **51**, 793–801.

Mills R. A. and Elderfield H. (1995) Hydrothermal activity and the geochemistry of metalliferous sediment. *Geophys. Monogr. (AGU)* **91**, 392–407.

Mills R. A., Elderfield H., and Thomson J. (1993) A dual origin for the hydrothermal component in a metalliferous sediment core from the Mid-Atlantic Ridge. *J. Geophys. Res.* **98**, 9671–9678.

Mottl M. J. (2003) Partitioning of energy and mass fluxes between mid-Ocean ridge axes and flanks at high and low temperature. In *Energy and Mass Transfer in Marine Hydrothermal Systems* (eds. P. Halbach, V. Tunnicliffe, and J. Hein), DUP, Berlin, pp. 271–286.

Mottl M. J. and McConachy T. F. (1990) Chemical processes in buoyant hydrothermal plumes on the East Pacific Rise near 21°N. *Geochim. Cosmochim. Acta* **54**, 1911–1927.

Mottl M. J. and Wheat C. G. (1994) Hydrothermal circulation through mid-ocean ridge flanks: fluxes of heat and magnesium. *Geochim. Cosmochim. Acta* **58**, 2225–2237.

Mottl M. J., Wheat G., Baker E., Becker N., Davis E., Feely R., Grehan A., Kadko D., Lilley M., Massoth G., Moyer C., and Sansome F. (1998) Warm springs discovered on 3.5 Ma oceanic crust, eastern flank of the Juan de Fuca Ridge. *Geology* **26**, 51–54.

Morton B. R., Taylor G. I., and Turner J. S. (1956) Turbulent gravitational convection from maintained and instantaneous sources. *Proc. Roy. Soc. London Ser. A.* **234**, 1–23.

Mullineaux L. S. and France S. C. (1995) Disposal mechanisms of deep-sea hydrothermal vent fauna. *Geophys. Monogr. (AGU)* **91**, 408–424.

Mullins T. D., Britschgi T. B., Krest R. L., and Giovannoni S. J. (1995) Genetic comparisons reveal the same unknown bacterial lineages in Atlantic and Pacific bacterioplankton communities. *Limnol. Oceanogr.* **40**, 148–158.

Murray J. and Renard A. F. (1891) *Deep-sea Deposits*. Report "Challenger" Expedition (1873–1876), London.

Nehlig P., Juteau T., Bendel V., and Cotten J. (1994) The root zones of oceanic hydrothermal systems—constraints from the Samail ophiolite (Oman). *J. Geophys. Res.* **99**, 4703–4713.

O'Grady K. M. (2001) The geochemical controls on hydrothermal vent fluid chemistry from two areas on the ultrafast spreading southern East Pacific Rise. MSc Thesis, University of New Hampshire, 134pp.

?A3B2 twb=0.24w?>Olivarez A. M. and Owen R. M. (1989) REE/Fe variation in hydrothermal sediments: implications for the REE content of seawater. *Geochim. Cosmochim. Acta* **53**, 757–762.

Oosting S. E. and Von Damm K. L. (1996) Bromide/chloride fractionation in seafloor hydrothermal fluids from 9–10°N East Pacific Rise. *Earth Planet. Sci. Lett.* **144**, 133–145.

Palmer M. R. and Edmond J. M. (1989) The strontium isotope budget of the modern ocean. *Earth Planet. Sci. Lett.* **92**, 11–26.

Palmer M. R. and Ernst G. G. J. (1998) Generation of hydrothermal megaplumes by cooling of pillow basalts at mid-ocean ridges. *Nature* **393**, 643–647.

Palmer M. R. and Ernst G. G. J. (2000) Comment on Lupton *et al.* (1999b). *Earth Planet. Sci. Lett.* **180**, 215–218.

Parson L. M., Murton B. J., Searle R. C., Booth D., Evans J., Field P., Keetin J., Laughton A., McAllister E., Millard N., Redbourne L., Rouse I., Shor A., Smith D., Spencer S., Summerhayes C., and Walker C. (1993) En echelon axial volcanic ridges at the Reykjanes ridge: a life cycle of volcanism and tectonics. *Earth Planet. Sci. Lett.* **117**, 73–87.

Parson L., Gracia E., Coller D., German C. R., and Needham H. D. (2000) Second order segmentation—the relationship between volcanism and tectonism at the MAR, 38°N–35°40' N. *Earth Planet. Sci. Lett.* **178**, 231–251.

Parsons B. (1981) The rates of plate creation and consumption. *Geophys. J. Roy. Astron. Soc.* **67**, 437–448.

Petersen S., Herzig P. M., and Hannington M. D. (2000) Third dimension of a presently forming VMS deposit: TAG hydrothermal mound, Mid-Atlantic Ridge, 26°N. *Mineralium Deposita* **35**, 233–259.

Peucker-Ehrenbrink B., Hofmann A. W., and Hart S. R. (1994) Hydrothermal lead transfer from mantle to continental crust—the role of metalliferous sediments. *Earth Planet. Sci. Lett.* **125**, 129–142.

Peucker-Ehrenbrink B., Ravizza G., and Hofmann A. W. (1995) The marine Os-187/Os-186 record of the past 180 million years. *Earth Planet. Sci. Lett.* **130**, 155–167.

Radford-Knoery J., German C. R., Charlou J.-L., Donval J.-P., and Fouquet Y. (2001) Distribution and behaviour of dissolved hydrogen sulfide in hydrothermal plumes. *Limnol. Oceanogr.* **46**, 461–464.

Ravizza G. (1993) Variations of the 187Os/186Os ratio of seawater over the past 28 million years as inferred from metalliferous carbonates. *Earth Planet. Sci. Lett.* **118**, 335–348.

Ravizza G., Martin C. E., German C. R., and Thompson G. (1996) Os isotopes as tracers in seafloor hydrothermal systems: a survey of metalliferous deposits from the TAG hydrothermal area, 26°N Mid-Atlantic Ridge. *Earth Planet. Sci. Lett.* **138**, 105–119.

Ravizza G., Blusztajn J., Von Damm K. L., Bray A. M., Bach W., and Hart S. R. (2001) Sr isotope variations in vent fluids from 9°46–54' N EPR: evidence of a non-zero-Mg fluid component at Biovent. *Geochim. Cosmochim. Acta* **65**, 729–739.

Rona P. A. and Trivett D. A. (1992) Discrete and diffuse heat transfer at ASHES vent field, axial volcano, Juan de Fuca Ridge. *Earth Planet. Sci. Lett.* **109**, 57–71.

Rona P. A., Bogdanov Y. A., Gurvich E. G., Rimskikorsakov N. A., Sagalevitch A. M., Hannington M. D., and Thompson G. (1993) Relict hydrothermal zones in the TAG hydrothermal field, Mid-Atlantic Ridge 26°N 45°W. *J. Geophys. Res.* **98**, 9715–9730.

Roth S. E. and Dymond J. (1989) Transport and settling of organic material in a deep-sea hydrothermal plume—evidence from particle-flux measurements. *Deep Sea Res.* **36**, 1237–1254.

Rubin K. H., MacDougall J. D., and Perfit M. R. (1994) $^{210}Po/^{210}Pb$ dating of recent volcanic eruptions on the seafloor. *Nature* **468**, 841–844.

Rudnicki M. D. and Elderfield H. (1992) Helium, radon and manganese at the TAG and SnakePit hydrothermal vent fields 26° and 23°N, Mid-Atlantic Ridge. *Earth Planet. Sci. Lett.* **113**, 307–321.

Rudnicki M. D. and Elderfield H. (1993) A chemical model of the buoyant and neutrally buoyant plume above the TAG vent field, 26 degrees N, Mid-Atlantic Ridge. *Geochim. Cosmochim. Acta* **57**, 2939–2957.

Rudnicki M. D., James R. H., and Elderfield H. (1994) Near-field variability of the TAG nonbuoyant plume 26°N Mid-Atlantic Ridge. *Earth Planet. Sci. Lett.* **127**, 1–10.

Ruhlin D. E. and Owen R. M. (1986) The rare earth element geochemistry of hydrothermal sediments from the East Pacific Rise: examination of a seawater scavenging mechanism. *Geochim. Cosmochim. Acta* **50**, 393–400.

Rüth C., Well R., and Roether W. (2000) Primordial ^{3}He in South Atlantic deep waters from sources on the Mid-Atlantic Ridge. *Deep-Sea Res.* **47**, 1059–1075.

Sauter D., Parson L., Mendel V., Rommevaux-Jestin C., Gomez O., Briais A., Mevel C., and Tamaki K. (2002) TOBI sidescan sonar imagery of the very slow-spreading southwest Indian Ridge: evidence for along-axis magma distribution. *Earth Planet. Sci. Lett.* **199**, 81–95.

Schaller T., Morford J., Emerson S. R., and Feely R. A. (2000) Oxyanions in metalliferous sediments: tracers for paleoseawater metal concentrations?. *Geochim. Cosmochim. Acta* **64**, 2243–2254.

Schultz A. and Elderfield H. (1997) Controls on the physics and chemistry of seafloor hydrothermal circulation. *Phil. Trans. Roy. Soc. London A* **355**, 387–425.

Schultz A., Delaney J. R., and McDuff R. E. (1992) On the partitioning of heat-flux between diffuse and point-source sea-floor venting. *J. Geophys. Res.* **97**, 12299–12314.

Seyfried W. E. and Ding K. (1995) Phase equilibria in subseafloor hydrothermal systems: a review of the role of redox, temperature, pH and dissolved Cl on the chemistry of hot spring fluids at mid-ocean ridges. *Geophys. Monogr. (AGU)* **91**, 248–272.

Shanks W. C., III (2001) Stable isotopes in seafloor hydrothermal systems: vent fluids, hydrothermal deposits, hydrothermal alteration, and microbial processes. In *Stable Isotope Geochemistry, Rev. Mineral. Geochem. 43* (eds. J.W. Valley and D.R. Cole), Mineralogical Society of America, pp. 469–525.

Sherrell R. M., Field M. P., and Ravizza G. (1999) Uptake and fractionation of rare earth elements on hydrothermal plume particles at 9°45' N, East Pacific Rise. *Geochim. Cosmochim. Acta* **63**, 1709–1722.

Simoneit B. R. T. (1991) Hydrothermal effects on recent diatomaceous sediments in Guaymas Basin—generation, migration, and deposition of petroleum. In *AAPG Memoir 47: The Gulf and Peninsular Province of the Californias*, American Association of Petroleum Geologists, Tulsa, OK, chap. 38, pp. 793–825.

Sohn R. A., Fornari D. J., Von Damm K. L., Hildebrand J. A., and Webb S. C. (1998) Seismic and hydrothermal evidence for a cracking event on the East Pacific Rise at 9°50' N. *Nature* **396**, 159–161.

Speer K. G. and Rona P. A. (1989) A model of an Atlantic and Pacific hydrothermal plume. *J. Geophys. Res.* **94**, 6213–6220.

Speer K. G., Maltrud M., and Thurnherr A. (2003) A global view of dispersion on the mid-oceanic ridge. In *Energy and Mass Transfer in Marine Hydrothermal Systems* (eds. P. Halbach, V. Tunnicliffe, and J. Hein), DUP, Berlin, pp. 287–302.

Spiess F. N., Ken C. M., Atwater T., Ballard R., Carranza A., Cordoba D., Cox C., Diaz Garcia V. M., Francheteau J., Guerrero J., Hawkins J., Haymon R., Hessler R., Juteau T., Kastner M., Larson R., Luyendyk B., Macdongall J. D., Miller S., Normark W., Orcutt J., and Rangin C. (1980) East Pacific Rise: hot springs and geophysical experiments. *Science* **207**, 1421–1433.

Statham P. J., Yeats P. A., and Landing W. M. (1998) Manganese in the eastern Atlantic Ocean: processes influencing deep and surface water distributions. *Mar. Chem.* **61**, 55–68.

Statham P. J., Connelly D. P., and German C. R. (2003) Fe(II) oxidation in Indian Ocean hydrothermal plumes. *Nature* (submitted).

Staudigel H. and Hart S. R. (1983) Alteration of basaltic glass—mechanisms and significance for oceanic-crust seawater budget. *Geochim. Cosmochim. Acta* **47**, 337–350.

Stein C. A. and Stein S. (1994) Constraints on hydrothermal heat flux through the oceanic lithosphere from global heat flow. *J. Geophys. Res.* **99**, 3081–3095.

Stein C. A., Stein S., and Pelayo A. M. (1995) Heat flow and hydrothermal circulation. Geophys. *Monogr. (AGU)* **91**, 425–445.

Summit M. and Baross J. A. (1998) Thermophilic subseafloor microorganisms from the 1996 north Gorda ridge eruption. *Deep-Sea Res.* **45**, 2751–2766.

Tivey M. K. (1995) The influence of hydrothermal fluid composition and advection rates on black smoker chimney mineralogy—insights from modelling transport and reaction. *Geochim. Cosmochim. Acta* **59**, 1933–1949.

Trefry J. H. and Metz S. (1989) Role of hydrothermal precipitates in the geochemical cycling of vanadium. *Nature* **342**, 531–533.

Trocine R. P. and Trefry J. H. (1988) Distribution and chemistry of suspended particles from an active hydrothermal vent site on the Mid-Atlantic Ridge at 26°N. *Earth Planet. Sci. Lett.* **88**, 1–15.

Turner J. S. (1973) *Buoyancy Effects in Fluids.* Cambridge University Press, 368pp.

Van Dover C. L., German C. R., Speer K. G., Parson L. M., and Vrijenhoek R. C. (2002) Evolution and biogeography of deep-sea vent and seep invertebrates. *Science* **295**, 1253–1257.

Von Damm K. L. (1995) Controls on the chemistry and temporal variability of seafloor hydrothermal fluids. *Geophys. Monogr. (AGU)* **91**, 222–247.

Von Damm K. L. (2000) Chemistry of hydrothermal vent fluids from 9–10°N, East Pacific Rise: time zero the immediate post-eruptive period. *J. Geophys. Res.* **105**, 11203–11222.

Von Damm K. L. (2003) Evolution of the hydrothermal system at East Pacific Rise 9°50' N: geochemical evidence for changes in the upper oceanic crust. *Geophys. Monogr. (AGU)* (submitted).

Von Damm K. L. and Bischoff J. L. (1987) Chemistry of hydrothermal solutions from the southern Juan de Fuca Ridge. *J. Geophys. Res.* **92**, 11334–11346.

Von Damm K. L. and Lilley M. D. (2003) Diffuse flow hydrothermal fluids from 9°50' N East Pacific Rise: origin, evolution and biogeochemical controls. *Geophys. Monogr. (AGU)* (in press).

Von Damm K. L., Edmond J. M., Grant B., Measures C. I., Walden B., and Weiss R. F. (1985a) Chemistry of submarine hydrothermal solutions at 21°N, East Pacific Rise. *Geochim. Cosmochim. Acta* **49**, 2197–2220.

Von Damm K. L., Edmond J. M., Measures C. I., and Grant B. (1985b) Chemistry of submarine hydrothermal solutions at Guaymas Basin, Gulf of California. *Geochim. Cosmochim. Acta* **49**, 2221 2237.

Von Damm K. L., Oosting S. E., Kozlowski R., Buttermore L. G., Colodner D. C., Edmonds H. N., Edmond J. M., and Grebmeier J. M. (1995) Evolution of East Pacific Rise hydrothermal vent fluids following a volcanic eruption. *Nature* **375**, 47–50.

Von Damm K. L., Buttermore L. G., Oosting S. E., Bray A. M., Fornari D. J., Lilley M. D., and Shanks W. C., III (1997) Direct observation of the evolution of a seafloor black smoker from vapor to brine. *Earth Planet. Sci. Lett.* **149**, 101–112.

Von Damm K. L., Bray A. M., Buttermore L. G., and Oosting S. E. (1998) The geochemical relationships between vent fluids from the lucky strike vent field, Mid-Atlantic Ridge. *Earth Planet. Sci. Lett.* **160**, 521–536.

Von Damm K. L., Gallant R. M., Hall J. M., Loveless J., Merchant E., and Scientific party of R/V Knorr KN162-13

(2001) The Edmond hydrothermal field: pushing the envelope on MOR brines. *EOS Trans. AGU (abstr.)* **82**, F646.

Von Damm K. L., Parker C. M., Gallant R. M., Loveless J. P., and the AdVenture 9 Science Party, (2002) Chemical evolution of hydrothermal fluids from EPR 21°N: 23 years later in a phase separating world. *EOS Trans. AGU (abstr.)* **83**, V61B-1365.

Von Damm K. L., Lilley M. D., Shanks W. C., III, Brockington M., Bray A. M., O'Grady K. M., Olson E., Graham A., Proskurowski G., and the SouEPR Science Party (2003) Extraordinary phase separation and segregation in vent fluids from the southern East Pacific Rise. *Earth Planet. Sci. Lett.* **206**, 365–378.

Welhan J. and Craig H. (1983) Methane, hydrogen, and helium in hydrothermal fluids at 21°N on the East Pacific Rise. In *Hydrothermal Processes at Seafloor Spreading Centres* (eds. P. A. Rona, K. Boström, L. Laubier, L. Laubier, and K. L. Smith, Jr.). NATO Conference Series IV: 12, Plenum, New York, pp. 391–409.

White R. S., McKenzie D., and O'Nions R. K. (1992) Oceanic crustal thickness from seismic measurements and rare earth element inversions. *J. Geophys. Res.* **97**, 19683–19715.

Winn C. D., Cowen J. P., and Karl D. M. (1995) Microbiology of hydrothermal plumes. In *Microbiology of Deep-sea Hydrothermal Vent Habitats* (ed. D.M. Karl), CRC, Boca Raton.

You C.-F., Butterfield D. A., Spivack A. J., Gieskes J. M., Gamo T., and Campbell A. J. (1994) Boron and halide systematics in submarine hydrothermal systems: effects of phase separation and sedimentary contributions. *Earth Planet. Sci. Lett.* **123**, 227–238.

Readings from the Treatise on Geochemistry
ISBN: 978-0-12-381391-6

pp. 337–378

12
The Geologic History of Seawater

H. D. Holland
Harvard University, Cambridge, MA, USA and University of Pennsylvania, Philadelphia, PA

12.1 INTRODUCTION

Aristotle proposed that the saltness of the sea was due to the effect of sunlight on water. Robert Boyle took strong exception to this view and—in the manner of the Royal Society—laid out a program of research in the opening paragraph of his *Observations and Experiments about the Saltness of the Sea* (1674) (Figure 1):

> The Cause of the Saltness of the Sea appears by *Aristotle's* Writings to have busied the Curiosity of Naturalists before his time; since which, his Authority, perhaps much more than his Reasons, did for divers Ages make the Schools and the generality of Naturalists of his Opinion, till towards the end of the last Century, and the beginning of ours, some Learned Men took the boldness to question the common Opinion; since when the Controversie has been kept on foot, and, for ought I know, will be so, as long as 'tis argued on both sides but by Dialectical Arguments, which may be probable on both sides, but are not convincing on either. Wherefore I shall here briefly deliver some particulars about the Saltness of the Sea, obtained by my own trials, where I was able; and where I was not, by the best Relations I could procure, especially from Navigators.

Boyle measured and compiled a considerable set of data for variations in the saltness of surface seawater. He also designed an improved piece of

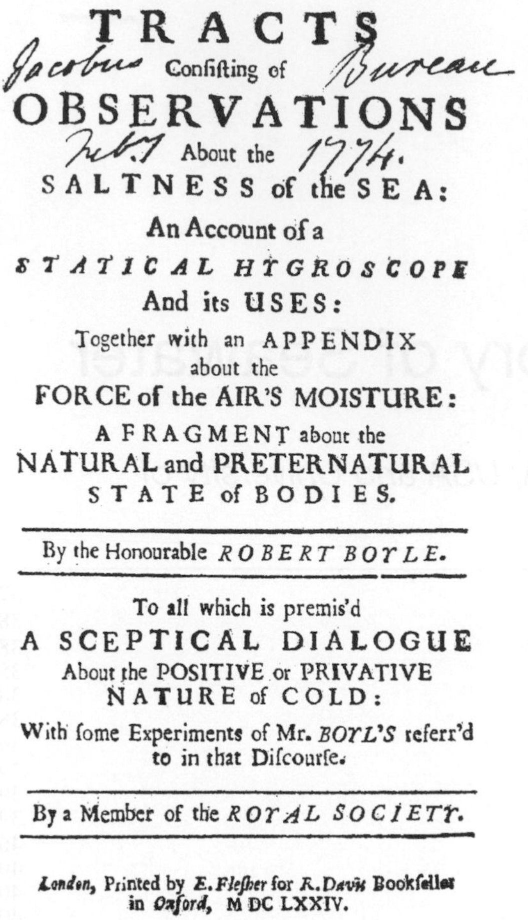
TRACTS
Jacobus Confifting of Bureau
OBSERVATIONS
Feb. 7 About the 1774.
SALTNESS of the SEA:
An Account of a
STATICAL HYGROSCOPE
And its USES:
Together with an APPENDIX
about the
FORCE of the AIR'S MOISTURE:
A FRAGMENT about the
NATURAL and PRETERNATURAL
STATE of BODIES.

By the Honourable ROBERT BOYLE.

To all which is premis'd
A SCEPTICAL DIALOGUE
About the POSITIVE or PRIVATIVE
NATURE of COLD:
With fome Experiments of Mr. BOYL'S referr'd
to in that Difcourfe.

By a Member of the ROYAL SOCIETY.

London, Printed by E. Flesher for R. Davis Bookfeller
in Oxford, M DC LXXIV.

Figure 1 Title page of Robert Boyle's Tracts consisting of Observations about the Saltness of the Sea and other essays (1674).

equipment for sampling seawater at depth, but the depths at which it was used were modest: 30 m with his own instrument, 80 m with another, similar sampler. However, the younger John Winthrop (1606–1676), an early member of the Royal Society, an important Governor of Connecticut, and a benefactor of Harvard College, was asked to collect seawater from the bottom of the Atlantic Ocean during his crossing from England to New England in the spring of 1663. The minutes of the Royal Society's meeting on July 20, 1663, give the following account of his unsuccessful attempt to do so (Birch, 1756; Black, 1966):

> Mr. Winthrop's letter written from Boston to Mr. Oldenburg was read, giving an account of the trials made by him at sea with the instrument for sounding of depths without a line, and with the vessel for drawing water from the bottom of the sea; both which proved successless, the former by reason of too much wind at the time of making soundings; the latter, on account of the leaking of the vessel. Capt. Taylor being to go soon to Virginia, and offering himself to make the same experiments, the society recommended to him the trying of the one in calm weather, and of the other with a stanch vessel.
>
> Mr. Hooke mentioning, that a better way might be suggested to make the experiment above-mentioned, was desired to think farther upon it, and to bring in an account thereof at the next meeting.

A little more than one hundred years later, in the 1780s, John Walker (1966) lectured at Edinburgh on the saltness of the oceans. He marshaled all of the available data and concluded that "these reasons seem all to point to this, that the water of the ocean in respect to saltness is pretty much what it ever has been."

In this opinion he disagreed with Halley (1715), who suggested that the salinity of the oceans has increased with time, and that the ratio of the total salt content of the oceans to the rate at which rivers deliver salt to the sea could be used to ascertain the age of the Earth. The first really serious attempt to measure geologic time by this method was made by Joly (1899). His calculations were refined by Clarke (1911), who inferred that the age of the ocean, since the Earth assumed its present form, is somewhat less than 100 Ma. He concluded, however, that "the problem cannot be regarded as definitely solved until all available methods of estimation shall have converged on one common conclusion." There was little appreciation in his approach for the magnitude of: (i) the outputs of salt from the oceans, (ii) geochemical cycles, and (iii) the notion of a steady-state ocean. In fact, Clarke's "age" of the ocean turns out to be surprisingly close to the oceanic residence time of Na^+ and Cl^-.

The modern era of inquiry into the history of seawater can be said to have begun with the work of Conway (1943, 1945), Rubey (1951), and Barth (1952). Much of the progress that was made between the appearance of these publications and the early 1980s was summarized by Holland (1984). This chapter describes a good deal of the progress that has been made since then.

12.2 THE HADEAN (4.5–4.0 Ga)

The broad outlines of Earth history during the Hadean are starting to become visible. The solar system originated 4.57 Ga (Allègre *et al.*, 1995). The accretion of small bodies in the solar nebula occurred within ~10 Myr of the birth of the solar system (Lugmaier and Shukolyukov, 1998). The Earth reached its present mass between 4.51 Ga and 4.45 Ga (Halliday, 2000; Sasaki and Nakazawa, 1986; Porcelli *et al.*, 1998). The core formed in <30 Ma (Yin *et al.*, 2002; Kleine *et al.*, 2002). The early Earth was covered by a magma ocean, but this must have cooled quickly at the end of the accretion process, and the first primitive crust must have formed shortly thereafter.

At present we do not have any rocks older than 4.03 Ga (Bowring and Williams, 1999). There are, however, zircons older than 4.03 Ga which were weathered out of their parent rocks and incorporated in 3 Ga quartzitic rocks in the Murchison District of Western Australia (Froude *et al.*, 1983; Compston and Pidgeon, 1986; Nutman *et al.*, 1991; Nelson *et al.*, 2000). A considerable number of 4.2–4.3 Ga zircon grains have been found in the Murchison District. A single 4.40 Ga zircon grain has been described by Wilde *et al.* (2001). The oxygen of these zircons has apparently retained its original isotopic composition. Mojzsis *et al.* (2001), Wilde *et al.* (2001), and Peck *et al.* (2001) have shown that the $\delta^{18}O$ values of the zircons which they have analyzed are significantly more positive than those of zircons which have crystallized from mantle magmas (see Figure 2). The most likely explanation for this difference is that the melts from which the zircons crystallized contained a significant fraction of material enriched in ^{18}O. This component was probably a part of the pre-4.0 Ga crust. Its enrichment in ^{18}O was almost certainly the result of subaerial weathering, which generates ^{18}O-enriched clay minerals (see, e.g., Holland, 1984, pp. 241–251). If this interpretation is correct, the data imply the presence of an active hydrologic cycle, a significant quantity of water at the Earth surface, and an early continental crust. However, as Halliday (2001) has pointed out, inferring the existence of entire continents from zircon grains in a single area requires quite a leap of the imagination. We need more zircon data from many areas to confirm the inferences drawn from the small amount of available data. Nevertheless, the inferences themselves are reasonable and fit into a coherent model of the Hadean Earth.

At present, little can be said with any degree of confidence about the composition of the proposed Hadean ocean, but it was probably not very different from that of the Early Archean ocean, about which a good deal can be inferred.

12.3 THE ARCHEAN (4.0–2.5 Ga)

12.3.1 The Isua Supracrustal Belt, Greenland

The Itsaq Gneiss complex of southern West Greenland contains the best preserved occurrences of ≥3.6 Ga crust. The gneiss complex had a complicated early history. It was added to and modified during several events starting ca. 3.9 Ga (Nutman *et al.*, 1996). Supracrustal, mafic, and ultramafic rocks comprise ~10% of the complex; these range in age from ≥3.87 Ga to ca. 3.6 Ga. A large portion of the Isua supracrustal belt (Figure 3) contains rocks that may be felsic volcanics and volcaniclastics, and abundant, diverse chemical sediments (Nutman *et al.*, 1997). The rocks are deformed, and many are substantially altered by metasomatism. However, transitional stages can be seen from units with relatively well-preserved primary volcanic and sedimentary features to schists in which all primary features have been obliterated.

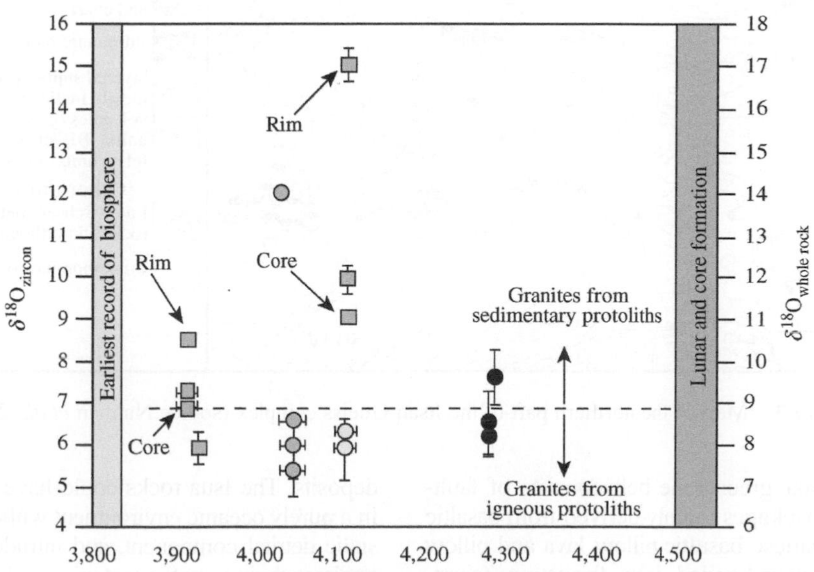

Figure 2 Ion microprobe $\delta^{18}O$ data for individual zircon spot analyses versus $^{207}Pb/^{206}Pb$ zircon age. The right vertical axis shows the estimated $\delta^{18}O$ data for the whole rock ($\delta^{18}O_{WR}$) from which the zircon crystallized (source Mojzsis *et al.*, 2001).

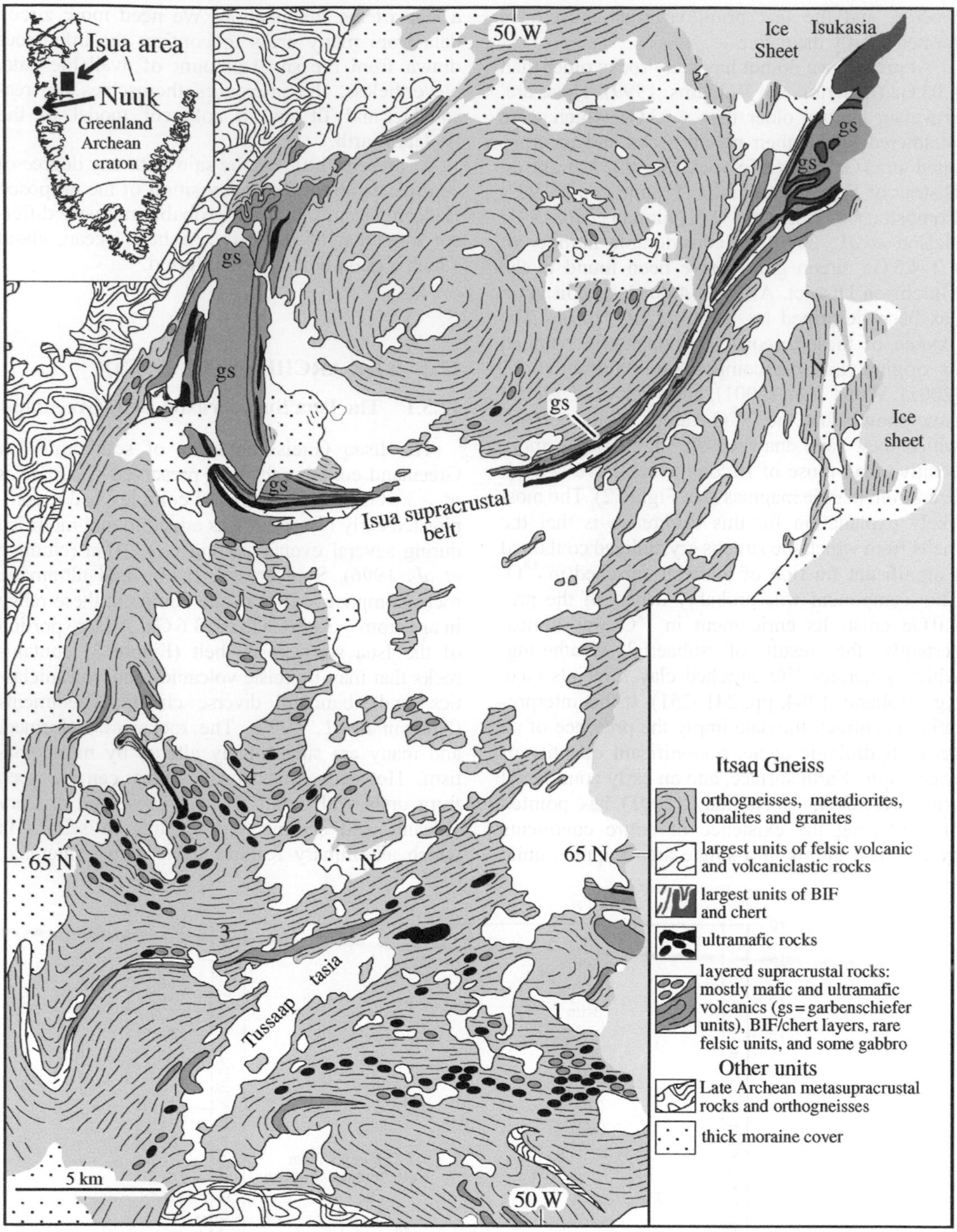

Figure 3 Map of the northern part of the Itsaq Gneiss complex (source Nutman *et al.*, 2002).

Most of the Isua greenstone belt consists of fault-bounded rock packages mainly derived from basaltic and high-manganese basaltic pillow lava and pillow lava breccia, chert-banded iron formation (chert-BIF), and a minor component of clastic sedimentary rocks derived from chert and basaltic volcanic rocks (Myers, 2001). The Isua sequence, as we now see it, resembles deep-sea sequences rather than platform deposits. The Isua rocks could have been deposited in a purely oceanic environment without a significant sialic detrital component, and intruded by the felsic gneisses during or after tectonic emplacement into the Amitsoq protocontinent (Rosing *et al.*, 1996).

The most compelling evidence for an ocean during the deposition of the Isua greenstone belt is provided by the BIF deposits which occur in this sequence. They are highly metamorphosed. Boak and Dymek (1982) have shown that the pelitic rocks in the sequence were exposed to temperatures of ~550°C at pressures of ~5 kbar. Temperatures during metamorphism could not have exceeded ~600 °C. The mineralogy and the chemistry of the iron formations have obviously been altered by metamorphism, but their major features are still preserved. Magnetite-quartz BIFs are particularly common. Iron enrichment can be extensive, as shown by the presence of a two-billion ton body at the very north-easternmost limit of the supracrustal belt (Bridgewater *et al.*, 1976).

In addition to the quartz-magnetite iron formation, Dymek and Klein (1988) described magnesian iron formation, aluminous iron formation, graphitic iron formation, and carbonate rich iron formation. They pointed out that the composition of the iron formation as a whole is very similar to that of other Archean and Proterozoic iron formations that have been metamorphosed to the amphibolite facies.

The source of the iron and probably much of the silica in the BIF was almost certainly seawater that had cycled through oceanic crust at temperatures of several hundred degree celsius. The low concentration of sulfur and of the base metals that are always present in modern solutions of this type indicates that these elements were removed, probably as sulfides of the base metals before the deposition of the iron oxides and silicates of the Isua iron formations.

The molar ratio of Fe_2O_3/FeO in the BIFs is less than 1.0 in all but one of the 28 analyses of Isua iron formation reported by Dymek and Klein (1988). In the one exception the ratio is 1.17. Unless the values of this ratio were reduced significantly during metamorphism, the analyses indicate that magnetite was the dominant iron oxide, and that hematite was absent or very minor in these iron formations. This, in turn, shows that some of the hydrothermal Fe^{2+} was oxidized to Fe^{3+} prior to deposition, but that not enough was oxidized to lead to the precipitation of Fe_2O_3 and/or Fe^{3+} oxyhydroxide precursors, or that these phases were subsequently replaced by magnetite.

The process(es) or processes by which the precursor(s) of magnetite in these iron formations precipitated is not well understood. A possible explanation involves the oxidation of Fe^{2+} by reaction with seawater to produce Fe^{3+} and H_2 followed by the precipitation of "green rust"—a solid solution of $Fe(OH)_2$ and $Fe(OH)_3$—and finally by the dehydration of green rust to magnetite (but see below). Figure 4 shows that the boundary between the stability field of $Fe(OH)_2$ and amorphous $Fe(OH)_3$ is at a rather low value of $p\varepsilon$. The field of green rust probably straddles this boundary. Saturation of solutions with siderite along this boundary at a total carbon concentration of 10^{-3}M and a total Fe concentration of 10^{-5}M lies within a reasonable range for the pH of

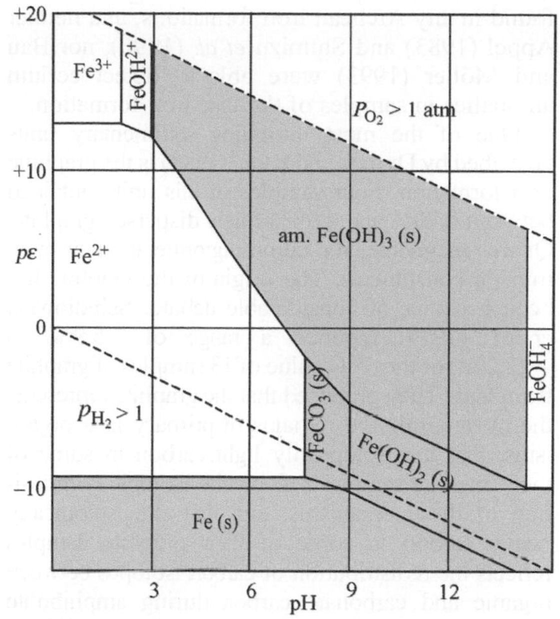

Figure 4 Diagram $p\varepsilon$ versus pH for the system $Fe–CO_2–H_2O$. The solid phases are $Fe(OH)_3$ (amorphous), $FeCO_3$ (siderite), $Fe(OH)_2$(s), and Fe(s); C_T=10^{-3}M. Lines are calculated for Fe(II) and Fe(III)=10^{-5}M at 25 °C. The possible conversion of carbonate to methane at low $p\varepsilon$ values was ignored (source Stumm and Morgan, 1996).

Archean seawater. However, the sequence of mineral deposition was probably more complex (see below).

The reaction of Fe^{2+} with H_2O is greatly accelerated by solar UV (see, e.g., Braterman *et al.*, 1984), and it has been suggested that solar UV played an important role in the deposition of oxide facies iron formation (Cairns-Smith, 1978; Braterman *et al.*, 1983; Sloper, 1983;François, 1986, 1987; Anbar and Holland, 1992). However, it now appears that solar UV played no more than a minor role in the deposition of oxide facies BIF. The reasons for this conclusion are detailed later in the chapter. The absence of significant quantities of hematite indicates that seawater from which the precursor of the magnetite-quartz iron formation was deposited was mildly reducing. This is corroborated by the mineralogy of the silicates, in which iron is present exclusively or nearly so in the divalent state. The only possible indication of a relatively high oxidation state is the presence of cerium anomalies reported in some of Dymek and Klein's (1988) rare earth element (REE) analyses. However, the validity of these anomalies is somewhat uncertain, because the analyses do not include praseodymium, and because neodymium measurements are lacking in a number of cases. Fryer (1983) had earlier observed that no significantly anomalous behavior of cerium has been found in any Archean iron formations, and neither Appel (1983) and Shimizu *et al.* (1990), nor Bau and Möller (1993) were able to detect cerium anomalies in samples of the Isua iron formation.

One of the most intriguing sedimentary units described by Dymek and Klein (1988) is the graphitic iron formation. Four samples of this unit contained between 0.70% and 2.98% finely dispersed graphite. Quartz, magnetite, and cummingtonite are their main mineral constituents. The origin of the graphite has been a matter of considerable debate. Schidlowski *et al.* (1979) reported a range of −5.9‰ to −22.2‰ for the δ^{13}C value of 13 samples of graphite from Isua. They proposed that the graphite represents the metamorphosed remains of primary Isua organisms, that the isotopically light carbon in some of their graphite samples reflects the isotopic composition of these organisms, and that the isotopically heavy carbon in some of their graphite samples reflects the redistribution of carbon isotopes between organic and carbonate carbon during amphibolite grade metamorphism. Perry and Ahmad (1977) found δ^{13}C values between −9.3‰ and −16.3‰ in Isua supracrustal rocks and pointed out that the fractionation of the carbon isotopes between siderite and graphite in their samples is consistent with inorganic equilibrium of these phases at ~400–500 °C. Oehler and Smith (1977) found graphite with δ^{13}C values of −11.3‰ to −17.4‰ in Isua metapelites (?) containing 150–4,800 ppm reduced carbon, and graphite with a δ^{13}C range between −21.4‰ and −26.9‰ in metasediments from the Isua iron formation which contain only trace amounts of graphite (4–56 ppm). The carbon in the latter samples is thought to be due to postdepositional contamination.

Since then Mojzsis *et al.* (1996) have used *in situ* ion microprobe techniques to measure the isotopic composition of carbon in Isua BIF and in a unit from the nearby Akilia Island that may be BIF. δ^{13}C in carbonaceous inclusions in the BIF ranged from −23‰ to −34‰. those in carbon inclusions occluded in apatite micrograins from the Akilia Island BIF ranged from −21‰ to −49‰. Since the carbon grains embedded in apatite were small and irregular, the precision and accuracy of the individual δ^{13}C measurements was typically ± 5‰ (1σ). These measurements tend to confirm the Schidlowski *et al.* (1979) interpretation, and suggest that the graphite in the Isua and Akilia BIFs could well be the metamorphosed remains of primitive organisms. However, other interpretations have been advanced very forcefully (see, e.g., Holland, 1997; Van Zuilen *et al.*, 2002).

Naraoka *et al.* (1996) have added additional δ^{13}C measurements, which fall in the same range as those reported previously. They emphasize that graphite with δ^{13}C values around −12‰ was probably formed by an inorganic, rather than by a biological process. Rosing (1999) reported δ^{13}C values ranging from −11.4‰ to −20.2‰ in 2–5 μm graphite globules in turbiditic and pelagic sedimentary rocks from the Isua supracrustal belt. He suggests that the reduced carbon in these samples represents biogenic detritus, which was perhaps derived from planktonic organisms.

This rather large database certainly suggests that life was present in the 3.7–3.9 Ga oceans, but it is probably best to treat the proposition as likely rather than as proven. One of the arguments against the presence of life before 3.8 Ga is based on the likelihood of large extraterrestrial impacts during a late heavy bombardment (LHB). The craters of the moon record an intense bombardment by large bodies, ending abruptly ca. 3.85 Ga (Dalrymple and Ryder, 1996; Hartmann *et al.*, 2000; Ryder, 1990). The Earth was probably impacted at least as severely as the moon, and there is a high probability that impacts large enough to vaporize the ocean's photic zone occurred as late as 3.8 Ga (Sleep *et al.*, 1989). The environment of the early Earth, therefore, may have been extremely challenging to life (Chyba, 1993; Appel and Moorbath, 1999). There has, however, been little direct examination of the Earth's surface environment during this period. The metasediments at Isua and on Akilia Island supply a small window on the effects of extraterrestrial bombardment between ca. 3.8 Ga and 3.9 Ga. Anbar *et al.* (2002) have determined the concentration of iridium and platinum in three samples of a ~5 m thick BIF chert unit and in three samples of mafic–ultramafic flows interposed with BIF in a relatively undeformed section on Akilia Island. The iridium and platinum concentrations in the Akilia metasediments are all extremely low. The Iridium content of only one of the BIF/chert samples is above 3 ppt, the detection limit of the techniques. ID-ICP-MS techniques. The concentration of platinum is below the detection limit (40 ppt) in nearly all of the BIF/chert samples. Both elements were readily detected in the mafic–ultramafic samples. The extremely low concentration of iridium and platinum in the BIF/chert samples shows that their composition was not significantly affected by extraterrestrial impacts. It is difficult, however, to extrapolate from these few analyses to other environments between 3.8Ga and 3.9 Ga, especially because it has been proposed that these rocks are not sedimentary (Fedo and Whitehouse, 2002). As Anbar *et al.* (2002) have pointed out, the large time gaps between the large, life-threatening impact events require extensive sampling of the rock record for the detection of the impact events. The slim evidence supplied by the Anbar *et al.* (2002) study encourages the view that conditions were sufficiently benign for the existence of life on Earth during the deposition of the sediments at Isua and on Akilia, but optimism on this point must surely be tempered, and it is premature to speculate on the effects of the potential biosphere on the state of the oceans 3.8–3.9 Ga.

12.3.2 The Mesoarchean Period (3.7–3.0 Ga)

Evidence for the presence of life is abundant during the later part of the Mesoarchean period. Unmetamorphosed carbonaceous shales with $\delta^{13}C$ values that are consistent with a biological origin of the contained carbon are reasonably common (see, e.g., Schidlowski *et al.*, 1983; Strauss *et al.*, 1992). The nature of the organisms that populated the Mesoarchean oceans is still hotly disputed. The description of microfossils in the 3.45 Ga Warrawoona Group of Western Australia by Schopf and Packer (1987) and Schopf (1983) suggested that some of these microfossils were probably the remains of cyanobacteria. If so, oxygenic photosynthesis is at least as old as 3.45 Ga. Brasier *et al.* (2002) have re-examined the type sections of the material described by Schidlowski et al. (1983) and have reinterpreted all of his 11 holotypes as artifacts formed from amorphous graphite within multiple generations of metalliferous hydrothermal vein chert and volcanic glass. However, Schopf *et al.* (2002) maintain that the laser Raman imagery of this material not only establishes the biogenicity of the fossils which they have studied, but also provides insight into the chemical changes that accompanied the metamorphism of these organics. Whatever the outcome of this debate, it is most likely that the oceans contained an upper, photic zone, that ~20% of the volcanic CO_2 added to the atmosphere was reduced and buried as a constituent of organic water, and that the remaining 80% were buried as a constituent of marine carbonates (see, e.g., Holland, 2002).

The influence of these processes on the composition of Mesoarchean seawater is still unclear. De Ronde *et al.* (1997) have studied the fluid chemistry of what they believe are Archean seafloor hydrothermal vents, and have explored the implications of their analyses for the composition of contemporary seawater. They estimate that seawater contained 920 mmol L^{-1} Cl, 2.25 mmol L^{-1} Br, 2.3 mmol L^{-1} SO_4, 0.037 mmol L^{-1} I, 789 mmol L^{-1} Na, 5.1 mmol L^{-1} NH_4, 18.9 mmol L^{-1} K, 50.9 mmol L^{-1} Mg, 232 mmol L^{-1} Ca, and 4.52 mmol L^{-1} Sr. This composition, if correct, implies that Archean seawater was rather similar to modern seawater. Unfortunately, there is considerable doubt about the correctness of these concentrations. First, the composition of seawater is altered significantly during passage through the oceanic crust, and the reconstruction of the composition of seawater from that of hydrothermal fluids is not straightforward. Second the charge balance of the proposed seawater is quite poor. Third, the quartz which contained the fluid inclusions analyzed by De Ronde *et al.* (1997) is intimately associated with hematite and goethite. The former mineral is most unusual as an ocean floor mineral at 3.2 Ga. The latter is also unusual, because these sediments passed through a metamorphic event at 2.7 Ga during which the temperature rose to >200 °C (De Ronde *et al.*, 1994). It is not unlikely that the inclusion fluids analyzed by De Ronde *et al.* were trapped more recently than 3.2 Ga, and that they are not samples of 3.2 Ga seawater (Lowe, personal communication, 2002).

The direct evidence for the composition of Mesoarchean seawater is, therefore, quite weak. For the time being it seems best to rely on indirect evidence derived from the mineralogy of sediments from this period, the composition of these minerals, and the isotopic composition of their contained elements. The carbonate minerals in Archean sediments are particularly instructive (see, e.g., Holland, 1984, chapter 5). Calcite, aragonite, and dolomite were the dominant carbonate minerals. Siderite was only a common constituent of BIFs. These observations imply that the Archean oceans were saturated or, more likely, supersaturated with respect to $CaCO_3$ and $CaMg(CO_3)_2$. Translating this observation into values for the concentration of Ca^{2+}, Mg^{2+}, HCO_3^-, and CO_3^{2-} in seawater is difficult in the absence of other information, but it can be shown that for the likely range of values of atmospheric P_{CO_2} ($\leq$0.03 atm; Rye *et al.*, 1995), the pH of seawater was probably $\geq$6.5. At saturation with respect to calcite and dolomite at 25 °C, the ratio $m_{Mg^{2+}}/m_{Ca^{2+}}$ in solutions is close to 1.0. This does not have to be the value of the ratio in Archean seawater. In Phanerozoic seawater (see below), the Mg^{2+}/Ca^{2+} ratio varied considerably from values as low as 1 up to its present value of 5.3 (Lowenstein *et al.*, 2001; Horita *et al.*, 2002).

A rough upper limit to the Fe^{2+}/Ca^{2+} ratio in Archean seawater can be derived from the scarcity of siderite except as a constituent of carbonate iron formations. At saturation with respect to siderite and calcite, the ratio $m_{Fe^{2+}}/m_{Ca^{2+}}$ is approximately equal to the ratio of the solubility product of siderite (Bruno *et al.*, 1992) and calcite (Plummer and Busenberg, 1982):

$$\frac{m_{Fe^{2+}}}{m_{Ca^{2+}}} \approx \frac{K_{sid}}{K_{cal}} = \frac{10^{-10.8}}{10^{-8.4}} = 4 \times 10^{-3} \qquad (1)$$

The absence of siderite from normal Archean carbonate sequences indicates that this is a reasonable upper limit for the Fe^{2+}/Ca^{2+} ratio in normal Archean seawater. An approximate lower limit can be set by the Fe^{2+} content of limestones and dolomites. These contain significantly more Fe^{2+} and Mn^{2+} than their Phanerozoic counterparts (see, e.g., Veizer *et al.*, 1989), a finding that is consistent with a much lower O_2 content in the atmosphere and in near-surface seawater than today.

The strongest evidence for no more than a few ppm O_2 in the Archean and Early Paleoproterozoic atmosphere is the evidence for mass-independent fractionation (MIF) of the sulfur isotopes in pre-2.47 Ga sulfides and sulfates (Farquhar *et al.*, 2000, 2001; Pavlov and Kasting, 2002; Bekker *et al.*, 2002). In the absence of O_2, solar UV interacts with SO_2 and generates MIF of the sulfur isotopes in the reaction products. The MIF signal is probably preserved in the sedimentary record, because this signal in elemental sulfur produced by this process differs from that of the gaseous products. The fate of the elemental sulfur and sulfur gases is not well understood, but S^0 may well be deposited largely as a constituent of sulfide minerals and the sulfur that is present as a constituent of sulfur gases in part as a constituent of sulfates. Today very little of the MIF signal is preserved, because all of the products are gases that become isotopically well mixed before the burial of their contained sulfur.

It is not surprising that the geochemical cycle of sulfur during the low-O_2 Archean differed from that of the present day. As shown in Figure 5, the mass-dependent fractionation of the sulfur isotopes in sedimentary sulfides was smaller prior to 2.7 Ga than in more recent times. Several explanations have been advanced for this observation. The absence of microbial sulfate reduction is one. However, the presence of microscopic sulfides in ca. 3.47 Ga barites from north pole, Australia with a maximum sulfur fractionation of 21.1% and a mean of 11.6% clearly indicates that microbial sulfate reduction was active during the deposition of these, probably evaporitic sediments (Shen *et al.*, 2001). A second explanation involves high temperatures in the pre-2.7 Ga oceans (Ohmoto *et al.*, 1993; Kakegawa *et al.*, 1998). However, Canfield *et al.* (2000) have shown that at both high and low temperatures large fractionations are expected during microbial sulfate reduction in the presence of abundant sulfate. A third possibility is that the concentration of sulfate in the Mesoarchean oceans was very much smaller than its current value of 28 mm. At present it appears that, until ca. 2.3 Ga, $m_{SO_4^{2-}}$ was $\leq 200\,\mu mol\,L^-$, the concentration below which isotopic fractionation during sulfate reduction is greatly reduced (Harrison and Thode, 1958; Habicht *et al.*, 2002). Such a low sulfate concentration in seawater prior to 2.3 Ga is quite reasonable. In the absence of atmospheric O_2, sulfide minerals would not have been oxidized during weathering, and this source of river SO_4^{2-} would have been extremely small. The other major source of river SO_4^{2-}, the solution of evaporite minerals, would also have been minimal. Constructing a convincing, quantitative model of the Archean sulfur cycle is still, however, very difficult. Volcanic SO_2 is probably disproportionated, at least partially, into H_2S and H_2SO_4 by reacting with H_2O at temperatures below 400 °C. SO_4^{2-} from this source must have cycled through the biosphere. Some of it was probably lost during passage through the oceanic crust at hydrothermal temperatures. Some was lost as a constituent of sulfides (mainly pyrite) and relatively rare sulfates (mainly barite).

Another potentially major loss of SO_4^{2-} may well have been the anaerobic oxidation of methane via the overall reaction

$$CH_4 + SO_4^{2-} \rightarrow HCO_3^- + HS^- + H_2O \quad (2)$$

(Iversen and Jørgensen, 1985; Hoehler and Alperin, 1996; Orphan *et al.*, 2001). It seems likely that the reaction is accomplished in part by a consortium of Archaea growing in dense aggregates of ~100 cells, which are surrounded by sulfate-reducing bacteria (Boetius *et al.*, 2000; De Long, 2000). In sediments rich in organic matter SO_4^{2-} is depleted rapidly. Below the zone of SO_4^{2-} depletion CH_4 is produced. The gas diffuses upward and is destroyed, largely in the transition zone, where the concentration of SO_4^{2-} in the interstitial water is in the range of 0.1–1 mmol kg^{-1} (Iversen and Jørgensen, 1985). The rate of CH_4 oxidation is highest where its concentration is equal to that of SO_4^{2-}. In two stations

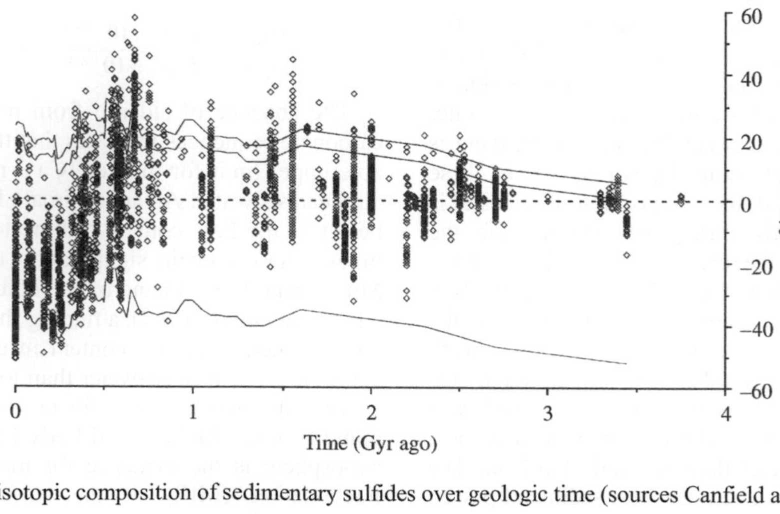

Figure 5 The isotopic composition of sedimentary sulfides over geologic time (sources Canfield and Raiswell, 1999).

studied by Iversen and Jørgensen, the total anaerobic methane oxidation was close to $1\,mmol\,m^{-2}\,d^{-1}$, of which 96% occurred in the sulfate–methane transition zone. If this rate were characteristic of the ocean floor as a whole, SO_4^{2-} reduction within marine sediments would occur at a rate of $\sim(1\times10^{14})\,mol\,yr^{-1}$, which exceeds the present-day input of volcanic SO_2 by ca. two orders of magnitude. The process is, therefore, potentially important for the global geochemistry of sulfur and carbon.

The anaerobic oxidation of CH_4 also occurs in anoxic water masses. In the Black Sea only ~2% of the CH_4 which escapes from sediments reaches the atmosphere. The remainder is largely lost by sulfate reduction in the anoxic parts of the water column (Reeburgh *et al.*, 1991). In an ocean containing $\leq 1\,mmol\,L\,SO_4^{2-}$ the rate of CH_4 loss in the water column would almost certainly be smaller than in the Black Sea today, and the flux of CH_4 to the atmosphere would almost certainly be greater than today. Pavlov *et al.* (2000) have shown that the residence time of CH_4 in an anoxic atmosphere is $\sim 3\times10^4$ yr, i.e., some 1,000 times longer than today. The combination of a higher rate of CH_4 input and a longer residence time in the atmosphere virtually assures that the partial pressure of CH_4 was much higher in the Archean atmosphere than its present value of $\sim(1\times10^{-6})$ atm. A CH_4 pressure of 10^{-4}–10^{-3} atm is not unlikely (Catling *et al.*, 2001). At these levels CH_4 generates a very significant greenhouse warming, enough to overcome the likely lower luminosity of the Sun during the early part of Earth history (Kasting *et al.*, 2001). The recent discovery of microbial reefs in the Black Sea fueled by the anaerobic oxidation of methane by SO_4^{2-} (Michaelis *et al.*, 2002) suggests that this process was important during the Archean, and that it can account for some of the organic matter generated in the oceans before the rise of atmospheric O_2.

12.3.3 The Neoarchean (3.0–2.5 Ga)

The spread in the $\delta^{34}S$ value of sulfides and sulfates increased significantly between 3.0 Ga and 2.5 Ga. The first major increase in the $\delta^{34}S$ range occurred ~2.7 Ga (see Figure 5). However, sulfate concentrations probably stayed well below the present value of $28\,mmol\,kg^{-1}$ until the Neoproterozoic. Grotzinger (1989) has reviewed the mineralogy of Precambrian evaporites and has shown that calcium sulfate minerals (or their pseudomorphs) are scarce before ~1.7–1.6 Ga. Bedded or massive gypsum/anhydrite formed in evaporitic environments is absent in the Archean and Paleoproterozoic record. A low concentration of SO_4^{2-} in the pre-1.7 Ga oceans is the most reasonable explanation for these observations (Grotzinger and Kasting, 1993).

Rather interestingly, the oldest usable biomarkers in carbonaceous shales date from the Neoarchean. Molecular fossils extracted from 2.5 Ga to 2.7 Ga shales of the Fortescue and Hamersley groups in the Pilbara Craton, Western Australia, if they are not due to contamination, indicate that the photic zone of the water column in the areas where these shales were deposited was probably weakly oxygenated, and that cyanobacteria were part of the microbial biota (Brocks *et al.*, 1999, 2002; Summons *et al.*, 1999). The similarity of the timing of the rise in the range of $\delta^{34}S$ in sediments and the earliest evidence for the presence of cyanobacteria may, however, be coincidental, because to date no sediments older than 2.7 Ga have been found that contain usable biomarker molecules (Brocks, personal communication, 2002).

Despite the biomarker evidence for the generation of O_2 at 2.7 Ga, the atmosphere seems to have contained very little or no O_2, and much of the ocean appears to have been anoxic. Pyrite, uraninite, gersdorffite, and, locally, siderite occur as unequivocally detrital constituents in 3,250–2,750 Ma fluvial siliciclastic sediments in the Pilbara Craton in Australia (Rasmussen and Buick, 1999). These sediments have never undergone hydrothermal alteration. Some grains of siderite display evidence of several episodes of erosion, rounding, and subsequent authigenic overgrowth (see Figure 6). Their frequent survival after prolonged transport in well-mixed and, therefore, well-aerated Archean rivers that contained little organic matter strongly implies that the contemporary atmosphere was much less oxidizing than at present. The paper by Rasmussen and Buick (1999) was criticized by Ohmoto (1999), but staunchly defended by Rasmussen *et al.* (1999).

These observations complement those made since the early 1990s on the gold–uranium ores of the Witwatersrand Basin in South Africa and on the uranium ores of the Elliot Lake District in Canada. The origin of these ores has been hotly debated (see, e.g., Phillips *et al.*, 2001). The rounded shape of many of the pyrite and uraninite grains (see Figures 7 and 8) are in a geologic setting appropriate for the placer accumulation of heavy minerals. Figure 7 shows some of the muffin-shaped uraninite grains described by Schidlowski (1966), and Figure 8 shows rounded grains of pyrite described by Ramdohr (1958). It is clear that some of the rounded pyrite grains are replacements of magnetite, ilmenite, and other minerals. The origin of any specific rounded pyrite grain if based on textural evidence alone is, therefore, somewhat ambiguous. However, the Re–Os age of some pyrite grains indicates that they are older than the depositional age of the sediments (Kirk *et al.*, 2001). The detrital origin of the uraninite muffins is essentially established by their chemical composition. As shown in Table 1 these

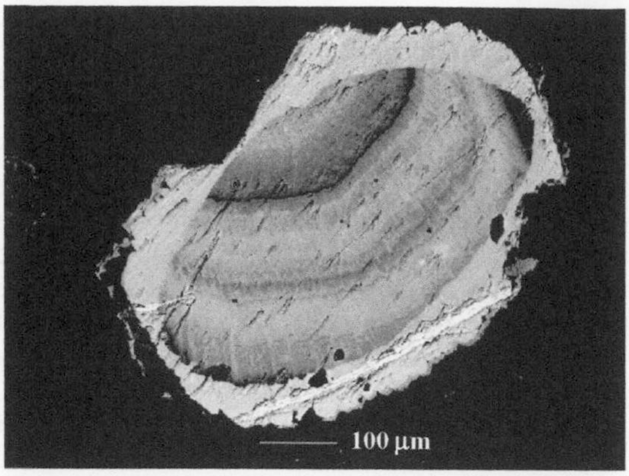

Figure 6 Rounded siderite grain, with core of compositionally banded siderite and gray to black syntaxial overgrowths (source Rasmussen and Buick, 1999).

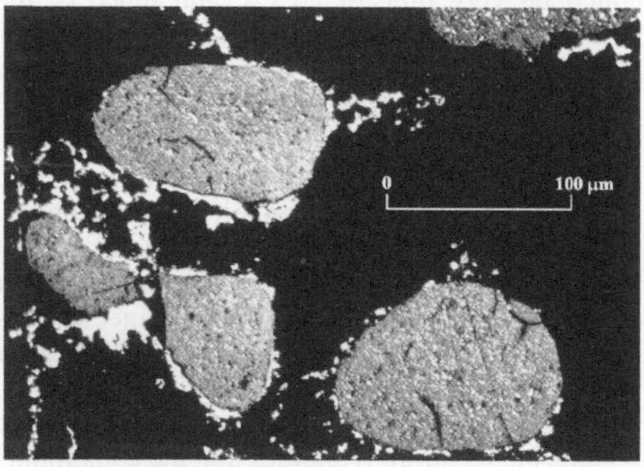

Figure 7 Detrital grains of uraninite with characteristic dusting of galena, partly surrounded by PbS overgrowths. The big grain displays a typical "muffin shape." Basal Reef, footwall; Loraine Gold Mines, South Africa oil immersion; 375× (source Schidlowski, 1966).

contain significant concentrations of ThO_2, which are characteristic of uraninite derived from pegmatites but not of hydrothermal pitchblende. It is, therefore, very difficult to assign anything but a detrital origin to the uraninite in the Witwatersrand ores (see, e.g., Hallbauer, 1986).

Experiments by Grandstaff (1976, 1980) and more recently by Ono (2002) on the oxidation and dissolution or uraninite can be used to set a rough upper limit of 10^{-2}–10^{-3} atm on the O_2 content of the atmosphere during the formation of the Au–U deposits of the Witwatersrand Basin (Holland, 1984, chapter 7). This maximum O_2 pressure is much greater than that permitted by the presence of MIF of sulfur isotopes during the last 0.5 Ga of the Archean; the observations do, however, complement each other.

The chemical composition of soils developed during the Late Archean and during the Paleoproterozoic also fit the pattern of a low- or no-O_2 atmosphere. During weathering on such an Earth, elements which are oxidized in a high-O_2 atmosphere remain in their lower valence states and behave differently within soils, in ground-waters, and in rivers. The theory connecting this qualitative statement to the expected behavior of redox sensitive elements has been developed in papers by Holland and Zbinden (1988), Pinto and Holland (1988), and Yang and Holland (2003). The available data for the chemical evolution of paleosols have been summarized by Rye and Holland (1998) and by Yang and Holland (2003). The composition of paleosols is consistent with a change from a low- or no-O_2 atmosphere to a highly oxygenated atmosphere between 2.3 Ga and 2.0 Ga; a different interpretation of the available data has been proposed by Ohmoto (1996) and by Beukes *et al.* (2002a,b).

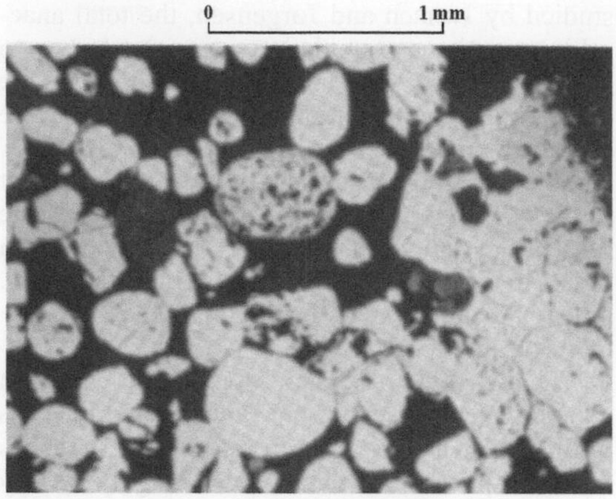

Figure 8 Conglomerate consisting of several types of pyrite together with zircon, chromite, and other heavy minerals. The large pyrite grain in the right part of the figure is a complex assemblage of older pyrite grains which have been cemented by younger pyrite (source Ramdohr, 1958).

One consequence of the proposed great oxidation event (GOE) of the atmosphere between 2.3 Ga and 2.0 Ga is that trace elements such as molybdenum, rhenium, and uranium, which are mobile during weathering in an oxidized environment, would have been essentially immobile before 2.3 Ga. Their concentration in seawater would then have been very much lower than today, and their enrichment in organic carbon-richshales would have been minimal. This agrees with the currently available data (Bekker *et al.*, 2002; Yang and Holland, 2002). Carbonaceous shales older than ca. 2.3 Ga are not enriched in molybdenum, rhenium, and uranium. A transition to highly enriched shales occurs ~2.1 Ga; by 1.6 Ga the enrichment of carbonaceous shales in these elements was comparable to that in their Phanerozoic counterparts (see, e.g., Werne *et al.*, 2002).

Table 1 Electron microprobe analyses of uraninite in some Witwatersrand ores.

Source	*Grain no.*	*UO_2* (%)	*ThO_2* (%)	*PbO_2* (%)	*FeO* (%)	*TiO_2* (%)	*CaO* (%)	*Total* (%)	*UO_2/ ThO_2*
Cristaalkop Reef (171) (Vaal Reefs South Mine)	T21	65.8	5.3	23.3	0.8	<0.01	0.6	95.8	12.4
	T22	66.5	6.1	21.9	0.5	0.02	0.7	95.7	10.9
	T23	70.5	1.7	23.8	0.5	0.04	0.8	97.3	41.5
	T23	62.4	6.1	24.7	0.4	<0.01	1.1	94.7	10.2
	T25	63.0	8.0	23.0	0.5	<0.01	1.0	95.5	7.9
	T27	66.6	3.2	23.3	1.0	0.04	0.9	95.0	20.8
	T28	66.3	1.4	28.0	0.6	0.02	0.9	97.2	47.4
	T29	63.5	3.9	28.5	0.6	<0.01	0.8	97.3	16.3
	T30	65.7	10.2	18.9	0.7	0.08	1.1	96.7	6.4
	Average	65.6	5.1	23.9	0.6	0.02	0.9	96.1	12.9
Carbon Leader (135) (Western Deep Levels Mine)	T13	69.6	2.7	26.1	0.2	0.08	1.0	99.7	25.8
	T14	66.4	2.5	27.8	0.2	0.12	0.7	97.7	26.6
	T15	63.8	2.0	30.3	0.2	0.06	0.7	97.1	31.9
	T16	62.6	7.0	24.5	0.2	0.04	0.8	95.1	8.9
	T17	67.1	5.2	21.1	0.2	0.06	0.7	94.4	12.9
	T18	69.2	2.1	28.0	0.2	0.10	0.7	100.3	33.0
	T19	67.9	7.2	18.3	0.2	0.06	0.6	94.3	9.4
	Average	66.7	4.1	25.2	0.2	0.07	0.7	97.0	16.3
Carbon Leader (167) (West Driefontein Mine)	B43	61.1	5.4	27.9	1.1	0.25	0.4	96.2	11.3
	B44	71.3	2.6	24.2	0.6	0.25	0.6	99.6	27.4
	B46	68.2	4.3	28.2	0.4	0.10	0.5	101.7	15.9
	B47	70.0	2.1	23.6	0.4	0.10	0.6	96.8	33.3
	B48	68.2	3.5	24.7	0.5	0.12	0.3	97.3	19.5
	B49	67.3	6.4	21.6	0.4	0.16	0.5	96.4	10.5
	B52	67.6	9.2	22.8	0.9	0.14	0.3	100.9	7.3
	Average	68.1	4.2	24.7	0.6	0.16	0.5	98.3	16.2
Main Reef (151) (SA Lands Mine)	B31	68.7	3.3	14.7	2.3	1.20	0.4	90.6	20.8
	B32	69.1	2.1	21.0	1.1	0.50	0.4	94.2	32.9
	B36	67.3	1.5	24.4	0.9	0.19	0.3	94.6	44.9
	B37	63.5	2.9	17.3	5.0	2.03	0.5	91.2	21.9
	Average	67.2	2.5	19.4	2.3	0.98	0.4	92.8	26.9
Basal Reef (184) (Welkom Mine)	U4	68.2	6.3	19.6	2.4	<0.1	0.4	96.9	10.8
	U5	70.5	3.3	16.8	4.1	0.2	0.4	95.3	21.4
	U7	70.1	5.2	24.8	1.1	0.1	0.5	101.8	13.5
	U8	66.6	4.5	25.2	0.9	<0.1	0.3	97.5	14.8
	U9	72.5	2.6	26.1	1.6	<0.1	0.5	103.3	27.9
	U10	64.8	3.3	23.7	5.2	0.2	0.5	97.7	19.6
	Average	68.8	4.2	22.7	2.6	0.1	0.4	98.7	16.4
Overall average:		67.2	3.9	23.6	1.0	0.16	0.6	93.8	17.2

Source: Feather (1980).

The data for the mineralogy of BIFs tell much the same story. These sediments provide strong evidence for the view that the deep oceans were anoxic throughout Archean time (James, 1992). Evidence regarding the oxidation state of the shallow parts of the Archean oceans is still very fragmentary. The shallow water facies of the 2.49 ± 0.03 Griquatown iron formation (Nelson *et al.*, 1999) in the Transvaal Supergroup of South Africa (Beukes, 1978, 1983; Beukes and Klein, 1990) were deposited on the $\sim$800 km$\times$800 km shelf shown in the somewhat schematic Figure 9. The stratigraphic relations are illustrated in the south–north cross-section of Figure 10. Several of the units in the Danielskuil Member of the Griquatown iron formation can be traced across the shallow platform from the subtidal, low-energy epeiric sea, through the high-energy zone of the shelf, into the lagoonal, near-shore parts of the platform. In the deeper parts of the shelf, siderite and iron silicates dominate the mineralogy of the iron formation. Siderite and minor (<10%)

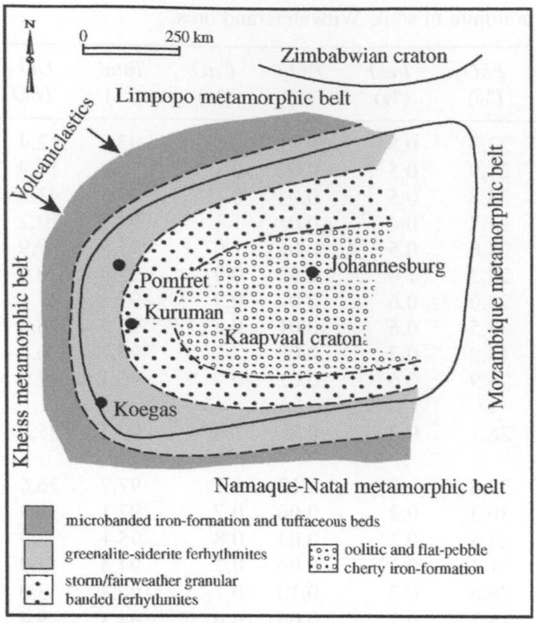

Figure 9 Depositional model for the Kuruman–Griquatown transition zone in a plan view, illustrating lithofacies distribution during drowning of the Kaapvaal craton (source Beukes and Klein, 1990).

hematite dominate the sediments of the high-energy zone. Greenalite and siderite lutites are most common in the platform lagoonal zone.

The dominance of Fe^{2+} minerals in even the shallowest part of the platform can only be explained if the O_2 content of the ambient atmosphere was very low. The half-life of Fe^{2+} oxidation in the Gulf Stream and in Biscayne Bay, Florida is only a few minutes (Millero *et al.*, 1987). The half-life of Fe^{2+} oxidation is similar in the North Sea, the Sargasso Sea, Narragansett Bay, and Puget Sound (for summary see Millero *et al.*, 1987).

The half-life of Fe^{2+} in the solutions from which the Griquatown iron formation was deposited was obviously many orders of magnitude longer than this, and it is useful to inquire into the cause for the difference. Stumm and Lee (1961) have shown that the rate of oxidation of Fe^{2+} in aqueous solutions is governed by the equation

$$-\mathrm{d}m_{\mathrm{Fe}^{2+}}/\mathrm{d}t = k m_{\mathrm{OH}^-}^2 m_{\mathrm{O}_2} m_{\mathrm{Fe}^{2+}} \tag{3}$$

where

m_{OH^-} = concentration of free OH^-

m_{O_2} = concentration of dissolved O_2

Integration of Equation (3) yields

$$\ln m_{\mathrm{Fe}^{2+}}/{}_{\mathrm{o}}m_{\mathrm{Fe}^{2+}} = -k m_{\mathrm{OH}^-}^2 m_{\mathrm{O}_2} t \tag{4}$$

where ${}_{\mathrm{o}}m_{\mathrm{Fe}^{2+}}$ is the initial concentration of Fe^{2+} in the solution. If we substitute a_{OH^-} for m_{OH^-}, the value of k for modern seawater is $\sim 0.9\times10^{15}\,\mathrm{min}^{-1}$ (Millero *et al.*, 1987). An upper limit for $\ln m_{\mathrm{Fe}^{2+}}/{}_{\mathrm{o}}m_{\mathrm{Fe}^{2+}}$ can be obtained from the field data for the Griquatown iron formation on the Campbellrand platform. Hematite accounts for $\leq 10\%$ of the iron in the near-shore iron formation. If all of the iron that was oxidized to Fe^{3+} was precipitated as a constituent of Fe_2O_3 during the passage of seawater across the platform,

$$m_{\mathrm{Fe}^{2+}}/{}_{\mathrm{o}}m_{\mathrm{Fe}^{2+}} \geq 0.9 \tag{5}$$

The time, t, required for the passage of water across the Campbellrand platform is uncertain. The modern Bahamas are probably a reasonable analogue for the Campbellrand platform. On the Bahama Banks tidal currents of $25\,\mathrm{cm\,s}^{-1}$ are common, and velocities of $1\,\mathrm{m\,s}^{-1}$ have been recorded in channels (Sellwood, 1986). At a rate of $25\,\mathrm{cm\,s}^{-1}$ it would have taken seawater $\sim$1 month (4×10^4 min) to traverse the $\sim$800 km diameter of the Campbellrand platform. This period is much longer than the time required to precipitate Fe^{3+} oxyhydroxide after the oxidation of Fe^{2+} to Fe^{3+} (Grundl and Delwiche, 1993). The best estimate of the residence time of seawater on the Grand Bahama Bank is $\sim$1 yr (Morse *et al.*, 1984; Millero, personal communication). The pH of the solutions from which the iron formations were deposited was probably less than that of seawater today, but probably not lower than 7.0.

If we combine all of these rather uncertain values for the terms in Equation (4), we obtain

$$\begin{aligned} m_{\mathrm{O}_2} &\sim (<0.10)/0.9\times10^{15}\times(\geq 10^{-14.0})\times 4 \\ &\quad \times 10^4\,\mathrm{mol\,kg}^{-1} \\ &<3\times10^{-7}\,\mathrm{mol\,kg}^{-1} \end{aligned} \tag{6}$$

In an atmosphere in equilibrium with seawater containing this concentration of dissolved O_2,

$$P_{\mathrm{O}_2} = 2.4\times10^{-4}\,\mathrm{atm}$$

The maximum value of atmospheric P_{O_2} estimated in this manner is consistent with inferences from the MIF of the sulfur isotopes during the deposition of the Griquatown iron formation that $P_{\mathrm{O}_2} \leq 1\times10^{-5}$ PAL.

A rather curious observation in the light of these observations is that in many unmetamorphosed oxide facies BIFs, the first iron oxide mineral precipitated was frequently hematite (Han, 1982, 1988). This phase was later replaced by magnetite. Klein and Beukes (1989) have reported the presence of hematite as a minor component in iron formations of the Paleoproterozoic Transvaal Supergroup, South Africa. The hematite occurs in two forms: as fine hematite dust and as very fine grained specularite. The former could well be a very early phase in these BIFs. The early deposition of hematite in BIFs followed by large-scale

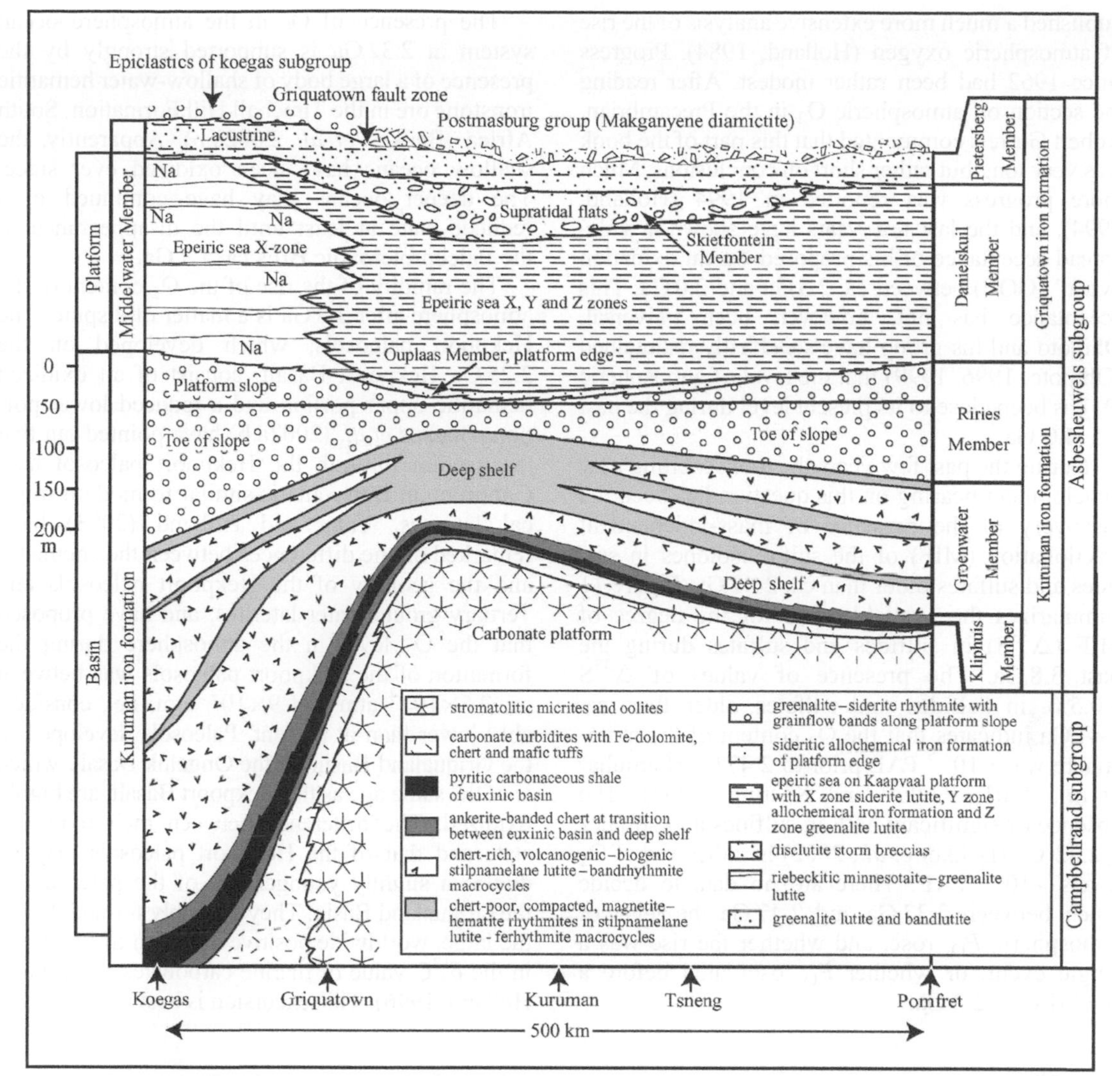

Figure 10 Longitudinal cross-section illustrating stratigraphic relationships and inferred palaeodepositional environments of the Asbesheuwels Subgroup in the Griqualand West basin (after Beukes, 1978).

replacement by magnetite could simply be the result of reactions in mixtures of hydrothermal vent fluids with ambient O_2-free seawater. High-temperature hydrothermal vent fluids are strongly undersaturated with respect to hematite. However, on mixing with ambient O_2-free seawater their pH rises, and path calculations indicate that they can become supersaturated with respect to hematite. During hematite precipitation P_{H_2} would increase, and the early hematite could well be replaced by magnetite during early diagenesis. The sequence of mineral deposition from such solutions depends not only on the composition of the vent fluids and that of the ambient seawater, but also on the kinetics of the precipitation mechanisms. These, in turn, could have been influenced by bacterial processes (Konhauser *et al.*, 2002). A thorough study of the effects of these parameters remains to be done.

12.4 THE PROTEROZOIC

12.4.1 The Paleoproterozoic (2.5–1.8 Ga)

In 1962 the author divided the evolution of the atmosphere into three stages (Holland, 1962). On the basis of rather scant evidence from the mineralogy of Precambrian sedimentary uranium deposits, it was suggested that free oxygen was not present in appreciable amounts until ca. 1.8 Ga, but that by the end of the Paleozoic the O_2 content of the atmosphere was already a large fraction of its present value. In a similar vein, Cloud (1968) proposed that the atmosphere before 1.8–2.0 Ga could have contained little or no free oxygen. In 1984 the author published a much more extensive analysis of the rise of atmospheric oxygen (Holland, 1984). Progress since 1962 had been rather modest. After reading the section on atmospheric O_2 in the Precambrian, Robert Garrels commented that this part of the book was very long but rather short on conclusions. Much more progress was reported in 1994 (Holland, 1994), and the last few years have shown a widespread acceptance of his proposed "great oxidation event" (GOE) between ca. 2.3 Ga and 2.0 Ga. This acceptance has not, however, been universal. Ohmoto and his group have steadfastly maintained (Ohmoto, 1996, 1999) that the level of atmospheric O_2 has been close to its present level during the past 3.5–4.0 Ga.

During the past few years the most exciting new development bearing on this question has been the discovery of the presence of mass-independent fractionation (MIF) of the sulfur isotopes in sulfides and sulfates older than ca. 2.47 Ga. Figure 11 summarizes the available data for the degree of MIF ($\Delta^{33}S$) in sulfides and sulfates during the past 3.8 Ga. The presence of values of $\Delta^{33}S$ >0.5‰ in sulfides and sulfates older than ca. 2.47 Ga indicates that the O_2 content of the atmosphere was $<10^{-5}$ PAL prior to 2.47 Ga (Farquhar *et al.*, 2001; Pavlov and Kasting, 2002). The absence of significant MIF in sulfides and sulfates $\leq$2.32 Ga (Bekker *et al.*, 2002) is indicative of O_2 levels $>10^{-5}$ PAL. There are no data to decide when between 2.32 Ga and 2.47 Ga the level of atmospheric P_{O_2} rose, and whether the rise was a single event, or whether P_{O_2} oscillated before a final rise by 2.3 Ga.

The presence of O_2 in the atmosphere–ocean system at 2.32 Ga is supported strongly by the presence of a large body of shallow-water hematitic ironstone ore in the Timeball Hill Formation, South Africa (Beukes *et al.*, 2002a,b). Apparently, the shallow oceans have been oxidized ever since. The deeper oceans may have continued in a reduced state at least until the disappearance of the Paleoproterozoic BIFs ca. 1.7 Ga.

The rapidity of the rise of the O_2 content of the atmosphere after 2.3 Ga is a matter of dispute. The Hekpoort paleosols, which developed on the 2.25 Ga Hekpoort Basalt, consist of an oxidized hematitic upper portion and a reduced lower portion. Beukes *et al.* (2002a,b) have pointed out that the section through the Hekpoort paleosol near Gaborone in Botswana is similar to modern tropical laterites. Yang and Holland (2003) have remarked on the differences between the chemistry and the geology of the Hekpoort paleosols and Tertiary groundwater laterites, and have proposed that the O_2 level in the atmosphere during the formation of the Hekpoort paleosols was between ca. 2.5×10^{-4} atm and 9×10^{-3} atm, i.e., considerably lower than at present. Paleosols developed in the Griqualand Basin on the Ongeluk Basalt, which is of the same age as the Hekpoort Basalt, are highly oxidized. The difference between their oxidation state and that of the Hekpoort paleosols may be due to a slightly younger age of the paleosols in the Griqualand Basin. They probably formed during the large, worldwide positive variation of Figure 12 in the $\delta^{13}C$ value of marine carbonates (Karhu and Holland, 1996). This excursion is best interpreted as

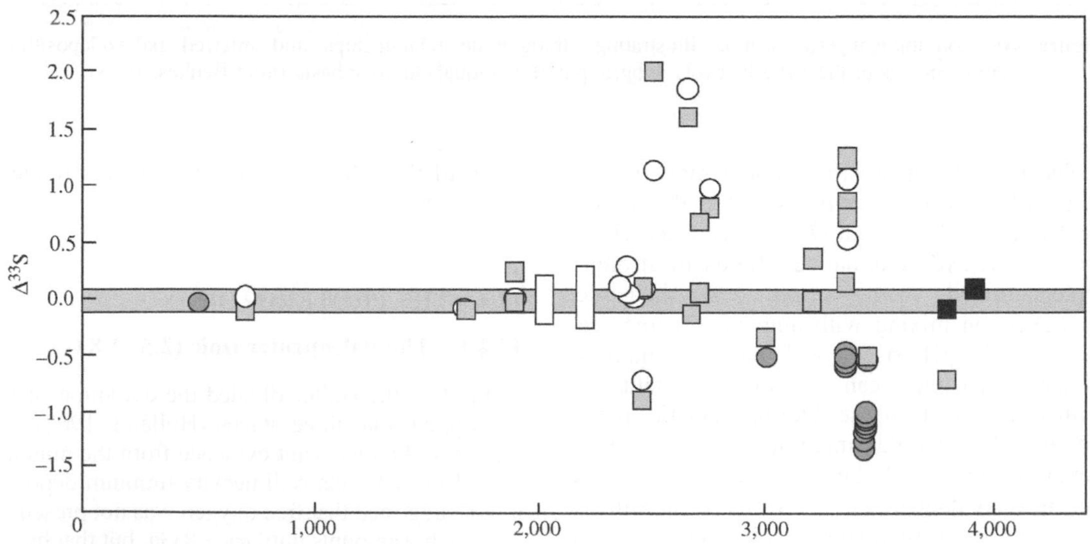

Figure 11 Summary of data for the degree of MIF of the sulfur isotopes in sulfides and sulfates. Data from Farquhar *et al.* (2000) (chemically defined sulfur minerals (▢) sulfides, (■) total sulfur, and (○) sulfates; (●) macroscopic sulfate minerals) with updated ages and from Bekker *et al.* (2002) ((□) range of values for pyrites in black shales). The gray band at $\Delta^{33}S\sim0$ represents the mean and 1 SD of recent sulfides and sulfates from Farquhar *et al.* (2000).

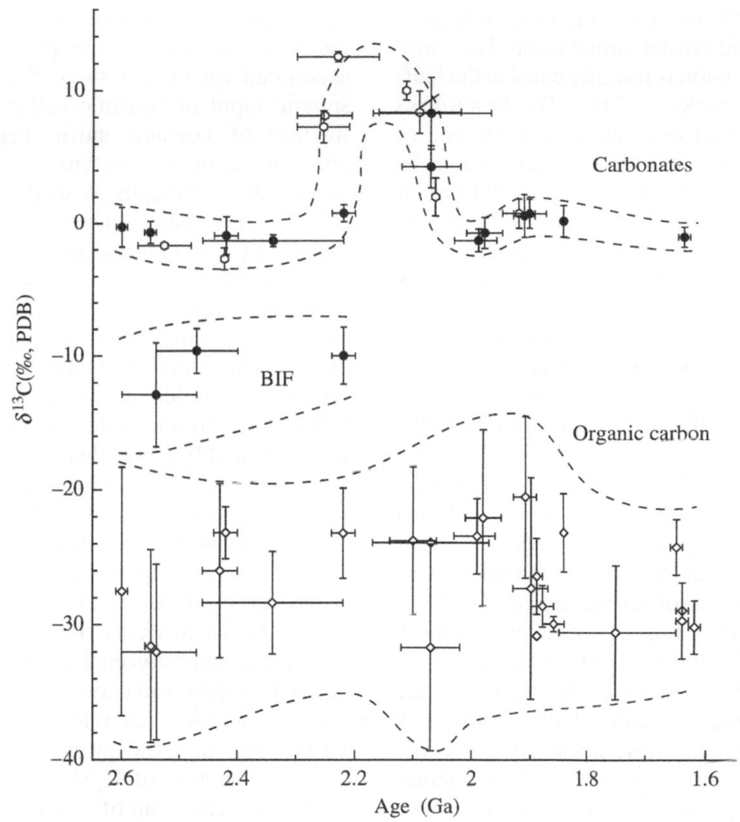

Figure 12 Variation in isotopic composition of carbon in sedimentary carbonates and organic matter during Paleoproperozoic time. Mean $\delta^{13}C$ values of carbonates from Fennoscandian Shield from Karhu (1993) are indicated by open circles. Vertical bars represent ± 1 SD of $\delta^{13}C$ values, and horizontal bars indicate uncertainty in age of each stratigraphic unit. Arrows combine dated formations that are either preceded or followed by major $\delta^{13}C$ shift. BIF denotes field for iron and manganese formations. Note that uncertainties given for ages do not necessarily cover uncertainties in entire depositional periods of sample groups. PDB—Peedee belemnite (source Karhu and Holland, 1996).

a signal of the production of a large quantity of O_2 between 2.22 Ga and 2.06 Ga. Estimates of this quantity are perforce very rough. The average isotopic composition of organic carbon during the excursion is somewhat uncertain, as is the total rate of carbon burial with organic matter and carbonate sediments during the $\delta^{13}C$ excursion, and there is the possibility that the $\delta^{13}C$ excursion in Figure 12 was preceded by another, shorter excursion (Bekker *et al.*, 2001; Young, 1969).It seems likely, however, that the total excess quantity of O_2 produced during the ca. 160 Ma of the $\delta^{13}C$ excursion was ~12–22 times the inventory of atmospheric oxygen (Karhu, 1993; Karhu and Holland, 1996). This large amount of O_2 must somehow have disappeared into the sedimentary record. The most likely sinks are crustal iron and sulfur. As indicated by the mineralogy of marine evaporites and the isotopic composition of sulfur in black shales, the SO_4^{2-} concentration of seawater probably remained very modest until well beyond 2.0 Ga (Grotzinger and Kasting, 1993; Shen *et al.*, 2002). Iron is, therefore, the most likely major sink for the O_2 produced during the $\delta^{13}C$ excursion between 2.22 Ga and 2.06 Ga. This is not unreasonable. Before the rise of atmospheric O_2, FeO was not oxidized to Fe_2O_3 during weathering, as shown by the record of the Fe_2O_3/FeO ratio in pre-2.3 Ga sedimentary rocks (Bekker *et al.*, 2003). The appearance of extensive red beds ca. 2.3 Ga indicates that a major increase in the Fe_2O_3/FeO ratio of sediments and sedimentary rocks occurred at that time. Shales before the GOE contained, on average, ~6.5% FeO and ~1.3% Fe_2O_3. Shales deposited between 2.3 Ga and 2.1 Ga contain, on average, ~4.1% FeO and 2.5% Fe_2O_3. There seems to have been little change between 2.1 Ga and 1.0 Ga (Bekker *et al.*, 2003). Approximately 2% of the FeO in pre-GOE rocks seem to have been converted to Fe_2O_3 during weathering in the course of the GOE. Since each mole of FeO requires 0.25 mol O_2 for conversion to Fe_2O_3, ~0.08 mol O_2 was used during the weathering of each kilogram of rock. If weathering rates during the GOE were comparable to current rates, some 1.6×10^{12} mol O_2 were used annually to convert FeO to Fe_2O_3 during weathering in the course of the $\delta^{13}C$ excursion. The total O_2 use was, therefore, 2.6×10^{20} mol, i.e., ~6 times the present atmospheric O_2 inventory.

Some of the excess O_2 was probably used to increase the redox state of the crustal sulfur cycle. The duration of the $\delta^{13}C$ excursion is roughly equal to the half-life of sedimentary rocks at 2 Ga. The Fe_2O_3/FeO ratio of rocks subjected to weathering at the end of the $\delta^{13}C$ excursion was, therefore, greater than at its beginning and was approaching a value typical of Mesoproterozoic sedimentary rocks. Post-$\delta^{13}C$ excursion weathering of sediments produced during the $\delta^{13}C$ excursion, therefore, required much less additional O_2 than the weathering of pre-GOE rocks.

Although this is a likely explanation for the fate of most of the "extra" O_2 generated during the $\delta^{13}C$ excursion between ca. 2.22 Ga and 2.06 Ga, it does not account for the GOE itself or for the cause of the $\delta^{13}C$ excursion. The appearance of O_2 in the atmosphere between ca. 2.47 Ga and 2.32 Ga could be explained easily if cyanobacteria evolved at that time. However, this explanation has been rendered very unlikely by the discovery of biomarkers that are characteristic of cyanobacteria and eukaryotes in 2.5–2.7 Ga sedimentary rocks (Brocks *et al.*, 1999). An alternative explanation involves a change in the redox state of volcanic gases as the trigger for the change of the oxidation state of the atmosphere (Kasting *et al.*, 1993). These authors pointed out that the loss of H_2 from the top of a reducing atmosphere into interplanetary space would have increased the overall oxidation state of the Earth as a whole, and almost certainly that of the mantle. This, in turn, would have led to an increase in the f_{O_2} of volcanic gases and to a change in the redox state of the atmosphere.

In a more detailed analysis of this mechanism, Holland (2002) showed that the change in the average f_{O_2} of volcanic gases required for the transition of the atmosphere from an anoxygenic to an oxygenic state is quite small. There is no inconsistency between the required change in f_{O_2} and the limits set on such changes by the data of Delano (2001) and Canil (1997, 1999, 2002) for the evolution of the redox state of the upper mantle during the past 4.0 Ga. The estimated changes in f_{O_2} due to H_2 loss are consistent with the likely changes in the redox state of the upper mantle if the major control on that state is exerted by the Fe_2O_3 /FeO buffer. In this explanation the average composition of volcanic gases before the GOE was such that 20% of their contained CO_2 could be reduced to CH_2O, and all of the sulfur gases to FeS_2. Excess H_2 present in the gases would have escaped from the atmosphere, possibly via the decomposition of CH_4 in the upper atmosphere. The loss of H_2 would have produced an irreversible oxidation of the early Earth (Catling *et al.*, 2001). The GOE began when the composition of volcanic gases had changed, so that not enough H_2 was present to convert 20% of the contained CO_2 to CH_2O and all of the sulfur gases to FeS_2. Before the GOE the only, or nearly the only, sulfate mineral deposited in sediments seems to have been barite. Since barium is a trace element, its precipitation as $BaSO_4$ accounted for only a small fraction of the atmospheric input of volcanic sulfur. After the GOE a fraction of volcanic sulfur began to leave the atmosphere–ocean system as a constituent of other sulfate minerals as well, largely as gypsum ($CaSO_4 \cdot 2H_2O$) and anhydrite ($CaSO_4$).

During the Phanerozoic close to half of the volcanic sulfur in volcanic gases has been removed as a constituent of FeS_2, the other half as a constituent of gypsum and anhydrite (see, e.g., Holland, 2002). The shift from the essentially complete removal of volcanic sulfur as a constituent of FeS_2 to the present state was gradual (see below). It was probably controlled by a feedback mechanism involving an increase in the sulfur content of volcanic gases. This was probably the result of an increase in the rate of subduction of $CaSO_4$ added to the oceanic crust by the cycling of sea water at temperatures above ca. 200 °C.

The burial of excess organic matter during the $\delta^{13}C$ excursion between 2.22 Ga and 2.06 Ga almost certainly required an excess of PO_4^{3-}. It seems likely that this excess was released from rocks during weathering due to the lower pH of soil waters related to the generation of H_2SO_4 that accompanied the oxidative weathering of sulfides. Toward the end of the $\delta^{13}C$ excursion, this excess PO_4^{3-} was probably removed by adsorption on the Fe^{3+} hydroxides and oxyhydroxides produced by the oxidative weathering of Fe^{2+} minerals (Colman and Holland, 2000). Although this sequence of events is reasonable, and although some parts of it can be checked semiquantitatively, the proposed process by which the anoxygenic atmosphere became converted to an oxygenic state should be treated with caution. Too many pieces of the puzzle are still either missing or of questionable shape.

A most interesting and geochemically significant change in the oceans may have occurred ca. 1.7 Ga. BIFs ceased to be deposited. They are apparently absent from the geologic record until their reappearance 1 Ga later in association with the very large Neoproterozoic ice ages (Beukes and Klein, 1992). Three explanations have been advanced for the hiatus in BIF deposition between 1.7 and 0.7 Ga. The first proposes that the deposition of BIF ended when the deep waters of the oceans became aerobic (Cloud, 1972; Holland, 1984). After 1.7 Ga, Fe^{2+} from hydrothermal vents was oxidized to Fe^{3+} close to the vents and was precipitated as Fe^{3+} oxides and/or oxyhydroxides on the floor of the oceans. The second explanation proposes that anoxic bottom waters persisted until well after the deposition of BIFs ceased, and that an increase in the concentration of H_2S rather than the advent of oxygen was responsible for removing iron from deep ocean water (Canfield, 1998). The sulfur isotope record indicates that the concentration of oceanic sulfate began to increase 2.3 Ga leading to increasing rates of sulfide production by bacterial sulfate reduction. Canfield (1998) has suggested that sulfide production became sufficiently intense ~1.7 Ga to precipitate the total hydrothermal flux of iron as a constituent of pyrite in the deep oceans. As a basis for this contention, he points out that the generation of aerobic deep ocean water would have required levels of atmospheric O_2 within a factor of 2 or 3 of the present level, a level which he believes was not attained until the Neoproterozoic. However, Canfield's (1998) analysis of his three-box model of the oceans assumes that the rate of sinking of organic matter into the deep ocean was the same during the Paleoproterozoic as at present. This is unlikely. Organic matter requires ballast to make it sink. Today most of the ballast is supplied by siliceous and carbonate tests (Logan *et al.*, 1995; Armstrong *et al.*, 2002; Iglesias-Rodriguez *et al.*, 2002; Sarmiento *et al.*, 2002). Clays and dust seem to be minor constituents of the ballast, although they may have been more important before the advent of soil-binding plants. There is no evidence for the production of siliceous or calcareous tests in the Paleoproterozoic oceans. Inorganically precipitated SiO_2 and/or $CaCO_3$ could have been important, but precipitation of these phases probably occurred mainly in shallow-water evaporitic settings. It is, therefore, likely that ballast was much scarcer during the Paleoproterozoic than today, and that the quantity of particulate organic matter (POC) transported annually from shallow water into the deep oceans was much smaller than today. This, in turn, implies that the amount of dissolved O_2 that was required to oxidize the rain of POC was much smaller than today. Evidence from paleosols suggests that atmospheric O_2 levels ca. 2.2 Ga were $\geq$15% PAL (Holland and Beukes, 1990). This implies that the proposal for the end of BIF deposition based on the development of oxygenated bottom waters ca. 1.7 Ga is quite reasonable. It does not, of course, prove that the proposal is correct. For one thing, too little is known about the mixing time of the Paleoproterozoic oceans. Data for the oxidation state of the deep ocean since 1.7 Ga are needed to settle the issue. The third explanation posits that no large hydrothermal inputs such as are required to produce BIFs occurred between 1.7 Ga and 0.7 Ga. This seems unlikely but not impossible.

12.4.2 The Mesoproterozoic (1.8–1.2 Ga)

Sedimentary rocks of the McArthur Basin in Northern Australia provide one of the best windows on the chemistry of the Mesoproterozoic ocean. Some 10 km of 1.6–1.7 Ga sediments accumulated in this intracratonic basin (Southgate *et al.*, 2000). In certain intervals, they contain giant strata-bound Pb–Zn–Ag mineral deposits (Jackson *et al.*, 1987; Jackson and Raiswell, 1991; Crick, 1992). The sediments have experienced only low grades of metamorphism.

Shen *et al.* (2002) have reported data for the isotopic composition of sulfur in carbonaceous shales of the lower part of the 1.72–1.73 Ga Wollogorang Formation and in the lower part of the 1.63–1.64 Ga Reward Formation of the McArthur Basin. These shales were probably deposited in a euxinic intracratonic basin with connection to the open ocean. The $\delta^{34}S$ of pyrite in black shales of the Wollogorang Formation ranges from −1‰ to +6.3‰ with a mean and SD of 4.0 ± 1.9‰ (n = 14). Donnelly and Jackson (1988) reported similar values. The $\delta^{34}S$ values of pyrite in the lower Reward Formation range from +18.2‰ to +23.4‰ with an average and SD of 18.4 ±1.8‰ (n=10). The spread of $\delta^{34}S$ values within each formation is relatively small. The sulfur is quite ^{34}S-enriched compared to compositions expected from the reduction of seawater sulfate with a $\delta^{34}S$ of 20−25‰ (Strauss, 1993). This is especially true of the sulfides in the Reward Formation. Shen *et al.* (2002) propose that the Reward data are best explained if the concentration of sulfate in the contemporary seawater was between 0.5 mmol kg^{-1} and 2.4 mmol kg^{-1}. Sulfate concentrations in the Mesoproterozoic ocean well below those of the present oceans have also been proposed on the basis of the rapid change in the value of $\delta^{34}S$ in carbonate associated sulfate of the 1.2 Ga Bylot Supergroup of northeastern Canada (Lyons *et al.*, 2002). However, the value of $m_{SO_4^2}$ in Mesoproterozoic seawater is still rather uncertain.

Somewhat of a cross-check on the SO_4^{2-} concentration of seawater can be obtained from the evaporite relics in the McArthur Group (Walker *et al.*, 1977). Up to 40% of the measured sections of the Amelia Dolomite consist of such relics in the form of carbonate pseudomorphs after a variety of morphologies of gypsum and anhydrite crystals, chert pseudomorphs after anhydrite nodules, halite casts, and microscopic remnants of original, unaltered sulfate minerals. Muir (1979) and Jackson *et al.* (1987) have pointed out the similarity of this formation to the recent sabkhas along the Persian Gulf coast. The pseudomorphs crosscut sedimentary features such as bedding and laminated microbial mats, suggesting that the original sulfate minerals crystallized in the host sediments during diagenesis.

Pseudomorphs after halite are common throughout the McArthur Group. The halite appears to have formed by almost complete evaporation of seawater in shallow marine environments and probably represents ephemeral salt crusts. The general lack of association of halite and calcium sulfate minerals in these sediments probably resulted in part from the dissolution of previously deposited halite during surface flooding, but also indicates that evaporation did not always proceed beyond the calcium sulfate facies.

This observation allows a rough check on the reasonableness of the Shen *et al.* (2002) estimate of the sulfate concentration in seawater during the deposition of the McArthur Group. On evaporating modern seawater, gypsum begins to precipitate when the degree of evaporation is ~3.8. As shown in Figure 13, the onset of gypsum and/or anhydrite precipitation occurs at progressively greater degrees of evaporation as the product $m_{Ca^{2+}}m_{SO_4^{2-}}$ in seawater decreases. Today $m_{Ca^{2+}}m_{SO_4^{2-}} = 280$ (mmol $kg^{-1})^2$. If this product is reduced to 23 (mmol $kg^{-1})^2$, anhydrite begins to precipitate simultaneously with halite at a degree of evaporation of 10.8. The presence of gypsum casts without halite in the sediments of the McArthur Group indicates that in seawater at that time $m_{Ca^{2+}}m_{SO_4^{2-}} > 23$ (mmol $kg^{-1})^2$ provided the salinity of seawater was the same as today. If $m_{SO_4^{2-}}$ was 2.4 mmol kg^{-1}, the upper limit suggested by Shen *et al.* (2002), $m_{Ca^{2+}}$, must then have been >10 mmol kg^{-1}, the concentration of Ca^{2+} in modern seawater. An SO_4^{2-} concentration of 2.4 mmol kg^{-1} is, therefore, permissible. Sulfate concentrations as low as 0.5 mmol kg^{-1} require what are probably unreasonably high concentrations of Ca^{2+} in seawater to account for the precipitation of gypsum before halite in the McArthur Group sediments.

The common occurrence of dolomite in the McArthur Group indicates that the $m_{Mg^{2+}}/m_{Ca^{2+}}$ ratio in seawater was >1 (see below). This is also indicated by the common occurrence of aragonite as the major primary $CaCO_3$ phase of sediments on Archean and Proterozoic carbonate platforms (Grotzinger, 1989; Winefield, 2000). Although these hints regarding the composition of Mesoproterozoic seawater are welcome, they need to be confirmed and expanded by analyses of fluid inclusions in calcite cements.

Perhaps the most interesting implication of the close association of gypsum, anhydrite, and halite relics in the McArthur Group is that the temperature during the deposition of these minerals was not much above 18 °C, the temperature at which gypsum, anhydrite, and halite are stable together (Hardie, 1967). At higher temperatures anhydrite is the stable calcium sulfate mineral in equilibrium with halite. The coexistence of gypsum and anhydrite with halite suggests that the temperature during their deposition was possibly lower but probably no higher than in the modern sabkhas of the Persian Gulf, where anhydrite is the dominant calcium sulfate mineral in association with halite (Kinsman, 1966).

In their paper on the carbonaceous shales of the McArthur Basin, Shen *et al.* (2002) comment that euxinic conditions were common in marine-connected basins during the Mesoproterozoic, and they suggest that low concentrations of seawater sulfate and reduced levels of atmospheric oxygen at this time are compatible with euxinic deep ocean waters. Anbar and Knoll (2002) echo this sentiment. They point out that biologically important trace metals would then have been scarce in most marine environments, potentially restricting the nitrogen cycle, affecting primary productivity, and limiting the ecological distribution of eukaryotic algae. However, some of the presently available evidence does not support the notion of a Mesoprotcrozoic euxinic ocean floor. Figure 14 shows that the redox sensitive elements molybdenum, uranium, and rhenium are well correlated with the organic carbon content of carbonaceous shales in the McArthur Basin. The slope of the correlation lines is close to that in many Phanerozoic black shales, suggesting that the concentration of these elements in McArthur Basin seawater

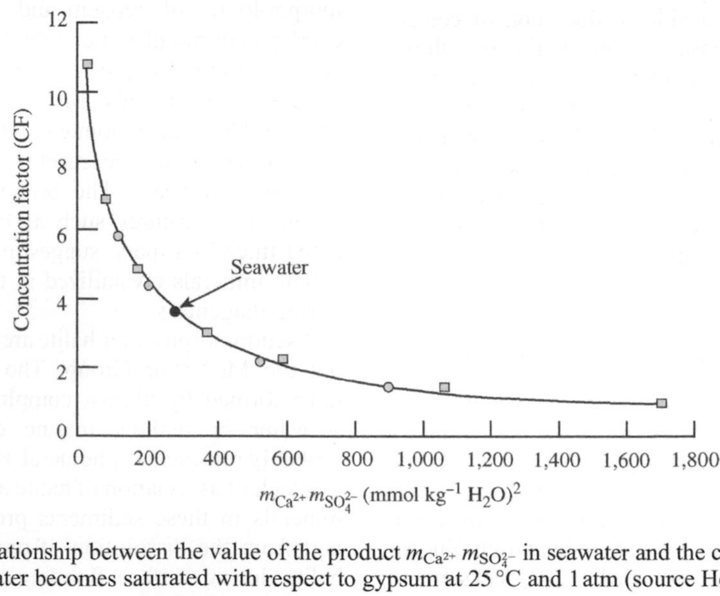

Figure 13 The relationship between the value of the product $m_{Ca^{2+}}\, m_{SO_4^{2-}}$ in seawater and the concentration factor at which seawater becomes saturated with respect to gypsum at 25 °C and 1 atm (source Holland, 1984).

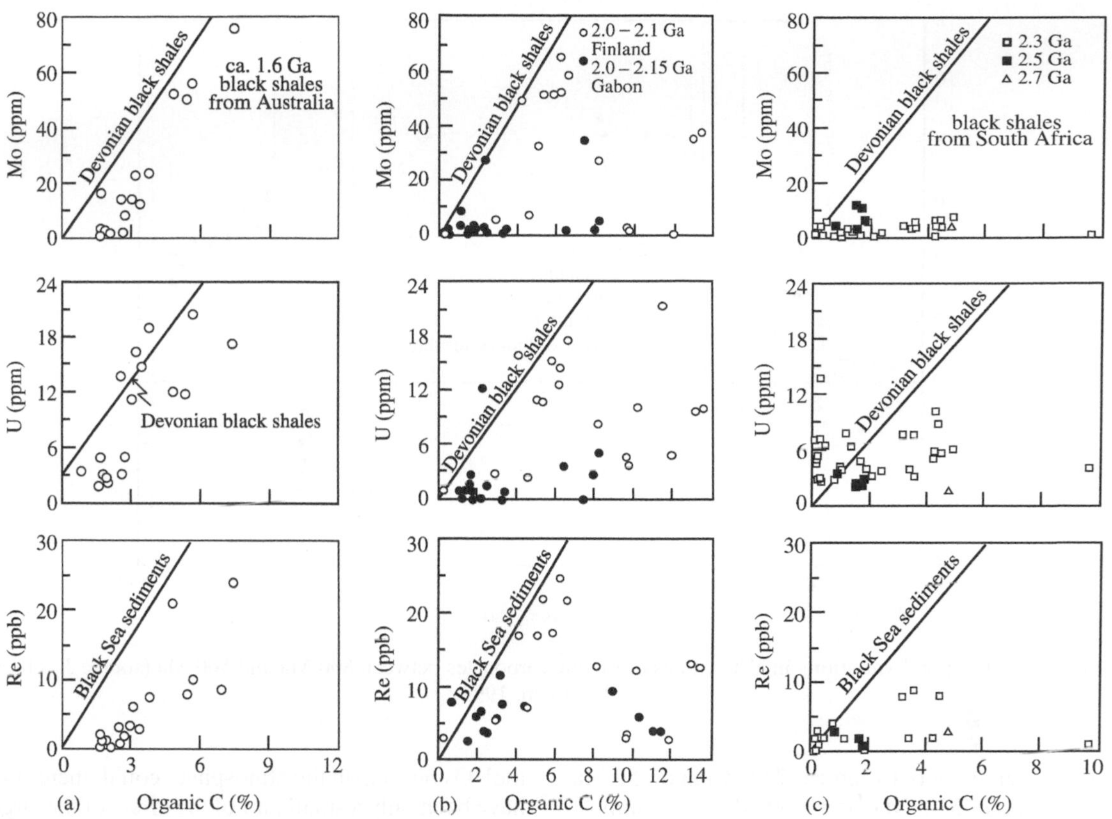

Figure 14 The concentration of Mo, U, and Re in carbonaceous shales: (a) McArthur Basin, Australia, 1.6 Ga; (b) Finland and Gabon, 2.0–2.15 Ga; and (c) South Africa, ≥2.3 Ga.

was comparable to their concentration in modern seawater. Preliminary data for the isotopic composition of molybdenum in the Wollogorang Formation of the McArthur Basin (Arnold et al., 2002) suggest somewhat more extensive sulfidic deposition of molybdenum in the Mesoproterozoic than in the modern oceans. Their data may, however, reflect a greater extent of shallow water euxinic basins rather than an entirely euxinic ocean floor. The good correlation of the concentration of sulfur and total iron in the McArthur Basin shales (Shen *et al*., 2002) confirms the euxinic nature of the Basin; the large value of the ratio of sulfur to total iron indicates that this basin cannot have been typical of the oceans as a whole. Additional data for the concentration of redox sensitive elements in carbonaceous shales and more data for the isotopic composition of molybdenum and perhaps of copper in carbonaceous shales will probably clarify and perhaps settle the questions surrounding the redox state of the deep ocean during the Mesoproterozoic.

12.4.3 The Neoproterozoic (1.2 – 0.54 Ga)

After what appears to have been a relatively calm and uneventful climatic, atmospheric, and marine history during the Mesoproterozoic, the Neoproterozoic returned to the turbulence of the Paleoproterozoic era. The last 300 Ma of the Proterozoic were times of extraordinary global environmental and biological change. Major swings in the $\delta^{13}C$ value of marine carbonates were accompanied by several very large glaciations, the sulfate content of seawater rose to values comparable to that of the modern oceans (Horita *et al.*, 2002), and the level of atmospheric O_2 probably attained modern values by the time of the biological explosion at the end of the Precambrian and the beginning of the Paleozoic. A great deal of research has been done on the last few hundred million years of the Proterozoic, stimulated in part by the discovery of the extensive glacial episodes of this period. Nevertheless, many major questions remain unanswered. The description of the major events and particularly their causes are still quite incomplete.

Figure 15 is a recent compilation of measurements of the $\delta^{13}C$ values of marine carbonates between 800 Ma and 500 Ma (Jacobsen and Kaufman, 1999). The $\delta^{13}C$ values experienced a major positive excursion interrupted by sharp negative spikes. The details of these spikes are still quite obscure (see, e.g., Melezhik *et al.*, 2001), but the negative excursions associated with

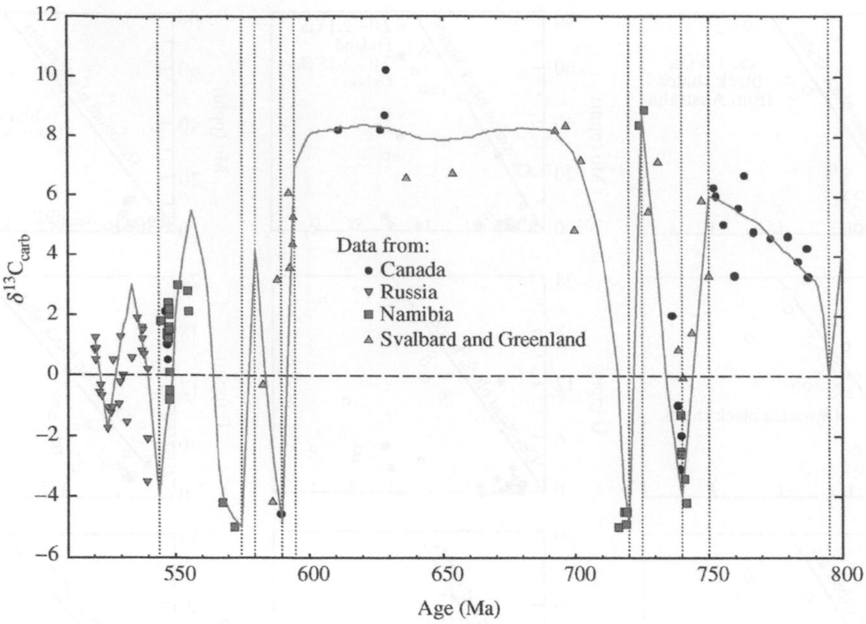

Figure 15 Temporal variations in $\delta^{13}C$ values of marine carbonates between 800 Ma and 500 Ma (source Jacobsen and Kaufman, 1999).

major glaciations between ca. 720 Ma and 750 Ma and between 570 Ma and 600 Ma were almost certainly separated by a 100 Ma plateau of very strongly positive $\delta^{13}C$ values (see, e.g., Walter *et al.*, 2000; Shields and Veizer, 2002; Halverson, 2003). In some ways the Late Neoproterozoic positive $\delta^{13}C$ excursion is reminiscent of the Paleoproterozoic excursion between 2.22 Ga and 2.06 Ga. Their duration and magnitude are similar. However, the Paleoproterozoic ice ages preceded the large positive $\delta^{13}C$ excursion, whereas in the Neoproterozoic they occur close to the beginning and close to the end of the excursion. Figure 15 can be used to estimate the "excess O_2" produced during the Neoproterozoic excursion. The average value of $\delta^{13}C$ between 700 Ma and 800 Ma was +3.0‰; between 595 Ma and 700 Ma it was ±8.0‰, and between 540 Ma and 595 Ma it was −0.4‰. The average $\delta^{13}C$ value for the period between 595 Ma and 800 Ma was +5.6‰. The variation of $\delta^{13}C$ during this time interval has been defined much more precisely by Halverson (2003). Although his $\delta^{13}C$ curve differs considerably from that of Jacobsen and Kaufman (1999), Halverson's (2003) average value of $\delta^{13}C$ between 595 Ma and 800 Ma is very similar to that of Jacobsen and Kaufman (1999). This indicates that ~40% of the carbon in the sediments of this period were deposited as a constituent of organic matter.

The rate of excess O_2 generation was probably $\sim 6\times 10^{12}$ mol yr^{-1}, and the total excess O_2 produced between 600 Ma and 800 Ma was $\sim 12\times 10^{20}$ mol. This quantity is ~30 times the O_2 content of the present atmosphere, 0.4×10^{20} mol. O_2 buildup in the atmosphere could, therefore, have been only a small part of the effect of the large $\delta^{13}C$ excursion. The excess O_2 is also much larger than 0.8×10^{20} mol, the quantity required to raise the SO_4^{2-} concentration of seawater from zero to its present value by oxidizing sulfide. Fortunately, additional sulfate sinks are available to account for the estimated excess O_2:$CaSO_4$ and $CaSO_4\cdot 2H_2O$ in evaporites, $CaSO_4$ precipitated in the oceanic crust close to MORs during the cycling of seawater at hydrothermal temperatures, and an increase in the Fe_2O_3/FeO ratio in sedimentary rocks. These sinks seem to be of the right order of magnitude to account for the use of the excess O_2. The magnitude of the $CaSO_4$ reservoir in sedimentary rocks during the last part of the Neoproterozoic has been estimated on the basis of models based on sulfur isotope data to be $(2\pm 0.5)\times 10^{20}$ mol (Holser *et al.*, 1989). The conversion of this quantity of sulfur from sulfide to sulfate requires $(4\pm 1)\times 10^{20}$ mol O_2.

At present the loss of $CaSO_4$ from seawater to the oceanic crust seems to be $\sim 1.0\times 10^{12}$ mol yr^{-1} (Holland, 2002). At this rate the loss of SO_4^{2-} to the oceanic crust between 600 Ma and 800 Ma would have been 2×10^{20} mol. The total O_2 sinks due to the sulfur cycle during this period might, therefore, have amounted to $\sim 6\times 10^{20}$ mol. The increase in the Fe_2O_3/FeO ratio in sedimentary rocks probably required $\sim 1\times 10^{20}$ mol O_2. Given all the rather large uncertainties and somewhat shaky assumptions which have been made in this mass balance calculation, the agreement between the estimated quantity of excess O_2 and the estimated quantity of

O_2 required to convert the sulfur cycle from its pre-1,200 Ma state to its state at the beginning of the Paleozoic is quite reasonable. The logic behind the change is also compelling. Carbon, iron, and sulfur are the three elements which dominate the redox state of the near-surface system. The carbon cycle seems to have been locked into its present state quite early in Earth history, probably by its linkage to the geochemical cycle of phosphorus. The iron cycle took on a more modern cast during the positive Paleoproterozoic $\delta^{13}C$ excursion. It is not unreasonable to propose that the positive $\delta^{13}C$ excursion during the Neoproterozoic was responsible for converting the sulfur cycle to its modern mode and for generating a further increase in the Fe_2O_3/FeO ratio.

Two questions now come to mind: (i) what triggered the Neoproterozoic $\delta^{13}C$ excursion? and (ii) Are the strong negative excursions due to instabilities inherent in the long positive excursion? The answers that have been given to both questions are still speculative, but it seems worthwhile to attempt a synthesis. The $\delta^{13}C$ excursion was accompanied by the reappearance of BIFs (Klein and Beukes, 1993), which are related to glacial periods but in a somewhat irregular manner (Young, 1969, 1976; James, 1983). They are widely distributed, and their tonnage is significant. Their reappearance virtually demands that the deeper parts of the oceans were anoxic. If, as suggested earlier, the deep oceans were oxidized during the Mesoproterozoic, they returned to their pre-1.7 Ga state during the last part of the Neoproterozoic. One possible cause for this return is the appearance of organisms which secreted SiO_2 or $CaCO_3$, that could serve as ballast for particulate organic matter. Recently discovered vase-shaped microfossils (VSMs) in the Chuar Group of the Grand Canyon could be members of one of these groups. The fossils appear to be testate amoebae (Porter and Knoll, 2000; Porter *et al.*, 2003). The structure and composition of testate amoebae tests are similar to those inferred for the VSMs. A number of testate amoebae have agglutinated tests; others have tests in which internally synthesized 1 μm to >10 μm scales of silica are arrayed in a regular pattern (Figure 16). The age of the VSM fossils in the Grand Canyon and in the Mackenzie Mountain Supergroup, NWT, is between 742 ± 7 Ma and ca. 778 Ma. Their presence at this time suggests that they or other SiO_2-secreting organisms could have supplied ballast for the transport of particulate organic matter into the deep ocean near the beginning of the Neoproterozoic $\delta^{13}C$ excursion. If the O_2 content of the atmosphere at that time was still significantly less than today, the flux of organic matter required to make the deep oceans anoxic would only need to have been a small fraction of the present-day flux. An increase in the flux of organic matter to the deep ocean probably followed the Cambrian explosion (Logan *et al.*, 1995).

Figure 16 The test of *Euglypha tuberculata*. Note regularly arranged siliceous scales: scale bar, 20 μm (courtesy of Ralf Meisterfeld).

Arguments can be raised against the importance of the evolution of SiO_2-secreting organisms as ballast for organic matter. The VSM organisms were shallow water and benthic, not open ocean and planktonic, and the remains of SiO_2-secreting organisms have not been found in Neoproterozoic deep-sea sediments. Alternative explanations have been offered for the burial of "excess" organic carbon between 800 Ma and 600 Ma. Knoll (1992) and Hoffman *et al.* (1998) pointed out that the Late Proterozoic was a time of unusual, if not unique, formation of rapidly subsiding extensional basins flooded by marine waters. Organic matter buried with rapidly deposited sediments is preserved more readily than in slowly deposited sediments (Suess, 1980). However, the consequences of this effect on the total rate of burial of organic matter are small unless additional PO_4^{3-} becomes available. Anoxia would tend to provide the required addition (Colman and Holland, 2000). However, the evidence for anoxia is still limited. The reappearance of BIFs demands deep-ocean anoxia during their formation, but the Neoproterozoic BIFs are associated with the major glaciations, which may indicate—but surely does not prove—that deep-water anoxia between 800 Ma and 550 Ma was restricted to these cold periods. Other indications of deep-water anoxia are needed to define the oxidation state of deep water and the causes of anoxia during the Neoproterozoic.

In a low-sulfate ocean, anoxia would probably have increased the rate of methane production in marine sediments. Some of this methane probably escaped into the water column. In the deeper, O_2-free parts of the oceans, it was oxidized in part by organisms using SO_4^{2-} as the oxidant. In the upper parts of the oceans, methane was partly oxidized by O_2. The remainder escaped into the atmosphere. There it was in part decomposed and oxidized inorganically to CO_2; in part it was returned to the ocean in nonmethanogenic areas and was oxidized there biologically. Some methane generated in marine sediments was probably sequestered at least temporarily in methane clathrates. The methane concentration in the Neoproterozoic atmosphere could have been as high as 100–300 ppm (Pavlov *et al.*, 2003). If methane concentrations were as high as this, it would have been a very significant greenhouse gas. At a concentration of 100 ppm, methane could well have increased the surface temperature by 12 °C (Pavlov *et al.*, 2003).

It seems strange, therefore, that the Late Neoproterozoic should have been the time of severe glaciations (Kirschvink, 1992). In their review of the snowball Earth hypothesis, Hoffman and Schrag (2002) point out that in some sections a steep decline in $\delta^{13}C$ values by 10–15‰ preceded any physical evidence of glaciation or sea-level fall. The most likely explanation for such a sharp drop involves a major decrease in the burial rate of organic carbon. The reason(s) for this are still obscure. It is intriguing that major periods of phosphogenesis coincided roughly with the glaciations between 750–800 Ma and ca. 620 Ma (Cook and McElhinny, 1979). The first of these appears to coincide with the appearance of Rapitan-type iron ores. The second seems to have commenced shortly after the latest Neoproteroizc glaciation (Hambrey and Harland, 1981), reached a peak during the Early Cambrian, then declined rapidly, and finally ended in the Late Mid-Cambrian (Cook and Shergold, 1984, 1986). However, the time relationship between the glacial and the phosphogenic events is still not entirely resolved. The rapid removal of phosphate from the oceans may have begun before the onset of glaciation. Perhaps continental extension and rifting during the Neoproterozoic and the creation of many shallow epicontinental seaways at low paleolatitudes created environments that were particularly favorable for the deposition of phosphorites (Donnelly *et al.*, 1990).

Some of the Neoproterozoic BIFs and associated rock units are also quite enriched in P_2O_5. For instance, the P_2O_5 content of the Rapitan iron formation in Canada and its associated hematitic mudstones ranges from 0.49% to 2.16% (Klein and Beukes, 1993). It can readily be shown, however, that the phosphate output from the oceans into the known phosphorites and BIFs is a small fraction of the phosphate metabolism of the oceans as a whole between 800 Ma and 600 Ma. This observation does not eliminate the possibility that phosphate removal into other sedimentary rocks was abnormally rapid during the two phosphogenic periods between 800 Ma and 600 Ma. Such abnormally rapid phosphate removal as a constituent of apatite would have decreased the availability of phosphate for deposition with organic matter. This would have produced a decrease in the $\delta^{13}C$ value of carbonates as observed before the onset of the snowball Earth glaciations. It would probably also have reduced the rate of methane generation, the methane concentration in the atmosphere, and the global temperature. If this, in turn, led to the onset of glaciation, the decrease in the rate of weathering would have further restricted the riverine flow of phosphate and thence to a decrease in $\delta^{13}C_{carb}$ and the global temperature. This scenario is highly speculative, but it does seem to account at least for the onset of the glaciations.

It seems very likely that the continents were largely ice covered during the Neoproterozoic glaciations. The state of the oceans is still a matter of debate. The Hoffman–Schrag Snowball Earth hypothesis posits that the oceans were completely, or nearly completely, ice covered. This seems unlikely. Leather *et al.* (2002) have pointed out that the sedimentology and stratigraphy of the Neoproterozoic glacials of Arabia were more like those of the familiar oscillatory glaciations of the Pleistocene than those required by the Snowball Earth hypothesis. Similarly, Condon *et al.* (2002), who studied the stratigraphy and sedimentology of six Neoproterozoic glaciomarine successions, concluded that the Neoproterozoic seas were not totally frozen, and that the hydrologic cycle was functioning during the major glaciations. This suggests that the tropical oceans were ice free or only partially ice covered. Perhaps an Earth in such a state might be called a frostball, rather than a snowball. Chemical weathering on the continents in this state would have been very minor. CO_2 released from volcanoes would have built up in the atmosphere and in the oceans until its partial pressure was high enough to overcome the low albedo of the Earth at the height of the glacial episodes. In the Hoffman–Schrag model, some 10 Ma of CO_2 buildup are needed to raise the atmospheric CO_2 pressure sufficiently to overcome the low albedo of a completely ice-covered Earth. The only test of this timescale has been provided by the Bowring *et al.*'s (2003) data for the duration of the Gaskiers glacial deposits in Newfoundland. This unit is often described as a Varanger-age glaciomarine deposit. It is locally overlain by a thin cap carbonate bed with a highly negative carbon isotopic signature. The U–Pb geochronology of zircons separated from ash beds below, within, and above the glacial deposits indicates that these glacial deposits accumulated in less than 1 Ma. The short duration of this episode may be more consistent with a frostball than with a snowball Earth.

At the very low stand of sea level during the heights of these glaciations, the release of methane from clathrates might have contributed significantly to the subsequent warming and to the negative value of $\delta^{13}C_{CARB}$ in the ocean–atmosphere system (Kennedy *et al.*, 2001). Intense weathering after the retreat of the glaciers would—among other products—have released large quantities of phosphate. This might have speeded the recovery of photosynthesis and the return of the $\delta^{13}C$ of marine carbonates to their large positive values along the course of the 800–600 Ma positive $\delta^{13}C$ excursion.

Between the end of the positive $\delta^{13}C$ excursion and the beginning of the Cambrian, several large marine evaporites were deposited. These are still preserved and offer the earliest opportunity to use the mineralogy of marine evaporites and the composition of fluid inclusions in halite to reconstruct the composition of the contemporaneous seawater. Horita *et al.* (2002) have used this approach to show that the sulfate concentration in latest Neoproterozoic seawater was $\sim$23 mmol kg^{-1}, i.e., only slightly less than in modern seawater, and significantly greater than during some parts of the Phanerozoic. By the latest Proterozoic, a new sulfur regimen had been installed. The cycling of seawater through MORs had probably reached present-day levels, the S/C ratio in volcanic gases had therefore risen, and the proportion of volcanic sulfur converted to constituents of sulfides and sulfates had approached unity as demanded by the composition of average Phanerozoic volcanic gases (Holland, 2002). The conversion of the composition of sedimentary rocks from their pre-GOE to their modern composition had been nearly completed.

The level of atmospheric O_2 seems to have been the last of the redox parameters to approach modern values. The evidence for this comes from the changes in the biota that occurred between the latest Proterozoic and the middle of the Cambrian period. These changes are discussed in the next section. Although a good deal of progress has been made since 1980 in our understanding of the atmosphere and oceans during the Proterozoic Era, we are still woefully ignorant of even the most basic oceanographic data for Precambrian seawater.

12.5 THE PHANEROZOIC

12.5.1 Evidence from Marine Evaporites

Our understanding of the chemical evolution of Phanerozoic seawater has increased enormously since the end of World War II. Rubey's (1951) presidential address to the Geological Society of America was aptly entitled "The geologic history of seawater, an attempt to state the problem." During the following year, Barth (1952) introduced the concept of the characteristic time in his analysis of the chemistry of the oceans, and this can, perhaps, be considered the beginning of the application of systems analysis to marine geochemistry. Attempts were made by the Swedish physical chemist Lars Gunnar Sillén to apply equilibrium thermodynamics to define the chemical history of seawater, but, as he pointed out, "practically everything that interests us in and around the sea is a symptom of nonequilibrium . . . What we can hope is that an equilibrium model may give a useful first approximation to the real system, and that the deviations of the real system may be treated as disturbances" (Sillén, 1967). Similar sentiments were expressed by Mackenzie and Garrels (1966) and by Garrels and Mackenzie (1971). They proposed that there has been little change in seawater composition since 1.5–2 Ga, although they were concerned by the discovery by Ault and Kulp (1959), Thode *et al.* (1961), Thode and Monster (1965), and Holser and Kaplan (1966) of very significant fluctuations in the isotopic composition of sulfur in seawater during the Phanerozoic.

A good deal of optimism regarding the constancy or near constancy of the composition of seawater during the Phanerozoic was, however, permitted by the Holser (1963) discovery that the Mg/Cl and Br/Cl ratios in brines extracted from fluid inclusions in Permian halite from Hutchinson, Kansas were close to those of modern brines. Holland (1972) published his analysis of the constraints placed by the constancy of the early mineral sequence in marine evaporites during the Phanerozoic. This paper showed that most of the seawater compositions permitted by the precipitation sequence $CaCO_3$–$CaSO_4$–NaCl in marine evaporites fall within roughly twice and half of the concentration of the major ions in seawater today. These calculations were extended by Harvie *et al.* (1980) and Hardie (1991) to include the later, more complex mineral assemblages of marine evaporites. Their calculations explained the mineral sequence in modern evaporites, and confirmed that the composition of Permian seawater was similar to that of modern seawater.

The development of a method to extract brines from fluid inclusions in halite and to obtain quantitative analyses by means of ion chromatography (Lazar and Holland, 1988) led to a study of the composition of trapped brines in halite from several Permian marine evaporites (Horita *et al.*, 1991). Their results together with those of Stein and Krumhansl (1988) confirmed that the composition of modern seawater is similar to that of Permian seawater, and suggested that the composition of seawater has been quite conservative during the Phanerozoic. This suggestion, however, has turned out to be far off the mark. Analyses of fluid inclusions in halite of the Late Silurian Salina Group of the Michigan Basin (Das *et al.*, 1990), the Middle Devonian Prairie Formation in the Saskatchewan Basin (Horita *et al.*, 1996), and a growing number of other marine evaporites (for a summary of the results of other groups, see Horita *et al.* (2002)) have shown that the similarity between Permian and modern seawater is the exception rather than the rule. Only the fluid inclusions in halite of the latest Neoproterozoic Ara Formation in Oman (Horita *et al.*, 2002) are similar to their Permian and modern equivalents. All of the other fluid inclusions contain brines which are very significantly depleted in Mg^{2+} and SO_4^{2-} relative to modern seawater. Their composition is consistent with the mineralogy of the associated evaporites. The difference between these inclusion fluids and their modern counterparts is either due to significant differences in the composition of the seawater from which they were derived or to reactions which depleted the Mg^{2+} and SO_4^{2-} content of the brines along their evaporation path. The Silurian and Devonian evaporites which we studied are associated with large carbonate platforms. The dolomitization of $CaCO_3$ followed by the deposition of gypsum and/or anhydrite during the passage of seawater across such platforms can deplete the evaporating brines in Mg^{2+} and SO_4^{2-}. Their composition can then become similar to that of the brines in the Silurian and Devonian halites. We opted for this interpretation of the fluid inclusion data, wrongly as it turned out. An obvious test of the proposition that the differences were due to dolomitization and $CaSO_4$ precipitation was to analyze fluid inclusions in halite from marine evaporites which are not associated with extended carbonate platforms. Zimmermann's (2000) work on Tertiary evaporites did just that. Her analyses showed that in progressively older Tertiary evaporites the composition of the seawater from which the brines in these evaporites developed was progressively more depleted in Mg^{2+} and SO_4^{2-} (Figure 17). As shown in Figure 18, this trend continued into the Cretaceous and was not reversed until the Triassic or latest Permian. The composition of fluid inclusion brines in marine halite is, therefore, a reasonably good guide to the composition of their parent seawater. However, changes due to the reaction of evaporating seawater with the sediments across which it passes en route to trapping have almost certainly occurred and cannot be neglected.

12.5.2 The Mineralogy of Marine Oölites

The proposed trend of the Mg/Ca ratio of seawater (Figure 17) during the Tertiary is supported by two independent lines of evidence: the mineralogy of marine oölites and the magnesium content of foraminifera. Sandberg (1983, 1985) discovered that the mineralogy of marine oölites has alternated several times between dominantly calcitic and dominantly aragonitic (see Figure 18). On this basis he divided the Phanerozoic into periods of calcitic and aragonitic seas. He suggested that the changes in mineralogy were related to changes in atmospheric P_{CO_2} or to changes in the Mg/Ca ratio of seawater. The experiments by Morse *et al.* (1997) have shown that changes in the Mg/Ca ratio are the most likely cause of the changes in oölite mineralogy. The most recent switch in oölite mineralogy occurred near the base of the Tertiary (see Figures 17 and 18). Unfortunately, the Mg/Ca ratio of seawater at the

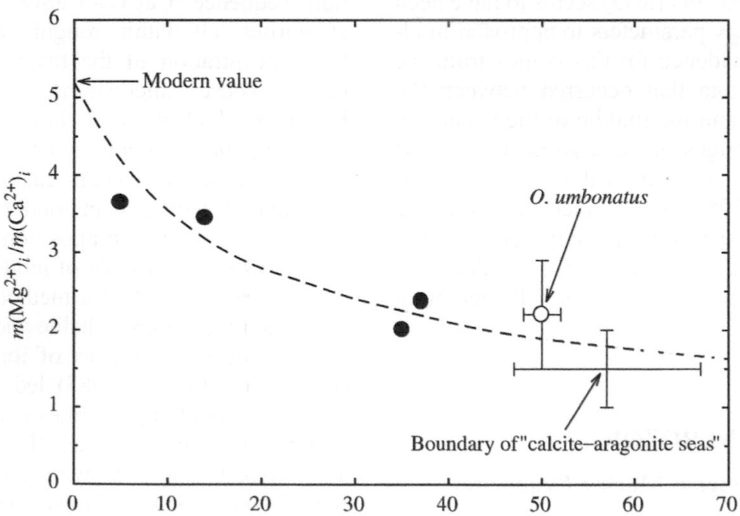

Figure 17 $m(Mg^{2+})_i/m(Ca^{2+})_i$ ratio in seawater during the Tertiary based on analyses of fluid inclusions in marine halite (● and dashed line), compared with data based on the Mg/Ca ratio of *O. umbonatus* (○) (Lear *et al.*, 2000) and the boundary of "aragonite-calcite seas" of Sandberg (1985) (source Horita *et al.*, 2002).

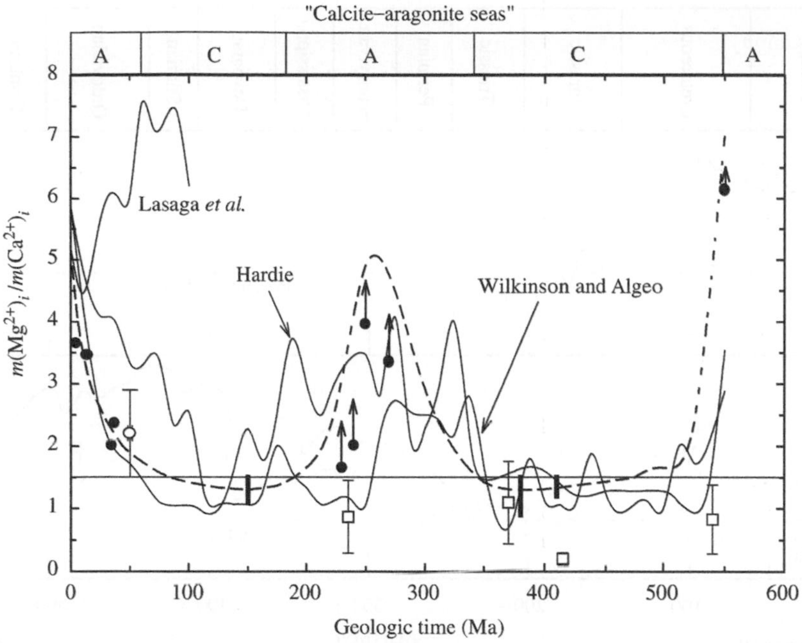

Figure 18 $m(Mg^{2+})_i/m(Ca^{2+})_i$ ratio in seawater during the Phanerozoic based on analyses of fluid inclusions in marine halite (solid symbols and dashed line), compared with the data based on Mg/Ca of *O. umbonatus* (○) (Lear *et al.*, 2000) and of abiogenic marine carbonate cements (□) (Cicero and Lohmann, 2001). Also shown are the results of modeling by Lasaga *et al.* (1985), Wilkinson and Algeo (1989), and Hardie (1996). "A" and "C" at the top indicate "aragonite seas" and "calcite seas" of Sandberg (1985) (source Horita *et al.*, 2002).

time of this switch can be defined only roughly, because the change in oölite mineralogy from calcite to aragonite depends on temperature as well as on the Mg/Ca ratio of seawater, and probably also on other compositional and kinetic factors.

12.5.3 The Magnesium Content of Foraminifera

The magnesium content of foraminifera supplies another line of evidence in support of a low Mg/Ca ratio in Early Tertiary seawater. Lear *et al.* (2000) found that the magnesium content of *O. umbonatus* decreased progressively with increasing age during the Tertiary. This change in composition is due in large part to changes in seawater temperature and in the Mg/Ca ratio of seawater, but it can also be overprinted by diagenetic processes. Much more data are needed to define the course of the Mg/Ca ratio of seawater on the basis of paleontologic data, but it is encouraging that its rather uncertain course during the Tertiary is consistent with that derived from the two other lines of evidence.

The correlation between the mineralogy of marine oölites and the Mg/Ca ratio of seawater as inferred from the composition of fluid inclusion brines extends throughout the Phanerozoic. This is also true for the correlation of the temporal distribution of the taxa of major calcite and aragonite reef builders and the temporal distribution of KCl-rich and $MgSO_4$-rich marine evaporites (Figure 19; Stanley and Hardie, 1998). There is no reason, therefore, to doubt that the composition of seawater has changed significantly during the Phanerozoic. As shown in Figures 20–23, only the concentration of potassium seems to have remained essentially constant. Lowenstein *et al.* (2001) have confirmed these trends and have shown that ancient inclusion fluids in halite had somewhat lower Na^+ concentrations and higher Cl^- concentrations during halite precipitation than present-day halite-saturated seawater brines.

Figures 20 – 23 compare the changes in the composition of seawater during the Phanerozoic that have been proposed by various authors since the early 1980s. Some of these differences are sizable. The Berner–Lasaga–Garrels (BLAG) box model developed by Berner *et al.* (1983) and Lasaga *et al.* (1985) included most of the major geochemical processes that affect the composition of seawater. Their changes in the concentration of Mg^{2+} and Ca^{2+} during the past 100 Ma are, however, much smaller than the estimates of Horita *et al.* (2002). The differences are due, in part, to the absence of dolomite as a major sink of Mg^{2+} in the BLAG model. However, Wallmann (2001) proposed concentrations of Ca^{2+} much higher than those of Horita *et al.* (2002), because he assumed, as an initial value in his model, that at 150 Ma the Ca^{2+} concentration in seawater was twice that of modern seawater.

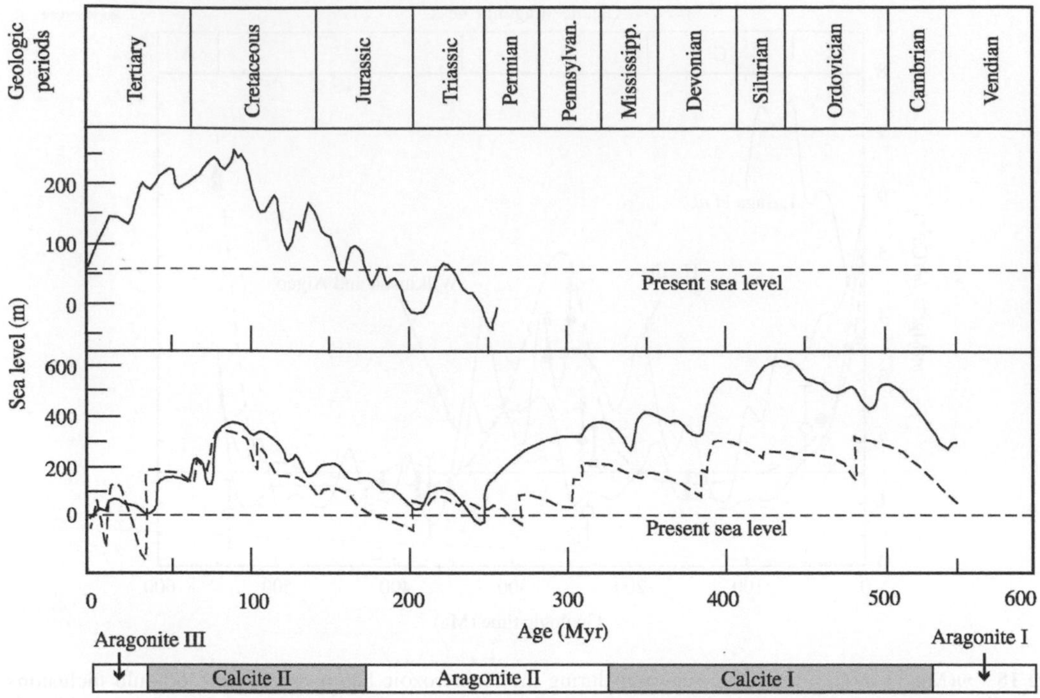

Figure 19 Sea-level changes and secular variations in the mineralogy of marine carbonates (Holland and Zimmermann, 2000): A—mean sea level during the Mesozoic and Cenozoic (Haq *et al.*, 1987); B—mean sea level during the Phanerozoic ((--) Vail *et al.*, 1977; (—) Hallam, 1984); C—secular variation in the mineralogy of Phanerozoic nonskeletal marine carbonates (source Stanley and Hardie, 1998).

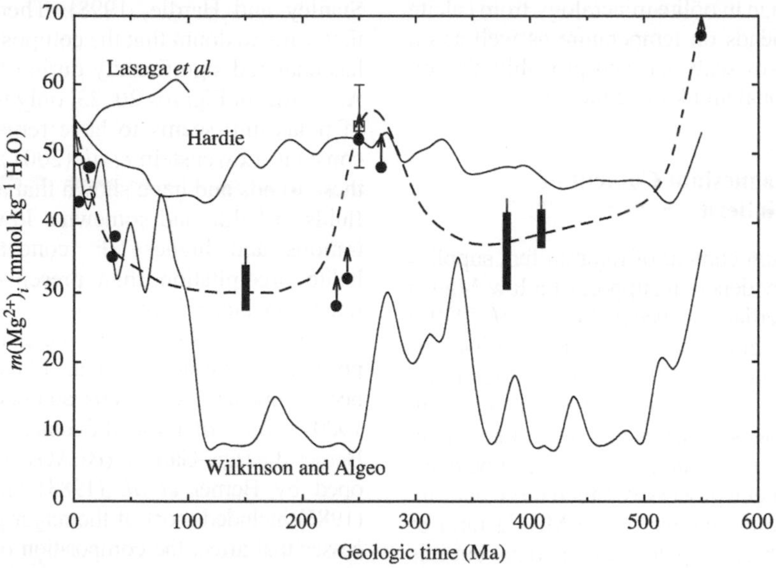

Figure 20 Concentration of Mg^{2+} in seawater during the Phanerozoic based on analyses of fluid inclusions in marine halite (solid symbols): thick and thin vertical bars are based on the assumption of different values for $m(Ca^{2+})_i/m(SO_4^{2-})_i$. Dashed line is our best estimate of age curve: (□)—Horita *et al.* (1991) and (○)—Zimmermann (2000). Also shown are the results of modeling by Lasaga *et al.* (1985), Wilkinson and Algeo (1989), and Hardie (1996) (source Horita *et al.*, 2002).

The effect of penecontemporaneous dolomite deposition on the chemical evolution of seawater was included in the model published by Wilkinson and Algeo (1989), which was based on Given and Wilkinson's (1987) compilation of the distribution of limestones and dolomites in Phanerozoic sediments.

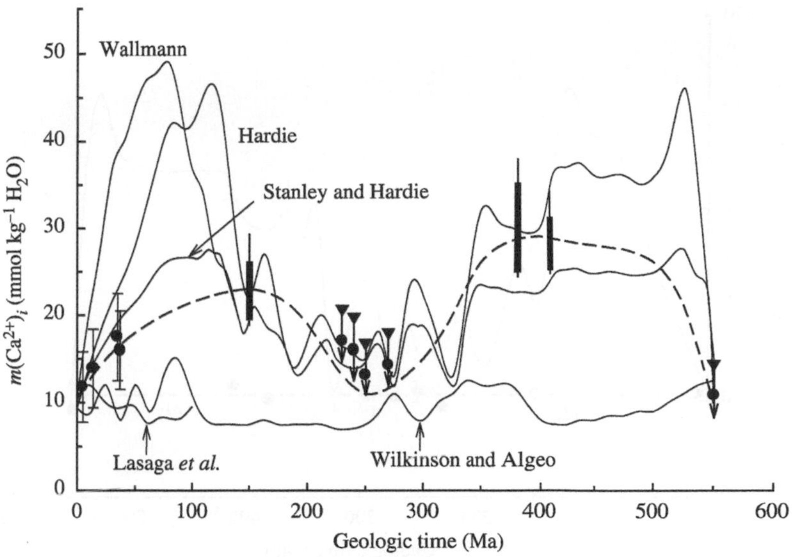

Figure 21 Concentration of Ca^{2+} in seawater during the Phanerozoic based on analyses of fluid inclusions in marine halite (solid symbols): circles–triangles and thick–thin vertical bars are based on the assumption of different values for $m(Ca^{2+})_i\ m(SO_4^{2-})_i$. Dashed line is our best estimate of age curve. Also shown are the results of modeling by Lasaga *et al.* (1985), Wilkinson and Algeo (1989), Hardie (1996), Stanley and Hardie (1998), and Wallmann (2001) (source Horita *et al.*, 2002).

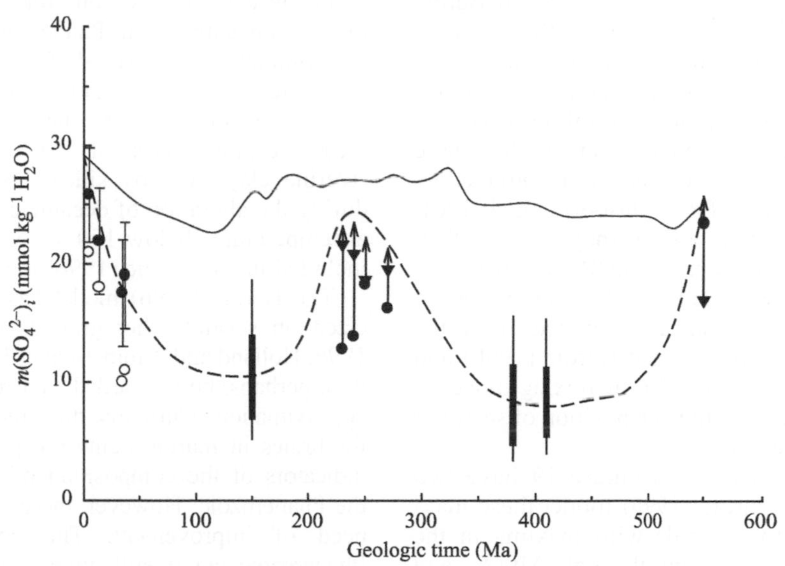

Figure 22 Concentration of SO_4^{2-} in seawater during the Phanerozoic based on analyses of fluid inclusions in marine halite (solid symbols): circles–triangles and thick–thin vertical bars are based on the assumption of different values for $m(Ca^{2+})_i m(SO_4^{2-})_i$. Dashed line is our best estimate of age curve. Data in open and filled circles are from Zimmermann (2000). Also shown are the results of modeling by Hardie (1996) (source Horita *et al.*, 2002).

Their calculations suggest that the concentration of Ca^{2+} in seawater remained relatively constant (±20%) during the Phanerozoic, but that the concentration of Mg^{2+} changed significantly, largely in phase with their proposed dolomite-age curve. The changes that they proposed for the concentration of both elements differ significantly from those of Horita *et al.* (2002), in part because of their incomplete compilation of Phanerozoic carbonate rocks (Holland and Zimmermann, 2000).

12.5.4 The Spencer–Hardie Model

A very different approach to estimating the composition of seawater during the Phanerozoic was

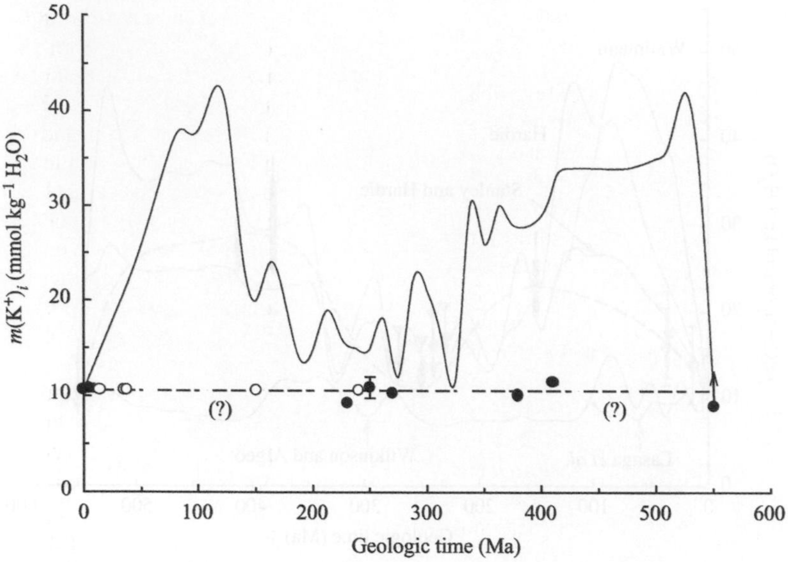

Figure 23 Concentration of K^+ in seawater during the Phanerozoic based on analyses of fluid inclusions in marine halite (solid symbols): circles–triangles and thick–thin vertical bars are based on the assumption of different values for $m(Ca^{2+})_i$ $m(SO_4^{2-})_i$. Dashed line is our best estimate of age curve. Open circles are from Zimmermann (2000). Also shown are the results of modeling by Hardie (1996) (source Horita *et al.,* 2002).

taken by Spencer and Hardie (1990) and Hardie (1996). These authors proposed that the composition of seawater was determined by the mixing ratio of river water and mid-ocean ridge solutions coupled with the precipitation of solid $CaCO_3$ and SiO_2 phases. They accepted Gaffin's (1987) curve for the secular variation of ocean crust production, which is based on the Exxon first-order global sea-level curve (Vail *et al.*, 1977). They assumed that, as a first approximation, the MOR flux of hydrothermal brines has scaled linearly with the rate of ocean crust production as estimated from the sea-level curves. Spencer and Hardie (1990) and Hardie (1996) then applied their mixing model to estimate the course of the composition of seawater during the Phanerozoic.

The sea-level curves in Figure 19 have two maxima. In the Hardie (1996) model these maxima are taken to coincide with maxima in the rate of seawater cycling through MORs and hence to minima in the concentration of Mg^{2+} and SO_4^{2-} in the contemporaneous seawater. To that extent the predictions of the Hardie (1996) model agree well with the fluid inclusion data and represented a distinct advance in understanding the chemical evolution of seawater during the Phanerozoic.

There are, however, rather serious discrepancies between the Hardie (1996) model and the fluid inclusion data of Horita *et al.* (2002). The most glaring is the difference between the large variations in the K^+ concentration predicted by Hardie (1996) and the essentially constant value of the K^+ concentration indicated by the fluid inclusion data. The reasons for the constancy of the K^+ concentration in Phanerozoic seawater are not completely understood. They must, however, be related to the mechanisms by which K^+ is removed from seawater. The most important of these are almost certainly the uptake of K^+ by riverine clays and by silicate phases produced during the alteration of oceanic crust by seawater at temperatures below 100 °C. Neither process is included in the Hardie (1996) model.

The Hardie (1996) model has also been criticized on several other grounds (Holland *et al.*, 1996; Holland and Zimmermann, 1998). At present it is, perhaps, best regarded as a rough, first-order approximation. Currently the fluid inclusion data for brines in marine halite are probably the best indicators of the composition of seawater during the Phanerozoic. However, these data are much in need of improvement. The coverage of the Phanerozoic era is still quite spotty, and a large number of additional measurements are needed before anything more can be claimed than a preliminary outline of the Phanerozoic history of seawater. Even the presently available data are not above suspicion. The assumptions that underlie the data points in Figures 17–23 were detailed by Horita *et al.* (2002). They are reasonable but not necessarily correct. The composition of brines trapped in halite is far removed from that of their parent seawater. Reconstructing the evolution of these brines is a considerable challenge, given the complexity of the precipitation, dissolution, reprecipitation, and mixing processes in evaporite basins.

12.5.5 The Analysis of Unevaporated Seawater in Fluid Inclusions

Analyses of unevaporated seawater are probably essential for defining precisely the evolution of seawater during the Phanerozoic. Johnson and Goldstein (1993) have described single-phase fluid inclusions in low-magnesium calcite cement of the Wilberns Formation in Texas. These almost certainly contain Cambro–Ordovician seawater. The salinity of the inclusion fluids ranges from 31‰ to 47‰. This is essentially identical to the range of seawater salinity observed in shallow-water marine settings today, and is consistent with the precipitation of the calcite cements within slightly restricted environments. Banner has reported a similar salinity range for fluid inclusions in low-magnesium calcite cements in the Devonian Canning Basin of Australia. The fluid inclusions in both areas have diameters $\leq 30\,\mu m$. They are large enough for making heating–freezing measurements but too small until now to serve for quantitative chemical analysis. The recent development of analytical techniques based on ICPMS technology brings the analysis of these fluid inclusions within reach. It will be important to determine whether their composition agrees with the composition of seawater that has been inferred from studies of the composition of inclusion fluids in halite from marine evaporites.

12.5.6 The Role of the Stand of Sea Level

The correlation between the composition of seawater and the sea-level curves in Figure 19 is quite striking. The reasons for the correlation are not entirely clear. Hardie (1996) suggested that the stand of sea level reflects the rate of ocean crust formation, that this determines the rate of seawater cycling through MORs, and hence the mixing ratio of hydrothermally altered seawater with average river water. However, the rate of seafloor spreading has apparently not changed significantly during the last 40 Ma (Lithgow-Bertelloni *et al.*, 1993), and Rowley's (2002) analysis of the rate of plate creation and destruction indicates that the rate of seafloor spreading has not varied significantly during the past 180 Ma. If this is correct, the rate of ridge production has been essentially constant since the Early Jurassic. Since the rate of seawater cycling through MORs is probably proportional to this rate, other changes in the Earth system have been responsible for the changes in sea level and in the composition of seawater during the past 180 Ma. Holland and Zimmermann (2000) have pointed out that marine carbonate sediments deposited during the past 40 Ma contain, on average, less dolomite than Proterozoic and Paleozoic carbonates. The lower dolomite content of the more recent carbonate sediments is due to the increase in the deposition of $CaCO_3$ in the deep sea, where dolomitization only takes place in unusual circumstances. The decrease in the rate of Mg^{2+} output from the oceans due to dolomite formation has been balanced by an increase in the output of oceanic Mg^{2+} by the reaction of seawater with clay minerals and with ocean-floor basalts, mainly at MORs. The increase in the output of Mg^{2+} into these reservoirs has been brought about by an increase in the Mg^{2+} concentration of seawater. A simple quantitative model of these processes (Holland and Zimmermann, 2000) can readily account for the observed increase in the Mg^{2+} and SO_4^{2-} concentration of seawater during the Tertiary.

This explanation cannot account for the changes in seawater chemistry before the development of abundant open-ocean $CaCO_3$ secreting organisms. Coccolithophores first appeared in the Jurassic and diversified tremendously during the Cretaceous. The foraminifera radiated explosively during the Jurassic and Cretaceous. Prior to the evolution of coccoliths and planktonic foraminifera, carbonate sediments were largely or entirely deposited on the continents and in shallow-water marine settings. It is likely, therefore, that changes in seawater composition before ca. 150 Ma were related either to changes in the rate of seafloor spreading or to the mineralogy of continental and near-shore carbonate sediments. There are not enough data to rule out the first alternative. However, the apparent near-constancy of the rate of ocean crust formation during the past 180 Ma (Rowley, 2002) is not kind to the notion of major changes in this rate during the first part of the Phanerozoic. The second alternative is more attractive. Flooding of the continents was extensive during high stands of sea level, and dolomitization, which is favored in warm, shallow evaporative settings, must have been widespread. During low stands of sea level, carbonate deposition on the continents was probably replaced by deposition on rims along continental margins, where dolomitization was kinetically less favored. One can imagine that this is the major reason for the correlation of the Mg^{2+} and SO_4^{2-} concentration of seawater and the stand of sea level. The relationship may, however, be more complicated. During the Tertiary, sea level fell, yet the shallow water, near-shore carbonates deposited during the last 40 Ma are strongly and extensively dolomitized. It is not clear why this should not have happened equally enthusiastically during the Permian and Late Neoproterozoic low stands of sea level.

12.5.7 Trace Elements in Marine Carbonates

Many attempts have been made to relate the trace element distribution in marine minerals, particularly in carbonates, to the rate of seawater cycling through MORs, and to tectonics in general. The concentration of lithium and the concentration and isotopic composition of strontium have been studied particularly intensively. In their paper on the lithium and strontium content of foraminiferal shells, Delaney and Boyle (1986) proposed that the Li/Ca ratio has varied rather little during the past 116 Ma, but that the Sr/Ca ratio increased significantly during this time period. More recent measurements by Lear *et al.* (2003) on a very large number of foraminifera have shown that the Sr/Ca ratio of benthic foraminifera has had a more complicated history during the past 75 Ma. Relating the Sr/Ca ratio of foraminiferal shells to the Sr/Ca ratio in seawater is complicated, because the Sr/Ca ratio in foraminifera is a rather strong function of temperature, shell size, and pressure, and because it is species specific (Elderfield et al., 2000, 2002). Lear *et al.* (2003) have taken all these factors into account, have combined their data for a variety of benthic foraminifera, and have proposed the course shown in Figure 24 for the Sr/Ca ratio in seawater during the last 75 Ma. The ratio decreased rapidly between 75 Ma and 40 Ma; during the last 40 Ma it climbed significantly, but the increase was interrupted by a decrease between 15 Ma and 7 Ma.

The data in Figure 24 can be combined with those in Figure 21 for the course of the calcium concentration in seawater to yield the course of the strontium concentration of seawater during the last 75 Ma. At 75 Ma, the strontium concentration was $\sim 22\times10^{-5}$ mol kg^{-1} H_2O. At 40 Ma, it was $\sim 11\times10^{-5}$ mol kg^{-1} H_2O. At present it is 8.5×10^{-5} mol kg^{-1} H_2O. The 60% decrease in the strontium concentration during the last 75 Ma exceeds the decrease of the calcium concentration (45%).

Reconstructing the course of the Sr/Ca concentration in seawater from the composition of fossils is limited by the effects of diagenesis. These are particularly disturbing in carbonates older than 100 Ma. Figure 25 shows Steuber and Veizer's (2002) data for the strontium content of Phanerozoic biological low-magnesium calcites. The averages of the strontium concentrations indicate a course similar to that of the changes in sea level during the Phanerozoic; but the scatter in their data is so large that the significance of their average curve is somewhat in doubt. Lear *et al.* (2003) have attempted to interpret the changes in the concentration and the isotopic composition of strontium in seawater during the last 75 Ma. The rise of the Himalayas and perhaps of other major mountain chains, the lowering of sea level, and the transfer of a significant fraction of marine $CaCO_3$ deposition from shallow to deep waters have all played a role in the changes in seawater composition during the Tertiary. A quantitative treatment of the available data for strontium is still quite difficult; there are still too many poorly defined parameters in the controlling equations.

12.5.8 The Isotopic Composition of Boron in Marine Carbonates

The isotopic composition of several elements in marine mineral phases is a much better indicator of oceanic conditions than their concentration in these phases. The $^{11}B/^{10}B$ ratio of living planktonic foraminifera is related to the pH of seawater (Sanyal *et al.*, 1996). This relationship has opened the possibility of using the $^{11}B/^{10}B$ ratio in foraminifera to infer the pH of seawater in the past (Spivack *et al.*, 1993) and thence the course of past P_{CO_2}. This approach has

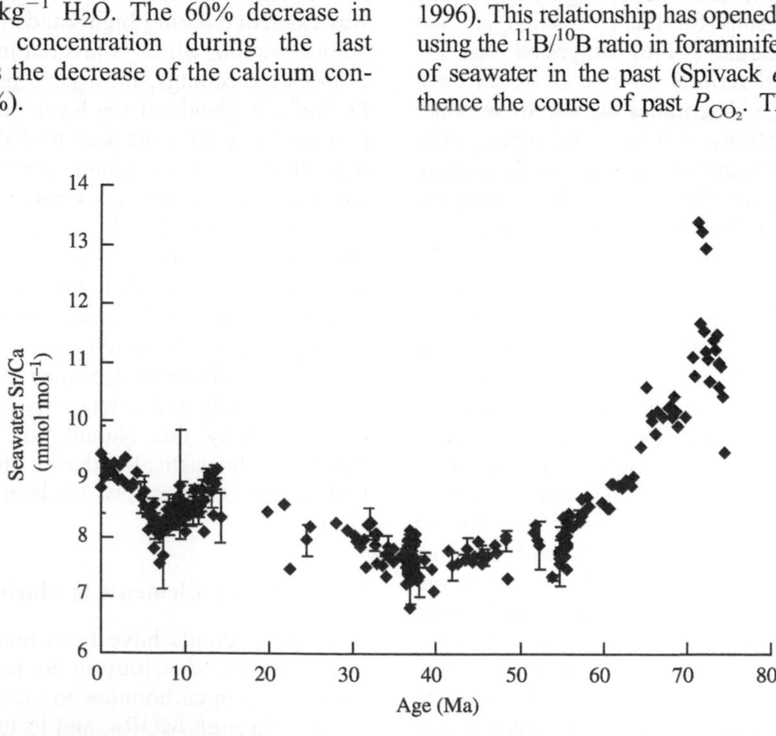

Figure 24 Seawater Sr/Ca record for the Cenozoic (source Lear *et al.*, 2003).

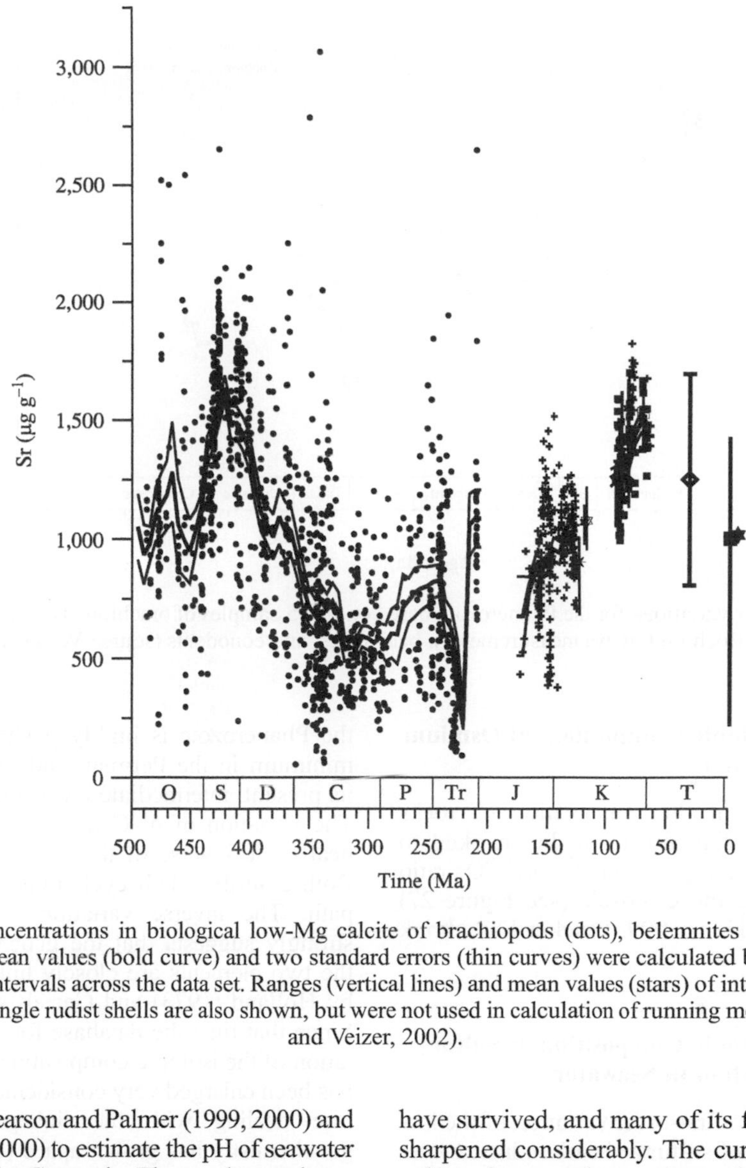

Figure 25 Sr concentrations in biological low-Mg calcite of brachiopods (dots), belemnites (crosses) and rudist bivalve (boxes). Mean values (bold curve) and two standard errors (thin curves) were calculated by moving a 20 Myr window in 5 Myr intervals across the data set. Ranges (vertical lines) and mean values (stars) of intrashell variations for concentrations in single rudist shells are also shown, but were not used in calculation of running means (source Steuber and Veizer, 2002).

been applied by Pearson and Palmer (1999, 2000) and by Palmer *et al.* (2000) to estimate the pH of seawater and P_{CO_2} during the Cenozoic. The results are, however, somewhat uncertain, because they depend rather heavily on the assumed value of the isotopic composition of boron in Cenozoic seawater, and because the fractionation of the boron isotopes during the uptake of this element in foraminifera is somewhat species specific (Sanyal *et al.*, 1996).

12.5.9 The Isotopic Composition of Strontium in Marine Carbonates

A very important contribution to paleoceanography has been made by measurements of the isotopic composition of strontium in Phanerozoic carbonates. The Burke *et al.* (1982) compilation of the isotopic composition of strontium in 744 Phanerozoic marine carbonates has now been expanded by a factor of about ~6. Figure 26 is taken from the summary of these data by Veizer *et al.* (1999). The major features of the Burke curve have survived, and many of its features have been sharpened considerably. The curve in Figure 26 is quite robust, and our view of variations in the $^{87}Sr/^{86}Sr$ ratio of seawater during the Phanerozoic is unlikely to change significantly. The $^{87}Sr/^{86}Sr$ ratio of seawater has fluctuated quite significantly. The overall decrease from its high value during the Cambrian to a minimum in the Jurassic and the return to its Early Paleozoic value during the Tertiary does not mirror the two megacycles of the sea-level curve. There is an obvious second-order correlation with orogenies, but their effect on the $^{87}Sr/^{86}Sr$ ratio of seawater has been overshadowed by changes in the isotopic composition and the flux of river strontium to the oceans. The very rapid rise of the $^{87}Sr/^{86}Sr$ ratio in seawater during the Tertiary must be related to the weathering of rocks of very high $^{87}Sr/^{86}Sr$ ratios, and changes in the $^{87}Sr/^{86}Sr$ ratio of rocks undergoing weathering have probably played a major role in determining the fluctuations in the $^{87}Sr/^{86}Sr$ ratio of seawater during the entire Phanerozoic.

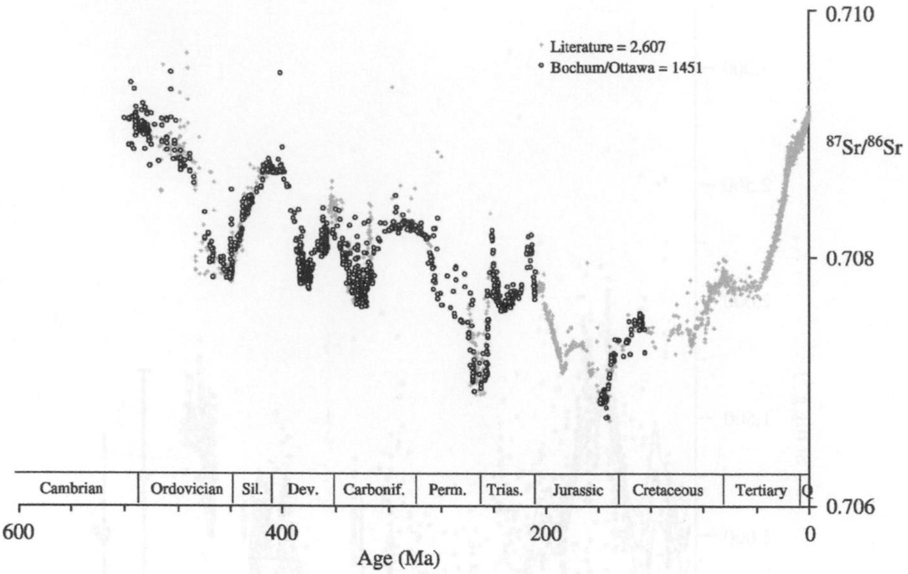

Figure 26 $^{87}Sr/^{86}Sr$ variations for the Phanerozoic based on 4,055 samples of brachiopods ("secondary" layer only for the new Bochum/Ottawa measurements), belemnites, and conodonts (source Veizer *et al.,* 1999).

12.5.10 The Isotopic Composition of Osmium in Seawater

Large changes in the composition of rocks undergoing weathering must also be invoked to explain the major changes in the $^{187}Os/^{186}Os$ ratio of seawater during the Cenozoic (see Figure 27) (Pegram and Turekian, 1999; Peucker-Ehrenbrink *et al.*, 1995).

12.5.11 The Isotopic Composition of Sulfur and Carbon in Seawater

Interestingly, the first-order variation of the isotopic composition of sulfur and carbon in seawater during the Phanerozoic is qualitatively similar to that of strontium. The value of $\delta^{34}S$ at the base of the Phanerozoic is highly positive. It drops to a minimum in the Permian and then rises again to its present, intermediate level (Figures 28 and 29). The variation of $\delta^{13}C$ during the Phanerozoic is nearly the inverse of the $\delta^{34}S$ curve (Figure 30). Both describe a half-cycle rather than a two-cycle path. The inverse variation of $\delta^{34}S$ and $\delta^{13}C$ strongly suggests that the geochemical cycles of the two elements are closely linked, as suggested by Holland (1973) and Garrels and Perry (1974). Since that time the database for assessing the variation of the isotopic composition of both elements has been enlarged very considerably, and the literature dealing with the linkage between their geochemistry has grown apace (e.g., Veizer *et al.*, 1980; Holser *et al.*, 1988, 1989; Berner, 1989; Kump, 1993; Carpenter and Lohmann, 1997;

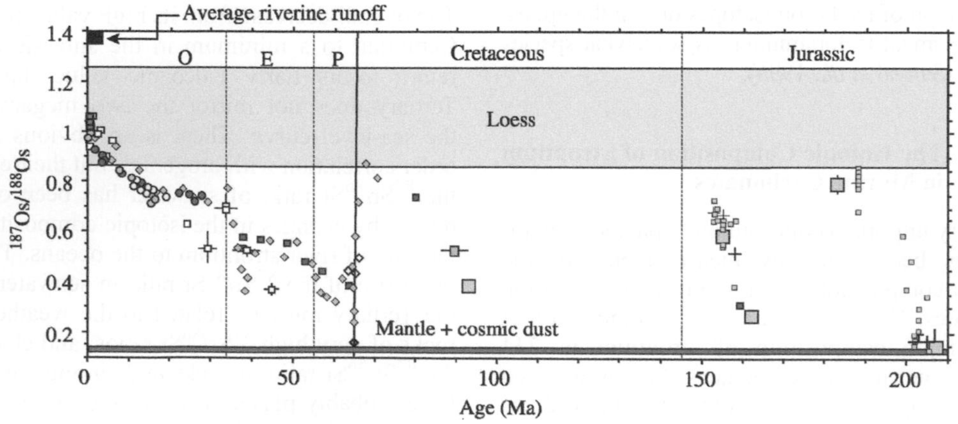

Figure 27 The marine Os isotope record during the last 200 Ma (source Peucker-Ehrenbrink and Ravizza, 2000.

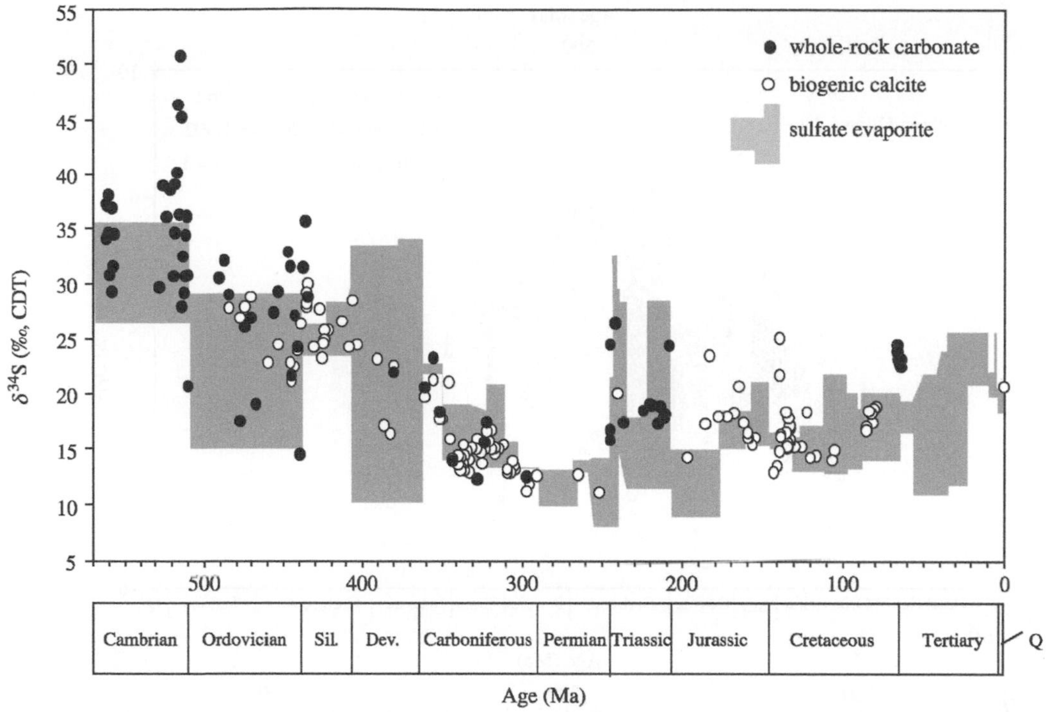

Figure 28 The sulfur isotopic composition of Phanerozoic seawater sulfate based on the analysis of structurally substituted sulfate in carbonates (Kampschulte and Strauss, in press) and evaporite based δ^{34} data (source Strauss, 1999).

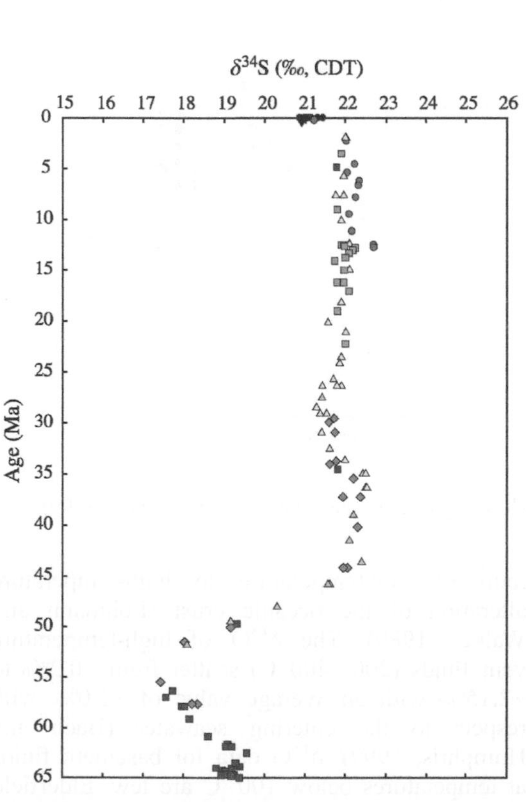

Figure 29 Isotopic composition of seawater sulfate during the past 65 Ma (Paytan *et al.*, 1998, 2002) with ages modified somewhat (sources Paytan, personal communication, 2003).

Petsch and Berner, 1998). Nevertheless, a truly quantitative understanding of the linkage has not yet been achieved. The inputs of sulfur and carbon to the oceans as well as the functional relationships that relate the composition of these reservoirs to their outputs into the rock record are still not very well defined. The large difference between the residence time of HCO_3^- and SO_4^{2-} in the oceans has invited deviations from steady state both in their concentration in seawater and in the isotopic composition of their constituents.

12.5.12 The Isotopic Composition of Oxygen in Seawater

The use of the isotopic composition of oxygen in marine carbonates as a means of reconstructing the course of the $\delta^{18}O$ value of seawater during the Phanerozoic has been a matter of considerable contention. The $\delta^{18}O$ value of carbonates and cherts tends to become more negative with increasing geologic age (see, e.g., Figure 31). In many instances this decrease is clearly due to diagenetic changes involving reactions with isotopically light water. However, the $\delta^{18}O$ of carbonates which have been chosen carefully to avoid overprinting by diagenetic alteration also tends to decrease with increasing geologic age.

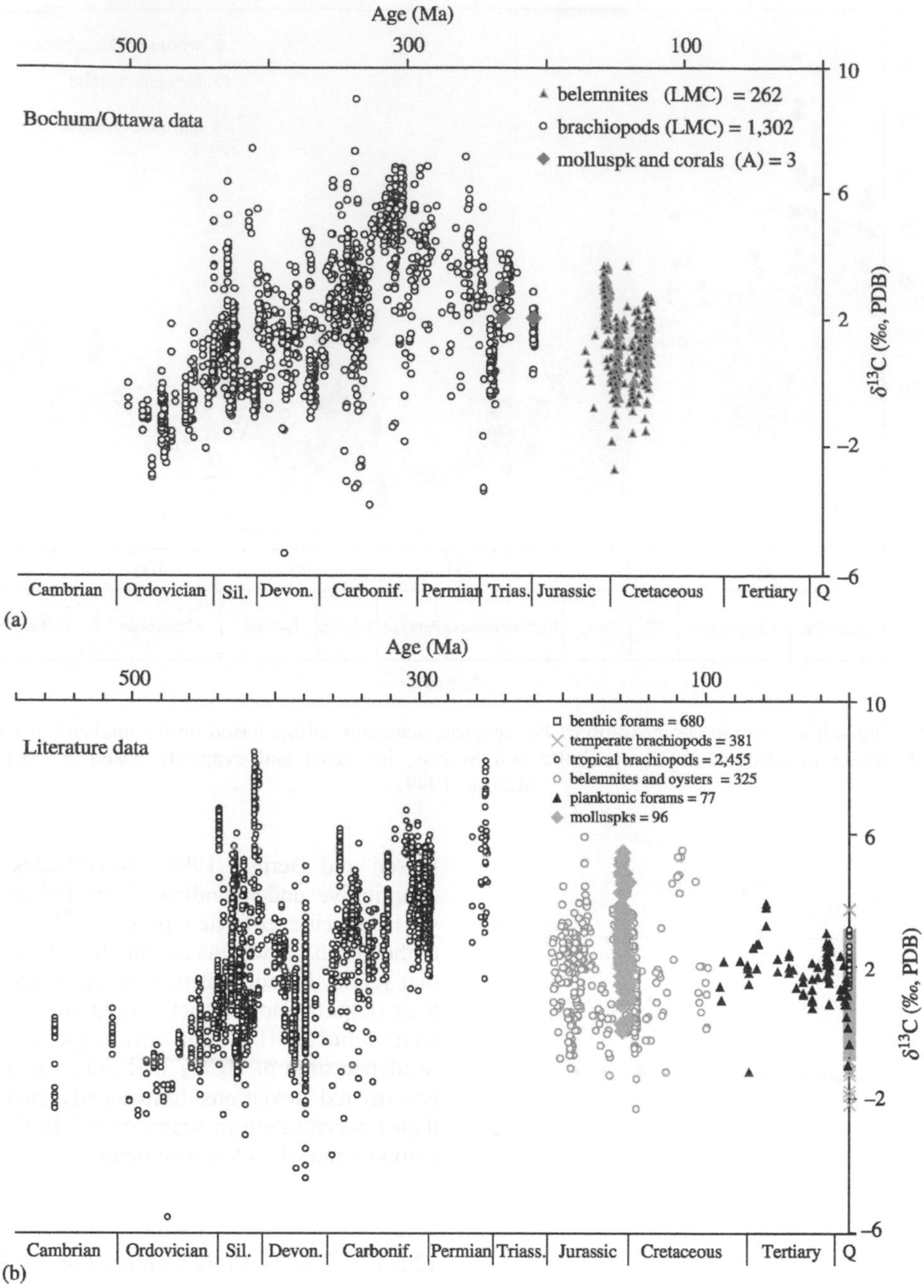

Figure 30 Carbon isotope composition of Phanerozoic low-Mg calcitic shells (source Veizer *et al.*, 1999).

The strongest argument in favor of a near-constant $\delta^{18}O$ value of seawater during much of Earth history is based on an analysis of the effects of seawater cycling through MORs at high temperatures (Muehlenbachs and Clayton, 1976; Holland, 1984; Muehlenbachs, 1986). Changes in the high-temperature cycling of seawater through mid-ocean ridges do not seem to be capable of accounting for major changes in the $\delta^{18}O$ of seawater. At present the only promising mechanism for explaining large changes in the $\delta^{18}O$ of seawater is a major change in the ratio of low-temperature to high-temperature alteration of the oceanic crust (Lohmann and Walker, 1989). The $\delta^{18}O$ of high-temperature vent fluids (200 – 400 °C) scatter from +0.2‰ to +2.15‰ with an average value of +1.0‰ with respect to the entering seawater (Bach and Humphris, 1999). $\delta^{18}O$ data for basement fluids at temperatures below 100 °C are few. Elderfield *et al.* (1999) have shown that thermally driven seawater in an 80 km transect across the eastern flank of the Juan de Fuca Ridge at 48° latitude has temperatures between 15.5 °C and 62.8 °C

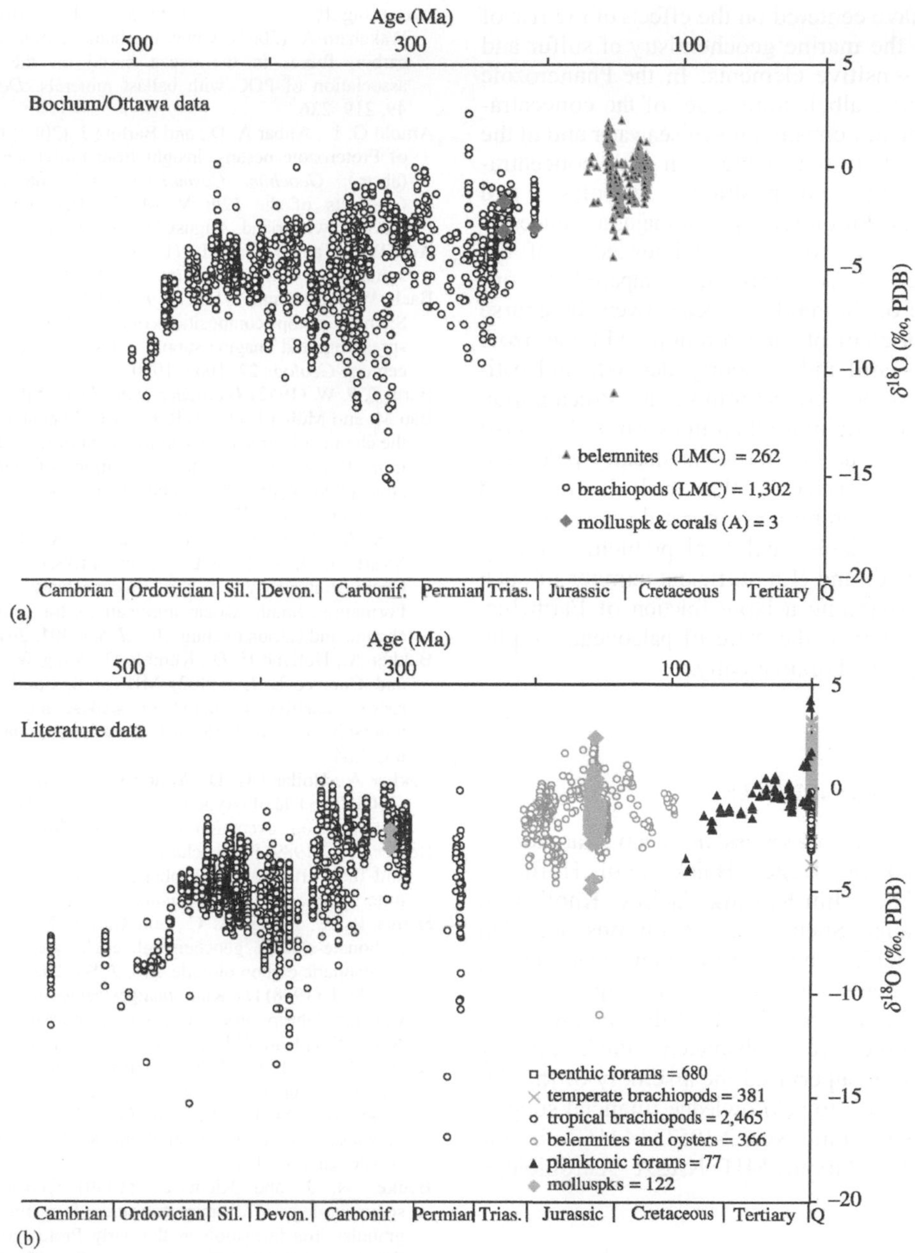

Figure 31 Oxygen isotope composition of Phanerozoic low-Mg calcitic shells (source Veizer *et al.,* 1999).

and $\delta^{18}O$ values, on average, −1.05‰ with respect to the entering seawater. About 5% of the global ridge flank heat flux would have to be associated with exchange of $\delta^{18}O$ in the ridge flanks to balance the enrichment in seawater $\delta^{18}O$ due to high-temperature hydrothermal activity. The estimate of 5% is similar to that proposed by Mottl and Wheat (1994) (see also Mottl, 2003). It remains to be seen whether the large changes in the balance between low-temperature and high-temperature alteration of the oceanic crust that seem to be required to shift the $\delta^{18}O$ values of seawater to −8‰ have actually occurred. Even if they have not, the Earth's total hydrologic system is so complex that major changes in the $\delta^{18}O$ of seawater during the history of the planet should not be dismissed out of hand.

12.6 A BRIEF SUMMARY

The large amount of research that has been completed since the early 1980s has done much to clarify our understanding of the chemical evolution of seawater. In the Precambrian, major advances have centered on the effects of the rise of oxygen on the marine geochemistry of sulfur and the redox sensitive elements. In the Phanerozoic the definition, albeit imprecise, of the concentration of the major constituents of seawater and of the relationship between changes in their concentration, isotopic composition, tectonics, and biological evolution represent a major advance.

Despite this progress our knowledge of the ancient oceans is miniscule compared to our knowledge of the modern ocean. Even the course of the major element concentrations in Phanerozoic seawater is still rather poorly defined, and estimates of their concentration in Precambrian seawater are little more than guesswork. The most promising avenue to a more satisfactory paleoceanography is probably the analysis of fluid inclusions containing unevaporated seawater. This presents formidable analytical problems. If they can be solved, and if a sufficient number of such inclusions spanning a large fraction of Earth history are analyzed, the state of paleoceanography will be improved dramatically.

ACKNOWLEDGMENTS

I would like to express my deep gratitude to Andrey Bekker, Roger Buick, Paul Hoffman, Stein Jacobsen, Jim Kasting, Andrew Knoll, Dan Schrag, Yanan Shen, Roger Summons, and Jan Veizer for taking the time to review this chapter. Their comments ranged from thoroughly enthusiastic to highly critical. All of the reviews were helpful; some were absolutely vital. Andrey Bekker kindly supervised the assembly of the figures. I also wish to acknowledge financial support from NASA Grant NCC2-1053 144167 for the work of the Harvard-MIT NASA Astrobiology Institute.

REFERENCES

Allègre C., Manhès G., and Göpel C. (1995) The age of the Earth. *Geochim. Cosmochim. Acta* **59**, 1445–1456.

Anbar A. D. and Holland H. D. (1992) The photochemistry of manganese and the origin of banded iron formations. *Geochim. Cosmochim. Acta* **56**, 2595–2603.

Anbar A. D. and Knoll A. H. (2002) Proterozoic ocean chemistry and evolution: a bioinorganic bridge? *Science* **297**, 1137–1142.

Anbar A. D., Zahnle K. J., Arnold G. L., and Mojzsis S. J. (2002) Extraterrestrial iridium, sediment accumulation and the habitability of the early Earth's surface. *J. Geophys. Res.* **106**, 3219–3236.

Appel P. W. U. (1983) Rare earth elements in the Early Archean Isua iron-formation, West Greenland. *Precamb. Res.* **20**, 243–258.

Appel P. W. U. and Moorbath S. (1999) Exploring Earth's oldest geological record in Greenland. *EOS, Trans., AGU* **80**, 257, 261, 264.

Armstrong R. A., Lee G. F., Hedges J. I., Honjo S., and Wakeham A. (2002) A new mechanistic model for organic carbon fluxes in the ocean based on the quantitative association of POC with ballast minerals. *Deep-Sea Res.* **49**, 219–236.

Arnold G. L., Anbar A. D., and Barling J. (2002) Oxygenation of Proterozoic oceans: insight from molybdenum isotopes (abstr.). *Geochim. Cosmochim. Acta* **66** (15A), A30 (abstracts of the 12th V. M. Goldschmidt Conference, Davos, Switzerland, August 18–23, 2002).

Ault W. U. and Kulp J. L. (1959) Isotopic geochemistry of sulfur. *Geochim. Cosmochim. Acta* **16**, 201–235.

Bach W. and Humphris S. E. (1999) Relationship between the Sr and O isotope compositions of hydrothermal fluids and the spreading and magma-supply rates at oceanic spreading centers. *Geology* **27**, 1067–1070.

Barth T. F. W. (1952) *Theoretical Petrology*. Wiley, New York.

Bau M. and Möller P. (1993) Rare earth element systematics of the chemically precipitated component in Early Precambrian iron formations and the evolution of the terrestrial atmosphere–hydrosphere–lithosphere system. *Geochim. Cosmochim. Acta* **57**, 2239–2249.

Bekker A., Kaufman A. J., Karhu J. A., Beukes N. J., Swart Q. D., Coetzee L. L., and Eriksson K. A. (2001) Chemostratigraphy of the Paleoproterozic Duitschland Formation, South Africa: implications for coupled climate change and carbon cycling. *Am. J. Sci.* **301**, 261–285.

Bekker A., Holland H. D., Rumble D., Yang W., Wang P.-L., and Coetzee L. L. (2002) MIF of S, oölitic ironstones, redox sensitive elements in shales, and the rise of atmospheric oxygen (abstr.). *Geochim. Cosmochim. Acta* **66**, A64.

Bekker A., Holland H. D., Young G. M., and Nesbitt H. W. (2003) The fate of oxygen during the early Paleoproterozoic carbon isotope excursion. *Astrobiology* **2**(4), 477.

Berner R. A. (1989) Biogeochemical cycles of carbon and sulfur and their effect on atmospheric oxygen over Phanerozoic time. *Paleogeogr. Paleoclimatol. Paleoecol.* **75**, 97–122.

Berner R. A., Lasaga A. C., and Garrels R. M. (1983) The carbonate-silicate geochemical cycle and its effect on atmospheric carbon dioxide. *Am. J. Sci.* **283**, 641–683.

Beukes N. J. (1978) Die Karbonaatgesteentes en Ysterformasies van die Ghaap-Groep van die Transvaal-Supergroep in Noord-Kaapland. PhD Thesis, Rand Afrikaans University.

Beukes N. J. (1983) Palaeoenvironmental setting of iron-formations in the depositional basin of the Transvaal Supergroup, South Africa. In *Iron Formation: Facts and Problems* (eds. A. F. Trendall and R. C. Morris). Elsevier, Amsterdam, pp. 131–209.

Beukes N. J. and Klein C. (1990) Geochemistry and sedimentology of a facies transition from microbanded to granular iron-formation in the early Proterozoic Transvaal Supergroup, South Africa. *Precamb. Res.* **47**, 99–139.

Beukes N. J. and Klein C. (1992) Time distribution, stratigraphy, and sedimentologic setting, and geochemistry of Precambrian iron-formations. In *The Proterozoic Biosphere* (eds. J. W. Schopf and C. Klein). Cambridge University Press, Cambridge, pp. 139–146.

Beukes N. J., Dorland H., and Gutzmer J. (2002a) Pisolitic ironstone and ferricrete in the 2.22–2.4 Ga Timeball Hill Formation, Transvaal Supergroup: implications for the history of atmospheric oxygen (abstr.). 2002 Annual Meeting of the Geological Society of America, 283p.

Beukes N. J., Dorland H., Gutzmer J., Nedachi M., and Ohmoto H. (2002b) Tropical laterites, life on land, and the history of atmospheric oxygen in the Paleoproterozoic. *Geology* **30**, 491–494.

Birch T. (1756) *A History of the Royal Society of London* (ed. A. Millar). London, vol. 1.

Black R. C. (1966) *The Younger John Winthrop*. Columbia University Press, New York.

Boak J. L. and Dymek R. F. (1982) Metamorphism of the ca. 3,800 Ma supercrustal rocks at Isua, West Greenland:

implications for early Archaean crustal evolution. *Earth Planet. Sci. Lett.* **59**, 155–176.

Boetius A., Ravenschlag K., Schubert C. J., Rickert D., Widdel F., Gieseke A., Amann R., Jørgensen B. B., Witte U., and Pfannkuche O. (2000) A marine microbial consortium mediating anaerobic oxidation of methane. *Nature* **407**, 623–626.

Bowring S. A. and Williams I. S. (1999) Priscoan (4.00–4.03) orthogneiss from northwestern Canada. *Contrib. Mineral. Petrol.* **134**, 3–16.

Bowring S. A., Landing E., Myrow P., and Ramenzavi J. (2003) Geochronological constraints on the terminal Neoproterozoic events and the rise of Metazoans. *Astrobiology* **2**(4), 457–458.

Brasier M. D., Green O. R., Steele A., Van Kranendonk M., Jephcoat A. P., Kleppe A. K., Lindsay J. F., and Grassineau N. V. (2002) Questioning the evidence for Earth's oldest fossils. In Astrobiology Science Conference, NASA Ames Research Center, April 7–11, 2002.

Braterman P. S., Cairns-Smith A. G., and Sloper R. W. (1983) Photo-oxidation of hydrated Fe^{+2}—significance for banded iron formations. *Nature* **303**, 163–164.

Braterman P. S., Cairns-Smith A. G., Sloper R. W., Truscott T. G., and Craw M. (1984) Photooxidation of iron (II) in water between pH 7.4 and 4.0. *J. Chem. Soc.: Dalton Trans.*, 1441–1445.

Bridgewater D., Keto L., McGregor V. R., and Myers J. S. (1976) Archaean gneiss complex in Greenland. In *Geology of Greenland*, Geological Survey of Greenland (eds. A. Escher and W. S. Watt). Copenhagen, pp. 20–75.

Bridgewater D., Keto L., McGregor V. R., and Myers J. S. (1976) Archaean gneiss complex in Greenland. In *Geology* Geology of Greenland, Geological Survey of (eds. A. Escher and W. S. Watt), Copenhagen, pp. 20–75.

Brocks J. J., Logan G. A., Buick R., and Summons R. E. (1999) Archean molecular fossils and the early rise of Eukaryotes. *Science* **285**, 1033–1036.

Brocks J. J., Summons R. E., Logan G. A., and Buick R. (2002) Molecular fossils in Archean Rocks: constraints on the oxygenation of the upper water column (abstr.). In *Astrobiology Science Conference, NASA Ames Research Center, April 7–11, 2002.*

Bruno J., Wersin P., and Stumm W. (1992) On the influence of carbonate in mineral dissolution: II. The stability of $FeCO_3$(s) at 25° and 1 atm. total pressure. *Geochim. Cosmochim. Acta* **56**, 1149–1155.

Burke W. H., Denison R. E., Heatherington E. A., Koepnick R. B., Nelson H. F., and Otto J. B. (1982) Variation of seawater $^{87}Sr/^{86}Sr$ throughout Phanerozoic time. *Geology* **10**, 516–519.

Cairns-Smith A. G. (1978) Precambrian solution photochemistry-inverse segregation and banded iron formations. *Nature* **276**, 807–808.

Canfield D. E. (1998) A new model for Proterozoic ocean chemistry. *Nature* **396**, 450–452.

Canfield D. E. (2001) Biogeochemistry of Sulfur Isotopes. In *Stable Isotope Geochemistry,* Reviews in Mineralogy and Geochemistry (eds. J. W. Valley and D. R. Cole). Washington, DC, vol. 43, pp. 607–636.

Canfield D. E. and Raiswell R. (1999) The evolution of the sulfur cycle. *Am. J. Sci. 299*, 697–729.

Canfield D. E., Habicht K., and Thamdrup B. (2000) The Archean sulfur cycle and the early history of atmospheric oxygen. *Science* **288**, 658–661.

Canfield D. E. and Raiswell R. (1999) The evolution of the sulfur cycle. *Am. J. Sci.* **299**, 697–729.

Canil D. (1997) Vanadium partitioning and the oxidation state of Archaean komatiite magmas. *Nature* **389**, 842–845.

Canil D. (1999) Vanadium partitioning between orthopyroxene, spinel and silicate melt and the redox states of mantle source regions for primary magmas. *Geochim. Cosmochim. Acta* **63**, 557–572.

Canil D. (1999) Vanadium partitioning between orthopyroxene, spinel and silicate melt and the redox states of mantle source regions for primary magmas. *Geochim. Cosmochim. Acta* **63**, 557–572.

Canil D. (2002) Vanadium in peridotites, mantle redox and tectonic environments: Archean to present. *Earth Planet. Sci. Lett.* **195**, 75–90.

Carpenter S. J. and Lohmann K. C. (1997) Carbon isotope ratios of Phanerozoic marine cements: re-evaluating the global carbon and sulfur systems. *Geochim. Cosmochim. Acta* **61**, 4831–4846.

Catling D. C., Zahnle K. J., and McKay C. P. (2001) Biogenic methane, hydrogen escape, and the irreversible oxidation of early Earth. *Science* **293**, 839–843.

Chyba C. F. (1993) The violent emergence of the origin of life: progress and uncertainties. *Geochim. Cosmochim. Acta* **57**, 3351–3358.

Cicero A. D. and Lohmann K. C. (2001) Sr/Mg variation during rock-water interaction: implications for secular changes in the elemental chemistry of ancient seawater. *Geochim. Cosmochim. Acta* **65**, 741–761.

Clarke F. W. (1911) *The Data of Geochemistry*, 2nd edn. The Government Printing Office, Washington, DC.

Cloud P. E., Jr. (1968) Atmospheric and hydrospheric evolution on the primitive Earth. *Science* **160**, 729–736.

Cloud P. E., Jr. (1972) A working model for the primitive Earth. *Am. J. Sci.* **272**, 537–548.

Colman A. S. and Holland H. D. (2000) The global diagenetic flux of phosphorus from marine sediments to the oceans: redox sensitivity and the control of atmospheric oxygen levels. In *Marine Authigenesis: From Global to Microbial*, SEPM Special Publication No. 66 (eds. C. R. Glenn, L. Prévôt-Lucas, and J. Lucas). Tulsa, UK, pp. 53–75.

Compston W. and Pidgeon R. T. (1986) Jack Hills, evidence of more very old detrital zircons in Western Australia. *Nature* **321**, 766–769.

Condon D. J., Prave A. R., and Benn D. I. (2002) Neoproterozoic glacial-rainout intervals: observations and implications. *Geology* **30**, 35–38.

Conway E. J. (1943) The chemical evolution of the ocean. *Proc. Roy. Irish Acad.* **48B**(8), 161–212.

Conway E. J. (1945) Mean losses of Na, Ca, etc. in one weathering cycle and potassium removal from the ocean. *Am. J. Sci.* **243**, 583–605.

Cook P. J. and McElhinny M. W. (1979) A reevaluation of the spatial and temporal distribution of sedimentary phosphate deposits in the light of plate tectonics. *Econ. Geol.* **74**, 315–330.

Cook P. J. and Shergold J. H. (1984) Phosphorus, phosphorites and skeletal evolution at the Precambrian–Cambrian boundary. . *Nature* **308**, 231–236.

Cook P. J. and Shergold J. H. (1986) Proterozoic and Cambrian phosphorites—nature and origins. In *Phosphate Deposits of the World: Volume 1. Proterozoic and Cambrian Phosphorites* (eds. P. J. Cook and J. H. Shergold). Cambridge University Press, Cambridge, pp. 369–386.

Crick I. H. (1992) Petrological and maturation characteristics of organic matter from the Middle Proterozoic McArthur Basin, Australia. . *Austral. J. Earth Sci.* **39**, 501–519.

Dalrymple G. B. and Ryder G. (1996) Argon-40/argon-39 age spectra of Apollo 17 highlands breccia samples by laser step heating and the age of the Serenitatis basin. . *J. Geophys. Res.* **101**, 26069–26084.

Das N., Horita J., and Holland H. D. (1990) Chemistry of fluid inclusions in halite from the Salina Group of the Michigan Basin: implications for Late Silurian seawater and the origin of sedimentary brines. *Geochim. Cosmochim. Acta* **54**, 319–327.

Delaney M. L. and Boyle F. A. (1986) Lithium in foraminiferal shells: implications for high-temperature hydrothermal circulation fluxes and oceanic crustal generation rates. *Earth Planet. Sci. Lett.* **80**, 91–105.

Delano J. W. (2001) Redox history of the Earth's interior: implications for the origin of life. . *Origins Life Evol. Biosphere* **31**, 311–334.

De Long E. F. (2000) Resolving a methane mystery. *Nature* **407**, 577–579.

De Ronde C. E. J., de Wit M. J., and Spooner E. T. C. (1994) Early Archean (>3.2 Ga) iron-oxide-rich hydrothermal discharge vents in the Barberton greenstone belt. *South Africa . Geol. Soc. Am. Bull.* **106**, 86–104.

De Ronde C. E. J., Channer R. M., Faure de R., Bray C. J., and Spooner E. T. C. (1997) Fluid chemistry of Archean seafloor hydrothermal vents: implications for the composition of circa 3.2 Ga seawater. *Geochim. Cosmochim. Acta* **61**, 4025–4042.

Donnelly T. H. and Jackson M. J. (1988) *Sedimentology and geochemistry of a mid-Proterozoic lacustrine unit from Northern Australia. Sedim. Geol.* **58**, 145–169.

Donnelly T. H., Shergold J. H., Southgate P. N., and Barnes C. J. (1990) Events leading to global phosphogenesis around the Proterozoic–Cambrian boundary. In *Phosphorite Research and Development*, Geological Society Special Publication No. 52 (eds. A. J. G. Notholt and I. Jarvis). London, pp. 273–287.

Dymek R. F. and Klein C. (1988) Chemistry, petrology and origin of banded iron-formation lithologies from the 3,800 Ma Isua supracrustal belt. *West Greenland. Precamb. Res.* **39**, 247–302.

Elderfield H., Wheat C. G., Mottle M. J., Monnin C., and Spiro B. (1999) Fluid and geochemical transport through oceanic crust: a transect across the eastern flank of the Juan de Fuca Ridge. *Earth Planet. Sci.* **172**, 151–165.

Elderfield H., Cooper M., and Ganssen G. (2000) Sr/Ca in multiple species of planktonic foraminifera: implications for reconstructions of seawater Sr/Ca. *Geochem. Geophys. Geosys.* **1**, paper number 1999GC000031.

Elderfield H., Vautravers M., and Cooper M. (2002) The relationship between cell size and Mg/Ca, Sr/Ca, $\delta^{18}O$ and $\delta^{13}C$ of species of planktonic foraminifera. *Geochem. Geophys. Geosys.* **3**(8), paper number 10.1029/2001GC000194.

Farquhar J., Bao H., and Thiemens M. (2000) Atmospheric influence of Earth's earliest sulfur cycle. *Science* **289**, 756–758.

Farquhar J., Savarino J., Airieau S., and Thiemens M. (2001) Observation of wavelength-sensitive mass-independent sulfur isotope effect during SO_2 photolysis: implications for the early atmosphere. *J. Geophys. Res.* **106**, 32829–32839.

Feather C. E. (1980) Some aspects of Witwatersrand mineralization with special reference to uranium minerals: Prof. Paper. *US Geol. Surv.,* 1161-M.

Fedo C. M. and Whitehouse M. J. (2002) Metasomatic origin of quartz-pyroxene rock, Akilia, Greenland, and implications for Earth's earliest life. *Science* **296**, 1448–1452.

François L. M. (1986) Extensive deposition of banded iron formations was possible without photosynthesis. *Nature* **320**, 352–354.

Francçois L. M. (1987) Reducing power of ferrous iron in the Archean ocean: 2. Role of $Fe(OH)^+$ photo-oxidation. *Paleoceanography* **2**, 395–408.

Froude D. O. (1983) Ion microprobe identification of 4,100–4,200 Myr old terrestrial zircons. Nature **304**, 616–618.

Fryer B. J. (1983) (1983) Rare-earth elements. In *Iron Formation: Facts and Problems* (eds. A. F. Trendall and R. C. Morris). Elsevier, Amsterdam, pp. 345–358.

Gaffin S. (1987) Ridge volume dependence on sea floor generation rate and inversion using long-term sea level change. *Am. J. Sci.* **287**, 596–611.

Garrels R. M. and Mackenzie F. T. (1971) *Evolution of Sedimentary Rocks*. W.W. Norton, New York.

Garrels R. M. and Perry E. C., Jr. (1974) Cycling of carbon, sulfur and oxygen through geologic time. In *The Sea* (ed. E. D. Goldberg). Wiley, New York, vol. 5, pp. 303–336.

Gellatly A. M. and Lyons T. W. (2005) Trace sulfate in mid-Proterozoic carbonates and the sulfur isotope record of biospheric evolution. *Geochim. Cosmochim. Acta* **69**, 3813–3829.

Given R. K. and Wilkinson B. H. (1987) Dolomite abundance and stratigraphic age: constraints on rates and mechanisms of Phanerozoic dolostone formation. *J. Sedim. Petrol.* **57**, 1068–1078.

Grandstaff D. E. (1976) A kinetic study of the dissolution of uraninite. *Econ. Geol.* **71**, 1493–1506.

Grandstaff D. E. (1980) Origin of uraniferous conglomerates at Elliott Lake, Canada and Witwatersrand, South Africa: implications for oxygen in the Precambrian atmosphere. *Precamb. Res.* **13**, 1–26.

Grotzinger J. P. (1989) Facies and evolution of Precambrian carbonate depositional systems: emergence of the modern platform archetype. In *Controls on Carbonate Platform and Basin Development,* Special Publication No. 44 (eds. P. D. Crevello, J. L. Wilson, J. F. Sarg, and J. F. Read). The Society of Economic Paleontologists and Mineralogists, Tulsa, UK, pp. 79–106.

Grotzinger J. P. and Kasting J. F. (1993) New constraints on Precambrian ocean composition. *J. Geol.* **101**, 235–243.

Grundl T. and Delwiche J. (1993) Kinetics of ferric oxyhydroxide precipitation. *J. Contam in. Hydrol.* **14**, 71–87.

Habicht K. S., Gade M., Thamdrup B., Berg P., and Canfield D. E. (2002) Calibration of sulfate levels in Archean ocean. *Science* **298**, 2372–2374.

Hallam A. (1984) Pre-quaternary sea level change. *Ann. Rev. Earth Planet. Sci.* **12**, 205–243.

Hallbauer D. K. (1986) The mineralogy and geochemistry of Witwatersrand pyrite, gold, uranium and carbonaceous matter. In *Mineral Deposits of Southern Africa* (eds. C. R. Anhaeusser and S. Maske). Geological Society of South Africa, Johannesburg. SA, pp. 731–752.

Halley E. (1715) A short account of the cause of the saltness of the ocean, and of the several lakes that emit no rivers; with a proposal, by help thereof, to discover the Age of the World. *Phil. Trans. Roy. Soc. London* **29**, 296–300.

Halliday A. N. (2000) Terrestrial accretion rates and the origin of the Moon. *Earth Planet. Sci. Lett.* **176**, 17–30.

Halliday A. N. (2001) In the beginning. . *Nature* **409**, 144–145.

Halverson G. P. (2003) Towards an integrated stratigraphic and carbon isotopic record for the Neoproterozoic. Doctoral Dissertation, Department of Earth and Planetary Sciences, Harvard University, Cambridge, MA.

Hambrey M. J. and Harland W. B. (1981) *Earth's Pre-Pleistocene Glacial Record*. Cambridge University Press, Cambridge.

Han T.-M. (1982) Iron formations of Precambrian age: hematite–magnetite relationships in some Proterozoic iron deposits—a microscopic observation. In *Ore Genesis—The State of the Art* (eds. G. C. Amstutz, A. E. Goresy, G. Frenzel, C. Kluth, G. Moh, A. Wauschkuhn, and R. A. Zimmermann). Springer, Berlin, pp. 451–459.

Han T.-M. (1988) Origin of magnetite in iron-formations of low metamorphic grade. In *Proceedings of the 7th Quadrennial IAGOD Symposium*, pp. 641–656.

Haq B. U., Hardenbol J., and Vail P. R. (1987) Chronology of fluctuating sea levels since the Triassic. *Science* **235**, 1156–1167.

Hardie L. A. (1967) The gypsum-anhydrite equilibrium at one atmosphere pressure. *Am. Mineral.* **52**, 171–200.

Hardie L. A. (1991) On the significance of evaporites. *Ann. Rev. Earth Planet. Sci.* **19**, 131–168.

Hardie L. A. (1996) Secular variation in seawater chemistry: an explanation for the coupled variation in the mineralogies of marine limestones and potash evaporites over the past 600 my. *Geology* **24**, 279–283.

Harrison A. G. and Thode H. G. (1958) Mechanisms of the bacterial reduction of sulfate from isotope fractionation studies. *Trans. Faraday Soc.* **53**, 84–92.

Hartmann W. K., Ryder G., Grinspoon D., and Dones L. (2000) The time-dependent intense bombardment of the primordial Earth/Moon system. In *Origin of the Earth and Moon* (eds.

R. M. Righter and K. Righter). University of Arizona Press, Tucson, pp. 493–512.

Harvie C. E., Weare J. H., Hardie L. A., and Eugster H. P. (1980) Evaporation of seawater: calculated mineral sequences. *Science* **208**, 498–500.

Hoehler T. M. and Alperin M. J. (1996) Anaerobic methane oxidation by a methanogen-sulfate reducer consortium: geochemical evidence and biochemical considerations. In *Microbial Growth in C1 Compounds* (eds. M. E. Lindstrom and F. R. Tabita). Kluwer Academic, San Diego.

Hoffman P. F. and Schrag D. P. (2002) The snowball earth hypothesis: testing the limits of global change. *Terra Nova* **14**, 129–155.

Hoffman P., Kaufman A. J., Halverson G. P., and Schrag D. P. (1998) A Neoproterozoic snowball Earth. *Science* **281**, 146–1342.

Holland H. D. (1962) Model for the evolution of the Earth's atmosphere. In *Petrologic Studies: A Volume to Honor A. F. Buddington* (eds. A. E. J. Engel, H. L. James, and B. F. Leonard). Geological Society of America, pp. 447–477.

Holland H. D. (1972) The geologic history of seawater: an attempt to solve the problem. *Geochim. Cosmochim. Acta* **36**, 637–651.

Holland H. D. (1973) Systematics of the isotopic composition of sulfur in the oceans during the Phanerozoic and its implications for atmospheric oxygen. *Geochim. Cosmochim. Acta* **37**, 2605–2616.

Holland H. D. (1984) *The Chemical Evolution of the Atmosphere and Oceans*. Princeton University Press, Princeton, NJ, 582p.

Holland H. D. (1994) Early Proterozoic atmospheric change. In *Early Life on Earth*, Nobel Symposium 84 (ed. S. Bengtson). Columbia University Press, New York, pp. 237–244.

Holland H. D. (1997) Evidence for life on Earth more than 3,850 million years ago. *Science* **275**, 38–39.

Holland H. D. (2002) Volcanic gases, black smokers, and the Great Oxidation Event. *Geochim. Cosmochim. Acta* **66**, 3811–3826.

Holland H. D. and Beukes N. J. (1990) A paleoweathering profile from Griqualand West, South Africa: evidence for a dramatic rise in atmospheric oxygen between 2.2 and 1.9 BYBP. *Am. J. Sci.* **290**, 1–34.

Holland H. D. and Zbinden E. A. (1988) Paleosols and the evolution of the atmosphere: Part I. In *Physical and Chemical Weathering in Geochemical Cycles* (eds. A. Lerman and M. Meybeck). Kluwer Academic, San Diego, pp. 61–82.

Holland H. D. and Zimmermann H. (1998) On the secular variations in the composition of Phanerozoic marine potash evaporites: Comment and reply. *Geology* **26**, 91–92.

Holland H. D. and Zimmermann H. (2000) The dolomite problem revisited. *Int. Geol. Rev.* **42**, 481–490.

Holland H. D., Horita J., and Seyfried W. E. (1996) On the secular variations in the composition of Phanerozoic marine potash evaporites. *Geology* **24**, 993–996.

Holser W. T. (1963) Chemistry of brine inclusions in Permian salt from Hutchinson, Kansas. In *Symposium on Salt (First)* (ed. A. C. Bersticker). Northern Ohio Geol. Soc., Cleveland, OH, pp. 86–95.

Holser W. T. and Kaplan I. R. (1966) Isotope geochemistry of sedimentary sulfates. *Chem. Geol.* **1**, 93–135.

Holser W. T., Schidlowski M., Mackenzie F. T., and Maynard J. B. (1988) Biogeochemical cycles of carbon and sulfur. In *Chemical Cycles in the Evolution of the Earth* (eds. C. B. Gregor, R. M. Garrels, F. T. Mackenzie, and J. B. Maynard). Wiley-Interscience, New York, pp. 105–173.

Holser W. T., Maynard J. B., and Cruikshank K. M. (1989) Modelling the natural cycle of sulphur through Phanerozoic time. In *Evolution of the Global Biogeochemical Sulphur Cycle* (eds. P. Brimblecombe and A. Y. Lein). Wiley, New York, chap 2, pp. 21–56.

Horita J., Friedman T. J., Lazar B., and Holland H. D. (1991) The composition of Permian seawater. *Geochim. Cosmochim. Acta* **55**, 417–432.

Horita J., Weinberg A., Das N., and Holland H. D. (1996) Brine inclusions in halite and the origin of the Middle Devonian Prairie Evaporites of Western Canada. *J. Sedim. Res.* **66**, 956–964.

Horita J., Zimmermann H., and Holland H. D. (2002) The chemical evolution of seawater during the Phanerozoic: implications from the record of marine evaporites. *Geochim. Cosmochim. Acta* **66**, 3733–3756.

Iglesias-Rodriguez M. D., Armstrong R. A., Feely R., Hood R., Kleypas J., Milliman J. D., Sabine C., and Sarmiento J. L. (2002) Progress made in study of ocean's calcium carbonate budget. *EOS* **83**, 365–375.

Iversen N. and Jørgensen B. B. (1985) Anaerobic methane oxidation rates at the sulfate-methane transition in marine sediments from Kattegat and Skagerrak (Denmark). *Limnol. Oceanogr.* **30**, 944–955.

Jackson M. J. and Raiswell R. (1991) Sedimentology and carbonsulfur geochemistry of the Velkerry Formation, a mid-Proterozoic potential oil source in Northern Australia. *Precamb. Res.* **54**, 81–108.

JacksonM. J., Muir M. D., and Plumb K. A. (1987) Geology of the southern McArthur Basin, Northern Territory. *Bureau of Mineral Resources, Geology and Geophysics*. Bulletin 220.

Jackson M. J., Muir M. D., and Plumb K. A. (1987) Geology of the southern McArthur Basin, Northern Territory. *Bureau of Mineral Resources, Geology and Geophysics*. Bulletin 220.

Jacobsen S. B. and Kaufman A. J. (1999) The Sr, C, and O isotopic evolution of Neoproterozoic seawater. *Chem. Geol.* **161**, 37–57.

James H. L. (1983) Distribution of banded iron-formation in space and time. In *Iron Formation: Facts and Problems* (eds. A. F. Trendall and R. C. Morris). Elsevier, Amsterdam, pp. 471–490.

James H. L. (1992) Precambrian iron-formations: nature, origin, and mineralogical evolution from sedimentation to metamorphism. In *Diagenesis III: Developments in Sedimentology* (eds. K. H. Wolf and G. V. Chilingarian). Elsevier, Amsterdam, pp. 543–589.

Johnson W. J. and Goldstein R. H. (1993) Cambrian seawater preserved as inclusions in marine low-magnesium calcite cement. *Nature* **362**, 335–337.

Joly J. (1899) An estimate of the geological age of the Earth. *Sci. Trans. Roy. Dublin Soc.* **7**(II), 23–66.

Kakegawa T., Kawai H., and Ohmoto H. (1998) Origins of pyrite in the .5 Ga Mt. McRae Shale, the Hamersley District, Western Australia. *Geochim. Cosmochim. Acta* **62**, 3205–3220.

Karhu J. A. (1993) Paleoproterozoic evolution of the carbon isotope ratios of sedimentary carbonates in the Fennoscandian Shield. *Geol. Soc. Finland Bull.* 371, 87.

Karhu J. A. and Holland H. D. (1996) Carbon isotopes and the rise of atmospheric oxygen. *Geology* **24**, 867–870.

Kasting J. F., Eggler D. H., and Raeburn S. P. (1993) Mantle redox evolution and the state of the Archean atmosphere. *J. Geol.* **101**, 245–257.

Kasting J. F., Pavlov A. A., and Siefert J. L. (2001) A coupled ecosystem-climate model for predicting the methane concentration in the Archean atmosphere. *Origin Life Evol. Biosphere* **31**, 271–285.

Kennedy M. J., Christie-Blick N., and Sohl L. E. (2001) Are Proterozoic cap carbonates and isotopic excursions a record of gas hydrate destabilization following Earth's coldest intervals? *Geology* **29**, 443–446.

Kinsman D. J. J. (1966) Gypsum and anhydrite of Recent age, Trucial Coast, Persian Gulf. (ed. J.L. Rau). The Northern Ohio Geological Society, Cleveland, OH, vol. 1, pp. 302–326.

Kirk J., Ruiz J., Chesley J., Titley S., and Walshe J. (2001) A detrital model for the origin of gold and sulfides in the Witwatersrand basin based on Re–Os isotopes. *Geochim. Cosmochim. Acta* **65**, 2149–2159.

Kirschvink J. L. (1992) Late Proterozoic low-latitude glaciation: the snowball earth. In *The Proterozoic Biosphere* (eds. J. W.

Schopf and C. Klein). Cambridge University Press, Cambridge, pp. 51–52., pp. 51–52.

Klein C. and Beukes N. J. (1989) Geochemistry and sedimentology of a facies transition from limestone to iron-formation deposition in the Early Proterozoic Transvaal Supergroup, South Africa. *Econ. Geol.* **84**, 1733–1774.

Klein C. and Beukes N. J. (1993) Sedimentology and geochemistry of the glaciogenic Late Proterozoic Rapitan iron-formation in Canada. *Econ. Geol.* **88**, 542–565.

Kleine T., Münker C., Mezger K., and Palme H. (2002) Rapid accretion and early core formation on asteroids and the terrestrial planets from Hf–W chronometry. *Nature* **418**, 952–955.

Knoll A. H. (1992) Biological and biogeochemical preludes to the Edeacaran radiation. In *Origin and Early Evolution of the Metazoa* (eds. J. H. Lipps and P. W. Signor). Plenum, New York, chap. 4.

Konhauser K. O., Hamade T., Raiswell R., Morris R. C., Ferris F. G., Southam G., and Canfield D. E. (2002) Could bacteria have formed the Precambrian banded iron formations? *Geology* **30**, 1079–1082.

Kump L. R. (1993) The coupling of the carbon and sulfur biogeochemical cycles over Phanerozoic time. In *Interactions of C, N, P, and S Biogeochemical Cycles and Global Change* (eds. R. Wollast, F.T. Mackenzie, and L. Chou). Springer, pp. 475–490., pp. 475–490.

Lasaga A. C., Berner R. A., and Garrels R. M. (1985) An improved geochemical model of atmospheric CO_2 fluctuations over the past 100 million years. In *The Carbon Cycle and Atmospheric CO_2, Natural Variations Archean to Present,* Geophysical Monograph 32 (eds. E. T. Sundquist and W. S. Broecker). American Geophysical Union, Washington, DC, pp. 397–411.

Lazar B. and Holland H. D. (1988) The analysis of fluid inclusions in halite. *Geochim. Cosmochim. Acta* **52**, 485–490.

Lear C. H., Elderfield H., and Wilson P. A. (2000) Cenozoic deep-sea temperatures and global ice volumes from Mg/Ca in benthic foraminiferal calcite. *Science* **287**, 269–272.

Lear C. H., Elderfield H., and Wilson P. A. (2003) A Cenozoic seawater Sr/Ca record from benthic foraminiferal calcite and its application in determining global weathering fluxes. *Earth Planet. Sci. Lett.* **208**, 69–84.

Leather J., Allen P. A., Brasier M. D., and Cozzi A. (2002) Neoproterozoic snowball Earth under scrutiny: evidence from the Fig glaciation of Oman. *Geology* **30**, 891–894.

Lithgow-Bertelloni C., Richards M. A., Ricard Y., O'Connell R. J., and Engebretson D. C. (1993) Toroidal–poloidal partitioning of plate motions since 120 Ma. *Geophys. Res. Lett.* **20**, 375–378.

Logan G. A., Hayes J. M., Hieshima G. B., and Summons R. E. (1995) Terminal Proterozoic reorganization of biogeochemical cycles. *Nature* **376**, 53–56.

Lohmann K. C. and Walker C. G. (1989) The $\delta^{18}O$ record of Phanerozoic abiotic marine calcite cements. *Geophys. Res. Lett.* **16**, 319–322.

Lowenstein T. K., Timofeeff M. N., Brennan S. T., Hardie L. A., and Demicco R. V. (2001) Oscillations in Phanerozoic seawater chemistry: evidence from fluid inclusions in salt deposits. *Science* **294**, 1086–1088.

Lugmaier G. W. and Shukolyukov A. (1998) Early solar system timescales according to $^{53}Mn–^{53}Cr$ systematics. *Geochim. Cosmochim. Acta* **62**, 2863–2886.

Lyons T. W., Gellatly A. M., and Kah L. C. (2002) Paleoenvironmental significance of trace sulfate in sedimentary carbonates. In *Abstracts Volume, 6th International Symposium on the Geochemistry of the Earth's Surface, May 20–24, 2002, Honolulu*, Hawaii, pp. 162–165.

Mackenzie F. T. and Garrels R. M. (1966) Chemical mass balance between rivers and oceans. *Am. J. Sci.* **264**, 507–525.

Melezhik V. A., Gorokhov I. M., Kuznetsov A. B., and Fallick A. E. (2001) Chemostratigraphy of neoproterozoic carbonates: implications for "blind" dating. *Terra Nova* **13**, 1–11.

Melezhik V. A., Fallick A. E., Rychanchik D. V., and Kuznetsov A. B. (2005) Palaeoproterozoic evaporites in Fennoscandia: implications for seawater sulphate, the rise of atmospheric oxygen and local amplification of the $\delta^{13}C$ excursion. *Terra Nova* **17**, 141–148.

Michaelis W., Seifert R., Nauhaus K., Treude T., Thiel V., Blumenberg M., Knittel K., Gieseke A., Peterknecht K., Pape T., Boetius A., Amann R., Jørgensen B. B., Widdel F., Peckmann J., Pimenov N., and Gulin M. (2002) Microbial reefs in the Black Sea fueled by anaerobic oxidation of methane. *Science* **297**, 1013–1015.

Millero F. J., Sotolongo S., and Izaguirre M. (1987) The oxidation kinetics of Fe(II) in seawater. *Geochim. Cosmochim. Acta* **51**, 793–801.

Mojzsis S. J., Arrhenius G., McKeegan K. D., Harrison T. M., Nutman A. P., and Friend C. R. L. (1996) Evidence for life on Earth before 3,800 million years ago. . *Nature* **384**, 55–59.

Mojzsis S. J., Harrison T. M., and Pidgeon R. T. (2001) (2001) Oxygen-isotope evidence from ancient zircons for liquid water at the Earth's surface 4,300 Myr ago. *Nature* **409**, 178–181.

Morse J., Millero F. J., Thurmond V., Brown E., and Ostlund H. G. (1984) The carbonate chemistry of Grand Bahama Bank waters: after 18 years another look. *J. Geophys. Res.* **89**, 3604–3614.

Morse J. W., Wang Q., and Tsio M.-Y. (1997) Influences of temperature and Mg: Ca ratio on $CaCO_3$ precipitates from seawater. *Geology* **25**, 85–87.

Mottl M. J. (2003) Partitioning of energy and mass fluxes between mid-ocean ridge axes and flanks at high and low temperature. In *Energy and Mass Transfer in Marine Hydrothermal Systems* (eds. P. E. Halbach, V. Tunnicliffe, and J. R. Hein). Dahlem University Press, Berlin, pp. 271–286.

Mottl M. J. and Wheat C. G. (1994) Hydrothermal circulation through mid-ocean ridge flanks: fluxes of heat and magnesium. *Geochim. Cosmochim. Acta* **58**, 2225–2237.

Muehlenbachs K. (1986) Alteration of the oceanic crust and the ^{18}O history of seawater. In *Stable Isotopes in High Temperature Geological Processes*, Reviews in Mineralogy (eds. J. W. Valley, H. P. Taylor, Jr., and J. R. O'Neil). Mining Society of America, Chelsea, MI, vol. 16, chap. **12**, pp. 425–444.

Muehlenbachs K. and Clayton R. N. (1976) Oxygen isotope composition of the oceanic crust and its bearing on seawater. *J. Geophys. Res.* **81**, 4365–4369.

Muir M. D. (1979) A sabkha model for deposition of part of the Proterozoic McArthur Group of the Northern Territory, and implications for mineralization. *BMR J. Austral. Geol. Geophys.* **4**, 149–162.

Myers J. S. (2001) Protoliths of the 3.8–3.7 Ga Isua greenstone belt, West Greenland. *Precamb. Res.* **105**, 129–141.

Naraoka H., Ohtake M., Maruyama S., and Ohmoto H. (1996) Non-biogenic graphite in 3.8-Ga metamorphic rocks from the Isua district, Greenland. *Chem. Geol.* **133**, 251–260.

Nelson D. R., Trendall A. F., and Altermann W. (1999) Chronological correlations between the Pilbara and Kaapvaal cratons. *Precamb. Res.* **97**, 165–189.

Nelson D. R., Robinson B.W., and Myers J. S. (2000) Complex geological histories extending for $\geq$4.0 Ga deciphered from xenocryst zircon microstructures. *Earth Planet. Sci.* **181**, 89–102.

Nutman A. P., Kinny P. D., Compston W., and Williams J. S. (1991) Shrimp U–Pb zircon geochronology of the Narryer Gneiss Complex, Western Australia. *Precamb. Res.* **52**, 275–300.

Nutman A. P., McGregor V. R., Friend C. R. L., Bennett V. C., and Kinny P. D. (1996) The Itsaq Gneiss complex of southern West Greenland: the world's most extensive record of early crustal evolution (3,900–3,600 Ma). *Precamb. Res.* **78**, 1–39.

Nutman A. P., Bennett V. C., Friend C. R. L., and Rosing M. T. (1997) 3,710 and >3,790 Ma volcanic sequences in the Isua (Greenland) supracrustal belt: structural and Nd isotope implications. *Chem. Geol. (Isotope Geosci.)* **141**, 271–287.

Nutman A. P., Friend C. R. L., and Bennett V. (2002) Evidence for 3,650–3,600 Ma assembly of the northern end of the Itsaq Gneiss Comples, Greenland: implication for early Archaean tectonics. *Tectonics* **21**, 1–28.

Oehler D. Z. and Smith J. W. (1977) Isotopic composition of reduced and oxidized carbon in early Archaean rocks from Isua, Greenland. *Precamb. Res.* **5**, 221–228.

Ohmoto H. (1996) Evidence in pre-2.2 Ga paleosols for the early evolution of the atmospheric oxygen and terrestrial biota. *Geology* **24**, 1135–1138.

Ohmoto H. (1999) Redox state of the Archean atmosphere: evidence from detrital heavy minerals in ca. 3,250–2,750 Ma sandstones from the Pilbara Craton, Australia: Comment. *Geology* **27**, 1151–1152.

Ohmoto H., Kakegawa T., and Lowe D. R. (1993) 3.4-billion-year-old biogenic pyrites from Barberton, South Africa: sulfur isotope evidence. *Science* **262**, 555–557.

Ono S. (2002) Detrital uraninite and the early Earth's atmosphere: SIMS analyses of uraninite in the Elliot Lake district and the dissolution kinetics of natural uraninite. Doctoral Dissertation, Pennsylvania State University, State College, PA.

Orphan V. J., House C. H., Hinrichs K. U., McKeegan K. D., and DeLong E. F. (2001) Methane-consuming archaea revealed by directly coupled isotopic and phylogenetic analysis. *Science* **293**, 484–487.

Palmer H. R., Pearson P. N., and Cobb S. J. (2000) Reconstructing past ocean pH-depth profiles. *Science* **282**, 1468–1471.

Pavlov A. A. and Kasting J. F. (2002) Mass-independent fractionation of sulfur isotopes in Archean sediments: strong evidence for an anoxic Archean atmosphere. *Astrobiology* **2**, 27–41.

Pavlov A. A., Kasting J. F., Brown L. L., Rages K. A., and Freedman R. (2000) Greenhouse warming by CH_4 in the atmosphere of early Earth. *J. Geophys. Res.* **105**, 11981–11990.

Pavlov A. A., Hurtgen M., Kasting J. F., and Arthur M. A. (2003) A methane-rich Proterozoic atmosphere? *Geology* **31**, 87–90.

Paytan A., Kastner M., Campbell D., and Thiemens M. (1998) Sulfur isotopic composition of Cenozoic seawater sulfate. *Science* **282**, 1459–1462.

Paytan A., Mearon S., Cobb K., and Kastner M. (2002) Origin of marine barite deposits: Sr and S characterization. *Geology* **30**, 747–750.

Pearson P. N. and Palmer M. R. (1999) Middle Eocene seawater pH and atmospheric carbon dioxide concentrations. *Science* **284**, 1824–1826.

Pearson P. N. and Palmer M. R. (2000) Atmospheric carbon dioxide concentrations over the past 60 million years. *Nature* **406**, 695–699.

Peck W. H., Valley J. W., Wilde S. A., and Geraham C. M. (2001) Oxygen isotope ratios and rare earth elements in 3.3 to 4.4 Ga zircons: ion microprobe evidence for high $\delta^{18}O$ continental crust and oceans in the early Archean. *Geochim. Cosmochim. Acta* **65**, 4215–4229.

Pegram W. J. and Turekian K. K. (1999) The Osmium isotopic composition change of Cenozoic seawater as inferred from a deep-sea core corrected for meteoritic contributions. *Geochim. Cosmochim. Acta* **63**, 4053–4058.

Perry E. C., Jr. and Ahmad S. N. (1977) Carbon isotope composition of graphite and carbonate minerals from the 3.8-AE metamorphosed sediments, Isukasia, Greenland. *Earth Planet. Sci. Lett.* **36**, 281–284.

Petsch S. T. and Berner R. A. (1998) Coupling the geochemical cycles of C, P, Fe, and S: the effect on atmospheric O_2 and the isotopic records of carbon and sulfur. *Am. J. Sci.* **298**, 246–262.

Peucker-Ehrenbrink B. and Ravizza G. (2000) The marine osmium isotope record. *Terra Nova* **12**, 205–219.

Peucker-Ehrenbrink B., Ravizza G., and Hofmann A. W. (1995) The marine $^{187}Os/^{186}Os$ record of the past 80 million years. *Earth Planet. Sci. Lett.* **130**, 155–167.

Phillips G. N., Law J. D. M., and Myers R. E. (2001) Is the redox state of the Archean atmosphere constrained? *Soc. Econ. Geologists SEG Newslett.* **47**(1), 9–18.

Pinto J. P. and Holland H. D. (1988) Paleosols and the evolution of the atmosphere: Part II. In *Paleosols and Weathering Through Geologic Time,* Geol. Soc. Am. Spec. Pap. 216 (eds. J. Reinhardt and W. Sigleo), pp. 21–34.

Plummer L. N. and Busenberg E. (1982) The solubilities of calcite, aragonite and vaterite in CO_2–H_2O solutions between 0° and 90°, and an evaluation of the aqueous model for the system $CaCO_3$–CO_2–H_2O. *Geochim. Cosmochim. Acta* **46**, 1011–1040.

Porcelli D., Cassen P., Woolum, D., and Wasserburg G. J. (1998) Acquisition and early losses of rare gases from the deep Earth. In *Origin of the Earth and Moon: Lunar Planetary Institute Contribution, Report 957,* pp. 35–36.

Porter S. M. and Knoll A. H. (2000) Testate amoebae in the Neoproterozoic era: evidence from vase-shaped microfossils in the Chuar Group, Grand Canyon. *Paleobiology* **26**, 360–385.

Porter S. M., Meisterfeld R., and Knoll A. H. (2003) Vase-shaped microfossils from the Neoproterozoic Chuar Group, Grand Canyon: a classification guided by modern Testate amoebae. *J. Paleontol.* **77**, 205–255.

Ramdohr P. (1958) Die Uran-und Goldlagerstätten Witwatersrand-Blind River District-Dominion Reef-Serra de Jacobina: erzmikroskopische Untersuchungen und ein geologischer Vergleich. *Abh. Deutschen Akad. Wiss Berlin* **3**, 1–35.

Rasmussen B. and Buick R. (1999) Redox state of the Archean atmosphere: evidence from detrital heavy minerals in ca. 3,250–2,750 Ma sandstones from the Pilbara Craton, Australia. *Geology* **27**, 115–118.

Rasmussen B., Buick R., and Holland H. D. (1999) Redox state of the Archean atmosphere: evidence from detrital heavy minerals in ca. 3250–2750Ma sandstones from the Pilbara Craton, Australia: Reply. *Geology* **27**, 1152.

Reeburgh W. S., Ward B. B., Whalen S. C., Sandbeck K. A., Kilpatrick K. A., and Kerkhof L. J. (1991) Black Sea methane geochemistry. *Deep Sea Res.* **38**(suppl. 2), 1189–1210.

Rosing M. T. (1999) ^{13}C-depleted carbon microparticles in >3,700-Ma sea-floor sedimentary rocks from West Greenland. *Science* **283**, 674–676.

Rosing M. T., Rose N. M., Bridgewater D., and Thomsen H. S. (1996) Earliest part of Earth's stratigraphic record: a reappraisal of the >3.7 Ga Isua (Greenland) supracrustal sequence. *Geology* **24**, 43–46.

Rowley D. B. (2002) Rate of plate creation and destruction: 180 Ma to present. *Geol. Soc. Am. Bull.* **114**, 927–933.

Rubey W. W. (1951) Geologic history of seawater, an attempt to state the problem. *Bull. Geol. Soc. Am.* **62**, 1111–1147.

Ryder G. (1990) Lunar samples, lunar accretion and the early bombardment of the Moon. *EOS, Trans., AGU,* **71**(10), 313, 322–323.

Rye R. and Holland H. D. (1998) (1998) Paleosols and the evolution of the atmosphere: a critical review. *Am. J. Sci.* **298**, 621–672.

Rye R., Kuo P. H., and Holland H. D. (1995) Atmospheric carbon dioxide concentration before 2.2 billion years ago. *Nature* **378**, 603–605.

Sandberg P. A. (1983) An oscillating trend in Phanerozoic non-skeletal carbonate mineralogy. *Nature* **305**, 19–22.

Sandberg P. A. (1985) Nonskeletal aragonite and pCO_2 in the Phanerozoic and Proterozoic. In *The Carbon Cycle and Atmospheric CO_2, Natural Variations Archean to Present,* Geophysical Monograph 32 (eds. E. T. Sundquist and W. S. Broecker) American Geophysical Union, Washington, DC, pp. 585–594.

Sanyal A., Hemming N. G., Broecker W. S., Lea D. W., Spero H. J., and Hanson G. N. (1996) Oceanic pH control on the boron isotopic composition of foraminifera: evidence from culture experiments. *Paleoceanography* **11**, 513–517.

Sarmiento J. L., Dunne J., Gnanadesikan A., Key R. M., Matsumoto K., and Slater R. (2002) A new estimate of the $CaCO_3$ to organic carbon export ratio. *Global Biogeochem. Cycles* **16**(4), 54-1–54-12.

Sasaki S. and Nakazawa K. J. (1986) Metal–silicate fractionation in the growing Earth: energy source for the terrestrial Magma Ocean. *J. Geophys. Res.* **91**, B9231–B9238.

Schidlowski M. (1966) Beiträge zur Kenntnis der radioactiven Bestandteile der Witwatersrand-Konglomerate. I Uranpecherz in den Konglomeraten des Oranje-Freistaat-Goldfeldes. *N. Jb. Miner Abh.* **105**, 183–202.

Schidlowski M., Appel P. W. U., Eichmann R., and Junge C. E. (1979) Carbon isotope geochemistry of the 3.7×10^9 yr-old Isua sediments, West Greenland: implications for the Archaean carbon and oxygen cycles. *Geochim. Cosmochim. Acta* **43**, 189–199.

Schidlowski M., Hayes J. M., and Kaplan I. R. (1983) Isotopic inferences of ancient biochemistries: carbon, sulfur, hydrogen, and nitrogen. In *Earth's Earliest Biosphere, Its Origin and Evolution* (ed. J. W. Schopf). Princeton University Press, Princeton, NJ, chap. 7, pp. 149–186.

Schopf J. W. (1983) Microfossils of the Early Archean Apex Chert: new evidence for the antiquity of life. *Science* **260**, 640–646.

Schopf J. W. and Packer B. M. (1987) Early Archean (3.3-billion to 3.5-billion-year-old) microfossils from Warrawoona Group, Australia. *Science* **237**, 70–73.

Schopf J. W., Kudryatsev A. B., Agresti D. G., Czaja A. D., and Widowiak T. J. (2002) Laser-Raman Imagery of the oldest fossils on Earth. In *Astrobiology Science Conference, NASA Ames Research Center, April* 7–11, 2002.

Sellwood B. W. (1986) Shallow marine carbonate environments. In *Sedimentary Environments and Facies* (ed. H. G. Reading). Blackwell, UK, chap. 10, pp. 283–342.

Shen Y., Buick R., and Canfield D. E. (2001) Isotopic evidence for microbial sulphate reduction in the early Archaean era. *Nature* **410**, 77–81.

Shen Y., Canfield D. E., and Knoll A. H. (2002) Middle Proterozoic ocean chemistry: evidence from the McArthur Basin, Northern Australia. *Am. J. Sci.* **302**, 81–109.

Shields G. and Veizer J. (2002) Precambrian marine carbonate isotope database: version 1.1. *Geochem. Geophys. Geosys.* **3**, doi:10.1029/2001GC000266.

Shimizu H., Umemoto N., Masuda A., and Appel P. W. U. (1990) Sources of iron-formations in the Archean Isua and Malene supracrustals, West Greenland: evidence from La–Ce and Sm–Nd isotopic data and REE abundances. *Geochim. Cosmochim. Acta* **54**, 1147–1154.

Sillén L. G. (1967) The ocean as a chemical system. *Science* **156**, 1189–1197.

Sleep N. H., Zahnle K. J., Kasting J. F., and Morowitz H. J. (1989) Annihilation of ecosystems by large asteroid impacts on the early Earth. *Nature* **342**, 139–142.

Southgate P. N., Bradshaw B. E., Domagala J., Jackson M. J., Idnurm M., Krassay A. A., Page R. W., Sami T. T., Scott D. L., Lindsay J. F., McConachie B. A., and Tarlowski C. (2000) Chronostratigraphic basin framework for Paleoproterozoic rocks (1,730–1,575 Ma) in northern Australia and implications for base-metal mineralization. *Austral. J. Earth Sci.* **47**, 461–483.

Spencer R. J. and Hardie L. A. (1990) Control of seawater composition by mixing of river waters and mid-ocean ridge hydrothermal brines. In *Fluid–Mineral Interactions: A Tribute to H. P. Eugster.* Special Publication 2 (eds. R. J. Spencer and I.-M. Chou). Geochemical Society, San Antonio, TX, pp. 409–419.

Spivack A. J., You C.-F., and Smith H. J. (1993) Foraminiferal boron isotope ratios as a proxy for surface ocean pH over the past 21 Myr. *Nature* **363**, 149–151.

Stanley S. M. and Hardie L. A. (1998) Secular oscillations in the carbonate mineralogy of reef-building and sediment-producing organisms driven by tectonically forced shifts in seawater chemistry. *Paleogeogr. Paleoclimatol. Paleoecol.* **144**, 3–19.

Stein C. L. and Krumhansl J. L. (1988) A model for the evolution of brines in salt from the lower Salado Formation, southeastern New Mexico. *Geochim. Cosmochim. Acta* **52**, 1037–1046.

Steuber T. and Veizer J. (2002) Phanerozoic record of plate tectonic control of seawater chemistry and carbonate sedimentation. *Geology* **30**, 1123–1126.

Strauss H. (1993) The sulfur isotopic record of Precambrian sulfates: new data and a critical evaluation of the existing record. *Precamb. Res.* **63**, 225–246.

Strauss H. (1999) Geological evolution from isotope proxy signals-sulfur. *Chem. Geol.* **161**, 89–101.

Strauss H., DesMarais D. J., Hayes J. M., and Summons R. E. (1992) The carbon-isotopic record. In *The Proterozoic Biosphere* (eds. J. W. Schopf and C. Klein). Cambridge University Press, Cambridge, chap. 3, pp. 117–127.

Stumm W. and Lee G. F. (1961) Oxygenation of ferrous iron. *Ind. Eng. Chem.* **53**, 143–146.

Stumm W. and Morgan J. J. (1966) *Aquatic Chemistry; Chemical Equilibria and Rates in Natural Waters*. Wiley-Interscience, New York.

Suess E. (1980) Particulate organic carbon flux in the oceans—surface productivity and oxygen utilization. *Nature* **288**, 260–263.

Summons R. E., Jahnke L. L., Hope J. M., and Logan G. A. (1999) 2-methylhopanoids as biomarkers for cyanobacterial oxygenic photosynthesis. *Nature* **400**, 554–557.

Thode H. G. and Monster J. (1965) Sulfur isotope geochemistry of petroleum, evaporites, and ancient seas. In *Fluids in Subsurface Environments, Mem. 4*. Am. Assoc. Petrol. Geol., Tulsa, OK, pp. 367–377.

Thode H. G., Monster J., and Sunford H. B. (1961) Sulfur isotope geochemistry. *Geochim. Cosmochim. Acta* **25**, 159–174.

Vail P. R., Mitchum R. W., and Thompson S. (1977) Seismic stratigraphy and global changes of sea level 4, Global cycles of relative changes of sea level. *AAPG Memoirs* **26**, 82–97.

Van Zuilen M. A., Lepland A., and Arrhenius G. (2002) Reassessing the evidence for the earliest traces of life. *Nature* **418**, 627–630.

Veizer J., Holser W. T., and Wilgus C. K. (1980) Correlation of $^{13}C/^{12}C$ and $^{34}S/^{32}S$ secular variations. *Geochim. Cosmochim. Acta* **44**, 579–587.

Veizer J., Hoefs J., Lowe D. R., and Thurston P. C. (1989) Geochemistry of Precambrian carbonates: II. Archean greenstone belts and Archean seawater. *Geochim. Cosmochim. Acta* **53**, 859–871.

Veizer J., Ala D., Azmy K., Bruckschen P., Buhl D., Bruhn F., Carden G. A. F., Diener A., Ebneth S., Godderis Y., Jasper T., Korte C., Pawellek F., Podlaha O. G., and Strauss H. (1999) $^{87}Sr/^{86}Sr$, $\delta^{13}C$ and $\delta^{18}O$ evolution of Phanerozoic seawater. *Chem. Geol.* **161**, 59–88.

Walker J. (1966) *Lectures on Geology.* The University of Chicago Press, Chicago, IL.

Walker R. N., Muir M. D., Diver W. L., Williams N., and Wilkins N. (1977) Evidence of major sulphate evaporite deposits in the Proterozoic McArthur Group, Northern Teritory, Australia. *Nature* **265**, 526–529.

Wallmann K. (2001) Controls on the Cretaceous and Cenozoic evolution of seawater composition, atmospheric CO_2 and climate. *Geochim. Cosmochim. Acta* **65**, 3005–3025.

Walter M. R., Veevers J. J., Calver C. R., Gorjan P., and Hill A. C. (2000) Dating the 840–544 Neoproterozoic interval by isotopes of strontium, carbon, and sulfur in seawater and some interpretative models. *Precamb. Res.* **100**, 371–433.

Werne J. P., Sageman B. B., Lyons T. W., and Hollander D. J. (2002) An intergrated assessment of a "type euxinic" deposit: evidence for multiple controls on black shale deposition in the Middle Devonian Oatka Creek formation. *Am. J. Sci.* **302**, 110–143.

Wilde S. A., Valley J. W., Peck W. H., and Graham C. M. (2001) Evidence from detrital zircons for the existence of continental crust and oceans on the Earth 4.4 Gyr ago. *Nature* **409**, 175–178.

Wilkinson B. H. and Algeo T. J. (1989) Sedimentary carbonate record of calcium-magnesium cycling. *Am. J. Sci.* **289**, 1158–1194.

Winefield P. R. (2000) Development of late Paleoproterozoic aragonitic sea floor cements in the McArthur Group, Northern Australia. In *Carbonate Sedimentation and Diagenesis in the Evolving Precambrian World*, SEPM Special Publication 67 (eds. J. P. Grotzinger and N. P. James). pp. 145–159.

Yang W. and Holland H. D. (2002) The redox sensitive trace elements Mo, U, and Re in Precambrian carbonaceous shales: indicators of the Great Oxidation Event (abstr.). *Geol. Soc. Am. Ann. Mtng.* **34**, 382.

Yang W. and Holland H. D. (2003) The Hekpoort paleosol profile in Strata 1 at Gaborone, Botswana: soil formation during the Great Oxidation Event. *Am. J. Sci.* **303**, 187–220.

Yin Q., Jacobsen S. B., Yamashita K., Blichert-Toft J., Télouk P., and Albarède F. (2002) A short timescale for terrestrial planet formation from Hf–W chronometry of meteorites. *Nature* **418**, 852–949.

Young G. M. (1969) Geochemistry of Early Proterozoic tillites and argillites of the Gowganda Formation, Ontario, Canada. *Geochim. Cosmochim. Acta* **33**, 483–492.

Young G. M. (1976) Iron-formation and glaciogenic rocks of the Rapitan Group, Northwest Territories, Canada. *Precamb. Res.* **3**, 137–158.

Zimmermann H. (2000) Tertiary seawater chemistry—implications from fluid inclusions in primary marine halite. *Am. J. Sci.* **300**, 723–767.

Readings from the Treatise on Geochemistry
ISBN: 978-0-12-381391-6

pp. 379–422

13
Geochemistry of Fine-grained Sediments and Sedimentary Rocks

B. B. Sageman
Northwestern University, Evanston, IL, USA
and
T. W. Lyons
University of California, Riverside, CA, USA

13.1 INTRODUCTION

The nature of detrital sedimentary (siliciclastic) rocks is determined by geological processes that occur in the four main Earth surface environments encountered over the sediment's history from source to final sink: (i) the site of sediment production (provenance), where interactions among bedrock geology, tectonic uplift, and climate control weathering and erosion processes; (ii) the transport path, where the medium of transport, gradient, and distance to the depositional basin may modify the texture and composition of weathered material; (iii) the site of deposition, where a suite of physical, chemical, and biological processes control the nature of sediment accumulation and early burial modification; and (iv) the conditions of later burial, where diagenetic processes may further alter the texture and composition of buried sediments. Many of these geological processes leave characteristic geochemical signatures, making detrital sedimentary rocks one of the most important archives of geochemical data available for reconstructions of ancient Earth surface environments. Although documentation of geochemical data has long been a part of the study of sedimentation (e.g., Twenhofel, 1926, 1950; Pettijohn, 1949; Trask, 1955), the development and application of geochemical methods specific to sedimentary geological problems blossomed in the period following the Second World War (Degens, 1965; Garrels and Mackenzie, 1971) and culminated in recent years, as reflected by the publication of various texts on marine geochemistry (e.g., Chester, 1990, 2000), biogeochemistry (e.g., Schlesinger, 1991; Libes, 1992), and organic geochemistry (e.g., Tissot and Welte, 1984; Engel and Macko, 1993).

Coincident with the growth of these subdisciplines a new focus has emerged in the geological sciences broadly represented under the title of "Earth System Science" (e.g., Kump *et al.*, 1999). Geochemistry has played the central role in this revolution (e.g., Berner, 1980; Garrels and Lerman, 1981; Berner *et al.*, 1983; Kump *et al.*, 2000), with a shifting emphasis toward sophisticated characterization of the linkages among solid Earth, oceans, biosphere, cryosphere, atmosphere, and climate, mediated by short- and long-term biogeochemical cycles. As a result, one of the primary objectives of current geological inquiry is improved understanding of the interconnectedness and associated feedback among the cycles of carbon, nitrogen, phosphorous, oxygen, and sulfur, and their relationship to the history of Earth's climate. This "Earth System" approach involves uniformitarian extrapolations of knowledge gained from modern environments to proxy-based interpretations of environmental change recorded in ancient strata. The strength of modern data lies with direct observations of pathways and products of physical, chemical, and biological processes, but available time-series are short relative to the response times of many of the biogeochemical systems under study. By contrast, stratigraphically constrained geological data offer time-series that encompass a much fuller range of system response. But with the enhanced breadth of temporal resolution and signal amplitude provided by ancient sedimentary records comes a caveat—we must account for the blurring of primary paleo-environmental signals by preservational artifacts and understand that proxy calibrations are extended from the modern world into a nonsubstantively uniformitarian geological past.

Fortunately, detrital sedimentary rocks preserve records of multiple proxies (dependent and independent) that illuminate the processes and conditions of sediment formation, transport, deposition, and burial. An integrated multiproxy approach offers an effective tool for deconvolving the history of biogeochemical cycling of, among other things, carbon and sulfur, and for understanding the range of associated paleo-environmental conditions (e.g., levels of atmospheric oxygen and carbon dioxide, oceanic paleoredox, and paleosalinity). Authors of a single chapter can hope, at best, to present a cursory glance at the many biogeochemical proxies currently used and under development in sedimentary studies. Our goal, instead, is to focus on a selected suite of tools of particular value in the reconstruction of paleoenvironments preserved in fine-grained siliciclastic sedimentary rocks.

Fine-grained, mixed siliciclastic–biogenic sedimentary facies—commonly termed hemipelagic (mainly calcareous or siliceous mudrocks containing preserved organic matter (OM))—are ideal for unraveling the geological past and are thus the focus of this chapter. These strata accumulate in predominantly low-energy basinal environments where the magnitude (and frequency) of lacunae is diminished, resulting in relatively continuous, though generally condensed sequences. Fortunately, condensation tends to benefit geochemical analysis as it helps to amplify some subtle environmental signals. Because hemipelagic facies include contributions from both terrigenous detrital and pelagic biogenic systems, as well as from authigenic components reflecting the burial environment (Figure 1), they are rich archives of geochemical information. In this chapter we present a conceptual model linking the major processes of detrital, biogenic, and authigenic accumulation in fine-grained hemipelagic settings. This model is intended to be a fresh synthesis of decades of prior research on the geochemistry of modern and ancient mudrocks, including our own work.

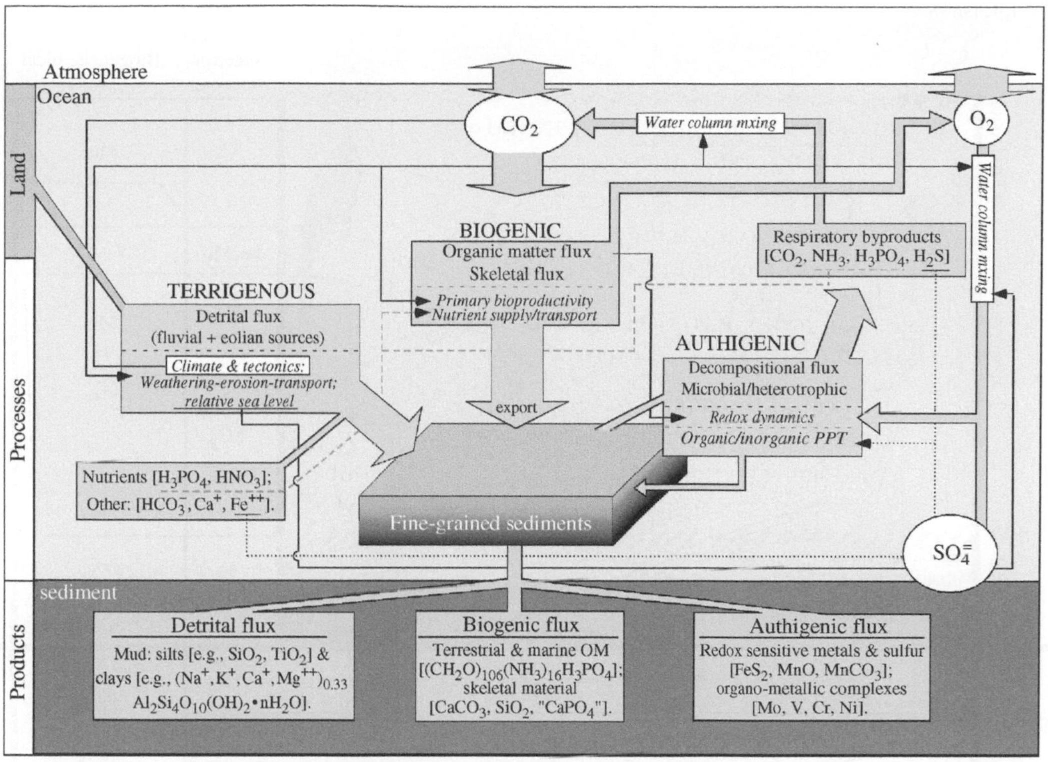

Figure 1 Conceptual model for the origin of mixed detrital–biogenic facies relating the three major inputs to the processes that control them. The major inputs are shown in boxes with bold-type labels. Controlling factors are shown in italics. Large and medium scale arrows represent fluxes of key components involved in sedimentation and the biogeochemical cycles of carbon, sulfur, and oxygen. Thin arrows illustrate relationships between major controlling factors and depositional processes and/or feedback. Dashed thin arrows apply to major nutrient fluxes only. Dotted thin arrows apply to major authigenic fluxes only. See text for further explanation.

In the sections that follow we will first develop and illustrate (Figures 1 and 2) the conceptual model and the proxy methods incorporated within this model—with additional details provided in the cited literature. The remainder of the chapter is devoted to demonstrating the utility of the model through a series of case studies ranging from the modern Black Sea to shales and argillites of the Precambrian. Based on these case studies, we conclude with a summary of similarities and differences in proxy application.

13.2 CONCEPTUAL MODEL—PROCESSES

A schematic representation of our model for the major inputs and feedback involved in the formation of fine-grained, mixed siliciclastic–biogenic facies in marine basins is shown in Figure 1. The major inputs include terrigenous detritus, such as material derived from weathering of continental crust or from volcanic sources, biogenic components (both OM and mineralized microskeletal remains) derived from primary photosynthetic production and heterotrophic processes on land and in the sea, and authigenic material precipitated at or near the sediment-water interface as a consequence of Eh-pH-controlled organic and inorganic reactions. Although these major inputs have been recognized for many years (e.g., see reviews by Potter *et al.*, 1980; Gorsline, 1984; Arthur and Sageman, 1994; Wignall, 1994; Hedges and Keil, 1995; Tyson, 1995; Schieber *et al.*, 1998a,b; Chester, 2000, and references therein), the synthesis in Figure 1 is novel in that it integrates physical (sedimentologic and oceanographic) and biogeochemical processes, relates major inputs to proximate and broader paleo-environmental controls, illustrates important linkages between causes and effects in the model (i.e., feedback), and, lastly, identifies and tracks key components of the major biogeochemical cycles (C, O, S, N, and P) involved in regulating conditions at the Earth's surface.

Climate and plate tectonics are the master controlling factors for the system represented in Figure 1. Climate includes a complex set of phenomena (temperature, evaporation, precipitation, and wind) and interactions among the atmosphere, land surface, ocean surface, biosphere, and cryosphere that are driven largely by variations in the amount and distribution of incoming solar

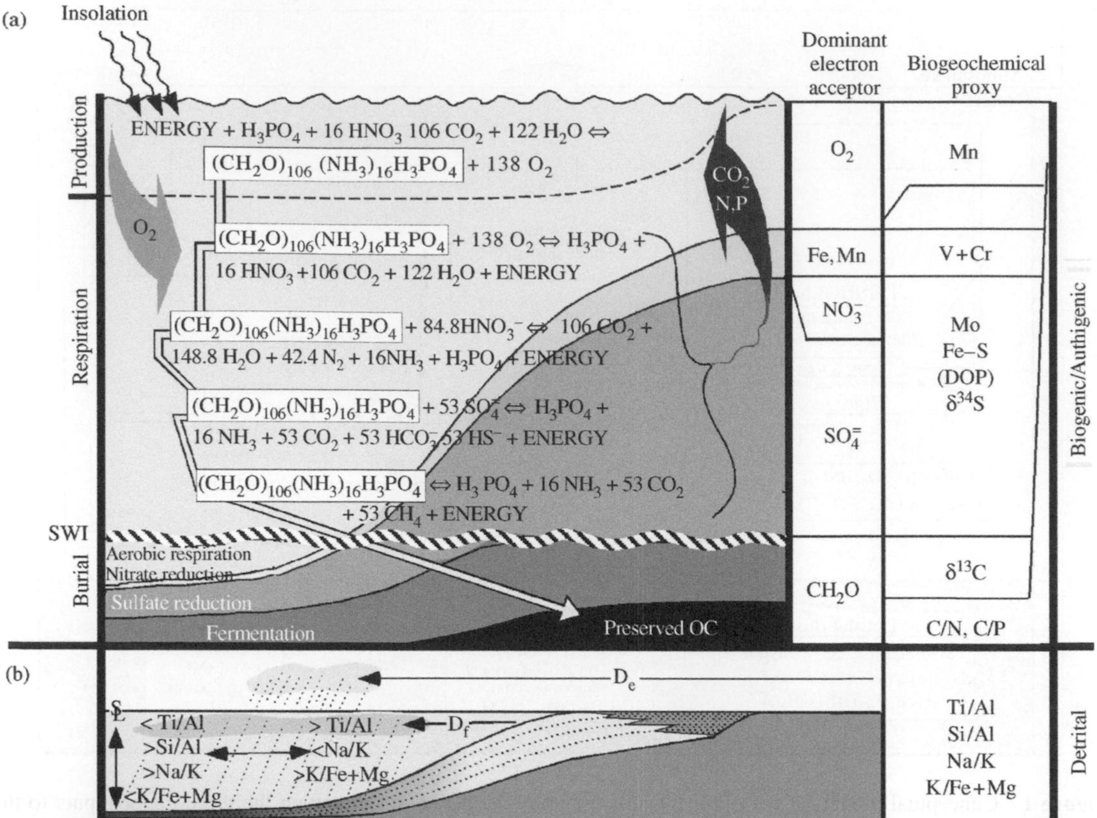

Figure 2 (a) Details of the interrelationship between biogenic and authigenic processes in Figure 1 are shown. Sequential steps in the remineralization of photosynthetically produced OM (represented by gray arrow) are represented by simplified chemical reactions. For each step in which a given electron acceptor is dominant, one or more characteristic biogeochemical proxies may accumulate (in some cases proxies accumulate across a range of conditions; in others the range is narrower than the text font: such cases are represented by slanted borders). The rate at which OM moves through the steps depends on availability of electron acceptors and bulk sedimentation rate, which may be evaluated using detrital proxies. (b) Cartoon illustrating sources of terrigenous flux and proxies for their detection. See text for further discussion.

radiation. Tectonic processes, by contrast, can be simplified to two parameters. These are vertical uplift, which creates crustal source areas for weathering and erosion, and subsidence, which together with eustasy acts to control the accommodation space available for accumulation of sediments (e.g., Sloss, 1962).

13.2.1 Detrital Flux

Climate and tectonics, particularly the interaction between uplift of continental crust and physical/chemical weathering, controls the generation of terrigenous detritus (sand, silt, and clay), as well as dissolved species (e.g., cations and anions such as Na^+, Ca^{2+}, SO_4^{2-}, Cl^-, and HCO_3^-), many of which play important roles in biogeochemical cycling (Figure 1). The chemical composition of the detrital fraction is largely determined by the mineralogy of the source rocks, the weathering regime, and the reactivity of weathering products during transport. Regional climatic conditions determine weathering regimes (temperature, precipitation, and runoff), and for a given weathering environment, uplift rate and drainage area determine net detrital flux or sediment yield into a basin of deposition (Milliman and Syvitski, 1992; Perlmutter *et al.*, 1998). Tectonic activity drives source area uplift to create relief, and on a global scale also influences long-term eustasy, which together with local subsidence will determine the bathymetric profile of a given sedimentary basin, as well as its relative sea-level history. Geography (latitude and basin orientation) and bathymetric profile combine with climate to determine the oceanographic character of a basin (e.g., circulation), which, in turn, controls delivery of the detrital fine fraction to distal depositional sites and strongly influences biogenic and authigenic processes. The ratio of accommodation (whether controlled by eustasy, local tectonics, or both) to sediment supply is one of the most important regulators of particulate detrital flux to a basin (e.g., Loutit *et al.*, 1988; Pasley *et al.*, 1991) because it controls the extent to which river-transported materials are stored in estuaries or transported beyond the "littoral energy fence" (Allen, 1970; Swift and Thorne, 1991), where they are dispersed by the shelfal transport system.

In general, transport of weathered material is directly mediated by runoff and wind-driven waves and currents. Both of these are controlled by climatic parameters and exhibit broadly predictable latitudinal patterns. Although runoff is the most volumetrically significant agent of detrital and chemical transport to the oceans, under certain circumstances windblown components can have a major influence on sedimentary composition and geochemical interpretations of detrital fluxes (e.g., Bertrand *et al.*, 1996). Once fine-grained material enters a marine basin, for example, from a riverine source, its transport is controlled by surface circulation, and deposition occurs by fall-out from suspension (Gorsline, 1984) often mediated by biologically and chemically induced aggregation. Large sediment plumes associated with major rivers characterize supply-dominated systems (Swift and Thorne, 1991) and are well known from modern observations (e.g., the Amazon).

Although good modern analogues for ancient epicratonic basins do not exist today, it is assumed that turbid plumes of muddy sediment blanketed such ancient settings (e.g., Pratt, 1984). There is, however, relatively little known about the dispersal of fine-sediment in ancient depositional systems beyond what can be implied by trends of increasing thickness, grain size, etc., toward dominant sediment sources (e.g., Elder, 1985). Sediment gravity processes, when bathymetric topography is sufficient (Gorsline, 1984), or impingement of storm wave base during large storms or episodes of decreased relative sea level can redistribute fine-grained sediment into thicker and thinner accumulations on the basin floor. These events can result in winnowing of clays and progressive concentration of coarser grained pelagic and benthic skeletal components (Sageman, 1996) or eolian-derived particles, thus altering the geochemical proportions of the detrital fraction. It is also well known (e.g., in the modern Black Sea; Lyons, 1991) that fine sediment is transported over large lateral distances along isopycnal surfaces within well-stratified water columns, often as a result of internal wave activity along the density interfaces.

The major elemental components of the net particulate flux from modern fluvial sources, in order of decreasing magnitude, are Si, Al, Fe, Ca, K, Mg, Na, Ti, P, Mn, and Ba (Martin and Whitfield, 1983; see also Bewers and Yeats, 1979; Martin and Maybeck, 1979; Chester and Murphy, 1990; Chester *et al.*, 1997). Similar proportions among the most abundant components (e.g., Si, Al, and Fe) generally characterize the mineralogy of the eolian dust flux (e.g., Lantzy and Mackenzie, 1979), but there can be significant regional variations in elemental concentrations depending on the composition of source areas (e.g., Chester, 2000). For example, titanium concentrations can be significantly elevated over average crustal values in eolian fluxes from arid regions (Bertrand *et al.*, 1996; Yarincik *et al.*, 2000b). Deviations from the crustal mean also characterize areas with a high volcanigenic contribution to the atmospheric flux (Weisel *et al.*, 1984; Arthur and Dean, 1991). Silicon is often involved in biogenic processes (see below), but aluminum and titanium are regarded as the most refractory products of crustal weathering (Taylor and McLennan, 1985), except under extreme weathering conditions (Young and Nesbitt, 1998).

13.2.2 Biogenic Flux

Photosynthetic production of OM and associated skeletal material defines the second major input to fine-grained, mixed siliciclastic–biogenic facies. Sedimentary organic carbon (C_{org}) has two major sources: terrestrial and marine. About 65% of the total terrestrial carbon flux survives as recalcitrant OM and can be incorporated in marine sediments (Ittekkot, 1988). In general, the concentration of this material varies as a function of proximity of the fluvial source and the dispersion processes. However, some recent studies have shown that much of this material is transported as sorbed phases on mineral surfaces, which can be shed in the nearshore zone and replaced by marine OM (e.g., Keil and Hedges, 1993; Mayer, 1994; Leithold and Blair, 2001). This marine C_{org} is far more labile and thus provides the dominant substrate for microbial decomposers, and through most of the marine realm its export to surface sediments is controlled by surface-water bioproductivity, heterotrophic "repackaging" (as fecal pellets with rapid down-column transit rates; Suess, 1980; Degens and Ittekkot, 1987), and water depth, which regulates the extent of degradation during export through the water column (e.g., Suess, 1980). In ancient epicratonic seaways, water depths probably varied from less than 100 m to not significantly more than 300 m. Consequently, decomposition during down-column transport was appreciably lower than for open-ocean settings (e.g., 61% versus 99% loss; Suess, 1980). Distal hemipelagic deposits in these basins commonly received minimal terrigenous OM, with marine phytoplanktonic and bacterial fractions dominating the sedimentary C_{org}.

The fundamental control on the production of marine OM and associated skeletal material

($CaCO_3$ and SiO_2) is the concentration and temporal continuity of biolimiting nutrients in surface waters (e.g., nitrate, phosphate, and micronutrients such as iron). In general, these are produced by chemical weathering and biogenic activities on land and transported via riverine or eolian processes to the marine realm, or they may be produced by biological processes within the water column (i.e., N-fixation). Although phosphate is transported in both dissolved (orthophosphate) and particulate (organic and inorganic) phases, only the dissolved form is readily bioavailable. However, preferential remineralization of particulate nutrients due to, for example, microbially mediated reactions under oscillating redox conditions (Aller, 1994; Ingall and Jahnke, 1997) may enhance nutrient recycling (Tyson and Pearson, 1991). In general, nitrogen is thought to be effective as a biolimiting nutrient on short (ecological) timescales, while phosphorus availability limits marine bioproductivity on geological timescales (Broecker and Peng, 1982; Berner and Canfield, 1989; Van Cappellen and Ingall, 1994; Falkowski, 1997; Tyrrell, 1999). In modern open-ocean settings the surface-water nutrient inventory is generally low, and upwelling zones represent major sites of enhanced bioproductivity. Nutrients remineralized in the deeper water column are actively recycled to the surface, thus driving new production. Climate exerts control on the biogenic flux by regulating nutrient supply, either through its effect on chemical weathering, soil biology, and rates of riverine transport, or by controlling water-column stability and wind-driven (Ekman-transport-induced) upwelling.

The pelagic biogenic flux includes carbon, nitrogen, and phosphorus in OM, which under conditions of exceptional preservation may be buried in proportions approximating the algal Redfield ratio (Redfield *et al.*, 1963; Murphy *et al.*, 2000a) but more commonly is enriched in carbon relative to nitrogen and phosphorus due to preferential nutrient release (Van Cappellen and Ingall, 1994; Aller, 1994; Ingall and Jahnke, 1997). Similarly, $CaCO_3$ accumulation can directly reflect primary productivity. Because epicratonic settings were sufficiently shallow to allow carbonate deposition with little or no dissolution, relations between $CaCO_3$ accumulation and total organic production in modern oceans (i.e., Broecker, 1982) may provide a means to reconstruct paleoproductivity in ancient hemipelagic deposits (Meyers *et al.*, in review).

13.2.3 Authigenic Flux

The third major input of material to fine-grained, mixed siliciclastic–biogenic facies is associated with authigenic processes that occur as a consequence of OM remineralization (bacterial and macrofaunal heterotrophy) in uncompacted sediments (Figure 1). In cases where the supply of O_2 exceeds the demand of microbial and macrofaunal respiration, most OM is efficiently oxidized by aerobes to dissolved inorganic C (ΣCO_2) and other byproducts, such as PO_4^{3-} and ultimately NO_3^-, which can return to surface waters and/or the atmosphere during water-column mixing events (Figures 1 and 2). If such conditions predominate on short to intermediate geological timescales (centuries to 1 Myr), the global carbon cycle remains in relative equilibrium. When O_2 supply fails to meet respiratory demand (e.g., due to diminished O_2 advection in a stratified water column, excess O_2 demand in the water column and sediments from enhanced production, or limitations in diffusional O_2 replenishment), oxygen is depleted and alternate terminal electron acceptors are employed in the remineralization of OM by a series of dysaerobic to anaerobic microbial communities (Figure 2); the dominant metabolic processes are nitrate reduction, manganese and iron reduction, sulfate reduction, and fermentation (methanogenesis), which represent progressively less efficient (i.e., less energy yielding per mole of C_{org} oxidized) forms of respiration (Froelich *et al.*, 1979; Berner, 1980; Canfield and Raiswell, 1991). While the relative net efficiencies of aerobic and anaerobic remineralization are debated (as discussed below), some fraction of the C_{org} is buried, and its removal from the short-term part of the carbon cycle may have direct consequences for climate change (i.e., Arthur *et al.*, 1988).

Nutrient release during OM remineralization appears to depend on the dominant type of bacterial decomposition, as well as its frequency and duration (e.g., Ingall *et al.*, 1993) and is thus related to the redox state of the system. Decoupled elemental release from OM (enhanced phosphorus regeneration relative to carbon) has been shown to occur during decomposition in sediments overlain by O_2-deficient bottom waters (e.g., Ingall and Jahnke, 1997) and may be particularly pronounced for nitrate under oscillating bottom-water redox conditions (Aller, 1994). In modern oceans, preferential nutrient release under normal aerobic conditions results in OM being buried with organic C : P ratios of ~250 : 1, compared to the assimilation ratio of ~106 : 1 (Van Cappellen and Ingall, 1994). The ratios increase dramatically under anoxic depositional conditions (Ingall *et al.*, 1993). Information about the extent to which nitrogen is preferentially regenerated from OM in natural settings is not abundant, as $C_{org} : N_{total}$ ratios are typically interpreted in terms of OM source rather than preservation state (e.g., Meyers, 1994). However, there is evidence that $C_{org} : N_{total}$ ratios increase with depth in sediments independent of source variation (e.g., Stevenson and Cheng, 1972), suggesting that postdepositional release of nitrogen may be enhanced by prolonged exposure to active microbial degradation.

In modern deep oceans, the aerobic zone dominates water columns and often the substrates, and most OM is remineralized by aerobic respiration during export, with less than 1–2% reaching the seafloor (Suess, 1980). This loss reflects the great depth of the oceanic water column, as well as vigorous resupply of O_2 due to thermohaline circulation. By contrast, up to 40% of primary production reaches substrates in shelfal areas where depths are ±100 m (and by inference shallow epicontinental seaways), even though water columns are fully oxic (Suess, 1980). However, most of this OM is subsequently remineralized by aerobic and anaerobic bacteria and other heterotrophic organisms in the sediments. Consequently, net carbon burial remains very low, but the percent of the flux to the sediment–water interface that becomes buried and preserved beyond early diagenetic remineralization also varies sympathetically with rates of bulk sediment accumulation (Stein, 1986; Henrichs and Reeburgh, 1987; Canfield, 1989, 1994; Betts and Holland, 1991). In open oceanic water columns, nitrate reduction may occur at mid-water depths where the decomposition of descending particulate OM depletes available oxygen (Falkowski, 1997), but thermohaline circulation tends to resupply oxygen to bottom waters, ensuring the near-complete remineralization of pelagic OM. Upwelling zones with high fluxes of recycled nutrients and enhanced production are characterized by benthic anoxia and high burial fluxes of C_{org}, but they tend to be nonsulfidic—suggesting that oxygen and nitrate are resupplied by currents at levels just sufficient to prevent sulfate reduction (Arthur *et al.*, 1998). Well-documented occurrences of water-column sulfate reduction and high sulfide concentrations are known to occur today only in highly restricted (silled) basins with relatively isolated bottom waters and predominantly stratified water columns (Black Sea, Fjords), and/or within restricted settings beneath fertile surface waters (Cariaco Basin, California borderland basins) (e.g., Rhoads and Morse, 1971; Demaison and Moore, 1980; Pedersen and Calvert, 1990; Werne *et al.*, 2000; Lyons *et al.*, 2003).

Ultimately, the redox state of a depositional system represents a dynamic balance between supply of electron acceptors and their consumption during OM remineralization. The key controls include water-column mixing rate, which is influenced by relative sea level and climate, and OM production and export rates (plus feedback expressed in water-column redox and associated effects on OM preservation—see Section 13.2.4) controlled mainly by nutrient supply and the biology of the planktic biota. Because the redox state of a depositional system regulates various organic and inorganic reactions by which dissolved constituents are precipitated as geologically preservable minerals (e.g., metal oxyhydroxides versus pyrite) and added to the burial flux, it is possible to reconstruct the history of changes in redox. Examples of these constituents include Fe, Mn, V, Cr, Mo, and U.

13.2.4 Carbon Cycle and Climate Feedback

In addition to the intensity of incoming solar radiation, the surface temperature on Earth is regulated by the warming effect of greenhouse gases in the atmosphere, in particular CO_2, and this relationship is influenced by several negative and positive feedback loops (e.g., Berner, 1999). The inorganic part of the carbon cycle (weathering of silicates by carbonic acid to produce Ca^{2+} and HCO_3^-, deposition of $CaCO_3$ in carbonate rocks, and metamorphic recycling of CO_2 from subducted $CaCO_3$) stores and transfers the largest masses of carbon over very long timescales (Berner, 1999). However, the focus of this chapter is fine-grained detrital sediments and sedimentary rocks. Although $CaCO_3$ can be volumetrically important in these fine-grained deposits, these dominantly siliciclastic facies are the primary reservoir for burial of C_{org} in both modern and ancient marine environments. Unlike inorganic (carbonate) carbon, this organic carbon pool has the potential to produce large and rapid perturbations in the carbon cycle, with concomitant climatic effects (e.g., Arthur *et al.*, 1988; Kump and Arthur, 1999; Berner and Kothavala, 2001).

Viewed in the context of Figure 1, when C_{org} fixation by photoautotrophs (biogenic flux) is not matched by sediment and water-column decomposition by bacteria and other heterotrophs, significant masses of C_{org} can be buried in detrital-biogenic and pelagic sediments. If this occurs over sufficiently large areas, geologically rapid and/or prolonged changes in atmospheric p_{CO2} may result (e.g., Arthur *et al.*, 1988; Berner and Canfield, 1989; Kump and Arthur, 1999; Berner *et al.*, 2000; Berner, 2001; Berner and Kothavala, 2001). Interpreted episodes of widespread carbon burial, such as the Miocene Monterey event (Vincent and Berger, 1985), Cretaceous oceanic anoxic events (Schlanger and Jenkyns, 1976; Schlanger *et al.*, 1987), the Permo-Triassic "superanoxia" event (Isozaki, 1997), and the Late Devonian Kellwasser events (Joachimski and Buggisch, 1993), span the geological record and are thought to be responsible for major changes in climate and biotic extinction/evolution. Because of this, the factors controlling C_{org} burial have long been topics of intense interest.

Many past studies have attributed enhanced OM burial to the relative inefficiency of anaerobic bacterial respiration (i.e., Demaison and Moore, 1980). In this model the driving mechanism of carbon burial is pervasive anoxia in paleo-oceans and shallow seas, attributed respectively to sluggish thermohaline circulation during warm climate intervals (e.g., Schlanger and Jenkyns, 1976) or enhanced salinity stratification because of high freshwater flux to shallow epicontinental basins (e.g., Seilacher, 1982; Ettensohn, 1985a,b). More recently, the notion of permanent stratification in shallow epeiric seas associated with the "stagnant basin" model has been challenged by a model invoking dynamic, seasonal (thermal?) stratification (e.g., Oschmann, 1991; Sageman and Bina, 1997; Murphy *et al.*, 2000a; Sageman *et al.*, 2003) based, in part, on observations of modern shallow marine systems (Tyson and Pearson, 1991). At about the same time that this new "dynamic stratification" model was being developed, evidence suggesting that increase in the biogenic flux (i.e., greater productivity) played a significant role in enhanced OM burial was accumulating (e.g., Arthur *et al.*, 1987). Ultimately, Pedersen and Calvert (1990) cited evidence for similar rates of aerobic and anaerobic degradation (e.g., unexceptional C_{org} preservation in the modern Black Sea; Calvert *et al.*, 1991), as well as patterns of accumulation in modern high-productivity zones that argue for enhanced production as the main driver of ancient OM burial. In their view anoxic conditions were a consequence rather than a cause of OM accumulation.

Following another decade of research, neither the "preservation" nor "production" mechanism has emerged as a comprehensive paradigm. Publication of new observational and experimental data has suggested that differential degradation of OM does occur under variable redox conditions (Hartnett *et al.*, 1998; Van Mooy *et al.*, 2002), providing support for the preservation end-member. More recently, secondary C_{org} loss in Mediterranean sapropels associated with downward advancing oxidation fronts—following transitions from anoxic to oxic bottom water conditions—has indicated the role of O_2 exposure in diminished OM preservation (Thomson *et al.*, 1999; Slomp *et al.*, 2002; and reference therein). There are, however, also cases in which increased production without coeval evidence for significant anoxicity appears to account for enhanced OM burial (Meyers *et al.*, 2001), thus providing support for the production end-member. Whereas both production and preservation factors appear to be critical components, depending on circumstances to be explored herein, the role of varying bulk sedimentation rate (i.e., Johnson-Ibach, 1982; Henrichs and Reeburgh, 1987; Mueller and Suess, 1979) has received comparatively less attention. Yet, this process regulates OM concentration and controls the rate at which OM is transported from the oxic zone into underlying suboxic to anoxic zones, where the efficiency of decomposition is reduced, and thus ultimately impacts the extent of OM preservation (Toth and Lerman, 1977; Canfield, 1994).

13.3 CONCEPTUAL MODEL: PROXIES

Two fundamental types of geochemical proxies are employed in the reconstruction of ancient depositional environments and geological processes. Broadly defined, these are elemental and compound concentration/accumulation data and stable isotopic data. Depending on the type of data used and the process under investigation, there are specific limitations that characterize each proxy.

13.3.1 Limitations of Proxy Data

Ideally, each proxy employed in the study of ancient fine-grained mixed siliciclastic–biogenic facies should be based on a known or inferred relationship between a primary geological process and the corresponding flux of a geochemical component to the sediment—or an isotopic fractionation expressed in elements involved in the process. The most useful proxies are those for which a single or predominant controlling factor can be identified based on direct modern observations and for which preserved signals are particularly sensitive to changes in the primary process. Sources of error in the development of such proxy concepts include: (i) those associated with quantifying the geochemical relationship in modern systems (sampling and instrument error, as well as uncertainties/simplifications about the modern processes—e.g., Popp *et al.*, 1998); (ii) those associated with preservation (diagenetic alteration); (iii) those associated with inferences, assumptions, or extrapolations made when the spatial and temporal scale of an ancient data set exceeds that of modern data; and (iv) those associated with measurement of components in ancient rock samples (also including sampling and instrument error).

Elemental concentration data suffer from the inherent limitations of reciprocal dilution. The risk of spurious suggestions of covariance driven by mutual dilution (and thus false interpretations of coupled delivery) provides a major impediment for unambiguous determinations of elemental fluxes within ancient sediments. Conversion of concentration values to flux terms (mass area^{-1}

time^{-1}) may be accomplished in cases where age–depth relationships are adequately known to allow the calculation of sedimentation rates, and factors like bulk density and porosity can be measured or estimated (e.g., Bralower and Thierstein, 1987; Park and Herbert, 1987; Meyers *et al.*, 2001). However, timescales of sufficient resolution are relatively rare in most pre-Pleistocene sequences, and sedimentary geochemists have devised other means of resolving relative trends in elemental fluxes. These include the calculation of elemental ratios (e.g., Turekian and Wedepohl, 1961; Brumsack, 1989; Arthur *et al.*, 1988; Arthur and Dean, 1991; Calvert and Pedersen, 1993; Piper and Isaacs, 1995; Davis *et al.*, 1999; Hofmann *et al.*, 2003; Lyons *et al.*, 2003), which serve to normalize components to major dilutants and thus allow contributions to flux variability to be better isolated. Absolute concentrations and ratios of elemental constituents can also be evaluated relative to mean values within a stratigraphic section (Sageman *et al.*, 2003) or relative to mean values for a given lithotype (e.g., world average shale or WAS: Turekian and Wedepohl, 1961; North American shale composite or NASC: Gromet *et al.*, 1984; Post-Archean average Australian shales or PAAS: Taylor and McLennan, 1985; mid-continent shales: Cullers, 1994). However, some normalization techniques (e.g., X/Al : Y/Al cross-plots, where X and Y are trace metals) are not without pitfalls and must be applied with caution (Van der Weijden, 2002).

Fractionations among stable isotopes of carbon, nitrogen, and sulfur are by-products of both kinetic and equilibrium effects associated with inorganic and biologically mediated reactions, and many of these have been well characterized in modern experimental and environmental studies (Hayes, 1993; Altabet and Francois, 1993, Canfield, 2001). Although trends in sedimentary bulk isotopic data reflect a net fractionation in the paleoreservoir being studied, the fact that many elements have multiple sources in the paleo-environment requires a mass balance (or at least species-based) approach to distinguish the major drivers of the fractionation (e.g., marine versus terrestrial C_{org} or pyrite versus organically bound sulfur). Compound-specific isotopic methods (e.g., Hayes *et al.*, 1989, 1990) have made possible the direct assessment of isotopic ratios in molecules that can be linked to specific primary sources.

In the sections below each of the processes introduced in Figure 1 (detrital, biogenic, and authigenic) is related to a corresponding set of proxies. In each case, specific strengths and limitations are reviewed, and working hypotheses for interpretation are evaluated. Given the low ratio of sedimentation rate-to-sample size (duration) that is typical for geochemical analysis in fine-grained facies, all proxy data are subject to a certain range of time averaging. The highest resolution data are commonly derived from 1 cm-thick (or less) samples taken through sequences with effective sedimentation rates (not corrected for compaction) between 0.5 cm kyr^{-1} and 2.0 cm kyr^{-1}, which therefore average, respectively, from 2 kyr to 0.5 kyr of depositional history into a single data point. However, recent undisturbed varved sequences at sites of comparatively rapid sedimentation (e.g., Black Sea and Cariaco Basin) provide a template for the very high level of temporal resolution possible in some settings (Hughen *et al.*, 1996, 1998).

13.3.2 Detrital Proxies

The objectives of detrital proxy analyses include: (i) identification of sources (e.g., weathered crust versus extrusive igneous), transport paths (fluvial, eolian, and ice-rafted), and depositional modes (suspension fallout versus gravity flow) of the terrigenous detrital constituents; (ii) determination of bulk and component fluxes of these constituents relative to biogenic and authigenic inputs; and (iii) determination of the controls on terrigenous fluxes (Figures 1 and 2). Determining the overall rate of terrigenous accumulation is important not only because it influences concentrations of biogenic and authigenic components through dilution/condensation (Johnson-Ibach, 1982) but more so because it actively modulates pore- and bottom-water redox conditions and organic carbon preservation by controlling the rate at which labile OM is transported through the successive decompositional zones shown in Figure 2 (Canfield, 1989; Meyers *et al.*, (in review)). While bulk sedimentation reflects the overall behavior of depositional systems tracts (Pasley *et al.*, 1991; Creaney and Passey, 1993), relative changes in the fluxes of individual components provide evidence for variation in dominant transport paths, which can be related to climate history (e.g., Arthur *et al.*, 1985).

Detrital geochemical proxies employed for reconstruction of source rock composition or provenance include rare-earth element suites and neodymium isotopes (e.g., McLennan *et al.*, 1993; McDaniel *et al.*, 1994; Cullers, 1994a,b; Weldeab *et al.*, 2002). Methods employing strontium and osmium isotopes (e.g., DePaolo, 1986; Ravizza, 1993) and Ge/Si ratios (Froelich *et al.*, 1992) have been used to reconstruct source area weathering and uplift history. Although such analyses are important in our understanding of the source-to-sink history of hemipelagic deposits, in this review we focus on the use of major-, minor-, and trace-element data to determine relative fluxes of detrital (weathered riverine or eolian) and/or volcanigenic fractions to bulk sedimentation (e.g., Arthur *et al.*, 1985; Pye and Krinsley, 1986; Arthur and Dean, 1991). These proxies track the relative proportions of detrital and volcanigenic mineral grains through their signature elemental compositions. Variations in these fluxes and the corresponding elemental signatures are conferred by differences in source, mode of transport (sorting), and rate of deposition and thus must be viewed in the appropriate geological context (paleogeography, tectonic framework, climate, etc.).

13.3.2.1 Physical methods

Hemipelagic facies generally include finely laminated to bioturbated to massive mudstones with variable concentrations of silt-sized particles (dominantly composed of quartz and authigenic pyrite), carbonate or siliceous content (<10 wt.% to >90 wt.%), and organic carbon (ranging from <1 wt.% to >20 wt.%). Variations in the proportion of a dominant biogenic component, such as calcium carbonate, relative to insoluble residue provide the basis for lithologic terms such as claystone (<10% $CaCO_3$), calcareous shale or mudstone (10–50% $CaCO_3$), marly shale or marlstone (50–75% $CaCO_3$), and limestone (>75% $CaCO_3$). The detrital component of these facies can be described in terms of its specific clay mineral composition, trends in grain size or optical characteristics of grains, and changes in other parameters directly controlled by the concentration of detrital grains relative to pelagic carbonate, such as magnetic susceptibility.

The major clay minerals include discrete illite, kaolinite, and chlorite, which reflect continental weathering and riverine discharge (Pratt, 1984; Leckie *et al.*, 1991), and mixed-layer illite/smectite, which results from postdepositional alteration of volcanic ash (Pollastro, 1980; Arthur *et al.*, 1985) and is an important background constituent of hemipelagic facies deposited near active volcanic belts. Variation in the proportions of different clays may reflect relative changes in dominance of riverine versus volcanigenic inputs (Arthur *et al.*, 1985; Dean and Arthur, 1998). Changes in the relative proportions of clays versus detrital grains larger than clay (>63 μm), including quartz, feldspar, biotite, and heavy mineral grains, are interpreted to reflect changes in bulk detrital flux by either mode of transport. Methods to quantify these changes include analysis of trends in grain size (e.g., Leithold, 1994; Rea and Hovan, 1995; Hassold *et al.*, 2003), quantification of weight percentages of detrital particles (e.g., DeMenocal, 1995), and in cases where sufficient time resolution is available, calculation of accumulation rates for the insoluble residue (Harris *et al.*, 1997) or individual detrital elements such as titanium (Meyers *et al.*, 2001). Petrographic and SEM analysis of grains can also assist in the determination of source and mode of transport (Schieber, 1996; Werne *et al.*, 2002). Finally, magnetic susceptibility—which measures changes in the proportion of magnetizable minerals and shows marked contrast between paramagnetic grains such as clays, ferromagnesian silicates, and iron sulfides versus diamagnetic components such as calcite—has been used as an indicator of changes in detrital flux (e.g., Ellwood *et al.*, 2000).

13.3.2.2 Elemental proxies

Elemental proxies of detrital flux are also based on changing proportions of mineral constituents and can be used to track subtle changes in grain size (e.g., Bertrand *et al.*, 1996). In hemipelagic rocks aluminum is generally regarded as the main conservative proxy for clay minerals, which dominate the terrigenous insoluble residue (Arthur *et al.*, 1985; Arthur and Dean, 1991; Calvert *et al.*, 1996). Although aluminum scavenging by sinking biogenic particles documented in the deep equatorial Pacific (Murray *et al.*, 1993; Murray and Leinen, 1996, Dymond *et al.*, 1997) would impair its use as a conservative tracer, this process is unlikely to have been significant in the comparatively shallow water columns of epicontinental basins where the studies described herein are focused. Thus, changes in detrital flux in these settings can be detected via: (i) variations in elements associated with coarser fractions relative to the aluminum proxy, such as changes in silicon related to detrital quartz silt, or in titanium and zirconium related to heavy mineral grains such as zircon, rutile, sphene, titanite, and ilmenite (Arthur *et al.*, 1985; Pye and Krinsley, 1986; Calvert *et al.*, 1996; Bertrand *et al.*, 1996; Davis *et al.*, 1999; Wortmann *et al.*, 1999; Yarincik *et al.*, 2000b; Haug *et al.*, 2003); (ii) changes in elements indicative of detrital clays, such as potassium associated with discrete illite, relative to background aluminum (Pratt, 1984, Arthur *et al.*, 1985; Pye and Krinsley, 1986; Yarincik *et al.*, 2000b; Hofmann *et al.*, 2003); and (iii) changes in elements indicative of altered volcanic ash, such as sodium and iron + magnesium—which reflect a background of eolian delivery—relative to indicators of hemipelagic detrital flux, such as potassium input as discrete illite (Dean and Arthur, 1998).

In each case, changes in the elemental ratios are caused by relative changes in bulk sedimentation—dilution or condensation results from changes in the terrigenous clay flux relative to eolian or other inputs. Correct interpretation of proxies depends on distinguishing elemental sources and transport modes. For example, some increases in Ti/Al are interpreted to reflect increased eolian flux relative to the hemipelagic background (Bertrand *et al.*, 1996; Yarincik *et al.*, 2000b), whereas in other cases the same signal is interpreted to indicate enhanced delivery of riverine detritus (Arthur *et al.*, 1985; Murphy *et al.*, 2000a; Meyers *et al.*, 2001). In situations where biotite is a dominant constituent of volcanic ash, changes in Ti/Al may track bentonite content. Similarly, enrichments in Si/Al that reflect proportions of silicon in excess of the aluminosilicate (mudrock) background may reflect enhanced quartz delivery due to eolian inputs (Pye and Krinsley, 1986; Werne *et al.*, 2002) or enhanced input of biogenic silicon and thus a productivity signal (Davis *et al.*, 1999). Geological context (e.g., proximity to arid source regions or deltaic systems), optical or SEM identification of grain types and surface textures (Schieber *et al.*, 2000; Werne *et al.*, 2002), and the use of multiple complementary proxies can help address these questions of source and transport mode.

13.3.3 Biogenic Proxies

The objectives of biogenic proxy analyses include: (i) identification of the sources of OM (terrigenous, marine algal, and bacterial) and biogenic skeletal material, (ii) determination of bulk and component fluxes of biogenic constituents, and (iii) determination of controls on ancient biogenic fluxes (Figures 1 and 2). Due to several factors, bulk organic carbon or pelagic skeletal material in ancient mudrocks cannot necessarily be viewed as reliable quantitative proxies for primary production in overlying surface waters. Firstly, OM and skeletal material such as $CaCO_3$ may derive from multiple sources. Second, as described above, concentrations of C_{org} and other constituents are influenced by bulk sedimentation rate, including relative dilution/condensation, as well as the possible influence of changes in mineral surface area and consequent sorption capacity with changing clay properties (Kennedy *et al.*, 2002). Therefore, identification of relative contributions of marine phytoplankton versus other biogenic components and, ideally, calculation of accumulation rates for individual components is required before primary and export production can be accurately determined. However, because the efficiency of OM remineralization, which depends on factors such as water depth and transport time, redox state of the water column and pore waters, and bulk sedimentation rates, exerts significant control on C_{org} burial flux (e.g., Emerson, 1985; Emerson and Hedges, 1988; Canfield, 1994; Meyers *et al.*, in review), reconstruction of productivity based on OM accumulation may be biased (similar arguments can be made for carbonate if dissolution is significant). Therefore, even if accurate accumulation rates can be determined, they must be calibrated against, among other things, proxies of redox history (see below; note that these arguments are the basis for the complementary multiproxy approach illustrated in Figures 1and 2).

In most pre-Pleistocene deposits, where timescales typically do not allow high-resolution accumulation rate estimates, assessment of paleoproduction relies mostly on indirect, qualitative, or semiquantitative methods. Nongeochemical approaches include, for example, analyses of changes in planktic bioassemblages (e.g., Watkins, 1989; Burns and Bralower, 1998; Peterson *et al.*, 1991). Among the host of geochemical techniques investigated in the literature we will review a subset that have produced consistent and complementary results for the different time intervals we have studied. These include methods to assess OM and skeletal sources, stable isotopic techniques, analyses of elemental ratios, compound specific approaches, and accumulation rate calculations for biogenic components (e.g., C_{org}, $CaCO_3$, or SiO_2).

13.3.3.1 Organic matter sources

Methods employed to determine OM sources in the hemipelagic facies described in our case studies include organic petrography, which allows delineation of relative proportions of terrigenous versus marine algal macerals (Durand, 1980; Pratt, 1984), and calculation of hydrogen and oxygen indices from OM pyrolysis (Rock Eval), which allows characterization of kerogen types and can distinguish between terrestrial and marine, as well as oxidized/thermally mature and well-preserved kerogen OM (Pratt, 1984; Kuhnt *et al.*, 1990). Other methods that contribute to recognition of OM source include C : N ratios (Meyers, 1994) and organic compounds or biomarkers (de Leeuw *et al.*, 1995).

13.3.3.2 Stable carbon isotopes of OM ($\delta^{13}C$)

Despite the wide diversity of controls on the carbon isotope composition of preserved OM and biogenic $CaCO_3$ (e.g., Hayes, 1993; Kump and Arthur, 1999), it is possible under certain circumstances to argue that a few variables are dominant and thus to relate changes in $\delta^{13}C_{org}$ to trends in paleoproduction. As argued originally by Scholle and Arthur (1980), Lewan (1986), Arthur *et al.* (1988), and others, changes in the $\delta^{13}C$ of preserved marine algal OM and biogenic $CaCO_3$ reflect changes in the isotopic composition of dissolved inorganic carbon in surface waters, which on short to intermediate timescales

(<1 Myr) may be controlled by the balance between net respiration and net burial of OM in sediments. For example, positive shifts in the $\delta^{13}C$ of organic carbon dominantly sourced from marine photoautotrophs have been interpreted to reflect elevated burial fluxes of OM related to global increase in primary productivity (Arthur *et al.*, 1987, 1988), whereas negative shifts have been interpreted to reflect recycling of respired CO_2 in a more localized reservoir (e.g., Saelen *et al.*, 1998; Murphy *et al.*, 2000a; Rohl *et al.*, 2001). For more detailed recent reviews of the controls on C-isotope fractionation see Kump and Arthur (1999), Hayes *et al.* (1999), and papers in Valley and Cole (2001).

13.3.3.3 Elemental ratios

A common indirect approach to reconstructing paleoproductivity is the analysis of components that reflect changes in nutrients. Among the numerous methods described in the literature, we have focused on the ratios of carbon, nitrogen, and phosphorous, which have been employed in studies of Paleozoic, Mesozoic, and Cenozoic black shales (Ingall *et al.*, 1993; Murphy *et al.*, 2000a; Slomp *et al.*, 2002; Filippelli *et al.*, 2003; Sageman *et al.*, 2003). Additional work has demonstrated the utility of barium (e.g., Ba/Al or Ba/Ca) as a proxy for paleoproductivity in some settings (Dymond *et al.*, 1992; Francois *et al.*, 1995; Paytan *et al.*, 1996; Van Santvoort *et al.*, 1996; cf. McManus *et al.*, 1999), but has also highlighted the limitations of this method under reducing conditions where low sulfate concentrations can lead to barite undersaturation (McManus *et al.*, 1998). Although promising in oceanic settings, use of the barium proxy in ancient epeiric deposits may be difficult.

Based on the observation that suboxic to anoxic decompositional processes favor the strongly preferential release of nitrogen and phosphorus from OM (Aller, 1994; Van Cappellen and Ingall, 1994; Ingall and Jahnke, 1997), increases in the ratios of C : N : P in preserved OM may imply increasing nitrogen and phosphorus bioavailability. First, preserved OM must be determined to be predominantly marine algal in origin (e.g., by organic petrography, biomarker abundance and isotopic composition, etc.). If so, it is reasonable to assume that its C : N : P content originally approximated the modern Redfield ratio of 106 : 16 : 1 (Redfield *et al.*, 1963). However, remineralized phosphorus is not necessarily released to the overlying water column but may instead become immobilized in sediments in inorganic form (e.g., Filippelli, 1997). Changes in total sedimentary phosphorus can be evaluated as an indication of the amount of phosphorus that was remineralized and not subsequently precipitated and thus potentially bioavailable. Although there may be sources of phosphorus in excess of that associated with sedimented OM (Schenau and DeLange, 2001), such additions would tend to reduce C : P anomalies, suggesting that C_{org} to total phosphorus values provide a minimum estimate of the phosphorus released by the sediment during OM remineralization.

The amount of OM that survives to become sedimentary OC is a function of the rate of production and down-column export, as well as the dominant type of decomposition (metazoan and/or microbial) and its duration relative to bulk sedimentation rate (burial). It should be noted that these generalizations, derived from the study of modern oceans, must be viewed in the context of shallow epicontinental settings where total depth probably did not exceed ± 300 m during maximum highstands. Thus, the transit time of OM in the water column was comparatively short, and the relative percentage of production to reach the sediment surface (export production) was high (Suess, 1980). As a consequence, respiratory demand in the bottom waters was likely to have been intense, especially during warm seasons when thermal stratification of the water column prevented downward advection of dissolved O_2. If decompositional release of nutrients occurred during establishment of seasonal (or longer-term) thermoclines in these shallow seas, and such nutrients were recycled to surface waters when the thermocline dissipated, an effective mechanism was available to drive primary production and increase the burial flux of carbon to the sedimentary reservoir (Murphy *et al.*, 2000a). This simplified view omits discussion of many issues, such as sedimentation rate controls on phosphorus burial (Tromp *et al.*, 1995) and dynamics of N-cycling in the water column (e.g., Altabet *et al.*, 1991; Holmes *et al.*, 1997). However, since studies of ancient stratigraphic sequences tend to span large time intervals based on analysis of significantly time-averaged samples, and since phosphorus is regarded as the more significant limiting nutrient on such geological timescales, we argue that the approach described here, in combination with other proxies, provides a reasonable first-order approximation of nutrient–productivity dynamics.

13.3.3.4 Biomarkers

Changes in abundance and isotopic composition of biomarker compounds can help delineate relative mass contributions from terrestrial, marine algal, and bacterial OM sources, as well as provide indications of specific processes and environmental conditions in ancient water columns (e.g., Hayes *et al.*, 1989; Sinninghe Damsté *et al.*, 1993; Silliman *et al.*, 1996; Kuypers *et al.*, 2001; Simons and Kenig, 2001; Pancost *et al.*, 2002). In some of our studies discussed below these methods have been used in a limited fashion, mainly to help constrain the relative contributions of different OM sources to observed $\delta^{13}C_{org}$ variations (e.g., Murphy *et al.*, 2000a).

13.3.3.5 Accumulation rates

With high-resolution timescales, calculation of accumulation rates (mass area^{-1} time^{-1}) for C_{org}, $CaCO_3$, and SiO_2 (e.g., Bralower and Thierstein, 1984; Archer, 1991; Pedersen *et al.*, 1991; Sancetta *et al.*, 1992; Arthur *et al.*, 1994; Calvert and Karlin, 1998) can provide a direct measure of the net burial flux of biogenic components. However, relating burial fluxes to primary production may be complicated by factors described above. For example, significant remineralization of OM can occur even under the most anoxic conditions (Canfield, 1989). Although the dissolution of $CaCO_3$ provides a similar impediment to its use as a linear proxy for paleoproduction, in shallow seas where surface-water productivity was dominated by calcareous nanoplankton, $CaCO_3$ accumulation rates may provide a reasonable first-order estimate of production. Broecker (1982) estimated a 1 : 4 relationship between $CaCO_3$ and C_{org} fixation rates in modern open-ocean settings. Meyers *et al.* (in review) evaluated this hypothesis using a compilation of modern data representing a range of values for $CaCO_3$ accumulation and C_{org} production and confirmed that the 1 : 4 ratio applied only under the highest rates of $CaCO_3$ accumulation. Although timescales of sufficient resolution to calculate accumulation rates are quite difficult to establish in pre-Pleistocene stratigraphic sequences where datable horizons may be few and disconformities many, in an increasing number of studies orbital timescales developed from analysis of rhythmically bedded hemipelagic facies are being employed for this purpose (Herbert and Fischer, 1986; Herbert *et al.*, 1986; Park and Herbert, 1987; Meyers *et al.*, 2001).

13.3.4 Authigenic Proxies

The chemical behavior of various minor and trace elements is relatively well characterized for particular redox conditions, and there has been significant effort directed at the development of geochemical proxies for paleo-oxygenation in black shale sequences (see reviews in Calvert and Pedersen, 1993; Arthur and Sageman, 1994; Jones and Manning, 1994; Wignall, 1994; Schieber *et al.*, 1998a,b). Elements of proven paleoredox utility include Mo (Coveney *et al.*, 1991; Dean *et al.*, 1999; Meyers *et al.*, in review), V–Ni (Lewan and Maynard, 1982; Lewan, 1984; Breit and Wanty, 1991), U (Wignall and Myers, 1988), Mn (Calvert and Pedersen, 1993), Re (Crusius *et al.*, 1996), and rare-earth elements or, more specifically, the Ce anomaly (Wright *et al.*, 1987; Wilde *et al.*, 1996; cf. Bright *et al.*, submitted; see German and Elderfield, 1990, for review). In addition, proxies that directly assess HS^- (ΣH_2S) availability in ancient water columns include: (i) degree of pyritization, which is a measure of the extent to which reactive iron has been transformed to pyrite, and related iron approaches (Berner, 1970; Raiswell *et al.*, 1988; Canfield *et al.*, 1996; Raiswell *et al.*, 2001); (ii) sulfur isotope relationships (Jørgensen, 1979; Goldhaber and Kaplan, 1974, 1980; Anderson *et al.*, 1987; Fisher and Hudson, 1987; Beier and Hayes, 1989; Habicht and Canfield, 1997; Lyons, 1997); (iii) bacterial pigments indicative of anoxygenic photosynthesis in the presence of hydrogen sulfide (e.g., Repeta, 1993; Sinninghe Damsté *et al.*, 1993; Koopmans *et al.*, 1996; Huang *et al.*, 2000); and (iv) pyrite framboid size distributions (Wilkin *et al.*, 1996; Wignall and Newton, 1998).

Studies in modern oxygen-deficient basins have helped to calibrate these methods (Spencer and Brewer, 1971; Jacobs and Emerson, 1982; Jacobs *et al.*, 1985, 1987; de Baar *et al.*, 1988; Anderson *et al.*, 1989a,b; German and Elderfield, 1989; German *et al.*, 1991; Lewis and Landing, 1992; Repeta, 1993; Sinninghe Damsté *et al.*, 1993; Van Cappellen *et al.*, 1998; Morford and Emerson, 1999), including those associated with iron and molybdenum cycling and resulting patterns of (sulfide-driven) mineralization that can preserve deep into the geological record (Francois, 1988; Emerson and Huested, 1991; Lewis and Landing, 1991; Canfield *et al.*, 1996; Calvert *et al.*, 1996; Crusius *et al.*, 1996; Helz *et al.*, 1996; Lyons, 1997; Wilkin *et al.*, 1997; Raiswell and Canfield, 1996, 1998; Dean *et al.*, 1999; Yarincik *et al.*, 2000a; Zheng *et al.*, 2000; Adelson *et al.*, 2001; Wijsman *et al.*, 2001; Wilkin and Arthur, 2001; Lyons *et al.*, 2003). Rather than providing cursory background for each of a large number of proxies, we will emphasize a few approaches of particular value in recent black shale research, while also providing some historical perspective.

13.3.4.1 C–S relationships

The C–S (organic carbon–pyrite sulfur) paleoenvironmental method—which has long been applied to ancient shale-bearing sequences—is based on the observation that different factors generally limit sedimentary pyrite formation in normal marine (oxic bottom waters), euxinic marine (anoxic and H_2S-containing), and freshwater to brackish settings (Berner, 1984). In normal marine Phanerozoic environments, the supply of dissolved sulfate for bacterial reduction to H_2S is typically not limiting because of the ample concentrations in seawater (contrasting low marine sulfate availability during the Precambrian is summarized in Lyons *et al.*, in press). As the result of reactions between the bacterially generated H_2S and detritally delivered reactive iron, appreciable pyrite sulfur (Spy) is produced during diagenesis if sufficient C_{org} is also available. The bacteria oxidize C_{org} in the process of reducing sulfate to H_2S. In normal marine settings, the formation of pyrite can be limited by the availability of bacterially metabolizable OM. This linkage is expressed as a positive linear relationship between concentrations of C_{org} and Spy—with a zero sulfur intercept. Normal marine sediments from a wide range of Holocene localities yield a linear trend with a mean C/S weight ratio of 2.8 (Berner, 1982; Morse and Berner, 1995). Expressions of the coupling between C_{org} and Spy and corresponding variations in C/S ratios in normal shales spanning the Phanerozoic are addressed in Raiswell and Berner (1986), although studies are increasingly demonstrating the potential for reactive-iron limitation within organic-rich sediments beneath oxic water columns (Canfield *et al.*, 1992; see Lyons *et al.*, 2003).

Euxinic marine environments are complex in that limitations in iron, rather than C_{org}, are the norm, and the presence of sulfide within the water column can decouple pyrite formation from the local burial flux of C_{org} (Raiswell and Berner, 1985; Lyons and Berner, 1992; Lyons, 1997). Regardless of the complexity, these anoxic marine systems are noted for their relatively high abundances of Spy and C_{org}. By contrast, sediments deposited under the generally sulfate-deficient conditions of natural freshwater to brackish settings display low amounts of Spy despite the potential for high concentrations of C_{org} and thus plot close to the C_{org} axis on a C–S plot (Figure 3; Berner, 1984; Berner and Raiswell, 1984). In practice, specific paleo-environmental variations within the marine realm are often difficult to delineate uniquely using the C–S technique because of the complex interplay among primary and secondary controlling factors (e.g., Lyons and Berner, 1992). As a result, distinctions between fully marine and low salinity freshwater-to-brackish settings are the least ambiguous application of the C–S method. Furthermore, primary C–S relationships can be masked by weathering, metamorphism, and both low-temperature and hydrothermal secondary overprints (Figure 3; Leventhal, 1995; Lyons *et al.*, 2000, 2003; Petsch *et al.*, 2000, 2002).

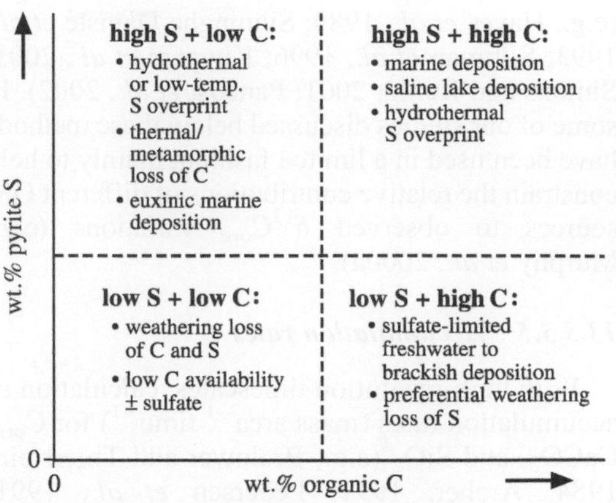

Figure 3 Schematic plot emphasizing the principal environmental controls, both primary and secondary, on C_{org} versus S_{py} distributions in fine-grained siliciclastic sediments and sedimentary rocks (after Lyons *et al.*, 2000). Preferential loss of sulfur during weathering is addressed in Petsch *et al.* (2000, 2002); however, weathering effects are minimized through the use of drill core. Low- and high-temperature sulfur overprints can be distinguished from primary S_{py} signals by their $\delta^{34}S$ characteristics (Lyons *et al.*, 2000, 2003). Background is provided in Section 13.3.4.1 and Lyons *et al.* (2000).

13.3.4.2 Sulfur isotope relationships

Details of sulfur isotope geochemistry are presented elsewhere in this volume) and are only highlighted here as related to paleo-environmental interpretations of fine-grained siliciclastic sequences. Formation of sedimentary pyrite initiates with bacterial sulfate reduction (BSR) under conditions of anoxia within the water column or sediment pore fluids. The kinetic isotope effect associated with bacterial sulfate reduction results in hydrogen sulfide (and ultimately pyrite) that is depleted in ^{34}S relative to the $^{34}S/^{32}S$ ratios of residual sulfate (Goldhaber and Kaplan, 1974). The balance between net burial versus oxidative weathering of pyrite controls the $^{34}S/^{32}S$ ratio in the global oceanic sulfate reservoir and, along with the redox cycling of organic carbon, is the principal modulator of PO_2 in the atmosphere over geological time (Claypool *et al.*, 1980; Berner and Petsch, 1998).

Dissimilatory BSR under pure-culture laboratory conditions can produce sulfide depleted in ^{34}S by ~2–46‰ relative to the parent sulfate (Chambers *et al.*, 1975; Canfield, 2001; Detmers *et al.*, 2001). Although this range is generally accepted, controls on the magnitude of this fractionation are subjects of recent debate. For example, contrary to a long-held assumption, the isotopic offset between parent sulfate and HS-produced during BSR ($\Delta^{34}S$) may not vary with a simple inverse relationship to the rate of sulfate reduction (cf. Kaplan and Rittenberg, 1964; Canfield, 2001; Detmers *et al.*, 2001; Habicht and Canfield, 2001). Furthermore, in light of the significantly smaller isotope effects attributable to BSR under pure-culture conditions and by inference in natural settings (cf. Wortmann *et al.*, 2001), recent studies have addressed the fractionations of up to and exceeding 60‰ that abound in the Phanerozoic record. One model invokes bacterial disproportionation of elemental sulfur and other sulfur mediates as a means of exacerbating the ^{34}S depletions observed in HS- and pyrite (Canfield and Thamdrup, 1994; Habicht and Canfield, 2001).

Ultimately, net isotopic fractionations preserved in geological systems reflect both the magnitudes of bacterial fractionations and the properties of the sulfate reservoir—as recorded in the integrated history of pyrite formation (Zaback *et al.*, 1993). Even in the presence of large fractionations during BSR and coupled disproportionation, comparatively high $\delta^{34}S_{sulfide}$ values can occur in environments where renewal of sulfate is restricted relative to the rate of bacterial consumption (e.g., within sediments under conditions of rapid accumulation). Conversely, low $\delta^{34}S$ values are typical of marine systems where sulfate availability does not limit BSR. Because of these multiple controlling factors, bacteriogenic pyrite can display a broad range of $\delta^{34}S$ values that are often very low (^{34}S-depleted) relative to coeval seawater sulfate. In euxinic settings (i.e., those with persistently anoxic and sulfidic bottom waters such as the modern Black Sea and Cariaco Basin), much (often most) of the pyrite forms (syngenetically) within the sulfidic water column and shows the light and uniform $\delta^{34}S$ values expected under conditions of large sulfate and sulfide reservoirs. Conversely, under oxic depositional conditions all of the pyrite forms diagenetically and is thus vulnerable to the ^{34}S enrichments and wide $\delta^{34}S$ variability expected in a restricted pore-water reservoir. At euxinic sites of very rapid sediment accumulation, such as coastal Fjords and marginal locations within larger anoxic basins (Lyons, 1997; Hurtgen *et al.*, 1999), iron sulfides form both diagenetically and syngenetically, which results in ^{34}S-enrichment relative to distal, largely syngenetic pyrite pools. These facies-dependent sulfur isotope trends are well expressed in a number of ancient marine sequences (Figure 4; Anderson *et al.*, 1987; Fisher and Hudson, 1987; Beier and Hayes, 1989).

In the two Mesozoic examples shown in Figure 4, the light and uniform $\delta^{34}S$ values that typify water-column pyrite formation under euxinic conditions, and are further favored by the comparatively slow rates of sediment accumulation, are well expressed. The Cretaceous data of Gautier (1986, 1987) also show the classic broad range of $\delta^{34}S$ values possible for oxic deposition. Under the diffusion-controlled sulfate fluxes of diagenesis, rapid sedimentation can favor ^{34}S enrichment in pyrite by enhancing the rates of BSR and reactive iron availability below surface sediment layers and by facilitating protracted transformations of iron monosulfide precursors to pyrite (Hurtgen *et al.*, 1999). These pore-water reservoir effects occur beneath both oxic and anoxic bottom waters. Conversely, the openness of pore waters beneath oxic bottom waters can be maintained by infaunal mixing and bio-irrigation. These processes

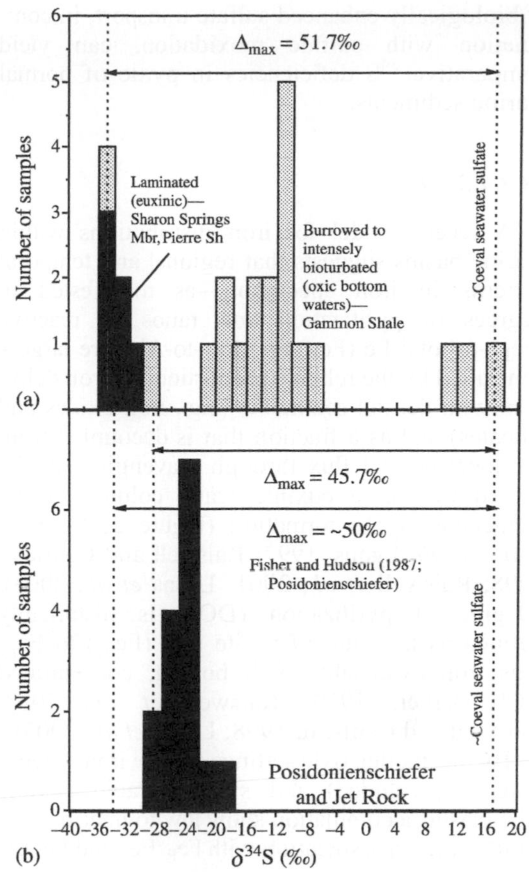

Figure 4 Sulfur isotope summary for black shales from the Pierre Shale of the Cretaceous Western Interior, North America (Gautier, 1986, 1987), and the Jurassic Posidonienschiefer and Jet Rock (Raiswell *et al.*, 1993). For comparison, the maximum fractionation observed in the Posidonienschiefer by Fisher and Hudson (1987) is also shown. The isotopically uniform and strongly ^{34}S-depleted pyrites of the Jurassic shales and the Cretaceous Sharon Springs Member of the Pierre Shale—like the sediments of the modern Black Sea and Cariaco Basin (Figure 7)—are diagnostic of euxinic (water-column) pyrite formation (see Section 13.3.4.2). By contrast, the Cretaceous Gammon Shale shows the ^{34}S enrichments and broad range of $\delta^{34}S$ values possible under oxic depositional conditions (Gautier, 1986, 1987).

of biologically enhanced sulfate transport, in combination with sulfide reoxidation, can yield comparative ^{34}S deficiencies in pyrite of normal marine sediments.

13.3.4.3 Iron

A recent model for iron distributions within marine basins suggests that regional and temporal patterns in iron speciation—as manifested in degrees of pyritization and ratios of reactive (Fe_R)-to-total Fe (Fe_T) and Fe_T-to-Al—are largely controlled by the relative proportions of iron delivered with detrital sediments (e.g., iron oxides and silicates) and as a fraction that is decoupled from the local detrital flux through scavenging of dissolved iron in a euxinic water column during syngenetic pyrite formation (Figure 5; Canfield *et al.*, 1996; Lyons, 1997; Raiswell and Canfield, 1998; Raiswell *et al.*, 2001; Lyons *et al.*, 2003). (Degree of pyritization (DOP) is historically defined as the ratio of pyrite iron (Fe_{py}) to Fe_{py} plus iron extractable with boiling, concentrated HCl—Berner, 1970; Raiswell *et al.*, 1988; Raiswell and Canfield, 1998; Lyons *et al.*, 2003).

By this model, oxic settings (where iron scavenging is precluded) and sites of rapid euxinic siliciclastic accumulation show low to intermediate DOP values in association with Fe_R/Fe_T and Fe_T/Al ratios that are typical of the riverine (continental) flux, while more-condensed euxinic settings can show dramatic augmentation of the iron reservoir (via Fe_{scav})—and thus diagnostic elevation of the three iron-based paleoredox proxies (Raiswell *et al.*, 1988; Raiswell and Canfield, 1998; Raiswell *et al.*, 2001; Lyons *et al.*, 2003). Because the iron paleoredox proxies are a function of (i) the presence or absence of hydrogen sulfide in the water column, (ii) the intensity of water-column pyrite formation and associated iron scavenging, and (iii) the relative siliciclastic flux (Figure 6), iron distributions at euxinic sites near the basin margin can look like those of oxic sediments because the scavenged iron is swamped by the high rates of siliciclastic accumulation (Raiswell *et al.*, 2001; Lyons *et al.*, 2003). As a result, high values of DOP (and Fe_T/Al ratios elevated above the local detrital flux) point uniquely toward strong and likely persistent euxinic conditions in the basin, while very low DOP values are suggestive of oxic deposition under conditions of low OM accumulation. Intermediate DOP values, by contrast, can reflect either oxic deposition associated with appreciable organic accumulation—wherein pyrite formation can be iron limited—or euxinic deposition at sites of rapid siliciclastic influx (Lyons *et al.*, 2003).

While the details are almost certainly more complex than the model presented here and elsewhere (e.g., Raiswell *et al.*, 2001; Lyons *et al.*, 2003), nearshore–offshore gradients in DOP (and overall iron patterns of enrichment as recorded in

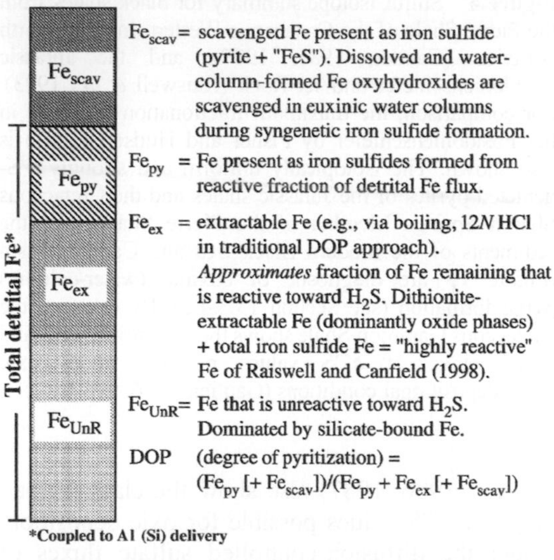

Fe_scav
Fe_py
Fe_ex
Fe_UnR
Total detrital Fe*
Fe_scav = scavenged Fe present as iron sulfide (pyrite + "FeS"). Dissolved and water-column-formed Fe oxyhydroxides are scavenged in euxinic water columns during syngenetic iron sulfide formation.
Fe_py = Fe present as iron sulfides formed from reactive fraction of detrital Fe flux.
Fe_ex = extractable Fe (e.g., via boiling, 12N HCl in traditional DOP approach). Approximates fraction of Fe remaining that is reactive toward H2S. Dithionite-extractable Fe (dominantly oxide phases) + total iron sulfide Fe = "highly reactive" Fe of Raiswell and Canfield (1998).
Fe_UnR = Fe that is unreactive toward H2S. Dominated by silicate-bound Fe.
DOP (degree of pyritization) = (Fe_py [+ Fe_scav])/(Fe_py + Fe_ex [+ Fe_scav])
*Coupled to Al (Si) delivery

Figure 5 Summary of iron speciation in fine-grained siliciclastic sediments and sedimentary rocks. Total Fe (Fe_T) is equal to the sum of all these fractions. DOP increases in oxic sediments through the conversion of Fe_{ex} to Fe_{py}, although the HCl procedure generally overestimates the readily reactive iron available. In euxinic settings, high DOP values (Fe_T/Al ratios) result from scavenging of dissolved iron during pyrite formation in the water column. See Section 13.3.4.3 and Lyons *et al.* (2003) for further discussion and background.

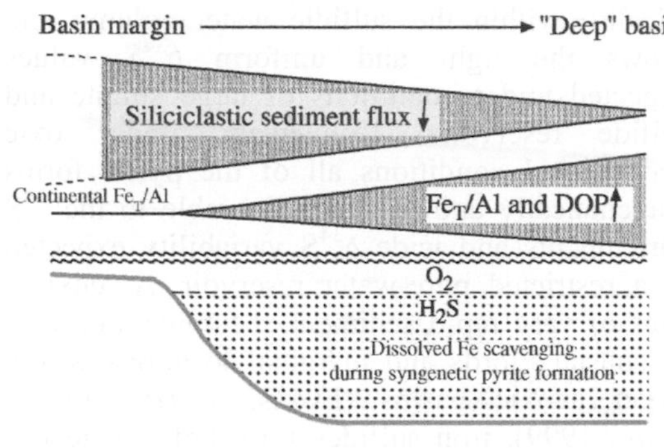

Figure 6 Schematic representation of the spatial gradients in Fe_T/Al ratios and DOP values in the bottom sediments of euxinic basins. Nearshore-to-offshore increases in these parameters derive from corresponding decreases in the siliciclastic flux and thus increases in the proportion of (i) iron scavenged during syngenetic pyrite formation in the euxinic water column to (ii) iron delivered with the local siliciclastic flux. The scavenged iron is initially present as dissolved iron and as oxyhydroxides formed near the chemocline and is therefore decoupled from the local siliciclastic (aluminum) flux. Details and other supporting references are provided in Section 13.3.4.3 and Lyons *et al.* (2003).

Fe_R/Fe_T and Fe_T/Al ratios) cannot simply be attributed to transport-related phenomena independent of depositional redox (e.g., sorting by grain size, mineralogy, or grain type—such as fluxes dominated by riverine versus eolian inputs). For example, the Orca Basin—a small anoxic brine pool on the northern slope of the Gulf of Mexico—shows a transition from Fe_T/Al ratios of 0.40–0.50 for oxic sediments (much like average continental crust; Taylor and McLennan, 1985) to values of ~0.55–0.75 within the anoxic zone. Ratios of ~0.85–1.00 occur in the transition zone where the redox interface (chemocline), with its associated iron cycling and particulate iron maximum (Van Cappellen *et al.*, 1998), impinges on the substrate (Lyons, unpublished data). The essential point here is that the iron pattern is linked to depositional redox, and, from the standpoint of siliciclastic transport and associated fractionation of the iron reservoir, the oxic, transitional, and anoxic sites are all equidistant from the shoreline.

13.3.4.4 Trace metals

In our research (e.g., Werne *et al.*, 2002; Lyons *et al.*, 2003), molybdenum has emerged as one of the empirically most useful, but mechanistically least understood proxies for depositional redox (cf. Meyers *et al.*, in review). It is clear to us and others that molybdenum enrichments in organic-rich facies are linked to the presence of appreciable hydrogen sulfide and are coupled, directly or indirectly, to C_{org} accumulation, where the type of OM present may also play a critical role (Coveney *et al.*, 1991; Helz *et al.*, 1996). What is less clear, however, are the specific mineralization mechanisms by which molybdenum accumulates and whether such enrichment can speak specifically to the presence or absence of sulfide in the water column, in contrast to the sediment pore waters. Molybdenum (dominantly as molybdate) is about two orders of magnitude more abundant in oxic seawater than iron. Consequently, the iron enrichments observed in euxinic sediments must reflect scavenging within anoxic water columns where dissolved iron is greatly elevated, and diffusion of dissolved iron into sulfidic sediments beneath an oxic water column is quantitatively negligible. Molybdenum, however, because of its greater abundance in oxic waters can diffuse into sulfide-rich pore waters and become appreciably enriched in the sediments relative to detritally delivered molybdenum.

The absence of elevated molybdenum concentrations in particulate samples from sediment traps deployed in the water columns of modern anoxic basins (e.g., Saanich Inlet, Cariaco Basin, and Black Sea) has suggested to some that molybdenum enrichment occurs only during diagenesis by reaction at or through diffusion across the sediment–water interface (Francois, 1988; Emerson and Huested, 1991; Crusius *et al.*, 1996) and that the link between high molybdenum concentration and accumulation and independently indicated euxinicity is the abundance of organic- and sulfide-rich sediments accumulating in oxygen-deficient settings (Zheng *et al.*, 2000). If true, a euxinic setting is not required for molybdenum enrichment. It should be noted, however, that the hydrogen sulfide concentrations in these anoxic water columns (e.g., <100 μM to 300–400 μM for the Cariaco Basin and Black Sea, respectively) may be below the critical threshold described by Helz *et al.* (1996) for the switch to particle-reactive thiomolybdate and thus may not effectively represent all sites of ancient anoxia. Alternatively, the availability of sediment-trap data remains small and potentially ambiguous, particularly for modern anoxic sites of comparatively high and persistent sulfide concentrations—such as Framvaren Fjord, Norway. It is also interesting to note that molybdenum is enriched significantly within Unit 1 sediments in the Black Sea—even at sites with present-day pore-water concentrations of dissolved sulfide that are not appreciably elevated above bottom-waters values (Lyons and Berner, 1992; Lyons, unpublished data).

In the case studies provided below we see evidence for molybdenum enrichments that are precisely coincident with the onset of euxinic conditions (Werne *et al.*, 2002; Lyons *et al.*, 2003), despite evidence for appreciable diagenetic dissolved sulfide in the underlying sediments deposited under conditions ranging from oxic to perhaps dominantly anoxic/nonsulfidic. In this review we cannot hope to resolve the role water-column sulfide plays in molybdenum accumulation and enrichment other than to suggest that (i) high concentrations of sedimentary molybdenum appear to be diagnostic of depositional environments with euxinic water columns, although high rates of molybdenum accumulation are not necessarily indicative of euxinicity (Meyers *et al.*, in review), and (ii) reactive iron availability may ultimately control molybdenum concentration and accumulation rate by modulating the amount of sulfide buildup in sediments and in the water column through iron sulfide formation (Meyers *et al.*, in review). Also of interest is the commonly observed, strongly positive covariance between the concentrations and accumulation rates of molybdenum and those of C_{org}. We are currently exploring the mechanistic underpinnings of this relationship and how it might constrain paleoenvironments over geological time. Currently, we favor two models: (i) reactions between OM and molybdenum in the presence of dissolved sulfide that yield a systematic (stoichiometric?) relationship between the two (e.g., Coveney *et al.*, 1991) and (ii) the possibility that sulfide alone is the link, with increasing OM yielding increasing H_2S via BSR and thus parallel molybdenum sequestration (see review in Lyons *et al.*, 2003). It seems unlikely, however, that there will be a consistent slope for this relationship across time or among basins—as evidenced by the two distinct trends that are present for sediments deposited since the last oxic–anoxic transition in the Cariaco Basin (Lyons *et al.*, 2003).

Recently, mass-dependent isotopic variation for molybdenum has surfaced as a tool of high paleo-environmental potential (Barling *et al.*, 2001; Barling and Anbar, submitted). Specifically, relatively inefficient scavenging of molybdenum from seawater by oxide phases under oxic marine conditions—compared to efficient removal under anoxic settings—yields patterns of molybdenum isotope variability in seawater (recorded in the black shales) that may track the global proportion of oceanic anoxia over geological time.

Other metals that occur as minor and trace elements in marine waters and clastic sediments, but may become concentrated by precipitation from seawater under appropriate redox conditions and thus have redox proxy potential, include Mn, V, Cr, Ni, Co, Cu, U, and Th (Emerson and Huested, 1991; Jones and Manning, 1994; Piper, 1994; Calvert and Pedersen, 1996). Among these we have investigated Mn, V, and Cr relationships in some of our studies, in addition to the proxies described above. The rationale for the V + Cr proxy is as follows: V and Cr, both of which have a variety of common valence states and correspondingly complex chemistries, are precipitated from seawater as hydroxides or hydrated oxides at Eh conditions that correspond to the range in which denitrification occurs (Piper, 1994). Relative increases in the sedimentary concentration of V + Cr may, therefore, be indicative of a proportional increase in the significance of denitrification relative to aerobic respiration. This is an important parameter to constrain for two reasons: (i) denitrification is a dysaerobic to anaerobic metabolism, whose increased prominence suggests the relatively reduced availability of oxygen in the system (Froelich *et al.*, 1979), but unlike sulfate-reduction, does not produce a by-product that is toxic to aerobes; and (ii) denitrification leads to loss of the biolimiting nutrient nitrogen from the system as the chemically reduced N_2, which is insoluble and escapes to the atmosphere.

Below Eh values of ~500 mV, manganese occurs as the soluble Mn^{2+} ion in seawater, whereas Eh values >500 mV favor insoluble Mn^{4+} oxides (Hem, 1981). Thus, under oxidizing conditions insoluble Mn-oxyhydroxides precipitate, while under reducing conditions manganese is maintained as a dissolved species (e.g., Piper, 1994), especially through bacterially mediated MnO_2 reduction (e.g., Stumm and Morgan, 1996; Van Cappellen *et al.*, 1998). Calvert and Pedersen (1993) argued that sediments in oxic depositional settings can show manganese enrichments relative to the continental flux due to diagenetic remobilization under reducing pore-water conditions and corresponding reprecipitation at the redox interface in the uppermost layers of the sediment column. Through a repeated sequence of burial, dissolution, remobilization, and reprecipitation, a manganese pump is established wherein pore waters can locally reach supersaturation with respect to manganese carbonate. Calvert and Pedersen (1993) suggested that in the absence of oxyhydroxide precipitation at the sediment–water interface, such a pump does not develop beneath anoxic waters, and instead manganese can be lost to the water column. They further argued that anoxic water columns generally fail to reach the saturation states necessary for water-column Mn-carbonate precipitation—thus a scavenging mechanism analogous to that for iron (via syngenetic Fe-sulfide formation; Section 13.3.4.3) may not operate. By their model, sediment manganese enrichments record bottom-water oxygenation (see also Yarincik *et al.*, 2000a). This pattern of manganese cycling, however, may be most relevant for suboxic to moderately sulfidic settings because it ignores the possible sediment immobilization or water-column precipitation (scavenging) of manganese as a sulfide phase (Lewis and Landing, 1991). Nevertheless, manganese concentrations in ancient hemipelagic sediments can provide a useful redox proxy companion to Mo and V+Cr data and are particularly effective for identifying the oxic–anoxic transition. For example, Lyons *et al.* (1993) reported manganese enrichments in sediments of the Black Sea outer shelf, at sites that are presently under oxic bottom waters, as evidence for the dramatic, short-term vertical (tens of meters) excursions of the chemocline that have been widely reported and debated in the literature (e.g., Anderson *et al.*, 1994).

13.4 GEOCHEMICAL CASE STUDIES OF FINE-GRAINED SEDIMENTS AND SEDIMENTARY ROCKS

The data presented for each case study are largely abstracted from prior publications, where they are explained in greater detail (see references below). Rather than repeat these details, discussions here focus on similarities and differences in proxy interpretation among different hemipelagic systems in an effort to develop a more comprehensive understanding of the geochemistry of fine-grained sediments and sedimentary rocks.

13.4.1 Modern Anoxic Environments of OM Burial—Black Sea and Cariaco Basin

Voluminous recent work in the Black Sea and Cariaco Basin, the world's first and second largest modern anoxic basins, respectively, has centered on patterns and pathways of carbon–sulfur cycling and sequestration, and the comparative behaviors of redox-sensitive metals and their paleo-environmental implications. At the same time, many workers are challenging the validity of the Black Sea paradigm (e.g., Rhoads and Morse, 1971; Demaison and Moore, 1980)—i.e., deposition under deep, highly stratified water-column conditions—as a universally relevant model for ancient anoxic marine deposits (e.g., Murphy *et al.*, 2000a). Nevertheless, research in these modern anoxic basins continues to illuminate general geochemical pathways (e.g., metal cycling) under oxygen-deficient conditions that are independent of the specifics of basin hydrography, water depths, etc. This approach is allowing us to define, refine, and calibrate proxies of broad paleo-environmental relevance.

Studies since the 1990s have consistently demonstrated that the majority of the pyrite accumulating within the uppermost laminated intervals in the Black Sea and Cariaco Basin formed within the euxinic water column. This prevalence of syngenetic pyrite is supported by (i) sedimentary sulfur isotope data that show uniform $\delta^{34}S$ values matching the isotopic composition of the hydrogen sulfide within the present-day water column (Figure 7; Calvert *et al.*, 1996; Lyons, 1997; Wilkin and Arthur, 2001; Lyons *et al.*, 2003; Werne *et al.*, 2003), (ii) pyrite grains with size distributions controlled by the settling velocities of framboids formed within euxinic waters (Wilkin *et al.*, 1997; Wilkin and Arthur, 2001), and (iii) iron enrichments that are most consistent with iron scavenging within a sulfidic water column during syngenetic pyrite formation (Canfield *et al.*, 1996; Raiswell and Canfield, 1998; Raiswell *et al.*, 2001; Lyons *et al.*, 2003). Ratios of Fe_T/Al and, correspondingly, DOP values in the anoxic portions of the Black Sea increase from nearshore to central basin localities by a factor of roughly two in response to a siliciclastic flux that decreases by two orders of magnitude across the same transect (Lyons *et al.*, 2003). At the nearshore, rapidly accumulating sites in the Black Sea, $\delta^{34}S_{py}$ values are enriched by ~10‰ relative to the deep basin in response the diagenetic effects discussed above (Section 13.3.4.2, Figure 7). As expected from their intermediate rates of siliciclastic accumulation relative to the end-members in the Black Sea, microlaminated sediments of the Cariaco Basin show intermediate values for the iron proxies (Figure 8). Each of these expressions of the sulfidic waters of the Black Sea and Cariaco Basin translates into a signature of euxinicity that would readily preserve into the deep geological record—and when viewed collectively, would provide a clear picture of paleoredox.

Despite the Black Sea's long status as the quintessential euxinic basin—the term euxinic derives from the basin's ancient name, *Pontus Euxinus*—it is not presently a site of anomalous C_{org} accumulation or preservation. Calvert *et al.* (1991) demonstrated that rates of C_{org} accumulation in the deep anoxic basin are not meaningfully elevated relative to oxic sites when normalized to primary production, water depth, and sedimentation rate (also Arthur *et al.*, 1994). Furthermore, the transition from the organic-rich (Unit 2) sapropel, with C_{org} concentrations of up to ~20 wt.% (Arthur *et al.*, 1994), to the overlying, carbonate-rich Unit 1 deposit where C_{org} is less abundant and averages ~5.3 wt.% (Lyons and Berner, 1992) is driven by carbonate dilution rather than shifts in production or preservation. Both intervals record euxinic conditions in the water column (Repeta, 1993; Sinninghe Damsté *et al.*, 1993; Huang *et al.*, 2000; Wilkin and Arthur, 2001), which is consistent with the persistence of lamination throughout. The onset of Unit 1 deposition specifically marks the attainment of a threshold salinity for coccolith (*E. huxleyi*) production as the basin evolved from the essentially freshwater conditions of the last glacial to the brackish marine environment of today—but it does not reflect a significant change in the rate of C_{org} accumulation relative to the upper part of Unit 2 (Arthur *et al.*, 1994; Arthur and Dean, 1998; Calvert and Karlin, 1998). Restricted marginal marine basins, such as the Black Sea and Cariaco Basin, are particularly susceptible to secondary sulfur overprints that correspond to a temporal evolution from lacustrine or oxic marine conditions during the sea-level lowstand of the last glacial to the anoxic marine settings now present (Middelburg *et al.*, 1991; Lyons *et al.*, 2003). Such overprints can complicate paleo-environmental interpretations based solely on C–S relationships (Figure 3), although sulfur isotopes assist in the recognition and characterization of secondary signals (see Figures 7 and 9; Lyons *et al.*, 2003).

The modern Black Sea also highlights the impact of siliciclastic dilution on C_{org} concentrations. At the euxinic sites on the basin margin characterized by extremely high rates of

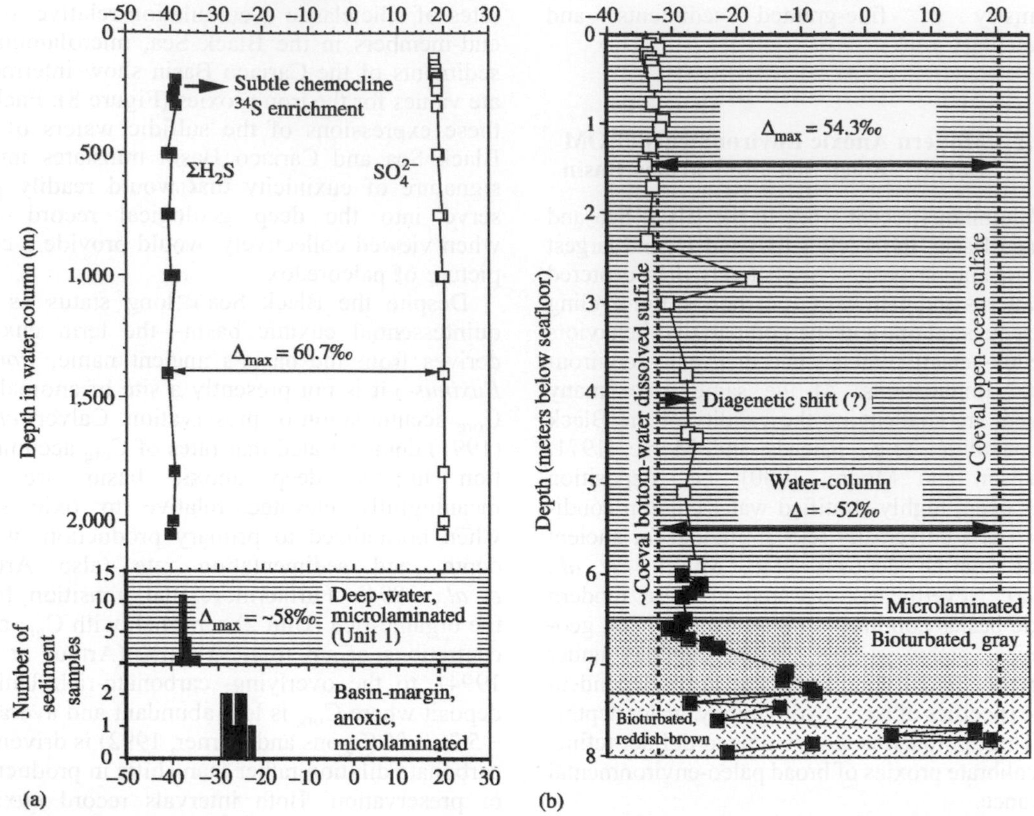

Figure 7 Summary of $\delta^{34}S$ values for pyrite in the sediments of the modern Black Sea (a) and Cariaco Basin (b). Data and discussions for the Black Sea and Cariaco sediments are available in Lyons (1997) and Lyons *et al.* (2003), respectively. $\delta^{34}S$ values from the basin-margin anoxic site in the Black Sea include pyrite and iron monosulfide (Lyons, 1997). Water-column data for the Black Sea and Cariaco, showing net fractions of 50–60‰, are from Sweeney and Kaplan (1980) and Fry *et al.* (1991), respectively. Note the strong correspondence between the data from the water column and the uniform $\delta^{34}S$ values of the bottom sediments—suggesting that most of the pyrite accumulating in the sediments formed (syngenetically) within the water column. Diagenetic effects at the basin margin in the Black Sea and possibly deep within the microlaminated zone of the Cariaco Basin enrich the iron sulfides in ^{34}S relative to the water-column signal. The basal, oxic portion of the Cariaco profile shows the $\delta^{34}S$ signature of the secondary sulfur overprint (see Figure 9 for S_{py} concentrations). Further discussions are available in Sections 13.3.4.2 and 13.4.1.

siliciclastic accumulation, mean C_{org} concentrations are 1.6 wt.% compared to 5.3 wt.% in the central basin despite $CaCO_3$ contents averaging ~52 wt.%, while C_{org} accumulation rates are ~30.9 $gC_{org}\,m^{-2}\,y^{-1}$ and 1.9 $gC_{org}\,m^{-2}\,y^{-1}$, respectively (Figure 10; Calvert and Karlin, 1991; Calvert *et al.*, 1991; Lyons and Berner, 1992; Lyons, 1997). Ancient facies analogous to these euxinic upper-slope environments along the Black Sea margin provide a unique challenge for the paleoredox proxies outlined here. However, these complications can also work to our benefit. For example, a multiproxy approach, including benthic ecologies, can speak to the presence or absence of sulfide in the water column, the persistence of such conditions, and the relative clastic fluxes within and among oxygen-deficient basins.

A recurring theme in this paper is the challenge of deducing detailed environmental change from ancient sequences lacking adequate age models (compare Meyers *et al.*, 2001). By contrast, the Black Sea and Cariaco Basin provide high-resolution ^{14}C-, ^{210}Pb- and varve-based chronological control under conditions of comparatively rapid sedimentation (Calvert *et al.*, 1991; Crusius and Anderson, 1992; Anderson *et al.*, 1994; Arthur *et al.*, 1994; Arthur and Dean, 1998; Hughen *et al.*, 1996) and are thus ideally suited to paleoceanographic study—including the principal controls on C_{org} accumulation. Among these important findings, recent work in the Cariaco Basin has yielded a clear low-latitude record of high C_{org} accumulation during the Younger Dryas cold event, which is readily attributable to high productivity during enhanced, trade-wind-driven

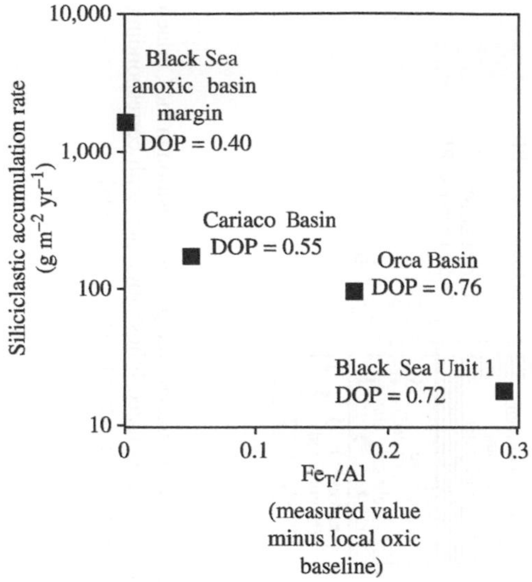

Figure 8 Approximate data showing the general inverse relationship between rate of siliciclastic accumulation and Fe_T/Al ratios as predicted by our model (Sections 13.3.4.3 and 13.4.1; Lyons *et al.*, 2003). The chronological framework for the Black Sea is from Calvert *et al.* (1991) and Anderson *et al.* (1994) (see also Lyons *et al.*, 2003), and details for the Cariaco Basin are provided in Lyons *et al.* (2003). The plotted Fe_T/Al ratio is the difference between the measured mean at a given euxinic site and a mean value for oxic sediments within the basin (analogous to the "GMS" introduced in Section 13.4.3). The Fe_T/Al ratios for the Orca Basin are unpublished, but other details are available in Hurtgen *et al.* (1999). Our unpublished results from Effingham Inlet, British Columbia—an anoxic Fjord where Fe_T/Al ratios are significantly elevated despite very high siliciclastic accumulation rates (see Hurtgen *et al.*, 1999)—have already suggested that this relationship is not universally valid. These complexities are the subjects of ongoing research.

upwelling (Peterson *et al.*, 1991; Hughen *et al.*, 1996; Werne *et al.*, 2000).

Figure 9 shows the relationships among Mo/Al, C_{org}, and pyrite sulfur for the euxinic interval of the last ~14.5 kyr in the Cariaco Basin and the uppermost oxic deposition, which correspond to the last glacial lowstand. Although the gray sediments at the top of the oxic interval bear a strong secondary sulfur overprint marking the onset of euxinic conditions, Mo/Al within the bioturbated gray and reddish-brown clays occurs at roughly average shale (continental) values (Lyons *et al.*, 2003). By contrast, molybdenum is enriched within the laminated euxinic interval, as expressed in Mo/Al ratios, by roughly two orders of magnitude beyond the oxic sediments, and the ratios vary sympathetically with C_{org} concentration. Despite the potential for ambiguity in molybdenum relationships in organic-rich sediments, as outlined in Section 13.3.4.4, Mo/Al ratios do sharply delineate the onset of euxinic deposition in the Cariaco Basin and suggest a strong mechanistic linkage between molybdenum and C_{org} accumulation (Figure 9; Lyons *et al.*, 2003; Meyers *et al.*, in review). Our recent work (see Section 13.4.2 and Meyers *et al.*, in review) is showing, however, that iron may play a central in controlling patterns of molybdenum enrichment. Specifically, the buildup of HS^-, which seems essential for molybdenum mineralization, is ultimately modulated by the balance between rates of sulfide production through BSR and consumption through iron sulfide formation. Therefore, in the Cariaco Basin, the strong diffusional sulfide overprint linked to the transition from oxic marine to euxinic marine conditions (Figure 9) resulted in extensive pyrite formation but likely little accumulation of dissolved sulfide because of the high availability of reactive iron. This relationship was compounded by the very low levels of OM in the oxic sediments and thus low potential for in situ sulfide production.

13.4.2 Cretaceous Western Interior Basin

During the Mesozoic, greenhouse melting of polar ice caps combined with global tectonic processes resulted in eustatic highstands that flooded many continental regions with epeiric seas, significantly expanding the shallow coastal zone of the oceans. These sites not only preserved key records of oceanic and climatic events but also acted as reservoirs of carbon burial and thus played important role in the linkage between biogeochemical cycles and oceanographic/climatic events. One of the best preserved records of a Cretaceous epeiric sea is the Western Interior seaway, a mid- to high-latitude, meridional retroarc foreland basin that extended from present day Arctic Canada to the Gulf of Mexico and connected the northern Boreal Sea with the circum-equatorial Tethys Sea (Kauffman, 1984). The basin is bounded on the west by the Sevier Orogenic Belt, which supplied most of its sedimentary fill and was responsible for a significant portion of its subsidence history, especially in the western foredeep (Kauffman, 1984; Kauffman and Caldwell, 1993). This subsidence, in combination with long-term eustatic rise, resulted in accumulation of over 5 km of sediment in the basin spanning Albian through Maastrichtian time (e.g., Dyman *et al.*, 1994). These deposits, which represent mixing of two main sedimentary sources (Sevier-derived siliciclastics and OM, and pelagic-derived carbonate and OM), have been studied in great detail through the years, culminating in what is surely one of the most highly refined and comprehensive chronostratigraphic frameworks and geological databases of any

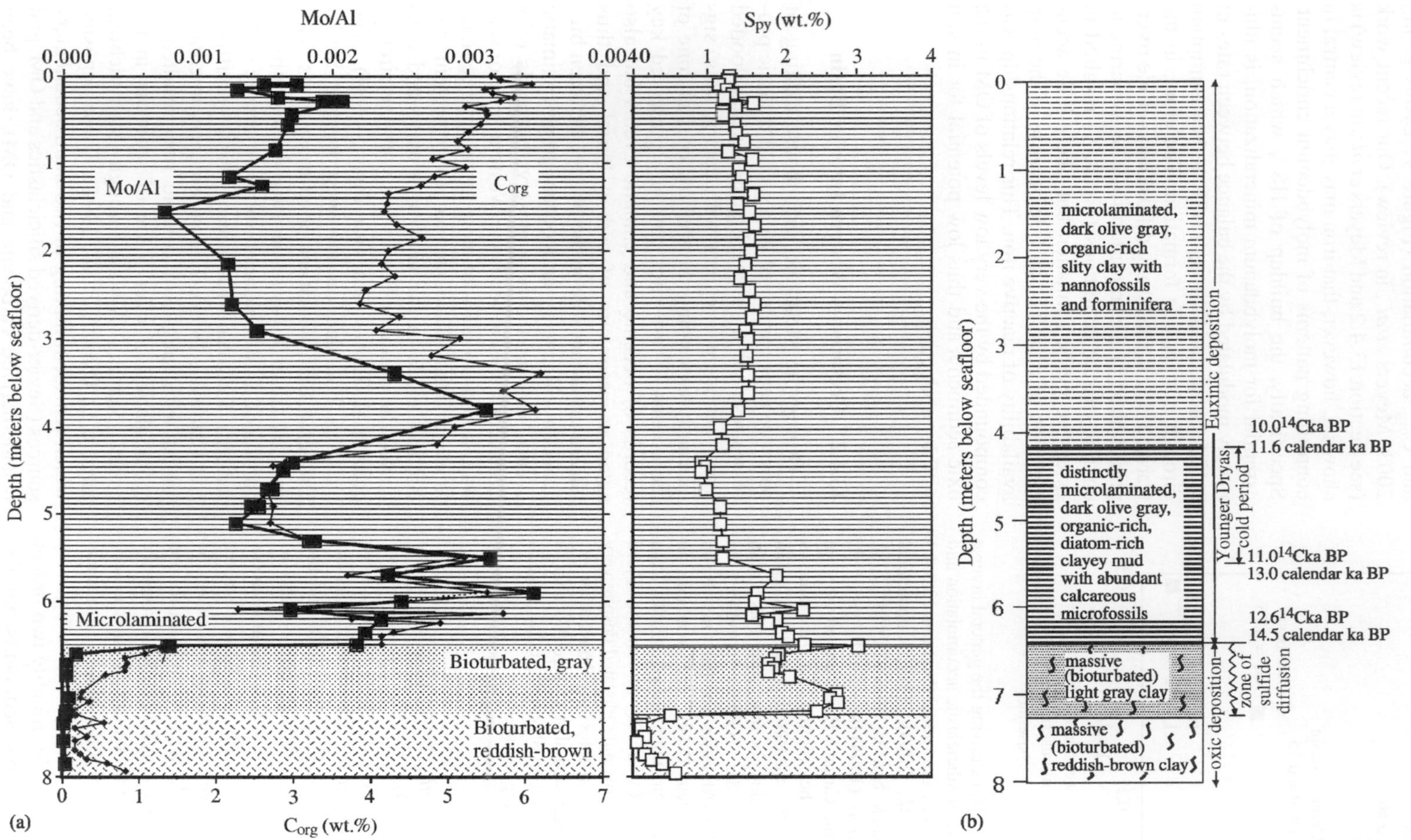

Figure 9 (a) Depth profiles for total Mo/Al ratios and wt.% pyrite sulfur (S_{py}) for both oxic and euxinic sediments spanning the most recent glacial–interglacial transition in the Cariaco Basin, Venezuela (Ocean Drilling Program Site 1002). Corresponding data for C_{org} concentrations are provided for comparison. (b) Generalized lithostratigraphy with corresponding chronology. Note general covariance between C_{org} and Mo/Al within the euxinic interval. See Sections 13.3.4.4 and 13.4.1 (after Lyons *et al.*, 2003).

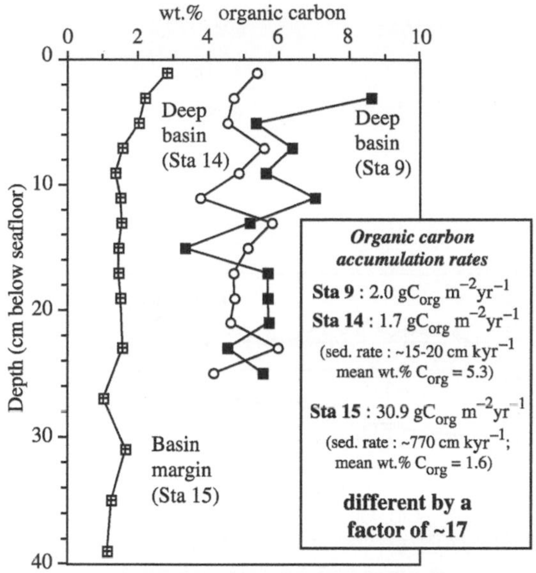

Figure 10 Summary of C_{org} concentration and accumulation rate data for modern euxinic sediments in the Black Sea. Stations 9 and 14 in the deep basin are characterized by microlaminated, carbonate-rich Unit 1 deposits, and Station 15 represents the rapidly accumulating, soupy, black (iron-monosulfide-rich), dominantly siliciclastic sediments on the anoxic upper slope (Calvert and Karlin, 1991; Lyons, 1991, 1997; Calvert *et al.*, 1991; Lyons and Berner, 1992; Anderson *et al.*, 1994; see also Lyons *et al.*, 2003). The sediments at all three sites are laminated. The ^{210}Pb and ^{137}Cs data of Moore and O'Neill (1991) and Anderson *et al.* (1994) for Station 15 confirm that the radiocarbon accumulation rates of Calvert *et al.* (1991) are spurious (see Lyons, 1997) and are related to reworking of older terrestrial and/or marine $CaCO_3$ and OM in the upper slope setting (Lyons, unpublished data). In the deep basin, however, ^{14}C-based chronologies (e.g., Calvert *et al.*, 1991) are consistent with independent measures of sedimentation rate, including varve counts (Arthur *et al.*, 1994). This figure is included to highlight the potential for low C_{org} concentrations as a result of rapid clastic dilution—even under conditions of high C_{org} accumulation (Section 13.4.1).

ancient sedimentary basin. Using this framework, numerous geochemical investigations of Cretaceous hemipelagic sedimentation have been conducted in the Western Interior (e.g., Pratt, 1984, 1985; Arthur *et al.*, 1985; Hayes *et al.*, 1989; Arthur and Dean, 1991; Pratt *et al.*, 1993; Sageman *et al.*, 1997, 1998; Dean and Arthur, 1998; Meyers *et al.*, 2001, submitted). The data shown in Figure 11 represent a distillation of these studies organized using the conceptual model proposed above.

The study interval discussed here includes the Middle to Late Cenomanian and Early Turonian Lincoln Limestone (LLM), Hartland Shale (HSM), and Bridge Creek Limestone Members (BCMs) of the Greenhorn Formation (Figure 11). These units were penetrated in the USGS #1 Portland core as part of a continental drilling study (Dean and Arthur, 1998), sampled at high resolution, and analyzed for many of the proxies described above (Sageman *et al.*, 1997, 1998; Meyers *et al.*, 2001; Sageman, unpublished data). The data series in Figure 11(a) include concentration data (5–10 cm resolution and 2 m moving average), and isotopic data (from Pratt, 1985). In Figure 11(b) the same concentration data are plotted with calculated mass accumulation rates (MAR: 2 m moving average). In the BCM a high-resolution orbital timescale was used for accumulation rate calculations, whereas MARs in the underlying units are based on Ar–Ar dating of bentonite layers that occur within the lowermost Bridge Creek, upper Lincoln, and basal Lincoln Members (Obradovich, 1993); analytical methods used to generate concentration and accumulation rate data are described in detail in Meyers *et al.* (2001). Although parts of some of the plots in Figure 11 are repeated from earlier publications, this compilation of high-resolution data comparing all the members of the Greenhorn Formation has not been presented before. Our main purpose in discussing this data set is to simply highlight the strengths and weaknesses of selected detrital, biogenic, and authigenic proxies.

The data set is excellent for this purpose because paleo-environmental conditions and sea-level history for the units are well known from regional high-resolution chronostratigraphic correlation and detailed lithic and paleobiologic data sets. The lithic character of the Greenhorn Formation reflects a significant part of the transgressive phase of the Greenhorn cyclothem, culminating in peak flooding in the Early Turonian (Kauffman and Caldwell, 1993). The LLM is predominantly characterized by weakly to moderately calcareous shale with intercalated skeletal limestone and bentonite beds in the upper half (Sageman and Johnson, 1985). Based on correlation to progradational units in nearshore western settings, the presence of submarine (wave base) unconformities on the eastern cratonic margin, and evidence of storm influence in the basin center, Sageman (1985, 1996) interpreted upper LLM skeletal limestones to reflect winnowing during a relative sea level fall, followed by condensation during subsequent sea level rise and transgression. The HSM is dominated by well-laminated calcareous shales and a restricted benthic fauna (Sageman, 1985; Sageman and Bina, 1997) and also includes skeletal limestone and bentonite beds in its middle section that correlate to a western prograding clastic wedge, suggesting a relative sea-level oscillation (Sageman, 1985, 1996). However, this sea-level event was likely of lower amplitude

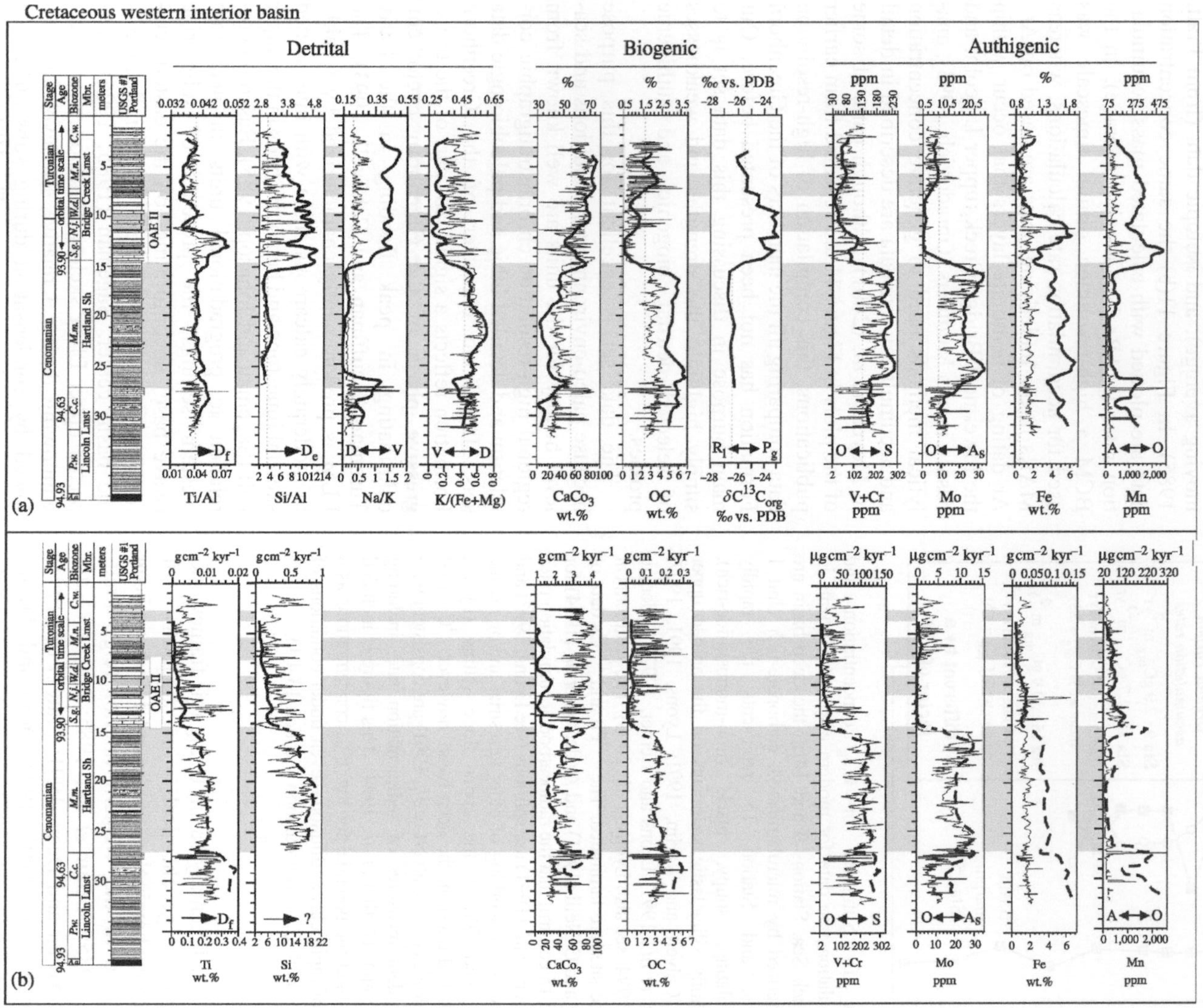

Cretaceous western interior basin
Detrital
Biogenic
Authigenic
(a)
(b)
Stage
Age
Biozone
Mbr.
meters
USGS #1 Portland
Turonian
Cenomanian
orbital time scale
93.90
94.63
94.93
Bridge Creek Lmst
Hartland Sh
Lincoln Lmst
OAE II
Ti/Al
Si/Al
Na/K
K/(Fe+Mg)
CaCo3 wt.%
OC wt.%
δC13Corg ‰ vs. PDB
V+Cr ppm
Mo ppm
Fe wt.%
Mn ppm
Ti wt.%
Si wt.%
g cm−2 kyr−1
μg cm−2 kyr−1

than the Lincoln event. Throughout most of HSM deposition the basin was dominated by oxygen-deficient conditions as indicated by faunal data (Sageman and Bina, 1997). The overlying BCM is marked by a conspicuous change to rhythmically interbedded limestone and calcareous shale or marlstone beds with interbedded bentonites and some skeletal limestones. This change is associated with a major increase in faunal diversity and abundance and a shift to extensively burrowed substrates (Elder, 1985, 1991; Sageman *et al.*, 1997). Faunal diversity decreases in a stepwise fashion up to the Cenomanian–Turonian boundary and this biotic decline reflects a major extinction event (Elder, 1989). This event is characterized by global C_{org} burial and oxygen deficient conditions and has been termed Oceanic Anoxic Event II (Schlanger *et al.*, 1987; Arthur *et al.*, 1987); it is marked in Figure 11 by a positive $\delta^{13}C_{org}$ excursion originally documented in Western Interior strata by Pratt (1985). The BCM retains its lithologic character over thousands of square kilometers in the basin (Kansas to western Colorado; South Dakota to New Mexico), and the transition into this facies can be traced to a major flooding surface on the western margin (Elder *et al.*, 1994). This lateral relationship suggests that Bridge Creek deposition was initiated by a rapid pulse of relative sea-level rise and transgression resulting in clastic sediment starvation over a large portion of the basin (Meyers *et al.*, 2001).

The proxies for balance between detrital and volcanigenic input (Na/K, K/(Fe+Mg)), which are interpreted to reflect increase/decrease in riverine siliciclastic flux relative to (±constant) background volcanic ashfall, are consistent with higher detrital flux during HSM deposition, relative condensation during the upper LLM, and a significant decrease in terrigenous flux at the base of the BCM. These data agree with Pratt's (1984) observations for trends in discrete illite. Anomalous increase in Si/Al ratios above the average for hemipelagic clays could reflect enhanced biogenic silicon input (Eicher and Diner, 1989) or greater eolian inputs of quartz silt, suggesting relative condensation (Arthur *et al.*, 1985; Arthur and Dean, 1991). For the Si/Al data shown in Figure 11 we favor the latter interpretation based on the concentration maxima of silicon in BCM limestone beds, which we interpret to reflect times of decreased siliciclastic flux (see also Arthur and Dean, 1991; Dean and Arthur, 1998), as well as the accumulation rate data for biogenic components, which suggest higher overall levels of primary production during HSM deposition. The Ti/Al ratio is weakly indicative of these same trends, showing greater than average values below the lower BCM and less than average values above. The accumulation rate data for titanium are more illustrative, broadly tracking the Greenhorn transgression (Figure 11). Interestingly, the largest pulse in Ti/Al occurs in the lower BCM following the major transition to dominance of carbonate dilution over insoluble residue. This pulse is matched by other indicators of detrital input, and it correlates to a progradational interval in SW Utah (Laurin and Sageman, 2001). However, the progradation is a relatively minor clastic wedge, suggesting that the sensitivity of the Ti/Al proxy may vary as a function of the proportion of insoluble residue relative to carbonate.

The most prominent indicator of biogenic processes in the C–T interval is the positive excursion in $\delta^{13}C_{org}$ (Figure 11), which has been interpreted to reflect a global increase in primary productivity (e.g., Arthur *et al.*, 1988). Interpretations of the isotopic record in the study interval are constrained by the determination that preserved OM in the distal basin is dominated by marine algal input (Pratt, 1984; Pancost *et al.*, 1998). As discussed by Meyers *et al.* (2001), evidence for relative increase in production in the BCM (increased MAR-$CaCO_3$ and MAR-C_{org}) does not coincide with the initial isotopic shift but occurs instead in the latter part of the excursion, suggesting that the onset of environmental conditions characteristic of OAE II in the Western Interior basin lagged behind

Figure 11 Geochemical data series and lithologic log for Cretaceous Greenhorn Formation in USGS #1 Portland core, modified from Sageman *et al.* (1997, 1998), Meyers *et al.* (2001; submitted), and Sageman (unpublished data). Data series include: (a) high-resolution concentration data (thin lines) with 2 m moving averages (thick lines) and (b) high-resolution concentration data (thin lines) with 2 m moving averages for accumulation rate. In both (a) and (b) units on the upper scales refer to thick lined plots. Thin vertical dashed lines represent mean values for the data set ($N = 309$ for most proxies). Horizontal shaded areas mark the organic-rich intervals. Data were generated using analytical methods and orbital timescale described in Meyers *et al.* (2001), except MAR values below BCM (dashed), which employ 2 m moving averages of concentration data, bulk density values of 2.6 g cm^{-3}, and linear effective sedimentation rates based on radiometric ages of Obradovich (1993). Also shown are $\delta^{13}C_{org}$ data from Pratt (1985). Location of OAE II, defined by the $\delta^{13}C_{org}$ excursion, is indicated next to the lithologic log. Abbreviations for proxy interpretation terms include: D—detrital source; V—volcanogenic source; D_f—detrital fluvial source; D_e—detrital eolian source; R_l—respiration/local reservoir-dominated; P_g—production/global reservoir-dominant; O—oxic; S—suboxic; A—anoxic; A_s—anoxic/sulfidic.

the global record. In addition, although the $\delta^{13}C_{org}$ excursion interval is marked by indicators of euxinic conditions in the North Atlantic basin (Brumsack and Thurow, 1986; Sinninghe Damsté and Koster, 1998), trace-metal evidence for pervasive occurrence of such conditions in the BCM occurs only after the C-isotope excursion—during maximum highstand and maximum sediment starvation (see also Simons and Kenig, 2001). Also interesting is the fact that MAR values for $CaCO_3$ and C_{org} a re much higher in the underlying HSM than in the BCM (Figure 11). The higher burial flux of C_{org} in this interval may be explained by enhanced preservation, as authigenic proxies for redox conditions suggest common transitions through the suboxic zone (maximum concentration and MAR of V + Cr) and at least intermittent dominance of anoxic–sulfidic conditions (maximum concentration and MAR of molybdenum). Yet, the HSM water column was never permanently sulfidic, as indicated by the frequency of benthic colonization events (Sageman, 1985, 1989), but probably experienced strong seasonal thermal stratification with only weak mixing by winter storms, perhaps punctuated by 100-year-storm ventilation events (Sageman and Bina, 1997). Consistent with this interpretation is the observation that average values of $\delta^{13}C_{org}$ in the HSM are slightly depleted relative to the post-OAE II background (Figure 11), suggesting that recycled CO_2 may have more strongly influenced the HSM isotopic signal. Manganese data also support these interpretations, showing enrichment (concentration and MAR) during the upper LLM and lower BCM, likely reflecting precipitation of Mn-oxyhydroxides or Mn-carbonates during oxygenation events, and depletion during the HSM, recording time averaged dominance of oxygen deficient conditions and prevalence of soluble Mn^{2+}.

Based on a comparison of modern chemical oceanographic data and the results described above, Meyers *et al.* (submitted) have attempted to link the production, terrigenous dilution, sedimentation rate, and redox processes illustrated graphically in Figure 2 into a quantitative model (Figure 12). This model depends on the fact that OM decomposition in sediments may be approximated using first-order rate equations (see Figure 12(a)), where the degree of OM degradation (ΔG) in each remineralization regime is dependent upon: (i) the amount of time OM resides in each regime (t), (ii) the initial concentration of metabolizable OM at the top of each regime (G), and (iii) the first-order OM degradation rate constants for each regime (k). The rate of organic carbon remineralization at a given time may be expressed as $-kG$, where the concentration of metabolizable organic carbon (G) is dependent upon the balance between export production (OM-export), dilution (MAR-dilutant), and the degree of degradation that has occurred (ΔG). Meyers *et al.* (submitted) demonstrate that the integration of this conceptual framework for OM diagenesis with high-resolution geochemical proxy burial fluxes in ancient organic-rich strata (Figure 12(a)) provides a unique opportunity to (i) evaluate the role of anaerobic remineralization on OM export to the lithosphere during OM burial events, and (ii) to deconvolve the paleo-environmental mechanisms that control OM delivery to these anaerobic microbial zones over geological time.

Application of this framework to evaluate anaerobic remineralization during ancient organic burial events requires a proxy for the rate of anaerobic OM degradation within sediments ($-kG$). The sensitivity of molybdenum accumulation to the rate of hydrogen sulfide production via sulfate reduction (Helz *et al.*, 1996) provides a promising proxy for $-k_sG_s$. Achievement of the critical pore water H_2S levels necessary for molybdenum accumulation depends upon a balance between the processes that source H_2S (*in situ* sulfate reduction rate and the rate of diffusion/advection of hydrogen sulfide into the pore water in the case of euxinic basins), the processes that deplete H_2S (the rate of formation of Fe-sulfides and OM sulfurization and the rate of diffusion/advection of H_2S out of the pore water), and the volume of water in the connected pore space. When the rate of sulfate reduction is high, production of hydrogen sulfide can outpace depletion within pore waters (which is largely controlled by reactive-Fe availability and pyrite formation), and hydrogen sulfide may amass to the levels necessary for molybdenum scavenging. Such high rates of sulfate reduction may be a consequence of shallowing of the sediment sulfate reduction zone ($d_a + d_n$), increased export production (OM-export), changes in inorganic dilutant flux (MAR-dilutant), or changes in bulk sedimentation rate (ω) (Figure 12(a)). Based on these lines of reasoning, Meyers *et al.* (in review) employed molybdenum accumulation as a proxy for the rate of sulfate reduction during and immediately following Oceanic Anoxic Event II (OAE II) and evaluated it in tandem with iron accumulation rates (primarily reflecting pyrite formation) to assess changes in the relative demand for H_2S within the sediment pore water (Figure 12(b)). When compared to C_{org} accumulation rates through the same interval, these data suggest that the highest rates of sulfate reduction and the highest rates of C_{org} accumulation within the sediments occur not during OAE II but following this global event. Iron accumulation rate data from the same interval suggest that lower C_{org} and molybdenum accumulation during OAE II was due to buffering of H_2S levels via reactive-Fe delivery and pyrite formation. Decreased dilution of labile OM and decreased delivery of reactive iron in the post-OAE II interval served to (i) increase the rates of

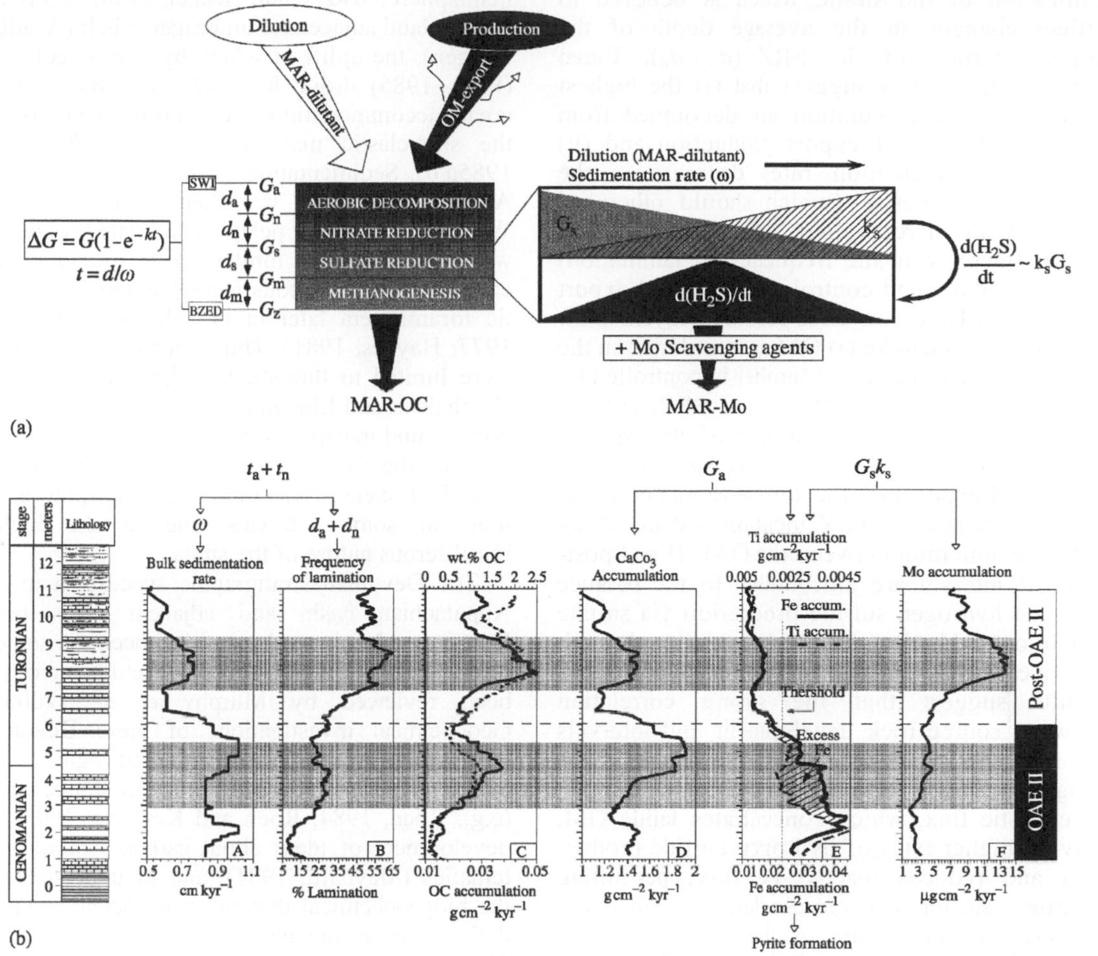

Figure 12 (a) Theoretical model for OM degradation in the sediment, and the linkage between OM burial and molybdenum accumulation. Subscript "a" refers to aerobic decomposition, subscript "n" refers to nitrate reduction, subscript "s" refers to sulfate reduction, and subscript "m" refers to methanogenesis. G = concentration of OC at the top of each remineralization zone; ΔG = wt.% OC change due to decomposition in each remineralization zone; ω = bulk sedimentation rate in cm/kyr; k = decomposition rate constant in kyr^{-1}; t = time spent in decomposition zone in kyr; d = thickness of decomposition zone in cm; (dH_2S/dt) = rate of hydrogen sulfide production via sulfate reduction; MAR-dilutant = dilutant accumulation in g cm^{-2} kyr^{-1}; OM-export = organic carbon export in g cm^{-2} kyr^{-1}; MAR-Mo = authigenic molybdenum accumulation in g cm^{-2} kyr^{-1}; MAR-OC = organic carbon accumulation in g cm^{-2} kyr^{-1}; G_z = final wt.% OC; SWI = sediment–water interface; BZED = base of zone of early diagenesis. (b) Bulk sedimentation rates, 2 m moving-average geochemical proxy accumulation rates, 2 m moving average wt.% OC data, and percent lamination data from the Bridge Creek Limestone Member (#1 Portland core). Percent lamination was calculated using the ORI rank data from Savrda (1998) and a 2 m moving-window, and is employed to assess the average depth of the upper interface of the SRZ. $CaCO_3$ accumulation is employed as a proxy for export production, and titanium accumulation is employed as a proxy for siliciclastic dilution.

hydrogen sulfide generation and (ii) decrease the buffering capacity of the system, driving the sediments to a more sulfidic state. High levels of reactive-Fe delivery during OAE II were a consequence of greater detrital flux, and this iron source was supplemented by an additional dissolved reactive-Fe source. The timing of this excess (dissolved) iron accumulation (Figure 12(b)) correlates with the hypothesized initiation of Tethyan oxygen minimum zone (OMZ) influence within the basin (Leckie *et al.*, 1998; Meyers *et al.*, 2001).

Two additional factors that must be considered when assessing the mechanism for enhanced molybdenum and OC accumulation in the post-OAE II interval are the rate of export production (OM-export) and the amount of time this export production spends prior to entering the sediment SRZ ($t_a + t_n$). Based on calibration with modern oceanic sites, Meyers *et al.* (submitted) employ $CaCO_3$ accumulation as a proxy for export production. The term $t_a + t_n$ is estimated based on the bulk sedimentation rate (ω) and frequency of lamination of the strata, which is believed to reflect changes in the average depth of the upper interface of the SRZ ($d_a + d_n$). Taken together, these data suggest that (i) the highest rates of C_{org} accumulation are decoupled from the highest rates of export production and (ii) although sedimentation rates decrease in the post-OAE II interval (which should otherwise result in an increase in $t_a + t_n$), a decrease in $d_a + d_n$ (increase in the frequency of lamination) exerts the dominant control on labile OM export to the SRZ. Based on these results, Meyers *et al.* (submitted) conclude: (i) OM accumulation in the Bridge Creek Limestone Member is controlled by the rate of export of OM to the SRZ; (ii) the location of the upper boundary of the SRZ is the first-order control (OAE II versus post-OAE II) on OM export into the sulfate reduction zone; and (iii) changes in SRZ location and molybdenum accumulation between the OAE II and post-OAE II interval are attributable to the balance between hydrogen sulfide production via sulfate reduction and hydrogen sulfide depletion through reactive-Fe delivery and pyrite formation. These results suggest that the strong correlation between source rock development and intervals of transgression in the geological record is the biogeochemical consequence of a decrease in the siliciclastic flux, which concentrates labile OM, driving higher rates of hydrogen sulfide production and reduces reactive-Fe flux, permitting hydrogen sulfide levels to escalate and enhancing the preservational state of the system. As the burial flux data for the Bridge Creek demonstrate, the increased rate of export of labile OM into a shallower SRZ resulted in elevated C_{org} accumulation rates, even under lower rates of primary production.

13.4.3 Devonian Appalachian Basin

During a significant portion of the Paleozoic, conditions were broadly similar to the Mesozoic greenhouse: elevated p_{CO2} levels, warm temperatures, decreased equator to pole temperature gradients, and widespread marine flooding of continental areas due to reduced ice volumes and possibly tectonoeustatic effects (e.g., Woodrow, 1985; Johnson *et al.*, 1985; Berner and Kothavala, 2001). Within these greenhouse times, the Devonian is of particular biogeochemical interest due to events such as the rise of vascular plants, widespread black shale deposition (Algeo *et al.*, 1995), and the Frasnian–Famennian mass extinction (McGhee, 1982). One of the best-preserved records of Devonian faunal history and environmental conditions is found in deposits of the Appalachian basin. Like the Western Interior, this basin formed in mid-latitudes (southern hemisphere; Witzke and Heckel, 1988) as a retro-arc foreland adjacent to an orogenic belt (Acadian Orogen), the uplift of which by terrane collision (Faill, 1985) drove load-induced subsidence to create accommodation space and sourced most of the siliciclastic material to fill it (Ettensohn, 1985a,b). Sedimentation in the distal part of the Appalachian basin was hemipelagic like the Western Interior, but pelagic carbonate production was comparatively limited prior to significant expansion of calcareous nanoplankton and planktic foraminifera later in the Mesozoic (Gartner, 1977; Haynes, 1981). Thus, sources of carbonate were limited to thin-shelled planktic styliolinids (Yochelson and Lindemann, 1986), allodapic carbonate mud transported from shallow areas to the west of the foredeep (Werne *et al.*, 2002), and reworked skeletal material, which is quite common in some intervals due to the highly fossiliferous nature of the strata.

The Devonian stratigraphic succession in the Appalachian basin and adjacent areas (e.g., Illinois and Michigan basins) has been the subject of much study through the years, and has recently been reviewed by Murphy *et al.* (2000b). Geochemical investigations of these Devonian fine-grained facies have contributed significantly to the characterization of petroleum source rocks (e.g., Roen, 1984; Roen and Kepferle, 1993), to development of ideas about carbon isotope systematics (Maynard, 1981), and to understanding the biogeochemical dynamics of ancient oxygen-deficient environments (e.g., Ingall *et al.*, 1993). The data presented in Figure 13 are abstracted from results of a recent study undertaken by the authors and their students (Murphy *et al.*, 2000a,c; Werne *et al.*, 2002; Sageman *et al.*, 2003). The project's main objective was to develop a high-resolution, continuous geochemical database from analysis of core samples spanning Eifelian through Famennian strata in western New York State and to use it to delineate the controls on C_{org} burial based on the conceptual approach illustrated in Figures 1 and 2. The complete data set, with descriptions of analytical methods and detailed interpretations, was summarized recently by Sageman *et al.* (2003). Here we focus on four of the studied intervals that include transitions between black and gray shale facies, but omit data from the thick intervening intervals of relatively homogenous gray shale. Our objective is to highlight the strengths and weaknesses of selected detrital, biogenic, and authigenic proxies.

The four study intervals are, in ascending order, the Marcellus subgroup (including Union Springs Member, Bakoven Formation and Oatka Creek Formation: US and OC), the Geneseo Formation (GS), the Pipe Creek Formation (PC), and a thin section of organic-rich facies in the uppermost Hanover Formation (UH) (Figure 13);

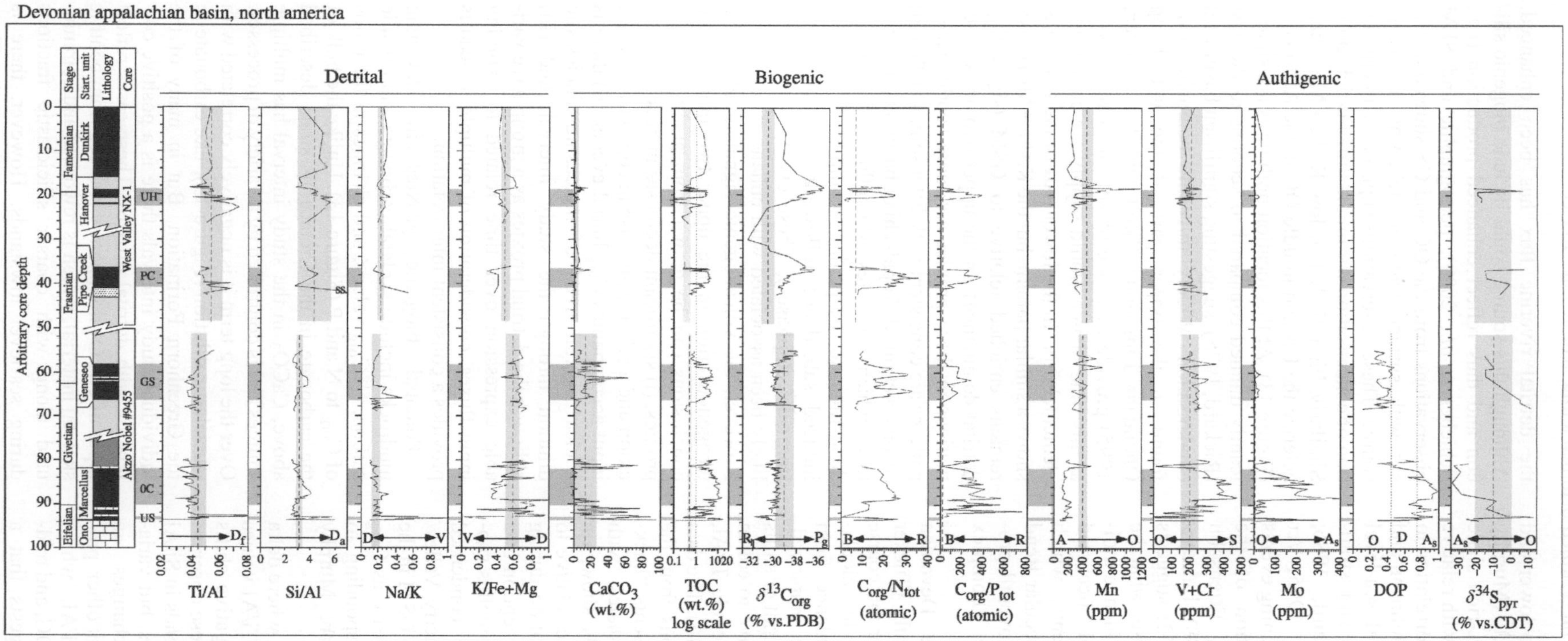

Figure 13 Geochemical data series and lithologic log for Devonian Marcellus subgroup, Geneseo, Pipe Creek, and Hanover–Dunkirk Formations in the Akzo Nobel #9455 and West Valley NX-1 cores, modified from Murphy *et al.* (2000a), Werne *et al.* (2002), and Sageman *et al.* (2003). Data series include concentration data (thin lines) with gray shale mean (GSM) indicated by vertical dashed line and $\pm 1\sigma$ indicated by vertical gray shading. For calculation of GSM in Akzo core, $N = 227$ for most proxies; for West Valley $N = 141$ for most proxies. In some cases vertical dotted lines mark threshold values, such as 1% C_{org} (gray shale cutoff), or 0.45 and 0.75 DOP (divisions for oxic, dysoxic, and anoxic: Raiswell *et al.*, 1988; Canfield *et al.*, 1996). Horizontal shaded areas mark the organic-rich intervals and are labeled with stratigraphic unit abbreviations used in text. Data were generated using analytical methods described in Murphy *et al.* (2000a) and Werne *et al.* (2002). Abbreviations for proxy interpretation terms as in Figure 11, except: B—buried nutrients; R—recycled nutrients; D_x—dysoxic.

the latter two intervals correspond to lower and upper Kellwasser horizons, respectively (Murphy *et al.*, 2000c). Each of these organic-rich shale units has been interpreted to coincide with relative deepening based on biostratigraphic correlation to proposed global eustatic events (i.e., Johnson and Sandberg, 1988), as well as regional sea-level reconstructions based on lithofacies and stratigraphic architecture (Brett and Baird; 1986; House and Kirchgasser, 1993; Brett, 1995; Ver Straeten and Brett, 1995). These deepening events may be related to tectonic evolution of the Appalachian basin as suggested by Ettensohn (1985a,b), and Ettensohn *et al.* (1988), but evidence discussed in Werne *et al.* (2002) suggests that eustasy also played a role. No matter what the mechanism, the bulk of geological evidence indicates that organic carbon burial occurred preferentially during deepening events, which are superimposed upon a long-term secular trend of relative shallowing reflecting the overall progradation of the Catskill Delta complex (Ettensohn, 1985a).

Although geochronology for the Devonian study interval has been refined recently (Tucker *et al.*, 1998), control points are still far too coarse to allow meaningful accumulation rate calculations. Instead, multiple lines of evidence from concentration and isotopic data are employed. Proxies reflecting changes in detrital input or shifts in the proportion of detrital flux relative to eolian flux (siliciclastic or volcanogenic) include Ti/Al, Si/Al, Na/K, and K/(Fe + Mg) (Figure 13). The data are best evaluated relative to deviations from the gray shale mean (GSM), which represents "background" values averaged from shales with C_{org}<1%. The C_{org}-enriched intervals show, on average, Ti/Al values that are depleted relative to GSM, consistent with sediment starvation during transgression (Figure 13). With the exception of a major excursion at the base of the Marcellus subgroup, which represents the highly condensed distal record of a lowstand event (Cherry Valley/Halihan Hill bed), Ti/Al values are sub-GSM in the US/OC and GS black shales but increase as C_{org} levels decrease (e.g., late GS, corresponding to progradation and progressive dilution; Murphy *et al.*, 2000a).

In the younger two study intervals, where delta progradation was more pronounced, Ti/Al values are higher overall but still show negative shifts associated with organic-rich units. These observations are generally corroborated by trends in Si/Al and the Na/K and K/(Fe+Mg) proxies, but variations up-section may reflect changes in background volcanogenic fluxes and other processes. For example, increased Si/Al, which clearly characterizes the US, middle OC, and middle GS intervals (Figure 13), suggests that a source of silicon other than that associated with the detrital riverine flux has been enhanced. Although this source could include biogenic silicon and thus reflect enhanced production (i.e., Schieber *et al.*, 2000), petrographic and SEM observations from the OC and GS showed quartz silt with surface features indicating an eolian source. These observations support the interpretation of condensation within the organic-rich units. Similarly, high Na/K and low K/(Fe + Mg) characterize the US and middle OC in the Akzo core (Figure 13). At this location both units are significantly thinned compared to sections to the east (Rickard, 1975), suggesting significant basinward condensation. Widely traceable bentonite beds found in the upper part of the underlying Onondaga Limestone (Brett and Ver Straeten, 1995) provide evidence of volcanogenic activity at this time, which would supply the contrasting elemental flux. In the lower part of the GS, Na/K shows a similar pattern, but the K/(Fe + Mg) proxy remains enriched relative to GSM (suggesting detrital dominance). In the upper two units, fluctuations in these proxies are more subtle. These patterns are consistent with Ettensohn's (1985a,b) reconstruction of Appalachian basin tectonic history, which attributes Marcellus and Geneseo deposition to deepening resulting from load-induced subsidence following collision events (termed tectophases). These events would have likely been associated with enhanced volcanism. However, the volcanogenic signal is diluted by terrigenous flux in the upper two units. Three conclusions can be drawn from analysis of the proxies: (i) significant decreases in bulk sedimentation and a relative increase in eolian over fluvial input accompanies C_{org} burial events, (ii) there is a clear overall trend of increasing siliciclastic dilution through the study interval, and (iii) although the detrital proxies as a group have variable expression over time related to evolving input fluxes, consideration of multiple indicators provides a consistent interpretation.

Potential biogenic proxies in Devonian mudrocks include skeletal carbonate and organic carbon, $\delta^{13}C_{org}$ values, and changes in the ratios of C_{org} to N and P (Figure 13). Unlike the bulk of the carbonate in the Cretaceous strata described above, $CaCO_3$ in the study interval has multiple sources and is controlled by a variety of processes. Over the long term it is negatively correlated with siliciclastic dilution (Figure 13), like carbonate in the Greenhorn Formation. But in many of the individual study intervals there is a positive correlation with detrital proxies. These cases reflect input of $CaCO_3$ dominantly sourced from benthic skeletal material, which is reworked and comminuted along with coarser siliciclastic fractions during sea-level lowstands. However, there is also evidence for ecologic and diagenetic effects on $CaCO_3$ production and dissolution (see discussion in Werne *et al.*, 2002).

Since no clear relationship between $CaCO_3$ concentration and pelagic production can be established, C_{org} is the main potential product of biogenic primary production in surface waters (following confirmation of marine algal source dominance, which was based on organic petrography and compound specific isotopic analysis; Murphy *et al.*, 2000a). As described above, however, the relationship between production and C_{org} concentration may be masked by variations in detrital dilution and OM remineralization. The observation that C_{org} enrichment in the studied units is definitely associated with condensation, and thus probably decreased bulk sedimentation rates, is consistent with evidence for transgression. As for elevated production, both the US/OC and the GS intervals have $\delta^{13}C_{org}$ values that are depleted relative to the GSM, which could reflect dominance of respired CO_2 in a restricted local reservoir (Lewan, 1986; Rohl *et al.*, 2001). Based on the parallel observation in each C_{org}-rich unit of significant deviations in C:N:P ratios from values expected for marine algal sources (Figure 13), Murphy *et al.* (2000a) proposed a model of fluctuating anoxia, nutrient remineralization, and relative eutrophication based on arguments summarized earlier (e.g., Aller, 1994; Ingall and Jahnke, 1997). The driving force for this model is establishment and breakdown of thermal stratification on timescales sufficient to allow nutrient build up and release and to cycle respired CO_2 back to surface waters with regenerated nutrients. In contrast to the US/OC and GS, the upper two organic-rich units (PC and UH) are characterized by major positive $\delta^{13}C_{org}$ excursions (Figure 6) that have been identified in localities around the world (e.g., Kellwasser horizons; Joachimski and Buggisch, 1993). These events are hypothesized to reflect, like the OAE II scenario described above, times when the extent of global net primary production and C_{org} burial exceeded respiration and recycling at a scale sufficient to overprint local reservoir effects.

Redox dynamics play a critical role in the hypothesis described above, and trends in the authigenic proxies support the model. For example, manganese concentrations are generally depleted throughout organic-rich intervals but commonly show enrichments at the top of these intervals, marking transitions from predominantly dysoxic–anoxic to predominantly oxic states. This pattern is particularly pronounced in the UH interval (Figure 13). The other indicators of reducing conditions show enrichments in the organic-rich intervals, but they clearly fall into two groups. The US/OC interval—characterized by the largest enrichments in V+Cr, Mo, and reactive-Fe (expressed in Fe_T/Al and DOP) and the most depleted values of $\delta^{34}S_{py}$ (Figure 13)—is the best candidate for a euxinic system of any Phanerozoic sequence studied by the authors (see detailed discussions in Werne *et al.*, 2002, and the paleoredox proxy details provided above). Although the other organic-rich intervals show depleted manganese values, small enrichments in V+Cr and Mo and shifts toward more depleted $\delta^{34}S_{py}$ values relative to GSM values, these changes are comparatively modest. Collectively, these observations suggest that most of the studied black shales were deposited in a shallow epeiric sea that was dynamically balanced between production, which led to increased oxygen utilization, and water-column mixing, which effectively recycled limiting nutrients to drive production but also bottom waters that were sufficiently oxygenated to ensure that sulfidic conditions could not be maintained for significant periods. The euxinic conditions of the US/OC interval were actually the exception rather than the norm (contrary to prior views—Byers, 1977; Ettensohn, 1985a,b) and resulted from a combination of maximum basin subsidence and eustatic rise, which resulted in water depths sufficient to limit effective mixing of the water column (Werne *et al.*, 2002). Notably, the most euxinic conditions in the studied units, although correlative with evidence for significant oxygen deficiency in other parts of the world (e.g., Truyols-Massoni *et al.*, 1990), do not correlate with a positive C-isotope excursion.

13.4.4 Recent Precambrian Advances

Precambrian studies are among the most promising yet least explored frontiers in shale geochemistry. In light of new analytical techniques, an improved understanding of sulfur microbiology, and vast additions to the geochemical database, however, shales are starting to play a central role in the development and validation of important new models for evolution of the early ocean–atmosphere–biosphere (e.g., Kakegawa *et al.*, 1998; Lyons *et al.*, 2000; Shen *et al.*, 2002; Strauss, 2002; see Lyons *et al.*, in press). One model argues for a globally euxinic Proterozoic ocean (Canfield, 1998), reflecting increases in seawater sulfate concentrations (linked to oxic continental weathering) in combination with a deep ocean that remained largely oxygen free. With the buildup of sulfate in an anoxic deep ocean, BSR drove extensive hydrogen sulfide production and accumulation, which may have ultimately led to the disappearance of banded iron formations through the corresponding decrease in iron solubility. (The Archean ocean, although anoxic, would have lacked the sulfate necessary for euxinic conditions

(Canfield, 1998; Habicht *et al.*, 2002). Under these widely sulfidic conditions of the Proterozoic, sequestration of bioessential metals such as molybdenum and iron, which are necessary for the production and utilization of bioavailable nitrogen, may have limited the ecological range and ultimately the evolution of eukaryotic algae (Anbar and Knoll, 2002).

Despite assertions made elsewhere in this report, delineation of an oxygen-deficient, sulfide-rich Proterozoic ocean is particularly elusive. For example, it is generally regarded that sulfate concentrations were substantially lower than those of the present ocean (perhaps only 10% of present levels during the Mesoproterozoic; Canfield, 1998; Hurtgen *et al.*, 2002; Shen *et al.*, 2002; Lyons *et al.*, in press; Kah *et al.*, submitted). Under such conditions, the light and uniform $\delta^{34}S_{py}$ values that are diagnostic of euxinic environments during the Phanerozoic (Section 13.3.4.2) are replaced by an abundance of ^{34}S-enriched pyrite that records local and likely global sulfate limitation (Lyons *et al.*, 2000; Luepke and Lyons, 2001; Shen *et al.*, 2002). By contrast, iron-based proxies, such as DOP, show more promise for the recognition of euxinicity on scales of individual basins (Shen *et al.*, 2002), and molybdenum isotope data may speak to the global distribution of such settings (Arnold *et al.*, 2002). Pyritic sulfur in fine-grained sediments spanning the ~2.3 Ga "great oxidation event" of the Precambrian atmosphere (e.g., Canfield, 1998; Holland, 2002; as reviewed in Lyons *et al.*, in press) also records the telltale disappearance of strong mass-independent sulfur isotope fractionation expected with shifting global redox (Farquhar *et al.*, 2000). Similarly, patterns of mass-dependent fraction expressed in these pyrites faithfully track the evolution of oxygen availability in the Precambrian ocean–atmosphere (Canfield and Teske, 1996; Canfield, 1998; Canfield *et al.*, 2000; Habicht *et al.*, 2002; Hurtgen *et al.*, 2002; reviewed in Lyons *et al.*, in press), and trace amounts of sulfate within carbonate rocks are yielding continuous, high-resolution records of the evolving $\delta^{34}S$ of seawater (Burdett *et al.*, 1989; Hurtgen *et al.*, 2002; Gellatly and Lyons, submitted; Kah *et al.*, submitted; Lyons *et al.*, in press).

13.5 DISCUSSION: A UNIFIED VIEW OF THE GEOCHEMISTRY OF FINE-GRAINED SEDIMENTS AND SEDIMENTARY ROCKS

In this chapter we first outlined a conceptual model for the relationship between primary geological processes involved in the formation of fine-grained, siliciclastic–biogenic sediments and sedimentary rocks and a selected set of geochemical proxies particularly useful for the reconstruction of these processes in ancient deposits (Figures 1 and 2). We then illustrated applications of this model through four case studies representing Precambrian, Paleozoic, Mesozoic, and modern marine environments. The conceptual model encompasses a formidable array of interrelated processes and proxies and draws on a very broad body of scholarship, which we endeavored to cite as fully as space would allow. Our rationale for pursuing such a broad-based approach is that fine-grained, siliciclastic–biogenic sediments provide perhaps the most fully integrated record of past Earth surface processes of any sedimentary facies and thus offer the opportunity to open a clearer window into the workings of the Earth system *if the interdependent parts of the depositional system can be adequately assessed.*

Although the general aspects of our conceptual model have their roots in the broad base of literature cited throughout the chapter, the specifics were refined in the course of research conducted by ourselves, our students, and others on the intervals described in the case studies. We argue that the model illustrated in Figures 1 and 2 provides the unifying framework for understanding these depositional systems, and analysis of key similarities and differences between the case studies reveals the constraints on ranges of proxy expression across space and time. We conclude with a summary of the key similarities and differences in proxy expression and a prospectus for future work.

- *Applicability of the mode*—A common attribute of the case studies is that all three factors in the tripartite model (Figure 1) play important roles in the origin of fine-grained, siliciclastic–biogenic facies. There are variations in the dominance of one control over another through geological history (e.g., authigenic processes in the anoxic–sulfidic oceans of the Precambrian versus the predominantly oxic waters of Phanerozoic seas) and through the individual histories of each study interval (e.g., changing roles of production and benthic redox in controlling OM accumulation during BCM deposition). End-member models, as exemplified by the "productivity" versus "preservation" dichotomy, are more clearly viewed as process continua to which additional axes must be added (bulk sedimentation rates) to fully assess depositional processes and their feedback on paleo-environmental conditions. It is the changes in process dominance, placed within appropriate geological contexts (paleogeography, paleoclimate, etc.) that provide the most penetrating information about Earth system dynamics.

- *Detrital fluxes and bulk sedimentation*—In hemipelagic systems oscillations in the detrital flux play a critical role in modulating the dynamic interplay between biogenic fluxes and authigenic modification of those fluxes. As a general rule, relative condensation (decreased dilution by detrital or other sediment sources) accompanies most cases of elevated C_{org} concentrations. However, the sources and pathways of dilutants vary widely through space and time. There is no single unambiguous proxy for distinguishing terrigenous dilutants, and even the most conservative tracers may vary in their transport paths (e.g., fluvial versus eolian), resulting in significant differences of interpretation. In each case, proxies must be calibrated to the geological context of a depositional system.
- *Productivity and nutrients*—One of the most interesting linkages represented in the model in Figure 1 is that between OM remineralization and nutrient recycling, a process with significant potential to affect rates of primary production. The key factor for this interpretation is the physical oceanography of the system, which we argue was characterized by strong seasonal thermal stratification and intermittent mixing in ancient shallow epeiric seas. This model provides the best explanation for observations of Cretaceous and Devonian black shales described herein. Only rarely were such basins deep and/or isolated enough to experience near-permanent stratification and euxinicity (Devonian US/OC interval), such as that present in the modern Black Sea. Further, in cases where C_{org} accumulation and paleoproductivity estimates are possible (e.g., Greenhorn Formation), these estimates suggest elevated production during black shale deposition, even at maximum highstands when the distance to terrigenous nutrient sources would be greatest. These observations indirectly support the hypothesis of nutrient recycling.
- *Productivity and C-isotope records*—Attributing trends in bulk $\delta^{13}C_{org}$ only to the trade-off between dominance of respiration-influenced local carbon reservoirs versus dominance of production-influenced global carbon reservoirs is clearly a gross oversimplification, as a number of other factors may influence C-isotope fractionation in the biogenic and aqueous reservoirs of the carbon cycle. However, this hypothesis finds support in the observations of Cretaceous and Devonian basins summarized herein. In each of these case studies, relative depletion of $\delta^{13}C_{org}$ values are observed in association with units interpreted to represent examples of the anoxia–nutrient feedback mechanism, and recycling of respired CO_2 to surface waters along with regenerated nutrients would be a plausible scenario. Similarly, each case study includes examples of major positive shifts in $\delta^{13}C_{org}$ documented from sites on distant continents that have been interpreted to reflect global increases in C_{org} burial fluxes sufficient to well exceed local reservoir effects. Considering the global carbon isotopic record of OAE II, the positive excursion is very well constrained by biostratigraphy and thus precisely correlated. Comparing numerous pelagic and hemipelagic sites where preserved OM is dominated by marine algal material, we observe variation of up to 4‰ in the pre-OAE II mean of $\delta^{13}C_{org}$ among different sites and a similar variation of up to 4‰ in the magnitude of the excursion (e.g., Arthur *et al.*, 1988; Pratt *et al.*, 1993; Kuypers, 2001). There are similar variations in the $\delta^{13}C_{org}$ values from the Frasnian–Famennian Kellwasser horizons at different localities (Murphy *et al.*, 2000c), although the OM sources are less well constrained. We hypothesize that these variations largely reflect the dynamic interplay between a local balance of OM respiration and burial and large-scale reservoir effects forced by global environmental perturbations.
- *Mo, Fe, and redox control*—Oxygen deficiency has been recognized as a key factor influencing the formation of Phanerozoic organic-rich hemipelagic facies and is clearly a critical boundary condition in the Precambrian. Reviewing the literature since the early 1980s on this topic, we note a transition to progressively more quantitative approaches to the chemistry of marine redox dynamics. We believe that one of the most exciting advances is the integration of first-order rate equations for OM degradation, for example, via sulfate reduction, developed based on modern observations, to paleoredox proxies such as molybdenum—the accumulation of which appears to reflect sulfide generation (Meyers *et al.*, in review). As described above, reactive-Fe availability plays a key role in this process, and may in fact be a primary regulator in the OM burial process. This idea needs to be tested in more sequences where accumulation rates for key proxies can be calculated.
- *Elemental ratios versus accumulation rates*—In most sections of the geological record, timescales are insufficiently resolved to calculate sedimentation rates at a scale capable of yielding meaningful accumulation rate data. In recent years, however, orbital timescales have been developed for this purpose in some ancient stratigraphic sequences (Park and Herbert, 1987; Meyers *et al.*, 2001). In this paper, the presentation of accumulation rates with the more common concentration and elemental ratio data serve to highlight the strengths of each. Because interpretations are commonly

built upon recognition of *relative changes* among interrelated components, multiproxy concentration and ratio data can be very useful. However, integration of the two types of information clearly represents the optimal approach. In the case of the Greenhorn Formation, accumulation rate data allow actualistic comparisons to be made—whereas average MAR-C_{org} values in the HSM and BCM (0.16 g cm^{-2} kyr^{-1} and 0.03 g cm^{-2} kyr^{-1}, respectively) exceed modern open-ocean oligotrophic sites by at least two orders of magnitude, they are conversely at least one order of magnitude lower than modern high-productivity upwelling sites (Bralower and Thierstein, 1987). Notably, the Late Cenomanian HSM experienced higher MAR C_{org} rates (by a factor of 5) than the overlying BCM, which contains the putative production pulse of OAE II. Although there is certainly a relative change in production that can be linked to this event, these data again underscore the importance of recognizing local records of global events. With the exception of some Cretaceous sites that are analogous to modern upwelling zones (e.g., Tarfaya Basin), it is possible that most areas experienced relatively modest production levels during OAE II (e.g., Thurow *et al.*, 1988; Kuhnt *et al.*, 1990), a conclusion that highlights the importance of changes in bulk sedimentation and redox conditions to explain the increased global C_{org} burial flux at this time. As originally suggested by Arthur *et al.* (1987, 1988), sea-level rise and its effect on bulk terrigenous sediment flux and water depth, as well as reactive-Fe delivery to distal hemipelagic sites (Meyers *et al.*, in review), was likely a master variable. This type of model appears to have similar explanatory power for OM burial in the Devonian Appalachian basin (Sageman *et al.*, 2003).

- *Future prospects*—Some of the topics of future importance not covered in detail in this report include advances in the analysis of OM sources, transport dynamics, and reactivities. These studies center on: (i) identification of terrestrial OM and weathered kerogen in the terrigenous flux and the fate of these components during transport and (re)burial (Leithold and Blair, 2001; Petsch *et al.*, 2000, 2002); (ii) further analysis of the role of mineral surface area in fine-grained, organic-rich facies (Kennedy *et al.*, 2002); and (iii) better quantification of decomposition rates for different components of sedimentary OM (e.g., Van Mooy *et al.*, 2002). Other emerging directions in carbon-cycle research include models that rigorously incorporate coupled nutrient cycles (e.g., Kump and Arthur, 1999), quantitative molecular analysis of bacterial and archeal biomass and the new information yielded about metabolic mechanisms (such as anaerobic methane oxidation) over time and space (e.g., Kuypers *et al.*, 2001), and methane storage in gas hydrates and release as a mechanism for driving abrupt climate change (Froelich *et al.*, 1993; Dickens *et al.*, 1995; Dickens, 1999).

ACKNOWLEDGMENTS

The authors wish to acknowledge financial support of the National Science Foundation (e.g., EAR-9725441, EAR-0001093), as well as Northwestern University and the University of Missouri, which made possible some of the projects upon which this compilation is based. The thesis work of our students Adam Murphy, Josef Werne, Matt Hurtgen, and Steve Meyers contributed greatly to the progression of ideas developed herein, and Steve Meyers is specifically thanked for providing significant scientific and editorial input to the manuscript. Lastly, thanks go to the other members of the NU—Sedimentary Research Group (Joniell Borges, Jason Flaum, Michael Fortwengler, Rob Locklair, Petra Pancoskova, and Ramya Sivaraj), who all helped in compiling parts of the bibliography.

REFERENCES

Adelson J. M., Helz G. R., and Miller C. V. (2001) Reconstructing the rise of recent coastal anoxia: molybdenum in Chesapeake Bay sediments. *Geochim. Cosmochim. Acta* **65**, 237–252.

Algeo T. J., Berner R. A., Maynard J. B., and Scheckler S. E. (1995) Late Devonian oceanic anoxic events and biotic crises: "rooted" in the evolution of vascular land plants? *GSA Today* **5**(45), 64–66.

Allen J. R. L. (1970) Studies in fluviatile sedimentation: a comparison of fining-upwards cyclothems, with special reference to coarse-member composition and interpretation. *J. Sedim. Petrol.* **40**, 298–323.

Aller R. C. (1994) Bioturbation and remineralization of sedimentary organic matter: effects of redox oscillation. *Chem. Geol.* **114**, 331–345.

Altabet M. A. and Francois R. (1993) The use of nitrogen isotopic ratio for reconstruction of past changes in surface ocean nutrient utilization. In *Carbon Cycling in the Glacial Ocean: Constraints on the Ocean's Role in Global Change: Quantitative Approaches in Paleoceanography, Series I, Global Environmental Change*, 17 NATO ASI Series (eds. R. Zahn, T. F. Pedersen, M. A. Kaminski, and L. Labeyrie). Springer, Berlin, pp. 281–306.

Altabet M. A., Deuser W. G., Honjo S., and Stienen C. (1991) Seasonal and depth-related changes in the source of sinking particles in the North Atlantic. *Nature* **354**, 136–139.

Anbar A. D. and Knoll A. H. (2002) Proterozoic ocean chemistry and evolution: a bioinorganic bridge? *Science* **297**, 1137–1142.

Anderson R. F., Fleisher M. Q., and Le Huray A. P. (1989a) Concentration, oxidation state, and particulate flux of

uranium in the Black Sea. *Geochim. Cosmochim. Acta* **53**, 2215–2224.
Anderson R. F., Le Huray A. P., Fleisher M. Q., and Murray R. W. (1989b) Uranium depletion in Saanich Inlet sediments, Vancouver Island. *Geochim. Cosmochim. Acta* **53**, 2205–2213.
Anderson R. F., Lyons T. W., and Cowie G. L. (1994) Sedimentary record of a shoaling of the oxic/anoxic interface in the Black Sea. *Mar. Geol.* **116**, 373–384.
Anderson T. F., Kruger J., and Raiswell R. (1987) C–S–Fe relationships and the isotopic composition of pyrite in the New Albany Shales of the Illinois Basin USA. *Geochim. Cosmochim. Acta* **51**, 2795–2805.
Archer D. E. (1991) Equatorial Pacific calcite preservation cycles: production or dissolution? *Paleoceanography* **6**, 561–571.
Arnold G. L., Anbar A. D., and Barling J. (2002) Oxygenation of Proterozoic oceans: insight from molybdenum isotopes. *Geochim. Cosmochim. Acta* **66**, A30 Goldschmidt Conference Abstracts.
Arthur M. A. and Dean W. E. (1991) An holistic geochemical approach to cyclomania: examples from Cretaceous pelagic limestone sequences. In *Cycles and Events in Stratigraphy* (eds. G. Einsele, W. Ricken, and A. Seilacher). Springer, Berlin, pp. 126–166.
Arthur M. A. and Dean W. E. (1998) Organic-matter production and preservation and evolution of anoxia in the Holocene Black Sea. *Paleoceanography* **13**, 395–411.
Arthur M. A. and Sageman B. B. (1994) Marine black shales: a review of depositional mechanisms and environments of ancient deposits. *Ann. Rev. Earth Planet. Sci.* **22**, 499–552.
Arthur M. A., Dean W. E., Pollastro R., Scholle P. A., and Claypool G. E. (1985) A comparative geochemical study of two transgressive pelagic limestone units, Cretaceous Western Interior basin US. In *Fine-Grained Deposits and Biofacies of the Cretaceous Western Interior Seaway: Evidence of Cyclic Sedimentary Processes*. (eds. L. M. Pratt, E. G. Kauffman, and F. B. Zelt). Society Economic Paleontologists and Minerologists, Tulsa, pp. 16–27.
Arthur M. A., Schlanger S. O., and Jenkyns H. C. (1987) The Cenomanian-Turonian Oceanic Anoxic Event: II. Paleoceanographic controls on organic matter production and preservation. In *Marine Petroleum Source Rocks.* Geological Society of London Special Publication 26 (eds. J. Brooks and A. J. Fleet). pp. 401–420.
Arthur M. A., Dean W. E., and Pratt L. M. (1988) Geochemical and climatic effects of increased marine organic carbon burial at the Cenomanian/Turonian boundary. *Nature* **335**, 714–717.
Arthur M. A., Dean W. E., Neff E. D., Hay B. J., Jones G., and King J. (1994) Late Holocene (0–2000 yBP) organic carbon accumulation in the Black Sea. *Global Biogeochem. Cycles* **8**, 195–217.
Arthur M. A., Dean W. E., and Laarkamp K. (1998) Organic carbon accumulation and preservation in surface sediments on the Peru margin. *Chem. Geol.* **152**, 273–283.
Barling J. and Anbar A. D. Molybdenum isotope fractionation during adsorption by manganese oxides. *Earth Planet. Sci. Lett.* (submitted).
Barling J., Arnold G. L., and Anbar A. D. (2001) Natural mass-dependent variations in the isotopic composition of molybdenum. *Earth Planet. Sci. Lett.* **193**, 447–457.
Beier J. A. and Hayes J. M. (1989) Geochemical and isotope evidence for paleoredox conditions during deposition of the Devonian-Mississippian New Albany Shale, southern Indiana. *Geol. Soc. Am. Bull.* **101**, 774–782.
Berner R. A. (1970) Sedimentary pyrite formation. *Am. J. Sci.* **268**, 1–23.
Berner R. A. (1980) *Early Diagenesis—A Theoretical Approach.* Princeton University Press, Princeton, NJ, 241pp.
Berner R. A. (1982) Burial of organic carbon and pyrite sulfur in the modern ocean: its geochemical and environmental significance. *Am. J. Sci.* **282**, 451–473.
Berner R. A. (1984) Sedimentary pyrite formation: an update. *Geochim. Cosmochim. Acta* **48**, 605–615.
Berner R. A. (1999) A new look at the long-term carbon cycle. *GSA Today* **9**, 1–6.
Berner R. A. (2001) Modeling atmospheric O_2 over Phanerozoic time. *Geochim. Cosmochim. Acta* **65**, 685–694.
Berner R. A. and Canfield D. E. (1989) A new model for atmospheric oxygen over Phanerozoic time. *Am. J. Sci.* **289**, 333–360.
Berner R. A. and Kothavala Z. (2001) GEOCARB III: a revised model of atmospheric CO_2 over Phanerozoic time. *Am. J. Sci.* **301**, 182–204.
Berner R. A. and Petsch S. T. (1998) The sulfur cycle and atmospheric oxygen. *Science* **282**, 1426–1427.
Berner R. A. and Raiswell R. (1984) C/S method for distinguishing freshwater from marine sedimentary rocks. *Geology* **12**, 365–368.
Berner R. A., Lasaga A. C., and Garrels R. M. (1983) The carbonate-silicate geochemical cycle and its effect on atmospheric carbon dioxide over the past 100 million years. *Am. J. Sci.* **283**, 641–683.
Berner R. A., Petsch S. T., Lake J. A., Berling D. J., Popp B. N., Lane R. S., Laws E. A., Westley M. B., Cassar N., Woodward F. I., and Quick W. P. (2000) Isotope fractionation and atmospheric oxygen: implications for Phanerozoic O_2 evolution. *Science* **287**, 1630–1633.
Bertrand P., Shimmield G., Martinez P., Grousset F., Jorissen F., Paterne M., Pujol C. J., Bouloubassi I., Buat Menard P., Peypouquet J. P., Beaufort L., Sicre M. A., Lallier-Verges E., Foster J. M., and Ternois Y. (1996) The glacial ocean productivity hypothesis: the importance of regional temporal and spatial studies. *Mar. Geol.* **130**, 1–9.
Betts J. N. and Holland H. D. (1991) The oxygen content of bottom waters, the burial efficiency of organic carbon, and the regulation of atmospheric oxygen. *Global Planet. Change* **5**, 5–18.
Bewers J. M. and Yeats P. A. (1979) The behavior of trace metals in estuaries of the St. Lawrence Basin. *Le Naturaliste Canadien* **106**, 149–160.
Bralower T. J. and Thierstein H. R. (1984) Low productivity and slow deep water circulation in mid-Cretaceous oceans. *Geology* **12**, 614–618.
Bralower T. J. and Thierstein H. R. (1987) Organic-carbon and metal accumulation in Holocene and Mid-Cretaceous marine sediments: palaeoceanographic significance. In *Marine Petroleum Source Rocks.* Geological Society of London Special Publication 26 (eds. J. Brooks and A. J. Fleet), pp. 345–369.
Breit G. N. and Wanty R. B. (1991) Vanadium accumulation in carbonaceous rocks: a review of geochemical controls during deposition and diagenesis. *Chem. Geol.* **91**, 83–97.
Brett C. E. (1995) Sequence stratigraphy, biostratigraphy, and taphonomy in shallow marine environments. *Palaios* **10**, 597–616.
Brett C. E. and Baird G. C. (1986) Symmetrical and upward shallowing cycles in the Middle Devonian of New York State and their implications for the punctuated aggradational cycle hypothesis. *Paleoceanography* **1**, 431–445.
Brett C. E. and Ver Straeten C. A. (1995) Stratigraphy and facies relationships of the Eifelian Onondaga Limestone (Middle Devonian) in western and west central New York State. In *67th Annual Meeting Field Trip Guidebook* (eds. J. I. Garver and J. A. Smith). New York State Geological Association, Albany, pp. 221–269.
Bright C. A., Lyons T. W., MacLeod K. G., Glascock M. D., Rexroad C. B., Brown L. M., and Ethington R. L. Arguments against preservation of primary seawater signals in the rare-earth element compositions of biogenic (conodont) apatite. *J. Sedim. Res.* (submitted).
Broecker W. S. (1982) Ocean chemistry during glacial time. *Geochim. Cosmochim. Acta* **4**, 1689–1705.
Broecker W. S. and Peng T. H. (1982) *Tracers in the Sea.* Eldigio Press, Palisades, New York, 690pp.

Brumsack H. J. (1989) Geochemistry of recent TOC-rich sediment from the Gulf of California and the Black Sea. *Geol. Rundsch.* **78**, 851–882.

Brumsack H. J. and Thurow J. (1986) The geochemical facies of black shales from the Cenomanian Turonian boundary event (CTBE). In *Biogeochemistry of Black shales—Case Studies* (eds. E. T. Degens, P. A. Meyers, and S. C. Brassell). Mitteilungen aus dem Geologisch-Palaeontologischen Institut der Universitaet Hamburg, Vol. 60, pp. 247–265.

Burdett J. W., Arthur M. A., and Richardson M. (1989) A Neogene seawater sulfate isotope age curve from calcareous pelagic microfossils. *Earth Planet. Sci. Lett.* **94**, 189–198.

Burns C. E. and Bralower T. J. (1998) Upper Cretaceous nannofossil assemblages across the Western Interior seaway: implications for the origin of lithologic cycles in the Greenhorn and Niobrara Formations. In *Stratigraphy and Paleoenvironments of the Cretaceous Western Interior Seaway.* Concepts in Sedimentology and Paleontology no. **6** (eds. M. A. Arthur and W. E. Dean). Society Economic Paleontologists Minerologists, Tulsa, pp. 35–58.

Byers C. W. (1977) Biofacies patterns in euxinic basins: a general model. In *Deep-water Carbonate Environments.* Special Publication, no. 25 (eds. H. E. Cook and P. Enos). Society Economic Paleontologists Minerologists, Tulsa, pp. 5–17.

Calvert S. E. and Karlin R. E. (1991) Relationships between sulphur, organic carbon, and iron in the modern sediments of the Black Sea. *Geochim. Cosmochim. Acta* 55, 2483–2490.

Calvert S. E. and Karlin R. E. (1998) Organic carbon accumulation in the Holocene sapropel of the Black Sea. *Geology* **26**, 107–110.

Calvert S. E. and Pedersen T. F. (1993) Geochemistry of recent oxic and anoxic marine sediments: implications for the geological record. *Mar. Geol.* **113**, 67–88.

Calvert S. E. and Pedersen T. F. (1996) Sedimentary geochemistry of manganese: implications for the environment of formation of manganiferous black shales. *Econ. Geol.* **91**, 36–47.

Calvert S. E., Karlin R. E., Toolin L. J., Donahue D. J., Southon J. R., and Vogel J. S. (1991) Low organic carbon accumulation rates in Black Sea sediments. *Nature* **350**, 692–695.

Calvert S. E., Thode H. G., Yeung D., and Karlin R. E. (1996) A stable isotope study of pyrite formation in the late Pleistocene and Holocene sediments of the Black Sea. *Geochim. Cosmochim. Acta* **60**, 1261–1270.

Canfield D. E. (1989) Sulfate reduction and oxic respiration in marine sediments: implications for organic carbon preservation in euxinic sediments. *Deep-Sea Res.* **36**, 121–138.

Canfield D. E. (1994) Factors influencing organic carbon preservation in marine sediments. *Chem. Geol.* **114**, 315–329.

Canfield D. E. (1998) A new model for Proterozoic ocean chemistry. *Nature* **396**, 450–453.

Canfield D. E. (2001) Isotope fractionation by natural populations of sulfate-reducing bacteria. *Geochim. Cosmochim. Acta* **65**, 1117–1124.

Canfield D. E. and Raiswell R. (1991) Pyrite formation and fossil preservation. In *Taphonomy: Releasing the Data Locked in the Fossil Record.* Topics in Geobiology 9 (eds. P. A. Allison and D. E. G. Briggs). Plenum, New York, pp. 411–453.

Canfield D. E. and Teske A. (1996) Late Proterozoic rise in atmospheric oxygen concentration inferred from phylogenetic and sulphur-isotope studies. *Nature* **382**, 127–132.

Canfield D. E. and Thamdrup B. (1994) The production of ^{34}S-depleted sulfide during bacterial disproportionation of elemental sulfur. *Science* **266**, 1973–1975.

Canfield D. E., Raiswell R., and Bottrell S. H. (1992) The reactivity of sedimentary iron minerals toward sulfide. *Am. J. Sci.* **292**, 659–683.

Canfield D. E., Lyons T. W., and Raiswell R. (1996) A model for iron deposition to euxinic Black Sea sediments. *Am. J. Sci.* **296**, 818–834.

Canfield D. E., Habicht K. S., and Thamdrup B. (2000) The Archean sulfur cycle and the early history of atmospheric oxygen. *Science* **288**, 658–661.

Chambers L. A., Trudinger P. A., Smith J. W., and Burns M. S. (1975) Fractionation of sulfur isotopes by continuous cultures of *Desulfovibrio desulfuricans. Can. J. Microbiol.* **21**, 1602–1607.

Chester M. A., Guymer I., and Freestone R. (1997) Managing water quality in the tidal Ouse (UK), problems associated with suspended sediment oxygen demand. In *Environmental and Coastal Hydraulics: Protecting the Aquatic Habitat.* Proceedings of Congress of International Association for Hydraulic Research 27, Theme B (eds. F. M. Holly, Jr., A. Alsaffar, S. S. Wang, and T. Carstens). International Association for Hydraulic Research Congress pp. 665–670.

Chester R. (1990) *Marine Geochemistry.* Harper Collins, NY, 690pp.

Chester R. (2000) *Marine Geochemistry* 2nd edn. Blackwell, Oxford, UK, 506pp.

Chester R. and Murphy K. J. T. (1990) Metals in the marine atmosphere. In *Heavy Metals in the Marine Environment* (eds. R. Furnace and P. Rainbow). CRC Press, Boca Raton, pp. 27–49.

Claypool G. E., Holser W. T., Kaplan I. R., Sakai H., and Zak I. (1980) The age curves of sulfur and oxygen isotopes in marine sulfate and their mutual interpretations. *Chem. Geol.* **28**, 199–260.

Coveney R. M., Jr, Watney W. L., and Maples C. G. (1991) Contrasting depositional models for Pennsylvanian black shale discerned from molybdenum abundances. *Geology* **19**, 147–150.

Creaney S. and Passey Q. R. (1993) Recurring patterns of total organic carbon and source rock quality within a sequence stratigraphic framework. *Am. Assoc. Petrol. Geol. Bull.* **77**, 386–401.

Crusius J. and Anderson R. F. (1992) Inconsistencies in accumulation rates of Black Sea sediments inferred from records of laminae and ^{210}Pb. *Paleoceanography,* **7**, 215–227.

Crusius J., Calvert S., Pedersen T., and Sage D. (1996) Rhenium and molybdenum enrichments in sediments as indicators of oxic, suboxic and sulfidic conditions of deposition. *Earth Planet. Sci. Lett.* **145**, 65–78.

Cullers R. L. (1994a) The controls on the major and trace element variation of shales, siltstones, and sandstones of Pennsylvanian-Permian age from uplifted continental blocks in Colorado to platform sediment in Kansas, USA. *Geochim. Cosmochim. Acta* **58**, 4955–4972.

Cullers R. L. (1994b) The chemical signature of source rocks in size fractions of Holocene stream sediment derived from metamorphic rocks in the Wet Mountains region, Colorado, USA. *Chem. Geol.* **113**, 327–343.

Davis C., Pratt L. M., Sliter W. V., Mompart L., and Murat B. (1999) Factors influencing organic carbon and trace metal accumulation in the Upper Cretaceous La Luna Formation of the western Maracaibo Basin, Venezuela. In *Evolution of the Cretaceous Ocean-Climate System.* Geological Society of America Special Paper 332 (eds. E. Barrera and C. Johnson). Geological Society of America, Boulder, pp. 203–230.

de Baar H. J. W., German C. R., Elderfield H., and van Gaans P. (1988) Rare earth element distributions in anoxic waters of the Cariaco Trench. *Geochim. Cosmochim. Acta* **52**, 1203–1219.

de Leeuw J. W., Frewin N. L., van Bergen P. F., Sinninghe Damsté J. S., and Collinson M. E. (1995) Organic carbon as a palaeoenvironmental indicator in the marine realm. In *Marine Paleoenironmental Analysis from Fossils.* Geological Society of London Special Publication 83 (eds. D. W. J. Bosence and P. A. Allison). Geological Society of London, London, pp. 43–72.

DeMenocal P. B. (1995) Plio-Pleistocene African climate. *Science* **270**, 53–59.

Dean W. E. and Arthur M. A. (1998) Geochemical expression of cyclicity in Cretaceous pelagic limestone sequences: Niobrara Formation, Western Interior Seaway. In *Stratigraphy and Paleoenvironments of the Cretaceous Western Interior Seaway USA*. Concepts in Sedimentology and Paleontology, no 6 (eds. W. E. Dean and M. A. Arthur). Society Economic Paleontologists Minerologists, Tulsa, pp. 227–255.

Dean W. E., Piper D. Z., and Peterson L. C. (1999) Molybdenum accumulation in Cariaco basin sediment over the past 24 k.y: a record of water-column anoxia and climate. *Geology* **27**, 507–510.

Degens E. T. (1965) *Geochemistry of Sediments: A Brief Survey.* Prentice Hall, Englewood Cliffs, NJ, 342pp.

Degens E. T. and Ittekkot V. (1987) The carbon cycle: tracking the path of organic particles from sea to sediment. In *Marine Petroleum Source Rocks.* Geological Society of London Special Publication 26 (eds. J. Brooks and A. J. Fleet). Geological Society of London, London, pp. 121–135.

Demaison G. I. and Moore G. T. (1980) Anoxic marine environments and oil source bed genesis. *Am. Assoc. Petrol. Geol.Bull.* **64**, 1179–1209.

DePaolo D. J. (1986) Detailed record of the Neogene Sr isotopic evolution of seawater from DSDP Site 590B. *Geology* **14**, 103–106.

Detmers J., Brüchert V., Habicht K. S., and Kuever J. (2001) Diversity of sulfur isotope fractionations by sulfate-reducing prokaryotes. *Appl. Environ. Microbiol.* **67**, 888–894.

Dickens G. R. (1999) Carbon cycle: the blast in the past. *Nature* **401**, 752–755.

Dickens G. R., O'Neil J. R., Rea D. K., and Owen R. M. (1995) Dissociation of oceanic methane hydrate as a cause of the carbon-isotope excursion at the end of the Paleocene. *Paleoceanography* 10, 965–971.

Durand B. (1980) Sedimentary organic matter and kerogen: definition and quantitative importance of kerogen. In *Kerogen: Insoluble Organic Matter from Sedimentary Rocks* (ed. B. Durand). Technip, Paris, pp. 13–34.

Dyman T. S., Cobban W. A., Fox J. E., Hammond R. H., Nichols D. J., Perry W. J., Jr., Porter K. W., Rice D. D., Setterholm D. R., Shurr G. W., Tysdal R. G., Haley J. C., and Campen E. B. (1994) Cretaceous rocks from southwestern Montana to southwestern Minnesota, Northern Rocky Mountains and Great Plains region. In *Perspectives on the Eastern Margin of the Cretaceous Western Interior Basin.* Geological Society of America Special Paper 287 (eds. G. W. Shurr, G. A. Ludvigson, and R. H. Hammond). Geological Society of America, Boulder, pp. 5–26.

Dymond J., Suess E., and Lyle M. (1992) Barium in deep-sea sediment: a geochemical proxy for paleoproductivity. *Paleoceanography* **7**, 163–181.

Dymond J., Collier R., McManus J., Honjo S., and Manganini S. (1997) Can the aluminum and titanium contents of ocean sediments be used to determine the paleoproductivity of the oceans? *Paleoceanography* **12**, 586–593.

Eicher D. L. and Diner R. (1989) Origin of the Cretaceous Bridge Creek cycles in the western interior, United States. *Palaeogeogr. Palaeoclimatol. Palaeoecol.* **74**, 127–146.

Elder W. P. (1985) Biotic patterns across the Cenomanian-Turonian extinction boundary near Pueblo, Colorado. In *Fine-Grained Deposits and Biofacies of the Cretaceous Western Interior Seaway: Evidence of Cyclic Sedimentary Processes* (eds. L. M. Pratt, E. G. Kauffman, and F. B. Zelt). Society Economic Paleontologists and Minerologists, Tulsa, pp. 157–169.

Elder W. P. (1989) Molluscan extinction patterns across the Cenomanian-Turonian stage boundary in the Western Interior of the United States. *Paleobiology* **15**, 299–320.

Elder W. P. (1991) Molluscan paleoecology and sedimentation patterns of the Cenomanian-Turonian extinction interval in the southern Colorado Plateau region. In *Stratigraphy, Depositional Environments, and Sedimentary Tectonics of the Western Margin, Cretaceous Western Interior Seaway.* Geological Society of America Special Paper 260 (eds. J. D. Nations and J. G. Eaton). Geological Society of America, Boulder, pp. 113–137.

Elder W. P., Gustason E. R., and Sageman B. B. (1994) Basinwide correlation of parasequences in the Greenhorn Cyclothem, Western Interior US. *Geol. Soc. Am. Bull.* **106**, 892–902.

Ellwood B. B., Crick R. E., El Hassani A., Benoist S. L., and Young R. H. (2000) Magnetosusceptibility event and cyclostratigraphy method applied to marine rocks: detrital input versus carbonate productivity. *Geology* **28**, 1135–1138.

Emerson S. (1985) Organic carbon preservation in marine sediments. In *The Carbon Cycle and Atmospheric CO_2: Natural Variations Archean to Present.* American Geophysical Union, Geophysical Monograph 32 (eds. E. T. Sunquist and W. S. Broecker). American Geophysical Union, Washington, DC, pp. 78–87.

Emerson S. and Hedges J. I. (1988) Processes controlling the organic carbon content of open ocean sediments. *Paleoceanograpy* **53**, 1233–1240.

Emerson S. R. and Huested S. S. (1991) Ocean anoxia and the concentrations of molybdenum and vanadium in seawater. *Mar. Chem.* **34**, 177–196.

Engel M. H. and Macko S. (1993) *Organic Geochemistry: Principles and Applications.* Plenum, New York, 861pp.

Ettensohn F. R. (1985a) The Catskill Delta complex and the Acadian Orogeny: a model. In *The Catskill Delta.* Geological Society of America Special Paper 201 (eds. D. L. Woodrow and W. D. Sevon). Geological Society of America, Boulder, pp. 39–49.

Ettensohn F. R. (1985b) Controls on development of Catskill Delta complex basin-facies. In *The Catskill Delta.* Geological Society of America Special Paper 201 (eds. D. L. Woodrow and W. D. Sevon). Geological Society of America, Boulder, pp. 65–77.

Ettensohn F. R., Miller M. L., Dillman S. B., Elam T. D., Geller K. L., Swager D. R., Markowitz G., Woock R. D., and Barron L. S. (1988) Characterization and implications of the Devonian-Mississippian black-shale sequence, eastern and central Kentucky, USA: Pycnoclines, transgression, regression and tectonism. In *Devonian of the World: Proceedings of the Second International Symposium on the Devonian System, Volume II, Sedimentation.* (eds. N. J. McMillan, A. F. Embry, and D. J. Glass) Canadian Society of Petroleum Geologists, Memoir 14, pp. 323–345.

Faill R. T. (1985) The Acadian Orogeny and the Catskill Delta. In *The Catskill Delta.* Geological Society of America Special Paper 201 (eds. D. L. Woodrow and W. D. Sevon). Geological Society of America, Boulder, pp. 15–37.

Falkowski P. G. (1997) Evolution of the nitrogen cycle and its influence on the biological sequestration of CO_2 in the ocean. *Nature* **387**, 272–275.

Farquhar J., Bao H. M., and Thiemens M. (2000) Atmospheric influence of Earth's earliest sulfur cycle. *Science* **289**, 756–758.

Filippelli G. (1997) Controls on the phosphorus concentration and accumulation in marine sediments. *Mar. Geol.* **139**, 231–240.

Filippelli G. M., Sierro F. J., Flores J. A., Vazquez A., Utrilla R., Perez-Folgado M., and Latimer J. C. (2003) A sediment-nutrient-oxygen feedback responsible for productivity variations in Late Miocene sapropel sequences of the western Mediterranean. *Palaeogeogr. Palaeoclimatol. Palaeoecol.* **190**, 335–348.

Fisher I. J. St. and Hudson J. D. (1987) Pyrite formation in Jurassic shales of contrasting biofacies. In *Marine Petroleum Source Rocks.* Geological Society of London Special Publications 26 (eds. J. Brooks and A. J. Fleet). Geological Society of London, London, pp. 69–78.

Francois R. (1988) A study on the regulation of the concentrations of some trace metals (Rb, Sr, Zn, Pb, Cu, V, Cr, Ni, Mn and Mo) in Saanich Inlet sediments, British Columbia, Canada. *Mar. Geol.* **83**, 285–308.

Francois R., Honjo S., Manganini S. J., and Ravizza G. E. (1995) Biogenic barium fluxes to the deep sea: implications for paleoproductivity reconstruction. *Global Biogeochem. Cycles* **9**, 289–303.

Froelich P. N., Klinkhammer G. P., Bender M. L., Luedtke N., Heath G. R., Cullen D., Dauphin P., Hammond D., Hartman B., and Maynard V. (1979) Early oxidation of organic matter in pelagic sediments of the eastern equatorial Atlantic: suboxic diagenesis. *Geochim. Cosmochim. Acta* **43**, 1075–1090.

Froelich P. N., Blanc V., Mortlock R. A., Chilrud S. N., Dunstan W., Udomkit A., and Peng T. H. (1992) River fluxes of dissolved silica to the ocean were higher during glacials: Ge/Si in diatoms rivers, and oceans. *Paleoceanography* **7**, 739–767.

Froelich P. N., Kvenvolden K. A., and Torres M. (1993) Evidence for gas hydrate in the accretionary prism near the Chile triple junction: ODP Leg 141. *Eos. Trans., AGU* **74**, 369.

Fry B., Jannasch H. W., Molyneaux S. J., Wirsen C. O., Muramoto J. A., and King S. (1991) Stable isotope studies of the carbon, nitrogen and sulfur cycles in the Black Sea and the Cariaco Trench. *Deep-Sea Res.* **38**, S1003–S1019.

Garrels R. M. and Lerman A. (1981) Phanerozoic cycles of sedimentary carbon and sulfur. *Proc. Natl. Acad. Sci. USA* **78**, 4652–4656.

Garrels R. M. and Mackenzie F. T. (1971) *Evolution of Sedimentary Rocks*. W.W. Norton and Company, New York, 397pp.

Gartner S. (1977) Nannofossils and biostratigraphy: an overview. *Earth Sci. Rev.* **13**, 227–250.

Gautier D. L. (1986) Cretaceous shales from the Western Interior of North America: sulfur/carbon ratios and sulfur-isotope composition. *Geology* 14, 225–228.

Gautier D. L. (1987) Isotopic composition of pyrite: relationship to organic matter type and iron availability in some North American Cretaceous shales. *Chem. Geol.: Isotope Geosci. Sect.* 65, 293–303.

Gellatly A. M. and Lyons T. W. Trace sulfate in Mesoproterozoic carbonates: implications for seawater sulfate and oxygen availability. *Geochim. Cosmochim. Acta.* (submitted).

German C. R. and Elderfield H. (1989s) Rare earth elements in Saanich Inlet, British Columbia, a seasonally anoxic basin. *Geochim. Cosmochim. Acta* 53, 2561–2571.

German C. R. and Elderfield H. (1990) Application of the Ce anomaly as a paleoredox indicator: the ground rules. *Paleoceanography* 5, 823–833.

German C. R., Holliday B. P., and Elderfield H. (1991) Redox cycling of rare earth elements in the suboxic zone of the Black Sea. *Geochim. Cosmochim. Acta* 55, 3553–3558.

Goldhaber M. B. and Kaplan I. R. (1974) The sulfur cycle. In *The Sea, vol. 5* (ed. E. D. Goldberg). Wiley, New York, pp. 569–655.

Goldhaber M. B. and Kaplan I. R. (1980) Mechanisms of sulfur incorporation and isotope fractionation during early diagenesis in sediments of the Gulf of California. *Mar. Chem.* **9**, 95–143.

Gorsline D. S. (1984) A review of fine-grained sediment origins, characteristics, transport and deposition. In *Finegrained Sediments: Deep-water Processes and Facies*, Geological Society of America Special Publication 5 (eds. D. A. V. Stow and D. J. W. Piper). Geological Society of America, Boulder, pp. 17–34.

Gromet L. P., Dymek R. F., Haskin L. A., and Kortev R. L. (1984) The "North American shale composite" its compilation, major and trace element characteristics. *Geochim. Cosmochim. Acta* **48**, 2469–2482.

Habicht K. S. and Canfield D. E. (1997) Sulfur isotope fractionation during bacterial sulfate reduction in organic-rich sediments. *Geochim. Cosmochim. Acta* **6**, 5351–5361.

Habicht K. S. and Canfield D. E. (2001) Isotope fractionation by sulfate-reducing natural populations and the isotopic composition of sulfide in marine sediments. *Geology* **29**, 555–558.

Habicht K. S., Gade M., Thamdrup B., Berg P., and Canfield D. E. (2002) Calibration of sulfate levels in the Archean Ocean. *Science* **298**, 2372–2374.

Harris S. E., Mix A. C, and King T. (1997) Biogenic and terrigenous sedimentation at Ceara Rise, western tropical Atlantic, supports Pliocene-Pleistocene deep-water linkage between hemispheres. *Proc. ODP Sci. Results* **154**, 331–345.

Hartnett H. E., Keil R. G., Hedges J. I., and Devol A. H. (1998) Influence of oxygen exposure time on organic carbon preservation in continental margin sediments. *Nature* **391**, 572–574.

Hassold N., Rea D. K., and Meyers P. A. (2003) Grain size evidence for variations in delivery of terrigenous sediments to a Middle Pleistocene interrupted sapropel from ODP Site 969, Mediterranean Ridge. *Palaeogeogr. Palaeoclimatol. Palaeoecol.* **190**, 211–219.

Haug G. H., Gunther D., Peterson L. C., Sigman D. M., Hughen K. A., and Aeschlimann B. (2003) Climate and the collapse of Maya Civilization. *Science* **299**, 1731–1735.

Hayes J. M. (1993) Factors controlling ^{13}C contents of sedimentary organic compounds: principles and evidence. *Mar. Geol.* **113**, 111–125.

Hayes J. M., Popp B. N., Takigiku R., and Johnson M. W. (1989) An isotopic study of biogeochemical relationships between carbonates and organic carbon in the Greenhorn Formation. *Geochim. Cosmochim. Acta* **53**, 2961–2972.

Hayes J. M., Freeman K. H., Popp B. N, and Hoham C. H. (1990) Compound-specific isotopic analyses: a novel tool for reconstruction of ancient biogeochemical processes. *Org. Geochem.* **16**, 1115–1128.

Hayes J. M., Strauss H., and Kaufman A. J. (1999) The abundance of ^{13}C in marine organic matter and isotopic fractionation in the global biogeochemical cycle of carbon during the past 800 Ma. *Chem. Geol.* **161**, 103–125.

Haynes J. R. (1981) *Foraminifera*. Wiley, New York, 433pp.

Hedges J. I. and Keil R. G. (1995) Sedimentary organic matter preservation: an assessment and speculative synthesis. *Mar. Chem.* **49**, 81–115.

Helz G. R., Miller C. V., Charnock J. M., Mosselmans J. F. W., Pattrick R. A. D., Garner C. D., and Vaughan D. J. (1996) Mechanism of molybdenum removal from the sea and its concentration in black shales: EXAFS evidence. *Geochim. Cosmochim. Acta* **60**, 3631–3642.

Henrichs S. M. and Reeburgh W. S. (1987) Anaerobic mineralization of marine sediment organic matter: rates and the role of anaerobic processes in the oceanic carbon economy. *Geomicrobiol. J.* **5**, 191–237.

Herbert T. D. and Fischer A. G. (1986) Milankovitch climatic origin of mid-Cretaceous black shale rhythms in central Italy. *Nature* **321**, 739–743.

Herbert T. D., Stallard R. F., and Fischer A. G. (1986) Anoxic events, productivity rhythms, and the orbital signature in a mid-Cretaceous deep-sea sequence from central Italy. *Paleoceanography* **1**, 495–506.

Hem J. D. (1981) Rates of manganese oxidation in aqueous systems. *Geochim. Cosmochim. Acta* 45, 1369–1374.

Hofmann P., Wagner T., and Beckmann B. (2003) Millennial-to centennial-scale record of African climate variability and organic carbon accumulation in the Coniacian-Santonian eastern tropical Atlantic (Ocean Drilling Program Site 959, off Ivory Coast and Ghana). *Geology* **31**, 135–138.

Holland H. D. (2002) Volcanic gases, black smokers, and the Great Oxidation Event. *Geochim. Cosmochim. Acta* **66**, 3811–3826.

Holmes M. E., Schneider R. R., Mueller P. J., Segl M., and Wefer G. (1997) Reconstruction of past nutrient utilization in the eastern Angola Basin based on sedimentary $^{15}N/^{14}N$ ratios. *Paleoceanography* **12**, 604–614.

House M. R. and Kirchgasser W. T. (1993) Devonian goniatite biostratigraphy and timing of facies movements in the Frasnian of eastern North America. In *High Resolution Stratigraphy.* Geological Society of London Special Publications 70 (eds. E. A. Hailwood and R. B. Kidd). Geological Society of London, London, pp. 267–292.

Huang Y., Freeman K. H., Wilkin R. T., Arthur M. A., and Jones A. D. (2000) Black Sea chemocline oscillations during the Holocene: molecular and isotopic studies of marginal sediments. *Org. Geochem.* **31**, 1525–1531.

Hughen K., Overpeck J. T., Peterson L. C., and Trumbore S. (1996) Rapid climate changes in the tropical Atlantic region during the last deglaciation. *Nature* **380**, 51–54.

Hughen K. A., Overpeck J. T., Lehman S. J., Kashgarian M., Southon J., Peterson L. C., Alley R., and Sigman D. M. (1998) Deglacial changes in ocean circulation from an extended radiocarbon calibration. *Nature* **391**, 65–68.

Hurtgen M. T., Lyons T. W., Ingall E. D., and Cruse A. M. (1999) Anomalous enrichments of iron monosulfide in euxinic marine sediments and the role of H_2S in iron sulfide transformations: examples from Effingham Inlet, Orca Basin, and the Black Sea. *Am. J. Sci.* **299**, 556–588.

Hurtgen M. T., Arthur M. A., Suits N. S., and Kaufman A. J. (2002) The sulfur isotopic composition of Neoproterozoic seawater sulfate: implications for a snowball Earth? *Earth Planet. Sci. Lett.* 203(1), 413–429.

Ingall E. D. and Jahnke R. (1997) Influence of water-column anoxia on the elemental fractionation of carbon and phosphorus during sediment diagenesis. *Mar. Geol.* **139**, 219–229.

Ingall E. D., Bustin R. M., and Van Cappellen P. (1993) Influence of water column anoxia on the burial and preservation of carbon and phosphorus in marine shales. *Geochim. Cosmochim. Acta* **57**, 303–316.

Isozaki Y. (1997) Permo-Triassic boundary superanoxia and stratified superocean: records from lost deep sea. *Science* **276**, 235–238.

Ittekkot V. (1988) Global trends in the nature of organic matter in river suspensions. *Nature* **332**, 436–438.

Jacobs L. and Emerson S. (1982) Trace metal solubility in an anoxic Fjord. *Earth Planet. Sci. Lett.* **60**, 237–252.

Jacobs L., Emerson S., and Skei J. (1985) Partitioning and transport of metals across the O_2/H_2S interface in a permanently anoxic basin: Framvaren Fjord, Norway. *Geochim. Cosmochim. Acta* **49**, 1433–1444.

Jacobs L., Emerson S., and Huested S. S. (1987) Trace metal geochemistry in the Cariaco Trench. *Deep-Sea Res.* 34, 965–981.

Joachimski M. M. and Buggisch W. (1993) Anoxic events in the late Frasnian: causes of the Frasnian-Famennian faunal crisis? *Geology* **21**, 675–678.

Johnson-Ibach L. E. (1982) Relationship between sedimentation rate and total organic carbon content in ancient marine sediments. *Am. Assoc. Petrol. Geol. Bull.* **66**, 170–188.

Johnson J. G. and Sandberg C. A. (1988) Devonian eustatic events in the Western United States and their biostratigraphic responses. In *Devonian of the World: Proceedings of the Second International Symposium on the Devonian System, Volume III, Paleontology, Paleoecology and Biostratigraphy* (eds. N. J. McMillan, A. F. Embry, and D. J. Glass) Canadian Society of Petroleum Geologists, Memoir **14**, pp. 171–178.

Johnson J. G., Klapper G., and Sandberg C. A. (1985) Devonian eustatic fluctuations in Euramerica. *Geol. Soc. Am. Bull.* **96**, 567–587.

Jones B. and Manning D. A. C. (1994) Comparison of geochemical indices used for the interpretations of paleoredox conditions in ancient mudstones. *Chem. Geol.* **111**, 111–129.

Jørgensen B. B. (1979) A theoretical model of the stable sulfur isotope distribution in marine sediments. *Geochim. Cosmochim. Acta* **43**, 363–374.

Kah L. C., Lyons T. W., and Frank T. D. Mesoproterozoic marine sulfate: Evidence for a changing biosphere. *Science* (submitted).

Kakegawa T., Kawai H., and Ohmoto H. (1998x) Origins of pyrites in the approximately 2.5 Ga Mt. McRae Shale, the Hamersley District, Western Australia. *Geochim. Cosmochim. Acta* **62**, 3205–3220.

Kaplan I. R. and Rittenberg S. C. (1964) Microbial fractionation of sulphur isotopes. *J. Gen. Microbiol.* **34**, 195–212.

Kauffman E. G. (1984) Paleobiogeography and evolutionary response dynamic in the Cretaceous Western Interior Seaway of North America. In *Jurassic-Cretaceous Biochronology and Paleogeography of North America.* Geological Association of Canada Special Paper 27 (ed. G. E. G. Westermann). Geological Association of Canada, St. John's, pp. 273–306.

Kauffman E. G. and Caldwell W. G. E. (1993) The Western Interior Basin in space and time. In *Evolution of the Western Interior Basin.* Geological Association Canada Special Paper 39 (eds. W. E. Caldwell and E. G. Kauffman). Geological Association of Canada, St. John's, pp. 1–30.

Keil R. G. and Hedges J. I. (1993) Sorption of organic matter to mineral surfaces and the preservation of organic matter in coastal marine sediments. *Chem. Geol.* **107**, 385–388.

Kennedy M. J., Pevear D. R., and Hill R. J. (2002) Mineral surface control of organic carbon in black shale. *Science* **295**, 657–660.

Koopmans M. P., Koester J., van Kaam-Peters H. M. E., Kenig F., Schouten S., Hartgers W. A., de Leeuw J. W., and Sinninghe Damste J. S. (1996) Diagenetic and catagenetic products of isorenieratene; molecular indicators for photic zone anoxia. *Geochim. Cosmochim. Acta* **60**, 4467–4496.

Kuhnt W., Herbin J. P., Thurow J., and Wiedmann J. (1990) Distribution of Cenomanian-Turonian organic facies in the western Mediterranean and along the adjacent Atlantic margin. *Am. Assoc. Petrol. Geol. Stud. Geol.* **30**, 133–160.

Kump L. R. and Arthur M. A. (1999) Interpreting carbon-isotope excursions: carbonates and organic matter. *Chem. Geol.* **161**, 181–198.

Kump L. R., Arthur M. A., Patzkowsky M. E., Gibbs M. T., Pinkus D. S., and Sheehan P. M. (1999) A weathering hypothesis for glaciation at high atmospheric pCO_2 during the Late Ordovician. *Palaeogeogr. Palaeoclimatol. Palaeoecol.* **152**, 173–187.

Kump L. R., Brantley S. L., and Arthur M. A. (2000) Chemical weathering, atmospheric CO_2, and climate. *Ann. Rev. Earth Planet. Sci.* **28**, 611–667.

Kuypers M. M. M (2001) Mechanisms and biogeochemical implications of the mid-Cretaceous global organic carbon burial events. PhD Dissertation, Universitiet Utrecht, 135pp (unpublished).

Kuypers M. M. M., Blokker P., Erbacher J., Kinkel H., Pancost R. D., Schouten S., and Sinninghe Damsté J. S. (2001) Massive expansion of marine Archea during a Mid-Cretaceous oceanic anoxic event. *Science* **293**, 92–94.

Lantzy R. J. and Mackenzie F. T. (1979) Atmospheric trace metals: global cycles and assessment of man's impact. *Geochim. Cosmochim. Acta* **43**, 511–526.

Laurin J. and Sageman B. (2001) Tectono-sedimentary evolution of the western margin of the Colorado Plateau during the latest Cenomanian and early Turonian. In *The Geologic Transition: High Plateaus to Great Basin.* Utah Geological Association Publication 30 (eds. M. C. Erskin, J. E. Faulds, J. M. Bartley, and P. D. Rowley). Utah Geological Association, Salt Lake City, pp. 57–74.

Leckie R. M., Schmidt M. G., Finkelstein D., and Yuretich R. (1991) Paleoceanographic and paleoclimatic interpretations of the Mancos Shale (Upper Cretaceous), Black Mesa Basin, Arizona. In *Stratigraphy, Depositional Environments, and Sedimentary Tectonics of the Western Margin, Cretaceous*

Western Interior Seaway. Geological Society of America Special Paper 260 (eds. J. D. Nations and J. G. Eaton). Geological Society of America, Boulder, pp. 139–152.

Leckie R. M., Yuretich R. F., West O., Finkelstein D. B., and Schmidt M. G. (1998) Paleoceanography of the southwestern Western Interior sea during the time of Cenomanian-Turonian boundary (Late Cretaceous). In *Stratigraphy and Paleoenvironments of the Cretaceous Western Interior Seaway USA.* Concepts in Sedimentology and Paleontology, no. 6 (eds. W. E. Dean and M. A. Arthur). Society Economic Paleontologists Minerologists, Tulsa, pp. 101–126.

Leithold E. L. (1994) Stratigraphical architecture at the muddy margin of the Cretaceous Western Interior Seaway, southern Utah. *Sedimentology* **41**, 521–542.

Leithold E. L. and Blair N. E. (2001) Watershed control on the carbon loading of marine sedimentary particles. *Geochim. Cosmochim. Acta* **65**, 2231–2240.

Leventhal J. S. (1995) Carbon-sulfur plots to show diagenetic and epigenetic sulfidation in sediments. *Geochim. Cosmochim. Acta* **59**, 1207–1211.

Lewan M. D. (1984) Factors controlling the proportionality of vanadium to nickel in crude oils. *Geochim. Cosmochim. Acta* **48**, 2231–2238.

Lewan M. D. (1986) Stable carbon isotopes of amorphous kerogens from Phanerozoic sedimentary rocks. *Geochim. Cosmochim. Acta* **50**, 1977–1987.

Lewan M. D. and Maynard J. B. (1982) Factors controlling enrichment of vanadium and nickel in the bitumen of organic sedimentary rocks. *Geochim. Cosmochim. Acta* **46**, 2547–2560.

Lewis B. L. and Landing W. M. (1991) The biogeochemistry of manganese and iron in the Black Sea. *Deep-Sea Res.* **38**, S773–S803.

Lewis B. L. and Landing W. M. (1992) The investigation of dissolved and suspended-particulate trace metal fractionation in the Black Sea. *Mar. Chem.* **40**, 105–141.

Libes S. M. (1992) *An Introduction to Marine Biogeochemistry.* Wiley, New York, 734pp.

Loutit T. S., Hardenbol J., Vail P. R., and Baum G. R. (1988) Condensed sections: the key to age determination and correlation of continental margin sequences. In *Sea-Level Changes: An Integrated Approach.* Special Publication, no. 42 (eds. C. K. Wilgus, B. S. Hastings, C. A. Ross, H. Posamentier, J. Van Wagoner, and C. G. St. C Kendall). Society Economic Paleontologists Minerologists, Tulsa, pp. 183–213.

Luepke J. J. and Lyons T. W. (2001) Pre-Rodinian (Mesoproterozoic) supercontinental rifting along the western margin of Laurentia: geochemical evidence from the Belt-Purcell Supergroup. *Precamb. Res.* **111**, 79–90.

Lyons T. W. (1991) Upper Holocene sediments of the Black Sea: summary of leg 4 box cores (1988 Black Sea Oceanographic Expedition). In *Black Sea Oceanography.* NATO ASI Series (eds. E. Izdar and J. W. Murray). Kluwer, pp. 401–441.

Lyons T. W. (1997) Sulfur isotopic trends and pathways of iron sulfide formation in upper Holocene sediments of the anoxic Black Sea. *Geochim. Cosmochim. Acta* **61**, 3367–3382.

Lyons T. W. and Berner R. A. (1992) Carbon-sulfur-iron systematics of the uppermost deep-water sediments of the Black Sea. *Chem. Geol.* **99**, 1–27.

Lyons T. W., Berner R. A., and Anderson R. F. (1993) Evidence for large pre-industrial perturbations of the Black Sea chemocline. *Nature* **365**, 538–540.

Lyons T. W., Luepke J. J., Schreiber M. E., and Zieg G. A. (2000) Sulfur geochemical constraints on Mesoproterozoic restricted marine deposition: Lower Belt Supergroup, Northwestern United States. *Geochim. Cosmochim. Acta* **64**, 427–437.

Lyons T. W., Werne J. P., Hollander D. J., and Murray R. W. (2003) Contrasting sulfur geochemistry and Fe/Al and Mo/Al ratios across the last oxic-to-anoxic transition in the Cariaco Basin, Venezuela. *Chem. Geol.* **195**, 131–157.

Lyons T. W., Kah L. C., and Gellatly A. M. The Precambrian sulfur isotope record of evolving atmospheric oxygen. In *Tempos and Events in Precambrian Time* (eds. Eriksson *et al.).* Developments in Precambrian Geology Series, Elsevier (in press).

Martin J. M. and Maybeck M. (1979) Elemental mass-balance of material carried by major world rivers. *Mar. Chem.* **7**, 173–206.

Martin J. M. and Whitfield M. (1983) The significance of the river input of chemical elements to the ocean, NATO Conference Series, IV. *Mar. Sci.* **9**, 265–296.

Mayer L. M. (1994) Relationships between mineral surfaces and organic carbon concentrations in soils and sediments. *Chem. Geol.* **114**, 347–363.

Maynard J. B. (1981) Carbon isotopes as indicators of dispersal patterns in Devonian-Mississippian shales of the Appalachian Basin. *Geology* **9**, 262–265.

McDaniel D. K., Hemming S. R., McLennan S. M., and Hanson G. N. (1994) Resetting of neodymium isotopes and redistribution of REEs during sedimentary processes; the early Proterozoic Chelmsford Formation, Sudbury Basin, Ontario, Canada. *Geochim. Cosmochim. Acta* **58**, 931–941.

McGhee G. R., Jr (1982) The Frasnian-Famennian extinction event: a preliminary analysis of Appalachian marine ecosystems. In *Geological Implications of Impacts of Large Asteroids and Comets on the Earth.* Geological Society of America Special Paper (eds. L. T. Silver and P. H. Schultz). Geological Society of America, Boulder, **190**, pp. 491–500.

McLennan S. M., Hemming S., McDaniel D. K., and Hanson G. N. (1993) Geochemical approaches to sedimentation, provenance, and tectonics. In *Processes Controlling the Composition of Clastic Sediments.* Geological Society of America Special Paper (eds. M. J. Johnsson and A. Basu). Geological Society of America, Boulder, **284**, pp. 21–40.

McManus J., Berelson W. M., Klinkhammer G. P., Johnson K. S., Coale K. H., Anderson R. F., Kumar N., Burdige D. J., Hammond D. E., Brumsack H. J., McCorkle D. C., and Rusdi A. (1998) Geochemistry of barium in marine sediments: implications for its use as a paleoproxy. *Geochim. Cosmochim. Acta* **62**, 3453–3473.

McManus J., Berelson W. M., Hammond D. E., and Klinkhammer G. P. (1999) Barium cycling in the North Pacific: implications for the utility of Ba as a paleoproductivity and paleoalkalinity proxy. *Paleoceanography* **14**, 53–61.

Meyers P. A. (1994) Preservation of elemental and isotopic source identification of sedimentary organic matter. *Chem. Geol.* **114**, 289–302.

Meyers S., Sageman B., and Hinnov L. (2001) Integrated quantitative stratigraphy of Cenomanian-Turonian Bridge Creek Limestone Member using Evolutive Harmonic Analysis and stratigraphie modeling. *J. Sedim. Res.* **71**, 628–644.

Meyers S., Sageman B., and Lyons T. The role of sulfate reduction in organic matter degradation and molybdenum accumulation: theoretical framework and application to a Cretaceous organic matter burial event, Cenomanian-Turonian OAE II. *Paleoceanography* (in review).

Middelburg J. J., Calvert S. E., and Karlin R. (1991x) Organic-rich transitional facies in silled basins: response to sea-level change. *Geology* **19**, 679–682.

Milliman J. D. and Syvitski J. P. M. (1992) Geomorphic/tectonic control of sediment discharge to the ocean: the importance of small mountainous rivers. *J. Geology* **100**, 524–544.

Morford J. L. and Emerson S. (1999) The geochemistry of redox sensitive trace metals in sediments. *Geochim. Cosmochim. Acta* **63**, 1735–1750.

Moore W. S. and O'Neill D. J. (1991) Radionuclide distributions in recent Black Sea sediments. In *Black Sea Oceanography.*

NATO ASI Series (eds. E. Izdar and J. W. Murray). Kluwer, pp. 343–359.

Morse J. W. and Berner R. A. (1995) What determines sedimentary C/S ratios? *Geochim. Cosmochim. Acta* **59**, 1073–1077.

Mueller P. J. and Suess E. (1979) Productivity, sedimentation rate and sedimentary organic matter in the oceans: I. *Organic carbon preservation. Deep-Sea Res.* **26**, 1347–1362.

Murray R. W. and Leinen M. (1996) Scavenged excess aluminum and its relationship to bulk titanium in biogenic sediment from the central Equatorial Pacific Ocean. *Geochim. Cosmochim. Acta* **60**, 3869–3878.

Murray R. W., Leinen M., and Isern A. R. (1993) Biogenic flux of Al to sediment in the central Equatorial Pacific Ocean: evidence for increased productivity during glacial periods. *Paleoceanography* **8**, 651–670.

Murphy A. E., Sageman B. B., Hollander D. J., Lyons T. W., and Brett C. E. (2000a) Black shale deposition in the Devonian Appalachian Basin: siliciclastic starvation, episodic water-column mixing, and efficient recycling of biolimiting nutrients. *Paleoceanography* **15**, 280–291.

Murphy A. E., Sageman B. B., and Hollander D. J. (2000b) Organic carbon burial and faunal dynamics in the Appalachian basin during the Devonian (Givetian-Famennian) greenhouse: an integrated paleoecological/biogeochemical approach. In *Warm Climates in Earth History* (eds. B. Huber, K. MacLeod, and S. Wing). Cambridge University Press, Cambridge, pp. 351–385.

Murphy A. E., Sageman B. B., and Hollander D. J. (2000c) Eutrophication by decoupling of the marine biogeochemical cycles of C, N, and P: a mechanism for the Late Devonian mass extinction. *Geology* **28**, 427–430.

Obradovich J. (1993) A cretaceous time scale. In *Evolution of the Western Interior Basin.* Geological Association Canada Special Paper 39 (eds. W. G. E. Caldwell and E. G. Kauffman). Geological Association Canada, St. John's, pp. 379–396.

Oschmann W. (1991) Anaerobic-poikiloaerobic-aerobic: a new facies zonation for modern and ancient neritic redox facies. In *Cyclic and Event Stratification* (eds. G. Einsele and A. Seilacher). Springer, Berlin, pp. 565–571.

Pancost R. D., Freeman K. H., Patzkowsky M. E., Wavrek D. A., and Collister J. W. (1998) Molecular indicators of redox and marine photoautotroph composition in the late Middle Ordovician of Iowa USA. *Org. Geochem.* **29**, 1649–1662.

Pancost R. D., Baas M., van Geel B., and Sinninghe Damsté J. S. (2002) Biomarkers as proxies for plant inputs to peats: an example from a sub-boreal ombrotrophic bog. *Org. Geochem.* **33**, 675–690.

Park J. and Herbert T. (1987) Hunting for paleoclimatic periodicities in a geologic time series with an uncertain time scale. *J. Geophys. Res.* **92**, 14027–14040.

Pasley M. A., Gregory W. A., and Hart G. F. (1991) Organic matter variations in transgressive and regressive shales. *Org. Geochem.* **17**, 483–509.

Paytan A., Kastner M., and Chavez F. (1996) Glacial to interglacial fluctuations in productivity in the equatorial Pacific as indicated by marine barite. *Science* **274**, 1355–1357.

Pedersen T. F. and Calvert S. E. (1990) Anoxia vs. productivity: What controls the formation of organic-carbon-rich sediments and sedimentary rocks? *Am. Assoc. Petrol. Geol. Bull.* **74**, 454–466.

Pedersen T. F., Nielsen B., and Pickering M. (1991) Timing of late Quaternary productivity pulses in the Panama Basin and implications for atmospheric CO_2. *Paleoceanography* **6**, 657–677.

Perlmutter M. A., Radovich B. J., Matthews M. D., and Kendall C. G. St. c (1998) The impact of high-frequency sedimentation cycles on stratigraphic interpretation. In *Sequence Stratigraphy—Concepts and Application.* Special Publication 8 (eds. F. M. Gradstein, K. O. Sandvik, and N. J. Milton). Norwegian Petroleum Society, Oslo, pp. 141–170.

Peterson L. C., Overpeck J. T., Kipp N. G., and Imbrie J. (1991) A high-resolution late Quaternary upwelling record from the anoxic Cariaco Basin, Venezuela. *Paleoceanography* **6**, 99–119.

Petsch S. T., Berner R. A., and Eglinton, T. I. (2000) A field study of the chemical weathering of ancient sedimentary organic matter. *Org. Geochem.* **31**, 475–487.

Petsch S. T., Edwards K. J. and Eglington T. (2002) Interactions of chemical, biological and physical processes during weathering of black shales. In *Proceedings, 6th International Symposium on the Geochemistry of the Earth's Surface (GES-6):* Honolulu, 47–52.

Pettijohn F. J. (1949) *Sedimentary rocks.* Harpers Inc., New York, 526pp.

Piper D. Z. (1994) Seawater as the source of minor elements in black shales, phosphorites and other sedimentary rocks. *Chem. Geol.* **114**, 95–114.

Piper D. Z. and Isaacs C. M. (1995) Minor elements in Quaternary sediment from the Sea of Japan: a record of surface-water productivity and intermediate-water redox conditions. *Geol. Soc. Am. Bull.* **107**, 54–67.

Pollastro R. M. (1980) Mineralogy and diagenesis of gas-bearing reservoirs in Niobrara Chalk. *US Geol. Surv. Prof. Pap.* **P1175**, 36–37.

Popp B. N., Laws E. A., Bidigare R. R., Dore J. E., Hanson K. L., and Wakeham S. G. (1998) Effect of phytoplankton cell geometry on carbon isotopic fractionation. *Geochim. Cosmochim. Acta* **62**, 69–77.

Potter P. E., Maynard B. J., and Pryor W. A. (1980) *Sedimentology of Shale: Study Guide and Reference Source.* Springer, New York, 270pp.

Pratt L. M. (1984) Influence of paleoenvironmental factors on preservation of organic matter in the Middle Cretaceous Green Formation, Pueblo, CO. *Am. Assoc. Petrol. Geol. Bull.* **68**, 1146–1159.

Pratt L. M. (1985) Isotopic studies of organic matter and carbonate in rocks of the Greenhorn Marine Cycle. In *Fine-Grained Deposits and Biofacies of the Cretaceous Western Interior Seaway: Evidence of Cyclic Sedimentary Processes.* (eds. L. M. Pratt, E. G. Kauffman, and F. B. Zelt). Society Economic Paleontologists and Minerologists, Tulsa, pp. 38–48.

Pratt L., Arthur M., Dean W., and Scholle P. (1993) Paleoceanographic cycles and events during the late Cretaceous in the Western Interior Seaway of North America. In *The Evolution of the Western Interior Basin.* Geological Association of Canada Special Paper 39 (eds. W. G. E. Caldwell and E. G. Kauffman) Geological Association of Canada, St. John's, pp. 333–353.

Pye K. and Krinsley D. H. (1986) Diagenetic carbonate and evaporite minerals in Rotliegend aeolian sandstones of the southern North Sea: their nature and relationship to secondary porosity. *Clay Min.* **21**, 443–457.

Raiswell R. and Berner R. A. (1985) Pyrite formation in euxinic and semi-euxinic sediments. *Am. J. Sci.* **285**, 710–724.

Raiswell R. and Berner R. A. (1986) Pyrite and organic matter in Phanerozoic normal marine shales. *Geochim. Cosmochim. Acta* **50**, 1967–1976.

Raiswell R. A. and Canfield D. E. (1996) Rates of reaction between silicate iron and dissolved sulfide in Peru margin sediments. *Geochim. Cosmochim. Acta* **60**, 2777–2787.

Raiswell R. and Canfield D. E. (1998) Sources of iron for pyrite formation in marine sediments. *Am. J. Sci.* **298**, 219–245.

Raiswell R., Buckley F., Berner R. A., and Anderson T. F. (1988) Degree of pyritization of iron as a paleoenvironmental indicator of bottom-water oxidation. *J. Sedim. Petrol.* **58**, 812–819.

Raiswell R., Bottrell S. H., Al-Biatty H. J., and Tan M. M. D. (1993) The influence of bottom water oxygenation and reactive iron content on sulfur incorporation into bitumens from Jurassic marine shales. *Am. J. Sci.* **293**, 569–596.

Raiswell R., Newton R., and Wignall P. B. (2001) An indicator of water-column anoxia: resolution of biofacies variations in the Kimmeridge Clay (Upper Jurassic UK). *J. Sedim. Res.* **71**, 286–294.

Ravizza G. (1993) Variations of the $^{187}Os/^{186}Os$ ratio of seawater over the past 28 million years as inferred from metalliferous carbonates. *Earth Planet. Sci. Lett.* **118**, 335–348.

Rea D. R. and Hovan S. A. (1995) Grain size distribution and depositional processes of the mineral component of abyssal sediments: lessons from the North Pacific. *Paleoceanography* **10**, 251–258.

Redfield A. C., Ketchum B. H., and Richards F. A. (1963) The influence of organisms on the composition or seawater. In *The Sea, vol. 2* (ed. M. N. Hill). Wiley, New York, pp. 26–77.

Repeta D. J. (1993) A high resolution historical record of Holocene anoxygenic primary production in the Black Sea. *Geochim. Cosmochim. Acta* **57**, 4337–4342.

Rhoads D. C. and Morse J. W. (1971) Evolutionary and ecologic significance of oxygen-deficient marine basins. *Lethaia* **4**, 413–428.

Rickard, L.V. (1975) *Correlation of the Devonian Rocks in New York State.* New York Museum and Science Service, Map and Chart Series, No. 24.

Roen J. B. (1984) Geologic framework and hydrocarbon evaluation of Devonian and Mississippian black shales in the Appalachian Basin. *AAPG Eastern Section meeting, Am. Assoc. Petrol. Geol. Bull.* **68**, 1927.

Roen J. B. and Kepferle R. C. (1993) Petroleum geology of the Devonian and Mississippian black shale of eastern North America. *US Geol. Surv. Bull.* **B1909**, A1–A8.

Rohl H. J., Schmid-Rohl A., Oschmann W., Frimmel A., and Schwark L. (2001) The Posidonia Shale (Lower Toarcian) of SW-Germany: an oxygen-depleted ecosystem controlled by sea level and palaeoclimate. *Palaeogeogr. Palaeoclimatol. Palaeoecol.* **165**, 27–52.

Saelen G., Tyson R. V., Talbot M. R., and Telnaes N. (1998) Evidence of recycling of isotopically light $CO_{2(aq)}$ in stratified black shale basins: contrasts between the Whitby Mudstone and Kimmeridge Clay formations, United Kingdom. *Geology* **26**, 747–750.

Sageman B. B. (1985) High-resolution stratigraphy and paleobiology of the Hartland Shale Member: analysis of an oxygen-deficient epicontinental sea. In *Fine-Grained Deposits and Biofacies of the Cretaceous Western Interior Seaway: Evidence of Cyclic Sedimentary Processes.* (eds. L. M. Pratt, E. G. Kauffman, and F. B. Zelt). Society Economic Paleontologists and Minerologists, Tulsa, pp. 110–121.

Sageman B. B. (1989) The benthic boundary biofacies model: Hartland Shale Member, Greenhorn Formation (Cenomanian), Western Interior, North America. *Palaeogeogr. Palaeoclimatol. Palaeoecol.* **74**, 87–110.

Sageman B. B. (1996) Lowstand tempestites: depositional model for cretaceous skeletal limestones, Western Interior US. *Geology* **24**, 888–892.

Sageman B. B. and Bina C. (1997) Diversity and species abundance patterns in Late Cenomanian black shale biofacies: western interior US. *Palaios* **12**, 449–466.

Sageman B. B. and Johnson C. C. (1985) Stratigraphy and paleobiology of the Lincoln Limestone Member, Greenhorn Limestone, Rock Canyon Anticline, Colorado. In *Fine-Grained Deposits and Biofacies of the Cretaceous Western Interior Seaway: Evidence of Cyclic Sedimentary Processes.* (eds. L. M. Pratt, E. G. Kauffman, and F. B. Zelt). Society Economic Paleontologists and Minerologists, Tulsa, pp. 100–109.

Sageman B. B., Rich J., Arthur M. A., Birchfield G. E., and Dean W. E. (1997) Evidence for Milankovitch periodicities in Cenomanian-Turonian lithologic and geochemical cycles, Western Interior USA. *J. Sedim. Res.* **67**, 286–302.

Sageman B. B., Murphy A. E., Werne J. P., Ver Straeten C. A., Hollander D. J., and Lyons T. W. (2003) A tale of shales: the relative roles of production, decomposition, and dilution in the accumulation of organic-rich strata, Middle-Upper Devonian, Appalachian basin. *Chem. Geol.* **195**, 229–273.

Sageman B., Rich J., Savrda C. E., Bralower T., Arthur M. A., and Dean W. E. (1998) Multiple Milankovitch cycles in the Bridge Creek Limestone (Cenomanian-Turonian), Western Interior basin. In *Stratigraphy and Paleoenvironments of the Cretaceous Western Interior Seaway USA.* Concepts in Sedimentology and Paleontology no. 6 (eds. W. E. Dean and M. A. Arthur). Society Economic Paleontologists Minerologists, Tulsa, pp. 153–171.

Sancetta C., Lyle M., Heusser L., Zahn R., and Bradbury J. P. (1992) Late-glacial to Holocene changes in winds, upwelling, and seasonal production of the Northern California current system. *Quat. Res.* **38**, 359–370.

Savrda C. E. (1998) Ichnology of the Bridge Creek Limestone: evidence for chemical and spatioal variations in paleo-oxygenation in the Western Interior Seaway. In *Stratigraphy and Paleoenvironments of the Cretaceous Western Interior Seaway USA.* Concepts in Sedimentology and Paleontology, no. 6 (eds. W. E. Dean and M. A. Arthur). Society Economic Paleontologists Minerologists, Tulsa, pp. 127–136.

Schenau S. J. and de Lange G. J. (2001) Phosphorous regeneration vs. burial in sediments of the Arabian Sea. *Mar. Chem.* **75**, 201–217.

Schieber J. (1996) Early diagenetic silica deposition in algal cysts and spores: a source of sand in black shales? *J. Sedim. Res.* **66**, 175–183.

Schieber J., Krinsley D., and Riciputi L. (2000) Diagenetic origin of quartz silt in mudstones and implications for silica cycling. *Nature* **406**, 981–985.

Schieber J., Zimmerle W., and Sethi P. S. (1998) *Shales and Mudstones.* vol. 1, E. Schweizerbart'sehe Verlagsbuchhandlung, Stuttgart, 384pp.

Schieber J., Zimmerle W., and Sethi P. S. (1998) *Shales and Mudstones.* vol. 2, E. Schweizerbart'sehe Verlagsbuchhandlung, Stuttgart, 296pp.

Schlanger S. O. and Jenkyns H. C. (1976b) Cretaceous Oceanic Anoxic Events: causes and consequences. *Geologie En Mijnbouw* **55**, 179–184.

Schlanger S. O., Arthur M. A., Jenkyns H. C., and Scholle P. A. (1987) The Cenomanian-Turonian oceanic anoxic event: I. Stratigraphy and distribution of organic carbon-rich beds and the marine d C excursion. In *Marine Petroleum Source Rocks.* Geological Society of London Special Publication 26 (eds. J. Brooks and A. J. Fleet). Geological Society of London, London, pp. 371–399.

Schlesinger W. H. (1991) *Biogeochemistry: An Analysis of Global Change.* Academic Press, San Diego, 443pp.

Scholle P. and Arthur M. A. (1980) Carbon isotope fluctuations in Cretaceous pelagic limestones: potential stratigraphic and petroleum exploration tool. *Am. Assoc. Petrol. Geol. Bull.* **64**, 67–87.

Seilacher A. (1982) Posidonia Shales (Toarcian S. Germany)—Stagnant Basin Model Revalidated. In *Proceedings of the First International Meeting on "Paleontology, Essential of Historical Geology"* (ed., E. M. Gallitelli). Venice, Italy, pp. 25–55.

Shen Y. N., Canfield D. E., and Knoll A. H. (2002) Middle Proterozoic ocean chemistry: evidence from the McArthur Basin, northern Australia. *Am. J. Sci.* **302**, 81–109.

Silliman J. E., Meyers P. A., and Bourbonniere R. A. (1996) Record of post-glacial organic matter delivery and burial in sediments of Lake Ontario. *Org. Geochem.* **24**, 463–472.

Simons D.-J. H. and Kenig F. (2001) Molecular fossil constraints on water column structure of the Cenomanian-Turonian Western Interior Seaway, USA. *Palaeogeogr. Palaeoclimatol. Palaeoecol.* **169**, 129–152.

Sinninghe Damsté J. S. and Koester J. (1998) A euxinic southern North Atlantic Ocean during the Cenomanian Turonian oceanic anoxic event. *Earth Planet. Sci. Lett.* **165**, 173.

Sinninghe Damsté J. S., Wakeham S. G., Kohnen M. E. L., Hayes J. M., and de Leeuw J. W. (1993) A 6000-year sedimentary molecular record of chemocline excursions in the Black Sea. *Nature* **827**, 829.

Slomp C. P., Thomson J., and de Lange G. J. (2002) Enhanced regeneration of phosphorous during formation of the most recent eastern Mediterranean sapropel (S1). *Geochim. Cosmochim. Acta* **66**, 1171–1184.

Sloss L. L. (1962) Stratigraphic models in exploration. *J. Sedim. Petrol.* **32**, 415–462.

Spencer D. W. and Brewer P. G. (1971) Vertical advection diffusion and redox potentials as controls on the distribution of manganese and other trace metals dissolved in waters of the Black Sea. *J. Geophys. Res.* **76**, 5877–5892.

Stein R. (1986) Organic carbon and sedimentation rate, further evidence for anoxic deep-water conditions in the Cenomanian/Turonian Atlantic Ocean. *Mar. Geol.* **72**, 199–209.

Stevenson F. J. and Cheng C.-N. (1972) Organic geochemistry of the Argentine Basin sediments: carbon-nitrogen relationships and Quarternary correlations. *Geochim. Cosmochim. Acta* **36**, 653–671.

Strauss H. (2002) The isotopic composition of Precambrian sulphide—Seawater chemistry and biological evolution. In *Precambrian Sedimentary Environments: A Modern Approach to Ancient Depositional Systems.* International Association of Sedimentologists Special Publication no. 33 (eds. W. Altermann and P. L. Corocoran). Blackwell, Oxford, pp. 67–105.

Stumm W. and Morgan J. J. (1996) *Aquatic Chemistry.* Wiley, New York, 1022pp.

Suess E. (1980) Particulate organic carbon flux in the oceans–surface productivity and oxygen utilization. *Nature* **288**, 260–263.

Sweeney R. E. and Kaplan I. R. (1980) Stable isotope composition of dissolved sulfate and hydrogen sulfide in the Black Sea. *Mar. Chem.* **9**, 145–152.

Swift D. J. P. and Thorne J. A. (1991) Sedimentation on continental margins: I. A general model for shelf sedimentation. *International Association of Sedimentologists, (Special Publication)* **14**, 3–31.

Taylor S. R. and McLennan S. M. (1985) *The Continental Crust: Its Composition and Evolution.* Blackwell, Maiden, Massachusetts, 312pp.

Thomson J., Mercone D., de Lange G. J., and van Santvoort P. J. M. (1999) Review of recent advances in the interpretation of eastern Mediterranean sapropel S1 from geochemical evidence. *Mar. Geol.* **153**, 77–89.

Thurow J., Moullade M., Brumsack H. J., Masure E., Taugourdeau-Lantz J., and Dunham K. W. (1988) The Cenomanian/Turonian boundary event (CTBE) at Hole 641A, ODP Leg 103 (compared with the CTBE interval at Site 398). *Proc. ODP Sci. Results* **103**, 587–634.

Tissot B. P. and Welte D. H. (1984) *Petroleum Formation and Occurrence.* Springer, Berlin, 699pp.

Toth D. J. and Lerman A. (1977) Organic matter reactivity and sedimentation rates in the ocean. *Am. J. Sci.* **277**, 465–485.

Trask P. D. (ed.) (1955) *Recent Marine Sediments.* Society of Economic Paleontologists and Mineralogists, Tulsa, OK, 736pp.

Tromp T. K., Van Cappellen P., and Key R. M. (1995) A global model for the early diagenesis of organic carbon and organic phosphorus in marine sediments. *Geochim. Cosmochim. Acta* **59**, 1259–1284.

Truyols-Massoni M., Montesinos R., Garcia-Alcalde J. L., and Leyva F. (1990) The Kacak-Otomari Event and its characterization in the Palentine Domain (Cantabrian Zone, NW Spain). In *Extinction Events in Earth History: Proceedings of Project 216, Global Biological Events in Earth History, 3rd International Conference on Global Bio-events.* Lecture Notes in Earth Sciences (eds. E. G. Kauffman and O. H. Walliser) 30, pp. 133–144.

Tucker R. D., Bradley D. C., Ver Straeten C. A., Harris A. G., Ebert J. R., and McCutcheon S. R. (1998) New U–Pb zircon ages and the duration and division of Devonian time. *Earth Planet. Sci. Lett.* **158**, 175–186.

Turekian K. K. and Wedepohl K. H. (1961) Distribution of the elements in some major units of the Earth's crust. *Geol. Soc. Am. Bull.* **72**, 175–192.

Twenhofel W. H. (1926) *Treatise on sedimentation.* Williams and Wilkins, Baltimore, 661pp.

Twenhofel W. H. (1950) *Principles of sedimentation* 2nd edn. McGraw Hill, New York, 673pp.

Tyrrell T. (1999) The relative influences of nitrogen and phosphorus on oceanic primary production. *Nature* **400**, 525–531.

Tyson R. V. (1995) *Sedimentary Organic Matter: Organic Facies and Palynofacies.* Chapman and Hall, London, 615pp.

Tyson R. V. and Pearson T. H. (1991) Modern and ancient continental shelf anoxia: an overview. In *Modern and Ancient Continental Shelf Anoxia.* Geological Society of London, Special Publication, no. 58 (eds. R. V. Tyson and T. H. Pearson). Geological Society of London, London, pp. 1–24.

Valley J. and Cole D. R. (2001) Stable Isotope Geochemistry, Rev. Mineral. Geochem. Mineralogical Society of America, Washington, DC, vol. 43, 662p.

Van Cappellen P. and Ingall E. D. (1994) Benthic phosphorous regeneration, net primary production, and ocean anoxia: a model of the coupled biogeochemical cycles of carbon and phosphorous. *Paleoceanography* **9**, 667–692.

Van Cappellen P., Viollier E., Roychoudhury A., Clark L., Ingall E., Lowe K., and DiChristina T. (1998) Biogeochemical cycles of manganese and iron at the oxic-anoxic transition of a stratified marine basin (Orca Basin, Gulf of Mexico). *Environ. Sci. Technol.* **32**, 2931–2939.

Van der Weijden C. H. (2002) Pitfalls of normalization of marine geochemical data using a common divisor. *Mar. Geol.* **184**, 167–187.

Van Mooy B. A. S., Keil R. G., and Devol A. H. (2002) Impact of suboxia on sinking particulate organic carbon; enhanced carbon flux and preferential degradation of amino acids via denitrification. *Geochim. Cosmochim. Acta* **66**, 457–465.

Van Santvoort P. J. M., de Lange G. J., Thomson J., Cussen H., Wilson T. R. S., Krom M. D., and Ströhle K. (1996) Active post-depositional oxidation of the most recent sapropel (S1) in sediments of the eastern Mediterranean Sea. *Geochim. Cosmochim. Acta* **60**, 4007–4024.

Ver Straeten C. A. and Brett C. E. (1995) Lower and Middle Devonian foreland basin fill in the Catskill Front: stratigraphic synthesis, sequence stratigraphy, and the Acadian Orogeny. In *67th Annual Meeting Field Trip Guidebook* (eds. J. I. Garver and J. A. Smith). New York State Geological Association, Albany, pp. 313–356.

Vincent E. and Berger W. H. (1985) Carbon dioxide and polar cooling in the Miocene: the monterey hypothesis. In *The Carbon Cycle and Atmospheric CO2: Natural Variations Archean to Present.* American Geophysical Union, Geophysical Monograph, no. 32 (eds. E. T. Sunquist and W. S. Broecker). American Geophysical Union, Washington, DC, pp. 455–468.

Watkins D. K. (1989) Nanoplankton productivity fluctuations and rhythmically bedded pelagic carbonates of the Greenhorn Limestone (Upper Cretaceous). *Palaeogeogr. Palaeoclimatol. Palaeoecol.* **74**, 75–86.

Weisel C. P., Duce R. A., Fasching J. L., and Heaton R. W. (1984) Estimates of the transport of trace metals from the ocean to the atmosphere. *J. Geophys. Res. D Atmos.* **89**, 11607–11618.

Weldeab S., Emeis K., Hemleben C., and Siebel W. (2002) Provenance of lithogenic surface sediments and pathways of riverine suspended matter in the eastern Mediterranean Sea: evidence from $^{143}Nd/^{144}Nd$ and $^{87}Sr/^{86}Sr$ ratios. *Chem. Geol.* **186**, 139–149.

Werne J. P., Hollander D. J., Lyons T. W., and Peterson L. C. (2000) Climate-induced variations in the productivity and planktonic ecosystem structure from the Younger Dryas to Holocene in the Cariaco Basin, Venezuela. *Paleoceanography* **15**, 19–29.

Werne J. P., Sageman B. B., Lyons T., and Hollander D. J. (2002) An integrated assessment of a "type euxinic" deposit: evidence for multiple controls on black shale deposition in the Middle Devonian Oatka Creek Formation. *Am. J. Sci.* **302**, 110–143.

Werne J. P., Lyons T. W., Hollander D. J., Formolo M. J., Formolo M. J., and Sinninghe Damsté J. S. (2003) Reduced sulfur in euxinic sediments of the Cariaco Basin: sulfur isotope constraints on organic sulfur formation. *Chem. Geol.* **195**, 159–179.

Wignall P. B. (1994) *Black Shales.* Claredon, Oxford, 127pp.

Wignall P. B. and Myers K. J. (1988) Interpreting benthic oxygen levels in mudrocks: a new approach. *Geology* **16**, 452–455.

Wignall P. B. and Newton R. (1998) Pyrite framboid diameter as a measure of oxygen deficiency in ancient mudrocks. *Am. J. Sci.* **298**, 537–552.

Wijsman J. W. M., Middelburg J. J., and Heip C. H. R. (2001) Reactive iron in Black Sea sediments: implications for iron cycling. *Mar. Geol.* **172**, 167–180.

Wilde P., Quinby-Hunt M. S., and Erdtmann B.-D. (1996) The whole-rock cerium anomaly: a potential indicator of eustatic sea-level changes in shales of the anoxic facies. *Sedim.Geol.* **101**, 43–53.

Wilkin R. T. and Arthur M. A. (2001) Variations in pyrite texture, sulfur isotope composition, and iron systematics in the Black Sea: evidence for late Pleistocene to Holocene excursions of the O_2-H_2S redox transition. *Geochim. Cosmochim. Acta* **65**, 1399–1416.

Wilkin R. T., Barnes H. L., and Brantley S. L. (1996) The size distribution of framboidal pyrite in modern sediments: an indicator of redox conditions. *Geochim. Cosmochim. Acta* **60**, 3897–3912.

Wilkin R. T., Arthur M. A., and Dean W. E. (1997) History of water-column anoxia in the Black Sea indicated by pyrite framboid size distributions. *Earth Planet. Sci. Lett.* **148**, 517–525.

Witzke B. J. and Heckel P. H. (1988) Paleoclimatic indicators and inferred Devonian paleolatitudes of Euramerica. In *Devonian of the World: Proceedings of the Second International Symposium on the Devonian System, Volume II, Sedimentation*, Canadian Society of Petroleum Geologists, Memoir 14 (eds. N. J. McMillan, A. F. Embry, and D. J. Glass), Canadian Society of Petroleum Geologists, Calgary, pp. 49–63.

Woodrow D. L. (1985) Paleogeography, paleoclimate, and sedimentary processes of the Late Devonian Catskill Delta. In *The Catskill Delta.* Geological Society of America Special Paper 201 (eds. D. L. Woodrow and W. D. Sevon). Geological Society of America, Boulder, pp. 51–63.

Wortmann U. G., Hesse R., and Zacher W. (1999) Major-element analysis of cyclic black shales: paleoceanographic implications for the early cretaceous deep western Tethys. *Paleoceanography* **114**, 525–541.

Wortmann U. G., Bernasconi S. M., and Boettcher M. E. (2001) Hypersulfidic deep biosphere indicates extreme sulfur isotope fractionation during single-step microbial sulfate reduction. *Geology* **29**, 647–650.

Wright J., Schrader H., and Holser W. T. (1987) Paleoredox variations in ancient oceans recorded by rare earth elements in fossil apatite. *Geochim. Cosmochim. Acta* **51**, 631–644.

Yarincik K. M., Murray R. W., Lyons T. W., Peterso L. C., and Haug G. H. (2000a) Oxygenation history of bottom waters in the Cariaco Basin, Venezuela, over the past 578,000 years: results from redox-sensitive metals (Mo, V, Mn, and Fe). *Paleoceanography* **15**, 593–604.

Yarincik K. M., Murray R. W., and Peterson L. C. (2000b) Climatically sensitive eolian and hemipelagic deposition in the Cariaco Basin, Venezuela, over the past 578,000 years: results from Al/Ti and K/Al. *Paleoceanography* **15**, 210–228.

Yochelson E. L. and Lindemann R. H. (1986) Considerations on the systematic placement of the Styliolines *(incertae sedis:* Devonian). In *Problematic Fossil Taxa.* Oxford Monographs in Geology and Geophysics, no. 5 (eds. Hoffman and Nitecki). Oxford University Press, Oxford, pp. 45–58.

Young G. M. and Nesbitt H. W. (1998) Processes controlling the distribution of Ti and Al in weathering profiles, siliciclastic sediments. *J. Sedim. Res.* **68**, 448–455.

Zaback D. A., Pratt L. M., and Hayes J. M. (1993) Transport and reduction of sulfate and immobilization of sulfide in marine black shales. *Geology* **21**, 141–144.

Zheng Y., Anderson R. F., van Geen A., and Kuwabara J. S. (2000) Authigenic molybdenum formation in marine sediments: a link to pore water sulfide in the Santa Barbara Basin. *Geochim. Cosmochim. Acta* **64**, 4165–4178.

Published by Elsevier Ltd.

Readings from the Treatise on Geochemistry
ISBN: 978-0-12-381391-6

pp. 423–466

14
Evolution of Sedimentary Rocks

J. Veizer

Ruhr University, Bochum, Germany and University of Ottawa, ON, Canada

and

F. T. Mackenzie

University of Hawaii, Honolulu, HI, USA

14.1 INTRODUCTION

For almost a century, it has been recognized that the present-day thickness and areal extent of Phanerozoic sedimentary strata increase progressively with decreasing geologic age. This pattern has been interpreted either as reflecting an increase in the rate of sedimentation toward the present (Barrell, 1917; Schuchert, 1931; Ronov, 1976) or as resulting from better preservation of the younger part of the geologic record (Gilluly, 1949; Gregor, 1968; Garrels and Mackenzie, 1971a; Veizer and Jansen, 1979, 1985).

Study of the rocks themselves led to similarly opposing conclusions. The observed secular (=age) variations in relative proportions of lithological types and in chemistry of sedimentary rocks (Daly, 1909; Vinogradov *et al.*, 1952; Nanz, 1953; Engel, 1963; Strakhov, 1964, 1969; Ronov, 1964, 1982) were mostly given an evolutionary interpretation. An opposing, uniformitarian, approach was proposed by Garrels and Mackenzie (1971a). For most isotopes, the consensus favors deviations from the present-day steady state as the likely cause of secular trends.

This chapter attempts to show that recycling and evolution are not opposing, but complementary, concepts. It will concentrate on the lithological and chemical attributes of sediments, but not deal with the evolution of sedimentary mineral deposits (Veizer *et al.*, 1989) and of life (Sepkoski, 1989), both well amenable to the outlined conceptual treatment. The chapter relies heavily on Veizer (1988a) for the sections dealing with general recycling concepts, on Veizer (2003) for the discussion of isotopic evolution of seawater, and on Morse and Mackenzie (1990) and Mackenzie and Morse (1992) for discussion of carbonate rock recycling and environmental attributes.

14.2 THE EARTH SYSTEM

The lithosphere, hydrosphere, atmosphere, and biosphere, or rocks, water, air, and life are all part of the terrestrial exogenic system that is definable by the rules and approaches of general system science theory, with its subsets, such as population dynamics and hierarchical structures.

14.2.1 Population Dynamics

The fundamental parameters essential for quantitative treatment of population dynamics are the population size (A_0) and its recycling rate. A_0 is normalized in the subsequent discussion to one population (or 100%) and the rates of recycling relate to this normalized size. Absolute rates can be established by multiplying this relative recycling rate (parameter b below) by population size.

A steady state natural population, characterized by a continuous generation/destruction (birth/death) cycle, is usually typified by an age structure similar to that in Figure 1, the cumulative curve defining all necessary parameters of a given population. These are its *half-life* τ_{50}, *mean age* τ_{mean}, and *oblivion age* or life expectancy $\tau_{\max}$.

For steady state first-order (=single population) systems, the survival rate of constituent units can be expressed as

$$A_{t^*} = A_0\, e^{-kt^*} \quad (1)$$

where A_{t^*} is the cumulative fraction of the surviving population older than t^*, $A_0 = 1$ (one population), t^* is age (not time), and k is the rate constant for the recycling process. In the subsequent discussion, the recycling rate is often considered in the form of a *recycling*

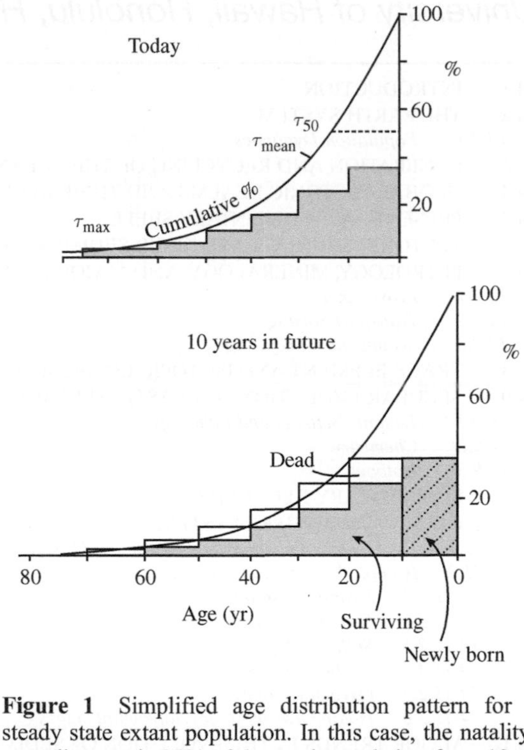

Figure 1 Simplified age distribution pattern for a steady state extant population. In this case, the natality/mortality rate is 35% of the total population for a 10-yr interval ($b = 35 \times 10^{-3}\,\text{yr}^{-1}$). $\tau_{\max}$ is defined as the fifth percentile. In practice $\tau_{\max}$ is the age at which the resolution of the database becomes indistinguishable from the background. The b value for the same population is inversely proportional to the available time resolution T (Equation (2)). Today's "instantaneous" rates of deposition and erosion of sediments exceed those calculated from the geological record based on time resolution of 10^6–10^7 yr (Sadler, 1981). It is therefore essential to stipulate the resolution T in consideration of rates (after Veizer and Jansen, 1985).

proportionality constant b, which is related to the above equation through formalism:

$$b = 1 - e^{-kT} \quad (2)$$

where T is the time resolution or duration of recycling (cf. Veizer and Jansen, 1979, 1985). In general, the larger the b value—i.e., the faster the rate of recycling—the steeper the slope of the cumulative curve and the shorter the τ_{50} and τ_{max} of the population. For a steady state extant population, generation per unit time must equal combined destruction for all age groups during the same time interval. Consequently, the cumulative slope remains the same but propagates into the future (Figure 1).

The above terminology is applicable to internally (cannibalistically) recycling populations. In an external type of recycling, the influx and efflux cause a similar age structure, but the terminology differs. In this case, the average duration an individual unit resides within the population is termed the *residence time* τ_{res}. Mathematically, τ_{res} is similar to the cannibalistic τ_{mean} and it relates to the above parameters as $\tau_{max} \geq \tau_{res} \geq \tau_{50}$. It is this alternative—populations interconnected by external fluxes—that is usually referred to as the familiar box model by natural scientists. Frequently, box models are nothing more than one possible arrangement for propagation of cyclic populations.

In the subsequent discussion, the terminology of the cannibalistic populations is employed. Note, however, that the age distribution patterns and the recycling rates calculated from these patterns are a consequence of both cannibalistic and external recycling. At this stage, we lack the data and the criteria for quantification of their relative significance. Nevertheless, from the point of view of preservation probability, it may be desirable, but not essential, to know whether the constituent units (geologic entities) have been created and destroyed by internal, external, or combined phenomena.

Among natural populations, two major deviations from the ideal pattern are ubiquitous (Pielou, 1977; Lerman, 1979). The first deviation consists of populations with excessive proportions of young units (e.g., planktonic larval stages) because their destruction rate is very high, but chances for survival improve considerably with maturation (type II in Figure 2). The mathematical formalism for such populations (e.g., Lerman, 1979) is a power-law function,

$$A_{t^*} = A_0(1 + kt^*)^{-z} \quad (3)$$

where the exponent z increases for populations with progressively larger destruction rates of young units.

The other common exception consists of populations with suppressed destruction of young units

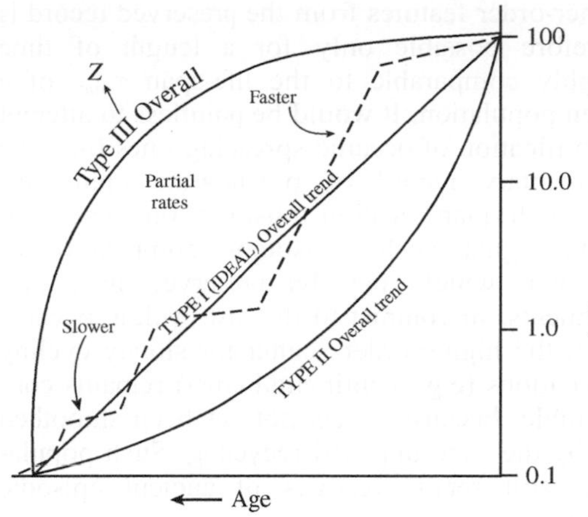

Figure 2 Cumulative age distribution functions for various populations plotted on a semilogarithmic scale (after Veizer, 1988a).

(type III, Figure 2). In these instances, the destruction rates increase rapidly as the life span τ_{max} is approached. Mathematical expression for this relation is

$$A_{t^*} = A_0[1 - (kt^*)^z] \quad (4)$$

where the exponent z increases for populations with progressively larger mortality rates of old units.

These relationships are valid for populations of constant size. For non-steady-state populations with stable age structures—i.e., those with overall rates of growth or decline much slower than the rates of recycling of their constituents—the age distributions approach the pattern of the constant-size populations. The calculated recycling rates are therefore identical. For populations where the overall growth (decline) approaches the rate of recycling of constituent units, independent criteria are required to differentiate recycling from the growth (decline) component.

The above discussion assumed a quasi-continuous generation/destruction process for a first-order system, but natural variability causes oscillations, at a hierarchy of frequencies and amplitudes, around the smooth overall patterns. Furthermore, geological processes are usually discrete and episodic phenomena. Because of all these factors, partial intervals have generation/destruction rates that deviate from the smooth average rates (Figure 2), with the connecting tangents having either shallower or steeper slopes. Note again that a given partial slope may reflect deviations in generation, in destruction, or in their combined effect. Usually, the problem is not resolvable, but the combined effect is the most likely alternative. As the population ages, the magnitude of this higher-order scatter diminishes to the level of uncertainties in the database (Figure 2). Quantitative interpretation of such higher-order features from the preserved record is therefore possible only for a length of time roughly comparable to the life-span τ_{max} of a given population. It would be pointless to attempt quantification of oceanic spreading rates from the fragmentary record of pre-Jurassic ophiolites. Any such quantification must rely on some derivative signal, such as isotopic composition of seawater, which may be preserved in coeval sediments. In contrast to the fast cycling oceanic crust, the higher-order scatter for slowly cycling populations (e.g., continental crust) remains considerable, because it has not yet been smoothed out by the superimposed recycling. Such populations still retain vestiges of ancient episodic events.

For geologic entities (e.g., crustal segments, mineral deposits, tectonic domains, and fossils), the age distribution patterns can be extracted from their stratigraphic and geochronologic assignments. At present, only major features of the record can be quantitatively interpreted, because the database is usually not of the desired reliability.

14.3 GENERATION AND RECYCLING OF THE OCEANIC AND CONTINENTAL CRUST

The concept of global tectonics (Dietz, 1961; Hess, 1962; Morgan, 1968; Le Pichon *et al.*, 1973) combined the earlier proposals of continental drift and seafloor spreading into a unified theory of terrestrial dynamics. It introduced the notion of continual generation and destruction of oceanic crust and implied similar consequences for other tectonic realms.

The present-day age distribution pattern of the *oceanic crust* is well known (Sprague and Pollack, 1980; Sclater *et al.*, 1981; Rowley, 2002), and the plate tectonic concept of ocean floor generation/subduction well established. The age distribution pattern (Figure 3) conforms to systematics with a half-life (τ_{50}) of ~60 Myr, translating to generation/destruction of ~3.5 km^2 (or ~20 km^3) of oceanic crust per year. The maximal life-span τ_{max} (oblivion age) for its tectonic settings and their associated sediments is therefore less than 200 Myr.

The situation for continental crust is more complex and still dogged by the controversy (e.g., Sylvester, 2000) that pits the proponents of its near-complete generation in the early planetary history, followed only by crust/mantle recycling (Armstrong, 1981 and the adherents) against those advocating an incremental growth in the course of geologic evolution (see Taylor and McLennan, 1985). The latest summary of the volume/age estimates for the continental crust

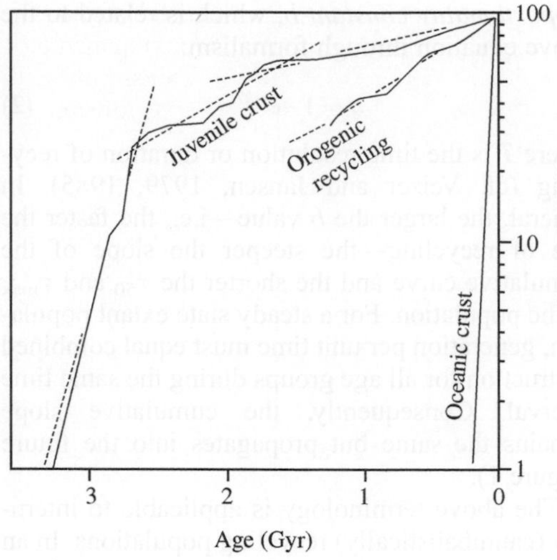

Figure 3 Cumulative age distribution of oceanic and continental crusts. Continental crust based on Condie (2001), oceanic crust after Sclater *et al.* (1981), and Rowley (2002).

(Condie, 2001) provides a more definitive constraint for the discussion of the issue. The total volume of the continental crust is estimated at 7.177×10^9 km^3 (Cogley, 1984), and the age distribution pattern of its juvenile component (Figure 3) suggests a tripartite evolution, with ~25% ~4.0–2.6 Gyr old, another 35% added between ~2.6 Gyr and 1.7 Gyr, and the remaining 40% subsequently. The observed growth pattern is of a sinusoidal (logistic) type (Veizer and Jansen, 1979), with commencement of large-scale crustal generation at ~4 Gyr, accelerating growth rate that culminated in major phases of crustal generation and cratonisation during the ~2.6–1.7 Gyr time span, and a declining rate subsequently.

The above crustal generation pattern is only a minimal estimate, based on the assumption that no continental crust was recycled into the mantle. The other limiting alternative can be based on the proposition that the continental crust attained its present-day steady state ~1.75 Gyr ago. If this were the case, today's preserved post 1.75 Gyr crust is only about one-half of that generated originally, with the equivalent amount recycled into the mantle. This recycling may go hand in hand with orogenic activity that has a τ_{50} of ~800 Myr (Figure 3), a value in good agreement with the previous estimates for the low- and high- grade metamorphic reworking rates (τ_{50} of 673 Myr and 987 Myr, respectively) by Veizer and Jansen (1985). Furthermore, in the post-1.6 Ga record, the orogenic segments are composed, on average, of about equal amounts of juvenile (2.7×10^9 km^3) and recycled (2.6×10^9 km^3) crust (figure 4 in Condie, 2001). This would suggest that each orogenic episode results in incorporation of about one-half of the juvenile crust into the reworked crustal segment, with the other half being subducted into the mantle.

Considering the above scenarios as limiting alternatives, the long-term average rate of continental crust generation would be $\sim 1.7 \pm 0.1$ $km^3\,yr^{-1}$ (1.1 ± 0.5 $km^3\,yr^{-1}$ by Reymer and Schubert, 1984) for no mantle recycling and about twice that much for the alternative where about one-half of the juvenile crust contributes to the orogenic buildup of the continents, while the other half is recycled into the mantle.

14.4 GLOBAL TECTONIC REALMS AND THEIR RECYCLING RATES

Tectonic setting is the principal controlling factor of lithology, chemistry, and preservation of sediment accumulations in their depocenters, the sedimentary basins. The latest classification of Ingersoll and Busby (1995) assigns sedimentary basins into five major groups based on their relationship to plate boundaries (Figure 4). It groups together the basins that are associated with the divergent, interplate, convergent, transform, and hybrid settings, and recognizes 26 basin types. This classification, while it cannot take into account the entire complexity of natural systems, implies that basins associated with transform and convergent boundaries are more prone to destruction than basins associated with divergent and intraplate settings, particularly those developed on continental crust (Figure 4).

This qualitative observation is consistent with the prediction of oblivion ages for specific tectonic realms based on the concept of population dynamics (Veizer and Jansen, 1985). Theoretically, if τ_{max} is taken as the fifth percentile, the oblivion ages for specific tectonic realms should be a factor of ~4.5 times the respective half-lives, but empirically, due to deviations from the ideal type I age pattern, the τ_{max} is usually some ~3.0–3.5 times τ_{50}. This qualification notwithstanding, the short-lived basins are erased faster from the geologic record and the degree of tectonic diversity must be a function of time. The diversity diminishes as the given segment of the solid earth ages, and the rate of memory loss is inversely proportional to recycling rates of the constituent tectonic realms. For a steady state system, the calculated theoretical preservational probabilities are depicted in Figure 5. This reasoning shows that the realms of the oceanic domain (basins of active margins to immature orogenic belts) should have only ≤5% chance of survival in crustal segments older than ~100–300 Myr, while the platformal and intracratonic basins can survive for billions of years.

Due to rather poor inventories, the proposed systematics should be viewed as nothing more than a conceptual framework, but it nevertheless helps to visualize the probability of preservation of sedimentary packages in the geologic record.

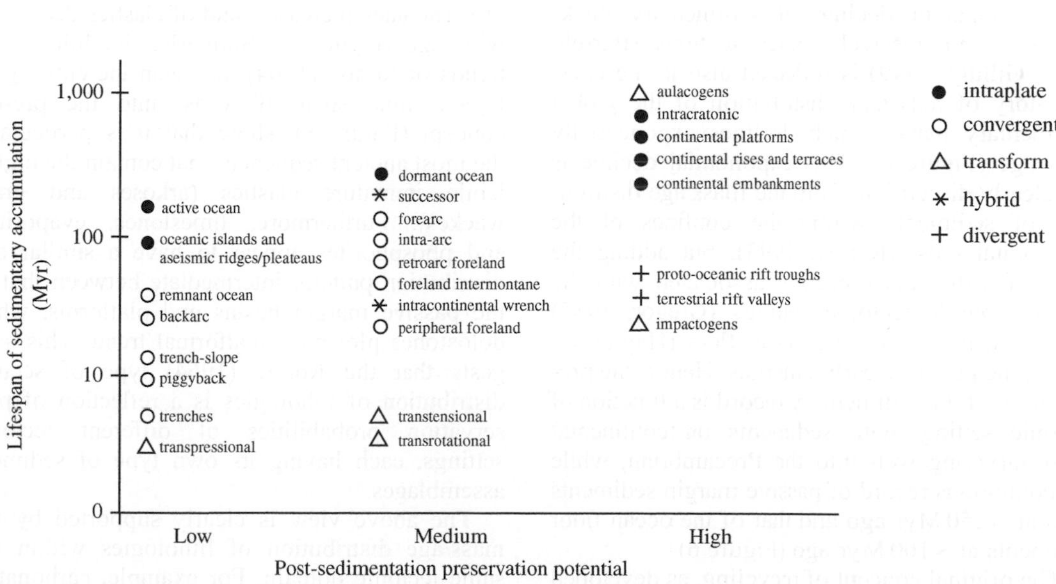

Figure 4 Typical life spans for sedimentary basins versus their post-sedimentation preservation potential. "Preservation potential" refers to average amount of time during which basins will not be uplifted and eroded, or be tectonically destroyed during and following sedimentation. Sedimentary or volcanic fill may be preserved as accretionary complexes during and after basin destruction (true of all strata deposited on oceanic crust). Basins with full circles, particularly intraplate continental margins, are "preserved" in the sense of retaining their basement, but they are likely to be subcreted beneath or within suture belts, and are difficult to recognize in the ancient record in such settings (after Ingersoll and Busby, 1995).

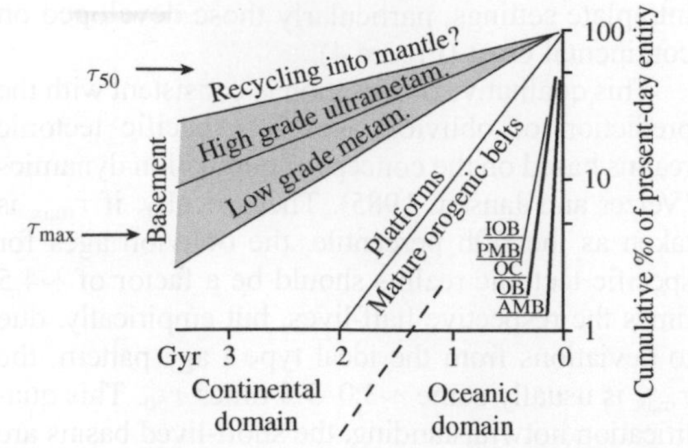

Figure 5 Preservation probabilities for major global tectonic realms. AMB = active margin basins, OB = oceanic intraplate basins, OC = oceanic crust, PMB = passive margin basins, IOB = immature orogenic belts. Rates were derived on the assumption that all deviations from the ideal pattern of the type I age distributions were a consequence of the poor quality of the available database (after Veizer, 1988b).

14.5 PRESENT-DAY SEDIMENTARY SHELL

The present-day mass of global sediments is $\sim 2.7 \times 10^{24}$ g (Ronov, 1982; Hay *et al.*, 2001). Of these, ~72.6% are situated within the confines of the present-day continents (orogenic belts 51.9%, platforms 20.7%), 12.9% at passive margin basins, 5.5% at active margin basins, and the sediments covering the ocean floor account for ~8.3% of the total (Ronov, 1982; Gregor, 1985; Veizer and Jansen, 1985).

The apparent decline of sedimentary thicknesses in progressively older sections (Barrell, 1917; Gilluly, 1949) is reflected also in the latest inventory of mass/age distribution of the global sedimentary mass, which declines exponentially with age (Figure 6). This exponential decline is not clearly discernible from the mass/age distribution of sediments within the confines of the continental crust (Ronov, 1993), but adding the mass of sediments presently associated with the passive margin tectonic settings (Gregor, 1985) and the sediments on the ocean floor (Hay *et al.*, 1988), the pattern clearly emerges. Hence, the preservation of the sedimentary record is a function of tectonic setting, with sediments on continental crust surviving well into the Precambrian, while the continuous record of passive margin sediments ends at ~250 Myr ago and that of the ocean floor sediments at ~100 Myr ago (Figure 6).

The original concept of recycling, as developed by Garrels and Mackenzie (1971a), was based solely on the "continental" database assembled by the group at the Vernadsky Institute of Geochemistry in Moscow (Ronov, 1949, 1964, 1968, 1976, 1982, 1993). The former authors proposed that the present-day mass/age distribution of global sedimentary mass is consistent with a half-mass age of ~600 Myr, resulting in deposition and destruction of about five sedimentary masses over the entire geologic history. Furthermore, the observed temporal relationship of clastics/carbonates/evaporites led Garrels and Mackenzie (1971a) to propose a concept of differential recycling rates for different lithologics based on their susceptibility to chemical weathering, with clastics having half-mass age of ~600 Myr, carbonates ~200 Myr, and evaporites ~100 Myr.

Subsequently, Veizer (1988a) and Hay *et al.* (1988) pointed out that the concept of differential recycling, although valid, can only partially be based on the susceptibility to chemical weathering. In a layer cake stratigraphy, the removal of carbonate strata, for instance, would necessarily result in a collapse of all the overlying strata, regardless of their lithology. On a macroscale, therefore, the sediments are removed en-masse, with chemical weathering rates coupled to the physical ones as (Millot *et al.*, 2002)

$$\text{Chem} = 0.39(\text{Phy})^{0.66} \qquad (5)$$

This point of view is supported also by the fact that the particulate load accounts for ~3/4 of the present-day fluvial sediment flux (Garrels and Mackenzie, 1971a).

Indeed, the detailed consideration of temporal lithological trends does not follow the pattern anticipated from recycling based on their susceptibility to chemical weathering. For example, during the Phanerozoic, the relative proportion of carbonates increases, and of clastics decreases, with age (Figure 7). Similarly, the lithological trends of Ronov (1964) that span the entire geological time span, if recast into the present concept (Figure 8), show that it is particularly the most ancient sequences that contain the highly labile immature clastics (arkoses and graywackes). Furthermore, limestones, evaporites, and phosphorites appear to have a similar age distribution pattern, intermediate between that of the passive margin basins and platforms, while dolostones plot on a platformal trend. This suggests that the Ronov (1964) type of secular distribution of lithologies is a reflection of preservation probabilities of different tectonic settings, each having its own type of sediment assemblages.

The above view is clearly supported by the mass/age distribution of lithologies within the same tectonic domain. For example, carbonates, chert, red clay, and terrigeneous sediments on the ocean floor (Hay *et al.*, 1988) all have the same type of age distribution pattern that is controlled by a single variable, the rate of spreading and subduction of the ocean floor. This sedimentary mass also differs lithologically from its "continental" counterpart, because it is comprised of

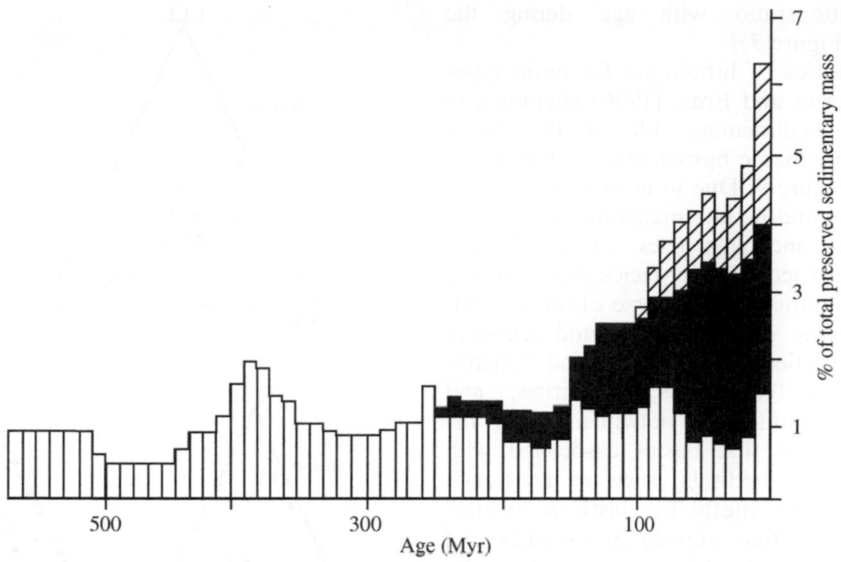

Figure 6 The mass/age distribution of preserved sedimentary mass deposited on continental crust (vertical lines), basin of passive margins (black), and on the oceanic floor (cross-hatched) (courtesy of W. W. Hay).

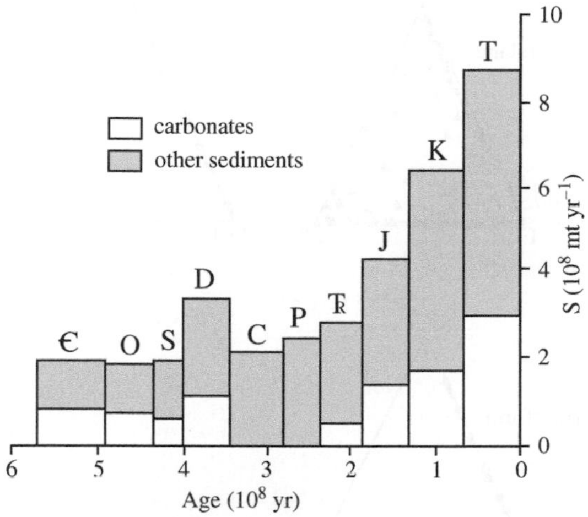

Figure 7 Surviving amounts of epicontinental terrigenous-clastic and marine-carbonates during the Phanerozoic (after Morse and Mackenzie, 1990).

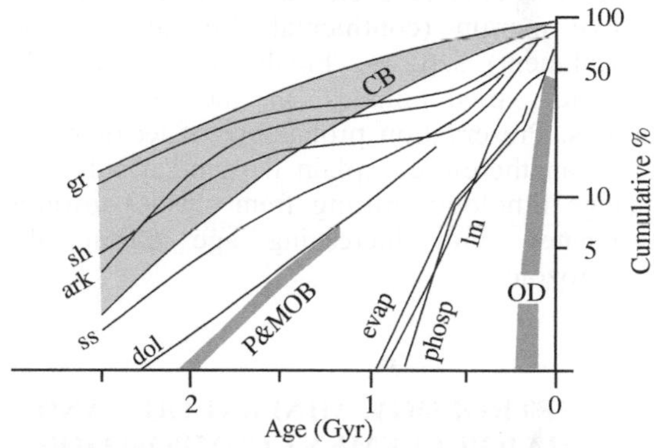

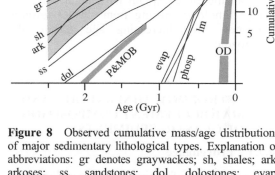

Figure 8 Observed cumulative mass/age distributions of major sedimentary lithological types. Explanation of abbreviations: gr denotes graywackes; sh, shales; ark, arkoses; ss, sandstones; dol, dolostones; evap, evaporites; lm, limestones; phosp, phosphorites; CB, continental basement; P, platforms; MOB, mature orogenic belts; OD, oceanic domain (after Veizer, 1988c).

76 wt. % terrigeneous material, 7% calcium carbonate, 10% opal, and 7% mineral bound water (Plank and Langmuir, 1998). Based on its age distribution pattern (Figure 6), the average rate of pelagic sediment subduction is $\sim 1\times 10^{21}$ g Myr^{-1} (Hay *et al.*, 1988), an estimate in good agreement with an upper limit of $1.1\pm 0.5\times 10^{21}$ g Myr^{-1} for sediment subduction based on Sm/Nd isotopic constraints (Veizer and Jansen, 1985) discussed later in the text. The estimates based on direct measurements, however, suggest a sediment subduction rate of $\sim$1.4 to 1.8×10^{21} g Myr^{-1} (Rea and Ruff, 1996; Plank and Langmuir, 1998), the difference likely accounted for by material entering the trenches from the adjacent accretionary wedge. This material can be scraped off the subducting slab, uplifted, eroded, and rapidly recycled (Hay *et al.*, 2001).

14.6 TECTONIC SETTINGS AND THEIR SEDIMENTARY PACKAGES

Is the proposition that differential recycling of sediments is controlled by tectonic settings supported by the observational data? Could this explain, for example, the paradox of increasing carbonate/clastic ratio with age during the Phanerozoic (Figure 7)?

The inventories of lithologies for basin types are sparse. Veizer and Ernst (1996) attempted to quantify the sedimentary fill of the North American Phanerozoic basins, and the results are presented in Figure 9. Due to inherent limitations of the database and its interpretation, and particularly due to inconsistencies in lithological descriptions, the sedimentary facies were grouped into three categories only: coarse clastics (sandstones, siltstones, conglomerates, and arkoses), fine clastics (shales, graywackes), and "chemical" sediments (carbonates, evaporites, and cherts). These limitations notwithstanding, the compilation shows that basins associated with immature tectonic settings, such as arc-trench systems, are filled chiefly by clastic sediments, mostly immature fine grained graywackes and shales (forearc). Foreland basins, situated usually on the continental side of the continental-margin/arch-trench system, contain a higher proportion of coarse, often mature, clastics and some chemical sediments. This is even more the case for the passive margin (continental rise, terrace, and embankment) settings. Finally, carbonate sedimentation predominates in the intracratonic settings. Preservation probability of tectonic settings can therefore explain the tendency for the average lithology shifting from clastics towards carbonates with increasing age during the Phanerozoic.

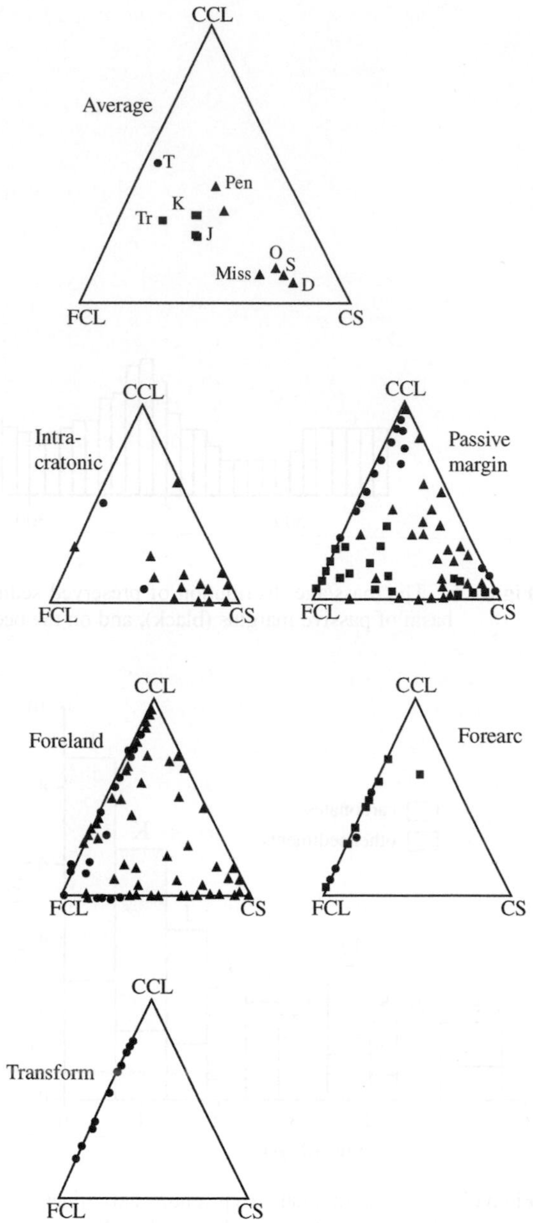

Figure 9 Relative proportions of lithological types within intracratonic to transform basins and the averages for geological periods. Based on the North American cross-sections in Cook and Bally (1975). An independent compilation by Berry and Wilkinson (1994) from the same source, but based on different criteria, yielded comparable temporal patterns. CCL coarse clastics; FCL fine clastics; CS "chemical" sediments. A global review based on the Ronov (1982) database also yields comparable patterns (after Veizer and Ernst, 1996).

14.7 PETROLOGY, MINERALOGY, AND MAJOR ELEMENT COMPOSITION OF CLASTIC SEDIMENTS

14.7.1 Provenance

The petrology of *coarse clastic* (conglomerate-size) *sediments* is controlled in the first instance by their provenance that, in turn, is a function of tectonic setting. Advancing tectonic stability is accompanied by an increasingly mature composition of the clasts (Figure 10). For first cycle sediments (Cox and Lowe, 1995), the early arc stage of tectonic evolution is dominated by volcanic clasts, from the growing and accreting volcanic pile. Subsequently, plutonic and metamorphic lithologies dominate the orogenic and uplift stages. The post-tectonic conglomerates contain a high proportion of recycled sedimentary clasts. Mineralogically, the evolution is from clasts with abundant plagioclase, to K-feldspar rich, and finally to quartz (chert) dominated clasts.

A similar provenance control relates also the *sandstone* petrology to tectonic setting, as expressed in the Q (quartz)—F (feldspar)—L (unstable lithic fragments) ternary diagrams (Figure 11) of Dickinson *et al.* (1983). Again, increasing tectonic maturity shifts the mode of sandstone petrology from L towards the F/Q tangent, terminating in the recycled Q mode.

Fine clastic sediments, mostly *mudrocks*, in contrast to their coarser counterparts, are either derived by first cycle weathering of silicate minerals or glass, or from recycling of older mudrocks. Physical comminution plays only a secondary role. The average shale is composed of ~40–60% clay minerals, 20–30% quartz, 5–10% feldspar and minor iron oxide, carbonate, organic matter, and other components (Yaalon, 1962; Shaw and Weaver, 1965). Granitic source rocks produce

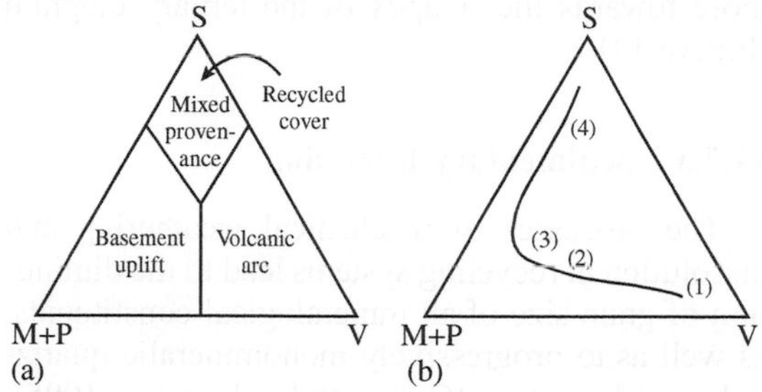

Figure 10 Idealized evolution of conglomerate clast composition. S = sedimentary clasts; M + P = metamorphic + plutonic clasts; V = volcanic clasts: (a) Approximate compositional fields for conglomerate clast populations of differing provenance. (b) Generalized stages in the evolution of conglomerate clast compositions on a crustal block: (1) volcanic arc stage; (2) dissected arc/accretion orogen stage; (3) post-tectonic granite/basement uplift stage; and (4) sediment-recycling-dominated stage (after Cox and Lowe, 1995).

shales richer in kaolinite and illite, the mafic ones richer in smectites (Cox and Lowe, 1995).

Geochemical processes associated with weathering and soil formation are dominated by alteration of feldspars (and volcanic glass), feldspars accounting for 70% of the upper crust, if the relatively inert quartz is discounted (Taylor and McLennan, 1985). Advancing weathering leads to a shift towards an aluminum rich composition that can be approximated by the chemical index of alteration (CIA) of Nesbitt and Young (1984)

$$\mathrm{CIA} = 100(\mathrm{Al_2O_3}/(\mathrm{Al_2O_3} + \mathrm{CaO^*} + \mathrm{Na_2O} + \mathrm{K_2O})) \qquad (6)$$

The suspended sediments of major rivers clearly reflect this alteration trend and plot on a tangent between the source (upper continental crust (UCC)) and the clays, the end-products of weathering (Figure 12).

The overall outcome is a depletion of the labile Ca–Na plagioclases in the sediments of progressively more stable tectonic settings, the trend being more pronounced in the fine-grained muddy sediments than in their coarser counterparts (Figure 13).

14.7.2 Transport Sorting

Transport processes, involving first cycle as well as recycled components, result in further

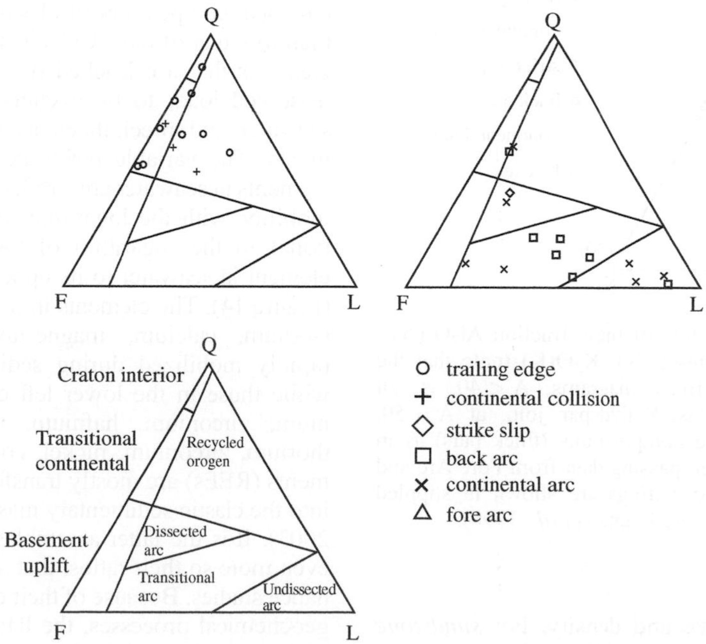

Figure 11 Ternary diagrams of framework quartz (Q)–feldspar (F) –unstable lithic fragments (L) for sands. Provenance fields from Dickinson *et al.* (1983) (after McLennan *et al.*, 1990).

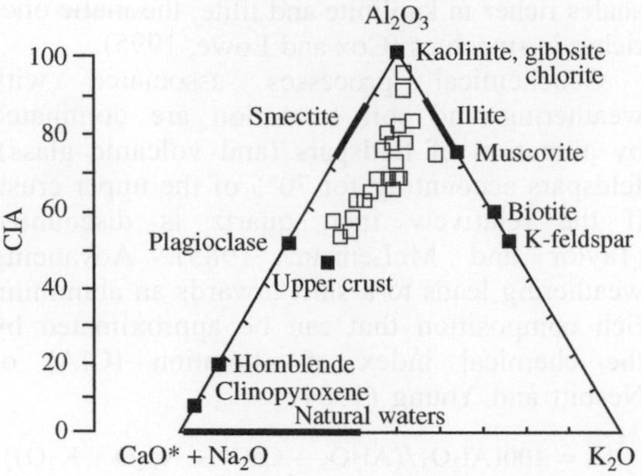

Figure 12 Ternary plots of molecular proportions of Al_2O_3–($Na_2O + CaO^*$)–K_2O with the Chemical Index of Alteration (CIA) scale shown on the left. Also plotted are selected idealized igneous and sedimentary minerals and the range of typical natural waters. Squares are suspended sediments from major rivers throughout the world representing a variety of climatic regimes. CaO^* is the silicate bound concentration only (after McLennan *et al.*, 2003).

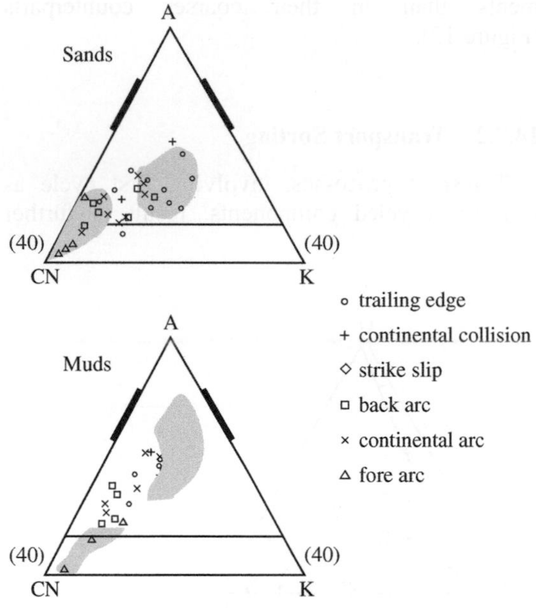

Figure 13 Ternary plots of mole fraction Al_2O_3(A) – $CaO + Na_2O$ in silicates (CN) – K_2O(K) (note that the lower part of the ternary diagrams, $A < 40$, is not shown). The plagioclase-K-feldspar join, at $A = 50$, and illite and smectite compositions (thick bars) as in Figure 12. Fields encompassing data from Fore Arc and Trailing Edge tectonic settings are shown in stippled patterns (after Mc Lennan *et al.*, 1990).

sorting by grain size and density. For *sandstone* components, the higher stability of quartz, compared to feldspar and lithic grains, results in an increasing SiO_2/Al_2O_3 ratio and a decrease in concentration of trace elements that were associated chiefly with the labile aluminosilicate minerals (McLennan *et al.*, 2003). Simultaneously, the labile nature of plagioclase relative to K-feldspar leads to a rise in the K_2O/Na_2O ratio. As for provenance, the overall shift in major element composition is towards the A–K tangent (Figure 13). More importantly, transport processes are the main factor that separates the sand- and the mud-size fractions. As for sandstones, *mudstones* also evolve towards the A/K tangent, but with increasing maturity they shift more towards the A apex of the ternary diagram (Figure 13).

14.7.3 Sedimentary Recycling

The processes of mechanical weathering and dissolution in recycling systems lead to the diminution of grain size of all mineralogical constituents, as well as to progressively monomineralic quartz-rich sediments (Cox and Lowe, 1995). Sedimentary recycling is particularly effective in redistributing the trace elements, a topic discussed in the next section.

14.8 TRACE ELEMENT AND ISOTOPIC COMPOSITION OF CLASTIC SEDIMENTS

As already pointed out, the suspended load of rivers falls on the tangent connecting the UCC and its clay rich weathering products (Figure 12). The chemical composition of clastic sediments reflects therefore that of the UCC, albeit depleted for those elements that are leached out, and transported as dissolved load, to be eventually concentrated in seawater and precipitated as (bio)chemical sediments. The variable residence times (τ) of these elements in seawater are a reflection of their relative mobility, with the logarithm of τ directly proportional to the logarithm of the ratio of a given element in seawater to its upper crustal abundance (Figure 14). The elements in the upper right corner (sodium, calcium, magnesium, strontium) are rapidly mobilized during sedimentary processes, while those in the lower left corner, such as titanium, zirconium, hafnium, niobium, tantalum, thorium, zirconium, nickel, cobalt, rare-earth elements (REEs) are mostly transferred from the UCC into the clastic sedimentary mass (McLennan *et al.*, 2003). It is the latter assemblage of elements, and even more so their ratios, that are useful for provenance studies. Because of their coherent behavior in geochemical processes, the REE and the elements such as thorium, zirconium, scandium, titanium... (Taylor and McLennan, 1985) are particularly suitable for tracing the ultimate provenence of clastic

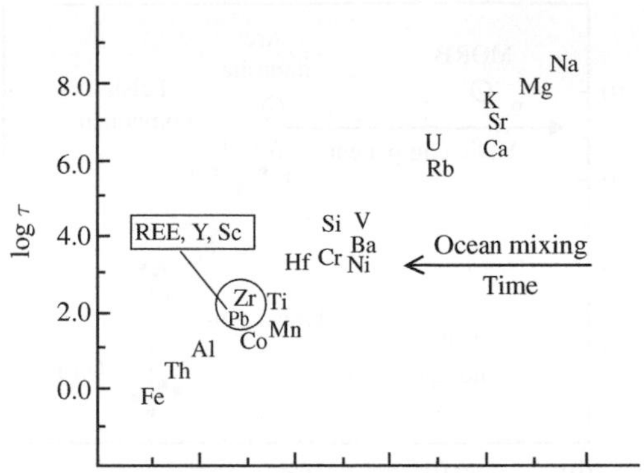

Figure 14 Plot of log τ (residence time in years) versus log K^{SW} (concentration in seawater/concentration in upper continental crust) for selected elements (after McLennan *et al.*, 2003).

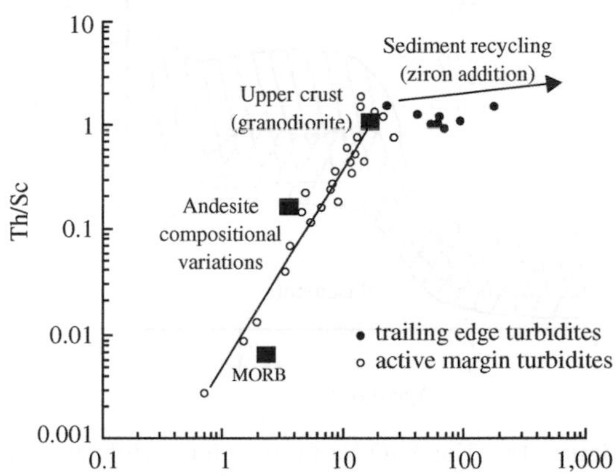

Figure 15 Th/Sc versus Zr/Sc for modern turbidites from active and passive margin setting (after McLennan *et al.*, 2003).

sediments (McLennan *et al.*, 1990). An example is given in Figure 15, where the modern active margin turbidites reflect directly the composition of the source, with Th/Sc and Zr/Sc ratios increasing in tandem with increasing igneous differentiation of the source rocks. The coherent trends break down in the trailing margin turbidites because these contain mostly recycled components of the older sediments. Repeated sedimentary recycling tends to enrich the sand-size fraction in heavy minerals. Heavy minerals, such as zircon (uranium, hafnium), monazite (thorium), chromite (chromium), titanium-minerals (ilmenite, titanite, rutile), or cassiterite (tin) are the dominant carriers of trace elements in sandstones. In trailing edge turbidites, the recycling tends to concentrate the much more abundant zircon (carrier phase of zirconium) than monazite (thorium), thus increasing the Zr/Sc but not so much the Th/Sc ratio.

Since a number of isotope systematics (U, Th/Pb, Lu/Hf, Sm/Nd) in clastic sediments is essentially controlled by the heavy mineral fraction, such considerations are of considerable importance for any geological interpretations.

Again, modern turbidites provide a classic example (Taylor and McLennan, 1985). Their Th/Sc ratio in active arc settings straddles the mafic to felsic join, with the bulk of samples reflecting the dominant andesitic component (Figure 16). Their ε_{Nd} of $\sim$+5 is that of modern oceanic crust. The turbidites in progressively more evolved tectonic settings contain increasing proportions of recycled sedimentary components, become more quartzose, have more negative ε_{Nd} and higher Th/Sc ratios. The observation that the more "evolved" tectonic settings incorporate recycled components from progressively older sources is confirmed by the neodymium model ages of these clastic sediments that increase from $\sim$250 Myr from fore arc settings to some 1.8 Gyr for the trailing age settings (Figure 17).

14.9 SECULAR EVOLUTION OF CLASTIC SEDIMENTS

14.9.1 Tectonic Settings and Lithology

Based on the tectonic concept of differential preservational probabilities, the progressively older segments of the continental crust should retain only the remnants of the most stable tectonic settings, that is they should be increasingly composed of basement and its platformal to intracratonic sedimentary cover. This is the case throughout the Phanerozoic and Proterozoic, but not for the oldest segment, the Archean. Compared to the Proterozoic, the Archean contains a disproportionate abundance of the perishable greenstones (e.g., Windley, 1984; Condie, 1989, 2000) with mid-ocean ridge basalt (MORB) and oceanic plateau basalt (OPB) affinities and immature clastic lithologies, particularly graywackes (Figure 8). Regardless of their precise present-day analogue, these greenstones are an expression of the ephemeral oceanic tectonic domain in the sense of Veizer and Jansen (1985) (Figure 5). The temporal distribution of greenstones is, therefore, entirely opposite to that expected from continuous recycling, regardless of its actual rate. The fact that so many of them survived to this day, despite this recycling, argues for their excessive original abundance and entrainment into the growing and stabilizing continents. How is this tectonic and lithological evolution, from an oceanic to continental domain

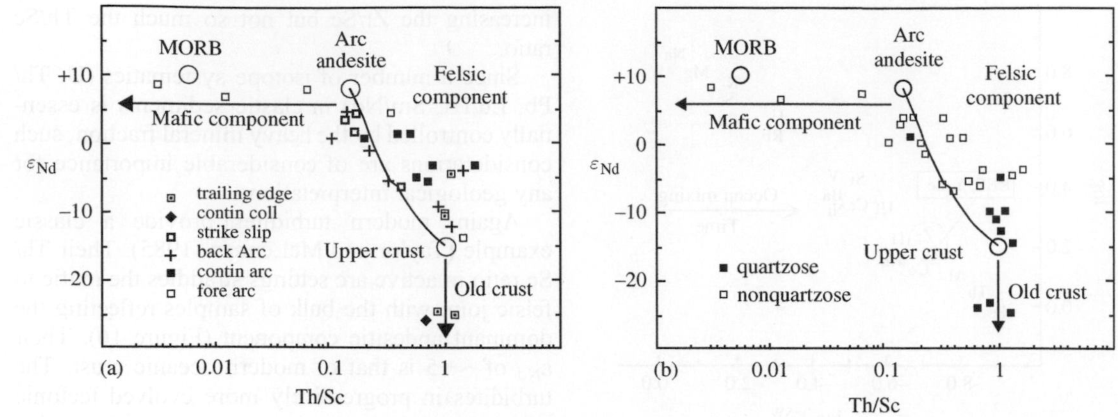

Figure 16 Plot of ε_{Nd} versus Th/Sc ratio for deep-sea turbidites according to tectonic setting of deposition (a) and according to quartz content (b). Also shown are compositions of various geochemical reservoirs (from Taylor and McLennan, 1985) and mixing relationships between average island arc andesite and upper continental crust (after McLennan *et al.*, 1990).

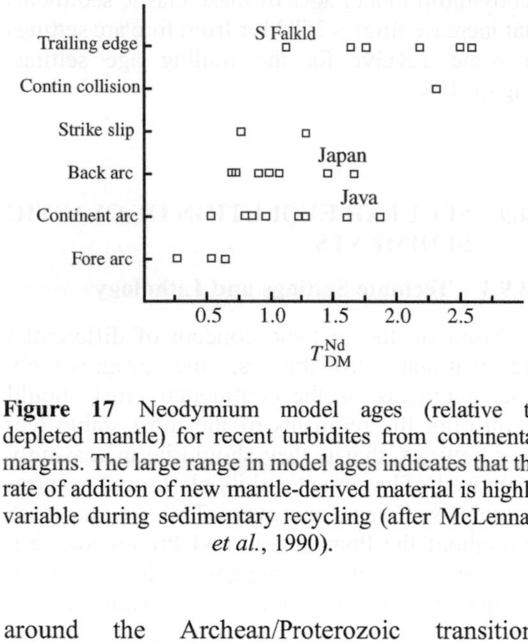

Figure 17 Neodymium model ages (relative to depleted mantle) for recent turbidites from continental margins. The large range in model ages indicates that the rate of addition of new mantle-derived material is highly variable during sedimentary recycling (after McLennan *et al.*, 1990).

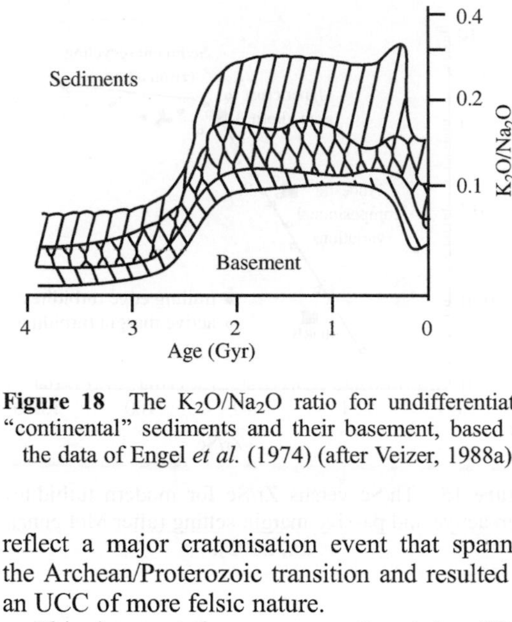

Figure 18 The K_2O/Na_2O ratio for undifferentiated "continental" sediments and their basement, based on the data of Engel *et al.* (1974) (after Veizer, 1988a).

around the Archean/Proterozoic transition, reflected in the chemistry of clastic sediments?

14.9.2 Chemistry

The K_2O/Na_2O ratio of sediments and continental basement increased considerably at about the Archean/Proterozoic transition (Figure 18), as already pointed out by Engel *et al.* (1974). Subsequently, Taylor (1979) and Taylor and McLennan (1985) emphasized that the REE showed similar trends in their overall concentrations, in LREE/HREE and Yb/La ratios, and in the decline of the size of the europium anomaly. The general interpretation of these data was based on the proposition that the compositions of the continental crust, and of "continental" sediments, reflect a major cratonisation event that spanned the Archean/Proterozoic transition and resulted in an UCC of more felsic nature.

This interpretation was questioned by Gibbs *et al.* (1988) and Condie (1993), who argued that the global averages for clastic rocks reflect only the variable proportions of facies associated with the predominant tectonic settings at any given time and not the change in the composition of the continental crust. This is undoubtedly the case, but—as discussed in the previous section—the types of the Archean tectonic settings are exactly the opposite to that expected from preservation probabilities based on the recycling concept. This feature must therefore reflect the fact that immature tectonic settings and lithologies were the norm in the Archean and they still dominate the preserved record, despite the high rate of recycling. Furthermore, the secular trends are present regardless whether the tectonic facies assemblages are considered together or separately (Figure 19). Combined, or separately, the Th/Sc

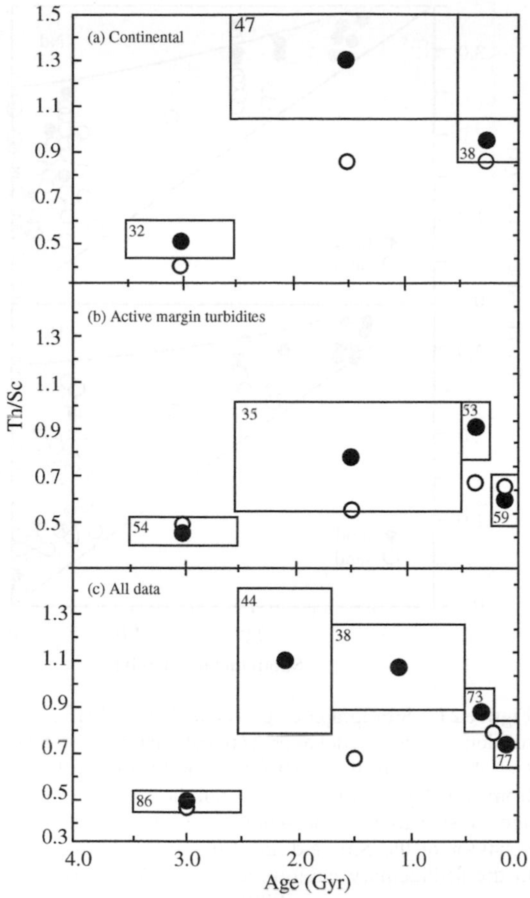

Figure 19 Secular variation in Th/Sc ratio for sedimentary rocks deposited in stable continental regions (shales) and tectonically active regions (shales/graywackes). Sample numbers are shown in or near boxes. Solid symbols and boxes are arithmetic means and 95% confidence intervals. Open circles are sediment averages reported by Condie (1993) (after McLennan *et al.*, in press).

secular trends show the same features as the K_2O/Na_2O parameter, with a rise in the ratio at the Archean/Proterozoic boundary and a reversal of the trend for the youngest portion of the record. Trends similar to that of Th/Sc one were observed also for Th, U, Th/U ratio (McLennan *et al.*, 2003) and La/Sc ratio (Condie, 1993). This is to a degree true also for the Eu/Eu* anomaly, although the magnitude is smaller than previously believed.

The Archean clastic rocks are also often enriched in chromium, nickel, and cobalt (Danchin, 1967; Condie, 1993), but this feature may only be of regional significance, mostly for the Kaapval Craton and the Pilbara block, where it may reflect the ubiquity of komatiites in the source regions of the sediments. Nevertheless, considering the frequent discrepancies in the anticipated Ni/Cr ratios, in lower than expected MgO content and, at times, high abundance of incompatible elements (Condie, 1993; McLennan *et al.*, 1990), the ultramafic provenance is not an unequivocal explanation and some secondary processes, such as weathering, may have played a role in repartitioning of these elements.

The apparent decline of the K_2O/Na_2O and Th/Sc ratios in the youngest segment of the secular trend is due to the fact that the youngest segments contain mostly the transient immature tectonic settings with their immature clastic assemblages. These are prone to destruction with advancing tectonic maturation (Figure 9) and their preservation into the Paleozoic and Proterozoic is therefore limited.

14.9.3 Isotopes

The discussion of the chemistry of clastic sediments suggested an overall mafic to felsic evolution of global sediments, and presumably of UCC, their ultimate source, in the course of geologic history, with a major evolutionary step across the Archean/Proterozoic transition. The response of isotopes to this evolutionary scenario can best be gauged by consideration of the REE isotope systematics, such as the Sm/Nd and Lu/Hf.

The major fractionation of REE is accomplished during igneous differentiation of rocks from the mantle, resulting in lower parent to daughter ratios of the crustal products for both, the $^{147}Sm/^{143}Nd$ (McCulloch and Wasserburg, 1978) and the $^{176}Lu/^{177}Hf$ (White and Patchett, 1982) systematics. Although exceptions do exist (e.g., McLennan *et al.*, 2003; Patchett, 2003), the subsequent igneous, metamorphic, and sedimentary history of the rocks usually does not affect their inherited parent/daughter ratios. For crustal rocks, including clastic sediments, it is therefore possible to calculate the time when the original material "departed" from the mantle, the latter approximated by the chondritic uniform reservoir (CHUR) evolutionary trend (Figure 20). The intercept with the CHUR is the model age. A similar reasoning applies also to the Lu/Hf systematics, with one exception. In clastic sediments, the usual carrier phase for hafnium is the heavy mineral zircon that tends to be fractionated into the sand fraction by the processes of sedimentary recycling. Mature sands therefore may contain an "excess" of hafnium and low Lu/Hf ratios (Patchett, 2003).

When model ages are plotted against stratigraphic ages of clastic sediments it becomes clear that the former are at best similar to, but mostly higher than their stratigraphic ages (Figure 21). Furthermore, the discrepancy increases from the Archean to today, from an average "excess" of ~250 Myr to ~1.8 Gyr. This is true for both isotope systematics and for muds as well as

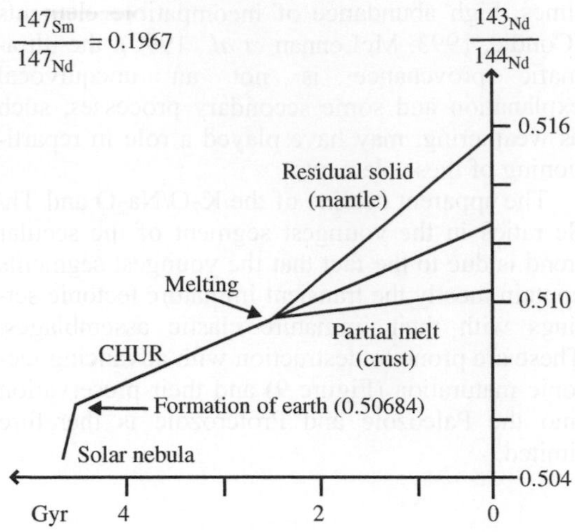

Figure 20 Theoretical evolution of Sm/Nd system in the course of planetary evolution.

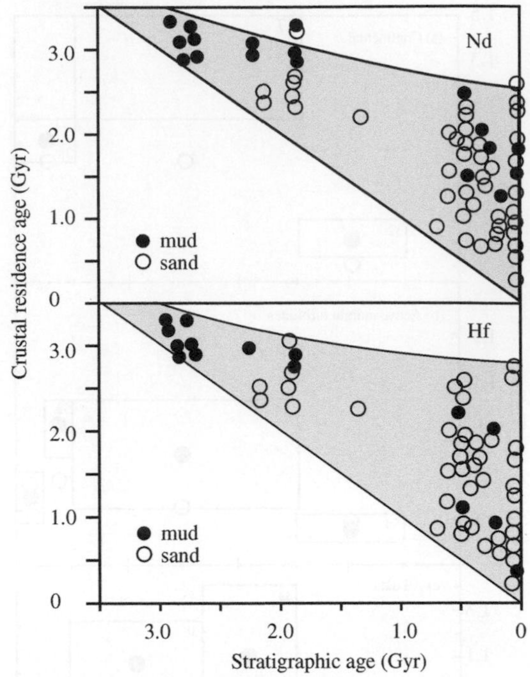

Figure 21 Stratigraphic age versus Nd- and Hf-crustal residence ages. Model ages were calculated using linear ε evolution from 0 to + 10 for Nd and 0 to + 16 for Hf, from 4.56 Gyr to present. The similarity of the model age systematics underscores the overall coherent behavior of the Sm–Nd and Lu–Hf isotopic systems in the sedimentary environment (after Vervoort *et al.*, 1999).

sands. These excess ages are usually interpreted as "crustal residence ages" (O'Nions *et al.*, 1983; Allègre and Rousseau, 1984), with an implication that they reflect the evolution of the ultimate crystalline crustal source. The average present-day modal excess of ~1.8 Gyr (Figure 21) is indeed a measure of the average ultimate provenance and thus of the mean age of the continental crust. In reality, however, sedimentary recycling is much faster than the metamorphism/erosion of the crystalline basement (Figure 5) and the excess model ages are therefore a consequence of the cannibalistic recycling of the ancient sedimentary mass (Veizer and Jansen, 1979, 1985). These relationships can be utilized for evaluation of the degree of cannibalistic recycling.

14.10 SEDIMENTARY RECYCLING

Sedimentary accumulations were ultimately derived from disintegration of the UCC and the global sedimentary mass should therefore have a chemical composition comparable to this part of the crust. This indeed is mostly the case (Figure 22). Compared with the UCC, the composition of present-day average global sediments (AS) for most elements does not deviate more than ±50% from that of the UCC. Exceptions are the enrichments in boron, calcium, vanadium, chromium, iron, cobalt, and nickel, and the depletion in sodium. In addition, the sediments (Goldschmidt, 1933; Rubey, 1951; Vinogradov, 1967; Ronov, 1968) are strongly enriched in excess volatiles and have a higher oxidation state.

The anomalous enrichment in calcium and depletion in sodium are a consequence of hydrothermal exchange between the ocean floor and seawater, processes discussed later in the text. Hydrothermal processes can also account for most of the other elements enriched in sediments. Vanadium, chromium, cobalt, nickel plus tin are even more enriched in the sediments of the ocean floor (global subducting sediment (GLOSS)) than the AS (Figure 22) due to stronger impact of the hydrothermal systems (Plank and Langmuir, 1998). Nevertheless, the general overall absence of large anomalies in the normalized average composition of sediments suggests that the exogenic ± endogenic inputs and sinks for most major elements—except possibly calcium and sodium—were balanced throughout most of geologic history, a proposition supported by the fact that the average composition of the subducting sediments (GLOSS; Plank and Langmuir, 1998) is also approaching that of the continental crust (Figure 22).

Because the continental sedimentary mass accounts for the bulk of the present-day global sediments, the growth of the global sedimentary mass should be a function of the growth of the continents. It is feasible, therefore, that at least

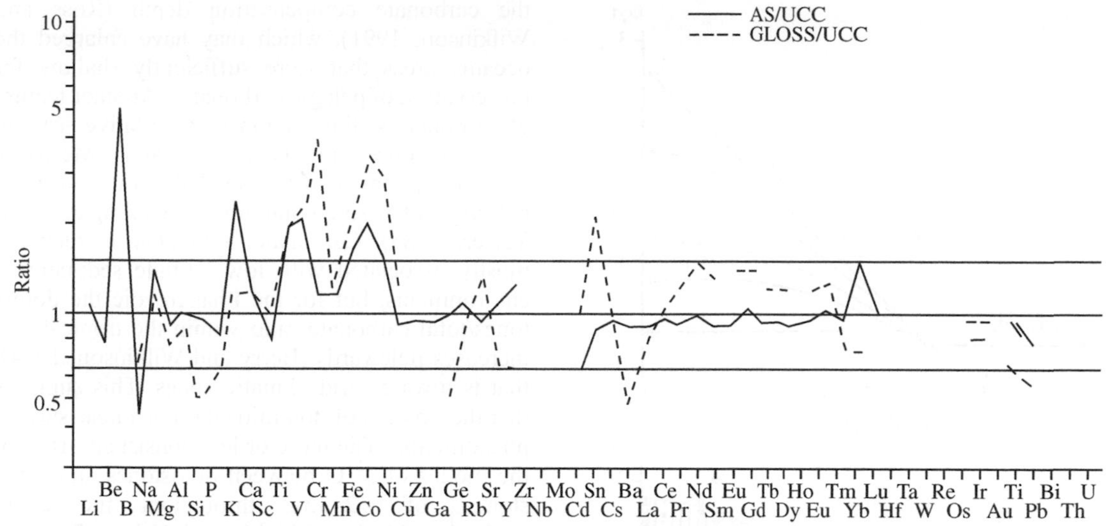

Figure 22 The elemental composition of average global sediment (AS) and global subducting sediment (GLOSS) normalized to upper continental crust (UCC). AS and UCC from McLennan and Murray (1999) and McLennan and Taylor (1999), respectively. Ti, Nb, Cs, and Ta are corrected as proposed in Plank and Langmuir (1998). GLOSS from Plank and Langmuir (1998).

some of this mass, particularly during the early stages of the earth's history and at the beginning of each tectonomagmatic cycle, evolved on oceanic or intermediate type crust. If so, it would have been derived from a source more mafic than the present-day UCC. In an entirely cannibalistic (closed) recycling system, this composition would have been perpetuated indefinitely regardless of the nature of the later continents.

In reality we must be dealing with a partially open system, because some sediments are being subducted while others are being formed at the expense of primary igneous and metamorphic rocks. Estimates based on Sm/Nd systematics (Figure 23) indicate that the sedimentary cycle is $\sim 90 \pm 5\%$ cannibalistic, attaining its near present-day steady state around the Archean/Proterozoic transition. The *first-order* features of secular trends, such as the K_2O/Na_2O, Th/Sc (Figures 18 and 19), REE, U, Th, and Th/U (Collerson and Kamber, 1999; McLennan *et al.*, 2003) can be explained, provided the Archean was the time when the sedimentary mass was mostly growing by addition of the first-cycle sediments from erosion of contemporaneous young ($\leq$250 Myr old) igneous precursors. Subsequent to the large-scale cratonization events, and subsequent to establishment of a substantial global sedimentary mass at $\sim 2.5 \pm 0.5$ Gyr, cannibalistic sediment—sediment recycling became the dominant feature of sedimentary evolution. The general absence of neodymium model ages much in excess of their stratigraphic ages in most Archean sediments (Figure 23) is consistent with the absence of the inherited old detrital components. This observation strongly argues against the presence of large continental landmasses prior to $\sim$3 Gyr.

14.11 OCEAN/ATMOSPHERE SYSTEM

The previous discussion dealt with the solid earth component of the exogenic cycle, a system that is recycled on 10^6–10^9 yr timescales. In contrast, the ocean, atmosphere, and life are recycled at much faster rates and the continuity of the past record is lost rapidly. For quantitative evaluations we have to rely therefore on proxy signals embedded in marine (bio)chemical sediments.

Other contributions in this Treatise (Volumes 6–8) deal with the lithological and chemical aspects of evolution of specific types of (bio)chemical sediments, such as cherts, phosphorites, hydrocarbons, and evaporites, and we will therefore concentrate on the most ubiquitous category, the carbonate rocks (see also Section 14.12). This carrier phase also contains the largest number of chemical and isotopic tracers.

As already pointed out in Figure 7, the relative proportion of carbonate rocks within the continental realm generally decreases in the course of the Phanerozoic, and the Mesozoic and Cenozoic "deficiency" of carbonates was attributed to a tectonic cause, the ubiquity of transient immature tectonic settings. Another reason is the general northward drift of continents, which resulted in a progressive decline in the shelf areas that fell within the confines of the tropical climatic belt

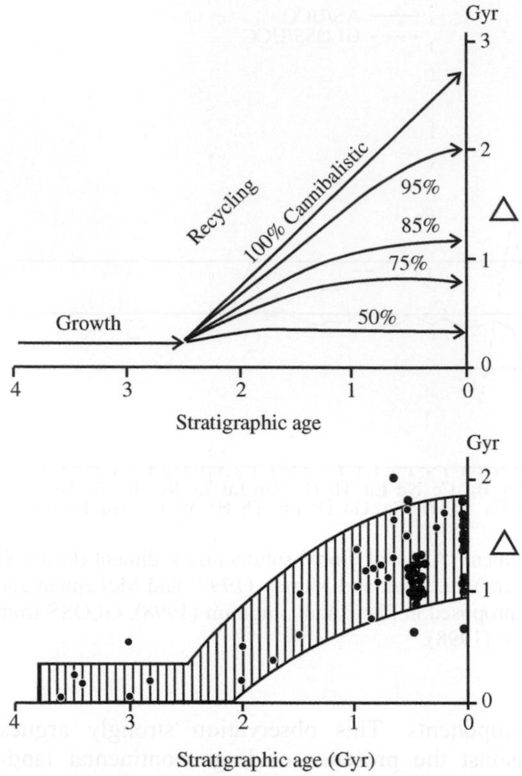

Figure 23 Models of Sm/Nd excess ages for sedimentary rocks. The Sm/Nd systematics dates the time of fractionation from the mantle. Regardless of whether most or only some of the sediments were generated during early terrestrial history, they would inherit Sm/Nd systematics from their igneous precursors. In a cannibalistic sedimentary recycling, these ancient systematics will be perpetuated and, as a consequence, Sm/Nd of all second-stage sediments should exceed their stratigraphic ages (Figure 21) and the Δ (Sm/Nd model age minus stratigraphic age) should increase toward the present, with a 45° slope being an upper limit for a completely closed system. In order to generate the observed smaller Δ, it is necessary to add sediments formed from a young source. The upper figure represents model calculations based on the assumption that prior to 2.5 Ga the sedimentary mass was growing through addition of first-cycle sediments. The post-Archean evolution assumes cannibalistic recycling of the steady state mass, and the slopes represent the degree of cannibalism for this recycling. The bottom part is a collation of experimental data (after Veizer and Jansen, 1985).

(Walker *et al.*, 2002; Bluth and Kump, 1991; Kiessling, 2002). Finally, in the course of the Mesozoic and Cenozoic, the locus of carbonate sedimentation migrated from the shelves to the pelagic realm, mirroring the role that calcareous shells of foraminiferans, pteropods, and coccolithophorids commenced to play in the carbonate budget (Kuenen, 1950). This environmental shift may have been accompanied by the deepening of the carbonate compensation depth (Ross and Wilkinson, 1991), which may have enlarged the oceanic areas that were sufficiently shallow for preservation of pelagic carbonates. Another feature of carbonate sedimentation is the relative scarcity of dolostones in Cenozoic and Mesozoic sequences, compared to their Paleozoic and particularly Proterozoic counterparts (Chilingar, 1956; Veizer, 1985). Carbonate sediments as such are mostly associated with low latitude sedimentary environments, but for the Phanerozoic the dolostones/total carbonate ratio within the tropical belt increases polewards (Berry and Wilkinson, 1994), that is towards arid climatic zones. This suggests that the process of dolomitization is a near-surface phenomenon. The more or less consistent offset of the modes of $\delta^{18}O$ between dolostones and limestones, of ~2–3‰, throughout the entire geological history (Shields and Veizer, 2002) is also consistent with such an interpretation, because the observed $\Delta^{18}O_{dol\text{-}cal}$ is a near-equilibrium value (Land, 1980). These carbonates are therefore either primary marine precipitates, or more likely they are early diagenetic products of stabilization and dolomitization of carbonate precursors. Pore waters at this stage were still in contact with the overlying seawater and/or contained an appreciable seawater component. The high frequency of dolostones in ancient sequences may again be principally a reflection of the ubiquity of shelf, epicontinental, and platformal tectonic settings preserved from the Paleozoic and Proterozoic times. Changes in seawater chemistry, such as Mg/Ca ratio, saturation state, or SO_4 content, may have been a complementary factor.

14.11.1 The Chemical Composition of Ancient Ocean

Earlier studies (e.g., Holland, 1978, 1984) assumed that the chemical composition of seawater during the Phanerozoic was comparable to the present-day one. Subsequently, experimental data on fluid inclusions in halite (Lowenstein *et al.*, 2001; Horita *et al.*, 1991, 2002) and on carbonate cements (Cicero and Lohmann, 2001) suggested that at least the magnesium, calcium, and strontium, and their ratios, in Phanerozoic seawater may have been variable (see Chapter 12). Steuber and Veizer (2002) assembled a continuous record of Sr/Ca variations for the Phanerozoic oceans (Figure 24) that covaries positively with the "accretion rate of the oceanic crust" (Gaffin, 1987) and negatively with the less well-known Mg/Ca ratio. Such a covariance would suggest that we are dealing with coupled phenomena and they proposed that all these variables are ultimately driven by tectonics, specifically seafloor spreading rates that, in turn, control the associated hydrothermal and

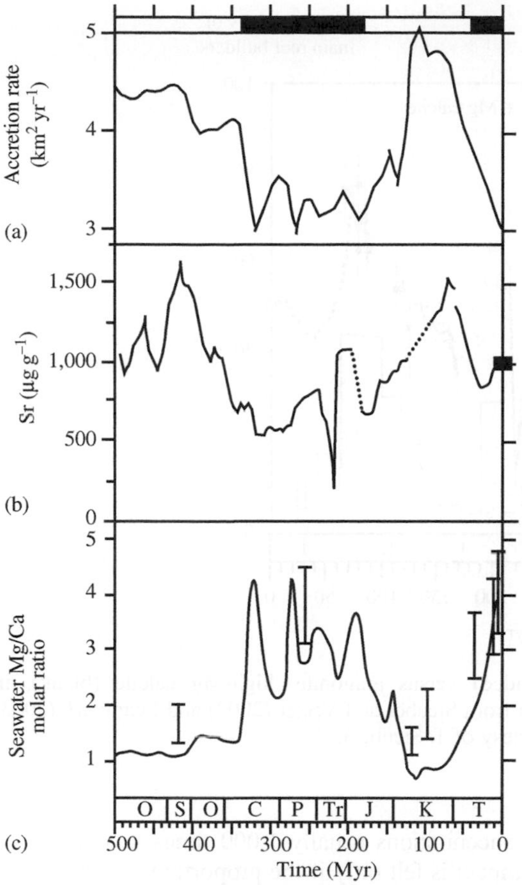

Figure 24 (a) "Accretion rate of oceanic crust" (Gaffin, 1987); (b) trends in Sr concentrations in biological LMC (Steuber and Veizer, 2002; Lear *et al.*, 2003); and (c) Mg/Ca ratio of seawater. Vertical bars are based on experimental data from fluid inclusions in halite (Lowenstein *et al.*, 2001), while the trend line results from model calculations (Stanley and Hardie, 1998). The bars at the top represent "calcitic" (blank) and "aragonitic" (black) seas (Sandberg, 1983) (after Steuber and Veizer, 2002).

low-temperature alteration processes. Since the hydrothermal alteration of young oceanic crust effectively exchanges magnesium for calcium (see Chapter 11), the accretion of the oceanic crust would modulate the Mg/Ca ratio of seawater. At high accretion rates, the low Mg/Ca ratio favors precipitation of calcite and, as a result, higher retention of strontium in seawater. At slow accretion rates, the high Mg/Ca ratio favors aragonite precipitation and a high rate of strontium removal from seawater.

The above scenario is consistent with the observation that calcite was the dominant mineralogy of carbonate skeletal components in the early to mid-Paleozoic and the mid-Jurassic to mid-Tertiary (Figure 25), the times of high Sr/Ca and low Mg/Ca ratio (Figure 24). Aragonite mineralogy, however, dominated the mid-Carboniferous to Jurassic and the Tertiary to Quaternary intervals (Sandberg, 1983; Kiessling, 2002) with opposite chemical attributes. The changing Mg/Ca ratio of seawater can also be at least in part responsible for the general scarcity of $MgSO_4$-bearing potash minerals (Hardie, 1996) in the Paleozoic and Mesozoic marine evaporites.

While all the above trends and their correlations are likely real, the proposed causative mechanism is being questioned lately due to the proposition that seafloor spreading rates have been about constant since at least 180 Myr (Rowley, 2002). If so, the sea-level stands cannot be inverted into "accretion rates of oceanic crust," as done by Gaffin (1987). A causative mechanism for all these covariant phenomena remains therefore enigmatic.

The reconnaissance studies of fluid inclusions (Horita *et al.*, 1991, 2002) suggest also that early Paleozoic seawater was ~2.5×depleted in SO_4, compared to its present-day counterpart. From model considerations, based on the mineralogy and volume of evaporites, claims have been made also for changes in the potassium concentration of Phanerozoic oceans (Hardie, 1996), and for an increase in the total salinity, from the modern 35 ppt to ~50 ppt in the Cambrian (Hay *et al.*, 2001). The experimental confirmation for all these theoretical assertions is presently not available (see Chapter 12).

On timescales of billion of years, ancient Precambrian carbonates appear to have been enriched in Fe^{2+} and Mn^{2+}, if compared to their Phanerozoic counterparts (Veizer, 1985). In part, this may be a reflection of diagenetic alteration processes that tend to raise the iron and manganese contents of successor phases (Brand and Veizer, 1980; Veizer, 1983). However, the Archean manganese concentrations, in the 10^3–10^4 ppm range, are likely not explained by diagenetic processes alone. Accepting that the redox state of the Archean and early Proterozoic oceans may have been lower than that of their Phanerozoic counterparts (Cloud, 1976), the high Fe^{2+} and Mn^{2+} content of contemporaneous carbonates may reflect higher concentrations of these elements in the ancient oceans.

14.11.2 Isotopic Evolution of Ancient Oceans

In contrast to chemistry, where the secular trends are still mostly obscured by the natural scatter in the database, the isotope evolution of seawater is better resolved.

14.11.2.1 Strontium isotopes

In modern oceans the concentration of strontium is ~8 ppm and its residence time is ~4–8 Myr

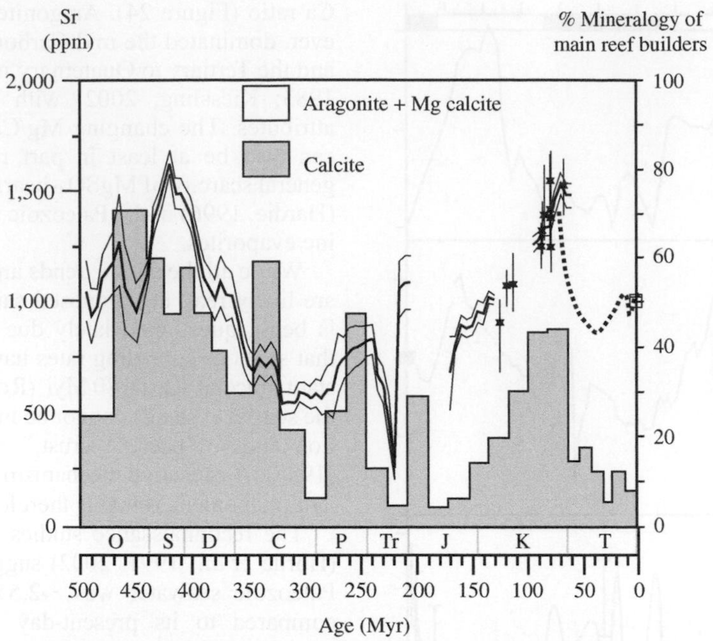

Figure 25 Predominant skeletal mineralogy, calcite (shaded) versus aragonite + high-Mg calcite (blank), in Phanerozoic reefs. Mineralogy from Kiessling (2002), Sr data from Steuber and Veizer (2002) and Lear *et al.* (2003) (see also Figure 24) (courtesy of T. Steuber).

(Holland, 1984). The present isotopic ratio $^{87}Sr/^{86}Sr$ is 0.7092 (McArthur, 1994), controlled essentially by two fluxes, the "mantle" and the "river" flux. The former represents strontium exchanged between seawater and oceanic crust ($^{87}Sr/^{86}Sr$ ~0.703) in hydrothermal systems on the ocean floor. The latter, reflecting the more fractionated composition of the continental crust, feeds into the oceans more radiogenic strontium, with an average isotope ratio for rivers of ~0.711 (Wadleigh *et al.*, 1985; Goldstein and Jacobsen, 1987; Palmer and Edmond, 1989). Note, nevertheless, that the latter may vary from 0.703 to 0.730 or more, depending on whether the river is draining a young volcanic terrane or an old granitic shield. The third input is the flux of strontium from diagenesis of carbonates, which results in expulsion of some strontium from the solid phase during precursor to product (usually aragonite to low-magnesium calcite) recrystallization, but this flux is not large enough to influence the isotopic composition of seawater. A simple balance calculation based on isotopes, therefore, shows that strontium in seawater originates ~3/4 from the "river" flux and ~1/4 from the "mantle" flux, generating the modern value of 0.7092. This value is uniform with depth and into marginal seas. Even water bodies such as Hudson Bay, with a salinity ~1/2 of the open ocean due to large riverine influx, have this same isotope ratio. This is because the rivers are very dilute relative to seawater, with strontium concentrations usually 1,000 times less, and their impact is felt only if the proportion of seawater in the mixtures is less than 10%.

The above considerations show that the strontium isotopic composition of seawater is controlled essentially by tectonic evolution, that is, by relative contributions from weathering processes on continents and from the intensity of submarine hydrothermal systems. Over geological time, however, the isotopic compositions of these two fluxes have evolved, because ^{87}Sr is a decay product of ^{87}Rb:

$$\left(\frac{^{87}Sr}{^{86}Sr}\right)_p = \left(\frac{^{87}Sr}{^{86}Sr}\right)_o + \left(\frac{^{87}Rb}{^{86}Sr}\right)(e^{\lambda t} - 1) \qquad (7)$$

where p = present, o = initial ratio at the formation of the Earth 4.5 Gyr ago (0.699), λ = decay constant (1.42×10^{-11} yr^{-1}), and t = time since the beginning (e.g., formation of the Earth 4.5 Gyr ago).

From Equation (7) it is evident that the term $(^{87}Sr/^{86}Sr)_p$ for coeval rocks originating from the same source $(^{87}Sr/^{86}Sr)_o$ depends only on their Rb/Sr ratios. Since this ratio is ~6 times larger for the more fractionated continental rocks than for the basalts (~0.15 to 0.027; Faure, 1986), the $^{87}Sr/^{86}Sr$ of the continental crust at any given time considerably exceeds that of the mantle and oceanic crust, increasing to the present-day values of ~0.730 for the average continental crust, as opposed to ~0.703 for the oceanic crust. The rivers draining the continents are less radiogenic

(~0.711) than the crust itself due to the fact that most riverine strontium originates from the weathering of carbonate rocks rather than from their silicate counterparts. The former, as marine sediments, inherited their strontium from seawater, which—as discussed above—contains also the less radiogenic strontium from hydrothermal sources.

The lower envelope of the strontium isotopic trend during the *Precambrian* (Figure 26) straddles the mantle values until about the Archean/Proterozoic transition, afterwards deviating towards more radiogenic values, reflecting the input from the continental crust. The large spread of values above the lower envelope is a consequence of several factors. First, it reflects the impact of secondary alteration that mostly results in resetting of the signal towards more radiogenic values (Veizer and Compston, 1974). Second, the scatter also includes higher order oscillations in the strontium isotope ratio of seawater which cannot be as yet resolved for the Precambrian due to poor stratigraphic resolution and inadequate geochronological control.

From the above discussion, it is clear that the primary control on strontium isotopic composition of the first order (billion years trend) for seawater will be exercised by the growth pattern of the continental crust. Two competing hypotheses dominate this debate:

(i) The generation of the entire continental crust was an early event and the present-day scarcity of older remnants is a consequence of their destruction (recycling) by subsequent tectonic processes (Armstrong, 1981; Sylvester *et al.*, 1997).

(ii) The continents were generated episodically over geologic history, with major phases of continent formation in the late Archean and early Proterozoic, and attainment of a near modern extent by ~1.8 Gyr (Veizer and Jansen, 1979; Taylor and McLennan, 1985; McCulloch and Bennett, 1994).

Model calculations by Goddéris and Veizer (2000) of seawater $^{87}Sr/^{86}Sr$ evolution for the two alternatives, and for the coeval mantle, show that the measured experimental data fit much better with the second pattern of continental growth. Note that this scenario is also fully compatible with the evolution of sediments and their chemistry discussed in the preceding sections of this chapter. The Archean oceans were "mantle buffered" (Veizer *et al.*, 1982) by vigorous circulation of seawater via submarine hydrothermal systems. With the exponential decline of internal heat dissipation, the vigor of the hydrothermal system also declined and at the same time the flux of radiogenic strontium from growing continents brought in by rivers started to assert itself. This tectonically controlled transition from "mantle" to "river buffered" oceans across the Archean/Proterozoic transition is a first order feature of terrestrial evolution, with consequences for other isotopic systematics, for redox state of the ocean/atmosphere system and for other related phenomena (see Chapter 12).

The resolution of the database is considerably better for the *Phanerozoic* than for the Precambrian due to higher quality of samples and to much better biostratigraphic resolution, with duration of biozones from ~1 Myr in the Cenozoic to ~5 Myr in the early Paleozoic. The first data documenting the strontium isotopic variations in Phanerozoic seawater were published by Peterman *et al.* (1970), with subsequent advances by Veizer and Compston (1974) and Burke *et al.* (1982). The latest version, by Veizer *et al.* (1999),

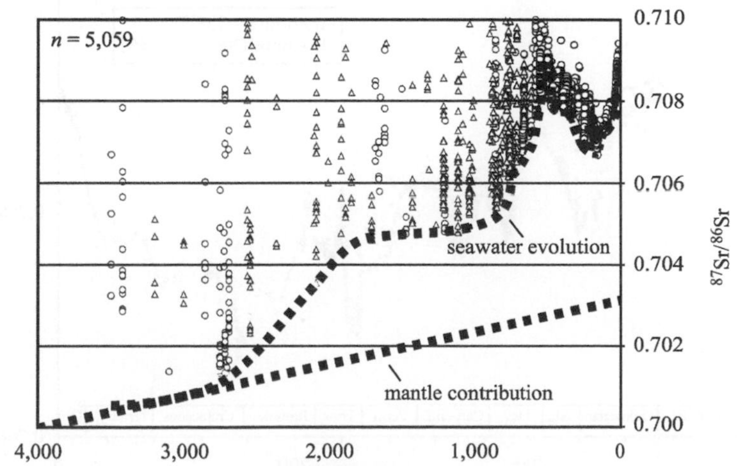

Figure 26 Strontium isotopic composition of sedimentary carbonate rocks during geologic history. Reproduced from Shields and Veizer (2002). Circles and triangles represent samples with good and poor age assignment, respectively.

is reproduced in Figure 27. Overall, this second order Phanerozoic trend shows a decline in $^{87}Sr/^{86}Sr$ values from the Cambrian to the Jurassic, followed by a steep rise to today's values, with superimposed third order oscillations at 10^7 yr frequency. Because of the better quality of samples, the experimental data indicate an existence of still higher order $^{87}Sr/^{86}Sr$ oscillations within biozones. However, since these samples often do not originate from the same profile, their relative ages within a biozone are difficult to discern. They have to be treated therefore as coeval and the secular trend thus becomes a band (Figure 28).

In general, it is again tectonics that is the cause of the observed Phanerozoic trend, with the "mantle" input of greater relative importance at times of the troughs and the "river" flux dominating in the Tertiary and early Paleozoic. Nevertheless, it is difficult to correlate the overall trend or the superimposed oscillations with specific tectonic events. The problem arises from the fact that model solutions do not produce unique answers. The "river" flux is likely the major reason for the observed $^{87}Sr/^{86}Sr$ oscillations, because the changes in seafloor spreading rates, apart from being disputed (Rowley, 2002), are relatively sluggish and the strontium isotope ratio of the "mantle" flux is relatively constant at ~0.703. The "river" flux, however, may vary widely in both strontium elemental flux and its isotope ratio. For example, the rapid Tertiary rise in $^{87}Sr/^{86}Sr$ (Figure 27) is commonly interpreted as reflecting the uplift of the Himalayas. Accepting this to be the case, it still remains an open question whether the rise is due to higher flux of "river" strontium (increased weathering rate), its more radiogenic nature (unroofing of older core complexes), or both (cf. the contributions in Ruddiman, 1997). For these reasons, it is difficult to utilize the strontium isotopic curve of seawater as a direct proxy for continental weathering rates in model considerations. Nevertheless, it is intriguing that the Phanerozoic seawater strontium isotope curve correlates surprisingly well with the estimated past sediment fluxes (Hay *et al.*, 2001) that were reconstructed by the "population dynamics" approach discussed in the introductory section of this chapter.

14.11.2.2 Osmium isotopes

The isotope ^{187}Os is generated by β decay of ^{187}Re. In many ways the systematics and the presently known secular evolution of $^{187}Os/^{186}Os$ in seawater is similar to that of $^{87}Sr/^{86}Sr$ (Peucker-Ehrenbring and Ravizza, 2000). The present-day $^{187}Os/^{186}Os$ of UCC is ~1.2 – 1.3, runoff ~1.4 (due likely to preferential weathering of radiogenic black shales), seawater, ~1.06 and that of meteorites and mantle, ~0.13. In contrast to strontium with a seawater residence time of 4 – 8 Myr, the residence time of osmium is on the order of 10^4 yr due to the effective scavenging of osmium by Fe/Mn crusts and organic-rich sediments. This enables tracing of short-term fluctuations in seawater composition, such as the Quaternary glacial/interglacial cycles, a feat difficult to replicate by the buffered strontium system.

Metalliferous sediments usually have low Re/Os ratios and reflect well the isotopic composition of seawater. Their disadvantage is the slow

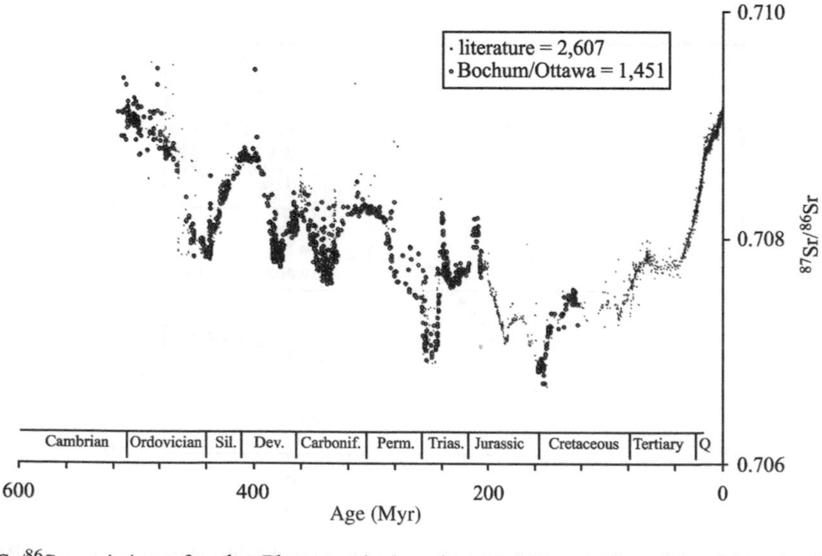

Figure 27 $^{87}Sr/^{86}Sr$ variations for the Phanerozoic based on 4,055 samples of brachiopods, belemnites, and conodonts. Normalized to NBS 987 of 0.710240 (after Veizer *et al.*, 1999).

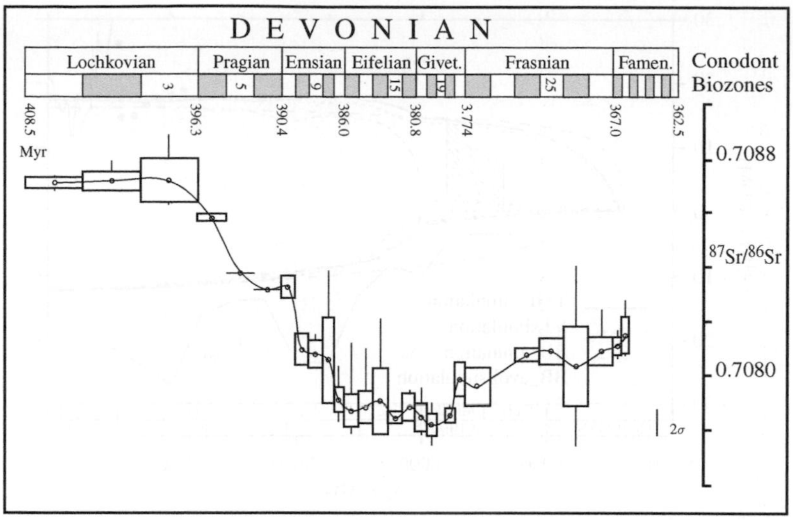

Figure 28 $^{87}Sr/^{86}Sr$ variations during the Devonian based on conodont biozones. Explanations: circle = mean; box = $\pm 1\sigma$; vertical line = minimum and maximum. The 2 σ in the lower right corner is an average 2 σ for the NBS 987 standard. Note that only brachiopods are included in this. Figures 1–37 are conodont biozones (after Veizer *et al.*, 1999).

accumulation rate that limits temporal resolution. Organic rich shales, alternatively, have high Re/Os ratios, requiring age correction by the isochron technique. Corals and carbonate sediments do not appear to preserve the hydrogenous (seawater) $^{187}Os/^{186}Os$ record.

The presently known data for seawater $^{187}Os/^{186}Os$ cover mostly the Cenozoic, with fragmentary results for the Cretaceous and the Jurassic. As for $^{87}Sr/^{86}Sr$, the $^{187}Os/^{186}Os$ declines from present-day value of ~1.06 to 0.15 at the K/T boundary. The sudden drop, from ~0.8 to 0.15 at the K/T boundary likely reflects the cosmic input from the meteoric impact. The $^{187}Os/^{186}Os$ ratio for the Mesozoic oscillates between 0.8 and 0.15, but the details are not yet resolved. For further discussion of Re/Os systematics.

14.11.2.3 Sulfur isotopes

In contrast to strontium and osmium isotopes, the isotopes of sulfur are strongly fractionated by biological processes, particularly during the dissimilatory bacterial reduction of sulfate to sulfide. The laboratory results for this step are anywhere from +4 to −46‰ (CDT), but even larger fractionations have been observed in natural systems (Harrison and Thode, 1958; Chambers and Trudinger, 1979; Habicht and Canfield, 1996).

The geologic record is characterized by a dearth of *Precambrian* evaporitic sequences, including their sulfate facies. Stratiform barites do exist, but they may be, at least in part, of hydrothermal origin. The scarcity of evaporites is partly due to their poor survival rates in the face of tectonic processes, but another reason may be low pO_2 levels in the contemporaneous ocean/atmosphere system, particularly during the Archean (Veizer, 1988a). In addition, most Archean sulfides, such as pyrites, contain $\delta^{34}S$ close to 0‰ CDT, the value typical of the mantle (Figure 29), rather than the expected highly negative ones characteristic of bacterial dissimilatory sulfate reduction. These observations were usually interpreted (e.g., Schidlowski *et al.*, 1983; Hayes *et al.*, 1992) as being due to biological evolution, where it is assumed that the invention of oxygen generating photosynthesis and of bacterial sulfate reduction were only later developments. In that case, most of the sulfur in the Archean host phases would have originated from mantle sources and carried its isotopic signature. Only with the onset of these two biological processes, in about the late Archean or early Proterozoic, was enough oxygen generated to stabilize sulfate in seawater and to initiate its bacterial reduction to H_2S, the latter eventually forming sulfide minerals, such as pyrite. This development resulted in the burial of large quantities of sulfides depleted in ^{34}S in the sediments, causing the residual sulfate in the ocean to shift towards heavier values. The result is the bifurcation of $\delta^{34}S$ sulfate/sulfide values at the time of "invention" of bacterial dissimilatory sulfate reduction (Figure 29).

The above scenario is appealing, but not mandatory. As shown by Goddéris and Veizer (2000), the same "logistic" scenario of continental growth that generated the strontium isotope trend can also generate the observed $\delta^{34}S$ pattern

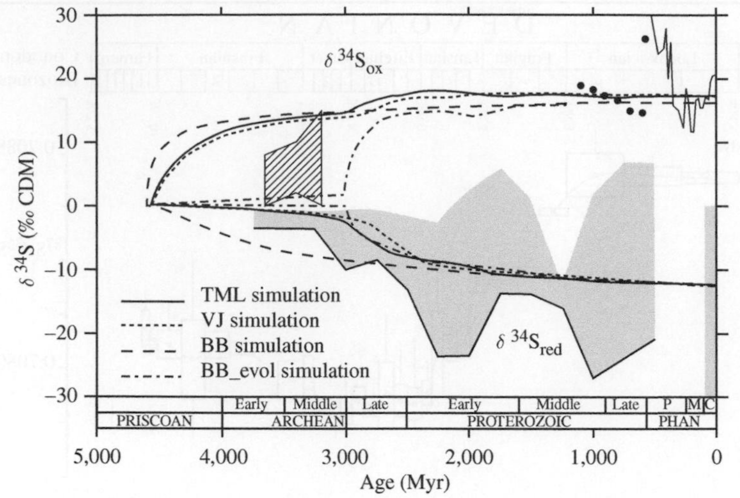

Figure 29 Model sulfur isotopic evolution in the course of geologic history. The lower trend (labeled $\delta^{34}S_{red}$) represents the $\delta^{34}S$ for sulfides. The upper curves (labeled $\delta^{34}S_{ox}$) are the $\delta^{34}S$ of marine sulfates. TML and VJ simulations assume a logistic type of continental growth as proposed by Taylor and McLennan (1985) and Veizer and Jansen (1979), respectively. BB simulation assumes an instantaneous generation of continental crust, BB-evol simulation assumes instantaneous continental generation, but with delayed "invention" of oxygen generating photosynthesis. The Phanerozoic trend as in Figure 30. Dots represent measurements of Precambrian sulfates (Claypool *et al.*, 1980) and the hatched field represents sulfates from Holser *et al.* (1988) (after Goddéris and Veizer, 2000).

(Figure 29) and the growth of sulfate in the oceans. This explanation has the advantage that a single scenario generates all these (and other) evolutionary patterns. In short, the early "mantle" buffered oceans (Veizer *et al.*, 1982) had a large consumption of oxygen in the submarine hydrothermal systems (Wolery and Sleep, 1988; see Chapter 12) because they operated at considerably higher rates than today. The capacity of this sink declined exponentially in the course of geologic history, reflecting the decay in the dissipation of the heat from the core and mantle. As a result, the buffering of the ocean was taken over by the continental "river" flux. In summary, the bacterial dissimilatory sulfate reduction must have been extant at the time of bifurcation of the $\delta^{34}S$ record, but the isotope data do not provide a definitive answer as to the timing of this invention. Tectonic evolution would override its impact even if established much earlier.

The Precambrian $\delta^{34}S$ record is spotty for sulfide-S and almost nonexistent for sulfate-S (Canfield and Teske, 1996; Strauss, 1993). A fragmentary record for the latter exists only for the latest Neoproterozoic (Strauss, 1993), suggesting a large shift from ~20 to 33 ± 2‰ at the transition into the Phanerozoic.

The $\delta^{34}S_{sulfate}$ variations in *Phanerozoic* oceans, based on evaporites (Holser and Kaplan, 1966; Claypool *et al.*, 1980), form an overall trough-like trend similar to strontium isotopes (Figure 30). Note, however, the large age uncertainties for, and the large gaps between, the studied evaporitic sequences. This is due to their episodic occurrence and uncertain chronology and is part of the reason for the large spread in the coeval $\delta^{34}S$ values despite the fact that the $\delta^{34}S_{sulfate}$ in seawater is spatially homogeneous (Longinelli, 1989). Another reason for this large spread in the $\delta^{34}S$ values is the evolution of sulfur isotopes in the course of the evaporative process, from sulfate to chloride to late salt facies. A recent development of the technique that enabled measurement of $\delta^{34}S$ in structurally bound sulfate in carbonates (Kampschulte and Strauss, 1998; Kampschulte, 2001) yielded a Phanerozoic secular curve with much greater temporal resolution (Figure 30).

The $\delta^{34}S_{sulfate}$ and $\delta^{13}C_{carbonate}$ secular curves correlate negatively (Veizer *et al.*, 1980), suggesting that it is the redox balance (peddling of oxygen between the carbon and sulfur cycles) that controls the $\delta^{34}S$ variations in Phanerozoic oceans. Note, nevertheless, that a physical geological scenario for this coupling is as yet not clarified. If redox balance is indeed a major control mechanism, it would suggest that the withdrawal of ^{32}S due to pyrite burial in sediments was twice as large as today in the early Paleozoic versus about one-half in the late Paleozoic (Kump, 1989).

In addition to the long-term 10^8 yr trends, shorter spikes, on 10^5–10^6 yr scales (e.g., Permian/Triassic transition; Holser, 1977), do exist, but their "catastrophic" geological causes are not as yet resolved.

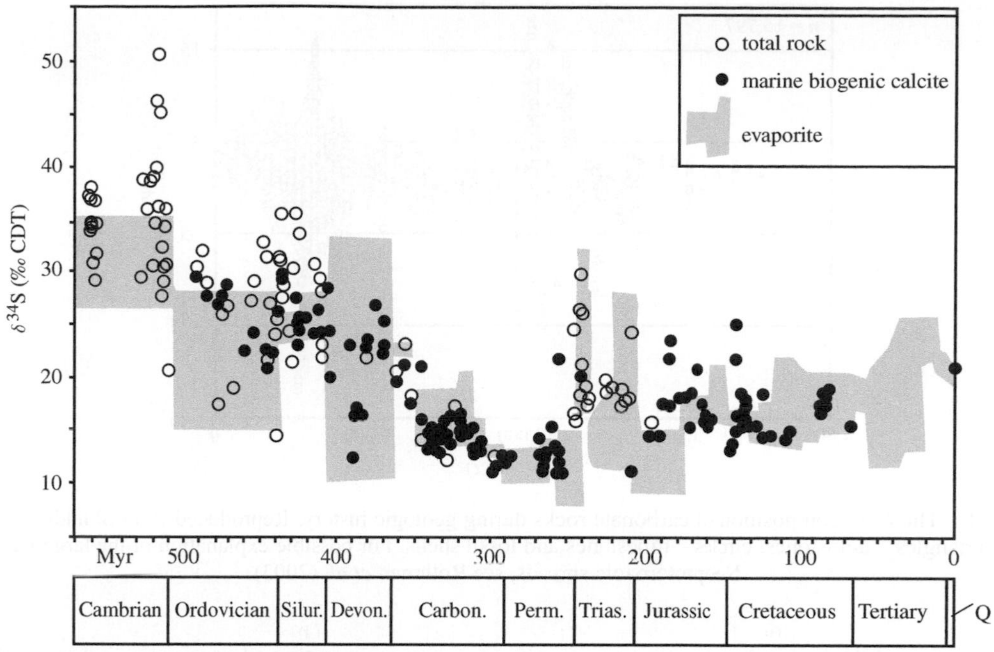

Figure 30 Sulfur isotopic composition of Phanerozoic seawater based on measurements of sulfur structurally bound in calcitic shells as well as evaporites. Note that the shell samples are mostly the same as those of the Sr, O, and C isotopes in Veizer *et al.* (1999) (after Kampschulte, 2001 by courtesy of the author).

14.11.2.4 Carbon isotopes

The two dominant exogenic reservoirs of carbon are carbonate rocks and organic matter in sediments. They are linked in the carbon cycle via atmospheric CO_2 and the carbon species dissolved in the hydrosphere. The $\delta^{13}C$ for the total dissolved carbon (TDC) in seawater is $\sim+1\pm0.5$‰ (PDB), with surficial waters generally heavier and deep waters lighter than this average (Kroopnick, 1980; Tan, 1988). Atmospheric CO_2 in equilibrium with TDC of marine surface water has a $\delta^{13}C$ of ~-7‰. CO_2 is preferentially utilized by photosynthetic plants for production of organic carbon causing further depletion in ^{13}C (Equation 8):

$$6CO_2 + 6H_2O \rightarrow C_6H_{12}O_6 + 6O_2 \qquad (8)$$

Most land plants utilize the so-called C_3, or Calvin pathway (O'Leary, 1988), that results in tissue with a $\delta^{13}C_{org}$ of ~-25‰ to -30‰. The situation for aquatic plants is somewhat different because they utilize dissolved and not gaseous CO_2. Tropical grasses, however, utilize the C_4 (Hatch-Slack or Kranz) pathway and have a $\delta^{13}C$ of some -10‰ to -15‰. A third group that combines these two pathways, the CAM plants (algae and lichens), has intermediate $\delta^{13}C$ values. In detail, the nature of the discussed variations is far more complex (Deines, 1980; Sackett, 1989) and depends on the type of organic compounds involved. For our purposes, however, it is only essential to realize that C_{org} is strongly depleted in ^{13}C. This organic matter, which is very labile, is easily oxidized into CO_2 that inherits the ^{13}C-depleted signal.

The $\delta^{13}C$ of mantle carbon is ~-5‰ PDB (Schidlowski *et al.*, 1983; Hayes *et al.*, 1992) and in the absence of life and its photosynthetic capabilities, this would also be the isotopic composition of seawater. Yet, as far back as 3.5 Gyr ago, and possibly as far as ~4 Gyr ago (Schidlowski *et al.*, 1983), the carbonate rocks (~seawater) had $\delta^{13}C$ at ~0‰ PDB (Figure 31). This suggests that a reservoir of reduced organic carbon that accounted for ~1/5 of the entire exogenic carbon existed already some 4 Gyr ago, "pushing up" the residual 4/5 of carbon, present in the oxidized form in the ocean/atmosphere system, from -5‰ to 0‰ PDB. This is an oxidized/reduced partitioning similar to that we have today. Stated in a simplified manner, life with its photosynthetic capabilities, and possibly of present-day magnitude, can be traced almost as far back as we have a rock record. This photosynthesis may or may not have been generating oxygen as its byproduct, but was essential in order to "lift" the seawater $\delta^{13}C$ to values similar to the present-day ones. In order to sustain seawater $\delta^{13}C$ at this level during the entire geologic history, it is necessary that the input and output in the carbon cycle have the same $\delta^{13}C$. Since the input from the mantle, via volcanism and hydrothermal systems, has a $\delta^{13}C$ of -5‰ and the subducted carbonates are 0‰, the subduction process must involve also a

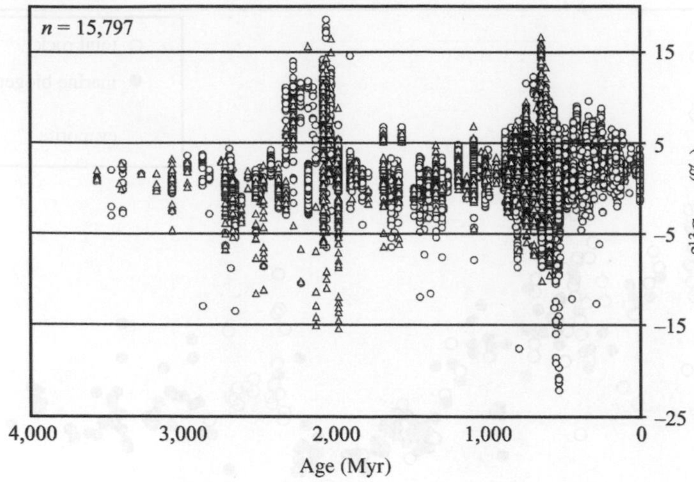

Figure 31 The $\delta^{13}C$ composition of carbonate rocks during geologic history. Reproduced from Shields and Veizer (2002). Triangles—dolostones, circles—limestones and fossil shells. For possible explanation of the large Paleo- and Neoproterozoic spreads, see Rothman *et al.* (2003).

complementary ^{13}C-depleted component, organic matter. This is possible to contemplate as long as oceanic waters were not fully oxygenated, such as may have been the case in the Archean. This is either because oxygen generating photosynthesis was "invented" as late as the late Archean or early Proterozoic (Cloud, 1976; Holland, 1984), or because tectonic evolution led to a progressive oxygenation of the ocean/atmosphere system due to a switchover from a "mantle"—to a "river"—buffered ocean system (Goddéris and Veizer, 2000). For the latter alternative, and in analogy to sulfur, it is possible to argue that oxygen generating photosynthesis (photosystem 2) may have been extant as far back as we have the geologic record, without necessarily inducing oxygenation of the early ocean/atmosphere system (but see Lasaga and Ohmoto, 2002). Whatever the cause, the oxygenation of the system in the early Proterozoic would have resulted in oxidation of organic matter that was settling down through the water column. Today only ~1% of organic productivity reaches the ocean floor and ~0.1% survives into sedimentary rocks. As a result, the addition of mantle carbon, coupled with the subduction loss of the ^{13}C-enriched limestone carbon, would slowly force the $\delta^{13}C$ of seawater back to mantle values. In order to sustain the near 0‰ PDB of seawater during the entire geologic history, it is necessary to lower the input of mantle carbon into the ocean/atmosphere system by progressively diminishing the impact of hydrothermal and volcanic activity over geologic time.

Superimposed on this invariant *Precambrian* $\delta^{13}C$ seawater trend are two intervals with very heavy (and very light) values, at ~2.2 Gyr and in the Neoproterozoic (Figure 31). The former has been interpreted as a result of the invention of oxygen generating photosynthesis that resulted in the sequestration of huge quantities of organic matter (Karhu and Holland, 1996) into coeval sediments and the Neoproterozoic interval was the time of the proposed "snowball earth" (Hoffman *et al.*, 1998). At this stage, the reasons for the high frequency of the anomalous $\delta^{13}C$ values during these two intervals are not well understood, but it is interesting that both were associated with large glaciations, as was the later discussed ^{13}C-enriched Permo/Carboniferous interval.

The sampling density and time resolution in the *Phanerozoic* enabled the delineation of a much better constrained secular curve (Figure 32), with a maximum in the late Permian, but even in this case we are dealing with a band of data, reflecting the fact that the $\delta^{13}C_{DIC}$ of seawater is not uniform in time and space, that organisms can incorporate metabolic carbon into their shells (vital effect), and that some samples may also contain a diagenetic overprint. Superimposed on the overall trend are higher oscillations, at 10^7yr and shorter timescales, but their meaning is not yet understood.

Frakes *et al.* (1992) proposed that the $\delta^{13}C_{carbonate}$ (seawater) becomes particularly heavy at times of glaciations, and that such times are also characterized by low CO_2 levels. The coincidences of the $\delta^{13}C$ peaks with the late Ordovician and Permocarboniferous glacial episodes appear to support this proposition, but the Mesozoic/Cenozoic record is divergent. Accepting the validity of the present $\delta^{13}C$ trend, it is possible to calculate the model p_{CO_2} levels of ancient atmospheres. Three Phanerozoic p_{CO_2} reconstructions exist (Berner and Kothavala, 2001; Berner and Streif, 2001; Rothman, 2002)

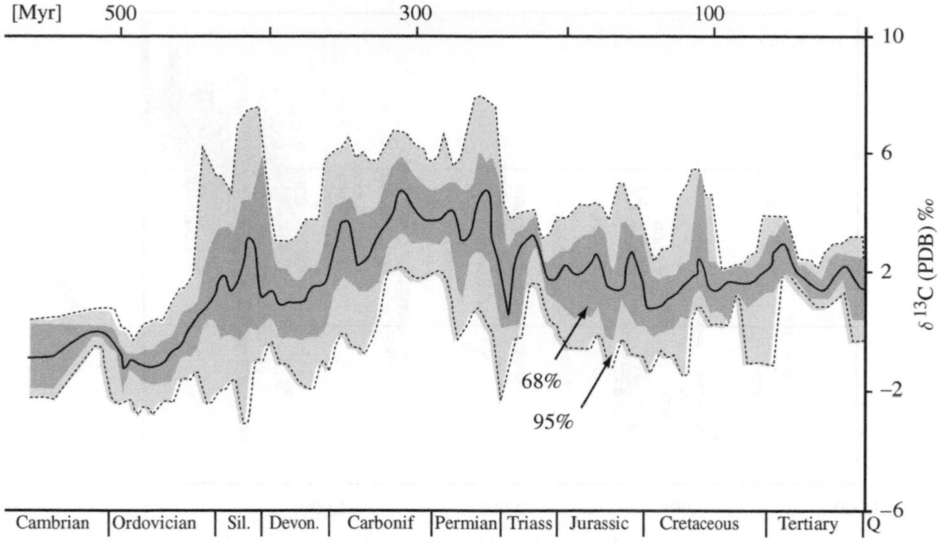

Figure 32 The Phanerozoic $\delta^{13}C$ trend for LMC shells. The running mean for ~4,500 samples is based on a 20 Ma window and 5 Ma forward step. The shaded areas around the running mean include the 68% ($\pm1\sigma$) and 95% ($\pm2\sigma$) of all data (after Veizer *et al.*, 1999).

that are internally inconsistent and not one of them shows any correlation with the paleoclimate deduced from sedimentological criteria (Veizer *et al.*, 2000; Boucot and Gray, 2001; Veizer, 2003; Figure 35). This led Veizer *et al.* (2000) to conclude that either the estimates of paleo-CO_2 were unreliable or there was no direct relationship between p_{CO_2} levels and climate for most of the Phanerozoic.

Higher order peaks, at a 10^6 yr resolution, have been observed in the geologic record, particularly in deep sea borehole sections and are discussed by Ravizza and Zachos (see Chapter 6.20).

14.11.2.5 Oxygen isotopes

The oxygen isotope record of some 10,000 limestones and low-magnesium calcitic fossils (Shields and Veizer, 2002) shows a clear trend of ^{18}O depletion with age of the rocks (Figure 33). This isotope record in ancient marine carbonates (but also cherts and phosphates) is one of the most controversial topics of isotope geochemistry. It centers on the issue of the primary versus post-depositional origin of the secular trend (e.g., Land, 1995 versus Veizer, 1995). Undoubtedly, diagenesis, and other post-depositional phenomena reset the $\delta^{18}O$, usually to more negative values, during stabilization of original metastable phases (e.g., aragonite, high-magnesium calcite), into the more stable phase, diagenetic low-magnesium calcite. Every carbonate rock is subjected to this stabilization stage and most, if not all, of its internal components are reset. The only exception can be the original low-magnesium calcitic shells of some organisms, such as brachiopods, belemnites, and foraminifera. Yet, the overall bulk rock depletions, relative to these stable phases, are ~2–3‰ (Veizer *et al.*, 1999) and not some 7‰ or more as is the case for the Precambrian limestones (Figure 33). The rocks, once diagenetically stabilized become relatively "inert" to further resetting. The retention of $\Delta^{18}O_{dolomite-calcite}$ of ~3‰ during the entire geologic history (Shields and Veizer, 2002) is also consistent with such an interpretation.

The observed *Precambrian* $\delta^{18}O$ secular trend is therefore real, albeit shifted by 2–3‰ to lighter values, and likely reflects the changing $\delta^{18}O$ of seawater. The exchange of oxygen at $T > 350\,°C$ between percolating seawater and oceanic crust results in ^{18}O enrichment of the water and ultimately oceans. The opposite happens at $T < 350\,°C$ (Muehlenbachs, 1998; Gregory, 1991; Wallmann, 2001). One interpretation could be that over geologic history this "isotopically neutral" crossover point migrated to shallower depths, thus reducing the profile of the low-T alteration relative to the deeper one, perhaps due to blanketing of the ocean floor by pelagic biogenic sediments during the Phanerozoic that sealed the off-ridge oceanic crust from seawater percolation.

The *Phanerozoic* trend (Figure 34) is based on ~4,500 samples of low-magnesium calcitic fossils from about 100 localities worldwide. The reasons for believing that it is essentially a primary trend were discussed in detail by Veizer *et al.* (1999).

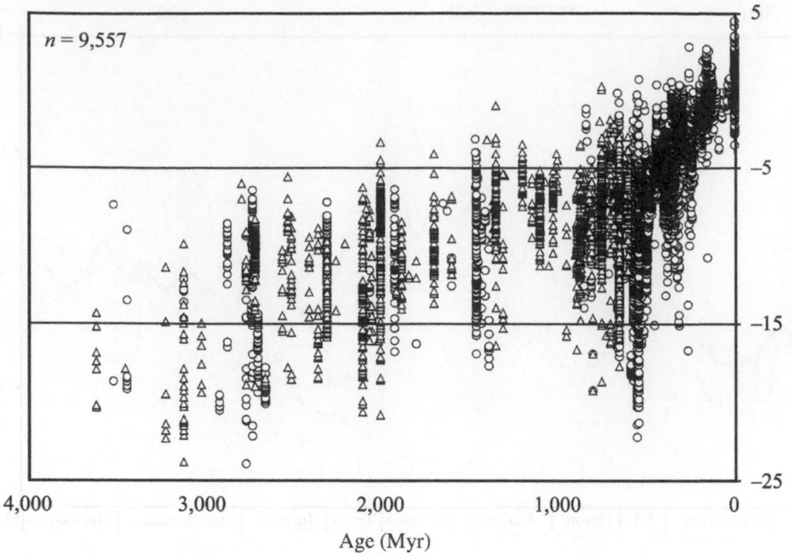

Figure 33 Oxygen isotopic composition of limestones and calcitic shells during geologic history. Triangles and circles represent samples with good and poor age assignment, respectively (after Shields and Veizer, 2002).

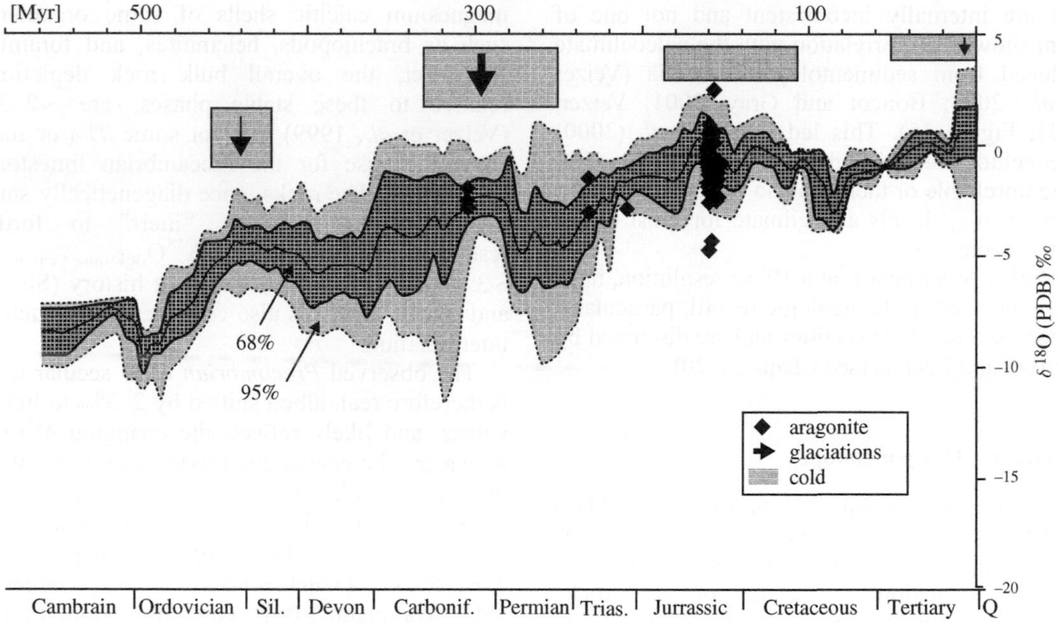

Figure 34 Phanerozoic $\delta^{18}O$ trend for low-Mg calcitic and aragonitic shells. Note that this is a trend for $CaCO_3$, offset by the fractionation factor α from that of seawater. See Figure 32 for further explanations (after Veizer *et al.*, 1999).

The pattern in Figure 34 has considerable implications for our understanding of past climate, but is still subject to debate. The models of Muehlenbachs (1998) and Gregory (1991) claimed that due to a balance of high and low temperature reactions during interaction of the water cycle with the lithosphere, the $\delta^{18}O$ of the oceans should have been buffered near its present-day value. If seawater always had $\delta^{18}O$ of ~0‰ SMOW (standard mean ocean water), the primary nature of the $\delta^{18}O$ record (Figure 34) would demand cooling oceans in the course of the Phanerozoic. With such an assumption, however, the early- to mid-Paleozoic ocean temperatures would have to have been in excess of 40 °C, even at times of glaciations. This is an unpalatable proposition, not only climatologically, but also in view of the similarity in faunal assemblages, in our case brachiopods, during this entire time span. Accepting that the $\delta^{18}O$ of past seawater was evolving towards ^{18}O-enriched values (Wallmann, 2001) and detrending the data accordingly (Figure 35), the superimposed second order structure of the curve correlates well with the Phanerozoic paleoclimatic record (cf. also Boucot and Gray, 2001). The observed structure, therefore, likely reflects paleotemperatures. If so, this would indicate that global climate swings were not confined to higher latitudes, but involved equatorial regions as well. As already pointed out, neither the $\delta^{18}O$ nor the paleoclimate record correlate with the model p_{CO_2} estimates for the ancient atmosphere.

14.11.2.6 *Isotope tracers in developmental stages*

The advances in instrumentation and particularly the arrival of the multicollector inductively coupled plasma mass spectrometers (MC-ICPMS) opened a window for a number of new tracers that were difficult to tackle with the old instrumentation. A pattern for seawater evolution is thus emerging for isotopes of boron and calcium.

The present-day boron concentration in seawater and its $\delta^{11}B$ are uniform at 4.5 ppm and −39.6‰, respectively (Lemarchand *et al.*, 2000) and its residence time is ~14 Myr. Boron is present in seawater as $B(OH)_4^-$ and $B(OH)_3$. The relative proportion of these species is a function of pH (Palmer and Swihart, 1996), with $B(OH)_3$ 19.8‰ enriched in ^{11}B relative to $B(OH)_4^-$. Boron incorporation into carbonate skeletons, at concentrations of ~10–60 ppm, is from $B(OH)_4^-$ (Hemming and Hanson, 1992) and their $\delta^{11}B$ can therefore be used for tracing the pH of ancient seawater. Pearson and Palmer (2000), utilizing foraminiferal calcite, argued that the pH of Cenozoic seawater increased from ~7.4 at 60 Ma to its present-day value of 8.1. In contrast, Lemarchand *et al.* (2000) argued that the $\delta^{11}B$ trend in these foraminifera reflects the changing $^{11}B/^{10}B$ composition of seawater at constant pH, a development largely due to scavenging of boron by an increasing flux of clastic sediments.

The residence time of *calcium*, ~1 Myr versus the mixing rates of ocean of ~1,000 yr, means that its isotopes are distributed homogeneously in seawater (Skulan *et al.*, 1997), with $^{40}Ca/^{44}Ca$ equal to 45.143 (Schmitt *et al.*, 2001). Modern carbonate shells show a variation of ~4‰. These shells are ~1‰ enriched relative to magmatic rocks, but compared to seawater they are depleted by 1–3‰, depending on the trophic level of the organism (Skulan *et al.*, 1997). Zhu and MacDougall (1998) showed that calcium isotope

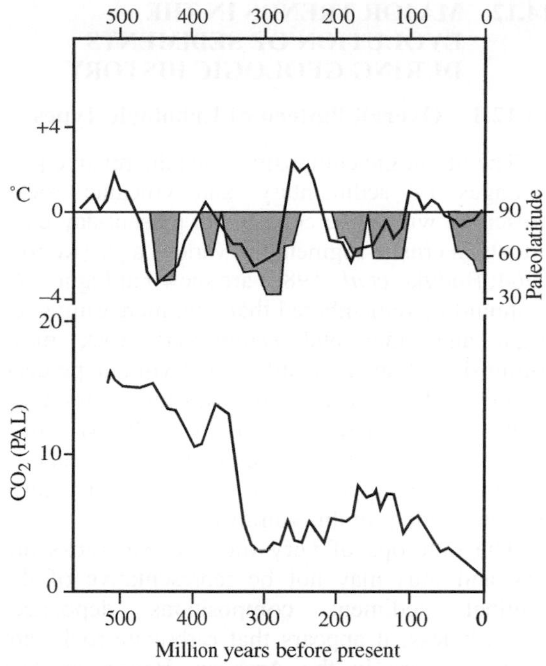

Figure 35 Reconstructed variations in mean temperature of shallow low-latitude seawater during the Phanerozoic based on the data in Figure 34. Note the good agreement of the cooling episodes with the extended latitudinal dispersion of ice rafted debris (shaded histograms). In the subsequent publication, Shaviv and Veizer (2003) showed that the proposed temperature variations correlated with the intensity of cosmic-ray flux reaching the Earth. The p_{CO2} (PAL—present-day atmospheric level) is that for the Geocarb model of Berner (1994).

composition of the shells was both species and temperature dependent and that river water is depleted by 2‰ relative to seawater. The temperature dependency of calcium isotope fractionation enables this tracer to be a potential paleotemperature proxy (Nägler *et al.*, 2000). The seawater $^{44}Ca/^{40}Ca$ secular variations are indicated by the results of De La Rocha and De Paolo (2000) for the last 160 Myr, and by the data of Zhu (1999) for the entire 3.4 Gyr of earth history. The latter indicate that, in analogy to strontium isotopes (Veizer and Compston, 1976), the Archean samples have $^{44}Ca/^{40}Ca$ ratios similar to the earth mantle, with the crustal-like values first appearing at the Archean/Proterozoic transition. However, in contrast to the strontium isotope trend, the calcium isotope ratios appear to dip towards mantle values also at ~1.6 Gyr ago.

In the near future, the isotopes of *silicon, iron, and magnesium* also will likely develop into useful paleoceanographic tracers, but at this stage their utility for pre-Quaternary studies is limited.

14.12 MAJOR TRENDS IN THE EVOLUTION OF SEDIMENTS DURING GEOLOGIC HISTORY

14.12.1 Overall Pattern of Lithologic Types

The lithologic composition and the relative percentages of sedimentary and volcanic rocks preserved within the confines of present-day continents in crustal segments of various ages (Ronov, 1964; Budyko *et al.*, 1985) are shown in Figure 36. It should be remembered that with increasing geologic age, the total sedimentary rock mass diminishes (Figure 6) and a given volume percentage of rock 3 Gyr ago represents much less mass than an equal percentage of rocks 200 Myr old. Despite this limitation, some general trends in lithologic rock types agreed on by most investigators are evident in this summary.

The outcrops of very old Archean rocks are few and thus may not be representative of the original sediment compositions deposited. Nevertheless, it appears that carbonate rocks are relatively rare in the Archean. Based on data from the limited outcrops, Veizer (1973) concluded that Archean carbonate rocks are predominantly limestones. During the early Proterozoic, the abundance of carbonates increases markedly, and for most of this Era the preserved carbonate rock mass is typified by the ubiquity of early diagenetic, and perhaps primary, dolostones (Veizer, 1973; Grotzinger and James, 2000). In the Phanerozoic, carbonates constitute ~30% of the total sedimentary mass, with sandstones and shales accounting for the rest. The Phanerozoic record of carbonates will be elaborated upon in the subsequent text.

Other highlights of the lithology-age distribution of Figure 36 are: (i) a marked increase in the abundance of submarine volcanogenic rocks and immature sandstones (graywackes) with increasing age; (ii) a significant percentage of arkoses in the early and middle Proterozoic, and an increase in the importance of mature sandstones (quartz-rich) with decreasing age; red beds are significant rock types of Proterozoic and younger age deposits; (iii) a significant "bulge" in the relative abundance of banded iron-ore formations, chert-iron associations, in the early Proterozoic; and (iv) a lack of evaporitic sulfate (gypsum, anhydrite) and salt (halite, sylvite) deposits in sedimentary rocks older than ~800 Myr. Marine evaporites owe their existence to a unique combination of tectonic, paleogeographic, and sea-level conditions. Seawater bodies must be restricted to some degree, but also must exchange with the open ocean to permit large volumes of seawater to enter these restricted basins and evaporate. Environmental settings of evaporite deposition may occur on cratons or in rifted basins. Figure 37 illustrates that because

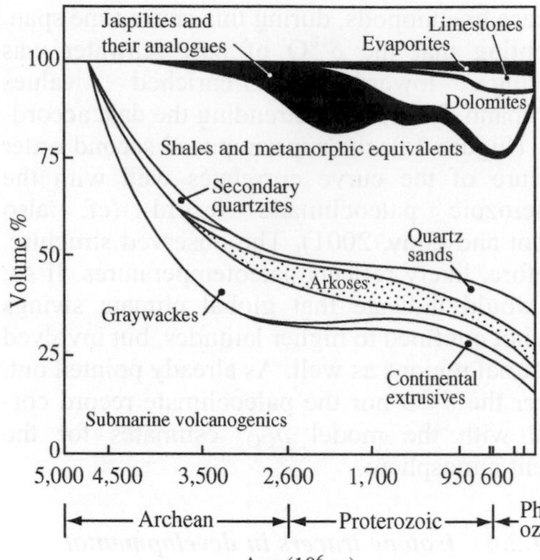

Figure 36 Volume percent of sedimentary rocks as a function of age. Extrapolation beyond ~3 Ga is hypothetical (after Ronov, 1964).

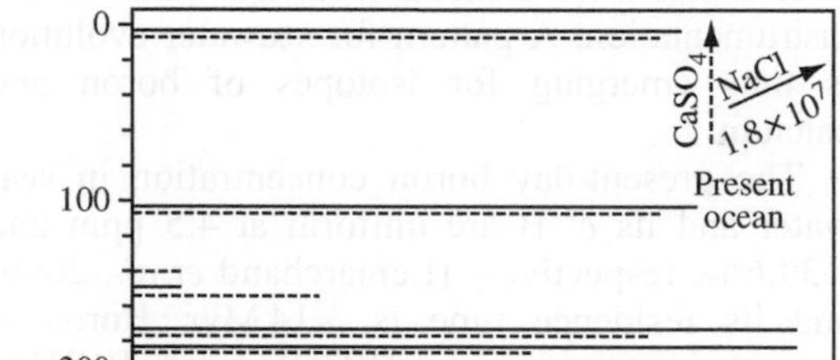

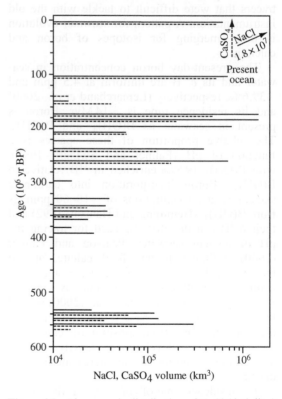

Figure 37 Phanerozoic distribution of NaCl (dark line) and $CaSO_4$ (dashed line) in marine evaporites (after Holser, 1984).

evaporite deposition requires an unusual combination of circumstances, a "geological accident" (Holser, 1984), the intensity of deposition has varied significantly during geologic time. This conclusion implies that the volume preserved of

NaCl and $CaSO_4$ per unit of time in the Phanerozoic reasonably reflects the volume deposited, and because of the lack of any secular trend in evaporite mass per unit time, there has been only minor differential recycling of these rocks relative to other lithologies. Because the oceans are important reservoirs for these components, such large variations in the rate of NaCl and $CaSO_4$ output from the ocean to evaporites imply changes in the salinity and chemistry of seawater (Hay *et al.*, 2001).

These trends in lithologic features of the sedimentary rock mass are a consequence of evolution of the surface environment of the planet as well as recycling and post-depositional processes, and both secular and cyclic processes have played a role in generating the lithology-age distribution we see today (Veizer, 1973, 1988a; Mackenzie, 1975). For the past 1.5–2.0 Gyr, the Earth has been in a near present-day steady state, and the temporal distribution of rock types since then has been controlled primarily by recycling in response to plate tectonic processes.

Because sedimentary carbonates are important rock types in terms of providing mineralogical, chemical, biological, and isotopic data useful in interpretation of the history of Earth's surface environment, the following sections discuss these rock types in some detail. The discussion is mainly limited to the Phanerozoic because of the more complete database for this Eon than for the Precambrian.

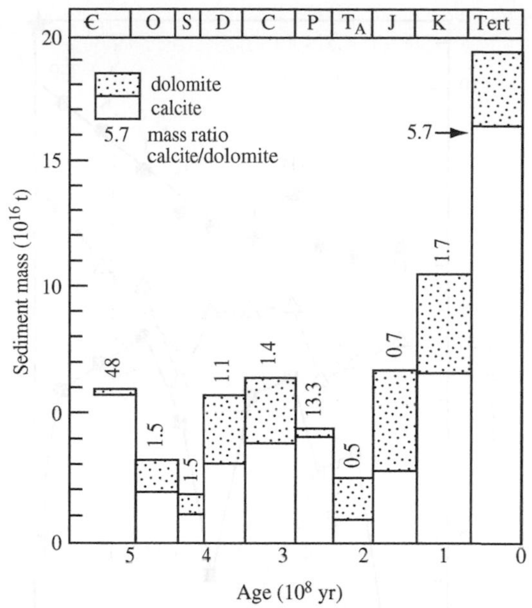

Figure 38 The Phanerozoic sedimentary carbonate mass distribution as a function of geologic age. Period masses of calcite and dolomite and the period mass ratios of calcite/dolomite are also shown (after Mackenzie and Morse, 1992).

14.12.2 Phanerozoic Carbonate Rocks

14.12.2.1 Mass-age distribution and recycling rates

As stated above, carbonate rocks comprise ~30% of the mass of Phanerozoic sediments. Given and Wilkinson (1987) reevaluated all the existing data on Phanerozoic carbonate rocks, their masses, and their relative calcite and dolomite contents (Figure 38). It can be seen that, as with the total sedimentary mass (Garrels and Mackenzie, 1971a,b), the mass of carbonate rock preserved is pushed toward the front of geologic time. The Tertiary, Carboniferous, and Cambrian periods are times of significant carbonate preservation, whereas the preservation of Silurian and Triassic carbonates is minimal.

The survival rates of the carbonate masses for different Phanerozoic systems are shown in Figure 39, together with the Gregor (1985) plot for the total sedimentary mass. The difference between the survival rate of the total carbonate mass and that of dolomite is the mass of limestone surviving per interval of time. The half-life of all the post-Devonian sedimentary mass is 130 Myr, and for a constant mass with a constant probability of destruction, the mean sedimentation rate is $\sim 100\times 10^{14}\,g\,yr^{-1}$. The modern global erosional flux is $\sim 200\times 10^{14}\,g\,yr^{-1}$, of which ~15% is particulate and dissolved carbonate. Although the data are less reliable for the survival rate of Phanerozoic carbonate sediments than for the total sedimentary mass, the half-life of the post-Permian carbonate mass is ~86 Myr. This gives a mean sedimentation rate of $\sim 35\times 10^{14}$ g carbonate per year, compared to the present-day carbonate flux of 30×10^{14} g yr^{-1} (Morse and Mackenzie, 1990). The difference in half-lives between the total sedimentary mass, which is principally sandstone and shale, and the carbonate mass probably is a consequence of the more rapid recycling of the carbonate mass at a rate ~1.5 times that of the total mass.

This is not an unlikely situation. With the advent of abundant carbonate-secreting, planktonic organisms in the Jurassic, the site of carbonate deposition shifted significantly from shallow-water areas to the deep sea. This gradual shift will increase still further the rate of destruction (by eventual subduction) of the global carbonate mass relative to the total sedimentary mass because the recycling rate of oceanic crust (the "b" values of Veizer and Jansen, 1985; Veizer, 1988a) exceeds that of the "continental" sediments by a factor of 6. Also, Southam and Hay (1981), using a half-life of 100 Myr for pelagic sediment, estimated that as much as 50% of all sedimentary rock formed by weathering of igneous rock may have been lost by subduction during the past 4.5 Gyr.

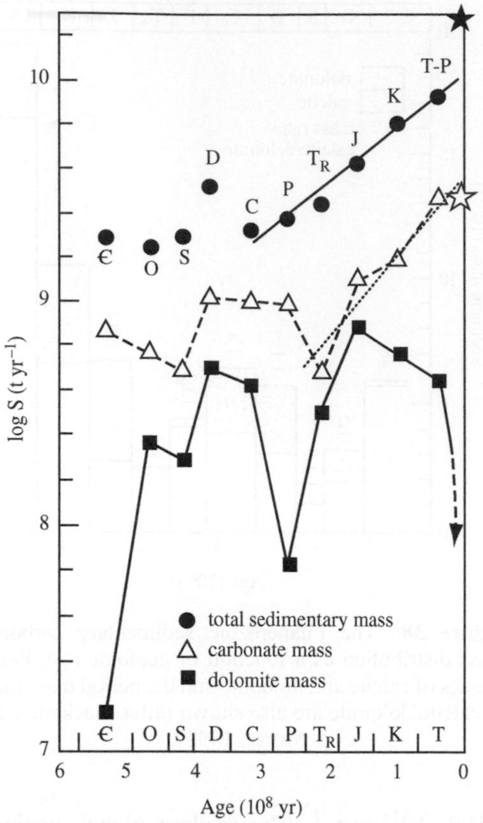

Figure 39 Phanerozoic sedimentary rock mass–age relationships expressed as the logarithm of the survival rate in tons per year versus time. The straight lines are best fits to the total mass data (solid line) and to the carbonate mass data (dotted line) for particular intervals of Phanerozoic time. The difference between the logarithm of S for the carbonate mass and that of the dolomite mass is the survival rate of the calcite mass. Filled star is present total riverine flux to the oceans, whereas open star is carbonate flux (after Mackenzie and Morse, 1992).

Thus, it appears, as originally suggested by Garrels and Mackenzie (1969, 1971a, 1972), that the carbonate component of the sedimentary rock mass may have a cycling rate slightly different from that of the total sedimentary mass. These authors argued that the differential recycling rates for different lithologies were related to their resistance to chemical weathering and transport. Evaporites are the most easily soluble, limestones are next, followed by dolostones, and shales and sandstones are the most inert. Although resistance to weathering may play a small role in the selective destruction of sedimentary rocks, it is likely, as argued previously, that differences in the recycling rates of different tectonic regimes in which sediments are deposited are more important.

14.12.2.2 Dolomite/calcite ratios

For several decades it has been assumed that the Mg/Ca ratio of carbonate rocks increases with increasing rock age (see Daly, 1909; Vinogradov and Ronov, 1956a,b; Chilingar, 1956; Figure 40). In these summaries, the magnesium content of North American and Russian Platform carbonates is relatively constant for the latest 100 Myr, and then increases gradually, very close to, if not the same as, the commencement of the general increase in the magnesium content of pelagic limestones (Renard, 1986). The dolomite content in deep sea sediments also increases erratically with increasing age back to ~125 Myr before present (Lumsden, 1985). Thus, the increase in magnesium content of carbonate rocks with increasing age into at least the early Cretaceous appears to be a global phenomenon, and to a first approximation, is not lithofacies related. In the 1980s, the accepted truism that dolomite abundance increases relative to limestone with increasing age has been challenged by Given and Wilkinson (1987). They reevaluated all the existing data and concluded that dolomite abundances do vary significantly throughout the Phanerozoic but may not increase systematically with age (Figure 40). Yet the meaning of these abundance curves, and indeed their actual validity, is still controversial (Zenger, 1989).

Voluminous research on the "dolomite problem" (see Hardie, 1987, for discussion) has shown that the reasons for the high magnesium content of carbonates are diverse and complex. Some dolomitic rocks are primary precipitates; others were deposited as $CaCO_3$ and then converted entirely or partially to dolomite before deposition of a succeeding layer; still others were dolomitized by migrating underground waters tens or hundreds of millions of years after deposition. It is therefore exceedingly important to know the distribution of the calcite/dolomite ratios of carbonate rocks through geologic time. This information has a bearing on the origin of dolomite, as well as on the properties of the coeval atmosphere–hydrosphere system (Given and Wilkinson, 1987; Wilkinson and Algeo, 1989; Berner, 1990; Morse and Mackenzie, 1990; Mackenzie and Morse, 1992; Arvidson and Mackenzie, 1999; Arvidson *et al.*, 2000; Holland and Zimmermann, 2000). For example, it could be argued that if the dolomite/calcite ratio progressively increases with age of the rock units (Figure 40), the trend principally reflects enhanced susceptibility of older rock units to processes of dolomitization. Such a trend would then be only a secondary feature of the sedimentary carbonate rock mass due to progressive diagenesis and seawater driven dolomitization (Garrels and Mackenzie, 1971a; Mackenzie, 1975; Holland

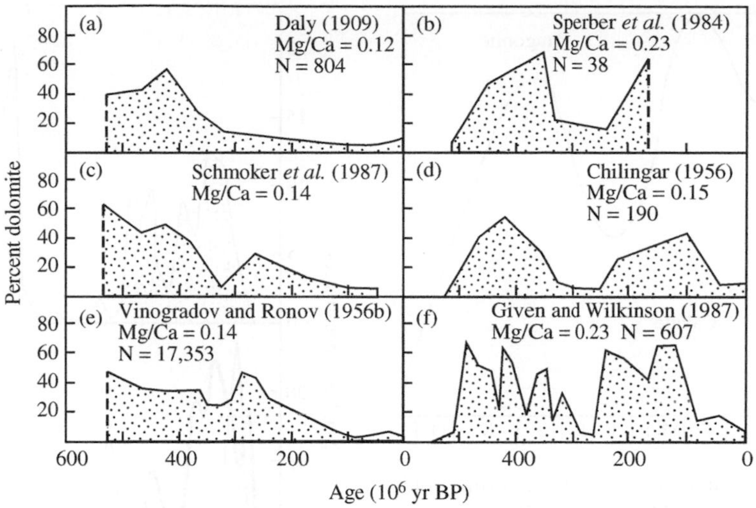

Figure 40 Estimates ((a) Daly (1909), (b) Sperber *et al.* (1984), (c) Schmoker *et al.* (1987), (d) Chilingar (1956), (e) Vinogradov and Ronov (1956b), and (f) Given and Wilkinson (1987)) of percent dolomite in Phanerozoic cratonic carbonate rocks as a function of age. Mg/Ca = average ratio (after Wilkinson and Algeo, 1989).

and Zimmermann, 2000). Alternatively, if the trend in the calcite/dolomite ratio is cyclic in nature, this cyclicity could be interpreted as representing environmental change in the ocean-atmosphere system (Wilkinson and Algeo, 1989). For the discussion here, we accepted the data of Given and Wilkinson (1987) on the Ca/Mg ratio of Phanerozoic sedimentary carbonates (Figure 38) to calculate the mass ratios of these carbonate components as a function of age, but do realize that such data are still a matter of controversy (Zenger, 1989; Holland and Zimmermann, 2000). In reality, it is most likely that both cyclical and secular compositional changes in the ocean-atmosphere-sediment system, as well as diagenesis, contribute to dolomite abundance during geologic time.

The period-averaged mass ratio of calcite to dolomite is anomalously high for the Cambrian and Permian System rocks (Figure 38). For the remainder of the Phanerozoic, it appears to oscillate within the 1.1 ± 0.6 range, except for the limestone-rich Tertiary. Comparison with the generalized Phanerozoic sea-level curves of Vail *et al.* (1977) and Hallam (1984) (Figure 42) hints that dolomites are more abundant at times of higher sea levels. Mackenzie and Agegian (1986, 1989) and Given and Wilkinson (1987) were the first to suggest this possible cyclicity in the calcite/dolomite ratio during the Phanerozoic, and Lumsden (1985) observed a secular decrease in dolomite abundance in deep marine sediments from the Cretaceous to Recent, corresponding to the general fall of sea level during this time interval. These cycles in calcite/dolomite ratios correspond crudely to the Fischer (1984) two Phanerozoic super cycles and to the Mackenzie and Pigott (1981) oscillatory and submergent tectonic modes.

14.12.2.3 Ooids and ironstones

Although still somewhat controversial (e.g., Bates and Brand, 1990), the textures of ooids appear to vary during Phanerozoic time. Sorby (1879) first pointed out the petrographic differences between ancient and modern ooids: ancient ooids commonly exhibit relict textures of a calcitic origin, whereas modern ooids are dominantly made of aragonite. Sandberg (1975) reinforced these observations by study of the textures of some Phanerozoic ooids and a survey of the literature. His approach, and that of others who followed, was to employ the petrographic criteria, among others, of Sorby: i.e., if the microtexture of the ooid is preserved, then the ooid originally had a calcite mineralogy; if the ooid exhibits textural disruption, its original mineralogy was aragonite. The textures of originally aragonitic fossils are usually used as checks to deduce the original mineralogy of the ooids. Sandberg (1975) observed that ooids of inferred calcitic composition are dominant in rocks older than Jurassic.

Following this classical work, Sandberg (1983, 1985) and several other investigators (Mackenzie and Pigott, 1981; Wilkinson *et al.*, 1985; Bates and Brand, 1990) attempted to quantify further this relationship. Figure 41 is a schematic diagram representing a synthesis of the inferred mineralogy of ooids during the Phanerozoic. This diagram is highly tentative, and more data are needed to document the trends. However, it appears that while originally aragonitic ooids are found throughout the

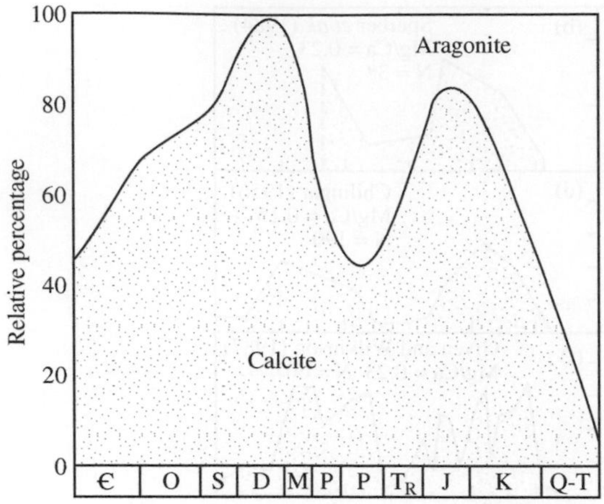

Figure 41 Inferred mineralogy of Phanerozoic ooids (after Morse and Mackenzie, 1990).

Phanerozoic, an oscillatory trend in the relative percentage of calcite versus aragonite ooids may be superimposed on a long-term evolutionary decrease in ooids with an inferred original calcitic mineralogy. Although the correlation is not strong, the two major maxima in the sea-level curves of Vail *et al.* (1977) and Hallam (1984) appear to coincide with times when calcite ooids were important seawater precipitates (Figure 42). Wilkinson *et al.* (1985) found that the best correlation between various data sets representing global eustasy and ooid mineralogy is that of inferred mineralogy with percentage of continental freeboard. Sandberg (1983) further concluded that the cyclic trend in ooid mineralogy correlates with cyclic trends observed for the inferred mineralogy of carbonate cements. Van Houten and Bhattacharyya (1982; and later Wilkinson *et al.*, 1985) showed that the distribution of Phanerozoic ironstones (hematite and chamosite oolitic deposits) also exhibits a definite cyclicity (Figure 42) that too appears to covary with the generalized sea-level curve. Minima appear to coincide with times of sea-level withdrawal from the continents.

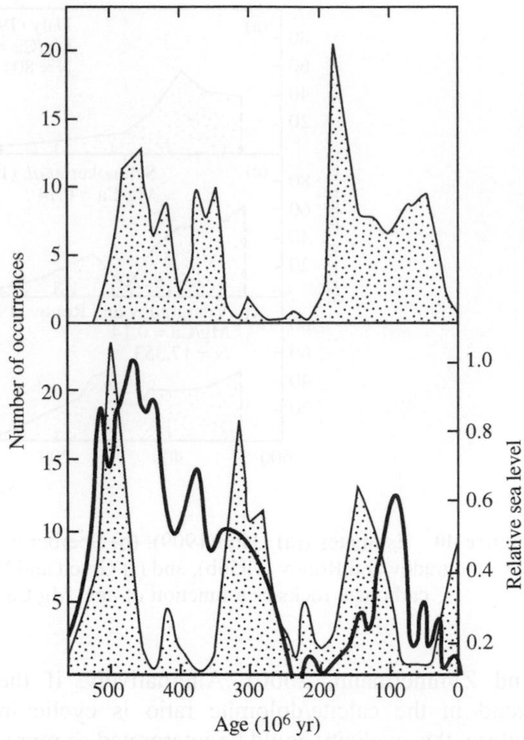

Figure 42 Number of occurrences of Phanerozoic ironstones (upper diagram, source Van Houten and Bhattacharyya, 1982) and oolitic limestones (lower diagram, source Wilkinson *et al.*, 1985) as a function of geologic age. The relative sea-level curve is that of Hallam (1984). Minima in occurrences appear to correlate with times of sea level withdrawal from the continents (after Morse and Mackenzie, 1990).

14.12.2.4 Calcareous shelly fossils

In some similarity to the trends observed for the inorganic precipitates of ooids and ironstones, the mineralogy of calcareous fossils during the Phanerozoic also shows a cyclic pattern (Figure 25) with calcite being particularly abundant during high sea levels of the early to mid-Paleozoic and the Cretaceous (Stanley and Hardie, 1998; Kiessling, 2002). Overall (Figure 43) there is a general increase in the diversity of major groups of calcareous organisms such as coccolithophorids, pteropods, hermatypic corals, and coralline algae. It is noteworthy that the major groups of pelagic and benthic organisms contributing to carbonate sediments in today's ocean first appeared in the fossil record during the middle Mesozoic and progressively became more abundant. What is most evident in Figure 43 is a long-term increase in the production of biogenic carbonates dominated by aragonite and magnesian calcite mineralogies. Because organo-detrital carbonates are such an important part of the Phanerozoic carbonate rock record, this increase in metastable mineralogies played an important role in the pathway of diagenesis of carbonate sediments. The ubiquity of low magnesian calcite skeletal organisms for much of the Paleozoic led to production of calcitic organo-detrital sediments whose original bulk chemical and mineralogical composition was closer to that of their altered and lithified counterparts of Cenozoic age.

14.12.2.5 The carbonate cycle in the ocean

The partitioning of carbonate burial between shoal-water and deep sea realms has varied in a cyclic pattern through Phanerozoic time

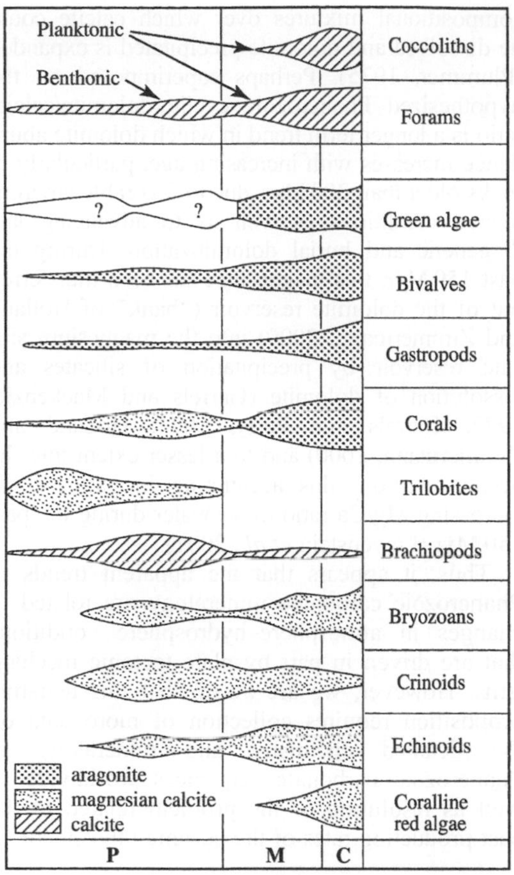

Figure 43 Mineralogical evolution of benthic and planktonic organism diversity during the Phanerozoic based on summaries of Milliken and Pigott (1977) and Wilkinson (1979). P is Paleozoic, M is Mesozoic, and C is Cenozoic.

(Morse and Mackenzie, 1990). The variation in the magnitudes of the fluxes of $(Ca,Mg)CO_3$ to the two environments through time is difficult to assess; even today's fluxes are probably not known within a factor of 2. To gain some impression of the fluxes involved, a tentative model of the carbonate carbon cycle in the world's oceans is shown in Figure 44. This is a representation of the mean state of the cycle during the most recent glacial to interglacial transition. About 18×10^{12} moles of calcium and magnesium (equivalent to 216×10^{6} metric tons of carbon) accumulate yearly as carbonate minerals (Morse and Mackenzie, 1990), mainly as biological precipitates. Of this flux $\sim6\times10^{12}$ moles are deposited as calcium and magnesium carbonates in shoal-water areas (Milliman, 1974; Smith, 1978; Wollast, 1993; Milliman *et al.*, 1999), and the remainder accumulates as calcareous oozes in the pelagic realm. The 12×10^{12} moles of carbonate accumulated annually in the deep sea are only $\sim$17% of the annual carbonate production rate of

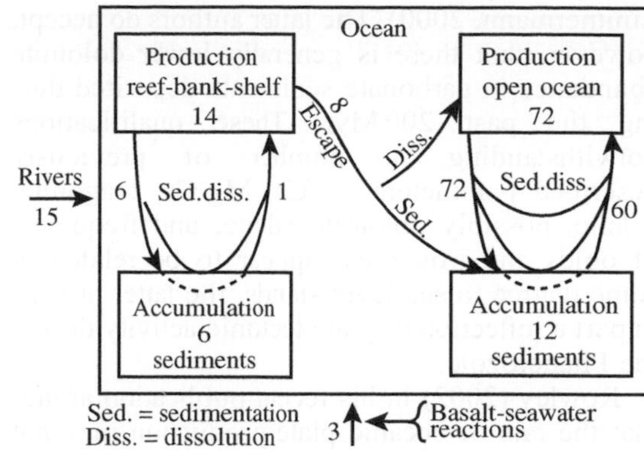

Figure 44 Tentative model of global ocean carbonate cycle. Fluxes are in units of 10^{12} mol C yr^{-1} as $(Ca,Mg)CO_3$ and represent estimates of fluxes averaged over the most recent glacial–interglacial transition (after Mackenzie and Morse, 1992).

72×10^{12} mol of the open ocean photic zone. This efficient recycling of carbonate carbon in the open ocean water column and at the sediment-water interface is a well-known feature of the marine carbon cycle (Broecker and Peng, 1982). It is important to note that much shoal-water carbonate production ends up in sediments of reefs, banks, etc., so that, in contrast to the pelagic realm, production rate more closely approximates sedimentation rate. However there is an escape of carbonate sediment from shoal-water areas to the deep sea, where it is deposited or dissolved (Land, 1979; Kiessling, 2002). The magnitude of this flux is poorly known but may affect the chemistry of open-ocean regions owing to dissolution of the carbonate debris (Droxler *et al.*, 1988; Agegian *et al.*, 1988; Sabine and Mackenzie, 1975). Furthermore, its accumulation on the slopes of banks may act as a record of paleoenvironmental change (Droxler *et al.*, 1983).

14.12.3 Geochemical Implications of the Phanerozoic Carbonate Record

The reasons for the mass-age relationships discussed previously are not totally clear. A number of investigators (Mackenzie and Pigott, 1981; Sandberg, 1985; Wilkinson *et al.*, 1985; Wilkinson and Given, 1986; Wilkinson and Algeo, 1989; Mackenzie and Agegian, 1989; Stanley and Hardie, 1998; Arvidson *et al.*, 2000) concluded that these observations are the result of changing atmosphere–hydrosphere environmental conditions through the Phanerozoic. Others argued that, for example, the ooid observations are not statistically significant (Bates and Brand, 1990) or that the Given and Wilkinson (1987) mass-age database is not valid (Holland and

Zimmermann, 2000). The latter authors do accept, however, that there is generally lower dolomite abundance in carbonate sediments deposited during the past 200 Myr. These qualifications notwithstanding, a number of previously described parameters (Sr/Ca, Mg/Ca, aragonite/calcite, possibly dolomite/calcite, and frequency of ooids and iron ores) appear to be related in some degree to sea-level stands, the latter at least in part a reflection of plate tectonic activity during the Phanerozoic.

Rowley (2002) in his recent publication argues that the rate of oceanic plate production may not have varied significantly for the latest 180 Myr. If so, this may have major implications on our understanding of the model linkage of tectonics to sea-level change, atmospheric CO_2, seawater chemistry and related phenomena. Nevertheless, while the 30–40% variations in seafloor spreading rates during the latest 100 Myr (Delaney and Boyle, 1986) are probably not justifiable, we cannot at present dismiss entirely the proposition that hydrothermal exchange between seafloor and ocean, and presumably the rate of plate generation, might still have varied somewhat. It may be possible therefore that the first-order changes in sea level can still be driven by the accretion of the ridges, but, at the same time, we should also search for alternative linkages of sea-level stands to the mineralogical and chemical properties of the Phanerozoic sedimentary cycle.

In the standard reasoning, the first-order changes in sea level are driven by the accretion of ridges: high accretion rate, high sea level; low accretion rate, low sea level. Regardless of the tie between oceanic plate production and sea level, extended times of global high sea level may have been times of enhanced atmospheric CO_2 levels (Berner *et al.*, 1983; Lasaga *et al.*, 1985; Berner, 2000), higher temperatures (not necessarily related solely to atmospheric CO_2 concentrations), probably lower seawater Mg/Ca ratios (Lowenstein *et al.*, 2001; Dickson, 2002), different saturation states of seawater (Arvidson *et al.*, 2000), and perhaps different seawater sulfate concentrations than at present. The converse is true for first-order global sea-level low stands. It appears that the environmental conditions for early dolomitization, formation of calcitic ooids and cements, and preponderance of calcitic reef-building organisms are best met during extended times of global high sea levels, with ubiquitous shallow-water and sabkha-like environments (calcite seas; Sandberg, 1983). Dolomitization of precursor calcite and aragonite phases, either in marine waters or in mixed continental-marine waters, would be enhanced under these conditions. Furthermore, the potentially lowered pH of marine waters during times of enhanced atmospheric carbon dioxide would favor syndepositional or later dolomitization in mixed marine-meteoric waters, because the range of seawater-meteoric compositional mixtures over which calcite could be dissolved and dolomite precipitated is expanded (Plummer, 1975). Perhaps superimposed on the hypothesized Phanerozoic cyclic dolomite/calcite ratio is a longer term trend in which dolomite abundance increases with increasing age, particularly in rocks older than 200 Myr, due to favorable environmental conditions as well as to advancing late diagenetic and burial dolomitization. During the past 150 Myr, this magnesium has been transferred out of the dolomite reservoir ("bank" of Holland and Zimmermann, 2000) into the magnesium silicate reservoir by precipitation of silicates and dissolution of dolomite (Garrels and Mackenzie, 1972; Garrels and Perry, 1974; Holland and Zimmermann, 2000) and to a lesser extent into the ocean reservoir, this accounting in part for the increasing Mg/Ca ratio of seawater during the past 150 Myr (Lowenstein *et al.*, 2001).

Thus, it appears that the apparent trends in Phanerozoic carbonate mineralogy are related to changes in atmosphere–hydrosphere conditions that are driven in part by plate tectonic mechanisms. However, we are aware that this tentative proposition requires collection of more data on the detailed chemistry and mineralogy of Phanerozoic carbonate sequences worldwide as well as resolution of the problem related to the past production rates of the oceanic lithosphere.

ACKNOWLEDGMENT

We would like to dedicate this chapter to Robert M. Garrels.

REFERENCES

Agegian C. R., Mackenzie F. T., Tribble J. S., and Sabine C. (1988) Carbonate production and flux from a mid-depth bank ecosystem. *Natl. Undersea Res. Prog. Res. Report* **88**(1), 5–32.

Allègre C. J. and Rousseau D. (1984) The growth of the continents through geological time studied by Nd isotope analysis of shales. *Earth Planet. Sci. Lett.* **67**, 19–34.

Armstrong R. L. (1981) Radiogenic isotopes: the case for crustal recycling on a near-steady-state no-continental-growth Earth. *Phil. Trans. Roy. Soc. London A Ser.* **301**, 443–472.

Arvidson R. S. and Mackenzie F. T. (1999) The dolomite problem: control of precipitation kinetics by temperature and saturation state. *Am. J. Sci.* **299**, 257–288.

Arvidson R. S., Mackenzie F. T., and Guidry M. W. (2000) Ocean/atmosphere history and carbonate precipitation rates: a solution to the dolomite problem. In *Marine Authigenesis: from Global to Microbial*, SEPM Spec. Publ. No. 65 (eds. C. R. Glenn, L. Prévôt-Lucas, and J. Lucas), SEPM, Tulsa, pp. 1–5.

Bates N. R. and Brand U. (1990) Secular variation of calcium carbonate mineralogy: an evaluation of oöid and micrite chemistries. *Geol. Rundsch.* **79**, 27–46.

Barrell J. (1917) Rhythms and the measurement of geological time. *Geol. Soc. Am. Bull.* **28**, 745–904.

Berner R. A. (1990) Atmospheric carbon dioxide levels over Phanerozoic time. *Science* **249**, 1382–1386.

Berner R. A. (1994) GEOCARB: II. A revised model of atmospheric CO_2 over Phanerozoic time. *Am. J. Sci.* **294**, 56–91.

Berner R. A. (2000) The effect of the rise of land plants on atmospheric CO_2 during the paleozoic. In *Plants Invade the Land: Evolutionary and Environmental Approaches* (eds. P. G. Gensel and D. Edwards). Columbia University Press, New York, pp. 173–178.

Berner R. A. and Kothavala Z. (2001) GEOCARB: III. A revised model of atmospheric CO_2 over phanerozoic time. *Am. J. Sci.* **301**, 182–204.

Berner U. and Streif H. (2001) *Klimafakten, 2001, Der Rückblick—Ein Schlüssel für die Zukunft*. Schweizerbart'sche Verlagsbuchhandlung, Science Publishers, Stuttgart.

Berner R. A., Lasaga A. C., and Garrels R. M. (1983) The carbonate silicate geochemical cycle and its effect on atmospheric carbon dioxide over the past 100 million years. *Am. J. Sci.* **283**, 641–683.

Berry J. P. and Wilkinson B. H. (1994) Paleoclimatic and tectonic control on the accumulation of North American cratonic sediment. *Geol. Soc. Am. Bull.* **106**, 855–865.

Bluth G. J. and Kump L. R. (1991) Phanerozoic paleogeology. *Am. J. Sci.* **291**, 284–308.

Boucot A. J. and Gray J. (2001) A critique of Phanerozoic climatic models involving changes in the CO_2 content of the atmosphere. *Earth Sci. Rev.* **56**, 1–159.

Brand U. and Veizer J. (1980) Chemical diagenesis of a multicomponent carbonate system: I. Trace elements. *J. Sedim. Petrol.* **50**, 1219–1236.

Broecker W. S. and Peng T. H. (1982) *Tracers in the Sea*. Eldigio Press, Palisades, NY.

Budyko M. I., Ronov A. B., and Yanshin A. L. (1985) *History of the Earth's Atmosphere*. Springer, Heidelberg.

Burke W. H., Denison R. F., Hetherington E. A., Koepnick R. F., Nelson H. F., and Otto J. B. (1982) Variation of seawater $^{87}Sr/^{86}Sr$ throughout Phanerozoic time. *Geology* **10**, 516–519.

Canfield D. E. and Teske A. (1996) Late Proterozoic rise in atmospheric oxygen concentration inferred from phylogenetic and sulphur-isotope studies. *Nature* **382**, 127–132.

Chambers L. A. and Trudinger P. A. (1979) Microbiological fractionation of stable sulfur isotopes: a review and critique. *Geomicrobiol. J.* **1**, 249–293.

Chilingar G. V. (1956) Relationship between Ca/Mg ratio and geologic age. *Am. Assoc. Petrol. Geol. Bull.* **40**, 2256–2266.

Cicero A. D. and Lohmann K. C. (2001) Sr/Mg variation during rock-water interaction: implications for secular changes in the elemental chemistry of ancient seawater. *Geochim. Cosmochim. Acta* **65**, 741–761.

Claypool G. E., Holser W. T., Kaplan I. R., Sakai H., and Zak I. (1980) The age curves of sulfur and oxygen isotopes in marine sulfate and their mutual interpretation. *Chem. Geol.* **28**, 199–260.

Cloud P. E. (1976) Major features of crustal evolution. *Trans. Geol. Soc. S. Afr.* **79**, 1–32.

Cogley J. G. (1984) Continental margins and the extent and number of continents. *Rev. Geophys. Space Phys.* **22**, 101–122.

Collerson K. D. and Kamber B. S. (1999) Evolution of the continents and the atmosphere inferred from the Th–U–Nb systematics of the depleted mantle. *Science* **283**, 1519–1522.

Condie K. C. (1989) *Plate Tectonics and Crustal Evolution*. Pergamon, London.

Condie K. C. (1993) Chemical composition and evolution of the upper continental crust: contrasting results from surface samples and shales. *Chem. Geol.* **104**, 1–37.

Condie K. C. (2000) Episodic continental growth models: afterthoughts and extensions. *Tectonophysics* **322**, 153–162.

Condie K. C. (2001) Continental growth during formation of Rodinia at 1.35–0.9 Ga. *Gondwana Res.* **4**, 5–16.

Cook T. D. and Bally A. W. (1975) *Shell Atlas: Stratigraphic Atlas of North and Central America*. Princeton University Press, Princeton, NJ.

Cox R. and Lowe D. R. (1995) A conceptual review of regional-scale controls on the composition of clastic sediment and the co-evolution of continental blocks and their sedimentary cover. *J. Sedim. Res.* **A65**, 1–12.

Daly R. A. (1909) First calcareous fossils and evolution of limestones. *Geol. Soc. Am. Bull.* **20**, 153–170.

Danchin R. V. (1967) Chromium and nickel in the Fig Tree shale from South Africa. *Science* **158**, 261–262.

Deines P. (1980) The isotopic composition of reduced organic carbon. In *Handbook of Environment Isotope Gechemistry: Vol. 1. The Terrestrial Environment* (eds. P. Fritz and J. C. Fontes). Elsevier, Amsterdam, pp. 329–406.

Delaney M. L. and Boyle E. A. (1986) Lithium in foraminiferal shells: implications for high-temperature hydrothermal circulation fluxes and oceanic generation rates. *Earth Planet. Sci. Lett.* **80**, 91–105.

De La Rocha C. L. and De Paolo J. (2000) Isotopic evidence for variations in the marine calcium cycle over the Cenozoic. *Science* **289**, 1176–1178.

Dickinson W. R., Beard L. S., Brakenridge G. R., Erjavec J. L., Ferguson R. C., Inman K. F., Knepp R. A., Lindberg F. A., and Ryberg P. T. (1983) Provenance of North American Phanerozoic sandstones in relation to tectonic setting. *Geol. Soc. Am. Bull.* **94**, 222–235.

Dickson J. A. D. (2002) Fossil echinoderms as monitor of the Mg/Ca ratio of Phanerozoic oceans. *Science* **298**, 1222–1224.

Dietz R. S. (1961) Continent and ocean basin evolution by spreading of the seafloor. *Nature* **190**, 854–857.

Droxler A. W., Schlager W., and Wallon C. C. (1983) Quaternary aragonite cycles and oxygen-isotopic records in Bahamian carbonate ooze. *Geology* **11**, 235–239.

Droxler A. W., Morse J. W., and Kornicker W. A. (1988) Controls on carbonate mineral accumulation in Bahamian basins and adjacent Atlantic ocean sediments. *J. Sedim. Petrol.* **58**, 120–130.

Engel A. E. J. (1963) Geologic evolution of North America. *Science* **140**, 143–152.

Engel A. E. J., Itson S. P., Engel C. G., Stickney D. M., and Cray E. J. (1974) Crustal evolution and global tectonics, a petrogenic view. *Bull. Geol. Soc. Am.* **85**, 843–858.

Faure G. (1986) *Principles of Isotope Geology*. Wiley, New York.

Fischer A. G. (1984) The two Phanerozoic super cycles. In *Catastrophes in Earth History* (eds. W. A. Berggren and J. A. Vancouvering). Princeton University Press, NJ, pp. 129–148.

Frakes L. A., Francis J. E., and Syktus J. I. (1992) *Climate Mode of the Phanerozoic: The History of the Earth's Climate over the Past 600 Million Years*. Cambridge University Press, Cambridge, UK.

Gaffin S. (1987) Ridge volume dependence of seafloor generation rate and inversion using long term sea level change. *Am. J. Sci.* **287**, 596–611.

Garrels R. M. and Mackenzie F. T. (1969) Sedimentary rock types: relative proportions as a function of geological time. *Science* **163**, 570–571.

Garrels R. M. and Mackenzie F. T. (1971a) *Evolution of Sedimentary Rocks*. Norton, New York.

Garrels R. M. and Mackenzie F. T. (1971b) Gregor's denudation of the continents. *Nature* **231**, 382–383.

Garrels R. M. and Mackenzie F. T. (1972) A quantitative model for the sedimentary rock cycle. *Mar. Chem.* **1**, 22–41.

Garrels R. M. and Perry E. A., Jr. (1974) Chemical history of the oceans deduced from postdepositional changes in sedimentary rocks. In *Studies in Paleo-Oceanography Special Publication Society Economic Paleontologists and Mineralogists*, **20**, (ed. W.W. Hay). SPEM, Tulsa, OK, pp. 193–204.

Gibbs A. K., Montgomery C. W., O'Day P. A., and Erslev E. A. (1988) Crustal evolution revisited: reply to comments by

S. M. McLennan et al., on "The Archean–Proterozoic transition: evidence from the geochemistry of metasedimentary rocks from Guyana and Montana". *Geochim. Cosmochim. Acta* **52**, 793–795.

Gilluly J. (1949) Distribution of mountain building in geologic time. *Geol. Soc. Am. Bull.* **60**, 561–590: **120**, 135–139.

Given R. K. and Wilkinson B. H. (1987) Dolomite abundance and stratigraphic age: constraints on rates and mechanisms of Phanerozoic dolostone formation. *J. Sedim. Petrol.* **57**, 1068–1079.

Goddéris Y. and Veizer J. (2000) Tectonic control of chemical and isotopic composition of ancient oceans: the impact of continental growth. *Am. J. Sci.* **300**, 434–461.

Goldschmidt V. M. (1933) Grundlagen der quantitativen Geochemie. *Fortschr. Mineral. Kristallog. Petrogr.* **17**, 1–112.

Goldstein S. J. and Jacobsen S. B. (1987) The Nd and Sr isotope systematics of river water dissolve material: implications for the source of Nd and Sr in seawater. *Chem. Geol.* **66**, 245–272.

Gregor C. B. (1968) The rate of denudation in post-Algonkian time. *Proc. Koninkl. Ned. Akad. Wetenschap. Ser. B: Phys. Sci.* **71**, 22–30.

Gregor C. B. (1985) The mass-age distribution of Phanerozoic sediments. In *Geochronology and the Geologic Record*, Mem. No. 10 (ed. N. J. Snelling). Geol. Soc. London, London, UK, pp. 284–289

Gregory R. T. (1991) Oxygen isotope history of seawater revisited: composition of seawater. In *Stable Isotope Geochemistry: a Tribute to Samuel Epstein*, Geochem. Soc. Spec. Publ. 3 (eds. H. P. Tailor, Jr., J. R. O'Neil, and I. R. Kaplan), Min. Soc. America, Washington, DC, pp. 65–76.

Grotzinger J. P. and James N. P. (2000) *Precambrian Carbonates, Evolution and Understanding: Carbonate Sedimentation and Diagenesis in the Evolving Precambrian World*. SEPM Special Publication # 67.

Habicht K. A. and Canfield D. E. (1996) Sulphur isotope fractionation in modern microbial mats and the evolution of the sulphur cycle. *Nature* **382**, 342–343.

Hallam A. (1984) Pre-Quaternary sea-level changes. *Ann. Rev. Earth Planet. Sci.* **12**, 205–243.

Hardie L. A. (1987) Perspectives on dolomitization: a critical view of some current views. *J. Sedim. Petrol.* **57**, 166–183.

Hardie L. A. (1996) Secular variation in seawater chemistry: an explanation of the coupled secular variations in the mineralogies of marine limestones and potash evaporites over the past 600 Myr. *Geology* **24**, 279–283.

Harrison A. G. and Thode H. G. (1958) Mechanism of the bacterial reduction of sulfate from isotope fractionation studies. *Faraday Soc. Trans.* **54**, 84–92.

Hay W. W., Sloan J. L. II, and Wold C. N. (1988) Mass/age distribution and composition of sediments on the ocean floor and the global rate of sediment subduction. *J. Geophys. Res.* **93**, 14933–14940.

Hay W. W., Wold C. N., Söding E., and Flügel S. (2001) Evolution of sediment fluxes and ocean salinity. In *Geologic Modelling and Simulations: Sedimentary Systems* (eds. D. F. Merriam and J. C. Davis). Kluwer Academic/Plenum, Dordrecht, pp. 153–167.

Hayes J. M., Des Marais D. J., Lambert J. B., Strauss H., and Summons R. E. (1992) Proterozoic biogeochemistry. In *The Proterozoic Biosphere: A Multidisciplinary Study* (eds. J. W. Schopf and C. Klein). Cambridge University Press, Cambridge, UK, pp. 81–543.

Hemming N. G. and Hanson G. N. (1992) Boron isotope composition and concentration in modern marine carbonates. *Geochim. Cosmochim. Acta* **56**, 537–543.

Hess H. H. (1962) History of ocean basins. In *Petrologic Studies* (eds. A. E. J. Engel, H. L. James, and B. F. Leonard). Geological Society of America, Bouldev, co., pp. 599–620.

Hoffman P. F., Kaufman A. J., Halverson G. P., and Schrag D. P. (1998) A Neoproterozoic snowball Earth. *Science* **281**, 1342–1346.

Holland H. D. (1978) *The Chemistry of the Atmosphere and Oceans*. Wiley, New York.

Holland H. D. (1984) *The Chemical Evolution of the Atmosphere and Oceans*. Princeton University Press, Princeton, NJ.

Holland H. D. and Zimmerman H. (2000) The dolomite problem revisited. *Int. Geol. Rev.* **2**, 481.

Holser W. T. (1977) Catastrophic chemical events in the history of the ocean. *Nature* **267**, 403–408.

Holser W. T. (1984) Gradual and abrupt shifts in ocean chemistry during Phanerozoic time. In *Patterns of Change in Earth Evolution* (eds. H. D. Holland and A. F. Trendall). Springer, Heidelberg, pp. 123–143.

Holser W. T. and Kaplan I. R. (1966) Isotope geochemistry of sedimentary sulfates. *Chem. Geol.* **1**, 93–135.

Holser W. T., Schidlowski M., McKenzie F. T., and Maynard J. B. (1988) Geochemical cycles of carbon and sulfur. In *Chemical Cycles in the Evolution of the Earth* (eds. C. B. Gregor, R. M. Garrels, F. T. Mackenzie, and J. B. Maynard). Wiley, New York, pp. 105–173.

Horita J., Friedman T. J., Lazar B., and Holland H. D. (1991) The composition of Permian seawater. *Geochim. Cosmochim. Acta* **55**, 417–432.

Horita J., Zimmermann H., and Holland H. D. (2002) The chemical evolution of seawater during the Phanerozoic: implications from the record of marine evaporites. *Geochim. Cosmochim. Acta* **66**, 3733–3756.

Ingersoll R. V. and Busby C. J. (1995) Tectonics of sedimentary basins. In *Tectonics of Sedimentary Basins* (eds. C. J. Busby and R. V. Ingersoll). Blackwell, Oxford, pp. 1–51.

Kampschulte A. (2001) Schwefelisotopenuntersuchungen an strukturell substituierten Sulfaten in marinen Karbonaten des Phanerozoikums—Implikationen für die geochemische Evolution des Meerwassers und Korrelation verschiedener Stoffkreisläufe. PhD Thesis, Ruhr Universität, Bochum.

Kampschulte A. and Strauss H. (1998) The isotopic composition of trace sulphates in Paleozoic biogenic carbonates: implications for coeval seawater and geochemical cycles. *Min. Mag.* **62A**, 744–745.

Karhu J. A. and Holland H. D. (1996) Carbon isotopes and the rise of atmospheric oxygen. *Geology* **24**, 867–870.

Kiessling W. (2002) Secular variations in the Phanerozoic reef ecosystem. *SEPM Spec. Publ.* **72**, 625–690.

Kroopnick P. (1980) The distribution of ^{13}C in the Atlantic ocean. *Earth Planet. Sci. Lett.* **49**, 469–484.

Kuenen Ph. H. (1950) *Marine Geology*. Wiley, New York.

Kump L. R. (1989) Alternative modeling approaches to the geochemical cycles of carbon, sulfur, and strontium isotopes. *Am. J. Sci.* **289**, 390–410.

Land L. S. (1979) The fate of reef-derived sediment on the north Jamaica island slope. *Mar. Geol.* **29**, 55–71.

Land L. S. (1980) The isotopic and trace element geochemistry of dolomite: the state of the art. *SEPM Spec. Publ.* **28**, 87–110.

Land L. S. (1995) Comment on "Oxygen and carbon isotopic composition of Ordovician brachiopods: implications for coeval seawater" by H. Qing and J. Veizer. *Geochim. Cosmochim. Acta* **59**, 2843–2844.

Lasaga A. C. and Ohmoto H. (2002) The oxygen geochemical cycle: dynamics and stability. *Geochem. Cosmochim. Acta* **66**, 361–381.

Lasaga A. C., Berner R. A., and Garrels R. M. (1985) An improved geochemical model of atmospheric CO_2 fluctuations over the past 100 million years. In *The Carbon Cycle and Atmospheric CO_2: Natural Variations Archean to Present*, Geophysical Monograph 32 (eds. E. T. Sundquist and W. S. Broecker). American Geophysical Union, Washington, DC, pp. 397–411.

Lear C. H., Elderfield H., and Wilson P. A. (2003) A Cenozoic seawater Sr/Ca record from benthic foraminiferal calcite and its application in determining global weathering fluxes. *Earth Planet. Sci. Lett.* **208**, 69–84.

Lemarchand D., Gaillardet J., Lewin E., and Allègre J. C. (2000) The influence of rivers on marine boron isotopes and implications for reconstructing past ocean pH. *Nature* **408**, 951–954.

Le Pichon X., Francheteau J., and Bonnin J. (1973) *Plate Tectonics*. Elsevier, Amsterdam.

Lerman A. (1979) *Geochemical Processes: Water and Sediment Environments*. Wiley, New York.

Longinelli A. (1989) Oxygen-18 and sulphur-34 in dissolved oceanic sulphate and phosphate. In *Handbook of Environmental Isotope Geochemistry* (eds. P. Fritz and J. C. Fontes). Elsevier, Amsterdam, pp. 219–255.

Lowenstein T. K., Timofeeft M. N., Brennan S. T., Hardie L. A., and Demicco R. V. (2001) Oscillations in Phanerozoic seawater chemistry: evidence from fluid inclusions. *Science* **294**, 1086–1088.

Lumsden D. N. (1985) Secular variations in dolomite abundance in deep marine sediments. *Geology* **13**, 766–769.

Mackenzie F. T. (1975) Sedimentary cycling and the evolution of seawater. In *Chemical Oceanography*, 2nd edn. (eds. J. P. Riley and G. Skirrow) Academic Press, London, vol. 1, pp. 309–364.

Mackenzie F. T. and Agegian C. (1986) Biomineralization, atmospheric CO_2 and the history of ocean chemistry. In *Proc. 5th Int. Conf. Biomineral. Department of Geology, University Texas, Arlington, 2*.

Mackenzie F. T. and Agegian C. (1989) Biomineralization and tentative links to plate tectonics. In *Origin, Evolution, and Modern Aspects of Biomineralization in Plants and Animals* (ed. R. E. Crick). Plenum, New York, pp. 11–28.

Mackenzie F. T. and Morse J. W. (1992) Sedimentary carbonates through Phanerozoic time. *Geochim. Cosmochim. Acta* **56**, 3281–3295.

Mackenzie F. T. and Pigott J. P. (1981) Tectonic controls of Phanerozoic sedimentary rock cycling. *J. Geol. Soc. London* **138**, 183–196.

McArthur J. M. (1994) Recent trends in strontium isotope stratigraphy. *Terra Nova* **6**, 331–358.

McCulloch M. T. and Bennett V. C. (1994) Progressive growth of the Earth's continental crust and depleted mantle: geochemical constraints. *Geochim. Cosmochim. Acta* **58**, 4717–4738.

McCulloch M. T. and Wasserburg G. J. (1978) Sm–Nd and Rb–Sr chronology of continental crust formation. *Science* **200**, 1003–1011.

McLennan S. M. and Murray R. W. (1999) Geochemistry of sediments. In *Encyclopedia of Geochemistry* (eds. C. P. Marshall and R. W. Fairbridge). Kluwer, Dordrecht, pp. 282–292.

McLennan S. M. and Taylor S. R. (1999) Earth's continental crust. In *Encyclopedia of Geochemistry* (eds. C. P. Marshall and R. W. Fairbridge). Kluwer, Dordrecht, pp. 145–151.

McLennan S. M., Taylor S. R., McCulloch M. T., and Maynard J. B. (1990) Geochemical and Nd–Sr isotopic composition of deep-sea turbidites: crustal evolution and plate tectonic associations. *Geochim. Cosmochim. Acta* **54**, 2015–2050.

McLennan S. M., Bock B., Hemming S. R., Hurowitz J. A., Lev S. M., and McDaniel D. K. (2003) The roles of provenance and sedimentary processes in the geochemistry of sedimentary rocks. In *Geochemistry of Sediments and Sedimentary Rocks: Evolutionary Considerations to Mineral Deposit-Forming Environments.* (ed. D. R. Lentz), Geol. Assoc. Canada GEOtext. St. John's, Nfld, vol. 5, pp. 1–31.

McLennan S. M., Taylor S. R., and Hemming S. R. Composition, differentiation and evolution of continental crust: constraints from sedimentary rocks and heat flow. In *Evolution and Differentiation of Continental Crust* (eds. M. Brown and T. Rushmer). Cambridge University Press (in press).

Milliken K. L. and Pigott J. D. (1977) Variation of oceanic Mg/Ca ratio through time-implications for the calcite sea. *Geol Soc. Am. South-Central Meet. (abstr.),* 64–65.

Milliman J. D. (1974) *Recent Sedimentary Carbonates: 1. Marine Carbonates*. Springer, Heidelberg.

Milliman J. D., Troy P. J., Balch W. M., Adams A. K., Li Y.-H., and Mackenzie F. T. (1999) Biologically mediated dissolution of calcium carbonate above the chemical lysocline? *Deep Res. I* **46**, 1653–1669.

Millot R., Gaillardet J., Dupre B., and Allègre J. C. (2002) The global control of silicate weathering rates and the coupling with physical erosion: new insights from rivers on the Canadian shield. *Earth Planet. Sci. Lett.* **196**, 83–98.

Morgan W. J. (1968) Rises, trenches, great faults, and crustal blocks. *J. Geophys. Res.* **73**, 1959–1982.

Morse J. W. and Mackenzie F. T. (1990) *Geochemistry of Sedimentary Carbonates*. Elsevier, Amsterdam.

Muehlenbachs K. (1998) The oxygen isotopic composition of the oceans, sediments and the seafloor. *Chem. Geol.* **145**, 263–273.

Nägler T. F., Eisenhauer A., Müller A., Hemleben C., and Kramers J. (2000) The δ^{44}Ca-temperature calibration on fossil and cultured *Globigerinoides sacculifer*: new tool for reconstruction of past sea surface temperatures. *Geochem. Geophys. Geosys.* **1 20009000091.**

Nanz R. H., , Jr. (1953) Chemical composition of Precambrian slates with notes on the geochemical evolution of lutites. *J. Geol.* **61**, 51–64.

Nesbitt H. W. and Young G. M. (1984) Predictions of some weathering trends of plutonic and volcanic rocks based on thermodynamic and kinetic considerations. *Geochim. Cosmochim. Acta* **48**, 1523–1534.

O'Leary M. H. (1988) Carbon isotopes in photosynthesis. *Bioscience* **38**, 328–336.

O'Nions R. K., Hamilton P. J., and Hooker P. J. (1983) A Nd isotope investigation of sediments related to crustal development in the British Isles. *Earth Planet. Sci. Lett.* **63**, 329–338.

Palmer M. R. and Edmond J. M. (1989) Strontium isotope budget of the modern ocean. *Earth Planet. Sci. Lett.* **92**, 11–26.

Palmer M. R. and Swihart G. H. (1996) Boron isotope geochemistry: an overview. *Rev. Mineral.* **33**, 709–744.

Patchett P. J. (2003) Provenance and crust-mantle evolution studies based on radiogenic isotopes in sedimentary rocks. In *Geochemistry of Sediments and Sedimentary Rocks: Evolutionary Considerations to Mineral Deposit—Forming Environments*, Geological Association of Canada GEOtext (ed. D. R. Lentz). St. John's, Nfld, vol 5, pp. 89–97.

Pearson P. N. and Palmer M. R. (2000) Atmospheric carbon dioxide concentrations over the past 60 million years. *Nature* **406**, 695–699.

Peterman Z. E., Hedge C. E., and Tourtelot H. A. (1970) Isotopic composition of strontium in seawater throughout Phanerozoic time. *Geochim. Cosmochim. Acta* **34**, 105–120.

Peucker-Ehrenbring B. and Ravizza G. (2000) The marine osmium isotope record. *Terra Nova* **12**, 205–219.

Pielou E. C. (1977) *Mathematical Ecology*. Wiley, New York.

Plank T. and Langmuir C. H. (1998) The chemical composition of subducting sediment and its consequences for the crust and mantle. *Chem. Geol.* **145**, 325–394.

Plummer L. N. (1975) Mixing of seawater with calcium carbonate ground water. *Geol. Soc. Am. Mem.* **142**, 219–236.

Rea D. K. and Ruff L. J. (1996) Composition and mass flux of sediment entering the world's subduction zones: implications for global sedimentary budgets, great earthquakes, and volcanism. *Earth Planet. Sci. Lett.* **140**, 1–12.

Renard M. (1986) Pelagic carbonate chemostratigraphy (Sr, Mg, ^{18}O, ^{13}C). *Mar. Micropaleontol.* **10**, 117–164.

Reymer A. and Schubert G. (1984) Phanerozoic additions to the continental crust and crustal growth. *Tectonics* **3**, 63–77.

Ronov A. B. (1949) A history of the sedimentation and epeirogenic movements of the European part of the USSR (based on the volumetric method). *AN SSSR Geofiz. Inst. Trudy* **3**, 1–390. (in Russian).

Ronov A. B. (1964) Common tendencies in the chemical evolution of the Earth's crust, ocean and atmosphere. *Geochem. Int.* **1**, 713–737.

Ronov A. B. (1968) Probable changes in the composition of seawater during the course of geological time. *Sedimentology* **10**, 25–43.

Ronov A. B. (1976) Global carbon geochemistry, volcanism, carbonate accumulation, and life. *Geokhimiya* **8**, 1252–1257; *Geochem. Int.* **13**, 175–196.

Ronov A. B. (1982) The Earth's sedimentary shell (quantitative patterns of its structure, compositions, and evolution). *Int. Geol. Rev.* **24**, 1313–1388.

Ronov A. B. (1993) *Stratisphere or Sedimentary Layer of the Earth*. Nauka, Russian.

Ross S. K. and Wilkinson B. M. (1991) Planktogenic/eustatic control on cratonic/oceanic carbonate accumulation. *J. Geol.* **99**, 497–513.

Rothman D. H. (2002) Atmospheric carbon dioxide levels for the last 500 million years. *Proc. Natl. Acad. Sci.* **99**, 4167–4171.

Rothman D. H., Hayes J. M., and Summons R. (2003) Dynamics of the Neoproterozoic carbon cycle. *Proc. Natl. Acad. Sci* **100**, 8124–8129.

Rowley D. B. (2002) Rate of plate creation and destruction: 180 Ma to present. *Geol. Soc. Am. Bull.* **114**, 927–933.

Rubey W. W. (1951) Geologic history of seawater: an attempt to state the problem. *Geol. Soc. Am. Bull.* **62**, 1111–1148.

Ruddiman W. F. (1997) *Tectonic Uplift and Climate Change*. Plenum, New York.

Sabine C. and Mackenzie F. T. (1975) Bank-derived carbonate sediment transport and dissolution in the Hawaiian Archipelago. *Aquat. Geochem.* **1**, 189–230.

Sackett W. M. (1989) Stable carbon isotope studies on organic matter in the marine environment. In *Handbook of Environmental Isotope Geochemistry* (eds. P. Fritz and J. C. Fontes). Elsevier, Heidelberg, vol. 3, pp. 139–169.

Sadler P. M. (1981) Sediment accumulation rates and the completeness of stratigraphic sections. *J. Geol.* **89**, 569–584.

Sandberg P. A. (1975) New interpretation of Great Salt Lake oöids and of ancient non-skeletal carbonate mineralogy. *Sedimentology* **22**, 497–538.

Sandberg P. A. (1983) An oscillating trend in non-skeletal carbonate mineralogy. *Nature* **305**, 19–22.

Sandberg P. A. (1985) Nonskeletal aragonite and pCO_2 in the Phanerozoic and Proterozoic. In *The Carbon Cycle and Atmospheric CO_2: Natural Variations Archean to Present*, Geophys. Monogr. Ser. 32 (eds. E. T. Sundquist and W. S. Broecker). American Geophysical Union, Washington, DC, pp. 585–594.

Schidlowski M., Hayes J. M., and Kaplan I. R. (1983) Isotopic inferences of ancient biochemistries: carbon, sulfur, hydrogen, and nitrogen. In *Earth's Earliest Biosphere* (ed. J. W. Schopf). Princeton University Press, Princeton, NJ, pp. 149–186.

Schmitt A.-D., Bracke G., Stille P., and Kiefel B. (2001) The calcium isotope composition of modern seawater determined by thermal ionisation mass spectrometry. *Geostand. Newslett.* **25**, 267–275.

Schuchert C. (1931) Geochronology of the age of the Earth on the basis of sediments and life. *Natl. Res. Council Bull.* **80**, 10–64.

Sclater J. G., Parsons B., and Jaupart C. (1981) Oceans and continents: similarities and differences in the mechanism of heat loss. *J. Geophys. Res.* **86**, 11535–11552.

Sepkoski J. J., , Jr. (1989) Periodicity in extinction and the problem of catastrophism in the history of life. *J. Geol. Soc. London* **146**, 7–19.

Shaviv N. J. and Veizer J. (2003) Celestial driver of Phanerozoic climate?. *GSA Today* **13**(7), 4–10.

Shaw D. B. and Weaver C. E. (1965) The mineralogical composition of shales. *J. Sedim. Petrol.* **35**, 213–222.

Shields G. and Veizer J. (2002) The Precambrian marine carbonate isotope database: version 1. *Geochem. Geophys. Geosys.* **3**(6), June 6, 2002, p. 12 (http://g-cubed.org/gc2002/2001GC000266).

Skulan J., De Paolo D. J., and Owens T. L. (1997) Biological control of calcium isotopic abundences in the global calcium cycle. *Geochim. Cosmochim. Acta* **61**, 2505–2510.

Smith S. V. (1978) Coral reef area and contributions of reefs to processes and resources of the world's oceans. *Nature* **273**, 225–226.

Sorby H. C. (1879) The structure and origin of limestones. *Proc. Geol. Soc. London* **35**, 56–95.

Southam J. R. and Hay W. W. (1981) Global sedimentary mass balance and sea level changes. In *The Oceanic Lithosphere*, The Sea (ed. C. Emiliani). Wiley, New York, vol. 7, pp. 1617–1684.

Sperber C. M., Wilkinson B. H., and Peacor D. R. (1984) Rock composition, dolomite stoichiometry, and rock/water reactions in dolomitic carbonate rocks. *J. Geol.* **92**, 609–622.

Sprague D. and Pollack H. N. (1980) Heat flow in the Mesozoic and Cenozoic. *Nature* **285**, 393–395.

Stanley S. M. and Hardie L. A. (1998) Secular oscillations in the carbonate mineralogy of reef-building and sediment-producing organisms driven by tectonically forced shifts in seawater chemistry. *Palaeogeogr. Palaeoclimat. Palaeoecol.* **144**, 3–19.

Steuber T. and Veizer J. (2002) A Phanerozoic record of plate tectonic control of seawater chemistry and carbonate sedimentation. *Geology* **30**, 1123–1126.

Strakhov N. M. (1964) States and development of the external geosphere and formation of sedimentary rocks in the history of the Earth. *Int. Geol. Rev.* **6**, 1466–1482.

Strakhov N. M. (1969) *Principles of Lithogenesis*. Oliver and Boyd, Edinburgh, vol. 2.

Strauss H. (1993) The sulfur isotopic record of Precambrian sulfates: new data and a critical evaluation of the existing record. *Precamb. Res.* **63**, 225–246.

Sylvester P. J. (2000) Continental formation, growth and recycling. *Tectonophysics* **322**, 163–190.

Sylvester P. J., Campbell I. H., and Bowyer D. A. (1997) Niobium/uranium evidence for early formation of the continental crust. *Science* **275**, 521–523.

Tan F. C. (1988) Stable carbon isotopes in dissolved inorganic carbon in marine and estuarine environments. In *Handbook of Environmental Isotope Geochemistry* (eds. P. Fritz and J. C. Fontes). Elsevier, Amsterdam, vol. 3, pp. 171–190.

Taylor S. R. (1979) Chemical composition and evolution of continental crust: the rare earth element evidence. In *The Earth: Its Origin, Structure and Evolution* (ed. M. W. McElhinny). Academic Press, New York, pp. 353–376.

Taylor S. R. and McLennan S. M. (1985) *The Continental Crust: its Composition and Evolution*. Blackwell, Oxford, UK.

Vail P. R., Mitchum R. W., and Thompson S. (1977) Seismic stratigraphy and global changes of sea level. 4, Global cycles of relative changes of sea level. *AAPG Mem.* **26**, 83–97.

Van Houten F. B. and Bhattacharyya D. P. (1982) Phanerozoic oölitic ironstones-geologic record and facies. *Ann. Rev. Earth Planet. Sci.* **10**, 441–458.

Veizer J. (1973) Sedimentation in geologic history: recycling versus evolution or recycling with evolution. *Contrib. Mineral. Petrol.* **38**, 261–278.

Veizer J. (1983) Trace element and isotopes in sedimentary carbonates. *Rev. Mineral.* **11**, 265–300.

Veizer J. (1985) Carbonates and ancient oceans: isotopic and chemical record on timescales of 10^7–10^9 years. *Geophys. Monogr. Am. Geophys. Union* **32**, 595–601.

Veizer J. (1988a) The evolving exogenic cycle. In *Chemical Cycles in the Evolution of the Earth* (eds. C. B. Gregor R. M. Garrels, F. T. Mackenzie, and J. B. Maynard). Wiley, New York, pp. 175–220.

Veizer J. (1988b) Continental growth: comments on "The Archean–Proterozoic transition: evidence from Guyana and Montana" by A. K. Gibbs C. W. Montgomery P. A. O'Day and E. A. Erslev. *Geochim. Cosmochim. Acta* **52**, 789–792.

Veizer J. (1988c) Solid Earth as a recycling system: temporal dimensions of global tectonics. In *Physical and Chemical*

Weathering in Geochemical Cycle (eds. A. Lerman and M. Meyback). Reidel, Dordrecht, pp. 357–372.

Veizer J. (1995) Reply to the comment by L. S. Land on "Oxygen and carbon isotopic composition of Ordovician brachiopods: implications for coeval seawater: discussion". *Geochim. Cosmochim. Acta* **59**, 2845–2846.

Veizer J. (2003) Isotopic evolution of seawater on geological timescales: sedimentological perspective. In *Geochemistry of Sedimentary Rocks: Secular Evolutionary Considerations to Mineral Deposit-forming Environments*, Geological Association of Canda, GEOtext (ed. D. R. Lentz). St. John's, Nfld, vol. 5, pp. 99–114.

Veizer J. and Compston W. (1974) $^{87}Sr/^{86}Sr$ composition of seawater during the Phanerozoic. *Geochim. Cosmochim. Acta* **38**, 1461–1484.

Veizer J. and Compston W. (1976) $^{87}Sr/^{86}Sr$ in Precambrian carbonates as an index of crustal evolution. *Geochim. Cosmochim. Acta* **40**, 905–914.

Veizer J. and Ernst R. E. (1996) Temporal pattern of sedimentation: Phanerozoic of North America. *Geochem. Int.* **33**, 64–76.

Veizer J. and Jansen S. L. (1979) Basement and sedimentary recycling and continental evolution. *J. Geol.* **87**, 341–370.

Veizer J. and Jansen S. L. (1985) Basement and sedimentary recycling: 2. Time dimension to global tectonics. *J. Geol.* **93**, 625–643.

Veizer J., Holser W. T., and Wilgus C. K. (1980) Correlation of $^{13}C/^{12}C$ and $^{34}S/^{32}S$ secular variations. *Geochim. Cosmochim. Acta* **44**, 579–587.

Veizer J., Compston W., Hoefs J., and Nielsen H. (1982) Mantle buffering of the early oceans. *Naturwissenschaften* **69**, 173–180.

Veizer J., Laznicka P., and Jansen S. L. (1989) Mineralization through geologic time: recycling perspective. *Am. J. Sci.* **289**, 484–524.

Veizer J., Ala D., Azmy K., Bruckschen P., Buhl D., Bruhn F., Carden G. A. F., Diener A., Ebneth S., Goddéris Y., Jasper T., Korte C., Pawellek F., Podlaha O. G., and Strauss H. (1999) $^{87}Sr/^{86}Sr$, $\delta^{13}C$ and $\delta^{18}O$ evolution of Phanerozoic seawater. *Chem. Geol.* **161**, 59–88.

Veizer J., Goddéris Y., and François L. M. (2000) Evidence for decoupling of atmospheric CO_2 and global climate during the Phanerozoic eon. *Nature* **408**, 698–701.

Vervoort J. D., Patchett P. J., Blichert-Toft J., and Albarede F. (1999) Relationship between Lu–Hf and Sm–Nd isotopic systems in the global sedimentary system. *Earth Planet. Sci. Lett.* **168**, 79–99.

Vinogradov A. P. (1967) The formation of the oceans. *Izv. Akad. Nauk SSSR, Ser. Geol.* **4**, 3–9.

Vinogradov A. P. and Ronov A. B. (1956a) Composition of the sedimentary rocks of the Russian platform in relation to the history of its tectonic movements. *Geochemistry* **6**, 533–559.

Vinogradov A. P. and Ronov A. B. (1956b) Evolution of the chemical composition of clays of the Russian platform. *Geochemistry* **2**, 123–129.

Vinogradov A. P., Ronov A. B., and Ratynskii V. Y. (1952) Evolution of the chemical composition of carbonate rocks. *Izv. Akad. Nauk SSSR Ser. Geol.* **1**, 33–60.

Wadleigh M. A., Veizer J., and Brooks C. (1985) Strontium and its isotopes in Canadian rivers: fluxes and global implications. *Geochim. Cosmochim. Acta* **49**, 1727–1736.

Walker L. J., Wilkinson B. H., and Ivany L. C. (2002) Continental drift and Phanerozoic carbonate accumulation in shallow-shelf and deep-marine settings. *J. Geol.* **110**, 75–87.

Wallmann K. (2001) The geological water cycle and the evolution of marine $\delta^{18}O$. *Geochim. Cosmochim. Acta* **65**, 2469–2485.

White W. M. and Patchett P. J. (1982) Hf–Nd–Sr and incompatible-element abundances in island arcs: implications for magma origin and crust-mantle evolution. *Earth Planet. Sci. Lett.* **67**, 167–185.

Wilkinson B. H. (1979) Biomineralization, paleoceanography, and the evolution of calcareous marine organisms. *Geology* **7**, 524–527.

Wilkinson B. H. and Algeo T. J. (1989) Sedimentary carbonate record of calcium–magnesium cycling. *Am. J. Sci.* **289**, 1158–1194.

Wilkinson B. H. and Given R. K. (1986) Secular variation in abiotic marine carbonates: constraints on Phanerozoic atmospheric carbon dioxide contents and oceanic Mg/Ca ratios. *J. Geol.* **94**, 321–334.

Wilkinson B. H., Owen R. M., and Carroll A. R. (1985) Submarine hydrothermal weathering, global eustasy and carbonate polymorphism in Phanerozoic marine oolites. *J. Sedim. Petrol.* **55**, 171–183.

Windley B. F. (1984) *The Evolving Continents*, 2nd edn. Wiley, New York.

Wolery T. J. and Sleep N. H. (1988) Interaction of the geochemical cycles with the mantle. In *Chemical Cycles in the Evolution of the Earth* (eds. C. B. Gregor R. M. Garrels, F. T. Mackenzie, and J. B. Maynard). Wiley, New York, pp. 77–104.

Wollast R. (1993) The relative importance of biomineralization and dissolution of $CaCO_3$ in the global carbon cycle. In *Past and Present Biomineralization Processes*. Bulletin de l'Institut Oceanographie, Monaco No. 13 (ed. Francoise Doumenge), pp. 13–35.

Yaalon D. H. (1962) Mineral composition of average shale. *Clay Mineral Bull.* **5**, 31–36.

Zenger D. H. (1989) Dolomite abundance and stratigraphic age: constraints on rates and mechanisms of Phanerozoic dolostone formation. *J. Sedim. Petrol.* **59**, 162–164.

Zhu P. (1999) *Calcium Isotopes in the Marine Environment*. PhD Thesis, University of California, San Diego.

Zhu P. and MacDougall J. D. (1998) Calcium isotopes in the marine environment and the oceanic calcium cycle. *Geochim. Cosmochim. Acta* **62**, 1691–1698.

Readings from the Treatise on Geochemistry
ISBN: 978-0-12-381391-6
pp. 467–506

15 Biogeochemistry of Primary Production in the Sea

P. G. Falkowski
Rutgers University, New Brunswick, NJ, USA

As the present condition of nations is the result of many antecedent changes, some extremely remote and others recent, some gradual, others sudden and violent, so the state of the natural world is the result of a long succession of events, and if we would enlarge our experience of the present economy of nature, we must investigate the effects of her operations in former epochs.

Charles Lyell,
Principles of Geology, 1830

15.1 INTRODUCTION

Earth is the only planet in our solar system that contains vast amounts of liquid water on its surface and high concentrations of free molecular oxygen in its atmosphere. These two features are not coincidental. All of the original oxygen on Earth arose from the photobiologically catalyzed splitting of water by unicellular photosynthetic organisms that have inhabited the oceans for at least 3 Gyr. Over that period, these organisms have used the hydrogen atoms from water and other substrates to form organic matter from CO_2 and its hydrated equivalents. This process, the *de novo* formation of organic matter from inorganic carbon, or primary production, is the basis for all life on Earth. In this chapter, we examine the evolution and biogeochemical consequences of primary production in the sea and its relationship to other biogeochemical cycles on Earth.

15.1.1 The Two Carbon Cycles

There are two major carbon cycles on Earth. The two cycles operate in parallel. One cycle is slow and abiotic. Its effects are observed on multi-million-year timescales and are dictated by tectonics and weathering (Berner, 1990). In this cycle, CO_2 is released from the mantle to the atmosphere and oceans via vulcanism and seafloor spreading, and removed from the atmosphere and ocean primarily by reaction with silicates to form carbonates in the latter reservoir. Most of the carbonates are subsequently subducted into the mantle, where they are heated, and their carbon is released as CO_2 to the atmosphere and ocean, to carry out the cycle again. The chemistry of this cycle is dependent on acid–base reactions, and would operate whether or not there was life on the planet (Kasting *et al.*, 1988). This slow carbon cycle is a critical determinate of the concentration of CO_2 in Earth's atmosphere and oceans on timescales of tens and hundreds of millions of years (Kasting, 1993).

The second carbon cycle is dependent on the biologically catalyzed reduction of inorganic carbon to form organic matter, the overwhelming majority of which is oxidized back to inorganic carbon by respiratory metabolism (Schlesinger, 1997). This cycle, which is observable on timescales of days to millenia, is driven by reduction—oxidation (redox) reactions that evolved over ~2 Gyr, first in microbes, and subsequently in multicellular organisms (Falkowski *et al.*, 1998). A very small fraction of the reduced carbon escapes respiration and becomes incorporated into the lithosphere. In so doing, some of the organic matter is transferred to the slow carbon cycle. In this chapter, we will focus primarily on this fast, biologically mediated carbon cycle in the sea, and the supporting biogeochemical processes and feedbacks.

15.1.2 A Primer on Redox Chemistry

The biologically mediated redox reactions cycle carbon through three mobile pools: the atmosphere, the ocean, and the biosphere. Of these, the ocean is by far the largest (Table 1); however, more than 98% of this carbon is found in its oxidized state as CO_2 and its hydrated equivalents, HCO_3^- and CO_3^{2-}. To form organic molecules, the inorganic carbon must be chemically reduced, a process that requires the addition of hydrogen atoms (not just protons, but protons plus electrons) to the carbon atoms. Broadly speaking, these biologically catalyzed reduction reactions are carried out by two groups of organisms, chemoautotrophs and photoautotrophs, which are collectively called primary

Table 1 Carbon pools in the major reservoirs on Earth.

Pools	*Quantity* ($\times 10^{15}$ g)
Atmosphere	720
Oceans	38,400
Total inorganic	37,400
Surface layer	670
Deep layer	36,730
Dissolved organic	600
Lithosphere	
Sedimentary carbonates	>60,000,000
Kerogens	15,000,000
Terrestrial biosphere (total)	2,000
Living biomass	600–1,000
Dead biomass	1,200
Aquatic biosphere	1–2
Fossil fuels	4,130
Coal	3,510
Oil	230
Gas	140
Other (peat)	250

producers. The organic carbon they synthesize fuels the growth and respiratory demands of the primary producers themselves and all remaining organisms in the ecosystem.

All redox reactions are coupled sequences. Reduction is accomplished by the addition of an electron or hydrogen atom to an atom or molecule. In the process of donating an electron to an acceptor, the donor molecule is oxidized. Hence, redox reactions require pairs of substrates, and can be described by a pair of partial reactions, or half-cells:

$$A_{ox} + n(e^-) \leftrightarrow A_{re} \quad (1a)$$

$$B_{red} - n(e^-) \leftrightarrow B_{ox} \quad (1b)$$

The tendency for a molecule to accept or release an electron is therefore "relative" to some other molecule being capable of conversely releasing or binding an electron. Chemists scale this tendency, called the redox potential, E, relative to the reaction

$$H_2 \leftrightarrow 2H^+ + 2e^- \quad (2)$$

which is arbitrarily assigned an E of 0 at pH 0, and is designated E_0. Biologists define the redox potential at pH 7, 298 K (i.e., room temperature) and 1 atm pressure (=101.3 kPa). When so defined, the redox potential is denoted by the symbols E_0' or sometimes E_{m_7}. The E_0' for a standard hydrogen electrode is −420 mV.

15.2 CHEMOAUTOTROPHY

Organisms capable of reducing sufficient inorganic carbon to grow and reproduce in the dark without an external organic carbon source are called chemoautotrophs (literally, "chemical self-feeders"). Genetic analyses suggest that chemoautotrophy evolved very early in Earth's history, and is carried out exclusively by prokaryotic organisms in both the Archea and Bacteria superkingdoms (Figure 1).

Early in Earth's history, the biological reduction of inorganic carbon may have been directly coupled to the oxidation of H_2. At present, however, free H_2 is scarce on the planet's surface. Rather, most of the hydrogen on the surface of Earth is combined with other atoms, such as sulfur or oxygen. Activation energy is required to break these bonds in order to extract the hydrogen. One source of energy is chemical bond energy itself. For example, the ventilation of reduced mantle gases along tectonic plate subduction zones on the seafloor provides hydrogen in the form of H_2S. Several types of microbes can couple the oxidation of H_2S to the reduction of inorganic carbon, thereby forming organic matter in the absence of light.

Ultimately all chemoautotrophs depend on a nonequilibrium redox gradient, without which there is no thermodynamic driver for carbon fixation. For example, the reaction involving the oxidation of H_2S by microbes in deep-sea vents described above is ultimately coupled to oxygen in

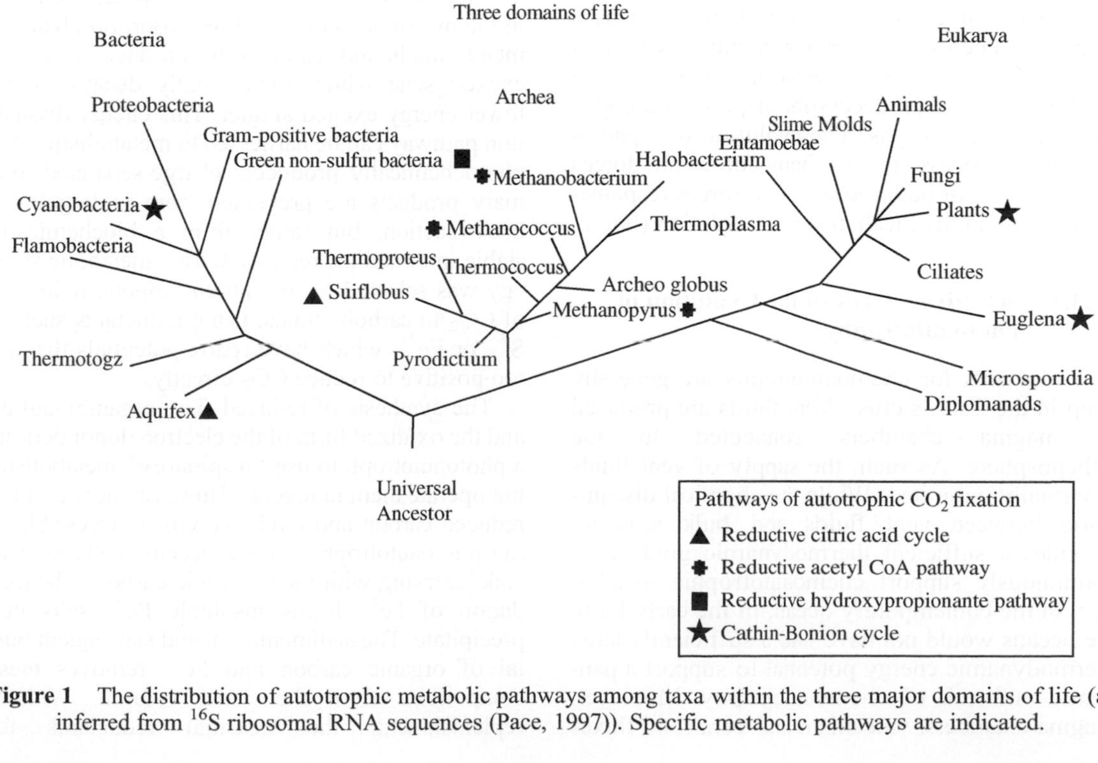

Figure 1 The distribution of autotrophic metabolic pathways among taxa within the three major domains of life (as inferred from [16]S ribosomal RNA sequences (Pace, 1997)). Specific metabolic pathways are indicated.

the ocean interior. Hence, this reaction is dependent on the chemical redox gradient between the ventilating mantle plume and the ocean interior that thermodynamically favors oxidation of the plume gases. Maintaining such a gradient requires a supply of energy, either externally, from radiation (solar or otherwise), or internally, via planetary heat and tectonics, or both.

The overall contribution of chemoautotrophy in the contemporary ocean to the formation of organic matter is relatively small, accounting for <1% of the total annual primary production in the sea. However, this process is critical in coupling reduction of carbon to the oxidation of low-energy substrates, and is essential for completion of several biogeochemical cycles.

15.3 PHOTOAUTOTROPHY

The oxidation state of the ocean interior is a consequence of a second energy source: light, which drives photosynthesis. Photosynthesis is a redox reaction of the general form:

$$2H_2A + CO_2 + \text{light} \rightarrow (CH_2O) + H_2O + 2A \quad (3)$$

where A is an atom, e.g., S. In this formulation, light is specified as a substrate, and a fraction of the light energy is stored as chemical bond energy in the organic matter. Organisms capable of reducing inorganic carbon to organic matter by using light energy to derive the source of reductant or energy are called photoautotrophs. Analyses of genes and metabolic sequences strongly suggest that the machinery for capturing and utilizing light as a source of energy to extract reductants was built on the foundation of chemoautotrophic carbon fixation; i.e., the predecessors of photoautotrophs were chemoautotrophs. The evolution of a photosynthetic process in a chemoautotroph forces consideration of both the selective forces responsible (why) and the mechanism of evolution (how).

15.3.1 Selective Forces in the Evolution of Photoautotrophy

Reductants for chemoautotrophs are generally deep in the Earth's crust. Vent fluids are produced in magma chambers connected to the Athenosphere. As such, the supply of vent fluids is virtually unlimited. While the chemical disequilibria between vent fluids and bulk seawater provides a sufficient thermodynamic gradient to continuously support chemoautotrophic metabolism in the contemporary ocean, in the early Earth the oceans would not have had a sufficiently large thermodynamic energy potential to support a pandemic outbreak of chemoautotrophy. Moreover, magma chambers, vulcanism and vent fluid fluxes are tied to tectonic subduction and spreading regions, which are transient features of Earth's crust and hence only temporary habitats for chemoautotrophs. In the Archean and early Proterozoic oceans, the chemoautotrophs would have already been dispersed throughout the oceans by physical mixing and helping to colonize new vent regions. This same dispersion process would have also helped ancestral chemoautotrophs exploit solar energy near the ocean surface.

Although the processes that selected the photosynthetic reactions as the major energy transduction pathway remain obscure, central hypotheses have emerged based on our understanding of the evolution of Earth's carbon cycle, the evolution of photosynthesis, biophysics, and molecular phylogeny. Photoautotrophs are found in all three major superkingdoms (Figure 1); however, there are very few known Archea capable of this form of metabolism. Efficient photosynthesis requires harvesting solar radiation, and hence the evolution of a light harvesting system. While some Archea and Bacteria use the pigment-protein rhodopsin, by far, the most efficient and ubiquitous light harvesting systems are based on chlorins. The metabolic pathway for the synthesis of porphyrins and chlorins is one of the oldest in biological evolution, and is found in all chemoautotrophs (Xiong *et al.*, 2000). Mulkidjanian and Junge (1997) proposed that the chlorin-based photosynthetic energy conversion apparatus originally arose from the need to prevent UV radiation from damaging essential macromolecules such as nucleic acids and proteins. The UV excitation energy could be transferred from the aromatic amino acid residues in the macromolecule to a blue absorption band of membrane-bound chlorins to produce a second excited state which subsequently decays to the lower-energy excited singlet. This energy dissipation pathway can be harnessed to metabolism if the photochemically produced, charge-separated, primary products are prevented from undergoing a back-reaction, but rather form a biochemically stable intermediate reductant. This metabolic strategy was selected for the photosynthetic reduction of CO_2 to carbohydrates, using reductants such as S^{2-} or Fe^{2+}, which have redox potentials that are too positive to reduce CO_2 directly.

The synthesis of reduced (i.e., organic) carbon and the oxidized form of the electron donor permits a photoautotroph to use "respiratory" metabolism, but operate them in reverse. However, not all of the reduced carbon and oxidants remain accessible to the photoautotrophs. In the oceans, cells tend to sink, carrying with them organic carbon. The oxidation of Fe^{2+} forms insoluble Fe^{3+} salts that precipitate. The sedimentation and subsequent burial of organic carbon and Fe^{3+} removes these components from the water column. Without replenishment, the essential reductants for anoxygenic photosynthesis would eventually become depleted in the surface waters. Thus, the necessity to regenerate reductants potentially prevented anoxygenic photoautotrophs from providing the major source of fixed carbon on Earth for eternity. Major net accumulation of reduced organic carbon in Proterozoic sediments implies local depletion of reductants such as S^{2-} and Fe^{2+} from the euphotic zone of the ocean. These limitations almost certainly provided the evolutionary selection pressure for an alternative electron donor.

15.3.2 Selective Pressure in the Evolution of Oxygenic Photosynthesis

H_2O is a potentially useful biological reductant with a vast supply on Earth relative to any redox-active solute dissolved in it. Liquid water contains ~100 kmol of H atoms per m^3, and, given $>10^{18}$ m^3 of water in the hydrosphere and cryosphere, $>10^{20}$ kmol of reductant are potentially accessible. Use of H_2O as a reductant for CO_2, however, requires a larger energy input than does the use of Fe^{2+} or S^{2-}. Indeed, to split water by light energy requires 0.82 eV at pH 7 and 298 K. Utilizing light at such high energy levels required the evolution of a new photosynthetic pigment, chlorophyll *a*, which has a red (lowest singlet) absorption band that is 200–300 nm blue shifted relative to bacteriochlorophylls. Moreover, stabilization of the primary electron acceptor to prevent a back-reaction necessitates thermodynamic inefficiency that ultimately requires two light-driven reactions operating in series. This sequential action of two photochemical reactions is unique to oxygenic photoautotrophs and presumably involved horizontal gene transfer through one or more symbiotic events (Blankenship, 1992).

In all oxygenic photoautotrophs, Equation (3) can be modified to:

$$2H_2O + CO_2 + \text{light} \overset{\text{Chl } a}{\rightarrow} (CH_2O) + H_2O + O_2 \quad (4)$$

where Chl *a* is the pigment chlorophyll *a* exclusively utilized in the reaction. Equation (4) implies that somehow chlorophyll *a* catalyzes a reaction or a series of reactions whereby light energy is used to oxidize water:

$$2H_2O + \text{light} \overset{\text{Chl } a}{\rightarrow} 4H^+ + 4e + O_2 \quad (5)$$

yielding gaseous, molecular oxygen. Hidden within Equation (5) are complex suites of biological innovations that have heretofore not been successfully mimicked *in vitro* by humans. At the core of the water splitting complex is a quartet of manganese atoms, that sequentially extract electrons, one at a time, from $2H_2O$ molecules, releasing gaseous O_2 to the environment, and storing the reductants on biochemical intermediates.

The photochemically produced reductants generated by the reactions schematically outlined in Equation (5) are subsequently used in the fixation (fixation is an archaic term meaning to make non-volatile, as in the chemical conversion of a gas to a solid phase) of CO_2 by a suite of enzymes that can operate *in vitro* in darkness and, hence, the ensemble of these reactions are called the dark reactions. At pH 7 and 25 °C, the formation of glucose from CO_2 requires an investment of 915 cal mol^{-1}. If water is the source of reductant, the overall efficiency for photosynthetic reduction of CO_2 to glucose is ~30%; i.e., 30% of the absorbed solar radiation is stored in the chemical bonds of glucose molecules.

15.4 PRIMARY PRODUCTIVITY BY PHOTOAUTOTROPHS

When we subtract the costs of all other metabolic processes by the chemoautotrophs and photoautotrophs, the organic carbon that remains is available for the growth and metabolic costs of heterotrophs. This remaining carbon is called *net* primary production (NPP) (Lindeman, 1942). From biogeochemical and ecological perspectives, NPP provides an upper bound for all other metabolic demands in an ecosystem. If NPP is greater than all respiratory consumption of the ecosystem, the ecosystem is said to be net autotrophic. Conversely, if NPP is less than all respiratory consumption, the system must either import organic matter from outside its bounds, or it will slowly run down—it is net heterotrophic.

It should be noted that NPP and photosynthesis are not synonymous. On a planetary scale, the former includes chemoautrophy, the latter does not. Moreover, photosynthesis *per se* does not include the integrated respiratory term for the photoautotrophs themselves (Williams, 1993). In reality, that term is extremely difficult to measure directly, hence NPP is generally approximated from measurements of photosynthetic rates integrated over some appropriate length of time (a day, month, season, or a year) and respiratory costs are either assumed or neglected.

15.4.1 What are Photoautotrophs?

In the oceans, oxygenic photoautotrophs are a taxonomically diverse group of mostly single-celled, photosynthetic organisms that drift with currents. In the contemporary ocean, these organisms, called phytoplankton (derived from Greek, meaning to wander), are comprised of $\sim 2 \times 10^4$ species distributed among at least eight taxonomic divisions or phyla (Table 2). By comparison, higher plants are comprised of $>2.5 \times 10^5$ species, almost all of which are contained within one class in one division. Thus, unlike terrestrial plants, phytoplankton are represented by relatively few species but they are phylogenetically diverse.

Table 2 The taxonomic classification and species abundances of oxygenic photosynthetic organisms in aquatic and terrestrial ecosystems. Note that terrestrial ecosystems are dominated by relatively few taxa that are species rich, while aquatic ecosystems contain many taxa but are relatively species poor

Taxonomic group	*Known species*	*Marine*	*Freshwater*
Empire: Bacteria (=Prokaryota)			
Kingdom: Eubacteria			
Subdivision: Cyanobacteria (*sensu strictu*) (=Cyanophytes, blue-green algae)	1,500	150	1,350
Subdivision: Chloroxybacteria (=Prochlorophyceae)	3	2	1
Empire: Eukaryota			
Kingdom: Protozoa			
Division: Euglenophyta	1,050	30	1,020
Class: Euglenophyceae			
Division: Dinophyta (Dinoflagellates)			
Class: Dinophyceae	2,000	1,800	200
Kingdom: Plantae			
Subkingdom: Biliphyta			
Division: Glaucocystophyta			
Class: Glaucocystophyceae	13		
Division: Rhodophyta			
Class: Rhodophyceae	6,000	5,880	120
Subkingdom: Viridiplantae			
Division: Chlorophyta			
Class: Chlorophyta	2,500	100	2,400
Prasinophyceae	120	100	20
Ulvophyceae	1,100	1,000	100
Charophyceae	12,500	100	12,400
Division: Bryophyta (mosses, liverworts)	22,000		1,000
Division: Lycopsida	1,228		70
Division: Filicopsida (ferns)	8,400		94
Division: Magnoliophyta (flowering plants)	(240,000)		
Subdivision: Monocotyledoneae	52,000	55	455
Subdivision: Dicotyledoneae	188,000		391
Kingdom: Chromista			
Subkingdom: Chlorechnia			
Division: Chlorarachniophyta			
Class: Chlorarachniophyceae	3–4	3–4	0
Subkingdom: Euchromista			
Division: Crytophyta			
Class: Crytophyceae	200	100	100
Division: Haptophyta			
Class: Prymensiophyceae	500	100	400
Division: Heterokonta			
Class: Bacillariophyceae (diatoms)	10,000	5,000	5,000
Chrysophyceae	1,000	800	200
Eustigmatophyceae	12	6	6
Fucophyceae (brown algae)	1,500	1,497	3
Raphidophyceae	27	10	17
Synurophyceae	250		250
Tribophyceae (Xanthophyceae)	600	50	500
Kingdom: Fungi			
Division: Ascomycontina (lichens)	13,000	15	20

Source: Falkowski (1997).

This deep taxonomic diversity is reflected in their evolutionary history and ecological function (Falkowski, 1997).

Within this diverse group of organisms, three basic evolutionary lineages are discernable (Delwiche, 2000). The first contains all prokaryotic oxygenic phytoplankton, which belong to one class of bacteria, namely, the cyanobacteria. Cyanobacteria are the only known oxygenic photoautotrophs that existed prior to ~2.5 Gyr BP (Ga) (Lipps, 1993; Summons *et al.*, 1999). These prokaryotes numerically dominate the photoautotrophic community in contemporary marine ecosystems and their continued success bespeaks an extraordinary adaptive capacity. At any moment in time, there are $\sim 10^{24}$ cyanobacterial cells in the contemporary oceans. To put that into perspective, the number of cyanobacterial cells in the oceans is two orders of magnitude more than all the stars in the sky.

The evolutionary history of cyanobacteria is obscure. The first microfossils assigned to this group were identified in cherts from 3.1 Ga by Schopf (Schopf, 1993). Macroscopic stromabolites, which are generally of biological (oxygenic photoautotrophic) origin, are first found in strata a few hundred million years younger. However, much of the fossil evidence provided by Schopf (e.g., Schopf, 1993) has been questioned (Brasier *et al.*, 2002), and many researchers believe in a later origin. The origin of this group is critical to establishing when net O_2 production (and hence, an oxidized atmosphere) first occurred on the planet. Although photodissociation of H_2O vapor could have provided a source of atmospheric O_2 in the Archean, the UV absorption cross-section of O_2 constrains the reaction, and theoretical calculations supported by geochemical evidence suggest that prior to ca. 2.4 Ga atmospheric O_2 was less than 10^{-5} of the present level (Holland and Rye, 1998; Pavlov and Kasting, 2002). There was a lag between the first occurrence of oxygenic photosynthesis and a global buildup of O_2 possibly due to the presence of alternative electron acceptors, especially Fe^{2+} and S^{2-} in the ocean. Indeed, the dating of oxidation of Earth's oceans and atmosphere is, in large measure, based on analysis of the chemical precipitation of oxidized iron in sedimentary rocks (the "Great Rust Event (Holland and Rye, 1998) and the mass-independent (Farquhar *et al.*, 2002) and mass-dependent (Habicht *et al.*, 2002) fractionation of sulfur isotopes. The ensemble of these analyses indicate that atmospheric oxygen rose sharply, from virtually insignificant levels, to between 1% and 10% of the present atmospheric concentration over a 100 Myr period beginning ca. 2.4 Ga. Thus, there may be as much as a 1 Gyr or as little as a 100 Myr gap between the origin of the first oxygenic photoautotrophs and oxygenation of Earth's oceans and atmosphere.

All other oxygen-producing organisms in the ocean are eukaryotic, i.e., they contain internal organelles, including a nucleus, one or more chloroplasts, one or more mitochondria, and, in some cases, a membrane-bound storage compartment, the vacuole. Within the eukaryotes, we can distinguish two major groups, both of which appear to have descended from a common ancestor thought to be the endosymbiotic appropriation of a cyanobacterium into a heterotrophic host cell (Delwiche, 2000). The appropriated cyanobacterium became a chloroplast.

15.4.1.1 The red and green lineages

In one group of eukaryotes, chlorophyll *b* was synthesized as a secondary pigment; this group forms the "green lineage," from which all higher plants have descended. The green lineage played a major role in oceanic food webs and the carbon cycle from ca. 1.6 Ga until the end-Permian extinction, ~250 Ma (Lipps, 1993). Since that time however, a second group of eukaryotes has risen to ecological prominence in the oceans; that group is commonly called the "red lineage" (Figure 2). The red lineage is comprised of several major phytoplankton divisions and classes, of which the diatoms, dinoflagellates, haptophytes (including the coccolithophorids), and the chrysophytes are the most important. All of these groups are comparatively modern organisms; indeed, the rise of dinoflagellates and coccolithophorids approximately parallels the rise of dinosaurs on land, while the rise of diatoms approximately parallels the rise of mammals in the Cenozoic. The burial and subsequent diagenesis of organic carbon produced primarily by members of the red lineage in shallow seas in the Mesozoic era provide the carbon source for many of the petroleum reservoirs that have been exploited for the past century by humans.

15.4.2 Estimating Chlorophyll Biomass

As implied in Equation (4), given an abundance of the two physical substrates, CO_2 and H_2O, primary production is, to first order, dependent on the concentration of the catalyst Chl *a* and light. The distribution of Chl *a* in the upper ocean can be discerned from satellite images of ocean color. The physical basis of the measurement is straightforward; making the measurements is technically challenging. Imagine two small parcels of water that are adjacent to each other. As photons from the sun enter the water column, they are either absorbed or scattered. Water itself absorbs red wavelengths of light, at shorter wavelengths of the visible spectrum, light is not as efficiently absorbed. However, because water molecules can randomly move from one adjacent parcel to another, there are continuous minor changes in

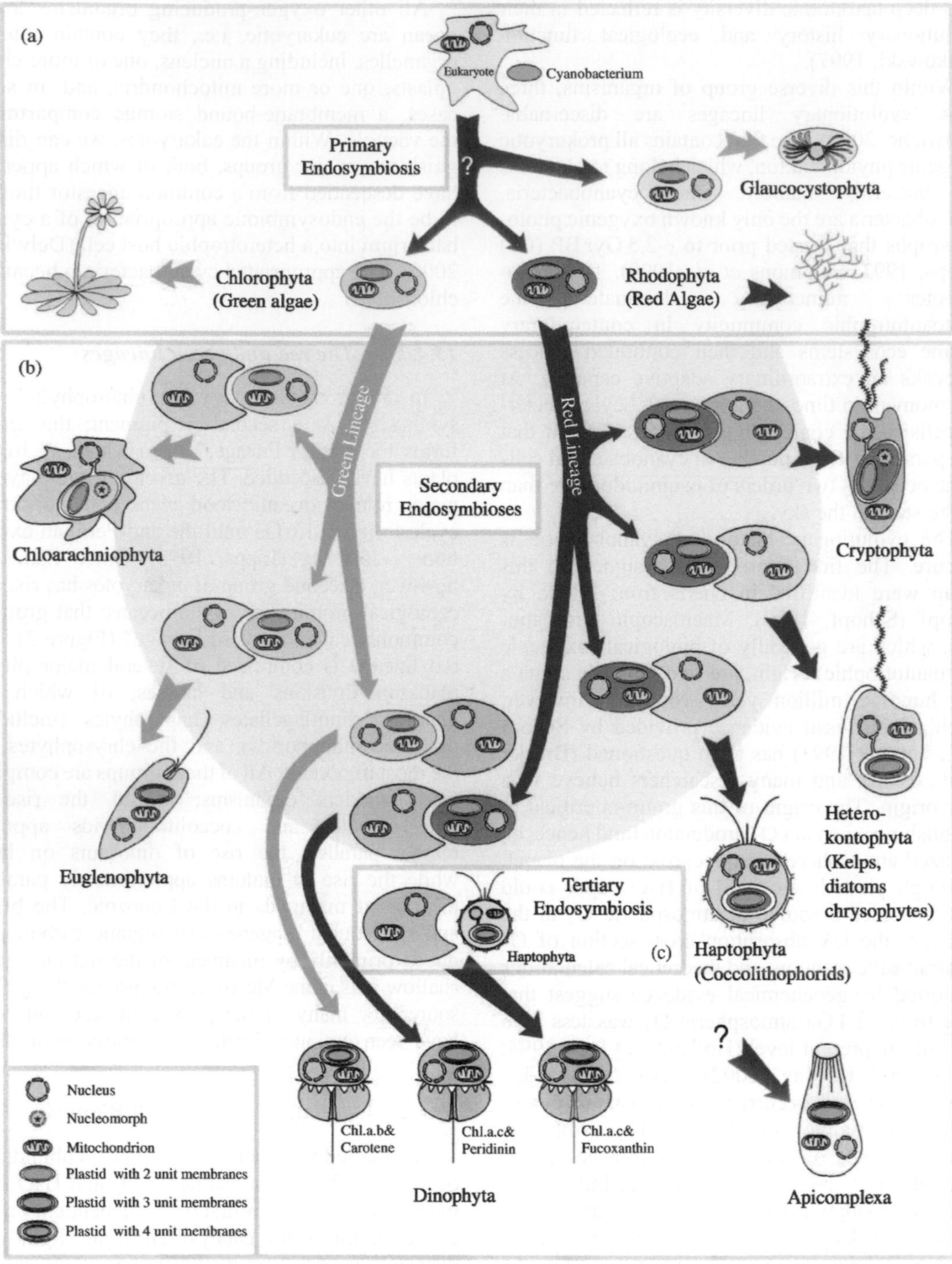

Figure 2 The basic pathway leading the evolution of eukaryotic algae. The primary symbiosis of a cyanobacterium with a apoplastidic host gave rise to both chlorophyte algae and red algae. The chlorophyte line, through secondary symbioses, gave rise to the "green" line of algae, one division of which was the predecessor of all higher plants. Secondary symbioses in the red line with various host cells gave rise to all the chromophytes, including diatoms, cryptophytes, and haptophytes (after Delwiche, 2000).

density and hence in the refractive index of the water parcels. These minor changes in refractive index lead to incoherence in the downwelling light stream. The incoherence, in turn, increases the probability of photon scattering (a process called "fluctuation density scattering"), such that light in the shorter wavelengths is more likely to be scattered back to space (Einstein, 1910; Morel, 1974). If the ocean contained sterile, pure seawater, an observer looking at the surface from space would see the oceans as blue. However, chlorophyll *a* has a prominent absorption band in the blue portion of the spectrum. Hence, in the presence of chlorophyll some of the downwelling and upwelling photons from the sun are absorbed by the phytoplankton themselves and the ocean becomes optically darker. As chlorophyll concentrations increase even further, blue wavelengths are largely eliminated from the outbound reflectance, and the ocean appears optically dark green (Morel, 1988).

15.4.2.1 Satellite based algorithms for ocean color retrievals

Empirically, satellite sensors that measure ocean color utilize a number of wavelengths. In addition to the blue and green region, red and far-red spectra are determined to derive corrections for scattering and absorption of the outbound or reflected radiation from the ocean by the atmosphere. In fact, only a very small fraction (~5%) of the light leaving the ocean is observed by a satellite; the vast majority of the photons are scattered or absorbed in the atmosphere. However, based on the ratio of blue-green light that is reflected from the ocean, estimates of photosynthetic pigments are derived. It should be pointed out that the blue-absorbing region of the spectrum is highly congested; it is virtually impossible to derive the fraction of absorption due solely to chlorophyll *a* as opposed to other photosynthetic pigments that absorb blue light. The estimation of chlorophyll *a* is based on empirical regression of the concentration of the pigment to the total blue-absorbing pigments (Gordon and Morel, 1983). Water-leaving radiances (L_W) at specific wavelengths are corrected for atmospheric scattering and absorption, and the concentration of chlorophyll is calculated from the ratios of blue and green light reflected from the water body. The calibration of the sensors is empirical and specifically derived for individual satellites. Examples of such algorithms for five satellites are given in Table 3.

One limitation of satellite images of ocean chlorophyll is that they do not provide information about the vertical distribution of phytoplankton. The water-leaving radiances visible to an observer outside of the ocean are confined to the upper 20% of the euphotic zone (which is empirically defined as the depth to which 1% of the solar radiation penetrates). In the open ocean there is almost always a subsurface chlorophyll maximum that is not visible to satellite ocean color sensors. A number of numerical models have been developed to estimate the vertical distribution of chlorophyll based on satellite color data (Berthon and Morel, 1992; Platt, 1986). The models rely on statistical parametrizations and require numerous *in situ* observations to obtain typical profiles for a given area of the world ocean (Morel and Andre, 1991; Platt and Sathyendranath, 1988). In addition, large quantities of phytoplankton associated with the bottom of ice flows in both the Arctic and Antarctic are not visible to satellite sensors but do contribute significantly to the primary production in the polar seas (Smith and Nelson, 1990). Despite these deficiencies, the satellite data allow high-resolution, synoptic observations of the temporal and spatial changes in phytoplankton chlorophyll in relation to the physical circulation of the atmosphere and ocean on a global scale.

The global distribution of phytoplankton chlorophyll in the upper ocean for winter and summer, derived from a compilation of satellite images, is shown in Figure 3. To a first order, the images reveal how the horizontal and temporal distribution of phytoplankton is related to the physical circulation of the oceans, especially the major features of the basin-scale gyres. For example, throughout most of the central ocean basins, between 30°N and 30°S, phytoplankton biomass is extremely low, averaging 0.1–0.2 mg chlorophyll *a* m^{-3} at the sea surface. In these regions the vertical flux of nutrients is generally extremely low, limited by eddy diffusion through the thermocline. Most of the chlorophyll biomass is associated with the thermocline. Because there is no seasonal convective overturn in this latitude band,

Table 3 Algorithms used to calculate chlorophyll *a* (*C*) from remote sensing reflectance. *R* is determined as the maximum of the values shown. Sensor algorithms are for Sea-viewing Wide Field of view Sensor (SeaWiFS), Ocean Color and Temperature Scanner (OCTS), Moderate Resolution Imaging Spectroradiometer (MODIS), Coastal Zone Color Scanner (CZCS), and Medium Resolution Imaging Spectrometer (MERIS).

Sensor	*Equation*	*R*
SeaWiFS/OC2	$C = 10.0^{(0.341-3.001R+2.811R^2-2.041R^3)} - 0.04$	490/555
OCTS/OC4O	$C = 10.0^{(0..405-2.900R+1.690R^2-0.530R^3-1.144R^4)}$	443 > 490 > 520/565
MODIS/OC3M	$C = 10.0^{(0.2830-2.753R+1.457R^2-0.659R^3-1.403R^4)}$	443 > 490/550
CZCS/OC3C	$C = 10.0^{(0.362-4.066R+5.125R^2-2.645R^3-0.597R^4)}$	443 > 520/550
MERIS/OC4E	$C = 10.0^{(0.368-2.814R+1.456R^2+0.768R^3-1.292R^4)}$	443 > 490 > 510/560
SeaWiFS/OC4v4	$C = 10.0^{(0.366-3.067R+1.930R^2+0.649R^3-1.532R^4)}$	443 > 490 > 510/555
SeaWiFS/OC2v4	$C = 10.0^{(0.319-2.336R+0.879R^2-0.135R^3)} - 0.071$	490/555

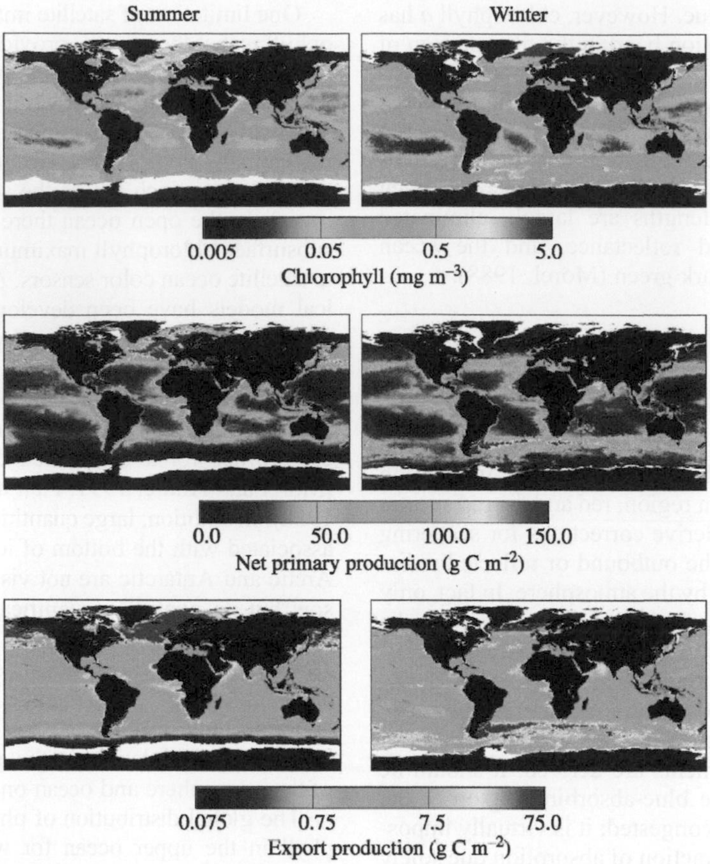

Figure 3 Composite global images for winter and summer of upper-ocean chlorophyll concentrations (top panels) derived from satellite-based observations of ocean color, net primary production (middle panels) calculated based on the algorithms of Behrenfeld and Falkowski (1997a), and export production (lower panels) calculated from the model of Laws *et al.* (2000).

there is no seasonal variation in phytoplankton chlorophyll. The chlorophyll concentrations are slightly increased at the equator in the Pacific and Atlantic Oceans, and south of the equator in the Indian Ocean. In the equatorial regions the thermocline shoals laterally as a result of long-range wind stress at the surface (Pickard and Emery, 1990). The wind effectively piles up water along its fetch, thereby inclining the upper mixed layer. This results in increased nutrient fluxes, shallower mixed layers, and higher chlorophyll concentrations on the eastern end of the equatorial band, and decreased nutrient fluxes, deeper mixed layers, and lower chlorophyll concentrations on the western end. This effect is most pronounced in the Pacific. The displacement of the band south of the equator in the Indian Ocean is primarily a consequence of basin scale topography.

At latitudes above ~30°, a seasonal cycle in chlorophyll can occur (Figure 3). In the northern hemisphere, areas of high chlorophyll are found in the open ocean of the North Atlantic in the spring and summer. The southern extent and intensity of the North Atlantic phytoplankton bloom are not found in the North Pacific. The North Atlantic bloom is associated with deep vertical convective mixing, which allows resupply of nutrients to the upper mixed layer of the ocean. This phenomenon does not occur in the Pacific due to a stronger vertical density gradient in that basin (driven by the hydrological cycle). The North Atlantic bloom leads to a flux of organic matter into the ocean interior that is observed even at the seafloor.

In the southern hemisphere, phytoplankton chlorophyll is generally lower at latitudes symmetrical with the northern hemisphere in the corresponding austral seasons. For example, in the austral summer (January–March), phytoplankton chlorophyll is slightly lower between 30°S and the Antarctic ice sheets than in the northern hemisphere in July to September (Yoder *et al.*, 1993).

15.4.3 Estimating Net Primary Production

15.4.3.1 Global models of net primary production for the ocean

Using satellite data to estimate upper-ocean chlorophyll concentrations, satellite-based observations of incident solar radiation, atlases of seasonally averaged sea-surface temperature, and models that incorporate a temperature response function for photosynthesis, it is possible to estimate global net photosynthesis in the world oceans (Antoine and Morel, 1996; Behrenfeld and Falkowski, 1997a; Longhurst *et al.*, 1995). Although estimates vary between models, based on how the parameters are derived, for illustrative purposes we use a model based on empirical parametrization of the daily integrated photosynthesis profiles as a function of depth. The physical depth at which 1% of irradiance incident on the sea surface remains is called the euphotic zone. This depth can be calculated from surface chlorophyll concentrations, and defines the base of the water column at which net photosynthesis can be supported. Given such information, net primary production can be calculated following the general equation:

$$PP_{eu} = C_{sat} \cdot Z_{eu} \cdot P^{b}_{opt} \cdot DL \cdot F \qquad (6)$$

where PP_{eu} is daily net primary production integrated over the euphotic zone, C_{sat} is the satellite-based (upper water column; i.e., derived from Table 3) chlorophyll concentration, P^{b}_{opt} is the maximum daily photosynthetic rate within the water column, Z_{eu} is the depth of the euphotic zone, DL is the photoperiod (Behrenfeld and Falkowski, 1997a), and F is a function describing the shape of the photosynthesis depth profile. This general model can be both expanded (differentiated) and collapsed (integrated) with respect to time and irradiance; however, the global results are fundamentally similar (Behrenfeld and Falkowski, 1997). The models predict that NPP in the world oceans amounts to 40–50 Pg per annum (Figure 3 and Table 4).

In contrast to terrestrial ecosystems, the fundamental limitation of primary production in the ocean is not irradiance *per se*, but temperature and the concentration of chlorophyll in the upper ocean. The latter is a negative feedback; i.e., the more the chlorophyll in the water column, the shallower is the euphotic zone. Hence, to double NPP requires nearly a fivefold increase in chlorophyll concentration.

15.4.4 Quantum Efficiency of NPP

The photosynthetically available radiation (400–700 nm) for the world oceans is 4.5×10^{18} mol of photons per annum, which is $\simeq 9.8 \times 10^{20}$ kJ yr^{-1}. The average energy stored by photosynthetic organisms amounts to ~39 kJ per gram of carbon fixed (Platt and Irwin, 1973). Given an annual net production of 40 Pg C for phytoplankton, and an estimated production of 4 Pg yr^{-1} by benthic photoautotrophs, the photosynthetically stored radiation is equal to $\sim 1.7 \times 10^{18}$ kJ yr^{-1}. The fraction of photosynthetically available solar energy conserved by photosynthetic reactions in the world oceans

Table 4 Annual and seasonal net primary production (NPP) of the major units of the biosphere.

	Ocean NPP		*Land NPP*
Seasonal			
April–June	10.9		15.7
July–September	13.0		18.0
October–December	12.3		11.5
January–March	11.3		11.2
Biogeographic			
Oligotrophic	11.0	Tropical rainforests	17.8
Mesotrophic	27.4	Broadleaf deciduous forests	1.5
Eutrophic	9.1	Broadleaf and needleleaf forests	3.1
Macrophytes	1.0	Needleleaf evergreen forests	3.1
		Needleleaf deciduous forest	1.4
		Savannas	16.8
		Perennial grasslands	2.4
		Broadleaf shrubs with bare soil	1.0
		Tundra	0.8
		Desert	0.5
		Cultivation	8.0
Total	48.5		56.4

Source: Field *et al.* (1998). After Field *et al.* (1998). All values in GtC. Ocean color data are averages from 1978 to 1983. The land vegetation index is from 1982 to 1990. Ocean NPP estimates are binned into three biogeographic categories on the basis of annual average C_{sat} for each satellite pixel, such that oligotrophic = C_{sat}<0.1 mg m^{-3}, mesotrophic = 0.1 < C_{sat}<1 mg m^{-3}, and eutrophic = C_{sat}>1 mg m^{-3} (Antoine *et al.*, 1996). This estimate includes a 1 GtC contribution from macroalgae (Smith, 1981). Differences in ocean NPP estimates between Behrenfeld and Falkowski (1997) and those in the global annual NPP for the biosphere and this table result from: (i) addition of Arctic and Antarctic monthly ice masks; (ii) correction of a rounding error in previous calculations of pixel area; and (iii) changes in the designation of the seasons to correspond with Falkowski *et al.* (1998).

amounts to 1.7×10^{18} kJ/$9.8 \times 10^{20} = 0.0017$ or 0.17%. Thus, on average, in the oceans, 0.0007 mol C is fixed per mole of incident photons; this is equivalent to an effective quantum requirement of 1,400 quanta per CO_2 fixed. This value is less than 1% of the theoretical maximum quantum efficiency of photosynthesis; the relatively small realized efficiency is due to the fact that photons incident on the ocean surface have a small probability of being absorbed by phytoplankton before they are either absorbed by water or other molecules (e.g., organic matter), or are scattered back to space.

15.4.4.1 Comparing efficiencies for oceanic and terrestrial primary production

The average chlorophyll concentration of the world ocean is $0.24\,\mathrm{mg\,m^{-3}}$ and the average euphotic zone depth is 54 m; thus the average integrated chlorophyll concentration is ~13 mg m^{-2}. Carbon to chlorophyll ratios of phytoplankton typically range between 40:1 and 100:1 by weight (Banse, 1977). Given the total area of the ocean of $3.1 \times 10^8\,\mathrm{km}^2$, the total carbon biomass in phytoplankton is 0.25–0.65 Pg. If NPP is ~40 Pg per annum, and assuming the ocean is in steady state (a condition we will discuss in more detail), the living phytoplankton biomass turns over 60–150 times per year, which is equivalent to a turnover time of 2–6 d. In contrast, terrestrial plant biomass amounts to ~600–1,000 Pg C, most of which is in the form of wood (Woodwell *et al.*, 1978). Estimates of terrestrial plant NPP are in the range of 50–65 Pg C per annum, which gives an average turnover time of ~12–20 yr (Field *et al.*, 1998). Thus, the flux of carbon through aquatic photosynthetic organisms is about 1,000-fold faster than terrestrial ecosystems, while the storage of carbon in the latter is about 1,000-fold higher than the former. Moreover, the total photon flux to terrestrial environments amounts to $\sim 2 \times 10^{18}$ mol yr^{-1}, which gives an effective quantum yield of ~0.002. In other words, on average one CO_2 molecule is fixed for every 500 incident photons. The results of these calculations suggest that terrestrial vegetation is approximately three times more efficient in utilizing *incident* solar radiation to fix carbon than are aquatic photoautotrophs. This situation arises primarily from the relative paucity of aquatic photoautotrophs in the ocean and the fact that they must compete with the media (water) for light.

This comparison points out a fundamental difference between the two ecosystems in the context of the global carbon cycle. On timescales of decades to centuries, carbon fixed in terrestrial ecosystems can be temporarily stored in organic matter (e.g., forests), whereas most of the carbon fixed by marine phytoplankton is rapidly consumed by grazers or sinks and is transferred from the surface ocean to the ocean interior. Upon entering the ocean interior, virtually all of the organic matter is oxidized by heterotrophic microbes, and in the process is converted back to inorganic carbon. Elucidating how this transfer occurs, what controls it, how much carbon is transferred via this mechanism, on what timescales, and whether the process is in steady state was a major focus for research in the latter portion of the twentieth century.

15.5 EXPORT, NEW AND "TRUE NEW" PRODUCTION

We can imagine that NPP produced by photoautotrophs in the upper, sunlit regions of the ocean (the euphotic zone) is consumed in the same general region by heterotrophs. In such a case, the basic reaction given by Equation (4) is simply balanced in the reverse direction due to respiration by heterotrophic organisms, and no organic matter leaves the ecosystem. This very simple "balanced state" model, also referred to as the microbial loop (e.g., Azam, 1998), accounts for the fate of most of the organic matter in the oceans (and on timescales of decades, terrestrial ecosystems as well). In marine ecology, this process is sometimes called "regenerated production"; i.e., organic matter produced by photoautotrophs is locally regenerated to inorganic nutrients (CO_2, NH_4^+, PO_3^{2-}) by heterotrophic respiration. It should be noted here that with the passage of organic matter from one level of a marine food chain to the next (e.g., from primary producer to heterotrophic consumer), a metabolic "tax" must be paid in the form of respiration, such that the net metabolic potential of the heterotrophic biomass is always less than that of the primary producers. This does not mean, *a priori*, that photoautotrophic biomass is always greater, as heterotrophs may grow slowly and accumulate biomass; however, as heterotrophs grow faster, their respiratory rates must invariably increase. The rate of production (i.e., the energy flux) of heterotrophic biomass is always constrained by NPP.

Let us imagine a second scenario. Some fraction of the primary producers and/or heterotrophs sink below a key physical gradient, such as a thermocline, and for whatever reason cannot ascend back into the euphotic zone. If the water column is very deep, sinking organic matter will most likely be consumed by heterotrophic microbes in the ocean interior. The flux of organic carbon from the euphotic zone is often called *export production*, a term coined by Wolfgang Berger. Export production is an important conduit for the exchange of carbon between the upper ocean and the ocean interior (Berger *et al.*, 1987). This conduit depletes the upper ocean of inorganic carbon and other nutrients essential for photosynthesis and the biosynthesis of organic matter. In the central ocean basins, export production is a relatively small fraction of total primary production, amounting to 5–10% of the total carbon fixed per annum (Dugdale and Wilkerson, 1992). At high latitudes and in nutrient-rich areas, however, diatoms and other large, heavy cells can form massive blooms and sink rapidly. In such regions, export production can account for 50% of the total carbon fixation (Bienfang, 1992; Campbell and Aarup, 1992; Sancetta *et al.*, 1991; Walsh, 1983). The subsequent oxidation and remineralization of the exported production enriches the ocean interior with inorganic carbon by ~200 μM in excess of that which would be supported solely by air–sea exchange (Figure 4 and Table 5). This enrichment is called the *biological pump* (Broecker *et al.*, 1980; Sarmiento and Bender, 1994; Volk and Hoffert, 1985). The biological pump is crucial to maintaining the steady-state levels of atmospheric CO_2 (Sarmiento *et al.*, 1992; Siegenthaler and Sarmiento, 1993).

15.5.1 Steady-state versus Transient State

The concepts of new, regenerated, and export production are central to understanding many aspects of the role of aquatic photosynthetic organisms in biogeochemical cycles in the oceans. In steady state, the globally averaged fluxes of new nutrients must match the loss of the nutrients contained in organic material. If this were not so, there would be a continuous depletion of nutrients in the euphotic zone and photoautotrophic biomass and primary production would slowly decline (Eppley, 1992). Thus, in the steady state, the sinking fluxes of organic nitrogen and the production of N_2 by denitrifying bacteria must equal the sum of the upward fluxes

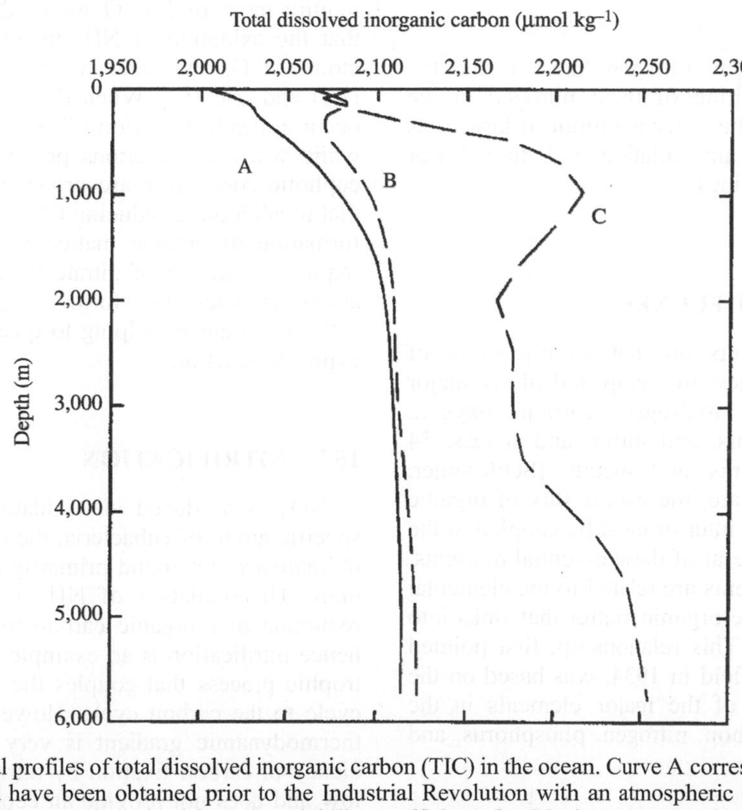

Figure 4 Vertical profiles of total dissolved inorganic carbon (TIC) in the ocean. Curve A corresponds to a theoretical profile that would have been obtained prior to the Industrial Revolution with an atmospheric CO_2 concentration of 280 μmol mol^{-1}. The curve is derived from the solubility coefficients for CO_2 in seawater, using a typical thermal and salinity profile from the central Pacific Ocean, and assumes that when surface water cools and sinks to become deep water it has equilibrated with atmospheric CO_2. Curve B corresponds to the same calculated solubility profile of TIC, but in the year 1995, with an atmospheric CO_2 concentration of 360 μmol mol^{-1}. The difference between these two curves is the integrated oceanic uptake of CO_2 from anthropogenic emissions since the beginning of the Industrial Revolution, with the assumption that biological processes have been in steady state (and hence have not materially affected the net influx of CO_2). Curve C is a representative profile of measured TIC from the central Pacific Ocean. The difference between curve C and B is the contribution of biological processes to the uptake of CO_2 in the steady state (i.e. the contribution of the "biological pump" to the TIC pool.) (courtesy of Doug Wallace and the World Ocean Circulation Experiment).

Table 5 Export production and ef ratios calculated from the model of Laws *et al.* (2000).

	Export ($GtC\,y^{-1}$)	*ef*
Ocean basin		
Pacific	4.3	0.19
Atlantic	4.3	0.25
Indian	1.5	0.15
Antarctic	0.62	0.28
Arctic	0.15	0.56
Mediterranean	0.19	0.24
Global	11.1	0.21
Total production		
Oligotrophic (chl a < 0.1 mg m^{-3})	1.04	0.15
Mesotrophic ($0.1 \leq$ chl a < 1.0 mg m^{-3})	6.5	0.18
Eutrophic (chl $a \geq 1.0$ mg m^{-3})	3.6	0.36
Ocean depth		
0 – 100 m	2.2	0.31
100 m–1 km	1.4	0.33
>1 km	7.4	0.18

of inorganic nitrogen, nitrogen fixation, and the atmospheric deposition of fixed nitrogen in the form of aerosols (the latter is produced largely as a consequence of air pollution and, to a lesser extent, from lightning).

15.6 NUTRIENT FLUXES

Primary producers are not simply sacks of organic carbon. They are composed of six major elements, namely, hydrogen, carbon, oxygen, nitrogen, phosphorus, and sulfur, and at least 54 other trace elements and metals (Schlesinger, 1997). In steady state, the export flux of organic matter to the ocean interior must be coupled to the upward flux of several of these essential nutrients. The fluxes of nutrients are related to the elemental stoichiometry of the organic matter that sinks into the ocean interior. This relationship, first pointed out by Alfred Redfield in 1934, was based on the chemistry of four of the major elements in the ocean, namely, carbon, nitrogen, phosphorus, and oxygen.

15.6.1 The Redfield Ratio

In the ocean interior, the ratio of fixed inorganic nitrogen (in the form of NO_{3-}) to PO_4 in the dissolved phase is remarkably close to the ratios of the two elements in living plankton. Hence, it seemed reasonable to assume that the ratio of the two elements in the dissolved phase was the result of the sinking and subsequent remineralization (i.e., oxidation) of the elements in organic matter produced in the open ocean. Further, as carbon and nitrogen in living organisms are largely found in chemically reduced forms, while remineralized forms are virtually all oxidized, the remineralization of organic matter was coupled to the depletion of oxygen. The relationship could be expressed stoichiometrically as

$$[106(CH_2O)16NH_31PO_4^{2-}] + 138O_2 \rightarrow 106CO_2 + 122H_2O + 16NO_3^- + PO_4^{2-} + 16H^+] \quad (7)$$

Hidden within this balanced chemical formulation are biochemical redox reactions, which are contained within specific groups of organisms. (In the oxidation of organic matter, there is some ambiguity about the stoichiometry of O_2/P. Assuming that the mean oxidation level of organic carbon is that of carbohydrate (as is the case in Equation (9)), then the oxidation of that carbon is equimolar with O. Alternatively, some organic matter may be more or less reduced than carbohydrate, and therefore require more or less O for oxidation. Note also that the oxidation of NH_3 to NO_3^- requires four atoms of O, and leads to the formation of one H_2O and one H^+.) When the reactions primarily occur at depth, Equation (7) is driven to the right, while when the reactions primarily occur in the euphotic zone, they are driven to the left. Note that in addition to reducing CO_2 to organic matter, formation of organic matter by photoautotrophs requires reduction of nitrate to the equivalent of ammonia. These two forms of nitrogen are critically important in helping to quantify "new" and export production.

15.7 NITRIFICATION

NO_3^- is produced via oxidation of NH_3 by a specific group of eubacteria, the nitrifiers, that are oblibate aerobes found primarily in the water column. The oxidation of NH_3 is coupled to the reduction of inorganic carbon to organic matter; hence nitrification is an example of a chemoautotrophic process that couples the aerobic nitrogen cycle to the carbon cycle. However, because the thermodynamic gradient is very small, the efficiency of carbon fixation by nitrifying bacteria is low and does not provide an ecologically significant source of organic matter in the oceans. In the contemporary ocean, global CO_2 fixation by marine nitrifying bacteria only amounts to ~0.2 Pg C per annum, or ~0.5% of marine photoautotrophic carbon fixation.

There are two major sources of nutrients in the euphotic zone. One is the local regeneration of simple forms of combined elements (e.g., NH_4^+, HPO_4^{2-}, SO_4^{2-}) resulting from the metabolic activity of metazoan and microbial degradation. The second is the influx of distantly produced, "new" nutrients, imported from the deep ocean, the atmosphere (i.e., nitrogen fixation, atmospheric pollution), or terrestrial runoff from streams, rivers, and estuaries (Dugdale and Goering, 1967). In the open ocean, these two sources can be usefully related to the form of inorganic nitrogen assimilated by phytoplankton. Because biological nitrogen fixation is relatively low in the ocean (see below) and nitrification in the upper mixed layer is sluggish relative to the assimilation of nitrogen by photoautotrophs, nitrogen supplied from local regeneration is assimilated before it has a chance to become oxidized. Hence, regenerated nitrogen is primarily in the form of ammonium or urea. In contrast, the fixed inorganic nitrogen in the deep ocean has sufficient time (hundreds of years) to become oxidized, and hence the major source of new nitrogen is in the form of nitrate. Using $^{15}NH_4$ and $^{15}NO_3$ as tracers, it is possible to estimate the fraction of new nitrogen that fuels phytoplankton production (Dugdale and Wilkerson, 1992). This approach provides an estimate of both the upward flux of nitrate required to sustain the $^{15}NO_3^-$ supported production, as well as the downward flux of organic carbon, which is required to maintain a steady-state balance (Dugdale and Wilkerson, 1992; Eppley and Peterson, 1979).

15.7.1 Carbon Burial

On geological timescales, there is one important fate for NPP, namely, burial in the sediments. By far the largest reservoir of organic matter on Earth is locked up in rocks (Table 1). Virtually all of this organic carbon is the result of the burial of exported marine organic matter in coastal sediments over literally billions of years of Earth's history. On geological timescales, the burial of marine NPP effectively removes carbon from biological cycles, and places most (not all) of that carbon into the slow carbon cycle. A small fraction of the organic matter escapes tectonic processing via the Wilson cycle and is permanently buried, mostly in continental rocks. The burial of organic carbon is inferred not from direct measurement, but rather from indirect means. One of the most common proxies used to derive burial on geological timescales is based on isotopic fractionation of carbonates. The rationale for this analysis is that the primary enzyme responsible for inorganic carbon fixation is ribulose 1,5-bisphosphate carboxylase/oxygenase (RuBisCO), which catalyzes the reaction between ribulose 1,5-bisphosphate and CO_2 (not HCO_3^-), to form two molecules of 3-phosphoglycerate. The enzyme strongly discriminates against ^{13}C, such that the resulting isotopic fractionation amounts to ~27‰ relative to the source carbon isotopic value. The extent of the actual fractionation is somewhat variable and is a function of carbon availability and of the transport processes for inorganic carbon into the cells, as well as the specific carboxylation pathway (Laws *et al.*, 1997). However, regardless of the quantitative aspects, the net effect of carbon fixation is an enrichment of the inorganic carbon pool in ^{13}C, while the organic carbon produced is enriched in ^{12}C.

15.7.2 Carbon Isotope Fractionation in Organic Matter and Carbonates

The isotopic fractionation in carbonates mirrors the relative amount of organic carbon buried. It is generally assumed that the source carbon, from vulcanism (the so-called "mantle" carbon) has an isotopic value of approximately −5‰. As mass balance must constrain the isotopic signatures of carbonate carbon and organic carbon with the mantle carbon, then

$$f_{org} = \frac{\delta_w - \delta_{carb}}{\Delta_B} \quad (8)$$

where f_{org} is the fraction of organic carbon buried, δ_w is the average isotopic content of the carbon weathered, δ_{carb} is the isotopic signature of the carbonate carbon, and Δ_B the isotopic difference between organic carbon and carbonate carbon deposited in the ocean. Equation (8) is a steady-state model that presumes the source of carbon from the mantel is constant over geological time. This basic model is the basis of nearly all estimates of organic carbon burial rates (Berner *et al.*, 1983; Kump and Arthur, 1999).

Carbonate isotopic analyses reveal positive excursions (i.e., implying organic carbon burial) in the Proterozioc, and more modest excursions throughout the Phanaerozic (Figure 5). Burial of organic carbon on geological timescales implies that export production must deviate from the steady state on ecological timescales. Such a deviation requires changing one or more of (i) ocean nutrient inventories, (ii) the utilization of unused nutrients in enriched areas, (iii) the average elemental composition of the organic material, or (iv) the "rain" ratios of particulate organic carbon to particulate inorganic carbon to the seafloor.

15.7.3 Balance between Net Primary Production and Losses

In the ecological theater of aquatic ecosystems, the observed photoautotrophic biomass at any moment in time represents a balance between the rate of growth and the rate of removal of that trophic level. The burial of organic carbon in the

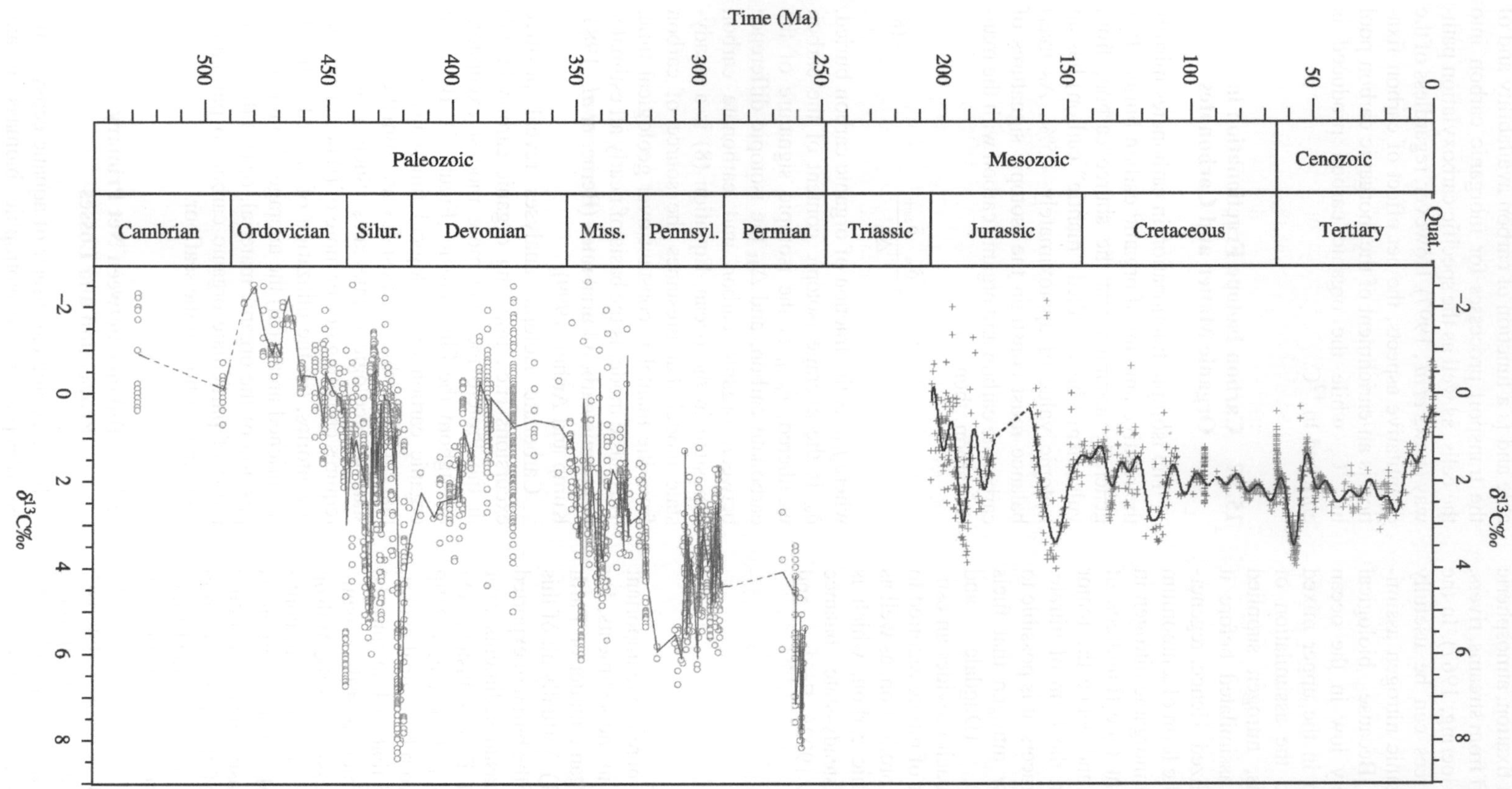

Figure 5 Phanerozoic $\delta^{13}C_{carb}$ record. The Jurassic through the Cenozoic record was generated from bulk sediment carbonates primarily from open ocean Atlantic Deep Sea Drilling Project boreholes; Lower Jurassic samples were used from the Mochras Borehole (Wales) (Katz *et al.*, in review). Dashed intervals indicate data gaps. Singular spectrum analysis was used to generate the Mesozoic–Cenozoic curve. The Paleozoic record was generated from brachiopods (Veizer *et al.*, 1999; note that the timescale has been adjusted from the original reference by interpolating between period boundaries). Data were averaged for each time slice to obtain the Paleozoic curve. The timescales of Berggren *et al.* (1995; Cenozoic), Gradstein *et al.* (1995; Mesozoic), and GSA (Paleozoic) were used.

lithosphere requires that the ecological balance between NPP and respiration diverge; i.e., the global ocean must be net autotropic. For simplicity, we can express the time-dependent change in photoautrophic biomass by a linear differential equation:

$$dP/dt = [P](\mu - m) \qquad (9)$$

where $[P]$ is photoautrophic biomass (e.g., organic carbon), μ is the specific growth rate (units of 1/time), and m is the specific mortality rate (units of 1/time). In this equation, we have lumped all mortality terms, such as grazing and sinking together into one term, although each of these loss processes can be given explicitly (Banse, 1994). Two things should be noted regarding Equation (9). First, μ and m are independent variables; i.e., changes in P can be independently ascribed to one or the other process. Second, by definition, a steady state exists when dP/dt is zero.

15.7.4 Carbon Burial in the Contemporary Ocean

The burial of organic carbon in the modern oceans is primarily confined to a few regions where the supply of sediments from terrestrial sources is extremely high. Such regions include the Amazon outfall and Indonesian mud belts. In contrast, the oxidation of organic matter in the interior of the contemporary ocean is extremely efficient; virtually no carbon is buried in the deep sea. Similarly, on most continental margins, organic carbon that reaches the sediments is consumed by microbes within the sediments, such that very little is actually buried in the contemporary ocean (Aller, 1998). The solution to Equation (9) must be close to zero; and consequently, in the absence of human activities, the oxygen content of Earth's atmosphere is very close to steady state and has been zso for tens, if not hundreds, of millions of years.

15.7.5 Carbon Burial in the Precambrian Ocean

In contrast, carbon burial in the Proterozoic ocean must have occurred as oxygen increased in the atmosphere; i.e., the global solution to Equation (9) must have been >0. Photoautotrophic biomass could have increased until some element became limiting. Thus, the original feedback between the production of photoautotrophic biomass in the oceans and the atmospheric content of oxygen was determined by an element that limited the crop size of the photoautotrophs in the Archean or Proterozoic ocean. What was that element, and why did it become limiting?

15.8 LIMITING MACRONUTRIENTS

A general feature of aquatic environments is that because the oxidation of organic nutrients to their inorganic forms occurs below the euphotic zone where the competing processes of assimilation of nutrients by photoautotrophs do not occur, the pools of inorganic nutrients are much higher at depth. As the only natural source of photosynthetically active radiation is the sun, the gradients of light and nutrients are from opposite directions. Thermal or salinity differences in the surface layers produce vertical gradients in density that effectively retard the vertical fluxes of soluble nutrients from depth. Thus, in the surface layers of a stratified water column, nutrients become depleted as the photoautotrophs consume them at rates exceeding their rate of vertical supply. Indeed, throughout most of the world oceans, the concentrations of dissolved inorganic nutrients, especially fixed inorganic nitrogen and phosphate, are exceedingly low, often only a few nM. One or the other of these nutrients can limit primary production. However, the concept of limitation requires some discussion.

15.8.1 The Two Concepts of Limitation

The original notion of limitation in ecology was related to the *yield* of a crop. A limiting factor was the substrate least available relative to the requirement for synthesis of the crop (Liebig, 1840). This concept formed a strong underpinning of agricultural chemistry and was used to design the elemental composition of fertilizers for commercial crops. This concept subsequently was embraced by ecologists and geochemists as a general "law" (Odum, 1971).

Nutrients can also limit the *rate* of growth of photoautotrophs (Blackman, 1905; Dugdale, 1967). Recall that if organisms are in balanced growth, the rate of uptake of an inorganic nutrient relative to the cellular concentration of the nutrient defines the growth rate (Herbert *et al.*, 1956). The uptake of inorganic nutrients is a hyperbolic function of the nutrient concentration and can be conveniently described by a hyperbolic expression of the general form

$$V = (V_{max}(U,\ldots))/(K_s + (U,\ldots)) \qquad (10)$$

where V is the instantaneous rate of nutrient uptake, V_{max} is the maximum uptake rate, $(U,\ldots)$ represent the substrate concentration of nutrient U, etc., and K_s is the concentration supporting half the maximum rate of uptake (Dugdale, 1967; Monod, 1942). There can be considerable variation between species with regard to K_s and V_{max} values and these variations are potential sources of competitive selection (Eppley *et al.*, 1969; Tilman, 1982).

It should be noted that Liebig's notion of limitation was not related *a priori* to the intrinsic rate of photosynthesis or growth. For example, photosynthetic rates can be (and often are) limited by light or temperature. The two concepts of limitation (yield and rate) are often not understood correctly: the former is more relevant to biogeochemical cycles, the latter is more critical to selection of species in ecosystems.

15.9 THE EVOLUTION OF THE NITROGEN CYCLE

Globally, nitrogen and phosphorus are the two elements that immediately limit, in a Liebig sense, the biologically mediated carbon assimilation in the oceans by photoautotrophs. It is frequently argued that since N_2 is abundant in both the ocean and the atmosphere, and, in principle, can be biologically reduced to the equivalent of NH_3 by N_2-fixing cyanobacteria, nitrogen cannot be limiting on geological timescales (Barber, 1992; Broecker *et al.*, 1980; Redfield, 1958). Therefore, phosphorus, which is supplied to the ocean by the weathering of continental rocks, must ultimately limit biological productivity. The underlying assumptions of these tenets should, however, be considered within the context of the evolution of biogeochemical cycles.

By far, the major source of fixed inorganic nitrogen for the oceans is via biological nitrogen fixation. Although in the Archean atmosphere, electrical discharge or bolide impacts may have promoted NO formation from the reaction between N_2 and CO_2, the yield for these reactions is extremely low. Moreover, atmospheric NH_3 would have photodissociated from UV radiation (Kasting, 1990), while N_2 would have been stable (Kasting, 1990; Warneck, 1988). Biological N_2 fixation is a strictly anaerobic process (Postgate, 1971), and the sequence of the genes encoding the catalytic subunits for nitrogenase is highly conserved in cyanobacteria and other eubacteria, strongly suggesting a common ancestral origin (Zehr *et al.*, 1995). The antiquity and homology of nitrogen fixation capacity also imply that fixed inorganic nitrogen was scarce prior to the evolution of diazotrophic organisms; i.e., there was strong evolutionary selection for nitrogen fixation in the Archean or early Proterozoic periods. In the contemporary ocean, N_2 is still catalyzed solely by prokaryotes, primarily cyanobacteria (Capone and Carpenter, 1982).

While apatite and other calcium-based and substituted solid phases of phosphate minerals precipitated in the primary formation of crustal sediments, secondary reactions of phosphate with aluminum and transition metals such as iron are mediated at either low salinity, low pH, or high oxidation states of the cations (Stumm and Morgan, 1981). Although these reactions would reduce the overall soluble phosphate concentration, the initial condition of the Archean ocean probably had a fixed low N:P ratio in the dissolved inorganic phase. As N_2 fixation proceeded, that ratio would have increased with a buildup of ammonium in the ocean interior. The accumulation of fixed nitrogen in the oceans would continue until the N:P ratio of the inorganic elements reached equilibrium with the N:P ratio of the sedimenting particulate organic matter (POM). Presumably, the latter ratio would approximate that of extant, nitrogen-fixing marine cyanobacteria, which is ~16:1 by atoms (Copin-Montegut and Copin-Montegut, 1983; Redfield, 1958; Quigg *et al.*, 2003) or greater (Letelier *et al.*, 1996) and would ultimately be constrained by the availability of phosphate (Falkowski, 1997; Tyrell, 1999).

The formation of nitrate from ammonium by nitrifying bacteria requires molecular oxygen; hence, nitrification must have evolved following the formation of free molecular oxygen in the oceans by oxygenic photoautotrophs. Therefore, from a geological perspective, the conversion of ammonium to nitrate probably proceeded rapidly and provided a substrate, NO_3^-, that eventually could serve both as a source of nitrogen for photoautotrophs and as an electron acceptor for a diverse group of heterotrophic, anaerobic bacteria, the denitrifiers.

In the sequence of the three major biological processes that constitute the nitrogen cycle, denitrification must have been the last to emerge. This process, which permits the reduction of NO_3^- to (ultimately) N_2, occurs in the modern ocean in three major regions, namely, continental margin sediments, areas of restricted circulation such as fjords, and oxygen minima zones of perennially stratified seas (Christensen *et al.*, 1987; Codispoti and Christensen, 1985; Devol, 1991; Nixon *et al.*, 1996). In all cases, the process requires hypoxic or anoxic environments and is sustained by high sinking fluxes of organic matter. Denitrification appears to have evolved independently several times; the organisms and enzymes responsible for the pathway are highly diverse from a phylogenetic and evolutionary standpoint.

With the emergence of denitrification, the ratio of fixed inorganic nitrogen to dissolved inorganic phosphate in the ocean interior could only be depleted in nitrogen relative to the sinking flux of the two elements in POM. Indeed, in all of the major basins in the contemporary ocean, the N:P ratio of the dissolved inorganic nutrients in the ocean interior is conservatively estimated at 14.7

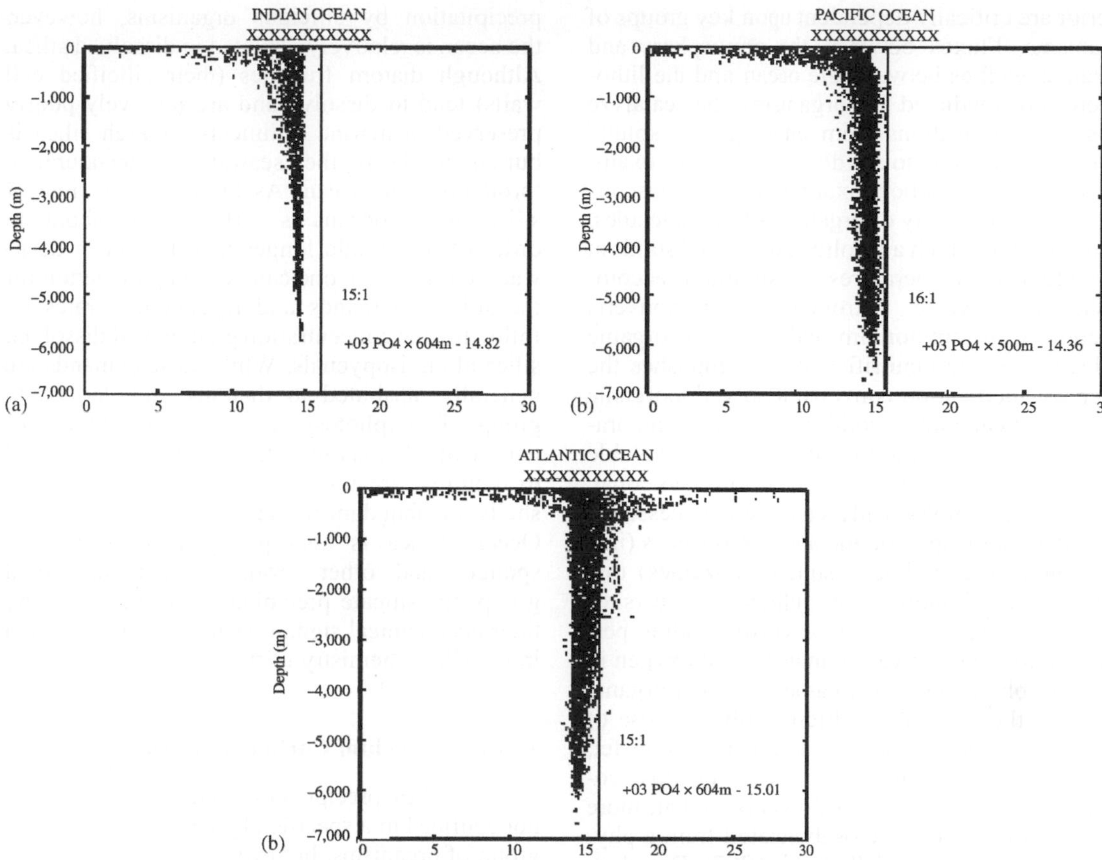

Figure 6 (a)–(c) Vertical profiles of NO_3^-/HPO_4^{3-} ratios in each of the three major ocean basins. The data were taken from the GEOSECS database. In all three basins, the N:P ratio converges on an average value that is significantly lower than the 16:1 ratio predicted by Redfield. The deficit in nitrogen relative to phosphorus is presumed to be a result of denitrification. Note that in the upper 500 m of the water column, NO_3^-/HPO_4^{3-} ratios generally decline except in a portion of the Atlantic that corresponds to the eastern Mediterranean.

by atoms (Fanning, 1992) or less (Anderson and Sarmiento, 1994) (Figure 6).

There are three major conclusions that may be drawn from the foregoing discussion:

(i) Because the ratio of the sinking flux of particulate organic nitrogen and particulate phosphorus exceeds the N:P ratio of the dissolved pool of inorganic nutrients in the ocean interior, the average upward flux of inorganic nutrients must be slightly enriched in phosphorus relative to nitrogen as well as to the elemental requirements of the photoautotrophs (Gruber and Sarmiento, 1997; Redfield, 1958). Hence, although there are some exceptions (Kromer, 1995; Wu *et al.*, 2000), dissolved, inorganic fixed nitrogen generally limits primary production throughout most of the world's oceans (Barber, 1992; Falkowski *et al.*, 1998).

(ii) The N:P ratio of the dissolved pool of inorganic nutrients in the ocean interior was established by biological processes, not vice versa (Redfield *et al.*, 1963, 1934). The elemental composition of marine photoautotrophs has been conserved since the evolution of the eukaryotic phytoplankton (Lipps, 1993). The Redfield N:P ratio of 16:1 for POM (Codispoti, 1995; Copin-Montegut and Copin-Montegut, 1983; McElroy, 1983; Redfield *et al.*, 1963, 1958) is an upper bound, which is not observed for the two elements in the dissolved inorganic phase in the ocean interior. The deficit in dissolved inorganic fixed nitrogen relative to soluble phosphate in the ocean represents a slight imbalance between nitrogen fixation and denitrification on timescales of $\sim 10^3$–10^4 yr (Codispoti, 1995).

(iii) If dissolved inorganic nitrogen rather than phosphate limits productivity in the oceans, then it follows that the ratio of nitrogen fixation/denitrification plays a critical role in determining the net biologically mediated exchange of CO_2 between the atmosphere and ocean (Codispoti, 1995).

15.10 FUNCTIONAL GROUPS

As we have implied throughout the foregoing discussion, the biologically mediated fluxes of elements between the upper ocean and the ocean interior are critically dependent upon key groups of organisms. Fluxes between the atmosphere and ocean, as well as between the ocean and the lithosphere, are mediated by organisms that catalyze phase state transitions from either gas to solute/solid or from solute to solid/gas phases. For example, autotrophic carbon fixation converts gaseous CO_2 to a wide variety of organic carbon molecules, virtually all of which are solid or dissolved solids at physiological temperatures. Respiration accomplishes the reverse. Nitrogen fixation converts gaseous N_2 to ammonium and thence to organic molecules, while denitrification accomplishes the reverse. Calcification converts dissolved inorganic carbon and calcium to solid phase calcite and aragonite, whereas silicification converts soluble silicic acid to solid hydrated amorphous opal. Each of these biologically catalyzed processes is dependent upon specific metabolic sequences (i.e., gene families encoding a suite of enzymes) that evolved over hundreds of millions of years of Earth's history, and have, over corresponding periods, led to the massive accumulation of oxygen in the atmosphere, and opal, carbonates, and organic matter in the lithosphere. Presumably, because of parallel evolution as well as lateral gene transfer, these metabolic sequences have frequently co-evolved in several groups of organisms that, more often than not, are not closely related from a phylogenetic standpoint (Falkowski, 1997). Based on their biogeochemical metabolism, these homologous sets of organisms are called functional groups or biogeochemical guilds; i.e., organisms that are related through common biogeochemical processes rather than a common evolutionary ancestor affiliation.

15.10.1 Siliceous Organisms

In the contemporary ocean, the export of particulate organic carbon from the euphotic zone is highly correlated with the flux of particulate silicate. Most of the silicate flux is a consequence of precipitation of dissolved orthosilicic acid by diatoms to form amorphous opal that makes up the cell walls of these organisms. These hard-shelled cell walls presumably help the organisms avoid predation, or if ingested, increase the likelihood of intact gut passage through some metazoans (Smetacek, 1999). In precipitating silicate, diatoms simultaneously fix carbon. Upon depleting the euphotic zone of nutrients, the organisms frequently sink *en masse*, and while some are grazed en route, many sink as intact cells. Ultimately, either fate leads to the gravitationally driven export flux of particulate organic carbon into the ocean interior.

Silica is supplied to the oceans from the weathering of continental rocks. Because of precipitation by silicious organisms, however, the ocean is relatively depleted in dissolved silica. Although diatom frustules (their silicified cell walls) tend to dissolve and are relatively poorly preserved in marine sediments, enough silica is buried to keep the seawater undersaturated, throughout the ocean. As the residence time of silica in the oceans is $\sim 10^4$ yr (i.e., about an order of magnitude longer than the mean deep-water circulation), one can get an appreciation for the silicate demands and regeneration rates by following the concentration gradients of dissolved silica along isopycnals. While these demands are generally attributed to diatoms, radiolarians (a group of nonphotosynthetic, heterotrophic protists with silicious tests that are totally unrelated to diatoms) are not uncommon, and radiolarian shells are abundant in the sediments of Southern Ocean. Silica is also precipitated by various sponges and other protists. As a functional group, the silicate precipitators are identified by their geochemical signatures in the sediments and in the silica chemistry of the oceans.

15.10.2 Calcium Carbonate Precipitation

Like silica precipitation, calcium carbonate is not confined to a specific phylogenetically distinct group of organisms, but evolved (apparently independently) several times in marine organisms. Carbonate sediments blanket much of the Atlantic basin, and are formed from the shells of both coccolithophorids and foraminifera (Milliman, 1993). (In the Pacific, the carbon compensation depth is generally higher than the bottom, and hence, in that basin carbonates tend to dissolve rather than become buried.) As the crystal structure of the carbonates in both groups is calcite (as opposed to the more diagenetically susceptible aragonite), the preservation of these minerals and their co-precipitating trace elements provides an invaluable record of ocean history. Although on geological timescales huge amounts of carbon are removed from the atmosphere and ocean and stored in the lithosphere as carbonates, on ecological timescales, carbonate formation leads to the formation of CO_2. This reaction can be summarized by the following:

$$2HCO_3^- + Ca^{2+} \rightarrow CaCO_3 + CO_2 + H_2O \quad (11)$$

Unlike silicate precipitation, calcium carbonate precipitation leads to strong optical signatures that can be detected both *in situ* and remotely (Balch *et al.*, 1991; Holligan and Balch, 1991). The basic principle of detection is the large, broadband (i.e., "white") scattering cross-sections of calcite. The high scattering cross-sections are detected by satellites observing the upper ocean as relatively highly reflective properties (i.e., a "bright" ocean). Using this detection scheme, one can reconstruct global maps of planktonic calcium carbonate precipitating organisms in the upper ocean. *In situ* analysis can be accompanied by optical rotation properties (polarization) to discriminate calcite from other scattering particles. *In situ* profiles of calcite can be used to construct the vertical distribution of calcium carbonate-precipitating planktonic organisms that would otherwise not be detected by satellite remote sensing because they are too deep in the water column.

Over geological time, the relative abundances of key functional groups change. For example, relative coccolithophorid abundances generally increased through the Mesozoic, and underwent a culling at the Cretaceous Tertiary (K/T) boundary, followed by a general waning throughout the Cenozoic. The changes in the coccolithophorid abundances appear to follow eustatic sea-level variations, suggesting that transgressions lead to higher calcium carbonate deposition. In contrast, diatom sedimentation increases with regressions and, since the K/T impact, diatoms have generally replaced coccolithophorids as ecologically important eukaryotic phytoplankton. On much finer timescales during the Pleistocene, it would appear that interglacial periods favor coccolithophorid abundance, while glacial periods favor diatoms. The factors that lead to glacial–interglacial variations between these two functional groups are relevant to elucidating their distributions in the contemporary ecological setting of the ocean (Tozzi, 2001).

15.10.3 Vacuoles

In addition to a silicic acid requirement, diatoms, in contrast to dinoflagellates and coccolithophores, have evolved a nutrient storage vacuole (Raven, 1987). The vacuole, which occupies ~35% of the volume of the cell, can retain high concentrations of nitrate and phosphate. Importantly, ammonium cannot be (or is not) stored in a vacuole. The vacuole allows diatoms to access and hoard pulses of inorganic nutrients, thereby depriving potentially competing groups of these essential resources. Consequently, diatoms thrive best under eutrophic conditions and in turbulent regions where nutrients are supplied with high pulse frequencies.

The competition between diatoms and coccolithophorids can be easily modeled by a resource acquisition model based on nutrient uptake (Equation (9)). In such a model, diatoms dominate under highly turbulent conditions, when their nutrient storage capacity is maximally advantageous, while coccolithophorids dominate under relatively quiescent conditions (Tozzi *et al.*, 2003).

The geological record during the Pleistocene reveals a periodicity of opal/calcite deposition corresponding to glacial/interglacial periods. Such alterations in mineral deposition are probably related to upper ocean turbulence; i.e., the sedimentary record is a "fax" machine of mixing (Falkowski, 2002). Glacial periods appear to be characterized by higher wind speeds and a stronger thermal contrast between the equator and the poles. These two factors would, in accordance with the simple nutrient uptake model, favor diatoms over coccolithophores. During interglacials, more intense ocean stratification, weaker winds, and a smaller thermal contrast between the equator and the poles would tend to reduce upper-ocean mixing and favor coccolithophores (Iglesias-Rodriguez *et al.*, 2002). While other factors such as silica availability undoubtedly also influenced the relative success of diatoms and coccolithophores on these timescales, we suggest that the climatically forced cycle, played out on timescales of 40 kyr and 100 kyr (over the past 1.9 Myr), can be understood as a long-term competition that never reaches an exclusion equilibrium condition (Falkowski *et al.*, 1998).

Can the turbulence argument be extended to even longer timescales to account for the switch in the dominance from coccolithophorids to diatoms in the Cenozoic? The fossil record of diatoms in the Mesozoic is obscured by problems of preservation; however several species are preserved in the late Jurassic (Harwood and Nikolaev, 1995), suggesting that the origins were in the early Jurassic or perhaps as early as the Triassic. It is clear, however, that this group did not contribute nearly as much to export production during Mesozoic times. We suggest that the ongoing successional displacement of coccolithophores by diatoms in the Cenozoic is, to first order, driven by tectonics (i.e., the Wilson cycle). The Mesozoic period was relatively warm and was characterized by a two-cell Hadley circulation, with obliquity greater than 37°, resulting in a thoroughly mixed atmosphere with nearly uniform temperatures over the surface of the Earth. The atmospheric meridional heat transport decreased the latitudinal thermogradients; global winds and ocean circulation were both sluggish (Huber *et al.*, 1995). This relatively quiescent period of Earth's history was ideal for coccolithophorids. Following the K/T impact, and more critically, the onset of polar ice caps about 32 Ma, the Hadley circulation changed dramatically. Presently, there are six Hadley cells, and the atmosphere has become drier. The net result is more intense thermohaline circulation, greater wind mixing and decreased stability (Barron *et al.*, 1995; Chandler *et al.*, 1992). Associated with this decreased stability is the rise of the diatoms.

Over the past 50 Myr, both carbon and oxygen isotopic records in fossil foraminifera suggest that there has been a long-term depletion of CO_2 in the ocean–atmosphere system and a decrease in temperature in the ocean interior. The result has been increased stratification of the ocean, which has, in turn, led to an increased importance for wind-driven upwelling and mesoscale eddy turbulence in providing nutrients to the euphotic zone. The ecological dominance of diatoms under sporadic mixing conditions suggests that their long-term success in the Cenozoic reflects an increase in event-scale turbulent energy dissipation in the upper ocean. But, was the Wilson cycle the only driver?

Although weathering of siliceous minerals by CO_2 (the so-called "Urey reactions"; but see (Berner and Maasch, 1996)) contributed to the long-term flux of silica to the oceans (Berner, 1990), and potentially fostered the radiation of diatoms in the Cenozoic, by itself, orogeny cannot explain the relatively sharp increase in diatoms at the Eocene/Oligocene boundary. Indeed, the seawater strontium isotope record does not correspond with these radiations in diatoms (Raymo and Ruddiman, 1992). We must look for other contributing processes.

Shortly after, or perhaps coincident with Paleocene thermal maximum (55 Ma), was a rise in true grasses (Retallack, 2001). This group, which rapidly radiated in the Eocene, rose to prominence in Oligocene, a period coincident with a global climatic drying. During this period, however, there was a rapid co-evolution of grazing ungulates that displaced browsers (Janis and Damuth, 1990). Grasses contain up to 10% dry weight of silica, which forms micromineral deposits in the cell walls; phytoliths (Conley, 2002). Indeed, the selection of hypsodont (high crown) dentition in ungulates from the brachydont (leaf eating) early-appearing browsing mammals, coincides with the widespread distribution of phytoliths and grit in grassland forage. It is tempting to suggest that the rise of grazing ungulates, which spurred the radiation of grasses, was, in effect, a biologically catalyzed silicate weathering process. The deep-root structure of Eocene grasses certainly facilitated silicate mobilization into rivers and groundwaters (Conley, 2002). Additionally, upon their annual death and decay, the phytoliths of many temperate grasses are potentially transported to the oceans via wind.

The feedback between the co-evolution of mammals and grasses and the supply of silicates to the ocean potentially explains the rapid radiation of diatoms, and their continued dominance in the Cenozoic. There is another potential feedback at play, however, which "locked in" the diatom preeminence. It is likely that the increase in diatom dominance, and the associated increase in the efficiency of carbon burial, played a key role in decreasing atmospheric CO_2 over the past 32 Myr. That biological selection may influence climate is clearly controversial; however, the trends in succession between taxa on timescales of tens of millions of years, and cycles in dominance on shorter geological timescales beg for explanation.

15.11 HIGH-NUTRIENT, LOW-CHLOROPHYLL REGIONS—IRON LIMITATION

On ecological timescales, the biologically mediated net exchange of CO_2 between the ocean and atmosphere is limited by nutrient supply and the efficiency of nutrient utilization in the euphotic zone. There are three major areas of the world ocean where inorganic nitrogen and phosphate are in excess throughout the year, yet the mixed-layer depth appears to be shallower than the critical depth; these are the eastern equatorial Pacific, the subarctic Pacific, and Southern (i.e., Antarctic) Oceans. In the subarctic North Pacific, it has been suggested that there is a tight coupling between phytoplankton production and consumption by zooplankton (Miller *et al.*, 1991). This grazer-limited hypothesis has been used to explain why the phytoplankton in the North Pacific do not form massive blooms in the spring and summer like their counterparts in the North Atlantic (Banse, 1992). In the mid-1980s, however, it became increasingly clear that the concentration of trace metals, especially iron, was extremely low in all three of these regions (Martin, 1991). Indeed, in the eastern equatorial Pacific, for example, the concentration of soluble iron in the euphotic zone is only 100–200 pM. Although iron is the most abundant transition metal in the Earth's crust, in its most commonly occurring form, Fe^{3+}, it is virtually insoluble in seawater. The major source of iron to the euphotic zone is Aeolian dust, originating from continental deserts. In the three major areas of the world oceans with high inorganic nitrogen in the surface waters and low chlorophyll concentrations, the flux of Aeolian iron is extremely low (Duce and Tindale, 1991). In experiments in which iron was artificially added on a relatively large scale to the waters in the equatorial Pacific, Southern Ocean, and subarctic Pacific, there were rapid and dramatic increases in photosynthetic energy conversion efficiency and phytoplankton chlorophyll (Abraham *et al.*, 2000; Behrenfeld *et al.*, 1996; Kolber *et al.*, 1994; Tsuda *et al.*, 2003). Beyond doubt, NPP and export production in all three regions are limited by the availability of a single micronutrient—iron.

15.12 GLACIAL–INTERGLACIAL CHANGES IN THE BIOLOGICAL CO_2 PUMP

In the modern (i.e., interglacial) ocean, two major factors affect iron fluxes. First, changes in land-use patterns and climate over the past several thousand years have had, and continue to have, marked effects on the areal distribution and extent of deserts. At the height of the Roman Empire some 2,000 years ago, vast areas of North Africa were forested, whereas today these same areas are desert. These changes were climatologically induced. Similarly, the Gobi Desert in North Central Asia has increased markedly in modern times. The flux of Aeolian iron from the Sahara Desert fuels photosynthesis for most of the North Atlantic Ocean; that from the Gobi is deposited over much of the North Pacific (Duce and Tindale, 1991). The primary source of iron for the Southern Ocean is Australia, but the prevailing wind vectors constrain the delivery of the terrestrial dust to the Indian Ocean. Consequently, the Southern Ocean is iron limited in the modern epoch (Martin, 1990).

The second factor in this climatological feedback is that the major wind vectors are driven by atmosphere–ocean heat gradients. Changes in thermal gradients between the equator and poles lead to changes in wind speed and direction. Wind vectors prior to glaciations appear to have supported high fluxes of iron to the Southern Ocean, thereby presumably stimulating phytoplankton production, the export of carbon to depth, and the drawdown of atmospheric CO_2 appears to have accompanied glaciations in the recent geological past (Berger, 1988).

15.13 IRON STIMULATION OF NUTRIENT UTILIZATION

The enhancement of *net* export production (i.e., "true" new production) requires the addition of a limiting nutrient to the ocean, an increase in the efficiency of utilization of preformed nutrients in the upper ocean, and/or a change in the elemental stoichiometry of primary producers (Falkowski *et al.*, 1998; Sarmiento and Bender, 1994). Indeed, an analysis of ice cores from Antarctica, reconstruction of Aeolian iron depositions and concurrent atmospheric CO_2 concentrations over the past 4.2×10^5 yr (spanning four glacial–interglacial cycles) suggests that, when iron fluxes were high, CO_2 levels were low and vice versa (Martin, 1990; Petit *et al.*, 1999). Variations in iron fluxes were presumably a consequence of the areal extent of terrestrial deserts and wind vectors. It is hypothesized that increased fluxes of iron to the high-nutrient, low chlorophyll Southern Ocean stimulated phytoplankton photosynthesis and led to a drawdown of atmospheric CO_2. Model calculations suggest that the magnitude of this drawdown could have been cumulatively significant, and accounted for the observed variations in atmospheric CO_2 recorded in gases trapped in the ice cores. However, the sedimentary records reveal large glacial fluxes of organic carbon in low- and mid-latitude regions; areas that are presumably nutrient impoverished. Was there another factor besides iron addition to high-nutrient, low-chlorophyll regions that contributed to a net export of carbon during glacial times?

Given that N:P ratios in the ocean interior are lower than for the sinking flux, an increase in the net delivery of fixed inorganic nitrogen to the ocean would also potentially contribute to a net drawdown of CO_2. The Antarctic ice core records suggest that atmospheric CO_2 declined from $\sim$290 μmol mol^{-1} to $\sim$190 μmol mol^{-1} over a period of $\sim8\times10^4$ yr between the interglacial and glacial maxima (Sigman and Boyle, 2000). Assuming a C:N ratio of about 6.5 by atoms for the synthesis of new organic matter in the euphotic zone, a simple equilibrium, three-box model calculation suggests that 600 Pg of inorganic carbon should have been fixed by marine photoautotrophs to account for the change in atmospheric CO_2. This amount of carbon is approximately threefold greater than that released to the atmosphere from the cumulative combustion of fossil fuels since the beginning of the Industrial Revolution. The calculated change in atmospheric CO_2 would have required an addition of $\sim$1.5 Tg fixed nitrogen per annum; this is $\sim$2% of the global mean value in the contemporary ocean.

15.14 LINKING IRON TO N_2 FIXATION

Iron also appears to exert a strong constraint on N_2 fixation in the modern ocean. Nitrogen-fixing cyanobacteria, like most cyanobacteria, require relatively high concentrations of iron (Berman-Frank *et al.*, 2001). The high-iron requirements come about because these organisms generally have high requirements for this element in their photosynthetic apparatus (Fujita *et al.*, 1990), as well as for nitrogen fixation and electron carriers that are critical for providing the reductants for CO_2 and N_2 fixation *in vivo*. Increased Aeolian flux of iron to the oceans during glacial periods may have therefore not only stimulated the utilization of nutrients in high-nutrient, low-chlorophyll regions, but also stimulated nitrogen fixation by cyanobacteria, and hence indirectly provided a significant source of new nitrogen. Both effects would have led to increased photosynthetic carbon fixation, and a net drawdown of atmospheric CO_2 (Falkowski, 1997; Falkowski *et al.*, 1998).

15.15 OTHER TRACE-ELEMENT CONTROLS ON NPP

Is iron the only limiting trace element in the sea? Prior to the evolution of oxygenic photosynthesis, the oceans contained high concentrations (~1 mM) of dissolved iron in the form of Fe^{2+} and manganese (>1 mM) in the form of Mn^{2+}, but essentially no copper or molybdenum; these and other elements would have been precipitated as sulfides. Thus, both iron and manganese were readily available to the early photoautotrophs. The availability of these two elements permitted the evolution of the photosynthetic apparatus and the oxygen-evolving system that ultimately became the genetic template for all oxygenic photoautotrophs (Blankenship, 1992). Indeed, the availability of these transition metals, which is largely determined by the oxidation state of the environment, appears to account for their use in photosynthetic reactions. However, over geological time, oxygenic photoautotrophs themselves altered the redox state of the ocean, and hence the availability of the elements in the soluble phase (Anbar and Knoll, 2002).

As photosynthetic oxygen evolution proceeded in the Proterozoic oceans, singlet oxygen ($^1O_2$), peroxide (H_2O_2), superoxide anion radicals ($O_2^{\cdot-}$), and hydroxide radicals ($^{\cdot}OH$) were all formed as by-products (Kasting, 1990; Kasting *et al.*, 1988). These oxygen derivatives can oxidize proteins and photosynthetic pigments as well as cause damage to reaction centers (Asada, 1994). A range of molecules evolved to scavenge or quench the potentially harmful oxygen by-products; these include superoxide dismutase (which converts $O_2^{\cdot-}$ to O_2 and H_2O_2), peroxidase (which reduces H_2O_2 to H_2O by oxidizing an organic cosubstrate for the enzyme), and catalase (which converts $2H_2O_2$ to H_2O and O_2). The oldest superoxide dismutases contained iron and/or manganese, while the peroxidases and catalases contained iron (Asada *et al.*, 1980). These transition metals facilitate the electron transfer reactions that are at the core of the respective enzyme activity, and their incorporation into the proteins undoubtedly occurred because the metals were readily available (Williams and Frausto da Silva, 1996). As O_2 production proceeded, the oxidation of Fe^{2+}, Mn^{2+}, and S^{2-} eventually led to the virtual depletion of these forms of the elements in the euphotic zone of the oceans. The depletion of these elements had profound consequences on the subsequent evolution of life. In the first instance, a number of enzymes were selected that incorporated alternative transition metals that were available in the oxidized ocean. For example, a superoxide dismutase evolved in the green algae, and hence higher plants (and many nonphotosynthetic eukaryotes) that utilized copper and zinc (Falkowski, 1997). Similar metal substitutions occurred in the photosynthetic apparatus and in mitochondrial electron transport chains.

The overall consequence of the co-evolution of oxygenic photosynthesis and the redox state of the ocean is a relatively well-defined trace-element composition of the bulk phytoplankton. Analogous to Redfield's relationship between the macronutrients, trace-element analyses of phytoplankton reveals a relation for trace elements normalized to cell phosphorus of $(C_{125}N_{16}P_1S_{1.3}K_{1.7}Mg_{0.56}Ca_{0.4})_{1000}Sr_{4.4}Fe_{7.5}Mn_{3.8}Zn_{0.80}Cu_{0.38}Co_{0.19}Cd_{0.21}Mo_{0.03}$. The relative composition of transition trace elements in phytoplankton is reflected in that of black shales (Figure 7). Thus, while there is a strong correlation between N:P ratios in phytoplankton and seawater (Anderson and Sarmiento, 1994; Lenton and Watson, 2000), there is no parallel correlation for trace elements (Whitfield, 2001; Morel and Hudson, 1985) (Figure 7 inset). There are two underlying explanations for the lack of a strong relationship. First, low abundance cations that have a valance state of two or higher are often assimilated with particles, whether or not they are metabolically required. Hence, while zinc, manganese, and cobalt are depleted in surface waters and are used in metabolic cycles, mercury, lanthanum, and other rare-earth elements are also depleted and have no known biological function. The profiles of these elements are dictated, to first order, by particle fluxes and their ligands. Second, the absolute abundance of transition metals is critically dependent on the solubility of the source minerals, which is, in turn, regulated by redox chemistry. Most key metabolic pathways evolved prior to the oxidation of Earth's atmosphere and ocean, and hence many of the transition metals selected for catalyzing biological redox reactions reflect their relative abundance under anoxic or suboxic conditions. For example, although the contemporary ocean is oxidized, no known metal can substitute for manganese in the water splitting complex in the photosynthetic apparatus. Similarly, the photosynthetic machinery has maintained a strict requirement for iron for over 2.8 Ga, and all nitrogenases require at least 16 iron atoms per enzyme complex. Some key biological processes do not have the flexibility to substitute trace elements based simply on their availability (Williams and Frausto da Silva, 1996). Hence, while oxygenic photoautotrophs indirectly determine the distribution of the major trace elements in the ocean interior, the distribution of these elements in the soluble phase does not reflect with the composition of the organisms.

On long timescales, seawater concentrations of trace elements reflect the balance between their sources and sinks. For trace elements, the sink

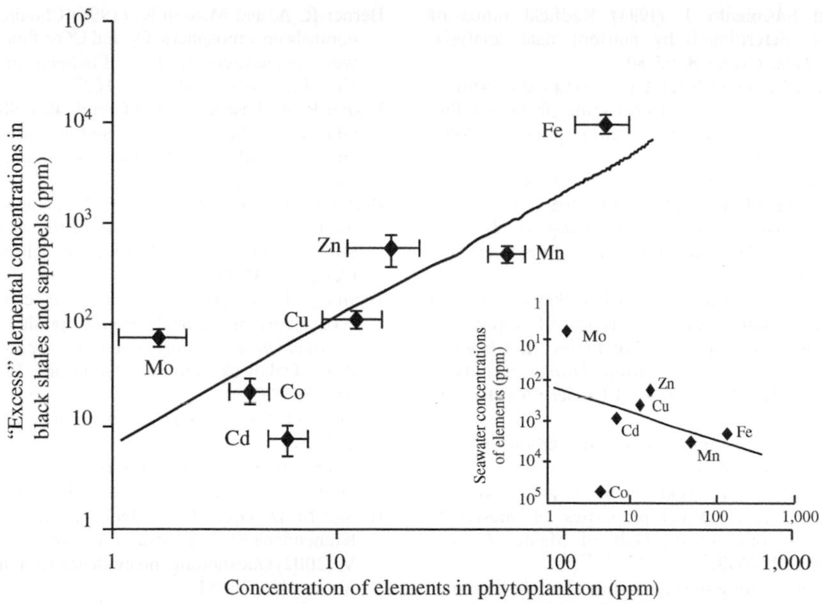

Figure 7 The excess trace element composition in black shales compared with that of calculated marine phytoplankton, and the relationship between trace elements in seawater and phytoplankton (inset) (source Quigg *et al.*, 2003).

term is tightly coupled with the extent of redox conditions throughout the ocean, where periods of extended reducing conditions result in greater partitioning of organic carbon into the deep ocean and sediments. This, in turn, leads to enhanced sequestering of redox-sensitive trace elements into sediments, thereby decreasing their seawater concentrations. These sedimentary rocks have a high content of marine fractions (i.e., organic matter, apatite, biogenic silica and carbonates), and so are enriched by 1–2 orders of magnitude in several trace elements: zinc, copper, nickel, molybdenum, chromium, and vanadium (Piper, 1994). The positive correlation between trace-element ratios in phytoplankton and sediments is consistent with the notion that phytoplankton have imprinted their activities on the lithosphere.

15.16 CONCLUDING REMARKS

The evolution of primary producers in the oceans profoundly changed the chemistry of the atmosphere, ocean, and lithosphere of Earth. The photosynthesis processes catalyzed by ensemble of these organisms not only influences the six major light elements, but directly and indirectly affect every major soluble redox-sensitive trace element and transition metal on Earth's surface. These processes continue to provide, primarily through the utilization of solar radiation, a disequilibrium in geochemical processes, such that Earth maintains an oxidized atmosphere and ocean. This disequilibrium prevents atmospheric oxygen from being depleted, maintains a lowered atmospheric CO_2 concentration, and simultaneously imprints on the ocean interior and lithosphere elemental composition that reflect those of the bulk biological material from which it is derived.

While primary producers in the ocean comprise only ~1% of Earth's biomass, their metabolic rate and biogeochemical impact rivals the much larger terrestrial ecosystem. On geological timescales, these organisms are the little engines that are essential to maintaining life as we know it on this planet.

ACKNOWLEDGMENTS

The author's research is supported by grants from the US National Science Foundation, the National Aeronautical and Space Administration, the US Department of Energy, and the US Department of Defense.

REFERENCES

Abraham E. R., Law C. S., Boyd P. W., Lavender S. J., Maldonado M. T., and Bowie A. R. (2000) Importance of stirring in the development of an iron-fertilized phytoplankton bloom. *Nature* **407**(6805), 727–730.

Aller R. C. (1998) Mobile deltaic and continental shelf muds as fluidized bed reactors. *Mar. Chem.* **61**, 143–155.

Anbar A. D. and Knoll A. H. (2002) Proterozoic ocean chemistry and the evolution: a bioinorganic bridge? *Science* **297**, 1137–1142.

Anderson L. and Sarmiento J. (1994) Redfield ratios of remineralization determined by nutrient data analysis. *Global Biogeochem. Cycles* **8**, 65–80.

Antoine D., Andre J. M., and Morel A. (1996) Oceanic primary production 2. Estimation at global-scale from satellite (coastal zone color scanner) chlorophyll. *Global Biogeochem. Cycles* **10**, 57–69.

Asada K. (1994) Mechanisms for scavenging reactive molecules generated in chloroplasts under light stress. In *Photoinhibition of Photosynthesis: from Molecular Mechanisms to the Field* (eds. N. Baker and J. Bowyer). Bios Scientific, Cambridge, pp. 129–142.

Asada K., Kanematsu S., Okada S., and Hayakawa T. (1980) Phytogenetic distribution of three types of superoxide dismutases in organisms and cell organelles. In *Chemical and Biochemical Aspects of Superoxide Dismutase* (eds. J. V. Bannister and H. A. O. Hill). Elsevier, Amsterdam, pp. 128–135.

Azam F. (1998) Micriobial control of oceanic carbon flux: the plot thickens. *Science* **280**, 694–696.

Balch W. M., Holligan P. M., Ackleson S. G., and Voss K. J. (1991) Biological and optical properties of mesoscale coccolithophore blooms in the Gulf of Maine. *Limnol. Oceanogr.* **36**(4), 629–643.

Banse K. (1977) Determining the carbon-to-chlorophyll ratio of natural phytoplankton. *Mar. Biol.* **41**, 199–212.

Banse K. (1992) Grazing, temporal changes of phytoplankton concentrations, and the microbial loop in the open sea. In *Primary Productivity and Biogeochemical Cycles in the Sea* (ed. P. G. Falkowski). Plenum, New York and London, pp. 409–440.

Banse K. (1994) Grazing and zooplankton production as key controls of phytoplankton production in the open ocean. *Oceanogr.* **7**, 13–20.

Barber R. T. (1992) Geological and climatic time scales of nutrient availability. In *Primary Productivity and Biogeochemical Cycles in the Sea* (eds. P. G. Falkowski and A. Woodhead). Plenum, New York, pp. 89–106.

Barron E. J., Fawcett P. J., Peterson W. H., Pollard D., and Thompson S. L. (1995) A simulation of midcretaceous climate. *Paleoceanography* **10**(5), 953–962.

Beaumont V. I., Jahnke L. L., and Des Marais D. J. (2000) Nitrogen isotopic fractionation in the synthesis of photosynthetic pigments in Rhodobacter capsulatus and Anabaena cylindrica. *Org. Geochem.* **31**(11), 1075–1085.

Behrenfeld M. J. and Falkowski P. G. (1997a) Photosynthetic rates derived from satellite-based chlorophyll concentration. *Limnol. Oceanogr.* **42**, 1–20.

Behrenfeld M. and Falkowski P. (1997b) A consumer's guide to phytoplankton productivity models. *Limnol. Oceanogr.* **42**, 1479–1491.

Behrenfeld M., Bale A., Kolber Z., Aiken J., and Falkowski P. (1996) Confirmation of iron limitation of phytoplankton photosynthesis in the equatorial Pacific. *Nature* **383**, 508–511.

Beja O., Spudich E. N., Spudich J. L., Leclerc M., and DeLong E. F. (2001) Proteorhodopsin phototrophy in the ocean. *Nature* **411**(6839), 786–789.

Berger A. (1988) Milankovitch theory and climate. *Rev. Geophys.* **26**, 624–657.

Berger W. H., Smetacek V. S., *et al.* (eds.) (1989) *Productivity of the Ocean: Present and Past.* Wiley, New York, 471pp.

Berggren W. A., Kent D. V., Swisher C. C., and Aubry M.-P. (1995) A revised Cenozoic geochronology and chronostrati-graphy. In *Geochronology, Time Scales, and Global Stratigraphic Correlations: A Unified Temporal Framework for an Historical Geology,* Spec. Vol. Soc. Econ. Paleontol. Mineral. 54 (eds. W. A. Berggren, D. V. Kent, and J. Hardenbol). pp. 129–212.

Berman-Frank I., Cullen J. T., Shaked Y., Sherrell R. M., and Falkowski P. G. (2001) Iron availability, cellular iron quotas, and nitrogen fixation in Trichodesmium. *Limnol. Oceanogr.* **46**, 1249–1260.

Berner R. A. (1990) Atmospheric carbon dioxide levels over phaneroic time. *Science* **249**, 1382–1386.

Berner R. A. and Maasch K. (1996) Chemical weathering and controls on atmospheric O_2 and CO_2: fundamental principles were enunciated by J. J. Ebelmen in 1845. *Geochem. Geophys. Geosys.* **60**, 1633–1637.

Berner R. A., Lasaga A., and Garrels R. (1983) The carbonate-silicate geochemical cycle and its effect of atmospheric carbon dioxide over the past 100 million years. *Am. J. Sci.* **283**, 641–683.

Berthon J. F. and Morel A. (1992) Validation of a spectral light-photosynthesis model and use of the model in conjunction with remotely sensed pigment observations. *Limnol. Oceanogr.* **37**(4), 781–796.

Bienfang P. K. (1992) The role of coastal high latitude ecosystems in global export production. In *Primary Productivity and Biogeochemical cycles in the Sea* (eds. P. G. Falkowski and A. Woodhead). Plenum, New York. pp. 285–297.

Blackman F. F. (1905) Optima and limiting factors. *Ann. Bot.* **19**, 281–298.

Blankenship R. E. (1992) Origin and early evolution of photosynthesis. *Photosyn. Res.* **33**, 91–111.

Brasier M. D., Green O. R., Jephcoat A. P., Kleppe A. K., Van Kranendonk M. J., Lindsay J. F., Steele A., and Grassinau N. V. (2002) Questioning the evidence for Earth's oldest fossils. *Nature* **416**, 76–81.

Bricaud A. and Morel A. (1987) Atmospheric corrections and interpretation of marine radiances in CZCS imagery: use of a reflectance model. In *Oceanography From Space: Proceedings Of The Atp Symposium On Remote Sensing, Brest, France* vol. **7**, pp. 33–50.

Broecker W. S., Peng T.-H., and Engh R. (1980) Modeling the carbon system. *Radiocarbon* **22**, 565–598.

Brumsack H.-J. (1986) The inorganic geochemistry of Cretaceous black shales (DSDP Leg 41) in comparison to modern upwelling sediments from the Gulf of California. In *North Atlantic Palaeoceanography,* Geological Society Special Publication. No. 21, (eds. C. P. Summerhayes and N. J. Shackleton), pp. 447–462.

Campbell J. W. and Aarup T. (1992) New production in the North Atlantic derived from the seasonal patterns of surface chlorophyll. *Deep-Sea Res.* **39**, 1669–1694.

Capone D. G. and Carpenter E. J. (1982) Nitrogen fixation in the marine environment. *Science* 217(4565), 1140–1142.

Chandler M. A., Rind D., and Ruedy R. (1992) Pangean climate during the Early Jurassic: GMC simulations and the sedimentary record of paleoclimate. *Geol. Soc. Am. Bull.* **104**, 543–559.

Christensen J. P., Murray J. W., Devol A. H., and Codispoti L. A. (1987) Denitrification in continental shelf sediments has major impact on the oceanic nitrogen budget. *Global Biogeochem. Cycles* **1**, 97–116.

Codispoti L. (1995) Is the ocean losing nitrate? *Nature* 376, 724.

Codispoti L. A. and Christensen J. P. (1985) Nitrification, denitrification and nitrous oxide cycling in the eastern tropical south Pacific Ocean. *Mar. Chem.* **16**, 277–300.

Conley D. J. (2002) Terrestrial ecosystems and the global biogeochemical cycle. *Glob. Biogeochem. Cycles* 16, doi: 10.129/2002GB001894.

Copin-Montegut C. and Copin-Montegut G. (1983) Stoichiometry of carbon, nitrogen, and phosporus in marine particulate matter. *Deep-Sea Res.* **30**, 31–46.

Delwiche C. (2000) Tracing the thread of plastid diversity through the tapestry of life. *Am. Nat.* **154**, S164–S177.

Des Marais D. J. (2000) When did photosynthesis emerge on Earth? *Science* **289**, 1703–1705.

Devol A. H. (1991) Direct measurement of nitrogen gas fluxes from continental shelf sediments. *Nature* **349**, 319–321.

Duce R. A. and Tindale N. W. (1991) Atmospheric transport of iron and its deposition in the ocean. *Limnol. Oceanogr.* **36**, 1715–1726.

Dugdale R. C. (1967) Nutrient limitation in the sea: dynamics, identification and significance. *Limnol. Oceanogr.* **12**, 685–695.

Dugdale R. C. and Goering J. J. (1967) Uptake of new and regenerated forms of nitrogen in primary productivity. *Limnol. Oceanogr.* **12**, 196–206.

Dugdale R. and Wilkerson F. (1992) Nutrient limitation of new production in the sea. In *Primary Productivity and Biogeochemical Cycles in the Sea.* (eds. P. G. Falkowski). Plenum, New York and London, pp. 107–122.

Einstein A. (1910) Therioe der Opaleszenz von homogenen Flussigkeiten un Flussigkeitsgemischen in der Nahe des kritischen Zustandes. *Ann. Physik.* **33**, 1275.

Eppley R. W. (1992) Towards understanding the roles of phytoplankton in biogeochemical cycles: personal notes. In *Primary Productivity and Biogeochemical Cycles in the Sea* (eds. P. G. Falkowski and A. D. Woodhead). Plenum, New York and London, pp. 1–7.

Eppley R. W. and Peterson B. J. (1979) Particulate organic matter flux and planktonic new production in the deep ocean. *Nature* **282**, 677–680.

Eppley R. W., Rogers J. N., and McCarthy J. J. (1969) Half-Saturation constant for uptake of nitrate and ammonium by marine phytoplankton. *Limnol. Oceanogr.* **14**, 912–920.

Falkowski P. (1997) Evolution of the nitrogen cycle and its influence on the biological sequestration of CO_2 in the ocean. *Nature* **387**, 272–275.

Falkowski P. G. (2002) On the evolution of the carbon cycle. In *Phytoplankton Productivity: Carbon Assimilation in Marine and Freshwater Ecosystems* (eds. P. I. B. Williams, D. Thomas, and C. Renyolds). Blackwell, Oxford, pp. 318–349.

Falkowski P. G. and Raven J. A. (1997) *Aquatic Photosynthesis.* Blackwell, Oxford, 375pp.

Falkowski P., Barber R., and Smetacek V. (1998) Biogeochemical controls and feedbacks on ocean primary production. *Science* **281**, 200–206.

Fanning K. (1992) Nutrient provinces in the sea: concentration ratios, reaction rate ratios, and ideal covariation. *J. Geophys. Res.* **97C**, 5693–5712.

Farquhar J., Wing B. A., McKeegan K. D., Harris J. W., Cartigny P., and Thiemens M. H. (2002) Mass-independent sulfur of inclusions in diamond and sulfur recycling on early Earth. *Science* **298**(1502), 2369–2372.

Field C., Behrenfeld M., Randerson J., and Falkowski P. (1998) Primary production of the biosphere: integrating terrestrial and oceanic components. *Science* **281**, 237–240.

Fujita Y., Murakami A., and Ohki K. (1990) Regulation of the stoichiometry of thylakoid components in the photosynthetic system of cyanophytes: model experiments showing that control of the synthesis or supply of Ch A can change the stoichiometric relationship between the two photosystems. *Plant. Cell. Physiol.* **31**, 145–153.

Gordon H. R. and Morel A. (1983) *Remote Sensing of Ocean Color for Interpretation of Satellie Visible Imagery: A Review.* Springer, New York.

Gradstein F. M., Agterberg F. P., Ogg J. G., Hardenbol H., van Veen P., Thierry J., and Huang Z. A. (1995) A Triassic, Jurassic, and Cretaceous time scale. In *Geochronology, Time Scales, and Global Stratigraphic Correlations: A Unified Temporal Framework for an Historical Geology* (eds. W. A. Berggren, D. V. Kent, and J. Hardenbol). Spec.Vol.- Soc. Econ. Paleontol. Mineral. **54**, 95–126.

Gruber N. and Sarmiento J. (1997) Global patterns of marine nitrogen fixation and denitrification. *Global Biogeochem. Cycles* **11**, 235–266.

Habicht K. S., Gade M., Thamdrup B., Berg P., and Canfield D. E. (2002) Calibration of sulfate levels in the Archean Ocean. *Science* **298**, 2372–2374.

Harrison W. G. (1980) Nutrient regeneration and primary production in the sea. In *Primary Production in the Sea* (ed. P. G. Falkowski). Plenum, New York, pp. 433–460.

Harwood D. M. and Nikolaev V. A. (1995) Cretaceous diatoms:morphology, taxonomy, biostratigraphy. In *Siliceous Microfossils* (eds. C. D. Blome, P. M. Whalen, and R. Katherine). Paleontological Society, Number 8, pp. 81–106.

Herbert D., Elsworth R., and Telling R. C. (1956) The continuous culture of bacterial a theoretical and experimental study. *J. Gen. Microbiol.* **14**, 601–622.

Holland H. and Rye R. (1998) *Am. J. Sci.* **298**, 621–672.

Holligan P. M. and Balch W. M. (1991) From the ocean to cells: coccolithophore optics and biogeochemistry. In *Particle Analysis in Oceanography* 27 (eds. S. Demers). Springer, Berlin, pp. 301–324.

Huber B. T., Hodell D., and Hamilton C. (1995) Middle–Late Cretaceous climate of the southern high Latitudes—stable isotopic evidence for minimal equator-to-pole thermal gradients. *Geol. Soc. Am. Bull.* **107**, 1164–1191.

Iglesias-Rodriguez D. M., Brown C. W., Doney S. C., Kleypas J. A., Kolber D., Kolber Z., Hayes P. K., and Falkowski G. P. (2002) Representing key phytoplankton functional groups in ocean carbon cycle models: coccolithophorids. *Global Biogeochem. Cycles* **16** (in press).

Janis N. and Damuth J. (1990) Mammals. In *Evolutionary Trends* (ed. K. McNammara). Belknap, London, pp. 301–345.

Johnston A. M. and Raven J. A. (1987) The C{-4}-like characteristics of intertidal macroalga Ascophyllum nedosum (Fucales, Phaeophyta). *Phycologia* **26**(2), 159–166.

Kasting J. F. (1990) Bolide impacts and the oxidation state of carbon in the Earth's early atmosphere. *Origins Life Evol. Biosphere* **20**, 199–231.

Kasting J. F. (1993) Earth's early atmosphere. *Science* **259**, 920–926.

Kasting J. F., Toon O. B., and Pollack J. B. (1988) How climate evolved on the terrestrial planets. *Sci. Am.* **258**, 90–97.

Katz M. E., Wright J. D., Miller K. G., Cramer B. S., Fennel K., and Falkowski P. G. Biological overprint of the geological carbon cycle. *Nature* (in review).

Kolber Z. S., Barber R. T., Coale K. H., Fitzwater S. E., Greene R. M., Johnson K. S., Lindley S., and Falkowski P. G. (1994) Iron limitation of phytoplankton photosynthesis in the Equatorial Pacific Ocean. *Nature* **371**, 145–149.

Kromer S. (1995) Respiration during photosynthesis. *Ann. Rev. Plant Physiol. Plant Mol. Biol.* **0046**, 00045–00070.

Kump L. and Arthur M. (1999) Interpreting carbon-isotope excursions: carbonates and organic matter. *Chem. Geol.* **161**, 181–198.

Laws E. A., Popp B. N., and Bidigare R. (1997) Effect of growth rate and CO_2 concentration on carbon siotoic fractionation by the marine diatom *Phaeodactylum tricornutum. Limnol. Oceanogr.* (in press).

Lenton T. and Watson A. (2000) Redfield revisited: 2. What regulates the oxygen content of the atmosphere. *Global Biogeochem. Cycles* **14**, 249–268.

Letelier R. M., Dore J. E., Winn C. D., and Karl D. M. (1996) Seasonal and interannual variations in photosynthetic carbon assimilation at Station ALOHA. *Deep-Sea Res.* **43**, 467–490.

Liebig J. (1840) *Chemistry and its Application to Agriculture and Physiology.* Taylor and Walton, London.

Lindeman R. (1942) The trophic-dynamic aspect of ecology. *Ecology* **23**, 399–418.

Lipps J. H. (ed.) (1993) *Fossil Prokaryotes and Protists.* Blackwell, Oxford.

Martin J. H. (1990) Glacial-interglacial CO_2 change: the iron hypothesis. *Paleoceangraphy* **5**, 1–13.

Martin J. H. (1991) Iron, Liebig's Law and the greenhouse. *Oceanography* **4**, 52–55.

McElroy M. (1983) Marine biological controls on atmospheric CO_2 and climate. *Nature* **302**, 328–329.

Miller C. B., Frost B. W., Wheeler P. A., Landry M. R.,Welschmeyer N., and Powell T. M. (1991) Ecological dynamics in the subarctic Pacific, a possibly iron-limited ecosystem. *Limnol. Oceanogr.* **36**, 1600–1615.

Milliman J. (1993) Production and accumulation of calcium carbonate in the ocean: budget of a nonsteady state. *Global Biogeochem. Cycles* **7**, 927–957.

Monod J. (1942) *Recheres sur la croissance des cultures bacteriennes.* Hermann & Cie, Paris.

Morel A. (1974) Optical properties of pure water and pure seawater. In *Optical Aspects of Oceanography* (eds. N. G. Jerlov and E. S. Nielsen). Academic Press, London, pp.1–24.

Morel A. (1988) Optical modeling of the upper ocean in relation to it's biogenous matter content (case one waters). *J. Geophys. Res.* **93**, 10749–10768.

Morel F. M. M. and Hudson R. J. (1985) The geobiological cycle of trace elements in aquatic systems: Redfield revisited. In *Chemical Processes in Lakes* (ed. W. Strumm). Wiley-Interscience, pp. 251–281.

Morel A. and Andre J. M. (1991) Pigment distribution and primary production in the western Mediterranean as derived and modeled from coastal zone color scanner observations. *J. Geophys. Res. C. Oceans* **96**(C7), 12685–12698.

Mulkidjanian A. and Junge W. (1997) On the origin of photosynthesis as inferred from sequence analysis—a primordial UV-protector as common ancestor of reaction centers and antenna proteins. *Photosyn. Res.* **51**, 27–42.

Nixon S. W., Ammerman J. W., Atkinson L. P., Berounsky V. M., Billen G., Boicourt W. C., Boynton W. R., Church T. M., Ditoro D. M., Elmgren R., Garber J. H., Giblin A. E., Jahnke R. A., Owens N. J. P., Pilson M. E. Q., and Seitzinger S. P. (1996) The fate of nitrogen and phosphorus at the land-sea margin of the North Atlantic Ocean. *Biogeochemistry* **35**, 141–180.

Odum E. P. (1971) *Fundamentals of Ecology.* Philadelphia.

Pace N. R. (1997) A molecular view of microbial diversity and the biosphere. *Science* **276**(5313), 734–740.

Pavlov A. and Kasting J. (2002) Mass-independent fractionation of sulfur isotopes in Archean sediments: strong evidence for an anoxic Archean atmosphere. *Astrobiology* **2**, 27–41.

Petit J. R., Jouzel J., Raynaud D., Barkov N. I., Barnola J.-M., and Basile I. (1999) Climate and atmospheric history of the past 420,000 years from the Vostok ice core, Antarctica. *Nature* **399**, 429–436.

Pickard G. L. and Emery W. J. (1990) *Descriptive Physical Oceanography.* Pergamon, Oxford.

Piper D. Z. (1994) Seawater as the source of minor elements in black shales, phosphorites and other sedimentary rocks. *Chem. Geol.* **114**, 95–114.

Platt T. (1986) Primary production of the ocean water column as a function of surface light intensity: algorithms for remote sensing. *Deep-Sea Res.* **33**, 149–163.

Platt T. and Irwin B. (1973) Caloric content of phytoplankton. *Limnol. Oceanogr.* **18**, 306–310.

Platt T. and Sathyendranath S. (1988) Oceanic primary production: estimation by remote sensing at local and regional scales. *Science* **241**, 1613–1620.

Postgate J. R. (ed.) (1971) *The Chemistry and Biochemistry of Nitrogen Fixation.* Plenum, New York.

Quigg A., Finkel Z. V., Irwin A. J., Rosenthal Y., Ho T.-Y., Reinfelder J. R., Schofield O., More F., and Falkowski P. (2003) Plastid inheritance of elemental stoichiometry in phytoplankton and its imprint on the geological record. *Nature* **425**, 291–294.

Ramus J. (1992) Productivity of seaweeds. In *Primary Productivity and Biogeochemical Cycles in the Sea* (ed. P. G. Falkowski). Plenum, New York and London, pp. 239–255.

Raven J. A. (1987) The role of vacuoles. *New Phytol.* **106**, 357–422.

Raymo M. and Ruddiman W. (1992) Tectonic forcing of the late Cenozoic climate. *Nature* **359**, 117–122.

Redfield A. C. (1934) *On the Proportions of Organic Derivatives in Sea Water and Their Relation to the Composition of Plankton.* Liverpool, James Johnstone Memorial Volume, Liverpool Univ. Press, pp. 176–192.

Redfield A. C. (1958) The biological control of chemical factors in the environment. *Am. Sci.* **46**, 205–221.

Redfield A., Ketchum B., and Richards F. (1963) The influence of organisms on the composition of sea-water. In *The Sea* (ed. M. Hill). Interscience, New York, **2**, 26–77.

Retallack G. (2001) Cenozoic expansion of grasslands and climatic cooling. *J. Geol.* **109**, 407–426.

Sancetta C., Villareal T., and Falkowski P. G. (1991) Massive fluxes of rhizosolenoid diatoms: a common occurrence? *Limnol. Oceanogr.* **36**, 1452–1457.

Sarmiento J. L. and Bender M. (1994) Carbon biogeochemistry and climate change. *Photosyn. Res.* **39**, 209–234.

Sarmiento J. L., Orr J. C., and Siegenthaler U. (1992) A perturbation simulation of CO_2 uptake in an ocean general circulation model. *J. Geophys. Res.* **94**, 3621–3645.

Schlesinger W. H. (1997) *Biogeochemistry: An Analysis of Global Change.* Academic Press, New York.

Schopf J. (1993) Microfossils of the early Archean Apex Chert: new evidence of the antiquity of life. *Science* **260**, 640–646.

Siegenthaler U. and Sarmiento J. L. (1993) Atmospheric carbon dioxide and the ocean. *Nature* **365**, 119–125.

Sigman D. and Boyle E. (2000) Glacial/interglacial variations in atmospheric carbon dioxide. *Nature* **407**, 859–869.

Smetacek V. (1999) Diatoms and the ocean carbon cycle.*Protist* **150**, 25–32.

Smith D. F. (1981) Tracer kinetic analysis applied to problems in marine biology. In *Physiological Bases of Phytoplankton Ecology* (ed. T. Platt) Can. Bull. Fish Aquat. Sci. **170**, Ottawa, pp. 113–129.

Smith W. O., Jr. and Nelson D. M. (1990) The importance of ice-edge phytoplankton production in the Southern Ocean. *BioScience* **36**, 151–157.

Stumm W. and Morgan J. J. (1981) *Aquatic Chemistry.* Wiley, New York.

Summons R., Jahnke L., Hope J., and Logan G. (1999) 2-Methylhopanoids as biomarkers for cyanobacterial oxygenic photosynthesis. *Nature* **400**, 55–557.

Tilman D. (1982) *Resource Competition and Community Structure.* Princeton University Press, Princeton.

Tozzi S. (2001) Competition and succession of key marine phytoplankton functional groups in a variable environment. *IMCS.* New Brunswick, Rutgers University, 79.

Tozzi S., Schofield O., and Falkowski P. (2003) Turbulence as a selective agent of two phytoplankton functional groups. *Global Change Biology* (in press).

Tsuda A. and Kawaguchi S., *et al.* (2003) Microzooplankton grazing in the surface-water of the Southern Ocean during an Austral Summer. *Polar Biology.*

Tyrell T. (1999) The relative influences of nitrogen and phosphorus on oceanic primary production. *Nature* **400**, 525–531.

Veizer J., Ala D., Azmy K., Bruckschen P., Buhl D., Bruhn F.,Carden G. A. F., Diener A., Ebneth S., Godderis Y., Jasper T., Korte C., Pawellek F., Podlaha O. G., and Strauss H. (1999) $^{87}Sr/^{86}Sr$, $\delta^{13}C$ and $\delta^{18}O$ evolution of Phanerozoic seawater. *Chem. Geol.* **161**, 59–88.

Volk T. and Hoffert M. I. (1985) Ocean carbon pumps: analysis of relative strengths and efficiencies in ocean-driven atmospheric CO_2 exchanges. In *The Carbon Cycle and Atmospheric CO_2: Natural Variations Archean to Present* (eds. E. T. Sunquist and W. S. Broeker). American Geophysical Union. **32**, Washington, DC, pp. 99–110.

Whitfield M. (2001) Interactions between phytoplankton and trace metals in the ocean. *Adv. Mar. Biol.* **41**, 3–128.

Walsh J. J. (1983) Death in the sea: enigmatic phytoplankton losses. *Prog. Oceanogr.* **12**, 1–86.

Warneck P. (1988) *Chemistry of the Natural Atmosphere.* Academic Press, New York.

Williams P. J. L. (1981) Incorporation of microheterotrophic processes into the classical paradigm of the planktonic food web. *Kieler Meeresforsch. Sonderh.* **5**, 1–27.

Williams P. J. L. (1993) On the definition of plankton production terms. *ICES Mar. Sci. Symp.* **197**, 9–19.
Williams R. and Frausto da Silva J. (1996) *The Natural Selection of the Chemical Elements*. Clarendon Press, Oxford.
Woodwell G. M., Whittaker R. H., Reiners W. A., Likens G. E., Delwiche C. C., and Botkin D. B. (1978) The biota and the world carbon budget. *Science* **199**, 141–146.
Wu J., Sunda W., Boyle E. A., and Karl D. M. (2000) Phosphate depletion in the western North Atlantic ocean. *Science* **289**(5480), 759–762.
Xiong J., Fischer W. M., Inoue K., Nakahara M., and Bauer C. E. (2000). Molecular evidence for the early evolution of photosynthesis. *Science* **289**(5485), 1724–1730.
Yoder J. A., McClain C. R., Feldman G. C., Esaias W. E., (1993) Annual cycles of phytoplankton chlorophyll concentrations in the global ocean: a satellite view. *Global Biogeochem. Cycles*, **7** 181–193.
Zehr J. P., Mellon M., Braun S., Litaker W., Steppe T., and Paerl H. W. (1995) Diversity of heterotrophic nitrogen-fixation genes in a marine cyanobacterial mat. *Appl. Environ. Microbiol.* **0061**, 02527–02532.

Readings from the Treatise on Geochemistry
ISBN: 978-0-12-381391-6

pp. 507–536

16 The Contemporary Carbon Cycle

R. A. Houghton

Woods Hole Research Center, MA, USA

16.1 INTRODUCTION

The global carbon cycle refers to the exchanges of carbon within and between four major reservoirs: the atmosphere, the oceans, land, and fossil fuels. Carbon may be transferred from one reservoir to another in seconds (e.g., the fixation of atmospheric CO_2 into sugar through photosynthesis) or over millennia (e.g., the accumulation of fossil carbon (coal, oil, gas) through deposition and diagenesis of organic matter). This chapter emphasizes the exchanges that are important over years to decades and includes those occurring over the scale of months to a few centuries. The focus will be on the years 1980–2000 but our considerations will broadly include the years ~1850–2100.

The carbon cycle is important for at least three reasons. First, carbon forms the structure of all life on the planet, making up ~50% of the dry weight of living things. Second, the cycling of carbon approximates the flows of energy around the Earth, the metabolism of natural, human, and industrial systems. Plants transform radiant energy into chemical energy in the form of sugars, starches, and other forms of organic matter; this energy, whether in living organisms or dead organic matter, supports food chains in natural ecosystems as well as human ecosystems, not the least of which are industrial societies habituated (addicted?) to fossil forms of energy for heating, transportation, and generation of electricity. The increased use of fossil fuels has led to a third reason for interest in the carbon cycle. Carbon, in the form of carbon dioxide (CO_2) and methane (CH_4), forms two of the most important greenhouse gases. These gases contribute to a natural greenhouse effect that has kept the planet warm enough to evolve and support life (without the greenhouse effect the Earth's average temperature would be −33°C). Additions of greenhouse gases to the atmosphere from industrial activity, however, are increasing the concentrations of these gases, enhancing the greenhouse effect, and starting to warm the Earth.

The rate and extent of the warming depend, in part, on the global carbon cycle. If the rate at which the oceans remove CO_2 from the atmosphere were faster, e.g., concentrations of CO_2 would have increased less over the last century. If the processes removing carbon from the atmosphere and storing it on land were to diminish, concentrations of CO_2 would increase more rapidly than projected on the basis of recent history. The processes responsible for adding carbon to, and withdrawing it from, the atmosphere are not well enough understood to predict future levels of CO_2 with great accuracy. These processes are a part of the global carbon cycle.

Some of the processes that add carbon to the atmosphere or remove it, such as the combustion of fossil fuels and the establishment of tree plantations, are under direct human control. Others, such as the accumulation of carbon in the oceans or on land as a result of changes in global climate (i.e., feedbacks between the global carbon cycle and climate), are not under direct human control except through controlling rates of greenhouse gas emissions and, hence, climatic change. Because CO_2 has been more important than all of the other greenhouse gases under human control, combined, and is expected to continue so in the future, understanding the global carbon cycle is a vital part of managing global climate.

This chapter addresses, first, the reservoirs and natural flows of carbon on the earth. It then addresses the sources of carbon to the atmosphere from human uses of land and energy and the sinks of carbon on land and in the oceans that have kept the atmospheric accumulation of CO_2 lower than it would otherwise have been. The chapter describes changes in the distribution of carbon among the atmosphere, oceans, and terrestrial ecosystems over the past 150 years as a result of human-induced emissions of carbon. The processes responsible for sinks of carbon on land and in the sea are reviewed from the perspective of feedbacks, and the chapter concludes with some prospects for the future.

Earlier comprehensive summaries of the global carbon cycle include studies by Bolin *et al.* (1979, 1986), Woodwell and Pecan (1973), Bolin (1981), NRC (1983), Sundquist and Broecker (1985), and Trabalka (1985). More recently, the Intergovernmental Panel on Climate Change (IPCC) has summarized information on the carbon cycle in the context of climate change (Watson *et al.*, 1990; Schimel *et al.*, 1996; Prentice *et al.*, 2001). The basic aspects of the global carbon cycle have been understood for decades, but other aspects, such as the partitioning of the carbon sink between land and ocean, are being re-evaluated continuously with new data and analyses. The rate at which new publications revise estimates of these carbon sinks and re-evaluate the mechanisms that control the magnitude of the sinks suggests that portions of this review will be out of date by the time of publication.

16.2 MAJOR RESERVOIRS AND NATURAL FLUXES OF CARBON

16.2.1 Reservoirs

The contemporary global carbon cycle is shown in simplified form in Figure 1. The four major reservoirs important in the time frame of decades to centuries are the atmosphere, oceans, reserves of fossil fuels, and terrestrial ecosystems, including vegetation and soils. The world's oceans contain ~50 times more carbon than either the atmosphere or the world's terrestrial vegetation, and thus shifts in the abundance of carbon among the major reservoirs will have a much greater significance for the terrestrial biota and for the atmosphere than they will for the oceans.

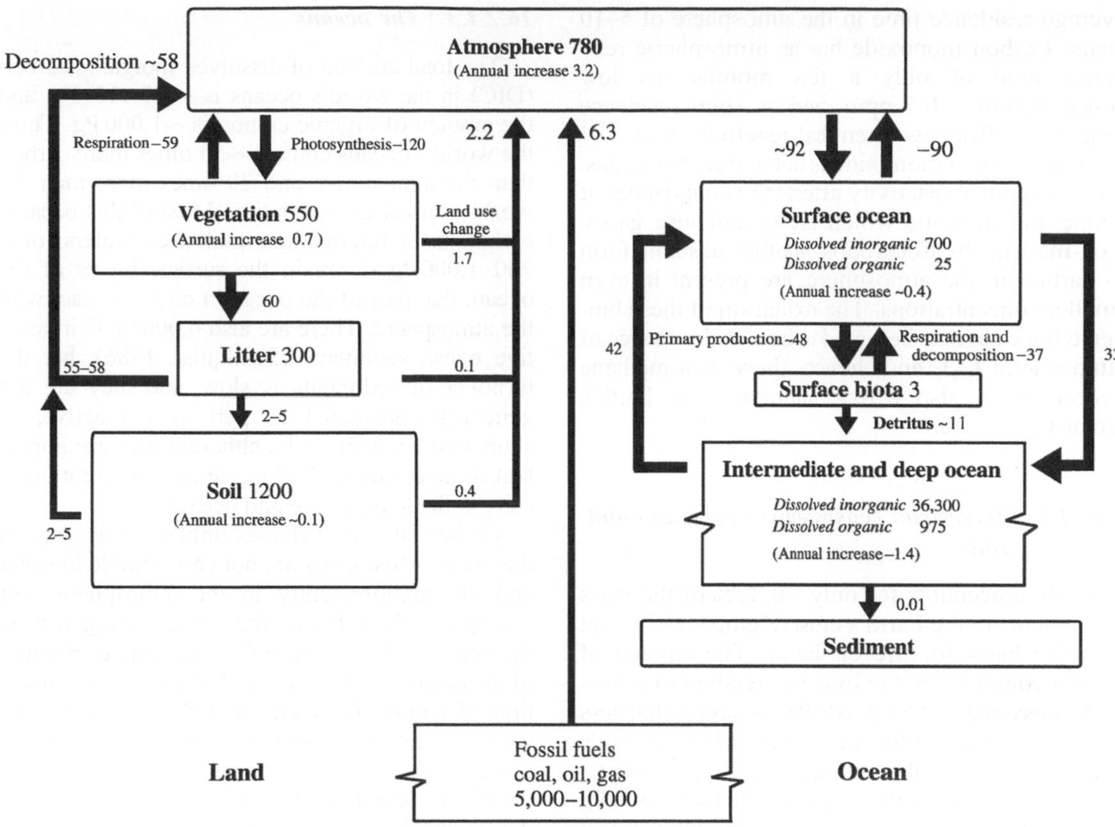

Figure 1 The contemporary global carbon cycle. Units are Pg C or Pg C yr^{-1}.

16.2.1.1 The atmosphere

Most of the atmosphere is made up of either nitrogen (78%) or oxygen (21%). In contrast, the concentration of CO_2 in the atmosphere is only $\sim$0.04%. The concentrations of CO_2 in air can be measured to within one tenth of 0.1 ppmv, or 0.00001%. In the year 2000 the globally averaged concentration was $\sim$0.0368%, or 368 ppmv, equivalent to $\sim$780Pg C (1 Pg = 1 petagram = 10^{15} g = 10^9 t) (Table 1).

The atmosphere is completely mixed in about a year, so any monitoring station free of local contamination will show approximately the same year-to-year increase in CO_2. There are at least 77 stations worldwide, where weekly flask samples of air are collected, analyzed for CO_2 and other constituents, and where the resulting data are integrated into a consistent global data set (Masarie and Tans, 1995; Cooperative Atmospheric Data Integration Project—Carbon Dioxide, 1997). The stations generally show the same year-to-year increase in concentration but vary with respect to absolute concentration, seasonal variability, and other characteristics useful for investigating the global circulation of carbon.

Most of the carbon in the atmosphere is CO_2, but small amounts of carbon exist in concentrations of CH_4, carbon monoxide (CO), and non-methane hydrocarbons. These trace gases are important because they modify the chemical and/or the radiative properties of the Earth's atmosphere. Methane is present at $\sim$1.7 ppm, two orders of magnitude more dilute than CO_2. Methane is a reduced form of carbon, is much less stable than CO_2, and has an

Table 1 Stocks and flows of carbon.

Carbon stocks (Pg C)	
Atmosphere	780
Land	2,000
Vegetation	500
Soil	1,500
Ocean	39,000
Surface	700
Deep	38,000
Fossil fuel reserves	10,000
Annual flows (Pg C yr^{-1})	
Atmosphere-oceans	90
Atmosphere-land	120
Net annual exchanges (Pg C yr^{-1})	
Fossil fuels	6
Land-use change	2
Atmospheric increase	3
Oceanic uptake	2
Other terrestrial uptake	3

average residence time in the atmosphere of 5–10 years. Carbon monoxide has an atmospheric residence time of only a few months. Its low concentration, ~0.1 ppm, and its short residence time result from its chemical reactivity with OH radicals. Carbon monoxide is not a greenhouse gas, but its chemical reactivity affects the abundances of ozone and methane which are greenhouse gases. Non-methane hydrocarbons, another unstable form of carbon in the atmosphere, are present in even smaller concentrations. The oxidation of these biogenic trace gases is believed to be a major source of atmospheric CO, and, hence, these non-methane hydrocarbons also affect indirectly the Earth's radiative balance.

16.2.1.2 Terrestrial ecosystems: vegetation and soils

Carbon accounts for only ~0.27% of the mass of elements in the Earth's crust (Kempe, 1979), yet it is the basis for life on Earth. The amount of carbon contained in the living vegetation of terrestrial ecosystems (550 ± 100 Pg) is somewhat less than that present in the atmosphere (780 Pg). Soils contain 2–3 times that amount (1,500–2,000 Pg C) in the top meter (Table 2) and as much as 2,300 Pg in the top 3 m (Jobbágy and Jackson, 2000). Most terrestrial carbon is stored in the vegetation and soils of the world's forests. Forests cover ~30% of the land surface and hold ~75% of the living organic carbon. When soils are included in the inventory, forests hold almost half of the carbon of the world's terrestrial ecosystems. The soils of woodlands, grasslands, tundra, wetlands, and agricultural lands store most of the rest of the terrestrial organic carbon.

16.2.1.3 The oceans

The total amount of dissolved inorganic carbon (DIC) in the world's oceans is ~3.7×10^4 Pg, and the amount of organic carbon is ~1,000 Pg. Thus, the world's oceans contain ~50 times more carbon than the atmosphere and 70 times more than the world's terrestrial vegetation. Most of this oceanic carbon is in intermediate and deep waters; only 700–1,000 Pg C are in the surface layers of the ocean, that part of the ocean in direct contact with the atmosphere. There are also 6,000 Pg C in reactive ocean sediments (Sundquist, 1986), but the turnover of sediments is slow, and they are not generally considered as part of the active, or short-term, carbon cycle, although they are important in determining the long-term concentration of CO_2 in the atmosphere and oceans.

Carbon dioxide behaves unlike other gases in the ocean. Most gases are not very soluble in water and are predominantly in the atmosphere. For example, only ~1% of the world's oxygen is in the oceans; 99% exists in the atmosphere. Because of the chemistry of seawater, however, the distribution of carbon between air and sea is reversed: 98.5% of the carbon in the ocean–atmosphere systems is in the sea. Although this inorganic carbon is dissolved, less than 1% of it is in the form of dissolved CO_2 (p_{CO_2}); most of the inorganic carbon is in the form of bicarbonate and carbonate ions (Table 3).

About 1,000 Pg C in the oceans (out of the total of 3.8×10^4 Pg) is organic carbon. Carbon in living organisms amounts to ~3 Pg in the sea, in comparison to ~550 Pg on land. The mass of animal life in the oceans is almost the same as on land, however, pointing to the very different trophic structures in the two environments. The ocean's plants are

Table 2 Area, carbon in living biomass, and net primary productivity of major terrestrial biomes

Biome	*Area* (10^9 ha)		*Global carbon stocks* (Pg C)						*Carbon stocks* (Mg C ha^{-1})				*NPP* (Pg C yr^{-1})	
	WBGU	*MRS*	*WGBU*			*MRS IGBP*			*WBGU*		*MRS IGPB*		*Ajtay*	*MRS*
			Plants	*Soil*	*Total*	*Plants*	*Soil*	*Total*	*Plants*	*Soil*	*Plants*	*Soil*		
Tropical forests	17.6	17.5	212	216	428	340	214	553	120	123	194	122	13.7	21.9
Temperate forests	1.04	1.04	59	100	159	139	153	292	57	96	134	147	6.5	8.1
Boreal forests	1.37	1.37	88	471	559	57	338	395	64	344	42	247	3.2	2.6
Tropical savannas and grasslands	2.25	2.76	66	264	330	79	247	326	29	117	29	90	17.7	14.9
Temperate grasslands and shrublands	1.25	1.78	9	295	304	23	176	199	7	236	13	99	5.3	7.0
Deserts and semi-deserts	4.55	2.77	8	191	199	10	159	169	2	42	4	57	1.4	3.5
Tundra	0.95	0.56	6	121	127	2	115	117	6	127	4	206	1.0	0.5
Croplands	1.60	1.35	3	128	131	4	165	169	2	80	3	122	6.8	4.1
Wetlands	0.35		15	225	240				43	643			4.3	
Total	15.12	14.93	466	2,011	2,477	654	1,567	2,221					59.9	62.6

Source: Prentice *et al.* (2001).

Table 3 The distribution of 1,000 CO_2 molecules in the atmosphere–ocean.

Atmosphere	15
Ocean	985
CO_2	5
HCO_3^-	875
CO_3^{2-}	105
Total	1,000

Source: Sarmiento (1993).

microscopic. They have a high productivity, but the production does not accumulate. Most is either grazed or decomposed in the surface waters. Only a fraction (~25%) sinks into the deeper ocean. In contrast, terrestrial plants accumulate large amounts of carbon in long-lasting structures (trees). The distribution of organic carbon between living and dead forms of carbon is also very different on land and in the sea. The ratio is ~1:3 on land and ~1:300 in the sea.

16.2.1.4 Fossil fuels

The common sources of energy used by industrial societies are another form of organic matter, so-called fossil fuels. Coal, oil, and natural gas are the residuals of organic matter formed millions of years ago by green plants. The material escaped oxidation, became buried in the Earth, and over time was transformed to a (fossil) form. The energy stored in the chemical bonds of fossil fuels is released during combustion just as the energy stored in carbohydrates, proteins, and fats is released during respiration.

The difference between the two forms of organic matter (fossil and nonfossil), from the perspective of the global carbon cycle, is the rate at which they are cycled. The annual rate of formation of fossil carbon is at least 1,000 times slower than rates of photosynthesis and respiration. The formation of fossil fuels is part of a carbon cycle that operates over millions of years, and the processes that govern the behavior of this long-term system (sedimentation, weathering, vulcanism, seafloor spreading) are much slower from those that govern the behavior of the short-term system. Sedimentation of organic and inorganic carbon in the sea, e.g., is ~0.2 Pg C yr^{-1}. In contrast, hundreds of petagrams of carbon are cycled annually among the reservoirs of the short-term, or active, carbon cycle. This short-term system operates over periods of seconds to centuries. When young (nonfossil) organic matter is added to or removed from the atmosphere, the total amount of carbon in the active system is unchanged. It is merely redistributed among reservoirs. When fossil fuels are oxidized, however, the CO_2 released represents a net increase in the amount of carbon in the active system.

The amount of carbon stored in recoverable reserves of coal, oil, and gas is estimated to be 5,000–10,000 Pg C, larger than any other reservoir except the deep sea, and ~10 times the carbon content of the atmosphere. Until ~1850s this reservoir of carbon was not a significant part of the short-term cycle of carbon. The industrial revolution changed that.

16.2.2 The Natural Flows of Carbon

Carbon dioxide is chemically stable and has an average residence time in the atmosphere of about four years before it enters either the oceans or terrestrial ecosystems.

16.2.2.1 Between land and atmosphere

The inorganic form of carbon in the atmosphere (CO_2) is fixed into organic matter by green plants using energy from the Sun in the process of photosynthesis, as follows:

$$6CO_2 + 6H_2O \rightleftharpoons C_6H_{12}O_6 + 6O_2$$

The reduction of CO_2 to glucose ($C_6H_{12}O_6$) stores some of the Sun's energy in the chemical bonds of the organic matter formed. Glucose, cellulose, carbohydrates, protein, and fats are all forms of organic matter, or reduced carbon. They all embody energy and are nearly all derived ultimately from photosynthesis.

The reaction above also goes in the opposite direction during the oxidation of organic matter. Oxidation occurs during the two, seemingly dissimilar but chemically identical, processes of respiration and combustion. During either process the chemical energy stored in organic matter is released. Respiration is the biotic process that yields energy from organic matter, energy required for growth and maintenance. All living organisms oxidize organic matter; only plants and some microbes are capable of reducing CO_2 to produce organic matter.

Approximately 45–50% of the dry weight of organic matter is carbon. The organic carbon of terrestrial ecosystems exists in many forms, including living leaves and roots, animals, microbes, wood, decaying leaves, and soil humus. The turnover of these materials varies from less than one year to more than 1,000 years. In terms of carbon, the world's terrestrial biota is almost entirely vegetation; animals (including humans) account for less than 0.1% of the carbon in living organisms.

Each year the atmosphere exchanges ~120 Pg C with terrestrial ecosystems through photosynthesis and respiration (Figure 1 and Table 1). The uptake of carbon through photosynthesis is gross primary production (GPP). At least half of this production is respired by the plants, themselves (autotrophic respiration (Rs_a)), leaving a net primary production (NPP) of ~60 Pg C yr^{-1}. Recent estimates of global terrestrial NPP vary between 56.4 Pg C $yr^{-1}$and 62.6 Pg C yr^{-1} (Ajtay *et al.*, 1979; Field *et al.*, 1998; Saugier *et al.*, 2001). The annual production of organic matter is what fuels the nonplant world, providing food, feed, fiber, and fuel for natural and human systems. Thus, most of the NPP is consumed by animals or respired by decomposer organisms in the soil (heterotrophic respiration (Rs_h)). A smaller amount (~4 Pg C yr^{-1} globally) is oxidized through fires. The sum of autotrophic and heterotrophic respiration is total respiration or ecosystem respiration (Rs_e). In steady state the net flux of carbon between terrestrial ecosystems and the atmosphere (net ecosystem production (NEP)) is approximately zero, but year-to-year variations in photosynthesis and respiration (including fires) may depart from this long-term balance by as much as 5–6 Pg C yr^{-1}. The annual global exchanges may be summarized as follows:

$$\mathrm{NPP} = \mathrm{GPP} - \mathrm{Rs_a}$$
$$(\sim 60 = 120 - 60\,\mathrm{Pg\ C\ yr^{-1}})$$

$$\mathrm{NEP} = \mathrm{GPP} - \mathrm{Rs_a} - \mathrm{Rs_h}$$
$$(\sim 0 = 120 - 60 - 60\,\mathrm{Pg\,C\,yr^{-1}})$$

$$\mathrm{NEP} = \mathrm{NPP} - \mathrm{Rs_h}$$
$$(\sim 0 = 60 - 60\,\mathrm{Pg\,C\,yr^{-1}})$$

Photosynthesis and respiration are not evenly distributed either in space or over the course of a year. About half of terrestrial photosynthesis occurs in the tropics where the conditions are generally favorable for growth, and where a large proportion of the Earth's land area exists (Table 2). Direct evidence for the importance of terrestrial metabolism (photosynthesis and respiration) can be seen in the effect it has on the atmospheric concentration of CO_2 (Figure 2(a)). The most striking feature of the figure is the regular sawtooth pattern. This pattern repeats itself annually. The cause of the oscillation is the metabolism of terrestrial ecosystems. The highest concentrations occur at the end of each winter, following the season in which respiration has exceeded photosynthesis and thereby caused a net release of CO_2 to the atmosphere. Lowest concentrations occur at the end of each summer, following the season in which photosynthesis has exceeded respiration and drawn CO_2 out of the atmosphere. The latitudinal variability in the amplitude of this oscillation suggests that it is driven largely by northern temperate and boreal ecosystems: the highest amplitudes (up to ~16 ppmv) are in the northern hemisphere with the largest land area. The phase of the amplitude is reversed in the southern hemisphere, corresponding to seasonal terrestrial metabolism there. Despite the high rates of production and respiration in the tropics, equatorial regions are thought to contribute little to this oscillation. Although there is a strong seasonality in precipitation throughout much of the tropics, the seasonal changes in moisture affect photosynthesis and respiration almost equally and thus the two processes remain largely in phase with little or no net flux of CO_2.

16.2.2.2 Between oceans and atmosphere

There is ~50 times more carbon in the ocean than in the atmosphere, and it is the amount of DIC in the ocean that determines the atmospheric concentration of CO_2. In the long term (millennia) the most important process determining the exchanges of carbon between the oceans and the atmosphere is the chemical equilibrium of dissolved CO_2, bicarbonate, and carbonate in the ocean. The rate at which the oceans take up or release carbon is slow on a century timescale, however, because of lags in circulation and changes in the availability of calcium ions. The carbon chemistry of seawater is discussed in more detail in the next section.

Two additional processes besides carbon chemistry keep the atmospheric CO_2 lower than it otherwise would be. One process is referred to as the solubility pump and the other as the biological pump. The solubility pump is based on the fact that CO_2 is more soluble in cold waters. In the ocean, CO_2 is ~2 times more soluble in the cold mid-depth and deep waters than it is in the warm surface waters near the equator. Because sinking of cold surface waters in Arctic and Antarctic regions forms these mid-depth and deep waters, the formation of these waters with high CO_2 keeps the CO_2 concentration of the atmosphere lower than the average concentration of surface waters.

The biological pump also transfers surface carbon to the intermediate and deep ocean. Not all of the organic matter produced by phytoplankton is respired in the surface waters where it is produced; some sinks out of the photic zone to deeper water. Eventually, this organic matter is decomposed at depth and reaches the surface again through ocean circulation. The net effect of the sinking of organic matter is to enrich the deeper waters relative to surface waters and thus to reduce the CO_2 concentration of the atmosphere. Marine photosynthesis and the sinking of organic matter out of the surface water are estimated to keep the concentration of CO_2 in air ~30% of what it would be in their absence.

Together the two pumps keep the DIC concentration of the surface waters ~10% lower than at depth. Ocean models that simulate both carbon

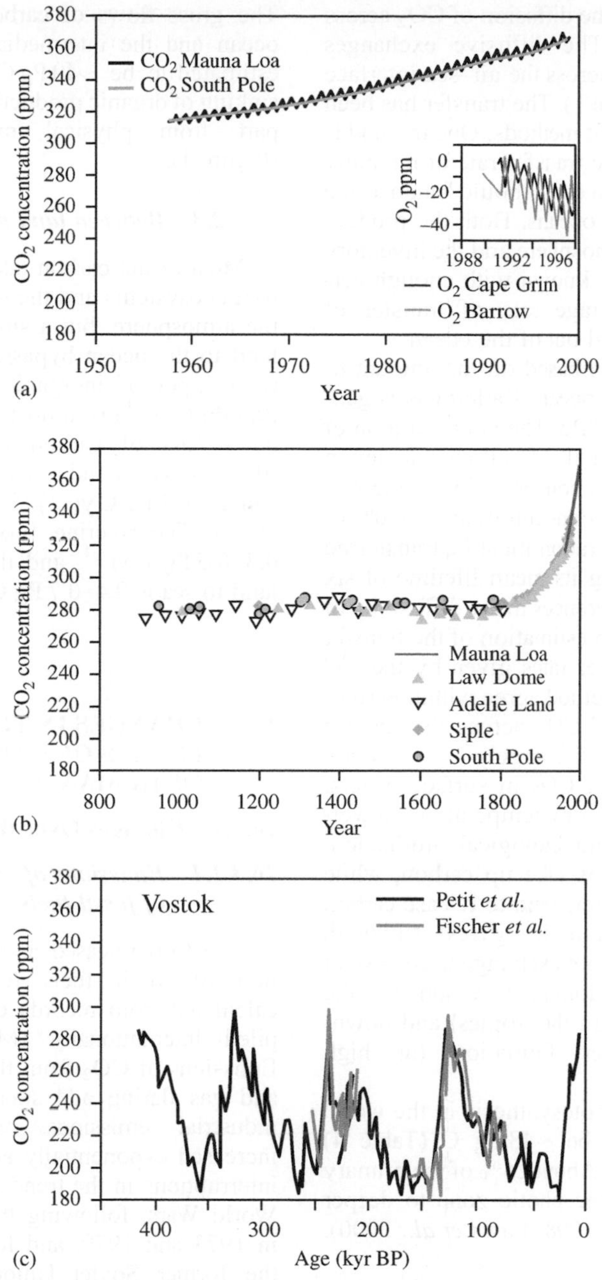

Figure 2 Concentration of CO_2 in the atmosphere: (a) over the last 42 years, (b) over the last 1,000 years, and (c) over the last $\sim 4\times10^5$ years (Prentice *et al.*, 2001) (reproduced by permission of Intergovernmental Panel on Climate Change from *Climate Change 2001: The Scientific Basis*, **2001**, pp. 183–237).

chemistry and oceanic circulation show that the concentration of CO_2 in the atmosphere (280 ppmv pre-industrially) would have been 720 ppmv if both pumps were turned off (Sarmiento, 1993).

There is another biological pump, called the carbonate pump, but its effect in reducing the concentration of CO_2 in the atmosphere is small. Some forms of phytoplankton have $CaCO_3$ shells that, in sinking, transfer carbon from the surface to deeper water, just as the biological pump transfers organic carbon to depth. The precipitation of $CaCO_3$ in the surface waters, however, increases the partial pressure of CO_2, and the evasion of this CO_2 to the atmosphere offsets the sinking of carbonate carbon.

Although ocean chemistry determines the CO_2 concentration of the atmosphere in the long term and the solubility and biological pumps act to modify this long-term equilibrium, short-term exchanges of carbon between ocean and atmosphere result from the diffusion of CO_2 across the air–sea interface. The diffusive exchanges transfer ~90 Pg C yr^{-1} across the air–sea interface in both directions (Figure 1). The transfer has been estimated by two different methods. One method is based on the fact that the transfer rate of naturally produced ^{14}C into the oceans should balance the decay of ^{14}C within the oceans. Both the production rate of ^{14}C in the atmosphere and the inventory of ^{14}C in the oceans are known with enough certainty to yield an average rate of transfer of ~100 Pg C yr^{-1}, into and out of the ocean.

The second method is based on the amount of radon gas in the surface ocean. Radon gas is generated by the decay of ^{226}Ra. The concentration of the parent ^{226}Ra and its half-life allow calculation of the expected radon gas concentration in the surface water. The observed concentration is ~70% of expected, so 30% of the radon must be transferred to the atmosphere during its mean lifetime of six days. Correcting for differences in the diffusivity of radon and CO_2 allows an estimation of the transfer rate for CO_2. The transfer rates given by the ^{14}C method and the radon method agree within ~10%.

The net exchange of CO_2 across the air–sea interface varies latitudinally, largely as a function of the partial pressure of CO_2 in surface waters, which, in turn, is affected by temperature, upwelling or downwelling, and biological production. Cold, high-latitude waters take up carbon, while warm, lower-latitude waters tend to release carbon (outgassing of CO_2 from tropical gyres). Although the latitudinal pattern in net exchange is consistent with temperature, the dominant reason for the exchange is upwelling (in the tropics) and downwelling, or deep-water formation (at high latitudes).

The annual rate of photosynthesis in the world oceans is estimated to be ~48 Pg C (Table 4) (Longhurst *et al.*, 1995). About 25% of the primary production sinks from the photic zone to deeper water (Falkowski *et al.*, 1998; Laws *et al.*, 2000).

Table 4 Annual net primary production of the ocean.

Domain or ecosystem	*NPP* (Pg C yr^{-1})
Trade winds domain (tropical and subtropical)	13.0
Westerly winds domain (temperate)	16.3
Polar domain	6.4
Coastal domain	10.7
Salt marshes, estuaries, and macrophytes	1.2
Coral reefs	0.7
Total	48.3

Source: Longhurst *et al.* (1995).

The gross flows of carbon between the surface ocean and the intermediate and deep ocean are estimated to be ~40 Pg C yr^{-1}, in part from the sinking of organic production (11 Pg C yr^{-1}) and in part from physical mixing (33 Pg C yr^{-1}) (Figure 1).

16.2.2.3 Between land and oceans

Most of the carbon taken up or lost by terrestrial ecosystems and the ocean is exchanged with the atmosphere, but a small flux of carbon from land to the ocean bypasses the atmosphere. The river input of inorganic carbon to the oceans (0.4 Pg C yr^{-1}) is almost balanced in steady state by a loss of carbon to carbonate sediments (0.2 Pg C yr^{-1}) and a release of CO_2 to the atmosphere (0.1 Pg C yr^{-1}) (Sarmiento and Sundquist, 1992). The riverine flux of organic carbon is 0.3–0.5 Pg C yr^{-1}, and thus, the total flux from land to sea is 0.4–0.7 Pg C yr^{-1}.

16.3 CHANGES IN THE STOCKS AND FLUXES OF CARBON AS A RESULT OF HUMAN ACTIVITIES

16.3.1 Changes Over the Period 1850–2000

16.3.1.1 Emissions of carbon from combustion of fossil fuels

The CO_2 released annually from the combustion of fossil fuels (coal, oil, and gas) is calculated from records of fuel production compiled internationally (Marland *et al.*, 1998). Emissions of CO_2 from the production of cement and gas flaring add small amounts to the total industrial emissions, which have generally increased exponentially since ~1750. Temporary interruptions in the trend occurred during the two World Wars, following the increase in oil prices in 1973 and 1979, and following the collapse of the former Soviet Union in 1992 (Figure 3). Between 1751 and 2000, the total emissions of carbon are estimated to have been ~275 Pg C, essentially all of it since 1860. Annual emissions averaged 5.4 Pg C yr^{-1} during the 1980s and 6.3 Pg C yr^{-1} during the 1990s. Estimates are thought to be known globally to within 20% before 1950 and to within 6% since 1950 (Keeling, 1973; Andres *et al.*, 1999).

The proportions of coal, oil, and gas production have changed through time. Coal was the major contributor to atmospheric CO_2 until the late 1960s, when the production of oil first exceeded that of coal. Rates of oil and gas consumption grew rapidly until 1973. After that their relative rates of growth declined dramatically, such that emissions of carbon from coal were, again, as large as those

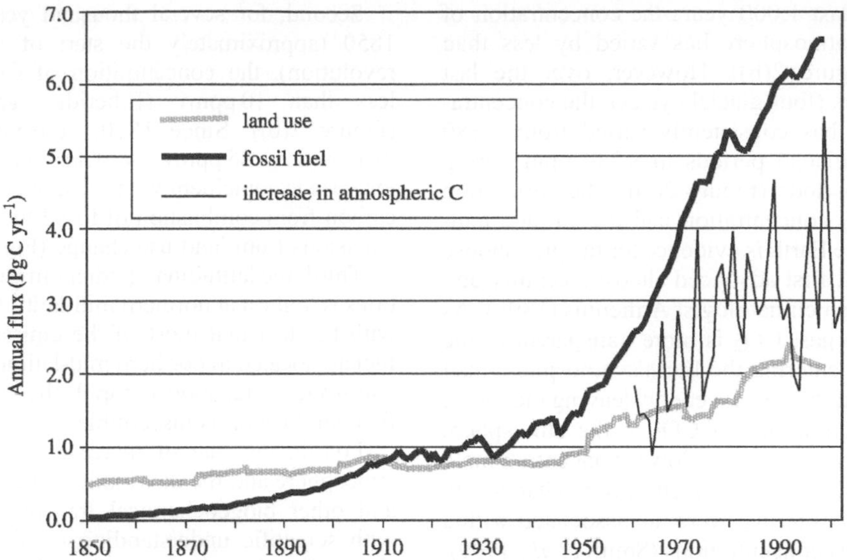

Figure 3 Annual emissions of carbon from combustion of fossil fuels and from changes in land use, and the annual increase in atmospheric CO_2 (in Pg C) since ~1750 (interannual variation in the growth rate of atmospheric CO_2 is greater than variation in emissions).

from oil during the second half of the 1980s and in the last years of the twentieth century.

The relative contributions of different world regions to the annual emissions of fossil fuel carbon have also changed. In 1925, the US, Western Europe, Japan, and Australia were responsible for ~88% of the world's fossil fuel CO_2 emissions. By 1950 the fraction contributed by these countries had decreased to 71%, and by 1980 to 48%. The annual rate of growth in the use of fossil fuels in developed countries varied between 0.5% and 1.4% in the 1970s. In contrast, the annual rate of growth in developing nations was 6.3% during this period. The share of the world's total fossil fuel used by the developing countries has grown from 6% in 1925, to 10% in 1950, to ~20% in 1980. By 2020, the developing world may be using more than half of the world's fossil fuels annually (Goldemberg *et al.*, 1985). They may then be the major source of both fossil fuel and terrestrial CO_2 to the atmosphere (Section 16.3.1.5).

Annual emissions of CO_2 from fossil fuel combustion are small relative to the natural flows of carbon through terrestrial photosynthesis and respiration (~120 Pg C yr^{-1}) and relative to the gross exchanges between oceans and atmosphere (~90 Pg C yr^{-1}) (Figure 1). Nevertheless, these anthropogenic emissions are the major contributor to increasing concentrations of CO_2 in the atmosphere. They represent a transfer of carbon from the slow carbon cycle to the active carbon cycle.

16.3.1.2 The increase in atmospheric CO_2

Numerous measurements of atmospheric CO_2 concentrations were made in the nineteenth century (Fraser *et al.*, 1986), and Callendar (1938) estimated from these early measurements that the amount of CO_2 had increased by 6% between 1900 and 1935. Because of geographical and seasonal variations in the concentrations of CO_2, however, no reliable measure of the rate of increase was possible until after 1957 when the first continuous monitoring of CO_2 concentrations was begun at Mauna Loa, Hawaii, and at the South Pole (Keeling *et al.*, 2001). In 1958 the average concentration of CO_2 in air at Mauna Loa was ~315 ppm. In the year 2000 the concentration had reached ~368 ppm, yielding an average rate of increase of ~1 ppm yr^{-1} since 1958. However, in recent decades the rate of increase in the atmosphere has been ~1.5 ppm yr^{-1} (~3 Pg C yr^{-1}).

During the early 1980s, scientists developed instruments that could measure the concentration of atmospheric CO_2 in bubbles of air trapped in glacial ice. Ice cores from Greenland and Antarctica show that the pre-industrial concentration of CO_2 was between 275 ppm and 285 ppm (Neftel *et al.*, 1985; Raynaud and Barnola, 1985; Etheridge *et al.*, 1996) (Figure 2(b)). The increase between 1700 and 2000, therefore, has been ~85 ppm, equivalent to ~175 Pg C, or 30% of the pre-industrial level.

Over the last 1,000 years the concentration of CO_2 in the atmosphere has varied by less than 10 ppmv (Figure 2(b)). However, over the last 4.2×10^5 years (four glacial cycles), the concentration of CO_2 has consistently varied from ~180 ppm during glacial periods to ~280 ppm during interglacial periods (Figure 2(c)). The correlation between CO_2 concentration and the surface temperature of the Earth is evidence for the greenhouse effect of CO_2, first advanced almost a century ago by the Swedish climatologist Arrhenius (1896). As a greenhouse gas, CO_2 is more transparent to the Sun's energy entering the Earth's atmosphere than it is to the re-radiated heat energy leaving the Earth. Higher concentrations of CO_2 in the atmosphere cause a warmer Earth and lower concentrations a cooler one. There have been abrupt changes in global temperature that were not associated with a change in CO_2 concentrations (Smith *et al.*, 1999), but never in the last 4.2×10^5 years have concentrations of CO_2 changed without a discernible change in temperature (Falkowski *et al.*, 2000). The glacial–interglacial difference of 100 ppm corresponds to a temperature difference of ~10°C. The change reflects temperature changes in the upper troposphere and in the region of the ice core (Vostoc, Antarctica) and may not represent a global average. Today's CO_2 concentration of 368 ppm represents a large departure from the last 4.2×10^5 years, although the expected increase in temperature has not yet occurred.

It is impossible to say that the increase in atmospheric CO_2 is entirely the result of human activities, but the evidence is compelling. First, the known sources of carbon are more than adequate to explain the observed increase in the atmosphere. Balancing the global carbon budget requires additional carbon sinks, not an unexplained source of carbon (see Section 16.3.1.4). Since 1850, ~275 Pg C have been released from the combustion of fossil fuels and another 155 Pg C were released as a result of net changes in land use, i.e., from the net effects of deforestation and reforestation (Section 16.3.1.5). The observed increase in atmospheric carbon was only 175 Pg C (40% of total emissions) over this 150-year period (Table 5).

Table 5 The global carbon budget for the period 1850 to 2000 (units are Pg C).

Fossil fuel emissions	275
Atmospheric increase	175
Oceanic uptake	140
Net terrestrial source	40
Land-use net source	155
Residual terrestrial sink	115

Second, for several thousand years preceding 1850 (approximately the start of the industrial revolution), the concentration of CO_2 varied by less than 10 ppmv (Etheridge *et al.*, 1996) (Figure 2(b)). Since 1850, concentrations have increased by 85 ppmv (~30%). The timing of the increase is coincident with the annual emissions of carbon from combustion of fossil fuels and the net emissions from land-use change (Figure 3).

Third, the latitudinal gradient in CO_2 concentrations is highest at northern mid-latitudes, consistent with the fact that most of the emissions of fossil fuel are located in northern mid-latitudes. Although atmospheric transport is rapid, the signal of fossil fuel combustion is discernible.

Fourth, the rate of increase of carbon in the atmosphere and the distribution of carbon isotopes and other biogeochemical tracers are consistent with scientific understanding of the sources and sinks of carbon from fossil fuels, land, and the oceans. For example, while the concentration of CO_2 has increased over the period 1850–2000, the ^{14}C content of the CO_2 has decreased. The decrease is what would be expected if the CO_2 added to the system were fossil carbon depleted in ^{14}C through radioactive decay.

Concentrations of other carbon containing gases have also increased in the last two centuries. The increase in the concentration of CH_4 has been more than 100% in the last 100 years, from background levels of less than 0.8 ppm to a value of ~1.75 ppm in 2000 (Prather and Ehhalt, 2001). The temporal pattern of the increase is similar to that of CO_2. There was no apparent trend for the 1,000 years before 1700. Between 1700 and 1900 the annual rate of increase was ~1.5 ppbv, accelerating to 15 ppb yr^{-1} in the 1980s. Since 1985, however, the annual growth rate of CH_4 (unlike CO_2) has declined. The concentration is still increasing, but not as rapidly. It is unclear whether sources have declined or whether atmospheric sinks have increased.

Methane is released from anaerobic environments, such as the sediments of wetlands, peatlands, and rice paddies and the guts of ruminants. The major sources of increased CH_4 concentrations are uncertain but are thought to include the expansion of paddy rice, the increase in the world's population of ruminants, and leaks from drilling and transport of CH_4 (Prather and Ehhalt, 2001). Atmospheric CH_4 budgets are more difficult to construct than CO_2 budgets, because increased concentrations of CH_4 occur not only from increased sources from the Earth's surface but from decreased destruction (by OH radicals) in the atmosphere as well. The increase in atmospheric CH_4 has been more significant for the greenhouse effect than it has for the carbon budget. The doubling of CH_4 concentrations since 1700 has amounted to only ~1 ppm, in comparison to the CO_2 increase of almost 90 ppm. Alternatively, CH_4 is, molecule for molecule, ~15 times more effective than CO_2 as a greenhouse gas. Its atmospheric lifetime is only 8–10 years, however.

Carbon monoxide is not a greenhouse gas, but its chemical effects on the OH radical affect the destruction of CH_4 and the formation of ozone. Because the concentration of CO is low and its lifetime is short, its atmospheric budget is less well understood than budgets for CO_2 and CH_4. Nevertheless, CO seems to have been increasing in the atmosphere until the late 1980s (Prather and Ehhalt, 2001). Its contribution to the carbon cycle is very small.

16.3.1.3 Net uptake of carbon by the oceans

As discussed above, the chemistry of carbon in seawater is such that less than 1% of the carbon exists as dissolved CO_2. More than 99% of the DIC exists as bicarbonate and carbonate anions (Table 3). The chemical equilibrium among these three forms of DIC is responsible for the high solubility of CO_2 in the oceans. It also sets up a buffer for changes in oceanic carbon. The buffer factor (or Revelle factor), ξ, is defined as follows:

$$\xi = \frac{\Delta p_{CO_2}/p_{CO_2}}{\Delta \Sigma CO_2/\Sigma CO_2}$$

where p_{CO_2} is the partial pressure of CO_2 (the atmospheric concentration of CO_2 at equilibrium with that of seawater), ΣCO_2 is total inorganic carbon (DIC), and Δ refers to the change in the variable. The buffer factor varies with temperature, but globally averages ~10. It indicates that p_{CO_2} is sensitive to small changes in DIC: a change in the partial pressure of CO_2 (p_{CO_2}) is ~10 times the change in total CO_2. The significance of this is that the storage capacity of the ocean for excess atmospheric CO_2 is a factor of ~10 lower than might be expected by comparing reservoir sizes (Table 1). The oceans will not take up 98% of the carbon released through human activity, but only ~85% of it. The increase in atmospheric CO_2 concentration by ~30% since 1850s has been associated with a change of only ~3% in DIC of the surface waters. The other important aspect of the buffer factor is that it increases as DIC increases. The ocean will become increasingly resistant to taking up carbon (see Section 16.4.2.1).

Although the oceans determine the concentration of CO_2 in the atmosphere in the long term, in the short term, lags introduced by other processes besides chemistry allow a temporary disequilibrium. Two processes that delay the transfer of anthropogenic carbon into the ocean are: (i) the transfer of CO_2 across the air–sea interface and (ii) the mixing of water masses within the sea. The rate of transfer of CO_2 across the air–sea interface was discussed above (Section 16.2.2.2). This transfer is believed to have reduced the oceanic absorption of CO_2 by ~10% (Broecker *et al.*, 1979).

The more important process in slowing the oceanic uptake of CO_2 is the rate of vertical mixing within the oceans. The mixing of ocean waters is determined from measured profiles of natural ^{14}C, bomb-produced ^{14}C, bomb-produced tritium, and other tracers. Profiles of these tracers were obtained during extensive oceanographic surveys: one called Geochemical Ocean Sections (GEOSECS) carried out between 1972 and 1978), a second called Transient Tracers in the Ocean (TTO) carried out in 1981, and a third called the Joint Global Ocean Flux Study (JGOSFS) carried out in the 1990s. The surveys measured profiles of carbon, oxygen, radioisotopes, and other tracers along transects in the Atlantic and Pacific Oceans. The differences between the profiles over time have been used to calculate directly the penetration of anthropogenic CO_2 into the oceans (e.g., Gruber *et al.*, 1996, described below). As of 1980, the oceans are thought to have absorbed only ~40% of the emissions (20–47%, depending on the model used; Bolin, 1986).

Direct measurement of changes in the amount of carbon in the world's oceans is difficult for two reasons: first, the oceans are not mixed as rapidly as the atmosphere, so that spatial and temporal heterogeneity is large; and, second, the background concentration of dissolved carbon in seawater is large relative to the change, so measurement of the change requires very accurate methods. Nevertheless, direct measurement of the uptake of anthropogenic carbon is possible, in theory if not practically, by two approaches. The first approach is based on measurement of changes in the oceanic inventory of carbon and the second is based on measurement of the transfer of CO_2 across the air–sea interface.

Measurement of an increase in oceanic carbon is complicated by the background concentration and the natural variability of carbon concentrations in seawater. The total uptake of anthropogenic carbon in the surface waters of the ocean is calculated by models to have been ~40 μmol kg^{-1} of water. Annual changes would, of course, be much smaller than 40 μmol kg^{-1}, as would the increase in DIC concentrations in deeper waters, where less anthropogenic carbon has penetrated. By comparison, the background concentration of DIC in surface waters is 2,000 μmol kg^{-1}. Furthermore, the seasonal variability at one site off Bermuda was 30 μmol kg^{-1}. Against this background and variability, direct measurement of change is a challenge. Analytical techniques add to uncertainties, although current techniques are capable of a precision of 1.5 μmol kg^{-1} within a laboratory and 4 μmol kg^{-1} between laboratories (Sarmiento, 1993).

A second method for directly measuring carbon uptake by the oceans, measurement of the air–sea exchange, is also made difficult by spatial and temporal variability. The approach measures the concentration of CO_2 in the air and in the surface mixed layer. The difference defines the gradient, which, together with a model that relates the exchange coefficient to wind speed, enables the rate of exchange to be calculated. An average air–sea difference (gradient) of 8 ppm, globally, is equivalent to an oceanic uptake of 2 Pg C yr^{-1} (Sarmiento, 1993), but the natural variability is greater than 10 ppm. Furthermore, the gas transfer coefficient is also uncertain within a factor of 2 (Broecker, 2001).

Because of the difficulty in measuring either changes in the ocean's inventory of carbon or the exchange of carbon across the air–sea interface, the uptake of anthropogenic carbon by the oceans is calculated with models that simulate the chemistry of carbon in seawater, the air–sea exchanges of CO_2, and oceanic circulation.

Ocean carbon models. Models of the ocean carbon cycle include three processes that affect the uptake and redistribution of carbon within the ocean: the air–sea transfer of CO_2, the chemistry of CO_2 in seawater, and the circulation or mixing of the ocean's water masses.

Three tracers have been used to constrain models. One tracer is CO_2 itself. The difference between current distribution of CO_2 in the ocean and the distribution expected without anthropogenic emissions yields an estimate of oceanic uptake (Gruber *et al.*, 1996). The approach is based on changes that occur in the chemistry of seawater as it ages. With age, the organic matter present in surface waters decays, increasing the concentration of CO_2 and various nutrients, and decreasing the concentration of O_2. The hard parts ($CaCO_3$) of marine organisms also decay with time, increasing the alkalinity of the water. From data on the concentrations of CO_2, O_2, and alkalinity throughout the oceans, it is theoretically possible to calculate the increased abundance of carbon in the ocean as a result of the increased concentration in the atmosphere. The approach is based on the assumption that the surface waters were in equilibrium with the atmosphere when they sank, or, at least, that the extent of disequilibrium is known. The approach is sensitive to seasonal variation in the CO_2 concentration in these surface waters.

A second tracer is bomb ^{14}C. The distribution of bomb ^{14}C in the oceans (Broecker *et al.*, 1995), together with an estimate of the transfer of $^{14}CO_2$ across the air–sea interface (Wanninkhof, 1992) (taking into account the fact that $^{14}CO_2$ equilibrates ~10 times more slowly than CO_2 across this interface), yields a constraint on uptake. A third constraint is based on the penetration of CFCs into the oceans (Orr and Dutay, 1999; McNeil *et al.*, 2003).

Ocean carbon models calculate changes in the oceanic carbon inventory. When these changes, together with changes in the atmospheric carbon inventory (from atmospheric and ice core CO_2 data), are subtracted from the emissions of carbon from fossil fuels, the result is an estimate of the net annual terrestrial flux of carbon.

Most current models of the ocean reproduce the major features of oceanic carbon: the vertical gradient in DIC, the seasonal and latitudinal patterns of p_{CO_2} in surface waters, and the interannual variability in p_{CO_2} observed during El Niños (Prentice *et al.*, 2001). However, ocean models do not capture the spatial distribution of ^{14}C at depth (Orr *et al.*, 2001), and they do not show an interhemispheric transport of carbon that is suggested from atmospheric CO_2 measurements (Stephens *et al.*, 1998). The models also have a tight biological coupling between carbon and nutrients, which seems not to have existed in the past and may not exist in the future. The issue is addressed below in Section 16.4.2.2.

16.3.1.4 Land: net exchange of carbon between terrestrial ecosystems and the atmosphere

Direct measurement of change in the amount of carbon held in the world's vegetation and soils may be more difficult than measurement of change in the oceans, because the land surface is not mixed. Not only are the background levels high (~550 Pg C in vegetation and ~1,500 in soils), but the spatial heterogeneity is greater on land than in the ocean. Thus, measurement of annual changes even as large as 3 Pg C yr^{-1}, in background levels 100 times greater, would require a very large sampling approach. Change may be measured over short intervals of a year or so in individual ecosystems by measuring fluxes of carbon, as, e.g., with the eddy flux technique (Goulden *et al.*, 1996), but, again, the results must be scaled up from 1 km^2 to the ecosystem, landscape, region, and globe.

Global changes in terrestrial carbon were initially estimated by difference, i.e., by estimates of change in the other three reservoirs. Because the global mass of carbon is conserved, when three terms of the global carbon budget are known, the fourth can be determined by difference. For the period 1850–2000, three of the terms (275 Pg C released from fossil fuels, 175 Pg C accumulated in the atmosphere, and 140 Pg C taken up by the oceans) define a net terrestrial uptake of 40 Pg C (Table 5). Temporal variations in these terrestrial sources and sinks can also be determined through inverse calculations with ocean carbon models (see Section 16.3.1.3). In inverse mode, models calculate the annual sources and sinks of carbon (output) necessary to produce observed concentrations of CO_2 in the atmosphere (input). Then, subtracting known fossil fuel sources from the calculated sources and sinks yields a residual flux of carbon, presumably terrestrial, because the other terms have been accounted for (the atmosphere and fossil fuels directly, the oceans indirectly). One such inverse calculation or deconvolution (Joos *et al.*, 1999b) suggests that terrestrial ecosystems were a net source of carbon until ~1940 and then became a small net sink. Only in the early 1990s was the net terrestrial sink greater than 0.5 Pg C yr^{-1} (Figure 4).

16.3.1.5 Land: changes in land use

At least a portion of terrestrial sources and sinks can be determined more directly from the large changes in vegetation and soil carbon that result from changes in land use, such as the conversion of forests to cleared lands. Changes in the use of land affect the amount of carbon stored in vegetation and soils and, hence, affect the flux of carbon between land and the atmosphere. The amount of carbon released to the atmosphere or accumulated on land depends not only on the magnitude and types of changes in land use, but also on the amounts of carbon held in different ecosystems. For example, the conversion of grassland to pasture may release no carbon to the atmosphere because the stocks of carbon are unchanged. The net release or accumulation of carbon also depends on time lags introduced by the rates of decay of organic matter, the rates of oxidation of wood products, and the rates of regrowth of forests following harvest or following abandonment of agriculture land. Calculation of the net terrestrial flux of carbon requires knowledge of these rates in different ecosystems under different types of land use. Because there are several important forms of land use and many types of ecosystems in different parts of the world, and because short-term variations in the magnitude of the flux are important, computation of the annual flux requires a computer model.

Changes in terrestrial carbon calculated from changes in land use. Bookkeeping models (Houghton *et al.*, 1983; Hall and Uhlig, 1991; Houghton and Hackler, 1995) have been used to calculate net sources and sinks of carbon resulting from land-use change in all the world's regions.

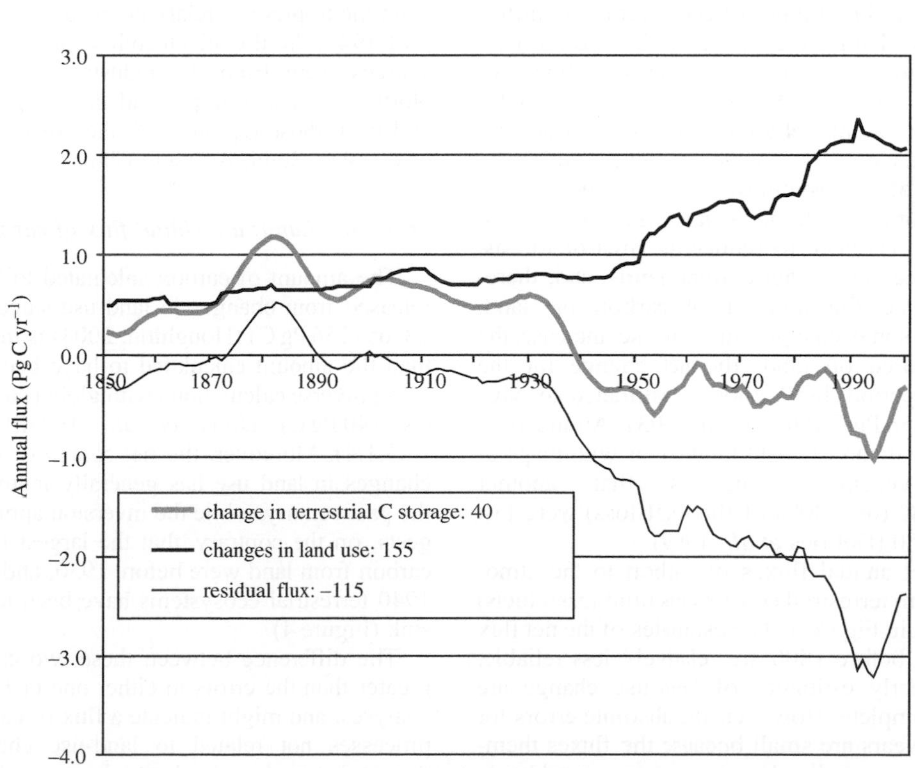

Figure 4 The net annual flux of carbon to or from terrestrial ecosystems (from inverse calculations with an ocean model (Joos *et al.*, 1999b), the flux of carbon from changes in land use (from Houghton, 2003), and the difference between the net flux and the flux from land-use change (i.e., the residual terrestrial sink). Positive values indicate a source of carbon from land and negative values indicate a terrestrial sink.

Calculations are based on two types of data: rates of land-use change and per hectare changes in carbon stocks that follow a change in land use. Changes in land use are defined broadly to include the clearing of lands for cultivation and pastures, the abandonment of these agricultural lands, the harvest of wood, reforestation, afforestation, and shifting cultivation. Some analyses have included wildfire because active policies of fire exclusion and fire suppression have affected carbon storage (Houghton *et al.*, 1999).

Bookkeeping models used to calculate fluxes of carbon from changes in land use track the carbon in living vegetation, dead plant material, wood products, and soils for each hectare of land cultivated, harvested, or reforested. Rates of land-use change are generally obtained from agricultural and forestry statistics, historical accounts, and national handbooks. Carbon stocks and changes in them following disturbance and growth are obtained from field studies. The data and assumptions used in the calculations are more fully documented in Houghton (1999) and Houghton and Hackler (2001).

The calculated flux is not the net flux of carbon between terrestrial ecosystems and the atmosphere, because the analysis does not consider ecosystems undisturbed by direct human activity. Rates of decay and rates of regrowth are defined in the model for different types of ecosystems and different types of land-use change, but they do not vary through time in response to changes in climate or concentrations of CO_2. The processes explicitly included in the model are the ecological processes of disturbance and recovery, not the physiological processes of photosynthesis and respiration.

The worldwide trend in land use over the last 300 years has been to reduce the area of forests, increase the area of agricultural lands, and, therefore, reduce the amount of carbon on land. Although some changes in land use increase the carbon stored on land, the net change for the 150-year period 1850–2000 is estimated to have released 156 Pg C (Houghton, 2003). An independent comparison of 1990 land cover with maps of natural vegetation suggests that another 58–75 Pg C (or ~30% of the total loss) were lost before 1850 (DeFries *et al.*, 1999).

The net annual fluxes of carbon to the atmosphere from terrestrial ecosystems (and fossil fuels) are shown in Figure 3. The estimates of the net flux from land before 1800 are relatively less reliable, because early estimates of land-use change are often incomplete. However, the absolute errors for the early years are small because the fluxes themselves were small. There were no worldwide economic or cultural developments in the eighteenth century that would have caused changes in land use of the magnitude that began in the nineteenth century and accelerated to the present day. The net annual biotic flux of carbon to the atmosphere before 1800 was probably less than 0.5 Pg and probably less than 1 Pg C until ~1950.

It was not until the middle of the last century that the annual emissions of carbon from combustion of fossil fuels exceeded the net terrestrial source from land-use change. Since then the fossil fuel contribution has predominated, although both fluxes have accelerated in recent decades with the intensification of industrial activity and the expansion of agricultural area.

The major releases of terrestrial carbon result from the oxidation of vegetation and soils associated with the expansion of cultivated land. The harvest of forests for fuelwood and timber is less important because the release of carbon to the atmosphere from the oxidation of wood products is likely to be balanced by the storage of carbon in regrowing forests. The balance will occur only as long as the forests harvested are allowed to regrow, however. If wood harvest leads to permanent deforestation, the process will release carbon to the atmosphere.

In recent decades the net release of carbon from changes in land use has been almost entirely from the tropics, while the emissions of CO_2 from fossil fuels were almost entirely from outside the tropics. The highest biotic releases were not always from tropical countries. The release of terrestrial carbon from the tropics is a relatively recent phenomenon, post-1945. In the nineteenth century the major sources were from the industrialized regions—North America, Europe, and the Soviet Union—and from those regions with the greatest numbers of people—South Asia and China.

16.3.1.6 *Land: a residual flux of carbon*

The amount of carbon calculated to have been released from changes in land use since the early 1850s (156 Pg C) (Houghton, 2003) is much larger than the amount calculated to have been released using inverse calculations with global carbon models (40 Pg C) (Joos *et al.*, 1999b) (Section 16.3.1.4). Moreover, the net source of CO_2 from changes in land use has generally increased over the past century, while the inversion approach suggests, on the contrary, that the largest releases of carbon from land were before 1930, and that since 1940 terrestrial ecosystems have been a small net sink (Figure 4).

The difference between these two estimates is greater than the errors in either one or both of the analyses, and might indicate a flux of carbon from processes not related to land-use change. The approach based on land-use change includes only the sources and sinks of carbon directly attributable to human activity; ecosystems not directly modified by human activity are left out of the analysis (assumed neither to accumulate nor release carbon). The approach based on inverse analyses with atmospheric data, in contrast, includes all ecosystems and all processes affecting carbon storage. It yields a net terrestrial flux of carbon. The difference between the two approaches thus suggests a generally increasing terrestrial sink for carbon attributable to factors other than land-use change. Ecosystems not directly cut or cleared could be accumulating or releasing carbon in response to small variations in climate, to increased concentrations of CO_2 in air, to increased availability of nitrogen or other nutrients, or to increased levels of toxins in air and soil resulting from industrialization. It is also possible that management practices not considered in analyses of land-use change may have increased the storage of carbon on lands that have been affected by land-use change. These possibilities will be discussed in more detail below (Section 16.4.1). Interestingly, the two estimates (land-use change and inverse modeling) are generally in agreement before 1935 (Figure 4), suggesting that before that date the net flux of carbon from terrestrial ecosystems was largely the result of changes in land use. Only after 1935 have changes in land use underestimated the net terrestrial carbon sink. By the mid-1990s this annual residual sink had grown to $\sim$3 Pg C yr^{-1}.

16.3.2 Changes Over the Period 1980–2000

The period 1980–2000 deserves special attention not because the carbon cycle is qualitatively different over this period, but because scientists have been able to understand it better. Since 1980 new types of measurements and sophisticated methods of analysis have enabled better estimates of the uptake of carbon by the world's oceans and terrestrial ecosystems. The following section addresses the results of these analyses, first at the global level, and then at a regional level. Attention focuses on the two outstanding questions that have concerned scientists investigating the global carbon cycle since the first carbon budgets were constructed in the late 1960s (SCEP, 1970): (i) How much of the carbon released to the atmosphere from combustion of fossil fuels and changes in land use is taken up by the oceans and by terrestrial ecosystems? (ii) What are the mechanisms responsible for the uptake of carbon? The mechanisms for a carbon sink in terrestrial ecosystems have received considerable attention, in part because different mechanisms have different implications for future rates of CO_2 growth (and hence future global warming).

The previous section addressed the major reservoirs of the global carbon cycle, one at a time. This section addresses the methods used to determine changes in the amount of carbon held on land and in the sea, the two reservoirs for which changes in carbon are less well known. In contrast, the atmospheric increase in CO_2 and the emissions from fossil fuels are well documented. The order in which methods are presented is arbitrary. To set the stage, top-down (i.e., atmospherically based) approaches are described first, followed by bottom–up (ground-based) approaches (Table 6). Although the results of different methods often differ, the methods are not entirely comparable. Rather, they are complementary, and discrepancies sometimes suggest mechanisms responsible for transfers of carbon (Houghton, 2003; House *et al.*, 2003). The results from each method are presented first, and then they are added to an accumulating picture of the global carbon cycle. Again, the emphasis is on, first, the fluxes of carbon to and from terrestrial ecosystems and the ocean and, second, the mechanisms responsible for the terrestrial carbon sink.

Table 6 Characteristics of methods use to estimate terrestrial sinks.

	Geographic limitations	*Temporal resolution*	*Attribution of mechanism(s)*	*Precision*
Inverse modeling: oceanic data	No geographic resolution	Annual	No	Moderate
Land-use models	Data limitations in some regions	Annual	Yes	Moderate
Inverse modeling: atmospheric data	Poor in tropics	Monthly to annual	No	High: North–South Low: East–West
Forest inventories	Nearly nonexistent in the tropics	5–10 years	Yes (age classes)	High for biomass; variable for soil carbon
CO_2 flux	Site specific (a few km^2); difficult to scale up	Hourly to annual	No	Some problems with windless conditions
Physiologically based models	None	Hourly to annual	Yes	Variable; difficult to validate

16.3.2.1 The global carbon budget

(i) Inferring changes in terrestrial and oceanic carbon from atmospheric concentrations of CO_2 and O_2. According to the most recent assessment of climate change by the IPCC, the world's terrestrial ecosystems were a net sink averaging close to zero (0.2 Pg C yr^{-1}) during the 1980s and a significantly larger sink (1.4 Pg C yr^{-1}) during the 1990s (Prentice *et al.*, 2001). The large increase during the 1990s is difficult to explain. Surprisingly, the oceanic uptake of carbon was greater in the 1980s than the 1990s. The reverse would have been expected because atmospheric concentrations of CO_2 were higher in the 1990s. The estimates of terrestrial and oceanic uptake were based on changes in atmospheric CO_2 and O_2 and contained a small adjustment for the outgassing of O_2 from the oceans.

One approach for distinguishing terrestrial from oceanic sinks of carbon is based on atmospheric concentrations of CO_2 and O_2. CO_2 is released and O_2 taken up when fossil fuels are burned and when forests are burned. On land, CO_2 and O_2 are tightly coupled. In the oceans they are not, because O_2 is not very soluble in seawater. Thus, CO_2 is taken up by the oceans without any change in the atmospheric concentration of O_2. Because of this differential response of oceans and land, changes in atmospheric O_2 relative to CO_2 can be used to distinguish between oceanic and terrestrial sinks of carbon (Keeling and Shertz, 1992; Keeling *et al.*, 1996b; Battle *et al.*, 2000). Over intervals as short as a few years, slight variations in the seasonality of oceanic production and decay may appear as a change in oceanic O_2, but these variations cancel out over many years, making the method robust over multiyear intervals (Battle *et al.*, 2000).

Figure 5 shows how the method works. The individual points show average annual global

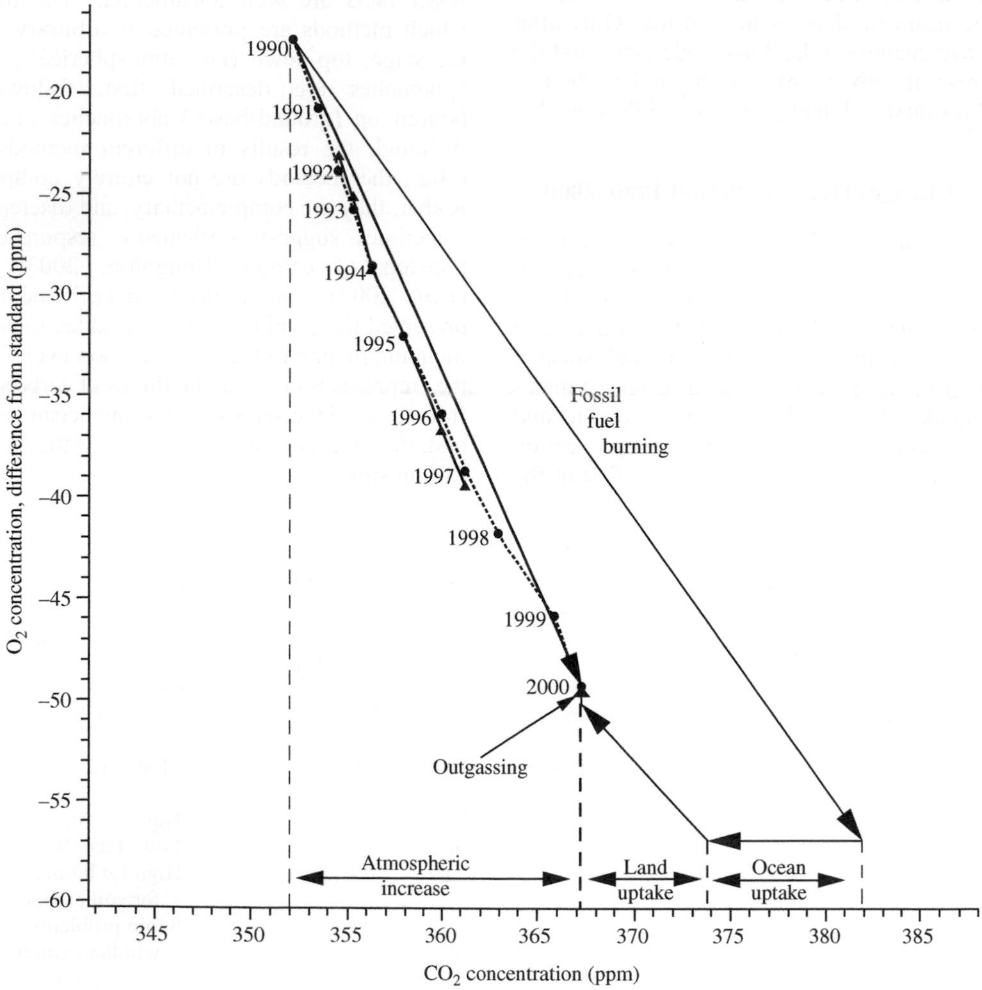

Figure 5 Terrestrial and oceanic sinks of carbon deduced from changes in atmospheric concentrations of CO_2 and O_2 (Prentice *et al.*, 2001) (reproduced by permission of the Intergovernmental Panel on Climate Change from *Climate Change 2001: The Scientific Basis*, **2001**, pp. 183–237).

CO_2/O_2 concentrations over the years 1990–2000. Changes in the concentrations expected from fossil fuel combustion (approximately 1:1) during this interval are drawn, starting in 1990. The departure of these two sets of data confirms that carbon has accumulated somewhere besides the atmosphere. The oceans are assumed not to be changing with respect to O_2, so the line for the oceanic sink is horizontal. The line for the terrestrial sink is approximately parallel to the line for fossil fuel, and drawn through 2000. The intersection of the terrestrial and the oceanic lines thus defines the terrestrial and oceanic sinks. According to the IPCC (Prentice *et al.*, 2001), these sinks averaged 1.4 Pg C yr^{-1} and 1.7 Pg C yr^{-1}, respectively, for the 1990s. The estimate also included a small correction for outgassing of O_2 from the ocean (in effect, recognizing that the ocean is not neutral with respect to O_2).

Recent analyses suggest that such outgassing is significantly larger than initially estimated (Bopp *et al.*, 2002; Keeling and Garcia, 2002; Plattner *et al.*, 2002). The observed decadal variability in ocean temperatures (Levitus *et al.*, 2000) suggests a warming-caused reduction in the transport rate of O_2 to deeper waters and, hence, an increased outgassing of O_2. The direct effect of the warming on O_2 solubility is estimated to have accounted for only 10% of the loss of O_2 (Plattner *et al.*, 2002). The revised estimates of O_2 outgassing change the partitioning of the carbon sink between land and ocean. The revision increases the oceanic carbon sink of the 1990s relative to that of the 1980s (average sinks of 1.7 Pg C yr^{-1} and 2.4 Pg C yr^{-1}, respectively, for the 1980s and 1990s). The revised estimates are more consistent with estimates from ocean models (Orr, 2000) and from analyses based on $^{13}C/^{12}C$ ratios of atmospheric CO_2 (Joos *et al.*, 1999b; Keeling *et al.*, 2001). The revised estimate for land (a net sink of 0.7 Pg C yr^{-1} during the 1990s) (Table 7) is half of that given by the IPCC (Prentice *et al.*, 2001). The decadal change in the terrestrial sink is also much smaller (from 0.4 Pg C yr^{-1} to 0.7 Pg C yr^{-1} instead of from 0.2 Pg C yr^{-1} to 1.4 Pg C yr^{-1}).

(ii) Sources and sinks inferred from inverse modeling with atmospheric transport models and atmospheric concentrations of CO_2, $^{13}CO_2$, and O_2. A second top-down method for determining oceanic and terrestrial sinks is based on spatial and temporal variations in concentrations of atmospheric CO_2 obtained through a network of flask air samples (Masarie and Tans, 1995; Cooperative Atmospheric Data Integration Project—Carbon Dioxide, 1997). Together with models of atmospheric transport, these variations are used to infer the geographic distribution of sources and sinks of carbon through a technique called inverse modeling.

Variations in the carbon isotope of CO_2 may also be used to distinguish terrestrial sources and sinks from oceanic ones. The ^{13}C isotope is slightly heavier than the ^{12}C isotope and is discriminated against during photosynthesis. Thus, trees have a lighter isotopic ratio (−22 ppt to −27 ppt) than does air (−7 ppt) (ratios are expressed relative to a standard). The burning of forests (and fossil fuels) releases a disproportionate share of the lighter isotope, reducing the isotopic ratio of $^{13}C/^{12}C$ in air. In contrast, diffusion of CO_2 across the air–sea interface does not result in appreciable discrimination, so variations in the isotopic composition of CO_2 suggest terrestrial and fossil fuels fluxes of carbon, rather than oceanic.

Spatial and temporal variations in the concentrations of CO_2, $^{13}CO_2$, and O_2 are used with models of atmospheric transport to infer (through inverse calculations) sources and sinks of carbon at the Earth's surface. The results are dependent upon the model of atmospheric transport (Figure 6; Ciais *et al.*, 2000).

The interpretation of variations in ^{13}C is complicated. One complication results from isotopic disequilibria in carbon pools (Battle *et al.*, 2000). Disequilibria occur because the $\delta^{13}C$ taken up by plants, e.g., is representative of the $\delta^{13}C$ currently in the atmosphere (allowing for discrimination), but the $\delta^{13}C$ of CO_2 released through decay represents not the $\delta^{13}C$ of the current atmosphere but of an atmosphere several decades ago. As long as the $\delta^{13}C$ of the atmosphere is changing, the $\delta^{13}C$ in pools will reflect a mixture of earlier and current conditions. Uncertainties in the turnover of various carbon pools add uncertainty to interpretation of

Table 7 The global carbon budget (Pg C yr^{-1}).

	1980s	*1990s*
Fossil fuel emissions[a]	5.4 ± 0.3	6.3 ± 0.4
Atmospheric increase[a]	3.3 ± 0.1	3.2 ± 0.2
Oceanic uptake[b]	−1.7 + 0.6 (−1.9 ± 0.6)	−2.4 + 0.7 (−1.7 ± 0.5)
Net terrestrial flux[b]	−0.4 + 0.7 (−0.2 ± 0.7)	−0.7 + 0.8 (−1.4 ± 0.7)
Land-use change[c]	2.0 ± 0.8	2.2 ± 0.8
Residual terrestrial flux	−2.4 + 1.1 (−2.2 ± 1.1)	−2.9 + 1.1(−3.6 ± 1.1)

[a] Source: Prentice *et al.* (2001). [b] Source: Plattner *et al.* (2002) (values in parentheses are from Prentice *et al.*, 2001). [c] Houghton (2003).

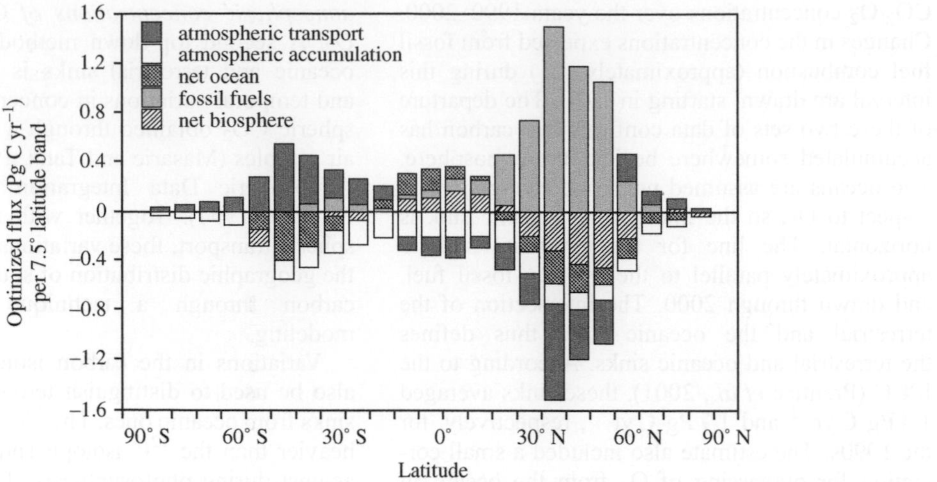

Figure 6 Terrestrial and oceanic sources and sinks of carbon inferred from inverse calculations with an atmospheric transport model and spatial and temporal variations in CO_2 concentrations. The net fluxes inferred over each region have been averaged into 7.5°-wide latitude strips (Ciais *et al.*, 2000) (reproduced by permission of the Ecological Society of America from *Ecol. Appl.*, **2000**, *10*, 1574–1589).

the $\delta^{13}C$ signal. Another complication results from unknown year-to-year variations in the photosynthesis of C_3 and C_4 plants (because these two types of plants discriminate differently against the heavier isotope). C_4 plants discriminate less than C_3 plants and leave a signal that looks oceanic, thus confounding the separation of land and ocean exchanges. These uncertainties of the $\delta^{13}C$ approach are most troublesome over long periods (Battle *et al.*, 2000); the approach is more reliable for reconstructing interannual variations in sources and sinks of carbon.

An important distinction exists between global approaches (e.g., O_2, above) and regional inverse approaches, such as implemented with ^{13}C. In the global top-down approach, changes in terrestrial or oceanic carbon *storage* are calculated. In contrast, the regional inverse method yields *fluxes* of carbon between the land or ocean surface and the atmosphere. These fluxes of carbon include both natural and anthropogenic components. Horizontal exchange between regions must be taken into account to estimate changes in storage. For example, the fluxes will not accurately reflect changes in the amount of carbon on land or in the sea if some of the carbon fixed by terrestrial plants is transported by rivers to the ocean and respired there (Sarmiento and Sundquist, 1992; Tans *et al.*, 1995; Aumont *et al.*, 2001).

An example of inverse calculations is the analysis by Tans *et al.* (1990). The concentration of CO_2 near the Earth's surface is ~3 ppm higher over the northern mid-latitudes than over the southern hemisphere. The "bulge" in concentration over northern mid-latitudes is consistent with the emissions of carbon from fossil fuel combustion at these latitudes. The extent of the bulge is also affected by the rate of atmospheric mixing. High rates of mixing would dilute the bulge; low rates would enhance it. By using the latitudinal gradient in CO_2 and the latitudinal distribution of fossil fuel emissions, together with a model of atmospheric transport, Tans *et al.* (1990) determined that the bulge was smaller than expected on the basis of atmospheric transport alone. Thus, carbon is being removed from the atmosphere by the land and oceans at northern mid-latitudes. Tans *et al.* estimated removal rates averaging between 2.4 Pg C yr^{-1} and 3.5 Pg C yr^{-1} for the years 1981–1987. From p_{CO_2} measurements in surface waters, Tans *et al.* calculated that the northern mid-latitude oceans were taking up only 0.2–0.4 Pg C yr^{-1}, and thus, by difference, northern mid-latitude lands were responsible for the rest, a sink of 2.0–3.4 Pg C yr^{-1}. The range resulted from uncertainties in atmospheric transport and the limited distribution of CO_2 sampling stations. Almost no stations exist over tropical continents. Thus, Tans *et al.* (1990) could not constrain the magnitude of a tropical land source or sink, but they could determine the magnitude of the northern sink relative to a tropical source. A large tropical source, as might be expected from deforestation, implied a large northern sink; a small tropical source implied a smaller northern sink.

The analysis by Tans *et al.* (1990) caused quite a stir because their estimate for oceanic uptake was only 0.3–0.8 Pg C yr^{-1}, while analyses based on ocean models yielded estimates of 2.0 ± 0.8 Pg C yr^{-1}. The discrepancy was subsequently reconciled (Sarmiento and Sundquist, 1992) by accounting for the effect of skin temperature on the calculated air–sea exchange, the effect of atmospheric transport and oxidation of CO on the carbon budget, and the effect of riverine transport of carbon on changes in carbon storage (see below). All of the adjustments increased the estimated oceanic uptake of carbon to values obtained by ocean models and lowered the estimate of the mid-latitude terrestrial sink.

Similar inverse approaches, using not only CO_2 concentrations but also spatial variations in O_2 and $^{13}CO_2$ to distinguish oceanic from terrestrial fluxes, have been carried out by several groups since 1990. An intercomparison of 16 atmospheric transport models (the TransCom 3 project) by Gurney *et al.* (2002) suggests average oceanic and terrestrial sinks of 1.3 Pg C yr^{-1} and 1.4 Pg C yr^{-1}, respectively, for the period 1992–1996.

The mean global terrestrial sink of 1.4 Pg C yr^{-1} for the years 1992–1996 is higher than that obtained from changes in O_2 and CO_2 (0.7 Pg C yr^{-1}) (Plattner *et al.*, 2002). However, the estimate from inverse modeling has to be adjusted to account for terrestrial sources and sinks of carbon that are not "seen" by the atmosphere. For example, the fluxes inferred from atmospheric data will not accurately reflect changes in the amount of carbon on land or in the sea if some of the carbon fixed by terrestrial plants or used in weathering minerals is transported by rivers to the ocean and respired and released to the atmosphere there. Under such circumstances, the atmosphere sees a terrestrial sink and an oceanic source, while the storage of carbon on land and in the sea may not have changed. Several studies have tried to adjust atmospherically based carbon budgets by accounting for the river transport of carbon. Sarmiento and Sundquist (1992) estimated a pre-industrial net export by rivers of 0.4–0.7 Pg C yr^{-1}, balanced by a net terrestrial uptake of carbon through photosynthesis and weathering. Aumont *et al.* (2001) obtained a global estimate of 0.6 Pg C yr^{-1}. Adjusting the net terrestrial sink obtained through inverse calculations (1.4 Pg C yr^{-1}) by 0.6 Pg C yr^{-1} yields a result (0.8 Pg C yr^{-1}) similar to the estimate obtained through changes in the concentrations of O_2 and CO_2 (Table 8). The two top-down methods based on atmospheric measurements yield similar global estimates of a net terrestrial sink (~0.7 ($\pm$0.8) Pg C yr^{-1} for the 1990s).

(iii) Land-use change. Another method, independent of those based on atmospheric data and models, that has been used to estimate terrestrial sources and sinks of carbon, globally, is a method based on changes in land use (see Section 16.3.1.5). This is a ground-based or bottom-up approach. Changes in land use suggest that deforestation, reforestation, cultivation, and logging were responsible for a carbon source, globally, that averaged 2.0 Pg C yr^{-1} during the 1980s and 2.2 Pg C yr^{-1} during the 1990s (Houghton, 2003). The approach includes emissions of carbon from the decay of dead plant material, soil, and wood products and sinks of carbon in regrowing ecosystems, including both vegetation and soil. Analyses account for delayed sources and sinks of carbon that result from decay and regrowth following a change in land use.

Other recent analyses of land-use change give results that bound the results of this summary, although differences in the processes and regions included make comparisons somewhat misleading. An estimate by Fearnside (2000) of a 2.4 Pg C yr^{-1} source includes only the tropics. A source of 0.8Pg C yr^{-1} estimated by McGuire *et al.* (2001) includes changes in global cropland area but does not include either the harvest of wood or the clearing of forests for pastures, both of which contributed to the net global source. The average annual release of carbon attributed by Houghton (2003) to changes in the area of croplands (1.2 Pg C yr^{-1} for the 1980s) is higher than the estimate found by McGuire *et al.* (0.8 Pg C yr^{-1}).

The calculated *source* of 2.2 ($\pm$0.8) Pg C yr^{-1} for the 1990s (Houghton, 2003) is very different from the global net terrestrial *sink* determined from top-down analyses (0.7 Pg C yr^{-1}) (Table 8). Are the methods biased? Biases in the inverse calculations may be in either direction. Because of the "rectifier effect" (the seasonal covariance between

Table 8 Estimates of the annual terrestrial flux of carbon (Pg C yr^{-1}) in the 1990s according to different methods. Negative values indicate a terrestrial sink

	O_2 and CO_2	*Inverse calculations CO_2, $^{13}CO_2$, O_2*	*Forest inventories*	*Land-use change*
Globe	−0.7 (±0.8)[a]	−0.8 (±0.8)[b]		2.2 (±0.6)[c]
Northern mid-latitudes		−2.1 (±0.8)[d]	−0.6 to −1.3[e]	−0.03 (±0.5)[c]
Tropics		1.5 (±1.2)[f]	−0.6 (±0.3)[g]	0.5 to 3.0[h]

[a] Plattner *et al.* (2002). [b] −1.4 (±0.8) from Gurney *et al.* (2002) reduced by 0.6 to account for river transport (Aumont *et al.*, 2001). [c] Houghton, 2003. [d] −2.4 from Gurney *et al.* (2002) reduced by 0.3 to account for river transport (Aumont *et al.*, 2001). [e] −0.65 in forests (Goodale *et al.*, 2002) and another 0.0–0.65 assumed for nonforests (see text). [f] 1.2 from Gurney *et al.* (2002) increased by 0.3 to account for river transport (Aumont *et al.*, 2001). [g] Undisturbed forests: −0.6 from Phillips *et al.* (1998) (challenged by Clark, 2002). [h] 0.9 (range 0.5–1.4) from DeFries *et al.* (2002) 1.3 from Achard *et al.* (2002) adjusted for soils and degradation (see text) 2.2 (±0.8) from Houghton (2003). 2.4 from Fearnside (2000).

the terrestrial carbon flux and atmospheric transport), inverse calculations are thought to underestimate the magnitude of a northern mid-latitude sink (Denning *et al.*, 1995). However, if the near-surface concentrations of atmospheric CO_2 in northern mid-latitude regions are naturally lower than those in the southern hemisphere, the apparent sink in the north may not be anthropogenic, as usually assumed. Rather, the anthropogenic sink would be less than 0.5 Pg C yr^{-1} (Taylor and Orr, 2000).

In contrast to the unknown bias of atmospheric methods, analyses based on land-use change are deliberately biased. These analyses consider only those changes in terrestrial carbon resulting directly from human activity (conversion and modification of terrestrial ecosystems). There may be other sources and sinks of carbon not related to land-use change (such as caused by CO_2 fertilization, changes in climate, or management) that are captured by other methods but ignored in analyses of land-use change. In other words, the flux of carbon from changes in land use is not necessarily the same as the net terrestrial flux from all terrestrial processes.

If the net terrestrial flux of carbon during the 1990s was 0.7 Pg C yr^{-1}, and 2.2 Pg C yr^{-1} were emitted as a result of changes in land use, then 2.9 Pg C yr^{-1} must have accumulated on land for reasons not related to land-use change. This residual terrestrial sink was discussed above (Table 7 and Figure 4). That the residual terrestrial sink exists at all suggests that processes other than land-use change are affecting the storage of carbon on land. Recall, however, that the residual sink is calculated by difference; if the emissions from land-use change are overestimated, the residual sink will also be high.

16.3.2.2 Regional distribution of sources and sinks of carbon: the northern mid-latitudes

Insights into the magnitude of carbon sources and sinks and the mechanisms responsible for the residual terrestrial carbon sink may be obtained from a consideration of tropical and extratropical regions separately. Inverse calculations show the tropics to be a moderate source, largely oceanic as a result of CO_2 outgassing in upwelling regions. Some of the tropical source is also terrestrial. Estimates vary greatly depending on the models of atmospheric transport and the years included in the analyses. The net global oceanic sink of 1.3 Pg C yr^{-1} for the period 1992–1996 is distributed in northern (1.2 Pg C yr^{-1}) and southern oceans (0.8 Pg C yr^{-1}), with a net source from tropical gyres (0.5 Pg C yr^{-1}) (Gurney *et al.*, 2002).

The net terrestrial sink of ~0.7 Pg C yr^{-1} is not evenly distributed either. The comparison by Gurney *et al.* (2002) showed net terrestrial sinks of 2.4 ± 0.8 Pg C yr^{-1} and 0.2 Pg C yr^{-1} for northern and southern mid-latitude lands, respectively, offset to some degree by a net tropical land source of 1.2 ± 1.2 Pg C yr^{-1}. Errors are larger for the tropics than the nontropics because of the lack of sampling stations and the more complex atmospheric circulation there.

River transport and subsequent oceanic release of terrestrial carbon are thought to overestimate the magnitude of the atmospherically derived northern terrestrial sink by 0.3 Pg C yr^{-1} and underestimate the tropical source (or overestimate its sink) by the same magnitude (Aumont *et al.*, 2001). Thus, the northern terrestrial sink becomes 2.1 Pg C yr^{-1}, while the tropical terrestrial source becomes 1.5 Pg C yr^{-1} (Table 8).

Inverse calculations have also been used to infer east–west differences in the distribution of sources and sinks of carbon. Such calculations are more difficult because east–west gradients in CO_2 concentration are an order of magnitude smaller than north–south gradients. Some estimates placed most of the northern sink in North America (Fan *et al.*, 1998); others placed most of it in Eurasia (Bousquet *et al.*, 1999a,b). More recent analyses suggest a sink in both North America and Eurasia, roughly in proportion to land area (Schimel *et al.*, 2001; Gurney *et al.*, 2002). The analyses also suggest that higher-latitude boreal forests are small sources rather than sinks of carbon during some years.

The types of land use determining fluxes of carbon are substantially different inside and outside the tropics (Table 9). As of early 2000s, the fluxes of carbon to and from northern lands are dominated by rotational processes, e.g., logging and subsequent regrowth. Changes in the area of forests are small. The losses of carbon from decay of wood products and slash (woody debris generated as a result of harvest) are largely offset by the accumulation of carbon in regrowing forests (reforestation and regrowth following harvest). Thus, the net flux of carbon from changes in land use is small: a source of 0.06 Pg C yr^{-1} during the 1980s changing to a sink of 0.02 Pg C yr^{-1} during the 1990s. Both the US and Europe are estimated to have been carbon sinks as a result of land-use change.

Inferring changes in terrestrial carbon storage from analysis of forest inventories. An independent estimate of carbon sources and sinks in northern mid-latitudinal lands may be obtained from forest inventories. Most countries in the northern mid-latitudes conduct periodic inventories of the growing stocks in forests. Sampling is designed to yield estimates of total growing stocks (volumes of merchantable wood) that are estimated with 95% confidence to within 1–5% (Powell *et al.*, 1993; Köhl and Päivinen, 1997; Shvidenko and Nilsson, 1997). Because annual changes due to growth and

Table 9 Estimates of the annual sources (+) and sinks (−) of carbon resulting from different types of land-use change and management during the 1990s (Pg C yr^{-1})

Activity	*Tropical regions*	*Temperate and boreal zones*	*Globe*
Deforestation	2.110[a]	0.130	2.240
Afforestation	−0.100	−0.080[b]	−0.190
Reforestation (agricultural abandonment)	0[a]	−0.060	−0.060
Harvest/management	0.190	0.120	0.310
Products	0.200	0.390	0.590
Slash	0.420	0.420	0.840
Regrowth	−0.430	−0.690	−1.120
Fire suppression[c]	0	−0.030	−0.030
Nonforests			
Agricultural soils[d]	0	0.020	0.020
Woody encroachment[c]	0	−0.060	−0.060
Total	2.200	0.040	2.240

[a] Only the net effect of shifting cultivation is included here. The gross fluxes from repeated clearing and abandonment are not included. [b] Areas of plantation forests are not generally reported in developed countries. This estimates includes only China's plantations. [c] Probably an underestimate. The estimate is for the US only, and similar values may apply in South America, Australia, and elsewhere. [d] These values include loss of soil carbon resulting from cultivation of new lands; they do not include accumulations of carbon that may have resulted from recent agricultural practices.

mortality are small relative to the total stocks, estimates of wood volumes are relatively less precise. A study in the southeastern US determined that regional growing stocks (m^3) were known with 95% confidence to within 1.1%, while changes in the stocks ($m^3\,yr^{-1}$) were known to within 39.7% (Phillips *et al.*, 2000). Allometric regressions are used to convert growing stocks (the wood contained in the boles of trees) to carbon, including all parts of the tree (roots, stumps, branches, and foliage as well as bole), nonmerchantable and small trees and nontree vegetation. Other measurements provide estimates of the carbon in the forest floor (litter) and soil. The precision of the estimates for these other pools of carbon is less than that for the growing stocks. An uncertainty analysis for 140×10^6 ha of US forests suggested an uncertainty of 0.028 Pg C yr^{-1} (Heath and Smith, 2000). The strength of forest inventories is that they provide direct estimates of wood volumes on more than one million plots throughout northern mid-latitude forests, often inventoried on 5–10-year repeat cycles. Some inventories also provide estimates of growth rates and estimates of mortality from various causes, i.e., fires, insects, and harvests. One recent synthesis of these forest inventories, after converting wood volumes to total biomass and accounting for the fate of harvested products and changes in pools of woody debris, forest floor, and soils, found a net northern mid-latitude terrestrial sink of between 0.6 Pg C yr^{-1} and 0.7 Pg C yr^{-1} for the years around 1990 (Goodale *et al.*, 2002). The estimate is ~30% of the sink inferred from atmospheric data corrected for river transport (Table 8). Some of the difference may be explained if nonforest ecosystems throughout the region are also accumulating carbon. Inventories of nonforest lands are generally lacking, but in the US, at least, nonforests are estimated to account for 40–70% of the net terrestrial carbon sink (Houghton *et al.*, 1999; Pacala *et al.*, 2001).

It is also possible that the accumulation of carbon below ground, not directly measured in forest inventories, was underestimated and thus might account for the difference in estimates. However, the few studies that have measured the accumulation of carbon in forest soils have consistently found soils to account for only a small fraction (5–15%) of measured ecosystem sinks (Gaudinski *et al.*, 2000; Barford *et al.*, 2001; Schlesinger and Lichter, 2001). Thus, despite the fact that the world's soils hold 2–3 times more carbon than biomass, there is no evidence, as of early 2000s, that they account for much of a terrestrial sink.

The discrepancy between estimates obtained from forest inventories and inverse calculations might also be explained by differences in the dates of measurements. The northern sink of 2.1 Pg C yr^{-1} from Gurney *et al.* (−2.4 + 0.3 for riverine transport) is for 1992–1996 and would probably have been lower (and closer to the forest inventory-based estimate) if averaged over the entire decade (see other estimates in Prentice *et al.*, 2001). Top-down measurements based on atmospheric data are sensitive to large year-to-year variations in the growth rate of CO_2 concentrations.

Both forest inventories and inverse calculations with atmospheric data show terrestrial ecosystems to be a significant carbon sink, while changes in land use show a sink near zero. Either the analyses of land-use change are incomplete, or other mechanisms besides land-use change must be responsible for the observed sink, or some combination of both. With respect to the difference between forest inventories and land-use change, a regional comparison suggests that the recovery of forests from land-use change (abandoned farmlands, logging, fire suppression) may either overestimate or underestimate the sinks measured in forest inventories (Table 10). In Canada and Russia, the carbon sink calculated for forests recovering from harvests (land-use change) is greater than the measured sink. The difference could be error, but it is consistent with the fact that fires and insect damage increased in these regions during the 1980s and thus converted some of the boreal forests from sinks to sources (Kurz and Apps, 1999). These sources would not be counted in the analysis of land-use change, because natural disturbances were ignored. In time, recovery from these natural disturbances will increase the sink above that calculated on the basis of harvests alone, but as of early 2000s the sources from fire and insect damage exceed the net flux associated with harvest and regrowth.

In the three other regions (Table 10), changes in land use yield a sink that is smaller than measured in forest inventories. If the results are not simply a reflection of error, the failure of past changes in land use to explain the measured sink suggests that factors not considered in the analysis have enhanced the storage of carbon in the forests of the US, Europe, and China. Such factors include past natural disturbances, more subtle forms of management than recovery from harvest and agricultural abandonment (and fire suppression in the US), and environmental changes that may have enhanced forest growth. It is unclear whether the differences between estimates (changes in land use and forest inventories) are real or the result of errors and omissions. The differences are small, generally less than 0.1 Pg C yr^{-1} in any region. The likely errors and omissions in analyses of land-use change include uncertain rates of forest growth, natural disturbances, and many types of forest management (Spiecker *et al.*, 1996).

16.3.2.3 Regional distribution of sources and sinks of carbon: the tropics

How do different methods compare in the tropics? Inverse calculations show that tropical lands were a net source of carbon, 1.2 ± 1.2 Pg C yr^{-1} for the period 1992–1996 (Gurney *et al.*, 2002). Accounting for the effects of rivers (Aumont *et al.*, 2001) suggests a source of 1.5 (±1.2) Pg C yr^{-1} (Table 8).

Forest inventories for large areas of the tropics are rare, although repeated measurements of permanent plots throughout the tropics suggest that undisturbed tropical forests are accumulating carbon, at least in the neotropics (Phillips *et al.*, 1998). The number of such plots was too small in tropical African or Asian forests to demonstrate a change in carbon accumulation, but assuming the plots in the neotropics are representative of undisturbed forests in that region suggests a sink of 0.62 (±0.30) Pg C yr^{-1} for mature humid neotropical forests (Phillips *et al.*, 1998). The finding of a net sink has been challenged, however, on the basis of systematic errors in measurement. Clark (2002) notes that many of the measurements of diameter included buttresses and other protuberances, while the allometric regressions used to estimate biomass were based on above-buttress relationships. Furthermore, these stem protuberances display disproportionate rates of radial growth. Finally, some of the plots were on floodplains where primary forests accumulate carbon. When plots with buttresses were excluded (and when recent floodplain (secondary) forests were excluded as well), the net increment was not statistically different from zero (Clark, 2002). Phillips *et al.* (2002) counter that the errors are minor, but the results remain contentious.

Thus, the two methods most powerful in constraining the northern net sink (inverse analyses and forest inventories) are weak or lacking in the tropics (Table 14), and the carbon balance of the tropics is less certain.

Direct measurement of CO_2 flux. The flux of CO_2 between an ecosystem and the atmosphere can be calculated directly by measuring the covariance

Table 10 Annual net changes in the living vegetation of forests (Pg C yr^{-1}) in northern mid-latitude regions around the year 1990. Negative values indicate an increase in carbon stocks (i.e., a terrestrial sink).

Region	*Land-use change*[a]	*Forest inventory*[b]	*Sink from land-use change relative to inventoried sink*
Canada	−0.025	0.040	0.065 (larger)
Russia	−0.055	0.040	0.095 (larger)
USA	−0.035	−0.110	0.075 (smaller)
China	0.075	−0.040	0.115 (smaller)
Europe	−0.020	−0.090	0.070 (smaller)
Total	−0.060	−0.160	0.100 (smaller)

[a] Houghton (2003). [b] From Goodale *et al.* (2002).

between concentrations of CO_2 and vertical wind speed (Goulden *et al.*, 1996). The approach is being applied at ~150 sites in North America, South America, Asia, and Europe. The advantage of the approach is that it includes an integrated measure for the whole ecosystem, not only the wood or the soil. The method is ideal for determining the short-term response of ecosystems to diurnal, seasonal, and interannual variations of such variables as temperature, soil moisture, and cloudiness. If measurements are made over an entire year or over a significant number of days in each season, an annual carbon balance can be determined. The results of such measured fluxes have been demonstrated in at least one ecosystem to be in agreement with independent measurements of change in the major components of the ecosystem (Barford, 2001).

As NEP is often small relative to the gross fluxes of photosynthesis and ecosystem respiration, the net flux is sometimes less than the error of measurement. More important than error is bias, and the approach is vulnerable to bias because both the fluxes of CO_2 and the micrometeorological conditions are systematically different day and night. Wind speeds below 17 cm s^{-1} in a temperate zone forest, e.g., resulted in an underestimate of nighttime respiration (Barford *et al.*, 2001). A similar relationship between nighttime wind speed and respiration in forests in the Brazilian Amazon suggests that the assumption that lateral transport is unimportant may have been invalid (Miller *et al.*, in press).

Although the approach works well where micrometeorological conditions are met, the footprint for the measured flux is generally less than 1 km^2, and it is difficult to extrapolate the measured flux to large regions. Accurate extrapolations require a distribution of tower sites representative of different flux patches, but such patches are difficult to determine *a priori*. The simple extrapolation of an annual sink of 1 Mg C ha yr^{-1} (based on 55 days of measurement) for a tropical forest in Brazil to all moist forests in the Brazilian Amazon gave an estimated sink of ~1 Pg C yr^{-1} (Grace *et al.*, 1995). In contrast, a more sophisticated extrapolation based on a spatial model of CO_2 flux showed a basin-wide estimate averaging only 0.3 Pg C yr^{-1} (Tian *et al.*, 1998). The modeled flux agreed with the measured flux in the location of the site; spatial differences resulted from variations in modeled soil moisture throughout the basin.

Initially, support for an accumulation of carbon in undisturbed tropical forests came from measurements of CO_2 flux by eddy correlation (Grace *et al.*, 1995; Malhi *et al.*, 1998). Results showed large sinks of carbon in undisturbed forests, that, if scaled up to the entire tropics, yielded sinks in the range of 3.9–10.3 Pg C yr^{-1} (Malhi *et al.*, 2001), much larger than the sources of carbon from deforestation. Tropical lands seemed to be a large net carbon sink. Recent analyses raise doubts about these initial results.

When flux measurements are corrected for calm conditions, the net carbon balance may be nearly neutral. One of the studies in an old-growth forest in the Tapajós National Forest, Pará, Brazil, showed a small net CO_2 source (Saleska *et al.*, in press). The results in that forest were supported by measurements of biomass (forest inventory) (Rice *et al.*, in press). Living trees were accumulating carbon, but the decay of downed wood released more, for a small net source. Both fluxes suggest that the stand was recovering from a disturbance several years earlier.

The observation that the rivers and streams of the Amazon are a strong source for CO_2 (Richey *et al.*, 2002) may help balance the large sinks measured in some upland sites. However, the riverine source is included in inverse calculations based on atmospheric data and does not change those estimates of a net terrestrial source (Gurney *et al.*, 2002).

Changes in land use in the tropics are clearly a source of carbon to the atmosphere, although the magnitude is uncertain (Detwiler and Hall, 1988; Fearnside, 2000; Houghton, 1999, 2003). The tropics are characterized by high rates of deforestation, and this conversion of forests to non-forests involves a large loss of carbon. Although rotational processes of land use, such as logging, are just as common in the tropics as in temperate zones (even more so because shifting cultivation is common in the tropics), the sinks of carbon in regrowing forests are dwarfed in the tropics by the large releases of carbon resulting from permanent deforestation.

Comparisons of results from different methods (Table 8) suggest at least two, mutually exclusive, interpretations for the net terrestrial source of carbon from the tropics. One interpretation is that a large release of carbon from land-use change (Fearnside, 2000; Houghton, 2003) is partially offset by a large sink in undisturbed forests (Malhi *et al.*, 1998; Phillips *et al.*, 1998, 2002). The other interpretation is that the source from deforestation is smaller (see below), and that the net flux from undisturbed forests is nearly zero (Rice *et al.*, in press; Saleska *et al.*, in press). Under the first interpretation, some sort of growth enhancement (or past natural disturbance) is required to explain the large current sink in undisturbed forests. Under the second, the entire net flux of carbon may be explained by changes in land use, but the source from land-use change is smaller than estimated by Fearnside (2000) or Houghton (2003).

A third possibility, that the net tropical source from land is larger than indicated by inverse calculations (uncertain in the tropics), is constrained by the magnitude of the net sink in northern mid-latitudes. The latitudinal gradient in CO_2 concentrations constrains the difference between the northern sink and tropical source more than it constrains the absolute fluxes. The tropical source can only be larger than indicated by inverse calculations if the northern mid-latitude sink is also larger. As discussed above, the northern mid-latitude sink is thought to be in the range of 1–2.6 Pg C yr^{-1}, but the estimates are based on the assumption that the pre-industrial north–south gradient in CO_2 concentrations was zero (similar concentrations at all latitudes). No data exist for the pre-industrial north–south gradient in CO_2 concentrations, but following Keeling *et al.* (1989), Tayor and Orr extrapolated the current CO_2 gradient to a zero fossil fuel release and found a negative gradient (lower concentrations in the north). They interpreted this negative gradient as the pre-industrial gradient, and their interpretation would suggest a northern sink larger than generally believed. In contrast, Conway and Tans (1999) interpret the extrapolated zero fossil fuel gradient as representing the current sources and sinks of carbon in response to fossil fuel emissions and other human activities, such as present and past land-use change. Most investigators of the carbon cycle favor this interpretation.

The second interpretation of existing estimates (a modest source of carbon from deforestation and little or no sink in undisturbed forests) is supported by satellite-based estimates of tropical deforestation. The high estimates of Fearnside (2000) and Houghton (2003) were based on rates of deforestation reported by the FAO (2001). If these rates of deforestation are high, the estimates of the carbon source are also high. Two new studies of tropical deforestation (Achard *et al.*, 2002; DeFries *et al.*, 2002) report lower rates than the FAO and lower emissions of carbon than Fearnside or Houghton. The study by Achard *et al.* (2002) found rates 23% lower than the FAO for the 1990s (Table 11). Their analysis used high resolution satellite data over a 6.5% sample of tropical humid forests, stratified by "deforestation hot-spot areas" defined by experts. In addition to observing 5.8×10^6 ha of outright deforestation in the tropical humid forests, Achard *et al.* also observed 2.3×10^6 ha of degradation. Their estimated carbon flux, including changes in the area of dry forests as well as humid ones, was 0.96 Pg C yr^{-1}. The estimate is probably low because it did not include the losses of soil carbon that often occur with cultivation or the losses of carbon from degradation (reduction of biomass within forests). Soils and degradation accounted for 12% and 26%, respectively, of Houghton's (2003) estimated flux of carbon for tropical Asia and America and would yield a total flux of 1.3 Pg C yr^{-1} if the same percentages were applied to the estimate by Archard *et al.*

A second estimate of tropical deforestation (DeFries *et al.*, 2002) was based on coarse resolution satellite data (8 km), calibrated with high-resolution satellite data to identify percent tree cover and to account for small clearings that would be missed with the coarse resolution data. The results yielded estimates of deforestation that were, on average, 54% lower than those reported by the FAO (Table 11). According to DeFries *et al.*, the estimated net flux of carbon for the 1990s was 0.9 (range 0.5–1.4) Pg C yr^{-1}.

If the tropical deforestation rates obtained by Archard *et al.* and DeFries *et al.* were similar, there would be little doubt that the FAO estimates are high. However, the estimates are as different from each other as they are from those of the FAO (Table 11). Absolute differences between the two studies are difficult to evaluate because Achard *et al.* considered only humid tropical forests, whereas DeFries *et al.* considered all tropical forests. The greatest differences are in tropical Africa, where the percent tree cover mapped by DeFries *et al.* is most unreliable because of the large areas of savanna. Both studies suggest that the FAO estimates of tropical deforestation are high, but the rates are still in question (Fearnside and Laurance, 2003; Eva *et al.*, 2003). The tropical emissions of carbon estimated by the two studies

Table 11 Annual rate of change in tropical forest area[a] for the 1990s.

	Tropical humid forests			*All tropical forests*		
	FAO (2001) (10^6 ha yr^{-1})	*Achard et al. (2002)*		*FAO (2001)* (10^6 hayr^{-1})	*DeFries et al. (2002)*	
		10^6 ha yr^{-1}	% lower than FAO		10^6 ha yr^{-1}	% lower than FAO
America	2.7	2.2	18	4.4	3.179	28
Asia	2.5	2.0	20	2.4	2.008	16
Africa	1.2	0.7	42	5.2	0.376	93
All tropics	6.4	4.9	23	12.0	5.563	54

[a] The net change in forest area is not the rate of deforestation but, rather, the rate of deforestation minus the rate of afforestation.

(after adjustments for degradation and soils) are about half of Houghton's estimate: 1.3 Pg C yr^{-1} and 0.9 Pg C yr^{-1}, as opposed to 2.2 Pg C yr^{-1} (Table 8).

16.3.2.4 Summary: synthesis of the results of different methods

Top-down methods show consistently that terrestrial ecosystems, globally, were a small net sink in the 1980s and 1990s. The sink was in northern mid-latitudes, partially offset by a tropical source. The northern sink was distributed over both North America and Eurasia roughly in proportion to land area. The magnitudes of terrestrial sinks obtained through inverse calculations are larger (or the sources smaller) than those obtained from bottom-up analyses (land-use change and forest inventories). Is there a bias in the atmospheric analyses? Or are there sinks not included in the bottom-up analyses?

For the northern mid-latitudes, when estimates of change in nonforests (poorly known) are added to the results of forest inventories, the net sink barely overlaps with estimates determined from inverse calculations. Changes in land use yield smaller estimates of a sink. It is not clear how much of the discrepancy is the result of omissions of management practices and natural disturbances from analyses of land-use change, and how much is the result of environmentally enhanced rates of tree growth. In other words, how much of the carbon sink in forests can be explained by age structure (i.e., previous disturbances and management), and how much by enhanced rates of carbon storage? The question is important for predicting future concentrations of atmospheric CO_2 (see below).

In the tropics, the uncertainties are similar but also greater because inverse calculations are more poorly constrained and because forest inventories are lacking. Existing evidence suggests two possibilities. Either large emissions of carbon from land-use change are somewhat offset by large carbon sinks in undisturbed forests, or lower releases of carbon from land-use change explain the entire net terrestrial flux, with essentially no requirement for an additional sink. The first alternative (large sources and large sinks) is most consistent with the argument that factors other than land-use change are responsible for observed carbon sinks (i.e., management or environmentally enhanced rates of growth). The second alternative is most consistent with the findings of Caspersen *et al.* (2000) that there is little enhanced growth. Overall, in both northern and tropical regions changes in land use exert a dominant influence on the flux of carbon, and it is unclear whether other factors have been important in either region. These conclusions question the assumption used in predictions of climatic change, the assumption that the current terrestrial carbon sink will increase in the future (see below).

16.4 MECHANISMS THOUGHT TO BE RESPONSIBLE FOR CURRENT SINKS OF CARBON

16.4.1 Terrestrial Mechanisms

Distinguishing between regrowth and enhanced growth in the current terrestrial sink is important. If regrowth is dominant, the current sink may be expected to diminish as forests age (Hurtt *et al.*, 2002). If enhanced growth is important, the magnitude of the carbon sink may be expected to increase in the future. Carbon cycle models used to calculate future concentrations of atmospheric CO_2 from emissions scenarios assume the latter (that the current terrestrial sink will increase) (Prentice *et al.*, 2001). These calculated concentrations are then used in general circulation models to project future rates of climatic change. If the current terrestrial sink is largely the result of regrowth, rather than enhanced growth, future projections of climate may underestimate the extent and rate of climatic change.

The issue of enhanced growth versus regrowth can be illustrated with studies from the US. Houghton *et al.* (1999) estimated a terrestrial carbon sink of 0.15–0.35 Pg C yr^{-1} for the US, attributable to changes in land use. Pacala *et al.* (2001) revised the estimate upwards by including additional processes, but in so doing they included sinks not necessarily resulting from land-use change. Their estimate for the uptake of carbon by forests, e.g., was the uptake measured by forest inventories. The measured uptake might result from previous land use (regrowth), but it might also result from environmentally enhanced growth, e.g., CO_2 fertilization (Figure 7). If all of the accumulation of carbon in US forests were the result of recovery from past land-use practices (i.e., no enhanced growth), then the measured uptake should equal the flux calculated on the basis of land-use change. The residual flux would be zero. The study by Caspersen *et al.* (2000) suggests that such an attribution is warranted because they found that 98% of forest growth in five US states could be attributed to regrowth rather than enhanced growth. However, the analysis by Houghton *et al.* (1999) found that past changes in land use accounted for only 20–30% of the observed accumulation of carbon in trees. The uptake calculated for forests recovering from agricultural abandonment, fire suppression, and earlier harvests was only 20–30% of the uptake measured by forest inventories (~40% if the uptake attributed to woodland "thickening" (0.26 Pg C yr^{-1}; Houghton, 2003) is included (Table 12)). The results are inconsistent

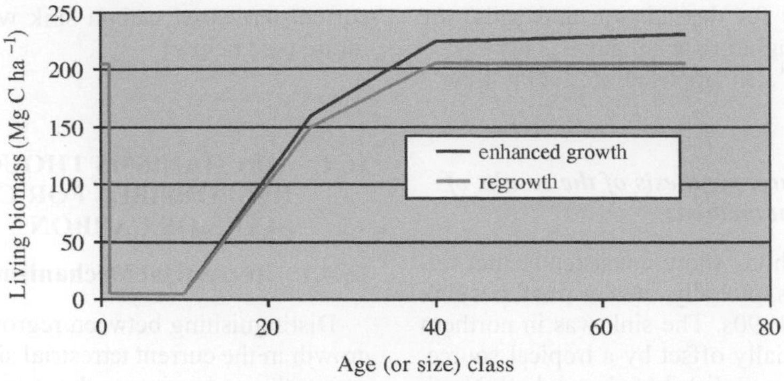

Figure 7 Idealized curves showing the difference between enhanced growth and regrowth in the accumulation of carbon in forest biomass.

Table 12 Estimated rates of carbon accumulation in the US ($PgC\,yr^{-1}$ in 1990).

	Pacala et al.[a] *(2001)*		*Houghton*[b] *et al. (1999)*	*Houghton*[b] *(2003)*	*Goodale et al. (2002)*
	Low	*High*			
Forest trees	−0.11	−0.15	−0.072[c]	−0.046[d]	−0.11
Other forest organic matter	−0.03	−0.15	0.010	0.010	−0.11
Cropland soils	0.00	−0.04	−0.138	0.00	NE
Woody encroachment	−0.12	−0.13	−0.122	−0.061	NE
Wood products	−0.03	−0.07	−0.027	−0.027	−0.06
Sediments	−0.01	−0.04	NE	NE	NE
Total sink	−0.30	−0.58	−0.35	−0.11	−0.28
% of total sink neither in forests nor wood products	43%	36%	74%	55%	NE

NE is "not estimated". Negative values indicate an accumulation of carbon on land.
[a] Pacala *et al.* (2001) also included the import/export imbalance of food and wood products and river exports. As these would create corresponding sources outside the US, they are ignored here. [b] Includes only the direct effects of human activity (i.e., land-use change and some management). [c] 0.020 $Pg\,C\,yr^{-1}$ in forests and 0.052 $Pg\,C\,yr^{-1}$ in the thickening of western pine woodlands as a result of early fire suppression. [d] 0.020 $Pg\,C\,yr^{-1}$ in forests and 0.026 $Pg\,C\,yr^{-1}$ in the thickening of western pine woodlands as a result of early fire suppression.

with those of Caspersen *et al.* (2000). Houghton's analysis requires a significant growth enhancement to account for the observed accumulation of carbon in trees; the analysis by Caspersen *et al.* suggests little enhancement.

Both analyses merit closer scrutiny. Joos *et al.* (2002) have pointed out, e.g., that the relationship between forest age and wood volume (or biomass) is too variable to constrain the enhancement of growth to between 0.001% and 0.01% per year, as Caspersen *et al.* claimed. An enhancement of 0.1% per year fits the data as well. Furthermore, even a small enhancement of 0.1% per year in NPP yields a significant sink (~2 $Pg\,C\,yr^{-1}$) if it applies globally (Joos *et al.*, 2002). Thus, Caspersen *et al.* may have underestimated the sink attributable to enhanced growth.

However, Houghton's analysis of land-use change (Houghton *et al.*, 1999; Houghton, 2003) most likely underestimates the sink attributable to regrowth. Houghton did not consider forest management practices other than harvest and subsequent regrowth. Nor did he include natural disturbances, which in boreal forests are more important than logging in determining the current age structure and, hence, rate of carbon accumulation (Kurz and Apps, 1999). Forests might now be recovering from an earlier disturbance. A third reason why the sink may have been underestimated is that Houghton used net changes in agricultural area to obtain rates of agricultural abandonment. In contrast, rates of clearing and abandonment are often simultaneous and thus create larger areas of regrowing forests than would be predicted from net changes in agricultural area. It is unclear how much of the carbon sink in the US can be attributed to changes in land use and management, and how much can be attributed to enhanced rates of growth.

The mechanisms responsible for the current terrestrial sink fall into two broad categories (Table 13 and Figure 7): (i) enhanced growth from physiological or metabolic factors that affect rates of photosynthesis, respiration, growth, and decay and (ii) regrowth from past disturbances, changes

Table 13 Proposed mechanisms for terrestrial carbon sinks.[a]

Metabolic or physiological mechanisms
CO_2 fertilization
N fertilization
Tropospheric ozone, acid deposition
Changes in climate (temperature, moisture)
Ecosystem mechanisms
Large-scale regrowth of forests following human disturbance (includes recovery from logging and agricultural abandonment)[b]
Large-scale regrowth of forests following natural disturbance[b]
Fire suppression and woody encroachment[b]
Decreased deforestation[b]
Improved agricultural practices[b]
Erosion and re-deposition of sediment
Wood products and landfills[b]

[a] Some of these mechanisms enhance growth; some reduce decomposition. In some cases these same mechanisms may also yield sources of carbon to the atmosphere. [b] Mechanisms included in analyses of land-use change (although not necessarily in all regions).

in land use, or management, affecting the mortality of forest stands, the age structure of forests, and hence their rates of carbon accumulation. What evidence do we have that these mechanisms are important? Consider, first, enhanced rates of growth.

16.4.1.1 Physiological or metabolic factors that enhance rates of growth and carbon accumulation

CO_2 fertilization. Numerous reviews of the direct and indirect effects of CO_2 on photosynthesis and plant growth have appeared in the literature (Curtis, 1996; Koch and Mooney, 1996; Mooney *et al.*, 1999; Körner, 2000), and only a very brief review is given here. Horticulturalists have long known that annual plants respond to higher levels of CO_2 with increased rates of growth, and the concentration of CO_2 in greenhouses is often deliberately increased to make use of this effect. Similarly, experiments have shown that most C_3 plants (all trees, most crops, and vegetation from cold regions) respond to elevated concentrations of CO_2 with increased rates of photosynthesis and increased rates of growth.

Despite the observed stimulative effects of CO_2 on photosynthesis and plant growth, it is not clear that the effects will result in an increased storage of carbon in the world's ecosystems. One reason is that the measured effects of CO_2 have generally been short term, while over longer intervals the effects are often reduced or absent. For example, plants often acclimate to higher concentrations of CO_2 so that their rates of photosynthesis and growth return to the rates observed before the

Table 14 Increases observed for a 100% increase in CO_2 concentrations.

Increased rates
60% increase in *photosynthesis* of young trees
33% average increase in net primary productivity(*NPP*) of crops
25% increase in *NPP* of a young pine forest
Increased stocks
14% average increase in *biomass* of grasslands and crops
~0% increase in the carbon content of mature forests

concentration was raised (Tissue and Oechel, 1987; Oren *et al.*, 2001).

Another reason why the experimental results may not apply to ecosystems is that most experiments with elevated CO_2 have been conducted with crops, annual plants, or tree seedlings. The few studies conducted at higher levels of integration or complexity, such as with mature trees and whole ecosystems, including soils as well as vegetation, suggest much reduced responses. Table 14 summarizes the results of experiments at different levels of integration. Arranged in this way (from biochemical processes to ecosystem processes), observations suggest that as the level of complexity, or the number of interacting processes, increases, the effects of CO_2 fertilization are reduced. This dampening of effects across ever-increasing levels of complexity has been noted since scientists first began to consider the effects of CO_2 on carbon storage (Lemon, 1977).

In other words, a CO_2-enhanced increase in photosynthesis is many steps removed from an increase in carbon storage. An increase in NPP is expected to lead to increased carbon storage until the carbon lost from the detritus pool comes into a new equilibrium with the higher input of NPP. But, if the increased NPP is largely labile (easily decomposed), then it may be decomposed rapidly with little net carbon storage (Davidson and Hirsch, 2001). Results from a loblolly pine forest in North Carolina suggest a very small increase in carbon storage. Elevated CO_2 increased litter production (with a turnover time of about three years) but did not increase carbon accumulation deeper in the soil layer (Schlesinger and Lichter, 2001). Alternatively, the observation that microbes seemed to switch from old organic matter to new organic matter after CO_2 fertilization of a grassland suggests that the loss of carbon may be delayed in older, more refractory pools of soil organic matter (Cardon *et al.*, 2001).

The central question is whether natural ecosystems will accumulate carbon as a result of elevated CO_2, and whether the accumulation will persist. Few CO_2 fertilization experiments have been carried out for more than a few years in whole ecosystems, but where they have, the results generally show an initial CO_2-induced increment in biomass that diminishes after a few years. The diminution of the initial CO_2-induced effect occurred after two years in an arctic tundra (Oechel *et al.*, 1994) and after three years in a rapidly growing loblolly pine forest (Oren *et al.*, 2001). Other forests may behave differently, but the North Carolina forest was chosen in part because CO_2 fertilization, if it occurs anywhere, is likely to occur in a rapidly growing forest. The longest CO_2 fertilization experiment, in a brackish wetland on the Chesapeake Bay, has shown an enhanced net uptake of carbon even after 12 years, but the expected accumulation of carbon at the site has not been observed (Drake *et al.*, 1996).

Nitrogen fertilization. Human activity has increased the abundance of biologically active forms of nitrogen (NO_x and NH_4), largely through the production of fertilizers, the cultivation of legumes that fix atmospheric nitrogen, and the use of internal combustion engines. Because the availability of nitrogen is thought to limit NPP in temperate-zone ecosystems, the addition of nitrogen through human activities is expected to increase NPP and, hence, terrestrial carbon storage (Peterson and Melillo, 1985; Schimel *et al.*, 1996; Holland *et al.*, 1997). Based on stoichiometric relations between carbon and nitrogen, many physiologically based models predict that added nitrogen should lead to an accumulation of carbon in biomass. But the extent to which this accumulation occurs in nature is unclear. Adding nitrogen to forests does increase NPP (Bergh *et al.*, 1999). It may also modify soil organic matter and increase its residence time (Fog, 1988; Bryant *et al.*, 1998). But nitrogen deposited in an ecosystem may also be immobilized in soils (Nadelhoffer *et al.*, 1999) or lost from the ecosystem, becoming largely unavailable in either case (Davidson, 1995).

There is also evidence that additions of nitrogen above some level may saturate the ecosystem, causing: (i) increased nitrification and nitrate leaching, with associated acidification of soils and surface waters; (ii) cation depletion and nutrient imbalances; and (iii) reduced productivity (Aber *et al.*, 1998; Fenn *et al.*, 1998). Experimental nitrogen additions have had varied effects on wood production and growing stocks. Woody biomass production increased in response to nitrogen additions to two New England hardwood sites, although increased mortality at one site led to a net decrease in the stock of woody biomass (Magill *et al.*, 1997, 2000). Several studies have shown that chronic exposure to elevated nitrogen inputs can inhibit forest growth, especially in evergreen species (Tamm *et al.*, 1995; Makipaa, 1995). Fertilization decreased rates of wood production in high-elevation spruce-fir in Vermont (McNulty *et al.*, 1996) and in a heavily fertilized red pine plantation (Magill *et al.*, 2000). The long-term effects of nitrogen deposition on forest production and carbon balance remain uncertain. Furthermore, because much of the nitrogen deposited on land is in the form of acid precipitation, it is difficult to distinguish the fertilization effects of nitrogen from the adverse effects of acidity (see below).

Atmospheric chemistry. Other factors besides nitrogen saturation may have negative effects on NPP, thus reducing the uptake of carbon in ecosystems and perhaps changing them from sinks to sources of carbon. Two factors that have received attention are tropospheric ozone and sulfur (acid rain). Experimental studies show leaf injury and reduced growth in crops and trees exposed to ozone. At the level of the ecosystem, elevated levels of ozone have been associated with reduced forest growth in North America (Mclaughlin and Percy, 2000) and Europe (Braun *et al.*, 2000). Acidification of soil as a result of deposition of NO_3^- and SO_4^{2-} in precipitation depletes the soils of available plant nutrients (Ca^{2+}, Mg^{2+}, K^+), increases the mobility and toxicity of aluminum, and increases the amount of nitrogen and sulfur stored in forest soils (Driscoll *et al.*, 2001). The loss of plant nutrients raises concerns about the long-term health and productivity of forests in the northeastern US, Europe, and southern China.

Although the effects of tropospheric ozone and sulfur generally reduce NPP, their actual or potential effects on carbon stocks are not known. The pollutants could potentially increase carbon stocks if they reduce decomposition of organic matter more than they reduce NPP.

Climatic variability and climatic change. Year-to-year differences in the growth rate of CO_2 in the atmosphere are large (Figure 3). The annual rate of increase ranged from 1.9 Pg C in 1992 to 6.0 Pg C in 1998 (Prentice *et al.*, 2001; see also Conway *et al.*, 1994). In 1998 the net global sink (ocean and land) was nearly 0 Pg C, while the average combined sink in the previous eight years was $\sim$3.5 Pg C yr^{-1} (Tans *et al.*, 2001). The terrestrial sink is generally twice as variable as the oceanic sink (Bousquet *et al.*, 2000). This temporal variability in terrestrial fluxes is probably caused by the effect of climate on carbon pools with short lifetimes (foliage, plant litter, soil microbes) through variations in photosynthesis, respiration, and possibly fire (Schimel *et al.*, 2001). Measurements in terrestrial ecosystems suggest that respiration, rather than photosynthesis, is the major contributor to variability (Valentini *et al.*, 2000). Annual respiration was almost twice as variable as photosynthesis over a five-year period in the Harvard Forest (Goulden *et al.*, 1996). Respiration is also more sensitive than photosynthesis to changes in both temperature and moisture. For example, during a dry year at the Harvard Forest, both photosynthesis and respiration were reduced, but the reduction in respiration was greater, yielding a greater than average net uptake of carbon for the year (Goulden *et al.*, 1996). A tropical forest in the Brazilian Amazon behaved similarly (Saleska *et al.*, in press).

The greater sensitivity of respiration to climatic variations is also observed at the global scale. An analysis of satellite data over the US, together with an ecosystem model, shows that the variability in NPP is considerably less than the variability in the growth rate of atmospheric CO_2 inferred from inverse modeling, suggesting that the cause of the year-to-year variability in carbon fluxes is largely from varying rates of respiration rather than photosynthesis (Hicke *et al.*, 2002). Also, global NPP was remarkably constant over the three-year transition from El Niño to La Niña (Behrenfeld *et al.*, 2001). Myneni *et al.* (1995) found a positive correlation between annual "greenness," derived from satellites, and the growth rate of CO_2. Greener years presumably had more photosynthesis and higher GPP, but they also had proportionately more respiration, thus yielding a net release of carbon from land (or reduced uptake) despite increased greenness.

Climatic factors influence terrestrial carbon storage through effects on photosynthesis, respiration, growth, and decay. However, prediction of future terrestrial sinks resulting from climate change requires an understanding of not only plant and microbial physiology, but the regional aspects of future climate change, as well. The important aspects of climate are: (i) temperature, including the length of the growing season; (ii) moisture; and (iii) solar radiation and clouds. Although year-to-year variations in the growth rate of CO_2 are probably the result of terrestrial responses to climatic variability, longer-term changes in carbon storage involve acclimation and other physiological adjustments that generally reduce short-term responses.

In cold ecosystems, such as those in high latitudes (tundra and taiga), an increase in temperature might be expected to increase NPP and, perhaps, carbon storage (although the effects might be indirect through increased rates of nitrogen mineralization; Jarvis and Linder, 2000). Satellite records of "greenness" over the boreal zone and temperate Europe show a lengthening of the growing season (Myneni *et al.*, 1997), suggesting greater growth and carbon storage. Measurements of CO_2 flux in these ecosystems do not consistently show a net uptake of carbon in response to warm temperatures (Oechel *et al.*, 1993; Goulden *et al.*, 1998), however, presumably because warmer soils release more carbon than plants take up. Increased temperatures in boreal forests may also reduce plant growth if the higher temperatures are associated with drier conditions (Barber *et al.*, 2000; Lloyd and Fastie, 2002). The same is true in the tropics, especially as the risk of fires increases with drought (Nepstad *et al.*, 1999; Page *et al.*, 2002). A warming-enhanced increase in rates of respiration and decay may already have begun to release carbon to the atmosphere (Woodwell, 1983; Raich and Schlesinger, 1992; Houghton *et al.*, 1998).

The results of short-term experiments may be misleading, however, because of acclimation or because the more easily decomposed material is respired rapidly. The long-term, or equilibrium, effects of climate on carbon storage can be inferred from the fact that cool, wet habitats store more carbon in soils than hot, dry habitats (Post *et al.*, 1982). The transient effects of climatic change on carbon storage, however, are difficult to predict, in large part because of uncertainty in predicting regional and temporal changes in temperature and moisture (extremes as well as means) and rates of climatic change, but also from incomplete understanding of how such changes affect fires, disease, pests, and species migration rates.

In the short term of seasons to a few years, variations in terrestrial carbon storage are most likely driven by variations in climate (temperature, moisture, light, length of growing season). Carbon dioxide fertilization and nitrogen deposition, in contrast, are unlikely to change abruptly. Interannual variations in the emissions of carbon from land-use change are also likely to be small (<0.2 Pg C yr^{-1}) because socioeconomic changes in different regions generally offset each other, and because the releases and uptake of carbon associated with a land-use change lag the change in land use itself and thus spread the emissions over time (Houghton, 2000). Figure 8 shows the annual net emissions of carbon from deforestation and reforestation in the Brazilian Amazon relative to the annual fluxes observed in the growth rates of trees and modeled on the basis of physiological responses to climatic variation. Clearly, metabolic responses to climatic variations are more important in the short term than interannual variations in rates of land-use change.

Understanding short-term variations in atmospheric CO_2 may not be adequate for predicting longer-term trends, however. Organisms and populations acclimate and adapt in ways that generally diminish short-term responses. Just as increased rates of photosynthesis in response to elevated levels of CO_2 often, but not always, decline within months or years (Tissue and Oechel, 1987), the same diminished response has been observed for higher temperatures (Luo *et al.*, 2001). Thus, over decades and centuries the factors most important in influencing concentrations of atmospheric CO_2 (fossil fuel emissions, land-use change, oceanic uptake) are probably different from those factors important in determining the short-term variations in atmospheric CO_2 (Houghton, 2000). Long-term changes in climate, as opposed to climatic variability, may eventually lead to long-term changes in

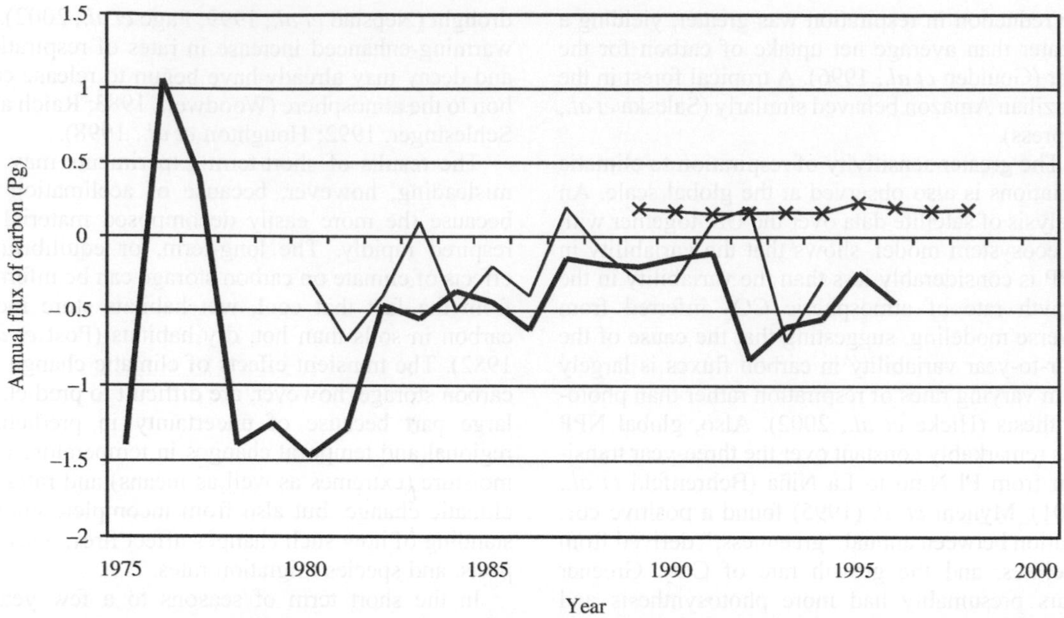

Figure 8 Net annual sources (+) and sinks (−) of carbon for the Brazilian Amazon, as determined by three different methods: (×) land-use change (Houghton *et al.*, 2000) (reproduced by permission of the American Geophysical Union from *J. Geophys. Res.*, **2000**, *105*, 20121–20130); (—) tree growth (Phillips *et al.*, 1998); and (—) modeled ecosystem metabolism (Tian *et al.*, 1998).

carbon storage, but probably not at the rates suggested by short-term experiments.

One further observation is discussed here. Over the last decades the amplitude of the seasonal oscillation of CO_2 concentration increased by ~20% at Mauna Loa, Hawaii, and by ~40% at Point Barrow, Alaska (Keeling *et al.*, 1996a). This winter–summer oscillation in concentrations seems to be largely the result of terrestrial metabolism in northern mid-latitudes. The increase in amplitude suggests that the rate of processing of carbon may be increasing. Increased rates of summer photosynthesis, increased rates of winter respiration, or both would increase the amplitude of the oscillation, but it is difficult to ascertain which has contributed most. Furthermore, the increase in the amplitude does not, by itself, indicate an increasing terrestrial sink. In fact, the increase in amplitude is too large to be attributed to CO_2 fertilization or to a temperature-caused increase in winter respiration (Houghton, 1987; Randerson *et al.*, 1997). It is consistent with the observation that growing seasons have been starting earlier over the last decades (Randerson *et al.*, 1999). The trend has been observed in the temperature data, in decreasing snow cover (Folland and Karl, 2001), and in the satellite record of vegetation activity (Myneni *et al.*, 1997).

Synergies among physiological "mechanisms." The factors influencing carbon storage often interact nonadditively. For example, higher concentrations of CO_2 in air enable plants to acquire the same amount of carbon with a smaller loss of water through their leaves. This increased water-use efficiency reduces the effects of drought. Higher levels of CO_2 may also alleviate other stresses of plants, such as temperature and ozone. The observation that NPP is increased relatively more in "low productivity" years suggests that the indirect effects of CO_2 in ameliorating stress may be more important than the direct effects of CO_2 on photosynthesis (Luo *et al.*, 1999).

Another example of synergistic effects is the observation that the combination of nitrogen fertilizer and elevated CO_2 concentration may have a greater effect on the growth of biomass in a growing forest than the expected additive effect (Oren *et al.*, 2001). The relative increase was greater in a nutritionally poor site. The synergy between nitrogen and CO_2 was different in a grassland, however (Hu *et al.*, 2001). There, elevated CO_2 increased plant uptake of nitrogen, increased NPP, and increased the carbon available for microbes; but it reduced microbial decomposition, presumably because the utilization of nitrogen by plants reduced its availability for microbes. The net effect of the reduced decomposition was an increase in the accumulation of carbon in soil.

Relatively few experiments have included more than one environmental variable at a time. A recent experiment involving combinations of four variables shows the importance of such work.

Shaw *et al.* (2003) exposed an annual grassland community in California to increased temperature, precipitation, nitrogen deposition, and atmospheric CO_2 concentration. Alone, each of the treatments increased NPP in the third year of treatment. Across all multifactor treatments, however, elevated CO_2 decreased the positive effects of the other treatments. That is, elevated CO_2 increased productivity under "poor" growing conditions, but reduced it under favorable growing conditions. The most likely explanation is that some soil nutrient became limiting, either because of increased microbial activity or decreased root allocation (Shaw *et al.*, 2003).

The expense of such multifactor experiments has led scientists to use process-based ecosystem models (see the discussion of "terrestrial carbon models" below) to predict the response of terrestrial ecosystems to future climates. When predicting the effects of CO_2 alone, six global biogeochemical models showed a global terrestrial sink that began in the early part of the twentieth century and increased (with one exception) towards the year 2100 (Cramer *et al.*, 2001). The maximum sink varied from $\sim$4 Pg C yr^{-1} to $\sim$10 Pg C yr^{-1}. Adding changes in climate (predicted by the Hadley Centre) to these models reduced the future sink (with one exception), and in one case reduced the sink to zero near the year 2100.

Terrestrial carbon models. A number of ecosystem models have been developed to calculate gross and net fluxes of carbon from environmentally induced changes in plant or microbial metabolism, such as photosynthesis, plant respiration, decomposition, and heterotrophic respiration (Cramer *et al.*, 2001; McGuire *et al.*, 2001). For example, six global models yielded net terrestrial sinks of carbon ranging between 1.5 Pg C yr^{-1} and 4.0 Pg C yr^{-1} for the year 2000 (Cramer *et al.*, 2001). The differences among models became larger as environmental conditions departed from existing conditions. The magnitude of the terrestrial carbon flux projected for the year 2100 varied between a source of 0.5 Pg C yr^{-1} and a sink of 7 Pg C yr^{-1}. Other physiologically based models, including the effects of climate on plant distribution as well as growth, projected a net source from land as tropical forests were replaced with savannas (White *et al.*, 1999; Cox *et al.*, 2000).

The advantage of such models is that they allow the effects of different mechanisms to be distinguished. However, they may not include all of the important processes affecting changes in carbon stocks. To date, e.g., few process-based terrestrial models have included changes in land use.

Although some processes, such as photosynthesis, are well enough understood for predicting responses to multiple factors, other processes, such as biomass allocation, phenology, and the replacement of one species by another, are not. Even if the physiological mechanisms and their interactions were well understood and incorporated into the models, other nonphysiological factors that affect carbon storage (e.g., fires, storms, insects, and disease) are not considered in the present generation of models. Furthermore, the factors influencing short-term changes in terrestrial carbon storage may not be the ones responsible for long-term changes (Houghton, 2000) (see next section). The variability among model predictions suggests that they are not reliable enough to demonstrate the mechanisms responsible for the current modest terrestrial sink (Cramer *et al.*, 2001; Knorr and Heimann, 2001).

16.4.1.2 *Demographic or disturbance mechanisms*

Terrestrial sinks also result from the recovery (growth) of ecosystems disturbed in the past. The processes responsible for regrowth include physiological and metabolic processes, but they also involve higher-order or more integrated processes, such as succession, growth, and aging. Forests accumulate carbon as they grow. Regrowth is initiated either by disturbances or by the planting of trees on open land. Disturbances may be either natural (insects, disease, some fires) or human induced (management and changes in land use, including fire management). Climatic effects—e.g., droughts, storms, or fires—thus affect terrestrial carbon storage not only through physiological or metabolic effects on plant growth and respiration, but also through effects on stand demography and growth.

In some regions of the world—e.g., the US and Europe—past changes in land use are responsible for an existing sink (Houghton *et al.*, 1999; Caspersen *et al.*, 2000; Houghton, 2003). Processes include the accumulation of carbon in forests as a result of fire suppression, the growth of forests on lands abandoned from agriculture, and the growth of forests earlier harvested. In tropical regions carbon accumulates in forests that are in the fallow period of shifting cultivation. All regions, even countries with high rates of deforestation, have sinks of carbon in recovering forests, but often these sinks are offset by large emissions (Table 9). The sinks in tropical regions as a result of logging are nearly the same in magnitude as those outside the tropics.

Sinks of carbon are not limited to forests. Some analyses of the US (Houghton *et al.*, 1999; Pacala *et al.*, 2001) show that a number of processes in nonforest ecosystems may also be responsible for carbon sinks. Processes include the encroachment of woody vegetation into formerly herbaceous ecosystems, the accumulation of carbon in agricultural soils as a result of conservation tillage or other practices, exportation of wood and food, and the riverine export of carbon from land to the sea (Table 12). At least a portion of these last two processes (import/export of food and wood and river export) represents an export of carbon from the US (an apparent sink) but not a global sink because these exports presumably become sources somewhere else (either in ocean waters or in another country).

Which terrestrial mechanisms are important? Until recently, the most common explanations for the residual carbon sink in the 1980s and 1990s were factors that affect the physiology of plants and microbes: CO_2 fertilization, nitrogen deposition, and climatic variability (see Table 13). Several findings have started to shift the explanation to include management practices and disturbances that affect the age structure or demography of ecosystems. For example, the suggestion that CO_2 fertilization may be less important in forests than in short-term greenhouse experiments (Oren *et al.*, 2001) was discussed above. Second, physiological models quantifying the effects of CO_2 fertilization and climate change on the growth of US forests could account for only a small fraction of the carbon accumulation observed in those forests (Schimel *et al.*, 2000). The authors acknowledged that past changes in land use were likely to be important. Third, and most importantly, 98% of recent accumulations of carbon in US forests can be explained on the basis of the age structure of trees without requiring growth enhancement due to CO_2 or nitrogen fertilization (Caspersen *et al.*, 2000). Either the physiological effects of CO_2, nitrogen, and climate have been unimportant or their effects have been offset by unknown influences. Finally, the estimates of sinks in the US (Houghton *et al.*, 1999; Pacala *et al.*, 2001; Table 12) are based, to a large extent, on changes in land use and management, and not on physiological models of plant and soil metabolism.

To date, investigations of these two different classes of mechanisms have been largely independent. The effects of changing environmental conditions have been largely ignored in analyses of land-use change (see Section 16.3.1.5), and physiological models have generally ignored changes in land use (see Section 16.4.1.1).

As of early 2000s, the importance of different mechanisms in explaining known terrestrial carbon sinks remains unclear. Management and past disturbances seem to be the dominant mechanisms for a sink in mid-latitudes, but they are unlikely to explain a large carbon sink in the tropics (if one exists). Recovery from past disturbances is unlikely to explain a large carbon sink in the tropics, because both the area of forests and the stocks of carbon within forests have been declining. Rates of human-caused disturbance have been accelerating. Clearly there are tropical forests recovering from natural disturbances, but there is no evidence that the frequency of disturbances changed during the last century, and thus no evidence to suggest that the sink in recovering forests is larger or smaller today than in previous centuries. The lack of systematic forest inventories over large areas in the tropics precludes a more definitive test of where forests are accumulating carbon and where they are losing it.

Enhanced rates of plant growth cannot be ruled out as an explanation for apparent sinks in either the tropics or mid-latitude lands, but it is possible that the current sink is entirely the result of recovery from earlier disturbances, anthropogenic and natural.

How will the magnitude of the current terrestrial sink change in the future? Identifying the mechanisms responsible for past and current carbon sinks is important because some mechanisms are more likely than others to persist into the future. As discussed above, physiologically based models predict that CO_2 fertilization will increase the global terrestrial sink over the next 100 years (Cramer *et al.*, 2001). Including the effects of projected climate change reduces the magnitude of projected sinks in many models but turns the current sink into a future global source in models that include the longer-term effects of climate on plant distribution (White *et al.*, 1999; Cox *et al.*, 2000). Thus, although increased levels of CO_2 are thought to increase carbon storage in forests, the effect of warmer temperatures may replace forests with savannas and grasslands, and, in the process, release carbon to the atmosphere. Future changes in natural systems are difficult to predict.

To the extent the current terrestrial sink is a result of regrowth (changes in age structure), the future terrestrial sink is more constrained. First, the net effect of continued land-use change is likely to release carbon, rather than store it. Second, forests that might have accumulated carbon in recent decades (whatever the cause) will cease to function as sinks if they are turned into croplands. Third, the current sink in regrowing forests will diminish as forests mature (Hurtt *et al.*, 2002).

Despite the recent evidence that changes in land use are more important in explaining the current terrestrial carbon sink than physiological responses to environmental changes in CO_2, nitrogen, or climate, most projections of future rates of climatic change are based on the assumption that the current terrestrial sink will not only continue, but grow in proportion to concentrations of CO_2. Positive biotic feedbacks and changes in land use are not included in the general circulation models (GCMs) used to predict future rates of climate change. The GCMs include physical feedbacks such as water vapor, clouds, snow, and polar ice, but not biotic feedbacks (Woodwell and Mackenzie, 1995). Thus, unless negative feedbacks in the biosphere become more important in the future, through physiological or other processes, these climate projections underestimate the rate and extent of climatic change. If the terrestrial sink were to diminish in the next decades, concentrations of CO_2 by the year 2100 might be hundreds of ppm higher than commonly projected.

16.4.2 Oceanic Mechanisms

16.4.2.1 Physical and chemical mechanisms

Increasing the concentration of CO_2 in the atmosphere is expected to affect the rate of oceanic uptake of carbon through at least eight mechanisms, half of them physical or chemical, and half of them biological. Most of the mechanisms reduce the short-term uptake of carbon by the oceans.

The buffer factor. The oceanic buffer factor (or Revelle factor), by which the concentration of CO_2 in the atmosphere is determined, increases as the concentration of CO_2 increases. The buffer factor is discussed above in Section 16.3.1.3. Here, it is sufficient to describe the chemical equation for the dissolution of CO_2 in seawater.

$$2HCO_3 + Ca \rightleftharpoons CaCO_3 + CO_2 + H_2O$$

Every molecule of CO_2 entering the oceans consumes a molecule of carbonate as the CO_2 is converted to bicarbonate. Thus, as CO_2 enters the ocean, the concentration of carbonate ions decreases, and further additions of CO_2 remain as dissolved CO_2 rather than being converted to HCO_3^-. The ocean becomes less effective in taking up additional CO_2. The effect is large. The change in DIC for a 100 ppm increase above 280 ppm (preindustrial) was 40% larger than a 100 ppm increase would be today. The change in DIC for a 100 ppm increase above 750 ppm will be 60% lower than it would be today (Prentice *et al.*, 2001). Thus, the fraction of added CO_2 going into the ocean decreases and the fraction remaining in the atmosphere increases as concentrations continue to increase.

Warming. The solubility of CO_2 in seawater decreases with temperature. Raising the ocean temperature 1°C increases the equilibrium p_{CO_2} in seawater by 10–20 ppm, thus increasing the atmospheric concentration by that much as well. This mechanism is a positive feedback to a global warming.

Vertical mixing and stratification. If the warming of the oceans takes place in the surface layers first, the warming would be expected to increase the stability of the water column. As discussed in Section 16.3.1.3, the bottleneck for oceanic uptake of CO_2 is largely the rate at which the surface oceans exchange CO_2 with the intermediate and deeper waters. Greater stability of the water column, as a result of warming, might constrict this bottleneck further. Similarly, if the warming of the Earth's surface is greater at the poles than at the equator, the latitudinal gradient in surface ocean temperature will be reduced; and because that thermal gradient plays a role in the intensity of atmospheric mixing, a smaller gradient might be expected to subdue mixing and increase stagnation. Alternatively, the increased intensity of the hydrologic cycle expected for a warmer Earth will probably increase the intensity of storms and might, thereby, increase oceanic mixing. Interactions between oceanic stability and biological production might also change the ocean's carbon cycle, with consequences for the oceanic uptake of carbon that are difficult to predict (Sarmiento *et al.*, 1998; Matear and Hirst, 1999).

One aspect of the ocean's circulation that seems particularly vulnerable to climate change is the thermohaline circulation, which is related to the formation of North Atlantic Deep Water (NADW). Increased warming of surface waters may intensify the hydrologic cycle, leading to a reduced salinity in the sea surface at high latitudes, a reduction (even collapse) of NADW formation, reduction in the surface-to-deep transport of anthropogenic carbon, and thus a higher rate of CO_2 growth in the atmosphere. In a model simulation, modest rates of warming reduced the rate of oceanic uptake of carbon, but the reduced uptake was largely compensated by changes in the marine biological cycle (Joos *et al.*, 1999a). For higher rates of global warming, however, the NADW formation collapsed and the concentration of CO_2 in the atmosphere was 22% (and global temperature 0.6 °C) higher than expected in the absence of this feedback.

Rate of CO_2 emission. High rates of CO_2 emissions will increase the atmosphere–ocean gradient in CO_2 concentrations. Although this gradient drives the uptake of carbon by surface waters, if the rate of CO_2 emissions is greater than the rate of CO_2 uptake, the fraction of emitted CO_2 remaining in the atmosphere will be higher. Under the business-as-usual scenario for future CO_2 emissions, rates of emissions increase by more than a factor of 3, from approximately 6 Pg C yr^{-1} in the 1990s to 20 Pg C yr^{-1} by the end of the twenty-first century.

16.4.2.2 Biological feedback/processes

Changes in biological processes may offset some of the physical and chemical effects described above (Sarmiento *et al.*, 1998; Joos *et al.*, 1999a), but the understanding of these processes is incomplete, and the net effects far from predictable. Potential effects fall into four categories (Falkowski *et al.*, 1998).

(i) Addition of nutrients limiting primary production. Nutrient enrichment experiments and observations of nutrient distributions throughout the oceans suggest that marine primary productivity is often limited by the availability of fixed inorganic nitrogen. As most of the nitrogen for marine production comes from upwelling, physical changes in ocean circulation might also affect oceanic primary production and, hence, the biological pump. Some nitrogen is made available through nitrogen fixation, however, and some is lost through denitrification, both of which are biological processes, limited by trace nutrients and the concentration of oxygen. The two processes are not coupled, however, and differential changes in either one would affect the inventory of fixed nitrogen in the ocean.

(ii) Enhanced utilization of nutrients. One of the mysteries of ocean biology today is the observation of "high nutrient, low chlorophyll (HNLC) regions." That is, why does primary production in major regions of the surface ocean stop before all of the available nitrogen and phosphorous have been used up? It is possible that grazing pressures keep phytoplankton populations from consuming the available nitrogen and phosphorous, and any reduction in grazing pressures might increase the export of organic matter from the surface. Another possibility that has received considerable attention is that iron may limit production (Martin, 1990). In fact, deliberate iron fertilization of the ocean has received serious attention as a way of reducing atmospheric CO_2 (see Section 16.5.2, below). Iron might also become more available naturally as a result of increased human eutrophication of coastal waters, or it might be less available as a result of a warmer (more strongly stratified) ocean or reduced transport of dust (Falkowski *et al.*, 1998). The aeolian transport of iron in dust is a major source of iron for the open ocean, and dust could either increase or decrease in the future, depending on changes in the distribution of precipitation.

(iii) Changes in the elemental ratios of organic matter in the ocean. The elemental ratio of C:N:P in marine organic particles has long been recognized as conservative (Falkowski *et al.*, 1998). The extent to which the ratios can depart from observed concentrations is not known, yet variations could reduce the limitation of nitrogen and thus act in the same manner as the addition of nitrogen in affecting production, export, and thus oceanic uptake of CO_2.

(iv) Increases in the organic carbon/carbonate ratio of export production. The biological and carbonate pumps are described above (Section 16.2.2.2). Both pumps transport carbon out of the surface waters, and the subsequent decay at depth is responsible for the higher concentration of carbon in the intermediate and deep ocean. The formation of carbonate shells in the surface waters has the additional effect of increasing the p_{CO_2} in these waters, thus negating the export of the carbonate shells out of the surface. Any increase in the organic carbon/carbonate ratio of export production would enhance the efficiency of the biological pump.

16.5 THE FUTURE: DELIBERATE SEQUESTERING OF CARBON (OR REDUCTION OF SOURCES)

Section 16.4 addressed the factors thought to be influencing current terrestrial and oceanic sinks, and how they might change in the future. It is possible, of course, that CO_2 fertilization will become more important in the future as concentrations of CO_2 increase. Multiyear, whole ecosystems experiments with elevated CO_2 do not uniformly support this possibility, but higher concentrations of CO_2, together with nitrogen deposition or increases in moisture, might yet be important. Rather than wait for a more definitive answer, a more cautious approach to the future, besides reducing emissions of CO_2, would consider strategies for withdrawing carbon from the atmosphere through management. Three general options for sequestering carbon have received attention: terrestrial, oceanic, and geological management.

16.5.1 Terrestrial

Even if CO_2 fertilization and other environment effects turn out to be unimportant in enhancing terrestrial carbon storage, terrestrial sinks can still be counted on to offset carbon emissions or to reduce atmospheric concentrations of CO_2. Increasing the amount of carbon held on land might be achieved through at least six management options (Houghton, 1996; Kohlmaier *et al.*, 1998): (i) a reduction in the rate of deforestation (current rates of deforestation in the tropics are responsible for an annual release of 1–2 Pg C (Section 16.3.2.3); (ii) an increase in the area of forests (afforestation); (iii) an increase in the stocks of carbon within existing forests; (iv) an increase in the use of wood (including increased efficiency of wood harvest and use); (v) the substitution of wood fuels for fossil fuels; and (vi) the substitution of wood for more energy-intensive materials, such as aluminum, concrete, and steel. Estimates of the amount of carbon that might be sequestered on land over the 55-year period 1995–2050 range between 60 Pg C and 87 Pg C (1–2 Pg C yr^{-1} on average) (Brown, 1996). Additional carbon might also be sequestered in agricultural soils through conservation tillage and other agricultural management practices, and in grassland soils (Sampson and Scholes, 2000). An optimistic assessment, considering all types of ecosystems over the Earth, estimated a potential for storing 5–10 Pg C yr^{-1} over a period of 25–50 years (DOE, 1999).

The amount of carbon potentially sequestered is small relative to projected emissions of CO_2 from business-as-usual energy practices, and thus the terrestrial options for sequestering carbon should be viewed as temporary, "buying time" for the development and implementation of longer-lasting measures for reducing fossil fuel emissions (Watson *et al.*, 2000).

16.5.2 Oceanic

Schemes for increasing the storage of carbon in the oceans include stimulation of primary production with iron fertilization and direct injection of CO_2 at depth. As pointed out in Section 16.4.2.2., there are large areas of the ocean with high nutrient, low chlorophyll, concentrations. One explanation is that marine production is limited by the micronutrient iron. Adding iron to these regions might thus increase the ocean's biological pump, thereby reducing atmospheric CO_2 (Martin, 1990; Falkowski *et al.*, 1998). Mesoscale fertilization experiments have been carried out (Boyd *et al.*, 2000), but the effects of large-scale iron fertilization of the ocean are not known (Chisholm, 2000).

The direct injection of concentrated CO_2 (probably in liquid form) below the thermocline or on the seafloor might sequester carbon for hundreds of years (Herzog *et al.*, 2000). The gas might be dissolved within the water column or held in solid, ice-like CO_2 hydrates. The possibility is receiving attention in several national and international experiments (DOE, 1999). Large uncertainties exist in understanding the formation and stability of CO_2 hydrates, the effect of the concentrated CO_2 on ocean ecosystems, and the permanence of the sequestration.

16.5.3 Geologic

CO_2 may be able to be sequestered in geological formations, such as active and depleted oil and gas reservoirs, coalbeds, and deep saline aquifers. Such formations are widespread and have the potential to sequester large amounts of CO_2 (Herzog *et al.*, 2000). A model project is underway in the North Sea off the coast of Norway. The Sleipner offshore oil and natural gas field contains a gas mixture of natural gas and CO_2 (9%). Because the Norwegian government taxes emissions of CO_2 in excess of 2.5%, companies have the incentive to separate CO_2 from the natural gas and pump it into an aquifer 1,000 m under the sea. Although the potential for sequestering carbon in geological formations is large, technical and economic aspects of an operational program require considerable research.

16.6 CONCLUSION

We are conducting a great geochemical experiment, unlike anything in human history and unlikely to be repeated again on Earth. "Within a few centuries we are returning to the atmosphere and oceans the concentrated organic carbon stored in sedimentary rocks over hundreds of millions of years" (Revelle and Suess, 1957). During the last 150 years (~1850–2000), there has been a 30% increase in the amount of carbon in the atmosphere. Although most of this carbon has come from the combustion of fossil fuels, an estimated 150–160 Pg C have been lost during this time from terrestrial ecosystems as a result of human management (another 58–75 Pg C were lost before 1850). The global carbon balance suggests that other terrestrial ecosystems have accumulated ~115 Pg C since about 1930, at a steadily increasing rate. The annual net fluxes of carbon appear small relative to the sizes of the reservoirs, but the fluxes have been accelerating. Fifty percent of the carbon mobilized over the last 300 years (~1700–2000) was mobilized in the last 30–40 of these years (Houghton and Skole, 1990) (Figure 3). The major drivers of the geochemical experiment are reasonably well known. However, the results are uncertain, and there is no control. Furthermore, the experiment would take a long time to stop (or reverse) if the results turned out to be deleterious.

In an attempt to put some bounds on the experiment, in 1992 the nations of the world adopted the United Nations Framework Convention on Climate Change, which has as its objective "stabilization of greenhouse gas concentrations in the atmosphere at a level that would prevent dangerous anthropogenic interference with the climate system" (UNFCCC, 1992). The Convention's soft commitment suggested that the emissions of greenhouse gases from industrial nations in 2000 be no higher than the emissions in 1990. This commitment has been achieved, although more by accident than as a result of deliberate changes in policy. The "stabilization" resulted from reduced emissions from Russia, as a result of economic downturn, balanced by increased emissions almost everywhere else. In the US, e.g., emissions were 18% higher in the year 2000 than they had been in 1990. The near-zero increase in industrial nations' emissions between 1990 and 2000 does not suggest that the stabilization will last.

Ironically, even if the annual rate of global emissions were to be stabilized, concentrations of the gases would continue to increase. Stabilization of concentrations at early 2000's levels, e.g., would require reductions of 60% or more in the emission of long-lived gases, such as CO_2. The 5% average reduction in 1990 emissions by 2010, agreed to by the industrialized countries in the Kyoto Protocol (higher than 5% for the participating countries now that the US is no longer participating), falls far short of stabilizing atmospheric concentrations. Such a stabilization will require nothing less than a switch from fossil fuels to renewable forms of energy (solar, wind, hydropower, biomass), a switch that would have salubrious economic, political, security, and health consequences quite apart from limiting climatic change. Nevertheless, the geophysical experiment seems likely to continue for at least the near future, matched by a socio-political experiment of similar proportions, dealing with the consequences of either mitigation or not enough mitigation.

REFERENCES

Aber J. D., McDowell W. H., Nadelhoffer K. J., Magill A. H., Berntson G., Kamakea M., McNulty S., Currie W., Rustad L., and Fernandex I. (1998) Nitrogen saturation in temperate forest ecosystems. *BioScience* **48**, 921–934.

Achard F., Eva H. D., Stibig H.-J., Mayaux P., Gallego J., Richards T., and Malingreau J.-P. (2002) Determination of deforestation rates of the world's humid tropical forests. Science **297**, 999–1002.

Ajtay G. L., Ketner P., and Duvigneaud P. (1979) Terrestrial primary production and phytomass. In *The Global Carbon Cycle* (eds. B. Bolin, E. T. Degens, S. Kempe, and P. Ketner). Wiley, New York, pp. 129–182.

Andres R. J., Fielding D. J., Marland G., Boden T. A., Kumar N., and Kearney A. T. (1999) Carbon dioxide emissions from fossil-fuel use, 1751–1950. *Tellus* **51B**, 759–765.

Arrhenius S. (1896) On the influence of carbonic acid in the air upon the temperature of the ground. *Phil. Magazine J. Sci.* **41**, 237–276.

Aumont O., Orr J. C., Monfray P., Ludwig W., Amiotte-Suchet P., and Probst J.-L. (2001) Riverine-driven interhemispheric transport of carbon. *Global Biogeochem. Cycles* **15**, 393–405.

Barber V., Juday G. P., and Finney B. (2000) Reduced growth of Alaskan white spruce in the twentieth century from temperature-induced drought stress. *Nature* **405**, 668–673.

Barford C. C., Wofsy S. C., Goulden M. L., Munger J. W., Hammond Pyle E., Urbanski S. P., Hutyr L., Saleska S. R., Fitzjarrald D., and Moore K. (2001) Factors controlling long- and short-term sequestration of atmospheric CO_2 in a mid-latitude forest. *Science* **294**, 1688–1691.

Battle M., Bender M., Tans P. P., White J. W. C., Ellis J. T., Conway T., and Francey R. J. (2000) Global carbon sinks and their variability, inferred from atmospheric O_2 and $\delta^{13}C$. *Science* **287**, 2467–2470.

Behrenfeld M. J., Randerson J. T., McClain C. R., Feldman G. C., Los S. O., Tucker C. J., Falkowski P. G., Field C. B., Frouin R., Esaias W. W., Kolber D. D., and Pollack N. H. (2001) Biospheric primary production during an ENSO transition. *Science* **291**, 2594–2597.

Bergh J., Linder S., Lundmark T., and Elfving B. (1999) The effect of water and nutrient availability on the productivity of Norway spruce in northern and southern Sweden. *Forest Ecol. Manage.* **119**, 51–62.

Bolin B. (ed.) (1981) *Carbon Cycle Modelling.* Wiley, New York.

Bolin B. (1986) How much CO_2 will remain in the atmosphere? In *The Greenhouse Effect, Climatic Change, and Ecosystems* (eds. B. Bolin, B. R. Doos, J. Jager, and R. A. Warrick). Wiley, Chichester, England, pp. 93–155.

Bolin B., Degens E. T., Kempe S., and Ketner P. (eds.) (1979) *The Global Carbon Cycle.* Wiley, New York.

Bolin B., Doos B. R., Jager J., and Warrick R. A. (1986) *The Greenhouse Effect, Climatic Change, and Ecosystems.* Wiley, New York.

Bopp L., Le Quéré C., Heimann M., and Manning A. C. (2002) Climate-induced oceanic oxygen fluxes: implications for the contemporary carbon budget. *Global Biogeochem. Cycles* **16**, doi: 10.1029/2001GB001445.

Bousquet P., Ciais P., Peylin P., Ramonet M., and Monfray P. (1999a) Inverse modeling of annual atmospheric CO_2 sources and sinks: 1. Method and control inversion. *J. Geophys. Res.* **104**, 26161–26178.

Bousquet P., Peylin P., Ciais P., Ramonet M., and Monfray P. (1999b) Inverse modeling of annual atmospheric CO_2 sources and sinks: 2. Sensitivity study. *J. Geophys. Res.* **104**, 26179–26193.

Bousquet P., Peylin P., Ciais P., Le Quérè C., Friedlingstein P., and Tans P. P. (2000) Regional changes in carbon dioxide fluxes of land and oceans since 1980. *Science* **290**, 1342–1346.

Boyd P. W., Watson A. J., Cliff S., Law C. S., Abraham E. R., Trulli T., Murdoch R., Bakker D. C. E., Bowie A. R., Buesseler K. O., Chang H., Charette M., Croot P., Downing K., Frew R., Gall M., Hadfield M., Hall J., Harvey M., Jameson G., Laroche J., Liddicoat M., Ling R., Maldonado M.T., Mckay R.M., Nodder S., Pickmere S., Pridmore R., Rintoul S., Safi K., Sutton P., Strzepek R. T., Tanneberger K., Turner S., Waite A., and Zeldis J. (2000) A mesoscale phytoplankton bloom in the polar Southern Ocean stimulated by iron fertilization. *Nature* **407**, 695–702.

Braun S., Rihm B., Schindler C., and Fluckiger W. (2000) Growth of mature beech in relation to ozone and nitrogen deposition: an epidemiological approach. *Water Air Soil Pollut.* **116**, 356–364.

Broecker W. S. (2001) A Ewing Symposium on the contemporary carbon cycle. *Global Biogeochem. Cycles* **15**, 1031–1032.

Broecker W. S., Takahashi T., Simpson H. H., and Peng T.-H. (1979) Fate of fossil fuel carbon dioxide and the global carbon budget. *Science* **206**, 409–418.

Broecker W. S., Sutherland S., Smethie W., Peng T. H., and Ostlund G. (1995) Oceanic radiocarbon: separation of the natural and bomb components. *Global Biogeochem. Cycles* **9**, 263–288.

Brown S. (1996) Management of forests for mitigation of greenhouse gas emissions. In *Climatic Change 1995. Impacts, Adaptations and Mitigation of Climate Change: Scientific-technical Analyses* (eds. J. T. Houghton, G. J. Jenkins, and J. J. Ephraums). Cambridge University Press, Cambridge, pp. 773–797.

Bryant D. M., Holland E. A., Seastedt T. R., and Walker M. D. (1998) Analysis of litter decomposition in an alpine tundra. *Canadian J. Botany* **76**, 1295–1304.

Callendar G. S. (1938) The artificial production of carbon dioxide and its influence on temperature. *Quarterly J. Roy. Meteorol. Soc.* **64**, 223–240.

Cardon Z. G., Hungate B. A., Cambardella C. A., Chapin F. S., Field C. B., Holland E. A., and Mooney H. A. (2001) Contrasting effects of elevated CO_2 on old and new soil carbon pools. *Soil Biol. Biochem.* **33**, 365–373.

Caspersen J. P., Pacala S. W., Jenkins J. C., Hurtt G. C., Moorcroft P. R., and Birdsey R. A. (2000) Contributions of land-use history to carbon accumulation in US forests. *Science* **290**, 1148–1151.

Chisholm S. W. (2000) Stirring times in the Southern Ocean. *Nature* **407**, 685–687.

Ciais P., Peylin P., and Bousquet P. (2000) Regional biospheric carbon fluxes as inferred from atmospheric CO_2 measurements. *Ecol. Appl.* **10**, 1574–1589.

Clark D. (2002) Are tropical forests an important carbon sink? Reanalysis of the long-term plot data. *Ecol. Appl.* **12**, 3–7.

Conway T. J. and Tans P. P. (1999) Development of the CO_2 latitude gradient in recent decades. *Global Biogeochem. Cycles* **13**, 821–826.

Conway T. J., Tans P. P., Waterman L. S., Thoning K. W., Kitzis D. R., Masarie K. A., and Zhang N. (1994) Evidence for interannual variability of the carbon cycle from the National Oceanic and Atmospheric Administration/Climate Monitoring and Diagnostics Laboratory Global Air Sampling Network. *J. Geophys. Res.* **99**, 22831–22855.

Cooperative Atmospheric Data Integration Project—Carbon Dioxide (1997) *GLOBALVIEW—CO_2.* National Oceanic and Atmospheric Administration, Boulder, CO (CD-ROM).

Cox P. M., Betts R. A., Jones C. D., Spall S. A., and Totterdell I. J. (2000) Acceleration of global warming due to carbon-cycle feedbacks in a coupled climate model. *Nature* **408**, 184–187.

Cramer W., Bondeau A., Woodward F. I., Prentice I. C., Betts R. A., Brovkin V., Cox P. M., Fisher V., Foley J. A., Friend A. D., Kucharik C., Lomas M. R., Ramankutty N., Sitch S., Smith B., White A., and Young Molling C. (2001) Global response of terrestrial ecosystem structure and function to CO_2 and climate change: results from six dynamic global vegetation models. *Global Change Biol.* **7**, 357–373.

Curtis P. S. (1996) A meta-analysis of leaf gas exchange and nitrogen in trees grown under elevated carbon dioxide. *Plant Cell Environ.* **19**, 127–137.

Davidson E. A. (1995) Linkages between carbon and nitrogen cycling and their implications for storage of carbon in terrestrial ecosystems. In *Biotic Feedbacks in the Global Climatic System: Will the Warming Feed the Warming?* (eds. G. M. Woodwell and F. T. Mackenzie). Oxford University Press, New York, pp. 219–230.

Davidson E. A. and Hirsch A. I. (2001) Fertile forest experiments. *Nature* **411**, 431–433.

DeFries R. S., Field C. B., Fung I., Collatz G. J., and Bounoua L. (1999) Combining satellite data and biogeochemical models to estimate global effects of human-induced land cover change on carbon emissions and primary productivity. *Global Biogeochem. Cycles* **13**, 803–815.

DeFries R. S., Houghton R. A., Hansen M. C., Field C. B., Skole D., and Townshend J. (2002) Carbon emissions from tropical deforestation and regrowth based on satellite observations for the 1980s and 90s. *Proc. Natl. Acad. Sci.* **99**, 14256–14261.

Denning A. S., Fung I. Y., and Randall D. A. (1995) Latitudinal gradient of atmospheric CO_2 due to seasonal exchange with land biota. *Nature* **376**, 240–243.

Department of Energy (DOE) (1999) *Carbon Sequestration Research and Development.* National Technical Information Service, Springfield, Virginia (www.ornl.gov/carbon_sequestration/).

Detwiler R. P. and Hall C. A. S. (1988) Tropical forests and the global carbon cycle. *Science* **239**, 42–47.

Drake B. G., Muche M. S., Peresta G., Gonzalez-Meler M. A., and Matamala R. (1996) Acclimation of photosynthesis, respiration and ecosystem carbon flux of a wetland on Chesapeake Bay, Maryland to elevated atmospheric CO_2 concentration. *Plant Soil* **187**, 111–118.

Driscoll C. T., Lawrence G. B., Bulger A. J., Butler T. J., Cronan C. S., Eagar C., Lambert K. F., Likens G. E., Stoddard J. L., and Weathers K. C. (2001) Acidic deposition in the northeastern United States: sources and inputs, ecosystem effects, and management strategies. *BioScience* **51**, 180–198.

Etheridge D. M., Steele L. P., Langenfelds R. L., Francey R. J., Barnola J. M., and Morgan V. I. (1996) Natural and anthropogenic changes in atmospheric CO_2 over the last 1000 years from air in Antarctic ice and firn. *J. Geophys. Res.* **101**, 4115–4128.

Eva H. D., Achard F., Stibig H. J., and Mayaux P. (2003) Response to comment on Achard *et al.* (2002). *Science* **299**, 1015b.Falkowski P. G., Barber R. T., and Smetacek V. (1998) Biogeochemical controls and feedbacks on ocean primary production. *Science* **281**, 200–206.

Falkowski P., Scholes R. J., Boyle E., Canadell J., Canfield D., Elser J., Gruber N., Hibbard K., Högberg P., Linder S., Mackenzie F. T., Moore B., Pedersen T., Rosenthal Y., Seitzinger S., Smetacek V., and Steffen W. (2000) The global carbon cycle: a test of our knowledge of earth as a system. *Science* **290**, 291–296.

Fan S., Gloor M., Mahlman J., Pacala S., Sarmiento J., Takahashi T., and Tans P. (1998) A large terrestrial carbon sink in North America implied by atmospheric and oceanic CO_2 data and models. *Science* **282**, 442–446.

FAO (2001) *Global Forest Resources Assessment 2000.* Main Report, FAO Forestry Paper 140, Rome.Fearnside P. M. (2000) Global warming and tropical land-use change: greenhouse gas emissions from biomass burning, decomposition and soils in forest conversion, shifting cultivation and secondary vegetation. *Climat. Change* **46**, 115–158.

Fearnside P. M. and Laurance W. F. (2003) Comment on Achard *et al.* (2002). *Science* **299**, 1015a.

Fenn M. E., Poth M. A., Aber J. D., Baron J. S., Bormann B. T., Johnson D. W., Lemly A. D., McNulty S. G., Ryan D. F., and Stottlemyer R. (1998) Nitrogen excess in North American ecosystems: predisposing factors, ecosystem responses and management strategies. *Ecol. Appl.* **8**, 706–733.

Field C. B., Behrenfeld M. J., Randerson J. T., and Falkowski P. G. (1998) Primary production of the biosphere: integrating terrestrial and oceanic components. *Science* **281**, 237–240.

Fog K. (1988) The effect of added nitrogen on the rate of decomposition of organic matter. *Biol. Rev. Cambridge Phil. Soc.* **63**, 433–462.

Folland C. K. and Karl T. R. (2001) Observed climate variability and change. In *Climate Change 2001: The Scientific Basis. Contribution of Working Group I to the 3rd Assessment Report of the Intergovernmental Panel on Climate Change* (eds. J. T. Houghton, Y. Ding, D. J. Griggs, M. Noguer, P. J. van der Linden, X. Dai, K. Maskell, and C.A. Johnson). Cambridge University Press, Cambridge, UK and New York, pp. 99–181.

Fraser P. J., Elliott W. P., and Waterman L. S. (1986) Atmospheric CO_2 record from direct chemical measurements during the 19 century. In *The Changing Carbon Cycle. A Global Analysis* (eds. J. R. Trabalka and D. E. Reichle). Springer, New York, pp. 66–88.

Gaudinski J. B., Trumbore S. E., Davidson E. A., and Zheng S. (2000) Soil carbon cycling in a temperate forest: radiocarbon-based estimates of residence times, sequestration rates and partitioning of fluxes. *Biogeochemistry* **51**, 33–69.

Goldemberg J., Johansson T. B., Reddy A. K. N., and Williams R. H. (1985) An end-use oriented global energy strategy. *Ann. Rev. Energy* **10**, 613–688.

Goodale C. L., Apps M. J., Birdsey R. A., Field C. B., Heath L. S., Houghton R. A., Jenkins J. C., Kohlmaier G. H., Kurz W., Liu S., Nabuurs G.-J., Nilsson S., and Shvidenko A. Z. (2002) Forest carbon sinks in the northern hemisphere. *Ecol. Appl.* **12**, 891–899.

Goulden M. L., Munger J. W., Fan S.-M., Daube B. C., and Wofsy S. C. (1996) Exchange of carbon dioxide by a deciduous forest: response to interannual climate variability. *Science* **271**, 1576–1578.

Goulden M. L., Wofsy S. C., Harden J. W., Trumbore S. E., Crill P. M., Gower S. T., Fries T., Daube B. C., Fau S., Sulton D. J., Bazzaz A., and Munger J. W. (1998) Sensitivity of boreal forest carbon balance to soil thaw. *Science* **279**, 214–217.

Grace J., Lloyd J., McIntyre J., Miranda A. C., Meir P., Miranda H. S., Nobre C., Moncrieff J., Massheder J., Malhi Y., Wright I., and Gash J. (1995) Carbon dioxide uptake by an

undisturbed tropical rain forest in southwest Amazonia, 1992 to 1993. *Science* **270**, 778–780.
Gruber N., Sarmiento J. L., and Stocker T. F. (1996) An improved method for detecting anthropogenic CO_2 in the oceans. *Global Biogeochem. Cycles* **10**, 809–837.
Gurney K. R., Law R. M., Denning A. S., Rayner P. J., Baker D., Bousquet P., Bruhwiler L., Chen Y.-H., Ciais P., Fan S., Fung I. Y., Gloor M., Heimann M., Higuchi K., John J., Maki T., Maksyutov S., Masarie K., Peylin P., Prather M., Pak B. C., Randerson J., Sarmiento J., Taguchi S., Takahashi T., and Yuen C.-W. (2002) Towards robust regional estimates of CO_2 sources and sinks using atmospheric transport models. *Nature* **415**, 626–630.
Hall C. A. S. and Uhlig J. (1991) Refining estimates of carbon released from tropical land-use change. *Canadian J. Forest Res.* **21**, 118–131.
Heath L. S. and Smith J. E. (2000) An assessment of uncertainty in forest carbon budget projections. *Environ. Sci. Policy* **3**, 73–82.
Herzog H., Eliasson B., and Kaarstad O. (2000) Capturing greenhouse gases. *Sci. Am.* **282**(2), 72–79.
Hicke J. A., Asner G. P., Randerson J. T., Tucker C., Los S., Birdsey R., Jenkins J. C., Field C., and Holland E. (2002) Satellite-derived increases in net primary productivity across North America, 1982–1998. *Geophys. Res. Lett.,* 10.1029/2001GL013578.
Holland E. A., Braswell B. H., Lamarque J.-F., Townsend A., Sulzman J., Muller J.-F., Dentener F., Brasseur G., Levy H., Penner J. E., and Roelots G.-J. (1997) Variations in the predicted spatial distribution of atmospheric nitrogen deposition and their impact on carbon uptake by terrestrial ecosystems. *J. Geophys. Res.* **102**, 15849–15866.
Houghton R. A. (1987) Biotic changes consistent with the increased seasonal amplitude of atmospheric CO_2 concentrations. *J. Geophys. Res.* **92**, 4223–4230.
Houghton R. A. (1996) Converting terrestrial ecosystems from sources to sinks of carbon. *Ambio* **25**, 267–272.
Houghton R. A. (1999) The annual net flux of carbon to the atmosphere from changes in land use 1850–1990. *Tellus* **51B**, 298–313.
Houghton R. A. (2000) Interannual variability in the global carbon cycle. *J. Geophys. Res.* **105**, 20121–20130.
Houghton R. A. (2003) Revised estimates of the annual net flux of carbon to the atmosphere from changes in land use and land management 1850–2000. *Tellus* **55B**, 378–390.
Houghton R. A. and Hackler J. L. (1995) *Continental Scale Estimates of the Biotic Carbon Flux from Land Cover Change: 1850–1980.*
ORNL/CDIAC-79, NDP-050, Oak Ridge National Laboratory, Oak Ridge, TN, 144pp.Houghton R. A. and Hackler J. L. (2001) *Carbon Flux to the Atmosphere from Land-use Changes: 1850–1990.*
ORNL/ CDIAC-131, NDP-050/R1, US Department of Energy, Oak Ridge National Laboratory, Carbon Dioxide Information Analysis Center, Oak Ridge, TN.Houghton R. A. and Skole D. L. (1990) Carbon. In *The Earth as Transformed by Human Action* (eds. B. L. Turner, W. C. Clark, R. W. Kates, J. F. Richards, J. T. Mathews, and W. B. Meyer). Cambridge University Press, Cambridge, pp. 393–408.
Houghton R. A., Hobbie J. E., Melillo J. M., Moore B., Peterson B. J., Shaver G. R., and Woodwell G. M. (1983) Changes in the carbon content of terrestrial biota and soils between 1860 and 1980: a net release of CO_2 to the atmosphere. *Ecol. Monogr.* **53**, 235–262.
Houghton R. A., Davidson E. A., and Woodwell G. M. (1998) Missing sinks, feedbacks, and understanding the role of terrestrial ecosystems in the global carbon balance. *Global Biogeochem. Cycles* **12**, 25–34.
Houghton R. A., Hackler J. L., and Lawrence K. T. (1999) The US carbon budget: contributions from land-use change. *Science* **285**, 574–578.
Houghton R. A., Skole D. L., Nobre C. A., Hackler J. L., Lawrence K. T., and Chomentowski W. H. (2000) Annual fluxes of carbon from deforestation and regrowth in the Brazilian Amazon. *Nature* **403**, 301–304.
House J. I., Prentice I. C., Ramankutty N., Houghton R. A., and Heimann M. (2003) Reconciling apparent inconsistencies in estimates of terrestrial CO_2 sources and sinks. *Tellus* **(55B)**, 345–363.
Hu S., Chapin F. S., Firestone M. K., Field C. B., and Chiariello N. R. (2001) Nitrogen limitation of microbial decomposition in a grassland under elevated CO_2. *Nature* **409**, 188–191.
Hurtt G. C., Pacala S. W., Moorcroft P. R., Caspersen J., Shevliakova E., Houghton R. A., and Moore B., III. (2002) Projecting the future of the US carbon sink. *Proc. Natl. Acad. Sci.* **99**, 1389–1394.
Jarvis P. and Linder S. (2000) Constraints to growth of boreal forests. *Nature* **405**, 904–905.
Jobbágy E. G. and Jackson R. B. (2000) The vertical distribution of soil organic carbon and its relation to climate and vegetation. *Ecol. Appl.* **10**, 423–436.
Joos F., Plattner G.-K., Stocker T. F., Marchal O., and Schmittner A. (1999a) Global warming and marine carbon cycle feedbacks on future atmospheric CO_2. *Science* **284**, 464–467.
Joos F., Meyer R., Bruno M., and Leuenberger M. (1999b) The variability in the carbon sinks as reconstructed for the last 1000 years. *Geophys. Res. Lett.* **26**, 1437–1440.
Joos F., Prentice I. C., and House J. I. (2002) Growth enhancement due to global atmospheric change as predicted by terrestrial ecosystem models: consistent with US forest inventory data. *Global Change Biol.* **8**, 299–303.
Keeling C. D. (1973) Industrial production of carbon dioxide from fossil fuels and limestone. *Tellus* **25**, 174–198.
Keeling C. D., Bacastow R. B., Carter A. F., Piper S. C., Whorf T. P., Heimann M., Mook W. G., and Roeloffzen H. (1989) A three-dimensional model of atmospheric CO_2 transport based on observed winds: 1. Analysis of observational data. In *Aspects of Climate Variability in the Pacific and the Western Americas.* Geophysical Monograph 55 (ed. D. H. Peterson). American Geophysical Union, Washington, DC, pp. 165–236.
Keeling C. D., Chin J. F. S., and Whorf T. P. (1996a) Increased activity of northern vegetation inferred from atmospheric CO_2 observations. *Nature* **382**, 146–149.
Keeling C. D., Piper S. C., Bacastow R. B., Wahlen M., Whorf T. P., Heimann M., and Meijer H. A. (2001) *Exchanges of Atmospheric CO_2 and $^{13}CO_2$ with the Terrestrial Biosphere and Oceans from 1978 to 2000: I. Global Aspects.* Scripps Institution of Oceanography, Technical Report SIO Reference Series, No. 01-06 (Revised from SIO Reference Series, No. 00-21), San Diego.
Keeling R. F. and Garcia H. (2002) The change in oceanic O_2 inventory associated with recent global warming. *Proc. US Natl. Acad. Sci.* **99**, 7848–7853.
Keeling R. F. and Shertz S. R. (1992) Seasonal and interannual variations in atmospheric oxygen and implications for the global carbon cycle. *Nature* **358**, 723–727.
Keeling R. F., Piper S. C., and Heimann M. (1996b) Global and hemispheric CO_2 sinks deduced from changes in atmospheric O_2 concentration. *Nature* **381**, 218–221.
Kempe S. (1979) Carbon in the rock cycle. In *The Global Carbon Cycle* (eds. B. Bolin, E. T. Degens, S. Kempe, and P. Ketner). Wiley, New York, pp. 343–377.
Knorr W. and Heimann M. (2001) Uncertainties in global terrestrial biosphere modeling: 1. A comprehensive sensitivity analysis with a new photosynthesis and energy balance scheme. *Global Biogeochem. Cycles* **15**, 207–225.
Koch G. W. and Mooney H. A. (1996) Response of terrestrial ecosystems to elevated CO_2: a synthesis and summary. In *Carbon Dioxide and Terrestrial Ecosystems* (eds. G. W. Koch and H. A. Mooney). Academic Press,San Diego, pp. 415–429.

Köhl M. and Päivinen R. (1997). *Study on European Forestry Information and Communication System.* Office for Official Publications of the European Communities, Luxembourg, Volumes 1 and 2, 1328pp.

Kohlmaier G. H., Weber M., and Houghton R. A. (eds.) (1998) *Carbon Dioxide Mitigation in Forestry and Wood Industry.* Springer, Berlin.

Körner C. (2000) Biosphere responses to CO_2-enrichment. *Ecol. Appl.* **10**, 1590–1619.

Kurz W. A. and Apps M. J. (1999) A 70-year retrospective analysis of carbon fluxes in the Canadian forest sector. *Ecol. Appl.* **9**, 526–547.

Laws E. A., Falkowski P. G., Smith W. O., Ducklow H., and McCarthy J. J. (2000) Temperature effects on export production in the open ocean. *Global Biogeochem. Cycles* **14**, 1231–1246.

Lemon E. (1977) The land's response to more carbon dioxide. In *The Fate of Fossil Fuel CO_2 in the Oceans* (eds. N. R. Andersen and A. Malahoff). Plenum Press, New York, pp. 97–130.

Levitus S., Antonov J. I., Boyer T. P., and Stephens C. (2000) Warming of the world ocean. *Science* **287**, 2225–2229.

Lloyd A. H. and Fastie C. L. (2002) Spatial and temporal variability in the growth and climate response of treeline trees in Alaska. *Climatic Change* **52**, 481–509.

Longhurst A., Sathyendranath S., Platt T., and Caverhill C. (1995) An estimate of global primary production in the ocean from satellite radiometer data. *J. Plankton Res.* **17**, 1245–1271.

Luo Y. Q., Reynolds J., and Wang Y. P. (1999) A search for predictive understanding of plant responses to elevated [CO_2]. *Global Change Biol.* **5**, 143–156.

Luo Y., Wan S., Hui D., and Wallace L. L. (2001) Acclimatization of soil respiration to warming in a tall grass prairie. *Nature* **413**, 622–625.

Magill A. H., Aber J. D., Hendricks J. J., Bowden R. D., Melillo J. M., and Steudler P. A. (1997) Biogeochemical response of forest ecosystems to simulated chronic nitrogen deposition. *Ecol. Appl.* **7**, 402–415.

Magill A., Aber J., Berntson G., McDowell W., Nadelhoffer K., Melillo J., and Steudler P. (2000) Long-term nitrogen additions and nitrogen saturation in two temperate forests. *Ecosystems* **3**, 238–253.

Makipaa R. (1995) Effect of nitrogen input on carbon accumulation of boreal forest soils and ground vegetation. *Forest Ecol. Manage.* **79**, 217–226.

Malhi Y., Nobre A. D., Grace J., Kruijt B., Pereira M. G. P., Culf A., and Scott S. (1998) Carbon dioxide transfer over a central Amazonian rain forest. *J. Geophys. Res.* **103**, 31593–31612.

Malhi Y., Phillips O., Kruijt B., and Grace J. (2001) The magnitude of the carbon sink in intact tropical forests: results from recent field studies. In *6th International Carbon Dioxide Conference, Extended Abstracts.* Tohoku University, Sendai, Japan, pp. 360–363.

Marland G., Andres R. J., Boden T. A., and Johnston C. (1998) *Global, Regional and National CO_2 Emission Estimates from Fossil Fuel Burning, Cement Production, and Gas Flaring: 1751–1995* (revised January 1998). ORNL/CDIACNDP-030/R8, http://cdiac.esd.ornl.gov/ndps/ndp030.html

Martin J. H. (1990) Glacial–interglacial CO_2 change: the iron hypothesis. *Paleoceanography* **5**, 1–13.

Masarie K. A. and Tans P. P. (1995) Extension and integration of atmospheric carbon dioxide data into a globally consistent measurement record. *J. Geophys. Res.* **100**, 11593–11610.

Matear R. J. and Hirst A. C. (1999) Climate change feedback on the future oceanic CO_2 uptake. *Tellus* **51B**, 722–733.

McGuire A. D., Sitch S., Clein J. S., Dargaville R., Esser G.,Foley J., Heimann M., Joos F., Kaplan J., Kicklighter D. W.,Meier R. A., Melillo J. M., Moore B., Prentice I. C., Ramankutty N., Reichenau T., Schloss A., Tian H., Williams L. J., and Wittenberg U. (2001) Carbon balance of the terrestrial biosphere in the twentieth century: Analyses of CO_2, climate and land use effects with four process-based ecosystem models. *Global Biogeochem. Cycles* **15**,183–206.

Mclaughlin S. and Percy K. (2000) Forest health in North America: some perspectives on actual and potential roles of climate and air pollution. *Water Air Soil Pollut.* **116**, 151–197.

McNeil B. I., Matear R. J., Key R. M., Bullister J. L., and Sarmiento J. L. (2003) Anthropogenic CO_2 uptake by the ocean based on the global chlorofluorocarbon data set. *Science* **299**, 235–239.

McNulty S. G., Aber J. D., and Newman S. D. (1996) Nitrogen saturation in a high elevation spruce-fir stand. *Forest Ecol. Manage.* **84**, 109–121.

Miller S. D., Goulden M. L., Menton M. C., da Rocha H. R., Freitas H. C., Figueira A. M., and Sousa C. A. D. Tower-based and biometry-based measurements of tropical forest carbon balance. *Ecol. Appl.* (in press).

Mooney H. A., Canadell J., Chapin F. S., Ehleringer J., Körner C., McMurtrie R., Parton W. J., Pitelka L., and Schulze E.-D. (1999) Ecosystem physiology responses to global change. In *Implications of Global Change for Natural and Managed Ecosystems: A Synthesis of GCTE and Related Research* (ed. B. H. Walker, W. L. Steffen, J. Canadel, and J. S. I. Ingram). Cambridge University Press, Cambridge, pp. 141–189.

Myneni R. B., Los S. O., and Asrar G. (1995) Potential gross primary productivity of terrestrial vegetation from 1982–1990. *Geophys. Res. Lett.* **22**, 2617–2620.

Myneni R. B., Keeling C. D., Tucker C. J., Asrar G., and Nemani R. R. (1997) Increased plant growth in the northern high latitudes from 1981 to 1991. *Nature* **386**, 698–702.

Nadelhoffer K. J., Emmett B. A., Gundersen P., Kjønaas O. J., Koopmans C. J., Schleppi P., Teitema A., and Wright R. F. (1999) Nitrogen deposition makes a minor contribution to carbon sequestration in temperate forests. *Nature* **398**, 145–148.

National Research Council (NRC) (1983) *Changing Climate.* National Academy Press, Washington, DC.

Neftel A., Moor E., Oeschger H., and Stauffer B. (1985) Evidence from polar ice cores for the increase in atmospheric CO_2 in the past two centuries. *Nature* **315**, 45–47.

Nepstad D. C., Verissimo A., Alencar A., Nobre C., Lima E., Lefebvre P., Schlesinger P., Potter C., Moutinho P., Mendoza E., Cochrane M., and Brooks V. (1999) Large-scale impoverishment of Amazonian forests by logging and fire. *Nature* **398**, 505–508.

Oechel W. C., Hastings S. J., Vourlitis G., Jenkins M., Riechers G., and Grulke N. (1993) Recent change of arctic tundra ecosystems from a net carbon dioxide sink to a source. *Nature* **361**, 520–523.

Oechel W. C., Cowles S., Grulke N., Hastings S. J., Lawrence B., Prudhomme T., Riechers G., Strain B., Tissue D., and Vourlitis G. (1994) Transient nature of CO_2 fertilization in Arctic tundra. *Nature* **371**, 500–503.

Oren R., Ellsworth D. S., Johnsen K. H., Phillips N., Ewers B. E., Maier C., Schäfer K. V. R., McCarthy H., Hendrey G., McNulty S. G., and Katul G. G. (2001) Soil fertility limits carbon sequestration by forest ecosystems in a CO_2-enriched atmosphere. *Nature* **411**, 469–472.

Orr J. C. (2000) OCMIP carbon analysis gets underway. *Research GAIM* **3**(2), 4–5.

Orr J. C. and Dutay J.-C. (1999) OCMIP mid-project workshop. *Res. GAIM Newslett.* **3**, 4–5.

Orr J., Maier-Reimer E., Mikolajewicz U., Monfray P., Sarmiento J. L., Toggweiler J. R., Taylor N. K., Palmer J., Gruber N., Sabine C. L., Le Quéré C., Key R. M., and Boutin J. (2001) Estimates of anthropogenic carbon uptake from four 3-D global ocean models. *Global Biogeochem. Cycles* **15**, 43–60.

Pacala S. W., Hurtt G. C., Baker D., Peylin P., Houghton R. A., Birdsey R. A., Heath L., Sundquist E. T., Stallard R. F.,Ciais P., Moorcroft P., Caspersen J. P., Shevliakova E., Moore B.,

Kohlmaier G., Holland E., Gloor M., Harmon M. E., Fan S.-M., Sarmiento J. L., Goodale C. L., Schimel D., and Field C. B. (2001) Consistent land- and atmosphere-based US carbon sink estimates. *Science* **292**, 2316–2320.

Page S. E., Siegert F., Rieley L. O., Boehm H.-D. V., Jaya A., and Limin S. (2002) The amount of carbon released from peat and forest fires in Indonesia during 1997. *Nature* **420**, 61–65.

Peterson B. J. and Melillo J. M. (1985) The potential storage of carbon by eutrophication of the biosphere. *Tellus* **37B**, 117–127.

Phillips D. L., Brown S. L., Schroeder P. E., and Birdsey R. A. (2000) Toward error analysis of large-scale forest carbon budgets. *Global Ecol. Biogeogr.* **9**, 305–313.

Phillips O. L., Malhi Y., Higuchi N., Laurance W. F., Núñez P. V., Vásquez R. M., Laurance S. G., Ferreira L. V., Stern M., Brown S., and Grace J. (1998) Changes in the carbon balance of tropical forests: evidence from land-term plots. *Science* **282**, 439–442.

Phillips O. L., Malhi Y., Vinceti B., Baker T., Lewis S. L., Higuchi N., Laurance W. F., Vargas P. N., Martinez R. V., Laurance S., Ferreira L. V., Stern M., Brown S., and Grace J. (2002) Changes in growth of tropical forests: evaluating potential biases. *Ecol. Appl.* **12**, 576–587.

Plattner G.-K., Joos F., and Stocker T. F. (2002) Revision of the global carbon budget due to changing air-sea oxygen fluxes. *Global Biogeochem. Cycles* **16(4)**, 1096, doi: 10.1029/2001GB001746.

Post W. M., Emanuel W. R., Zinke P. J., and Stangenberger A. G. (1982) Soil carbon pools and world life zones. *Nature* **298**, 156–159.

Powell D. S., Faulkner J. L., Darr D. R., Zhu Z., and MacCleery D. W. (1993) Forest resources of the US, 1992. General Technical Report RM-234,. USDA Forest Service, Rocky Mountain Forest and Range Experiment Station, Fort Collins, CO.

Prather M. and Ehhalt D. (2001) Atmospheric chemistry and greenhouse gases. In *Climate Change 2001: The Scientific Basis. Contribution of Working Group I to the 3rd Assessment Report of the Intergovernmental Panel on Climate Change* (eds. J. T. Houghton, Y. Ding, D. J. Griggs, M. Noguer, P. J. van der Linden, X. Dai, K. Maskell, and C. A. Johnson). Cambridge University Press, Cambridge, UK and New York, pp. 239–287.

Prentice I. C., Farquhar G. D., Fasham M. J. R., Goulden M. L., Heimann M., Jaramillo V. J., Kheshgi H. S., Le Quéré C., Scholes R. J., and Wallace D. W. R. (2001) The carbon cycle and atmospheric carbon dioxide. In *Climate Change 2001: The Scientific Basis. Contribution of Working Group I to the 3rd Assessment Report of the Intergovernmental Panel on Climate Change* (eds. J. T. Houghton, Y. Ding, D. J. Griggs, M. Noguer, P. J. van der Linden, X. Dai, K. Maskell, and C. A. Johnson). Cambridge University Press, Cambridge, UK and New York, pp. 183–237.

Raich J. W. and Schlesinger W. H. (1992) The global carbon dioxide flux in soil respiration and its relationship to vegetation and climate. *Tellus* **44B**, 81–99.

Randerson J. T., Thompson M. V., Conway T. J., Fung I. Y., and Field C. B. (1997) The contribution of terrestrial sources and sinks to trends in the seasonal cycle of atmospheric carbon dioxide. *Global Biogeochem. Cycles* **11**, 535–560.

Randerson J. T., Field C. B., Fung I. Y., and Tans P. P. (1999) Increases in early season ecosystem uptake explain recent changes in the seasonal cycle of atmospheric CO_2 at high northern latitudes. *Geophys. Res. Lett.* **26**, 2765–2768.

Raynaud D. and Barnola J. M. (1985) An Antarctic ice core reveals atmospheric CO_2 variations over the past few centuries. *Nature* **315**, 309–311.

Revelle R. and Suess H. E. (1957) Carbon dioxide exchange between atmosphere and ocean and the question of an increase of atmospheric CO_2 during the past decades. *Tellus* **9**, 18–27.

Rice A. H., Pyle E. H., Saleska S. R., Hutyra L., de Camargo P. B., Portilho K., Marques D. F., and Wofsy S. C. Carbon balance and vegetation dynamics in an old-growth Amazonian forest. *Ecol. Appl.* (in press).

Richey J. E., Melack J. M., Aufdenkampe A. K., Ballester V. M., and Hess L. L. (2002) Outgassing from Amazonian rivers and wetlands as a large tropical source of atmospheric CO_2. *Nature* **416**, 617–620.

Saleska S. R., Miller S. D., Matross D. M., Goulden M. L., Wofsy S. C., da Rocha H., de Camargo P. B., Crill P. M., Daube B. C., Freitas C., Hutyra L., Keller M., Kirchhoff V., Menton M., Munger J. W., Pyle E. H., Rice A. H., and Silva H. Carbon fluxes in old-growth Amazonian rainforests: unexpected seasonality and disturbance-induced net carbon loss (in press).

Sampson R. N. and Scholes R. J. (2000) Additional human-induced activities—article 3.4. In *Land Use, Land-use Change, and Forestry. A Special Report of the IPCC* (eds. R. T. Watson, I. R. Noble, B. Bolin, N. H. Ravindranath, D. J. Verardo, and D. J. Dokken). Cambridge University Press, New York, pp. 181–281.

Sarmiento J. L. (1993) Ocean carbon cycle. *Chem. Eng. News* **71**, 30–43.

Sarmiento J. L. and Sundquist E. T. (1992) Revised budget for the oceanic uptake of anthropogenic carbon dioxide. *Nature* **356**, 589–593.

Sarmiento J. L., Hughes T. M. C., Stouffer R. J., and Manabe S. (1998) Simulated response of the ocean carbon cycle to anthropogenic climate warming. *Nature* **393**, 245–249.

Saugier B., Roy J., and Mooney H. A. (2001) Estimations of global terrestrial productivity: converging toward a single number? In *Terrestrial Global Productivity* (eds. J. Roy, B. Saugier, and H. A. Mooney). Academic Press, San Diego, California, pp. 543–557.

SCEP (Study of Critical Environmental Problems) (1970) *Man's Impact on the Global Environment.* The MIT Press, Cambridge, Massachusetts.

Schimel D. S., Alves D., Enting I., Heimann M., Joos F., Raynaud D., and Wigley T. (1996) CO_2 and the carbon cycle. In *Climate Change 1995* (eds. J. T. Houghton, L. G. M. Filho, B. A. Callendar, N. Harris, A. Kattenberg, and K. Maskell). Cambridge University Press, Cambridge, pp. 76–86.

Schimel D., Melillo J., Tian H., McGuire A. D., Kicklighter D., Kittel T., Rosenbloom N., Running S., Thornton P., Ojima D., Parton W., Kelly R., Sykes M., Neilson R., and Rizzo B. (2000) Contribution of increasing CO_2 and climate to carbon storage by ecosystems in the United States. *Science* **287**, 2004–2006.

Schimel D. S., House J. L., Hibbard K. A., Bousquet P., Ciais P., Peylin P., Braswell B. H., Apps M. J., Baker D., Bondeau A., Canadell J., Churkina G., Cramer W., Denning A. S., Field C. B., Friedlingstein P., Goodale C., Heimann M., Houghton R. A., Melillo J. M., Moore B., III, Murdiyarso D, Noble I., Pacala S. W., Prentice I. C., Raupach M. R., Rayner P. J., Scholes R. J., Steffen W. L., and Wirth C. (2001) Recent patterns and mechanisms of carbon exchange by terrestrial ecosystems. *Nature* **414**, 169–172.

Schlesinger W. H. and Lichter J. (2001) Limited carbon storage in soil and litter of experimental forest plots under increased atmospheric CO_2. *Nature* **411**, 466–469.

Shaw M. R., Zavaleta E. S., Chiariello N. R., Cleland E. E., Mooney H. A., and Field C. B. (2003) Grassland responses to global environmetal changes suppressed by elevated CO_2. *Science* **298**, 1987–1990.

Shvidenko A. Z. and Nilsson S. (1997) Are the Russian forests disappearing? *Unasylva* **48**, 57–64.

Smith H. J., Fischer H., Wahlen M., Mastroianni D., and Deck B. (1999) *Nature* **400**, 248–250.

Spiecker H., Mielikainen K., Kohl M., Skovsgaard J. (eds.) (1996) *Growth Trends in European Forest—Studies from 12 Countries.* Springer, Berlin.

Stephens B. B., Keeling R. F., Heimann M., Six K. D., Murnane R., and Caldeira K. (1998) Testing global ocean carbon cycle models using measurements of atmospheric O_2 and CO_2 concentration. *Global Biogeochem. Cycles* **12**, 213–230.

Sundquist E. T. (1986) Geologic analogs: their value and limitation in carbon dioxide research. In *The Changing Carbon Cycle. A Global Analysis* (eds. J. R. Trabalka and D. E. Reichle). Springer, New York, pp. 371–402.

Sundquist E. T. and Broecker W. S. (eds.) (1985) *The Carbon Cycle and Atmospheric CO2: Natural Variations Archean to Present,* Geophysical Monograph 32. American Geophysical Union, Washington, DC.

Tamm C. O., Aronsson A., and Popovic B. (1995) Nitrogen saturation in a long-term forest experiment with annual additions of nitrogen. *Water Air Soil Pollut.* **85**, 1683–1688.

Tans P. P., Fung I. Y., and Takahashi T. (1990) Observational constraints on the global atmospheric CO_2 budget. *Science* **247**, 1431–1438.

Tans P. P., Fung I. Y., and Enting I. G. (1995) Storage versus flux budgets: the terrestrial uptake of CO_2 during the 1980s. In *Biotic Feedbacks in the Global Climatic System.* Will the Warming Feed the Warming (eds. G. M. Woodwell and F. T. Mackenzie). Oxford University Press, New York, pp. 351–366.

Tans P. P., Bakwin P. S., Bruhwiler L., Conway T. J., Dlugokencky E. J., Guenther D. W., Kitzis D. R., Lang P. M., Masarie K. A., Miller J. B., Novelli P. C., Thoning K. W., Vaughn B. H., White J. W. C., and Zhao C. (2001) Carbon cycle. In *Climate Monitoring and Diagnostics Laboratory Summary Report No. 25 1998–1999* (eds. R. C. Schnell, D. B. King, and R. M. Rosson). NOAA, Boulder, CO, pp. 24 46.

Taylor J. A. and Orr J. C. (2000) The natural latitudinal distribution of atmospheric CO_2. *Global Planet. Change* **26**, 375–386.

Tian H., Melillo J. M., Kicklighter D. W., McGuire A. D., Helfrich J. V. K., Moore B., and Vorosmarty C. J. (1998) Effect of interannual climate variability on carbon storage in Amazonian ecosystems. *Nature* **396**, 664–667.

Tissue D. T. and Oechel W. C. (1987) Response of *Eriophorum vaginatum* to elevated CO_2 and temperature in the Alaskan tussock tundra. *Ecology* **68**, 401–410.

Trabalka J. R. (ed.) (1985) *Atmospheric Carbon Dioxide and the Global Carbon Cycle.* DOE/ER-0239, US Department of Energy, Washington, DC.

UNFCCC (1992) *Text of the United Nations Framework Convention on Climate Change* (UNEP/WMO Information Unit on Climate Change.), Geneva, Switzerland, 29pp.

Valentini R., Matteucci G., Dolman A. J., Schulze E.-D., Rebmann C., Moors E. J., Granier A., Gross P., Jensen N. O., Pilegaard K., Lindroth A., Grelle A., Bernhofer C., Grünwald T., Aubinet M., Ceulemans R., Kowalski A. S., Vesala T., Rannik Ü., Berbigier P., Loustau D., Gudmundsson J., Thorgeirsson H., Ibrom A., Morgenstern K., Clement R., Moncrieff J., Montagnani L., Minerbi S., and Jarvis P. G. (2000) Respiration as the main determinant of European forests carbon balance. *Nature* **404**, 861–865.

Wanninkhof R. (1992) Relationship between wind-speed and gas-exchange over the ocean. *J. Geophys. Res.* **97**, 7373–7382.

Watson R. T., Rodhe H., Oeschger H., and Siegenthaler U. (1990) Greenhouse gases and aerosols. In *Climate Change,* The IPCC Scientific Assessment (eds. J. T. Houghton, G. J. Jenkins, and J. J. Ephraums). Cambridge University Press, Cambridge, pp. 1–40.

Watson R. T., Noble I. R., Bolin B., Ravindranath N. H., Verardo D. J., and Dokken D. J. (eds.) (2000) *Land Use, Land-Use Change, and Forestry.* A Special Report of the IPCC, Cambridge University Press, New York.

White A., Cannell M. G. R., and Friend A. D. (1999) Climate change impacts on ecosystems and the terrestrial carbon sink: a new assessment. *Global Environ. Change* **9**, S21–S30.

Woodwell G. M. (1983) Biotic effects on the concentration of atmospheric carbon dioxide: a review and projection. In *Changing Climate.* National Academy Press, Washington, DC, pp. 216–241.

Woodwell G. M. and Mackenzie F. T. (eds.) (1995) *Biotic Feedbacks in the Global Climatic System. Will the Warming Feed the Warming?* Oxford University Press, New York.

Woodwell G. M. and Pecan E. V. (eds.) (1973) *Carbon and the Biosphere,* US Atomic Energy Commission, Symposium Series 30. National Technical Information Service, Springfield, Virginia.

Readings from the Treatise on Geochemistry
ISBN: 978-0-12-381391-6
pp. 537–578

17 Environmental Geochemistry of Radioactive Contamination

M. D. Siegel and C. R. Bryan

Sandia National Laboratories, Albuquerque, NM, USA

NOMENCLATURE

a_{H^+}	activity of the hydrogen ion in solution
$a_{UO_2^{2+}}$	activity of uranyl in solution
$A'_{colloid}$	specific surface area of the colloid, in a colloid/rock system
A'_{IP}	specific surface area of the rock matrix, in a colloid/rock system
C	colloid concentration in solution
C_i	initial concentration of contaminant in solution
C_f	final concentration of contaminant in solution
C_L	concentration of a contaminant in solution
C_S	concentration of a contaminant sorbed onto solid
e	fundamental electrical charge
F	the ratio of the specific surface areas of the colloid and the rock matrix, in a colloid/rock system
k	Boltzmann's constant
K_d	distribution coefficient for a contaminant in a water/rock system
$K_{i,j,k}$	intrinsic formation constant for an aqueous species (*i, j, k* describe stoichiometry)
m	mass of substrate present in a batch system
R	retardation factor for a contaminant
R_d	sorption coefficient for a contaminant in a water/rock system
$R_{F,eff}$	effective retardation factor (including the effects of transport on colloids)
T	absolute temperature, in K
V_i	initial solution volume in a batch system
V_f	final solution volume in a batch system
β^{cat}	intrinsic surface-complexation constant for a cationic species
ρ	bulk density of a porous medium
ϕ	bulk porosity of a porous medium
ψ_0	electrical potential of the inner (o) surface plane

17.1 INTRODUCTION

Psychometric studies of public perception of risk have shown that dangers associated with radioactive contamination are considered the most dreaded and among the least understood hazards (Slovic, 1987). Fear of the risks associated with nuclear power and associated contamination has had important effects on policy and commercial decisions in the last few decades. In the US, no new nuclear power plants were ordered between 1978 and 2002, even though it has been suggested that the use of nuclear power has led to significantly reduced CO_2 emissions and may provide some relief from the potential climatic changes associated with fossil fuel use. The costs of the remediation of sites contaminated by radioactive materials and the projected costs of waste disposal of radioactive waste in the US dwarf many other environmental programs. The cost of disposal of spent nuclear fuel at the proposed repository at Yucca Mountain will likely exceed \$10 billion. The estimated total life cycle cost for remediation of US Department of Energy (DOE) weapons production sites ranged from \$203–247 billion dollars in constant 1999 dollars, making the cleanup the largest environmental project on the planet (US DOE, 2001). Estimates for the cleanup of the Hanford site alone exceeded \$85 billion through 2046 in some of the remediation plans.

Policy decisions concerning radioactive contamination should be based on an understanding of the potential migration of radionuclides through the geosphere. In many cases, this potential may have been overestimated, leading to decisions to clean up contaminated sites unnecessarily and exposing workers to unnecessary risk. It is important for both the general public and the scientific community to be familiar with information that is well established, to identify the areas of uncertainty and to understand the significance of that uncertainty to the assessment of risk.

17.1.1 Approach and Outline of Chapter

This chapter provides an applications-oriented summary of current understanding of environmental radioactive contamination by addressing three major questions:

(i) What are the major sources of radioactive contamination on the planet?

(ii) What controls the migration of radioactive contaminants in the environment?

(iii) How can an understanding of radionuclide geochemistry be used to facilitate environmental remediation or disposal of radioactive wastes, and how can we assess the associated risks?

The chapter starts with an overview of the nature of major sites of radioactive environmental contamination. A brief summary of the health effects associated with exposure to ionizing radiation and radioactive materials follows. The remainder of the chapter summarizes current knowledge of the properties of radionuclides as obtained and applied in three interacting spheres of inquiry and analysis: (i) experimental studies and theoretical calculations, (ii) field studies, and (iii) predictions of radionuclide behavior for remediation and waste disposal. Recent studies of radionuclide speciation, solubility, and sorption are reviewed, drawing upon the major US and European nuclear waste and remediation programs. Examples are given of the application of that information to understanding the behavior of radionuclides as observed at sites of natural and anthropogenic radioactive contamination. Finally, the uses of that information in remediation of radioactive contamination and in predicting the potential behavior of radionuclides released from proposed nuclear waste repositories are described.

17.1.2 Previous Reviews and Chapter Scope

Information describing the environmental geochemistry of radionuclides is being gathered at a rapid rate due to the high interest and pressing need to dispose of nuclear wastes and to remediate radioactively contaminated areas. Several important books have been written on the subject of actinide chemistry in the last few decades. Notable ones include Seaborg and Katz (1954), Katz *et al.* (1986), Ivanovich (1992), and Choppin *et al.* (1995). This chapter updates similar reviews that have been published in the last few decades, such as those by Allard (1983), Krauskopf (1986), Choppin and Stout (1989), Hobart (1990), Fuger (1992), Kim (1993), Silva and Nitsche (1995), and chapters in Barney *et al.* (1984), Langmuir (1997a), and Zhang and Brady (2002). National symposia dealing with the disposal of nuclear waste and remediation of radioactive environmental contamination have been held annually by the Material Research Society (e.g., McGrail and Cragnolino, 2002) and Waste Management Symposia (e.g., WM Symposia, 2001) since the 1980s. Proceedings of these conferences should continue to be valuable sources of the results of current research and work carried out after the publication of this chapter.

This chapter focuses on the interactions of radionuclides with geomedia in near-surface low-temperature environments. Due to the limitations on the chapter length, this review will not describe the mineralogy or economic geology of uranium deposits; the use of radionuclides as environmental tracers in studies of the atmosphere, hydrosphere, or lithosphere, the nature of the nuclear fuel cycle or processes involved in nuclear weapons production. Likewise, radioactive contamination associated with the use of atomic weapons during World War II, the contamination of the atmosphere, hydrosphere, or lithosphere related to nuclear weapons testing, and concerns over the contamination of the Arctic Ocean by the Soviet nuclear fleet are not discussed. The interested reader is advised to turn to other summaries of these topics and included references. The nuclear fuel cycle is summarized in publications of the US DOE (1997a,b). Eisenbud (1987) provides a comprehensive overview of environmental radioactivity from natural, industrial, and military sources. A recent publication of the Mineralogical Society of America included a series of review articles describing the mineralogy and paragenesis of uranium deposits and the environmental geochemistry of uranium and its decay products (Burns and Finch, 1999). Mahara and Kudo (1995) and Kudo *et al.* (1995) review the environmental behavior of plutonium released by the Nagasaki atomic bomb blast. A recent summary of the extent of contamination from atmospheric nuclear testing is found in Beck and Bennett (2002). Descriptions of radioactive contamination of the Arctic Ocean are found in Salbu *et al.* (1997), Aarkrog *et al.* (1999), and in publications of the Arctic Monitoring and Assessment Programme (AMAP, 2002; http://www.amap.no/).

17.2 THE NATURE AND HAZARDS OF RADIOACTIVE ENVIRONMENTAL CONTAMINATION

The relationships between radioactive contamination and the risks to human health have scientific and regulatory dimensions. From the scientific perspective, the risks associated with radioactive contamination depend systematically on the magnitude of the source, the type of radiation, exposure routes, and biological susceptibility to the effects of radiation damage. The effects of radiation on human health can be ascertained from the health status of people exposed to different levels of radiation in nuclear explosions or occupational settings. The most extreme exposures include those experienced by survivors of the atomic bomb blasts in Hiroshima and Nagasaki in World War II. Researchers have not observed an increase in cancer frequency for Japanese bomb survivors below an external dose of 0.2 Gy (20 rad), thus providing a lower limit for human biological susceptibility to the effects of radiation. In contrast, the regulatory perspective addresses estimates of an upper limit to biological susceptibility and involves a *di minimus* approach to exposure limits. Radiation protection standards are based on a "zero threshold" dose-response relationship and exposure for the public is limited to 1 mSv yr^{-1} (0.1 rem yr^{-1}). The following sections summarize basic information about the scientific and regulatory aspects of this issue. The different types of radioactivity are characterized, natural and anthropogenic sources of radioactivity are described, levels of exposure to these sources are estimated and compared to regulatory exposure limits, and the effects of radiation on human health are discussed.

17.2.1 Sources of Radioactivity

17.2.1.1 Radioactive processes

Only certain combinations of protons and neutrons result in stable atomic nuclei. Figure 1 shows a section of the chart of the nuclides, on which nuclides are plotted as a function of their proton number (*Z*) and neutron number (*N*). The radioactive decay chain for ^{238}U is indicated; only ^{206}Pb has a stable combination of protons and neutrons. At low atomic numbers (below $Z=20$), isotopes with proton: neutron ratios of $\sim$1 are stable, but a progressively higher proportion of neutrons is required to produce stability at higher atomic numbers. Unstable nuclei undergo radioactive decay—spontaneous transformations involving emission of particles and/or photons, resulting in changes in *Z* and *N*, and transformation of that atom into another element. Several types of radioactive decay may occur:

- β^--decay—a negatively charged beta particle (electron) is emitted from the nucleus of the atom, and one of the neutrons is transformed into a proton. *Z* increases by 1 and *N* decreases by 1.
- β^+-decay—a positively charged beta particle (positron) is emitted from the nucleus, and a proton is transformed into a neutron. *Z* decreases by 1 and *N* increases by 1.
- Electron capture—an unstable nucleus may capture an extranuclear electron, commonly a K-shell electron, resulting in the transformation of a proton to a neutron. This results in the same change in *Z* and *N* as β^+ decay; commonly, nuclides with a deficiency of neutrons can decay by either mechanism.
- α-decay—nuclei of high atomic number (heavier than cerium), and a few light nuclides, may decay by emission of an α-particle, a ^{4}He nucleus consisting of two protons and two neutrons. *Z* and *N* both decrease by 2.

In each case, the daughter nucleus is commonly left in an excited state and decays to the ground state by emission of γ-rays. If there is a significant delay between the two processes, the γ-emission is considered a separate event. Decay by γ-emission, resulting in no change in *Z* or *N*, is called an isomeric transition (e.g., decay of ^{99m}Tc to ^{99}Tc).

Many radioactive elements decay to produce unstable daughters. The radioactivity of many forms of radioactive contamination is due primarily to daughter products with short half-lives. The longest such decay chains that occur naturally are those for ^{238}U, ^{235}U, and ^{232}Th, which decay through a series of intermediate daughters to ^{206}Pb, ^{207}Pb, and ^{208}Pb, respectively (see Table 1). The decay of ^{238}U to ^{206}Pb results in the production of eight α-particles and six β-particles; that of ^{235}U to ^{207}Pb, seven α- and four β-particles; and that of ^{232}Th to ^{208}Pb, six α- and four β-particles. Thus, understanding the geochemistry of radioactive contamination requires consideration of the chemistries of both the abundant parents and of the transient

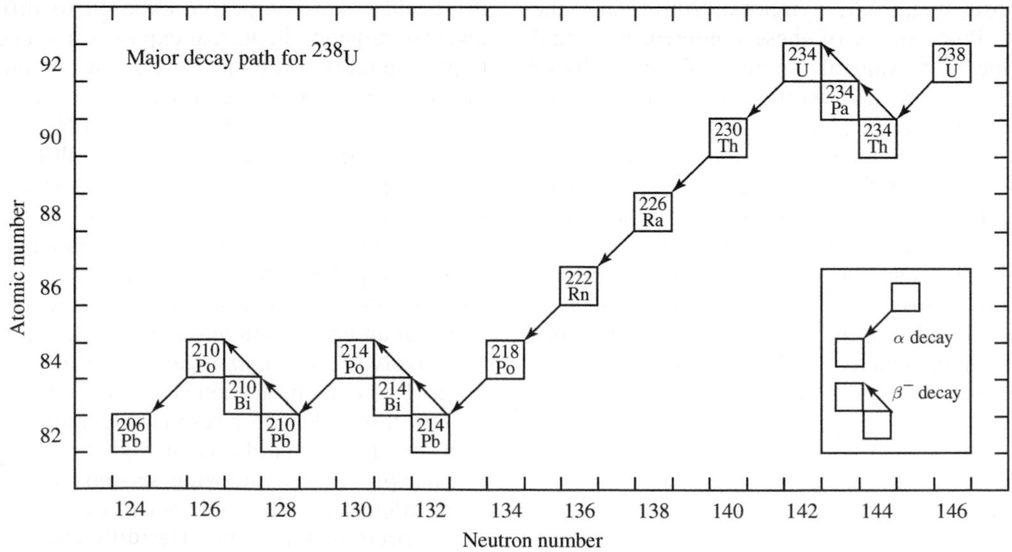

Figure 1 The major decay path for ^{238}U. A small fraction of decays will follow other possible decay paths—for example, decaying by β^- emission from ^{218}Po to ^{218}At, and then by α emission to ^{214}Bi—but ^{206}Pb is the stable end product in all cases.

Table 1 The major decay paths for several important actinides (isotope, half-life, and decay mode).

^{238}U decay series			^{235}U decay series			^{232}Th decay series		
^{238}U	*4.47×10^{9} yr*	α	*^{235}U*	*7.04×10^{8} yr*	α	*^{232}Th*	*1.40×10^{10} yr*	α
^{234}Th	24.1 d	β^-	^{231}Th	1.06 d	β^-	^{228}Ra	5.76 yr	β^-
^{234}Pa	1.17 m	β^-	^{231}Pa	3.28×10^{4} yr	α	^{228}Ac	6.15 h	β^-
^{234}U	2.46×10^{5} yr	α	^{227}Th	21.8 yr	β^-	^{228}Th	1.91 yr	α
^{230}Th	7.54×10^{4} yr	α	^{223}Ra	18.7 d	α	^{224}Ra	3.66 d	α
^{226}Ra	1.6×10^{3} yr	α	^{219}Rn	11.4 d	α	^{220}Rn	55.6 s	α
^{222}Rn	3.82 d	α	^{215}Po	4.0 s	α	^{216}Po	0.15 s	α
^{218}Po	3.10 m	α	^{211}Pb	1.78×10^{-3} s	α	^{212}Pb	10.6 h	β^-
^{214}Pb	27.0 m	β^-	^{214}Pb	36.1 m	β^-	^{212}Bi	60.6 m	β^-
^{214}Bi	19.9 m	β^-	^{211}Bi	2.14 m	α	^{212}Po	45 s	α
^{214}Po	1.64×10^{-4} s	α	^{207}Tl	4.78 m	β^-	^{208}Pb	Stable	
^{210}Pb	22.6 yr	β^-	^{207}Pb	Stable				
^{210}Bi	51.01 d	β^-						
^{210}Po	138 d	α						
^{206}Pb	Stable							

^{237}Np decay series			^{239}Pu decay series			^{238}Pu decay series			^{241}Am decay series		
^{237}Np	*2.14×10^{6} yr*	α	*^{239}Pu*	*2.44×10^{4} yr*	α	*^{238}Pu*	*87.7 yr*	α	*^{241}Am*	*433 yr*	α
^{233}Pa	27.0 d	β^-	^{235}U	As above		^{234}U	As above		^{237}Np	As above	
^{233}U	1.59×10^{5} yr	α									
^{229}Th	7.30×10^{3} yr	α									
^{225}Ra	14.9 d	β^-									
^{224}Ac	10.0 d	α									
^{221}Fr	4.8 m	α									
^{217}At	32 ms	α									
^{213}Bi	45.6 m	β^-									
^{213}Po	4 μs	α									
^{209}Pb	3.25 h	β^-									
^{209}Bi	Stable										

Source: Parrington *et al.* (1996).

daughters, which have much lower chemical concentrations. After several half-lives of the longest-lived intermediate daughter, a radioactive parent and its unstable daughters will reach secular equilibrium; the contribution of each nuclide to the total activity will be the same. Thus, a sample of ^{238}U will, after about a million and a half years, have a total α-activity that is ~8 times that of the uranium alone.

Heavy nuclei can also decay by *fission*, by splitting into two parts. Although some nuclei can spontaneously fission, most require an input of energy. This is most commonly accomplished by absorption of neutrons, although α-particles, γ-rays and even X-rays may also induce fission. Fission is usually asymmetric—two unequal nuclei, or *fission products*, are produced, with atomic weights ranging from 66 to 172. Fission product yields vary with the energy of the neutrons inducing fission; under reactor conditions, fission is highly asymmetric, with production maxima at masses of ~95 and ~135. High-energy neutrons result in more symmetric fission, with a less bimodal distribution of products.

Fission products generally contain an excess of neutrons and are radioactive, decaying by successive β^- emissions to stable nuclides. The high radioactivity of spent nuclear fuel and of the wastes generated by fuel reprocessing for nuclear weapons production is largely due to fission products, and decreases rapidly over the first few tens of years. In addition to the daughter nuclei, neutrons are released during fission (2.5–3.0 per fission event for thermal neutrons), creating the potential for a fission chain reaction—the basis for nuclear power and nuclear weapons.

The major decay paths for the naturally occurring isotopes of uranium and thorium are shown in Table 1. Other actinides of environmental importance include ^{237}Np, ^{238}Pu, ^{239}Pu, and ^{241}Am. These have decay series similar to and overlapping those of uranium and thorium. Neptunium-237 ($t_{1/2}=2.14\times10^{6}$ yr, α) decays to ^{209}Bi through a chain of intermediates, emitting seven α- and four β^--particles. Plutonium-238 ($t_{1/2}=86$ yr, α) decays into ^{234}U, an intermediate daughter on the ^{238}U decay series. Plutonium-239 ($t_{1/2}=2.44\times10^{4}$ yr, α) decays into ^{235}U. Americium-241 ($t_{1/2}=458$ yr, α) decays into ^{237}Np.

The basic unit of measure for radioactivity is the number of atomic decays per unit time. In the SI system, this unit is the becquerel (Bq), defined as one decay per second. An older, widely used measure of activity is the curie (Ci). Originally defined as the activity of 1 g of ^{226}Ra (1 Ci $=3.7\times10^{10}$ Bq). The units used to describe the dose, or energy absorbed by a material exposed to radiation, are dependent upon the type of radiation and the material. X-ray or γ-radiation absorbed by air is measured in Roentgens (R). The dose absorbed by any material, by any radiation, is measured in rad (radiation-absorbed-dose), where 1 rad corresponds to 100 erg g^{-1} (1 erg = 10^{-7} J) of absorbed energy. The SI equivalent is the gray (Gy), which is equal to 100 rad.

Different types of radiation affect biological materials in different ways, so a different unit is needed to describe the dose necessary to produce an equivalent biological damage. Historically, this unit is the rem (roentgen-equivalent-man). The dose in rem is equal to the dose in rad multiplied by a quality factor, which varies with the type of radiation. For β^--, γ-, and X-ray radiation, the quality factor is 1; for neutrons, it is 2–11, depending upon the energy of the particle; and for α-particles, the quality factor is 20. The SI unit for equivalent dose is the sievert (Sv), which is equivalent to 100 rem.

17.2.1.2 Natural sources of radioactivity

Radioactive materials have been present in the environment since the accretion of the Earth. The decay of radionuclides provides an important source of heat that drives many large-scale planetary processes. The most abundant naturally occurring radionuclides are ^{40}K, ^{232}Th, and ^{238}U and ^{235}U. The bulk of the natural global inventory of actinide radioactivity in the upper 100 m of the lithosphere ($\sim 10^{22}$ Bq or 2.7×10^{11} Ci) is due to activity of uranium and thorium isotopes (Santschi and Honeyman, 1989; Ewing, 1999). This is about equal to the total activity of ^{40}K in the world ocean. Average crustal concentrations of uranium (mostly 238) and thorium (mostly 232) are 2.7 μg g^{-1} and 9.6 μg g^{-1}, respectively. Both elements are enriched in silica-rich igneous rocks (4.4 μg g^{-1} and 16 μg g^{-1}, respectively in granites) and are highly enriched in zircons (2,000 μg g^{-1} and 2,500 μg g^{-1}, respectively).

In groundwater, average uranium concentrations range from <0.1 μg L^{-1} (reducing) to 100 μg L^{-1} (oxidizing) (Langmuir, 1997a). The average thorium concentration in groundwater is <1 μg L^{-1} and is not affected by solution redox conditions. Other naturally occurring radionuclides include actinium, technetium, neptunium, and protactinium. Small amounts of actinides (^{237}Np and ^{239}Pu) are present from neutron capture reactions with ^{238}U. Natural ^{99}Tc is a product of ^{238}U spontaneous fission (Curtis *et al.*, 1999). For comparison, Table 2 provides examples of large-scale sources of natural and anthropogenic radioactivity in the environment throughout the world.

17.2.1.3 Nuclear waste

It is estimated that the inventory of nuclear reactor waste in the US will reach 1.3×10^{21} Bq by 2020 (Ewing, 1999). Decay of the radionuclides from a reference inventory over 10^{10} yr is shown in Figure 2. In this figure, the change of ingestion toxicity of radionuclides important for disposal of high-level wastes (HLWs) is shown. Ingestion toxicity for a given radioisotope is defined as the isotopic quantity (in microcuries) divided by the maximum permissible concentration in water (in microcuries per cubic meters) for that isotope (Campbell *et al.*, 1978). Plots of time-dependent

Table 2 Examples of sources of radioactivity in the environment.

Location	*Source of radioactivity*	*Major radionuclides*	*Amount of radioactivity*	*Ref.*
Global	Top 100 m of lithosphere	^{238,235}U, ^{232}Th	1.0×10^{22} Bq	1,2
HLW geologic repository	70 kt spent fuel (proposed)	^{137}Cs, ^{90}Sr	1.0×10^{22} Bq	1,2
Atmospheric testing	220 Megaton yield	^{131}I and ^{3}H	2.0×10^{20} Bq	1,2
		^{239,240}Pu	1.0×10^{17} Bq	
Mayak, Russia	Nuclear production	Various HLW	3.6×10^{19} Bq	4
		^{90}Sr, ^{137}Cs	2.1×10^{19} Bq	
US weapons complex	High level waste/ 100 million gallons (3.8×10^{5} m^3)	Short-lived ($t_{1/2}$ < 50 yr): ^{137}Cs, ^{90}Sr, ^{90}Y, ^{137m}Ba, ^{241}Pu	3.3×10^{19} Bq	6
		Longer lived ($t_{1/2}$ = 50–500 yr): ^{238}Pu, ^{131}Sm, ^{241}Am	1.1×10^{17} Bq	
		Long lived ($t_{1/2}$ = 500–50,000 yr): ^{239}Pu, ^{240}Pu, ^{14}C	3.3×10^{15} Bq	
		Longest lived ($t_{1/2}$ > 50,000 yr): ^{99}Tc, ^{135}Cs, ^{233}U	2.0×10^{15} Bq	
Chernobyl	Reactor accident in 1986	^{131}I, ^{134,137}Cs, ^{103,106}Ru	1.2×10^{19} Bq	1,2,5
US	U mining and milling	^{226}Ra	1.9×10^{15} Bq	1

References: 1. Ewing (1999), 2. Santschi and Honeyman (1989), 3. NRC (2001), 4. Cochran *et al.* (1993), 5. IAEA (1996), and 6. US DOE (1997a).

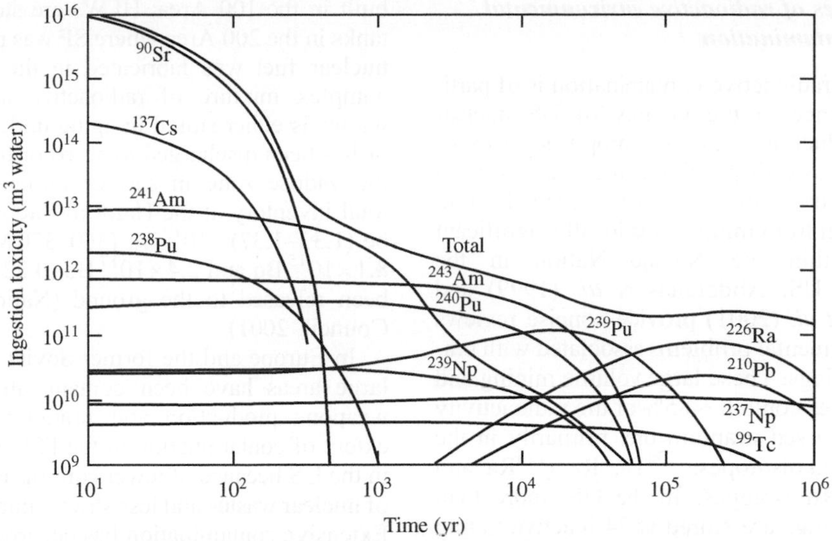

Figure 2 Ingestion toxicity for HLW as a function of decay time (Campbell *et al.* (1978); reproduced by permission of Nuclear Regulatory Commission from *Risk Methodology for Geologic Disposal of Nuclear Waste*, **1978**).

thermal output from the radionuclides show similar trends. It can be seen that initially the bulk of the radioactivity is due to short-lived radionuclides ^{137}Cs and ^{90}Sr. After 1,000 yr, the bulk of the hazard is due to decay of ^{241}Am, ^{243}Am, $^{239,240}Pu$, and ^{237}Np, and during the longest time periods, a mixture of the isotopes ^{99}Tc, ^{210}Pb, and ^{226}Ra dominates the small amount of radioactivity that remains. US Department of Energy (1980) shows that the relative toxicity (hazard) index (toxicity associated with ingesting a given weight of material) of spent fuel (SF) is about the same as uranium ore (0.2% U ore) after 10^5 yr.

Most regulations focus on the time period up to 10^4 yr and 10^5 yr after emplacement when radioactivity is dominated by the decay of americium, neptunium, and plutonium. Disposal of nuclear waste in the US is regulated by the Environmental Protection Agency (EPA) and the Nuclear Regulatory Commission (NRC). There are several classes of nuclear waste; each type is regulated by specific environmental regulations and each has a preferred disposal option, as described below.

Spent Fuel (SF) consists of irradiated fuel elements removed from commercial reactors or special fuels from test reactors. It is highly radioactive and generates a lot of heat; therefore, remote handling and heavy shielding are required. It is considered a form of HLW because of the uranium, fission products, and transuranics that it contains. HLW includes highly radioactive liquid, calcined or vitrified wastes generated by reprocessing of SF. Both SF and HLW from commercial reactors will be entombed in the geological repository at Yucca Mountain ~100 mile (1 mile = 1.609344 km) northwest of Las Vegas, Nevada. Disposal of spent nuclear fuel and HLW in the US is regulated by 40 CFR Part 191 (US EPA, 2001) and 10 CFR Part 60 (US NRC, 2001). It is discussed in more detail in a later section of this chapter.

Transuranic waste (TRU) is defined as waste contaminated with α-emitting radionuclides of atomic number greater than 92 and half-life greater than 20 yr in concentrations greater than 100 nCi g^{-1} (3.7×10^3 Bq g^{-1}). TRU is primarily a product of the reprocessing of SF and the use of plutonium in the fabrication of nuclear weapons. In the US, the disposal of TRU at the Waste Isolation Pilot Plant in southeastern New Mexico is regulated by 40 CFR Part 194 (US EPA, 1996). It is also discussed in more detail in a later section of this chapter.

Uranium mill tailings are large volumes of radioactive residues that result from the processing of uranium ore. In the US, the DOE has the responsibility for remediating mill tailing surface sites and associated groundwater under the Uranium Mill Tailings Radiation Control Act (UMTRCA) of 1978 and its modification in 1988. *Low-level wastes* (LLWs) are radioactive wastes not classified as HLW, TRU, SF, or uranium mill tailings. They are generated by institutions and facilities using radioactive materials and may include lab waste, towels, and lab coats contaminated during normal operations. Disposal of LLW is governed by agreements between states through state compacts at several facilities in the continental US. Geochemical data, conceptual models, and performance assessment methodologies relevant to LLW are summarized in Serne *et al.* (1990).

17.2.1.4 Sites of radioactive environmental contamination

In the US, radioactive contamination is of particular importance in the vicinity of US nuclear weapons production sites, near proposed or existing nuclear waste disposal facilities and in areas where uranium mining was carried out. Contamination from mining was locally significant in areas within the Navajo Nation in the southwestern US. Abdelouas *et al.* (1999) and Jove-Colon *et al.* (2001) provide concise reviews of the environmental problems associated with uranium mill tailings. These large volume mining and milling residues contain ~85% of the radioactivity of the unprocessed uranium ore, primarily in the form of U radioisotopes, ^{230}Th, Ra (^{226}Ra and ^{222}Ra), and Rn isotopes. In the US, more than 230 Mt of tailings are stored at 24 inactive tailing sites; in Canada more than 300 Mt of tailings have accumulated (Abdelouas *et al.*, 1999). By 1999, remediation of all 24 Title I UMTRA uranium mill tailings sites in the US had been completed through either in-place stabilization or relocation to more favorable sites. Other UMTRA sites (Title II sites) still had their mill tailings in place as of 2002 and have relatively long plumes that could release uranium to nearby aquifers (Brady *et al.*, 2002).

Both surface and subsurface processes are important for environmental contamination from uranium mill tailings. Surface soil/water contamination occurs by erosion and wind dispersion of contaminated soil; air pollution occurs by radon emission. Leaching and subsequent leaking of radioactive and hazardous metals (cadmium, copper, arsenic, molybdenum, lead, and zinc) from mill tailings contaminate groundwater. Concentrations of uranium in mill tailings and tailings pore waters in deposits reviewed by Abdelouas *et al.* (1999) were as high as 1.4 g kg^{-1} and 0.5 mg L^{-1}, respectively. Abdelouas *et al.* (1999) provide good examples of the relationships among environmental contamination, radiological hazards, and remediation activities in their discussion of the tailings piles sites at Tuba City, Arizona, at Rayrock, Northwest Territories, Canada, and at the complex of solution mining sites, open pits and tailings piles managed by the WIZMUT company in the uranium mining districts of East Germany.

The US DOE estimates that nuclear weapons production activities have led to radioactive contamination of ~63 Mm^3 of soil and 1,310 Mm^3 of groundwater in the US (US DOE, 1997a,b). The contamination is located at 64 DOE environmental management sites in 25 states. The Hanford Reservation in Washington State provides a good example of the diverse sources of radioactive contamination associated with weapons production. At this site, nine plutonium production reactors were built in the 100 Area; HLWs are stored in buried tanks in the 200 Area where SF was processed; and nuclear fuel was fabricated in the 300 Area. A complex mixture of radioactive and hazardous wastes is either stored in aging underground tanks or has been discharged to seeps or trenches and to the vadose zone in surface impoundments. The total inventory at the Hanford Site is estimated to be $(1.33–1.37)\times10^{19}$ Bq (360–370 MCi); between 8.1×10^{15} Bq and 2.4×10^{17} Bq (0.22–6.5 MCi) has been released to the ground (National Research Council, 2001).

In Europe and the former Soviet Union (FSU), large areas have been contaminated by nuclear weapons production and uranium mining. The extent of contamination in the FSU is greater than in the US because of fewer controls on the disposal of nuclear wastes and less strict mining regulations. Extensive contamination has occurred due to solution mining of uranium ore deposits. In solution mining, complexing agents (lixiviants) are added to groundwater and reinjected into a uranium ore body. In the US, nearly all of uranium solution mining was carried out using carbonated, oxidizing solutions to produce mobile uranyl carbonato complexes. These practices did not lead to significant contamination. In contrast, more aggressive acid leach techniques using large amounts of sulfuric and nitric acids were used in solution mining operations in the FSU. These operations created a legacy of widespread contamination comprised of large volumes of groundwater contaminated with acid and leached metals. In addition, a large number of open pit and underground mines, tailing ponds, waste rock, and low-grade ore piles are sources of potential radioactive contamination in Central and eastern European countries.

The Straz deposit in the Hamr District of the Czech Republic provides a good example of this problem (Slezak, 1997). Starting in 1968, more than 4 Mt of sulfuric acid, 3×10^5 t of nitric acid, and 1.2×10^5 t of ammonia were injected into the subsurface to mine uranium ore. Now, ~266 Mm^3 in the North Bohemian Cretaceous Cenomanian and Turonian aquifers are contaminated with uranium, radium, and manganese and other solutes. The contaminated area is more than 24 km^2 and threatens the watershed of the Plucnice River.

Large-scale contamination in regions of the Ural Mountains of Russia is severe due to weapons production at the Chelyabinsk-65 complex near the city of Kyshtym. Cochran *et al.* (1993) and Shutov *et al.* (2002) provide overviews of the extent of the contamination at the site, drawing upon a number of Russian source documents. The main weapons production facility, the Mayak Chemical Combine, produced over 5.5×10^{19} Bq of radioactive wastes from 1948 through 1992. Over 4.6×10^{18} Bq of long-lived radioactivity were discharged into open lake storage and other sites. During the period 1948–1951, $\sim 9.3\times10^{16}$ Bq of medium level β^--activity liquid waste was discharged directly into the nearby Techa River; from March 1950 to November 1951, the discharge averaged 1.6×10^{14} Bq d^{-1}. During this period, $\sim 1.24\times10^5$ people living downstream from the facility were exposed to elevated levels of radioactivity. In 1951, the medium level waste was discharged into nearby Lake Karachay. Through 1990, the lake had accumulated 4.4×10^{18} Bq of long-lived radionuclides (primarily ^{137}Cs (3.6×10^{18} Bq) and ^{90}Sr (7.4×10^{17} Bq)). By 1993, seepage of radionuclides from the lake had produced a radioactive groundwater plume that extended 2.5–3.0 mile from the lake. In 1993, the total volume of contaminated groundwater was estimated to be more than 4 Mm^3 containing at least 1.8×10^{14} Bq of long-lived (>30 yr half-life) fission products.

Several large-scale nuclear disasters have occurred at the Chelyabinsk-65 complex. Two of these events led to the spread of radioactive contamination over large areas, exposure to large numbers of people and to relocation of entire communities. In 1957, during the so-called "Kyshtym Disaster," an explosion of an HLW storage tank released 7.4×10^{17} Bq of radioactivity into the atmosphere (Medvedev, 1979; Trabalka *et al.*, 1980). About 90% of the activity fell out in the immediate vicinity of the tank. Approximately 7.8×10^{16} Bq formed a kilometer-high radioactive cloud that contaminated an area greater than 2.3×10^4 km^2 to levels above 3.7×10^9 Bq km^{-2} of ^{90}Sr. This area was home to $\sim 2.7\times10^5$ people; many of the inhabitants were evacuated after being exposed to radioactive contamination for several years.

A second disaster at the site occurred in 1967, after water in Lake Karachay evaporated during a hot summer that followed a dry winter. Dust from the lakeshore sediments was blown over a large area, contaminating 1,800–2,700 km^2 with a total of 2.2×10^{13} Bq of ^{137}Cs and ^{90}Sr. Approximately 4.1×10^4 people lived in the area contaminated at levels of 3.7×10^9 Bq km^{-2} of ^{90}Sr or higher. According to Botov (1992), releases of radioactive dust from the area continued through at least 1972. Since 1967, a number of measures have been taken to reduce the dispersion of radioactive contamination from Lake Karachay as described in Cochran *et al.* (1993) and source documents cited therein.

17.2.2 Exposure to Background and Anthropogenic Sources of Radioactivity

The amounts of radioactivity listed in Table 2 provide some idea of the maximum amount of radioactivity potentially available for exposure to the public. However, such inventories by themselves provide little information about the actual risk associated with the radioactive hazards. Although the exposure has been very high for those unfortunate few that have been involved in nuclear accidents, the exposure of the general public to ionizing radiation is relatively low. The 2×10^5 workers involved in the cleanup of Chernobyl nuclear power plant after the 1986 accident received average total body effective doses of 100 mSv (10 rem) (Ewing, 1999). Current US regulations limit whole-body dose exposures (internal + external exposures) of radiological workers to 50 mSv yr^{-1} (5 rem yr^{-1}).

The calculated effective whole-body γ-radiation doses on tailings piles at the Rayrock mill tailings site in Northwest Territories, Canada, ranged from 42 mSv yr^{-1} to 73 mSv yr^{-1}. This is ~ 24 times the annual dose from average natural radiation in Canada (Abdelouas *et al.*, 1999). Exposure to natural sources of radon (produced in the decay chain of crustal ^{238}U) averages 2 mSv yr^{-1} while other natural sources account for 1 mSv yr^{-1} (National Research Council, 1995). Total exposure to anthropogenic sources averages ~ 0.6 mSv yr^{-1}, with medical X-ray tests accounting for $\sim 2/3$ of the total. The average exposure related to the nuclear fuel cycle is estimated to be less than 0.01 mSv yr^{-1} and is comparable to that associated with the release of naturally occurring radionuclides from the burning of coal in fossil-fuel plants (Ewing, 1999; McBride *et al.*, 1978).

As shown previously, radioactive hazards associated with SF and HLW decrease exponentially over time (see Figure 2). After 10^4–10^5 yr, the risk to the public of a nuclear waste disposal vault approaches that of a high-grade uranium ore deposit and is less than the time invariant toxicity risk of ore deposits of mercury and lead (Langmuir, 1997a). The new EPA standard for nuclear waste repositories seeks to limit exposures from all exposure pathways for the reasonably maximally exposed individual living 18 km from a nuclear waste repository to 0.15 mSv yr^{-1} (15 mrem yr^{-1}) (US EPA, 2001). For comparison, the radiation dose on the shores of Lake Karachay near the Chelyabinsk-65 complex is ~ 0.2 Sv h^{-1} (Cochran *et al.*, 1993).

17.2.3 Health Effects and Radioactive Contamination

17.2.3.1 *Biological effects of radiation damage*

A summary of the nature of radioactive contamination would be incomplete without some mention of the human health effects related to radioactivity and radioactive materials. Several excellent reviews at a variety of levels of detail have been written and should be consulted by the reader

(ATSDR, 1990a,b,c, 1999, 2001; Harley, 2001; Cember, 1996; BEIR V, 1988). The subject is extremely complex, with a number of important controversies that are beyond the scope of this chapter. Some general principles, however, are summarized below.

Ionizing radiation loses energy by producing ion pairs when passing through matter. These can damage biological material directly or produce reactive species (free radicals) that can subsequently react with biomolecules. External and internal exposure pathways are important for human health effects. External exposure occurs from radiation sources outside the body such as soil particles on the ground, surface water, and from particles dispersed in the air. It is most important for γ-radiation that has high penetrating potential; however, except for large doses, most of the radiation will pass through the body without causing significant damage. External exposure is less important for β- and α-radiation, which are less penetrating and will deposit their energy primarily on the outer layer of the skin.

Internal exposures of α- and β-particles are important for ingested and inhaled radionuclides. Dosimetry models are used to estimate the dose from internally deposited radioactive particles. The amount and mode of entry of radionuclides into the body, the movement and retention of radionuclides within various parts of the body, and the amount of energy absorbed by the tissues from radioactive decay are all factors in the computed dose (BEIR IV, 1988). The penetrating power of α-radiation is low; therefore, most of the energy from α-decay is absorbed in a relatively small volume surrounding an ingested or inhaled particle in the gut or lungs. This means that the chance that damage to DNA or other cellular material will occur is greater and the associated human health risk is higher for radioactive contamination composed of α-emitters than for the other forms of radioactive contamination. As mentioned above, weighting parameters that take into account the radiation type, the biological half-life, and the tissue or organ at risk are used to convert the physically absorbed dose in units of gray (or rad) to the biologically significant committed equivalent dose and effective dose, measured in units of Sv (or rem).

There is considerable controversy over the shape of the dose-response curve at the chronic low dose levels important for environmental contamination. Proposed models include linear models, nonlinear (quadratic) models, and threshold models. Because risks at low dose must be extrapolated from available data at high doses, the shape of the dose-response curve has important implications for the environmental regulations used to protect the general public. Detailed description of dosimetry models can be found in Cember (1996), BEIR IV (1988), and Harley (2001).

The health effect of radiation damage depends on a combination of events on the cellular, tissue, and systemic levels. Exposure to high doses of radiation (>5 Gy) can lead to direct cell death before division due to interaction of free radicals with macromolecules such as lipids and proteins. At acute doses of 0.1–5.0 Gy, damage to organisms can occur on the cellular level through single strand and double strand DNA breaks. These led to mutations and/or cellular reproductive death after one or more divisions of the irradiated parent cell. The dose level at which significant damage occurs depends on the cell type. Cells that reproduce rapidly, such as those found in bone marrow or the gastrointestinal tract, will be more sensitive to radiation than those that are longer lived, such as striated muscle or nerve cells. The effect of high radiation doses on an organ depends on the various cell types that it contains.

Cancer is the major effect of low radiation doses expected from exposure to radioactive contamination. Laboratory studies have shown that α-, β-, and γ-radiation can produce cancer in virtually every tissue type and organ in animals that have been studied (ATSDR, 2001). Cancers observed in humans after exposure to radioactive contamination or ionizing radiation include cancers of the lungs, female breast, bone, thyroid, and skin. Different kinds of cancers have different latency periods; leukemia can appear within 2 yr after exposure, while cancers of the breast, lungs, stomach, and thyroid have latency periods greater than 20 yr. Besides cancer, there is little evidence of other human health effects from low-level radiation exposure (ATSDR, 2001; Harley, 2001).

17.2.3.2 Epidemiological studies

The five large epidemiological studies that provide the majority of the data on the effects of radiation on humans are reviewed by Harley (2001). These include radium exposures by radium dial painters, atom bomb survivors, patients irradiated with X-rays for ankylosing spondylitis and ringworm (tinea capitis), and uranium miners exposed to radon. The first four studies examine health effects due to external exposures of high doses of ionizing radiation. The studies of uranium miners are more relevant to internal exposures. There have been 11 large follow-up studies of underground miners who were exposed to high concentrations of radon (^{222}Rn) and radon decay products. The carcinogens are actually the short-term decay products of ^{222}Rn (^{218}Po, ^{214}Po), which are deposited on the bronchial airways during inhalation and exhalation. Because of their short range, the α-particles transfer most of their energy to the thin layer of bronchial epithelium cells. These cells are known to be involved in induction of cancer, and it is clear that even relatively short exposures to the high levels possible in mines lead to excess lung cancers.

The results of the studies on miners have been used as a basis for estimating the risks to the general public from exposures to radon in homes. There is considerable controversy over this topic. Although the health effects due to the high radon exposures experienced by the miners have been well established, the risks at the lower exposure levels in residences are difficult to establish due to uncertainties in the dose-response curve and the confounding effects of smoking and urbanization. The reader is referred to extensive documentation by the National Academy of Sciences (1998) and the National Institute of Health (1994) for more information.

Epidemiological studies of populations in the FSU exposed to fallout from the 1986 nuclear reactor explosion at Chernobyl and releases from the Chelyabinsk-65 complex demonstrate the health effects associated with exposure to radioactive iodine, strontium, and caesium. A study of 2.81×10^4 individuals exposed along the Techa River, downstream from Chelyabinsk-65, revealed that a statistically significant increase in leukemia mortality arose between 5 yr and 20 yr after the initial exposure (37 observed deaths versus 14–23 expected deaths; see Cochran *et al.* (1993) and cited references and comments). There has been a significant increase of thyroid cancers among children in the areas contaminated by fallout from the Chernobyl explosion (Harley, 2001; UNSCEAR, 2000). The initial external exposures from Chernobyl were due to ^{131}I and short-lived isotopes. Subsequently, external exposures to ^{137}Cs and ^{134}Cs and internal exposures to radiocaesium through consumption of contaminated foodstuffs were important.

17.2.3.3 Toxicity and carcinogenicity

Toxicity and carcinogenicity of radioactive materials are derived from both the chemical properties of the radioelements and the effects of ionizing radiation. The relative importance of the radiological and chemical health effects are determined by the biological and radiological half-lives of the radionuclide and the mechanism of chemical toxicity of the radioelement. Ionizing radiation is the main source of carcinogenicity of the radionuclides. Many of the damaging effects of radiation can be repaired by the natural defenses of cells and the body. However, the longer a radionuclide is retained by the body or localized in a specific organ system, the greater the chances that the damage to DNA or proteins will not be repaired and a cancer will be initiated or promoted. The biological half-life of a radioelement, therefore, is an important determinant of the health risk posed by a radionuclide. This is determined by the chemical and physical form of the radioelement when it enters the body, the metabolic processes that it participates in and the routes of elimination from the body. For example, ^{90}Sr substitutes for calcium and accumulates on the surfaces of bones. It has a long biological half-life (50 yr), because it is recycled within the skeletal system. In young children, it is incorporated into growing bone where it can irradiate both the bone cells and bone marrow. Consequently, high exposures to ^{90}Sr can lead to bone cancer and cancers of the blood such as leukemia.

In contrast, the chemical toxicity of uranium is more important than its radiological hazard. In body fluids, uranium is present as soluble U(VI) species and is rapidly eliminated from the body (60% within 24 h; Goyer and Clarkson (2001)). It is rapidly absorbed from the gastrointestinal tract and moves quickly through the body. The uranyl carbonate complex in plasma is filtered out by the kidney glomerulus, the bicarbonate is reabsorbed by the proximule tubules, and the liberated uranyl ion is concentrated in the tubular cells. This produces systemic toxicity in the form of acute renal damage and renal failure.

Because there are few data on the results of human exposure to actinides, the health effects of these radioelements are more uncertain than those discussed above for ionizing radiation, radon, and fission products. Americium accumulates in bones and will likely cause bone cancer due to its radioactive decay. Animal studies suggest that plutonium will cause effects in the blood, liver, bone, lung, and immune systems. Other potential mechanisms of chemical toxicity and carcinogenicity of the actinides are similar to those of heavy metals and include: (i) disruption of transport pathways for nutrients and ions; (ii) displacement of essential metals such as Cu^{2+}, Zn^{2+}, and Ni^{2+} from biomolecules; (iii) modification of protein conformation; (iv) disruption of membrane integrity of cells and cell organelles; and (v) DNA damage. More details can be found in the references cited at the beginning of this section.

17.3 EXPERIMENTAL AND THEORETICAL STUDIES OF RADIONUCLIDE GEOCHEMISTRY

Analysis of the risk from radioactive contamination requires consideration of the rates of release and dispersion of the contaminants through potential exposure pathways. Prediction of the release and dispersion of radionuclides from nuclear waste sites and contaminated areas must consider a series of processes including: (i) contact of the waste with groundwater, degradation of the waste, and release of radioactive aqueous species and particulate matter; (ii) transport of aqueous species and colloids through the saturated and vadose zone; and (iii) uptake of radionuclides by exposed populations or ecosystems. The geochemistry of the radionuclides will control migration through the geosphere by determining solubility, speciation, sorption, and the extent of transport by colloids. These are strong functions of the compositions of the groundwater and geomedia as well as the atomic structure of the radionuclides. These topics are the main focus of the sections that follow.

General predictions of radionuclide mobility are difficult to make; instead, site-specific measurements and thermodynamic calculations for the site-specific conditions are needed to make meaningful statements about radionuclide behavior. However, for the purposes of this chapter, some underlying themes are described. Trends in solubility and mobility are described from several perspectives: (i) by identity of the radionuclide and its dominant oxidation state under near-surface environmental conditions; (ii) by composition of the groundwater or pore waters (i.e., Eh, pH, and nature of complexing ligands and competing solutes); and (iii) by composition of the geomedia in contact with the solutions (i.e., mineralogy and organic matter).

17.3.1 Principles and Methods

17.3.1.1 Experimental methods

A wide variety of experimental techniques are used in radiochemical studies; a review of this subject is beyond the scope of this chapter. The interested reader should refer to the reviews and textbooks of actinide chemistry listed in the Section 17.1.2 above. Some general points, which should be considered in evaluating available data relevant to environmental radioactive contamination, are made below.

Solubility and speciation. Minimum requirements for reliable thermodynamic solubility studies include: (i) solution equilibrium conditions; (ii) effective and complete phase separation; (iii) well-defined solid phases; and (iv) knowledge of the speciation/oxidation state of the soluble species at equilibrium. Ideally, radionuclide solubilities should be measured in both "oversaturation" experiments, in which radionuclides are added to a solution until a solid precipitates, and "undersaturation" experiments, in which a radionuclide solid is dissolved in aqueous media. Due to the difference in solubilities of crystalline versus amorphous solids and different kinetics of dissolution, precipitation, and recrystallization, the results of these two types of experiments rarely agree. In some experiments, the maximum concentration of the radionuclide source term in specific water is of interest, so the solid that is used may be SF or nuclear waste glass rather than a pure radionuclide solid phase.

In addition, the maximum concentrations measured in laboratory experiments and the solubility-limiting solid phases identified are often not in agreement with the results of theoretical thermodynamic calculations. This discrepancy could be due to differences in the identity or the crystallinity of solubility-limiting solids assumed in the calculation or to errors in the thermodynamic property values used in the calculations. Thus, although theoretical thermodynamic calculations are useful in summarizing available information and in performing sensitivity analyses, it is important also to review the results of empirical experimental studies in site-specific solutions.

Good summaries of accepted experimental techniques can be found in the references that are cited for individual radionuclides in the sections below. Nitsche (1991) provides a useful general summary of the principles and techniques of solubility studies. A large number of techniques have been used to characterize the aqueous speciation of radionuclides. These include potentiometric, optical absorbance, and vibrational spectroscopy. Silva and Nitsche (1995) summarize the use of conventional optical absorption and laser-based photothermal spectroscopy for detection and characterization of solution species and provide an extensive citation list. A recent review of the uses of Raman and infrared spectroscopy to distinguish various uranyl hydroxy complexes is given by Runde *et al.* (2002b).

Extraction techniques to separate oxidation states and complexes are often combined with radiometric measurements of various fractions. A series of papers by Choppin and co-workers provides good descriptions of these techniques (e.g., Caceci and Choppin, 1983; Schramke *et al.*, 1989). Cleveland and co-workers used a variety of extraction techniques to characterize the speciation of plutonium, neptunium, and americium in natural waters (Rees *et al.*, 1983; Cleveland *et al.*, 1983a,b; and Cleveland and Rees, 1981).

A variety of methods have been used to characterize the solubility-limiting radionuclide solids and the nature of sorbed species at the solid/water interface in experimental studies. Electron microscopy and standard X-ray diffraction techniques can be used to identify some of the solids from precipitation experiments. X-ray absorption spectroscopy (XAS) can be used to obtain structural information on solids and is particularly useful for investigating noncrystalline and polymeric actinide compounds that cannot be characterized by X-ray diffraction analysis (Silva and Nitsche, 1995). X-ray absorption near edge spectroscopy (XANES) can provide information about the oxidation state and local structure of actinides in solution, solids, or at the solution/solid interface. For example, Bertsch *et al.* (1994) used this technique to investigate uranium speciation in soils and sediments at uranium processing facilities. Many of the surface spectroscopic techniques have been reviewed recently by Bertsch and Hunter (2001) and Brown *et al.* (1999). Specific recent applications of the spectroscopic techniques to radionuclides are described by Runde *et al.* (2002b). Rai and co-workers have carried out a number of experimental studies of the solubility and speciation of plutonium, neptunium, americium, and uranium that illustrate combinations of various solution and spectroscopic techniques (Rai *et al.*, 1980, 1997, 1998; Felmy *et al.*, 1989, 1990; Xia *et al.*, 2001).

Sorption studies. Several different approaches have been used to measure the sorption of radionuclides by geomedia. These include: (i) the laboratory batch method, (ii) the laboratory flow-through (column) method, and (iii) the *in situ* field batch sorption method. Laboratory batch tests are the simplest experiments; they can be used to collect distribution coefficient (K_d) values or other partitioning coefficients to parametrize sorption and ion-exchange models. The different sorption models are summarized in Section 17.3.1.2. The term *sorption* is often used to describe a number of surface processes including adsorption, ion exchange, and co-precipitation that may be included in the calculation of a K_d. For this reason, some geochemists will use the term sorption ratio (R_d) instead of distribution coefficient (K_d) to describe the results of batch sorption experiments. In this chapter, both terms are used in order to be consistent with the terminology used in the original source of information summarized.

Batch techniques. Descriptions of the batch techniques for radionuclide sorption and descriptions of calculations used to calculate distribution coefficients can be found in ASTM (1987), Park *et al.* (1992), Siegel *et al.* (1995a), and US EPA (1999a). The techniques typically involve the following steps: (i) contacting a solution with a known concentration of radionuclide with a given mass of solid; (ii) allowing the solution and solid to equilibrate; (iii) separating the solution from the solid; and (iv) measuring the concentration of radionuclide remaining in solution.

In batch systems, the distribution or sorption coefficient (K_d or R_d) describes the partitioning of a contaminant between the solid and liquid phases. The K_d is commonly measured under equilibrium or at least steady-state conditions, unless the goal of the experiment is to examine the kinetics of sorption. It is defined as follows:

$$K_d(\mathrm{mL\,g^{-1}}) = \frac{C_S}{C_L} \tag{1}$$

where C_S is the concentration of the contaminant on the solid and C_L is the concentration in solution. In practice, the concentration of the contaminant on the solid is rarely measured. Rather, it is calculated from the initial and final solution concentrations, and the operative definition for the K_d becomes

$$K_d(\mathrm{mL\,g^{-1}}) = \frac{(C_i V_i - C_f V_f)/m}{C_f} \tag{2}$$

where C_i and C_f are the initial and final concentrations of contaminant in solution, respectively; V_i and V_f are the initial and final solution volumes; and m is the mass of the substrate added to the system.

Measured batch K_d values can be used to calculate a retardation factor (R), which describes the ratio of the groundwater velocity ν_m to the velocity of radionuclide movement ν_r:

$$R = \frac{\nu_m}{\nu_r} = 1 + \frac{K_d \rho}{\phi} \tag{3}$$

where ρ is the bulk density of the porous medium and ϕ is the porosity. This equation can be rearranged, and contaminant retardation values measured from column breakthrough curves can be used to calculate K_d values.

Many published data from batch sorption measurements are subject to a number of limitations as described by Siegel and Erickson (1984, 1986), Serne and Muller (1987), and by US EPA (1999a). These include a solution: solid ratio that is much higher than that present in natural conditions, an inability to account for multiple sorbing species, an inability to measure different adsorption and desorption rates and affinities, and an inability to distinguish between adsorption and co-precipitation.

Batch methods are also used to collect data to calculate equilibrium constants for the surface-complexation models (SCMs) which are described in more detail in the next section. Commonly for these models, sorption is measured as function of pH and data are presented as pH–sorption edges similar to the one shown in Figure 3. Sorption is strongly affected by the surface charge of the geomedia. The proton is the surface potential determining ion (PDI) in metal oxyhydroxides and of the high-energy edge sites in aluminosilicates. In the absence of significant sorption of metal ions (i.e., low surface loading or site occupancy), the surface charge is determined primarily by the difference between the surface concentrations of positively charged, protonated sites, and negatively charged, deprotonated sites.

Considerable data have been collected describing the influence of pH on actinide sorption. Sorption edges are most commonly (and usefully) measured for single oxidation states of the radionuclide. Figure 3 shows a plot of a

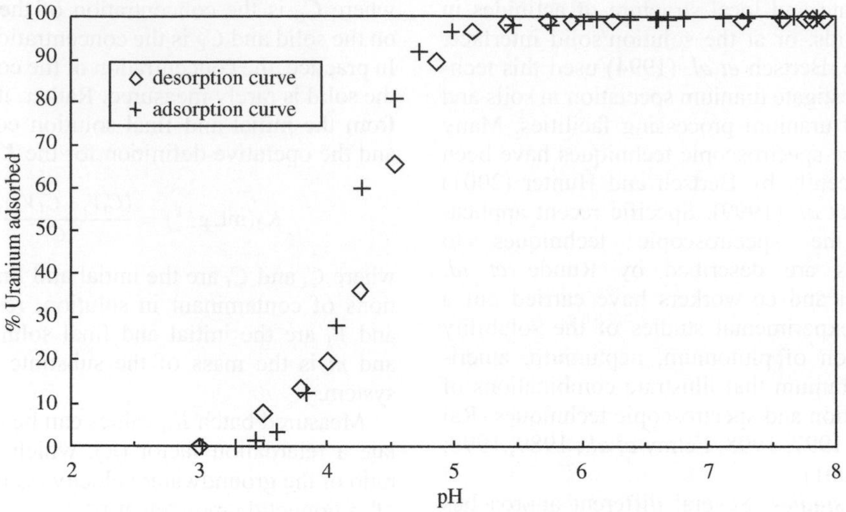

Figure 3 Sorption and desorption edges for uranyl on goethite. Each sample contained 60 $m^2 L^{-1}$ goethite and 100 μg mL^{-1} uranium. The sorption edge was measured 2 d after uranium addition; desorption samples were contacted with the uranyl solution at pH 7 for 5 d, the pH was adjusted to cover the range of interest, and the samples were re-equilibrated for 2 d prior to sampling (see also Bryan and Siegel, 1998).

sorption/ desorption edge for uranium under oxidizing conditions and illustrates reversible sorption. The effect of competition between protons and the radionuclide is illustrated by the sorption edge. Appreciable sorption can occur at a relatively low pH for radionuclides that form strong bonds with the surface, resulting in a low pH_{50} value (the pH at which 50% of the radionuclide is adsorbed). Radionuclides that form weaker surface complexes can only sorb appreciably when the concentration of competing protons is low (high pH) and therefore have high pH_{50} values. Comparisons of sorption edges for different radionuclides on the same substrate or for a single radionuclide on several substrates can be made by referring to their pH_{50} values. For example, Kohler *et al.* (1992) show that goethite strongly sorbs Np(V) from $NaClO_4$ solutions, while quartz only weakly sorbs Np(V). The pH_{50} sorption values for Np(V) increase in the order goethite < hematite < gibbsite < albite < quartz.

Other techniques. *Laboratory column tests* are more difficult to perform but overcome some of the limitations of the batch tests. Proper design, descriptions of experimental procedures, and methods of data interpretation for column tests can be found in US EPA (1999a), Relyea (1982), Van Genuchten and Wierenga (1986), Triay *et al.* (1992, 1996, 1997), Torstenfelt (1985a,b), Siegel *et al.* (1995b), Sims *et al.* (1996), and Gabriel *et al.* (1998). In these experiments, a solution containing a known concentration of radionuclide is introduced into a column of packed soil or rock at a specified flow rate. The concentration of the radionuclide in the column effluent is monitored to obtain a breakthrough curve; the shape of the curve provides information about sorption equilibrium and kinetics and other properties of the crushed rock or intact rock column. Limitations of this technique are: (i) complex, time-consuming and expensive experimental procedures are required; (ii) symmetric breakthrough curves are rarely obtained and a number of ad hoc assumptions may be required to interpret them; and (iii) the sorption parameters are dependent on the hydrodynamics of the specific experiment and are not applicable to other conditions.

In situ (field) batch sorption tests use measurements of the radionuclide contents of samples of rock cores and consanguineous pore water obtained at a field site. The advantage of this approach is that the water and rock are likely to be in chemical equilibrium and that the concentrations of any cations or anions competing for sorption sites are appropriate for natural conditions. Disadvantages of this technique are associated with limitations in obtaining accurate measurements of the concentration of radionuclides on the rock surface in contact with the pore fluid. Applications of this technique are described in Jackson and Inch (1989), McKinley and Alexander (1993), Read *et al.* (1991), Ward *et al.* (1990), and Payne *et al.* (2001).

Surface analytical techniques. A variety of spectroscopic methods have been used to characterize the nature of adsorbed species at the solid–water interface in natural and experimental systems (Brown *et al.*, 1999). Surface spectroscopy techniques such as extended X-ray absorption fine structure spectroscopy (EXAFS) and attenuated total reflectance Fourier transform infrared spectroscopy (ATR-FTIR) have been used to characterize complexes of fission products, thorium, uranium, plutonium, and uranium sorbed onto silicates, goethite, clays, and microbes (Chisholm-Brause *et al.*, 1992, 1994; Dent *et al.*, 1992; Combes *et al.*, 1992; Bargar *et al.*, 2000; Brown and Sturchio, 2002). A recent overview of the theory and applications of synchrotron radiation to the analysis of the surfaces of soils, amorphous materials, rocks, and organic matter in low-temperature geochemistry and environmental science can be found in Fenter *et al.* (2002).

Before the application of these techniques became possible, the composition and predominance of adsorbed species was generally inferred from information about aqueous species. In some cases, these models for adsorbed species have been shown to be incorrect. For example, spectroscopic studies demonstrated that U(VI)-carbonato complexes were the predominant adsorbed species on hematite over a wide pH range (Bargar *et al.*, 2000). This was contrary to expectations based on analogy to aqueous U(VI)-carbonato complexes, which are present in very low concentration at near-neutral and acidic pH. As discussed later, the use of different sorption models will produce alternate stoichiometries for surface species when data are fit to sorption edges. The information obtained from surface spectroscopy helps to constrain the interpretation of the results of batch sorption tests by revealing the stoichiometry of the sorbed species. For example, Redden and Bencheikh-Latmar (2001) demonstrate how data from an EXAFS study helped to elucidate the interactions among UO_2^{2+}, citrate, and goethite over the pH range 3.5–5.5. Previous studies of sorption equilibria (Redden *et al.*, 1998) showed that whereas citric acid reduced the sorption of uranium by gibbsite and kaolinite, the presence of citrate led to enhanced uranyl sorption at high citrate: UO_2^{2+} ratios. The EXAFS spectra were consistent with the existence of two principal surface species: an inner-sphere uranyl–goethite complex and an adsorbed uranyl–citrate complex, which displaces the binary uranyl–geothite complex at high citrate: UO_2^{2+} ratios. Other examples of the use of these techniques to understand the nature of radionuclide sorption can be found in a recent review by Brown and Sturchio (2002).

17.3.1.2 Theoretical geochemical models and calculations

Aqueous speciation and solubility. Several geochemical codes are commonly used for calculations of radionuclide speciation and solubilities. Recent reviews of the codes can be found in Serne *et al.* (1990), Mangold and Tsang (1991), NEA (1996), and US EPA (1999a). Extensive databases of thermodynamic property values and kinetic rate constants are required for these codes. Several databases of thermodynamic properties of the actinides have been developed since the early 1970s. Of historical importance are the compilations and reviews of Lemire and Tremaine (1980), Phillips *et al.* (1988), and Fuger *et al.* (1990). The more recent comprehensive and consistent databases have been based on compilations produced by the Nuclear Energy Agency (NEA) for plutonium and neptunium (Lemire *et al.*, 2001), americium (Silva *et al.*, 1995), uranium (Grenthe *et al.*, 1992), and technetium (Rard *et al.*, 1999). These books contain suggested values for ΔG, ΔH, C_p, and $\log K_f$ for formation reactions of radionuclide species. More recent publications often use these compilations as reference and add more recent property values or correct errors. For example, Langmuir (1997a) updates the 1995 NEA database for uranium (Grenthe *et al.*, 1995; Silva *et al.*, 1995) and provides results of solubility, speciation, and sorption calculations using the MINTEQ2A code (Allison *et al.*, 1991) as described later in this chapter.

The more recent compilations have been *internally self-consistent.* The process of compiling an internally self-consistent database consists of several steps: (i) compilation of process values such as equilibrium constants for reactions involving the element of interest; (ii) extrapolation of equilibrium constants to reference conditions (usually zero ionic strength and 25 °C), and (iii) calculation of property values such as free energies of formation of the products through reaction networks (Wagman *et al.*, 1982; Grenthe *et al.*, 1992). Calculated thermodynamic constants from different experimental studies will be incompatible if different reference states or reaction networks are used. Caution must be exercised in combining constants from different compilations, because they are dependent on the methods used for ionic strength correction and the values used for auxiliary species such as OH^- or SO_4^{2-}. Thus, strictly speaking, as new species are identified or suspect ones eliminated and as constants of previously recognized species are revised, the entire reaction network must be used to rederive all of the constants in order to maintain internal consistency. Software is vailable to recalculate the reaction networks to ensure that internal consistency is maintained with the NEA Thermochemical Database (TDB) Project (http://www.nea.fr/html/dbtdb/cgi-bin/tdbdoc proc.cgi). In practice, however, these recalculations are not commonly done, especially if only minor changes in values of equilibrium constants are expected.

For most solubility and speciation studies, calculations of the activity coefficients of aqueous species are required. For waters with relatively low ionic strength (0.01–0.1 *m*), simple corrections such as the Debye–Hückel equation are used (Langmuir, 1997a, p. 127). This model accounts for the electrostatic, nonspecific, long-range interactions between water and the solutes. At higher ionic strengths, short-range, nonelectrostatic interactions must be taken into account. The NEA has developed a database based on the specific interaction theory (SIT) approach of Bronsted (1922), Scatchard (1936), and Guggenheim (1966). In this model, activity coefficients are calculated from a set of virial coefficients obtained from experimental data in simple solutions (Ciavatta, 1980). This method is assumed to be valid for ionic strengths up to 3.0 m and had been used to extrapolate experimental data to zero ionic strength to obtain equilibrium constants and free energies for the NEA databases (Lemire *et al.*, 2001; Silva *et al.*, 1995; Grenthe *et al.*, 1992; Rard *et al.*, 1999).

The US DOE has adopted the more complex Pitzer model (Pitzer, 1973, 1975, 1979) for calculations of radionuclide speciation and solubility in its Nuclear Waste Management Programs. This model includes concentration-dependent interaction terms and is valid up to ionic strengths greater than 10 m. However, because it requires three parameters instead of the single interaction parameter of the SIT model, this method requires more extensive experimental data. Pitzer coefficients and coefficients from the Harvie–Moller–Weare model (Harvie *et al.*, 1984) are used in several geochemical codes such as PHRQPITZ (Plummer *et al.*, 1988), EQ3/6 (Wolery, 1992a), and REACT (Bethke, 1998). Considerably more experimental data are needed for the radionuclides, and the Pitzer coefficient database is currently being developed by the US DOE. Data are available for some radionuclides (Felmy and Rai, 1999) and have been used in the Waste Isolation Pilot Plant (WIPP) program as described in a later section.

Sorption overview. Both empirical and mechanistic approaches have emerged since the 1970s to describe interactions between radionuclides and geomedia. These are based on "conditional" constants, which are valid for specific experimental conditions, or more robust "intrinsic" constants, which are valid over a wider range of conditions. The "empirical approach" involves measurements of conditional radionuclide distribution or sorption coefficients (K_ds or R_ds) in site-specific water–rock systems using synthetic or natural ground waters and crushed rock samples. Mechanistic-based approaches produce intrinsic, thermodynamic surface-complexation constants for simple electrolyte solutions with pure mineral phases.

The different approaches to describing sorption can be discussed in order of increasing model complexity and the robustness of their associated constants:

(i) linear sorption (K_d or R_d);
(ii) nonlinear sorption (Freundlich and other isotherms);
(iii) constant-charge (ion-exchange) model;
(iv) constant-capacitance model;
(v) double layer or diffuse layer model (DLM); and
(vi) triple-layer model (TLM).

Figure 4 compares several of these models with respect to the nature of the constants that each uses. The simplest model (linear sorption or K_d) is the most empirical model and is widely used in contaminant transport models. K_d values are relatively easy to obtain using the batch methods described

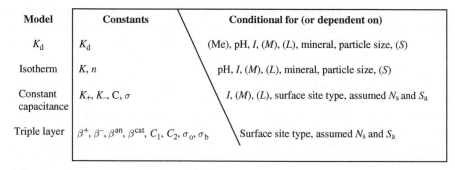

Model	Constants	Conditional for (or dependent on)
K_d	K_d	(Me), pH, I, (M), (L), mineral, particle size, (S)
Isotherm	K, n	pH, I, (M), (L), mineral, particle size, (S)
Constant capacitance	K_+, K_-, C, σ	I, (M), (L), surface site type, assumed N_s and S_a
Triple layer	β^+, β^-, β^{an}, β^{cat}, C_1, C_2, σ_o, σ_b	Surface site type, assumed N_s and S_a

Figure 4 Comparison of sorption models. Several commonly used sorption models are compared with respect to the independent *constants* they require. These *constants* are valid only under specific conditions, which must be specified in order to properly use them. In other words, the constants are *conditional* with respect to the *experimental* variables described in the third column of the figure. K_d is the radionuclide distribution constant; K and n are the Freundlich isotherm parameters; β^+ and β^- are surface complexation constants for protonation and deprotonation of surface sites; K_+, K_-, β^{an}, β^{cat} are surface complexation constants for sorption of cations and anions in the constant capacitance model and TLM, respectively; C, C_1, and C_2 are capacitances for the electrical double layers; σ, σ_o, and σ_b are surface charges at different surface planes; (Me) and (S) are concentrations of the sorbing ions and the surface sites, (M), (L) are concentrations of other cations and ligands in solution, respectively; I is the ionic strength of the background electrolyte; N_s and S_a are the site density and specific surface of the substrate, respectively. The requirements of the DLM are similar to those of the constant capacitance model.

above. The K_d model requires a single distribution constant, but the K_d value is conditional with respect to a large number of variables. Thus, even if a batch K_d experiment is carefully carried out to avoid introduction of extraneous effects such as precipitation, the K_d value that is obtained is valid only for the particular conditions of the experiment. As Figure 4 shows, the radionuclide concentration, pH, major and minor element composition, rock mineralogy, particle size and solid-surface-area/solution volume ratio must be specified for each K_d value.

Ion-exchange models are commonly used to describe radionuclide sorption onto the fixed-charged sites of materials like clays. Ion exchange will be strongly affected by competition with monovalent and divalent ions such as Na^+ and Ca^{2+}, whereas it will be less dependent on pH over the compositional ranges common for natural waters. Many studies of strontium and caesium sorption by aluminosilicates (e.g., Wahlberg and Fishman, 1962; Tamura, 1972) have been carried out within the framework of ion-exchange theory. Early mechanistic studies of uranium sorption were carried out within an ion-exchange framework (e.g., Tsunashima *et al.*, 1981); however, more recent studies relevant to environmental conditions have used SCMs (e.g., Davis, 2001).

The TLM (Davis and Leckie, 1978) is the most complex model described in Figure 4. It is an example of an SCM. These models describe sorption within a framework similar to that used to describe reactions between metals and ligands in solutions (Kent *et al.*, 1988; Davis and Kent, 1990; Stumm, 1992). Reactions involving surface sites and solution species are postulated based on experimental data and theoretical principles. Mass balance, charge balance, and mass action laws are used to predict sorption as a function of solution chemistry. Different SCMs incorporate different assumptions about the nature of the solid–solution interface. These include the number of distinct surface planes where cations and anions can attach (double layer versus triple layer) and the relations between surface charge, electrical capacitance, and activity coefficients of surface species.

Aqueous radionuclide species and other solutes can sorb to mineral surfaces by forming chemical bonds directly with the amphoteric sites or may be separated from the surface by a layer of water molecules and be bound through longer-range electrostatic interactions. In the TLM, complexes of the former type are often called "inner-sphere" complexes; those of the latter type are called "outer-sphere" complexes (Davis and Kent, 1990). The TLM includes an inner plane (o-plane), an outer plane (β-plane), and a diffuse layer that extends from the β-plane to the bulk solution. Sorption via formation of inner-sphere complexes is often referred to "chemisorption" or "specific sorption" to distinguish it from ion exchange at fixed charged sites or outer-sphere complexation that is dominantly electrostatic in nature.

The sorption of uranium to form the sorbed species ($>FeOH–UO_2^+$) at the o-plane of an iron oxyhydroxide surface (represented as $>FeOH$) can be represented by a surface reaction in a TLM as

$$>\mathrm{FeOH} + \mathrm{UO}_{2(s)}^{2+} \rightarrow >\mathrm{FeO} - \mathrm{UO}_2^+ + \mathrm{H}_{(s)}^+ \quad (4)$$

where $UO_{2(s)}^{2+}$ and $H_{(s)}^+$ are the aqueous uranyl ion and proton at the surface, respectively. The mass action law (equilibrium constant) for the reaction using the TLM is

$$\beta^{\mathrm{UO}_2^{2+}} = \frac{\{>\mathrm{FeO}-\mathrm{UO}_2^+\} \times a_{\mathrm{H}^+} \times \exp[e\psi_0/kT]}{\{>\mathrm{FeOH}\} \times a_{\mathrm{UO}_2^{2+}}} \quad (5)$$

where $\beta^{UO_2^{2+}}$ is the intrinsic surface-complexation constant for the uranyl cation; $\{>FeOH\}$ and $\{>FeO–UO_2^+\}$ are the activities of the uncomplexed and complexed surface sites, respectively; a_{H^+} and $a_{UO_2^{2+}}$ are activities of the aqueous species in the bulk solution; ψ_0 is the electrical potential for the inner (o) surface plane; and k, T, and e are the Boltzmann constant, absolute temperature, and the fundamental charge, respectively. The exponential term describes the net change in electrostatic energy required to exchange the divalent uranyl ion for the proton at the mineral surface. (The activities of the uranyl ion and proton at the surface differ from their activities in the bulk solution: $\{UO_{2(s)}^{2+}\} = a_{UO_2^{2+}} \times \exp(-2e\psi_0/kT)$ and $\{H_{(s)}^+\} = a_{H^+} \times \exp(-e\psi_0/kT)$. Equation (5) can be derived from the equilibrium constant for Equation (4), $\beta^{UO_2^{2+}} = [\{>FeO–UO_2^+\}/\{>FeOH\}] \times [\{H_{(s)}^+\}/\{UO_{2(s)}^{2+}\}]$, by substitution.)

In natural waters, other surface reactions will be occurring simultaneously. These include protonation and deprotonation of the $>FeOH$ site at the inner o-plane and complexation of other cations and anions to either the inner (o) or outer (β) surface planes. Expressions similar to Equation (5) above can be written for each of these reactions. In most studies, the activity coefficients of surface species are assumed to be equal to unity; thus, the activities of the surface sites and surface species are equal to their concentrations. Different standard states for the activities of surface sites and species have been defined either explicitly or implicitly in different studies (Sverjensky, 2003). Sverjensky (2003) notes that the use of a hypothetical 1.0 M standard state or similar convention for the activities of surface sites and surface species leads to surface-complexation constants that are directly dependent on the site density and surface area of the sorbent. He defines a standard state for surfaces sites and species that is based on site occupancy and produces equilibrium constants independent of these properties of the solids. For more details about the properties of the electrical double layer, methods to calculate surface speciation and alternative models for activity coefficients for surface sites, the reader should refer to the reference cited above and other works cited therein.

The TLM contains eight adjustable constants (identified in the caption of Figure 4) that are valid over the ranges of pH, ionic strength, solution composition, specific areas, and site densities of the experiments used to extract the constants. The surface-complexation constants, however, must be determined for each type of surface site of interest and should not be extrapolated outside the original experimental conditions. Although the TLM constants are valid under a wider range of conditions than are K_ds, considerably more experimental data must be gathered to obtain the adjustable parameters. An important advantage of surface-complexation constant models is that they provide a structured way to examine experimental data obtained in batch sorption studies. Application of such models may ensure that extraneous effects such as precipitation have not been introduced into the sorption experiment.

Between the simplicity of the K_d model and the complexity of the TLM, there are several other sorption models. These include various forms of isotherm equations (e.g., Langmuir and Freundlich isotherms) and models that include kinetic effects. The generalized two-layer model (Dzombak and Morel, 1990) (also referred to as the DLM) recently has been used to model radionuclide sorption by several research groups (Langmuir, 1997a; Jenne, 1998; Davis, 2001). Constants used in this model are dependent upon the concentration of background electrolytes and are thus less robust than those of the TLM. Reviews by Turner (1991), Langmuir (1997b), and US EPA (1999a) provide concise descriptions of many of these models.

Several researchers have illustrated the interdependence of the adjustable parameters and the nonunique nature of the SCM constants by fitting the same or similar sorption edges to a variety of alternate SCM models (Westall and Hohl, 1980; Turner, 1995; Turner and Sassman, 1996). Robertson and Leckie (1997) systematically examined the effects of SCM model choice on cation binding predictions when pH, ionic strength, cation loading, and proposed surface complex stoichiometry were varied. They show that although different models can be used to obtain comparable fits to the same experimental data set, the stoichiometry of the proposed surface complex will vary considerably between the models. In the near future, it is possible that the actual stoichiometry of adsorbed species can be determined using combinations of the spectroscopic techniques discussed in a previous section and molecular modeling techniques similar to those described in Cygan (2002).

There is no set of reference surface-complexation constants corresponding to the reference thermodynamic property values contained in the NEA thermodynamic database described in the previous section (Grenthe *et al.*, 1992; Silva *et al.*, 1995; Rard *et al.*, 1999; Lemire *et al.*, 2001). Wang *et al.* (2001a,b) used the DLM with original experimental data to obtain a set of internally consistent surface-complexation constants for Np(V), Pu(IV), Pu(V), and Am(III), I^-, IO_3^-, and TcO_4^- sorption by a variety of synthetic oxides and geologic materials in low-ionic-strength waters (<0.1 M). Turner and Sassman (1996) and Davis (2001) provide databases for uranium sorption also using the DLM. Langmuir (1997b) compiles surface-complexation constants for actinides and fission products based on several different SCMs. Other compilations are based on the TLM; these include those of Hsi and Langmuir (1985), Tripathi (1983), McKinley *et al.* (1995), Turner *et al.* (1996), and Lenhart and Honeyman (1999) for uranium; Girvin *et al.* (1991) and Kohler *et al.* (1999) for neptunium; Laflamme and Murray (1987), Quigley *et al.* (1996), and Murphy *et al.* (1999) for thorium; and Sanchez *et al.* (1985) for plutonium. Langmuir (1997b), Davis (2001), Wang *et al.* (2001a,b), and Turner *et al.* (2002) provide references to a large number of other surface-complexation studies of radionuclides.

Representation of sorption of radionuclides under natural conditions. Several approaches have been used to represent variability of sorption under natural conditions. These include: (i) sampling K_d values from a probability distribution function (PDF); (ii) calculating a K_d using empirical relations based on measurements over a range of experimental conditions, solution compositions, and mineral properties; and (iii) calculating aqueous and surface speciation using a thermodynamically based surface-complexation model. The first approach is used most commonly in risk assessment and remediation design calculations. Because of the diversity of solutions, minerals, and radionuclides that could be present at contaminated sites and potential waste disposal repository sites, a large body of empirical radionuclide sorption data has been generated. Databases of K_d values that can be used to estimate PDFs for various geologic media are summarized by Barney (1981a,b), Tien *et al.* (1985), Bayley *et al.* (1990), US DOE (1988), McKinley and Scholtis (1992), Triay *et al.* (1997), the US EPA (1999a,b), and Krupka and Serne (2000). Methods used to specify PDFs for K_ds for use with sampling techniques such as Latin Hypercube Sampling (Iman and Shortencarier, 1984; Helton and Davis, 2002) have been described by Siegel *et al.* (1983, 1989), Wilson *et al.* (1994), and Rechard (1996).

Serne and Muller (1987) describe attempts to find statistical empirical relations between experimental variables and the measured sorption ratios (R_ds). Mucciardi and Orr (1977) and Mucciardi (1978) used linear (polynomial regression of first-order independent variables) and nonlinear (multinomial quadratic functions of paired independent variables, termed the Adaptive Learning Network) techniques to examine effects of several variables on sorption coefficients. The dependent variables considered included cation-exchange capacity (CEC) and surface area (SA) of the solid substrate, solution variables (Na, Ca, Cl, HCO_3), time, pH, and Eh. Techniques such as these allow modelers to construct a narrow probability density function for K_ds.

The dependence of a K_d on the composition of the groundwater can also be described in terms of more fundamental thermodynamic parameters. This can be illustrated by considering the sorption of uranyl (UO_2^{2+}) onto a generic surface site (>SOH) of a mineral

$$>SOH + UO_2^{2+} \rightarrow SO - UO_2^{+} + H^{+} \quad (6)$$

with an equilibrium sorption binding constant β^{cat} defined for the reaction. The concentration of UO_2^{2+} available to complex with the surface site will be affected by complexation reactions with other ligands such as carbonate. The K_d in a system containing the uranyl ion and its hydroxo and carbonato complexes can be calculated as

$$K_d = \frac{\beta^{cat} \times \{SOH\} \times C}{\{H^+\} \times [1 + \sum_{ijk} K_{i,j,k}\{UO_2^{2+}\}^{i-1}\{OH\}^{j}\{CO_3^{2-}\}^{k}]} \quad (7)$$

For simplicity, in this equation, we have assumed that activities are equal to concentrations and brackets refer to activities. C is a units conversion constant $= V_v\,m^{-1}$, relating void volume V_v (mL) in the porous media and the mass m (g) of the aquifer material in contact with the volume V_v; $K_{i,j,k}$ is the formation constant for an aqueous uranyl complex, and the superscripts i, j, k describe the stoichiometry of the complex. The form that the sorption binding constant β^{cat} takes is different for the different sorption models shown in Figure 4 (e.g., see Equation (5)). Leckie (1994) derives similar expressions for more complex systems in which anionic and cationic metal species form polydentate surface complexes. Equation (7) can be derived from the following relationships for this system:

(i) K_d = total sorbed uranium/total uranium in solution;

(ii) total sorbed uranium $= \{>SO–UO_2^{2+}\} = \beta^{cat} \times \{>SOH\} \times \{UO_2^{2+}\}/\{H^+\}$;

(iii) total uranium in solution $= \{UO_2^{2+}\} + \{UO_2CO_3\}$ + other uranyl complexes;

(iv) $\{UO_2CO_3\} = K_{UO_2CO_3} \times UO_2^{2+} \times CO_3^{2-}$; and

(v) similar expressions can be written for other uranyl species.

Substituting (v) and (iv) into (iii), and then substituting (iii) and (ii) into (i), yields Equation (7) after some manipulation. Note that activity coefficients of all species are assumed to be equal to 1.0.

Expressions like Equation (7) can be solved using computer programs such as HYDRAQL. Using a spreadsheet program for postprocessing of the results, K_d values can easily be calculated over ranges of solution compositions. Using this approach, the effects of relatively small changes in the composition of the groundwaters can be shown to result in order-of-magnitude changes in the K_d. Figure 5 shows that the calculated K_d of uranium in systems containing several competing ligands can

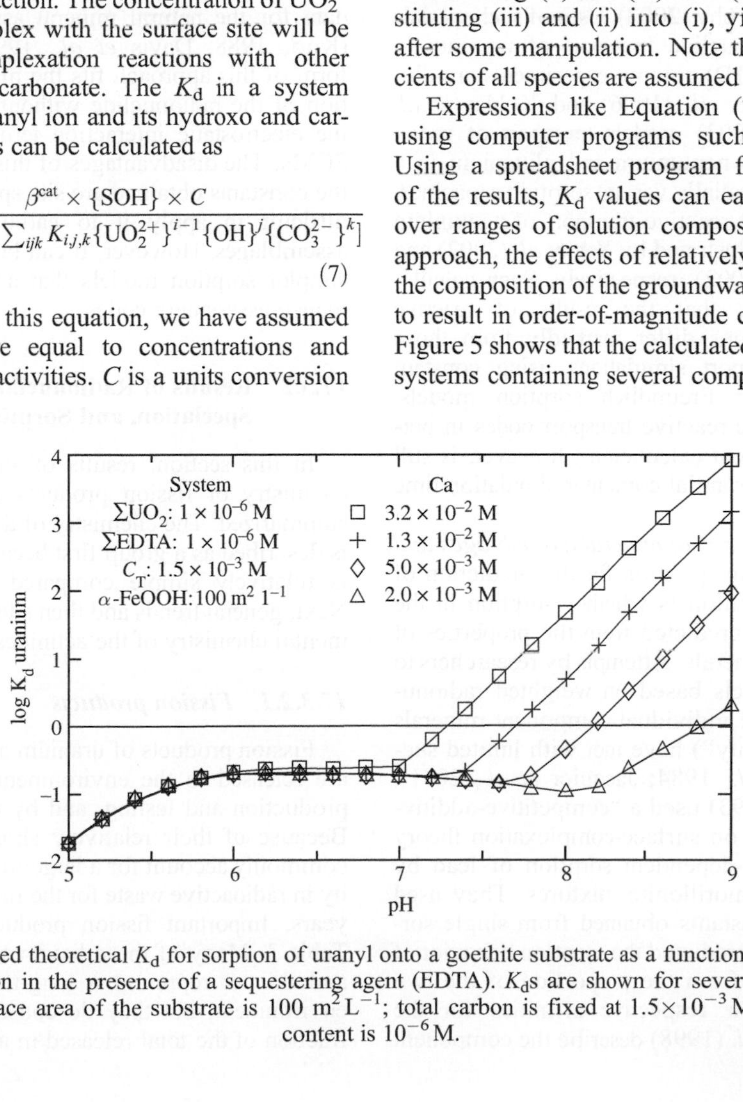

Figure 5 Calculated theoretical K_d for sorption of uranyl onto a goethite substrate as a function of pH at fixed total carbon concentration in the presence of a sequestering agent (EDTA). K_ds are shown for several levels of calcium concentration. Surface area of the substrate is 100 $m^2\,L^{-1}$; total carbon is fixed at 1.5×10^{-3} M and total uranium content is 10^{-6} M.

be sensitive to the concentration of other cations such as Ca^{2+}. Leckie (1995) provides examples of this methodology and produces multidimensional K_d response surfaces. Approaches to using thermodynamic sorption models to predict, interpret, or guide the collection of K_d data are summarized by the NEA (2001).

K_ds, whether sampled from probability distribution functions or calculated by regression equations or surface-complexation models, can be used in many contaminant transport models. Alternate forms of the retardation factor equation that use a K_d (Equation (3)) and are appropriate for porous media, fractured porous media, or discrete fractures have been used to calculate contaminant velocity and discharge (e.g., Erickson, 1983; Neretnieks and Rasmuson, 1984). An alternative approach couples chemical speciation calculations to transport equations. Such models of *reactive transport* have been developed and demonstrated by a number of researchers including Parkhurst (1995), Lichtner (1996), Bethke (1997), Szecsody *et al.* (1998), Yeh *et al.* (1995, 2002), and others reviewed in Lichtner *et al.* (1996), Steefel and Van Cappellen (1998), and Browning and Murphy (2003). Uses of such models to simulate radionuclide transport of uranium in one-dimensional (1D) column experiments are illustrated by Sims *et al.* (1996) and Kohler *et al.* (1996). Glynn (2003) models transport of redox sensitive elements neptunium and plutonium in a 1D domain with spatially variant sorption capacities. Simulations of 2D reactive transport of neptunium and uranium are illustrated by Yeh *et al.* (2002) and Criscenti *et al.* (2002), respectively. Such calculations demonstrate that the results of reactive transport simulations differ markedly from those obtained in transport simulations using constant K_d, Langmuir or Freundlich sorption models. Routine use of the reactive transport codes in performance assessment calculations, however, is still limited by the substantial computer simulation time requirements.

Sorptive properties of mineral assemblages and soils. An important question for the prediction of radionuclide migration is whether sorption in the geomedia can be predicted from the properties of the constituent minerals. Attempts by researchers to use sorption models based on weighted radionuclide K_d values of individual component minerals ("sorptive additivity") have met with limited success (Meyer *et al.*, 1984; Jacquier *et al.*, 2001). Tripathi *et al.* (1993) used a "competitive-additivity" model based on surface-complexation theory to model the pH-dependent sorption of lead by goethite/Ca-montmorillonite mixtures. They used complexation constants obtained from single sorbent systems and predicted the sorption behavior of mineral mixtures from the proportion of the two sorbents and their respective affinities for the metals. Davis *et al.* (1998) describe the component additivity (CA) model, a similar approach in which the wetted surface of a complex mineral assemblage is assumed to be composed of a mixture of one or more reference minerals. The surface properties of the individual phases are obtained from independent studies in monomineralic model systems and then are applied to the mineral assemblage without further fitting, based on the contributions of the individual minerals to the total surface area of the mixture. Applications of this approach to radionuclides are described by McKinley *et al.* (1995), Waite *et al.* (2000), Prikryl *et al.* (2001), Arnold *et al.* (2001), Davis (2001), and Davis *et al.* (2002). Strongly sorbent minerals such as clays or goethite are produced by the alteration of host rocks and line the voids of porous geomedia. In these cases, the sorption behavior of the mineral assemblage can be approximated by using the properties of one or two of minerals even though they constitute a small fraction of the rock mass (Davis and Kent, 1990; Ward *et al.*, 1994; Barnett *et al.*, 2002).

The generalized composite (GC) approach is an alternative approach in which surface-complexation constants are obtained by fitting experimental data for the natural mineral assemblage directly (Koβ, 1988; Davis *et al.*, 1998). A simplified form of this approach fits the pH-dependent sorption of the radionuclide without representation of the electrostatic interaction terms found in other SCMs. The disadvantages of this approach are: (i) the constants obtained are site specific and (ii) it is difficult to apply it to carbonate-rich mineral assemblages. However, it can be used to calibrate simpler sorption models that are used in performance assessment codes.

17.3.2 Results of Radionuclide Solubility, Speciation, and Sorption Studies

In this section, results of studies of the geochemistry of fission products and actinides are summarized. The chemistry of the fission products is described as a group first because their behavior is relatively simple compared to the actinides. Next, general trends and then site-specific environmental chemistry of the actinides are summarized.

17.3.2.1 Fission products

Fission products of uranium and other actinides are released to the environment during weapons production and testing, and by nuclear accidents. Because of their relatively short half-lives, they commonly account for a large fraction of the activity in radioactive waste for the first several hundred years. Important fission products are shown in Table 3. Many of these have very short half-lives and do not represent a long-term hazard in the environment, but they do constitute a significant fraction of the total released in a nuclear accident.

Table 3 Environmentally important fission products.

Fission product	$t_{1/2}$ (yr)
^{79}Se	6.5×10^5
^{90}Sr	28.1
^{93}Zr	1.5×10^6
^{99}Tc	2.12×10^5
^{103}Ru	0.11
^{106}Ru	0.56
^{110m}Ag	0.69
^{125}Sb	2.7
^{129}I	1.7×10^7
^{134}Cs	2.06
^{137}Cs	30.2
^{144}Ce	0.78

Only radionuclides with half-lives of several years or longer represent a persistent environmental or disposal problem. Of primary interest are ^{90}Sr, ^{99}Tc, ^{129}I, and ^{137}Cs, and to a lesser degree, ^{79}Se and ^{93}Zr; all are β^--emitters.

While fission product mobility is mostly a function of the chemical properties of the element, the initial physical form of the contamination can also be important. For radioactive contaminants released as particulates—"hot particles"—radionuclide transport is initially dominated by physical processes, namely, transport as aerosols (Wagenpfeil and Tschiersch, 2001) or as bedload/suspended load in river systems. At Chernobyl, the majority of fission products were released in fuel particles and condensed aerosols. Fission products were effectively sequestered—for example, little downward transport in soil profiles and little biological uptake—until dissolution of the fuel particles occurred and the fission products were released (Petryaev *et al.*, 1991; Konoplev *et al.*, 1992; Baryakhtar, 1995; Konoplev and Bulgakov, 1999). Thus, fuel particle dissolution kinetics controlled the release of fission products to the environment (Kruglov *et al.*, 1994; Kashparov *et al.*, 1999, 2001; Uchida *et al.*, 1999; Sokolik *et al.*, 2001).

^{90}Sr. Strontium occurs in only one valence state, (II). It does not form strong organic or inorganic complexes and is commonly present in solution as Sr^{2+}. The concentration is rarely solubility limited in soil or groundwater systems because the solubility of common strontium phases is relatively high (Lefevre *et al.*, 1993; US EPA, 1999b). The concentration of strontium in solution is commonly controlled by sorption and ion-exchange reactions with soil minerals. Parameters affecting strontium transport are CEC, ionic strength, and pH (Sr^{2+} sorption varies directly with pH, presumably, due to competition with H^+ for amphoteric sites). Clay minerals—illite, montmorillonite, kaolinite, and vermiculite—are responsible for most of the exchange capacity for strontium in soils (Goldsmith and Bolch, 1970; Sumrall and Middlebrooks, 1968). Zeolites (Ames and Rai, 1978) and manganese oxides/hydroxides also exchange or sorb strontium in soils. Because of the importance of ion exchange, strontium K_ds are strongly influenced by ionic strength of the solution, decreasing with increasing ionic strength (Mahoney and Langmuir, 1991; Nisbet *et al.*, 1994); calcium and natural strontium are especially effective at competing with ^{90}Sr. Strontium in soils is largely exchangeably bound; it does not become fixed with time (Serne and Gore, 1996). However, co-precipitation with calcium sulfate or carbonate and soil phosphates may also contribute to strontium retardation and fixation in soils (Ames and Rai, 1978).

^{137}Cs. Caesium, like strontium, occurs in only one valence state, (I). Caesium is a very weak Lewis acid and has a low tendency to interact with organic and inorganic ligands (Hughes and Poole, 1989; US EPA, 1999b); thus, Cs^+ is the dominant form in groundwater. Inorganic caesium compounds are highly soluble, and precipitation/co-precipitation reactions play little role in limiting caesium mobility in the environment. Retention in soils and groundwaters is controlled by sorption/desorption and ion-exchange reactions.

Caesium is sorbed by ion exchange into clay interlayer sites, and by surface complexation with hydroxy groups comprised of broken bonds on edge sites, and the planer surfaces of oxide and silicate minerals. CEC is the dominant factor in controlling caesium mobility. Clay minerals such as illite, smectites, and vermiculite are especially important, because they exhibit a high selectivity for caesium (Douglas, 1989; Smith and Comans, 1996). The selectivity is a function of the low hydration energy of caesium; once it is sorbed into clay interlayers, it loses its hydration shell and the interlayer collapses. Ions, such as magnesium and calcium, are unable to shed their hydration shells and cannot compete for the interlayer sites. Potassium is able to enter the interlayer and competes strongly for exchange sites. Because it causes collapse of the interlayers, caesium does not readily desorb from vermiculite and smectite and may in fact be irreversibly sorbed (Douglas, 1989; Ohnuki and Kozai, 1994; Khan *et al.*, 1994). Uptake by illitic clay minerals does not occur by ion exchange but rather by sorption onto frayed edge sites (Cremers *et al.*, 1988; Comans *et al.*, 1989; Smith *et al.*, 1999), which are highly selective for caesium. Although illite has a higher selectivity for caesium, it has a much lower capacity than smectites because caesium cannot enter the interlayer sites.

Caesium mobility increases with ionic strength because of competition for exchange sites (Lieser and Peschke, 1982). Potassium competes more effectively than calcium or magnesium. Since caesium is rapidly and strongly sorbed by soil and sediment particles, it does not migrate downward rapidly through soil profiles, especially forest soils (Bergman, 1994; Rühm *et al.*, 1996; Panin *et al.*, 2001). Estimated downward migration rates for Cs released by the Chernobyl accident are on the order of 0.2–2 cm yr^{-1} in soils in Bohemia (Hölgye and Malú, 2000), Russia (Sokolik *et al.*, 2001), and Sweden (Rosén *et al.*, 1999; Isaksson *et al.*, 2001).

^{99}Tc. Technetium occurs in several valence states, ranging from −1 to +7. In groundwater systems, the most stable oxidation states are (IV) and (VII) (Lieser and Peschke, 1982). Under oxidizing conditions, Tc(VII) is stable as pertechnetate, TcO_4^-. Pertechnetate compounds are highly soluble, and being anionic, pertechnetate is not sorbed onto common soil minerals and/or readily sequestered by ion exchange. Thus, under oxidizing conditions, technetium is highly mobile. Significant sorption of pertechnetate has been seen in organic-rich soils of low pH (Wildung *et al.*, 1979), probably due to the positive charge on the organic fraction and amorphous iron and aluminum oxides, and possibly coupled with reduction to Tc(IV).

Under reducing conditions, Tc(IV) is the dominant oxidation state because of biotic and abiotic reduction processes. Technetium(IV) is commonly considered to be essentially immobile, because it readily precipitates as low-solubility hydrous oxides and forms strong surface complexes on iron and aluminum oxides and clays.

Technetium(IV) behaves like other tetravalent heavy metals and occurs in solution as hydroxo and hydroxo-carbonato complexes. In carbonate-containing groundwaters, $TcO(OH)_{2(aq)}$ is dominant at neutral pH; at higher pH values, $Tc(OH)_3CO_3^-$ is more abundant (Erikson *et al.*, 1992). However, the solubility of Tc(IV) is low and is limited by precipitation of the hydrous oxide, $TcO_2{\cdot}nH_2O$. The number of waters of hydration is traditionally given as $n=2$ (Rard, 1983) but has more recently been measured as 1.63 ± 0.28 (Meyer *et al.*, 1991)). In systems containing H_2S or metal sulfides, the solubility-limiting phase for technetium may be Tc_2S_7 or TcS_2 (Rard, 1983).

Retention of pertechnetate in soil and groundwater systems usually involves reduction and precipitation as Tc(IV)-containing hydroxide or sulfide phases. Several mineral phases have been shown to fix pertechnetate through surface-mediated reduction/co-precipitation. These include magnetite (Haines *et al.*, 1987; Byegård *et al.*, 1992; Cui and Erikson, 1996) and a number of sulfides, including chalcocite, bournonite, pyrrhotite, tetrahedrite, and, to a lesser extent, pyrite and galena (Strickert *et al.*, 1980; Winkler *et al.*, 1988; Lieser and Bauscher, 1988; Huie *et al.*, 1988; Bock *et al.*, 1989). Sulfides are most effective at reducing technetium if they contain a multivalent metal ion in the lower oxidation state (Strickert *et al.*, 1980). Technetium sorption by iron oxides is minimal under near-neutral, oxidizing conditions but is extensive under mildly reducing conditions, where Fe(III) remains stable. It is minimal on ferrous silicates (Vandergraaf *et al.*, 1984).

In addition, technetium may be fixed by bacterially mediated reduction and precipitation. Several types of Fe(III)- and sulfate-reducing bacteria have been shown to reduce technetium, either directly (enzymatically) or indirectly through reaction with microbially produced Fe(II), native sulfur, or sulfide (Lyalikova and Khizhnyak, 1996; Lloyd and Macaskie, 1996; Lloyd *et al.*, 2002).

^{129}I. Iodine can exist in the oxidation states −1, 0, +1, +5, and +7. However, the +1 state is not stable in aqueous solutions and disporportionates into −1 and +5. In surface- and groundwaters at near-neutral pH, IO_3^- (iodate) is the dominant form in solution, while under acidic conditions, I_2 can form. Under anoxic conditions, iodine is present as I^- (iodide) (Allard *et al.*, 1980; Liu and van Gunten, 1988).

Iodide forms low-solubility compounds with copper, silver, lead, mercury, and bismuth, but all other metal iodides are quite soluble. As these metals are not common in natural environments, they have little effect on iodine mobility (Couture and Seitz, 1985). Retention by sorption and ion exchange appears to be minor (Lieser and Peschke, 1982). However, significant retention has been observed by the amorphous minerals imogolite and allophane (mixed Al/Si oxides–hydroxides, with SiO_2/Al_2O_3 ratios between 1 and 2). These minerals have high surface areas and positive surface charge at neutral pH and contribute significantly to the anion-exchange capacity in soils (Gu and Schultz, 1991). At neutral pH, aluminum and iron hydroxides are also positively charged and contribute to iodine retention, especially if iodine is present as iodate (Couture and Seitz, 1985). Sulfide minerals containing the metal ions which form insoluble metal iodides strongly sorb iodide, apparently through sorption and surface precipitation of the metal iodide. Iodate is also sorbed, possibly because it is reduced to iodide on the metal/sulfide surfaces (Allard *et al.*, 1980; Strickert *et al.*, 1980). Lead, copper, silver, silver chloride, and lead oxides/hydroxides and carbonates can also fix iodine through surface precipitation (Bird and Lopato, 1980; Allard *et al.*, 1980). None of these minerals are likely to be important in natural soils but may be useful in immobilizing iodine for environmental remediation.

Organic iodo compounds are not soluble and form readily through reaction with I_2 and, to a lesser extent, I^- (Lieser and Peschke, 19821; Couture and Seitz, 1985); retention of iodine in soils is mostly associated with the organic matter (Wildung *et al.*, 1974; Muramatsu *et al.*, 1990; Gu and Schultz, 1991; Yoshida *et al.*, 1998; Kaplan *et al.*, 2000). Several studies have suggested that fixation of iodine by organic soil compounds appears to be dependent upon microbiological activity, because sterilization by heating or radiation commonly results in much lower iodine retention (Bunzl and Schimmack, 1988; Koch *et al.*, 1989; Muramatsu *et al.*, 1990; Bors *et al.*, 1991; Rädlinger and Heumann, 2000).

17.3.2.2 Uranium and other actinides [An(III), An(IV), An(V), An(VI)]

General trends in solubility, speciation, and sorption. Actinides are hard acid cations (i.e., comparatively rigid electron clouds with low polarizability) and form ionic species as opposed to covalent bonds (Silva and Nitsche, 1995; Langmuir, 1997a). Several general trends in their chemistry can be described (although there are exceptions). Due to similarities in ionic size, coordination number, valence, and electron structure, the actinide elements of a given oxidation state have either similar or systematically varying chemical properties (David, 1986; Choppin, 1999; Vallet *et al.*, 1999). For a given oxidation state, the relative stability of actinide complexes with hard base ligands can be divided into three groups in the order: CO_3^{2-}, $OH^- > F^-$, HPO_4^{2-}, $SO_4^{2-} > Cl^-$, NO_3^-. Within these ligand groups, stability constants generally decrease in the order $An^{4+} > An^{3+} \approx AnO_2^{2+} > AnO_2^+$ (Lieser and Mohlenweg, 1988; Silva and Nitsche, 1995). In addition, the same order describes the decreasing stability (increasing solubility) of actinide solids formed with a given ligand (Langmuir, 1997a).

These trends have allowed the use of an oxidation analogy modeling approach, in which data for the behavior of one actinide can be used as an analogue for others in the same oxidation state. An oxidation state analogy was used for the WIPP to evaluate the solubility of some actinides and to develop a more complete set of modeling parameters for actinides included in the repository performance calculations. The results are assumed to be either similar to the actual case or can be shown to vary systematically (Fanghänel and Kim, 1998; Neck and Kim, 2001; Wall *et al.*, 2002). The similarities in chemical behavior extend beyond the actinides to the lanthanides—Nd(III) is commonly used as a nonradioactive analogue for the +III actinides. For instance, complexation and hydrolysis constants and Pitzer ion interaction parameters used in modeling Am(III) speciation and solubility for the WIPP were extracted from a suite of published experimental studies involving not only Am(III) but also Pu(III), Cm(III), and Nd(III) (US DOE, 1996).

Oxidation state. Differences among the potentials of the redox couples of the actinides account for much of the differences in their speciation and environmental transport. Detailed information about the redox potentials for these couples can be found in numerous references (e.g., Hobart, 1990; Silva and Nitsche, 1995; Runde, 2002). This information is not repeated here, but a few general points should be made. Important oxidation states for the actinides under environmental conditions are described in Table 4. Depending on the actinide, the potentials of the III/IV, IV/V, V/VI, and/or IV/VI redox couples can be important under near-surface environmental conditions. When the redox potentials between oxidation states are sufficiently different, then one or two redox states will predominate; this is the case for uranium, neptunium, and americium (Runde, 2002). The behavior of uranium is controlled by the predominance of U(VI) species under oxidizing conditions and U(IV) under reducing conditions. In the intermediate Eh range and neutral pH possible under many settings, the solubility of neptunium is controlled primarily by the Eh of the aquifer and will vary between the levels set by $Np^{IV}(OH)_{4(s)}$ (10^{-8} M under reducing conditions) and $Np_2^VO_{5(s)}$ (10^{-5} M under oxidizing conditions). Redox potentials of plutonium in the III, IV, V, and VI states are similar (~1.0 V); therefore, plutonium can coexist in up to four oxidation states in some solutions (Langmuir, 1997a; Runde, 2002). However, Pu(IV) is most commonly observed in environmental conditions and sorption of plutonium is strongly influence by reduction of Pu(V) to Pu(IV) at the mineral–water interface. More discussions of these behaviors will be found in the individual sections for each actinide that follow.

Complexation and solubility. In dilute aqueous systems, the dominant actinide species at neutral to basic pH are hydroxy and carbonato complexes. Similarly, solubility-limiting solid phases are commonly oxides, hydroxides, or carbonates. The same is generally true in high-ionic-strength brines, because common brine components—Na^+, Ca^{2+}, Mg^{2+}, Cl^-, SO_4^{2-}—do not complex as strongly with actinides. However, weak mono-, bis-, and tris-chloro complexes with hexavalent actinides

Table 4 Important actinide oxidation states in the environment.

Actinide element	*Oxidation states*			
Thorium		IV		
Uranium		IV		VI
Neptunium		IV	V	
Plutonium		IV	V	VI
Americium	III			
Curium	III			

(U(VI) and Pu(VI)) can contribute significantly to the solubility of these actinides in chloride-rich brines. Runde *et al.* (1999) measured shifts in the apparent solubility product constants for uranyl and plutonyl carbonate of nearly one log unit as chloride concentrations increased to 0.5 M. Carbonate complexes are important for radionuclides; thorium, plutonium, neptunium, and uranium all have strong carbonate complexes under environmental conditions. Carbonate complexation also leads to decreased sorption by forming strong anionic complexes that will not sorb to negatively charged mineral surfaces. The potential importance of carbonate complexes with respect to increasing actinide solubility and decreasing sorption influenced a decision by the DOE to use MgO as the engineered barrier in the WIPP repository. MgO and its hydration products sequester CO_2 through formation of carbonates and hydroxycarbonates, as well as buffering the pH at neutral to moderately basic values, where actinide solubilities are at a minimum.

Dissolved organic carbon may be present as strong complexing ligands that increase the aqueous concentration limits of actinides (Olofsson and Allard, 1983). In environments with high organic matter from natural or anthropogenic sources, complexation of actinides with ligands such as EDTA and other organic ligands may decrease the extent of sorption onto rocks. Langmuir (1997a) suggests that to a first approximation, complexation of An^{4+}, AnO_2^+, and AnO_2^{2+} with humic and fulvic acids can be ignored, because the actinides form such strong hydroxyl and carbonato-complexes in natural waters. In contrast, however, although An^{3+} species form OH^- and or CO_3^{2-} complexes, important actinide/humic–fulvic complexation does occur. Conditions under which actinide–humic interactions are important are discussed in more detail in Section 17.3.3.1.

Sorption. In general, actinide sorption will decrease in the presence of ligands that complex with the radionuclide (most commonly humic or fulvic acids, CO_3^{2-}, SO_4^{2-}, F^-) or cationic solutes that compete with the radionuclide for sorption sites (most commonly Ca^{2+}, Mg^{2+}). In general, sorption of the (IV) species of actinides (Np, Pu, U) is greater than of the (V) species.

As discussed previously (Section 17.3.1.1), plots of pH sorption edges (see Figure 3) are useful in summarizing the sorption of radionuclide by substrates that have amphoteric sites (i.e., SOH, SO^-, SOH_2^+). The pH sorption edges of actinides are similar for different aluminosilicates (quartz, α-alumina, clinoptilolite, montmorillonite, and kaolinite). For example, Np(V) and U(VI) exhibit similar pH-dependent sorption edges that are independent of specific aluminosilicate identity (Bertetti *et al.*, 1998; Pabalan *et al.*, 1998). Under similar solution conditions, the amount of radionuclide adsorbed is primarily a function of the surface area. This observation has led several workers to propose that the amount of actinide sorption onto natural materials can be predicted from the surface site density and surface area rather the specific molecular structure of the surface (Davis and Kent, 1990; Turner and Pabalan, 1999).

Carroll *et al.* (1992), Stout and Carroll (1993), Van Cappellen *et al.* (1993), Meece and Benninger (1993), Brady *et al.* (1999), and Reeder *et al.* (2001) summarize empirical data and theoretical models of actinide-carbonate mineral interactions. The surface PDI on carbonate minerals may be Ca^{2+} or Mg^{2+}. Increased solution concentration of Ca^{2+} will lead to decreased actinide sorption, which then leads to complex sorption behavior if the carbonate concentration and pH of the solution are varied. Carroll *et al.* (1992) studied the uptake of Nd(III), U(VI), and Th(VI) by pure calcite in dilute $NaHCO_3$ solutions using a combination of surface analysis techniques. They found that U(VI) uptake was limited to monolayer sorption and uranium–calcium solid solution was minimal even in solutions supersaturated with rutherfordine (UO_2CO_3). In contrast, they found that surface precipitation and carbonate solid solution was extensive for thorium and neodymium. Similarly, irreversible sorption and surface precipitation of americium onto carbonates were observed by Shanbhag and Morse (1982) and Higgo and Rees (1986).

Attempts to propose representative K_d values for actinides have met with controversy. For example, Silva and Nitsche (1995) suggested average K_d values for actinides in the order $An^{4+} > An^{3+} > AnO_2^{2+} > AnO_2^+$, as 500, 50, 5, and 1, respectively. This order corresponds to the order of the pH_{50} values of sorption edges for Th(IV), Am(III), Np(V), and Pu(V) in studies of sorption by γ-Al_2O_3 (Bidoglio *et al.*, 1989); and of Pu(IV), U(VI), and Np(V) in studies of sorption by α-FeOOH (Turner, 1995). Calculated or measured element-specific K_ds for natural soils and geomedia for many environmental sites, however, are quite different from these values. For example, a recent compilation listed the following suggested general ranges for soil/mineral K_ds, in mL g^{-1}: Pu, 11–300,000; U, 10.5–4,400; Am,1–47,230 (Krumhansl *et al.*, 2002). These wide ranges exist because sorption of radionuclides is very dependent on the radionuclide oxidation state, groundwater composition, and nature of rock surface, all of which may be variable and/or poorly characterized along the flow path. The databases of K_d values described previously (Section 17.3.1.2) should be used to obtain K_ds for site-specific conditions instead of using broad "generic" ranges whenever possible.

17.3.2.3 Site-specific geochemistry of the actinides

Introduction: actinide solubilities in reference waters. In this section, the environmental chemistry of the actinides is examined in more detail by considering three different geochemical environments. Compositions of groundwater from these environments are described in Tables 5 and 6. These include: (i) low-ionic-strength reducing waters from crystalline rocks at nuclear waste research sites in Sweden; (ii) oxic water from the J-13 well at Yucca Mountain, Nevada, the site of a proposed repository for high-level nuclear waste in tuffaceous rocks; and (iii) reference brines associated with the WIPP, a repository for TRU in the Permian Salt beds of SE New Mexico. These last brines are model solutions produced by the reaction of the Permian formation waters with the components of the engineered barrier at the WIPP as discussed below.

Table 5 Compositions of low ionic strength reference waters used in speciation and solubility calculations.

Component	*SKI-90 Stripa* (mM)[a]	*J-13 YM* (mM)[b]
Na^+	1.39	1.96
K^+	0.0256	0.14
Ca^{2+}	0.5	0.29
Mg^{2+}	0.0823	0.07
Fe (total)	0.00179	
SiO_2	0.0682	1.07
Cl^-	0.423	0.18
SO_4^{2-}	0.417	0.19
F^-	0.142	0.11
PO_4^{3-}	3.75e−5	
HCO_3^-	2.0	2.81
PH[c]	8.2	6.9
Eh (mV)	−0.3	0.34–0.7[c]

[a] SKI (1991). [b] Ogard and Kerrisk (1984). [c] Range of Eh used in different works.

Table 6 Compositions of brines used in WIPP speciation and solubility calculations (US DOE, 1996).

Component	*Salado brine* (mM)	*Castile brine* (mM)
Na^+	1,830	4,870
K^+	770	97
Ca^{2+}	20	12
Mg^{2+}	1,440	19
Fe (total)		
SiO_2		
Cl^-	5,350	4,800
SO_4^{2-}	40	170
F^-		
PO_4^{3-}		
Br^-	10	11
$B_4O_7^{2-}$	5	16
pH[a]	8.7	9.2
pC_{H^+}[a,b]	9.4	9.9
P_{CO_2}, atm[a]	$10^{-5.5}$	$10^{-5.5}$
Eh		
Total dissolved solids	306,000	330,000
Ionic strength	6,990	5,320

[a] In equilibrium with brucite and hydromagnesite.
[b] pC_{H^+} = negative log of the molar concentration of H^+.

The Swedish repository science program has investigated crystalline rock as a host rock for the disposal of radioactive waste and has measured the composition of granitic groundwaters Andersson, (1990). Much of this was done at the Stripa site, an abandoned iron mine located in a granitic intrusion in south-central Sweden. At Stripa, shallow groundwaters are dilute, carbonate-rich, pH neutral, oxidizing waters of meteoric origin; naturally occurring uranium is present in concentrations of 10–90 ppb. Once below this zone, the waters are slightly more saline (up to 1.3 g L^{-1} total dissolved solids), more basic (up to pH 10.1), and the Eh is lower—groundwater uranium concentrations are less than 1 ppb (Andrews *et al.*, 1989; Nordstrom *et al.*, 1989). The trace amounts of sulfide and ferrous iron in the groundwater have little capacity for maintaining reducing conditions, and groundwater interactions with radioactive waste, waste containers, or repository backfill materials are likely to govern the redox conditions in a real repository (Nordstrom *et al.*, 1989). The dilute, near-neutral, mildly reducing groundwater composition given in Table 5 is a composite of analyses from several Swedish sites and is a suggested reference composition (Andersson, 1990) for deep granitic groundwaters.

The proposed nuclear waste repository at Yucca Mountain, Nevada, would be located in a thick sequence of Tertiary volcanic tuffs. The range of groundwater compositions sampled at the site is discussed by Perfect *et al.* (1995). Numerous geochemical studies have been carried out in high-Eh waters from the alluvium and tuffaceous rocks (e.g., UZ-TP-7) from the unsaturated zone, high-Eh waters from the saturated zone (e.g., J-13) within tuffaceous rocks, and in lower-Eh waters from a deeper Paleozoic carbonate aquifer (e.g., UE252p-1) (Tien *et al.*, 1985; Triay *et al.*, 1997). Table 5 describes the composition of water from the J-13 well, which has been used as a reference water in systematic studies of sorption, transport, and solubility (Nitsche *et al.*, 1992). Its composition is controlled by a number of processes including dissolution of vitric and devitrified tuff, precipitation of secondary minerals, and ion exchange (Tien *et al.*, 1985; Triay *et al.*, 1997).

Table 7 contains the results of actinide solubility and speciation calculations for the J-13 and SKI-90 reference waters carried out using the MINTEQ2A

Table 7 Solubility-limiting solids and range of solubility-limited concentrations (M concentration) for low-ionic strength geochemical environments.[a]

Element	*Environment*	
	YMP[b]	*Stripa*[c]
Tc	None	$TcO_2{\cdot}2H_2O \leq 3.3\times10^{-8}$
Th	$Th(OH)_{4(am)} \leq 6.0\times10^{-7}$	$Th(OH)_{4(am)} \leq 5.7\times10^{-7}$
U	$Ca(H_3O)_2(UO_2)_2(SiO_4)_2{\cdot}3H_2O_{(cr)}\ 5.4\times10^{-9}$	$UO_{2(am)} \leq 1.4\times10^{-8}$
Np	$NaNpO_2CO_3{\cdot}3.5H_2O_{(cr)}\ 8.9\times10^{-4}$	$Np(OH)_{4(am)} \leq 1.6\times10^{-9}$
Pu	$Pu(OH)_{4(am)} \leq 6.6\times10^{-8}$	$Pu(OH)_{4(am)} \leq 1.7\times10^{-9}$
Am	$AmOH(CO_3)_{(cr)}5.6\times10^{-8}$	$AmOH(CO_3)_{(cr)} \leq 1.4\times10^{-7}$

[a]Based on Langmuir (1997a, table 13.11). [b]J-13 reference water (Ogard and Kerrisk, 1984). [c]SKI-90 reference water (SKI, 1991).

code (Allison *et al.*, 1991) as described by Langmuir (1997a). The MINTEQ2A thermodynamic database of Turner *et al.* (1993) was used with revised data for americium (Silva *et al.*, 1995) and modifications for uranium described in Langmuir (1997a). Langmuir (1997a) used formation constants that effectively eliminated the influence of $Np(OH)_5^-$ and $Pu(OH)_5^-$ complexes and assumed that the most soluble amorphous hydroxides and mixed carbonato-hydroxide phases controlled the solubility. These concentrations should be considered maximum soluble concentrations that might be important for short-term behavior of the radionuclides. Over longer time periods, the solubilities are likely controlled by more crystalline phases at levels that are several orders of magnitude lower than those listed in Table 7. These calculations should be considered as a set of baseline calculations for low-ionic-strength solutions; they illustrate the effect of redox potential on speciation and solubility.

WIPP is an underground repository for the permanent disposal of defense-related TRU wastes (NAS, 1996). The facility is located in the US in southeastern New Mexico in a thick, bedded salt, the Salado Formation, at a depth of 655 m. The Castile Formation is an evaporite sequence below the Salado that may serve as a brine source if the repository is breached by human activities in the future. Brines from both formations are a mixture of Na^+, Mg^{2+}, K^+, Ca^{2+}, Cl^-, and SO_4^{2-} (see Table 6) and are saturated with respect to halite (NaCl) and anhydrite ($CaSO_4$). For the discussion below, the pH, P_{CO_2} and radionuclide solubilities and speciation were calculated assuming that the brines were in equilibrium with halite, anhydrite, and minerals produced by hydration and carbonation of the MgO engineered barrier (brucite and hydromagnesite). The Eh of the reference brines was assumed to be controlled by the metallic iron in the waste and waste packages. Calculated solubilities in the two brines are grossly similar. The dominant aqueous species and solubility-limiting phases contain hydroxide and/or carbonate, and solubility differences in the two brines are largely due to differences in the pH (8.7 for the Salado, and 9.2 for the Castile) and CO_3^{2-} activities (3.81×10^{-7} M and 4.82×10^{-6} M, for the Salado and Castile brines, respectively).

Because of the high ionic strength of the brines, the calculations were carried out using a Pitzer ion interaction model (US DOE, 1996) for the activity coefficients of the aqueous species (Pitzer, 1987, 2000). Pitzer parameters for the dominant nonradioactive species present in WIPP brines are summarized in Harvie and Weare (1980), Harvie *et al.* (1984), Felmy and Weare (1986), and Pitzer (1987, 2000). For the actinide species, the Pitzer parameters that were used are summarized in the WIPP Compliance Certification Application (CCA) (US DOE, 1996). Actinide interactions with the inorganic ions H^+, Na^+, K^+, Mg^{2+}, Cl^-, and HCO_3^-/CO_3^{2-} were considered.

Americium. The low solubilities and high sorption affinity of thorium and americium severely limit their mobility under environmental conditions. However, because each exists in a single oxidation state—Th(IV) and Am(III)—under environmentally relevant conditions, they are relatively easy to study. In addition, their chemical behaviors provide valuable information about the thermodynamic properties of trivalent and tetravalent species of uranium, neptunium, and plutonium.

Silva *et al.* (1995) provide a detailed summary of experimental and theoretical studies of americium chemistry as well as a comprehensive, self-consistent database of reference thermodynamic property values. Solubility and speciation experiments with Am(III) indicate that the mixed hydroxy-carbonate $AmOHCO_{3(cr)}$ is the solubility-limiting solid phase under most surface and subsurface conditions. At neutral pH, $AmOH^{2+}$ or $AmCO_3^+$ can be the dominant solution species depending on the carbonate concentration. Langmuir (1997a) calculated a solubility of 5.6×10^{-8} M for J-13 water with the MINTEQA2 code using a revised formation constant $\log K_{sp}=7.2$ for $AmOHCO_3$ (compared to NEA $\log K_{sp}=8.605$ of Silva *et al.* (1995)); see Table 7.

This is similar to the value of 1.2×10^{-9} M measured by Nitsche *et al.* (1993) in solubility experiments in J-13 water. The americium solubility calculated by Langmuir (1997a) for reducing water from crystalline rock in Table 7 ($\leq1.4\times10^{-7}$ M) is similar to the range calculated by Bruno *et al.* (2000) using the EQ3NR (Wolery, 1992b) code for slightly basic, reducing groundwaters in granite at Äspö and Gideå, Sweden.

Fanghänel and Kim (1998) evaluated the solubility of trivalent actinides in brines, using Cm(III) as a representative analogue, and found that An(III) hydroxy and carbonato complexes are the most stable aqueous complexes. Multiple-ligand complexes with a high negative charge are more stable in brines than in dilute solutions, apparently because of the high cation concentrations. Chloride and sulfate complexes, although very weak, may be important aqueous species in some brines, especially at low pH.

In the WIPP speciation and solubility calculations, the solubility-controlling solid phase for americium, and by analogy, for all +III actinides under WIPP conditions, was $Am(OH)CO_{3(cr)}$ (Novak, 1997; US EPA, 1998a,b,c,d; Wall *et al.*, 2002). $Am(OH)_2^+$ was the most abundant aqueous species, and estimated americium solubilities in the reference Salado and Castile brines (Table 8) were 9.3×10^{-8} M and 1.3×10^{-8} M, respectively (Novak, 1997; US EPA, 1998d).

Americium is strongly sorbed by tuffaceous rocks from Yucca Mountain in waters of low ionic strength (Triay *et al.*, 1997). In a compilation by Tien *et al.* (1985), americium K_ds obtained with tuff in J-13 water ranged from 130 mL g^{-1} to 13,000 mL g^{-1}. Average values for devitrified, vitric, and zeolitized tuff were 2,975 mL g^{-1}, 1,430 mL g^{-1}, and 1,513 mL g^{-1}, respectively. Turin *et al.* (2002) measured K_ds ranging from 410 mL g^{-1} to 510 mL g^{-1} using similar waters and tuffaceous rocks from Busted Butte on the Nevada Test Site. They also provide Freundlich isotherm parameters from the sorption measurements. A K_d range of 500–50,000 ml g^{-1} is reported for crystalline rocks by McKinley and Scholtis (1992); a value of 5,000 mL g^{-1} is recommended for performance assessment.

Data are sparse for americium sorption in high-ionic-strength solutions. In experimental studies with near-surface sediments from the Gorleben site, Lieser *et al.* (1991) showed that americium sorption did not vary ($K_d\sim1,000$ mL g^{-1}) over a range of NaCl concentrations of 0–2 M, at a pH 7.5. They concluded that americium sorption was not sensitive to ionic strength, because at this pH americium is nearly completely hydrolyzed. Thus, ion-exchange reactions did not contribute to americium sorption, and competing ion concentrations had little effect on sorption K_ds. *In situ* studies of radionuclide transport through brackish bay sediments in Sweden (~seawater solution compositions) measured K_ds of 10^3–10^4 mL g^{-1} (Andersson *et al.*, 1992).

Thorium. Experimental and theoretical studies of thorium speciation, solubility, and sorption in low-ionic-strength waters are described by Langmuir and Herman (1980), Laflamme and Murray (1987), Östhols *et al.* (1994), Östhols (1995), and Quigley *et al.* (1996). Langmuir and Herman (1980) provide a critically evaluated thermodynamic database for natural waters at low temperature that is widely used. However, it does not contain information about important thorium carbonate complexes, and the stability of phosphate complexes may be overestimated (US EPA, 1999b).

In both low-ionic-strength groundwaters and in the WIPP brines, the solubility limiting phase is $ThO_{2(am)}$. In seawater, waters from Yucca Mountain, and reducing waters in crystalline rocks, the dominant aqueous species are $Th(OH)_{4(aq)}$ and mixed hydroxy carbonato complexes. In alkaline lakes and other environments with high carbonate concentrations, thorium carbonate complexes are dominant (Laflamme and Murray, 1987; Öthols *et al.*, 1994). In organic-rich stream waters, swamps, soil horizons, and sediments, organic thorium complexes may predominate (Langmuir and Herman, 1980). Calculated thorium solubilities in waters from Yucca Mountain and crystalline rocks in Table 7 are

Table 8 Calculated actinide solubilities in WIPP brines.

Actinide	*Solubility* (M)		*Solubility-limiting phase*	*Dominant aqueous phases*
	Salado	*Castile*		
An(VI)[a]	8.7×10^{-6}	8.8×10^{-6}		
An(V)	1.2×10^{-7}	4.8×10^{-7}	$KnpO_2CO_3\cdot2H_2O_{(s)}$	$NpO_2CO_3^-$
An(IV)	1.2×10^{-8}	4.1×10^{-8}	$ThO_{2(am)}$	$Th(OH)_3CO_3^-$
An(III)	9.3×10^{-8}	1.3×10^{-8}	$Am(OH)CO_{3(cr)}$	$Am(OH)_2^+$

Source: US EPA (1998d).
[a] Estimated from literature data (values as listed in US DOE, 1996).

similar (~6.0×10^{-7} M) but are much higher than those calculated by Bruno *et al.* (2000) for waters from Äspö and Gideå using EQ3NR (~2×10^{10} M).

The solubility of $ThO_{2(am)}$ increases with increasing ionic strength; above pH 7 in 3.0 M NaCl solutions, the solubility is approximately three orders of magnitude higher than that measured in 0.1 M $NaClO_4$ solution (Felmy *et al.*, 1991). Rai *et al.* (1997) describe solubility studies and a thermodynamic model for Th(IV) speciation and solubility in concentrated NaCl and $MgCl_2$ solutions. A Pitzer ion-interaction model was used to obtain a solubility product of $\log K_{sp} = -45.5$ for $ThO_{2(am)}$. In the speciation and solubility calculations for the WIPP performance assessment (Table 8), the only important aqueous species was $Th(OH)_3(CO_3)^-$; the corresponding estimated Th(IV) solubilities were 1.2×10^{-8} M in the Salado brine and 4.1×10^{-8} M in the Castile brine (Novak, 1997; US EPA 1998d).

Thorium sorbs strongly to iron oxyhydroxides and humic matter (Nash and Choppin, 1980; Hunter *et al.*, 1988; Murphy *et al.*, 1999) and weakly to silica at neutral to basic pH (Östhols, 1995). Thorium sorption is sensitive to carbonate alkalinity due to the formation of negatively charged aqueous mixed hydroxy-carbonato complexes (Laflamme and Murray, 1987); at alkalinities of 100 meq L^{-1}, thorium sorption by goethite decreases markedly. However, at the relatively low alkalinities measured at Yucca Mountain, this effect is not important for the proposed repository site (Triay *et al.*, 1997). Measured thorium sorption ratios in J-13 water from Yucca Mountain for devitrified, vitric, and zeolitized tuff ranged from 140 mL g^{-1} to 2.38×10^4 mL g^{-1} (Tien *et al.*, 1985; Thomas, 1987). Other compilations contain representative K_d values for thorium in crystalline rock that range from 100 mL g^{-1} to 5,000 mL g^{-1} (McKinley and Scholtis, 1992) and from 20 mL g^{-1} to 3×10^5 mL g^{-1} for low-temperature geochemical environments (US EPA, 1999b).

Thorium sorption at high ionic strength was examined using uranium-series disequilibrium techniques by Laul (1992). Laul measured thorium retardation in saline groundwaters from the Palo Duro Basin, Texas, and determined sorption K_ds of ~2,100 mL g^{-1}. Because tetravalent actinides are strongly sorbed by mineral colloids and have a strong tendency to form intrinsic colloids, increases in ionic strength may have more effect on An(IV) transport through destabilization and flocculation of colloidal particles (Lieser and Hill, 1992), rather than through changes in the degree of sorption.

Uranium. Uranium, neptunium, and plutonium are probably the most important actinides in assessment of the environmental risks posed by radioactive contamination. Uranium contamination is present at numerous sites contaminated by uranium mining, milling, and solution mining as described in previous sections. It is highly mobile and soluble under near-surface oxidizing conditions and thus presents an exposure hazard to humans and ecosystems.

Under oxidizing near-surface conditions, U(VI) is the stable oxidation state. Figure 6 shows the aqueous speciation of U(VI) and the solubility of crystalline schoepite (β-$UO_2(OH)_2$) under atmospheric conditions ($P_{CO_2} = 10^{-3.5}$) over the pH 4–9. The calculations were carried out with the

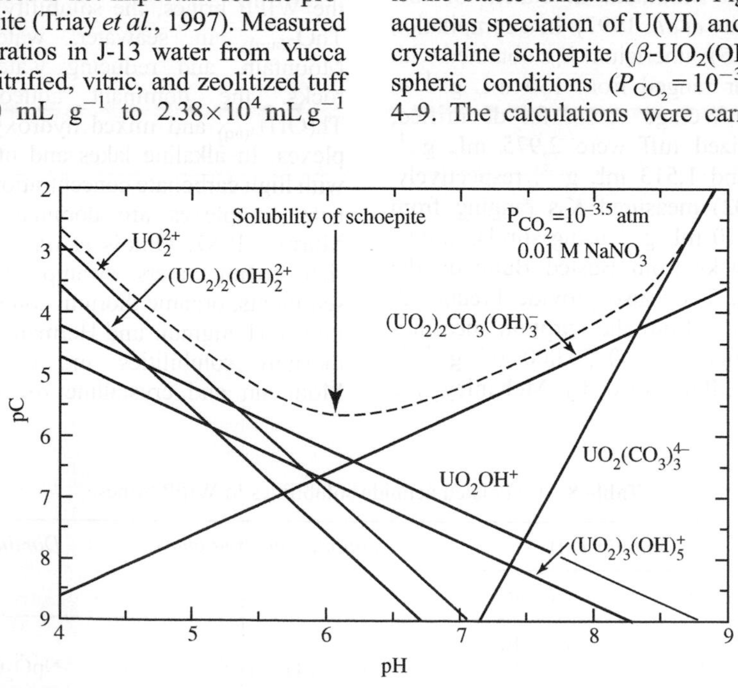

Figure 6 Aqueous speciation of uranium(VI) and the solubility of crystalline schoepite (β-$UO_2(OH)_2$) under atmospheric conditions ($P_{CO_2} = 10^{-3.5}$atm (Davis (2001); reproduced by permission of Nuclear Regulatory Commission from *Complexation Modeling of Uranium(VI) Adsorption on Natural Mineral Assemblages NUREG/ CR- 6708*, **2001**, p.12).

HYDRAQL code (Papelis *et al.*, 1988) using a thermodynamic database described by Davis *et al.* (2001), which is based primarily on the compilation of Grenthe *et al.* (1992). The figure shows that over the pH 6–8, the mixed hydroxy-carbonato binuclear complex $(UO_2)_2CO_3(OH)_3^-$ is predicted to predominate; that at lower pH, the uranyl ion (UO_2^{2+}) is most important; and that at higher pH, the polycarbonate $UO_2(CO_3)_3^{4-}$ has the highest concentration. In Figure 6, the total concentration of uranium is limited by the solubility of schoepite; therefore, it varies as a function of pH and can exceed 10^{-3} M. The relative importance of the multinuclear uranyl complexes are different from those shown in Figure 6 at different fixed total uranium concentrations. For example, if the total uranium concentration is limited to $<10^{-8}$ M (e.g., by slow leaching of uranium from a nuclear waste form), then the species $UO_2(OH)_{2(aq)}$ predominates at pH 6–7. At higher total uranium concentrations (e.g., 10^{-4} M), the multinuclear uranyl hydroxy complex $(UO_2)_3(OH)_5^+$ predominates at pH = 5–6 (Davis *et al.*, 2001).

Uncertainties in the identity and solubility product of the solubility-limiting uranium solid in laboratory studies lead to considerable uncertainty in estimates of the solubility under natural conditions. If amorphous $UO_2(OH)_2$ is assumed to limit the solubility in solutions open to the atmosphere, then the value of the minimum solubility and the pH at which it occurs change. If crystalline schoepite (β-$UO_2(OH)_2$) controls the solubility, then the minimum solubility of 2×10^{-6} M occurs at about pH 6.5. If amorphous $UO_2(OH)_2$ controls the solubility, the solubility minimum of 4×10^{-5} M occurs closer to pH 7.0 if the solubility product of Tripathi (1983) is assumed (Davis *et al.*, 2001).

In water of low Eh, such as crystalline rock environments studied in the European nuclear waste programs, uranium solubility is controlled by saturation with UO_2 and coffinite ($USiO_4$) (Langmuir, 1997a). Langmuir (1997a, pp. 501–502) describes some of the controversy surrounding estimation of the solubility of UO_2. Estimates for the log K_{sp} of UO_2 range from −51.9 to −61.0, corresponding to soluble concentrations (as $U(OH)_{4(aq)}$) ranging from 10^{-8} M (measured by Rai *et al.*, 1990) to $10^{-17.1}$ M (computed by Grenthe *et al.* (1992)). The reason for the wide range lies in the potential contamination of the experimental systems by O_2 and CO_2 and the varying crystallinity of the solid phase. Contamination and the presence of amorphous rather than crystalline UO_2 would lead to higher measured solubilities.

In the WIPP performance assessment calculations, it was assumed that redox conditions would be controlled by the presence of metallic iron and that U(VI) would be reduced to U(IV). An upper bound for the solubility of U(IV) was estimated from that of Th(IV), using the oxidation state analogy. Calculations by Wall *et al.* (2002) suggest that this is a conservative assumption—that the solubility of U(IV) is much lower than that of Th(IV). This is because the solubility product constant of $ThO_{2(am)}$, the solubility limiting phase in the An(IV) model, is several orders of magnitude greater than that of $UO_{2(am)}$.

Reed *et al.* (1996) examined An(VI) stability in WIPP brines under anoxic conditions (1 atm H_2 gas) and found that U(VI) was stable as a carbonate complex in Castile brine at pH 8–10. Xia *et al.* (2001) also observed that U(VI) may be stable under some WIPP-relevant conditions, finding that while U(VI) was rapidly reduced to U(IV) by Fe^0 in water and 0.1 M NaCl, it was not reduced in the Castile brine, at pC_H^+ 8–13, over the course of a 55-day experiment. The possible occurrence of U(VI) was considered in WIPP performance assessment calculations—solubility-limited concentrations for U(VI) in the Salado and Castile brines were estimated from literature values to be 8.7×10^{-6} M and 8.8×10^{-6} M, respectively (US DOE, 1996).

In areas affected by uranium solution mining using sulfuric acid, $UO_2SO_{4(aq)}$ will be important. In alkaline waters, carbonate complexes will dominate. Bernhard *et al.* (1998) studied uranium speciation in water from uranium mining districts in Germany (Saxony) using laser spectroscopy, and found that $Ca_2UO_2(CO_3)_{3(aq)}$ was the dominant species in neutral pH carbonate- and calcium-rich mine waters; $UO_2(CO_3)_3^{4-}$ was the dominant aqueous species in basic (pH = 9.8), carbonate-rich, calcium-poor mine waters; and $UO_2SO_{4(aq)}$ dominated in acidic (pH = 2.6), sulfate-rich mine waters.

A large number of studies of uranium sorption have been carried out in support of the nuclear waste disposal programs and the uranium mill tailings program (UMTRA). Park *et al.* (1992) and Prasad *et al.* (1997) describe studies of sorption of uranyl ion by corrensite, the clay mineral lining many fractures in the fractured Culebra Dolomite member of the Rustler Formation above the WIPP in SE New Mexico. The studies were carried out in dilute and concentrated NaCl (0.1–3 M) solutions in the presence of Ca^{2+}, Mg^{2+}, carbonate, and citrate. Binding constants for the TLM were fit to the sorption edges. They found that the adsorption edges were typical of cation adsorption on mineral surfaces; the uranium was nearly completely bound to the surface at neutral and near-neutral pH values. Neither the background electrolyte (NaCl) nor Ca^{2+} or Mg^{2+} ions (at 0.05 M) influenced the adsorption, suggesting that uranyl binds at pH-dependent edge sites on the corrensite surface as an inner-sphere complex. Both carbonate and citrate reduced the adsorption of uranyl on corrensite in near neutral solutions. Redden *et al.* (1998) carried out similar studies of uranium sorption by goethite, kaolinite, and gibbsite in the presence of citric acid. Davis (2001) and Jenne (1998) provide good summaries of studies of uranium sorption by synthetic and natural aluminosilicates and iron oxyhydroxides. Qualitative features of the sorption edges for these minerals are similar: U(VI) sorption at higher pH is typically low and likely is controlled by the predominance of the negatively charged uranyl-carbonate solution species. By analogy, sorption of U(VI) by aluminosilicates is predicted to be low in waters sampled at Yucca Mountain (Turner *et al.*, 1998; Turner and Pabalan, 1999).

Luckscheiter and Kienzler (2001) examined uranyl sorption onto corroded HLW glass simulant in deionized water, 5.5 M NaCl and 5.0 M $MgCl_2$, and found that sorption was greatly inhibited by the magnesium-rich brine, while the NaCl brine had little effect. Uranyl sorption at high ionic strength was also studied by Vodrias and Means (1993), who examined uranyl sorption onto crushed impure halite and limestone from the Palo Duro Basin, Texas, in a synthetic Na–K–Mg–Ca–Cl brine (I = 10.7 M). They measured K_ds of 1.3 mL g^{-1} on the halite and 4–7 mL g^{-1} on the limestone. This is in contrast to a K_d of 2,100 mL g^{-1} determined from uranium-series disequilibrium measurements on formation brines from the same region (Laul, 1992). The isotopic ratios suggest that naturally occurring uranium was more strongly sorbed because it was present as U(IV).

Neptunium. Neptunium and plutonium are the radioelements of primary concern for the disposal of nuclear waste at the proposed repository at Yucca Mountain. This is due to their long half-lives, radiotoxicity, and transport properties. Neptunium is considered to be the most highly mobile actinide because of its high solubility and low potential for sorption by geomedia. Its valence state (primarily Np(V) or Np(IV)) is the primary control of its environmental geochemistry. Oxide, hydroxide, and carbonate compounds are the most important solubility-limiting phases in natural waters. In low-ionic-strength, carbonate-free systems, $NpO_2(OH)$ and Np_2O_5 are stable Np(V) solids, while in brines, Np(V) alkaline carbonate solids are stable. Under reducing conditions, $Np(OH)_{4\ am}$ and NpO_2 are the stable Np(IV) solids. Under most near-surface environmental conditions, the dominant complexes of neptunium are those of the pentavalent neptunyl species (NpO_2^+). Neptunium(IV) aqueous species may be important under reducing conditions possible at some underground nuclear waste research facilities such as the WIPP and Stripa.

Kaszuba and Runde (1999)compiled thermodynamic data for neptunium relevant to Yucca Mountain. They updated the database of Lemire (1984) with recent experimental data and used the SIT to calculate ion activity coefficients. Their report has an extensive reference list and list of interaction parameters. Kaszuba and Runde (1999) used the EQ3NR (Wolery, 1992b) and the Geochemist Workbench (Bethke, 1998) codes to calculate solubility and speciation in the J-13 and UE25p#1 well waters that span the expected geochemical conditions for the proposed HLW repository at Yucca Mountain. They predicted that $Np(OH)_{4(aq)}$ is the dominant aqueous complex in neutral solutions at Eh < 0 mV, while under oxidizing conditions, NpO_2^+ and $NpO_2CO_3^-$ are predominant at pH < 8 and pH 8–13, respectively.

Although the calculations of Kaszuba and Runde (1999) indicate that $NpO_{2(s)}$ is the thermodynamically stable solid for most Eh–pH conditions of environmental interest, that phase has never been observed to precipitate in solubility experiments in natural waters; instead, $Np_2O_{5(s)}$ and amorphous $Np(OH)_{4(s)}$ precipitate. Figure 7 shows that if $Np_2O_{5(s)}$ controls the solubility under oxidizing conditions (Eh > 0.25 V), then the calculated solubility of neptunium decreases from $\sim 10^{-3.5}$ M at pH = 6 to 10^{-5} M at pH = 8. If amorphous $Np(OH)_{4(s)}$ controls the solubility under reducing conditions (Eh < −0.10 V), the solubility is $\sim 10^{-8}$ M over the same pH range. In the intermediate Eh range and neutral pH conditions possible under many environmental settings, the solubility of neptunium is controlled primarily by the Eh of the aquifer and will vary between the levels set by the solubilities of $Np(OH)_{4(s)}$ and $Np_2O_{5(s)}$ (Figure 7). The inset in the figure illustrates that at pH = 6.8, at Eh = −0.10 V, the concentration of neptunium in solution is approximately equal to that of the Np(IV) species and is controlled by the solubility of $Np(OH)_{4(s)}$. As the redox potential increases, Np(IV) in solution is oxidized to Np(V) and the aqueous concentration of neptunium increases. Phase transformation of $Np(OH)_{4(s)}$ to $Np_2O_{5(s)}$ occurs at about Eh = 0.25 V and then the solubility of the Np(V) oxide controls the aqueous neptunium concentration at higher Eh values.

Neptunium is expected to be present in the WIPP in either the IV or V oxidation state. For the WIPP CCA speciation and solubility calculations, an upper bound for the solubility of Np(IV) was estimated from that of Th(IV), using the oxidation state analogy. As with U(IV), calculations suggest that this assumption is conservative (Wall *et al.*, 2002). Modeling for the WIPP project suggested that Np(V) solubility in the reference Salado and Castile brines is limited by $KNpO_2CO_3 \cdot 2H_2O_{(s)}$ and is 1.2×10^{-7} M and 4.8×10^{-7} M, respectively (Novak, 1997; US EPA, 1998d). The most abundant aqueous species in both brines is $NpO_2(CO_3)^-$. Experimental measurements of the solubility of Np(V) in laboratory solutions representing unaltered Salado brine yielded a value of 2.4×10^{-7} M, after allowing the brine systems to

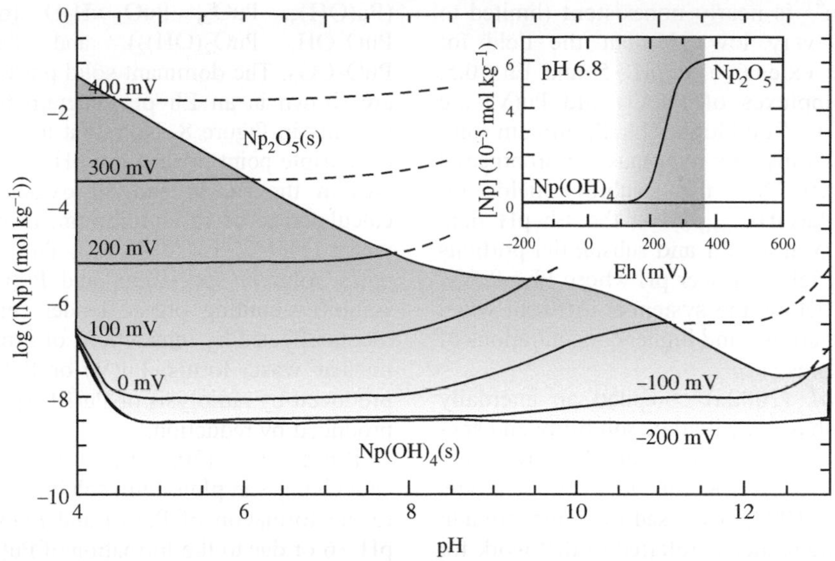

Figure 7 Calculated Np solubilities as a function of pH and Eh in J-13 groundwater variants (Table 5). $Np_2O_{5(s)}$ and $Np(OH)_{4(s)}$ were assumed to be the solubility-limiting phases. Inset shows regions of solubility control versus redox control (shaded area) (Kaszuba and Runde, 1999) (reproduced by permission of American Chemical Society from *Environmental Science and Technology* **1999**, *33*, 4433).

equilibrate for up to 2 yr (Novak *et al.*, 1996). The solubility-limiting phase was identified as $KNpO_2CO_3{\cdot}nH_2O_{(s)}$, in agreement with the results of the WIPP performance assessment modeling.

Aqueous neptunium species including bishydroxo and mixed hydroxy-carbonato species may be important, though not dominant, at higher pH and carbonate concentrations. Such conditions may exist at the Hanford Waste tanks, where $MNpO_2CO_3{\cdot}nH_2O$ and $M_3NpO_2(CO_3)_2$ ($M = Na^+$, K^+) are predicted to be stable phases. The solubilities are two to three orders of magnitude higher than in waters in which Np_2O_5 is stable. Under conditions expected in the near field of HLW geologic repositories in saline groundwater environments such as the salt domes and the bedded salts in Europe, Np(VI) species like $NpO_2(CO_3)_3^{4-}$ might be important due to radiolysis.

In general, sorption of Np(V) by aluminosilicates is expected to be low in waters at Yucca Mountain (Turner and Pabalan, 1999; Turner *et al.*, 1998). K_ds for sorption of neptunium by zeolites and tuff particles were typically less than 10 mL g^{-1} in waters from that site (Tien *et al.*, 1985; Runde, 2002). The low neptunium sorption is due to the relative dominance of the poorly sorbed hydrolyzed species $NpO_2(OH)_{(aq)}$ and the anionic $NpO_2CO_3^-$ species in solution. In contrast, the average K_ds for Np(V) uptake by colloidal hematite, montmorillonite, and silica were 880 mL g^{-1}, 150 mL g^{-1}, and 550 mL g^{-1}, respectively, in Yucca Mountain J-13 water (Efurd *et al.* (1998), probably due to the high surface area of the particles.

Similarly, McCubbin and Leonard (1997) reported neptunium K_ds of 10^3–10^4 mL g^{-1} for particulates in seawater, but the oxidation state was uncertain. Like other tetravalent actinides, Np(IV) has a strong tendency to polymerize and form colloids and is strongly sorbed. Neptunium(IV) migration is likely to occur as intrinsic colloids or sorbed species on pseudocolloids, and changes in ionic strength are likely to impact mobility mostly through destabilization of colloidal particles. Neptunium(V) intrinsic colloids are not expected at neutral pH (Tanaka *et al.*, 1992) and uptake by carrier colloids occurs by ion-exchange and surface complexation. Competition for sorption sites between Np(V) species and other ions, especially Ca^{2+} and Mg^{2+}, could be significant (Tanaka and Muraoka, 1999; McCubbin and Leonard, 1997).

Plutonium. Plutonium chemistry is complicated by the fact that it can exist in four oxidations states over an Eh range of −0.6–1.2 V and a pH range of 0–14. In the system $Pu–O_2–H_2O$, four triple points exist (Eh–pH where three oxidation states may coexist) and thus disproportionation reactions can occur in response to radiolysis or changes in Eh, pH, or the concentrations of other chemical species (Langmuir, 1997a). The most important of these reactions are disproportionation of PuO_2^+ to PuO_2^{2+} and Pu^{4+} or disproportionation of plutonium facilitated by humic acid (Guillaumont and Adloff, 1992) and radiolysis (Nitsche *et al.*, 1995).

Langmuir's Eh–pH calculations (1997a) show that in systems containing only Pu, H_2O, and carbonate/bicarbonate (10^{-2} M), the stability field for the species Pu^{4+} is nearly nonexistent (limited to high Eh and very low pH) but the field for $Pu(OH)_{4(aq)}$ is extensive at pH > 5 and Eh < 0.5. Carbonato complexes of Pu(VI) and Pu(V) are important at pH > 5 and higher Eh. Plutonium solubilities are generally low over most environmental conditions ($<10^{-8}$ M); the solubility fields for $PuO_{2(cr)}$ and $Pu(OH)_{4(am)}$ cover the Eh–pH field over the pH > 5 at all Eh and substantial portions of the Eh–pH field at lower pH where Eh > 0.5 V. As discussed below, the system is different when other ligands, cations, and higher concentrations of carbonate are present.

Runde *et al.* (2002a) compiled an internally consistent database to calculate solubility and speciation of plutonium in more complex low-ionic-strength waters. A specific interaction model (Grenthe *et al.*, 1992) was used for ionic strength corrections. The reader is referred to that work for details of the data sources and methods used for extrapolation and interpolation. Where reliable data for plutonium species were unavailable, thermodynamic constants were estimated from data for analogous americium, curium, uranium, and neptunium species. The most important solution species of plutonium are the aqueous ions, hydroxides, carbonates, and fluoride complexes. Important solids include oxides and hydroxides ($Pu(OH)_3$, PuO_2, $PuO_2{\cdot}nH_2O$ [or $Pu(OH)_4$], PuO_2OH, $PuO_2(OH)_2$), and the carbonate PuO_2CO_3. The dominant solid phases and species are shown in an Eh–pH diagram for J-13 water variants in Figure 8. Note that in this system, the only triple point occurs at a pH of 2.4 where species in the IV, V, and VI oxidation states are calculated to be in equilibrium. In reference J-13 water (pH = 7, Eh = 0.43 V), $Pu(OH)_{4(aq)}$ dominates solution speciation, and $Pu(OH)_{4(s)}$ is the solubility-limiting phase. Under certain environments affected by interactions of groundwater and nuclear waste forms, Pu(V) or Pu(VI) could be produced by radiolysis or Pu(III) species could be produced by reduction.

Runde *et al.* (2002a) demonstrate that significant changes in plutonium solubility can occur due to the formation of Pu(V) and Pu(VI) species at pH > 6 or due to the formation of Pu(III) species at pH < 6. Using either $Pu(OH)_{4(s)}$ or the more crystalline $PuO_{2(s)}$ as the solubility controlling solids, Runde *et al.* (2002a) calculated plutonium solubilities over ranges of pH (3–10), Eh (0–0.6 V), and total carbonate concentration (0.1–2.8 mmol). For conditions typical of groundwater environments (pH 6–9 and Eh 0.05 – 0.45 V), $Pu(OH)_{4(aq)}$ is the dominant aqueous species. Under alkaline conditions, solubility increases with Eh due to formation

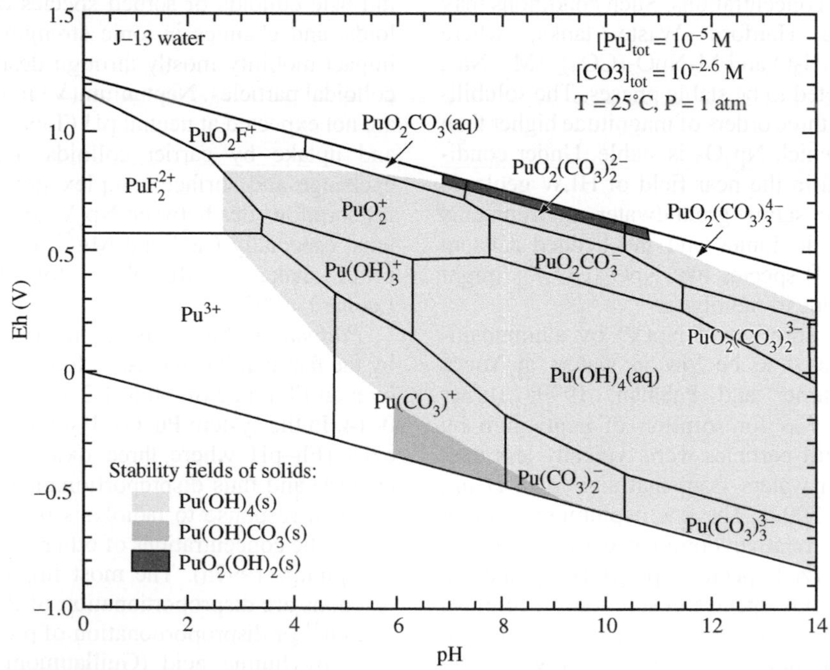

Figure 8 Eh–pH diagram showing dominant plutonium solid phases and species for J-13 water variants at 25 °C as calculated by Runde *et al.* (2002a). Solid lines indicate dominant solution species and shaded areas indicate solids supersaturated in 10^{-5} M Pu solutions. Precipitation of PuO_2 was suppressed in the calculations; see Runde *et al.* (2002a) for details. (reproduced by permission of Elsevier from *Appl. Geochem.*, **2002**, *17*, 844).

of Pu(V) and Pu(VI) solution species. At pH > 8 and Eh > 0.4 V, carbonate species are dominant. At pH < 7, the solubility increases with decreasing Eh due to the stability of $Pu(OH)_3^+$. The calculated solubilities and speciation are sensitive to changes in both Eh and pH. For systems in which $Pu(OH)_{4(s)}$ is the stable solid, they ranged from $>10^{-1}$ M at pH = 4 and Eh = 0 V to 10^{-11} M at pH > 5 and Eh < 0.4 V. Calculated solubilities were about four orders of magnitude lower when $PuO_{2(s)}$ was the stable solid.

Experimentally measured solubilities over this range of solution compositions are typically two orders of magnitude higher than calculated values (Runde *et al.*, 2002a), presumably due to the presence of Pu(IV) colloids (Capdevila and Vitorge, 1998; Efurd *et al.*, 1998; Knopp *et al.*, 1999). In experimental studies in J-13 water, at ambient temperature, plutonium solubility decreased from 5×10^{-8} M at pH = 6, to 9×10^{-9} M at pH = 9 (Efurd *et al.*, 1998).

Because of the presence of Fe^0 and Fe(II), Pu(VI) is not expected to be stable under WIPP repository conditions. Plutonium(IV) is expected to be the dominant oxidation state, although Pu(III) was also considered to be a possibility in the WIPP CCA. In the WIPP performance assessment speciation and solubility calculations, Th(IV) was used as an analogue for Pu(IV) (US DOE, 1996). Wall *et al.* (2002) evaluated the appropriateness of the analogy and found that this assumption was highly conservative and that predicted solubilities for Pu(IV) in Salado and Castile brines were 10–11 orders of magnitude lower than those for Th(IV). Similarly, Am(III) was used to estimate the solubility of Pu(III).

Studies of the sorption of plutonium are complicated by the high redox reactivity of plutonium. Sorption of Pu(V) by pure aluminosilicates and oxyhydroxide phases is usually characterized by initial rapid uptake followed by slow irreversible sorption and may represent a reductive uptake mechanism catalyzed by the electrical double layer of the mineral surface (Turner *et al.*, 1998; Runde *et al.*, 2002a). In Yucca Mountain waters, the K_d ranges for Pu(V) uptake by hematite, montmorillonite, and silica colloids were 4.9×10^3 mL g^{-1} to 1.8×10^5 mL g^{-1}; 5.8×10^3 mL $^{-1}$; and 8.1×10^3 mL g^{-1}, respectively. These are much higher than those observed for Np(V) in the same waters as described previously. High surface redox reactivity for plutonium and possible disproportionation of Pu(V) to Pu(VI) and Pu(IV) were observed in sorption studies using goethite by Keeney-Kennicutt and Morse (1985) and Sanchez *et al.* (1985). Desorption by plutonium was typically less from hematite than from aluminosilicates in studies with J-13 water described by Runde *et al.* (2002a).

17.3.3 Other Topics

17.3.3.1 Colloids

Introduction. Colloidal suspensions are defined as suspensions of particles with a mean diameter less than 0.45 μm, or a size range from 1 nm to 1 μm. They represent potentially important transport vectors for highly insoluble or strongly sorbing radionuclides in the environment. Colloids are important in both experimental systems and natural settings. In the former, unrecognized presence of colloids may lead to overestimation of the solubility and underestimation of the sorption of radionuclides if they are included in the estimation of the concentration of radionuclide solution species. In natural systems, they may provide an important transport mechanism for radionuclides not filtered out by the host rock. In fractured rock, local transport of radionuclides by colloids may be important. Useful reviews of the behavior of colloids in natural systems and their potential role in transporting contaminants include those of Moulin and Ouzounian (1992), Ryan and Elimelech (1996), Kretzschmar *et al.* (1999), and Honeyman and Ranville (2002).

Two types of colloids are recognized in the literature. Intrinsic colloids (also called "true" colloids, type I colloids, precipitation colloids, or "Eigencolloids") consist of radioelements with very low solubility limits. Carrier colloids (also known as "pseudocolloids," type II colloids or "Fremdkolloides") consist of mineral or organic phases (in natural waters primarily organic complexes, silicates and oxides) to which radionuclides are sorbed. Both sparingly soluble and very soluble radionuclides can be associated with this type of colloid. In addition, radionuclides can be associated with microbial cells and be transported as biocolloids.

Natural carrier colloids exist in most groundwaters; they include mineral particles, alteration products of mineral coatings, humic substances, and bacteria. In nuclear waste repositories, carrier colloids will be produced by degradation of engineered barrier materials and waste components: iron-based waste package materials can produce iron oxyhydroxide colloids, degradation of bentonite backfills can produce clay colloids, and alteration of HLW glass can produce a variety of silicate particulates. Intrinsic colloids potentially could be produced by direct degradation of the nuclear waste or by remobilization of precipitated actinide compounds (Avogadro and de Marsily, 1984; Bates *et al.*, 1992; Kim, 1994).

Two different processes could be important for the initiation of radionuclide transport by carrier colloids: (i) reversible sorption of radionuclides from solution onto pre-existing colloids and (ii) detachment of colloids from the host rock with high concentrations of previously sorbed irreversibly bound colloids. In both cases, radionuclides that sorb strongly to the rock matrix and would normally migrate very slowly will travel at a rapid rate while they are bound to colloids.

Naturally occurring colloids. Degueldre *et al.* (2000) and Honeyman and Ranville (2002) summarize modern techniques used to sample colloids from groundwater and to characterize particle concentration and size distributions. Naturally occurring colloids and radionuclide-colloid associations have been characterized at several natural analogue sites for nuclear waste repositories. These include: the Cigar Lake Uranium deposit in altered sandstone (Vilks *et al.*, 1993); the altered schist at the Koongarra Uranium deposit (Payne *et al.*, 1992); altered volcanic rock sites in Pocos de Caldas, Brazil (Miekeley *et al.*, 1991); shallow freshwater aquifers above the salt-hosted Gorleben repository test site in Germany (Dearlove *et al.*, 1991); the Grimsel test site in the Swiss Alps (Degueldre *et al.*, 1989); the Whiteshell Research area in fractured granite in Canada (Vilks *et al.*, 1991); the El Borrocal site in weathered fractured granite near Madrid, Spain (Gomez *et al.*, 1992); and 24 springs and wells near or within the Nevada Test Site (Kingston and Whitbeck, 1991). Major international studies of the occurrence of natural colloids and their potential importance to the European nuclear waste disposal program were carried out by the MIRAGE 2 (Migration of Radionuclides in the Geosphere) project and the Complex Colloid Group of the Commission of European Communities; these are reviewed in Moulin and Ouzounian (1992).

Although locally and globally there are wide variations in colloid concentration and size distribution, several general trends can be observed. Many of the observed particle concentrations fall within the range 0.01–5 mg L^{-1}; however, concentrations of >200 mg L^{-1} have been observed. There is an inverse correlation between particle concentration and particle size. Degueldre (1997) summarized the occurrence of colloids in groundwater from 17 different sites. In a marl aquifer near a proposed Swiss repository site for low-level nuclear waste, the concentration was found to be independent of flow rate, and colloid generation was caused only by resuspension/detachment of the rock clay fraction. Degueldre *et al.* (2000) expanded the scope of that study and found that colloid stability can be parsimoniously described as a function of groundwater chemistry while colloid composition is a function of rock composition. They found that at certain ionic strengths, the concentration of colloids was inversely correlated to the concentration of alkali metals and alkaline earth elements (below about 10^{-4} M and 10^{-2} M, respectively). Large concentrations of organics and the process of water mixing enhance colloid stability and concentration.

Experimental studies. Sorption of radionuclides by colloids is affected by the same solution composition parameters discussed in the previous section on sorption processes. The important parameters include pH, redox conditions, the concentrations of competing cations such as Mg^{2+} and K^+, and the concentrations of organic ligands and carbonate. The high surface area of colloids leads to relatively high uptake of radionuclides compared to the rock matrix. This means that a substantial fraction of mobile radionuclides could be associated with carrier colloids in some systems. The association of radionuclides with naturally occurring colloids and studies of radionuclide uptake by colloids in laboratory systems give some indication of the potential importance of colloid-facilitated radionuclide transport in the environment as discussed below.

For example, Kim (1994) summarizes evidence for strong sorption of americium by silica and alumina colloids in simple $NaClO_4$ solutions ($K_d > 10^4$ at pH = 8 and is independent of ionic strength, temperature, and concentrations of both americium and colloids). He also reports that significant sorption of Np(V) by alumina colloids occurs at pH > 7 at high colloid concentrations (>100 ppm) under the same conditions. Lieser *et al.* (1990) studied partitioning of strontium, caesium, thorium, and actinium between molecular (<0.002 μm) and particle-bound (0.002 μm to >0.45 μm) fractions in batch systems at low concentrations (several orders of magnitude below estimated solubility limits). Sediment–water systems comprised several particle-size fractions and natural waters of different salinities from the aquifers above the Gorleben salt dome were examined. In the groundwaters, colloids consisted primarily of clays, amorphous silica, and iron hydroxide. Appreciable fractions of the total concentrations of the cations in the groundwaters was associated with the large (>0.45 μm) and fine (0.002–0.45 μm) particles (in the order Ac = Th > Cs > Sr). When the groundwaters were passed through columns filled with sediments associated with the groundwaters, most of the caesium (99%) was retained (presumably by ion exchange) and the eluted caesium was primarily associated with fine-grained pseudocolloids. About 99% of the actinium and thorium were retained in the columns (presumably by chemisorption) and the colloidal fraction of the radionuclides dominated the effluent. In contrast, retention of strontium in the columns was less effective due to its lower potential for ion exchange.

Runde *et al.* (2002a) characterized plutonium precipitates and examined neptunium and plutonium uptake by inorganic colloidal particulates in J-13 water from the Yucca Mountain site.

Plutonium solubilities determined experimentally at pH 6, 7, and 8.5 were about two orders of magnitude higher than those calculated using the existing thermodynamic database, indicating the influence of colloidal Pu(IV) species. Solid-phase characterization using X-ray diffraction revealed primarily Pu(IV) in all precipitates formed at pH 6, 7, and 8.5. As discussed previously, hematite, montmorillonite, and silica colloids were used for uptake experiments with ^{239}Pu(V) and ^{237}Np(V). The capacity of hematite to sorb plutonium significantly exceeded that of montmorillonite and silica. A low desorption rate was indicative of highly stable plutonium–hematite colloids, which may facilitate plutonium transport to the accessible environment. Plutonium(V) uptake on all mineral phases was far greater than Np(V) uptake, suggesting that a potential Pu(V)–Pu(IV) reductive sorption process was involved.

Microbial and humic colloids. The transport of radionuclides and metals adsorbed to microbes has been considered by a number of researchers including McCarthy and Zachara (1989), Han and Lee (1997), and Gillow *et al.* (2000). Because of their small size (<10 μm diameter), these colloids can be transported rapidly through fractured media and either filtered out or be transported through porous media. Microbes can sorb to geologic media, thereby retarding transport. Alternatively, under conditions of low nutrient concentrations, the microbes can reduce their size and adhesion capabilities and become more easily transported. Studies performed in support of WIPP compliance certification indicated that, under relevant redox conditions, microbially bound actinides contributed significantly to the concentration of mobile actinides in WIPP brines (Strietelmeyer *et al.*, 1999; Gillow *et al.*, 2000). For performance assessment calculations, the concentration of each actinide sorbed onto microbial colloids was estimated to be 3.1, 0.0021, 12.0, 0.3, and 3.6 times the dissolved concentration for thorium, uranium, neptunium, plutonium, and americium, respectively (US DOE, 1996).

Evidence for strong sorption of actinides and fission products by humic substances, both in dilute and high-ionic-strength media, is provided by experimental studies and thermodynamic calculations. Humic colloids are stable and occur in concentrations up to 0.4 g L^{-1} in dilute, shallow groundwaters overlying the Gorleben test site in Germany (Buckau *et al.*, 2000). Humic substances have experimentally been shown to strongly complex the trivalent actinides (Czerwinski *et al.*, 1996; Artinger *et al.*, 1998; Morgenstern *et al.*, 2000): Th(IV) (Nash and Choppin, 1980); U(VI) and probably U(IV) (Czerwinski *et al.*, 1994; Zeh *et al.*, 1997); and Np(V) (Kim and Sekine, 1991; Rao and Choppin, 1995; Marquardt *et al.*, 1996; Marquardt and Kim, 1998) at mildly acidic to neutral pH. Under basic conditions, actinide–humic substance complexation is strongly a function of the carbonate concentration, because carbonate competes effectively with the humic acid as a ligand (Zeh *et al.*, 1997; Unsworth *et al.*, 2002). Little data are available for tetravalent actinides, but Tipping (1993) suggests, based on thermodynamic modeling, that these should be even more strongly complexed by humic substances than other oxidation states.

Several studies have shown that actinide–humic acid complexes are thermodynamically stable in high-ionic-strength solutions (Czerwinski *et al.*, 1996; Marquardt *et al.*, 1996; Labonne-Wall *et al.*, 1999). However, destabilization of humic colloids at high ionic strength (Buckau *et al.*, 2000) and competition for humic acid sites by divalent metal cations (Tipping, 1993; Marquardt *et al.*, 1996) may limit the importance of colloidal transport of actinides in brines. In the WIPP performance assessment, the estimated contribution of actinide sorbed onto humic substances to the total mobile concentration was >0.01 times the dissolved fraction for An(V), 0.1–1.4 times the dissolved fraction for An(III) and An(VI), and 6.3 times the dissolved fraction for An(IV) (US DOE, 1996).

Transport of radionuclides by colloids. Several numerical models have been developed to assess the potential magnitude of colloidal-facilitated transport of radionuclides compared to the transport of dissolved species. Vilks *et al.* (1998) proposed a simple modification to the standard equation for the retardation factor. The equation applies to the ideal case where colloids are not trapped by the rock matrix and the composition of the colloids and the rock matrix are the same:

$$R_{F,\text{eff}} = 1 + \frac{(1-\phi)\rho K_\text{d}}{\phi(1 + CFK_\text{d})} \tag{8}$$

where $R_{F,\text{eff}}$ is the effective retardation factor for radionuclides, including the effect of reversible sorption onto and transport by colloids. In the equation, ρ is the bulk density of the porous medium, ϕ is the porosity, K_d is the distribution coefficient for both the colloid and the rock matrix (mL g^{-1}), and C is the colloid concentration (mg L^{-1}). F is defined as

$$F = (A'_\text{colloid}/A'_{IP}) \tag{9}$$

where A'_colloid and A'_IP are the specific surface areas of the colloid and rock matrix, respectively.

When the above conditions are not met, more complex models that account for colloid generation, irreversible sorption, differences between the sorptive capacity of colloids and rock matrix (i.e., different K_ds), and colloid filtration should be used. Avogadro and de Marsily (1984) developed a simple model involving colloid filtration under field conditions. Nuttall *et al.* (1991) developed a 2D population-balance model for radiocolloid transport that includes production and filtration of colloids under saturated and unsaturated conditions. van der Lee *et al.* (1992) and Smith and Degueldre (1993) developed numerical models for colloid-facilitated transport through fractured media that incorporate a finite number of sorption sites, Langmuir isotherms, and irreversible sorption. Finally, the incorporation of colloid transport in systems with slow desorption, contrasts between the sorptive capacity of colloids and the rock matrix, and variable K_ds as functions of solution composition are simulated by the LEHGC reactive transport model as described by Yeh *et al.* (1995) and Honeyman and Ranville (2002).

There is considerable debate concerning the potential importance of colloid-facilitated transport of radionuclides for the design and performance assessment of nuclear waste repositories and for risk assessments of radioactively contaminated sites. Honeyman and Ranville (2002) develop a framework to determine the conditions under which colloid-facilitated contaminant transport will be important compared to the transport of solution species. They conclude that such conditions will be relatively rare in the environment. In contrast, Penrose *et al.* (1990) and Nuttall *et al.* (1991) suggest that colloidal transport of radionuclides in the unsaturated zone can be important. They describe field evidence, laboratory results, and computer simulations that suggest that colloidal transport of strongly sorbing actinides such as plutonium and americium is potentially significant in the unsaturated zone and in shallow aquifers near Los Alamos, New Mexico. Similarly, Kersting *et al.* (1999) provide evidence that measurable amounts of plutonium and perhaps cobalt, europium, and caesium produced by nuclear weapons tests (1956–1992) at the Nevada Test site have been transported at least 1.3 km from the blast sites by colloids. They argue that models that do not include colloid-facilitated transport may significantly underestimate the extent of radionuclide migration. In contrast, Vilks (1994) proposes that colloids do not have to be considered in the safety assessment for the Canadian repository in granite. He argues that the clay-based buffer to be used in the repository will filter out any colloids produced by degradation of the waste package. In addition, the concentration of naturally occurring colloids is too low to provide a substantial transport vector for radionuclides that escape to the far field of the repository.

Four types of colloids were considered in the WIPP program: intrinsic actinide colloids, mineral colloids, microbes, and humic acid colloids (US DOE, 1996). Intrinsic actinide colloids, consisting of polymerized hydrated actinide hydroxides, are not stable in the neutral to moderately basic pH conditions expected in the WIPP, and were assumed not to contribute to the total actinide concentrations in solution. Mineral colloids are destabilized and tend to flocculate in the high-ionic-strength WIPP brines (Kelly *et al.*, 1999). In the performance assessment calculations for the WIPP, a highly conservative value of 2.6×10^{-8} mol actinide per liter, for each actinide, was assumed to be bound to mineral colloids and to contribute to the mobile fraction. Actinides sorbed onto microbes and humic acids were estimated to contribute significantly to the concentration of mobile actinides in WIPP brines as discussed above (Section 17.3.2.2).

Contardi *et al.* (2001) used an SCM to examine the potential effect of colloidal transport on the effective retardation factors for americium, thorium, uranium, neptunium, and plutonium in waters from the proposed repository site at Yucca Mountain. They found that colloidal transport reduced the effective retardation of strongly sorbed radionuclides such as americium and thorium by several orders of magnitude compared to simulations in which such transport was ignored. Uranium, neptunium, and Pu(V) are less strongly sorbed by colloids and therefore were relative unaffected by colloidal transport. They also described performance assessment calculations of the effect of colloid-facilitated radionuclide transport on the peak mean annual total effective dose equivalent (TEDE) from the proposed nuclear water repository. The colloid transport simulations showed no increase in the TEDE within a compliance period of 10^4 yr; however, at longer simulation times, as the waste container failures increased, the TEDE from the colloid models are up to 60 times that of the base (noncolloid) case. Such simulations are strongly dependent on scenario assumptions and can be used to provide conservative estimates of the potential importance of colloids for radionuclide transport. They are also useful in demonstrating the relative importance of processes included in the performance assessment models.

17.3.3.2 Microbe–actinide interactions

In addition to possible transport of radionuclides by microbial colloids, microbe–actinide chemical interactions are important for the genesis of uranium ore bodies, dissolution of radioactive waste, and remediation of contaminated sites. Chapelle (1993) provides a recent comprehensive treatment of microbial growth, metabolism, and ecology for geoscience applications. Suzuki and Banfield (1999) and Abdelouas *et al.* (1999) provide well-documented overviews of geomicrobiology of uranium with discussion of applications to environmental transport and remediation of sites contaminated with uranium and actinides. Microbial–uranium interactions have been studied for technological applications such as bioleaching, which is an important method for uranium extraction (Bertheolet, 1997). *In situ* stabilization of uranium plumes by microbial reduction is an important method for remediation (Barton *et al.*, 1996). Microbial ecology in high uranium environments has been studied in uranium mill tail wastes, and the effects of microbes on the stability of radioactive wastes buried in geological repositories are discussed by Francis (1994) and Pedersen (1996).

Microbes can control the local geochemical environment of actinides and affect their solubility and transport. Francis *et al.* (1991) report that oxidation is the predominant mechanism of dissolution of UO_2 from uranium ores. The dominant oxidant is not molecular oxygen but Fe(III) produced by oxidation of Fe(II) in pyrite in the ore by the bacteria *Thiobacillus ferroxidans*. The Fe(III) oxidizes the UO_2 to UO_2^{2+}. The rate of bacterial catalysis is a function of a number of environmental parameters including temperature, pH, TDS, f_{O_2}, and other factors important to microbial ecology. The oxidation rate of pyrite may be increased by five to six orders of magnitude due to the catalytic activity of microbes such as *Thiobacillus ferroxidans* (Abdelouas *et al.*, 1999).

Suzuki and Banfield (1999) classify methods of microbial uranium accumulation as either *metabolism dependent* or *metabolism independent*. The former consists of precipitation or complexation with metabolically produced ligands, processes induced by active cellular pumping of metals, or enzyme-mediated changes in redox state. Examples include precipitation of uranyl phosphates due the activity of enzymes such as phosphatases, formation of chelating agents in response to metal stress, and precipitation of uraninite through enzymatic uranium reduction.

Metabolism-independent processes involve physicochemical interactions between ionic actinide species and charged sites in microorganisms; these can occur with whole living cells or cell fragments. Uranium can be accumulated in the cells by passive transport mechanisms across the cell membrane or by biosorption, a term that includes nondirected processes such as, absorption, ion exchange, or precipitation. Suzuki and Banfield (1999) describe the effects of pH and concentrations of other cations and anions on uranium uptake by a variety of organisms. Uptake of uranium by microbes can be described using the techniques and formalisms used to analyze the sorption of metals by metal hydroxide surfaces such as the Freundlich, Langmuir, and surface-complexation sorption models (Fein *et al.*, 1997; Fowel and Fein, 2000).

A large number of species of bacteria, algae, lichen, and fungi have been shown to accumulate high levels of uranium through these processes. Suzuki and Banfield provide examples of organisms, whose uranium uptake capacities range from approximately $50\,mg\,U\,g^{-1}$ to $500\,mg\,U\,g^{-1}$ dry cell weight. Maximum uptake occurs from pH 4–5. Microorganisms can develop resistance to the chemical and radioactive effects of actinides through genetic adaptation. In contaminated environments such as uranium mill tailings and mines, uranium accumulation levels exceed those observed in laboratory experiments with normal strains of bacteria. Suzuki and Banfield (1999) describe several mechanisms used by microbes to detoxify uranium.

Suzuki and Banfield (1999) discuss the similarities between the uranium–microbe interactions and transuranic–microbe interactions. Macaskie (1991) notes that it is possible to extrapolate the data for microbial uranium accumulation to other actinides. Hodge *et al.* (1973) observe that the biological behavior of uranium, thorium, and plutonium resemble that of ferric iron. Microbes can also affect the speciation and transport of multivalent fission products. For example, Fe^{3+}-reducing bacteria and sulfate-reducing bacteria can reduce soluble pertechnetate to insoluble Tc(IV), as discussed by Lloyd *et al.* (1997). For additional information about these topics, the reader is referred to the references cites above. Applications of these principles are described in the section on bioremediation later in this chapter.

17.4 FIELD STUDIES OF RADIONUCLIDE BEHAVIOR

17.4.1 Introduction

Studies of field sites establish the link between theoretical calculations, laboratory studies, and the behavior of radionuclides in the environment. The field observations may complement or in some cases conflict with information obtained in laboratory studies. Important field sites include the areas of anthropogenic radioactive contamination described in previous sections, such as the areas surrounding the Chernobyl reactor, the Hanford reservation, the Chleyabinsk-65 complex, the Hamr uranium mining district of the Czech Republic, the Nevada test site, releases from the Nagasaki atomic bomb detonation, and the Konegstein mine in Germany. Important sites of natural radioactivity (natural analogues) include the Pena Blanca deposit, the Alligator River Region, Cigar Lake, and the Oklo natural reactor. The Pena Blanca deposit in northern Mexico is an analogue for the proposed HLW repository in the unsaturated tuffs at Yucca Mountain (Pearcy *et al.*, 1994; Murphy, 2000). The Koongarra uranium deposit in the Alligator Rivers Region, Australia, has been widely studied in a coordinated international program (Payne *et al.*, 1992; Duerden *et al.*,

1992; van Maravic and Smellie, 1992; Davis, 2001). The Cigar Lake uranium deposit in altered sandstone in Saskatchewan, Canada is another analogue for nuclear waste repositories located below the groundwater table (Vilks *et al.*, 1993; Bruno *et al.*, 1997; Curtis, 1999). Natural analogue studies have also been carried out in the uranium, thorium, and REE ore bodies in altered volcanic rock at Pocos de Caldas, Brazil (Miekeley *et al.*, 1991) and in a uranium-bearing quartz dike/breccia complex in weathered fractured granite at the El Borrocal site near Madrid, Spain (Gomez *et al.*, 1992). The interested reader is referred to the sources cited above for detailed descriptions of how the geochemical and hydrologic characteristics of those sites are used to validate or calibrate hydrogeochemical models for radionuclide behavior.

The following section provides detailed information concerning the transport of radionuclides associated with two very different field analogues: the Chernobyl reactor accident and the Oklo Natural Reactor. These examples span wide temporal and spatial scales and include the rapid geochemical and physical processes important to nuclear reactor accidents or industrial discharges as well as the slower processes important to the geologic disposal of nuclear waste.

17.4.2 Short-term Behavior of Radionuclides in the Environment—Contamination from the Chernobyl Reactor Accident

On April 26, 1986, an explosion at the Chernobyl Nuclear Power Plant, and the subsequent fire in the graphite reactor core, released ~3% (6–8 Mt) of the total fuel inventory of the reactor core. During the initial explosion, most radionuclide release occurred as fragments of unoxidized uranium dioxide fuel, which were deposited mostly in a plume extending 100 km to the west of the plant. The core fire lasted 10 days and releases were again dominated by fuel particles. However, because the core was exposed, these particles were partially or completely oxidized. For the first four days, high temperatures in the reactor resulted in the release of volatile elements (xenon, krypton, iodine, tellurium, and caesium), much of which was deposited as condensed particles in plumes to the northwest, west, and northeast of the plant, extending as far as Scandinavia. Over the next six days, temperatures in the reactor decreased, and the release of volatile fission products decreased. The lower temperatures (600–1,200 K) favored oxidation of the nuclear fuel; therefore, fuel particles released during this phase were more heavily oxidized. Radionuclides released during this phase were deposited mostly in a southern plume, extending as far south as Greece.

A major part of the radionuclides released at Chernobyl were deposited as "hot particles," either fuel particles with an average median diameter of 2–3 μm or condensed particles. Within the exclusion zone, extending 30 km from the plant, more than 90% of the radioactive contamination was in the form of fuel particles (Kashparov *et al.*, 1999). This included ~80% of the ^{90}Sr and ~50–75% of the ^{137}Cs contamination. Particle size decreased with distance from the Chernobyl plant, and the proportion of condensed particles relative to the fuel particles increased. Other radionuclides released include the fission products ^{134}Cs, ^{144}Ce, ^{125}Sb, ^{106}Ru, ^{103}Ru, and ^{95}Zr, and the neutron activation products ^{110m}Ag and ^{54}Mn (Petropoulos *et al.*, 2001).

Considerable attention has been given to fuel particle behavior in the environment because of their importance to the total radionuclide release from Chernobyl. The transport properties of the radionuclides present in the fuel particles and type of hazard that they represented changed as the particles weathered. Initially, particle (and radionuclide) transport was governed by physical processes. Particles were transported both as aerosols in the atmosphere and as suspended load in runoff and rivers. The primary hazards represented by the particles were inhalation and dermal exposure. Particle α-activities were similar to that of the original nuclear fuel (2.5×10^8 Bq cm^{-3}; Boulyga *et al.* (1999)), while β- and γ-activities were somewhat lower due to loss of volatile fission products. The median fuel particle size released at Chernobyl (2–3 μm) falls within the respirable fraction of aerosols (defined as <7 μm), and such particles are easily resuspended by anthropogenic or natural processes that disturb the soil. Fuel particles are not readily transported downward through the soil column, and soil sampling in the exclusion zone, carried out 10 yr after the accident, showed that the particles were still concentrated in the upper 5 cm (Kashparov *et al.*, 1999). Thirteen years after the accident, agricultural activities in the exclusion zone resulted in local airborne concentrations of hot particles of 50–200 particles m^{-3} (Boulyga *et al.*, 1999).

Radionuclides sequestered in the fuel particles are not immediately available to the biosphere, because the particles generally have a low solubility in water, simulated lung fluids, and HCl solutions (Chamberlain and Dunster, 1958; Oughton *et al.*, 1993; Salbu *et al.*, 1994). Release of biologically important radionuclides such as ^{90}Sr and ^{137}Cs into the biosphere requires weathering and dissolution of the fuel particles. Particle weathering rates vary with several source-related particle characteristics (particle size, oxidation state, and structure) and depend on environmental parameters such as soil pH and redox conditions. Field studies have shown that the fraction of exchangeable ^{90}Sr in soil increases as fuel particles dissolve (Petryaev *et al.*, 1991; Konoplev *et al.*, 1992; Baryakhtar, 1995; Konoplev and Bulgakov, 1999). Kashparov *et al.* (1999) and Konoplev and Bulgakov (1999) used these observations to estimate the fraction of undissolved fuel particles present in soils in the Chernobyl 30 km exclusion zone. They found that the unoxidized fuel particles, deposited from the initial explosion in a plume to the west of Chernobyl, are more resistant to leaching and dissolution than the oxidized fuel particles released during the subsequent fire and deposited to the north, northeast, and south of the plant. Dissolution rates for the unoxidized fuel particles were about one-third those of oxidized particles, with 27–79% remaining undissolved in 1995, while only 2–30% of the oxidized fuel remained. Fuel particle dissolution rates also increased with increasing soil acidity and with decreasing particle size (Kashparov *et al.*, 1999; Konoplev and Bulgakov, 1999).

These results are consistent with laboratory measurements of the effects of pH and oxidation state on the leachability of nuclear fuel particles (Kashparov *et al.*, 2000). The increased leachability of the oxidized fuel particles may be due to: (i) an increased solubility because of the change in oxidation state; (ii) the higher surface area of the highly fractured oxidized particles; or (iii) the diffusion of radionuclides (strontium, caesium) to grain boundaries and particle surfaces during the heating and oxidation process.

Once released from the fuel particles, strontium and caesium are available to the biosphere and can be taken up by plants or can be transferred down the soil profile. At present, the relative rates of each process are not well enough constrained to predict future levels of ^{90}Sr and ^{137}Cs in vegetation in the exclusion zone (Kashparov *et al.*, 1999).

In more distal areas, caesium and strontium concentrations in soils, lakes, and rivers were initially high from direct fallout but have progressively dropped as these elements move downward into the soil profile and are flushed or sedimented out of bodies of water. The estimated ecological half-life for ^{137}Cs in German forest soils is 2.8 ± 0.5 yr for the L horizon and 7.7 ± 4.9 yr for the Ah horizon (Rühm *et al.*, 1996). ^{90}Sr concentrations in the Black Sea had dropped to pre-Chernobyl levels by 1994, and ^{137}Cs is predicted to reach pre-accident levels by 2025–2030 (Kanivets *et al.*, 1999). The main causes of the decreases are radioactive decay and loss through the Bosporus Strait. However, the relative proportion of ^{90}Sr entering the Black Sea as river input is increasing as fuel particles in the major watersheds weather and release sequestered radionuclides. Smith *et al.* (1999, 2000) have shown that caesium removal from lakes and rivers is dominantly by lake outflow and by sedimentation. Caesium is strongly sorbed onto the frayed edge sites of illitic clay minerals, and caesium removal rates from lakes correlate with aqueous K^+ concentrations.

17.4.3 Natural Analogues for the Long-term Behavior of Radionuclides in the Environment—The Oklo Natural Reactor

Naturally occurring uranium deposits have been an important source of information on the long-term behavior of actinides and fission products in the environment. Of special interest are the Oklo and Bangombe deposits of Gabon, which hosted natural fission reactors ~2 Gyr ago (Pourcelot and Gauthier-Lafaye, 1999; Jensen and Ewing, 2001). The Gabon deposits occur in the Francevillian series, a 2.1 Ga sedimentary series consisting of sandstones, conglomerates, black shales, and volcaniclastic sediments. All uranium mineralization occurs in the basal sandstone formation near the upper contact with overlying black shales. The ore deposits have been interpreted as classic uranium roll-front deposits, which formed in oil traps where uranium-rich oxidizing fluids met reducing conditions in the hydrocarbon accumulations. The uranium is present primarily as uraninite, and uranium concentrations in the normal ore vary from 0.1 wt.% to 10 wt.%. In the reactor zones, the sandstone was highly fractured, and the mineralized stockwork ore initially contained up to 20 wt.% uranium.

As redox conditions continued to concentrate uranium in the ore zone, the conditions required to initiate and sustain criticality were achieved. With initiation of criticality, hydrothermal circulation cells formed in the sandstone, resulting in migration of silica and other components out of the reactor. Uraninite was concentrated in the core; uranium grades as high as 80 wt.% have been reported. In some reactors, the volume lost from the core was sufficient to cause slumping and collapse of the overlying beds. The reactor cores are generally 10–50 cm thick and are commonly overlain by a clay-rich hydrothermal gangue, consisting of magnesium chlorite and illite, known as the "reactor clays" or "argile de pile." During criticality and cooling, aluminum-rich chlorites formed in the reactor core (Pourcelot and Gauthier-Lafaye, 1999).

During operation, the reactor cores reached temperatures of 200–450 °C (Brookins, 1990; Pourcelot and Gauthier-Lafaye, 1999), perhaps as high as 1,000 °C (Holliger and Derillers, 1981). The reactors operated for $(1\text{–}8)\times 10^5$ yr (Gauthier-Lafaye *et al.*, 1996) and in the largest, consumed up to 1,800 kg of ^{235}U. Because of this, uranium from the reactor cores is depleted in ^{235}U, with this isotope constituting as little as 0.29% of the total (in "normal" ore, it would be 0.72%).

Analysis of the isotopic concentrations of uranium, thorium, fission products, and their daughters gives information on element mobility during and after criticality. There is evidence for some actinide and fission product migration during criticality, during dolerite dike intrusion, and due to recent supergene weathering. Element retentivity in the Oklo natural reactors has application to nuclear waste disposal, because the uranium oxide reactor core and the reactor clays have been described as analogous to spent nuclear fuel that is embedded in a clay-rich backfill (Menet *et al.*, 1992).

A summary of actinide and fission product behaviors at Oklo is given by Gauthier-Lafaye *et al.* (1996). In general, the core of the reactor consists mainly of uraninite with varying amounts of clay. Grains of metal/metal-oxide/sulfide are present in the uraninite, similar to metal/metal oxide grains found in depleted uranium fuel. As the reactors have seen little oxidation, actinides, and fission products that are compatible with the uraninite structure have largely been retained by the core. Evidence that uranium and plutonium did not migrate during criticality is given by the presence of large amounts of ^{232}Th in the reactor cores. ^{232}Th is formed by decay of ^{236}U and ^{240}Pu, both of which were produced by neutron capture processes during criticality. Thorium is of low abundance in the surrounding rocks, and its presence in the reactor zones is taken to indicate that uranium and plutonium were mostly retained in the reactor core. Plutonium was also retained by clays in the "argile de pile," as is evidenced by enrichments of ^{235}U, the long-lived daughter of ^{239}Pu, in some reactor clay samples (Pourcelot and Gauthier-Lafaye, 1999). Concentrations of ^{209}Bi, the stable daughter of ^{237}Np, suggest that neptunium was also mostly retained by the reactor core.

Many environmentally important fission products have short half-lives; hence their behavior in the reactors is determined by proxy, by examining the distribution of stable daughters. Such parent/daughter pairs include $^{90}Sr/^{90}Zr$, $^{137}Cs/^{137}Ba$, $^{135}Cs/^{135}Ba$, $^{129}I/^{129}Xe$, and $^{99}Tc/^{99}Ru$. The daughters commonly have a significantly different chemistry than the parents and may have migrated during criticality, the 800 Ma dike intrusion, or during supergene weathering. Despite this uncertainty, comparisons between measured concentrations and theoretical fission yields have provided valuable information on fission product mobility (Hidaka *et al.*, 1992). These studies suggest that retention of fission products by the reactors was variable. Gaseous and volatile fission products—xenon, krypton, iodine, cadmium, and caesium—were largely lost from the reactor. Highly mobile alkali metals and alkali earths—rubidium, strontium, and barium—were also lost. Fission products that are compatible with the uraninite crystal structure—the REE, yttrium, neodymium, and zirconium—were largely retained in the uraninite core, the reactor clays, minor phosphate phases, and uranium and zirconium silicate phases (Gauthier-Lafaye *et al.*, 1996). Lighter REE—lanthanum, cerium, and praseodymium—were partially lost from the reactor. Finally, molybdenum, technetium, ruthenium, rhodium, and other metallic elements were retained in the metal/metal oxide inclusions and arsenide/sulfide inclusions in the core, and in the reactor clays (Hidaka *et al.*, 1993; Jensen and Ewing, 2001).

17.5 APPLICATIONS: DEALING WITH RADIONUCLIDE CONTAMINATION IN THE ENVIRONMENT

The previous sections of this chapter have briefly described the nature and locations of the most serious radioactive environmental contamination on the planet and have established the geochemical foundations for understanding the behavior of radionuclides in the environment. Applications of this information to remediating or assessing the risk posed by the contamination is the subject of this section of the review.

17.5.1 Remediation

In many remediation programs, simple excavation of contaminated soil and removal of contaminated groundwater by pumping are the preferred techniques. These techniques may be practical for removal of relatively small volumes of contaminated soils and water; however, after these source terms have been removed, large volumes of soil and water with low but potentially hazardous levels of contamination may remain. For poorly sorbing radionuclides, capture of contaminated water and removal of radionuclides may be possible using permeable reactive barriers and bioremediation. Alternatively, radionuclides could be immobilized in place by injecting agents that lead to reductive precipitation or irreversible sorption.

For strongly sorbing radionuclides, contaminant plumes will move very slowly and likely pose no potential hazards to current populations (Brady *et al.*, 2002). However, regulations may require cleanup of sites to protect present and future populations under a variety of future-use scenarios. In these cases, it may be necessary to use soil-flushing techniques to mobilize the radionuclides and then to collect them. Alternatively, it may be possible to demonstrate that contaminant plumes will not reach populations and that monitoring networks and contingency remedial plans are in place to protect populations if the plume moves more rapidly than predicted. This approach is called monitored natural attenuation (MNA) and is described in a later section.

17.5.1.1 Permeable reactive barriers

Permeable reactive barriers include reactive filter beds containing materials such as zero-valent iron (Fe^0), phosphate rock (apatite), silica sand, organic materials, or combinations of these materials (US EPA, 1999c). The barriers can be installed by digging a trench in the flow path of a contaminated groundwater plume and backfilling with reactive material or by injecting either a suspension of colloidal material or a solution containing a strong reductant (Cantrell *et al.*, 1995, 1997; Abdelouas *et al.*, 1999). The reactive filter material is used to reduce and precipitate the contaminant from solution while allowing the treated water to flow through the reactive bed. In some installations, the drain field can be designed to have removable cells should replacement and disposal of the reactive material be required. A classical funnel and gate arrangement can be used for this purpose. In this system, impermeable subsurface walls are used to direct the flow of the plume through a narrow opening where the reactive material is emplaced. A number of techniques applicable for remediation of radionuclide plumes are described in a collection of articles edited by Looney and Falta (2000) and in publications of the Federal Remediation Technologies Roundtable (e.g., US EPA, 2000a).

The use of Fe^0 to reduce and precipitate uranium out of solution has been shown to be effective by Gu *et al.* (1998) and Fiedor *et al.* (1998). The technique has been deployed in permeable reactive barriers at the Rocky Flats site in Colorado (Abdelaous *et al.*, 1999), the Y-12 Plant near Oak Ridge National Laboratories (Watson *et al.*, 1999), and other DOE sites. In this method, the Fe^0 reduces the U(VI) species to U(IV) aqueous species, which then precipitates as U(IV) solids (Gu *et al.*, 1998; Fiedor *et al.*, 1998). Reduction of U(VI) to U(IV) usually results in the precipitation of poorly crystalline U(IV) (e.g., uraninite, compositions ranging from UO_2 to $UO_{2.25}$) or mixed U(IV)/U(VI) solids (e.g., U_4O_9).

Uranium(VI) readily precipitates in the presence of phosphate to form a number of sparingly soluble U-phosphate phases (U phases, such as saleeite, meta-autunite, and autunite) and also is removed by sorption and co-precipitation in apatite. Several studies have shown that hydroxyapatite is extremely effective at removing heavy metals, uranium, and other radionuclides from solution (Gauglitz *et al.*, 1992; Arey and Seaman, 1999). Apatite was shown to be effective at removing a number of metals including uranium at Fry Canyon, Utah (US EPA, 2000b). Krumhansl *et al.* (2002) reviews the sorptive properties of a number of other materials for backfills around nuclear waste repositories and permeable reactive barriers.

Injection of a reductant such as sodium dithionite creates a reducing zone that may be effective in immobilizing uranium and other redox-active radionuclides. The technique is known as *in situ* redox manipulation (ISRM). It has been shown to be moderately effective for chromium and is proposed for use at the Hanford site for remediation of a uranium groundwater plume (Fruchter *et al.*, 1996).

17.5.1.2 Bioremediation

A number of remediation techniques based on biological processes are in use at contaminated sites. Examples include use of microbes to sequester uranium and phytoremediation of a number of metals. The former method involves reductive reactions by bacteria, particularly those of sulfate reduction (Lovely and Phillips, 1992a) and direct reduction (Lovely and Phillips, 1992b; Truex *et al.*, 1997). Several techniques have been employed to generate high organic loading by growth of plant and algal biomass. Injection of nutrients into the subsurface and subsequent microbial bloom leads to the low redox conditions favorable for reductive reactions, a significant decrease in the solubility and consequently, removal of the metal onto the geomedia.

Abdelouas *et al.* (1999) provide a good review of remediation techniques for uranium mill tailings and groundwater plumes. Biological processes used in bioremediation include biosorption, bioaccumulation, and bioreduction. Biosorption includes uptake of uranium by ion-exchange or surface complexation by living microbes or the cell membranes of dead organisms. In bioaccumulation, the radionuclides are precipitated with enzymatic reactions (Macaskie *et al.* (1996) and Abdelouas *et al.* (1999) and references therein). Bioreduction includes both direct reduction of radionuclides by organisms and indirect reduction. The latter involves creation of reducing conditions by the activity of sulfate and iron-reducing microbes and the subsequent reduction of the radionuclides by produced reductants such as H_2 and H_2S. Abdelouas *et al.* (1999) provide an excellent review of laboratory and field studies of microbes such as various *Desulfovibrio* species that have been shown to be effective in reducing hexavalent uranium by both of these processes.

Phytoremediation has been used to remove uranium and strontium from groundwaters and surface waters. Studies have been conducted on the uptake of heavy metals, uranium, and other radionuclides (Cornish *et al.*, 1995; Abdelouas *et al.* (1999); both the uptake rates and the phytoconcentration of the radionuclides are high. The plants can be harvested and the volume of the residuals minimized by combusting the plant material.

17.5.1.3 Monitored natural attenuation

Natural attenuation encompasses processes that lead to reduction of the mass, toxicity, mobility, or volume of contaminants without human intervention. The US EPA has recently published guidelines for the use of MNA for a variety of contaminated sites (US EPA, 1997). For inorganic constituents, the most potentially important processes include dispersion and immobilization (reversible and irreversible sorption, co-precipitation, and precipitation) (Brady *et al.*, 1998). Studies of remediation options at UMTRA sites (Jove-Colon *et al.*, 2001) and the Hanford Site (Kelley *et al.*, 2002) have addressed the viability of adopting an MNA approach for uranium and strontium, respectively. As discussed below, different approaches are required to establish the viability of MNA for these radioelements.

Laboratory experiments, transport modeling, field data, and engineering cost analysis provide complementary information to be used in an assessment of the viability of an MNA approach for a site. Information from kinetic sorption/desorption experiments, selective extraction experiments, reactive transport modeling, and historical case analyses of plumes at several UMTRA sites can be used to establish a framework for evaluation of MNA for uranium contamination (Brady *et al.*, 1998, 2002; Bryan and Siegel, 1998; Jove-Colon *et al.*, 2001). The results of a recent project conducted at the Hanford 100-N site provided information for evaluation of MNA for a ^{90}Sr plume that has reached the Columbia River (Kelley *et al.*, 2002). The study included strontium sorption–desorption studies, strontium transport and hydrologic modeling of the near-river system, and evaluation of the comparative costs and predicted effectiveness of alternative remediation strategies.

It is likely that it will be easier to gain acceptance for an MNA approach for radionuclides such as ^{90}Sr compared to $^{235/238}$U. This is because ^{90}Sr has a short half-life and uniformly strong sorption, whereas uranium isotopes have very long half-lives and complex sorption behavior. Strontium transport merely needs to be slowed enough to allow radioactive decay to remove the strontium, whereas demonstrating sorption irreversibility might be a key component of an MNA remedy for uranium. MNA may be acceptable for uranium only if it can be shown that an appreciable fraction of the uranium is irreversibly sequestered on mineral sorption sites or physically occluded and cannot be leached out in the foreseeable future. Institutional controls are also important for MNA. Monitoring programs and contingency remediation plans are required as part of an overall MNA strategy (US EPA, 1997; Brady *et al.*, 1998).

17.5.2 Geochemical Models in Risk Assessment

17.5.2.1 Overview

In performance assessment models, simplified process models and sampling techniques are linked to provide a description of the release of radionuclides from idealized source terms, transport through engineered barriers and surrounding geomedia, and finally uptake by potentially exposed populations. The resulting doses are compared to environmental and health regulatory standards to estimate the risk posed by the releases. A basic overview of the process of risk assessment is presented by Fjeld and Compton (1999). Probabilistic performance assessment methods have been developed to provide a basis for evaluation of the risk associated with nuclear waste disposal in geological repositories (Cranwell *et al.*, 1987; Rechard, 1996, 2002; Wilson *et al.*, 2002). Similar approaches are used for LLW disposal and uranium mill tailings (Serne *et al.*, 1990). Development of risk assessment models by the European community is summarized in NEA (1991). The current status of risk assessment programs in several countries was reviewed in a session devoted to performance assessment at the *2001 Materials Research Society Symposium on the Scientific Basis for Nuclear Waste Management* (McGrail and Cragnolino, 2002).

Abstraction of the hydrogeochemical properties of real systems into simple models is required for risk assessment. Heterogeneities in geochemical properties along potential flow paths, uncertainties in or lack of thermodynamic and kinetic parameter values, and the lack of understanding of geochemical processes all necessitate the use of a probabilistic approach to risk assessment. System complexity and limitations in computer technology preclude precise representation of geochemical processes in risk assessment calculations. Uncertainties in properties of the engineered and natural barriers are incorporated into the risk assessment by using ranges and probability distributions for the parameter values (K_ds and maximum aqueous radionuclide concentrations) in Monte Carlo simulations, by regression equations to calculate sorption and solubility limits from sampled geochemical parameter ranges, and by the use of alternative conceptual models. Representation of the probabilistic aspects of geochemical processes in risk assessment is discussed in Siegel *et al.* (1983, 1992), Chen *et al.* (2002), and Turner *et al.* (2002). Simplifications in solubility and sorption models used in performance assessment calculations for the WIPP and the proposed HLW repository at Yucca Mountain, respectively, are described below.

17.5.2.2 Solubility calculations for the waste isolation pilot plant

Performance assessment calculations of actinide speciation and solubility, and of the potential releases that could result if the repository is breached, were carried out as part of the CCA) for the waste isolation pilot plant (WIPP) (US DOE, 1996; US EPA, 1998a,b,c,d). The calculations modeled actinide behavior in a reference Salado brine and a less magnesium-rich brine from the Castile Formation as described previously (see Tables 6 and 8). The performance assessment calculations will be periodically repeated with updated parameter sets as part of site recertification.

Predicted repository conditions placed several constraints on the WIPP actinide speciation and solubility model. These conditions include: (i) high ionic strength, requiring the use of a Pitzer ion interaction model for calculating activity coefficients; (ii) the presence of magnesium oxide backfill, which will buffer P_{CO_2} and pH in the repository; and (iii) the presence of large amounts of iron and organics in the waste, establishing reducing conditions in the repository and constraining the actinides to their lower oxidation states.

The predicted waste inventory for the repository indicates that potentially significant quantities of the organic ligands—acetate, citrate, oxalate, and EDTA—will be present (US DOE, 1996). Actinide interactions with these compounds were not considered in the speciation and solubility modeling, as calculations suggested that they would be mostly complexed by transition metal ions (Fe^{2+}, Ni^{2+}, Cr^{2+}, V^{2+}, and Mn^{2+}) released by corrosion of the steel waste containers and waste components. A thermodynamic model of actinide–ligand interactions appropriate to brines will be included in solubility calculations for WIPP recertification.

Under many experimental conditions, it is difficult to maintain plutonium and some other actinides in a single oxidation state (Choppin, 1999; Neck and Kim, 2001). Depending on pH and solution composition, as many as four plutonium oxidation states may coexist, not necessarily in equilibrium. This leads to uncertainty in the oxidation state(s) present in experimental solutions. For this reason, there is little reliable speciation and solubility data available for Pu(III) and Pu(IV). WIPP solubility models were only developed for Am(III), Th(IV), and Np(V). Results of Am(III) calculations were used, through an oxidation state analogy, to predict the speciation of Pu(III) and to place an upper bound on its solubility. Similarly, results of the Th(IV) calculations were used to predict and bound the speciation and solubilities of Pu(IV), Np(IV), and U(IV). Recently, Wall *et al.* (2002) evaluated the appropriateness of the analogy and found that the predicted behavior of Am(III) was reasonably similar to that of Pu(III), while predicted solubilities for Th(IV) in Salado and Castile brines were 10–11 orders of magnitude higher than those for Pu(IV). Thus, Th(IV) is a highly conservative analogue for Pu(IV).

The An(V) model was developed using Np(V) but was not used for other actinides, because none are expected to be present in the +V oxidation state. Although U(VI) may be present, there were insufficient experimental data to develop an An(VI) model for the WIPP, and solubilities for this oxidation state were estimated from literature data rather than using a Pitzer model.

17.5.2.3 Models for radionuclide sorption at Yucca mountain

Since the 1970s, the US DOE has evaluated the suitability of Yucca Mountain, Nevada, as a potential site for a geologic repository for high-level nuclear waste. The proposed site is ~170 km northwest of Las Vegas, Nevada, and occupies a portion of the Nevada Test Site where nuclear weapons testing has been carried out since approximately 1945. The geochemical setting of the proposed repository site was described in Section 17.3.2.2 above. The results of performance assessment calculations for the proposed repository at Yucca Mountain suggest that the most significant contributors to risk are radionuclides that are highly soluble or poorly sorbing (^{99}Tc, ^{129}I, and ^{237}Np) in the oxidizing, sodium-bicarbonate-rich waters at the site. In addition, other radionuclides such as ^{239}Pu, ^{241}Am, ^{238}U, and ^{230}Th may be important due to colloidal transport or high dose-conversion factors.

Sorption of a radionuclide may vary drastically over postulated flow paths and over time at Yucca Mountain. Changes in mineralogy and transient concentrations of competing and complexing ligands may cause the sorption at any point along the flow path to change. One approach to represent this variability in performance assessment calculations is to sample K_ds for transport equations from a probability distribution based on experimental measurements as discussed above (Siegel *et al.*, 1983; Wilson *et al.*, 2002). Another approach is to calculate a range of K_ds from thermodynamic data for a range of groundwater compositions. Turner and Pabalon (1999) and Turner *et al.* (2002) outline two methods by which surface-complexation models can be used to obtain reasonable bounds on K_ds for stochastic performance assessment calculations for nuclear waste repositories. The authors represented the mineral surfaces of rocks at the Yucca Mountain site as two sorption sites (>SiOH and >AlOH). They used a DLM in the MINTEQA code (Allison *et al.*, 1991) to calculate ranges and distributions of K_ds for Am(III), Th(IV), Np(V), Pu(V), and U(VI) in a suite of groundwater compositions sampled at the site. The contours in calculated K_ds due to spatial variations in hydrochemistry were presented on maps of Yucca Mountain. In an alternate approach, Turner *et al.* (2002) used a simplified SCM to calculate neptunium sorption behavior over a wide range of pH and P_{CO_2}. K_ds can be sampled from such calculated response surfaces for use in simple transport codes in Monte Carlo simulations for performance assessment.

17.6 SUMMARY—CHALLENGES AND FUTURE RESEARCH NEEDS

Disposal of nuclear waste in geological repositories remains a topic of bitter controversy 30 yr after the nuclear waste program was initiated. Public opposition to geologic disposal occurs in all countries with active waste disposal programs. Many people have not accepted geologic disposal of nuclear waste and many environmental groups and some scientists call for monitored retrievable storage until additional information is gathered. The ability of the public to accept the risks associated with disposal and cleanup of nuclear contamination depends, in part, on our ability to predict the behavior of radionuclides in the environment. Although much data and model development have occurred since the 1970s, active research in a number of areas could potentially enhance our ability to predict nuclide migration accurately. These include: (i) better characterization of the chemical interactions between radionuclides and geomedia; (ii) better characterization of the sorptive media along potential flow paths between radionuclide sources and exposed populations; and (iii) cost-effective monitoring of potential radionuclides releases from waste sites. These areas are being pursued by advances in: (i) spectroscopic techniques and molecular simulation models; (ii) geostatistical models and computer simulations of radionuclide transport; (iii) improved geophysical and drilling techniques for characterizing the properties of geomedia; and (iv) improved monitoring technologies for potential radionuclide releases and exposures.

A large body of empirical sorption (K_d) data has been generated since the 1970s. One of the accomplishments of the 1990s has been the widespread awareness of the limitations of much of the data as discussed previously. It has been recognized that many of these data do not describe reversible equilibrium sorption. When precipitation or other mass transfer processes influence measurements of sorption, the resulting K_d will not provide accurate estimates of radionuclide velocities when used in transport equations. In addition, even when the K_d values represent only reversible sorption, they are valid only for the specific conditions of their experiment. Critically evaluated data sets that are useful for performance assessment of waste repositories have been assembled by several workers. It is hoped that when used with sampling schemes in Monte Carlo calculations, the data can be used to provide order-of-magnitude estimates and reasonable confidence intervals for radionuclide migration velocities that reflect uncertainties in radionuclide sorptive properties along flow paths.

Some workers argue that a more fundamental approach to sorption and solubility will lead to more accurate estimates of radionuclide transport behavior. Since the 1970s, much sorption data have been collected within the framework of SCMs. These data can be critically evaluated to determine if processes other than adsorption or ion exchange have occurred. They can also be applied over much wider ranges of solution compositions than empirical K_d measurements. Most of the available surface-complexation data have been collected in simple electrolyte solutions for single mineral phases. Application of SCMs to natural mineral assemblages is difficult due to the existence of multiple sorption sites. New experimental and theoretical methods for determination of the surface-complexation constants would improve our ability to apply the SCMs to natural systems. The resulting capacity to predict radionuclide sorption on natural materials based on fundamental properties could increase confidence in risk assessment and in the design of remediation of contaminated sites.

Cooperative international efforts such as those carried out by the NEA/OECD Thermodynamic Database Project allow sharing of the results of basic research, standardization of techniques for experiments, and establishment of reference values for thermodynamic and kinetic calculations. A comparable effort would be useful for sorption data modeling. A number of alternative SCMs are used in the literature. They are all based on conditional constants that are obtained by fitting equations or curves to experimental data obtained in solutions. The surface-complexation constants from different models cannot be combined. This is because the constants are dependent on the assumed stoichiometry of the surface species, the identity and properties of species present in solution, and properties of the electrical double layer. Recent advances in surface spectroscopy can remove some of the conditional nature of the sorption constants by providing direct information about the stoichiometry of the sorbing species.

Molecular modeling could also be used to constrain the likely stoichiometry of surface species. These techniques would allow establishment of a set of reference surface species. High-quality existing sorption data sets could then be reinterpreted in light of the reference surface species and the chosen SCM. This effort would result in a much larger internally consistent set of thermodynamic data than currently exists.

Prediction of radionuclide transport in the environment is complicated by the heterogeneity of rocks along radionuclide migration paths. Estimates of expected values or ranges of radionuclide discharges at exposed populations can benefit from improvements in computer technology and applications of geostatistical methods to the geochemical properties of the flow field. Surface-complexation constants have been used to calculate sorption ratios for a suite of groundwater compositions and plotted on a map of the Yucca Mountain site. This method provides a framework to assess the range of sorption behavior at the proposed waste site. The approach could be combined with geostatistical simulation of the sorptive properties of the site (site density, surface areas, identity of sorbing sites) based on samples from boreholes. When used as input parameters to reactive transport codes, geostatistical simulation of the compositions of coexisting water and rock could be used to produce multiple realizations of radionuclide transport at a site. These multiple simulations of radionuclide transport will lead to greater confidence that reasonable upper limits for the release of radionuclides to the environment have been calculated.

Improved monitoring techniques that are cheaper and more robust with respect to the environment will allow networks of monitoring wells to be placed between sources of radionuclides such as repositories or disposal sites and potentially exposed populations. This will improve the acceptance of MNA. With improved modeling capabilities and better understanding of radionuclide interactions, public confidence in predictions of the risk associated with radioactive waste management will increase.

This overview of the geochemistry of radioactive contamination in the environment has included a summary of available data and conceptual models, descriptions of experimental and computational methods, and examples of applications of the information to address environmental problems. Historical improvements in our ability to predict the migration of radionuclides in the environment and the hazards they pose to humans and ecosystems have benefited from advances in the field of geochemistry in general. Geochemical conceptual models have progressed from early thermodynamic models through kinetic models, *ab initio* molecular models, and models incorporating the molecular biology of microbes. Future advances in these areas will lead to an improved understanding of radionuclide geochemistry and an improved ability to manage radioactive contamination in the environment.

ACKNOWLEDGMENTS

The authors would like to thank our colleagues who reviewed this document for technical content, including Pat Brady, Carlos Jove-Colon, Louise Criscenti, James Krumhansl, Yifeng Wang, Donald Wall, Lawrence Brush, and Natalie Wall. Amy Rein provided library support; Judy Campbell assisted in technical editing; Mona Aragon drafted several figures. James Leckie (Stanford University) did the calculations that are plotted in Figure 5. A special thanks to Arielle Siegel for helping to create the database used to manage the citation list. This work was supported in part by the US Department of Energy, Office of Science and Technology. Sandia National Laboratories is a multi-program laboratory operated by the Sandia Corporation, a Lockheed Martin Company, for the US Department of Energy Under Contract DE-AC04-94AL85000.

REFERENCES

Aarkrog A., Dahlgaard H., and Nielsen S. P. (1999) Marine radioactivity in the Arctic: a retrospect of environmental studies in Greenland waters with emphasis on transport of Sr-90 and Cs-137 with the East Greenland Current. *Sci. Tot. Environ.* **238**, 143–151.

Abdelouas A., Lutze W., and Nuttall E. (1999) Uranium contamination in the subsurface: characterization and remediation. In *Uranium, Mineralogy, Geochemistry, and the Environment* (eds. P. Burns and R. Finch). Mineralogical Society of America, vol. 38, pp. 433–473.

Allard B. (1983) Actinide chemistry in geologic systems. *Kemia* **10**(2), 97–102.

Allard B., Torstenfelt B., Andersson K., and Rydberg J. (1980) Possible retention of iodine in the ground. In *International Symposium on the Scientific Basis for Waste Management.* Plenum, New York, vol. 2, pp. 673–680.

Allison J., Brown D., and Novo-Gradac K. (1991) *MINTEQA2/PRODEFA2, a Geochemical Assessment Model for Environmental Systems.* Environmental Protection Agency.

AMAP (2002) *Arctic Monitoring and Assessment Program.* http://www. amap/no/.

Ames L. and Rai D. (1978) Radionuclide interactions with soil and rock media: Volume I. Processes influencing radionuclide mobility and retention, element chemistry and geochemistry, conclusions and evaluation. Pacific Northwest National Laboratory.

Andersson K. (1990) *Natural Variability in Deep Groundwater Chemistry and Influence on Transport Properties of Trace Radionuclides.* SKTR 90:17, Swedish Nuclear Power Inspectorate.

Andersson K., Evans S., and Albinsson Y. (1992) Diffusion of radionuclides in sediments—*in situ* studies. *Radiochim. Acta* **58/59**, 321–327.

Andrews J. N., Ford D. J., Hussain N., Trivedi D., and Youngman M. J. (1989) Natural radioelement solution by circulating groundwaters in the Stripa granite. *Geochim. Cosmochim. Acta* **53**, 1791–1802.

Arey J. and Seaman J. (1999) Immobilization of uranium in contaminated sediments by hydroxyapatite addition. *Environ. Sci. Technol.* **33**(2), 337–342.

Arnold T., Zorn T., Zanker H., Berhard G., and Nitsche H. (2001) Sorption behavior of U(VI) on phyllite: experiments and modeling. *J. Contamin. Hydrol.* **47**, 219–231.

Artinger R., Kienzler B., Schussler W., and Kim J. I. (1998) Effects of humic substances on the Am-241 migration in a sandy aquifer: column experiments with Gorleben ground-water/sediment systems. *J. Contamin. Hydrol.* **35**(1–3), 261–275.

ASTM (1987) 24-hour batch-type measurement of contaminant sorption by soils and sediments. In *Annual Book of ASTM Standards Water and Environmental Technology.* Philadelphia, Pennsylvania, vol. 11.04, pp. 163–167.

ATSDR (1990a) *Toxicological Profile for Thorium.* US Department of Heath and Human Services. Public Health Service. Agency for Toxic Substances and Disease Registry.

ATSDR (1990b) *Toxicological Profile for Radium.* US Department of Heath and Human Services. Public Health Service. Agency for Toxic Substances and Disease Registry.

ATSDR (1990c) *Toxicological Profile for Radon.* US Department of Heath and Human Services. Public Health Service. Agency for Toxic Substances and Disease Registry.

ATSDR (1999) *Toxicological Profile for Uranium.* US Department of Heath and Human Services. Public Health Service. Agency for Toxic Substances and Disease Registry.

ATSDR (2001) Appendix D. Overview of basic radiation physics, chemistry and biology. In *Draft Toxicological Profile for Americium—Draft for Public Comment—July 2001.* US Department of Health and Human Services, Public Health Service, Agency for Toxic Substances and Disease Registry, Atlanta, GA, pp. D.1–D.11.

Avogadro A. and de Marsily G. (1984) The role of colloids in nuclear waste disposal. In *Scientific Basis for Nuclear Waste Management VII*, Materials Research Society Proceedings (ed. G. L. McVay). North-Holland, New York, vol. 26, pp. 495–505.

Bargar J. R., Reitmeyer R., Lenhart J. J., and Davis J. A. (2000) Characterization of U(VI)-carbonato ternary complexes on hematite: EXAFS and electrophoretic mobility measurements. *Geochim. Cosmochim. Acta* **64**(16), 2737–2749.

Barnett M. O., Jardine P. M., and Brooks S. C. (2002) U(VI) adsorption to heterogeneous subsurface media: application of a surface complexation model. *Environ. Sci. Technol.* **36**, 937–942.

Barney G. S. (1981a) *Radionuclide Reactions with Groundwater and Basalts from Columbia River Basalt Formations.* Rockwell Hanford Operations.

Barney G. S. (1981b) *Evaluation of Methods for Measurement of Radionuclide Distribution in Groundwater/Rock Systems.* Rockwell Hanford Operations.

Barney G. S., Navratil J. D., and Schulz W. W. (1984) *Geochemical Behavior of Disposed Radioactive Waste.* American Chemical Society Symposium Ser. 246, Washington, DC.

Barton L. L., Choudhury K., Thompson B., Steenhoudt K., and Groffman A. R. (1996) Bacterial reduction of soluble uranium: the first step of *in situ* immobilization of uranium. *Radioact. Waste Manage. Environ. Restor.* **20**, 141–151.

Baryakhtar V. G. (1995) *Chernobyl Catastrophe.* Export Publishing House, Kiev, Russia.

Bates J. K., Bradley J. P., Teetsov A., Bradley C. R., and Buchholtz M. (1992) Colloid formation during waste form reaction: implications for nuclear wasted disposal. *Science* **256**, 649–651.

Bayley S. E., Siegel M. D., Moore M., and Faith S. (1990) *Sandia Sorption Data Management System Version 2 (SSDMS II) User's Manual.* Sandia National Laboratories.

Beck K. M. and Bennett B. G. (2002) Historical overview of atmospheric nuclear weapons testing and estimates of fall out in the continental United States. *Health Phys.* **82**(5), 591–608.

BEIR IV (1988) *Health Risks of Radon and Other Internally Deposited Alpha Emitters.* Committee on the Biological Effects of Ionizing Radiations, National Research Council, National Academy Press.

BEIR V (1988) *The Effects of Exposure to Low Levels of Ionizing Radiation.* Committee on the Biological Effects of Ionizing Radiations, National Research Council, National Academy Press.

Bergman R. (1994) The distribution of radioactive caesium in boreal forest ecosystems. In *Nordic Radioecology, the Transfer of Radionuclides through Nordic Ecosystems to Man: Studies in Environmental Science.* Studies in Environmental Science (ed. H. Dahlgaard). Elsevier, New York, vol. 62, pp. 335–379.

Bernhard G., Geipel G., Brendler V., and Nitsche H. (1998) Uranium speciation in waters of different uranium mining areas. *J. Alloy. Comp.* **271**, 201–205.

Bertetti F., Pabalan R. T., and Almendarez M. (1998) Studies of neptunium^V sorption on quartz, clinoptilolite, montmorillonite, and alumina. In *Adsorption of Metals by Geomedia* (ed. E. Jenne). Academic Press, San Diego, CA, pp. 131–148.

Berthelot D., Leudc L. G., and Ferroni G. D. (1997) Iron-oxidizing autotrophs and acidophilic heterotrophs from uranium mine environments. *Geomicrobiol. J.* **14**, 317–324.

Bertsch P. M. and Hunter D. B. (2001) Applications of synchrotron-based x-ray microprobes. *Chem. Rev.* **101**, 1809–1842.

Bertsch P. M., Hunter D. B., Sutton S. R., Bajt S., and Rivers M. L. (1994) Situ chemical speciation of uranium in soils and sediments by micro x-ray absorption spectroscopy. *Environ. Sci. Technol.* **28**, 980–984.

Bethke C. M. (1997) Modelling transport in reacting geochemical systems. *Comptes Rendus de l'Academie des Sciences. Sciences de la Terre et des Planetes* **324**, 513–528.

Bethke C. M. (1998) *The Geochemist's Workbench Release 3.0.* University of Illinois at Urbana-Champaign.

Bidoglio G., De Plano A., and Righetto L. (1989) Interactions and transport of plutonium-humic acid particles in groundwater environments. In *Scientific Basis for Nuclear Waste Management XII*, Materials Research Society Proceedings (eds. W. Lutze and R. Ewing). Materials Research Society, Pittsburgh, PA, vol. 127, pp. 823–830.

Bird G. W. and Lopato V. J. (1980) Solution interaction of nuclear waste anions with selected geological materials. *Int. Symp. Sci. Basis Waste Manage.*, 419–426.

Bock W. D., Bruhl H., Trapp T., and Winkler A. (1989) Sorption properties of natural sulfides with respect to technetium. *Int. Symp. Sci. Basis Waste Manage. VII*, 973–977.

Bors J., Martens H., and Kühn W. (1991) Sorption studies of radioiodine on soils with special references to soil microbial biomass. *Radiochim. Acta* **52/53**, 317–325.

Botov N. G. (1992) ALWP-67: A little-known big nuclear accident. In *High Level Radioactive Waste Management, Proceedings of the Third International Conference, April 12–16, 1992. Las Vegas, Nevada.* American Society of Civil Engineers, New York, pp. 2331–2338.

Boulyga S. F., Lomonosova E. M., Zhuk J. V., Yaroshevich O. I., Kudrjashov V. P., and Mironov V. P. (1999) Experimental study of radioactive aerosols in the vicinity of the Chernobyl nuclear power plant. *Radiat. Measure.* **30**, 703–707.

Brady P. V., Brady M. V., and Borns D. J. (1998) *Natural Attenuation, CERCLA, RBCA's and the Future of Environmental Remediation.* Lewis Publishers, New York.

Brady P. V., Papenguth H. W., and Kelly J. W. (1999) Metal sorption to dolomite surfaces. *Appl. Geochem.* 14(5), 569–579.

Brady P. V., Jove-Colon C., Carr G., and Huang F. (2002) Soil Radionuclide Plumes. In *Geochemistry of Soil Radionuclides*, SSSA Special Publication Number 59 (eds. P. Zhang and P. Brady). Soil Science Society of America, Madison, Wisconsin, pp. 165–190.

Bronsted J. N. (1922) Studies on solubility: IV. The principals of the specific interactions of ions. *J. Am. Chem. Soc.* **44**, 877–898.

Brookins D. G. (1990) Radionuclide behaviour at the Oklo nuclear reactor, Gabon. *Waste Manage.* **10**, 285–296.

Brown G. E., Jr. and Sturchio N. C. (2002) An overview of synchrotron radiation applications to low temperature geochemistry and environmental science. In *Applications of Synchrotron Radiation in Low Temperature Geochemistry and Environmental Sciences*, Reviews in Mineralogy and Geochemistry (eds. P. A. Fenter, M. L. Rivers, N. C. Sturchio, and S. R. Sutton). Mineralogical Society of America, Washington, DC, vol. 49, pp. 1–33.

Brown G. E., Jr., and Henrich V. E. (1999) Metal oxide surfaces and their interactions with aqueous solutions and microbial organisms. *Chem. Rev.* **99**, 77–174.

Brown G. E., Jr., Henrich Z. E., Casey W. H., Clark D. L., Eggleston C., Felmy A., Goodman D. W., Grätzel M., Maciel G., McCarthy M. I., Nealson K. H., Sverjensky D. A., Toney M. F., and Zachara J. M. (1999) Metal oxide surfaces and their interactions with aqueous solutions and microbial organisms. *Chem. Rev.* **99**, 77–174.

Browning L. and Murphy W. M. (2003) Reactive transport modeling in the geosciences. *Comput. Geosci.* **29**(3), 245–411.

Bruno J., Casas I., Cera E., and Duro L. (1997) Development and application of a model for the long-term alteration of UO_2 spent nuclear fuel: test of equilibrium and kinetic mass transfer models in the Cigar Lake ore deposit. *J. Contamin. Hydrol.* **26**, 19–26.

Bruno J., Cera E., Grive M., Pablo J. D., Sellin P., and Duro L. (2000) Determination and uncertainties of radioelement solubility limits to be used by SKB in the SR 97' performance assessment exercise. *Radiochim. Acta* **88**, 823–828.

Bryan C. R. and Siegel M. D. (1998) Irreversible adsorption of uranium onto iron oxides; a mechanism for natural attenuation at uranium contaminated sites. In *The Eighth Annual West Coast Conference on Contaminated Soils and Groundwater Abstracts and Supplemental Information.* Association for the Environmental Health of Soils, Oxnard, CA, 201pp.

Buckau G., Artinger R., Fritz P., Geyer S., Kim J. I., and Wolf M. (2000) Origin and mobility of humic colloids in the Gorleben aquifer system. *Appl. Geochem.* **15**(2), 171–179.

Bunzl K. and Schimmack W. (1988) Distribution coefficients of radionuclides in the soil: analysis of the field variability. *Radiochim. Acta* **44**(5), 355–360.

Burns P. and Finch R. (1999) In *Uranium: Mineralogy, Geochemistry, and the Environment.* Reviews in Mineralogy 38 (ed. P. H. Ribbe). Mineralogical Society of America, Washington, DC.

Byegård J., Albinsson Y., Skarnemark G., and Skålberg M. (1992) Field and laboratory studies of the reduction and sorption of technetium(VII). *Radiochim. Acta* **58/59**, 239–244.

Caceci M. S. and Choppin G. R. (1983) The first hydrolysis constant of uranium(VI). *Radiochim. Acta* **33**, 207–212.

Campbell J. E., Dillon R. T., Tierney M. S., Davis H. T., McGrath P. E., Pearson F. J. J., Shaw H. R., Helton J. C., and Donath F. A. (1978) *Risk Methodology for Geologic Disposal of Nuclear Waste: Interim Report, NUREG/CR-0458; SAND78-0029.* Sandia National Laboratories, Albuquerque, NM.

Cantrell K. J., Kaplan D. I., and Wietsma T. W. (1995) Zero-valent iron for the *in situ* remediation of selected metals in groundwater. *J. Hazard. Mater.* **42**, 201–212.

Cantrell K. J., Kaplan D. I., and Gilmore T. J. (1997) Injection of colloidal Fe^0 particles in sand with shear-thinning fluids. *J. Environ. Eng.* **123**, 786–791.

Capdevila H. and Vitorge P. (1998) Solubility product of $Pu(OH)_{4(am)}$. *Radiochim. Acta* **82**, 11–16.

Carroll S. A., Bruno J., Petit J. C., and Dran J. C. (1992) Interactions of U(VI), Nd and Th(IV) at the calcite-solution interface. *Radiochim. Acta* **58/59**, 245–252.

Cember H. (1996) *Introduction to Health Physics.* McGraw Hill, New York.

Chamberlain A. C. and Dunster J. (1958) Deposition of radioactivity in north-west England from the accident at Windscale. *Nature* **182**, 629–630.

Chapelle F. H. (1993) *Ground-Water Microbiology and Geochemistry.* Wiley, New York.

Chen Y., Loch A. R., Wolery T. J., Steinborn T. L., Brady P. V., and Stockman C. T. (2002) Solubility evaluation for Yucca Mountain TSPA-SR. In *Scientific Basis for Nuclear Waste Management XXV,* Materials Research Society Symposium Proceedings (eds. B. P. McGrail and G. A. Cragnolino). Materials Research Society, Pittsburgh, PA, vol. 713, pp. 775–782.

Chisholm-Brause C., Conradson S. D., Eller P. G., and Morris D. E. (1992) Changes in U(VI) speciation upon sorption onto montmorillonite from aqueous and organic solutions. In *Scientific Basis for Nuclear Waste Management XVI,* Materials Research Society Symposium Proceedings (eds. C. G. Intcrrante and R. T. Pabalan). Materials Research Society, Pittsburgh, PA, vol. 294, pp. 315–322.

Chisholm-Brause C., Conradson S. D., Buscher C. T., Eller P. G., and Morris D. E. (1994) Speciation of uranyl sorbed at multiple binding sites on montmorillonite. *Geochim. Cosmochim. Acta* **58**(17), 3625–3631.

Choppin G. R. (1999) Utility of oxidation state analogs in the study of plutonium behavior. *Radiochim. Acta* 85, 89–95.

Choppin G. R. and Stout B. E. (1989) Actinide behavior in natural waters. *Sci. Tot. Environ.* **83**, 203–216.

Choppin G. R., Liljenzin J. O., and Rydberg J. (1995) *Radiochemistry and Nuclear Chemistry.* Butterworth-Heinemann Ltd., Oxford.

Ciavatta L. (1980) The specific interactions theory in evaluation ionic equilibria. *Ann. Chim.* **70**(11-1), 551–567.

Cleveland J. M. and Rees T. F. (1981) Characterization of plutonium in Maxey Flats radioactive trench leachates. *Science* **212**, 1506–1509.

Cleveland J. M., Rees T. F., and Nash K. L. (1983a) Plutonium speciation in water from Mono Lake, California. *Science* **222**, 1323–1325.

Cleveland J. M., Rees T. F., and Nash K. L. (1983b) Plutonium speciation in selected basalt, granite, shale, and tuff groundwaters. *Nuclear Technol.* **62**, 298–310.

Cochran T. B., Norris R. S., and Suokko K. L. (1993) Radioactive contamination at Chelyabinsk-65, Russia. *Ann. Rev. Energy Environ.* **18**, 507–528.

Comans R. N. J., Middleburg J. J., Zonderhuis J., Woittiez J. R.W., De Lange G. J., Das H. A., and Van der Weijden C. H. (1989) Mobilization of radiocaesium in pore water in lake sediments. *Nature* **339**(6223), 367–369.

Combes J., Chisholm-Brause D., Brown G. E. J., Parks G., Conradson S. D., Eller P. G., Triay I. R., Hobart D., and Meijer A. (1992) EXAFS spectroscopic study of neptunium (V) sorption at the a-FeOOH/water interface. *Environ. Sci. Technol.* **26**, 376–383.

Contardi J. S., Turner D. R., and Ahn T. M. (2001) Modeling colloid transport for performance assessment. *J. Contamin. Hydrol.* **47**(2–4), 323–333.

Cornish J. E., Goldberg W. C., Levine R. S., and Benemann J. R. (1995) Phytoremediation of soils contaminated with toxic elements and radionuclides. *Bioremediat. Inorg.* **3**, 55–62.

Couture R. A. and Seitz M. G. (1985) Sorption of anions and iodine by iron oxides and kaolinite. *Nuclear Chem. Waste Manage.* **4**, 301–306.

Cranwell R. M., Campbell J. E., Helton J. C., Iman R. L., Longsine D. E., Ortiz N. R., Runkle G. E., and Shortencarier M. J. (1987) *Risk Methodology for Geologic Disposal of Radioactive Waste: Final Report.* Sandia National Laboratories.

Cremers A., Elsen A., De Preter P., and Maes A. (1988) Quantitative analysis of radiocaesium retention in soils. *Nature* **335**, 247–249.

Criscenti L. J., Cygan R. T., Eliassi M., and Jove-Colon C. F. (2002) *Effects of Adsorption Constant Uncertainty on Contaminant Plume Migration.* US Nuclear Regulatory Commission.

Cui D. and Erikson T. (1996) Reactive transport of Sr, Cs, and Tc through a column packed with fracture-filling material. *Radiochim. Acta* **82**, 287–292.

Curtis D. (1999) Nature's uncommon elements: plutonium and technetium. *Geochim. Cosmochim. Acta* 63(2), 275–285.

Czerwinski K., Buckau G., Scherbaum F., and Kim J. I. (1994) Complexation of the uranyl-ion with aquatic humic-acid. *Radiochim. Acta* **65**(2), 111–119.

Czerwinski K., Kim J. I., Rhee D. S., and Buckau G. (1996) Complexation of trivalent actinide ions (Am^{3+}, Cm^{3+}) with humic acid: the effect of ionic strength. *Radiochim. Acta* **72**(4), 179–187.

Cygan R. (2002) Molecular models of radionuclide interaction with soil minerals. In *Geochemistry of Soil Radionuclides* (eds. P. Zhang and P. Brady). Soil Science Society of America, Madison, Wisansin, pp. 87–110.

David F. (1986) Thermodynamic properties of lanthanide and actinide ions in aqueous solution. *J. Less-Common Metals* **121**, 27–42.

Davis J. and Kent D. (1990) Surface complexation modeling in aqueous geochemistry. In *Mineral-Water Interface Chemistry*, Reviews in Mineralogy 23 (eds. M. Hochella and A. White). Mineralogical Society of America, Washington, DC, pp. 177–260.

Davis J. A. (2001) *Surface Complexation Modeling of Uranium (VI) Adsorption on Natural Mineral Assemblages NUREG/CR-6708.* US Nuclear Regulatory Commission, Washington, DC.

Davis J. A. and Leckie J. O. (1978) Surface ionization and complexation at the oxide/water interface: II. Surface properties of amorphous iron oxyhydroxide and adsorption on metal ions. *J. Colloid Interface Sci.* **67**, 90–107.

Davis J. A., Coston J., Kent D., and Fuller C. (1998) Application of the surface complexation concept to complex mineral assemblages. *Environ. Sci. Technol.* **32**, 2820–2828.

Davis J. A., Kohler M., and Payne T. E. (2001) Uranium (VI) aqueous speciation and equilibrium chemistry. In *Surface Complexation Modeling of Uranium (VI) Adsorption on Natural Mineral Assemblages NUREG/CR-6708.* US Nuclear Regulatory Commission, Washington, DC, pp. 11–18.

Davis J. A., Payne T. E., and Waite T. D. (2002) Simulation of the pH and pCO_2 dependence of Uranium(VI) adsorption by a weathered schist with surface complexation models. In *Geochemistry of Soil Radionuclides* (eds. P. Zhang and P. Brady). Soil Science Society of America, pp. 61–86.

Dearlove J. P., Longworth G., Ivanovich M., Kim J. I., Delakowitz B., and Zeh P. (1991) A study of groundwater-colloids and their geochemical interactions with natural radionuclides in Gorleben aquifer systems. *Radiochim. Acta* **52/53**, 83–89.

Degueldre C. (1997) Groundwater colloid properties and their potential influence on radionuclide transport. *Mater. Res. Soc. Symp. Proc.* **465**, 835–846.

Degueldre C., Baeyens B., Goerlich W., Riga J., Verbist J., and Stadelmann P. (1989) Colloids in water from a subsurface fracture in granitic rock, Grimsel test site. *Geochim. Cosmochim. Acta* **53**, 603–610.

Degueldre C., Triay I. R., Kim J. I., Vilks P., Laaksoharju M., and Miekeley N. (2000) Groundwater colloid properties: a global approach. *Appl. Geochem.* **15**, 1043–1051.

Dent A. J., Ramsay J. D. F., and Swanton S. W. (1992) An EXAFS study of uranyl ion in solution and sorbed onto silica and montmorillonite clay colloids. *Coll. Interface Sci.* **150**, 45–60.

Douglas L. A. (1989) Vermiculites. In *Minerals in Soil Environments* (eds. J. B. Dixon and S. B. Week). Soil Science Society of America.

Duerden P., Lever D. A., Sverjensky D., and Townley L. R. (1992) *Alligator Rivers Analogue Project Final Report.* Sandia National Laboratories.

Dzombak D. and Morel F. (1990) *Surface Complexation Modeling: Hydrous Ferric Oxide.* Wiley, New York.

Efurd D. W., Runde W., Banar J. C., Janecky D. R., Kaszuba J. P., Palmer P. D., Roesnsch F. R., and Tait C. D. (1998) Neptunium and plutonium solubilities in a Yucca Mountain groundwater. *Environ. Sci. Technol.* **32**, 3893–3900.

Eisenbud M. (1987) *Environmental Radioactivity from Natural, Industrial, and Military Sources.* Academic Press, New York.

Erickson K. L. (1983) Approximations for adapting porous media radionuclide transport models to analysis of transport in jointed porous rocks. In *Scientific Basis for Nuclear Waster Management VI*, Materials Research Society Symposium Proceedings 15 (ed. D. Brookins). Elsevier, Amsterdam, pp. 473–480.

Erikson T. E., Ndalamba P., Bruno J., and Caceci M. (1992) The solubility of $TcO_2 \cdot nH_2O$ in neutral to alkaline solutions under constant P_{CO_2}. *Radiochim. Acta* **58/59**, 67–70.

Ewing R. (1999) Radioactivity and the 20th century. In *Uranium: Mineralogy, Geochemistry and the Environment*, Reviews in Mineralogy 38 (eds. P. C. Burns and R. J. Finch). Mineralogical Society of America, Washington, DC, pp. 1–22.

Fanghänel T. and Kim J. I. (1998) Spectroscopic Evaluation of Thermodynamics of Trivalent Actinides in Brines. *J. Alloy. Comp.* **271–273**, 728–737.

Fein J. B., Daughney J., Yee N., and Davis T. A. (1997) A chemical equilibrium model for the sorption onto bacterial surfaces. *Geochim. Cosmochim. Acta* **61**, 3319–3328.

Felmy A. R. and Rai D. (1999) Application of Pitzer's equations for modeling aqueous thermodynamics of actinide species in natural waters: a review. *J. Solut. Chem.* **28**(5), 533–553.

Felmy A. R. and Weare J. H. (1986) The prediction of borate mineral equilibria in natural waters: application to Searles Lake, California. *Geochim. Cosmochim. Acta* **50**(12), 2771–2783.

Felmy A. R., Dhanpat R., Schramke J. A., and Ryan J. L. (1989) The solubility of plutonium hydroxide in dilute solution and in high-ionic-strength chloride brines. *Radiochim. Acta* **43**, 29–35.

Felmy A. R., Rai D., and Fulton R. W. (1990) The solubility of $AmOHCO_{3(C)}$ and the aqueous thermodynamics of the system Na^+-Am^{3+}-HCO_3 – -CO_3^2 – -OH – -H_2O. *Radiochim. Acta* **50**(4), 193–204.

Felmy A. R., Rai D., and Mason M. J. (1991) The solubility of hydrous thorium(IV) oxide in chloride media: development of an aqueous ion-interaction model. *Radiochim. Acta* **55**(4), 177–185.

Fenter P. A., Rivers M. L., Sturchio N. C., and Sutton S. R. (2002) *Applications of Synchrotron Radiation in Low-Temperature Geochemistry and Environmental Science*, Reviews in Mineralogy and Geochemistry 49 (eds. J. J. Rosso and P. H. Ribbe). Mineralogical Society of America, Washington, DC.

Fiedor J. N., Bostic W. D., Jarabek R. J., and Farrell J. (1998) Understanding the mechanism of uranium removal from groundwater by zero-valent iron using x-ray photoelectron spectroscopy. *Environ. Sci. Technol.* **32**(10), 1466–1473.

Fjeld R. A. and Compton K. L. (1999) Risk assessment. In *Encyclopedia of Environmental Pollution and Cleanup* (eds. R. A. Meyers and D. K. Dittrick). Wiley, vol. 2, pp. 1450–1473.

Fowel D. A. and Fein J. B. (2000) Experimental measurements of the reversibility of metal adsorption reactions. *Chem. Geol.* **168**(1-2), 27–36.

Francis A. J. (1994) Microbiological treatment of radioactive wastes. In *Chemical Pretreatment of Nuclear Waste for Disposal* (eds. W. W. Schulz and E. P. Horwitz). Plenum, New York, pp. 115–131.

Francis A. J., Dodge C. J., Gillow J. B., and Cline J. E. (1991) Microbial transformations of uranium in wastes. *Radiochim. Acta* **52/53**, 311–316.

Fruchter J. S., Amonette J. E., Cole C. R., Gorby Y. A., Humphrey M. D., Istok J. D., Olsen K. B., Spane F. A., Szecsody J. E., Teel S. S., Vermeul V. R., Williams M. D., and Yabusaki S. B. (1996) *In situ* redox manipulation field injection test report—Hanford 100 H area. Pacific Northwest National Laboratory.

Fuger J. (1992) Thermodynamic properties of actinide aqueous species relevant to geochemical problems. *Radiochim. Acta* **58/59**, 81–91.

Fuger J., Khodakovsky I., Medvedev V., and Navratil J. (1990) *The Chemical Thermodynamics of Actinide Elements and Compounds: Part 12. The Actinide Aqueous Inorganic Complexes.* International Atomic Energy Agency, Vienna, Austria.

Gabriel U., Gaudet J. P., Spandini L., and Charlet L. (1998) Reactive transport of uranyl in a goethite column: an experimental and modeling study. *Chem. Geol.* **151**, 107–128.

Gauglitz R., Holterdorf M., Franke W., and Marx G. (1992) Immobilization of actinides by hydroxylapatite. In *Scientific Basis for Nuclear Waste, Management XV*, Materials Research Society Symposium Proceedings (ed. C. G. Sombret). Materials Research Society, Pittsburgh, PA, vol. 257, pp. 567–573.

Gauthier-Lafaye F., Holliger P., and Blanc P. L. (1996) Natural fission reactors in the Franceville basin, Gabon: a review of the conditions and results of a "critical event" in a geologic system. *Geochim. Cosmochim. Acta* **60**(23), 4831–4852.

Gillow J. B., Dunn M., Francis A. J., Lucero D. A., and Papenguth H. W. (2000) The potential for subterranean microbes in facilitating actinide migration at the Grimsel Test Site and Waste Isolation Pilot Plant. *Radiochim. Acta* **88**(9–11), 769–774.

Girvin D. C., Ames L. L., Schwab A. P., and McGarrah J. E. (1991) Neptunium adsorption on synthetic amorphous iron oxyhydroxide. *J. Colloid Interface Sci.* **141**(1), 67–78.

Glynn P. D. (2003) Modeling Np and Pu transport with a surface complexation model and spatially variant sorption capacities: implications for reactive transport modeling and performance assessments of nuclear waste disposal sites. *Comput. Geosci.* **29**(3), 331–349.

Goldsmith W. A. and Bolch W. E. (1970) Clay slurry sorption of carrier-free radiocations. *J. Sanitary Eng. Div. Am. Soc. Civil Eng.* **96**, 1115–1127.

Gomez P., Turrero M. J., Moulin V., and Magonthier M. C. (1992) Characterization of natural colloids in groundwaters of El Berrocal, Spain. In *Water-Rock Interaction* (eds. Y. K. Kharaka and A. S. Maest). Balkema, Rotterdam, pp. 797–800.

Goyer R. A. and Clarkson T. W. (2001) Toxic effects of metals. In *Casarett and Doull's Toxicology: The Basic Science of Poisons* (ed. C. D. Klaassen). McGraw-Hill, New York, pp. 811–867.

Grenthe I., Fuger J., Konings R. J. M., and Lemire R. J. (1992) *Chemical Thermodynamics of Uranium.* North-Holland, Amsterdam.

Grenthe I., Puigdomenech I., Sandino M. C., and Rand M. H. (1995) Chemical thermodynamics of uranium. In *Chemical Thermodynamics of Americium* (eds. R. J. Silva, G. Bidoglio,M. H. Rand, P. Robouch, H. Wanner, and I. Puigdomenech).Elsevier, Amsterdam, pp. 342–374.

Gu B. and Schultz R. K. (1991) *Anion Retention in Soil: Possible Application to Reduce Migration of Buried T echnetium and Iodine: A Review.* US Nuclear Regulatory Commission, 32pp.

Gu B., Liang L., Dickey M. J., Yin X., and Dai S. (1998) Reductive precipitation of uranium(VI) by zero-valent iron. *Environ. Sci. Technol.* **32**, 3366–3373.

Guggenheim E. A. (1966) *Applications of Statistical Mechanics.* Claredon Press, Oxford.

Guillaumont R. and Adloff J. P. (1992) Behavior of environmental pollution at very low concentration. *Radiochim. Acta* **58**(59), 53–60.

Haines R. I., Owen D. G., and Vandergraaf T. T. (1987) Technetium-iron oxide reactions under anaerobic conditions: a Fourier transform infrared, FTIR study. *Nuclear J. Can.* **1**, 32–37.

Han B. S. and Lee K. J. (1997) The effect of bacterial generation on the transport of radionuclide in porous media. *Ann. Nuclear Energy* **24**(9), 721–734.

Harley N. H. (2001) Toxic effects of radiation and radioactive materials. In *Casarett and Doull's Toxicology*, 6th edn. (ed. C. D. Klaassen). McGraw-Hill, New York, pp. 917–944.

Harvie C. E. and Weare J. H. (1980) The prediction of mineral solubilities in natural waters: The Na–K–Mg–Ca–Cl–SO_4–H_2O system from zero to high concentration at 25 °C. *Geochim. Cosmochim. Acta* **44**(7), 981–997.

Harvie C. E., Möller N., and Weare J. H. (1984) The prediction of mineral solubilities in natural waters: the Na–K–Mg–Ca–H–Cl–SO_4–OH–HCO_3–CO_3–CO_2–H_2O system to high ionic strengths at 25 °C. *Geochim. Cosmochim. Acta* **48**, 723–751.

Helton J. C. and Davis F. J. (2002) *Latin Hypercube Sampling and the Propagation of Uncertainty in Analyses of Complex Systems.* Sandia National Laboratiories.

Hidaka H., Konishi T., and Masuda A. (1992) Reconstruction of cumulative fission yield curve and geochemical behaviors of fissiogenic nuclides in the Oklo natural reactors. *Geochem. J.* **26**, 227–239.

Hidaka H., Shinotsuka K., and Holliger P. (1993) Geochemical behaviour of Tc in the Oklo natural fission reactors. *Radiochim. Acta* **63**, 19–22.

Hobart D. (1990) Actinides in the environment. In *Fifty Years with Transuranium Elements, Robert A. Welch Foundation Conference on Chemical Research 34*, Robert A. Welch Foundation, Houston, TX, pp. 379–436.

Hodge H. C., Stannard J. N., and Hursh J. B. (1973) Uranium, plutonium, and transuranic elements. In *Handbook of Experimental Pharmacology XXXVI.* Spriner, New York, pp. 980–995.

Higgo J. J. W. and Rees L. V. C. (1986) Adsorption of actinides by marine sediments: effect of the sediment/seawater ratio on the measured distribution ratio. *Environ. Sci. Technol.* **20**, 483–490.

Hölgye Z. and Malú M. (2000) Sources, vertical distribution, and migration rates of $^{239,240}Pu$, ^{238}Pu, and ^{137}Cs in grassland soil in three localities of central Bohemia. *J. Environ. Radioact.* **47**, 135–147.

Holliger P. and Devillers C. (1981) Contribution to study of temperature in Oklo fossil reactors by measurement of lutetium isotopic ratio. *Earth Planet. Sci. Lett.* **52**(1), 76–84.

Honeyman B. D. and Ranville J. F. (2002) Colloid properties and their effects on radionuclide transport through soils and groundwater. In *Geochemistry of Soil Radionuclides*, SSSA Special Publication Number 59 (eds. P. Zhang and P. Brady). Soil Science Society of America, Madison, Wisconsin, pp. 131–164.

Hsi C.-K. and Langmuir D. (1985) Adsorption of uranyl onto ferric oxyhydroxides: application of the surface complexation site-binding model. *Geochim. Cosmochim. Acta* **49**, 1931–1941.

Hughes M. N. and Poole R. K. (1989) *Metals and Microorganisms.* Chapman and Hall, London.

Huie Z., Zishu Z., and Lanying Z. (1988) Sorption of radionuclides technetium and iodine on minerals. *Radiochim. Acta* **44/45**, 143–145.

Hunter K. A., Hawke D. J., and Choo L. K. (1988) Equilibrium adsorption of thorium by metal oxides in marine electrolytes. *Geochim. Cosmochim. Acta* **52**, 627–636.

Iman R. L. and Shortencarier M. J. (1984) *A Fortran 77 Program and User's Guide for the Generation of Latin Hypercube and Random Samples for Use with Computer Models, Nureg/Cr-3264, Sand83-2365.* Sandia National Laboratories.

Isaksson M., Erlandsson B., and Mattsson S. (2001) A 10-year study of the ^{137}Cs distribution in soil and a comparison of Cs soil inventory with precipitation-determined deposition. *J. Environ. Radioact.* **55**, 47–59.

Ivanovich M. (1992) *The Phenomenon of Radioactivity.* Oxford University Press.

Jackson R. E. and Inch K. J. (1989) The *in-situ* absorption of Sr-90 in a sand aquifer at the Chalk River nuclear laboratories. *J. Contamin. Hydrol.* **4**, 27–50.

Jacquier P., Meier P., and Ly J. (2001) Adsorption of radioelements on mixtures of minerals-experimental study. *Appl. Geochem.* **16**(1), 85–93.

Jenne E. A. (1998) *Adsorption of Metals by Geomedia: Variables, Mechanisms, and Model Applications.* Academic Press, Richland, Washington, 583pp.

Jensen K. and Ewing R. (2001) The Okelobondo natural fission reactor, southeast Gabon: geology, mineralogy, and retardation of nuclear-reaction products. *Geol. Soc. Am. Bull.* **113**(1), 32–62.

Jove-Colon C. F., Brady P. V., Siegel M. D., and Lindgren E. R. (2001) Historical case analysis of uranium plume attenuation. *Soil Sedim. Contamin.* **10**, 71–115.

Kanivets V. V., Voitsekhovitch O. V., Simov V. G., and Golubeva Z. A. (1999) The post-Chernobyl budget of ^{137}Cs and ^{90}Sr in the Black Sea. *J. Environ. Radioact.* **43**, 121–135.

Kaplan D., Serne R. J., Parker K. E., and Kutnyakov I. V. (2000) Iodide sorption to subsurface sediments and illitic minerals. *Environ. Sci. Technol.* **34**, 399–405.

Kashparov V. A., Lundin S. M., Khomutinin Y. V., Kaminsky S. P., Levchuk S. E., Protsak V. P., Kadygrib A. M., Zvarich S. I., Yoschenko V. I., and Tschiersch J. (2001) Soil contamination with ^{90}Sr in the near zone of the Chernobyl accident. *J. Environ. Radioact.* **56**, 285–298.

Kashparov V. A., Protsak V. P., Ahamdach N., Stammose D., Peres J. M., Yoschenko V. I., and Zvarich S. I. (2000) Dissolution kinetics of particles of irradiated Chernobyl nuclear fuel: influence of pH and oxidation state on the release of radionuclides in the contaminated soil of Chernobyl. *J. Nuclear Mater.* **279**, 225–233.

Kashparov V. A., Oughton D. H., Zvarich S. I., Protsak V. P., and Levchuk S. E. (1999) Kinetics of fuel particle weathering and ^{90}Sr mobility in the Chernobyl 30-km exclusion zone. *Health Phys.* **76**(3), 251–259.

Kaszuba J. P. and Runde W. (1999) The aqueous geochemistry of Np: dynamic control of soluble concentrations with applications to nuclear waste disposal. *Environ. Sci.Technol.* **33**, 4427–4433.

Katz J. J., Seaborg G. T., and Morse L. R. (1986) *The Chemistry of the Actinide Elements.* Chapman Hall, London. Keeney-Kennicutt W. L. and Morse J. W. (1985) The redox chemistry of $Pu(V)O_2^+$ interaction with common mineral surfaces in dilute solutions and seawater. *Geochim. Cosmo-chim. Acta* **49**, 2577–2588.

Kelly J. W., Aguilar R., and Papenguth H. W. (1999) Contribution of mineral-fragment type pseudo-colloids to the mobile actinide source term of the Waste Isolation Pilot Plant (WIPP). In *Actinide Speciation in High Ionic Strength Media* (eds. D. T. Reed, S. B. Clark, and L. Rao). Kluwer/Plenum, New York, pp. 227–237.

Kelley M., Maffit L., McClellan Y., Siegel M. D., and Williams C. V. (2002) Hanford 100-N area remediation options evaluation summary report. Sandia National Laboratories.

Kent D., Tripathi V., Ball N., Leckie J., and Siegel M. (1988) Surface-complexation modeling of radionuclide adsorption in subsurface environments. US Nuclear Regulatory Commission.

Kersting A. B., Efurd D. W., Finnegan D. L., Rokop D. J.,Smith D. K., and Thompson J. L. (1999) Migration of plutonium in ground water at the Nevada test site. *Nature* **397**, 56–59.

Khan S. A., Riaz-ur-Rehman, and Kahn M. (1994) Sorption of cesium on bentonite. *Waste Manage.* **14**(7), 629–642.

Kim J. I. (1993) The chemical behavior of transuranium elements and barrier functions in natural aquifer systems. In *Scientific Basis for Nuclear Waste Manage* (eds. C. G. Interrante and R. T. Pabalan). Materials Research Society, vol. 294, pp. 3–21.

Kim J. I. (1994) Actinide colloids in natural aquifer systems. *MRS Bull.* **19**(12), 47–53.

Kim J. I. and Sekine T. (1991) Complexation of neptunium(V) with humic-acid. *Radiochim. Acta* **55**(4), 187–192.

Kingston W. L. and Whitbeck M. (1991) Characterization of colloids found in various groundwater environments in central and southern Nevada. Water Resources Center Publication #45083. Desert Research Institute, University of Nevada System.

Knopp R., Neck V., and Kim J. I. (1999) Solubility, hydrolysis, and colloid formation of plutonium (IV). *Radiochim. Acta* **86**, 101–108.

Koβ V. (1988) Modelling of U(VI) sorption and speciation in a natural sediment-groundwater system. *Radiochim. Acta* **44/45**, 403–406.

Koch J. T., Rachar D. B., and Kay B. D. (1989) Microbial participation in iodide removal from solution by organic soils. *Can. J. Soil Sci.* **69**, 127–135.

Kohler M., Weiland E., and Leckie J. O. (1992) Metal-ligand-surface interactions during sorption of uranyl and neptunyl on oxides and silicates. In *Proceedings of 7th International Symposium Water–Rock Interaction*, pp. 51–54.

Kohler M., Curtis G., Kent D., and Davis J. A. (1996) Experimental investigation and modeling of uranium (VI) transport under variable chemical conditions. *Water Resour. Res.* **32**(12), 3539–3551.

Kohler M., Honeyman B. D., and Leckie J. O. (1999) Neptunium(V) sorption on hematite (α-Fe_2O_3) in aqueous suspension: the effect of CO_2. *Radiochim. Acta* **85**, 33–48.

Konoplev A., Bulgakov A., Popov V.E., and Bobovnikova T. I. (1992) Behaviour of long-lived radionuclides in a soil-water system. *Analyst* **117**, 1041–1047.

Konoplev A. V. and Bulgakov A. A. (1999) Kinetics of the leaching of Sr-90 from fuel particles in soil in the near zone of the Chernobyl power plant. *Atomic Energy* **86**(2), 136–141.

Krauskopf K. B. (1986) Aqueous geochemistry of radioactive waste disposal. *Appl. Geochem.* **1**, 15–23.

Kretzschmar R., Borovec M., Grollimund D., and Elimelech M. (1999) Mobile subsurface colloids and their role in contaminant transport. *Adv. Agronomy* **66**, 121–194.

Kruglov S. V., Vasil'eva N. A., Kurinov A. D., and Aleksakhin R. M. (1994) Leaching of radionuclides in Chernobyl fallout from soil by mineral acids. *Radiochemistry* **36**(6), 598–602.

Krumhansl J. L., Brady P. V., and Zhang P. (2002) Soil mineral backfills and radionuclide retention. In *Geochemistry of Soil Radionuclides.* SSSA Special Publication Number **59** (eds. P. Zhang and P. Brady). Soil Science Society of America, Madison, Wisconsin, pp. 191–210.

Krupka K. M. and Serne R. J. (2000) *Understanding Variation in Partition Coefficient, Kd, Values, Volume III: Review of Geochemistry and available Kd values for Americium, Arsenic, Curium, Iodine, Neptunium, Radium, and Technetium.* Pacific Northwest National Laboratory, Richland, WA.

Kudo A., Mahara Y., Santry D. C., Suzuki T., Miyahara S., Sugahara M., Zheng J., and Garrec J. (1995) Plutonium mass-balance released from the Nagasaki a-bomb and the applicability for future environmental-research. *Appl. Radiat. Isotopes* 46(11), 1089–1098.

Labonne-Wall N., Choppin G. R., Lopez C., and Monsallier J. M. (1999) Interaction of uranyl with humic and fulvic acids at high ionic strength. In *Actinide Speciation in High Ionic Strength Media* (eds. D. T. Reed, S. B. Reed, and L. Rao). Kluwer/Plenum, New York, pp. 199–211.

Laflamme B. D. and Murray J. W. (1987) Solid/solution interaction: the effect of carbonate alkalinity on adsorbed thorium. *Geochim. Cosmochim. Acta* **51**, 243–250.

Langmuir D. (1997a) *Aqueous Environmental Chemistry.* Prentice Hall, Upper Saddle River, NJ.

Langmuir D. (1997b) The use of laboratory adsorption data and models to predict radionuclide releases from a geological repository: a brief history. *Mater. Res. Soc. Symp. Proc.* **465**, 769–780.

Langmuir D. and Herman J. S. (1980) The mobility of thorium in natural waters at low temperatures. *Geochim. Cosmochim. Acta* **44**, 1753–1766.

Laul J. C. (1992) Natural radionuclides in groundwaters. *J. Radioanalyt. Nuclear Chem.* **156**(2), 235–242.

Leckie J. O. (1994) Ternary complex formation at mineral/ solution interfaces. In *Binding Models Concerning Natural Organic Substances in Performance Assessment: Proceedings of an NEA Workshop.* Nuclear Energy Agency, Bad Surzach, Switzerland, pp. 181–211.

Lefevre F., Sardin M., and Schweich D. (1993) Migration of Sr in clayey and calcareous sandy soil: precipitation and ion exchange. *J. Contamin. Hydrol.* **13**(4), 215–229.

Lemire R. J. (1984) *An Assessment of the Thermodynamic Behavior of Neptunium in Water and Model Groundwaters from 25 to 150°C.* AECL-7817, Atomic Energy of Canada Limited, Pinawa, Manitoba, Canada.

Lemire R. J. and Tremaine P. R. (1980) Uranium and plutonium equilibria in aqueous solutions to 200 °C. *J. Chem. Eng. Data* **25**, 361–370.

Lemire R. J., Fuger J., Nitsche H., and Potter P. (2001) *Chemical Thermodynamics of Neptunium and Plutonium.* Elsevier, Amsterdam.

Lenhart J. J. and Honeyman B. D. (1999) Uranium(VI) sorption to hematite in the presence of humic acid. *Geochim. Cosmochim. Acta* **63**(19/20), 2891–2901.

Lichtner P. (1996) Continuum formulation of multicomponent-multiphase reactive transport. In *Reactive Transport in Porous Media*, Reviews in Mineralogy (eds. P. Lichtner, C. I. Steefel, and E. H. Oelkers). Mineralogical Society of America, Washington, DC, vol. 34, pp. 1–81.

Lichtner P. C., Steefel C. I., and Oelkers E. H. (1996) *Reactive Transport in Porous Media*, Reviews in Mineralogy (ed. P. H. Ribbe). Mineralogical Society of America, Washington, DC.

Lieser K. H. and Bauscher C. (1988) Technetium in the hydrosphere and in the geosphere: II. Influence of pH, of complexing agents, and of some minerals on the sorption of technetium. *Radiochim. Acta* **44/45**, 125–128.

Lieser K. H. and Hill R. (1992) Hydrolysis and colloid formation of thorium in water and consequences for its migration behavior: comparison with uranium. *Radiochim.Acta* **56(1)**, 37–45.

Lieser K. H. and Mohlenweg U. (1988) Neptunium in the hydrosphere and in the geosphere. *Radiochim. Acta* **43**, 27–35.

Lieser K. H. and Peschke S. (1982) The geochemistry of fission products (Cs, I, Tc, Sr, Zr, Sm). In *NEA/OECD Workshop on Geochemistry and Waster Disposal.* Geneva, Switzerland, pp. 67–88.

Lieser K. H., Ament A., Hill R. N., Singh U., and Thybusch B. (1990) Colloids in groundwater and their influence onmigration of trace elements and radionuclides. *Radiochim. Acta* **49**, 83–100.

Lieser K. H., Hill R., Mühlenweg U., Singh R. N., Shu-De T.,and Steinkopff T. (1991) Actinides in the environment. *J. Radioanalyt. Nuclear Chem.* **147**(1), 117–131.

Liu Y. and van Gunten H. R. (1988) Migration chemistry and behavior of iodine relevant to geological disposal of radioactive wastes: a literature review with a compilation of sorption data. Paul Scherrer Institute.

Lloyd J. R. and Macaskie L. E. (1996) A novel phosphor-imager-based technique for monitoring the microbial reduction of technetium. *Appl. Environ. Microbiol.* **62**(2), 578–582.

Lloyd J. R., Cole J. A., and Macaskie L. E. (1997) Reduction of technetium from solution by *Escherichia coli. J. Bacteriol.* **179**, 2014–2021.

Lloyd J. R., Chenses J., Glasauer S., Bunker D. J., Livens F. R.,and Lovely D. R. (2002) Reduction of actinides and fission products by Fe(III)-reducing bacteria. *Geomicrobiol. J.* **19**, 103–120.

Looney B. B. and Falta R. W. (2000). *Vadose Zone Science andTechnology Solutions.* Battelle Press, Columbus, OH.

Lovely D. R. and Phillips E. J. P. (1992a) Reduction of uraniumby *Desulfovibrio desulfuricans. Appl. Environ. Microbiol.* **58**, 850–856.

Lovely D. R. and Phillips E. J. P. (1992b) Bioremediation of uranium contamination with enzymatic uranium reduction. *Environ. Sci. Technol.* **26**, 2228–2234.

Luckscheiter B. and Kienzler B. (2001) Determination of sorption isotherms for Eu, Th, U, and Am on the gel layer of corroded HLW glass. *J. Nuclear Mater.* **298**, 155–162.

Lyalikova N. N. and Khizhnyak T. V. (1996) Reduction of heptavalent technetium by acidophilic bacteria on the genus *Thiobacillus. Microbiology* **65**(4), 468–473.

Macaskie L. E. (1991) The application of biotechnology to the treatment of water produced from the nuclear fuel cycle:biodegradation and the bioaccumulation as a means of treating radionuclide-containing streams. *Critical Rev. Biotechnol.* **11**, 41–112.

Macaskie L. E., Lloyd J. R., Thomas R. A. P., and Tolley M. R. (1996) The use of micro-organisms for the remediation of solutions contaminated with actinide elements, other radionuclides, and organic contaminants generated by nuclear fuel cycle activities. *Nuclear Technol.* **35**, 257–271.

Mahara Y. and Kudo A. (1995) Plutonium released by the Nagasaki a-bomb: mobility in the environment. *Appl. Radiat. Isotopes* **46**(11), 1191–1201.

Mahoney J. J. and Langmuir D. (1991) Adsorption of Sr on kaolinite, illite, and montmorillonite at high ionic strengths. *Radiochim. Acta* **54**, 139–144.

Mangold D. C. and Tsang C. F. (1991) A summary of subsurface hydrological and hydrochemical models. *Rev. Geophys.* **29**, 51–79.

Marquardt C. and Kim J. I. (1998) Complexation of Np(V) with fulvic acid. *Radiochim. Acta* **81**(3), 143–148.

Marquardt C., Herrmann G., and Trautmann N. (1996) Complexation of neptunium(V) with humic acids at very low metal concentrations. *Radiochim. Acta* **73**(3), 119–125.

McBride J. P., Moore R. E., Witherspoon J. P., and Blanco R. E. (1978) Radiological impact of airborne effluents of coal and nuclear power plants. *Science* **202**, 1045–1050.

McCarthy J. F. and Zachara J. M. (1989) Surface transport of contaminants. *Environ. Sci. Technol.* **23**, 496–502.

McCubbin D. and Leonard K. S. (1997) Laboratory studies to investigate short-term oxidation and sorption behavior of neptunium in artificial and natural seawater solutions. *Mar. Chem.* **56**, 107–121.

McGrail B. P. and Cragnolino G. A. (2002) *Scientific Basis for Nuclear Waste Management XXV.* Materials Research Society Symposium Proceedings 713, Materials Research Society, Pittsburgh, PA.

McKinley I. G. and Alexander J. L. (1993) Assessment of radionuclide retardation: uses and abuses of natural analogue studies. *J. Contamin. Hydrol.* **13**, 727–732.

McKinley I. and Scholtis A. (1992) A comparison of sorption databases used in recent performance assessments. In *Disposal of Radioactive Waste: Radionuclide Sorption from the Safety Evaluation Perspective, Proceedings of an NEA Workshop, Interlaken, Switzerland, 16–18 October 1991*, NEA-OECD, Paris, France, pp. 21–55.

McKinley J., Zachara J. M., Smith S. C., and Turner G. (1995) The influence of hydrolysis and multiple site-binding reactions on adsorption of U(VI) to montmorillonite. *Clays Clay Mineral.* **43**, 586–598.

Medvedev Z. A. (1979) *Nuclear Disaster in Urals.* Norton, New York.

Meece D. E. and Benninger L. K. (1993) The coprecipitation of Pu and other radionuclides with $CaCO_3$. *Geochim. Cosmochim. Acta* **57**, 1447–1458.

Menet C., Ménager M. T., and Petit J. C. (1992) Migration of radioelements around the new nuclear reactors at Oklo: analogies with a high-level waste repository. *Radiochim. Acta* **58–59(2)**, 395–400.

Meyer R., Palmer D., Arnold W., and Case F. (1984) Adsorption of nuclides on hydrous oxides. Sorption isotherms on natural materials. *Geochem. Behavior Disp. Radioact. Waste*, 79–94.

Meyer R., Arnold W., Case F., and O'Kelley G. D. (1991) Solubilities of Tc(IV) Oxides. *Radiochim. Acta* **55**, 11–18.

Miekeley N., Coutinho de Jesus H, Porto da Siveira C. L., and Degueldre C. (1991) *Chemical and Physical Characterization of Suspended Particles and Colloids in Waters from the Osamu Utsumi and Morro de Ferro Analog Study Sites, Poco de Caldas, Brazil.* SKB Technical Report 90-18.

Morgenstern M., Lenze R., and Kim J. I. (2000) The formation of mixed-hydroxo complexes of Cm(III) and Am(III) with humic acid in the neutral pH range. *Radiochim. Acta* **88**(1), 7–16.

Moulin V. and Ouzounian G. (1992) Role of colloids and humic substances in the transport of radio-elements through the geosphere. *Appl. Geochem.* (1), 179–186.

Mucciardi A. N. (1978) Statistical investigation of the mechanics controlling radionuclide sorption: Part II. Task 4. In *Second Contractor Information Meeting.* Battelle Northwest Laboratory, Richland, WA, vol. II, pp. 333–425.

Mucciardi A. N. and Orr E. C. (1977) Statistical investigation of the mechanics controlling radionuclide sorption. In *Waste Isolation Safety Assessment Program, Task 4, Contractor Information Meeting Proceedings.* Battelle Northwest Laboratory, Richland, WA, pp. 151–188.

Muramatsu Y., Uchida S., Sriyotha P., and Sriyotha K. (1990) Some considerations on the sorption and desorption phenomena of iodide and iodate on soil. *Water Air Soil Pollut.* **49**, 125–138.

Murphy R. J., Lenhart J. L., and Honeyman B. D. (1999) The sorption of thorium(IV) and uranium(VI) to hematite in the presence of natural organic matter. *Physicochem. Eng. Aspects* **157**, 47–62.

Murphy W. M. (2000) Natural analogs and performance assessment for geologic disposal of nuclear water. In *Scientific Basis for Nuclear Waster Management XXII,* Materials Research Society Symposium Proceedings (eds. R. W. Smith and D. W. Shoesmith). Materials Research Society, Warrendale, PA, vol. 608, pp. 533–544.

Nash K. L. and Choppin G. R. (1980) Interaction of humic and fulvic acids with Th(IV). *J. Inorg. Nuclear Chem.* **42**, 1045–1050.

National Academy of Sciences (NAS) (1996) *The Waste Isolation Pilot Plant: A Potential Solution for the Disposal of Transuranic Waste.* National Academy Press, Washington, DC.

National Academy of Sciences (1998) *Health Effects of Exposure to Radon, National Academy of Sciences Report BEIR VI.* National Academy Press.

National Institute of Health (1994) *Radon and Lung Cancer Risk: a Joint Analysis of 11 Underground Miner Studies.* US Department of Health and Human Services, National Institutes of Health.

National Research Council (1995) *Technical Bases for Yucca Mountain Standards*, Washington, DC, 205pp.

National Research Council (2001) *Science and Technology for Environmental Cleanup at Hanford.* National Academy Press.

NEA (1991) *Disposal of Radioactive Waste: Review of Safety Assessment Methods.* Nuclear Energy Agency, OECD.

NEA Nuclear Science Committee (1996) Survey of thermodynamic and kinetic databases. http://www.nea.fr/html/science/chemistry/tdbsurvey.Html.

Neck V. and Kim J. L. (2001) Solubility and hydrolysis of tetravalent actinides. *Radiochim. Acta* **89**(1), 1–16.

Neretnieks I. and Rasmuson A. (1984) An approach to modeling radionuclide migration in a medium with strongly varying velocity and block sizes along the flow path. *Water Resour. Res.* **20**(12), 1823–1836.

Nisbet A. F., Mocanu N., and Shaw S. (1994) Laboratory investigation into the potential effectiveness of soil-based countermeasures for soils contaminated with radiocaesium and radiostrontium. *Sci. Tot. Environ.* **149**, 145–154.

Nitsche H. (1991) Solubility studies of transuranium elements for nuclear waste-disposal: principles and overview. *Radiochim. Acta* 52–**53**, 3–8.

Nitsche H., Muller A., Standifer E. M., Deinhammer R. S., Becraft K., Prussin T., and Gatti R. C. (1992) Dependence of actinide solubility and speciation on carbonate concentration and ionic strength in ground water. *Radiochim. Acta* 58/**59**, 27–32.

Nitsche H., Gatti R. C., Standifer E. M., Lee S. C., Müller A., Prussin T., Deinhammer R. S., Maurer H., Becraft K., Leung S., and Carpenter S. A. (1993) Measured solubilities and speciations of neptunium, plutonium and americium in a typical groundwater (J-13) from the Yucca Mountain region. Los Alamos National Laboratory.

Nitsche H., Roberts K., Becraft K., Prussin T., Keeney D., Carpenter S. A., and Hobart D. E. (1995) *Solubility and Speciation Results from Over- and Undersaturation Experiments on Np, Pu, and Am in Water from Yucca Mountain Region Well UE 25p#1.* LA-13017-MS, Los Alamos National Laboratories, Los Alamos, NM.

Nordstrom D. K., Ball J. W., Donahoe R. J., and Whittemore D. (1989) Groundwater chemistry and water–rock interactions at Stripa. *Geochim. Cosmochim. Acta* **53**, 1727–1740.

Novak C. F. (1997) Calculation of actinide solubilities in WIPP SPC and ERDA-6 brines under MgO backfill scenarios containing either nesquehonite or hydromagnesite as the Mg-CO_3 solubility-limiting phase. Sandia National Labs, WIPP Records Center.

Novak C. F., Nitsche H., Silber H. B., Roberts K., Torretto P. C., Prussin T., Becraft K., Carpenter S. A., Hobart D. E., and AlMahamid I. (1996) Neptunium(V) and neptunium(VI) solubilities in synthetic brines of interest to the Waste Isolation Pilot Plant (WIPP). *Radiochim. Acta* **74**, 31–36.

Nuclear Energy Agency (2001) *Using Thermodynamic Sorption Models for Guiding Radioelement Distribution Coefficient (Kd) Investigations. Radioactive Waste Management.* Nuclear Energy Agency, Paris, France.

Nuttall H. E., Jain R., and Fertelli Y. (1991) Radiocolloid transport in saturated and unsaturated fractures. In *2nd Annual International Conference on High Level Radioactive Waste Management*, Las Vegas, NV, pp. 189–196.

Ogard A. E. and Kerrisk J. F. (1984) Groundwater chemistry along the flow path between a proposed repository site and the accessible environment. Los Alamos National Laboratory.

Ohnuki T. and Kozai N. (1994) Sorption characteristics of radioactive cesium and strontium. *Radiochim. Acta* **66/67**, 327–331.

Olofsson U. and Allard B. (1983) Complexes of actinides with naturally occurring organic substances. Literature Survey Technical. KBS Report 83–09.

Östhols E. (1995) Thorium sorption on amorphous silica. *Geochim. Cosmochim. Acta* **59**(7), 1235–1249.

Östhols E., Bruno J., and Grenthe I. (1994) On the influence of carbonate on mineral dissolution: III. The solubility of microcrystalline ThO_2 in CO_2-H_2O media. *Geochim. Cosmochim. Acta* **58**(2), 613–623.

Oughton D. H., Salbu B., Brand T. L., Day J. P., and Aarkrog A. (1993) Underdetermination of Strontium-90 in soils containing particles of irradiated uranium oxide fuel. *Analyst* **118**, 1101–1105.

Pabalan R. T., Turner D. R., Bertetti F. P., and Prikryl J. (1998) Uranium(VI) sorption onto selected mineral surfaces: key geochemical parameters. In *Adsorption of Metals by Geomedia* (ed. E. Jenne). Academic Press, San Diego, CA, pp. 99–130.

Panin A. V., Walling D. E., and Golosov V. N. (2001) The role of soil erosion and fluvial processes in the post-fallout redistribution of Chernobyl-derived caesium-137: a case study of the Lapki Catchment, central Russia. *Geomorphology* **40**, 185–204.

Papelis C., Hayes K. F., and Leckie J. O. (1988) *HYDRAQL: A Program for the Computation of Chemical Equilibrium Composition of Aqueous Batch Systems Including Surface-complexation Modeling of Ion Adsorption at the Solution Oxide/Solution Interface. 306.* Environmental Engineering and Science, Department of Civil Engineering, Stanford University, Stanford, CA, 131pp.

Park S.-W., Leckie J. O., and Siegel M. D. (1992) *Surface Complexation Modeling of Uranyl Adsorption on Corrensite from the Waste Isolation Pilot Plant site.* Sandia National Laboratories.

Parkhurst D. L. (1995) *User's Guide to PHREEQC, a Computer Model for Speciation, Reaction-path, Advective-transport and Inverse Geochemical Calculations.* US Geological Survey Water-resources Investigations Report 95-4227, US Geological Survey, 143pp.

Parrington J. R., Knox H. D., Breneman S. L., Baum E. M., and Feiner F. (1996) *Nuclides and Isotopes: Wall Chart Information Booklet.* General Electric, Schenectady, NY.

Payne T., Edis R., and Seo T. (1992) Radionuclide transport by groundwater colloids at the Koongarra Uranium Deposit. *Sci. Basis Nuclear Waste Manage.* **XV**, 481–488.

Payne T. E., Fenton B. R., and Waite T. D. (2001) Comparison of "in-situ distribution coefficients" with experimental Rd values for uranium (VI) in the Koongarra Weathered Zone. In *Surface Complexation Modeling of Uranium (VI) Adsorption on Natural Mineral Assemblages* (ed. J. A. Davis). Nuclear Regulatory Commission, Washington, DC, pp. 133–142.

Pearcy E., Prikryl J. D., Murphy W. M., and Leslie B. W. (1994) Alteration of uraninite from the Nopal I deposit, Pena Bianca District, Chihuahua, Mexico, compared to degradation of spent nuclear fuel in the proposed US high-level nuclear waste repository at Yucca Mountain, Nevada. *Appl. Geochem.* **9**, 713–732.

Pedersen K. (1996) Investigations of subterranean bacteria in deep crystalline bedrock and their importance for the disposal of nuclear waste. *Can. J. Microbiol.* **42**, 382–400.

Penrose W. R., Polzer W. L., Essington E. H., Nelson D. M., and Orlandini K. A. (1990) Mobility of plutonium and americium through a shallow aquifer in a semiarid region. *Environ. Sci. Technol.* **24**, 228–234.

Perfect D. L., Faunt C. C., Steinkampf W. C., and Turner A. K. (1995) *Hydrochemical Database for the Death Valley Region, Nevada and California.* US Geological Survey Open-File Report 94-305, US Geological Survey.

Petropoulos N. P., Anagnostakis M. J., Hinis E. P., and Simopoulos S. E. (2001) Geographical mapping and associated fractal analysis of the long-lived Chernobyl fallout radionuclides in Greece. *J. Environ. Radioact.* **53**, 59–66.

Petryaev E. P., Ovsyannikova S. V., Rubinchik S., Lubkina I. J., and Sokolik G. A. (1991) Condition of Chernobyl fallout radionuclides in the soils of Belorussia. Proceedings of AS of BSSR. *Physico-Energent. Sci.* **4**, 48–55.

Phillips S., Hale F., Silvester L., and Siegel M. (1988) *Thermodynamic Tables for Nuclear Waste Isolation: Volume 1. Aqueous Solutions Database.* Sandia National Laboratories.

Pitzer K. S. (1973) Thermodynamics of electrolytes:I. Theoretical basis and general equations. *J. Phys. Chem. II*, **77**, 268–277.

Pitzer K. S. (1975) Thermodynamics of electrolytes: V. Effects of higher-order electrostatic terms. *J. Solut. Chem.* **4**, 249–265.

Pitzer K. S. (1979) Theory: Ion interaction approach. In *Activity Coefficients in Electrolyte Solutions* (ed. R. M. Pytkowicz). CRC Press, Boca Raton, FL, pp. 157–208.

Pitzer K. S. (1987) A thermodynamic model for aqueous solutions of liquid-like density. In *Thermodynamic Modeling of Geological Materials: Minerals, Fluids, and Melts.* Reviews in Mineralogy (eds. I. S. E. Carmichael and H. P.Eugster). Mineralogical Society of America, Washington, DC, vol.17, pp. 97–142.

Pitzer K. S. (2000) *Activity Coefficients in Electrolyte Solutions*, 2nd edn. CRC Press, Boca Raton, FL.

Plummer L., Parkhurst D., Fleming G., and Dunkle S. (1988) *A Computer Program Incorporating Pitzer's Equations for Calculation of Geochemical Reactions in Brines.* US Geological Survey. Pourcelot L. and Gauthier-Lafaye F. (1999) Hydrothermal and supergene clays of the Oklo natural reactors: conditions of radionuclide release, migration and retention. *Chem. Geol.***157**, 155–174.

Prasad A., Redden G., and Leckie J. O. (1997) *Radionuclide Interactions at Mineral/Solution Interfaces in the Wipp Site Subsurface Environment.* Sandia National Laboratories. Prikryl J. D., Jain A., Turner D. R., and Pabalan R. T. (2001) Uranium sorption behavior on silicate mineral mixtures. *J. Contamin. Hydrol.* 47(2–4), 241–253.

Quigley M. S., Honeyman B. D., and Santschi P. H. (1996) Thorium sorption in the marine environment: equilibrium partitioning at the hematite/water interface, sorption/desorption kinetics and particle tracing. *Aquat. Geochem.* **1**, 277–301.

Rädlinger G. and Heumann K. G. (2000) Transformation of iodide in natural and wastewater systems by fixation on humic substances. *Environ. Sci. Technol.* **34(18)**,3932–3936.

Rai D., Serne R. J., and Swanson J. L. (1980) Solution species of plutonium in the environment. *J. Environ. Qual.* **9**, 417–420.

Rai D., Felmy A. R., and Ryan J. L. (1990) Uranium(VI)hydrolysis constants and solubility product of $UO_2 \times H_2O$. *Inorg. Chem.* **29**, 260–264.

Rai D., Felmy A. R., Sterner S. M., Moore D. A., Mason M. J.,and Novak C. F. (1997) The solubility of Th(IV) and U(IV) hydrous oxides in concentrated NaCl and $MgCl_2$ solutions. *Radiochim. Acta* **79(4)**, 239–247.

Rai D., Felmy A. R., Hess H. J., and Moore D. A. (1998) A thermodynamic model of solubility of UO_2(am) in the aqueous $K^+–Na^+–HCO_3^-–CO_3^{2-}–OH^-–H_2O$. *Radiochim. Acta* **82**, 17–25.

Rao L. and Choppin G. R. (1995) Thermodynamic study of the complexation of neptunium(V) with humic acids. *Radiochim. Acta* **69(2)**, 87–95.

Rard J. A. (1983) *Critical Review of the Chemistry and Thermodynamics of Technetium and some of its Inorganic Compounds and Aqueous Species.* Lawrence Livermore National Laboratory, University of California, 86pp.

Rard J. A., Rand M. H., Anderegg G., and Wanner H. (1999) *Chemical Thermodynamics of Technetium.* North-Holland, Amsterdam, Holland.

Read D., Hooker P. J., Ivanovich M., and Milodowski A. E. (1991) A natural analogue study of an abandoned uranium mine in Cornwall, England. *Radiochim. Acta* **52/53**, 349–356.

Rechard R. P. (1996) *An Introduction to the Mechanics of Performance Assessment Using Examples of Calculations Done for the Waste Isolation Pilot Plant between 1990 and 1992.* Sandia National Laboratories.

Rechard R. P. (2002) General approach used in the performance assessment for the Waste Isolation Pilot Plant. In *Scientific Basis for Nuclear Waste Management XXV.* Materials Research Society Symposium Proceedings (eds. B. P. McGrail and G. A. Cragnolino). Materials Research Society, Pittsburgh, PA, vol. 713, pp. 213–228.

Redden G., Li J., and Leckie J. O. (1998) Adsorption of U(VI) and citric acid on goethite, gibbsite, and kaolinite. In

Adsorption of Metals by Geomedia (ed. E. Jenne). Academic Press, San Diego, CA, pp. 291–315.

Redden G. B. and Bencheikh-Latmar J. R. (2001) Citrate enhanced uranyl adsorption on goethite: an EXAFS analysis. *J. Colloid Interface Sci.* **244**(1), 211–219.

Reed D. T., Wygmans D. G., and Richman M. K. (1996) *Actinide Stability/Solubility in Simulated WIPP Brines: Interim Report under SNL WIPP Contract AP-2267.* Sandia National Laboratories.

Reeder R. J., Nugent M., Tait C. D., Morris D. E., Heald S. M., Beck K. M., Hess W. P., and Lanzirotti A. (2001) Coprecipitation of Uranium(VI) with calcite: XAFS, micro-XAS, and luminescence characterization. *Geochim. Cosmochim. Acta* **65**(20), 3491–3503.

Rees T. F., Cleveland J. M., and Nash K. L. (1983) The effect of composition of selected groundwaters from the basin and range province on plutonium, neptunium, and americium speciation. *Nuclear Technol.* **65**, 131–137.

Relyea J. F. (1982) Theoretical and experimental considerations for the use of the column method for determining retardation factors. *Radioact. Waste Manage. Nuclear Fuel Cycle* **3**(3), 151–166.

Robertson A. and Leckie J. O. (1997) Cation binding predictions of surface complexation models: effects of pH, ionic strength, cation loading, surface complex, and model fit. *J. Colloid Interface Sci.* **188**, 444–472.

Rosén K., Öborn I., and Lönsjö H. (1999) Migration of radiocaesium in Swedish soil profiles after the Chernobyl accident, 1987–1995. *J. Environ. Radioact.* **46**, 45–66.

Rühm W., Kammerer L., Hiersche L., and Wirth E. (1996) Migration of ^{137}Cs and ^{134}Cs in different forest soil layers. *J. Environ. Radioact.* **33**, 63–75.

Runde W. (2002) Geochemical interactions of actinides in the environment. In *Geochemistry of Soil Radionuclides* (eds. P. Zhang and P. Brady). Soil Science Society of America, pp. 21–44.

Runde W., Neu M. P., and Reilly S. D. (1999) Actinyl(VI) carbonates in concentrated sodium chloride solutions: characterization, solubility, and stability. In *Actinide Speciation in High Ionic Strength Media* (eds. D. T. Reed, S. B. Clark, and L. Rao). Kluwer/Plenum, New York, pp. 141–151.

Runde W., Conradson S. D., Efurd W., Lu N., VanPelt C. E., and Tait D. C. (2002a) Solubility and sorption of redox-sensitive radionuclides (Np, Pu) in J-13 water from Yucca Mountain site: comparison between experiment and theory. *Appl. Geochem.* **17**(6), 837–853.

Runde W., Neu M. P., Condrdson S. D., Li J., Lin M., Smith D. M., Van-Pelt C. E., and Xu Y. (2002b) Geochemical speciation of radionuclides in soil and solution. In *Geochemistry of Soil Radionuclides.* SSSA Special Publication Number 59 (eds. P. Zhang and P. Brady). Soil Science Society of America, Madison, Wisconsin, pp. 45–60.

Ryan J. N. and Elimelech M. (1996) Colloid mobilization and transport in groundwater. *Coll. Surf. A-Physicochem. Eng. Aspects* **107**, 1–56.

Salbu B., Krekling T., Oughton D. H., Ostby G., Kashparov V. A., Brand T. L., and Day J. P. (1994) Hot particles in accidental releases from Chernobyl and Windscale nuclear installations. *Analyst* **119**(1), 125–130.

Salbu B., Nikitin A. I., Strand P., Christensen G. C., Chumichev V. B., LInd B., Fjelldal H., Bergan T. D. S., Rudjord A. L., Sickel M., Valetova N. K., and Foyn L. (1997) Radioactive contamination from dumped nuclear waste in the Kara sea—results from the joint Russian–Norwegian expeditions in 1992–1994. *Sci. Tot. Environ.* **202**(1–3), 185–198.

Sanchez A. L., Murray J. W., and Sibley T. H. (1985) The adsorption of plutonium IV and V on goethite. *Geochim. Cosmochim. Acta* **49**, 2297–2307.

Santschi P. H. and Honeyman B. D. (1989) Radionuclides in aquatic environments. *Radiat. Phys. Chem.* **34**, 213–240.

Scatchard G. (1936) Concentrated solutions of strong electrolytes. *Phil. Mag.* **19**, 588–643.

Schramke J. A., Rai D., Fulton R. W., and Choppin G. R. (1989) Determination of aqueous plutonium oxidation states by solvent extractions. *J. Radioanalyt. Nuclear Chem.* **130**(2), 333–346.

Seaborg G. T. and Katz J. J. (1954) *The Actinide Elements.* McGraw-Hill, New York.

Serne R. J. and Gore V. L. (1996) Strontium-90 Adsorption-Desorption Properties and Sediment Characterization at the 100 N-area. Pacific Northwest National Laboratory.

Serne R. J. and Muller A. B. (1987) A perspective on adsorption of radionuclides onto geologic media. In *The Geological Disposal of High Level Radioactive Wastes* (ed. D. G. Brookins). Theophrastus Publications, pp. 407–443.

Serne R. J., Arthur R. C., and Krupka K. M. (1990) Review of Geochemical Processes and Codes for Assessment of Radionuclide Migration Potential at Commercial LLW Sites. 129pp., US Nuclear Regulatory Commission.

Shanbhag P. M. and Morse J. W. (1982) Americium interaction with calcite and aragonite sufaces in seawater. *Geochim. Cosmochim. Acta* **46**, 241–246.

Shutov V. N., Travnikova I. G., Brak G. Y., Golikov V. Y., and Balanov M. I. (2002) Current contamination by Cs-137 and Sr-9- of the inhabited part of the Techa river basin in the Urals. *J. Environ. Radioact.* **61**, 91–109.

Siegel M. D. and Erickson K. L. (1984) Radionuclide releases from a hypothetical nuclear waste repository: potential violations of the proposed EPA standard by radionuclides with multiple aqueous species. In *Waste Manage. 84* (ed. R. G. Post). University of Arizona, Tuscon, AZ, vol. 1, pp. 541–546.

Siegel M. D. and Erickson K. L. (1986) Geochemical sensitivity analysis for performance assessment of HLW repositories: effects of speciation and matrix diffusion. *Proceedings of the Symposium on Groundwater Flow and Transport Modeling for Performance Assessment of Deep Geologic Disposal of Radioactive Waste: A Critical Evaluation of the State of the Art.* Sandia National Laboratories, Albuquerque, NM, pp. 465–488.

Siegel M. D., Chu M. S., and Pepping R. E. (1983) Compliance assessments of hypothetical geological nuclear waster isolation systems with the draft EPA standard. In *Scientific Basis for Nuclear Waste Management VI.* Materials Research Society Symposium Proceedings 15 (ed. D. G. Brookins). Elsevier, Amsterdam, Holland, pp. 497–506.

Siegel M. D., Leckie J. O., Phillips S. L., and Kelly W. R. (1989) Development of a methodology of geochemical sensitivity analysis for performance assessment. In *Proceedings of the Conference on Geostatistical, Sensitivity, and Uncertainty Methods for Ground-Water Flow and Radionuclide Transport Modeling.* San Francisco, CA, Battdle, pp. 189–211.

Siegel M. D., Holland H. D., and Feakes C. (1992) Geochemistry. In *Techniques for Determining Probabilities of Geologic Events and Processes, Studies in Mathematical Geology 4.* Int. Nat. Assoc. Math. Geology (eds. R. L. Hunter and C. J. Mann). Oxford University Press, New York, pp. 185–206.

Siegel M. D., Ward D. B., Bryan C. R., and Cheng W. C. (1995a) *Characterization of Materials for a Reactive Transport Model Validation Experiment.* Sandia National Laboratories.

Siegel M. D., Ward D. B., Bryan C. R., and Cheng W. C. (1995b) *Batch and Column Studies of Adsorption of Li, Ni, and Br by a Reference Sand for Contaminant Transport Experiments.* Sandia National Laboratories.

Silva R. J. and Nitsche H. (1995) Actinide environmental chemistry. *Radiochim. Acta* 70/**71**, 377–396.

Silva R. J., Bidoglio G., Rand M. H., Rodouch P. B., Wanner H., and Puigdomenech I. (1995) *Chemical Thermodynamics of Americium.* Elsevier, New York.

Sims R., Lawless R., Alexander J., Bennett D., and Read D. (1996) Uranium migration through intact sandstone: effect of pollutant concentration and the reversibility of uptake. *J. Contamin. Hydrol.* **21**, 215–228.

SKI (1991) *SKI Project 90 Summary.* SKB Technical Report 91–23.

Slezak J. (1997) *National Experience on Groundwater Contamination Associated with Uranium Mining and Milling in the Czech Republic.* DIAMO s.p., Straz pod Ralskem, Czech Republic.

Slovic P. (1987) Perception of risk. *Science* **236**, 280–285.

Smith J. T. and Comans R. N. J. (1996) Modeling the diffusive transport and remobilization of ^{137}Cs in sediments: the effects of sorption kinetics and reversibility. *Geochim. Cosmochim. Acta* **60**, 995–1004.

Smith J. T., Comans R. N. J., and Elder D. G. (1999) Radiocaesium removal from European lakes and reservoirs: key processes determined from 16 Chernobyl-contaminated lakes. *Water Res.* **33**, 3762–3774.

Smith J. T., Comans R. N. J., Ireland D. G., Nolan L., and Hilton J. (2000) Experimental and *in situ* study of radiocaesium transfer across the sediment/water interface and mobility in lake sediments. *Appl. Geochem.* **15**(6), 833–848.

Smith P. A. and Degueldre C. (1993) Colloid-facilitated transport of radionuclides through fractured media. *J. Contamin. Hydrol.* **13**, 143–166.

Sokolik G. A., Ivanova T. G., Leinova S. L., Ovsiannikova S. V., and Kimlenko I. M. (2001) Migration ability of radionuclides in soil-vegetative cover of Belarus after Chernobyl accident. *Environ. Int.* **26**(3), 183–187.

Steefel C. I. and Van Cappellen P. (1998) Special Issue Reactive transport modeling of natural systems. *J. Hydrol.* 209(1–4), 1–7.

Stout D. L. and Carrol S. A. (1993) *A Literature Review of Actinide-carbonate Mineral Interactions.* Sandia National Laboratories.

Strickert R., Friedman A. M., and Fried S. (1980) The sorption of technetium and iodine radioisotopes by various minerals. *Nuclear Technol.* **49**, 253–266.

Strietelmeyer B. A., Gillow J. B., Dodge C. J., and Pansoy-Hjelvik M. E. (1999) Toxicity of actinides to bacterial strains isolated from the Waste Isolation Pilot Plant (WIPP) Environment. In *Actinide Speciation in High Ionic Strength Media* (eds. D. T. Reed, S. B. Clark, and L. Rao). Kluwer/Plenum, New York, pp. 261–268.

Stumm W. (1992) *Chemistry of the Solid–Water Interface.* Wiley, New York.

Sumrall C. L. I. and Middlebrooks E. J. (1968) Removal of radioisotopes from water by slurrying with Yazoo and Zilpha clays. *J. Am. Water Works Assoc.* **60**(4), 485–494.

Suzuki Y. and Banfield J. F. (1999) Geomicrobiology of uranium. In *Reviews in Mineralogy.* Uranium: Mineralogy, Geochemistry and the Environment (eds. P. Burns and R. Finch). Mineralogical Society of America, Washington, DC, vol. 38, pp. 393–432.

Sverjensky D. A. (2003) Standard states for the activities of mineral surface sites and species. *Geochem. Cosmochim. Acta* **67**(1), 17–28.

Szecsody J. E., Zachara J. M., Chilakapati A., Jardine P. M., and Ferrency A. C. (1998) Importance of flow and particle-size heterogeneity in $Co^{II/III}$ EDTA reactive transport. *J. Hydrol.* 209(1–4), 000.

Tamura T. (1972) Sorption phenomena significant in radioactive-waste disposal. In *Underground Waste Management and Environmental Implications*, American Association of Petroleum Geologists (ed. T. D. Cook). Tulsa, Oklahoma, pp. 318–330.

Tanaka S., Yamawaki M., Nagasaki S., and Moriyama H. (1992) Geochemical behavior of neptunium. *J. Nuclear Sci. Technol.* **29**(7), 706–718.

Tanaka T. and Muraoka S. (1999) Sorption characteristics of ^{237}Np, ^{238}Pu, ^{241}Am in sedimentary materials. *J. Radioanalyt. Nuclear Chem.* **240**(1), 177–182.

Thomas K. (1987) Summary of sorption measurements performed with Yucca Mountain, Nevada tuff samples and water from well J-13. Los Alamos National Laboratory.

Tien P.-L., Siegel M. D., Updegraff C. D., Wahi K. K., and Guzowski R. V. (1985) *Repository Site Data Report for Unsaturated Tuff, Yucca Mountain, Nevada.* US Nuclear Regulatory Commission.

Tipping E. (1993) Modeling the binding of europium and the actinides by humic substances. *Radiochim. Acta* **62**, 141–152.

Torstenfelt B. (1985a) Migration of the actinides thorium, protactinium, uranium, neptunium, plutonium, and americium in clay. *Radiochim. Acta* **39**, 105–112.

Torstenfelt B. (1985b) Migration of the fission products strontium, technetium, iodine and cesium in clay. *Radiochim. Acta* **39**, 97–104.

Trabalka J. R., Eyman L. D., and Auerbach S. I. (1980) Analysis of the 1957–1958 Soviet nuclear accident. *Science* **209**, 345–352.

Triay I. R., Mitchell A. J., and Ott M. A. (1992) *Radionuclide Migration Laboratory Studies for Validation of Batch-Sorption Data.* LA-12325-C, Los Alamos National Laboratory, Los Alamos.

Triay I. R., Furlano A. C., Weaver S. C., Chipera S. J., Bish D. L., Meijer A., and Canepa J. A. (1996) *Comparison of Neptunium Sorption Results Using Batch and Column Techniques.* Los Alamos National Laboratory.

Triay I. R., Meijer A., Conca J. L., Kung K. S., Rundberg R. S., Streitelmeier B. A., Tait C. D., Clark D. L., Neu M. P., and Hobart D. E. (1997) *Summary and Synthesis Report on Radionuclide Retardation for the Yucca Mountain Site Characterization Project LA-13262-MS.* Los Alamos National Laboratories.

Tripathi V. (1983) Uranium(VI) Transport modeling: geochemical data and sub-models. PhD Thesis, Stanford University.

Tripathi V. S., Siegel M. D., and Kooner Z. S. (1993) Measurements of metal adsorption in oxide-clay mixtures: "competitive-additivity" among mixture components. In *Scientific Basis for Nuclear Waste Management.* Materials Research Society, vol. XVI, pp. 791–796.

Truex M. J., Peyton B. M., Valentine N. B., and Gorby Y. A. (1997) Kinetics of U(VI) reduction by a dissimilatory Fe(III)-reducing bacterium under non-growth conditions. *Biotechnol. Bioeng.* **55**, 490–496.

Tsunashima A., Brindley W., and Bastovanov M. (1981) Adsorption of uranium from solutions by montmorillonite; compositions and properties of uranyl montmorillonites. *Clays Clay Min.* **29**(1), 10–16.

Turin H. J., Groffman A. R., Wolfsberg L. E., Roach J. L., and Strietelmeier B. A. (2002) Tracer and radionuclide sorption to vitric tuffs of Busted Butte. *Nevada. Appl. Geochem.* **17**(6), 825–836.

Turner D. and Sassman S. (1996) Approaches to sorption modeling for high-level waste performance assessment. *J. Contamin. Hydrol.* **21**, 311–332.

Turner D. R. (1991) *Sorption Modeling for High-level Waste Performance Assessment: a Literature Review.* Center for Nuclear Waste Regulatory Analyses.

Turner D. R. (1995) *A Uniform Approach to Surface Complexation Modeling of Radionuclide Sorption.* Center for Nuclear Waste Regulatory Analyses.

Turner D. R. and Pabalan R. T. (1999) Abstraction of mechanistic sorption model results for performance assessment calculations at Yucca Mountain. *Nevada. Waste Manage.* **19**, 375–388.

Turner D. R., Griffin T., and Dietrich T. (1993) Radionuclide sorption modeling using the MIN–TEQA2 speciation code. *Mater. Res. Soc. Symp. Proc.* 783–789.

Turner D. R., Pabalan R. T., and Bertetti F. P. (1998) Neptunium(V) sorption on montmorillonite: an experimental and surface complexation modeling study. *Clays Clay Min.* **46**, 256–269.

Turner G., Sachara J., McKinley J., and Smith S. C. (1996) Surface-charge properties and UO_2^{2+} adsorption on a surface smectite. *Geochim. Cosmochim. Acta* **60**, 3399–3414.

Turner D. R., Bertetti F. P., and Pabalan R. T. (2002) Role of radionuclide sorption in high-level waste performance assessment: approaches for the abstraction of detailed models. In *Geochemistry of Soil Radionuclides* (eds. P. Zhang and P. Brady). Soil Science Society of America, Madison, Wisconsin, pp. 211.

Uchida S., Tagami K., Rühm W., and Wirth E. (1999) Determination of ^{99}Tc deposited on the ground within the 30-km zone around the Chernobyl reactor and estimation of ^{99}Tc released into atmosphere by the accident. *Chemosphere* **39**(15), 2757–2766.

US Department of Energy (1980) *Project Review: Uranium Mill Tailings Remedial Action Project.* US Department of Energy, Washington, DC.

US Department of Energy (DOE) (1988) *Site Characterization Plan, Yucca Mountain Site, Nevada Research and Development Area, Nevada.* US Department of Energy, Office of Civilian Radioactive Waste Management.

US Department of Energy (DOE) (1996) *Title 40 CFR Part 191 Compliance Certification Application for the Waste Isolation Pilot Plant.* US Department of Energy, Carlsbad Area Office, vol. 1–21.

US Department of Energy (1997a) *Linking Legacies: Connecting the Cold War Nuclear Weapons Production Process to Their Environmental Consequences.* DOE/EM-0319, US Department of Energy, Washington, DC.

US Department of Energy (1997b) The Legacy Story. http://legacystory.apps.em.doe.gov/index.asp.US Department of Energy (2001) *Status Report on Paths to Closure.* DOE/EM-0526, US Department of Energy, Office of Environmental Management, Washington, DC.

US Environmental Protection Agency (1996) 40 CFR Part 194: criteria for the Certification and Re-certification of the Waste Isolation Pilot Plant's Compliance With the 40 CFR Part 191 Disposal Regulations; Final Rule. Federal Register,. (No. 28), Office of the Federal Register, National Archives and Records Administration, Washington, DC, vol. 61, pp. 5224–5245.

US Environmental Protection Agency (EPA) (1997) Use of monitored natural attenuation at superfund, RCRA corrective action and underground storage tank sites, directive 9200.4-17. US Environmental Protection Agency, Office of Solid Waste and Emergency Response.

US Environmental Protection Agency (EPA) (1998a) *Compliance Application Review Documents for the Criteria for the Certification and Recertification of the Waste Isolation Pilot Plant's Compliance with the 40 CFR Part 191 Disposal Regulations: Final Certification Decision. CARD 23: Models and Computer Codes.* US Environmental Protection Agency, Office of Radiation and Indoor Air.

US Environmental Protection Agency (EPA) (1998b) *Technical Support Document for Section 194.23: Models and Computer Codes.* US Environmental Protection Agency, Office of Radiation and Indoor Air.

US Environmental Protection Agency (EPA) (1998c) *Technical Support Document for Section 194.23: Parameter Justification Report.* US Environmental Protection Agency, Office of Radiation and Indoor Air.

US Environmental Protection Agency (EPA) (1998d) *Technical Support Document for Section 194.24: EPA's Evaluation of DOE's Actinide Source Term.* US Environmental Protection Agency, Office of Radiation and Indoor Air.

US Environmental Protection Agency (EPA) (1999a) *Understanding Variation in Parition Coefficient, Kd, Values Volume 1: the Kd Model Methods of Measurement and Application of Chemical Reaction Codes.* EPA-402-R-99-044A, United States Environmental Protection Agency Office of Air and Radiation, Washington, DC.

US Environmental Protection Agency (EPA) (1999b) Understanding variation in partition coefficient, Kd, values: Volume II. Review of geochemistry and available Kd values for cadmium, cesium, chromium, lead, plutonium, radon, strontium, thorium, tritium (3H) and uranium. Prepared for the EPA by Pacific Northwest National Laboratory.

US Environmental Protection Agency (EPA) (1999c) *Field Applications of in situ Remediation Technologies: Permeable Reactive Barriers.* US Environmental Protection Agency.

US Environmental Protection Agency (EPA) (2000a). *Abstracts of Remediation Case Studies.* Federal Remediation Technologies Roundtable, vol. 4.

US Environmental Protection Agency (EPA) (2000b) *Field Demonstration of Permeable Reactive Barriers to Remove Dissolved Uranium from Groundwater, Fry Canyon, Utah.* US Environmental Protection Agency.

US Environmental Protection Agency (EPA) (2001) 40 CFR Part 197: public Health and environmental radiation protection standards for Yucca Mountain, NV; final rule. *Federal Register* **66**, 32074–32135.

US Nuclear Regulatory Commission (2001) 10 CFR Parts 2, 19, 20, 21, etc. disposal of high-level radioactive wastes in a proposed geological repository at Yucca Mountain, Nevada; final rule. *Federal Register* **66**(213), 55732–55816.

UNSCEAR (2000) *Sources and Effects of Ionizing Radiation.* Report of the United Nations Scientific Committee on the Effects of Atomic Radiation. United Nations.

Unsworth E. R., Jones P., and Hill S. J. (2002) The effect of thermodynamic data on computer model predictions of uranium speciation in natural water systems. *J. Environ. Moniter.* **4**, 528–532.

Vallet V., Schimmelpfennig B., Maron L., Teichteil C., Leininger T., Gropen O., Grenthe I., and Wahlgren U. (1999) Reduction of uranyl by hydrogen: an *ab initio* study. *Chem. Phys.* 244(2–3), 185–193.

Van Cappellen P., Charlet L., Stumm W., and Wersin P. (1993) A surface complexation model of the carbonate mineral-aqueous solution interface. *Geochim. Cosmochim. Acta* **57**, 3505–3518.

van der Lee J., Ledoux E., and de Marsily G. (1992) Modeling of colloidal uranium transport in a fractured medium. *J.Hydrol.* **139**, 135–158.

Van Genuchten M. T. and Wierenga P. J. (1986) Solute dispersion coefficients and retardation factors. In *Methods of Soil Analysis, Part 1. Physical and Mineralogical Methods* (ed. A. Klute). American Society of Agronomy, Madison, WI, pp. 1025–1054.

van Maravic H. and Smellie J. (1992) *Fifth CEC Natural Analogue Working Group Meeting and Alligator Rivers Analogue Project (ARAP) Final Workshop.* Commission of the European Communities.

Vandergraaf T. T., Tichnor K. V., and George I. M. (1984) Reactions between technetium in solution and iron-containing minerals under oxic and anoxic conditions. In *Geochemical Behaviour of Disposed Radioactive Waste.* ACS Symposium Series 246 (eds. G. S. Barney, J. D. Navratil,, and W. W. Schultz). American Chemical Society, Washington, DC, pp. 25–44.

Vilks P. (1994) *The Role of Colloids and Suspended Particles in Radionuclide Transport in the Canadian Concept for Nuclear Fuel Waste Disposal.* AECL Research.

Vilks P., Miller H. G., and Doern D. C. (1991) Natural colloids and suspended particles in the Whiteshell Research area and their potential effect on radiocolloid formation. *Appl. Geochem.* **6**(5), 565–574.

Vilks P., Cramer J. J., Bachinski D. B., Doern D. C., and Miller H. G. (1993) Studies of colloids and suspended particles, Cigar Lake uranium deposit, Saskatchewan, Canada. *Appl. Geochem.* **8**, 605–616.

Vilks P., Caron F., and Haas M. (1998) Potential for the formation and migration of colloidal material from a near-surface waste disposal site. *Appl. Geochem.* **13**, 31–42.

Vodrias E. A. and Means J. L. (1993) Sorption of uranium by brine-saturated halite, mudstone, and carbonate minerals. *Chemosphere* **26**(10), 1753–1765.

Wagenpfeil F. and Tschiersch J. (2001) Resuspension of coarse fuel hot particles in the Chernobyl area. *J. Environ. Radioact.* **52(1)**, 5–16.

Wagman D. D., Evans W. H., Parker V. B., Schumm R. H., and Halow I. (1982) The NBS tables of chemical thermodynamic properties. *J. Phys. Chem. Ref. Data* **11**, 2-1–2-34.

Wahlberg J. S. and Fishman M. J. (1962) *Adsorption of Cesium on Clay Minerals, Geological Survey Bulletin 1140-A*. US Government Printing Office.

Waite T. D., Davis J. A., Fenton B. R., and Payne T. E. (2000) Approaches to modeling uranium(VI) adsorption on natural mineral assemblages. *Radiochim. Acta* **88**, 687–693.

Wall N. A., Giambalvo E. R., Brush L. H., and Wall D. E. (2002) *The Use of Oxidation-state Analogs for WIPP Actinide Chemistry.* Unpublished presentation at the 223rd American Chemical Society National Meeting, April 7–11, 2002. Sandia National Laboratories.

Wang P., Andrzej A., and Turner D. R. (2001a) Thermodynamic modeling of the adsorption of radionuclides on selected minerals: I. Cations. Indust. Eng. Chem. Res. **40**, 4428–4443.

Wang P., Andrzej A., and Turner D. R. (2001b) Thermodynamic modeling of the adsorption of radionuclides on selected minerals: II. Anions. *Indust. Eng. Chem. Res.* **40**, 4444–4455.

Ward D., Bryan C., and Siegel M. D. (1994) Detailed characterization and preliminary adsorption model for materials for an intermediate-scale reactive transport experiment. In *Proceedings of 1994 International Conference of High Level Radioactive Waste Management*, pp. 2048–2062.

Ward D. B., Brookins D. G., Siegel M. D., and Lambert S. J. (1990) Natural analog studies for partial validation of conceptual models of radionuclide retardation at the WIPP. In *Scientific Basis for Nuclear Waster Management XIV* (eds. T. A. Abrajano, Jr. and L. H. Johnson). Materials Research Society, Boston, MA, pp. 703–710.

Watson D., Gu B., Phillips D., and Lee S. Y. (1999) *Evaluation of Permeable Reactive Barriers for Removal of Uranium and other Inorganics at the Department of Energy Y-12 Plant, S-3 Disposal Ponds.* Oak Ridge National Laboratory, Environmental Sciences Division.

Westall J. and Hohl H. (1980) A comparison of electrostatic models for the oxide/solution interface. *Adv. Coll. Interface Sci.* **12**, 265–294.

Wildung R. E., Routson R. C., Serne R. J., and Garland T. R. (1974) Pertechnetate, iodide, and methyl iodide retention by surface soils. In *Pacific Northwest Laboratory Annual Report for 1974 to the USAEC Division of Biomedical and Environmental Research: Part 2. Ecological Sciences* (Manager, B. E. Vaughan), BNWL-1950 PT2, Pacific Northwest Laboratories, Richland, WA, pp. 37–40.

Wildung R. E., McFadden K. M., and Garland T. R. (1979) Technetium sources and behavior in the environment. *J. Environ. Qual.* **8**(2), 156–161.

Wilson M. L., Gauthier J. H., Barnard R. W., Barr G. E., Dockery H. A., Dunn E., Eaton R. R., Guerin D. C., Lu N., Martinez M. J., Nilson R., Rautman C. A., Robey T. H., Ross B., Ryder E. E., Schenker A. R., Shannon S. A., Skinner L. H., Halsey W. G., Gansemer J. D., Lewis L. C., Lamont A. D., Triay I. R. A. M., and Morris D. E. (1994) *Total-System Performance Assessment for Yucca Mountain-SNL. Second Iteration* vol. 2 *(TSPA-1993).* SAND93-2675. Sandia National Laboratories, Albuquerque, NM.

Wilson M. L., Swift P. N., McNeish J. A., and Sevougian S. D. (2002) Total-system performance assessment for the Yucca Mountain Site. In *Scientific Basis for Nuclear Waste Manage. XXV* (eds. B. P. McGrail and G. A. Cragnolino). Materials Research Society, vol. 713, pp. 53–164.

Winkler A., Bruhl H., Trapp C., and Bock W. D. (1988) Mobility of technetium in various rock and defined combinations of natural minerals. *Radiochim. Acta* 44/**45**, 183–186.

Wolery T. J. (1992a) *EQ3/EQ6, a Software Package for Geochemical Modeling of Aqueous Systems, Package Overview and Installation Guide (Version 7.0).* Lawrence Livermore National Laboratory.

Wolery T. J. (1992b) *EQ3NR, A Computer Program for Geochemical Aqueous Speciation-Solubility Calculations: Theoretical Manual, User's Guide, and Related Documentation (Version 7.0).* Lawrence Livermore National Laboratory.

WM Symposia (2001) WM 01 Proceedings, Feb. 24–28, 2001, Tucson, Arizona: HLW, LLW, mixed wastes and environmental restoration-working towards a cleaner environment. Waste Management, 2001.

Xia Y., Roa L., Rai D., and Felmy A. R. (2001) Determining the distribution of Pu, Np, and U oxidation states in dilute NaCl and synthetic brine solutions. *J. Radioanalyt. Nuclear Chem.* **250**(1), 27–37.

Yeh G. T., Carpenter S. L., Hopkins P. L., and Siegel M. D. (1995) *Users' Manual for LEHGC: A Lagrangian-Eulerian Finite-element Model of HydroGeoChemical Transport through Saturated-unsaturated Media-version 1.1.* Sandia National Laboratories.

Yeh G T., Li M. H., and Siegel M. D. (2002) Fluid flow and reactive chemical transport in variably saturated subsurface media. In *Environmental Fluid Mechanics* (eds. H. Shen, A. Cheng, K. Wang, M. Teng, and C. Liu). American Society of Civil Engineers, pp. 207–256.

Yoshida S., Muramatsu Y., and Uchida S. (1998) Soil-solution distribution coefficients, Kds of I^- and IO_3^- for 68 Japanese soils. *Radiochim. Acta* **82**, 293–297.

Zeh P., Czerwinski K. R., and Kim J. I. (1997) Speciation of uranium in Gorleben groundwaters. *Radiochim. Acta* **76**, 37–44.

Zhang P. and Brady P. V. (2002) *Geochemistry of Soil Radionuclides.* Soil Science Society of America.

Published by Elsevier Ltd.

Readings from the Treatise on Geochemistry
ISBN: 978-0-12-381391-6
pp. 579–636

INDEX

NOTES:

Cross-reference terms in italics are general cross-references, or refer to subentry terms within the main entry (the main entry is not repeated to save space). Readers are also advised to refer to the end of each article for additional cross-references - not all of these cross-references have been included in the index cross-references.

The index is arranged in set-out style with a maximum of three levels of heading. Major discussion of a subject is indicated by bold page numbers. Page numbers suffixed by *t* and *f* refer to Tables and Figures respectively. *vs.* indicates a comparison.

This index is in letter-by -letter order, whereby hyphens and spaces within index headings are ignored in the alphabetization. Prefixes and terms in parentheses are excluded from the initial alphabetization.

A

B

J

K

L

M

O

P

Q

R

S

T

U

V

Printed and bound by CPI Group (UK) Ltd, Croydon, CR0 4YY
03/10/2024
01040318-0014